Fourth Chinese International Turbomachinery Conference (CITC 2020)

IOP Conference Series: Materials Science and Engineering
Volume 1081

Guangzhou, China
30 October - 2 November 2020

Editors:

Matthias Meinke
Peter Childs
Xavier Ottavy

ISBN: 978-1-7138-4585-0
ISSN: 1757-8981

Printed from e-media with permission by:

Curran Associates, Inc.
57 Morehouse Lane
Red Hook, NY 12571

Some format issues inherent in the e-media version may also appear in this print version.

Institute of Physics
Dirac House, Temple Back
Bristol BS1 6BE UK

Phone: 44 1 17 929 7481
Fax: 44 1 17 920 0979

techtracking@iop.org

Additional copies of this publication are available from:

Curran Associates, Inc.
57 Morehouse Lane
Red Hook, NY 12571 USA
Phone: 845-758-0400
Fax: 845-758-2633
Email: curran@proceedings.com
Web: www.proceedings.com

TABLE OF CONTENTS

IOP Conf. Series: Materials Science and Engineering 1081 (2021) 011001 doi:10.1088/1757-899X/1081/1/011001

Preface of the CITC 2020

The CITC (Chinese International Turbomachinery Conference) has been established in 2013 and since then has become a reference event in the turbomachinery field for the Chinese and worldwide industry and research community. It is sponsored by the Chinese International Turbomachinery Alliance and Chinese Journal of Turbomachinery,

The CITC focuses on promoting both fundamental and engineering applications and is of primary interest to researchers, engineers, students and users in the field of turbomachinery. It is a key event for technology transfer through the presentation of the latest developments and best practices, with a specific focus on China. The CITC has become a trendsetting conference outlining the roadmap to the future and is known as one of the most cutting-edge meetings comprising all aspects of turbomachinery.

The fourth CITC was held in Nan Chang, China with a duration of two and half days from October 31 to November 2, 2020. More than 650 attendees were registered for the conference and exhibition, where 179 technical papers were presented in 26 sessions and 39 companies exhibited their products in the conference exhibition booths. Due to Covid 19 travel restrictions, international experts gave invited web based presentations in several sessions of the conference, also allowing questions from the audience in the conference room.

The CITC 2020 was a tremendous success in establishing a platform for exchanging ideas and solutions, encouraging partnerships across academia and industry.

CITC Committee:

Chu-hua Zhang Xi'an Jiaotong University, China

Cun-liang Liu Northwestern Polytechnical Univerisity, China

Gérard BOIS Arts et Metiers ParisTech, France

IOP Conf. Series: Materials Science and Engineering 1081 (2021) 011001 doi:10.1088/1757-899X/1081/1/011001

Grant Laidlaw Ingram, Durham University, UK
Guang Xi, Xi'an Jiao Tong University, China
Hong-wei Ma, Beihang University, China
Hong Yan, Northwestern Polytechnical Univerisity, China
Li Zhou, Northwestern Polytechnical Univerisity, China
Luca d'Agostino, University of Pisa, Italy
Luis San Andrés , Texas A&M University, US
Matthias Meinke, RWTH Aachen University, Germany
Mehdi Vahdati , Imperial College London, UK
Peter RN Childs, Imperial College London, UK
Pietro Zunino, University of Genova, Italy
Qun Zheng, Harbin Engineering University, China
Xavier Ottavy, Fluid Mechanics and Acoustics Laboratory (Ecole Centrale de Lyon), France

Editors:
Matthias Meinke, RWTH Aachen University, Germany
Peter RN Childs, Imperial College London, UK
Xavier Ottavy, Fluid Mechanics and Acoustics Laboratory (Ecole Centrale de Lyon) , France

IOP Conf. Series: Materials Science and Engineering 1081 (2021) 011002 doi:10.1088/1757-899X/1081/1/011002

Peer review declaration of the CITC 2020

All papers published in this volume of *IOP Conference Series: Materials Science and Engineering* have been peer reviewed through processes administered by the Editors. Reviews were conducted by expert referees to the professional and scientific standards expected of a proceedings journal published by IOP Publishing.

- Describe criteria used by reviewers:

CITC peer review focuses on innovation, originality, technical correctness, academic value, practicality, and significance of the topic deducted from the content of the paper. In addition, the organization of the paper, the presentation of results in form of figures, diagrams, and tables, the references and the language were evaluated.

Five categories for the evaluation of a paper were defined: 1. to be rejected, 2. weak, 3. satisfactory, 4. very good, and 5. excellent. The category of each paper was decided by the combination of scores for a list of cf quality criteria. The reviewer also had the task to provide comments on how to modify the paper to improve its quality concerning the theoretical contribution, accuracy of the applied methods, suitability of the concept, discussion of results, logical structure of the paper, clarity of the language, readability, etc..

If the paper was classified in a category lower than "satisfactory", the paper was rejected. If the paper was evaluated better than the category "very good", the paper was accepted, but the reviewer still required the paper authors to respond to his comments in a revised version of the paper to improve the paper quality. The author was asked to submit the revised version of the paper after a given period of time.

- Type of peer review: Double-blind
- Conference submission management system: CITC submission management system (http://conference.cftn.cn)
- Number of submissions received: 260
- Number of submissions sent for review: 140
- Number of submissions accepted: 57
- Acceptance Rate (Number of Submissions Accepted / Number of Submissions Received X 100): 20%
- Average number of reviews per paper: 2
- Total number of reviewers involved: 65
- Any additional info on review process:

Detailed description of the peer review process:

Authors needed to upload their papers to the CITC website. The conference organizers assigned at least two reviewers to each paper. Reviewers could accept or reject the invitation to review a paper. If the invitation was accepted, reviewers were obliged to provide review comments within 15 days. Peer-reviewers had three options: 1. reject the paper, 2. accept the paper, 3. send comments with requests for changes to the authors. The peer-reviewers rejected a paper, if the quality did not satisfy the requirements specified in the call for papers. If the quality of an article was sufficiently high, peer-reviewers could accept the paper without changes. In most of all cases, however, the peer-reviewers sent comments and requests for changes to the authors of the paper. Authors were obliged to provide a revised version of the paper according to the reviewer comments within 20 days. Once authors submitted the revised paper document, the editors checked whether all requested changes were addressed. If so, the article was sent to a reviewer to approve the changes. In case the reviewers approved the changes in the revised version of the paper, the paper was considered as accepted for the CITC2020. In case editors could not verify that the article was changed according to the reviewers' comments, they sent the paper back to authors with a second request to revise the paper.

- Contact person for queries: Hong Wang, Email: sfjjs@163.com.

IOP Conf. Series: Materials Science and Engineering 1081 (2021) 012001 doi:10.1088/1757-899X/1081/1/012001

STRUCTURAL PARAMETER ANALYSIS OF SELF-TIGHTENING METAL U-SHAPED SEAL RING

Tianfu Cheng Shuangxi Li

(Beijing University of Chemical Technology)

E-mail: 1246328540@qq.com

ABSTRACT. The effects of different structural parameters on the sealing performance of self-tightening metal U-shaped seal rings were studied. A two-dimensional axisymmetric model of a self-tightening U-shaped seal ring was established using finite element software. The influence of structural parameters such as side wall thickness, bottom wall thickness and boss radius on the sealing performance of the seal ring was analyzed by the control variable method. The results show that as the thickness of the side wall increases, the stress shows a downward parabolic trend, which is the smallest at 0.55 to 0.65 mm; as the bottom wall thickness increases, the stress first increases steadily, and then decreases slowly after 0.8 to 0.9 mm; When the inclination angle is from 0 to 9 °, the sealing ring can obtain better sealing performance. The results have guiding significance for the design, optimization and application of subsequent seal ring structures.

1. Introduction

Seals have a wide range of applications. About 90% of machinery requires seals. It involves many disciplines, including mechanics, tribology, materials science, heat transfer, fluid mechanics, and solid mechanics. It is an important guarantee for long-term efficient and safe operation of industrial equipment. In recent years, scholars at home and abroad have conducted a lot of research on various seals, such as labyrinth seals [1], dry gas seals [2-5], brush seals [6], and various seals on centrifugal compressors [7,8], etc. And made considerable progress. Rocket engines using liquid hydrogen and liquid oxygen as fuels are widely used in the aerospace industry, and their severe low temperature and high pressure conditions place high requirements on sealing [9]. The metal sealing ring is the most widely used static sealing structure in the industry. Although its structural size is small compared to the entire equipment, its sealing performance will directly affect the safety of the entire equipment and even the entire system. Metal sealing rings are used in the sealing of aerospace engines because of their advantages such as low temperature resistance and low pressure deformation. However, their disadvantages are high rigidity, difficulty in compression, high pre-tightening force, and high requirements on the surface roughness of the sealing [10-12]. Low temperature and high pressure self-tightening metal static seal, mainly refers to

the use of metal materials and has the characteristics of elastic compensation cold shrinkage deformation structural seals, commonly used structures are metal C-shaped, hollow O-shaped, K-shaped, X-shaped, E-shaped, W-shaped, V-shaped ,U-shaped, etc. [13-16]. The sealing features are: small cross-sectional area, low axial load requirements and elastic compensation capabilities. Elastic metal static seals can be manufactured by two methods: mechanical cutting and die spinning [17]. The working principle of the metal seal is that the elastoplastic deformation on the pre-pressure contact surface compensates the gap of the original seal contact surface, and seals the leakage channel to achieve medium sealing [12].
A series of researches on metal seals have been made at home and abroad since the 1970s [18-21]. The domestic research on self-tightening metal sealing rings started late, but the development speed is very fast and the results are significant [22].

2.ESTABLISHMENT OF FINITE ELEMENT MODEL OF SELF-TIGHTENING METAL U-SHAPED SEAL RING

In order to improve the shortcomings of the existing metal seal ring under pressure conditions, poor fatigue resistance, unreliable sealing performance under vibration conditions, combined with the advantages of metal U-shaped ring energy storage and good metal V-shaped ring resilience, A self-tightening metal U-shaped seal ring structure is improved and designed. Its structure is shown in Figure 1.

In recent decades, various metal seal structures have been newly developed in China, such as an improved Y-shaped seal ring that can be applied to both dynamic and static seals [23]; a non-contact type with progressive pressure reduction and throttling discharge functions static seal structure [24], which solves the problem of oil-water separation of the rotating parts of the hydrostatic bearing in a vertical turbine; a seal structure that does not flip during installation and has stable sealing performance during shearing; and an easy-to-install, bearing static sealing structure with strong compression ability and stable sealing performance [25];Dong Zhang[26] developed two types of static sealing structures with trapezoidal cross-sections to achieve the seal between the metal shell and the window, with a pressure higher than 150 good sealing effect at MPa and temperature higher than 100 ℃; Parker [27], as a pioneer in the sealing industry, has developed a variety of sealing structures, of which the BS type sealing structure has two sealing lips that are not sensitive to impact loads; AF type sealing uses the interference fit of the sealing groove and the metal to fix the dust ring in the open groove, and the lip can move with the movement of the piston rod. SSME uses a "U" -shaped pressure-acting sealing element made of silver-plated nickel-based material.

U-shaped seal rings are relatively rare in China. This paper uses finite element software to establish a two-dimensional axisymmetric model of U-shaped seal rings The model analyzes the structural parameters such as the side wall thickness, bottom wall thickness, and boss radius of the self-tightening metal U-shaped seal ring, and optimizes its structural parameters. The results have guiding significance for the design, optimization and application of seal rings in the future.

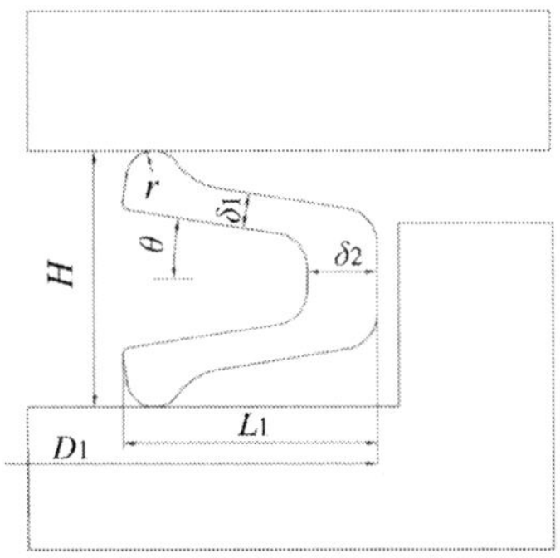

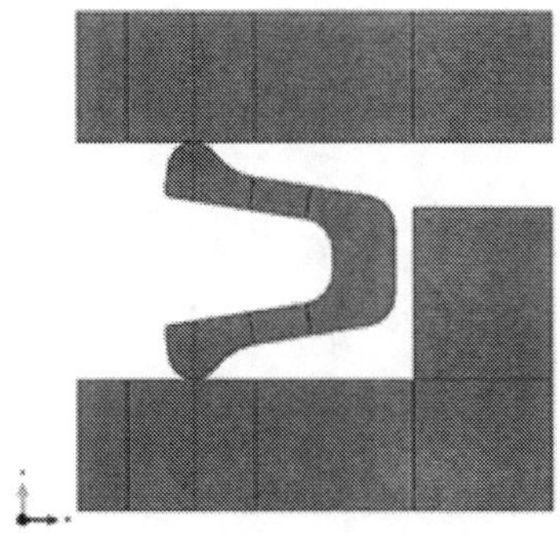

Fig.1 Geometry Size of U - shaped Seal Ring Fig.2 Simplified Model Diagram of Structure

The low-temperature and high-pressure self-tightening seal structure has axial symmetry. In order to ensure the accuracy of the structural analysis and reduce the calculation amount, a two-dimensional planar axisymmetric simplified model is established. This model includes a simplified upper flange, a metal seal ring, and a simplified lower flange. The assembly relationship is that the highest point of the upper boss of the metal seal ring contacts the lower boundary surface of the upper flange, and the lowest point of the lower boss and the lower flange. Contact the upper boundary face. Its axial section is shown in Figure 2.

Table 1 Integral Structure Parameter Table

CATEGORY	SIZE/mm	CATEGORY	SIZE/mm
Total Height H	3.64	Side Wall Thickness δ1	0.5
Diameter D1	Φ70	Bottom Wall Thickness δ2	1
Tilt Angle θ	9°	Boss Radius r	0.39
Section Length L1	3.64		

The mesh division of the two-dimensional axisymmetric model generally uses a quadrilateral mesh, which can improve the calculation efficiency and obtain accurate calculation results. In this paper, a four-node bilinear axisymmetric reduction and integration unit CAX4R is used to divide the grid. The grid division diagram is shown in Figure 3. The total number of grids is 16565. The grid independence proves that the number of grids is reliable.

IOP Conf. Series: Materials Science and Engineering 1081 (2021) 012001 doi:10.1088/1757-899X/1081/1/012001

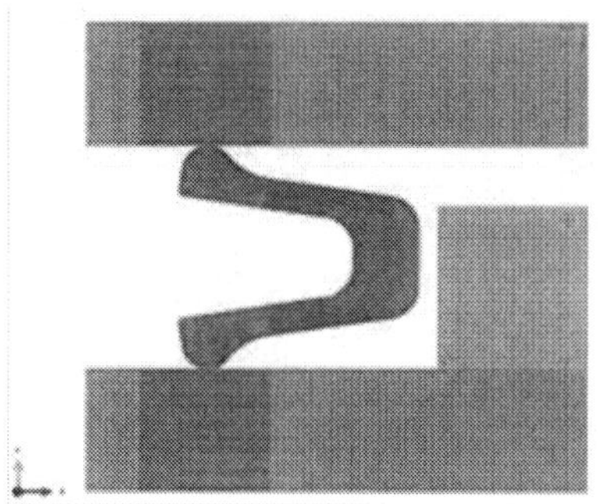

Fig.3 Grid Partition Diagram

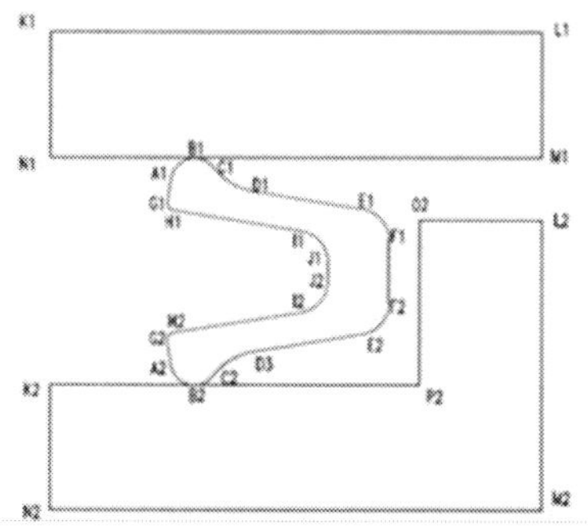

Fig.4 Boundary Condition Loading Diagram

In recent years, GH4169 [18] has been widely used in the field of aviation. After research, it has been found that the material has excellent elastoplastic properties under low temperature and high pressure conditions. It is a nickel-based high-temperature alloy with good fatigue resistance, radiation resistance, oxidation resistance, and corrosion resistance in the range of -253 ~ 700 ℃. Its approximate grade is Inceonel718, which has good processability. The specific parameters of GH4169 material are listed in Table 2.

Table 2 Material Parameters of GH4169

Parameter	数值	数值
Temperature /°C	-183	20
Density/ g.cm^{-3}	8.24	8.24
Elastic Modulus /MPa	213000	203000
Poisson's Ratio/μ	0.254	0.29
Yield Strength /MPa	1300	1030
Tensile Strength /MPa	1700	1270

The working parameters of the U-shaped seal ring are listed in Table 3. The boundary condition labels are shown in Figure 4, and the applied boundary conditions are shown in Table 4. The stress and displacement results obtained by loading in this paper are shown in Figure 5 and Figure 6.

Table 3 Working condition parameter

Pre-tightening Conditions		Operating Conditions	
Internal Pressure P_0/MPa	0	Internal Pressure P_0/MPa	20
Set Temperature T/°C	20	Set Temperature T/°C	-183
Compression △Y/mm	0.5	Compression △Y/mm	0.5

Table 4 Boundary condition（Coulomb friction)

Category	Boundary
Internal Pressure Boundary	N1B1、B1A1、A1G1、G1H1、H1I1、I1J1、J1J2、J2I2、I2H2、H2G2、G2A2、A2B2、B2K2、K2N2

IOP Conf. Series: Materials Science and Engineering 1081 (2021) 012001 doi:10.1088/1757-899X/1081/1/012001

Thermal Boundary	**Upper and Lower Flanges, Seal Rings**
Friction Boundary	**A1B1、B1C1、N1M1、A2B2、B2C2、K2P2**
Frictionless Boundary	**F1F2、O2P2**
Fully Constrained Boundary	**N2M2**
Displacement Constraint Y=-0.5mm	**K1L1**

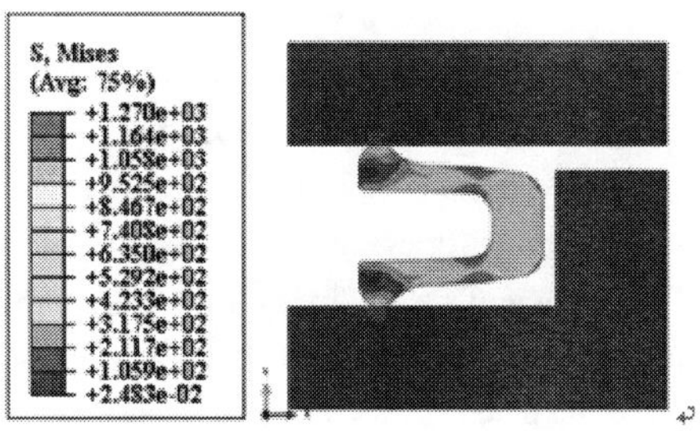

Fig.5 Structural stress nephogram

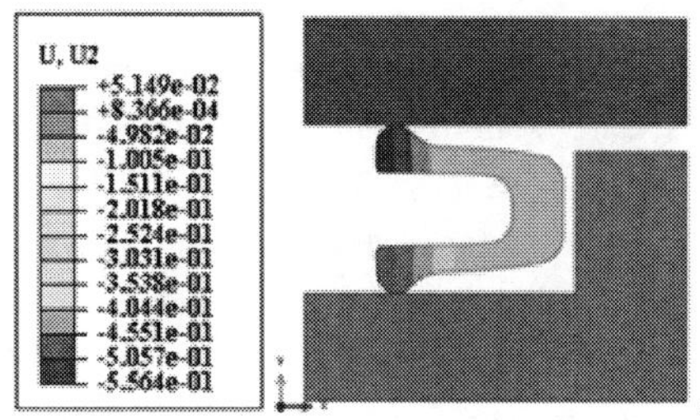

Fig. 6 Structural displacement nephogram

3.ANALYSIS OF THE INFLUENCE OF VARIOUS PARAMETERS ON THE RESULTS

3.1 Side Wall Thickness δ_1

The thickness of the side wall of the seal ring is mainly to provide the elastic force for the seal ring, but it is connected to the bottom wall thickness and is located at the discontinuity of the structure. It is easy to break when compressed. This paper analyzes the influence of the side wall thickness δ_1 on the mechanical properties of the seal ring as shown in Figure 7.

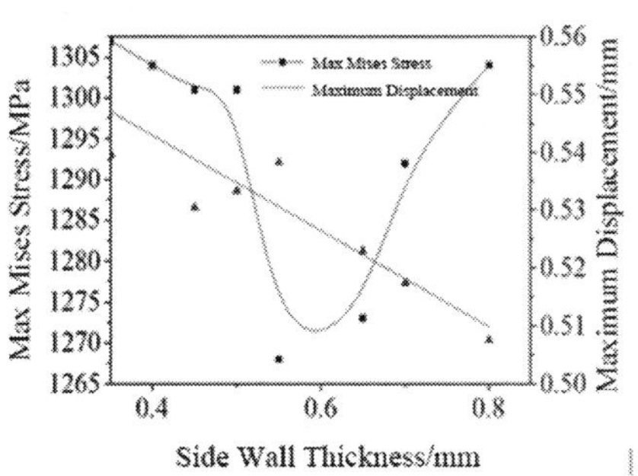

(a)

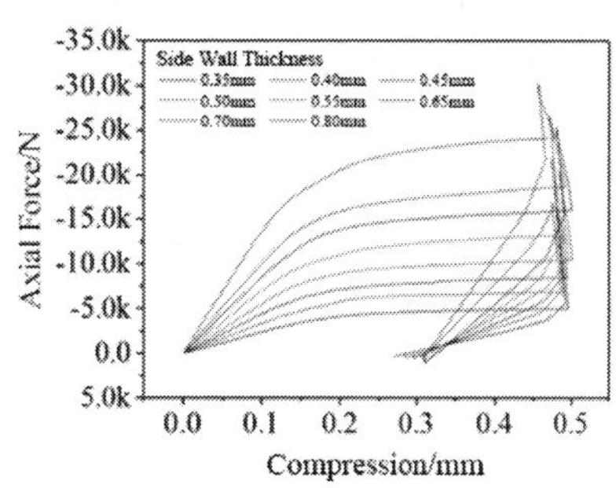

(b)

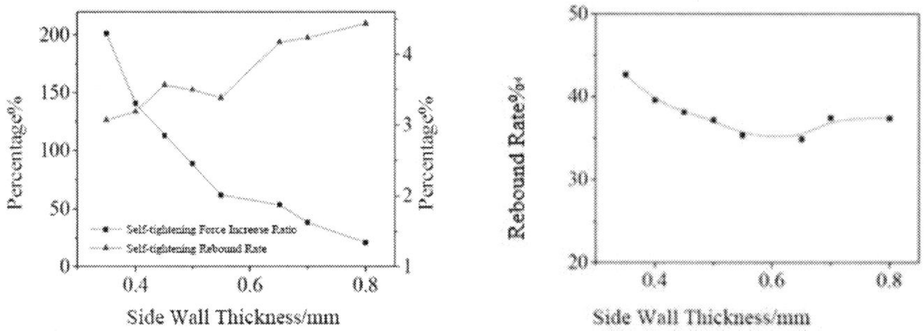

(c) (d)

Fig. 7 Mechanical Property Curve of Seal Ring when δ1 changes

It can be seen from Fig. 7 that as the sidewall thickness increases, the maximum Mises stress appears as a parabola, which is the smallest at 0.55 ~ 0.65mm; the maximum displacement decreases; from (b), it can be seen that as the sidewall thickness increases, the axial stiffness of the seal ring increases. And the graph (c ~ d) can be obtained from the compression rebound curve. The increase ratio of the self-tightening force decreases as the thickness of the side wall increases. As the thickness of the side wall increases, the stiffness becomes larger. The axial force increment of the position decreases, the self-tightening incremental ratio gradually decreases, and the self-tightening effect decreases. The total amount of self-tightening deformation increased, and the self-tightening effect increased. The rebound rate is reduced and the resilience of the seal ring is reduced. That is, as the thickness of the seal ring increases with the increase in the thickness of the side wall, the rigidity increases during pre-tightening, the self-tightening effect in the working state decreases, and unrecoverable deformation increases, resulting in a decrease in the elasticity of the seal ring. Therefore, the optimal value of the side wall thickness should be selected between 0.4 and 0.45 mm.

3.2 Bottom Wall Thickness δ2

When the bottom wall thickness is compressed, a large strain is generated. When the seal ring is compressed too much, the bottom wall thickness is easily broken, and it has the effect of absorbing energy and storing energy. In this paper, the influence of different bottom wall thickness on the mechanical properties of the seal ring is calculated, as shown in Figure 8.

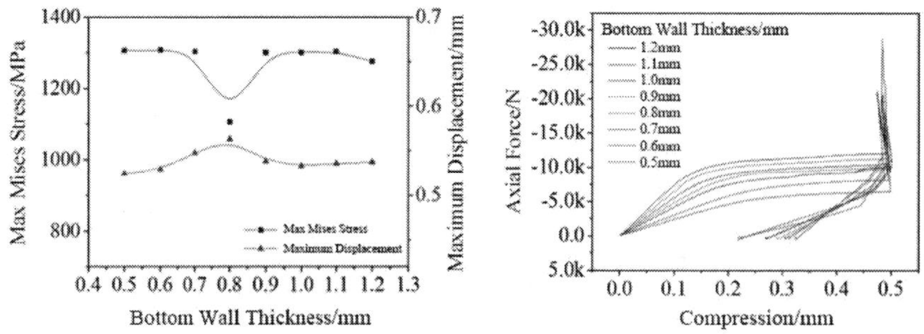

(a) (b)

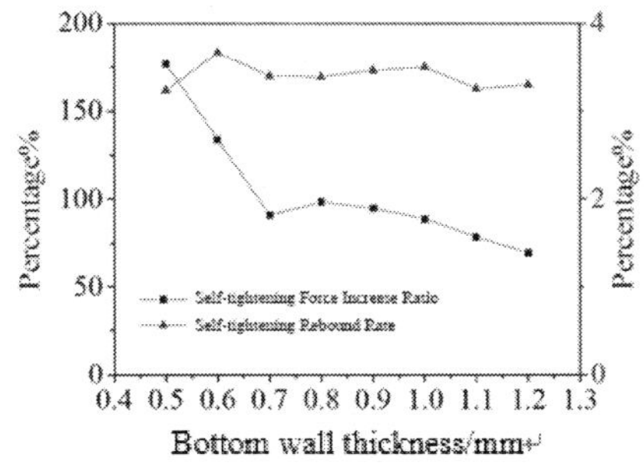

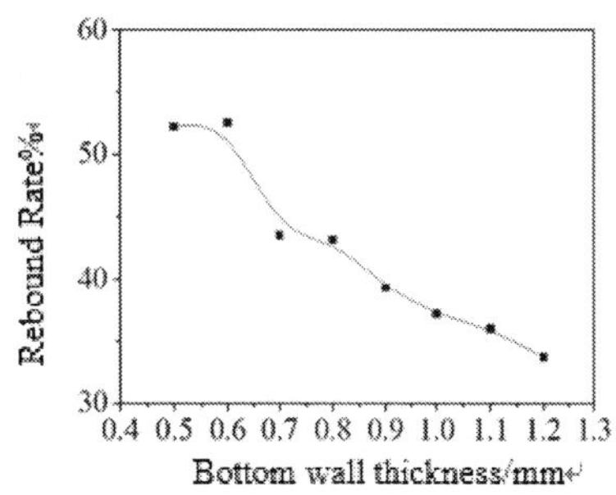

(c) **(d)**

Fig.8 Mechanical Property Curve of Seal Ring when δ2 changes

It can be seen from Fig. 8 that the maximum Mises stress increases steadily with the increase of the bottom wall thickness, and then decreases at 0.8 mm, and then decreases to 0.9 mm at the previous size. The maximum displacement has a maximum value at 0.8 mm. It can be seen from (b) that as the bottom wall thickness increases, the stiffness of the pre-tightened seal ring increases. From the compression rebound curve, we can get the graph (c ~ d), the self-tightening axial force increment decreases, and the self-tightening deformation rebound rate decreases. It can be seen that the stiffness of the seal ring under pressure increases, the elasticity decreases, and the self-tightening effect decreases. The rebound rate gradually decreases, and the elasticity of the seal loop decreases. The recommended value is 0.5 ~ 0.6mm.

3.3 Boss Radius r

In the design of the structure, a boss using spherical contact at the position of the contact point of the seal ring is selected. In order to ensure the line contact of the seal surface, when a smaller contact force is provided, a larger contact pressure can also be generated to increase the sealing effect. However, the radius of the boss cannot be increased indefinitely. It is directly connected to the thickness of the side wall of the seal ring. When the seal ring is too small, it is easy to detach the upper contact surface when the seal ring is rotated by the compressed side wall and make the flange directly with other contact at non-preset sealing surfaces. This not only causes a waste of the structure of the seal ring, but if other contact surfaces do not guarantee sufficient surface roughness of the contact surface, the seal will leak. However, the boss must not be too large. When its radius is too large, a weak point of stress may occur in the thick middle part of the sealing sidewall connected to it. During the compression process, the sealing ring is broken before it has sufficient sealing force. This paper calculates the influence of the boss radius on the mechanical properties of the seal ring, as shown in Figure 9.

IOP Conf. Series: Materials Science and Engineering 1081 (2021) 012001 doi:10.1088/1757-899X/1081/1/012001

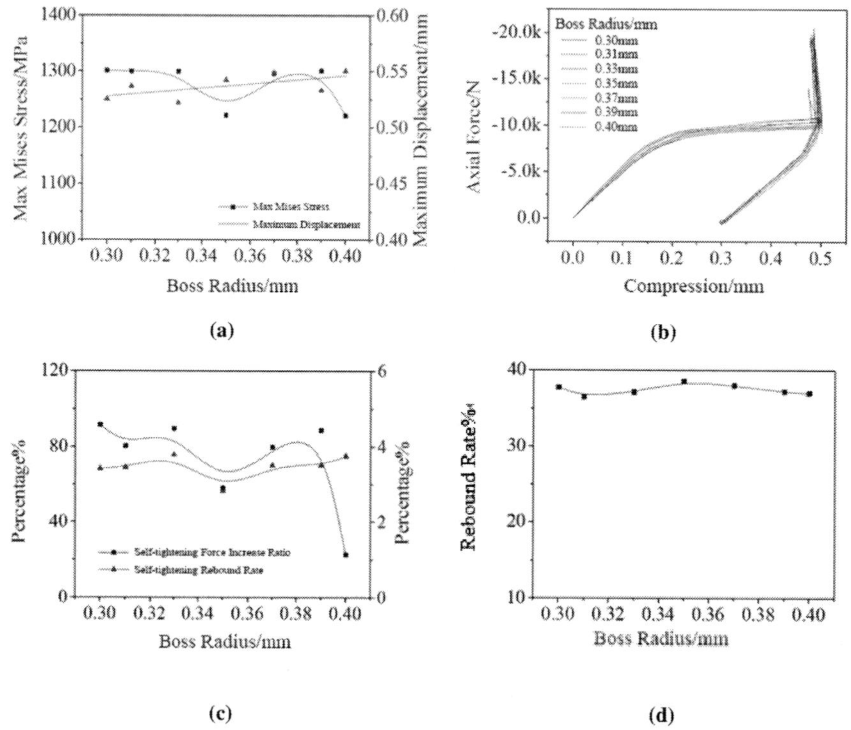

Fig.9 Mechanical Property Curve of Seal Ring when r changes

It can be seen from Fig. 9 that the maximum Mises stress of the seal ring is smallest at 0.39 mm with the increase of the boss radius, and the rest is basically near the yield limit. The overall trend of maximum displacement increases; the axial stiffness of the seal ring increases; from the compression rebound curve It can be obtained that (c ~ d), the self-tightening axial force increase of the sealing ring decreases slowly and then decreases sharply, while the rebound rate of self-tightening deformation slowly increases, and the rebound rate decreases first and then increases. Decreased, there is a maximum at 0.35mm. Considering the comprehensive mechanical properties, the best range is 0.31 ~ 0.33mm.

3.4 Tilt Angle θ

The inclination angle of the seal ring can directly affect the elasticity of the seal ring when it is compressed. This paper calculates the influence of the inclination angle on the mechanical properties of the seal ring, as shown in Figure 10.

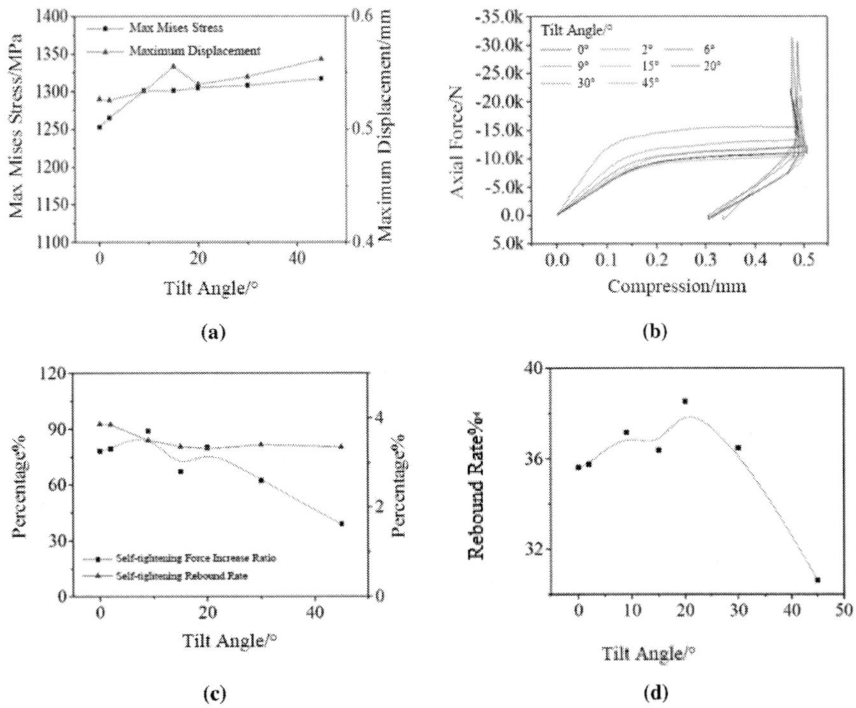

Fig.10 Mechanical Property Curve of Seal Ring when θ changes

From Figure 10, it can be seen that with the increase of the tilt angle, the maximum Mises stress increases faster and reaches the yield limit slowly after 9 °; the maximum displacement has a maximum abrupt value at 15 °, which generally increases; the axial stiffness increases ; From the compression rebound curve, we can get the graph (c ~ d). The self-tightening force increase ratio is generally reduced, and has a maximum value at 10 ° and 20 °, while the self-tightening deformation recovery amount gradually decreases, and at 20 ° there is the minimum value; the rebound rate increases first and then decreases rapidly, and the resilience performance increases first and then decreases. The best result is 0 ~ 9 °.

4 CONCLUSIONS

(1) By using the control variable method, a range of values for each structural parameter is obtained when the self-tightening metal U-shaped seal ring achieves better sealing performance.

(2) It can be known from the research results that with the increase of the thickness of the side wall, the stress shows a parabolic trend, which is the smallest at 0.55 ~ 0.65mm; as the bottom wall thickness increases, the stress increases steadily first, and then decreases slowly after 0.8 ~ 0.9mm ; As the radius of the boss increases, the stress of the seal ring is the smallest at 0.39mm, and the rest is basically near the yield limit; the sealing angle is better when the inclination angle is 0 ~ 9 °; the maximum Mises overall as the outer diameter increases showing a growing trend.

(3) According to the research results, it is known that the thickness of the side wall has a great

influence on the maximum Mises stress and the maximum displacement of the self-tightening metal U-shaped seal ring.

(4) According to the research results, it can be known that the bottom wall thickness and cross-section length have a greater impact on the rebound rate of the self-tightening U-shaped seal ring.

References

[1] Yaru Ma, Wenhao Huo, Jing Liu, Lijun Jiang, Zhi Fang, Zhigang Li.*Research on Wear Failure Leakage Characteristics of Labyrinth Seal and Design of Anti-bump Structure* [J] .Fan Technology, 2019,61 (05): 64-71.

[2] Liang *Han.Design and Application of Dry Gas Seal for Natural Gas Compressor* [J] .Fan Technology, 2017,59 (02): 76-82 + 86.

[3] Fengqin Li, Qiang Wang, Yuanhong Song.*Spiral Groove Dry Gas Seal Technology for Compressor* [J] .Fan Technology, 2008 (01): 33-35 + 54.

[4] Mingzhen *He.Failure Analysis and Treatment of Dry Gas Seal of Syngas Compressor* [J] .Fan Technology, 2013 (04): 87-90.

[5] Wei He, Ye Tao, Hailin Feng, Honghai Xu.*Cause analysis and measures of oil pollution in dry gas sealing system* [J] .Fan Technology, 2011 (05): 78-79.

[6] Guoqin Li, DanS un, Zhongze Jiao, Yanyu Du, Lu Bai.*Numerical Study on Flow and Heat Transfer Characteristics of Multi-stage Low Hysteresis Brush Seal* [J] .Fan Technology, 2019,61 (05): 56-63.

[7] Qingyu Shi, Huifeng Li, Nan Jin.*Fault diagnosis and characteristic analysis of friction between rotor and seal of centrifugal compressor* [J] .Fan Technology, 2017,59 (S1): 45-48.

[8] Xingyun Jia, Jie Zhou, Zhiqiang An, Qun Zheng, Hai Zhang.*Effects of Flange Sealing Arrangement on Mainstream Invasion in Transonic System (English)* [J] .Fan Technology, 2019,61 (02): 50-60 + 5.

[9] Yuansen *Zhu Hydrogen-oxygen rocket engine and its low temperature technology* [M] Beijing: National Defense Industry Press, 1995.

[10] Xiu Feng, Boqin Gu, Jianjun Sun, et al. *Research on Seal Mechanism of Metal Gaskets* [J]. Lubrication Engineering, 2007, 32 (10): 97-99

[11] Michel Lefrançois. Metal-to-metal seals: the innovative route in static sealing[J]. Sealing Technology, 2004:10-13

[12] Boqin Gu, Xinhua Li, Zheng Tian. *Static Seal Design Technology* [M]. Beijing: China Standard Press, 2004

[13] Renliang Cai, Boqin Gu, Pengyun Song. *Process Equipment Sealing Technology* [M]. Beijing: Chemical Industry Press, 2006

[14] Chuanjun Liao, Daolian Wang, Hongrui Wang, et al. *Characteristics and application of typical special-shaped metal seals* [J]. Cryogenic Engineering, 2014 (04): 55-60

[15] Changguo Huang. *High Pressure Static Seal Used on Spacecraft* [J]. Shanghai Aerospace, 1990 (04): 48-54

[16] Xuezheng Wu, Chao Hu, Deyou Qin. *Analysis of the sealing performance of a new type of seal ring* [J]. Natural Science, 2015: 265-266

[17] Jian Zhao, Yonghua Tan, Jianhua Chen, et al. *Study on sealing characteristics of self-tightening K-shaped metal seal assembly* [J]. Rocket Propulsion, 2013, 39 (6): 35-41

[18] Pingzhang Fan. *Space Shuttle Metal Static Seal Technology and Manufacturing Process* [J]. Aerospace Technology, 1999 (2): 116

[19] Kollmann K.Problems of the application of metal seals at very high pressures [C] // Proceeding of the 5th International Conference on Fluid Sealing.1971:53-67.

[20] Huang Changguo.High pressure static sealing using on spacecraft [J] Shanghai Aerospace,1990 (4) :48-54.

[21] Thomas T R.Influence of rough surfaces in static contact [C]//Proceeding of the 6th International Conference on Fluid Sealing.1973:33-46.

[22] Feiming Zhao, Yongxiang Xu, Shujun Wang. *Research on Low Temperature Static Sealing Technology* [J]. Cryogenic Engineering, 2001 (2): 23-26

[23] Xuezheng Wu, Chao Hu, Deyou Qin. *Analysis of the sealing performance of a new type of seal ring* [J]. Natural Science, 2015: 265-266

[24] Bing Wan, Yan Zheng, Tao Wang. *A non-contact static sealing structure* [P]. Chinese Patent, 201851667U, 2011-06-01

[25] De Peng. *A kind of static seal* [P]. Chinese Patent, 203374792U, 2014-01-01

[26] Dong Zhang, Chengyuan Lu, Zengzheng Lun, et al. *A kind of static sealing structure* [P]. Chinese Patent, 205578728U, 2016-09-14

[27] Parker company brochure.

Numerical analysis of magnetic grease seal pole teeth optimization based on pressure bearing capacity

Shuwen-Li[1], Rao-Zheng[1], Qiuxiang-Zhang[1], Shuangxi-Li[1]

[1] Beijing University of Chemical Technology, college of Electrical and Mechanical Engineering Beijing, 100029

buctlishuwen@163.com

Abstract. The static magnetic field model of the magnetic grease seal was established by the nonlinear magnetic field finite element method. The influence of three typical pole teeth with different seal gaps, namely rectangular teeth, single-side trapezoid, and symmetrical teeth, on pressure bearing capacity was analyzed. The results show that the maximum pressure bearing capacity of the optimized structure is from large to small: symmetric trapezoid teeth, single-sided trapezoid teeth, rectangular teeth. The seal gap increased by 0.1 mm and pressure bearing capacity decreased by 12 %. If the teeth width was doubled, the pressure bearing capacity can be increased by 17.8%, with an optimum range of 1.8 mm to 2.2 mm. The interval length of the teeth doubled, the pressure bearing capacity increased by 30.65 %, and the optimum range for the interval length of the teeth was 1.8 mm to 2.0 mm. There was no significant correlation between teeth height and pressure bearing capacity. The study on the optimization of the teeth structure based on pressure bearing capacity could provide a reference for technical personnel to improve the efficiency and rationality of the design of the teeth for the sealing of the magnetic grease.

1. Introduction

The sealing of the fan is of great significance for safe production, energy-saving, and environmental protection[1-5]. The common sealing form at the fan shaft end is a labyrinth seal, and there was still a small amount of gas leakage[6]. Magnetic grease seal is a new type of seal, which has the advantages that other seals do not have, such as no leakage, low friction, long life, and high reliability. Suitable for fan shaft end sealing. The magnetic fluid seal similar to the magnetic grease seal was applied to the shaft end of the centrifugal fan with a good effect[7]. Compared with magnetic fluid, the saturation magnetization and initial viscosity of magnetic grease are higher, and it could still maintain pressure bearing capacity under vibration or impurity conditions [8, 9]. Therefore, the magnetic grease seal could be applied to the fan shaft end seal, and it is suitable for the sealing of the above extreme working conditions at the fan shaft end.

Research on the magnetic grease seal showed that the magnetic grease seal had strong developability and application value. The magnetic grease seal at the shaft end of the evaporative cooling motor had been tested and had a strong sealing ability [10]. Combining the brush seal with the magnetic grease seal, the flexible pole piece magnetic grease gas-phase seal met engineering needs [11]. Reports on the numerical calculation of the pressure-pressure bearing capacity of magnetic grease seals were rare, and the current seal structure urgently needs to be optimized.

Szczech used the nonlinear magnetic field finite element method to calculate the influence of different pole teeth types on the pressure bearing capacity of the seal. It was found that with the same pole teeth end width, the maximum pressure bearing capacity was arranged in order of symmetrical trapezoidal pole teeth. Single-sided trapezoidal pole teeth, rectangular pole teeth. And compared with experiments, the calculated values were all larger than the experimental values [12-15]. In the literature, the end width of the pole teeth was the same, and the volume of the trapezoidal pole teeth was significantly larger than that of the rectangular pole teeth. The influence of the volume of the pole teeth on the result cannot be ruled out, and supplementary research is needed.

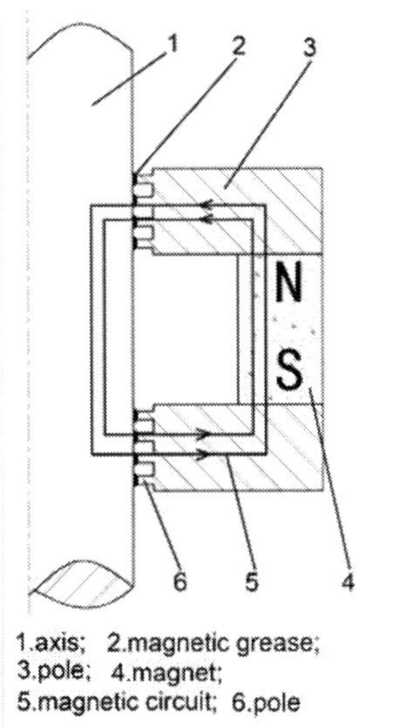

Figure 1. Schematic diagram of magnetic grease seals

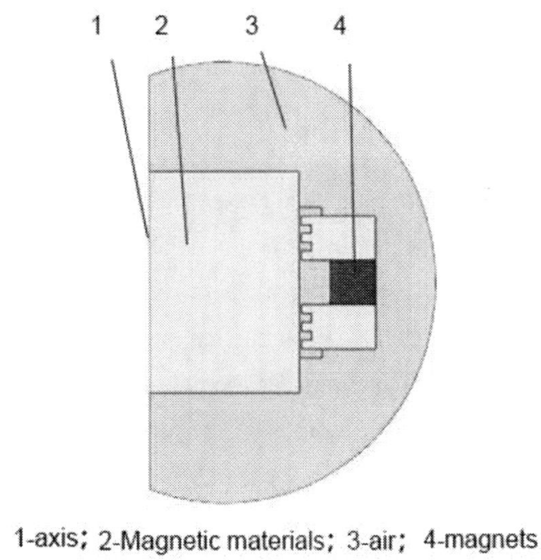

Figure 2. Magnetic field calculation model of magnetic grease seal

The pole teeth is an important structure of the magnetic grease seal. Under normal working conditions, the magnetic grease is maintained in the sealing gap formed by the pole teeth and the shaft. It is necessary to analyze the influence of the pole teeth structure on the pressure bearing capacity. This paper was based on the nonlinear magnetic field finite element method, based on the single factor principle, guided by the pressure-pressure bearing capacity, using the magnetic field force viewpoint to study the formation mechanism of the magnetic grease seal, and optimize the pole teeth structure concerning the numerical calculation results.

2. The principle of magnetic grease seal

As shown in Figure 1, the magnetic grease seal device is composed of a permanent magnet, magnetic pole, pole teeth, magnetic grease, and the rotation axis of the magnetic permeability, forming a magnetic circuit and having a high magnetic induction line density below the pole teeth. Under the action of non-uniform high gradient magnetic field force, magnetic fat is adsorbed on the end of pole teeth, forming a liquid "O" ring, and obtaining the sealing ability to resist pressure difference.

Compared with magnetic fluids, magnetic grease has higher saturation magnetization, higher initial viscosity, and obvious non-Newtonian fluid properties. Therefore, magnetic grease has higher pressure bearing capacity at a low rotation speed.

The pressure bearing capacity of the magnetic grease seal is related to the magnetization strength of magnetic grease, the difference of magnetic induction strength near the pole teeth, and the pressure resistance value of the carrier fluid. For the simplicity of the study, the liquid carrying pressure value of magnetic grease was ignored, and the effect of the magnetic field on pressure bearing capacity was considered only. Equation (1) represents the pressure bearing capacity of a single-pole tooth.

$$\Delta p = M\left(H_1\right)B_1 - M\left(H_2\right)B_2 \tag{1}$$

$$\Delta p = M_s \Delta B \tag{2}$$

$$\Delta p = \alpha M_s \sum_{i=1}^{N} \Delta B \tag{3}$$

In the formula,

α	Correction factor after superposition of seal pressure
N	The number of magnetic pole teeth
M_s	The saturation magnetization of magnetic lipid，kA/m
ΔB	The magnetic induction intensity difference of the sealing gap，T

The magnetic induction near the pole teeth far exceeds the saturation magnetization of the magnetic grease. That is, the position of the magnetic grease in contact with the sealing medium satisfies the relation formula: $M\left(H_1\right)=M\left(H_2\right)=M_S$. Therefore, the maximum pressure difference formula of single-stage magnetic grease seal can be simplified as Equation (2). The sealing ability of the single-stage magnetic grease seal is limited. The sealing pressure ability is enhanced by increasing the number of pole teeth. The maximum pressure bearing capacity of a multi-stage magnetic grease seal can be calculated by Formula (3). According to the experimental results, the correction coefficient of multi-stage rectangular tooth pressure bearing capacity was $\alpha = 0.88$ [15]。 To verify the method, the saturation magnetization of magnetic fat was selected according to the literature and specified as $M_s = 37.8KA/m$ [16].

3. Models and methods

3.1. Magnetic field models and materials

A magnetic grease seal magnetic field calculation model was established, as shown in Figure 2, including permanent magnet, magnetic pole, pole tooth, magnetic rotation axis, and external air. Under the action of the magnetic field, the magnetic induction intensity under the pole tooth was much higher than the saturation magnetization intensity of the magnetic grease, and its permeability was approximately equal to the vacuum permeability. Therefore, the air was applied to replace the magnetic fat in the model. Similarly, the air was also applied to replace the non-magnetic permeability of the sealed shell. To select the NdFeb magnet, brand N35, whose residual magnetic induction intensity was 1.2T, magnetized along the axis to the direction of the sealed cavity. The magnetic pole, pole tooth, and rotation axis were made of magnetic stainless steel material 12Cr13. The B-H curve of this material is shown in Figure 3.

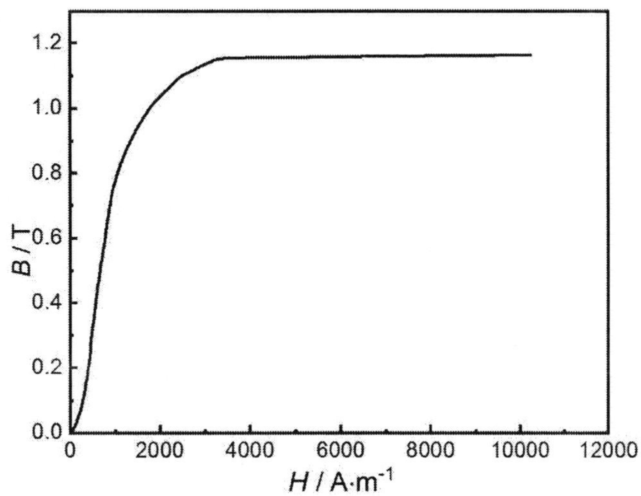

Figure 3. BH curve of magnetic material 12cr13

12Cr13 was an excellent magnetic stainless steel and a good material for magnetic grease seal fittings.Magnetic field intensity mode, magnetic induction intensity mode and relative permeability are obtained from the BH curve of the material.It was worth noting that the relative permeability refers to the slope of this curve, and the BH curve obtained by experimental measurement was usually a series of scatter points. Curve fitting was required to obtain an approximate function curve, and derivation of the slope of the curve was used as the relative permeability of this material. The process of curve fitting was omitted here. In fact, for the nonlinear static magnetic field finite element method, it is an important factor affecting convergence and should be paid attention to.

Table 1 listed the geometric parameters of the model. According to the principle of a single factor, the pressure bearing capacity of different seal clearance, the width of pole tooth, interval length of pole tooth, and height of pole tooth was calculated to optimize the structure of the pole tooth.

By using parametric modeling method, various sizes of polar teeth can be changed conveniently. The control variable method is used to study the effects of width, spacing distance, height and type of polar teeth on the bearing capacity of magnetic grease seal.

Table 1. Geometric parameters of the magnetic field model

IOP Conf. Series: Materials Science and Engineering 1081 (2021) 012002 doi:10.1088/1757-899X/1081/1/012002

Parameters	Values	Meanings
R_{axis}	35	The radius of the rotation axis
gap	0.1-1	Sealing gap
h_m	10.5	The radial thickness of the magnet
l_m	10	Axial thickness of the magnet
L_{axis}	50	Length of the axis
h_1	1.5-3.0	Height of pole teeth
b	1.4-2.8	Width of pole teeth
h_p	15	The radial thickness of the magnetic pole
t_1	1.4-2.8	Distance between pole teeth
l_p	$3 \cdot b + 2 \cdot t_1$	The thickness of pole teeth

3.2. Mesh independence verification

To reduce the calculation result error caused by the mesh, the mesh in Fig 4 was divided. Permanent magnet, magnetic pole, pole tooth, axis, and external air domain use triangular mesh, and set the maximum unit size as 1mm; Quadrilateral mesh is used in the area near the pole tooth to correlate the maximum unit size with the seal clearance and set as 1/x of the seal gap. The number of triangular meshes is 17540, and quadrilateral meshes are the core area of the calculation. Table 2 shows the relationship between the number of quadrilateral meshes and the extreme value of the magnetic induction intensity of pole teeth.

a b

Figure 4. meshes of calculation area (a) global mixed mesh (b) quadrilateral mesh near pole teeth

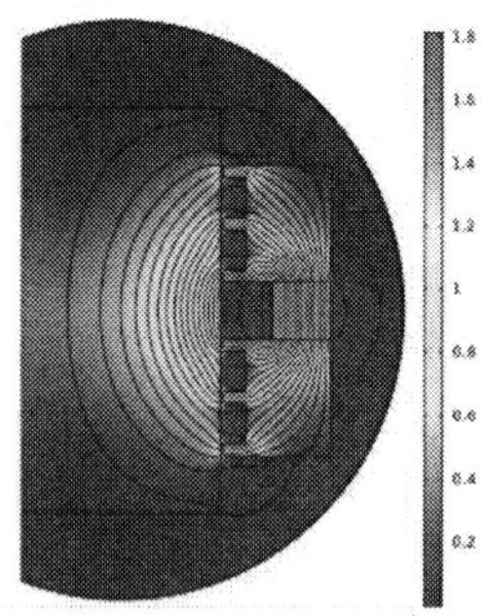

Figure 5. Calculation results of the magnetic field

The fraction of the minimum unit size in the seal clearance was no more than 0.2, the mesh precision met the calculation requirements, and there was no obvious correlation between the calculation results and the mesh.

Table 2. The corresponding relationship between the number of meshes and the magnetic induction intensity of the first pole teeth

Proportion	Maximum units size[mm]	Meshes numbers	Maximum of B[T]	Minimum of B[T]
1	0.5	508	0.815	0.509
0.8	0.4	809	1.063	0.502
0.6	0.3	1541	1.005	0.489
0.4	0.2	3409	0.985	0.490
0.2	0.1	12700	0.987	0.489
0.1	0.05	50800	0.987	0.489

3.3. Results validation

Fig. 5 showed the calculation results of the sealed static magnetic field. There was a strong magnetic induction intensity near the pole tooth. The black curve represented the magnetic induction lines, which were concentrated near the pole teeth, proving the magnetic gathering effect of the pole tooth. The maximum magnetic induction intensity near the pole tooth was about 1.2T, and the minimum magnetic induction intensity of the teeth groove was about 0.2T.The pressure bearing capacity of magnetic grease seal can be calculated by simplifying the pressure bearing capacity formula (3).

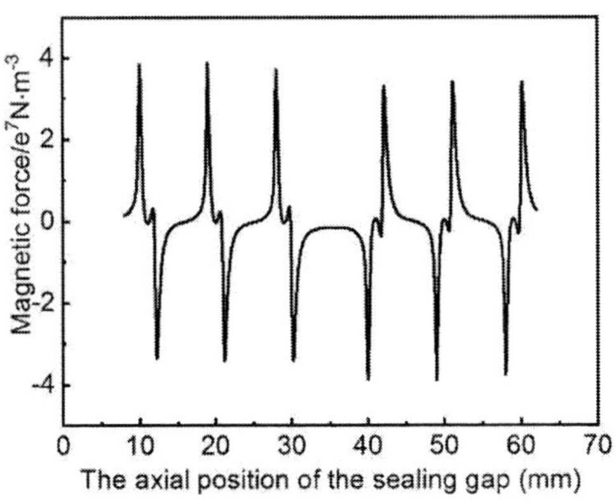

Figure 6. Axial magnetic force in sealing gap of the pole tooth

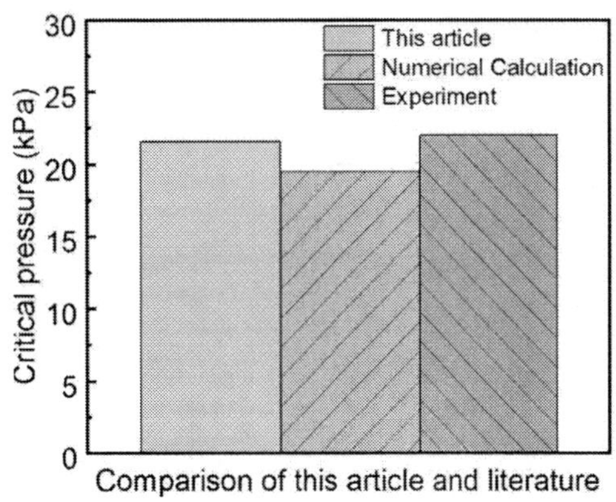

Figure 7. Comparison with existing literature data

The axial magnetic force of seal clearance was shown in Fig. 6. The magnetic force had obvious peaks and valleys, and the negative sign only represents the direction. The magnetic force is calculated according to Formula (4):

$$F_{m} = M_{s} \cdot \nabla B \tag{4}$$

The magnetic field was a potential field, with directional, magnetic field force, and magnetic induction intensity gradient was proportional, so there would be a sign of difference.
Fig 7 showed the difference of the pressure bearing capacity. Compared with the data given in the literature, the same material and structure parameters were used to calculate the sealing pressure capacity of magnetic grease. When the sealing clearance was 0.5mm, the pressure bearing capacity of the single-pole tooth calculated in this paper was 21.56kPa. Compared with 19.5kPa given in the literature[16], the result was 10.56% higher than the experimental data 0.022mpa, 2% lower[17]. The magnetic materials were nonlinear and had different permeability under different magnetic induction intensity, which might result in the difference between the calculated results and the literature data. Through comparative analysis, the numerical calculation method proposed in this paper was more accurate.

4. Result and discussion

4.1. Effect of pole tooth type on bearing capacity

The bearing capacity of single-pole teeth was studied when the pole tooth was a rectangular shape, single-side trapezoid, and symmetric trapezoid. The relationship between the bearing capacity and gap of different types of pole teeth is shown in Fig. 8.
The pressure bearing capacity of the three types of pole teeth was consistent with the changing trend of seal clearance, and the pressure bearing capacity decreased with the increase of the sealing gap. When the sealing gap changed from 0.1mm to 1mm, the change rate of pressure capacity was the same, which was about -12%, that was if the seal clearance increased by 0.1mm, the pressure capacity decreased by about 12%.

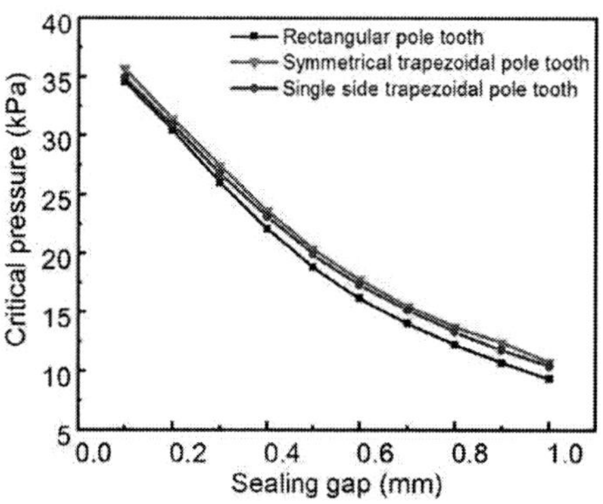

Figure 8. effect of pole teeth type on bearing capacity

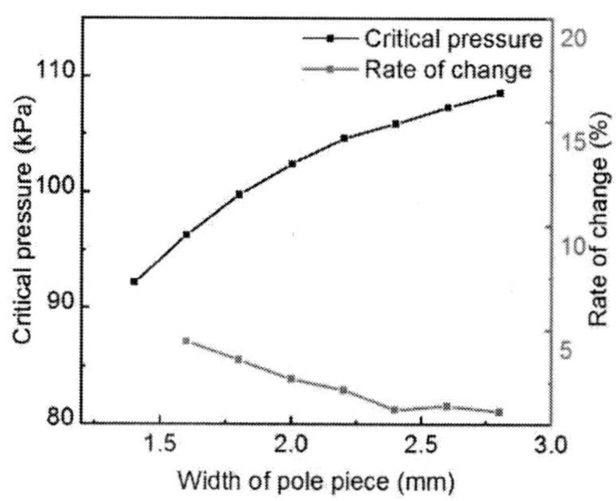

Figure 9. Effect of pole tooth width on bearing capacity

Among the three typical pole teeth, symmetrical trapezoidal pole teeth had the highest bearing capacity, followed by single-side trapezoidal pole teeth and rectangular pole teeth. With the same tooth root width, the end area of the trapezoidal pole tooth was less than the root area, and the magnetic induction line density was higher, that was, the greater the magnetic induction intensity was, the higher the bearing capacity was.

The order of bearing capacity with the same width of the tooth root was studied, which verified the correctness of the conclusion in the literature[12] between 0.1mm and 1mm in the sealing gap. No matter the width of the tooth root or the width of the tooth end was the same, the trapezoidal pole teeth had higher bearing capacity.

4.2. Effect of pole tooth width on bearing capacity

The rectangular pole tooth was selected and the remaining parameters were kept unchanged to study the change in the bearing capacity caused by the change in the width of the pole tooth. The comprehensive bearing capacity of the level-6 pole tooth was shown in Fig. 9.

With the increase of pole tooth width, the bearing capacity increased from 92.187kPa to 108.56kPa, increasing by 17.8%. With 0.2mm as the change interval, the change rate of bearing capacity at each pole tooth width was calculated, decreasing from 4.5% to 1%. This indicated that when the pole tooth width exceeded 2.2mm, the rate of change decreases significantly.

As the width of the pole tooth increases, the magnetic induction line was more likely to pass through the pole tooth, causing the magnetic induction intensity under the pole tooth to rise. According to formula 3, the sealing bearing capacity increased. However, the magnetic energy product of a permanent magnet was determined, so the rate of change of bearing capacity caused by increasing the width of the pole tooth decreased gradually. Compared with six single-stage, pressure bearing ability of multistage pole tooth was small, this was because the multistage pole tooth magnetic induction intensity distribution influenced each other, caused the magnetic induction intensity in the teeth groove minimum increased, according to the formula (3), the bearing capacity was not only related to the maximum, also related to the minimum value, so the bearing capacity to reduce. At the same time, it was noted that the increase of the width of the pole tooth would increase the volume of magnetic grease in the sealing gap. Considering the machining economy and bearing capacity, the optimal range of the width of the pole tooth was 1.8mm to 2.2mm.

4.3. Effect of interval length of pole tooth on bearing capacity

The rectangular pole teeth were selected and the remaining parameters were kept unchanged to study the change in the bearing capacity caused by the change in the spacing of the pole teeth. The comprehensive bearing capacity of the level-6 pole teeth was shown in Fig. 10.

With the increase of interval length of the pole tooth, the pressure bearing capacity increased from 87.318kPa to 114.08kPa by 30.65%. The mutation rate increased to 9.46% when t1=1.8mm, and then dropped to about 2.5%. The magnetic field of tooth groove position of was very small compared with the sealing gap magnetic field, increasing pole tooth interval length, caused by alveolar lines of magnetic induction, magnetic induction line easier to gather at the bottom of the pole tooth, tooth loss of magnetic induction intensity tooth at the bottom of the magnetic induction intensity increased, thereby increasing the bearing pressure ability. Due to the abrupt change in the rate of change, the bearing capacity was greatly improved after this pole tooth interval length. The pole tooth interval exceeding 1.8mm had no great influence on the overall sealing structure, and the optimal range of the pole tooth interval was 1.8mm to 2.0mm.

4.4. Effect of pole tooth height on bearing capacity

The rectangular pole tooth was selected and the remaining parameters were kept unchanged to study the change in the bearing capacity caused by the change in the height of the pole tooth. The comprehensive bearing capacity of the level-6 pole tooth was shown in Fig. 11.

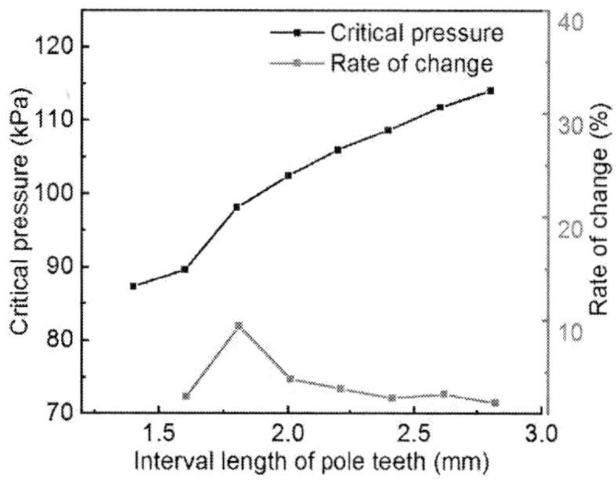

Figure 10. Effect of interval length of pole tooth on bearing capacity

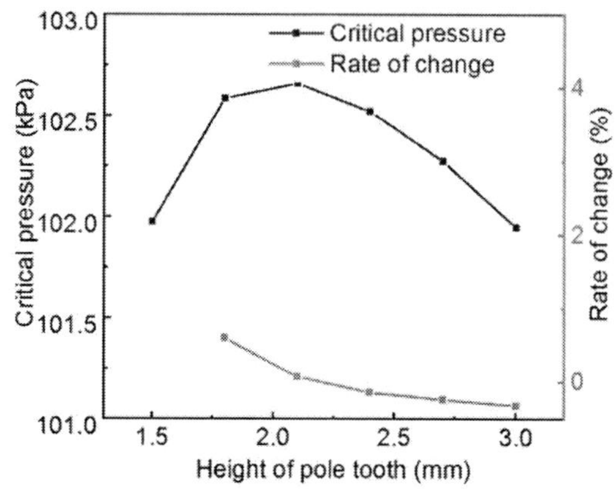

Figure 11. Effect of pole tooth height on bearing capacity

The influence of pole tooth height on the bearing capacity was relatively weak compared with the width and interval length of pole teeth. When the pole tooth height was increased, the bearing capacity was approximately equal to 102.3kPa, which only changed in a small range and the rate of change was always less than 1%. Therefore, it can be considered that the pole tooth height did not influence the bearing capacity. This was because the bearing capacity was related to the magnetic induction intensity and the difference of the magnetic induction intensity, and the pole tooth height did not significantly affect the magnetic induction intensity distribution near the pole tooth. The seal structure design selected many kinds of pole tooth depth, the bearing capacity does not change basically, had the larger design flexibility.

5. Conclusion

In this paper, the nonlinear magnetic field finite element method was adopted to obtain the magnetic induction intensity distribution of the sealing structure according to the single factor principle. Based on the magnetic field force point of view, the sealing mechanism was described, and the influence of the pole tooth structure on the bearing capacity of the magnetic grease seal was studied. The following findings were found:

The order of bearing capacity of three typical pole teeth from large to small was symmetrical trapezoidal pole teeth, single side trapezoidal pole teeth, and rectangular pole teeth. The structure of the pole tooth was able to generate a non-uniform magnetic force, and the maximum magnetic force of the rectangular pole tooth is located at the right angle of the pole tooth.

For every 0.1mm increase in the sealing gap, the pressure bearing capacity decreased by 12%; The pole tooth width increased from 1.5mm to 3mm, and the bearing capacity increased by 17.8%. The pole tooth interval length increased from 1.5mm to 3mm, and the bearing capacity increased by 30.65%. With the increase of pole tooth height, the pressure bearing capacity had little change.

Through the numerical calculation of pole tooth optimization, the better pole tooth parameters were obtained, and the sealing pressure capacity is improved.

This study could guide the design of the magnetic grease seal, optimize the structure of pole teeth, save the space of seal, and improve the bearing capacity of the seal.

Reference

[1] Mimi X and Lin C 2019 Numerical study on the effect of wheel cover seal and wheel disc seal on the performance of centrifugal compressor stages(In Chinese). *J. Chinese Journal of Turbomachinery*.**61**(02)61-68.

[2] Tiejun W 2020 Failure analysis of dry gas seal of centrifugal compressor (In Chinese) *J. Chinese Journal of Turbomachinery*.**62**(S1)44-48.

[3] Shen C, Jun L and Hao X 2020 Multistage centrifugal compressor performance Prediction Software Design (In Chinese) *J. Chinese Journal of Turbomachinery*.**62**(S1)1-4.

[4] Hongwang Z, Chen T, Huarong T and Jian X 2020 Case Analysis of abnormal noise of centrifugal fan (In Chinese) *J. Chinese Journal of Turbomachinery*.**62**(S1):66-68.

[5] Yaru M, Wenhao H and Jing L 2019 Failure and leakage characteristics of labyrinth seals and design of anti-friction structure(In Chinese) *J. Chinese Journal of Turbomachinery*. **61**(05)64-71.

[6] Yuhua L and Guoping S 2006 Analysis and design of H_2S fan seal structure (In Chinese).*J. Fluid Machinery*.**7** 28-31.

[7] Shifu G, Jinping G and Mingheng S 2006 Study on the magnetic fluid seal used at the shaft end of the centrifugal fan (In Chinese). *J. Fluid Machinery*. **2** 19-21.

[8] Yang X 2010 Numerical study on the magnetic field and heat transfer of magnetic grease seals(In Chinese).(Beijing: Beijing University of Chemical Technology Press)

[9] Sheng Z 2010 Experimental study and structural Optimization design of magnetic grease seal(In Chinese). (Beijing: Beijing University of Chemical Technology Press)

[10] Shangbin L, Yu Z and Shuangxi L 2016 Study on magnetic grease seal for the evaporative cooling motor(In Chinese) *J.Large Motor Technology*. **03** 19-23.

[11] Yu Z 2016 Development and Performance Research of magnetic grease seal for flexible pole boots(In Chinese). (Beijing: Beijing University of Chemical Technology Press)

[12] Horak W and Szczech M 2013 Experimental and numerical determination of the static critical pressure in ferrofluid seals. *J. Journal of Physics: Conference Series*. **412**.

[13] Szczech M and Horak W 2015 Tightness testing of rotary ferromagnetic fluid seal working in water environment. *J. Industrial Lubrication and Tribology*. **67**(5)455-9.

[14] Szczech M and Horak W 2017 Numerical Simulation and Experimental Validation of the Critical Pressure Value in Ferromagnetic Fluid Seals. *J. IEEE Transactions on Magnetics*. **53**(7) 1-5.

[15] Szczech M 2019 Magnetic fluid seal critical pressure calculation based on numerical simulations. *J. Simulation. 96*(4)403-13.

[16] Yang X, Qiuxiang Z and Jining C 2009 Pressure Resistance analysis of high viscosity non-Newtonian magnetic fluid seals (In Chinese). *J. .Lubrication and Sealing* **34**(08)42-4+9.

[17] Sheng Z, Qiuxiang Z and Jining C 2010 Experimental study on high viscosity non-Newtonian magnetic grease seal (In Chinese). *J. Lubrication and Sealing*. **35**(04)90-3.

IOP Conf. Series: Materials Science and Engineering 1081 (2021) 012003 doi:10.1088/1757-899X/1081/1/012003

Study on the Influence of High Temperature and High Speed Working Conditions on the Leakage Characteristics of Flexible Bristle Brush Seal

Y H Ma [1], S X Li [1], S Y Zhang [2], H R Liao [1], L X Sha [1]
[1] Mechanical and Electrical Engineering, Beijing University of Chemical Technology, Beijing, China
[2] China Aerospace Standardization Research Institute, Beijing, China
E-mail: buctlsx@126.com

Abstract. Compared with the metal bristles, the flexible bristles can reduce the heat and wear caused by the friction between the bristles bundle and the rotor, and improve the working stability and working life of the brush seal. In this paper, the L-shaped numerical mode is established to analyze leakage characteristics of flexible bristle brush seal. The flow field of brush seal is analyzed, and the effects of differential pressure, rotating speed and temperature on the leakage rate were studied as well. The leakage rate of the flexible bristles brush seal under different pressure and rotating speed was measured and compared with the numerical results. The results show that: Pressure drop mainly occurred in bristles pack; Small pressure difference, high temperature and high speed can improve the sealing performance, which, shows that the flexible bristles brush seal can well adapt to high temperature and high speed conditions; The leakage rate after running-in is slightly lower than that before running-in, and has better sealing performance, and the brush bristles arrangement is complete and tight after test, indicating that the flexible bristle brush seal has the ability of lasting and stable seal; The error between experimental results and numerical simulation results is less than 10%, and the variation trend is consistent, confirming the accuracy of the leakage rate obtained by numerical simulation.

Keywords. Flexible bristle brush seal, Porous media, Experimental investigation, Leakage characteristics

1. Introduction

Sealing technologies commonly used in rotating machinery such as fans and aircraft engines include brush seals, dry gas seals, floating ring seals, etc. [1-5] Brush seal is a flexible contact type dynamic seal device. As a new type of seal, it has developed rapidly in recent years. The flexible bristle of the brush seal can well adapt to the radial runout of the rotor, which can significantly improve the sealing performance and reduce leakage to 10%~20% of the traditional labyrinth seal [6].

The bristle materials used in brush seal are mostly cobalt-based and nickle-based high-temperature superalloy materials [6-9]. High-intensity friction between the metal bristle and the rotor will increase the temperature in the seal cavity and cause serious bristle wear, thus reduce the stability and operating life of brush seals. Brush seals are often applied in the conditions of high temperature and high speed. With the development of sealing technology, brush seals have also begun to choose some non-metallic bristle materials, such as aramid fibers [10-11]. Non-metallic bristle can reduce the overall weight of the brush seal, the temperature rise of the seal cavity, and friction and wear of the bristles, so it can meet the high requirements of modern engineering technology better. Therefore, it is particularly important to study the leakage flow characteristics and mechanism of flexible bristle brush seal under the conditions of high temperature and high speed.

Many scholars have done a lot of research on the leakage flow characteristics of brush seals. Bayley and Long [12] used a linear Darcian model of porous media which only considered the viscosity resistance of the bristle tow to the fluid, and numerically predicted the static leakage of the brush seal. Dogu et al. [13] applied the integral porous media model to numerically predict the leakage of brush seals with clearance and without clearance. Sun Dan et al. [14] numerically predicted the relationship between leakage and pressure ratio and velocity by using a fluid-solid coupling model considering the deformation of the bristle, and he verified it by experiments. Li Shuangxi et al. [15] comprehensively studied the leakage flow characteristics of end face brush seals by using numerical prediction of porous media model and combining it with experimental verification. Zhang yuanqiao et al. [16] used the porous media model to numerically analyze the leakage flow characteristics of brush seal with different back plate structures.

The predecessors mainly studied metal bristle brush seals, but they paid less attention to non-metallic bristle materials, and rarely considered the working conditions of high temperature and high speed of brush seal. In this paper, a carbon fiber flexible bristle material is selected, and the L-shaped numerical model of the leakage flow characteristics of the brush seal is established based on the modified RANS equation and energy equation which solve the Non-Darcian porous media model. This paper analyzes the effects of pressure difference, temperature and velocity on the leakage flow characteristics of carbon fiber brush seals. The leakage rates of carbon fiber brush seals under different pressure differences and rotational speeds are measured experimentally. The experimental results are compared with numerical simulation results to analyze the leakage performance of carbon fiber brush seals under the conditions of high temperature and high speed to meet the requirements of practical engineering applications better.

2. Structure and working principle

As is shown in Figure 1, the brush seal is composed of a front plate, a back plate, and a bristle pack. The front plate is mainly used to fix and protect the bristle pack, and the back plate is used to support the bristle pack. The two fit the bristle pack by interference fitting. The bristle contacts with the rotor to form a sealed interface. The irregularly arranged bristle generate irregular flow channels, causing the fluid flowing through the area of the bristle pack to flow irregularly, increasing the flow resistance, generating a self-sealing effect, and achieving a sealing effect.

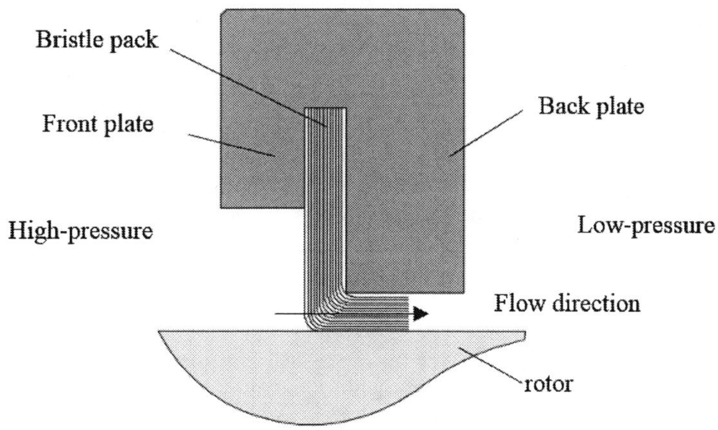

Fig.1 Schematic diagram of brush seal

The interference bristles will rub against the rotor, and the heat generated by the friction will increase the temperature in the seal cavity, affecting the stability of the brush seal operation. At the same time, the bristle wear caused by the friction between the bristle and the rotor will seriously reduce the operating life of the brush seal.

IOP Conf. Series: Materials Science and Engineering 1081 (2021) 012003 doi:10.1088/1757-899X/1081/1/012003

Fig.2 Carbon fiber brush seal

Figure 2 is a physical map of carbon fiber brush seal. The bristle is carbon fiber material. The carbon fiber bristle has the characteristics of high thermal conductivity, wear resistance and high temperature resistance. The flexible carbon fiber bristle can reduce the radial contact force with the rotor and reduce friction heat and friction wear. The flexible carbon fiber bristle has a strong recovery ability after deformation, which can well adapt to the radial runout of the rotor, alleviate the hysteresis effect and reduce leakage.

3. Numerical method and calculation model

3.1 Porous Media Model

The small diameter of bristle in brush seals and the uncertainty of the bristle arrangement make it difficult to solve the flow field inside the bristle pack. In the numerical study of the leakage characteristics of brush seals, the most widely used and reliable method is to treat the bristle pack area into a porous medium to simplify the flow of fluid inside the bristle pack. The upstream and downstream fluid in the bristle pack area is a compressible turbulent flow. The bristle pack area is processed into anisotropic materials, and the momentum equation and the continuity equation are calculated simultaneously. As shown in formulas (1) and (2):

$$\frac{\partial \rho}{\partial t}+\frac{\partial(\rho u_{i})}{\partial x_{i}}=0 \tag{1}$$

$$\frac{\partial(\rho u_{i})}{\partial t}+\frac{\partial(\rho u_{i}u_{j})}{\partial x_{j}}=-\frac{\partial p}{\partial x_{i}}+\frac{\partial \tau_{ij}}{\partial x_{ij}} \tag{2}$$

When the fluid flows through the bristle pack, it will be hindered by the bristle, causing a pressure drop in the fluid, and the density of the gas in the bristle pack will also change. Therefore, the fluid flowing inside the bristle pack is treated as an ideal compressible gas, and the state equation of ideal gas is as follows:

$$p=\rho RT \tag{3}$$

When the fluid flows through the bristle pack, the resistance of the bristle to the fluid is treated as an additional momentum source term. The momentum equation is amended as follows:

$$\frac{\partial(\rho u_{i})}{\partial t}+\frac{\partial(\rho u_{i}u_{j})}{\partial x_{j}}=-\frac{\partial p}{\partial x_{i}}+\frac{\partial \tau_{ij}}{\partial x_{ij}}+F_{i} \tag{4}$$

In the equation: $F_i=-(\sum_{j=1}^{2} D_{ij}\mu v_j+\sum_{j=1}^{2} C_{ij}\frac{1}{2}\rho|v|v_j)$, F_i is the momentum source term; D_{ij} is the viscous resistance coefficient matrix in the bristle pack region; C_{ij} is the inertial resistance coefficient matrix in the bristle pack region.

The viscous resistance coefficient αi and the inertia resistance coefficient β_i can be measured through experiments or can be obtained through the following empirical formulas [17]:

$$\alpha_i = \frac{150}{d^2}\frac{(1-\varepsilon)^2}{\varepsilon^3} \tag{5}$$

$$\beta_i = \frac{7}{2d}\frac{1-\varepsilon}{\varepsilon^3} \tag{6}$$

In the two equations, d is the diameter of the bristle; ε is the porosity of the bristle pack.

Porosity is defined as the ratio of the void volume in the porous medium to the total volume of the porous medium. The porosity of the bristle pack ε can be defined by the bristle arrangement density N. When the bristle arrangement does not have a radial tilt angle:

$$\varepsilon = 1 - \frac{\pi d^2 N}{4L} \tag{7}$$

In the equation, d is the diameter of the bristle ; N is the bristle arrangement density; L is the thickness of bristle pack.

3.2 Calculation Model

Figure 3 is the structural diagram of the L-shaped carbon fiber brush seal. It can be seen that the carbon fiber brush seal and its internal fluid are symmetrical in the circumferential direction. Therefore, the calculation model is simplified by selecting the arc of 1° in the circumferential direction to reduce computing time.

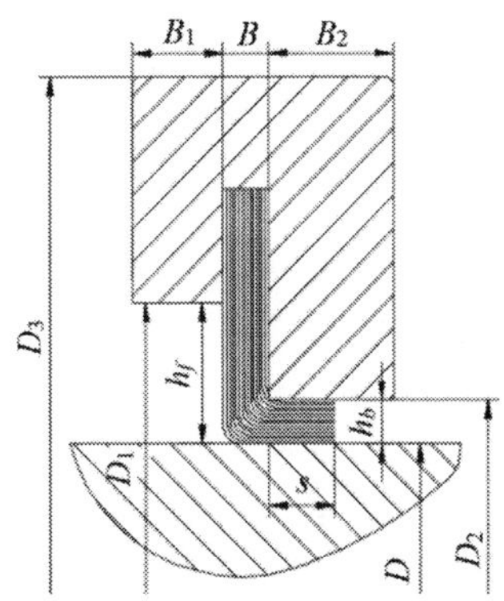

Fig.3 Structural diagram of L-shaped carbon fiber brush seal

The numerical calculation model of flexible bristle brush seal is established, as shown in Figure 4 (a). The model includes a common fluid region and a bristle pack region treated as a porous medium. There is a retracting effect when the fluid enters and exits the brush seal, and a part of the fluid domain needs to be added at the entrance and exit of the brush seal. The calculation model is meshed, as shown in Figure 4 (b). The area framed by red wire is the bristle area. Through mesh independence verification, as shown in Table 1, the optimal number of grids is determined to be 259470.

Table1 Mesh independence verification

Number of mesh	Leakage rate/g·s^{-1}
14777	0.24
31203	0.253
127470	0.367
259407	0.375
595150	0.375
2001700	0.376

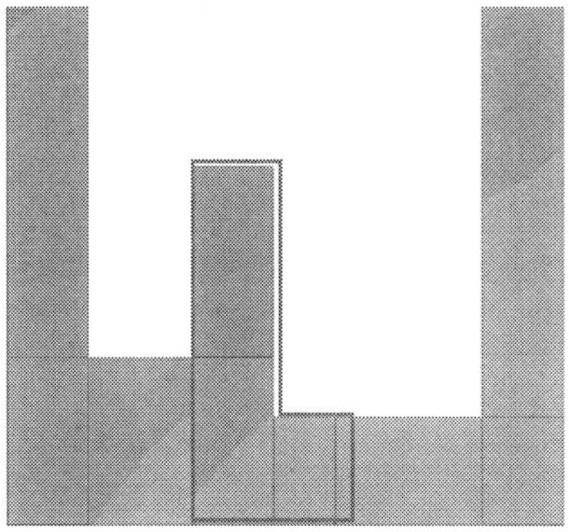

(a) Model diagram

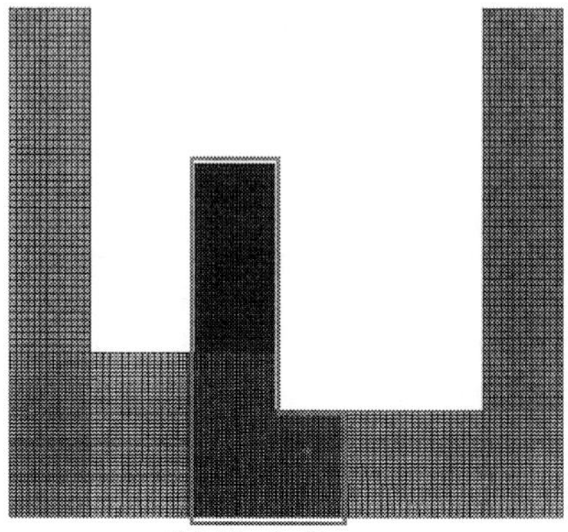

(b) Mesh diagram

Fig.4 Mesh map

The values of the geometric parameters for the numerical calculation of the brush seal and the attribute of the bristle material are shown in Table 2, where the bristle interference is reflected by the bristle length.

Table2 Main structural dimensions

Classification	Value/mm
Front plate clearance h_f	2.5
Back plate clearance h_b	0.8-1.4
Bristle diameter d	0.006
Bristle pack thickness B	1.0
Bristle length (interference) s	1.0
Sealing diameter D	80.6

Numerical calculation and the bristle material properties of the test piece are shown in Table 3.

Table3 Carbon fiber bristle material properties

Classification	Value
Thermal conductivity λ /W·(m·k)$^{-1}$	35.127
Density ρ /g·cm^{-3}	1.802
Specific heat capacity C /J·(kg·k)$^{-1}$	753.44
Tensile modulus M /GPa	226
Coefficient of thermal expansion γ /k^{-1}	-0.56×10^{-6}

3.3 Boundary setting

In this section, Fluent software is used for analysis and calculation, and the standard k-ε equation turbulence model is used to simulate ordinary fluid domains. The bristle pack adopts the laminar flow model and the porous medium model. The whole model is calculated by the simple algorithm, and the ideal compressible air is selected as the sealing medium. The wall surface in contact with the rotor is set as a rotation boundary. The wall surface rotating in the circumferential direction of the rotor is set as a periodic boundary. The remaining wall surfaces are set as non-slipping boundary. Table 4 shows the specific values of the boundary conditions of numerical simulation.

Table4 Boundary condition

Classification	Value
Inlet pressure P_j/MPa	0-1.0
Outlet pressure P_c /MPa	0

IOP Conf. Series: Materials Science and Engineering 1081 (2021) 012003 doi:10.1088/1757-899X/1081/1/012003

Inlet temperature T_j/K	300
Outlet temperature T_c/K	300
Rotor speed n/r·min^{-1}	0-38000

3.4 Numerical Analysis Results

3.4.1 Flow Field Analysis Results. The structural parameters and working condition parameters of flow field analysis numerical model of the carbon fiber brush seal are shown in Table 5.

Table 5 Flow field analysis parameters

Classification	Value
Front plate clearance h_f /mm	2.5
Back plate clearance h_b /mm	1.5
Bristle diameter d/mm	0.006
Bristle thickness B /mm	1.0
Bristle length (interference) s	1.0
Inlet pressure P_j/MPa	0.3
Outlet pressure Pc /MPa	0
Rotor speed n/r·min^{-1}	10000

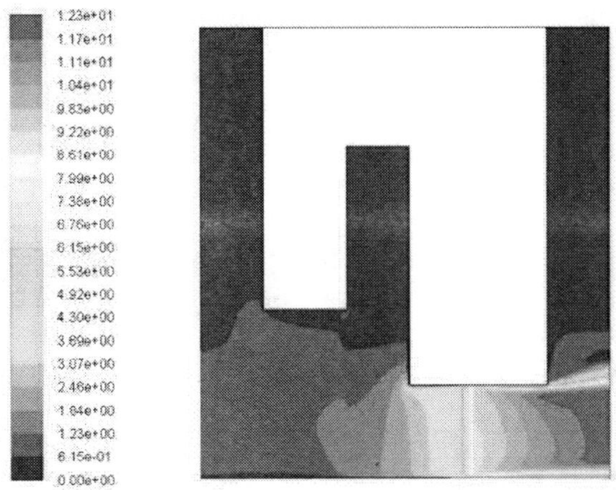

Fig.5 Velocity nephogram

The velocity nephogram of the fluid is shown in Figure 5. The velocity of the fluid hardly changes before entering the bristle pack region, and increases gradually after entering the bristle pack region. The velocity increases to the maximum at the exit of the back plate. After the fluid enters the bristle pack region, the flow channel suddenly narrows, which gradually increases the flow velocity. When the fluid reaches the exit of the back plate, the flow channel and flow direction suddenly change, further accelerating the fluid flow and making the flow velocity reach the maximum. The friction between the bristle and the rotor generate heat, which increases the temperature and accelerate the flow of fluid as well.

Figure 6 is a part of the velocity vector map of the fluid. When the fluid passes through the bristle pack, the flow velocity and direction of the fluid is deflected due to the sudden narrowing of the flow channel in the bristle pack, and the radial velocity increases rapidly. At this time, the radial force that the bristle pack receives increases, pressing the bristle toward the rotor, increasing the friction and wear of the bristle and the rotor, and causing the brush seal to have a "blow down effect". At the same time, the porosity of the bristle pack also decreases, reducing the leakage rate of the brush seal.

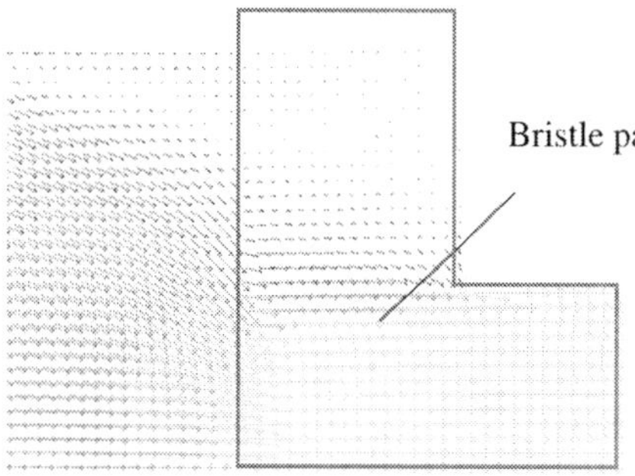

Fig.6 Velocity vector map

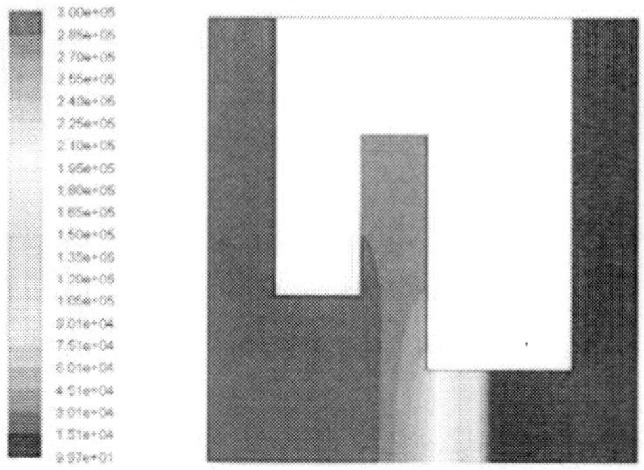

Fig.7 Pressure nephogram

It can be seen from the pressure nephogram in Figure 7, that the overall pressure decreases along the fluid flow direction. Because the bristle pack area has a large resistance to the fluid flow, there is a significant pressure drop. There is no flow resistance in the inlet area and the outlet area, so there is no significant pressure drop. This shows that the characteristics of the bristle pack have a very important influence on the sealing performance.

3.4.2 Leakage rate analysis results. Figure 8 is the varying curves of the leakage rate of brush seals with four different back plate gaps with pressure difference. As the pressure difference increases, the leakage rates increase approximately linearly. When the pressure difference increases from 0.05 MPa to 0.7 MPa, the curve of the brush seal leakage rate with a back plate gap of 0.8 mm varies significantly less than that of the brush seal with a back plate gap of 1.4 mm. It indicates that for the brush seal with a wider back plate gap, the leakage rate will increase faster when pressure difference increases. Reducing the pressure difference at both ends of the brush seal is an important measure to improve the sealing performance of the brush seal. Where conditions permit, multi-level brush seals can be arranged to share the pressure difference at different levels, so that each levle of the brush seal is in a differential environment with lower pressure.

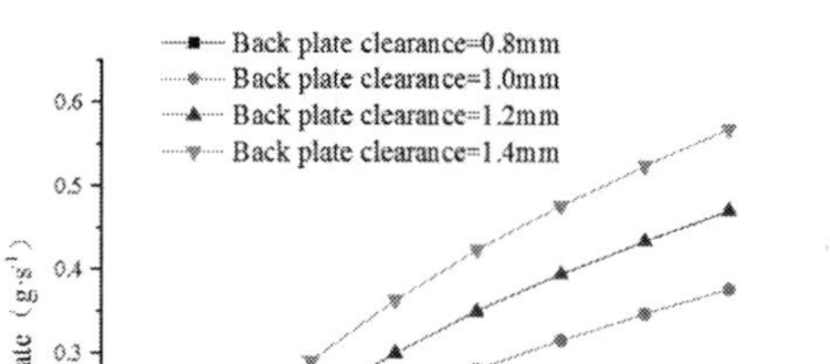

Fig.8 Effect of pressure difference on leakage rate

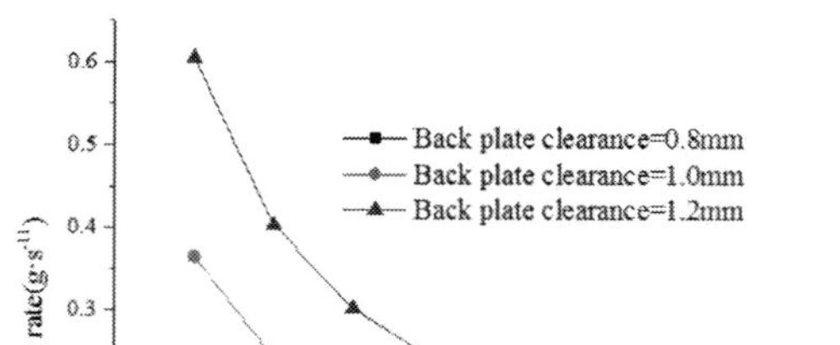

Fig.9 Effect of temperature on leakage rate

Figure 9 is the varying curves of the leakage rate of brush seals with three different back plate gaps with temperature. As the temperature increases, the leakage rates decrease approximately parabolically, because when the temperature increases, the gas density decreases and the viscosity increases, which increases the flow resistance in the bristle pack area and gradually reduces the leakage rate. When the temperature continues to rise, the density change and viscosity change of the gas will reach the peak, the flow resistance in the bristle pack area will no longer change, and the leakage rate will stabilize.

When the temperature increases from 200 K to 900 K, the curve of the brush seal leakage rate with a back plate gap of 0.8 mm changes significantly less than that of the brush seal with a back plate gap of 1.2 mm, indicating that for a brush seal with a wide back plate gap, the effect of the temperature change on the leakage rate will be greater.

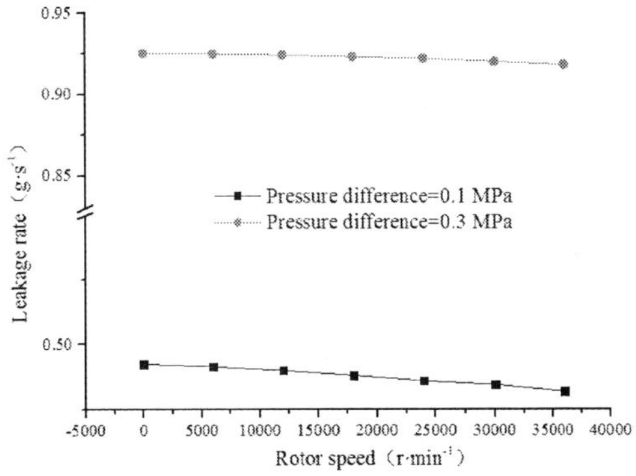

Fig. 10 Effect of rotor speed on leakage rate

Figure 10 is the changing curve of the leakage rate with the speed under two pressure differences. With the increase of the rotor speed, the leakage rates both decrease slightly, and finally stabilize, and the degree of the decline is not obvious. Increasing the rotor speed can reduce the leakage rate of the brush seal, but it has little effect.

4. Experimental research

4.1 Test device.

Figure 11 shows a rotating test bench of the brush seal, which is mainly composed of a sealed chamber system, a power system, a gas path system, and a measurement system. The sealed chamber system includes chamber on low-pressure side, carbon fiber brush seal, etc. The power system includes inverter, motor, rotor, etc. The gas system includes compressor, pressure regulator, gas pipeline, etc. The measurement system includes pressure gauge, flowmeter, temperature thermocouple, etc.

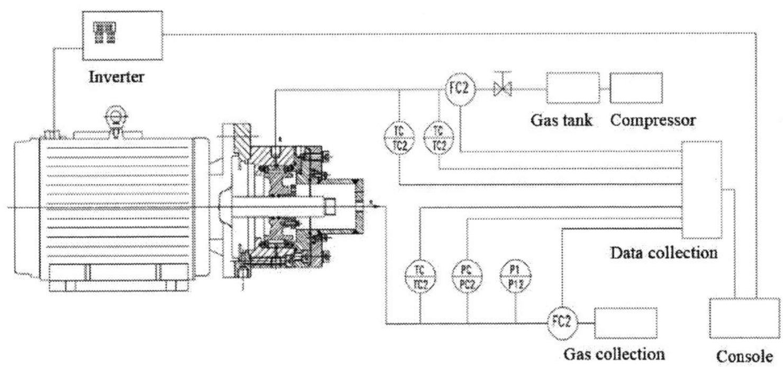

(a)

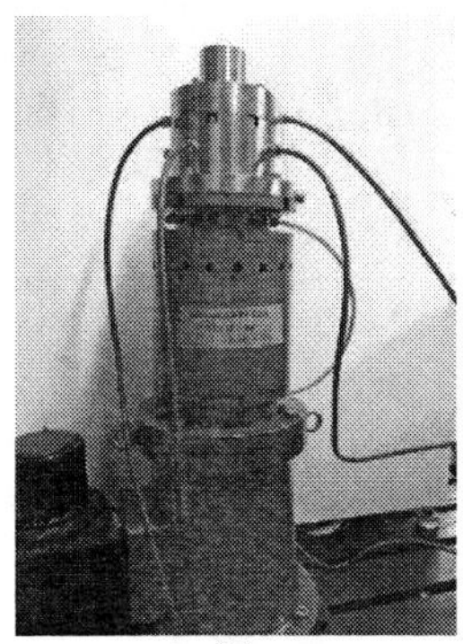

(b)

Fig.11 Brush seal test bench

4.2 Test process

The compressor presses the air into the surge tank, and the pressure regulator is adjusted to the required pressure, and the high-pressure gas is injected into the sealed test chamber (high-pressure side) through the air inlet pipe, and the high-pressure gas leaks to the sealed leakage chamber (low-pressure side)

IOP Conf. Series: Materials Science and Engineering 1081 (2021) 012003 doi:10.1088/1757-899X/1081/1/012003

through the bristle pack. The leaked gas is discharged into the atmosphere through the exhaust pipe connected to the flowmeter, and the leakage gas is measured by the flowmeter. The test chamber and the leakage chamber are connected with a pressure gauge and a temperature thermocouple to measure the inlet and outlet pressure and temperature. In dynamic tests, the inverter can be adjusted to change the frequency, and then the required motor speed is obtained.

4.3 Test piece

The test piece for this test is a carbon fiber brush seal designed and processed independently. The front plate and back plate are made of S30408 stainless steel, and the bristle is made of carbon fiber, as shown in Figure 12.

Fig.12 Brush seal test piece

The structural parameters of the carbon fiber brush seal for the test are shown in Table 6. The test measured the leakage rates of carbon fiber brush seals before and after running-in, and analyzed the influence of pressure difference and speed on the leakage of carbon fiber brush seals before and after running-in.

Table 6 Structural parameters of test piece

Classification	Value /mm
Front and back plate outer diameterD_3	93.0
Front and back plate inside diameter D_1/D_2	82.2
Front and back plate clearance h_f/h_b	0.8
Brush seal thickness	4.5
Bristle length (interference) s	1.2
Bristle pack thickness B	0.6

4.4 Test results

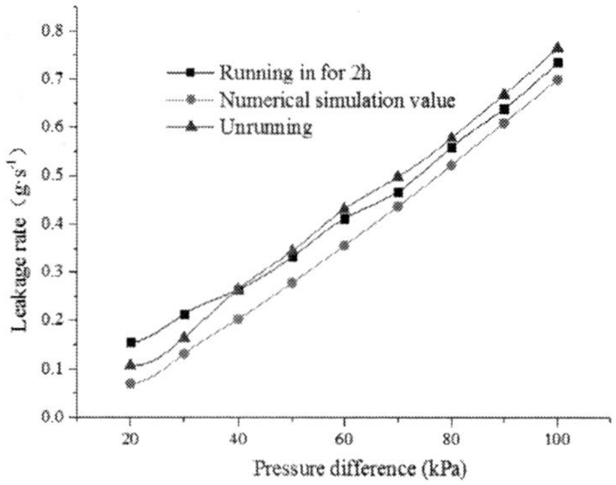

Fig.13 Effect of pressure difference on static leakage rate

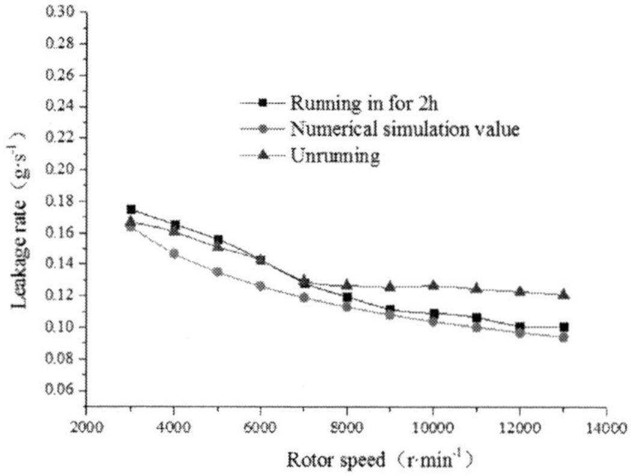

Fig.14 Effect of rotational speed on leakage rate

Figure 13 is the changing curve of the leakage rate with the pressure difference under the static condition. The trend of leakage rate of brush seals before and after the running-in is similar to that in numerical simulation, and both increase approximately linearly with the increase of the pressure difference.

Figure 14 is the varying curve of the leakage rate with the rotor speed under the pressure difference of 30 kPa. The change of the leakage rate before and after the running-in is approximately consistent with the numerical simulation, and both gradually decrease with the increase of the rotor speed, and finally stabilize.

In Figures 13 and 14, due to factors such as machining accuracy and mounting errors of the test pieces, the test values before and after the running-in are slightly larger than the simulated values. But the error remains within 10%, which proves the correctness of the theoretical simulation. Before the running-in of brush seals, the bristle does not adhere closely to the surface of the rotor, causing local gaps, resulting in a lower leakage rate and better sealing performance after running-in than before running-in, indicating that the flexible bristle brush seal has the potential for durable and stable sealing. However, before the pressure difference of 40 kPa in Figure 13 and the rotor speed of 5000 $r \cdot min^{-1}$ in Figure 14, the leakage rate after running-in is higher than that before running-in. This was caused by the fact that the brush seal was not reassembled and the deformed bristle was not recovered in time during the test after running-in.

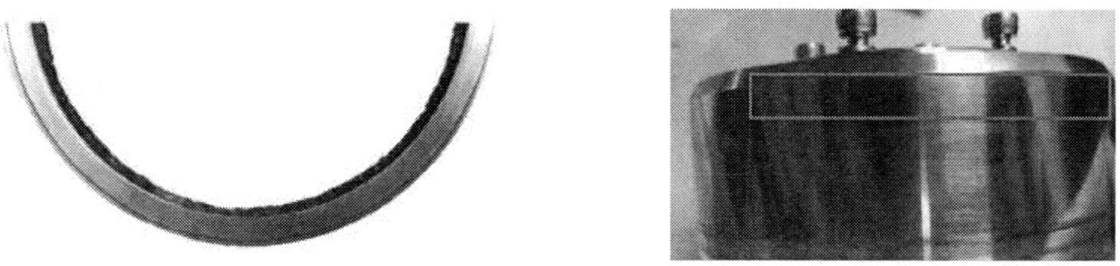

（a）Brush seal （b）Rotor

Fig.15 Brush seal after test

It can be seen from figure 15 that the bristle arrangement after the test is complete and tight, and there is no serious wear on the rotor surface, indicating that the brush seal has the ability to maintain good sealing performance for a long time.

5. Conclusion

1. The flow velocity of the fluid is low before entering the bristle pack region, and gradually increases after entering the bristle pack region, and the flow velocity reaches the maximum at the exit of the back plate. The pressure distribution in the upstream and downstream fluid regions is uniform, and the pressure drop mainly occurs in the bristle pack regions, which indicates that the bristle characteristics play an important role in sealing performance.

2. Small pressure difference, high temperature, and high rotor speed conditions can improve the sealing performance. The effect of rotor speed is lower, indicating that the flexible bristle brush seal can well adapt to high temperature and high rotor speed conditions and meet the requirements of practical engineering applications.

3. The test shows that the leakage rate after running-in is slightly smaller than that before running-in, and the sealing performance is better. The bristle arrangement is complete and tight after the test, which indicates that the flexible bristle brush seal has the ability of durable and stable sealing.

4. The variation trend of the measured leakage rate with pressure difference and rotor speed is the same as the numerical simulation result. The error between the test result and the numerical simulation result is less than 10%, and the changing trend is approximately the same, which fully confirms the accuracy of the leakage rate change law obtained by numerical simulation.

References

[1] Li G Q, Sun D, Jiao Z Z, Du C Y, Bai L 2019. Numerical Study on Inter-stage Pressure and Heat Transfer Characteristics of Multi-stage Low Hysteresis Brush Seal, *Chinese Journal of Turbomachinery*, **61**(05)56-63.

[2] Ma Y R et al. 2019. Investigations on the Leakage Performance of Wear Failure and Anti-rubbing Structure Design for a Labyrinth Seal, *Chinese Journal of Turbomachinery*, **61** (05) 64-71.

[3] He Z H, Xue Z H, Liu X F, Wang T Z 2019. The Rotor Dynamic Analysis of Centrifugal Pump Rotor System, *Chinese Journal of Turbomachinery*, **61**(S1)65-70.

[4] Xue M M, Cheng L 2019. Numerical Research on the Influence of Shroud Labyrinth Seal and Hub Labyrinth Seal on the Performance of Centrifugal Compressor Stage, *Chinese Journal of Turbomachinery*, **61** (02) 61-68.

[5] Jia X Y, Zhou J, An Z Q, Zheng Q, Zhang H 2019. Effect of the Rim Seal Configurations on the Hot Gas, *Chinese Journal of Turbomachinery,* **61**(02):50-60+5.

[6] He L D, Yuan X, Yin X 2001. The Advance in the Investigation of Brush Seal, *Proceedings of the Csee*, (12) 29-33+54.

[7] Wang Z S, Wang Z H 2005. Development and Prospect of Brush Seal in Aero-engine, *Lubrication Engineering*, (05) 203-205+209.

[8] Dogu Y, Aksit M F 2005. Brush seal temperature distribution analysis. *ASME Turbo Expo 2005: Power for Land, Sea, and Air*, June 6-9, Reno-Tahoe, USA.

[9] Stephen D, Hogg S I 2003. Development of brush seal technology for steam turbine retrofit applications. *Proceedings of IJPGC03*, June 16-19,2003, Atlanta, USA.

[10] Ruggiero E J, Allen J, Demiroglu M, et al. 2007. Heat generation characteristic of a Kevlar fiber brush seal.*43rd AIAA/ASME/SAE/A-SEE Joint Propulsion Conference & Exhibit*, July 8-11, Cincinnati, USA.

[11] Ruggiero E J, Allen J, Lusted M 2009. Experimental testing techniques for Kevlar fiber brush seal. *Proceeding of ASME Turbo Expo 2009: Power for Land, Sea and Air*, June 8-12, Orlando, USA.

[12] Bayley F J, Long C A 1993. A combined experimental and theoretical study of flow and pressure distributions in a brush seal. *ASME Journal of Engineering for Gas Turbines and Power*,**115**(2)404-410.

[13] Dogu Y, Aksit M F, Demiroglu M 2008. Evaluation of flow behavior clearance brush seals. *ASME Journal of Engineering for Gas Turbines and Power*,**130**(1)012507.

[14] Sun D, Ding H Y et al. 2018. Theory and experiment on the leakage characteristics of brush seals based on fluid-structure interaction. *Journal of Aerospace Power*,**34**(7)1519-1529.

[15] Li S X, Wang W W et al. 2012. Property Analysis of Face Brush Seal. *Fluid Machinery*, **40**(6)45-50+32

[16] Zhang T Q, Wang Y et al. 2017. Investigations on the Leakage and Heart transfer Characteristics of Brush Seal Part 1:Leakage Characteristics. *Journal of Engineering Thermophysics,* **38**(03):482-489.

[17] Wang W W 2011. Research on Brush Seal Characteristics and Analysis of Influencing Factors. *Beijing University of Chemical Technology.*

IOP Conf. Series: Materials Science and Engineering 1081 (2021) 012004 doi:10.1088/1757-899X/1081/1/012004

Experimental Study on the Performance of the Gas Centrifugal Seal with Continuous Injection and Discharge of Sealing Fluid

Chen Kunyi[1], Li Shuangxi[1]

1 College of Mechanical and Electrical Engineering, Beijing University of Chemical Technology, Beijing, China

E-mail: buctlsx@126.com

Abstract. The centrifugal seal has the advantages of simple structure, long service life and the cost is far lower than the dry gas seal. But under high-speed gas conditions, the temperature of the sealing fluid rises too high and low sealing pressure. This paper studies the gas centrifugal seal with continuous injection and discharge of sealing fluid, establishes a numerical model, simulates the gas-liquid two-phase flow field and temperature field in the sealing chamber. Revealing the effects of the speed and the sealing fluid flux on the sealing performance by comparing the changes in the sealing fluid pressure, temperature, and sealing capacity under different operating conditions. And the reliability of its application to gas mechanical seals is verified through experiments. The results show that at high speeds, the pumping compression effect of the sealing fluid is enhanced, and the sealing capability is improved. Within a certain range, increasing the flux of the sealing fluid can effectively reduce the temperature and prevent the sealing fluid from vaporizing, the continuous injection and discharge structure has a significant effect.

1. Introduction

Centrifugal seal is a device that pumps leaked liquid to high pressure area or forms a liquid barrier to prevent the leakage of high pressure gas through the rotation of rotating parts [1], so as to prevent the leakage of liquid or gas from high pressure to low pressure. As one of the traditional fluid dynamic seals, centrifugal seals have the characteristics of simple structure, no friction and wear of the main seals, and long life compared with commonly used dry gas seal [2-4] and labyrinth seal [5-7]. They have achieved good results in sealing liquids, have good sealing capabilities, and have been widely used in various seals similar to centrifugal pumps. However, when sealing the gas, due to the high-speed rotation of the impeller, the sealing fluid is driven to generate a large amount of stirring heat. This part of the heat cannot be taken away in time, which causes the temperature of the sealing fluid to rise rapidly, or even vaporize, leading to seal failure [8], which severely limits the further development of centrifugal seals in the field of gas seals. The centrifugal seal with continuous injection and discharge of sealing fluid can effectively remove the stirring heat, overcome the problem of the temperature rise of the sealing fluid, improve the sealing ability, and solve the disadvantages of the centrifugal seal application under high-speed gas conditions. Research on it has great significance for promoting the application of the centrifugal seal in the field of gas sealing and energy saving.

The research on centrifugal seals is mainly based on the engineering modification of seals on high-speed rotating equipment such as pump shafts and fans. In recent years, with the development of computational fluid dynamics and computer numerical simulation software, domestic research on the analysis mechanism of internal flow in centrifugal seals has gradually increased. Young C et al. [9] introduced the basic principle of the VOF method in CFD and its advantages in two-phase flow analysis, and used FLUENT software to analyze the two-phase flow field of the centrifugal seal in the steam turbine. Denecke J [10] then studied the geometric model of centrifugal seals in steam turbines, and compared the positions of oil and gas boundaries for centrifugal seals in steam turbines by theoretical calculations and numerical simulation analysis. At the same time, a test system was established and the above-mentioned tests were verified by experiments. But the author did not analyze the heat balance of the centrifugal seal in detail. Jianping Yuan et al. [11] numerically simulated the flow field inside the centrifugal pump with the auxiliary impeller seal, and used a steady-state model to analyze the sealing mechanism and internal flow conditions. The results show that the auxiliary impeller seal has a slight effect on the flow field in the main impeller and the volute. The slight effect, which shows that the centrifugal pump with the auxiliary impeller seal has a higher pressure head but a lower efficiency, and this change becomes more obvious as the flux increases. Long Cai [12] performed numerical simulations on four different sizes of back blades and analyzed the influence of the sealing gap on the sealing pressure. It was found that as the sealing gap decreases, the seal pressure of back blades with different sizes showed an increasing trend. Shujuan Li [13] used a combination of numerical simulation and orthogonal test to analyze the influence of the structural parameters of the auxiliary impeller centrifugal seal on the sealing performance and pump performance, and verified the feasibility of the orthogonal test in analyzing the sealing structure by experiments.

Based on this, this paper uses Fluent software to study the gas-liquid two-phase flow field and temperature field of the centrifugal seal with continuous injection and discharge of sealing fluid, so as to analyze the influence of operating parameters on the sealing performance parameters. Through research, it has been shown that the gas centrifugal seal with continuous injection and discharge of sealing fluid can effectively solve the problems of small sealing pressure and fast temperature rise of the centrifugal seal under high-speed gas conditions.

2. Working principle

The structure of the continuous injection and discharge centrifugal seal is shown in Figure 1. When the gas device is running, the rotating shaft drives the centrifugal impeller to rotate at high speed to promote the flux of the sealing fluid on both sides of the impeller. The sealing fluid with a higher pressure is injected into the sealing chamber from the inlet, flows through the centrifugal impeller through the water groove E on the high pressure side and flows out from the water groove F on the low pressure side. Due to the pressure difference between the two sides of the gas, the fluid in the gap between the impeller and the sealing chamber forms a rotary siphon because of rotation, thereby achieving a non-contact radial seal. Finally, the sealing fluid flowing from F will be recovered from the sealing fluid outlet to the liquid source in the control system.

When the main engine is stopped or running at low speed, a pneumatically controlled sliding valve type parking seal plays a role. The parking seal is an end face mechanical seal, and the axial movement of the spool valve is achieved through a spring and a control gas. At low speeds, the spring force closes the end face of the slide valve and the end face of the impeller to achieve the parking seal. When running at high speed, the spool valve is pushed open by introducing control gas to achieve centrifugal sealing.

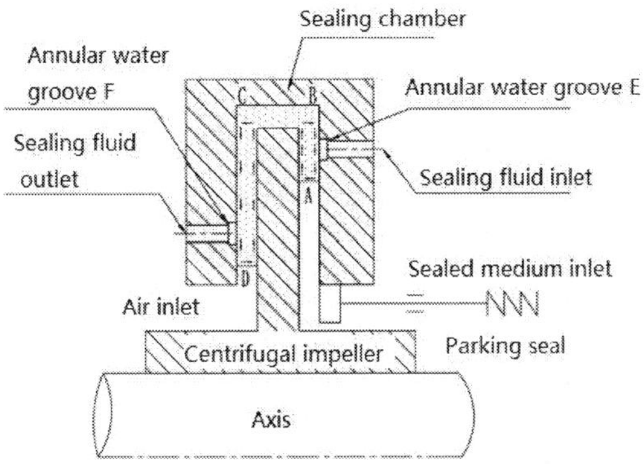

Figure 1. The structure diagram of the centrifugal seal with continuous injection and discharge of sealing fluid

In order to calculate the sealing ability of the performance of the centrifugal seal with continuous injection and discharge of sealing fluid, the sealing pressure formula of the seal is derived based on the momentum equation below. Take a fluid element at a radius R, from the momentum equation:

$$\frac{dp}{dR} = \rho R \omega^2 \tag{1-1}$$

Assuming the fluid is incompressible, integrate equation (1-1) from A to B and from D to C:

$$p_A - p_B = \frac{\rho}{2}\omega_2^2(R_A^2 - R_B^2) \tag{1-2}$$

$$p_D - p_C = \frac{\rho}{2}\omega_1^2(R_D^2 - R_C^2) \tag{1-3}$$

Among them, A and D respectively indicate the positions of the gas-liquid interface on the high-pressure medium side and the low-pressure atmospheric side. pA is the medium pressure, pD is the atmospheric pressure. B and C are the inner walls of the sealing chamber, and pB = pC. ω1 is the angular velocity of the liquid ring in the chamber 1, ω2 is the angular velocity of the liquid ring in the chamber 2. And ρ is the density of the blocking liquid. Generally, the difference between ω1 and ω2 can be ignored [14]. When ω1 = ω2, according to (1-2), (1-3) can get:

$$\Delta p = p_A - p_D = \frac{\rho}{2}\omega^2(R_A^2 - R_D^2) \tag{1-4}$$

Under normal circumstances, the speed and the density of the sealing fluid are known. It can be seen that determining the position of the gas-liquid interface is not only a measure of whether the seal has failed, but also a key factor in determining the sealing pressure value.

In the above derivation process, the assumption that the liquid ring angular velocity and the impeller angular velocity on both sides are equal is actually used. In fact, there is a certain difference between the liquid ring angular velocity and the impeller angle due to the viscosity of the liquid. In order to consider the impact of introducing this assumption, Introducing a opposite pressure coefficient k [15], the equation becomes:

$$\Delta p' = k \bullet \Delta p = k\frac{\rho}{2}\omega^2(R_A^2 - R_D^2) \tag{1-5}$$

In general, due to the viscosity of the liquid, there is a speed difference between the rotating part and the stationary part. The actual value of the sealing pressure is smaller than the theoretical value. Therefore, some scholars also refer to the opposite pressure coefficient k as the pressure reduction coefficient [16]. In fact, the opposite pressure coefficient k reflects the influence of the annular vortex inside the sealing chamber on the average angular velocity of the liquid ring. The larger the opposite pressure coefficient, the higher the degree that the impeller drive the liquid, the better the sealing effect.

Through the above analysis, the key performance parameters of the centrifugal seal with continuous injection and discharge of sealing fluid can be determined as the position of the air-liquid interface on the high and low pressure side, the temperature of the sealing fluid, the pressure of the medium, and the opposite pressure coefficient.

3. Numerical Simulation

The centrifugal seal with continuous injection and discharge of sealing fluid simultaneously achieves the purpose of preventing the leakage of the sealing medium and cooling the sealing chamber by continuously injecting and discharging the sealing fluid, but it also makes the fluid in the sealing chamber in a state of high-speed and high-pressure gas-liquid two-phase flow.

In order to simplify the calculation, the following assumptions are made [17]:

(1) The sealing fluid used in the calculation is incompressible fluid by default, and the sealing medium and the atmosphere are air of constant density by default.

(2) During the sealing process, the sealing fluid and the sealing medium will not undergo chemical reaction and physical phase change by default.

3.1. Geometric Model

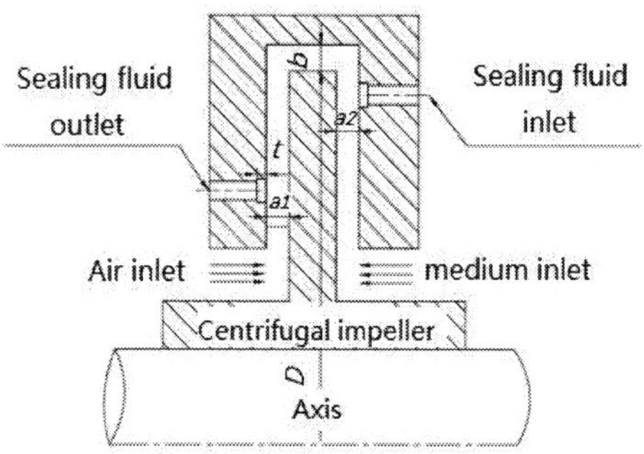

Figure 2. Fundamental diagram of the centrifugal seal with continuous injection and discharge of sealing fluid

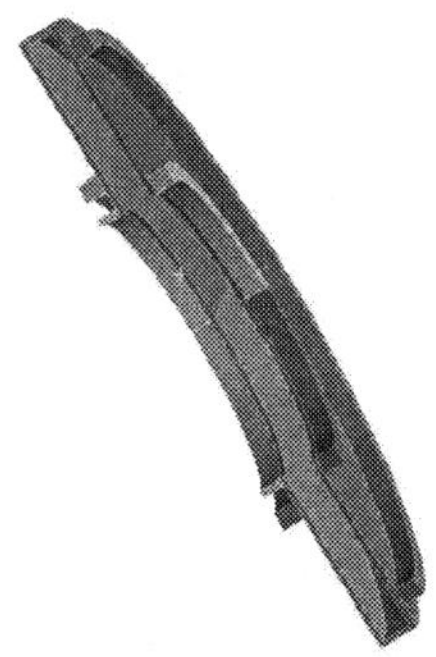

Figure 3. Calculation model

In order to reduce the calculation time and improve the calculation accuracy, the ring-shaped water groove on the cavities on both sides of the impeller are simplified into the inlet and outlet of the sealing fluid when performing the modeling.

Figure 3 is the calculation model of the centrifugal sealed flow channel. From left to right are the extended section of the low-pressure side air inlet, the extended section of the sealing fluid outlet, the axial gap 2, the radial gap, the axial gap 1, and the high-pressure side sealing medium. There are seven sections including the extension section of the inlet and the extension section of the sealing fluid inlet. Table 1 shows the main structural dimensions of the calculation domain.

Table 1. The main geometric parameters of the computed field.

The main geometric parameters	Value
Diameter of impeller D/mm	180
Axial clearance a_1/mm	1.5
Axial clearance a_2/mm	1.5
Radial clearance b/mm	1.5
Radial distance of sealing fluid inlet c/mm	30
Depth of annular water groove t/mm	2

IOP Conf. Series: Materials Science and Engineering 1081 (2021) 012004 doi:10.1088/1757-899X/1081/1/012004

3.2. Boundary condition and solution

In the calculation process, the gas phase mainly refers to air and sealed medium, and the liquid phase mainly refers to sealing fluid. The gas phase medium is selected as standard air, and the liquid phase medium is selected as water. The remaining operating parameters and boundary conditions are shown in Table 2. The physical properties of the sealing fluid and sealing medium are shown in Table 3.

Table 2. Operational parameters.

parameters	Boundary conditions	Value
Low-pressure side air inlet	Pressure-inlet	P_0=0MPa（G）
High-pressure side seal medium inlet	Pressure-inlet	P=0.15MPa（G）
Sealing fluid inlet	Flux-inlet	V_{in}=1.544m/s m_{in}= 4kg/s
Sealing fluid outlet	Pressure-outlet	P_{out}=0MPa（G）
Sealed chamber	Wall	0r/min
Centrifugal impeller	Rotation boundary	n=4000r/min

Since the default liquid phase is an incompressible fluid, the speed inlet in the sealing fluid inlet is equal to the mass inlet. Considering the annular area of the sealing fluid inlet, the calculated sealing fluid inlet flux is 4kg/s. In addition to the boundary conditions given in the table, the VOF model also needs to determine whether the medium at each inlet and outlet is a gas phase or a liquid phase. In FLUENT, the volume fraction of the second phase is used to represent this concept. In order to accelerate convergence, this paper uses the gas phase medium as the first phase and the liquid phase medium as the second phase. Therefore, the volume fraction of the low-pressure side air inlet and the high-pressure side seal medium inlet is set to 0, and the volume fraction of the sealing fluid inlet and outlet is 1.

Table 3. The physical properties of the sealing fluid.

Type	Density kg/m^3	Viscosity kg/m·s (Pa·s)	Thermal Conductivity w/m·K	Specific heat J/kg·K	Boiling point ℃
Water	1000	0.001003	0.5989	4183	100
Air/sealed medium	1.225	1.7894×10^{-5}	0.0242	1006.43	/

The solid boundary conditions in FLUENT mainly include the constant temperature boundary, constant heat flux density boundary, and convective heat transfer boundary. When the centrifugal impeller rotates at a fixed speed, on the one hand, the end surface of the impeller is used as the constant heat flux density boundary and the stirring heat flux density is given. On the other hand, the circumferential surface of the impeller is taken as the convective heat transfer boundary and the convective heat transfer coefficient is given. The wall surface of the sealing chamber is taken as the constant temperature boundary and its temperature is given.

3.3. Calculation results and analysis

3.3.1. Phase distribution and pressure and temperature distribution. When the gas centrifugal seal works, under the double action of external pressure and centrifugal force of impeller, the sealing fluid is continuously injected into the sealing chamber from the inlet and discharged from the outlet. On the other hand, under the action of centrifugal force of rotating impeller, it is thrown to the inner wall of the chamber, so after the flow is stable, a rotating siphon will be formed in the gap between the impeller and the inner wall of the chamber.

Figure 4 is a gas-liquid distribution cloud diagram of the half section of the sealing chamber at z = 0, where red represents the liquid phase and blue represents the gas phase. Figure 5 is a cloud diagram of the total pressure distribution of the half section of the sealing chamber at z = 0. From the figure, it can be seen that the high pressure area is close to the sealing medium side, and the low pressure area is close to the atmospheric side. Moreover, the pressure of the sealing fluid at the sealing fluid inlet is significantly higher than the pressure in other positions in the sealing chamber.

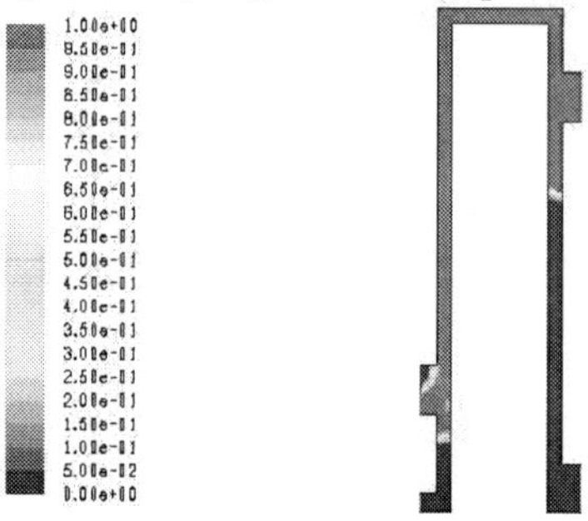

Figure 4. Two-phase distribution of the half section of the sealing chamber at Z=0

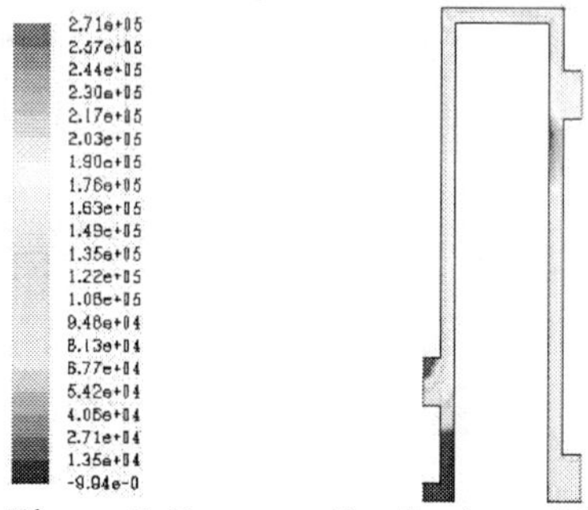

Figure 5. Pressure distribution of the half section of the sealing chamber at Z=0

Figure 6 shows the temperature distribution in a sealing chamber. It can be seen from the figure that the temperature is higher in the place full of medium and close to the impeller rotation boundary, which is caused by the stirring heat generated by the impeller rotation. And the temperature distribution in the other places full of sealing fluid is more average, which indicates that this method can effectively take away the stirring heat generated by the impeller rotation and reduce the temperature in the sealing chamber.

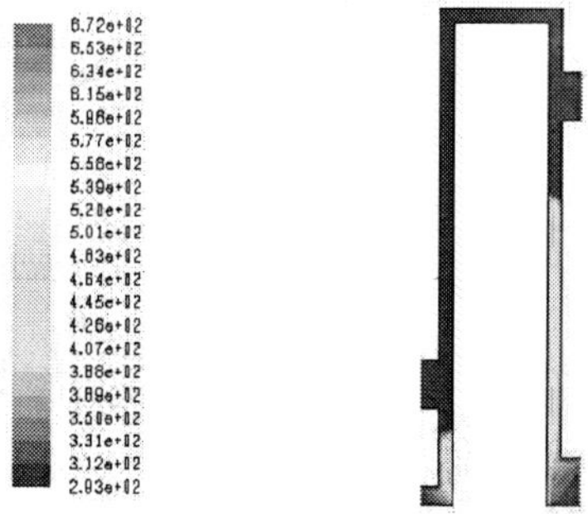

Figure 6. Temperature distribution of the half section of the sealing chamber at Z=0

3.3.2. Impact of rotational speed on performance. (1) Effect of rotational speed on pressure and opposite pressure coefficient of blocking fluid.

The change of the impeller speed will affect the centrifugal force generated by the rotation of the sealing fluid, which will cause the change of the air-liquid interface on both sides of the impeller, and will affect the compression of the liquid in the chamber, which will change the pressure field in the seal. Therefore, it is necessary to study the effect of rotational speed on the sealing performance under the condition that the flux of the sealing fluid is unchanged.

Figure 7 shows the change law of the sealing fluid pressure Pin with the impeller rotation n when the medium pressure is 0.15MPa and the flux of the sealing fluid is 4 kg/s. It can be seen that when the flux of the sealing fluid in and out of the sealing chamber is constant, as the impeller speed n increases, the required sealing fluid pressure Pin also gradually increases. The pumping compression effect of the sealing fluid is enhanced, which in turn causes an increase in pressure near the sealing fluid inlet. Therefore, in order to ensure a certain sealing fluid flow, a higher sealing fluid pressure is required.

Figure 8 is a curve of the opposite pressure coefficient k with the rotation speed n. It can be seen from the figure that when the flux is constant, the opposite pressure coefficient gradually increases with the increase of the rotation speed. This shows that the higher the rotation speed, the greater the impeller's driving degree on the sealing fluid, and the smaller the impact of the annular vortex in the sealing chamber on the sealing performance.

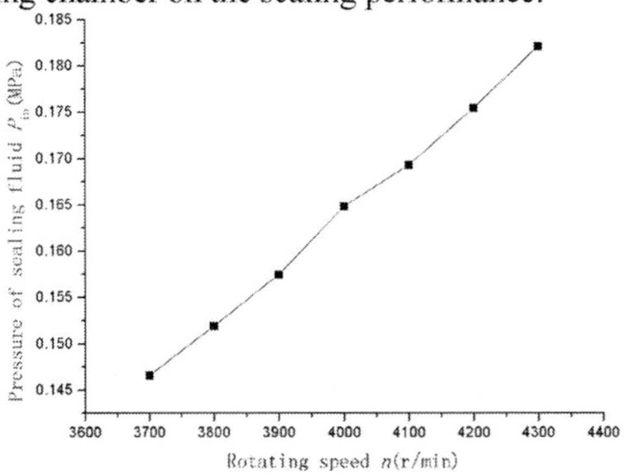

Figure 7. Effects of speed on the pressure of the sealing fluid

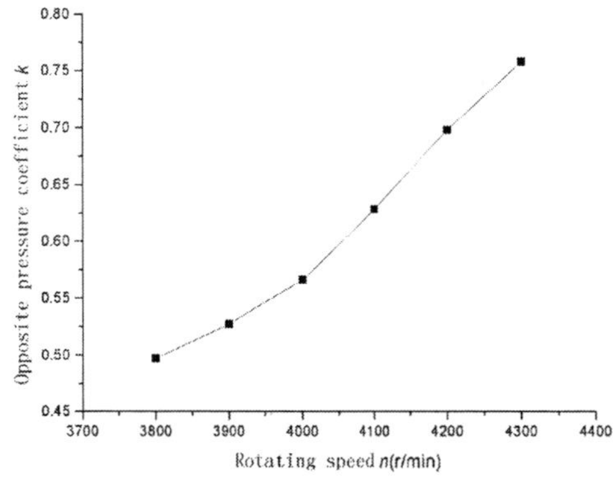

Figure 8. Effects of speed on the opposite pressure coefficient *k*

(2) Effect of rotational speed on the temperature of the sealing fluid

Figure 9 shows the change in the temperature of the sealing fluid T in the sealing chamber with the impeller speed n when the medium pressure is 0.15MPa and the flux of the sealing fluid is 4 kg/s. The increase in temperature T is relatively slow, which means that after the increase of the rotational speed, although the stirring heat generated by the impeller rotation will increase, the continuous injection and discharge of sealing fluid can remove the heat in the chamber in time, thereby effectively reducing the temperature rise of the fluid.

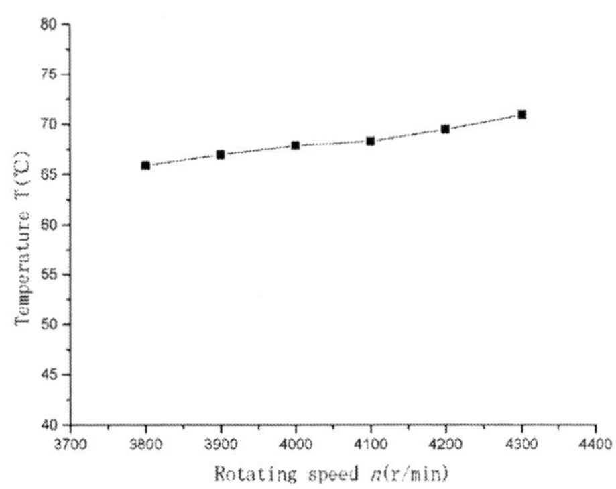

Figure 9. Effects of speed on the temperature of the sealing fluid in the chamber

(3) Effect of rotational speed on sealing ability

When the pressure of the medium changes, the centrifugal seal can automatically adjust the position of the air-liquid interface on both sides of the impeller to adapt to the seal of different pressure differences, which makes the sealing device have a certain pressure sealing ability at each speed.

In order to achieve continuous injection and discharge of sealing fluid while effectively sealing, investigate the sealing ability at different speeds when the flux of the sealing fluid is 4kg/s, and the changes of the corresponding pressure of sealing fluid and average temperature.

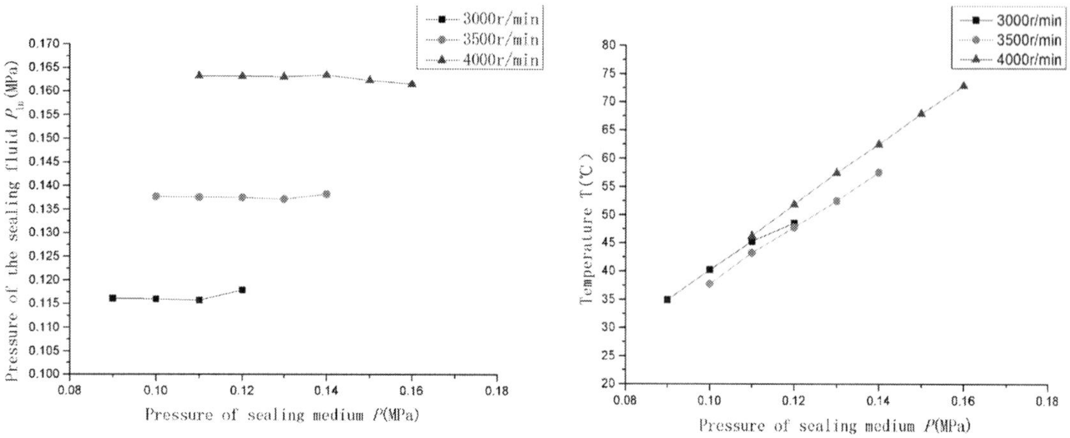

(a) Effects of the pressure of the sealing medium on the pressure of the sealing fluid (b) Effects of the pressure of the sealing medium on the temperature of the sealing fluid

Figure 10. Effects of speed on the range of the sealing pressure and correlation parameters

Figure 10 (a) shows the pressure of the sealing fluid at different speeds, and the sealing ability within the effective range is basically not affected by the change of the medium pressure. This shows that when the speed and the pressure of the sealing fluid are fixed, there is a certain range of pressure of the sealable medium, and the higher the speed, the greater the pressure of the sealing fluid, the larger the corresponding range.

Figure 10 (b) shows that the temperature of the sealing fluid increases with the increase of the medium pressure at different speeds, and the higher the speed, the higher the corresponding temperature. This shows that although increasing the rotational speed and the pressure of the sealing fluid can effectively increase the pressure of the sealing medium, it also easily causes the temperature of the sealing fluid to rise too high and cause vaporization.

In addition, it can be seen from Figure 10 that the sealing pressure range and sealing value increase when the rotational speed increases from 3000r/min to 4000r/min, which indicates that increasing the rotational speed can improve the sealing ability of the device.

3.3.3. Impact of sealing fluid flux on performance. The gas centrifugal sealing with continuous injection-discharge of sealing fluid requires constant renewal of the sealing fluid. Changes in the sealing fluid flux will cause changes in the blocking fluid pressure and temperature, which will affect the flow field and temperature field in the entire sealing chamber. It is necessary to study the influence of sealing fluid flux on sealing performance.

(1) Effect of sealing fluid flux on sealing fluid pressure and opposite pressure coefficient

Figure 11 shows the change rule of sealing fluid pressure pin with flux min when medium pressure is 0.15Mpa and rotating speed is 4000r/min. It can be seen that with the increase of the flux, the pressure of the sealing fluid increases gradually, which is due to the fact that the degree of compression of the impeller to the sealing fluid is basically unchanged when the rotating speed is fixed.

Figure 12 is a curve showing the change of the opposite pressure coefficient k with the flux of the sealing fluid, and it can be seen from the figure that the opposite pressure coefficient does not change basically with the flux of the sealing fluid. This shows that increasing or decreasing the flux of sealing fluid will not affect the annular vortex in the sealing chamber.

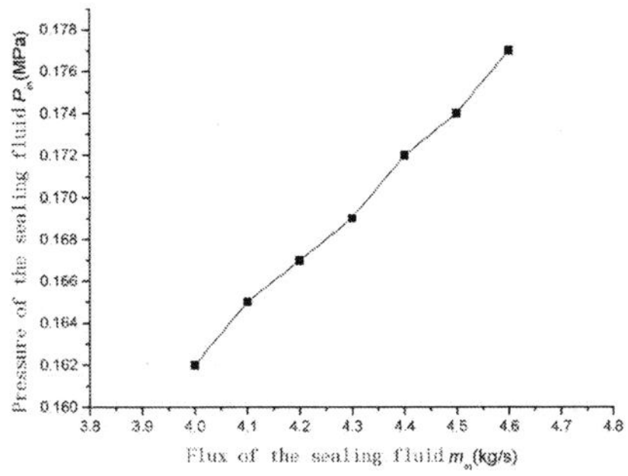

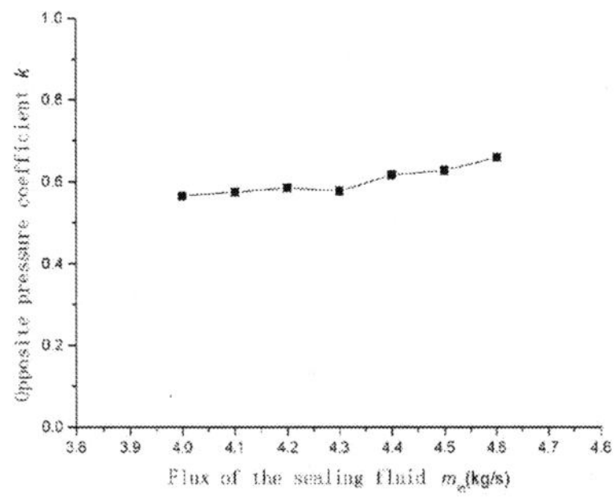

Figure 11. Effects of the flux of the sealing fluid on the pressure of the sealing fluid

Figure 12. Effects of the flux of the sealing fluid on the opposite pressure coefficient k

(2) Effect of sealing fluid flux on sealing fluid temperature

Figure13 shows the change law of the temperature T of the sealing fluid with the flux min when the medium pressure is 0.15MPa and the rotation speed is 4000 r/min. It can be seen from the figure that the temperature of the seal fluid decreases gradually with the increase of the flux. Because the stirring heat generated by the impeller is constant when the speed is constant, and the increase of the flux not only enhances the convection heat transfer between the seal fluid and the chamber, but also increases the heat taken away by the sealing fluid, thus reducing the temperature of the chamber.

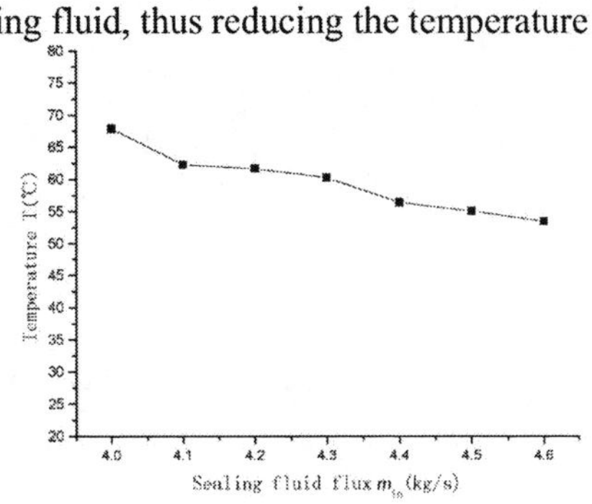

Figure 13. Effects of the flux of the sealing fluid on the temperature of the sealing fluid in the chamber

(3) Effect of sealing fluid flux on sealing ability

In order to analyze the effect of the sealing fluid flux on sealing ability at a fixed speed, investigating the sealing ability of the gas centrifugal seal at a speed of 4000r/min and a sealing fluid flux of 4, 4.2, 4.4kg/s. Further study the change curve of the sealing fluid pressure, opposite pressure coefficient and sealing fluid temperature under different medium pressure.

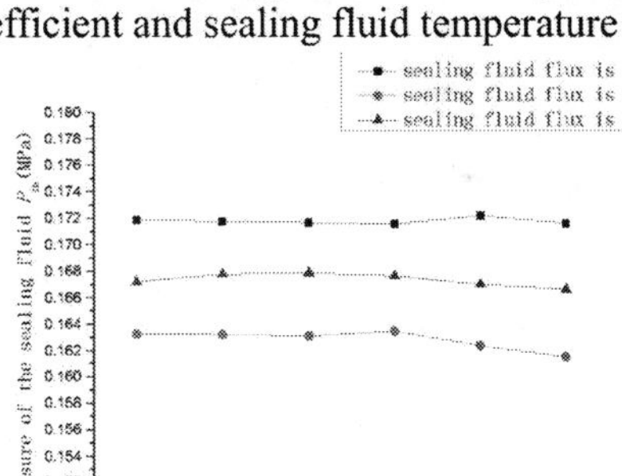

(a) Effects of the pressure of the sealing medium on the pressure of the sealing fluid

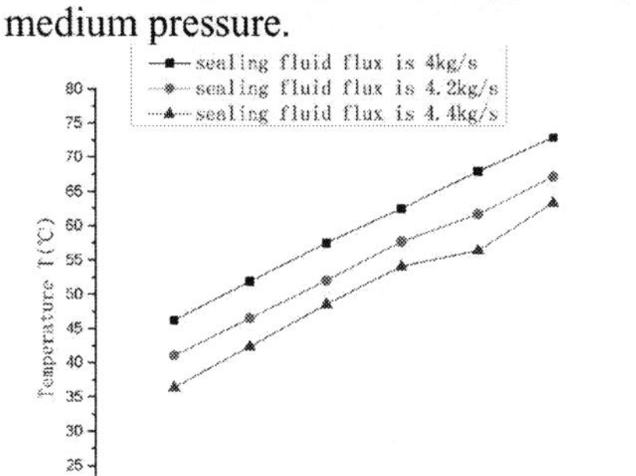

(b) Effects of the pressure of the sealing medium on the temperature of the sealing fluid in the chamber

Figure 14. Effects of mass flow rate on the range of the sealing pressure and correlation parameters

It can also be seen from figure 14 that the sealing ability of the gas centrifugal seal does not change substantially when the flux of the sealing fluid changes.

Figure 14 (a) shows that the pressure of the sealing fluid under different flux is basically not affected by the change in the pressure of the medium, and the pressure of the sealing fluid increases as the flux of the sealing fluid increases.

Figure 14 (b) shows the change curve of the temperature of the sealing fluid with the pressure of the medium under different flux. It can be seen from the figure that when the pressure of the medium increases, the temperature of the sealing fluid rises. The reason is that the increase of the pressure of the medium causes the gas-liquid interface on the high pressure side to rise, thereby reducing the cooling area of the sealing fluid. At the same time, the annular vortex flow intensify in the chamber, which generates more heat.

4. Test verification

The centrifugal seal test system is shown in Figure 15. Because the test needs to be performed at the shaft end of the test stand, the prototype uses a double-end structure as shown in Figure 16 (a), and the actual prototype of the test is shown in Figure 16 (b).

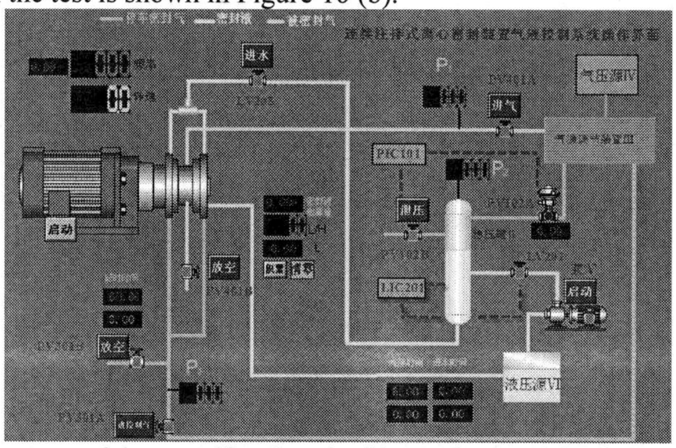

Figure 15. Testing system of the centrifugal seal with continuous injection and elimination of the sealing fluid

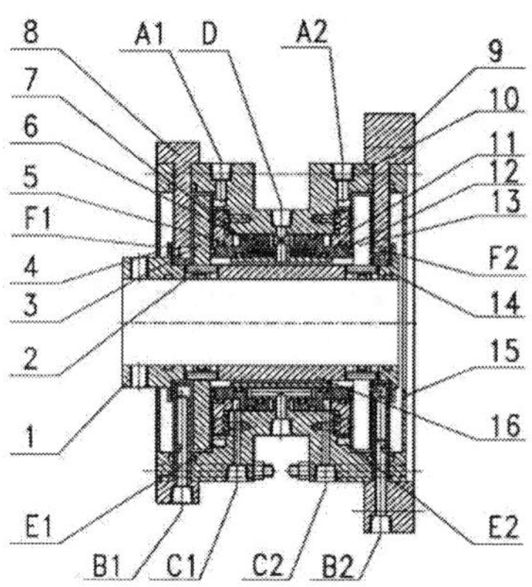

(a) structural drawing

(b) Prototype

1-Left flinging ring of water; 2-Drive pin; 3-Left centrifugal impeller; 4-Left Drain; 5-Left baffle; 6-Left slide valve; 7-spring; 8-Left end cap; 9-Right end cap; 10-Intermediate chamber; 11- Right gland; 12-Right slide valve; 13-Right centrifugal impeller; 14-Right flinging ring of water; 15-Right baffle; 16- Drive sleeve

Figure 16. Experimental Prototype of the centrifugal seal

In order to study the influence of the rotational speed on the sealing performance separately, in the test, the intake pressure was controlled by a high-precision automatic regulating valve, and the excessive pressure was discharged through the relief valve to stabilize the sealing fluid pressure. At the same time, the speed is adjusted by the inverter, so that the impact of the speed on the performance of the continuous injection and discharge centrifugal seal can be studied separately under the premise that the pressure of the sealing fluid is stable.

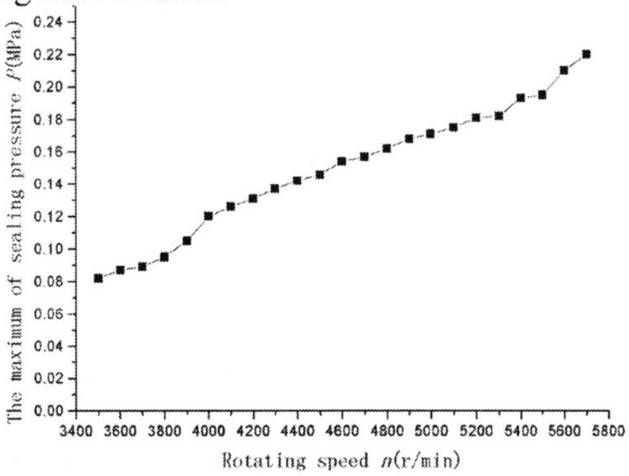

Figure 17. Effects of speed on the maximum of the sealing pressure

During the test, in order to obtain the maximum sealing pressure at each speed as much as possible, it is necessary to gradually increase the pressure of the sealing fluid and the medium until the zero leakage after stabilization is achieved. Recording the medium pressure at this time is the maximum sealing pressure at this speed. Figure 17 shows the relationship between the speed and the maximum sealing pressure. It can be seen from the figure that the maximum sealing pressure increases as the speed increases. This is because the pumping effect of the centrifugal impeller on the sealing fluid is increased when the speed is increased, resulting in the total pressure of the sealing fluid in the chamber is increased. And the medium with higher pressure can be sealed when the fluid level on both sides is not changed much. This test verifies the correctness of the increase in the pressure-sealing ability as the increase of the rotational speed proposed in 3.3.2 of this paper.

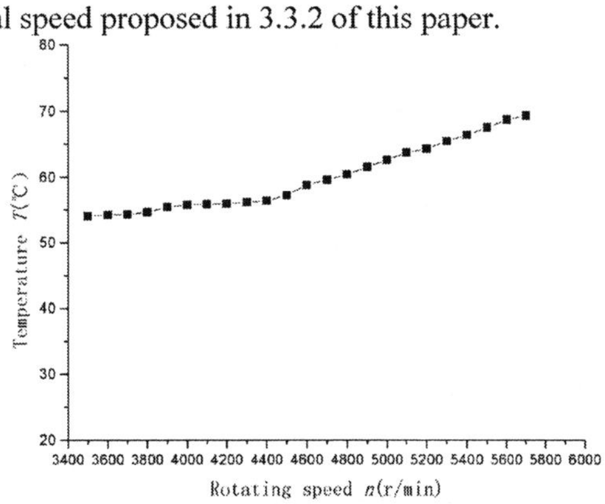

Figure 18. Effects of speed on the temperature of the sealing fluid at the exit

The temperature of the sealing fluid flowing out of the sealing chamber was then measured using a temperature sensor. Figure 18 is the curve of the temperature of the outlet of the sealing fluid when the medium pressure at each speed is the maximum. It can be seen from the figure that the temperature at the outlet of the sealing fluid rises slowly with the increase of the rotation speed in the range of 3500r/min-4300r/min, which is the same as the pattern shown in Figure 9. Then gradually increase the sealing speed to 6000r/min, and found that the temperature of the sealing fluid at the outlet gradually increased, which indicates that when the speed is around 4000r/min, the sealing fluid continuously injected and discharged can take away the stirring heat during the impeller rotation. Then, with the increase of the speed and the accumulation of heat during the test, the temperature of the sealing fluid at the outlet increased. This test verifies the feasibility of reducing the temperature rise by continuously injecting the sealing fluid.

According to the working parameters shown in Table 2, the rotational speed of the rotating shaft was fixed at 4000r/min, and the pressure in the surge tank was stabilized at about 0.165 MPa for testing. Then increase the medium pressure, record the medium pressure at this time and the flux and temperature of the sealing fluid at the corresponding outlet.

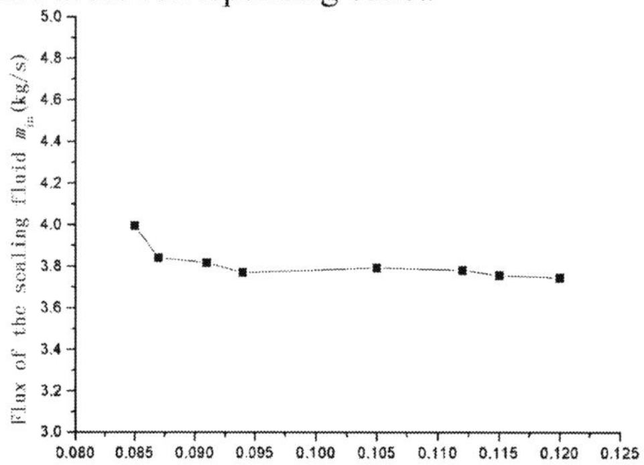

Figure 19. Effects of the pressure of the sealing medium on the mass flow rate of the sealing fluid at the exit

It can be seen from Figure 19 that under the same working conditions, the measured medium pressure is lower than the simulation result. The main reason for this phenomenon is the impact of the pressure of the medium and the pressure of the sealing fluid during the sealing operation. A small amount of axial movement will occur, causing the axial sealing gap on both sides to change, which will cause a reduction in sealing pressure.

Figure 19 is the change curve of the flux of sealing fluid at the outlet with the pressure of the medium. It can be seen from the Figure 19 that the flux is stable at about 4 kg/s when the pressure in the surge tank is 0.165 MPa. This verifies that the correspondence between the mass boundary and velocity boundary of the sealing fluid inlet set in the simulation calculation and the sealing fluid pressure calculated are correct.

5. Conclusion

Aiming at the problems of centrifugal seals in the field of gas seals, this paper studies the centrifugal seals with continuous injection and discharge of sealing fluid and draws the following conclusions:

(1) The continuous injection and discharge of the sealing fluid structure can effectively remove the stirring heat generated during the rotation of the seal, which solves the problems of low sealing pressure and fast temperature rise when the centrifugal seal is applied to the gas centrifuge machine;

(2) The higher the rotation speed and the greater the flux of the sealing fluid, the greater the sealing ability, but at the same time, the temperature of the sealing fluid in the chamber will rise higher, which will cause the vaporization of the sealing fluid. According to the analysis results, within the range of rated speed and sealing fluid flux, the effect of centrifugal seal on centrifugal machinery is significant;

(3) The correctness of the numerical simulation calculation was verified by experiments, and it was pointed out that there was a certain fluctuation in the flux of the sealing fluid limited by the test conditions, and the measured temperature was lower than that in the simulation due to the double-end seal structure used in the test.

References

[1] Yao R, Qin W, Jingdan L 2010. Design and numerical simulation of the hydrodynamical and merchanical sealing for slag pulp pumps, *Technology development*, **29** (7) 24-26, 54.

[2] Wu Zhi-qiang 2016. The Analysis and Solution of the Leakage Gas of Dry Gas Seal with High

IOP Conf. Series: Materials Science and Engineering 1081 (2021) 012004 doi:10.1088/1757-899X/1081/1/012004

Differential Pressure in Ammonia Compressor, *Chinese Journal of Turbomachinery*, **58** (01) 85-89.

[3] Han Liang 2017. Design and Application of a Dry Gas Seal on Gas for Natural Gas Compressor, *Chinese Journal of Turbomachinery*, **59** (02) 76-82.

[4] Wang Chunhua, Jiang Daren, 2008. The Application of Dry Gas Seal on Centrifugal Compressors, *Chinese Journal of Turbomachinery*, (01) 43-45.

[5] Ya-ru Ma et al. 2019. Investigations on the Leakage Performance of Wear Failure and Anti-rubbing Structure Design for a Labyrinth Seal, *Chinese Journal of Turbomachinery*, **61** (05) 64-71.

[6] Mi-mi Xue, Lin Cheng 2019. Numerical Research on the Influence of Shroud Labyrinth Seal and Hub Labyrinth Seal on the Performance of Centrifugal Compressor Stage, Chinese Journal of Turbomachinery, **61** (02) 61-68.

[7] Bian Shancheng et al. 2013. Study on Numerical Simulation of Leakage of Labyrinth Seals in Turbine Compressor, *Chinese Journal of Turbomachinery*, (02) 19-23.

[8] Wei Long, Feng Xiu 2011. Practical technology of chemical seal, *Chemical Industry Press*, Beijing.

[9] Banerjee R, Isaac KM 2003. Evaluation of Turbulence Closure Scheme for Stratified Two Phase Flow. *ASME 2003 International Mechanical Engineering Congress and Exposition.* Washington, American.

[10] Denecke J,Schramm V, Dullenkopf K,et al. 2006. Advanced Hydraulic Seal Design for High Temperature Environments. *ASME Turbo Expo: Power for Land, Sea, and Air*. Spain, American Society of Mechanical Engineers.

[11] Yuan JP, Zhang WJ, Jin R, et al. 2011. Flow numerical analysis within auxiliary impellers of centrifugal pumps. *The International Conference on Information Science and Technology*, China.

[12] Cai long 2010. The sealing ability of the centrifugal impeller back blade, *Pump Technology*, (2) 36-39

[13] Li Shujuan 2010. Numerical Simulation and Experimental Study on the Impeller Dynamic Seal of Centrifugal Pump. *Jiangsu University*, Nanjing.

[14] Zhou Zhian, Ding Keqin 1992.The explore and optimize of back blade sealing liquid sealing ability, *Chemical Equipment Technology*, **13**(5) 1-10.

[15] Liu Zhenduo 1986. Theoretical Analysis and Discussion on the *K* Value of the Back Pressure Coefficient of the Impeller Dynamic Seal, *Petro-Chemical Equipment*, (6) 18-25.

[16] Cai Renliang et al. 2006. Process equipment sealing technology, *Chemical Industry Press*, Beijing.

[17] Zi Dong, Li Shuangxi et al. 2012. Performance Analysis of the Gas Centrifugal Seal with Continuous Injection and Elimination of the Sealing Fluid, *Lubrication Engineering*, **37**(2) 40-44,48.

IOP Conf. Series: Materials Science and Engineering 1081 (2021) 012005 doi:10.1088/1757-899X/1081/1/012005

Design method and research of high temperature gas floating ring seal with slightly variable gap

Ren-jie Shi, Shuang-xi Li, Rao Zhen, Li-jun Ma, Jing-bo Zhang,

(Mechanical and Electrical Engineering, Beijing University of Chemical Technology, Beijing, China, 100029)

E-mail: shirenjie0630@163.com

Abstract. The problem of seal failure caused by rubbing between floating ring and shaft directly affects the actual production efficiency. In this paper, the mathematical model of floating ring seal was established, the expression of seal clearance under the working condition of floating ring was calculated, and the design method of "slightly variable gap" gas floating ring was proposed. The finite element model was established by using Workbench, and the correctness of the mathematical model was verified. The results show that the ideal insert ring material calculated by the "slightly variable gap" design method can keep the working clearance of floating ring seal stable. In practical application, the insert ring with the same material as the axle runway can also be selected to achieve a more stable working clearance. The design method can avoid the rubbing between the runway and the inside of the floating ring under harsh conditions.

1. Introduction

In today's era of rapid development of modern industry, sealing has become an essential technical product to ensure the efficient, long-term, safe and stable operation of modern industry. Common seals mainly include non-contact seals: floating ring seals, labyrinth seals, etc, contact seals: brush seals, packing seals, etc [1-3]. With the rapid development of science and technology, sealing conditions gradually develop to high temperature, ultra-low temperature, supercritical and other special conditions. The sealing effect also needs to meet the requirements of reducing leakage and environmental pollution [4]. These problems need people to constantly develop more superior sealing products [5,6]. Because of its simple structure, small axial width and reliable sealing performance, floating ring seal has been applied in many special conditions, so it plays an important role in modern sealing technology. According to the structure and principle of floating ring seal, the sealing effect is directly related to the clearance of

IOP Conf. Series: Materials Science and Engineering 1081 (2021) 012005 doi:10.1088/1757-899X/1081/1/012005

floating ring. From the point of view of reducing leakage and improving sealing efficiency, the clearance of floating ring should be reduced as much as possible. However, if the clearance is too small, the working condition of floating ring will be deteriorated, and even the floating ring will lock shaft [7,8]. If the floating ring clearance is too large, the leakage will increase [9~11]. Therefore, it is very important to choose reasonable clearance for sealing effect [12].

The initial clearance of floating ring is determined by the interference fit and initial size. The working time gap will change with the temperature, speed and other factors. The change of clearance may lead to the failure of the rubbing between the floating ring and the shaft, and eventually lead to machine failure and equipment shutdown. However, there are few studies and literatures on how to keep the floating ring clearance stable. Therefore, a design method of "slightly variable gap" is proposed in this paper. By analyzing the influence of different structural parameters and operating parameters on the clearance value, the mathematical modeling was used to quantify the clearance value, and the specific expression of floating ring seal clearance was obtained, and the method to achieve slightly variable gap by selecting materials was established. This paper also used Workbench software to verify the model, and the software data supports the conclusion.

2. Structure and working principle of gas floating ring seal with slightly variable gap

Floating ring seal, also known as clearance seal [6], its working principle is a flow resistance non-contact dynamic seal which combines radial clearance seal and axial face seal. The floating ring seal can minimize the radial clearance without friction, so as to minimize the leakage. Due to the existence of radial clearance, the mutual contact and friction between solids are avoided. It is suitable for principal axis of high speed, such as hydrogen turbine pump of liquid rocket engine, centrifugal compressor of high and medium pressure, steam turbine and centrifuge [13,14].

The structure of the floating ring is shown in Figure 1. The floating ring is located on the shaft and located in the sealing cavity, which has a small gap with the rotating shaft runway. After sealing oil or gas is injected into the sealing cavity of floating ring, it leaks along the clearance of floating ring. Due to the high-speed rotation of the rotor, the sealing oil or gas flowing into the floating ring gap forms a membrane with certain bearing capacity under the action of the rotating shaft. On the one hand, the film lifts the floating ring to realize liquid lubrication between the floating ring and the journal, so as to reduce friction and wear. On the other hand, because the oil film fills the whole floating ring gap, it can prevent the gas medium from leaking out, thus playing a role of sealing [15~17].

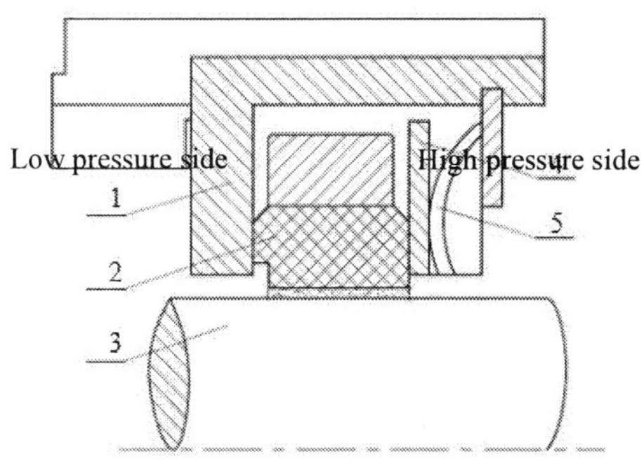

1-Floating ring seat;2-Floating ring;3-The runway;4-Spacer ring;5-Spring.

Fig.1 High Speed And High Temperature Gas Floating Ring Seal Structure Diagram.

3. Calculation basis and hypothesis of mathematical model

The calculation and analysis of the working clearance of floating ring seal is mainly based on the linear strain analysis of floating ring structure size under operating conditions, and the core part of calculation is to solve the contact surface stress of thick wall cylinder according to the stress of the microelement. The calculation model is that the interference structure is equivalent to the thick wall cylinder, the insert ring is equivalent to the thick wall cylinder under internal pressure, and the graphite ring is equivalent to the thick wall cylinder under external pressure. microelement stress solving equation [18~20]:.

(1) The equilibrium equation of microelement:

$$\sigma_\theta - \sigma_r = r\frac{d\sigma_r}{dr} \tag{1}$$

(2) Deformation compatibility equation:

$$\frac{1}{r}(\varepsilon_r - \varepsilon_\theta) = \frac{d\varepsilon_\theta}{dr} \tag{2}$$

(3) Comprehensive equations of equilibrium, geometry and physics:

$$\varepsilon_r - \varepsilon_\theta = \frac{1+\mu}{E}(\sigma_r - \sigma_\theta) \tag{3}$$

According to the above three calculation equations, the three-dimensional stress on the contact surface of the insert ring and graphite ring is solved, and the deformation of the floating ring under the operating condition is calculated by combining the interference inserting condition and the operating condition, and then the working clearance of the floating ring seal under the operating condition is calculated.

The following assumptions are made in the mechanical model calculation of the insert floating ring:

(1) The internal and external pressure of the insert ring and graphite ring are both axisymmetric, The elastic deformation of cylindrical shell is also axisymmetric without axial bending, Therefore, the stress and strain on the axial section are constant;

(2) The deformation of the insert ring and graphite ring on the contact surface is linear elastic deformation, and no plastic strain occurs.

4. Derivation of design method for gas floating ring seal with slightly variable gap

4.1. *Calculation Of Clearance Caused By Interference*

The floating ring is mainly composed of two parts. The outer ring is metal insert ring, and the inner ring is graphite ring. The two rings are assembled together by interference fit. The method of hot charging or cold pressing is widely used in interference connection. The interference of floating ring can be adjusted by modifying the design tolerance of inner circle of metal ring and outer circle of graphite ring, and the inside and seal end face of floating ring can be machined to the required size after inserting.

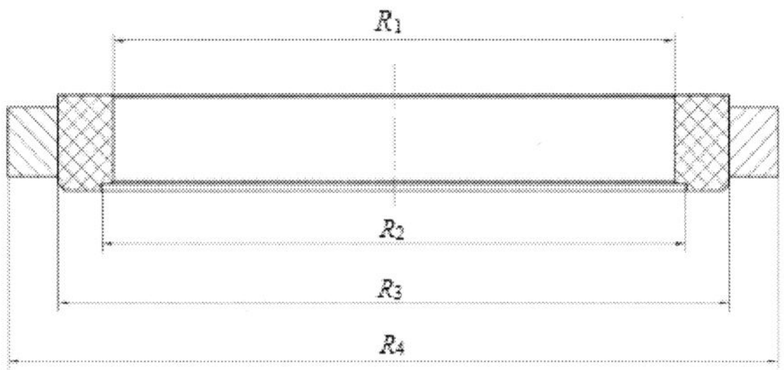

Fig.2 Insert Floating Ring Structure Diagram.

The symbols used are described as follows: σ_r - radial stress of contact surface, σ_θ - circumferential stress of contact surface, σ_z - axial stress, r -diameter of any point on the insert ring or graphite ring, P_m - the average contact stress on the contact surface of the insert ring, P_g - the average contact stress on the contact surface of graphite ring, R - diameter of the runway, E_1- elastic modulus of insert ring, E_2- elastic modulus of graphite ring, H_g - axial thickness of graphite ring, H_m - axial thickness of the insert ring.

Stress at any point on the insert ring:

$$\begin{cases} \sigma_\theta^m = \dfrac{P_m R_3^2}{R_4^2 - R_3^2}(1+\dfrac{R_4^2}{r^2}) \\ \sigma_r^m = \dfrac{P_m R_3^2}{R_4^2 - R_3^2}(1-\dfrac{R_4^2}{r^2}) \\ \sigma_z^m = \dfrac{P_m R_3^2}{R_4^2 - R_3^2} \end{cases} \quad (4)$$

Stress at any point on the graphite ring:

$$\begin{cases} \sigma_\theta^g = -\dfrac{P_g R_3^2}{R_3^2 - R_1^2}(1+\dfrac{R_1^2}{r^2}) \\ \sigma_r^g = -\dfrac{P_g R_3^2}{R_3^2 - R_1^2}(1-\dfrac{R_1^2}{r^2}) \\ \sigma_z^g = -\dfrac{P_g R_3^2}{R_3^2 - R_1^2} \end{cases} \quad (5)$$

According to the formula (4), when r=R_3, the preload of the metal ring on the contact surface can be obtained, and the three-dimensional stress on the contact surface of the metal ring is obtained as follows:

$$\begin{cases} \sigma_\theta^m \big|_{r=R_3} = \dfrac{P_m (R_4^2 + R_3^2)}{R_4^2 - R_3^2} \\ \sigma_r^m \big|_{r=R_3} = -P_m \\ \sigma_z^m \big|_{r=R_3} = \dfrac{P_m R_3^2}{R_4^2 - R_3^2} \end{cases} \quad (6)$$

According to the calculation formula (5), when $r=R_3$, the preload of graphite ring on the contact surface can be obtained, and the three-dimensional stress on the contact surface of graphite ring is obtained as follows:

$$\begin{cases} \sigma_\theta^g \big|_{r=R_3} = \dfrac{-P_g(R_3^2+R_1^2)}{R_3^2-R_1^2} \\ \sigma_r^g \big|_{r=R_3} = -P_g \\ \sigma_z^g \big|_{r=R_3} = -\dfrac{P_g R_3^2}{R_3^2-R_1^2} \end{cases} \tag{7}$$

According to the calculation hypothesis (2) of the mechanical model, the strain deformation of the sealing ring is linear elastic deformation. The interference between the insert ring and the graphite ring can be considered as the linear superposition of the deformation λ_1 of the inner circle of the insert ring and the deformation λ_2 of the outer circle of the graphite ring. The numerical calculation relationship is as follows:

$$\frac{\Delta}{2} = |\lambda_1| + |\lambda_2| \tag{8}$$

(1) Radial displacement λ_1 of the inner circle of the insert ring

Suppose that the radial displacement of the inner circle R_3 of the insert ring is λ_1 and the radial linear strain ε_3^m is:

$$\varepsilon_3^m = \frac{2\pi(R_3+\lambda_1)-2\pi R_3}{2\pi R_3} = \frac{\lambda_1}{R_3} \tag{9}$$

According to the generalized Hooke's Law:

$$\varepsilon_3^m = \frac{1}{E_1}[\sigma_r^m - \mu(\sigma_\theta^m + \sigma_z^m)] \tag{10}$$

Substituting (6) (9) into (10) shows that:

$$\varepsilon_3^m = \frac{\lambda_1}{R_3} = \frac{1}{E_1}[\sigma_r^m \big|_{r=R_3} - \mu(\sigma_\theta^m \big|_{r=R_3} + \sigma_z^m \big|_{r=R_3})] \tag{11}$$

After finishing, we can get the following results:

$$\lambda_1 = \frac{-P_m R_3}{E_1}(1+\mu_1 \frac{R_4^2+2R_3^2}{R_4^2-R_3^2}) \tag{12}$$

(2) Radial displacement λ_2 of the outer circle of the graphite ring:

Similarly, suppose the radial displacement of the outer circle R_3 of graphite ring is λ_2, After finishing, we can get the following results:

$$\lambda_2 = \frac{-P_g R_3}{E_2}(1-\mu_2 \frac{2R_3^2+R_1^2}{R_3^2-R_1^2}) \tag{13}$$

The diameter interference of the insert ring and graphite ring is Δ and the radial force of the insert ring and graphite ring is acting force and reaction force, and the force is assumed to be F_0. For the convenience of calculation, the following assumption is made on the contact surface of floating ring: the force of insert ring and graphite ring on the contact surface is evenly distributed on their respective contact surfaces. According to the balance of forces:

$$P_m A_m = P_g A_g \tag{14}$$

It can be obtained:

$$P_m = P_g \frac{H_g}{H_m} = P_g \gamma \tag{15}$$

Let $H_g/H_m=\gamma$, where γ is defined as the axial width ratio of graphite ring to metal ring, and γ is generally greater than 1. The radial displacement of the insert ring and graphite ring on the contact surface presents a linear distribution in the radial direction. According to the calculation hypothesis:

$$\frac{\Delta}{2} = |\lambda_1| + |\lambda_2| \tag{16}$$

Substituting the formula (12) and (13) into formula (16), we can get:

$$\frac{\Delta}{2} = \left|\frac{-P_m R_3}{E_1}(1+\mu_1 \frac{R_4^2 + 2R_3^2}{R_4^2 - R_3^2})\right| + \left|\frac{P_g R_3}{E_2}(1-\mu_2 \frac{2R_3^2 + R_1^2}{R_3^2 - R_1^2})\right| \tag{17}$$

The inserting stress P_g on the contact surface of graphite ring is obtained as:

$$P_g = \frac{\Delta/2}{\left|\frac{\gamma R_3}{E_1}(1+\mu_1 \frac{2R_3^2 + R_4^2}{R_4^2 - R_3^2})\right| + \left|\frac{R_3}{E_2}(1-\mu_2 \frac{2R_3^2 + R_1^2}{R_3^2 - R_1^2})\right|} \tag{18}$$

(3) Clearance C_0 caused by interference of insert:

According to formula (5), the stress of inner circle can be obtained when $r=R_1$. Stress distribution of the inner circle of graphite ring:

$$\begin{cases} \sigma_\theta^g = -\dfrac{2P_g R_3^2}{R_3^2 - R_1^2} \\ \sigma_r^g = 0 \\ \sigma_z^g = -\dfrac{P_g R_3^2}{R_3^2 - R_1^2} \end{cases} \tag{19}$$

Under this stress, the linear strain of the inner hole can be expressed by the following formula:

$$\varepsilon_1^g = \frac{\lambda_0}{R_1} = \frac{1}{E_2}[\sigma_r^g|_{r=R_1} - \mu(\sigma_\theta^g|_{r=R_1} + \sigma_z^g|_{r=R_1})] \tag{20}$$

After substituting (19) into (20), we can get the following results:After installation, the inner diameter displacement of floating ring is λ_0:

$$\lambda_0 = C_0 = \frac{R_1\mu_2}{E_2}\frac{3P_g R_3^2}{R_3^2 - R_1^2} \tag{21}$$

By substituting (18) into (21), the radial displacement of the inner hole of floating ring under interference condition can be obtained as:

$$C_0 = \frac{\frac{R_1\mu_2}{E_2}\frac{3R_3^2}{R_3^2 - R_1^2}\frac{\Delta}{2}}{\left|\frac{\gamma R_3}{E_1}(1+\mu_1 \frac{2R_3^2 + R_4^2}{R_4^2 - R_3^2})\right| + \left|\frac{R_3}{E_2}(1-\mu_2 \frac{2R_3^2 + R_1^2}{R_3^2 - R_1^2})\right|} \tag{22}$$

Let:

$$C_0 = k \cdot \frac{\Delta}{2} \tag{23}$$

Where k is the conversion factor:

$$k = \frac{\frac{R_1\mu_2}{E_2}\frac{3R_3^2}{R_3^2 - R_1^2}}{\left|\frac{\gamma R_3}{E_1}(1+\mu_1\frac{2R_3^2 + R_4^2}{R_4^2 - R_3^2})\right| + \left|\frac{R_3}{E_2}(1-\mu_2\frac{2R_3^2 + R_1^2}{R_3^2 - R_1^2})\right|} \tag{24}$$

4.2 Calculation Of Clearance Caused By Temperature Load

When the temperature is set at T_0, the diameter interference is Δ, while at T_1 is Δ_1, and $\Delta_1>0$. The radial displacement of the insert ring on the contact surface R_3 is ΔB, and that of the graphite ring on the contact surface R_3 is ΔC.

According to hypothesis (2) and contact surface deformation analysis, ΔB is approximately equal to the change of interference $\Delta - \Delta_1$ and thermal deformation ΔC of graphite ring outer circle, namely:

$$\Delta B = \Delta - \Delta_1 + \Delta C \tag{25}$$

Where:

$$\Delta C = R_3\beta_2(T_1 - T_0) \tag{26}$$

$$\Delta B = R_3\beta_1(T_1 - T_0) \tag{27}$$

It can be obtained by substituting the above formula into (25):

$$\Delta_1 = \Delta - R_3(\beta_2 - \beta_1)(T_1 - T_0) \tag{28}$$

Where β_1 and β_2 are the linear expansion coefficients of the insert ring and the graphite ring respectively. When there is contact stress and the operating temperature is T_1, the inner diameter displacement C_1 under interference condition is obtained:

$$\frac{C_1}{C_0} = \frac{\Delta_1}{\Delta} = 1 - \frac{R_3(\beta_2 - \beta_1)(T_1 - T_0)}{\Delta} \tag{29}$$

When the conversion factor k is brought in, the following results are obtained:

$$C_1 = C_0 - kR_3(\beta_2 - \beta_1)(T_1 - T_0) \tag{30}$$

The total deformation U_1 of floating ring inner circle at temperature T_1 consists of three parts:(1) At normal temperature, the radial displacement of floating ring inner circle caused by interference is C_0;(2) When the operating temperature is T_1, the radial displacement of floating ring inner circle is C_1 under interference condition;(3) When the operating temperature is from T_0 to T_1, The change of inner circle caused by the expansion of graphite ring is $R_1\beta_1(T_1 - T_0)$.

The numerical relationship is as follows:

$$U_1 = C_0 - C_1 + R_1\beta_1(T_1 - T_0) \tag{31}$$

$$U_1 = kR_3(\beta_2 - \beta_1)(T_1 - T_0) + R_1\beta_1(T_1 - T_0) \tag{32}$$

If the expansion coefficient of floating ring is defined as β_k, then:

$$U_1 = R_1\beta_k(T_1 - T_0) \tag{33}$$

By combining formula (32) and (33), we can get the following results:

$$\beta_k = k\frac{R_3}{R_1}(\beta_2 - \beta_1) + \beta_1 \tag{34}$$

Radial displacement of inner hole of floating ring at operating temperature:

$$\delta_{R1} = R_1\beta_k\frac{\Delta T}{2} \tag{35}$$

Radial displacement of runway at operating temperature:

$$\delta_{R2}^T = d_0\beta_3\frac{\Delta T}{2} \tag{36}$$

Where $d_0 \approx R_1$.

4.3 design method of gas floating ring seal with slightly variable gap

Assuming that the initial radial seal clearance of floating ring is δ_0, the radial displacement of ru δ_{R2}, δ_{R2} includes radial displacement δ_{R2}^n caused by rotation speed and radial displacement $\delta_R^{'}$ by temperature load. The actual radial seal clearance can be calculated by formula:

$$\delta = \delta_0 + R_1\beta_k\frac{\Delta T}{2} - \delta_{R2}^T - \delta_{R2}^n \tag{37}$$

To sum up, the actual radial working clearance of floating ring is as follows:

$$\delta = \delta_0 + R_1\beta_k\frac{\Delta T}{2} - R_1\beta_3\frac{\Delta T}{2} - \delta_{R2}^n \tag{38}$$

According to the result of formula (38), under the condition of not considering the influ rotating speed on the working clearance of floating ring temporarily, That is, $\delta_{R2}^n=0$, β_k is the ex coefficient of floating ring after inserted, see equation (34). The calculation formula of th working clearance of floating ring can be obtained as follows:

$$\delta = \delta_0 + [kR_3(\beta_2 - \beta_1) - R_1(\beta_3 - \beta_1)]\frac{\Delta T}{2} \tag{39}$$

The design concept of "slightly variable gap" is that the working clearance δ of floating r not change with the change of operating temperature. Therefore, the above formula can be co as $\delta=\delta_0$, after calculation, the condition to be satisfied by the linear expansion coefficient of tl ring is:

$$\beta_2 = \frac{R_1}{kR_3}(\beta_3 - \beta_1) + \beta_1 \tag{40}$$

Among them, the conversion factor is:

$$k = \frac{\dfrac{R_1\mu_2}{E_2}\dfrac{3R_3^2}{R_3^2 - R_1^2}}{\left|\dfrac{\gamma R_3}{E_1}(1+\mu_1\dfrac{2R_3^2+R_4^2}{R_4^2-R_3^2})\right| + \left|\dfrac{R_3}{E_2}(1-\mu_2\dfrac{2R_3^2+R_1^2}{R_3^2-R_1^2})\right|} \tag{41}$$

5. Finite element analysis and verification

In this section, the Workbench software is used to model and calculate the working clearance floating ring, and the correctness is verified by comparing the calculated value of th mathematical formula model with that of the model. The floating ring model established by the s is divided into two parts: insert ring and graphite ring. The grid division cell size is 0.5 mm, the of nodes is 361238, and the number of elements is 78013. After checking, the calculation results do not change with the grid density. Therefore, through the verification of grid independence, the meshing method meets the requirements of calculation accuracy. The floating ring structure diagram and grid division are shown in Figure 3.

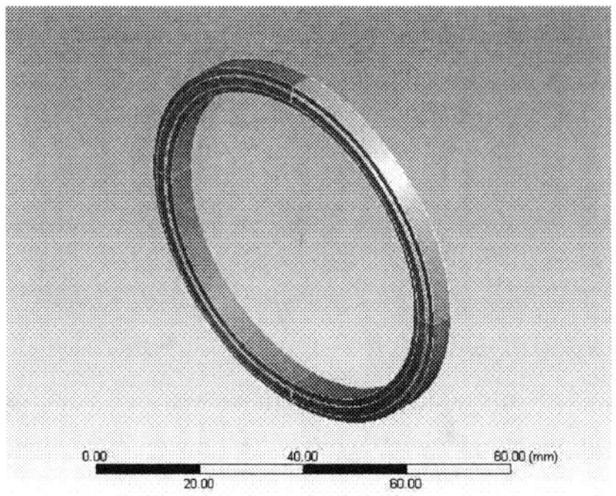

(a) Floating Ring Structure

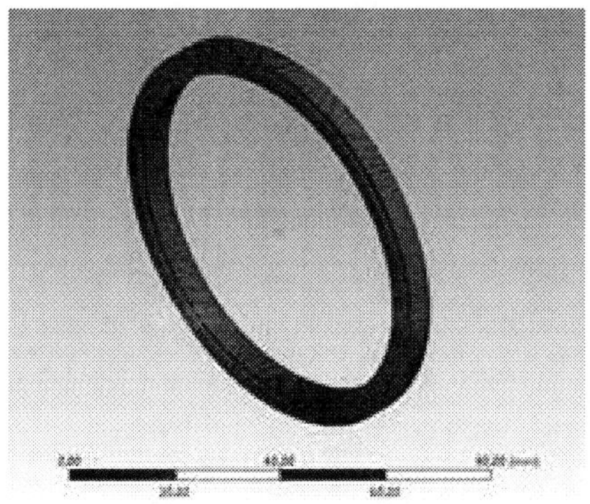

(b) Floating Ring Meshing

Fig.3 Numerical Calculation Model Of Insert Floating Ring

By changing the operating temperature and material of the insert ring, the calculation results of the working clearance between the mathematical model and the finite element floating ring were compared. Choose EY308 graphite ring and insert ring of different material, while material of runway is GH4169. The calculation results are arranged as shown in Figure 4. Through the comparison between the experimental results and the simulation value, it is known that the deviation between the measured working clearance and the simulation calculation is not large, and they are all within the allowable range of error. Therefore, it is feasible to use numerical simulation to calculate the inner diameter variation of insert floating ring.

According to Fig. 4, the calculation results show that the change trend of the two methods in the calculation of floating ring working clearance is almost the same. The mathematical model and finite element numerical calculation values of different material insert ring structures and different temperatures have good fitting, and the correctness of the mathematical model has been verified.

Furthermore, it can be deduced from the mathematical model that the conversion factor is only related to the floating ring structure. When the structural parameters are fixed, the ideal material parameters of the insert floating ring satisfying the design condition of "slightly variable gap" are

IOP Conf. Series: Materials Science and Engineering 1081 (2021) 012005 doi:10.1088/1757-899X/1081/1/012005

calculated as follows. Assuming that the elastic modulus of the insert ring is a constant value, the conversion factor k can be obtained by calculation. The linear expansion coefficient $\beta_2^{'}$ of the ideal material of insert ring is obtained by substituting the formula (40). The ideal material is compared with the other two materials, and the working clearance of the floating ring at fixed speed and different operating temperature are calculated respectively. The calculation results are shown in Fig. 5.

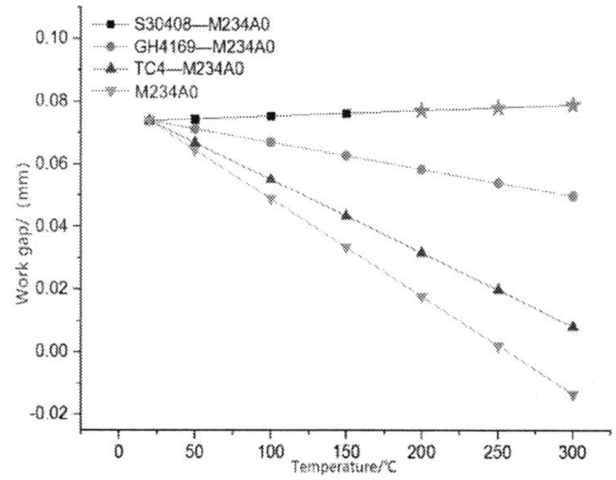

(a) The Results Of Working Clearance Are Calculated By Mathematical Formulas Under Different Temperatures And Materials

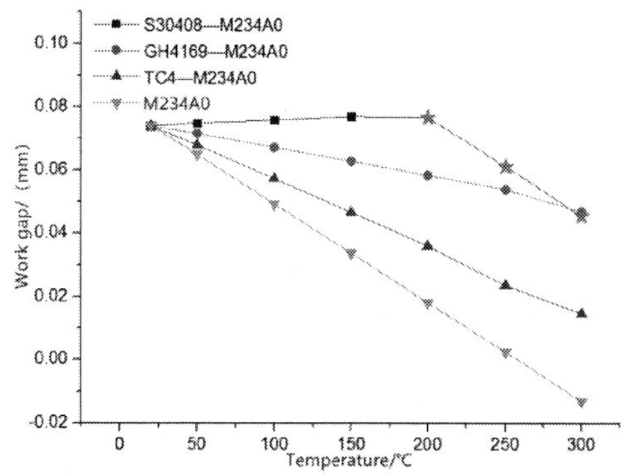

(b) Finite Element Numerical Calculation Of Working Gap Under The Corresponding Conditions

Note: the inflection point indicated by "★" is the calculated value of the working gap at the operating temperature after floating ring detangling, which is not considered in this paper

Fig.4 Comparison Between Mathematical Model And Finite Element Numerical Calculation Results

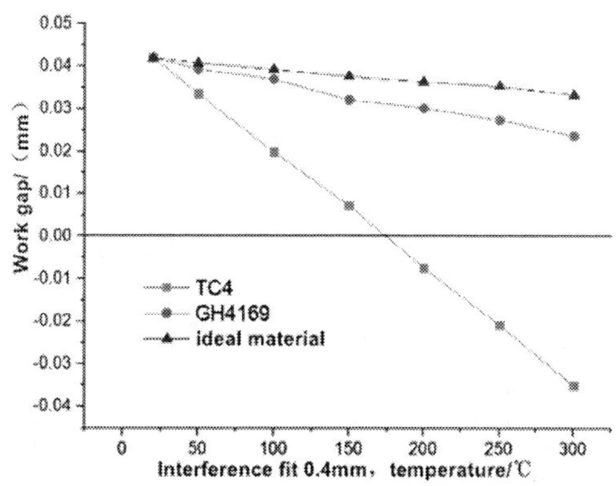

Fig.5 Working Clearance Comparison Under The Design Method Of "Slightly Variable Gap"

According to the calculation results, when the operating temperature rises to 300 °C, the clearance of floating ring of ideal material changes from 0.04196mm to 0.03336 mm. From the overall change trend, the working clearance of floating ring has little change, which indicates that it is feasible to calculate the linear expansion coefficient of ideal insert ring material by formula (40) in the structural design of "slightly variable gap" of insert floating ring. Therefore, in the design, the selection of insert ring material should be as close as possible to the linear expansion coefficient of the ideal material in the calculation of "slightly variable gap". According to the analysis, it can also be found that when the material of the insert ring is the same as that of the shaft runway like the GH4169 curve in Fig. 5 can also maintain a relatively stable slightly, which can be used in the actual production process.

Conclusions

(1) The calculation method of working clearance of gas floating ring seal is established, which provides reference for the research of floating ring seal clearance.

(2) The sealing clearance of floating ring can be kept stable by selecting suitable material of insert ring. The ideal floating ring material calculated by the design method of "slightly variable gap" in this paper can achieve better effect.

(3) In practical application, choosing the same material as the axis runway can have a small clearance change and keep the sealing performance stable.

References

[1] Yaru Ma, Wenhao Huo, etc. *Research on Wear Failure Leakage Characteristics of Labyrinth Seal and Design of Anti-bump Structure* [J] .Fan Technology, 2019, 61 (05): 64-71.

[2] Mimi Xue, Lin Chen. *Numerical study on the influence of wheel cover seal and disk seal on the performance of centrifugal compressor stage* [J]. Fan Technology, 2019, 61 (02): 61-68.

[3] Guoqing Li, Dan Sun, etc., *Numerical study on flow and heat transfer characteristics of multistage brush seal with low hysteresis* [J]. Fan Technology, 2019, (05): 56-63.

[4] Nan Yan. *Analysis and treatment of frequent oil leakage in bearing box of axial flow induced draft fan with adjustable moving blade in a power plant* [J]. Fan Technology, 2019,(07): 56-59.

[5] Rui Dai,Fengge Zhang, etc., *Characteristics and key technical problems of high speed motor* [J]. Fan Technology, 2019, 61(04): 59-66.

[6] Jangning Jia, Zhi Chen, etc. *Structure analysis and research status of floating ring seal* [J]. Mechanics, 2007, (10): 1.

[7] R.Gordon.Kirk, S.K.Baheti. *Evaluation of floating ring seals for centrifugal compressors using the finite element method* [J]. Transactions of the ASME, 1999: 121(1):131-136.

[8] W. Shapiro, C. Lee, H. Jones. *Analysis and design of a gas-lubricated sectored, floating ring seal* [J]. Transactions of the ASME, 1988: 110(7):525-532.

[9] Haiqing Wang. *Failure causes and preventive measures of floating ring seal* [J]. Fan Technology, 2000, (2).

[10] Yongfu Zhuang. *Failure analysis and improvement of floating ring seal of circulating gas compressor* [J]. Fluid machinery, 2006, (3): 48-51.

[11] Qipeng Liu. *Analysis and improvement of seal failure of rich gas compressor* [J]. Equipment management and maintenance, 2014, (6): 17-19.

[12] Tae-Woong Haa; Yong-Bok Lee etc. *Leakage and rotor dynamic analysis of a high pressure floating ring seal in the turbo pump unit of a liquid rocket engine* [J].Tribology International, 2002, Vol.35:153-161.

[13] Mnler.H.K, Nan.B.S, etc. *Fluid sealing technology: principle and Application* [M]. Beijing: Machinery Industry Press, 2002.

[14] Tae-Woong Ha, Yong-Bok Lee. etc. *Leakage and rotor dynamic analysis of high pressure floating ring seal in the turbo pump unit of a liquid rocket engine* [J]. Tribology International, 2002: 35(3):153-161.

[15] Qiang Wang, Yue Wang. *Comparison of application effect of floating ring seal at home and abroad* [J]. Petrochemical Equipment Technology, 2002, (06): 47-49.

[16] Wenbo Duan, Fulei Chu, etc. *A bulk-flow analysis of static and dynamic characteristics of floating ring seals* [J]. Tribology International, 2007:40(03):470-478.

[17] S. K. Baheti, R. G. Kirk. *Finite element thermos hydrodynamic analysis of a circumferentially grooved floating oil ring seal* [J]. Tribology Transactions, 1995:38(1):86-96.

[18] Jingyang Zheng,Zhifu Sang. *Process equipment design (4th Edition)* [M].Beijing: Chemical Industry Press.2015.11.

[19] Guohui Zhu, Jingyang Zheng. *New type of ribbon wound pressure vessel* [M]. Beijing: Chemical Industry Press.1995.

[20] Qiuer Zhu. *High voltage equipment design* [M].Shanghai: Shanghai Science and Technology Press, 1988.

Performance Analysis and Optimization of Internal Spiral Groove Dry Gas Seal Based on Fluid-solid-thermal Coupling Method

S X Li, H R Liao, R Zheng, S C Li, Y H Ma, K Y Chen

(Mechanical and Electrical Engineering, Beijing University of Chemical Technology, Beijing, 100029, China)

ABSTRACT: Internal spiral groove dry gas seal (ISGDGS) are used to seal special positions of rotating machinery. There is an urgent need to address the problems caused by contacting of rotating and static rings during high-speed rotation of ISGDGS. The main aim of this study is to investigate the influence of spiral groove structural parameters on ISGDGS sealing performance. Fluid-solid-thermal coupling method was used to analysis the relationship of sealing performance (opening force, leakage, friction power, and gas film stiffness) and spiral-groove structural parameters. The test analysis method is used to study the sealing performance under four different structural parameters, and the obtained trends are the same as the analysis results. According to the analysis results, the value range of structural parameters for optimal sealing performance is proposed This thesis has provided a deeper insight into the influence of the structural parameters of end face on the sealing performance, and provides a basis for its optimal design.

KEY WORDS :
Internal Spiral Groove, Fluid-solid-thermal Coupling, Sealing Performance

NOMENCLATURE

a Helix angle
r_i Sealing face inner radius
r_o Sealing face outer radius
r Spiral groove root radius
r_g Spiral groove outer radius
h_g depth of groove
h_o End flow field thickness (film thickness)
T Temperature distribution
N_g Number of grooves
R_w Ratio of groove to weir
R_d Ratio of groove to dam
r Radial position
θ Angular position

p Pressure
ρ Density
η Gas viscosity
ω Rotating speed
δ_g Position factor
δ_{ij}, δ_{id} Seal ring deformation components
h_i Initial thickness correction value
δ_{ij}', δ_{id}' New seal ring deformation components
δ_{kj}, δ_{kd} Thermal deformations
h_i' Modified thickness value
ε Convergence accuracy
F Opening force
q Leakage
P_n Friction power
s Film Stiffness
ψ Friction power

INTRODUCTION

Internal spiral groove dry gas seal (ISGDGS) are used to seal special positions of rotating machinery due to its advantages on pumping out medium and boosting pressure. ISGDGS are non-contact gas seals used in high-speed rotating gas machinery. If the end face parameters are not selected properly, the rotating and static rings cannot be opened at high speed. That scenario will lead to serious consequences, such as burning the seals which means that the sealing ring is malfunctioning [1]. ISGDGS are different from the traditional dry gas seals. There are grooves on the high-pressure inner diameter side, which can ensure the high-pressure medium to be pumped out with a small leakage rate avoiding medium pollution.

In actual working conditions, the state of the sealing end face is affected by structure, fluid and temperature. Recently, researchers have shown an increased interest in coupling analysis which is a research method that comprehensively considers the effects of various aspects. Researchers studied the deformation of the rotating and static rings of dry gas seals [2]. Hu Q [3] established a three-dimensional heat transfer model, and studied the effects of the spindle speed on the structure-thermal coupling deformation. Luo X [4] coupled the end face waviness and taper deformation to perform fluid calculation, and studied the influence of the groove shape on the non-contact sealing performance. Huang W F [5] coupled a multi-physics field to establish a fluid-solid-thermal simulation model to study the mechanical properties of upstream pumping, and revealed the weakening effect of temperature rise and deformation on upstream pumping capacity. Wang J H [6] established a thermo-mechanical coupling model to analyze the friction characteristics of friction pairs. Bai Y G [7] used the Gauss-Seidel block and iterative coupling method to analyze the influence of flow and heat transfer in the internal cavity of the sealed structure and external heat transfer on the seal. Chen H L [8] performed a fluid-solid-thermal coupling analysis on the seal end face, and obtained that the fluid film thickness change on the inner and outer diameter sides was 16%. Some researchers have established micro-gap flow field models based on micro-gap fluid flow characteristics [9-10], two-phase flow particle dispersion mechanism [11], and spiral groove end face lubrication state [12]. Li G Q [13] summarized common seal types such as dry gas seals. Wang R J [14] studied the characteristics of double-end mechanical seals and proposed common applications for double-end mechanical seals. Li F Q [15] summarized the dry gas sealing technology. The scholars have analyzed the dry gas sealing performance of end grooves such as herringbone grooves [16], spiral grooves [17,18], T-shaped grooves [19], double spiral grooves [20-22], and tree grooves [23] and obtain the influence of various structural parameters and working conditions on sealing performance.

Previous researches into dry gas seals mainly have focused on coupling method and the end face groove shape. Far too little attention has been paid to the analysis of the ISGDGS' end face fluid considering the heat and end face deformation. This study sets out to explore the influence of the structural parameters of the spiral grooves on the sealing performance using fluid-solid-thermal coupling analysis method. The effects of spiral angle, number of grooves, ratio of weir, ratio of groove to dam, and depth of groove on sealing performance such as opening force, leakage, friction power, and film stiffness were revealed. The findings shed new light on the optimal design of ISGDGS.

1 Structure and principle

Fig. 1 is a schematic diagram of ISGDGS' structure, which is mainly composed of a rotating ring and a static ring assembly. High pressure gas flows through the seal from right to left. There are micron-sized dynamic pressure groove on the rotating ring. The static ring component is composed of a static ring, a static ring seat, an O-ring, a wave spring, spring pad and snap ring.

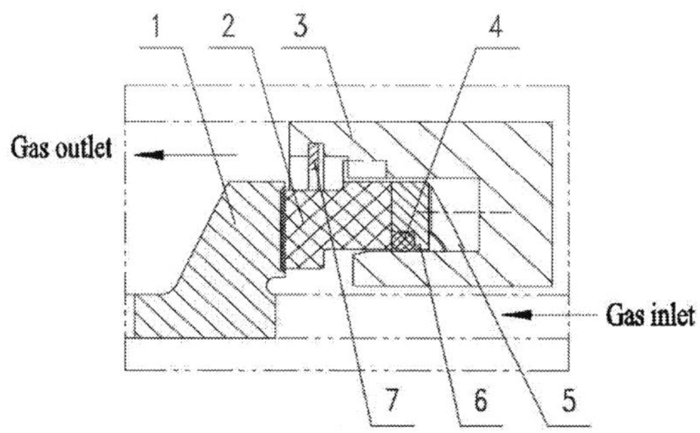

1- rotating ring; 2- static ring; 3- static ring seat; 4- O- ring; 5- wave spring; 6- spring pad; 7- Snap ring

Figure 1. Seal structure.

The gas film structure is shown in Fig. 2. The main leak location is at the contact surface of the rotating ring. When it is stationary, the static ring is tightly fitted with the rotating ring to seal the gas on both sides. At high-speed rotation, as the Figure 3 shows, high pressure gas generates dynamic pressure effect, as the flow channel becomes narrower in the groove. When the dynamic pressure effect is large enough, the end face divides, and a micron-scale air film with a certain rigidity will be formed between the end face. The rigid film is the reason for the sealing effect.

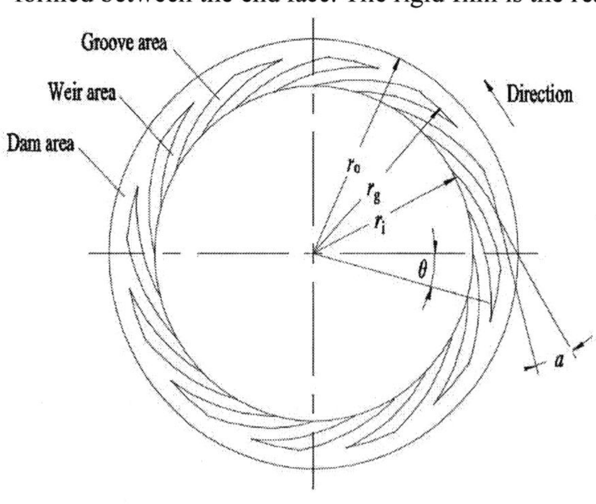

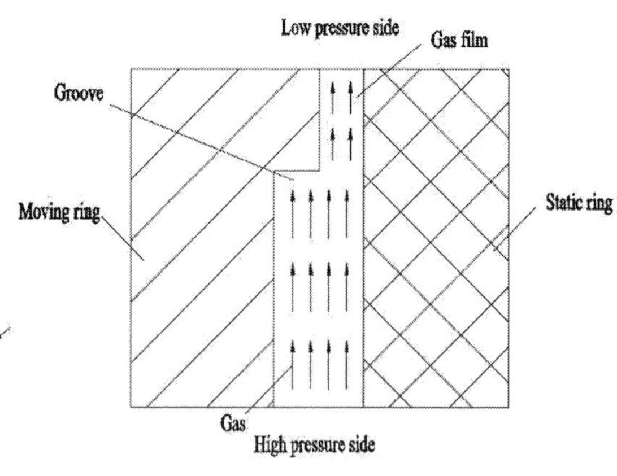

Figure 2. End face spiral groove structure. **Figure 3.** ISGDGS schematic diagram.

Dynamic pressure effects during rotation is related to the end face structural parameters. The structural parameters of the flow field analysis of the seal are shown in Tab. 1.

Table 1. Structural parameters of the flow field of the seal.

Structural parameters	Value
Sealing face inner radius r_i/ mm	28.5
Sealing face outer radius r_o/ mm	34.5
Spiral groove root radius r/ mm	27.5
Spiral groove outer radius r_g/ mm	32.5
Spiral groove curve equation	$R = re^{\theta \tan 15°}$
depth of groove h_g/ μm	5
End flow field thickness (film thickness)h_o/μm	3
Number of spiral grooves N_g	12

2 Analysis model

2.1 Gas Film Governing Equation and Solving Conditions

The basic equation describing the pressure distribution of fluid film between sealing faces of the dynamic pressure seal is the pressure control equation (Reynolds equation):

$$\frac{\partial}{r^2\partial\theta}\left(\frac{\rho h_o^3}{\eta}\frac{\partial p}{\partial\theta}\right)+\frac{\partial}{r\partial\theta}\left(\frac{r\rho h_o^3}{\eta}\frac{\partial p}{\partial r}\right)=6\omega\frac{\partial(\rho h_o)}{\partial\theta} \quad (1)$$

The solving method of governing equation (1) is mainly based on the variational equation of Galerkin method to discretize pressure governing equation and using the finite element method to solve total equation in the calculation domain so as to obtain the pressure value of each node of fluid film after dispersion p.

(1) Assumptions

In view of the actual characteristics of fluid film on the end face of the dynamic pressure seal, the following reasonable basic assumptions need to be made for fluid film between the seal faces:

① The fluid flows between the gap of end faces is laminar flow.

② The fluid film is thin enough, and dimensions between thickness direction of the working film and structural direction are thousands of times different. It can be assumed that the fluid density and pressure remain unchanged along the working film thickness direction;

③ Fluid does not cause relative sliding when it comes into contact with the rotating and static rings;

④ The speed gradient only considers $\partial u/\partial z$ and $\partial v/\partial z$ ignoring other speed gradients, since the working film thickness is micron level, but the speed changes greatly.

⑤ The temperature and viscosity of fluid during the flow process are constant, mainly because the pressure analyzed in this paper is not so high, and the temperature is normal.

⑥ This paper analyzes the steady-state performance, assuming that the fluid film is stable and will not rupture.

(2) Boundary conditions

Solving the pressure control equation must satisfy the following boundary conditions:

① When $r = r_i$,then $p = p_i$; when $r = r_o$,then $p = p_o$.

② The wall surfaces on both sides of the seal in the circumferential direction are periodic boundary conditions:

$$p(r,\theta) = p(r, \frac{2\pi}{N_g} + \theta) \quad (2)$$

2.2 Performance parameter solving equation

After obtaining the pressure distribution of the fluid film on the end surface, use the following equations to obtain performance parameters such as the opening force F, the film stiffness s, the leakage q, and the friction power P_n.

(1) Opening force F: The fluid opening force is calculated by integrating the pressure on the area. The opening force calculation formula is shown in (3).

$$F = \int_0^{2\pi} \int_{r_i}^{r_o} prdrd\theta \tag{3}$$

(2) Film stiffness s. The fluid film stiffness characterizes the ability of the sealing end face to resist interference. The film stiffness s calculation formula is shown in (4).

$$s = -\frac{\partial F}{\partial z}\Big|_{h_0} \tag{4}$$

(3) Leakage q. The radial mass flow through the film thickness h_o and width $rd\theta$ is:

$$q_r = \left(\int_0^h \rho u_r \mathrm{d}z\right) r\mathrm{d}\theta = \left(\int_0^h \frac{\rho}{2\mu}\frac{\mathrm{d}p}{\mathrm{d}r}\left(z^2 - hz\right)\mathrm{d}z\right) r\mathrm{d}\theta = -\left(\frac{\rho h^3}{12\mu}\frac{\mathrm{d}p_0}{\mathrm{d}r}\right) r\mathrm{d}\theta \tag{5}$$

Integrate it along the circumference of the leakage surface to get the radial leakage:

$$q = \int_0^{2\pi} q_r d\theta = -\int_0^{2\pi} \frac{\rho h^3}{12\mu}\frac{dp_0}{dr} rd\theta \tag{6}$$

(4) Friction power consumption P_n. The friction of the fluid on the end face of the seal mainly overcomes the friction torque m_θ generated by the shear force of the fluid viscosity. Obtain the friction torque m_θ as:

$$m_\theta = \int_0^{2\pi} \int_{r_i}^{r_o} \tau r^2 drd\theta = \int_0^{2\pi} \int_{r_i}^{r_o} \left(\frac{\partial p}{r\partial\theta}\frac{h}{2} + \mu\frac{\omega r}{h}\right) r^2 drd\theta \tag{7}$$

Further solving friction power:

$$P_n = m_\theta \omega \tag{8}$$

2.3 Fluid-solid-Thermal Analysis Method

Fig. 4 shows the process of fluid-solid-thermal analysis. Under the actual operating conditions, the end surface is deformed by the influence of centrifugal force, pressure and temperature, which causes the thickness of fluid film on the end surface to change, and then affect fluid film pressure distribution and sealing performance. The fluid- solid- thermal coupling method couples the calculation results of fluid, solid and heat, getting the actual fluid film pressure distribution. The pressure distribution contributes in calculating the fluid sealing performance. Since the pressure change of the fluid film has little effect on the temperature, for the sake of calculation, only the influence of temperature on the unidirectional of the fluid film is considered. The specific calculation steps are as follows:

(1) Divide the mesh according to the input parameters and calculate the deformation components δ_{ij} and δ_{id} of the seal ring;

(2) Correct the initial thickness of the fluid film to obtain the initial thickness correction value h_i for the fluid domain, where $h_i=h_o+\delta_{ij}+\delta_{id}$.

(3) The thermal deformations δ_{kj} and δ_{kd} are obtained through thermal calculation.

(4) Obtaining fluid pressure distribution p_1 through calculation of the fluid part

(5) The pressure p_1 is the boundary condition for the new round of solid deformation calculation, and the deformation components δ_{ij}' and δ_{id}' are obtained. The new fluid film thickness is $h_i'=h_o+\delta_{ij}'+\delta_{id}'+\delta_{kj}+\delta_{kd}$. Repeated iterative calculations. When the pressure difference meets the set accuracy (ε), the pressure distribution of the fluid film and the deformation of the seal ring reach a steady state.

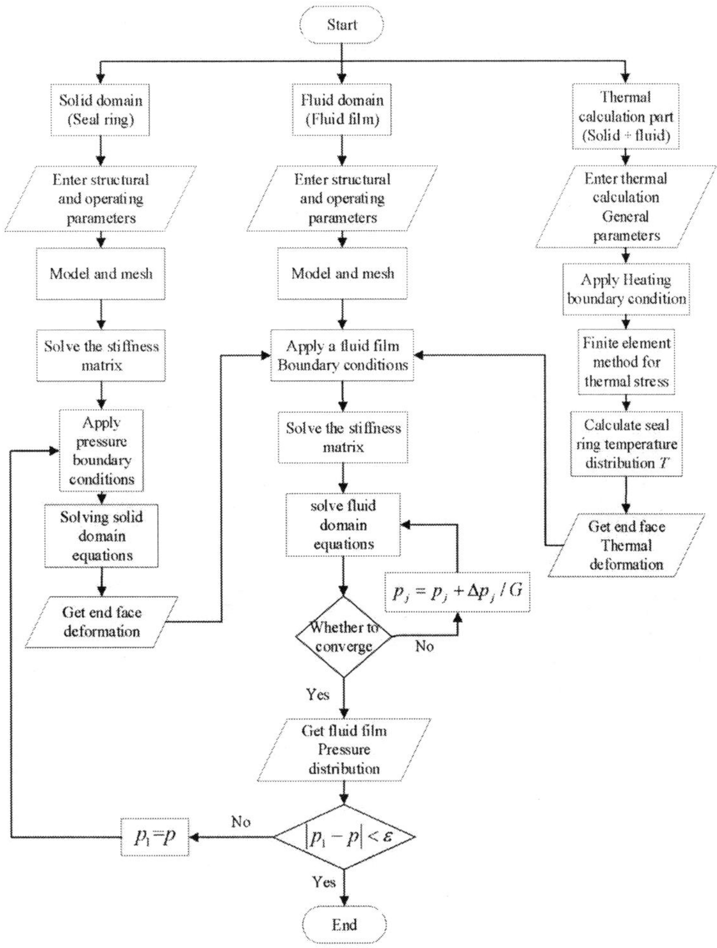

Figure 4. Flow chart of fluid-structure-thermal coupling analysis.

2.4 Verification calculation

With this method, the rotating ring end face deformation, the static ring end face deformation and end face fluid film pressure distribution are obtained showing in Fig. 5, Fig. 6, and Fig. 7.

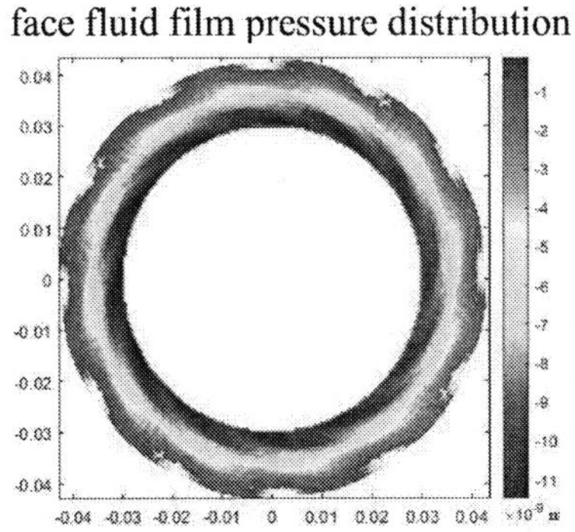

Figure 5. Deformation of the end face of the rotating ring after coupling.

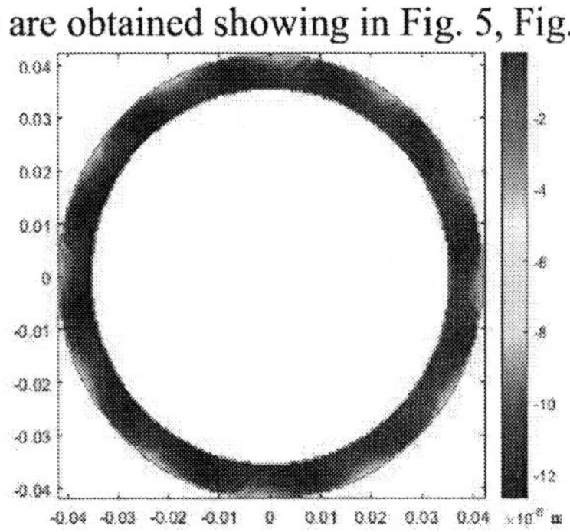

Figure 6. Deformation of the static ring end face after coupling.

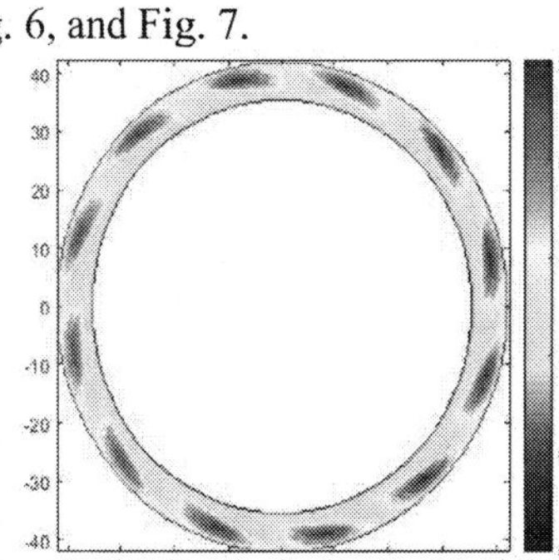

Figure 7. Fluid film pressure distribution after coupling.

It can be seen from the Fig.5 and Fig.6 that the outside of the end face is higher than the inside. It indicates that the end face gap is convergent. Convergence gap is conducive to the formation of fluid film, which is beneficial to the stability of the seal and the reduction of the leakage. Due to the existence of the dynamic pressure groove, the deformation of some parts of the end face changes. The figure also shows that the relative deformations of the rotating ring and static change are 1nm and 3nm, respectively, which has a certain impact on the micron-scale fluid film, demonstrating the necessity of fluid-structure-thermal coupling analysis. It can be known from the Fig. 7 that the seal can produce a considerable dynamic pressure effect under working conditions.

3 Test verification

3.1 Test device

The ISGDGS test bench is shown in Fig. 8, and the measuring device is shown in Fig. 9.

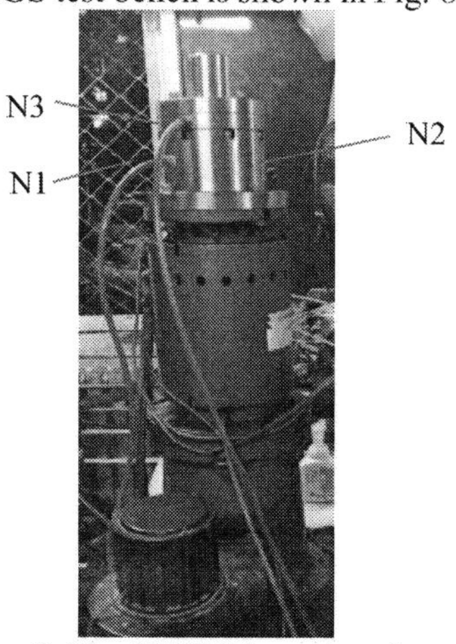

Figure 8. The ISGDGS test bench.

Figure 9. high-precision flow meter.

High pressure gas enters from the N1 port and exits from the N2 port. N3 is a leak measurement port opened on the side of the air atmosphere. N3 is connected to an electronic flow meter to measure the leakage of the seal. The test uses a high-precision flow meter to measure the leakage of the seal, and its measurement accuracy can reach 0.001 L/min.

3.2 Verification analysis

The static ring is softer than the rotating ring, so the wear marks caused by the friction are more obvious on the static ring. The micro-morphology of the static ring end surface before and after operation is shown in Fig. 10(a) and (b) respectively.

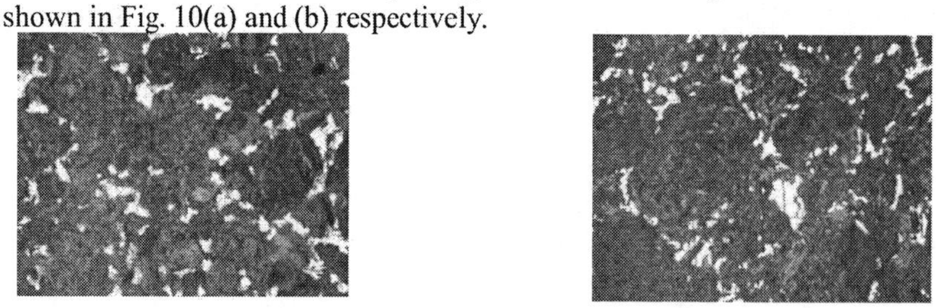

(a) (b)

Figure 10. Micro-morphology of static ring.

The flood white point in the picture is the metal antimony immersed in graphite. As can be seen from Fig. 10(a), the end face of the static ring is flat and free of scratches before operation. Fig. 10(b) shows the end face of the static ring after operation. There are regular scratches in the direction of the lower left corner to the upper right corner after the operation. Using the electronic scale with an accuracy of 0.1g, the static ring mass difference before and after operation is 0.0g. There is almost no wear on the end face at high speeds, which means that the ISGDGS design is feasible.

The test data and calculation data are shown in Fig. 11. What is striking in this figure is the growth of Leakage. The leakage increases with the increase of the rotation speed. The larger the pressure difference, the larger the leakage. The test leak and the calculated leak have the same trend with the rotation speed. But the numerical calculation assumes that the fluid film between the end faces is a stable gas film, the sealing dynamic pressure effect is weak at low speeds, the end face is not opened, and the fluid film is unstable. Therefore, at 2000 $r{\cdot}min^{-1}$, the test leakage of 0 $L{\cdot}min^{-1}$ is inconsistent with the calculated data. In the case of the seal opening, the deviation between the test and calculation data is very small so the seal analysis calculation is of reference significance.

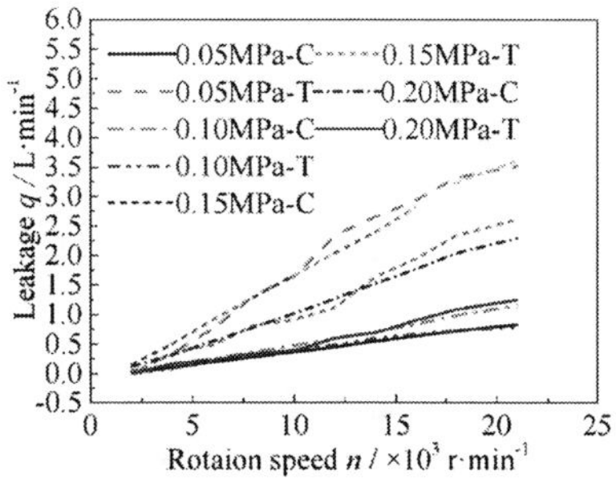

Tip: C for calculation data; T for test data.

Figure 11. Test and calculation of leakage under different pressures.

4 Structural parameter analysis results

The deformation of the rotating and static rings and the distribution of the film pressure will affect the sealing performance. The sealing performance is mainly achieved by adjusting the end face parameters such as the helix angle, the number of grooves, the ratio of groove to weir, the ratio of groove to dam, and the depth of groove. The analysis structural parameters are shown in Tab 1. Set the analysis external pressure to 0.15 MPa, internal pressure to 0 MPa, and the initial temperature of the medium to 20 ℃.

4.1 Helix Angle

The effect of the helix angle on the spiral groove dynamic pressure sealing performance is shown in Fig. 12. What stands out in this figure is the rapid increase in opening force with the increase of the helix angle. After the helix angle is 15-20°, the increasing trend slows down. The leakage increases with the increase of the helix angle, and the increase rate gradually slows down. This phenomenon demonstrates that within the scope of the study, the increase of the helix angle reduces the resistance of the liquid to enter the spiral groove, the dynamic pressure effect is enhanced, and the leakage and the opening force are large near 22-24 °. Friction power is not greatly affected by the helix angle, and the trend of friction power at high speeds is different from lower rotation speeds. This is mainly because friction power is related to the end surface state and deformation, and the end surface deformation is relatively large at high speed. Increasing helix angle leads to an increase in fluid film stiffness. After turning on, the rotation speed will not affect the process of fluid entering the spiral groove, and the influence of the spiral angle on the sealing performance under different rotation speeds is the same. Increasing the helix angle helps to improve the opening performance, but at the same time the increase in the leakage rate is detrimental to the normal operation of the seal. A smaller helix angle reduces the opening force, which in turn requires a greater rotational speed to balance the closing force, and the seal requires a greater rotational speed to open and ensure the rigidity of the gas film.

Therefore, considering the influence of the helix angle on the sealing performance, it is reasonable to choose a helix angle of 15°~20°.

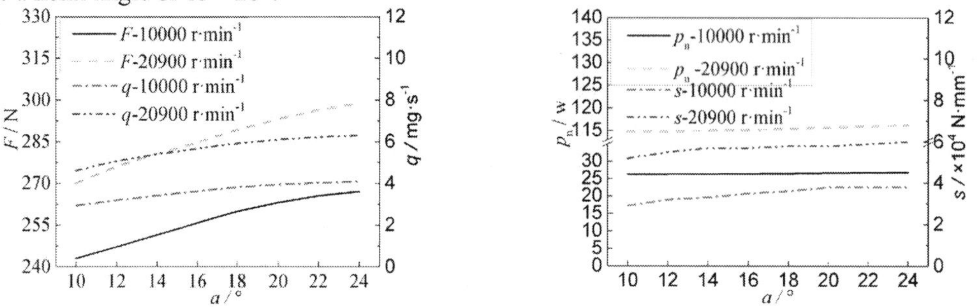

(a) (b)

Figure 12. Effect of helix angle on dynamic pressure sealing performance.

4.2 Number of Grooves

The effect of the number of grooves on the sealing performance is shown in Fig. 13.

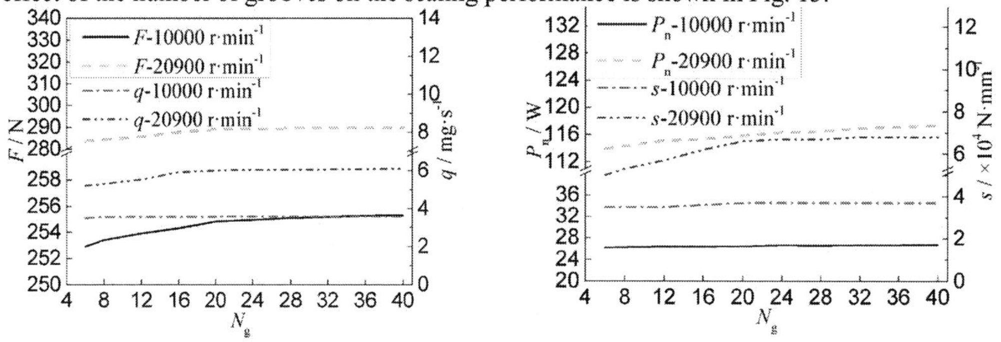

(a) (b)

Figure 13. Effect of the number of grooves on the dynamic pressure sealing performance.

Figure 13(a) reveals that the opening force increases with the increase of the number of grooves, and finally stabilizes. After the number of grooves is 16 to 20, the opening force reaches the maximum. This is because when the number of grooves is small, the dynamic pressure effect of the fluid entering the spiral groove is not obvious. Increased number of grooves mean increased leak paths so the greater the number of slots, the greater the leakage. The leakage reaches a maximum when the number of grooves is about 16. Due to the small overall leakage at low speeds, the variation of the leakage with the number of grooves is not obvious. The increase in the number of grooves makes the state of the end face worse and the frictional heat generation increases. So the friction power increases, there are more number of grooves. The increase of the number of grooves increases the rigidity of the air film, and the increase trend slows down when the number of grooves is 16-20. When the number of grooves exceeds 16, the opening force, leakage rate, stiffness, and friction power tend to stabilize. In order to ensure enough opening force and a small leakage, it is better to take 12-16 grooves.

4.3 Ratio of Groove to Weir

The influence of the ratio of groove to weir on the sealing performance is shown in Fig. 14. What can be clearly seen in Fig. 14 is the opening force and leakage rate have a maximum value within the ratio of groove to weir of 0.6 to 0.75. This is because a larger ratio is easier to enter the grooves area than a fluid, resulting in a dynamic pressure effect and a large leakage. When the ratio is too large, the flow channel is overdone, and the dynamic pressure effect is not obvious. The larger ratio, the larger the average gap between the end faces of the seal will be, which will cause the shear force between the end faces to decrease and the power consumption to decrease. When the ratio is 0.5, the groove area and the weir area is the same, the film pressure distribution is the most uniform, and the stiffness of the liquid film reaches the maximum, so the optimal range of stiffness is 0.4-0.6. Comprehensive analysis of the sealing performance, the ratio of groove to weir should be taken between 0.5-0.6.

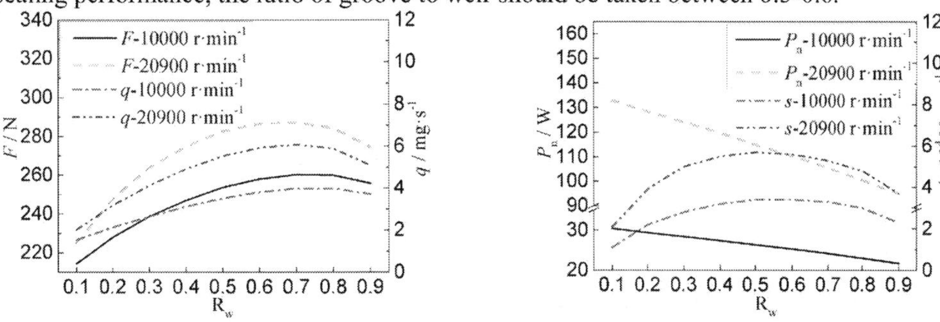

(a) (b)

Figure 14. Effect of the ratio of groove on the dynamic pressure sealing performance.

4.4 Ratio of Groove to Dam

The effect of the ratio of groove to dam on the sealing performance is shown in Fig. 15. As can be seen from the Fig. 15, there is a maximum opening force when the ratio is 0.5 to 0.7. The leakage increases with the increase of the ratio. This is because a small ratio is not conducive to the fluid entering the trough. A large ratio makes the dam area smaller, the fluid resistance and the dynamic pressure effect decreased, resulting in leakage increased. When the ratio of groove to dam increases, the difference in leakage between high and low rotation speeds becomes larger. This is because the gas is more likely to leak under the large ratio and the taper deformation of the end face. When the ratio is too large, the bearing capacity and throttling capacity of the film decrease, and the stiffness reaches the maximum between 0.55 and 0.75. These findings suggest that the best overall sealing performance is obtained when the ratio of groove to dam is between 0.65 and 0.75.

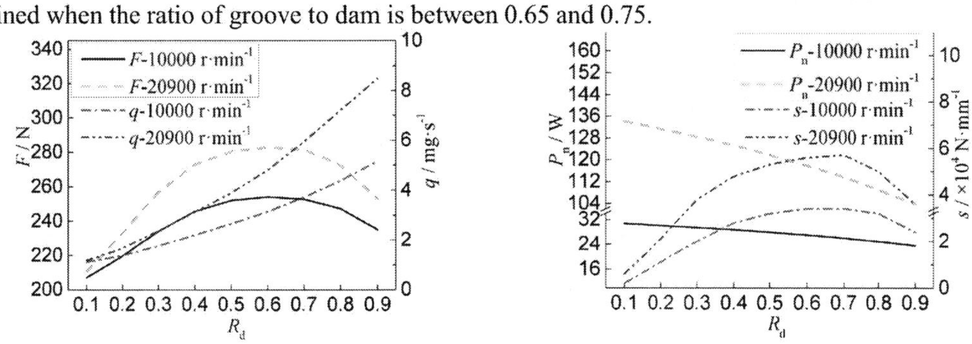

(a) (b)

Figure 15. Effect of the ratio of groove to dam on the dynamic pressure sealing performance.

4.5 Depth of Groove

The effect of groove of depth on sealing performance is shown in Fig. 16. When the groove of depth is less than 12μm, the opening force and leakage rate increase rapidly with the increase of the groove of depth, but when the groove of depth exceeds 12μm, the opening force and leakage rate both slowly decrease. This is because the resistance of the fluid in and out of the groove area is greater due to the large groove of depth, and the dynamic pressure effect decreases. When the groove depth excessive increase, pressure in the groove area decrease so that the pressure difference between groove areas and dam areas decrease causing the leakage reduced. Power consumption decreases with increasing groove of depth. When the groove of depth is 5-10μm, the gas film can produce a certain dynamic pressure effect and a better bearing capacity. At the same time, the thickness of the gas film is relatively small, and the stiffness reaches the maximum. In summary, when the groove of depth is about 7-10μm, it has better opening force, leakage rate and stiffness.

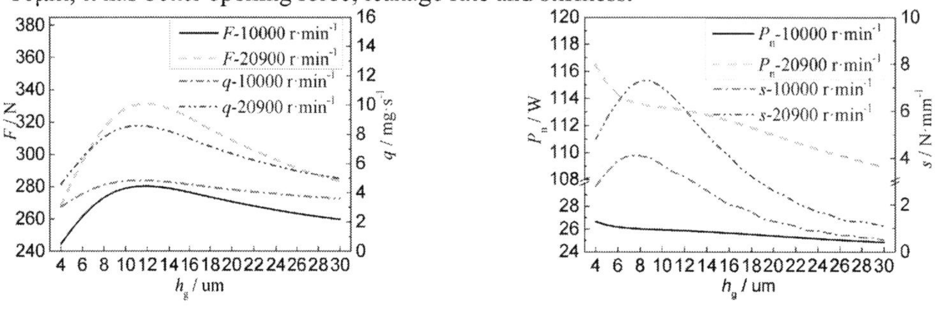

(a) (b)

Figure 16. Effect of the groove of depth on the dynamic pressure sealing performance.

4.6 Sealing performance comparison test analysis

According to the results of numerical analysis, four structures of T1~T4 were selected for experimental analysis. The specific parameters of T1~T4 are shown in Tab 2.

Table 2 Four kinds of spiral groove structure parameters.

structure type	r_o /mm	r_i /mm	N_g	h_g / μm	R_d	a/°
T1	34.5	28.5	12	5	0.7	15
T2	34.5	28.5	8	5	0.7	15
T3	34.5	28.5	12	10	0.7	15
T4	34.5	28.5	12	5	0.72	15

The test was carried out under three conditions of 0.05, 0.1MPa, and 0.15MPa, and the test data is shown in Figure 17(a), (b)and(c).

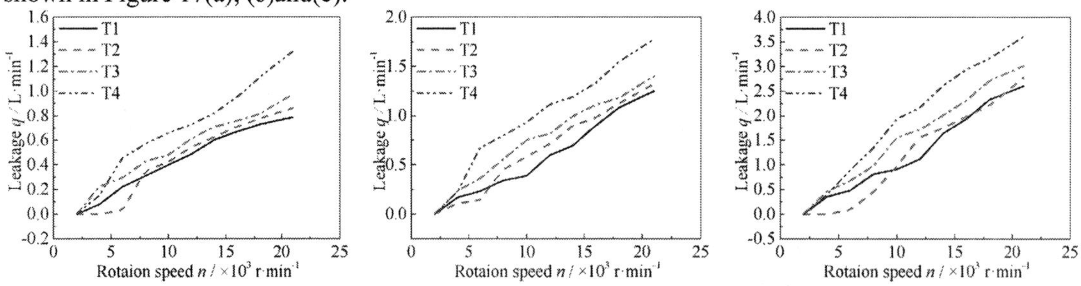

(a) (b) (c)

Figure 17. Test data.

The end face is easy to wear at low speed, so the 2000rpm working condition passed quickly in the test. Observe that there is no significant change in the leakage of the four types of seals under the three pressure conditions during this process. In order to observe the trend, temporarily replace the seal leakage at 2000 rpm with 0 leakage.

Under the three conditions of 0.05, 0.1MPa, and 0.15MPa, the leakage of the inner spiral groove dry gas seal increases with the increase of speed. This shows that the increase in rotation speed increases the film thickness and the leakage.

In the speed range of 0~8000rpm, the leakage of T1, T3, T4 three structural seals suddenly increases between 2000~6000rpm. The sealing of structure T2 suddenly increases in leakage between 6000 and 8000 rpm. The sudden increase in leakage means that the dynamic pressure effect generated by the end fluid pushes away the dynamic and static rings. This means that the end face opens when the leakage increases suddenly.

It can be seen from the leakage change trend that the order of opening performance from good to bad is T3, T4, T1, T2. It can be seen that the large depth of groove the opening performance is better. The reduction of the number of grooves has an obvious effect on the opening performance of the seal. The structure with a large ratio groove to dam has a larger leakage at a high speed, and a structure with a small number of grooves has a smaller leakage.It can be seen from the figure that when the structural parameter is T1, the seal opening speed is lower and the leakage is the smallest, which is similar to the numerical analysis result.

5 Conclusions

The study was designed to verify the feasibility of ISGDGS and analyze the impact of end face parameters on its performance. The findings suggest an important role for end face parameters in promoting ISGDGS' performance. When the helix angle is 15-20 °, the number of grooves is 12-16, the ratio of groove to weir is 0.5-0.6, the ratio of groove to dam is 0.65-0.75, and the depth of groove is 7-10 um, the opening force and air film stiffness are relatively large, the leakage and friction power are small, and the seal with the best comprehensive performance is obtained. The correctness of the analysis model established based on fluid-solid-thermal coupling has been verified by the experimental results, laying the groundwork for future research into the optimal design of ISGDGS.

References

[1] Shuang-xi Li, Wen-bo Song, Qiu-xiang Zhang, et al. *Opening Characteristics of Dry Gas Seal* [J]. CIESC Journal, 2011, 62(03): 766- 772.

[2] Liang Han. *Design and Application of a Dry Gas Seal on Gas for Natural Gas Compressor* [J]. Chinese Journal of Turbomachinery Chinese Journal of Turbomachinery, 2019, 59 (02): 76- 82.

[3] Qiong Hu, Kai Tao, Jian-jun Sun, et al. *Numerical Investigation on Heat Transfer and Thermal-structural Coupled Deformation of Split Mechanical Seals* [J]. Lubrication Engineering, 2018, 43 (08): 24- 31.

[4] Xian Luo, Kai Qin, Xin-yu Li. *A Numerical Analysis of the Grooved Surface with Waviness and Taper Effect on Sealing Performance* [J]. Lubrication Engineering, 2017, 42(04): 17-22.

[5] Wei-feng Huang, Xiao-bo Pan, Zi-xi Wang, et al. *Thermal-fluid-solid coupled analyses of upstream mechanical seals in pumps* [J]. Journal of Tsinghua University (Science and Technology): 1-8[2020-03-05].https://doi.org/10.16511/j.cnki.qhdxxb.2019. 26. 039.

[6] Ji-hui Wang, Zhi Chen, Can-hong Gu, et al. *Thermo-mechanical Coupling Analysis of Friction Interface of Mechanical Seals under Dry Friction* [J]. Tribology, 39(06): 737- 745.

[7] Yu-guang Bai, Guang-qing Xia, De-chuan Sun, et al. *Research on numerical method of heat-sealing structure for spacecraft hull based on multi-physics coupling* [J]. Journal of Solid Rocket Technology, 2014, 37(06): 756- 762.

[8] Hui-long Chen, Tong Li, Bing-juan Zhao, et al. *Analysis of flow field and performances in hydrodynamic mechanical seal based on fluid — solid — heat interaction* [J]. Journal of Drainage and Irrigation Machinery Engineering, 2017, 35(06): 502- 507.

[9] Pinel S I, Signer H R, Zaretsky E V. *Comparison between oil-mist and oil-jet lubrication of*

high-speed, small-bore, angular-contact ball bearings[J]. Tribology Transactions, 2001, 44(3): 327- 338

[10] PATHAK M, DASS A. K. and DEW ANA. *An investigation of turbulent rectangular jet discharged into a narrow channel weak crossflow*[J]. Journal of Hydrodynamics, 2008, 20(1): 154- 163

[11] Jing-yu Fan; Song-li Xu; Dao-zeng Wang. *PDA measurements of two-phase flow structure and particle dispersion for a particle-laden jet in crossflow*[J]. Journal of Hydrodynamics, 2010(1): 9- 18

[12] Y. Qiu, M.M. Khonsari. *Investigation of tribological behaviors of annular rings with groove spiral*[J]. Tribology International, 2011, 44(12): 1610- 1619

[13] Gui-qin Li, Yu-hua Wang. *Basic Principle and Use Analysis for Dry Gas Seal of Compressor* [J]. Chinese Journal of Turbomachinery Chinese Journal of Turbomachinery, 2000(01): 19- 23.

[14] Rui-jing Wang. *Application of Dry Gas Double Seal for Hydrogen Compressor* [J]. Chinese Journal of Turbomachinery Chinese Journal of Turbomachinery, 2002(01): 50- 52.

[15] Feng-qin Li, Qiang Wang, Yuan-hong Song. *Technology of Spiral Groove Dry Gas Seal for Compressors* [J]. Chinese Journal of Turbomachinery Chinese Journal of Turbomachinery, 2008(01): 33- 35.

[16] He Zhang, Mu-ming Hao, Jun-ma Yuan, et al. *Influence of Surface Roughness on Sealing Performance of Herringbone Groove under Different Lubrication Conditions*[J]. Lubrication Engineering, 2018, 43(06): 61- 66.

[17] Wen-bo Song, Shuang-xi Li, Ji-ning Cai. *Study on the Performance of Dry Gas Seal with Cone-shaped and Ladder-like Spiral Groove*[J]. Lubrication Engineering, 2010, 35(04): 80-84.

[18] Huan Li, Shuang-xi Li, Xiao-fen Li, et al. *Multi-Parameter Orthogonal Optimization and Experimental Study on The Structure of The Oil-Gas Two-Phase Dynamic Pressure Seal End Face* [J]. Journal of Beijing University of Chemical Technology (Natural Science), 2017, 44(01): 76- 84.

[19] Yan Wang, Jian-jun Sun, Kai Tao, et al. *Numerical Analysis of T — Groove Dry Gas Seal and Groove Optimization* [J]. Tribology, 2014, 34(04): 420- 427.

[20] Wen-qi Chen, Zhi-qiang Li, Yun-ke Wu, et al. *Optimization Design of Groove of Double Spiral Angle Groove Dry Gas Seal*[J]. Lubrication Engineering, 2018, 43(07): 23- 30.

[21] Peng-cheng He, Ye-lin Li. *The Comparison Analysis of Upstream Pumping Mechanical Seals with Double-row Spiral Grooves and Single-Row Spiral Grooves* [J]. Journal of Hunan University of Arts and Science (Science and Technology), 2019, 31(04): 25- 29.

[22] Yu-hua Wang, Da-ren Jiang. *The Brief Comparison between Spiral Groove Dry Gas Seal and Bidirectional Dry Gas Seal* [J]. Chinese Journal of Turbomachinery Chinese Journal of Turbomachinery, 2008(02): 35.

[23] Sun Xin-hui, Fang-qi Yan, Zhen-tao Li, et al. *Numerical Analysis on Steady Characteristics of Upstream Pumping Mechanical Seals with Fir-Tree Grooves* [J]. Lubrication Engineering, 2018, 43(12): 76- 80.

About the author: Li Shuangxi (1977-), male, associate professor, doctor
Research Interest: Fluid Sealing Technology.
E-mail: buctsx@126.com

IOP Conf. Series: Materials Science and Engineering 1081 (2021) 012007 doi:10.1088/1757-899X/1081/1/012007

Bifurcation Characteristics of Airfoil-NESs coupled System

Wenfan Zhang[1], Genbiao Zhou[1], Yafeng Zhou[1], Lei Wang[1], Jiazhong Zhang[2]

[1] Xi'an ShaanGu Power CO., LTD, Xi'an, Shaanxi 710611, P.R. China
[2] School of Energy and Power Engineering, Xi'an Jiaotong University, Xi'an, Shaanxi 710049, P.R. China

E-mail: sheen.z@163.com

Abstract. Aeroelastic instability has become a hot research topic. It has increasingly high-lighted its importance in the aircraft and fluid machinery industry and brings challenges to the safe flight of various aircraft and the safe operation of fluid machinery. Numerical simulations were performed to study the flow-induced oscillation of a two-dimensional airfoil with two non-linear energy sinks (NES). The main system has two degrees of freedom-pitch and heave. These two NESs are considered as subsystems, with the first NES at the leading edge and the second NES at the trailing edge. The bifurcation behaviour and the mechanism of the system leading to chaos of the coupled system under different parameters and initial conditions is studied by nonlinear analysis, and the relationship between large amplitude limit cycle oscillation and the chaotic occurrence point of the main system is revealed, this corresponding relationship has important theoretical significance for mastering and adjusting system parameters to make NES work within its ideal oscillation suppression parameters.

1. Introduction

The development of modern aircraft and fluid machinery have been gradually accelerated, and the aeroelastic problems always accompany it. In the classical theory of aeroelasticity, the aeroelasticity problem is simplified into a set of linear equations which are easy to solve. However, when the incoming velocity increases to high subsonic or transonic speeds, this assumption usually results in inaccurate results. In addition, the phenomena of flow separation and shock oscillation are beyond the scope of linear aeroelastic theory. In fact, due to the structural nonlinearity of the system, aeroelastic system may exhibit various phenomena, such as limit cycle oscillation and chaotic oscillation, etc., and the problems caused by structural nonlinearity have gradually attracted the attention of researchers in relative field.

Fung and Dowell et al. modified the basic linear model, studied several unstable oscillation responses caused by system instability [1, 2], and found that the amplitude increased exponentially. At the same time, due to the nonlinear characteristics of stiffness or damping in the structure, the exponential growth of this response is limited [3], and ultimately, the system maintains LCOs.

In fact, the LCO problem exists for A long time, such as F-16 and F/A-18 under high subsonic and transonic oscillation problem of LCOs, Robert discusses the specific situation to avoid the LCOs is the wing of the method [4], reveal some typical similarity and difference between flutter and LCOs, moreover, the LCOs in flight tests on aircraft external system and the ability of the whole task are emphatically studied. Later and Cattarius used numerical methods to study the interactions between the wing and its subsystems of the F-16 [5]. Denegri conducted several flight tests to measure the actual limit cycle vibration of these fighters and found three typical aeroelastic response types during the test, namely, classic flutter, typical LCOs and atypical LCOs [6]. Lee and LeBlanc used numerical methods to study the effect of cubic nonlinear stiffness on the flutter behavior of airfoils [7]. The research shows that when the system stiffness is small, it may have a subcritical LCO lower than the linear flutter speed. At this time, the system shows a dependence on the initial conditions; however, for the system with larger stiffness, the dependence on the initial conditions disappears. In addition, Lee and Desrochers consider and use different types of structural nonlinearity for flutter analysis [8]. Singh used the description function method to study the existence of LCO in the inflow excitation of wing section with pitch nonlinearity (including asymmetry) [9]. The results show that the amplitude of pitch LCO does not always increase with the increase of velocity when the stiffness center is determined. Generally, the amplitude of oscillation increases with the flow rate; however, for some positions of the elastic axis, the amplitude of the limit cycle can be reduced within a certain range of flow rate. Nonlinear LCOs in aeroelastic system was calculated and experimentally studied by O'Neil [10], the results show that linear analysis has been completely unable to predict the LCOs in the aerodynamic effect. The analysis tools provided by Neil show that the predicted LCO amplitude and frequency are very close to the experimental data, which proves that it is necessary to consider the nonlinear characteristics of fluid and structure modeling.

Gilliatt et al. studied the possibility of internal resonance of aeroelastic system (stall model) under nonlinear aerodynamic load [11], the selected control equation contains the cubic term of velocity, and the nonlinearity of the system shows a more complex form than the previous study. They studied the oscillation and bifurcation characteristics of the unstable limit cycle caused by different flow velocity, when the flow velocity is lower than the flutter velocity, it is found that the internal resonance with a frequency of 3:1 can lead to a large oscillation of the system. Lee used harmonic balance method to study the nonlinear aerodynamic characteristics of airfoils with strong cubic nonlinear stiffness. The results are in good agreement with the experimental results. In order to capture the secondary bifurcation phenomenon of the heave and pitch motion, the equation should contain at least the ninth order harmonic term [12]. Lind et al. simulated the structural nonlinearity of flight data by using wavelet transform and predicted the occurrence of LCOS by using numerical results [13]. They analyzed the transient response of the system under different flow velocities, and they predicted the initial conditions and occurrence time of limit cycle oscillation caused by Hopf bifurcation. The basic method used in the derivation can be used to analyze general nonlinear systems. As the nonlinear characteristics of the system may significantly affect the aeroelastic stability of some aircraft structures, Straganac et al. considered the structural nonlinear characteristics of the system to accurately predict and actively suppress the aeroelastic instability of the system [14]. In this paper, an active control strategy for the nonlinear oscillation is proposed to control the LCOS of the wing section with the nonlinear mechanism, At the same time, compared with the experimental results in the wind tunnel, the numerical results are in good agreement with the experimental results, which proves the effectiveness of this active control strategy. Li's work shows that the hysteresis of nonlinear oscillation has no effect on the aeroelastic response of the system. By using the bifurcation phenomenon of pitch freedom, the influence of the central position of the elastic axis, the ratio of airfoil to air mass and the structure preload on the chaotic motion of the system is studied. It is proved that when the position of the elastic axis moves from the middle chord to the front edge, the chaotic motion will disappear [15, 16].

The research of Nayfeh shows that the small imbalance in rotating machinery may also cause the solution to fall into resonance, and the rotating frequency of rotating parts and the frequency of system vibration fall into resonance capture phenomenon [17]. The resonance behavior in dynamic system will lead to complex dynamic behavior, even chaos [18-20]. Sethna has proved that the harmonic excitation of the system can lead to the coupled response among the vibration modes [21], which is caused by the nonlinearity of the system, also known as the mixed mode response. It is also proved that the internal resonance behavior in these systems can explain the interaction relationship between the vibration and the generating conditions of chaotic response [22]. Some other studies involve the dynamic response of arch structure [23, 24], tension chord [25, 26], and forced response of approximate square plate [27]. In these studies, the natural frequency of the internal vibration mode and the external driving frequency are fixed parameters. Subsequently, some scholars explore the way of resonance capture in the system by studying the phase space geometry of the system, which represents the internal resonance phenomenon in the slowly changing single degree of freedom system [28-30], and these theories have been applied to a series of studies on the rotational phase locking and reentry rolling resonance in the dual rotating spacecraft [31-36]. The limit cycle oscillation (LCO), which is to be suppressed by the NES, is studied from the viewpoint of the TET [37, 38]. The resonance capture (RC) in the coupled nonlinear system is also discussed by the means of the energy and spectrum analysis.

2. Modelling of the system

The rigid airfoil is used as the main system. As shown in Figure 1, the subsystem is a nonlinear energy sink located near the leading edge and trailing edge of the airfoil respectively.

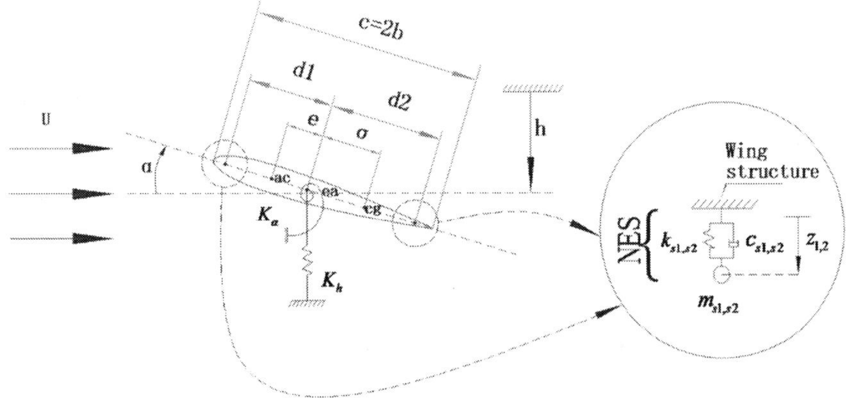

Fig .1 Two-DOF wing model coupled with two NESs

By using the virtual work principle, the governing equation of the coupled system is as follows,

$$\begin{cases} m\ddot{h} + S_\alpha \ddot{\alpha} + K_h(h + c_h h^3) + qSC_{L\alpha}(\alpha + \dot{h}/U) + c_{s1}(\dot{h} - d_1\dot{\alpha} - \dot{z}_1) + \\ k_{s1}(h - d_1\alpha - z_1)^3 + c_{s2}(\dot{h} - d_2\dot{\alpha} - \dot{z}_2) + k_{s2}(h - d_2\alpha - z_2)^3 = 0 \\ I_\alpha \ddot{\alpha} + S_\alpha \ddot{h} + K_\alpha(\alpha + c_p\alpha^3) - qeSC_{L\alpha}(\alpha + \dot{h}/U) + d_1 c_{s1}(d_1\dot{\alpha} + \dot{z}_1 - \dot{h}) + \\ d_1 k_{s1}(d_1\alpha + z_1 - h)^3 + d_2 c_{s2}(d_2\dot{\alpha} + \dot{z}_2 - \dot{h}) + d_2 k_{s2}(d_2\alpha + z_2 - h)^3 = 0 \\ m_{s1}\ddot{z}_1 + c_{s1}(\dot{z}_1 + d_1\dot{\alpha} - \dot{h}) + k_{s1}(z_1 + d_1\alpha - h)^3 = 0 \\ m_{s2}\ddot{z}_2 + c_{s2}(\dot{z}_2 + d_2\dot{\alpha} - \dot{h}) + k_{s2}(z_2 + d_2\alpha - h)^3 = 0 \end{cases} \tag{2.1}$$

And the non-dimensional form is as follows,

$$\begin{cases} y'' + x_\alpha \alpha'' + \Omega^2 y + \xi_y y^3 + \mu C_{L\alpha}\Theta(y' + \Theta\alpha) + \varepsilon_1\lambda_1(y' - \delta_1\alpha' - v_1') + \\ C_1(y - \delta_1\alpha - v_1)^3 + \varepsilon_2\lambda_2(y' - \delta_2\alpha' - v_2') + C_2(y - \delta_2\alpha - v_2)^3 = 0 \\ r_\alpha^2\alpha'' + x_\alpha y'' + r_\alpha^2\alpha + \xi_\alpha\alpha^3 - \gamma\mu C_{L\alpha}\Theta(y' + \Theta\alpha) + \delta_1\varepsilon_1\lambda_1(\delta_1\alpha' + v_1' - y') + \\ \delta_1 C_1(\delta_1\alpha + v_1 - y)^3 + \delta_2\varepsilon_2\lambda_2(\delta_2\alpha' + v_2' - y') + \delta_2 C_2(\delta_2\alpha + v_2 - y)^3 = 0 \\ \varepsilon_1 v_1'' + \varepsilon_1\lambda_1(v_1' + \delta_1\alpha' - y') + C_1(v_1 + \delta_1\alpha - y)^3 = 0 \\ \varepsilon_2 v_2'' + \varepsilon_2\lambda_2(v_2' + \delta_2\alpha' - y') + C_2(v_2 + \delta_2\alpha - y)^3 = 0 \end{cases} \tag{2.2}$$

3. Bifurcation in coupled systems

3.1. Instability of coupled systems

First, the parameters of the coupled system are taken as $\Theta = 0.9$, $\delta_1 = 0.9$, $\delta_2 = -0.9$, $\lambda_1 = \lambda_2 = 0.1$, $C_1 = C_2 = 10$, $\varepsilon_1 = \varepsilon_2 = 0.01$, the data after transient resonance at the initial stage of the system is captured.

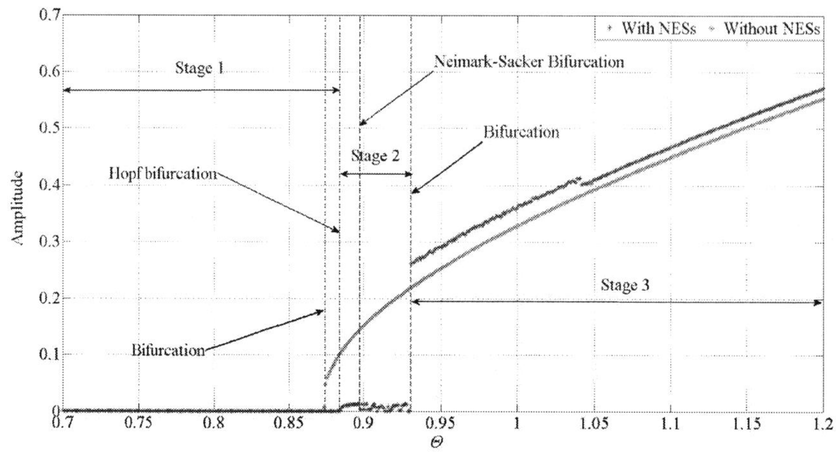

Fig .2 Bifurcation of coupled system

As in Fig .2, is used as abscissa, and vertical coordinate is the average value of the maximum amplitude of pitch oscillation. There are bifurcation point at in the system without NESs. At this point, the system changes from a stable point to a periodic oscillation with a large amplitude. It can be known that the main frequency of the heave oscillation of the system before and after the bifurcation point changes from lower frequency to high frequency, while the main frequency of the pitching oscillation is still low, the resonance mode of the two degrees of freedom of the system changes from 1:1 resonance to 3:1 resonance.

From the perspective of system bifurcation, as the front and rear NES are coupled to the airfoil as a subsystem, the mechanism of oscillation suppression and energy absorption is as follows:

First, the NES always stabilized the oscillation of the main system, and the stability of the system in this stage would not change with the increase of the incoming flow velocity.

Secondly, two types of bifurcation phenomena occurred in the system in stage 2： 1）Hopf bifurcation occurs in the system, and the system exhibits the characteristic of periodic oscillation, the oscillation of the system is always suppressed oscillation; 2）With the increase of the incoming flow velocity, the system experiences Neimark-Sacker bifurcation. The oscillation of the system reflects the quasi-periodic characteristics, and the oscillation of the system is a recurrent series of suppressed burst-out.

Finally, at the end of stage 2, with the third bifurcation of the system, the system returned to the periodic oscillation, and the amplitude of the main system suddenly jumped into the oscillation of the limit cycle of extreme large amplitude, at which point the oscillation suppression capacity of the front and rear NES to the main system failed.

3.2. Bifurcation and chaos of coupled systems

As the incoming velocity continues to increase, the system will exhibit more complex dynamic characteristics. The following two groups of representative parameters are taken to study the nonlinear characteristics of the system.

All parameters of the main system remain unchanged, and the parameters of NES1 and the NES2 are taken as $\delta_1 = 0.9$, $\delta_2 = -0.9$, $\lambda_1 = \lambda_2 = 0.1$, $C_1 = C_2 = 10$, $\varepsilon_1 = \varepsilon_2 = 0.01$. The bifurcation characteristics of the heave oscillation are shown in Fig 3:

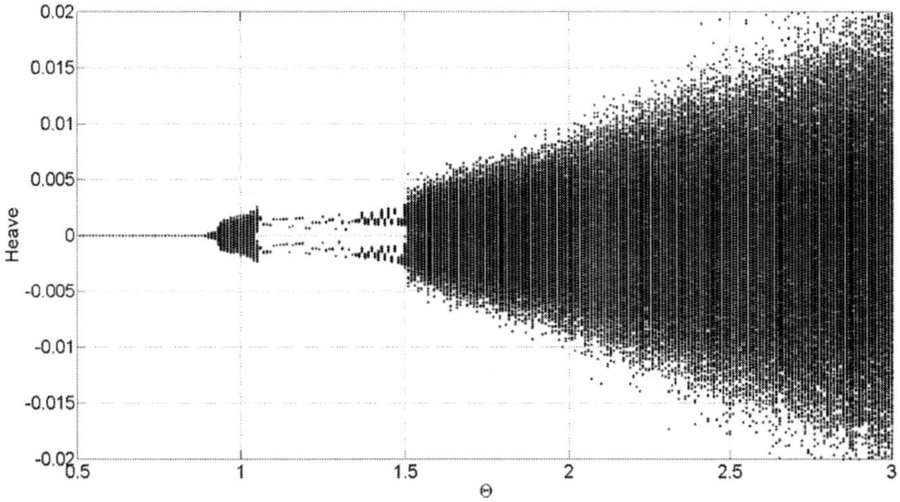

Fig .3 Bifurcation of heave mode

The horizontal sitting is still nondimensional incoming flow velocity, and the research scope is raised to $\Theta = 3.0$, The vertical coordinate is the point on the Poincare section of the heave oscillation.

According to the characteristics shown in Fig 3, the sections of $\Theta = 0.9 \sim 2.0$ are chosen to study.

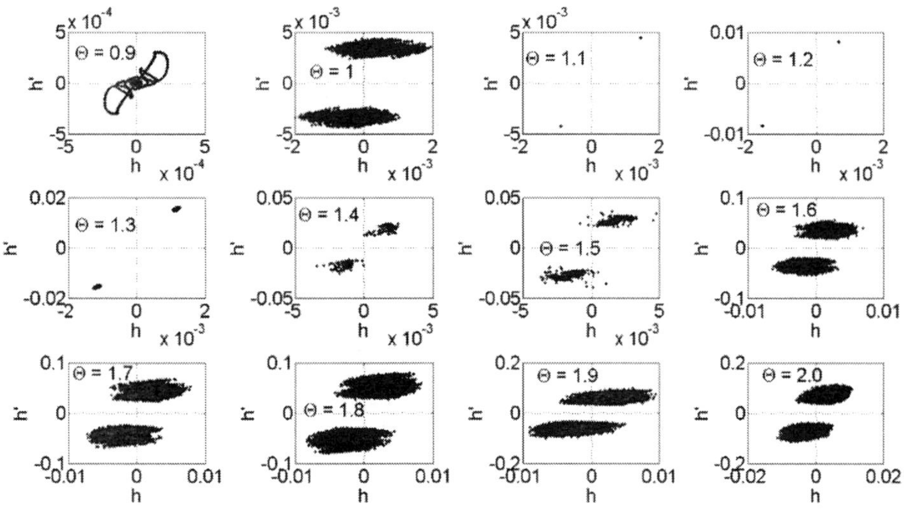

Fig .4 Trajectory change on the Poincare section

As it shows in Fig .4, first, when the non-dimensional incoming flow velocity is $\Theta = 0.9$, the oscillation of the system is quasi-periodic. As the velocity of the incoming flow increases, the oscillation state of the system suddenly develops into chaos, then again into periodic oscillation, and finally into permanent chaotic oscillation. The velocity at which a chaotic system occurs is between $\Theta = 0.9$ and $\Theta = 1$.

The specific location of chaos occurrence point in system 1 is studied. Take a point whose distance is near an exist solution. The sensitivity of the chaotic system to the initial value and the change of the distance between the final solutions are used to determine the location of the chaotic point.

IOP Conf. Series: Materials Science and Engineering 1081 (2021) 012007 doi:10.1088/1757-899X/1081/1/012007

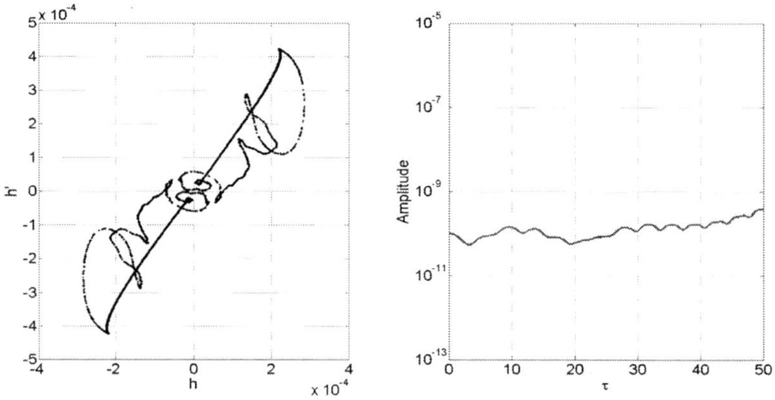

(a) Trajectories on Poincare section (b) Sensitivity to initial value

Fig .5 Trajectories on Poincare section in system 1 and sensitivity of system to initial value, $\Theta = 0.91$

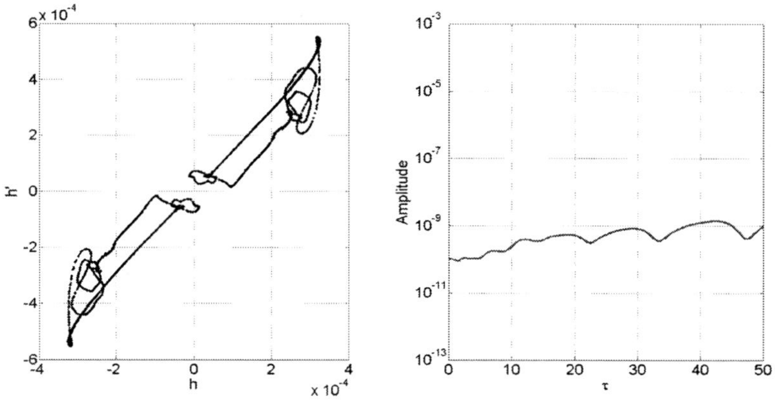

(a) Trajectories on Poincare section (b) Sensitivity to initial value

Fig .6 Trajectories on Poincare section in system 1 and sensitivity of system to initial value, $\Theta = 0.92$

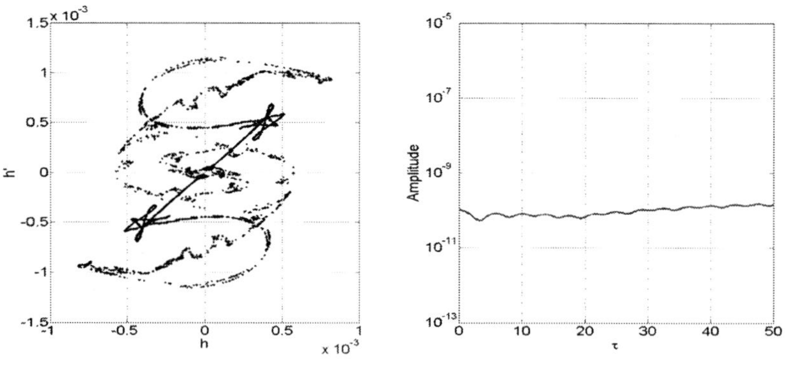

(a) Trajectories on Poincare section (b) Sensitivity to initial value

Fig .7 Trajectories on Poincare section in system 1 and sensitivity of system to initial value, $\Theta = 0.93$

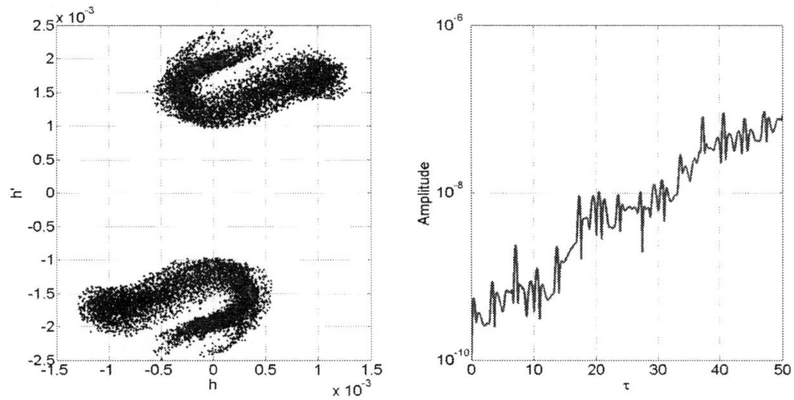

(a) Trajectories on Poincare section (b) Sensitivity to initial value

Fig .8 Trajectories on Poincare section in system 1 and sensitivity of system to initial value, $\Theta = 0.94$

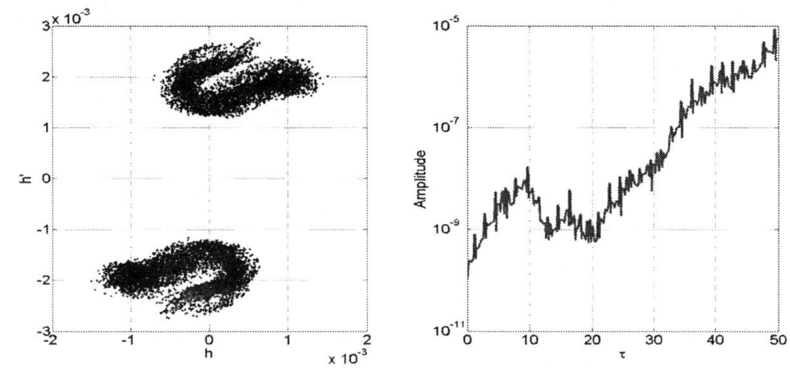

(a) Trajectories on Poincare section (b) Sensitivity to initial value

Fig .9 Trajectories on Poincare section in system 1 and sensitivity of system to initial value, $\Theta = 0.95$

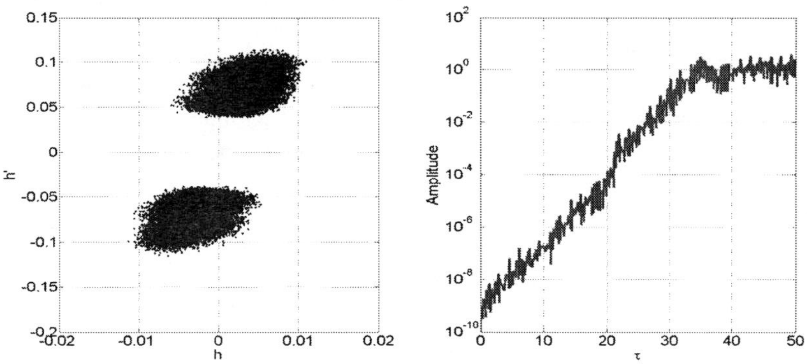

(a) Trajectories on Poincare section (b) Sensitivity to initial value

Fig .10 Trajectories on Poincare section in system 1 and sensitivity of system to initial value, $\Theta = 2.0$

From Fig. 5 to Fig. 7, it can be seen that the three systems under different inflow velocities are all quasi periodic oscillations, which have obvious quasi periodic behaviour on the Poincare section, and the system is not sensitive to the initial value, and the distance between the two adjacent points of the system as the initial value solution has been kept in a limited range. The other parameters remain unchanged, and the incoming flow speed continues to increase to 0.94, the oscillation state of the system changes into chaos. As can be seen from Fig .8, the track of the system on the Poincare section no longer keeps the closed curve. And the system becomes very sensitive to the initial value. When two adjacent points are taken as the initial value, the distance between the two system solutions will increase in order of magnitude in a short time and will increase with the development of time. The system represented in Fig .9 and Fig .10 have larger range of chaotic attractors, and the maximum distance between the solutions will be the same as the attractor diameter of the chaotic system. In this parameter system, the point of chaos is $\Theta = 0.94$.

4. Conclusions

The bifurcation characteristics of the system in different parameter range are studied, and it is concluded that the system will experience bifurcation twice before developing into large amplitude LCOs.

First, after the system experiences Hopf bifurcation, the system changes from stable point to periodic oscillation. With the increase of the incoming flow velocity, there occurs Neimark-Sacker bifurcation, the oscillation of the system changes from periodic oscillation to quasi-periodic oscillation. As the velocity of incoming flow continues to increase, the oscillation of the system is transformed into large amplitude LCOs. Finally, the system enters chaos when the incoming flow velocity is much higher.

The delay at the chaotic occurrence point of the system also corresponds to the delay at each bifurcation point of the system, that is, the higher the incoming flow velocity corresponding to the chaotic phenomenon, the larger the oscillation suppression parameter range of the NES for the main system.

References

[1] Fung YC 2008 An introduction to the theory of aeroelasticity[M]. New York: Courier Dover Publications.
[2] Dowell EH, Crawley EF Curtiss HC, et al. 2005 A modern course in aeroelasticity[M]. Netherlands: Springer.
[3] Nayfeh AH, Mook DT, Holmes P. Nonlinear oscillations[J]. Physics Today, 1980, 15(9): 63-65.
[4] Bunton RW, Denegri CM 2000 Limit cycle oscillation characteristics of fighter aircraft[J]. Journal of Aircraft, 37(5): 916-918.
[5] Cattarius J, Cochair JI, Douglas D, et al. 1999 Numerical wingstore interaction analysis of a parametric F-16 wing[D]. Virginia Tech.
[6] Denegri, Charles M. 2000 Limit cycle oscillation flight test results of a fighter with external stores[J]. Journal of Aircraft, 37(5): 761-769.
[7] Lee BH, Leblanc P 1986 Flutter analysis of a two-dimensional airfoil with cubic non-linear restoring force[J]. National Research Council Canada National Aeronautical Establishment.
[8] Lee BH 1987 Flutter analysis of a two-dimensional airfoil containing structural nonlinearities[R]. Canada: National Aeronautical Establishment, Aeronautical Report LR-618, National Research Council.
[9] Singh SN, Brenner M 2003Limit cycle oscillation and orbital stability in aeroelastic systems with torsional nonlinearity[J]. Nonlinear Dynamics, 31(4): 435-450.
[10] O'Neil T, Strganac TW 1998 Aeroelastic response of a rigid wing supported by nonlinear springs[J]. Journal of Aircraft, 35(4): 616-622.
[11] Gilliatt HC, Strganac TW 2003 Kurdila AJ. An investigation of internal resonance in aeroelastic systems[J]. Nonlinear Dynamics, 31(1): 1-22.
[12] Lee BH, Liu L, Chung KW 2005 Airfoil motion in subsonic flow with strong cubic nonlinear restoring forces[J]. Journal of Sound and Vibration, 281(3-5): 699-717.
[13] Lind R, Snyder K, Brenner M 2001 Wavelet analysis to characterise non-linearities and predict limit cycles of an aeroelastic system[J]. Mechanical Systems and Signal Processing, 15(2): 337-356.
[14] Strganac TW, Ko J, Thompson DE 2000 Identification and control of limit cycle oscillations in

aeroelastic systems[J]. Journal of Guidance Control and Dynamics, 23(6): 1127-33.

[15] Li D, Guo S, Xiang J, et al. 2010 Control of an aeroelastic system with control surface nonlinearity[C]. 51st AIAA/ASME/ASCE/AHS/ASC Structures, Structural Dynamics, and Materials Conference, Orlando, Florida: 2951.

[16] Li D, Guo S, Xiang J 2012 Study of the conditions that cause chaotic motion in a two-dimensional airfoil with structural nonlinearities in subsonic flow[J]. Journal of Fluids and Structures, 33: 109-126.

[17] Nayfeh AH, Mook DT 2008 Nonlinear oscillations[M]. New Jersey: John Wiley and Sons.

[18] Bajaj AK, Davies P, Chang SI 2018 On internal resonances in mechanical systems[J]. Nonlinear Dynamics and Stochastic Mechanics: 69-94.

[19] Nayfeh AH, Balachandran B 1989 Modal interactions in dynamical and structural systems[J]. Applied Mechanics Reviews, 42: 175-201.

[20] Sethna PR, Bajaj AK 1978 Bifurcations in dynamical systems with internal resonances[J]. Journal of Applied Mechanics, 45: 895-902.

[21] Sethna PR 1963 Coupled in certain classes of weakly nonlinear vibrating systems[C]. International Symposium on Nonlinear Differential Equations and Nonlinear Mechanics. Academic Press: 58-70.

[22] FengZC, SethnaPR 1990 Global bifurcation and chaos in parametrically forced systems with one-one resonance[J]. Dynamics and Stability of Systems, 5(4): 201-225.

[23] Tien WM, Namachchivaya N, Bajaj AK 1994 Nonlinear dynamics of a shallow arch under periodic excitation. part I: 1:2 internal resonance[J]. International Journal of Nonlinear Mechanics, 29(3): 349-366.

[24] Tien WM, Namachchivaya N, Bajaj AK 1994 Nonlinear dynamics of a shallow arch under periodic excitation. part II: 1:1 internal resonance[J]. International Journal of Nonlinear Mechanics, 29(3): 367-386.

[25] Johnson JM, Bajaj AK 1989 Amplitude modulated and chaotic dynamics in resonant motion of strings[J]. Journal of Sound and Vibration, 128: 87-107.

[26] Reilly O, Holmes PJ 1992 Non-linear non-planar and non-periodic vibrations of a string[J]. Journal of Sound and Vibration, 153: 413-435.

[27] YangXL, SethnaPR 1992 Nonlinear forced vibrations of a nearly square plate-Antisymmetric case[J]. Journal of Sound and Vibration, 155: 413-441.

[28] Henrard J 1982 Capture into resonance: An extension of the use of adiabatic invariants[J]. Celestial Mechanics, 27: 3-22.

[29] Neishtadt AI 1975 Passage through a separatrix in a resonance problem with a slowly-varying parameter[J]. Journal on Applied Mathematical Mechanics, 39(4): 594-605.

[30] Neishtadt AI 1990 Averaging and passage through resonances[C]. Proceedings of the International Congress of Mathematicians, Kyoto: 1271-1283.

[31] Hall CD, Rand RH 1994 Spinup dynamics of axial dual-spin spacecraft[J]. Journal of Guidance, Control, and Dynamics, 17(1): 30-37.

[32] Kinsey RJ, Mingori DL, Rand RH 1990 Spinup through resonance of rotating unbalanced systems with limited torque[C]. AIAA/AAS Astrodynamics Conference, Portland: 805-813.

[33] Rand RH, Kinsey RJ, Mingori DL 1992 Dynamics of spinup through resonance[J]. International Journal of Non-Linear Mechanics, 27(3): 489-502.

[34] Haberman R 1983 Energy bounds for the slow capture by a center in sustained resonance[J]. SIAM Journal on Applied Mathematics, 43(2): 244-256.

[35] Kath WL 1983 Necessary conditions for sustained reentry roll resonance[J]. SIAM Journal on Applied Mathematics, 43(2): 314-324.

[36] Kevorkian J 1974 On a model for re-entry roll resonance[J]. SIAM Journal on Applied Mathematics, 26: 638-669.

[37] Wenfan Zhang, Yan Liu, Shengli Cao, et al. 2017 Targeted energy transfer between 2-D wing and nonlinear energy sinks and their dynamic behaviors[J]. Nonlinear Dynamics, 90(3): 1841-

1850.

[38] Wenfan Zhang, Jiazhong Zhang, Le Wang, et al. 2018 Study on targted energy transfer and resonance captures in the 2D-wing and nonlinear energy sinks[J]. Journal of Vibration Testing and System Dynamics, 2(4): 297-306.

Study on Acid Corrosion Characteristics in Low-Pressure Stage Initial Condensation Zone of Supercritical H_2O/CO_2 Turbine

Zi-yue Ma, Xiao-fang Wang*, Tao Jiang, Wei Wang, Yang Xu

School of Energy and Power Engineering, Dalian University of Technology, Dalian 116024, China

Email: dlwxf@dlut.edu.cn

Abstract. Compared with the traditional steam turbine, the supercritical H_2O/CO_2 turbine has a relatively high CO_2 content in the working fluid. In the initial condensation zone, CO_2 dissolves in the condensed water to form carbonic acid, which intensifies the corrosion of blade. In order to study the acid corrosion characteristics in the initial condensation zone of supercritical H_2O/CO_2 turbine, the acid corrosion rate of the blades in the initial condensation zone of supercritical H_2O/CO_2 turbine was calculated and analyzed based on the numerical model of non-equilibrium condensation flow and a numerical model of CO_2 corrosion reaction. The results show that the temperature, pressure, proportion of CO_2 in the gas phase and thickness of the solution on blade surface are the main factors affecting the acid corrosion rate in the initial condensation zone of the turbine. The influence of temperature and pressure is more significant. Overall, the acid corrosion rate decreases stage by stage.

1. Introduction

Coal plays an important role in China's energy structure. However, the process of using coal causes nonnegligible waste and inevitable pollution. Therefore, in order to build a resource-conserving and environment-friendly society, China has adopted Clean Coal Technology as a main focus of the energy development strategy in the 21th century. The National Key Research and Development Project, Integrated Investigation on Supercritical Water Gasification of Coal for Hydrogen Production and Thermal Power Generation with H_2O/CO_2 Mixtures, proposed a new technology for clean coal utilization. Supercritical H_2O/CO_2 steam turbine is the key equipment of this technology to achieve thermal power conversion. The supercritical H_2O/CO_2 steam turbine has a relatively high CO_2 content in its working fluid(CO_2 accounts for about 10-12% in the main steam), much higher than the content of CO_2 in the working fluid of traditional steam turbines. In the initial condensation zone, CO_2 will dissolve in condensed water to form carbonic acid, which shows strong corrosiveness. This scenario is harmful to turbine blade and will reduce aerodynamic performance, safety, and reliability. In order to provide a design basis for supercritical H_2O/CO_2 steam turbines, it is necessary to investigate the corrosion characteristic of CO_2 in the initial condensation zone of supercritical H_2O/CO_2 steam turbines.

Many researchers have investigated the mechanism and influencing factors of CO_2 corrosion. For example, Tanupabrungsun[1] proposed a mechanism for the CO_2 corrosion of which the temperatures ranging from 25°C to 250°C. Nordsyeen et al.[2] studied the chemical reactions and mass transfer in

CO_2 corrosion process. Main influencing factors of CO_2 corrosion are pH, temperature, partial pressure of CO_2 and flow rate. Nordsveen et al.[2] found the corrosion rate decreases with the increase of pH. Many research[3-8] suggested that the CO_2 corrosion rate shows different characteristics in different range of temperature. Choi et al.[9] calculated the interaction of the pressure on the solubility of CO_2. Wei Liang[8] investigated the effect of flow on the corrosion rate.

Although the current research on CO_2 corrosion is relatively complete, working condition in the supercritical H_2O/CO_2 steam turbine is complex. The corrosion environment is different with most condition(due to the relative motion between blade and working fluid, changing high temperature and high pressure and stress on the blade). Therefore, it is significant to investigate the characteristics of CO_2 corrosion in the turbine internal environment.

2. Corrosion process in supercritical H_2O/CO_2 steam turbines initial condensation zone

2.1. An application of the CO_2 corrosion mechanism in the low-pressure stage initial condensation zone of steam turbines

The corrosion phenomenon of CO_2 to metal materials occurs in the initial condensation zone. The reason is that the steam starts to condense into droplets due to temperature and pressure factors as the working fluid flows, and the CO_2 dissolving in condensed water to form carbonic acid. Specifically, CO_2 dissolves in condensed water on turbine blades, then forms carbonic acid that corrodes the blades. From the electrochemical perspective, the corrosion reaction on the blade includes two half-reactions, including the oxidation reaction of iron at the anode and the reduction reaction of H^+, H_2CO_3, HCO_3^- and H_2O. This reactions are summarized in Table 1.

Table 1. Electrochemical reactions of carbon dioxide corrosion.

	Reaction Name	Reaction Formula
	H^+ reduction reaction	$2H^+(aq) + 2e^- \rightarrow H_2(g)$
Cathodic Reaction	H_2CO_3 reduction reaction	$2H_2CO_3(aq) + 2e^- \rightarrow 2HCO_3^-(aq) + H_2(g)$
	HCO_3^- reduction reaction	$2HCO_3^-(aq) + 2e^- \rightarrow CO_3^{2-}(aq) + H_2(g)$
	H_2O reduction reaction	$2H_2O(l) + 2e^- \rightarrow 2OH^-(aq) + H_2(g)$
Anode Reaction	Fe Oxidation reaction	$Fe(s) \rightarrow Fe_2^+(aq) + 2e^-$

2.2. Main factors of CO_2 corrosion in supercritical H_2O/CO_2 steam turbines

The internal environment of turbines is extremely complex. In order to build a simple and effective corrosion model, it is necessary to determine the main factors of CO_2 corrosion in turbines.

2.2.1. pH. The pH of condensed water in turbines depends on the dissolved CO_2 amount and temperature. Therefore, the effect of pH is essentially the effect of the partial pressure of CO_2 and temperature of condensed water.

2.2.2. Temperature. The temperature on CO_2 corrosion rate is another central factor as shown in previous works [1,3-6,11-12]. It is effected by two fundamental factors. On one hand, the corrosion rate increases with temperature due to the kinetic rate of electrochemical reactions being accelerated by temperature. On the other hand, the protective $FeCO_3$ forms on the steel surface with increasing temperature and the $FeCO_3$ will slow down both anodic and cathodic reactions. As a result, under the interaction of these two phenomena, the CO_2 corrosion rate reaches peak value when the temperature is ranging from 60 °C to 90 °C. However, its effect can be negligible as the temperature decreases further. Therefore, when temperature is relatively low, the corrosion rate is mainly effected by the reactant activity, will gradually decrease.

2.2.3. Partial pressure of CO_2. The effect of CO_2 partial pressure on the rate of carbon dioxide acid corrosion is mainly reflected in the solubility of CO_2 and the concentration of reactant in solution. In a steam turbine, the partial pressure of CO_2 decreases as the working fluid flows. However, the condensation of steam will increase the proportion of CO_2 in working fluid.

2.2.4. Flow rate. The flow rate mainly affects the corrosion rate by affecting the formation of the corrosion product film on the metal surface. Notably, the flow rate of working fluid in the steam turbine is much higher than most existing corrosion studies[9] and the working blade surface of steam turbines will be subject to great stress, so it is difficult to form a protective corrosion film on the blade surface. In this paper, the effect of corrosion products on corrosion rate is ignored. The effect of flow rate is reflected in the effect of mass transfer.

3. Numerical analysis of two-phase non-equilibrium condensation of wet steam in supercritical H_2O/CO_2 steam turbines initial condensation zone

In this study, the steady and viscous three-dimensional numerical simulation for a 5MW supercritical H2O/CO2 steam turbine initial condensation zone is carried out using the commercial software CFX. The Turbogrid is utilized to generate the mesh of stators and rotors. The geometry of blades is built based on one-dimensional design results. The SST turbulence model is used and a no-slip wall boundary condition is applied for the solid wall. The stage interface is set between stators and rotors. In order to simplify calculations, the steam based on IAPWS-IF97 is used as the working fluid. The total pressure of 200kPa and total temperature of 418 K are specified at the inlet of the domain and the static pressure of 5kPa is defined as outlet boundary condition with a rotating speed of 5000rpm. The distribution of thermal parameters is shown in Figure.1-3.

The pressure, velocity, and temperature at the surface of blades are obtained according to the calculation results, and these parameters will be used to calculate the CO_2 corrosion rate. The condensed water appears at the trailing edge of the second stage stator (as shown in Figure.4). Therefore, the parameters are collected after the second stage stator. The pressure, fluid velocity and condensed water temperature are collected at 10%, 50% and 90% heights of blades after the second stage stator for corrosion rate calculations.

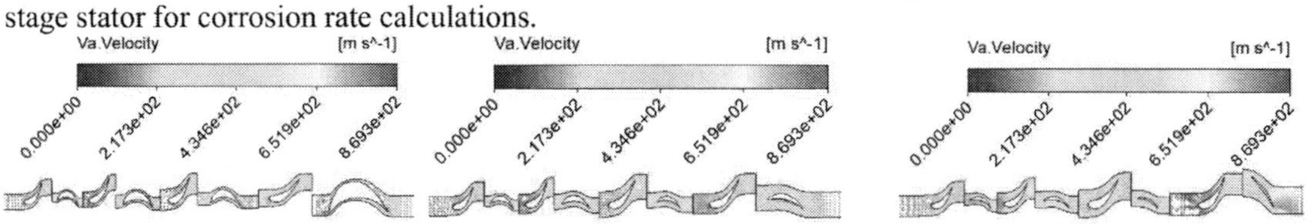

(a) 0.1 blade height (b) 0.5 blade height (c) 0.9 blade height

Figure 1. Speed distribution in blade-to-blade view

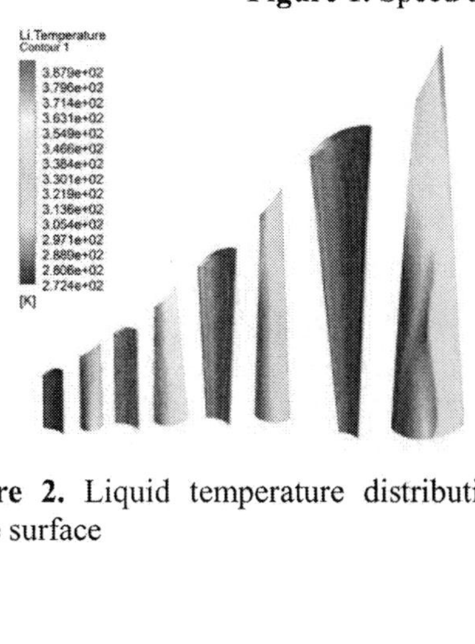

Figure 2. Liquid temperature distribution on blade surface

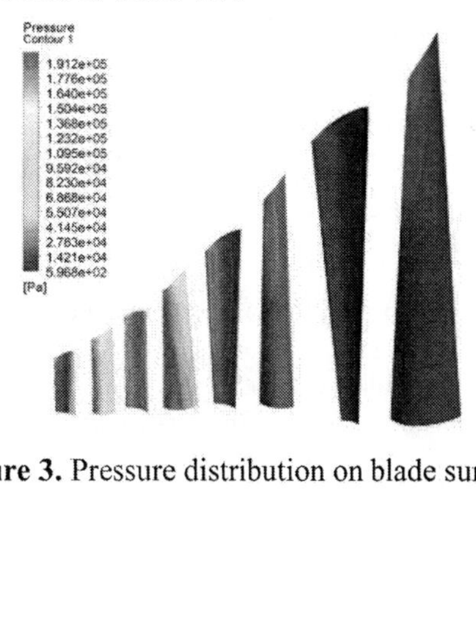

Figure 3. Pressure distribution on blade surface

IOP Conf. Series: Materials Science and Engineering 1081 (2021) 012008 doi:10.1088/1757-899X/1081/1/012008

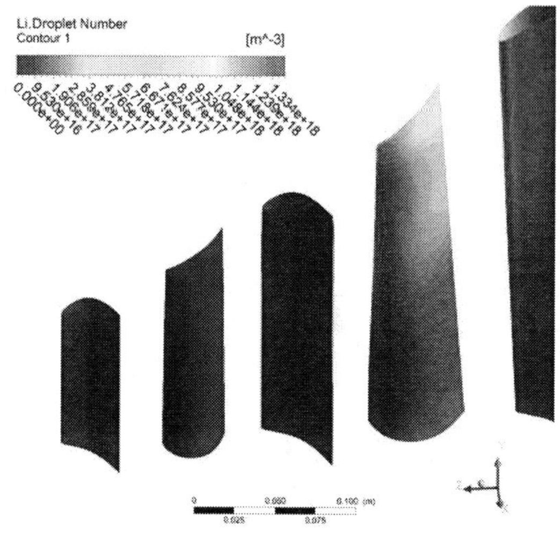

Figure 4. Droplet number distribution on blade surface

4. Numerical analysis of CO_2 Corrosion in supercritical H_2O/CO_2 steam turbines initial condensation zone

4.1. CO_2 Corrosion model in Supercritical H_2O/CO_2 steam turbines initial condensation zone
In this study, the CO_2 corrosion rate is calculated using the Corrosion Module from the commercial software Comsol Multiphysics. An one-dimensional acid corrosion model is used as shown in Fig.4. The thickness of the boundary layer depends on the Reynolds number. The diffusion and turbulent sublayers vary and are accounted for with the mass transfer parameters. All species are assumed to be diluted in water and the mass transfer is modeled by diffusion. CO_2 hydration, water dissociation, thress reduction reactions, and iron dissolution are accounted fro, resulting in seven species in the model. The species and diffusion coefficients are tabulated in Table.2.

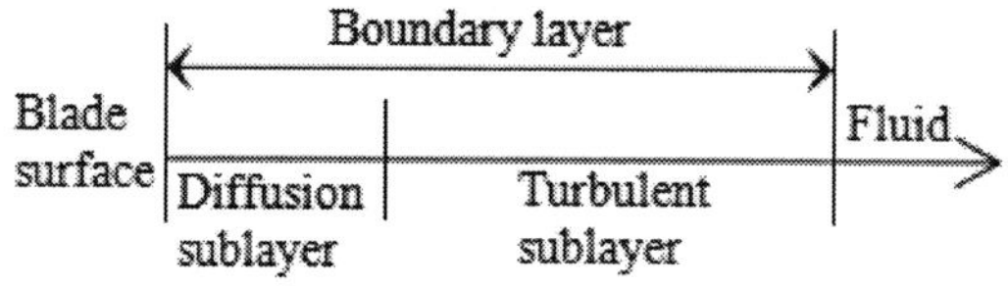

Figure 5. Description of the boundary layer adjacent to the steel surface

The turbulent sublayer is modeled by adding a turbulent diffusivity term to the diffusion coefficient. The term depends on the flow rate, viscosity, density of the liquid, and distance from the steel surface[2].

The electrode surface boundary feature is used to calculate the corrosion potential at the blade surface using the mixed potential theory. The total current of all electrochemical reactions is set to 0, which leads to the following equation:

$$\sum_{\mathrm{j}} i_{\mathrm{j}} = 0, \tag{1}$$

where i (SI unit: A/m^2) is the current density of j number of electrochemical reactions. The initial value is set to -0.5V around the free corrosion potential[10]. The outer point of the boundary layer is set to the bulk concentrations of the species. Fluxes of species converted in the electrochemical reactions are applied on the blade surface as i_j/F (Faraday's constant = 96485 C/mol).

Table 2. Modeled species with their respective diffusion coefficients

Species	Diffusion Coefficient $D(m^2/s)\cdot 10^9$
CO_2	1.96
H_2CO_3	2.00
HCO_3^-	1.11
CO_3^{2-}	0.92
H^+	9.31
OH^-	5.26
Fe^+	0.72

Except for the physical quantities calculated by CFD, some parameters involved in the reactions also need to be obtained from theoretical or empirical formulas.

The concentration of CO_2 in condensed water is calculated by Henry's Law. The Henry constant of CO_2 at different temperatures is estimated by the following formula:

$$K_{CO_2} = 0.0454\times[1.6616 - 5.736\times10^{-2}\times(T-273) + 1.031\times10^{-3}\times(T-273)^2 - 9.68\times10^{-6}\times(T-273)^3 + 4.471\times10^{-8}\times(T-273)^4 - 7.912\times10^{-11}\times(T-273)^5] \tag{2}$$

The ionization equilibrium constant of water is:

$$K_{H_2O} = 10^{-(29.3868 - 0.0737549\times T + 7.47881\times10^{-5}\times T^2)} \tag{3}$$

The first ionization constant of H_2CO_3 is:

$$K_{H_2CO_3} = 387.6\times10^{-(6.41 - 1.594\times10^{-3}T_F + 8.52\times10^{-6}T_F^2 - 3.07\times10^{-5}P - 0.4772I^{0.5} + 0.118I)} \tag{4}$$

The ionization equilibrium constant of HCO_3^- is:

$$K_{HCO_3^-} = 10^{-(10.61 - 4.97\times10^{-3}T_F + 1.331\times10^{-5}T_F^2 - 2.624\times10^{-5}P - 1.166I^{0.5} + 0.3466I)}, \tag{5}$$

where T_F (°F) is Fahrenheit temperature, and I (mol/L) is ionic strength. I is set to 0.185mol/L in this study.

According to Henry's Law, the concentration of dissolved CO_2 in the solution is:

$$c_{CO_{2(aq)}} = K_{CO_2}\times p_{CO_2}, \tag{6}$$

and the concentration of H2CO3 can be calculated by:

$$c_{H_2CO_3} = K_{CO_{2H}}\times c_{CO_{2(aq)}}, \tag{7}$$

where $K_{CO_{2H}}$ is the equilibrium constant of CO_2 hydration.

Steam is condensed into droplets as the working fluid flows, which makes the volume fraction and the partial pressure of CO_2 increase. In this study, the volume changing laws of CO_2 and steam under the effect of temperature and pressure are assumed to be the same. According to the dryness of each stage obtained from the one-dimensional design results, the proportion of steam reduction in gas phase is calculated to obtain the new volume fraction and partial pressure of CO_2. The dryness after each stage is listed in table. 3. The concentration of other substances can be calculated based on the concentration of H_2CO_3 and the equilibrium constant of each reaction.

The expression of electrode kinetics to control the progress of the reaction is:

$$i_{\mathrm{loc}} = i_0 \times 10^{\eta / A_\mathrm{a}}, \tag{8}$$

where i_{loc} (SI unit: A/m^2) is the local current density, i_0 (SI unit: A/m^2) is the initial current density. η (SI unit: V) is overpotential, which characterizes the difference between the disequilibrium potential and the equilibrium potential of this electrode reaction. A_a (SI unit: V) is the Tafel slope of the electrode reaction that characterizes the effect of current density on overpotential.
After consulting the literature, the Tafel slope of each substances involved in electron transfer are taken as follows:

$$A_{\mathrm{Fe}} = 0.04\ \mathrm{V},\ A_{\mathrm{H2O}} = 0.118\ \mathrm{V},\ A_{\mathrm{H2CO3}} = 0.12\ \mathrm{V},\ A_{\mathrm{H2}} = 0.118\ \mathrm{V}.$$

The equilibrium potential is necessary for overpotential calculation. The potential of each reduction reaction can be calculated by the Nernst equation:

$$E_{\mathrm{rev}} = 2.303\frac{RT}{nF}\cdot \lg c_{\mathrm{i}}, \tag{9}$$

where R is the standard gas constant, n is number of reaction transfer electrons, and F is Faraday's constant. For the reduction reactions involved in this study, H^+ gets electron to generate H_2, so the c_i is just the concentration of H^+. Therefore, the Nernst equation can can be written as:

$$E_{\mathrm{rev}} = -2.303\frac{RT}{F}\cdot \mathrm{pH} \tag{10}$$

Furthermore, the initial current density needs to be converted from the reference current density in standard state. The initial current density of each substance involved in the electron transfer is calculated by:

$$i_{0_{\mathrm{H_2O}}} = i_{\mathrm{ref}_{\mathrm{H_2O}}} \cdot \mathrm{e}^{[\frac{-30\times10^3\,(\mathrm{J/mol})}{R}\times(\frac{1}{T}-\frac{1}{T_{\mathrm{refH_2}}})]} \tag{11}$$

$$i_{0_{\mathrm{H_2CO_3}}} = i_{\mathrm{ref}_{\mathrm{H_2CO_3}}} \cdot \left(\frac{c_{\mathrm{H}}}{c_{\mathrm{HrefH_2CO_3}}}\right)^{a_{\mathrm{1H_2CO_3}}} \cdot \left(\frac{c_{\mathrm{H_2CO_3}}}{c_{\mathrm{H_2CO_3refH_2CO_3}}}\right)^{a_{\mathrm{3H_2CO_3}}} \cdot \mathrm{e}^{[-30\times10^3\,(\mathrm{J/mol})/R\times(\frac{1}{T}-\frac{1}{T_{\mathrm{refH_2CO_3}}})]} \tag{12}$$

$$i_{0_{\mathrm{H_2}}} = i_{\mathrm{ref}_{\mathrm{H_2}}} \cdot \left(\frac{c_{\mathrm{H}}}{c_{\mathrm{HrefH_2}}}\right)^{a_{\mathrm{1H_2}}} \cdot \mathrm{e}^{[\frac{-30\times10^3\,(\mathrm{J/mol})}{\mathrm{R}}\times(\frac{1}{T}-\frac{1}{T_{\mathrm{refH_2}}})]} \tag{13}$$

$$i_{0_{\mathrm{Fe}}} = i_{\mathrm{ref}_{\mathrm{Fe}}} \cdot \left(\frac{c_{\mathrm{H}}}{c_{\mathrm{HrefFe}}}\right)^{a_{\mathrm{1Fe}}} \cdot \left(\frac{c_{\mathrm{CO_2}}}{c_{\mathrm{CO_2refFe}}}\right)^{a_{\mathrm{2Fe}}} \cdot \mathrm{e}^{\frac{-37.5^3[\mathrm{J/mol}]}{\mathrm{R}}*(\frac{1}{T}-\frac{1}{T_{\mathrm{refFe}}})}, \tag{14}$$

where the values of constants are as follows:

$i_{\mathrm{refH_2O}} = 3\times10^{-5}\,\mathrm{A/m^2}, i_{\mathrm{refH_2CO_3}} = 0.06\mathrm{A/m^2}, i_{\mathrm{refH_2}} = 0.05\mathrm{A/m^2}, i_{\mathrm{Fe}} = 1\mathrm{A/m^2}, T_{\mathrm{refH_2}} = T_{\mathrm{refFe}} = 25°\mathrm{C}$,

$T_{\mathrm{refH_2CO_3}} = 20°\mathrm{C}, c_{\mathrm{HrefH_2CO_3}} = 0.01\mathrm{mol/m^3}, c_{\mathrm{HrefH_2}} = 0.1\mathrm{mol/m^3}, c_{\mathrm{H_2CO_3refH_2CO_3}} = 0.1\mathrm{mol/m^3}, c_{\mathrm{HrefFe}} = 0.1\mathrm{mol/m^3}$,

$c_{\mathrm{CO_2refFe}} = 0.0366\times10^3, a_{\mathrm{1H_2CO_3}} = -0.5, a_{\mathrm{3H_2CO_3}} = 1, a_{\mathrm{1H_2}} = 0.5, a_{\mathrm{1Fe}} = 1\times p_{\mathrm{CO_2}}$.

After calculating the corrosion current density of the Fe electrode, commercial software can be adopted in realizing this model to calculate the CO_2 corrosion rate.

4.2. Prediction and analysis of corrosion rate in the supercritical H_2O/CO_2 steam turbine initial condensation zone

The numerical calculation results of the internal flow field of turbine initial condensation zone are imported into the CO_2 corrosion model presented in part 4.1. The results of corrosion rate calculation are shown in the following figures (Figure.6 to 11) where the abscissa represents the non-dimensional distance along streamwise and the ordinate represents the corrosion rate.

First of all, as the steam flows, the temperature and pressure of the steam gradually decrease and part of steam condenses into water. However, the corrosion rate does not increase with the increase of humidity. In contrary, it gradually decreases. In this process, temperature and pressure are the dominant factors that affect the corrosion rate. The decrease of temperature will reduce the ionization of water and CO_2. And the ionization equilibrium reaction will be reversed, which will reduce the concentration of H^+, increase the pH, and reduce the reaction rate. When the pressure decreases greatly, the partial pressure of CO_2 will decrease and make CO_2 more difficult to be dissolved into water. As a result, the concentration of H_2CO_3 decreases, and pH eventually increase. Due to the comprehensive effect of temperature and pressure, the corrosion rate in the initial condensation zone of steam turbine gradually decreases stage by stage.

In the same way, as the steam flows, the temperature and pressure of steam gradually decreases, and the corrosion rate of stators is larger than that of rotors in the same stage. On the surface of a single blade, the corrosion rate of leading edge is larger than other position of the blade. This is because the fluid is at the stagnation state at leading edge, and the temperature and pressure of fluid decrease as the fluid flows in the cascade flow-path. At different height of a blade, the load of blade is different and the degrees of temperature and pressure reduction are diverse. The corrosion rate increases with the increase of blade height. However, in some positions of flow-path, the partial pressure of CO_2 increases due to the condensation of steam. It makes the corrosion rate larger than the other positions.

Another important influential factor of the corrosion rate is the thickness of liquid. At leading edge of the blade, the thickness of the boundary layer is minimal. The boundary layer thickens with the flow of fluid. The smaller the thickness of the liquid, the smaller the resistance of the mass transfer between main fluid area and the metal surface. It makes the progress of the corrosion reaction more conducive. Therefore, the corrosion rate at leading edge correspondingly increases. At the other position of the blade where the thickness of liquid is relatively large, the resistance of mass transfer between main fluid area and the metal surface is larger than the leading edge. For example, as shown in Figure.10 and 11, the thickness of the solution layer is small at leading edge point, and the difference in concentration of each substance on the metal surface and the outer surface of the solution is smaller than that at the throat of blade. A large concentration of the reaction product will hinder the reaction and decrease the corrosion rate.

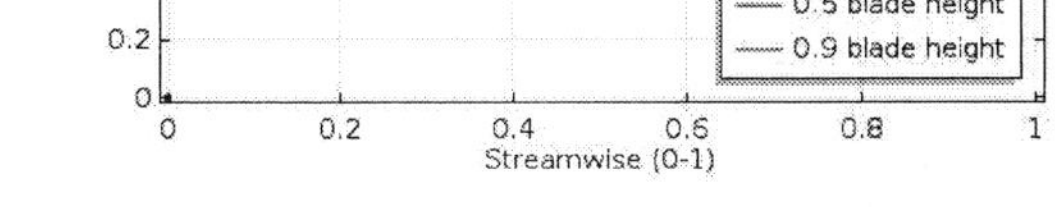

Figure 6. Corrosion rate of the third stator

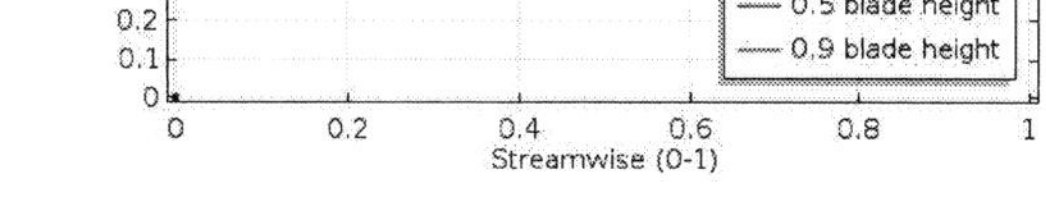

Figure 7. Corrosion rate of the third rotor

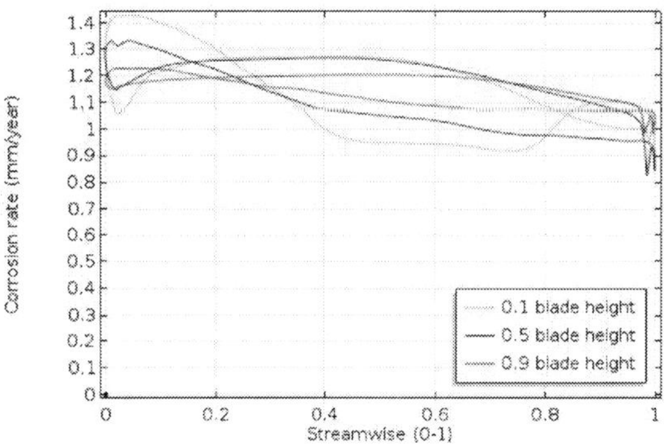

Figure 8. Corrosion rate of the forth stator

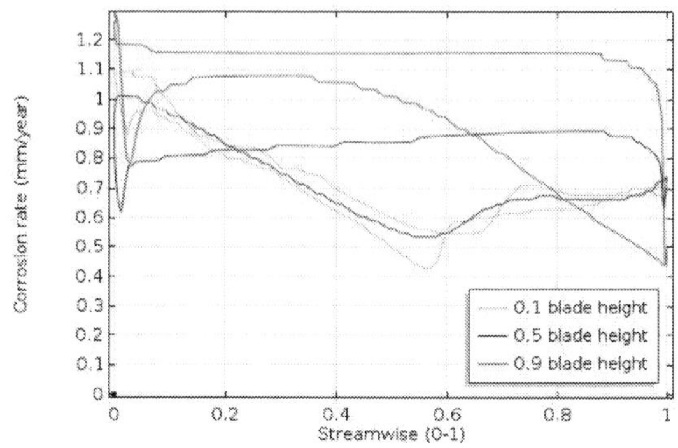

Figure 9. Corrosion rate of the forth rotor

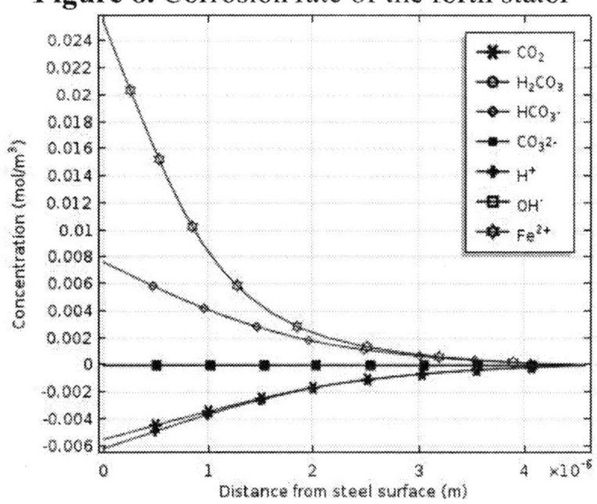

Figure 10. Deviation in concentration of the species compared to the bulk along the liquid boundary layer at the leading point, 50% blade height in the third rotor

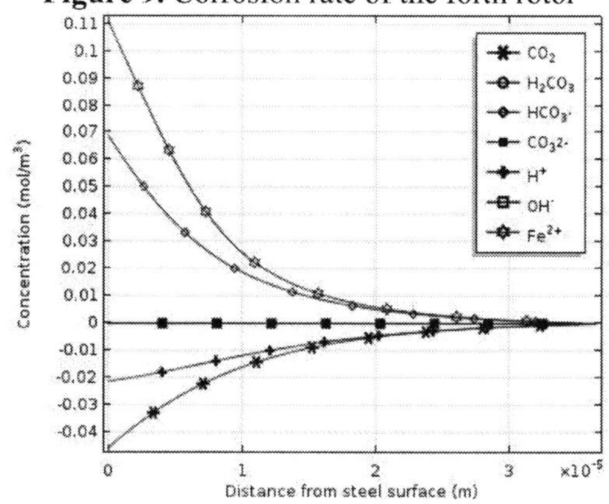

Figure 11. Deviation in concentration of the species compared to the bulk along the liquid boundary layer at the throat, 50% blade height in the third rotor

5. Conclusions

Supercritical H_2O/CO_2 steam turbine is the key equipment for realizing thermal conversion in supercritical water gasification of coal for hydrogen production and thermal power generation with H_2O/CO_2 mixtures technology. However, ensuring its safety and reliability might be of great significance. In this study, a novel CO_2 corrosion model is constructed for the environment of the initial condensation zone of a supercritical H_2O/CO_2 steam turbine, and the internal fluid field calculations of a 5MW supercritical H_2O/CO_2 steam turbine are used to analyze the corrosion characteristics. The main conclusions are:

1)In supercritical H_2O/CO_2 steam turbine, the main factor that affects the CO_2 corrosion rate is temperature and pressure of working fluid. Temperature affects the corrosion rate by changing equilibrium constants of reactions. Pressure is mainly related to the partial pressure of CO_2 in the working fluid, and then affects the dissolution of CO_2 in condensed water as well as the concentration of corrosive substances in solution. The corrosion rate decreases with temperature and pressure stage by stage.

2)For a single blade, the corrosion rate gradually decreases within the cascade flow-path while in some position the condensation of steam may make the corrosion rate larger than the other positions. Also, the corrosion rate will be larger at a larger blade height.

3)The thickness of the solution on blade surface also affects the CO_2 corrosion rate. Large distance between the blade surface and the main flow area (the greater the thickness of the solution)will lead to large resistance of mass transfer between the blade surface and the main fluid area, large concentration difference between product and reactant, and unfavorable the progress of the reaction. Eventually, the corrosion rate is reduced.

Reference

[1] Tanupabrungsun T "Thermodynamics and Kinetics of Carbon Dioxide Corrosion of Mild Steel at Elevated Temperatures" (PH.D. diss., Ohio University, 2012)
[2] Nordsveen M, Nesic S, Nyborg R and Stangeland A 2003 *Corrosion.* **59** 443
[3] Kahyarian A, Singer M and Nesic S 2016 *J. Nat. Gas. Sci. Eng.* **29** 530
[4] Waard D, Lotz U and Milliams E 1991 *Corrosion.* **47** 976
[5] Guo X and Tomoe Y 1999 *Corros. Sci.* **41** 1391
[6] Ikeda A, Ueda M and Mukai S 1983 *Corrosion* (Houston: NACE) p 45
[7] George K, Waard D and Nesic S 2004 *BMC Pediatr.* **13** 1
[8] Wei L "Study of Corrosion Mechanism of Steels in Supercritical CO_2 Environments" (PH.D. diss., University of Science and Technology Beijing, 2017)
[9] Choi Y and Nesic S 2011 *Int. J. Greenh. Gas Con.* **5** 788
[10] Nesic S, Postlethwaite J and Olsen S 1996 *Corrosion.* **52** 280
[11] Sa X, Liu m, Liu D 2016 *Chinese Journal of Turbomachinery* **58** 73
[12] Li Y, Wang S, Fan G, Wei Z, Xing W, Zhang Y 2020 *Chinese Journal of Turbomachinery* **62** 68

Acknowledgements

The research is supported by the National Key Research and Development Project (2016YFB0600104).

IOP Conf. Series: Materials Science and Engineering 1081 (2021) 012009 doi:10.1088/1757-899X/1081/1/012009

Investigation on Interaction between Geometry and Performance and Design of $S\text{-}CO_2$ Radial Inflow Turbine's Vaneless Inlet Volute

Chang Zhu[1],Wen-Yang Shao[1], Zi-Yue Ma[1], Guo-Chuan Lv[2], Xiao-Fang Wang[1*]

[1] School of Energy and Power Engineering, Dalian University of Technology, Dalian 116024, China
[2] NEIAS Commercial Aircraft Company Limited, AVIC, Nanjing 211100, China
[*] Corresponding Author

Email: dlwxf@dlut.edu.cn

Abstract. The inlet volute is one of the key components that guide the working fluid into the radial inflow turbine. It has a significant effect on the performance of the radial inflow turbine. The conventional design criteria for the vaneless inlet volute mainly focuses on the ideal gas, which does not apply to real gas cases, such as supercritical carbon dioxide($S\text{-}CO_2$). In this paper, an MW-class $S\text{-}CO_2$ radial inflow turbine is studied. Firstly, the inlet volute design of the radial inflow turbine with a vaned stator is completed, and the flow in the whole machine is calculated using CFD simulations. Then, based on the inlet boundary conditions of the impellers, the vaneless inlet volute with different throat areas and cross-sectional area distributions are investigated. The results indicate that the size of the counter rotating vortexes in the vaneless inlet volute increases with the throat area increasing, and is also influenced apparently by the cross-sectional area distribution. The volute throat area has little effect on the circumferential distribution of the outlet airflow angle but changes the average value of the airflow angle. The cross-sectional area distribution of the volute influences the circumferential distribution of the outlet flow angle. The result indicates that volute with a small-scale convex profile has a uniform outlet flow field. Compared to the turbine with the vaned stator, the radial dimension of the vaneless volute turbine obtained in the final design is significantly reduced, and the total-total efficiency at the design point is 85.62%, which meets the design requirements. Moreover, it also has good performance at off-design operating conditions.

Nomenclature

F	Area	c	Velocity
G	Mass flow rate	r	Radial direction
L	Volute outlet width	u	Tangential direction
P_t	Total pressure	θ	Circumferential azimuth angle
R	Radius	$\eta_{t\text{-}t}$	Total-total efficiency
V_m	Absolute radial velocity	ρ	Density
V_t	Absolute tangential velocity	φ	Velocity loss coefficient
V_{xyz}	Absolute velocity	1	Volute outlet
α	Airflow angle		

1. Introduction

In recent years, the Brayton cycle with S-CO_2 as the working fluid has become a research hot spot due to its high cycle heat efficiency, compactness and environmental protection[1-2]. When the cycle power level is lower than 10 MW, radial flow turbines are often used instead of the axial flow turbine due to the blade height limitation[3]. Compared to the turbine with the vaned stator, the radial dimension of the vaneless volute turbine is smaller, which is more conducive to the structural arrangement of the machine. Therefore, the vaneless volute turbine capable of reducing the radial dimension needs to be studied under the premise of ensuring efficiency. The vaneless inlet volute serves to connect the turbine directly to the outer pipeline, and whether the flow field uniformity at the volute outlet directly affects the performance of the impeller significantly.

Many researchers have investigated the vaneless volute. Yansheng Li et al. [4] considered the radial velocity variation of the flow in the vaneless volute, when integrating the mass flow over each volute cross-section, and proposed a simplified two-dimensional analysis method. Hara K et al. [5] conducted an experimental study on a vaneless volute and measured the flow field in the volute and the boundary layer of the sidewall. They found that the circumferential non-uniformity of the volute outlet flow field is greatly affected by the internal boundary layer of the volute. Spence S.W.T. et al. [6] conducted experimental tests on a series of radial inlet volutes with and without guide vanes. The vaneless volute turbine showed better performance and its measurement efficiency value had increased by 4.5% under the same conditions as the others. The above researches on the vaneless inlet volute mainly focus on incompressible fluid. However, there are few studies on real gas cases, such as S-CO_2 with high density and high specific heat capacity. For S-CO_2, the design method of the volute does not provide a clear range in the selection of the velocity loss coefficient. And the investigation of the interaction between geometry and performance of the S-CO_2 volute is also lacking. Moreover, the computational domain simulated in most researches on the S-CO_2 radial inflow turbine only consists of the vane and the impeller, and the simulation of the inlet volute is always neglected.

In this paper, an MW-class S-CO_2 radial inflow turbine is studied. Firstly, an inlet volute of the radial inflow turbine with a vaned stator is designed, and the performance of the whole machine is calculated using CFD simulations. Then, based on the inlet boundary conditions of the impeller, various vaneless inlet volute models are obtained by changing the throat area and cross-sectional area distribution. Then, the interaction between geometry and performance of the vaneless volute is investigated in detail. Finally, the vaneless inlet volute matching the target impeller is obtained.

2. Volute Design Method

The one-dimensional design method developed by Yansheng Li et al.[4] is adopted. The radial inflow turbine volute is designed based on the blade height and the estimated inlet airflow angle of the vaned stator, and the cross-section of the volute is taken as a circular cross-section. The main design parameters of the radial inflow turbine required for the design method are shown in Table.1.

Firstly, the flow is assumed to be steady and adiabatic. Approximately, the angular momentum of the flow in the inlet volute varies according to the constant circulation flow:

$$c_{u\theta}R_\theta\varphi = c_{u1}R_1 = const \tag{1}$$

The relationship between the mass flow rate G_θ and the azimuthal angle θ is as follows:

$$G_\theta = G(\theta_{\max} - \theta)/\theta_{\max} \tag{2}$$

Four additional relationships can be obtained from the continuity equation:

$$G = c_{r1}\rho_1 F_1 \tag{3}$$

$$F_1 = 2\pi R_1 L \tag{4}$$

$$G_\theta = c_{u\theta}\rho_\theta F_\theta \tag{5}$$

$$\tan\alpha_\theta = c_{u\theta}/c_{r\theta} \tag{6}$$

By solving equations (1-6) simultaneously, the distribution of the cross-sectional area F_θ with the azimuth angle θ can be obtained:

$$F_\theta / R_\theta = 2\pi L \varphi \rho_1 (\theta_{\max} - \theta) / \rho_\theta \theta_{\max} \tan\alpha_1 \quad (7)$$

The flow velocity is low in the scroll of the vaned stator's volute, and the density of S-CO_2 is very high. It can be considered that the density of S-CO_2 in the volute is constant, which is also in line with the calculation results of numerical simulations. Therefore, equation (7) can be simplified as follows:

$$F_\theta / R_\theta = 2\pi L \varphi (\theta_{\max} - \theta) / \theta_{\max} \tan\alpha_1 \quad (8)$$

It can be found from equation (8) that the ratio of the cross-sectional area F_θ to the core radius R_θ has a positive correlation with the azimuthal angle θ. Besides, the velocity loss coefficient φ at the volute inlet recommended by Baines [7] ranges from 0.85 to 0.95.

The preliminary design of the vaneless inlet volute is based on the two-dimensional design method [4], but it lacks the selection range of the velocity loss coefficient for the real gas. Therefore, the derivation of the method is not specifically described here. The geometric parameters of both the volute with vaned stator and the preliminary design of the vaneless inlet volute are shown in Table.2.

Table.1 Main parameters of an MW-class S-CO_2 radial inflow turbine.

Item	Unit	Parameter
Inlet total temperature	K	773.15
Inlet total pressure	MPa	20
Outlet total pressure	MPa	8.20
Mass flow rate	kg/s	16
Rotational speed	rpm	50000
Impeller inlet blade height	mm	5.20
Impeller inlet radius	mm	69
Total-total efficiency	%	>85

Table.2 Geometric parameters of the volute with vaned stator and the preliminary design of the vaneless inlet volute.

Item	Unit	Volute with vaned stator	Vaneless inlet volute
Volute outlet radius	mm	87	69
Volute outlet width	mm	5.20	5.20
F/R (Throat)	/	23.44	7.04
Throat centroid radius	mm	122.13	88.91
Throat area	mm^2	2862.80	625.80

3. Numerical Simulation

In this paper, the numerical simulation of the S-CO_2 radial inflow turbine with a vaned stator is carried out by the commercial CFD software NUMECA. Then, based on the inlet boundary conditions of the vaned turbine impeller, a series of vaneless inlet volute models with different volute throat areas and cross-sectional area distributions are designed and analyzed. Then, the interaction between geometry and performance of the vaneless inlet volute is investigated. The thermodynamic properties of S-CO_2 are interpolated from the physical property tables generated by the TABGEN during the numerical simulations. The SA model is used to simulate the turbulent flow. The computational grids of vanes and impellers are generated by AutoGrid5, and the grid of volute is generated by IGG. The resulting grids satisfy the turbulence model's requirement (SA model requires $y^+ \leq 10$) and the grid independence study is performed to obtain the proper mesh number. For the numerical simulation of the whole machine, the total temperature and the total pressure are imposed as the inlet boundary conditions, and the outlet condition is set with static pressure; for the numerical simulation of the individual volute model, the inlet conditions are specified by the total temperature and the total pressure, and the outlet condition is specified by the mass flow rate.

On basis of the geometric data of the SNL(Sandia National Laboratory) radial inflow turbine given in the literature [8], different turbulence models were adopted to simulate the turbine in the previous works of the authors, and the numerical simulations were compared with the SNL's experimental results. The results showed that the SA mode is accurate and reliable for the simulation of the S-CO_2 radial inflow turbine.

4. Results and Discussions

4.1. Analysis of the volute with the vaned stator

The results of three-dimensional simulations with or without the inlet volute are shown in Table.3. According to the authors' design experience, the total pressure of the vane inlet is adjusted to 19.85 MPa by considering the total pressure in the inlet volute. It can be found that the estimated total pressure of the vane inlet is slightly lower than the calculated CFD value. The loss caused by the inlet volute only reduces the total-total efficiency by 0.94%. In summary, the inlet volute designed by the one-dimensional method has little effect on the performance of the S-CO_2 radial inflow turbine under the design condition.

Table.3 Comparison of calculation results with or without inlet volute

Item	Unit	Vane+Impeller	Volute+Vane+ Impeller	Change rate
Vane inlet total pressure	MPa	19.85	19.94	0.45%
Mass flow rate	kg/s	16.06	16.10	0.25%
Impeller outlet total pressure	MPa	8.20	8.20	0
Total-total efficiency	%	89.05	88.21	-0.94%

Figure.1 shows the velocity field and the airflow angle distribution at the vaned stator mid-span section. It can be seen from the figure that the flow inside the vaned stator is relatively uniform, and the airflow angle at the volute outlet is well-matched with the vane inlet blade angle. The inlet volute with guide vane guides the flow well towards the impeller and no obvious flow separation can be seen. In summary, the inlet volute designed by the one-dimensional method matches well with the guide vane and the impeller under the design condition.

Figure.2 shows the performance curve of the radial inflow turbine total-total efficiency and mass flow rate as functions of the total-total expansion ratio. It can be seen that the total-total efficiency of the radial inflow turbine is higher than 0.8 under the working conditions of the study (The total-total expansion ratio ranges from 1.8 to 3). In the vicinity of the design point, the turbine efficiency is not significantly affected by the operating conditions. With the total-total expansion ratio increasing, the mass flow rate of the turbine increases gradually and finally keeps nearly constant. In summary, the radial inflow turbine composed of the designed inlet volute, guide vane and impeller has good performance under off-design operating conditions.

Figure.3 and Figure.4 show the velocity field and the airflow angle distribution of the vaned stator mid-span section under two off-design points with the total-total expansion ratio of 1.81 (small flow condition) and 2.97 (large flow condition). It can be seen that the flows inside the vaned stator under two off-design points are almost as uniform as under the design condition. The change of the velocity in the inlet volute is not obvious. This is because the internal velocity of the inlet volute is much

IOP Conf. Series: Materials Science and Engineering 1081 (2021) 012009 doi:10.1088/1757-899X/1081/1/012009

smaller compared with that in the vaned stator, which result in less obvious change of the velocity under off-design conditions. The same reason applies to the airflow angle at the outlet of the volute. The airflow angle distributions at the volute outlet are basically unchanged, matching well with the guide vane inlet blade angle.

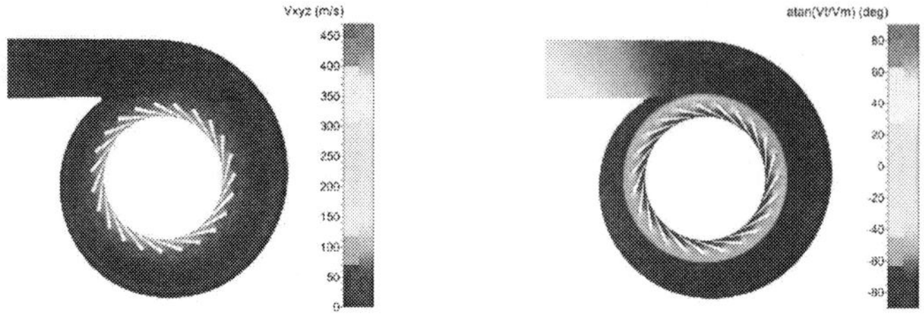

(a) Velocity field (b) Airflow angle

Figure.1 The velocity field and the airflow angle distribution at the vaned stator mid-span section

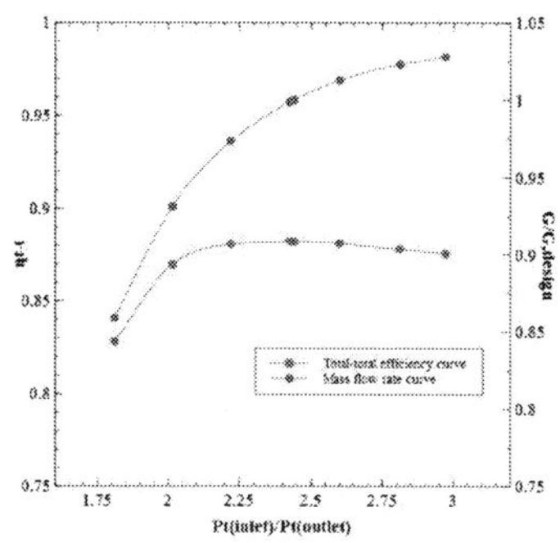

Figure.2 The performance curve of the radial inflow turbine total-total efficiency and mass flow rate as functions of the total-total expansion ratio

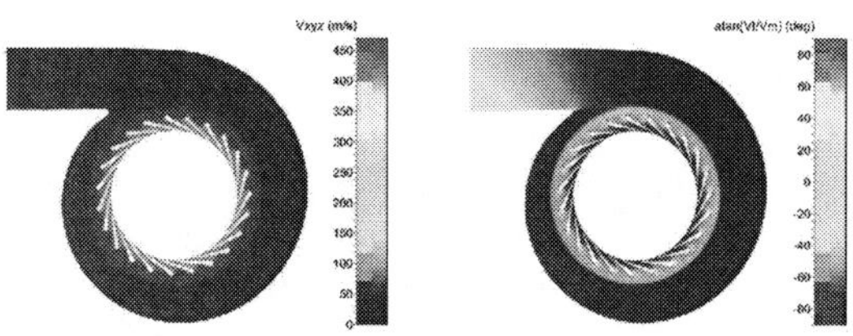

(a) Velocity field (b) Airflow angle

Figure.3 The velocity field and the airflow angle distribution at the vaned stator mid-span section with total-total expansion ratio of 1.81

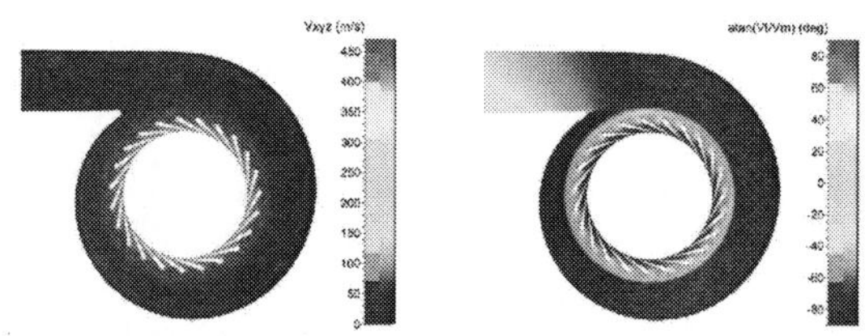

(a) Velocity field (b) Airflow angle

Figure.4 The velocity field and the airflow angle distribution at the vaned stator mid-span section with total-total expansion ratio of 2.97

4.2. Analysis of the vaneless inlet volute

The design of the vaneless inlet volute in the paper referenced to the two-dimensional design method proposed in the literature[4]. However, under the inlet and outlet conditions of the design point, the mass flow rate of the vaneless radial inflow turbine increases to 18.22 kg/s, which is 13.88% higher than the design mass flow rate. It means that the preliminary design of the vaneless inlet volute does not match the target impeller because of the lack of the determination of the real gas velocity loss coefficient. It has been determined that the section of the inlet volute is circular. On the basis of the preliminary design result, several models of vaneless inlet volute are obtained by changing the throat area and the cross-sectional area distribution. It should be noted that only the vaneless volute is simulated during the parameter exploration in this section in order to reduce the computation time. For the numerical simulation of the individual volute model, the inlet conditions are specified by the total temperature and the total pressure, and the outlet condition is specified by the mass flow rate.

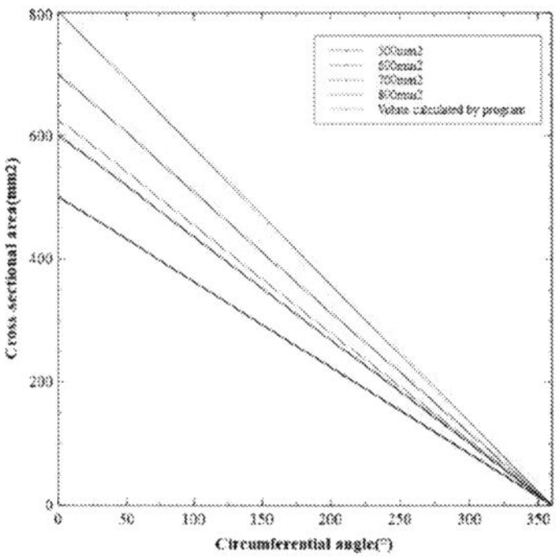

Figure.5 Cross-sectional area distributions curve of the vaneless inlet volute

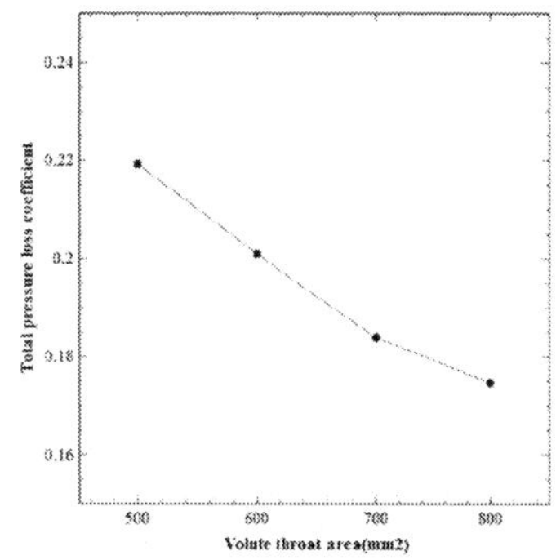

Figure.6 Total pressure loss coefficient of vaneless inlet volute as a function of the throat area

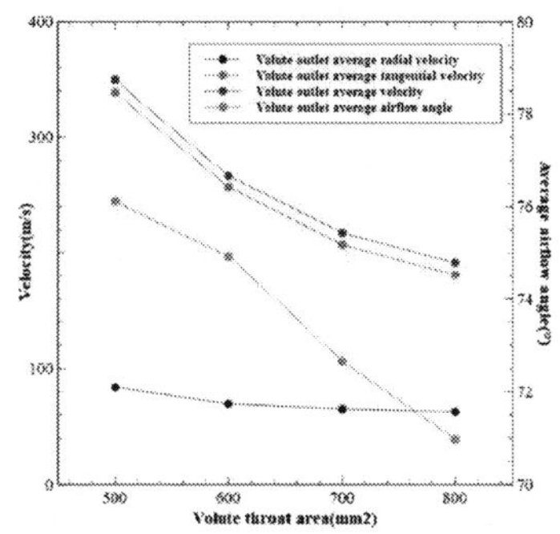

Figure.7 Outlet velocity of vaneless inlet volute as a function of the throat area

4.2.1. Influence of the vaneless inlet volute throat area. As shown in Figure.5, four new volute models are obtained by adjusting the throat area to 500-800 mm^2, with the same cross-sectional area distribution. Figure.6 shows the influence of the volute throat area on the total pressure loss coefficient. It can be found that the total pressure loss coefficient decreases with the volute throat area increasing. Figure.7 shows the variations of the volute outlet velocity as a function of the throat area. The axial velocity is very small and can be ignored. It can be found that the volute outlet velocity decreases with the throat area increasing. Because of the same cross-section area distribution applied, the change of the volute outlet radial velocity is not obvious. And the change of the volute outlet velocity caused by the throat area mainly results from the outlet tangential velocity. Therefore, the average airflow angle of the volute outlet also shows a trend of decrease with increasing the throat area.

Figure.8 shows the velocity field at the volute mid-span section with different throat areas. Figure.9 shows the static pressure contours at the volute mid-span section with different throat areas. It can be seen from the figures that the flow in the two different volutes is good, and there is no obvious flow separation phenomenon.

Figure.10 shows the streamlines of the secondary flow at several circumferential positions(45°to 225°) in the vaneless inlet volute with different throat areas. It can be found that the secondary flow in the volute is obvious. The results indicate that there are two pairs of counter rotating vortexes, one of which moves along the radial direction to the outlet of the volute, and the other pair of vortexes gradually reduces. The size of the counter rotating vortexes decreases in the circumferential direction and finally they almost disappear at the circumferential section of 225°. At the 45° circumferential sec-

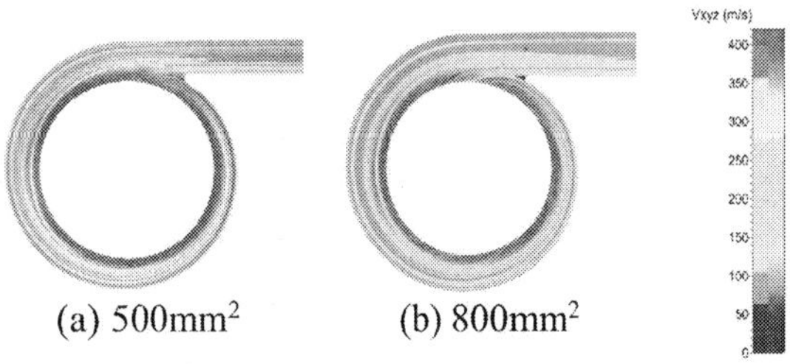

Figure.8 Velocity field at the volute mid-span section with different throat areas

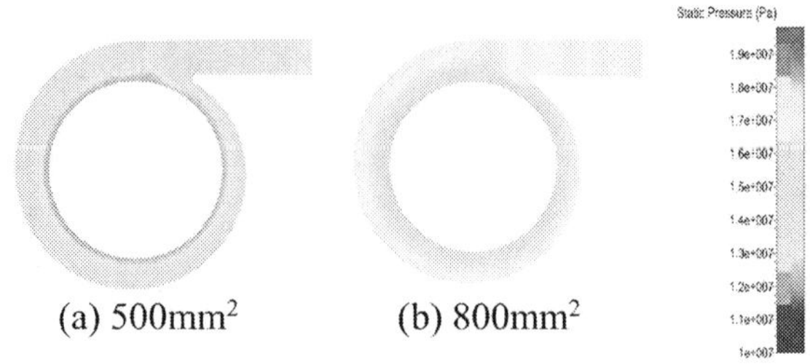

Figure.9 Static pressure contours at the volute mid-span section with different throat areas

tion, the volutes with a throat area of 500 mm^2 and 600mm^2 have two pairs of counter rotating vortexes, but there is only one pair of counter rotating vortexes in the other two volute models. At the 135° circumferential section, the volute with a throat area of 500 mm^2 still has two pairs of counter ro-

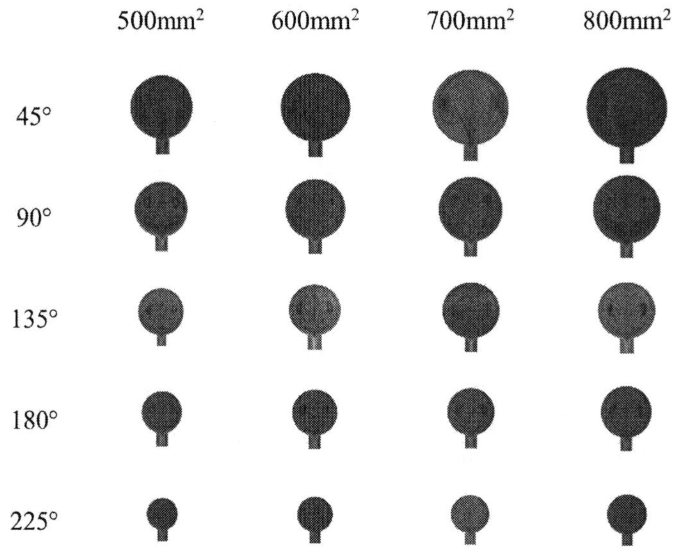

Figure.10 Streamlines of the secondary flow at different azimuthal positions in the vaneless inlet volute with different throat areas

tating vortexes, but in contrast, only one pair of counter rotating vortexes is left at the same position for the other three volute models. This indicates that the influence range of the two pairs of counter rotating vortexes is larger in the vaneless inlet volute with smaller throat area. In addition, at any circumferential section, the size of the vortexes increases with the volute throat area increasing. In summary, with the volute throat area increasing, the size of the counter rotating vortexes increases but the influence range of the two pairs of counter rotating vortexes, on the contrary, decrease.

Figure.11 shows the axial average airflow angle distributions at the outlet of the vaneless inlet volutes with different throat areas in the circumferential direction. It can be seen that volute throat area has a significant effect on the outlet flow angle, but almost does not affect the distribution trend of the flow angle in the circumference direction. With the circumferential position ranging from 30°to 140°, there is a fluctuation in the flow angle, and the fluctuation amplitude increases with the volute throat area increasing. In the range of the circumferential angle from 320°to 30°, there is a sharp change in the flow angle. This phenomenon is due to the mixing of high-speed flow and low-speed flow

IOP Conf. Series: Materials Science and Engineering 1081 (2021) 012009 doi:10.1088/1757-899X/1081/1/012009

downstream of the volute tongue. It can be found that the vaneless volute with a throat area of 600 mm^2 in Figure.11 has a relatively good flow angle distribution near the volute tongue. This shows that the uniformity of the flow angle near the volute tongue improves first and then deteriorates with increasing the volute throat area.

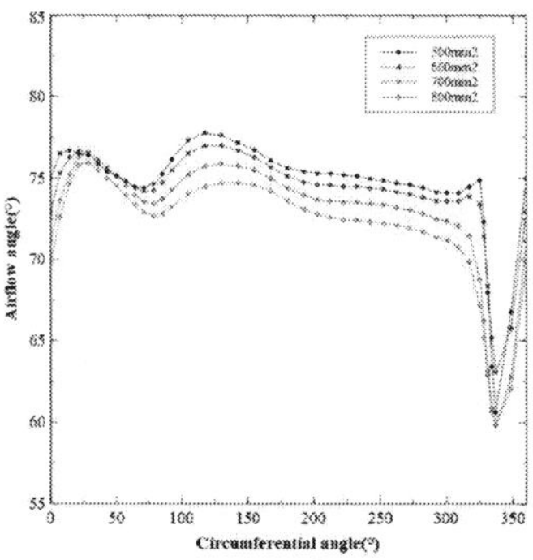

Figure.11 Distributions of the axial average airflow angle at the outlet of the vaneless inlet volutes with different throat areas in the circumferential direction

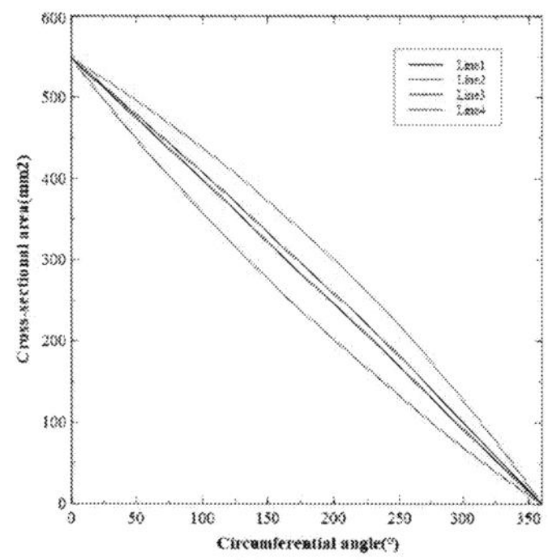

Figure.12 Cross-sectional area distributions of the vaneless inlet volutes

Table.4 The numerical simulations of the volutes with different cross-sectional area distributions

Item	Unit	Line1	Line2	Line3	Line4
Total pressure loss coefficient		0.198	0.213	0.228	0.235
Outlet average radial velocity	m/s	75.89	76.05	76.50	79.08
Outlet average tangential velocity	m/s	285.22	299.38	307.89	312.74
Outlet average velocity	m/s	295.95	308.92	317.53	323.00
Outlet average airflow angle	°	75.10	75.75	76.05	75.81

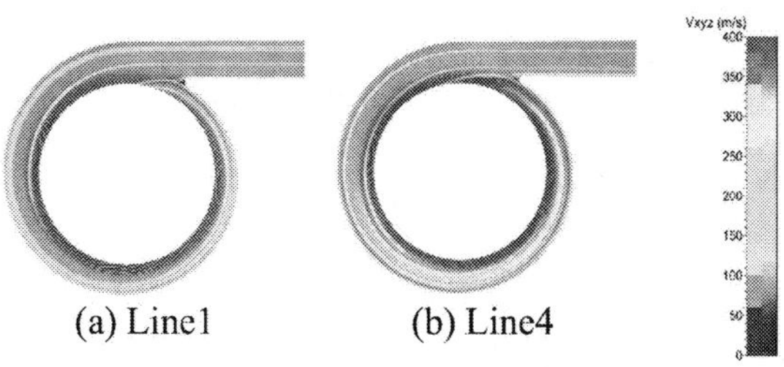

(a) Line1 (b) Line4

Figure.13 Velocity field at the volute mid-span section with different cross-sectional area distributions

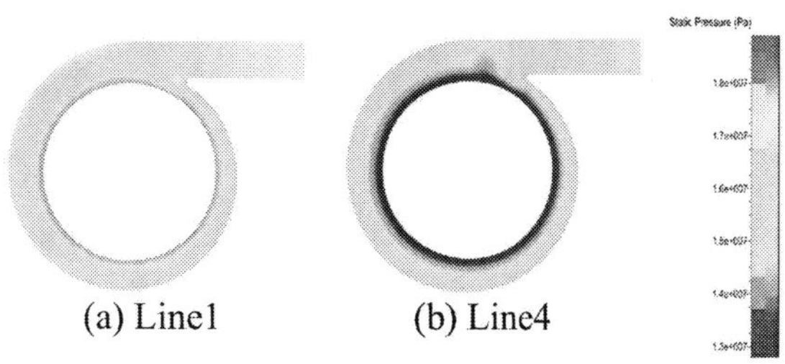

(a) Line1 (b) Line4

Figure.14 Static pressure contours at the volute mid-span section with different cross-sectional area distributions

4.2.2. Influence of the cross-sectional area distribution of the vaneless inlet volute. The optimum volute throat area is readjusted according to the parameter exploration results presented in the above section. In this section, the volute throat area is kept constant, and the effect of the cross-sectional area distribution is studied by comparing the performance of four volutes with different cross-sectional area distributions. Table.4 shows the numerical simulations of four vaneless inlet volutes. It can be found that the volute total pressure loss coefficient increases as the shape of the cross-sectional area distribution changes from concave to convex. The outlet average velocity also shows the same results. However, for the average flow angle, a maximum value could be expected as the shape of the cross-sectional area distribution changes from concave to convex.

Figure.13 and 14 show the velocity field and the static pressure contours at the mid-span section of the volutes with different cross-sectional area distributions. The results indicate that the flow in the two different volutes is good, and there is no obvious flow separation.

Figure.15 shows the streamlines of the secondary flow at the different azimuthal positions in the vaneless inlet volute with different cross-sectional area distributions. It can be found that the secondary flow inside the vaneless volute is basically the same as the results presented in the previous section. Apart from the circumferential position of 225°, the size of the counter rotating vortexes increases as the shape of the cross-sectional area distribution changes from concave to convex. At the 135° circumferential section, the vaneless inlet volute models corresponding to line3 and line4 still have two pairs of counter rotating vortexes, but only one pair of counter rotating vortexes is observed at the corresponding section of the other two volute models. This indicates that the influence range of the two pairs of counter rotating vortexes is larger in the vaneless inlet volute with a more convex profile. In summary, the size of the counter rotating vortexes increases as the shape of the cross-sectional area distribution changes from concave to convex, and it is the same case for the influence range of the two pairs of counter rotating vortexes.

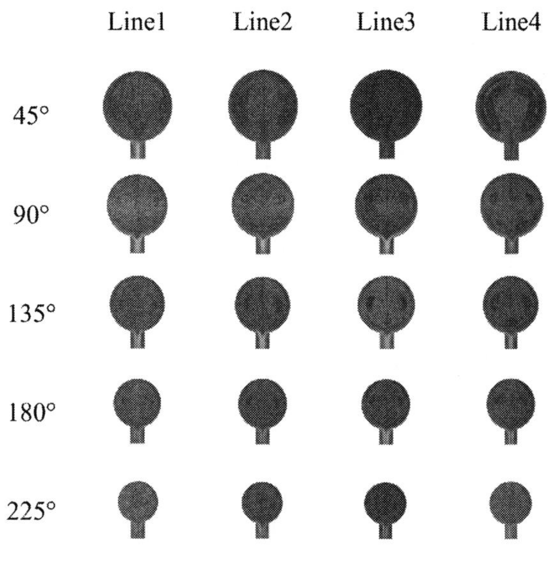

Figure.15 Streamlines of the secondary flow at different azimuthal positions in the vaneless inlet volute with different cross-sectional area distributions

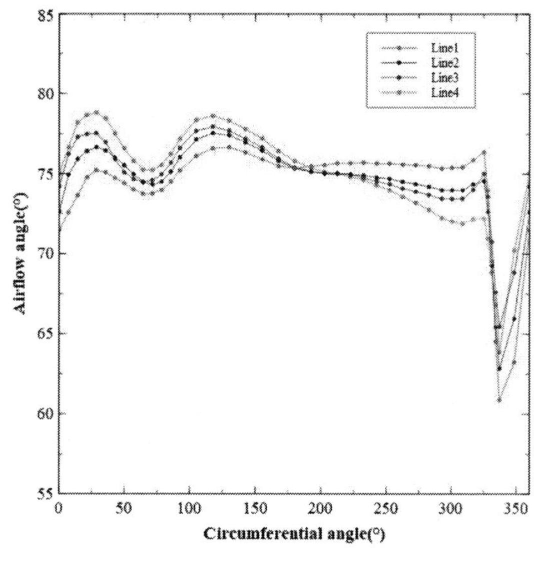

Figure.16 Effect of the cross-sectional area distribution on the axial average flow angle at the outlet of the volute

Figure.16 shows the effect of the cross-sectional area distribution on the axial average flow angle at the outlet of the volute. It can be seen that the uniformity of the outlet axial average flow angle distribution in the circumferential direction improves first and then deteriorates with the shape of the cross-sectional area distribution changes from concave to convex. For the spiral part, with the cross-sectional area distribution changes from concave to convex, the outlet axial average flow angle of the first half increases gradually, while the second half decreases gradually. For the flow near the volute tongue, the uniformity of outlet axial average flow angle distribution improves first and then deteriorates. According to the results in Table.4, the axial average flow angle at the vaneless volute outlet corresponding to line3 has the most uniform distribution in the circumferential direction, and the numerical result of the flow angle best matches the impeller in the four volute models. In summary, when the volute inlet area is constant, the vaneless inlet volute with a small-scale convex profile has a relatively uniform outlet flow angle distribution.

4.3. Analysis of the final vaneless inlet volute

The final design of the vaneless inlet volute is obtained by adjusting the inlet area and the cross-sectional area distribution. The performance comparison of the radial inflow turbine with guide vanes and the one with vaneless volute is shown in Table.5. It can be found that the total-total efficiency of the vaneless volute turbine is lower than the turbine with the vaned stator, but it still meets the design requirements (> 85%). The inlet total pressure of the vaneless volute turbine is 2.34% less than that of the turbine with the vaned stator, which shows the former's stator loss is higher than that of the latter. The mass flow rate of the vaneless volute turbine decreases with the lower total pressure of the impeller inlet. However, the radial dimension of the vaneless volute turbine is significantly smaller

Table.5 Performance comparison of the radial inflow turbine with guide vanes and the one with vaneless volute

Item	Unit	Vaned stator+ Impeller	Volute+ Impeller	Change rate
Impeller inlet total pressure	MPa	19.19	18.74	-2.34%
Mass flow rate	kg/s	16.10	15.95	-0.93%
Outlet total pressure	MPa	8.20	8.20	0
Total-total efficiency	%	88.21	85.62	-2.93%
Maximum radius	mm	152.33	101.22	-33.55%

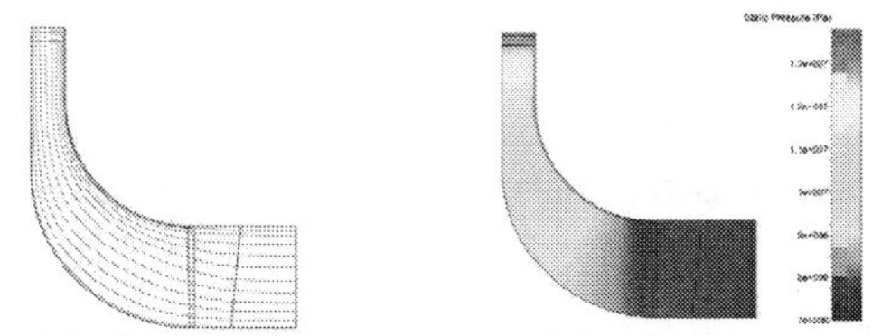

(a) Velocity field (b) Static pressure

Figure.17 Velocity field and the static pressure distribution on the impeller meridian surface

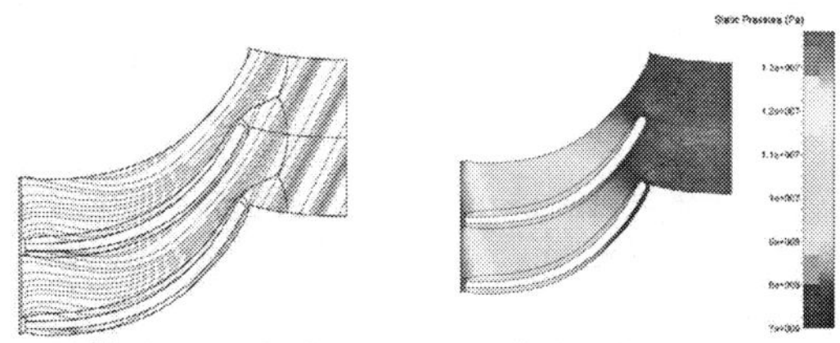

(a) Velocity field (b) Static pressure

Figure.18 Velocity field and the static pressure distribution at the impeller mid-span section

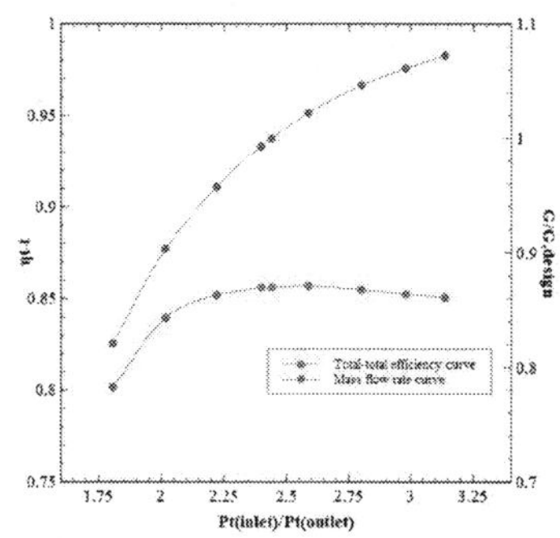

Figure.19 Performance curves of the vaneless volute turbine total-total efficiency and mass flow rate with total-total expansion ratio

than the turbine with the vaned stator, and the maximum radius is reduced by 33.55%. In summary, the final design of the vaneless volute meets the requirements at the design point.

Figure.17 and 18 show the velocity field and the static pressure contours on the impeller meridian surface and the mid-span section. It can be seen that no obvious flow separation occurs near the impeller leading edge, which indicates that S-CO_2 flowing through the vaneless volute flows well.

Figure.19 shows the performance curves of the vaneless volute turbine, total-total efficiency and mass flow rate with total-total expansion ratio. It can be seen that the total-total efficiency of the vaneless volute turbine is higher than 0.8 under the working conditions of the study. In the vicinity of the design point, the turbine efficiency is not significantly affected by the change of the operating conditions. With the increase of the total-total expansion ratio, the mass flow rate of the turbine increases gradually, but its increasing ratio decreases obviously. In summary, the vaneless volute turbine has good performance under off-design operating conditions.

5. Conclusions

In this paper, based on the two-dimensional design method, several vaneless inlet volute models are obtained by changing the volute throat area and the cross-sectional area distribution. Three dimensional CFD simulations are performed to investigate the interaction between the geometry and the performance of the vaneless inlet volute. Finally, the vaneless inlet volute matching with the target impeller is obtained. The conclusions are as follows:

(1) Under the same cross-sectional area distribution, the total pressure loss coefficient of the volute gradually decreases with the volute throat area increasing. With the volute throat area increasing, the size of the counter rotating vortexes increases but the influence range of the two pairs of counter rotating vortexes decreases.

(2) The volute throat area has little effect on the circumferential distribution trend of the outlet flow angle, but obviously affects the value of the flow angle. The uniformity of the flow angle near the volute tongue improves first and then deteriorates with increasing the volute throat area.

(3) With the same volute throat area, the total pressure loss coefficient of the volute gradually increases with the cross-sectional area distribution changes from concave to convex. The size of the counter rotating vortexes increases as the shape of the cross-sectional area distribution changes from concave to convex, and it is the same case for the influence range of two pairs of counter rotating vortexes.

(4) The uniformity of the outlet flow angle distribution in the circumferential direction improves first and then deteriorates with the shape of the cross-sectional area distribution changes from concave to convex, and the vaneless inlet volute with small-scale convex profile has a relatively uniform outlet flow angle distribution.

(5) The radial dimension of the vaneless volute turbine obtained in the final design is significantly reduced, and the total-total efficiency of the design point is 85.62%, which meets the design requirements and also has good performance under off-design operating conditions.

Acknowledgments

The research is supported by National Key Research and Development Project(2016YFB0600104) and Youth Program of National Natural Science Foundation of China(51606026).

References

[1] Crespi F, Gavagnin G, Sánchez D, et al. Supercritical carbon dioxide cycles for power generation: A review[J]. Applied Energy, 2017, 195:152-183.

[2] Garg P, Kumar P, Srinivasan K. Supercritical carbon dioxide Brayton cycle for concentrated solar power[J]. Journal of Supercritical Fluids, 2013, 76(4):54-60.

[3] Sienicki JJ, Moisseytsev A, Fuller RL, et al. Scale Dependencies of Supercritical Carbon Dioxide Brayton Cycle Technologies and the Optimal Size for a Next-Step Supercritical CO_2 Cycle Demonstration[C]//SCO2 Power Cycle Symp 2011.

IOP Conf. Series: Materials Science and Engineering 1081 (2021) 012009 doi:10.1088/1757-899X/1081/1/012009

[4] Yansheng Li. Simplified two-dimensional analysis method for radial inflow turbine's vaneless inlet volute[J]. Journal of University of Shanghai for Science and Technology, 1982,(4): 13-33.
[5] Hara K, Furukawa M, Inoue M. Behavior of three-dimensional boundary layers in a radial inflow turbine volute[J]. Journal of turbomachinery, 1994,116(3): 446-452.
[6] Spence SWT, Rosborough RSE, Artt DW, et al. A direct performance comparison of vaned and vaneless stators for radial turbines[J]. Journal of turbomachinery-transactions of the ASME, 2007, 129(1): 53–61.
[7] Moustapha H, Zelesky M F, Baines N C, et al. Axial and radial turbines[M]. White River Junction, VT: Concepts NREC, 2003.
[8] Vilim. Development and validation of a radial inflow turbine model for simulation of the SNL S-CO_2 split-flow loop.[J]. Office of Scientific & Technical Information Technical Reports, 2012.

IOP Conf. Series: Materials Science and Engineering 1081 (2021) 012010 doi:10.1088/1757-899X/1081/1/012010

Design and flow rate improvement of a small aeroengine turbine

Z Y Jiang[1], Z L Fan[1], M Qiu[1] and D W Yang[1]

[1] China Aerodynamics Research and Development Centre, 6 South Part of Second Ring Road, Mianyang, PRC

E-mail: jiangzhuyu@cardc.cn

Abstract. The need of small high thrust-to-weight ratio aeroengine is growing. One way to meet this is higher through flow rate of engine. Higher engine mass flow means higher turbine mass flow. In the paper, a small turbine is designed and then promoted with higher flow rate. First the definition of universal turbine through flow capability is discussed and formulated. Then the design and flow rate promotion process of a turbine is introduced. And three design results are obtained and compared, marked MI, MII, and MIII. Among them, MI is original design result with highest efficiency of 0.8899 but doesn't meet mass flow requirement. MII and MIII are two retrofitted turbines based on MI both with improved mass flow rate by about 3.40%. Different flow rate improvement methods are implemented for the two turbines. S shape endwall is applied to MII stator shroud and MIII turns down stator blade exit angle a bit. MII has higher efficiency than MIII but through flow capability is in fact lowered. Efficiency decline has been seen in both two redesigned turbines compared to MI. The reason possibly lies in the increased rotor shock loss found by detailed CFD flow analysis. Finally, MII is regarded as the best result of which efficiency is 0.8884.

1. Introduction

For the recent decades, the demand for small and light aerial vehicles has been growing gradually, like unmanned aerial vehicles (UAVs), rotorcrafts, combat drones, cruise missiles, and scaled aircraft models [1-3]. In the future aerial combat, small disposable drones and intelligent missiles would play a more important role in the field. And to improve a drone or missile's maneuverability and defense penetration, small high thrust-to-weight ratio aeroengine is needed. There hasn't been a rigorous definition of small engine somehow till now yet. Usually, as Buysschaert [4] and Huang [5] have concluded, a gas turbine engine of thrust class from 1 kN to 10 kN is called small turbine engine. Engine with lower thrust then is designated as micro turbine engine (MTE). And others whose thrust is higher than 10 kN are regarded as middle or large turbine engines. To increase engine thrust, one possible way is to increase mass flow rate or through flow of the engine given the constant exhaust velocity [6]. However, in small aeroengine development, turbine design work will meet new difficulties different from large engine [7, 8].

At present, the major difficulties to design a small turbine come from the size effect [9]. Since turbine blade height is constrained, to maintain the turbine output work ability, the spool rotation speed has to increase for sake of higher rim speed. Generally, the rotation speed can be up to 30,000 to 70,000 r/min in a small engine, some MTE's speed can even be more than 100,000 r/min [10]. Higher rotation speed results in larger disc and blade stress which presents a big challenge for structure and material, and causes complex rotor dynamics problems [11]. Meanwhile, present fabrication technology can't decrease rotor tip clearance, blade trailing thickness, and hub rounding radius anymore, so the leakage flow and flow blockage will be stronger [10, 12]. Furthermore, blade height to chord ratio of small turbine is usually close to 1, thus larger proportion of secondary flow will generate [10]. Also, during high altitude flight for some reconnaissance UAVs, low Reynolds number condition (approximately under 25,000) in small turbine makes the internal flow more unstable, and easy to separate [13, 14]. Hence engine efficiency will be reduced. And it is more difficult to design cooling configuration and place cooling holes as well [10].

In order to increase aeroengine through flow, it is inevitable to promote mass flow of turbomachinery component. In small aeroengine, radial and axial turbomachineries are both used in practice [15-18]. How to choose the type of turbine to increase mass flow rate is needed to be considered. Ewen [12] points out that axial turbine is preferred now than radial turbine in advanced small aeroengine. Radial turbine on the other hand is more suitable for low flow and high pressure ratio condition. Xia [18, 19] has compared one radial and two axial turbines for a MTE, and concluded that radial turbine has advantage of smaller size when engine flow rate is lower than 0.5kg/s, while on the contrast axial turbine will gain the advantage when flow rate is over 0.5kg/s. However, he thinks that design difficulty of micro turbine stands in the way to implement axial turbine. And Xie [20] as well says axial micro turbine is lighter, easier to realize multi-stage configuration with higher mass flow than radial turbine. But sometimes, as he suggests, turbine flow rate is constrained by compressor. Expect these findings, however, there has not been much more detailed information in open literature about small turbine through flow improvement methods till now. And turbine through flow capability itself is still a vague concept with no accurate and universal definition.

In this paper, the aerodynamic design and through flow improvement of axial turbine for a small short life aeroengine has been introduced in order to explore a way to improve turbine mass flow rate and maintain comprehensive performance as good as possible at the same time. Chapter 1 discusses the definition of universal turbine through flow capability. Chapter 2 introduces turbine design and improvement method. Chapter 3 is the analysis of flow field. Finally, Chapter 4 is conclusion.

2. Definition of universal turbine through flow capability

Before the presentation of turbine design and flow improvement process, it is necessary to find a common and standardized way of turbine mass flow capacity measurement at first. Since different turbines have different scale size, work condition, and configuration, the absolute mass flow rate (mass versus time) can't be a generalized standard to assess flow capacity. Therefore, the paper introduces a parameter of corrected mass flow rate divided by maximum turbine aerodynamic windward area or corrected flow per unit area, $\dot{M}\sqrt{T_0^*}(P_0^* A_{max})^{-1}$, to evaluate through flow capability. Note that the phrase "through flow capability" from now on becomes a specific term in this paper, which stands for this special normalized flow rate that contains the influences of turbine inlet condition and radial geometry. Through flow capability can better reflect turbine physical characteristics than mass flow rate (through flow rate). For convenience, the symbol $\widetilde{M}$ is used to stand for the parameter. Comparison of $\widetilde{M}$ of 8 different aeroengine turbines at design point is shown in Tab. 1 to display the function of the parameter. In the table, each turbine is represented by a sequence number. And as turbine number grows, the maximum tip radius of flow path rises. Here are some briefs of these turbines. No. 1 is the turbine of micro aeroengine MTE-A developed by Nanjing University of Aeronautics and Astronautics [19]. No. 2 is an ordnance turbojet engine turbine [21]. No. 3 is a subsonic experimental turbine developed by Pratt and Whitney Aircraft Division of United Aircraft Corporation [12]. No. 4 is a small low-cost turbofan engine turbine [22]. No. 5 is the high through flow turbine design result MII the paper introduces. No. 6 is the high pressure turbine of NASA/GE E^3 aeroengine [23]. No. 7 is a cold-air experimental turbine for high temperature aeroengine research [24]. Finally, No. 8 is the low pressure turbine of NASA/GE E^3 aeroengine [25]. Besides, number of turbine stages N and stage average

IOP Conf. Series: Materials Science and Engineering 1081 (2021) 012010 doi:10.1088/1757-899X/1081/1/012010

corrected specific work $L(NT_0^*)^{-1}$ are also presented in the table. Still, here use the symbol $\tilde{L}$ to represent $L(NT_0^*)^{-1}$.

Table 1. Turbine through flow capability.

	N	$R_{T,max}$ (m)	$\tilde{M}$ (kg$\sqrt{\text{K}}$(s · kPa · m^2)$^{-1}$)	$\tilde{L}$ (J(kg · K)$^{-1}$)
1. MTE-A turbine	1	0.033	9.452	139.13
2. Ordnance turbine	1	0.1235	8.589	146.28
3. Subsonic turbine	1	0.1259	1.923	153.22
4. Low-cost turbofan turbine	2	0.1260	6.578	137.06
5. MII turbine	1	0.15	6.629	256.05
6. E^3 HPT	2	0.15	12.234	111.27
7. Experimental turbine	1	0.381	6.764	132.35
8. E^3 LPT	5	0.59	3.731	63.64

Generally, it can be seen from the table that $\tilde{M}$ of most turbines is over 6.5 kg$\sqrt{\text{K}}$(s · kPa · m^2)$^{-1}$, and $\tilde{M}$ of different turbines can be quite different from each other with the highest of over 12 kg$\sqrt{\text{K}}$(s · kPa · m^2)$^{-1}$ and lowest below 2 kg$\sqrt{\text{K}}$(s · kPa · m^2)$^{-1}$. Even if two turbines have basically the same $R_{T,max}$, their $\tilde{M}$ can differ a lot in the table. So, it can be assured that some other factors than radial size possibly like engine overall requirement or compressor will limit turbine through flow. Another important factor that is connected with $\tilde{M}$ is $\tilde{L}$. As a matter of fact, stator throat area A_N decides turbine flow capacity and influences specific work at the same time. As A_N grows up (with unchanged number of blades), $\tilde{M}$ increases. However, flow turning and acceleration in stator cascade essentially becomes insufficient, and $\tilde{L}$ decreases in the meantime. Most turbines in the table have $\tilde{L}$ around 140 J(kg · K)$^{-1}$, while No. 5 turbine has a much higher $\tilde{L}$ over 250 J(kg · K)$^{-1}$. This high $\tilde{L}$ certainly brings profits. For instance, $\tilde{M}$ of No. 5 turbine is near to No. 4, but its $\tilde{L}$ is almost two times of the latter. Therefore, number of stages can be reduced and so can be the turbine weight. Thus thrust-to-weight ratio will be improved. Also, from this instance comparison, it can be found that N can be thought as the result of $\tilde{M}$ and $\tilde{L}$ combination. In all, with the help of $\tilde{M}$, turbine mass flow rate is translated to some intrinsic physical property.

3. Turbine design and through flow improvement

3.1. Design requirements

A turbine is asked to be developed mainly to be applied to a small short life aeroengine. Its biggest design difficult is to maintain mass flow rate under high specific output work requirement. Based on the discussion in Chapter 0, the turbine is decided to be axial with one stage. The design corrected flow rate is 0.460 kg$\sqrt{\text{K}}$(s · kPa)$^{-1}$, design total pressure ratio π^* is 2.895, and total efficiency η^* is required to be no less than 0.85. The stator blade needs cooling while rotor does not.

3.2. Overall design process

The whole design process herein still follows the traditional system from 1D, through flow, and blade design with the in-house code and CFD computation later [26]. And it takes several loops to finish the job. As Lu [27] advises, once overall performance has been achieved in lower dimensional job, detailed analysis and optimization effort should be spent in 3D phase. And the paper turbine design work basically agrees with this idea. The CFD simulation and local geometry adjustment have cost the majority of time during the design process.

In preliminary design, meridian flow path and three important parameters need to be decided, i.e. flow coefficient ψ, loading coefficient μ, and mean rotor reaction Ω. Definitions of them are shown from formula (1) to (3). Some initial geometry and aerodynamic design parameters are inputted manually into the in-house code. The 1D code uses energy loss ζ models (including profile loss ζ_p, secondary loss ζ_s, and tip clearance leakage loss ζ_k, shown in formula (4)) [28] to solve inviscid angular momentum equation based on mean flow theory. In formulas (4) to (7), β_1 and β_2 are stator and rotor exit relative flow angle, *h*/*b* is aspect ratio, and *k*/*b* is relative tip clearance height. Through flow code generally uses the same loss models to solve inviscid radial equilibrium equation with streamline curvature method neglecting blade force effect. Meanwhile, flow path is adjusted during through flow design for better performance. Some changes have been made on secondary and tip clearance leakage loss model. And consequently, ζ_s is divided into tip loss term ζ_{st} and hub loss term ζ_{sh} both varying with radius *r*, as formulas (9) to (11) show. And also becomes a function of *r*, shown in formula (12). This way radial flow loss distribution can be more accurate. Next, with energy loss coefficient known, blade row total pressure recovery σ can be calculated by formula (13). Since shock loss is not included in the loss model, some discrepancy may be produced when designing transonic turbines. Nevertheless, design practice shows that error won't be very large. Next with blade inlet and outlet angles, blade design code utilizes Pritchard 11 parameters profiling method to design blade sections. Leading and trailing edge circle radius and wedge angle, maximum thickness and its chord wise location, metal angle, and chord length need to be inputted into the code. The blade section profiles are represented by Bezier curve.

$$\psi = \frac{V_{1z}}{U_m} \tag{1}$$

$$\mu = \frac{\Delta h^*}{U_m^2} \tag{2}$$

$$\Omega = \frac{h_1 - h_2}{h_0^* - h_2} \tag{3}$$

$$\zeta = \zeta_p + \zeta_s + \zeta_k \tag{4}$$

$$\zeta_p = \frac{0.003}{\left(\frac{0.09k_{p1}}{\cos\beta_2} + 0.46\right)(\cos\beta_1 - \cos\beta_2) + 0.085} + 0.017 \tag{5}$$

$$k_{p1} = \begin{cases} \cos\beta_1 \cos(\beta_1 + \beta_2), \beta_1 + \beta_2 > 90^\circ \\ \frac{\cos\beta_1}{\cos(\beta_1 + \beta_2)}, \beta_1 + \beta_2 \le 90^\circ \end{cases} \tag{6}$$

$$\zeta_s = \begin{cases} 0.0474 \frac{\cos\beta_2}{\cos\beta_1}\left(\frac{\cot\beta_1 + \cot\beta_2}{\frac{h}{b}}\right) + 0.0118, \frac{h}{b} \ge 2 \\ 0.0474 \frac{\cos\beta_2}{\cos\beta_1}\left(\frac{\cot\beta_1 + \cot\beta_2}{\frac{h}{b}}\right)\left(1 - 0.25\sqrt{2 - \frac{h}{b}}\right) + 0.0118, \frac{h}{b} < 2 \end{cases} \tag{7}$$

$$\zeta_k = a_k \frac{k}{h} \tag{8}$$

$$\zeta_s = \zeta_{st} + \zeta_{sh} \tag{9}$$

$$\zeta_{st} = a_{st} r^2 \tag{10}$$

$$\zeta_{sh} = a_{sh} r^2 \tag{11}$$

$$\zeta_k = a_{ak1} \frac{k}{h} r^2 \tag{12}$$

$$\sigma = \left[\frac{1-\zeta\left(1+\frac{\gamma-1}{2}Ma^2\right)}{1-\zeta}\right]^{\frac{\gamma}{\gamma-1}} \tag{13}$$

3.3. CFD simulation method

After design work finished, CFD calculation plays a role in analysing design result with commercial software package NUMECA. Many researchers have used this software in turbine numerical simulation, and proven its reliability [14, 20, 27, 29]. The RANS method is a reliable and relatively simple tool in turbomachinery 3D calculation [30]. And steady RANS approach of FINE/Turbo is used to solve the flow field, and mixing plane method is used in stator/rotor interface flow solution. Spalart-Allmaras turbulence model is used, second order central difference scheme is used for spatial discretization, and fourth order Rung-Kutta scheme is used for time discretization. And O-H topology is adopted to construct computational mesh by AutoGrid with solid wall y^+ less than 5, minimum skewed angle greater than 30°, maximum aspect ratio less than 400, and maximum expansion ratio less than 2. Single turbine blade row channel mesh structure is show in figure 1, and number of grid points is around 2 million. Solver precision is double, and convergence criteria level is 10e-6 during computation. Inlet boundary condition is set according to total pressure and temperature, and outlet boundary condition is settled by static pressure imposed with radial equilibrium. To confirm the mesh level validation, independence test of mesh has been carried out with 8 different meshes and the paper design turbine MI. Among them, grid level varies from 440,000 to 4,200,000. The corrected mass flow of each meshes is displayed in figure 2. Clearly, very little change can be found of the mass flow after number of grids is higher than 1,600,000. Thus, 2,000,000 level mesh should be reliable for calculation.

Combustion gas property is used in CFD simulation which incorporates empirical functions of constant pressure specific heat C_p and heat conduction coefficient λ. Both C_p and λ are regarded as polynomial functions of temperature. Total efficiency η^* is then computed by equation (14) and (15), where T_{2i}^* is the isentropic rotor exit total temperature given the same total pressure ratio π^*. Yet, for simplification, some researchers still use equation (16) based on constant specific ratio γ to evaluate efficiency η_c^* even if some of them are using variant C_p during simulation at the same time [20, 27, 31]. Without a doubt, this will cause discrepancy. figure 3 exhibits an example of efficiency calculation result with varying γ. The η_c^* changes in a near linear fashion, and every time γ grows by 0.005, η_c^* will drop by about 0.01. Therefore, the error introduced by γ can be big.

$$\eta^* = \frac{\int_{T_2^*}^{T_0^*} C_p(T)\mathrm{d}T}{\int_{T_{2i}^*}^{T_0^*} C_p(T)\mathrm{d}T} \tag{14}$$

$$\int_{T_{2i}^*}^{T_0^*} \frac{C_p(T)}{T}\mathrm{d}T = R\ln\pi^* \tag{15}$$

$$\eta_c^* = \frac{T_1^*-T_2^*}{T_1^*\left(1-\frac{1}{\pi^{*\frac{\gamma-1}{\gamma}}}\right)} \tag{16}$$

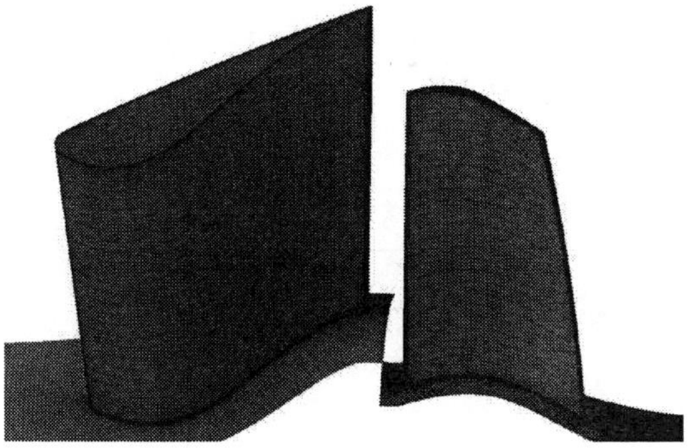

Figure. 1. Computational mesh structure.

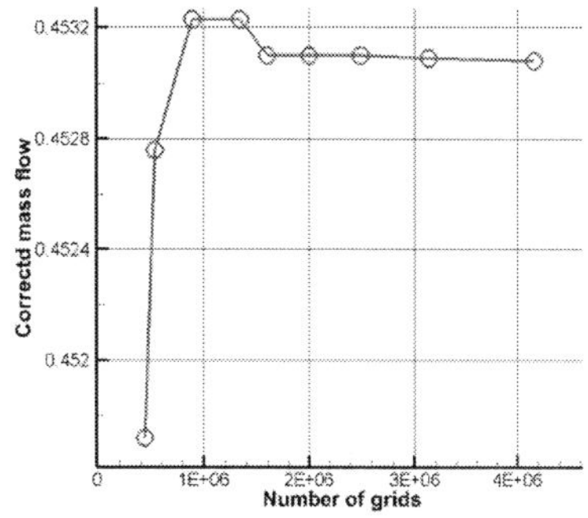

Figure. 2. Corrected mass flow of different meshes.

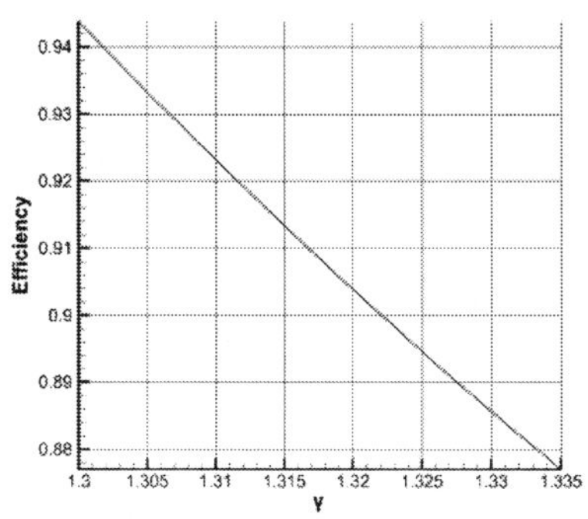

Figure. 3. γ impact on η^*.

4. Results comparison

4.1. Overall performance

After many trials, three design results have been accomplished, marked MI, MII, and MIII separately. Of them, MI is initial result which has in reality not met the mass flow rate requirement, even though it shows satisfactory pressure ratio and efficiency. MII and MIII, next, are two retrofitted plans of MI to raise mass flow. However, some performance characteristics are reduced as a sacrifice. MII has adjusted stator shroud endwall profile to S shape and MIII has reduced stator tip blade section exit angle by around 2° keeping axial chord length unchanged. Stator shroud S shape endwall is suggested as an easy and effective method to control endwall secondary, reduce loss, and improve efficiency [9, 11, 12, 32-34]. Also, it can increase flow rate [12, 33]. According to reference [34], annular cascade loss reduction is related to relative convergence, $(h_1 - h_2)/h_2$ and aspect ratio h_1/b (h_1 is inlet blade height, h_2 is exit blade height, and b is blade chord length). But in design practice as Shi [34] implies, $(h_1 - h_2)/h_2$ can be chosen within a relatively wide range. Hass [33] has tested two turbines with different contoured stators, and both turbines have higher efficiency compared to cylindrical wall stage due to lower stator loss. Due Jr. [35] has also conducted experimental survey on one cylindrical wall and three different contoured wall stators, and found different outer wall contouring could all improve cascade efficiency. Thus, basically almost any kind of S endwall can be beneficial. Finally, the $(h_1 - h_2)/h_2$ of MII is 0.156 and is 0.751. Referred to the picture provided by reference [34] (figure 4), it could be found that the match of these two parameters should be appropriate. And definitely smaller blade exit angle of MIII shall enlarge stator throat area, and then flow rate rises. MI and MIII share the same meridian flow path.

Flow path comparison of MI and MII is shown in figure 5. Stator tip blade section profile comparison of MII and MIII is shown in figure 6.

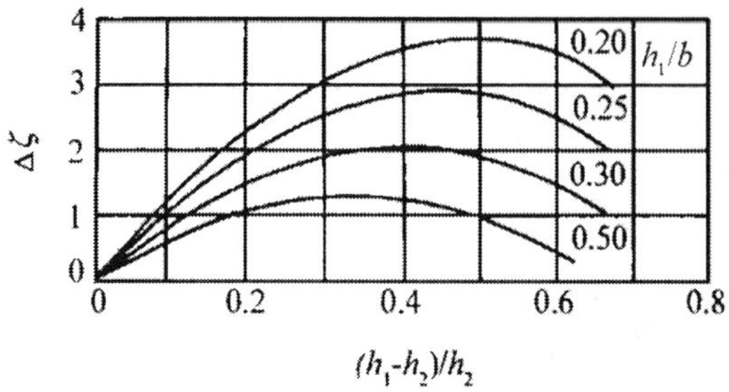

Figure. 4 $(h_1 - h_2)/h_2$ and h_1/b influence on loss reduction.

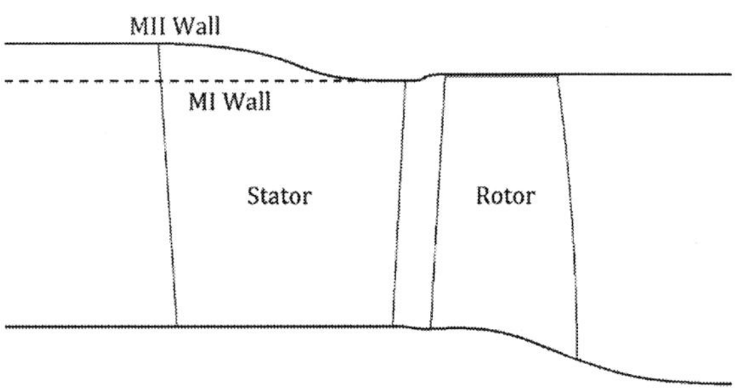

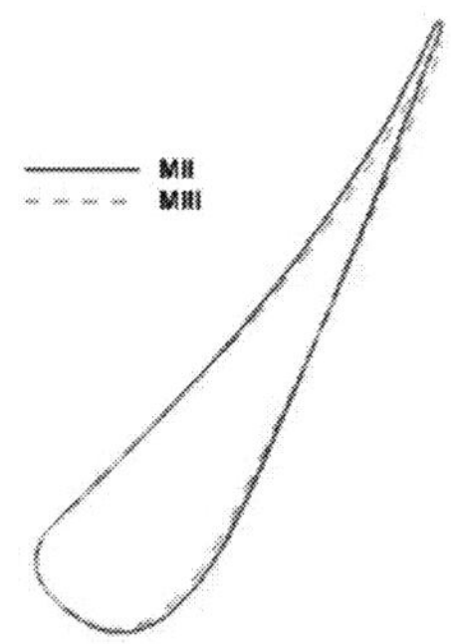

Figure. 5 Flow path comparison.

Figure. 6 Stator tip blade section profile comparison.

Because stator blade is in need of cooling, its thickness is designed to be higher. Stators blade profiles are designed at hub at tip section on the cylindrical surface, and rotor blade profiles are designed at hub, mean, and tip section. Stator blades sections are stacked along leading edge and rotor sections are stacked through centroids. Rotor blade stacking line has a small lean angle circumferentially to compensate some aerodynamic force on blade pressure surface to improve blade strength. For MII, when stacking stator blade sections, tip profile line is extrapolated radially to intersect with S endwall. Since the turbine is going to be used in short life engine, AN^2 is increased above conventional standard to improve through flow capability. As reference [9] indicates, usually AN^2 of high pressure turbine is within (19~32) $m^2(r/min)^2$. While the paper turbine's AN^2 is 36 $m^2(r/min)^2$. Empirically, it is believed this compromise could be acceptable. And meanwhile better material or fabrication can if necessary, make up for rotor strength flaw.

Overall aerodynamic characteristic parameters obtained by NUMECA of MI, MII, and MIII are displayed in Tab. 2. And three different total efficiency computation results of the turbines are demonstrated in Tab. 3. In the table, η^* is calculated by equation (14), η_{c1}^* is calculated with different γ at rotor/stator interface of each turbine respectively by equation (16), and η_{c2}^* is also obtained by equation (16) given the same γ (1.318) which is the mean value of the three turbines at rotor/stator interface. And among them, η^* is thought as the most accurate result.

Table 2. Aerodynamic characteristics.

	ψ	μ	Ω	$\bar{M}$ kg$\sqrt{K}$(s · kPa)$^{-1}$	$\tilde{M}$ (kg$\sqrt{K}$(s · kPa · m^2)$^{-1}$)	$\tilde{L}$ (J(kg · K)$^{-1}$)
MI	0.7846	1.675	0.2624	0.4531	6.859	256.51
MII	0.7794	1.672	0.2949	0.4685	6.629	256.06
MIII	0.7600	1.668	0.3040	0.4685	7.093	255.52

IOP Conf. Series: Materials Science and Engineering 1081 (2021) 012010 doi:10.1088/1757-899X/1081/1/012010

Table 3. Total efficiency.

	η^*	η^*_{c1}	η^*_{c2}
MI	0.8899	0.9084	0.9092
MII	0.8884	0.9086	0.9077
MIII	0.8866	0.9071	0.9058

From the first table above, it can be found that MII and MIII have effectively increased flow rate $\overline{M}$ by around 3.40% from MI, however both of them have suffered specific work deterioration at the same time. With consideration of manufacture error and blade corner rounding induced flow blockage, $\overline{M}$ of MII and MIII is a little higher than design value. $\widetilde{M}$ of MIII is the highest. Although MII has improved $\overline{M}$, its shroud radius goes up at the same time causing through flow capability to get lower. So, S endwall may not necessarily be effective in $\widetilde{M}$ promotion. Different design attempts have shown that incorporating low ψ and μ can be beneficial for turbine efficiency, but turbine radial size will increase at the same time. And to achieve high enough stator expansion work and required π^*, Ω should be low. Compared with MI, both MII and MIII has reduced ψ, and increased Ω. The reason of this change could be the larger stator throat of MII and MIII. And then flow acceleration in stator decreases as stated in Chapter 1. Consequently, V_{1z} and Δh turns smaller with higher bigger proportion of expansion work loaded on rotor. Higher Ω and increased stator exit gaging area attributed by S endwall is also found by Ewen [12].

Next from the second table, it shows that η^* of MI is the highest, MII is 0.0015 lower, and MIIII is 0.0033 lower when compared. The efficiency decline is the cause of $\widetilde{L}$ decrease. This does not accord with other researchers' findings where turbine efficiency is successfully promoted with implementation of S endwall [12, 32]. The reason will be explored the next section. Compared with η^*, η^*_{c1} and η^*_{c2} are higher by around 0.02. Since η^*_c is close to a linear function of γ as shown in figure 2, therefore the suitable value of γ might be higher by 0.01. And compared to η^*_{c1}, η^*_{c2} is more accurate and its change tendency from MI to MIII is very close to η^*. η^*_{c1} of MI is lower than MII, this is the result of different γ induced discrepancy. As η^*_c is very sensitive to γ, the small change of γ can lead to a large error of absolute value of efficiency covering the real physical quality. But equation (15) could still be used to compare relative efficiency among different turbines as soon as their operation conditions and configurations are similar, and the same γ is applied.

4.2. Flow field analysis

Since efficiency has dropped in two redesigned turbines, it is necessary to find out the reason. Secondary flow loss takes a big park of overall loss in small turbine, and alterations are made in stator of MII and MIII. Then first of all, secondary flow in stator is analyzed. figure 6 shows limiting streamlines at stator shroud and hub endwall respectively of MI, MII, and MIII. Clearly, S endwall has promoted the flow condition near stator shroud found from figure 7 (a). For MI and MIII, crossflow is stronger with pressures side separation line starting from saddle point moves toward and ends at the adjacent blade suction side. Pressure separation line of MII on the other hand moves downstream without intersection with the other blade. Thus, lower transverse pressure gradient should there be near stator tip of MII. Transverse pressure gradient is the source of secondary flow [34], therefore secondary flow is then effectively controlled. However, figure 7 (b) shows no obvious difference of flow near hub that can be seen among three turbines. Influence of S shape endwall is mainly focused on shroud area.

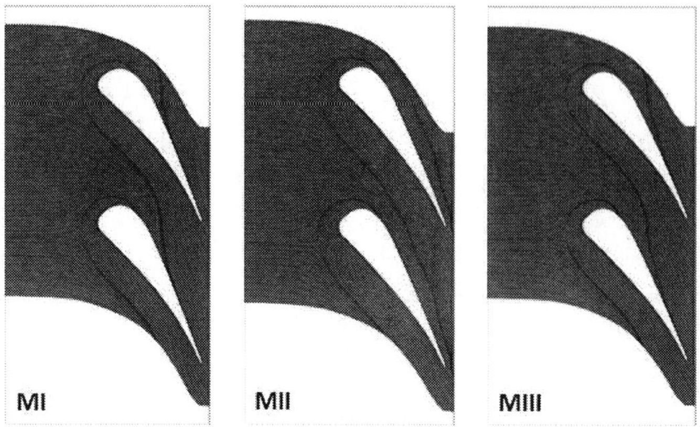

(a) Shroud.

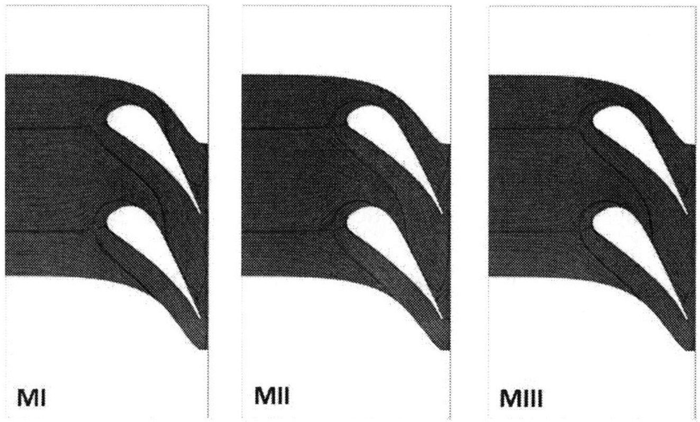

(b) Hub.

Figure. 7 Stator endwall limiting streamlines.

Then blade loading is analysed here. Stator and rotor blade section static pressure coefficient distribution P/P^* at different span locations of the three turbines is shown in figure 8 and 9. The benefit of S endwall on stator blade is remarkable as figure 8 shows. Stator blade loading distribution of MI and MIII are close. At 90% blade span location there is an abrupt pressure rise at suction side of MI and MIII from 0.35 to 0.55 axial position, behind the maximum curvature point of suction profile line. This is the result of strong acceleration on frontal suction side surface with pressurized waves formed, as depicted in figure 10. And at 10% and 50% span section, there is also a phenomenon of over speeding of flow on frontal half of suction side for MI and MIII. While for MII, stator blade pressure side P/P^* is raised especially after 0.5 axial position, and loading is successfully put back by reducing flow pressurization near high curvature profile area on suction side. In the meantime, figure 8 clearly displays the lower pressure difference between pressure and suction side of MII, which is the reason of weaker secondary flow as discussed before. And S shroud wall improvement is most effective at 90% span. However, the pressure rise near trailing edge is not restrained on all turbines.

As for rotor blade, MII and MIII have both managed to optimize the loading distribution at different levels compared to MI depicted by figure 9. Generally, P/P^* on pressure side is raised for MII and MIII. At 10% span location, there near the leading area of MI suction side exists reverse pressures gradient. MIII has reduced the pressure gradient and MII basically has eliminated it. Meanwhile, pressure rise after 0.7 axial position is also decreased in MII and MIII. The pressure distribution at 50% span section of MII and MIII are very close. And both two retrofitted turbines have improved suction side pressure gradient from near leading to trailing edge as well as eliminating the pressure rise at trailing edge. At 90% span, the pressurization of MI at trailing of suction side is also removed in MII and MIII. Compared to MI, positive pressure gradient along suction side surface is improved in MII and is further improved in MIII. Pressure side P/P^* rise is also increased more in MIII than MII.

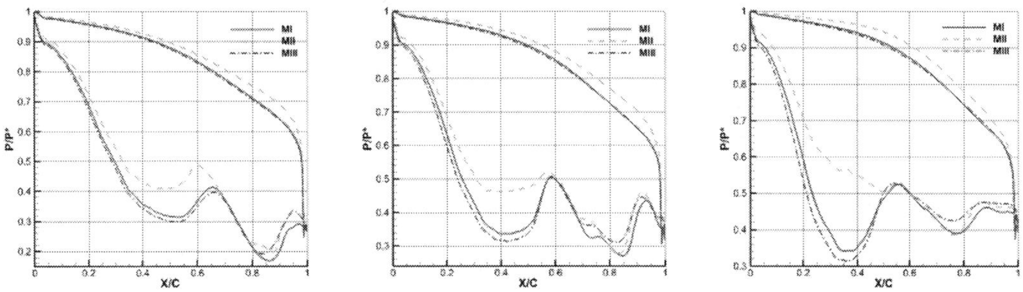

(a) 10% span. **(b)** 50% span. **(c)** 90% span.

Figure. 8 Stator blade section P/P^*

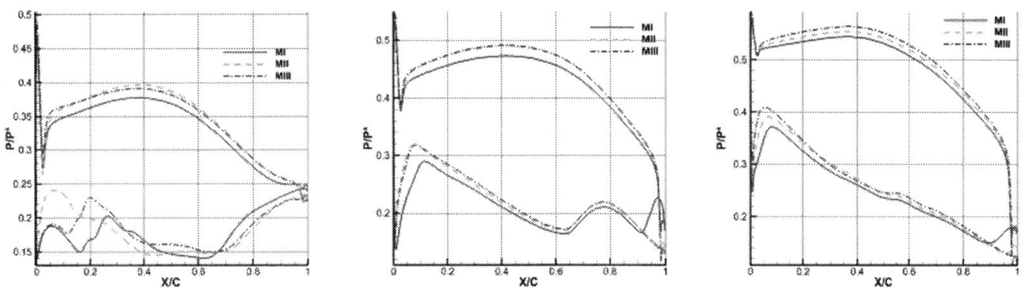

(a) 10% span. **(b)** 50% span. **(c)** 90% span.

Figure. 9 Rotor blade section P/P^*

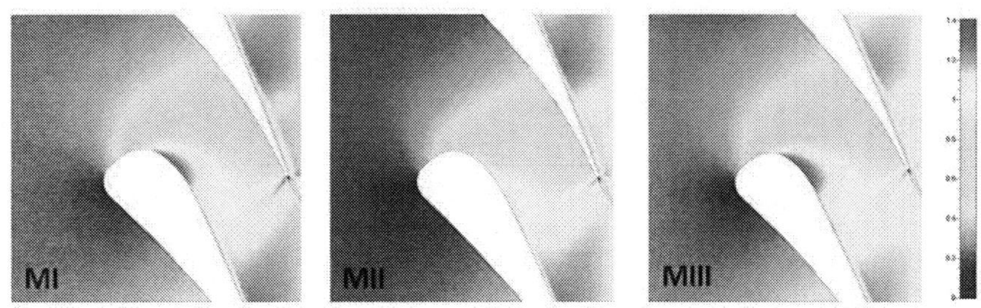

Figure. 10 Stator Mach number contour at 90%

Flow loss decides the efficiency of turbine, and the cause of efficiency drop of two redesigned turbines can be explored from loss analysis. Stator and rotor total pressure loss coefficient of three turbines is illustrated in figure 11. And blade inlet and exit Mach number is displayed in figure 12. It can be found that MIII stator loss is basically reduced compared to MI and MIII. This is possibly the result of effective control of secondary flow and blade suction side reverse pressures gradient. Also, it could be the consequence of decreased inlet and exit Mach number, shown in figure 12 (a). On the contrary, loss is hardly decreased in MIII, and though stator exit Mach number is reduced, inlet Mach number turns higher. In spite of the optimized rotor blade loading distribution, effective overall loss reduction is not seen in MII and MIII. Loss near rotor hub has been reduced in the two retrofitted turbines, and this could be the effect of endwall secondary flow control, since hub transverse pressure gradient of these turbines is lower. Also, the high Mach zone with more complex expansion and pressurized waves near leading edge are in MI rotor hub as figure 13 exhibits could bring extra flow loss. Lower rotor inlet Mach number could be part of the reason of lower hub loss in MII and MIII. But loss above 40% span area becomes higher in MII and MIII. The reason could be the higher rotor exit Mach number which produces stronger shock wave and more shock loss as shown in figure 12 (b) and figure 14. Therefore, increased rotor loss should possibly be blamed for the descended efficiency. Also, it can be explained why S shroud endwall does not promote stage efficiency like other researchers have found. Because in Ewen's experiment [12], both contoured and straight wall turbines testing S endwall function are subsonic. The three turbines tested in Hass' work [33] are also substantially subsonic, with only one contoured turbine having tip exit relative Mach number slightly over 1 (1.023) on design point. While in this study here, MII turbine is basically transonic with higher exit relative Mach number.

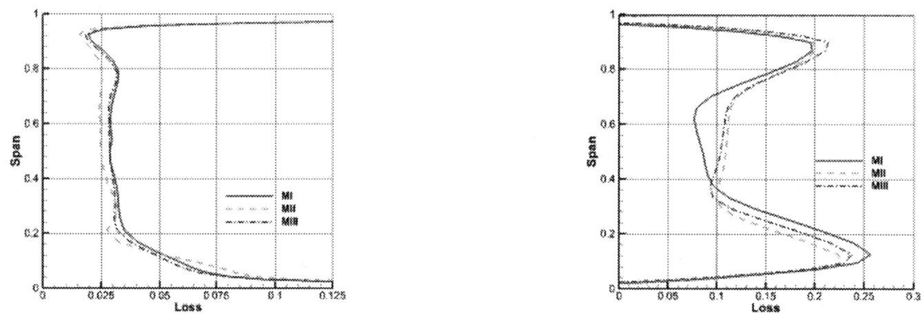

(a) Stator. (b) Rotor.

Figure. 11 Blade row total pressure loss.

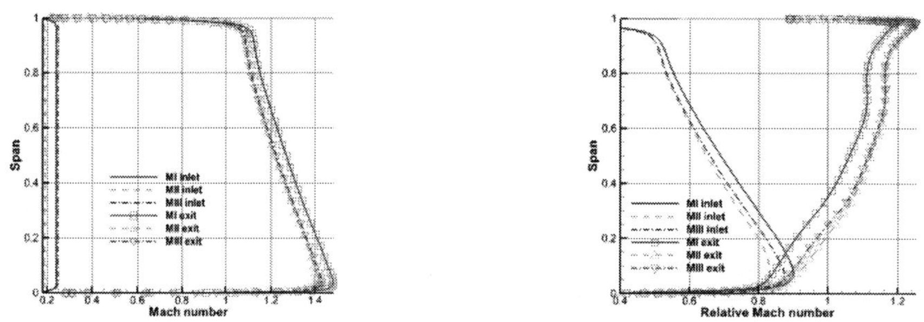

(a) Stator. (b) Rotor.

Figure. 12 Blade row inlet and exit Mach number.

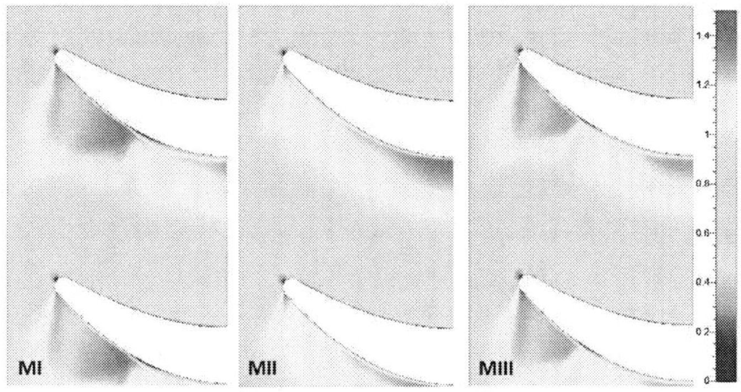

Figure. 13 Rotor relative Mach contour near inlet at 10% span.

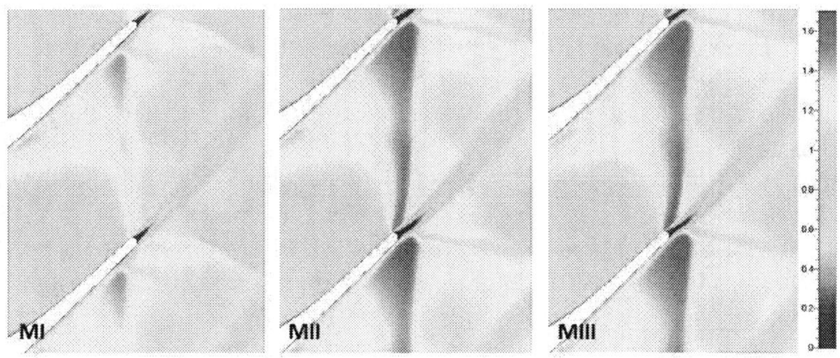

Figure. 14 Rotor exit relative Mach number contour at 50% span.

4.3. Design summary
After comparison of MI, MII, and MIII design results, it is decided that MII should be the best result. Because it meets the flow rate requirement and has higher efficiency than MIII. Although $\widetilde{M}$ is not promoted in MII but in MIII, it is still believed that efficiency improvement of 0.0018 from MIII is more important. The higher efficiency of MII among the two retrofitted turbines is achieved by S shape endwall. The S endwall is another way to enlarge stator throat area rather than turning down the blade exit angle. And it is more profitable in improving flow characteristics. Unfortunately, on the other hand, overall stage efficiency promotion has not been seen in MII. The reason could be in the larger rotor shock loss.

5. Conclusions
The paper discussed the definition of universal turbine through flow capability firstly, and then mainly introduces the design and flow rate improvement process of a turbine prospectively for application in small short life aeroengine. Original and two retrofitted design results are shown and compared, and finally turbine MII is decided as the best result. Conclusions can be summarized in 3 points.

(1) A precise and generalized assessment method of turbine through flow capability has been proposed with the introduction of the parameter, corrected flow per unit area $\widetilde{M}$. This way, through flow capability becomes a clear and special concept that can be measured quantitatively and better reflects intrinsic physical property.

(2) The final turbine design result MII is basically satisfactory with corrected mas flow of 0.4685 $\mathrm{kg}\sqrt{\mathrm{K}}(\mathrm{s}\cdot\mathrm{kPa})^{-1}$, through flow capability of 6.629 $\mathrm{kg}\sqrt{\mathrm{K}}(\mathrm{s}\cdot\mathrm{kPa}\cdot\mathrm{m}^2)^{-1}$, efficiency of 0.8884, and corrected specific work of 256.06 $\mathrm{J}(\mathrm{kg}\cdot\mathrm{K})^{-1}$ at design point and its exit flow is transonic. Its efficiency

is higher by 0.0016 than MIII. The biggest characteristic of the turbine is considerably large specific work much higher than many others with lowered strength requirement.

(3) Stator shroud S shape endwall is effective in stator secondary flow control, blade loading optimization and loss reduction. It can also improve loading distribution on rotor blade. It is as well a way to improve turbine mass flow rate. However, in transonic turbine, it might increase rotor exit Mach number and lower efficiency. And S endwall does not necessarily improve $\widetilde{M}$.

6. References

[1] Niedzwiecki R W and Meitner P L 1987 Small Gas Turbine Engine Technology *Preprint* (AVSCOM-TR-87-C-6)

[2] Su S M, Zhang L, Cai Y H and Tu Q Y 2009 Microturbo engine development and key technology analysis *J. Aerospace Power* **24** 11 pp 2410-4.

[3] Tian B l 2003 A survey of the development of engines for the unmanned aircraft and the cruise missile in the world *Aeroenigne* **29** 4 pp 51-4

[4] Buysschaert F and Hendrick P 2005 Efficient use of small gas turbine engine for UAV applications *41st AIAA/ASME/SAE/ASEE Joint Propulsion Conf. & Exhibit* Tucon Arizona

[5] Huang G P, Liang D W and He Z Q 2008 Comparison of technical characteristics of APU and micro turbine engines for large aircraft *J. Aerospace Power* **23** 2 pp 383-8

[6] Lian X C and Wu H 2005 *Aeroengine Principle* (Xi'an: Northwestern Polytechnical University Press) p 15

[7] Moustapha H 2003 Small gas turbine technology: evolution and challenges *AIAA/ICAS Int. Air and Space Symp. and Expo.: The Next 100 Y* Dayton Ohio

[8] Brooks V E 2008 Small turbine engine evolution *SAE Int. J. Aerospace* **1** 1 pp 915-28

[9] Shi J W, Du Q and Zhu J Q 2014 High-pressure turbine aerodynamics design of small turbofan engines *J. Engineering Thermophysics* **35** 6 pp 1068-74

[10] Zhang W H, Zou Z P, Liu H X and Li W 2010 Study on turbine aerodynamic design of small turbofan engines *J. Aerospace Power* **25** 1 pp 72-9

[11] Shi J and Han J Y 1991 Aerodynamical design for small gasturbines *J. Aerospace Power* **6** 1 pp 29-32

[12] Ewen J S, Huber F W and Mitchell J P 1973 Investigation of the aerodynamic performance of small axial turbine *J. Engineering for Power* **95** 4 pp 326-32

[13] Zou Z P, Ye J, Liu H X, Li W, Yang L and Feng T 2007 Research progress on low pressure turbine internal flows and related aerodynamic design *Advances in Mechanics* **37** 4 pp 511-62

[14] Wang X D, Liu X J, Liang L P and Kang S 2016 Comparisons of numerical simulation method of laminar flow separation on blade surface within a low speed axial turbine *Chinese J. Turbomachinery* **3** pp 25-30

[15] Zhun C Y, Ju Y P and Zhang C H 2016 Aerodynamic design and analysis of impeller and diffuser in a high flow coefficient centrifugal compressor *Chinese J. Turbomachinery* **5** pp 57-62

[16] Wang Y S, Lin F, Cao X J and Cao P 2016 Design of a single-stage high-speed, large flowrate, high pressure ratio centrifugal compressor *Chinese J. Turbomachinery* **1** pp 50-4

[17] Wang X, Li W, Zhang X H, Zhu Y L, Zuo Z T and Chen H S 2019 Tip profile optimization in a low aspect ratio CAES radial expander based on orthogonal design *Chinese J. Turbomachinery* **1** pp 11-21

[18] Xia C, Huang G P and Qiu J 2011 Comparison research on performance and size of micro axial turbine and radial turbine *J. Aerospace Power* **26** 3 pp 490-7

[19] Xia C 2012 *Design and Experiment Technique Researches of High Load Micro Radial Turbine* (Nanjing: Nanjing University of Aeronautics and Astronautics) pp 36-42

[20] Xie J, Xia C, Zhang Y S and Li C P 2015 Micro axial turbine bowed bladed design at low aspect radio *J. Nanjing University of Aeronautics & Astronautics* **47** 1 pp 160-6

[21] Kofskey M G, Nusbaum W J and Haas J E 1974 Turbine for Ordnance Turbojet Engine II – Cold Air Performance with Opened Stator *National Aeronautics and Space Administration Report*

(NASA TN-7626)
[22] Kofskey M G and Nusbaum W J 1972 Design and Cold-Air Investigation of a Turbine for a Small Low-Cost Turbofan Engine *National Aeronautics and Space Administration Report* (NASA TN D-6967)
[23] Halila E E 1982 Energy Efficient Engine High Pressure Turbine Test Hardware Detailed Design Report *National Aeronautics and Space Administration Report* (NASA CR-167955)
[24] Whitney W J, Szanca E M, Bider B and Monroe D E 1968 Cold-Air Investigation of a Turbine for High-Temperature-Engine Application III – Overall Stage Performance *National Aeronautics and Space Administration Report* (NASA TN D-4389)
[25] Chery D G, Gay C H and Lenahan D T 1982 Energy Efficient Engine Low Pressure Turbine Test Hardware Detailed Design Report *National Aeronautics and Space Administration Report* (NASA CR-167956)
[26] Zou Z P, Wang S T, Liu H X, Yang C and Zhang W H 2014 *Turbine Aerodynamics for Aero-engine: Flow Analysis and Aerodynamics Design* (Shanghai: Shanghai Jiao Tong University Press) p 242
[27] Lu S P, Wen F B, Liang C, Niu X Y, Cai L, Wang S T, Feng G T and Wang Z Q 2012 Investigation on multi-stage turbine aerodynamic design and optimization *J. Propulsion Technology* **33** 6 pp 888-96
[28] Huang Q N 2001 *Aeroengine Design Manual 10: Turbine* (Beijing: Aviation Industry Press) pp 63-7
[29] Han J, Wen F B and Zhao G B 2014 Curve optimization design of a turbine blade with low aspect ratio *J. Tsinghua University (Science and Technology)* **54** 1 pp 102-8
[30] Zhong L Y, Liu Y W and Lu L P 2019 Unsteady Behavior of Tip Leakage Vortex in an Axial Compressor with Different Rotor Tip-gap Sizes Using DDES *Chinese J. Turbomachinery* **61** 2 pp 1-9
[31] Afanas'ev I V and Granovskii A V 2018 Computational study on the effect of shroud shape on the efficiency of the gas turbine stage *Thermal Engineering* **65** 3 pp 136-42
[32] Haas J E 1982 Analytical and Experimental Investigation of Stator Endwall Contouring in a Small Axial-Flow Turbine I – Stator Performance *National Aeronautics and Space Administration Report* (NASA TP-2023)
[33] Haas J E and Boyle R J 1984 Analytical and Experimental Investigation of Stator Endwall Contouring in a Small Axial-Flow Turbine II – Stage Results *National Aeronautics and Space Administration Report* (NASA TP-2309)
[34] Shi J, Zhou Y, Zhou S P, Luo J Q and Lu C M 2002 Three dimensional aerodynamic investigation of a turbine nozzle guide van with combination design of tip endwall contouring and positive leaned airfoil *J. Aerospace Power* **17** 4 pp 385-91
[35] Due Jr H F, Kosier C L, Rogo C and Jasas G B 1977 Advanced Small Axial Turbine Technology *U.S. Army Air Mobility Research and Development Laboratory Report* (USAAMRDL-TR-77 -1)

Study on the influence of hot streak and swirl on the film cooling performance of the leading edge of the guide vane

ChenXiao[1], YI Wei-lin[1]

[1]Beijing Institute of Technology, Beijing

Abstract. In order to study the impacts of the non-uniform conditions of the combustor exit on the flow loss of the turbine guide blade and the cooling performance of the leading edge film, a hot streak and swirl distribution generation program has been developed to implement the parameterization specification of the non-uniform flow state. In addition, the calculation of this paper is based on GE-E3, the numerical simulation analysis is conducted for the swirl/hot streak of the positive blade on the leading edge of the film cooling guide blade. The results showed that the presence of hot streak increased the non-uniformity of vane heat load and had no effect on vane wall pressure distribution. The presence of swirl changes the pressure distribution on the vane wall, affects the migration form of hot streak in the passage, and reduces the cooling effect of the film cooling on the vane. The temperature at the outlet section of the guide vane is uniform.

Keywords: hot streak; swirl; turbine blade; leading edge film cooling; entropy distribution

1. Introduction

With the continuous improvement of the performance requirements of aero-engine like the thrust-to-weight ratio and thermodynamic efficiency, the turbine operating temperature is getting increasingly higher. The high temperature far exceeds the metal material tolerance range so that the turbine life is dramatically shortened. Due to the limited scientific development of turbine blade materials, film cooling technology [1] has been widely studied and applied as an effective cooling method. The leading edge of the turbine blade is completely exposed to high temperature fluids and is the most critical part. Giridhara [2] and Liu [3] respectively performed experimental and numerical simulations on the film cooling of the leading edge of the blade. Liu [3] studied the semi-cylindrical leading edge model with three vent film cooling holes on the leading edge, and analyzed its flow and heat transfer. Malkell [4] also used the cylindrical leading edge as the research object for numerical simulation calculations. It was found that due to the flow separation of the cooling flow, the film cooling efficiency was reduced at a higher blowing ratio. In the same amount of cooling flow, the film holes of the three rows are more efficient than the film holes of the single row. Li[5] et al. conducted a numerical simulation study on the cooling effect of fan-shaped film holes. Chen[6] et al. studied the interaction between the film cooling flow on the inner end wall of the turbine runner and the main secondary flow structure.The exploration of film cooling has been studied by many scholars. Most of the research is carried out under uniform conditions or on flat plates. However, the actual turbine blade surface will be affected by the structure of the combustion chamber, and the import conditions will be very complicated. To improve the fidelity of the simulations, realistic non-uniform inlet conditions must be included.

The complex combustor structure and working mode result in non-uniform exit temperature, and the local temperature is much higher than the ambient temperature. This local high temperature zone is called hot streak. Many scholars have conducted experimental measurements on the total temperature of combustor exit flow, such as Qureshi [7], Barringer [8] and others. Paolo [9] studied the change of the hot streak generated by the burner in the vane, and found that the temperature of the hot streak in the static vane has obvious attenuation, and the hot streak changes significantly to the leading edge. Sean [10] studied the effect of hot streak on the vane impact with film cooling by experimental methods, and found that the intensity of hot streak was reduced by 70% under the effect of film cooling. An [11] obtained numerical simulations that the presence of hot streak increased the unevenness of the thermal load of the stationary leaves, and the influence of hot streak at different circumferential positions on the thermal load of the turbine was given. Dong [12] performed hot streak analysis on the blades with film cooling, and found that the hot streak not only caused a large thermal shock to some areas of the blade, but also reduced the effect of film cooling.

In order to reduce NOx emissions, the combustion chamber tends to adopt Lean-Burn technology, and it is necessary to introduce a swirler to ensure flame combustion stability [13], so that the turbine inlet is further affected by the swirling flow while the hot streak is present. The influence of complex flow angle distribution on the turbine flow field and temperature field is difficult to ignore.Turrell [14] confirmed that there is a highly rotating vortex core at the exit of the combustion chamber through observing the trajectory of the film cooling jet.Tommaso Bacci [15] et al. simulated the existence of swirling flow at the leading edge of the vane through an experimental method, and the result pointed out that the swirl flow has a significant effect on the pressure distribution of the leading edge of the vane. Through the analysis of the blade adiabatic efficiency, it is also obtained The effect of swirl on film cooling cannot be ignored. Zhao [16] used two directions of rotation and two circumferential clock positions to simulate the swirl, and the effect of the inlet vortex on the pressure distribution was concentrated on the suction side. Stavros Pyliouras [17] studied the existence of swirl at the exit of the combustion chamber to reduce the efficiency of the turbine. Qureshi [18,19] added a swirl generator at the turbine inlet to simulate the swirl flow existing at the turbine inlet, and it was found that the blade load changed, and the hot streak caused a significant radial migration under the action of the swirl. The aerodynamic performance of the turbine has a large impact.

Most of the above literatures study the effects of hot streak or swirls on heat transfer and flow in the turbine alone, and less consideration is given to leading edge film cooling. The swirl flow affects the main flow near the leading edge of the vane of the high-pressure turbine, as well as the cold flowing out of the leading edge of the vane, which will have a great impact on film cooling.The first stage turbine vanes are facing the high temperature and high pressure gas discharged from the combustor, and the cooling performance of the film is not clear enough. In this paper, the E3 turbine is taken as the research object, and the aerodynamic and heat transfer effects generated by the combination of hot streak and eddy current are studied by numerical simulation.

2. Research objects and numerical methods

2.1. Computation mode

The experiment report for core machine high pressure turbine provided by GE Company is adopted for the computation model in this paper [20]. Only the vane cascade is considered,and the number of vanes is 46. The numerical caiculation model is shown in Figure 1. The position is measured according to the experimental results in [21]. P40 is the assumed combustor-turbine interface is, located 40% of the axial chord length in front of the leading edge of the vane. P1 represents the vane exit data measurement plane. Referring to the experimental cooling structure [22], three rows of cylindrical cooling holes row1, row2, and row3 are set at the leading edge of the blade, as shown in Figure 2, and the relevant parameters are shown in Table 1.

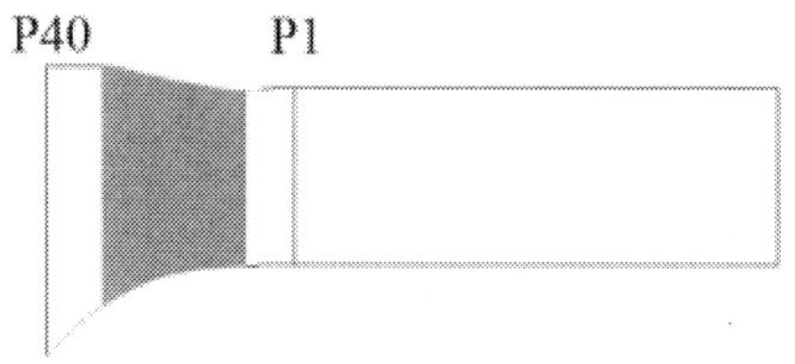

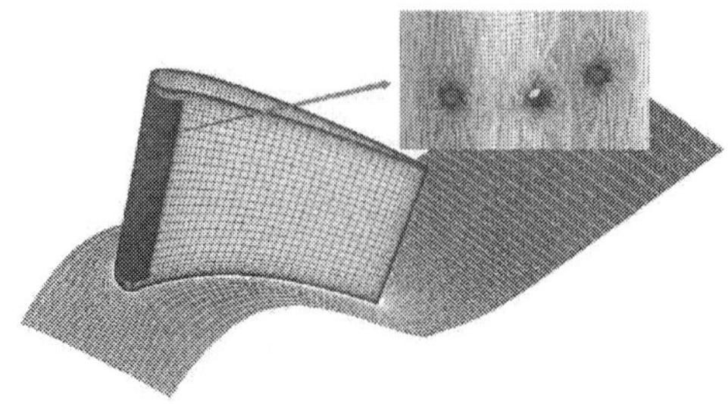

Figure 1. Computational domain of the E3 guide vane

Figure 2. E3 guide blade mesh and film cooling hole.

Table1. Size and location of film cooling hole

	Diameter(mm)	No.of holes on each Row	Span angle (degree)	Distance to the geometrical midline position of the leading edge (%)
Row1	0.019	12	30	3
Row2	0.019	12	30	0
Row3	0.019	12	30	-3

2.2. Numerical method

The calculation uses the Numeca AutoGrid5 module to generate structured grids. A total of 2.6 million, 3.8 million, and 5.2 million sets of grids were generated. The three types of grids were calculated separately. The calculation results were compared with those measured in the experimental report [21]. The distribution of isentropic Mach number distribution under cooling conditions is compared. The distribution of the isentropic Mach number distribution in the vane surface under the three kinds of grids is shown in Figure 3. The values on the pressure surface are significantly different from the experimental values. The distribution of film holes on the pressure surface is relatively dense in the experiment. In this article, we will not consider the distribution of air film holes on the pressure surface for the time being. Considering the calculation speed, it is considered that the calculation results under the 3.8 million grid are acceptable. Therefore, the number of grids used in the numerical simulation in this paper is 3.8 million.

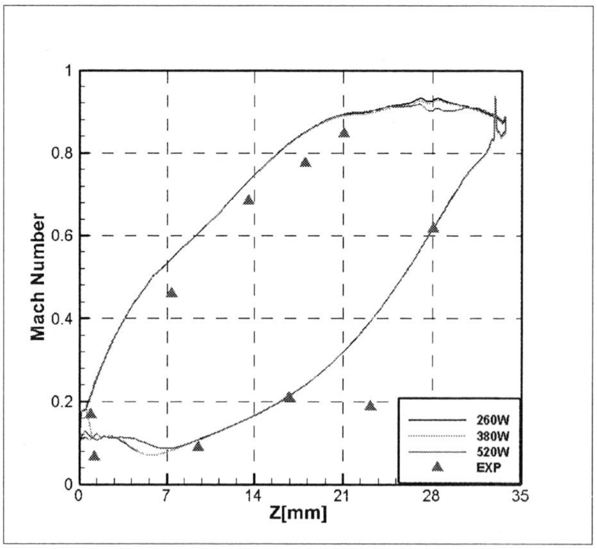

Figure3. Mach number distribution at 50% span under different mesh.

This paper used Ansys CFX for the computation. The ideal air gas is selected for the flow medium, the SST model is selected for the turbulence model, the boundary conditions specify the total inlet pressure, the total temperature, the inflow direction, the outlet specifies a given static pressure, the film hole specifies the inflow flow, the inflow temperature, and the vane adopts the adiabatic and no slip shift boundary. The boundary conditions for the computation are shown in Table 2.

Table2 .Setting of boundary conditions

Boundary conditions	**values**
Turbine inlet average temperature (K)	708.5[K]
Total inlet pressure (Pa)	343228 [Pa]
Imported turbulence intensity (%)	5%
Outlet average static pressure (Pa)	208865 [Pa]
Cooling massflow/inlet massflow (%)	1.7%
Cold air jet temperature (K)	378.56 [K]

2.3. Parameterization method for hot streak boundary condition

Before simulating the effects of hot streak, the actual distribution pattern of the combustion chamber outlet temperature is required to be considered to specify the appropriate turbine inlet hot streak conditions. Previous researchers have proposed a number of parameterization simulation strategies. In this paper, the two-dimensional Gaussian distribution function listed in Reference [22] is used to calculate the hot streak distribution of the turbine inlet, and the actual formula used to calculate the hot streak distribution of the turbine inlet is shown in Equation (1).

$$T_T = T_0 + \sum_{i=1}^{3} A\exp\left(-\frac{(m-m_i)^2}{2\sigma_m^2} - \frac{(n-n_i)^2}{2\sigma_n^2}\right) \tag{1}$$

The rectangular surface of the turbine inlet fan surface is simplified. The circumferential halved node is M=50, and the averaging node in the elevation direction is N=71. The coordinates represent the position of the current node and represent the core position of the first hot streak. The accumulation of the three Gaussian distributions is to ensure the periodicity of the temperature distribution in the circumferential direction. Therefore, the following relationship exists

$$\begin{cases} m_1+M=m_2=m_3-M \\ n_1=n_2=n_3 \end{cases} \tag{2}$$

In formula (1). T_0 is the base value before the Gaussian distribution is superimposed, A represents the fluctuation amplitude of the Gaussian distribution (A = 0.4), and controls the degree of contraction of the Gaussian distribution in the elevation and circumferential directions (σ_m=15, σ_n=18), n_2=40, respectively. In order to ensure that the average flow rate of the turbine inlet temperature does not change after the hot streak is introduced, the basic value T_0 of the Gaussian distribution is the difference between the total turbine inlet temperature of 1277.6R and the arithmetic mean of the right index portion of the formula (1).

2.4. Parameterization method for swirl boundary condition

In this paper, the non-uniform inlet airflow angle is set on the inlet boundary surface by setting the components in the axial, radial and tangential directions in the cylindrical coordinate system of the Ansys CFX software.

The configuration of the swirl flow refers to the Reference [23]. The position of the vortex core always overlaps with the core position of the hot streak. On this basis, the tangential velocity of the fluid on the combustor-turbine intersurface can be obtained by solving the Equation (3).

$$V^S(R) = d^S \frac{\Gamma}{2\pi R}\left[1-\exp\left(-\frac{4R^2}{(D^S_{core})^2}\right)\right] \quad (3)$$

The parameter in the equation is used to control the direction of the vortex. The parameter represents the size of the vortex core diameter, which was set to 16.8 mm in this study, approximately 36.3% of the length of the arc in the entrance surface. For the setting of the vortex intensity Γ , the maximum circumferential airflow angle should be referred to the actual combustion chamber exit conditions. Based on the test measurement data on the angular distribution of the turbine inlet airflow provided in the published references, the actual combustion chamber exit condition is about ±50°. Based on this, the vortex intensity Γ is set to 2.857 m^2/s , and the maximum circumferential airflow angle obtained is 41.8°.

The selected hot streak-swirl timing position is more representative of the aligning vane leading edge.The schematic diagram of hot streak and swirl was shown in Figure 4, which shows the total temperature distribution and the projection of the velocity vector on the inlet.

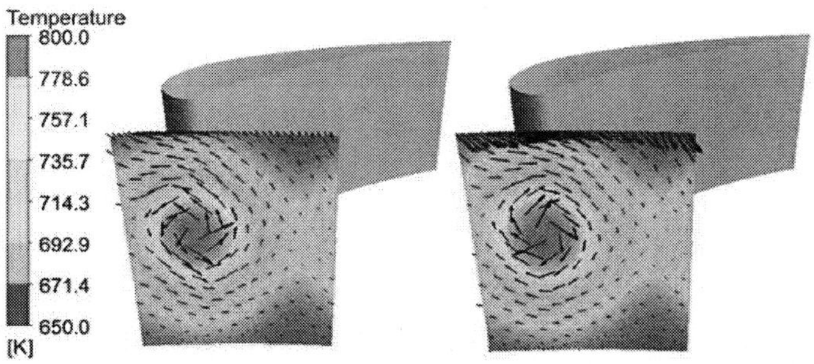

(a) Hot streak + Reverse swirl (b) Hot streak + Positive swirl

Figure4. Schematic diagram of hot streak and swirl.

In order to study the effects of uniform inlet conditions, hot streak in the inlet and swirl on the film cooling and vane, the final example settings are shown in Table 3.

Table3. Calculation example settings

	With or without hot streak	Swirl
Case1	No	None
Case2	Yes	None
Case3	Yes	Reverse swirl
Case4	Yes	Positive swirl

3. Results

3.1. Effects of hot streak

Figure 5 shows the vane temperature distribution for uniform inlet (Case1) and inlet with hot streak (Case2). It can be seen from the figure that the film cooling of the vane wall is different under the condition of Case1 and Case2, and the unevenness of the temperature distribution of the wall surface of Case2 is significantly greater than the unevenness of the wall temperature distribution of the blade under the uniform inlet condition of Case1. Case1 and Case2 have a low temperature zone at the leading edge of the high span, because the cooling holes have a radial span angle in jetting direction, causing the cooling airflow to concentrate at the high span. The wall temperature distribution of Case1 is generally uniform, and the maximum temperature is about 700K. In the case of Case2 with hot streaks, the temperature distribution in the leading edge region is unevenness. A high temperature region appears at the leading edge of 20%-50%, and this high temperature region always extends to the suction surface. The high temperature region appears here due to the location of the hot streak at

IOP Conf. Series: Materials Science and Engineering 1081 (2021) 012011 doi:10.1088/1757-899X/1081/1/012011

the entrance. The presence of hot streaks significantly boosts the unevenness of the thermal load of the vane.

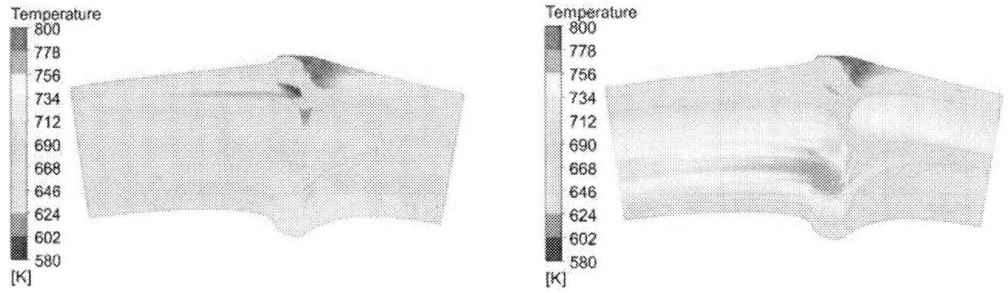

a)Case1(Only cooling) b)Case2(Cooling+Hot streak)

Figure5. Contour of vane temperature distribution

The pressure distribution on the vane wall of Case1 and Case2 was shown in Figure 6. It can be seen that there is no difference in the pressure distribution under the two inlet conditions, which is consistent with the conclusion of An et al. [10]. Figure 7 shows the pressure change at the position of chord length. There is no difference in the pressure in the channel under the two inlet conditions. That is, the hot streak does not affect the pressure change in the vane wall surface and the channel, so the flow situation of the main flow and cooling flow is the same. Through the above analysis, it can be concluded that the uneven thermal load of the vane caused by the presence of hot streak in the turbine inlet needs to be paid more attention.

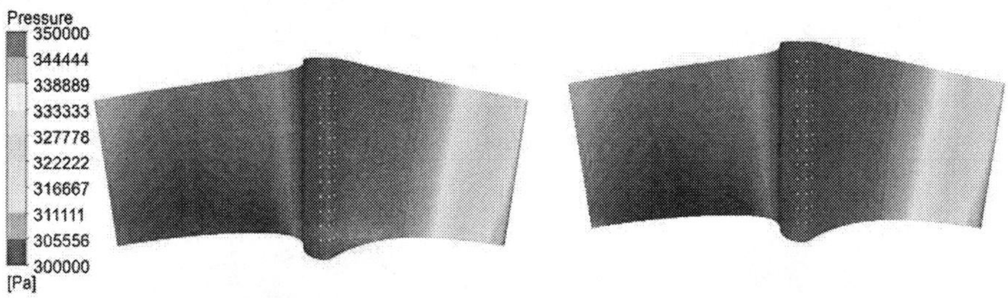

a)Case1(Only cooling) b)Case2(Cooling+Hot streak)

Figure 6. Contour of vane pressure distribution

a)Case1(Only cooling) b)Case2(Cooling+Hot streak)

Figure 7. Contour of passage pressure distribution

3.2. Effects of swirl

3.2.1. Effects of swirl on temperature

Under the influence of hot streak, the temperature distribution on the vane wall surface is uneven, and the influence of swirl at the turbine inlet will further affect the cooling effect of film cooling on the wall surface. Fig. 8 shows the temperature distribution on the vane under swirl flow. In In Figure 8-a. the contour on vane surface shows the temperature of the vane under the effect of positive swirl. There is a high temperature zone in the high span at the leading edge, and the temperature is up to 797K. Fig. 8-b shows the temperature distribution of the vane surface under the action of reverse swirl, which is more uniform than that of the leading edge of positive swirl. Under the action of positive and reverse swirl, high temperature zone appeared on the suction surface. Under the action of positive swirl, high temperature zone concentrated in the middle span zone, and under the reverse swirl, the distribution of cooling jet flow was opposite. In general, the temperature distribution on the wall surface of the vane under reverse swirling flow is more uniform than that under forward swirling flow.

a)Case3(Hot streak+Positive) b)Case4(Hot streak+Reverse)

Figure8. Contour of vane wall temperature distribution

3.2.2. Effects of swirl on pressure

The hot streak obtained from 2.1 has no effect on the static pressure of the vane wall. The change of the wall temperature is related to the flow structure of the main stream and the cooling jet. Figure 9 shows the pressure distribution of the vane wall under positive and reverse swirling, and the swirl has a significant effect on the vane surface pressure near the leading edge. Figure 9 shows the pressure distribution on the vane wall surface under the positive and reverse swirl. It can be seen that the swirl has a significant effect on the vane pressure near the leading edge. Under the action of positive swirl, the pressure stagnation line in the low-span area of the leading edge of the vane is shifted to the pressure side, and the pressure stagnation line in the high-span area is shifted to the suction side. Under the action of reverse swirl, The change of pressure stagnation line the leading edge is opposite to that under positive swirl. This change is also related to the flow mode of the cooling jet. Under the action of swirl flow, a tangential velocity is generated, which results in a change in the pressure gradient in this region and a decrease in static pressure, so the change of the pressure stagnation line will occur.

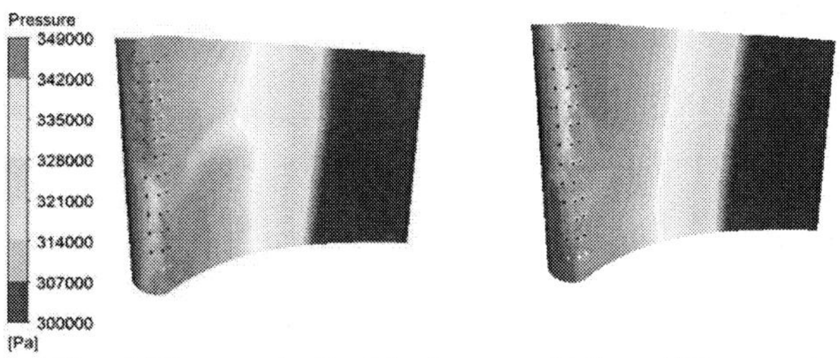

a)Case3(Hot streak+Positive) b)Case4(Hot streak+Reverse)

Figure9. Contour of vane wall pressure distribution

3.2.3. Effects of swirl on the shape of hot streak

Considering the influence of the swirl of the turbine inlet on the hot streak under the condition of film cooling in the leading edge, the position of the hot streak and the swirl is positively on the leading edge of the vane. The hot streak are blended at the leading edge with the cooling flow, and the range of hot streak spreads. The leading edge of the vane divides the hot streak into two parts, and the main flow with swirling is also divided into two parts by the vane. The two parts of the massflow still maintain the original direction of rotation, and the radial velocity component near the wall on both sides causes the flow to move along the span. Figure 10-a) shows the change of the hot streak in the chord length without swirling. The cooling jet is ejected from the film hole, which can form a gas film between the wall and the mainflow, thus preventing the vane temperature from being too high. The hot streak is divided into two parts by the vane, one part is on the pressure side, and the other part is on the suction side. After mainflow is mixed with the cooling gas sprayed from the film hole, the temperature of the hot streak is obviously lowered, and the highest temperature at the outlet is 706K. High temperature on both sides gathered together at the exit of the trailing edge, and the shape is approximately circular.

The change of hot streak along the chord length in Case3 and Case4 was shown in Figure 10-b) and Figure 10-c), respectively. On the 10% chord length section, there is a high temperature the pressure side of Case3 in 60%-80% span of the pressure side and the 40%-60% span of the suction side. Similarly,the high temperature main flow and the wall surface of the pressure surface in Case4 are separated by cooling film, and there is a high temperature zone near the wall surface of the suction surface 20%-50% span. At the 50% chord length section, the passage temperatures of Case3 and Case4 were lowered, and the temperature at the pressure side and the suction side near the wall did not change significantly. Under the action of the two direction swirls, the range of the hot streak in the radial direction becomes large. Under the action of positive swirling, the high temperature zone of the low span of the exit is "elongated" in the circumferential direction, and the high temperature zone of the mid-high span is "compressed". Under the action of reverse swirling, the high temperature zone of the mid-high span the exit is "elongated", and the high temperature zone of the mid-low span is "compressed".

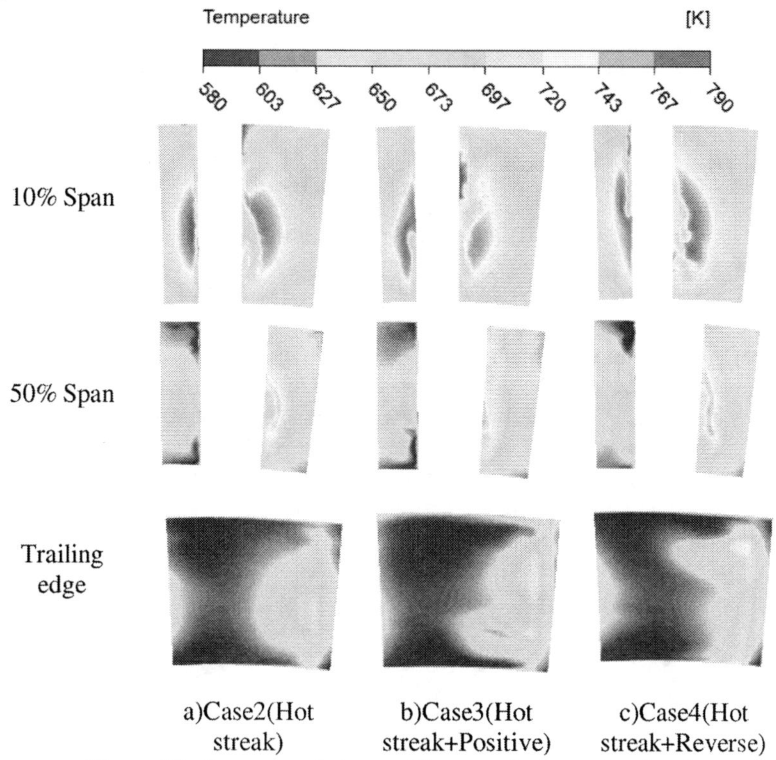

Figure10. Contours of shape of hot streak along the chord

3.2.4. Effects of swirl on heat tansfer

In order to study the heat transfer characteristics of the vane wall, the convective heat transfer coefficient h is defined as follows

$$h = \frac{q}{T_{aw} - T_w} \tag{4}$$

T_w is the temperature of the constant temperature wall;

q indicates the heat flux of the vane wall at a constant temperature 566.8K;

T_{aw} is the adiabatic wall temperature.

Figure 11 shows the convective heat transfer coefficient distribution of the vane wall of Case2, Case3 and Case4. It is a strong heat exchange zone at the leading edge of the film hole, where the cooling flow is strongly blended with the main flow, so that the convective heat transfer coefficient of the zone is large, and the heat transfer coefficient is gradually reduced to constant away from the area of the hole row. The convective heat transfer coefficient of Case2 and Case3 is uniform compared of Case4 in leading edge. In the three Cases, the heat transfer ratio of the pressure surface is lower than that of the suction surface. Although the cooling flow and the mainstream mixing are severe, the wall surface curvature of the pressure surface is large and the pressure difference in the channel is affected. There is a tendency to stay away from the pressure surface, so the coefficient of convective heat transfer coefficient is low at 0% - 50% of the chord length on the pressure side. Under the action of the positive swirl, the range of the high heat transfer zone at the proximal wall of Case3 is larger than that of the suction side of the Case2, and under the reverse swirl, the high heat transfer zone at the lower end wall disappears. In Case 4, the convective heat transfer coefficient in the 30%-40% span area of the leading edge becomes larger, and the convective heat transfer coefficient near the cooling film hole in the mid-high span is also larger than that in Case3 and Case4.

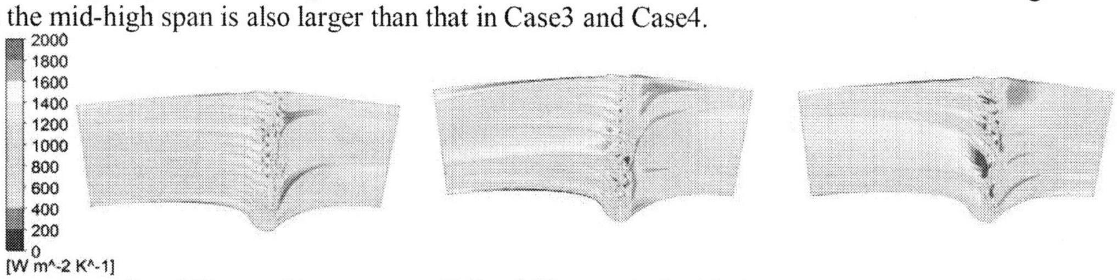

a)Case2(Hot streak) b)Case3(Hot streak+Positive) c)Case4(Hot streak+Reverse)

Figure 11. Contours of heat transfer coefficient

3.2.5. Effects of swirl on exit of vane

After introducing the hot streak and the swirl condition, the temperature and airflow angle of the vane exit will change, that is, the total temperature and airflow angle of the vane inlet are directly related to the turbine inlet temperature and the airflow direction. The total temperature distribution of the vane outlet is given in Figure 12. In Case2 without swirl, there is still hot streak at the exit of the vane, and the total temperature of the hot streak is 785K. In the presence of the swirling conditions of Case3 and Case4, the shape of the hot streak at the exit of the vane is destroyed. Under the influence of the radial velocity, the range of the hot streak in the span becomes larger, so that the distribution of total outlet temperature in the direction of developing height is more uniform.

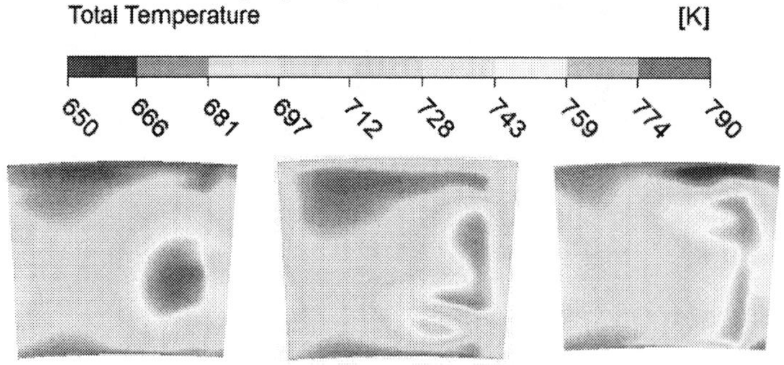

a)Case2(Hot streak) b)Case3(Hot streak+Positive) c)Case4(Hot streak+Reverse)

Figure 12. Contour of total temperature distribution of the vane exit

During the migration process, the high temperature fluid is affected by the vane splitting, the leading edge film cooling jet and the swirling flow. The jet angle of the vane exit changes, under the positive swirling flow. The jet angle varies widely, the amplitude is [162°, 173°], and the jet angle changes under reverse swirl is [160°, 167°].

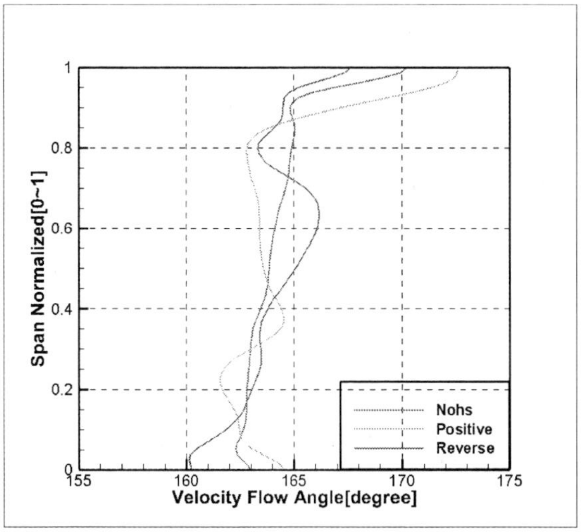

Figure 13. Flow angle of the vane exit

4. Conclusions

This paper develops a program that simulates hot streaks and swirls, which can easily adjust the hot streak temperature and swirl and its relative guide vane position. Taking the GE-E3 turbine guide vane as the research object, the hot streak and the swirl flow are set to the leading edge of the vane, and the influence of the hot streak and the swirl on the leading edge with the film cooling guide vane is analyzed by CFX. The following conclusions:

1) By comparing and analyzing the influence of hot streak on the film cooling in the leading edge, it is found that the hot streak mainly causes the temperature distribution of the leading edge to be uneven, which makes the heat load unevenness of the vane increase. The presence of hot streaks has little effect on the mainstream flow and the cooling flow, so more attention should be paid to thermodynamics.

2) The presence of swirl increases the unevenness of the thermal load on the surface of the vane. In general, the effect of the reverse swirl on its non-uniformity is smaller than that under the positive swirl, and under the action of swirl, the cooling flow is not evenly distributed on the pressure side and the suction side, resulting in the deviation of the pressure stagnation line at the leading edge of the guide blade.

3) The swirling process changes the migration process of the hot streak. Since the mainstream flow has a velocity in the radial direction, the effect of the hot streak on the wall surface is changed along the direction of the heightening, and the range of the hot streak in the channel becomes larger. At the trailing edge, the range of the high temperature zone becomes larger, and the change of the circumferential direction is affected by the swirling direction.

4) At the exit, the temperature range of the hot streak in the span becomes larger, increasing the unevenness of the vane outlet. The flow angle of the high, medium and low span also changes under the positive and reverse swirl.

References

[1] Bunker, Ronald S . Evolution of Turbine Cooling[J]. ASME Turbo Expo 2017: Turbomachinery Technical Conference and Exposition, Charlotte, North Carolina, USA, Monday 26 June 2017.

[2] Yepuri G B, Puttarangasetty A B T , Rajendran R, et al. Experimental Investigation of Adiabatic Film Cooling Effectiveness and Heat Transfer Coefficients over a Gas Turbine Blade Leading Edge Configuration[J]. 2014.

[3] Liu Z F, Wang Z D, Feng Z P. Study on the effect of inlet swirl on hot streak migration and rotor heat load in gas turbine[J]. Journal of Engineering Thermophysics, 2016.

[4] Maikell J, Bogard D, Piggush J, et al. Experimental Simulation of a Film Cooled Turbine Blade Leading Edge Including Thermal Barrier Coating Effects[J]. Journal of Turbomachinery, 2011, 133(1):011014.
[5] Xu Li Yan-jing Qu Liu-li Song. A Numerical Investigation on Cooling Effectiveness forAdvanced Fan-shaped Film Cooling[J]. Chinese Journal of Turbomachinery. 2019,03:72-77+5
[6] Chen Ziyu;Su Xinrong;Yuan Xin. Interaction Mechanism Between Coolant and Mainflow in Turbine Endwall Region[J]. Compressor Blower & Fan Technology. 2020,03:46-51
[7] Qureshi I, Smith A D, Chana K S , et al. Effect of Temperature Nonuniformity on Heat Transfer in an Unshrouded Transonic HP Turbine: An Experimental and Computational Investigation[J]. Journal of Turbomachinery, 2012, 134(1):011005.
[8] Barringer M D, Thole K A, Polanka M D, et al. Migration of Combustor Exit Profiles Through High Pressure Turbine Vanes[J]. Journal of Turbomachinery, 2009, 131(2).
[9] Paolo G, Giacomo P. Hot Streak Evolution in an Axial HP Turbine Stage[J]. International Journal of Turbomachinery, Propulsion and Power, 2017, 2(2):6-.
[10] Jenkins S, Varadarajan K, Bogard D G. The Effects of High Mainstream Turbulence and Turbine Vane Film Cooling on the Dispersion of a Simulated Hot Streak[J]. Journal of Turbomachinery, 2004, 126(1):203.
[11] An B T, Liu J J, Jiang H D. Numerical Investigation on Unsteady Effects of Hot Streak on Flow and Heat Transfer in a Turbine Stage[J]. Journal of Turbomachinery, 2009, 131(3):031015.
[12] Dong P, Wang Q, Guo Z, et al. Conjugate Calculation of Gas Turbine Vanes Cooled with Leading Edge Films[J]. Chinese Journal of Aeronautics, 2009, 22(2):145-152.
[13] Tsukagoshi K, Arimura H, Tanaka K, et al. Development of Air Cooled Combustor for Mitsubishi G Class Gas Turbine[C]. ASME Turbo Expo 2010: Power for Land, Sea, and Air. 2010.
[14] Turrell M D, Stopford P J. CFD Simulation of the Flow Within and Downstream of a High Swirl Lean Premixed Gas Turbine Combustor[J]. Proceedings of ASME Turbo Expo 2004, Vienna , 2004.
[15] Tommaso B, Riccardo B, Alessio P, et al. Adiabatic Effectiveness on High-Pressure Turbine Nozzle Guide Vanes under Realistic Swirling Conditions[J]. Journal of Turbomachinery, 2018, 141(1).
[16] Liu Z, Ye L, Wang C, et al. Numerical simulation on impingement and film composite cooling of blade leading edge model for gas turbine[J]. Applied Thermal Engineering, 2014, 73(2):1432-1443.
[17] Pyliouras S, Schiffer H P, Janke E, et al. Effects of Non-Uniform Combustor Exit Flow on Turbine Aerodynamics[C].Asme Turbo Expo: Turbine Technical Conference & Exposition. 2012:11.
[18] Qureshi I, Smith A D, Povey T. HP Vane Aerodynamics and Heat Transfer in the Presence of Aggressive Inlet Swirl[J]. Journal of Turbomachinery, 2013, 135(2):21040-21041.
[19] Qureshi I, Beretta A, Chana K, et al. Effect of Aggressive Inlet Swirl on Heat Transfer and Aerodynamics in an Unshrouded Transonic HP Turbine[J]. Journal of Turbomachinery, 2012, 134(6):061023.
[20] Timko L P. Energy Efficient Engine high pressure turbine component test performance report[R].CLEVELAND,OHIO:NASA,1984.
[21] Cha C M, Hong, S,et al. Experimental and Numerical Investigation of Combustor-Turbine Interaction Using an Isothermal , Nonreacting Tracer [J]. Eng Gas Turb Power, 2012,134(8): 081501.
[22] Prenter R, Ameri A, Bons J P. Computational Simulation of Deposition in a Cooled High-Pressure Turbine Stage With Hot Streaks[C] Asme Turbo Expo: Turbomachinery Technical Conference & Exposition. 2016.
[23] Schneider M, Schiffer H P, Lehmann K. Parameterised Model of 2D Combustor Exit Flow Conditions for High-Pressure Turbine Simulations[J].International Journal of Turbomachinery, Propulsion and Power, 2017, 2(4).

IOP Conf. Series: Materials Science and Engineering 1081 (2021) 012012 doi:10.1088/1757-899X/1081/1/012012

Study on Stiffness of Angular Contact Ball Bearing and Its Evolution Rule under Excessive Preload

Xianghe Yun[1], **Ning Li**[1], **Baogang Wen**[2] and **Qingkai Han**[1*]

[1] School of Mechanical Engineering, Dalian University of Technology, Dalian, PR China

[2] School of Mechanical Engineering and Automation, Dalian Polytechnic University, Dalian, PR China

E-mail:qk.han@hotmail.com

Abstract. There are many factors influencing the bearing stiffness variation, among which, preload has more significant effect. In addition, bearing stiffness is developing accompanying with its performance deterioration, which leads to vibration variation of bearing-rotor system. In this paper, system support stiffness model and system dynamics model of micro-turbine bearing-rotor system test rig are built to obtain the bearing stiffness identification method by model simplification based on modal theory. The identification method is validated adopting modal test. The stiffness evolution rule and rotor vibration of different stages are achieved via employing the identification method to extract bearing stiffness over the process of bearing run-to-failure. The results indicate that bearing stiffness is the main source of system support stiffness. Bearing stiffness will change during the whole life, which affects the vibration responses in time and frequency domain. Rotor stability is influenced by cage movement.

1. Introduction

Rolling bearing is main load-carrying component of rotary machinery, whose stiffness variation gives rise to the oscillation of bearing vibration[1], at the same time, it has the impression on safety and stable revolution of rotor. During the service course of bearing, its stiffness will alter accompanying with the function decay of bearing, which has an impact on the bearing-rotor system vibration. Some researches [2-7] demonstrated that the factors bringing about the variation of bearing stiffness are axial preload, inner radial clearance, speed and so on. Thereinto, axial preload not only has notable effect on bearing stiffness, but also on bearing life. The ball will skid and roll intermittently if the preload is too small, which will induces the instability of cage-ball. Then, the vibration amplitude of bearing is one or two orders of magnitude higher than that as bearing is in the normal operation[8]. However, the higher preload, the more heat production[9], it will cause the expansion of inner friction moment and shorten the bearing service life[10]. Therefore, appropriate preload not only extends the service life[11], but also strengthens the stiffness[2, 11, 12], depresses the vibration[2, 4], in addition, guarantees the stability of rotor. Consequently, the research on the influence of preload on bearing stiffness, the bearing stiffness evolution law of the whole life and rotor vibration characteristics should be carried out in detail, thus, it will explain bearing vibration mechanism fundamentally and provide guidance for bearing condition-based maintenance and stiffness design of rotor system.

In recent years, people pay their attention to study the effect of preload acting on bearing stiffness and vibration, bearing life, rotor stability and so on. Alfares et al. [4] advocated that bearing stiffness is increasing and vibration is decreasing with the strengthening of preload. However, bearing vibration keeps steady as preload exceeds to a certain value. Cao et al. [6] presented that bearing stiffness increases gradually along the axial and horizontal radial direction, while the stiffness along the vertical radial direction decreases firstly and then increases with the increment of preload. Gunduz et al. [13] proposed that the natural frequency of rotor system grows up along with the increment of preload. In addition, many researches [14, 15] introduced that the rigid preload mechanism is more efficient than constant preload mechanism in maintaining the dynamic stiffness of spindles.

People usually introduced the L_P fatigue life theory, which is firstly proposed by Lundberg and Palmgren[16], to gain a clear idea of the relationship between preload and bearing service life. Hagiu et al. [17, 18] determined the optimal preload by analyzing the relationship between bearing service life of shaft ball bearing and preload. Kaczor et al. [19] observed the influence of preload acting on the angular contact bearing service life. Zhang et al. [20] advocated a quasi-dynamic model to invest the load distribution of ball bearing standing arbitrary preload based on the contact state of ball-raceway, in the same time, it was employed to study the impact of external loads, rotating speed and preload on ball-raceway contact state and calculate the fatigue life of ball bearing in a running state. Some researchers took advantage of advanced algorithm to diagnose bearing initial fault[21] and prognose the remaining useful life[22].

In the aspect of rotor stability, inner radial clearance and axial preload have a notable effect on the rotor stability. Tiwari and Gupta et al. [3] firstly proposed the variation of inner radial clearance induces a third region of instability, moreover, it resulted in the appearance of regions of periodic, subharmonic and chaotic behavior in a large extent. Harsha et al. [23, 24] advocated that the rotational speed has the same effect with inner radial clearance. Bai et al. proved that the rotor stability is strongly dependent on the inner radial clearance, axial preload and radial force [25], for instance, inner radial clearance and rotational speed brought about three routes to instability, which are the period-doubling bifurcation, the secondary Hopf bifurcation and the boundary crisis [26], axial preload has a profound effect on bifurcation margins [27], the non-linear characteristics, hertz contact forces and internal clearance of ball bearings led to the occurrence of subharmonic resonance at the same time [28]. Zhang et al. [29] maintained that bearing contact angular changes notably in company with the axial preload, the contact angular significantly contributes to the ball passage frequency and its coupling frequency with rotational frequency. With considering the influence of contact angular, the main resonance region of bearing-rotor system switches to the lower speed ranges. Furthermore, Ashtekar et al. [30] analyzed that the rotor orbits affected by different preload. Allaire et al. [31] studied the influence of bearing parameter acting on synchronous vibration of turbine-gearbox-generator system. Xiao et al. [32] obtained the variety of bearing parameters and rotor stability affected by different preload. Long et al. [33] analyzed the critical speed of rotor influenced by static and dynamic stiffness, respectively.

From the above, people can draw a conclusion that the existing studies are mainly focused on discussing the influence of axial preload on bearing stiffness, life and rotor stability by analytic method. Nevertheless, people do not have an insight into bearing stiffness, its stiffness evolution law of the whole life and development of rotor axial orbit through these conclusions. For that reason, bearing stiffness identification method is proposed by means of simplification of support stiffness model and bearing-rotor system dynamics model of micro-turbine. Next, the bearing stiffness evolvement rule is achieved by introducing vibration data of bearing service period to the forementioned method. The rest of this paper are arranged as follows: In section 2, the theoretical background of bearing stiffness identification method, test equipment and location of sensors are illustrated. Section 3 presents bearing stiffness, its evolution law of the whole life and rotor vibration characteristics. The conclusions are made in Section 4.

2. Methodology of stiffness identification and vibration test

IOP Conf. Series: Materials Science and Engineering 1081 (2021) 012012 doi:10.1088/1757-899X/1081/1/012012

2.1. Construction of test rig

Micro-turbine test rig is shown in Figure 1(a), whose bearing-rotor system fulcrum construction and preload mechanism are illustrated in Figure 1(b). As we can see, the fulcrum consists of angular contact ball bearing, house and pedestal. In which, angular contact ball bearings are employed to support the rotor, which are treated as the main part of load-carrying and force-transferring, they are connected to tube-shaped house adopting end covers. The middle of the house is bolted to the pedestal, which forms an interference fit. The preload mechanism is rigid preload [14, 34], which can satisfy different tightness by the means of modifying the axial distance between end cover and house, it also alters the relative location of inner and outer rings. The relative distance between house and end cover is taken as e when the house is tight fit to the other one.

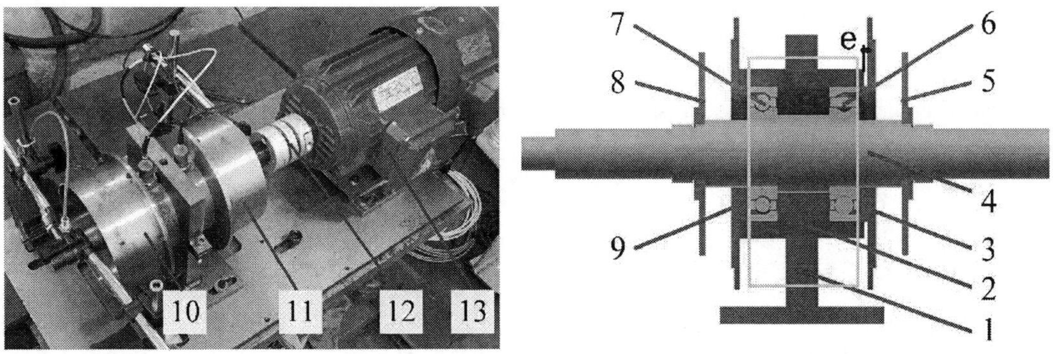

(a) (b)

Figure 1. Test Equipment: (b)(1)Pedestal (2)House (3)End Cover 1 (4)Shaft (5)Disc 1 (6)Tested Bearing (7)Contrast Bearing (8)Disc 2 (9)End Cover 2 (10)Drum 1 (11)Drum 2 (12)Elastic Couple (13)AC Motor.

2.2. Model of bearing-rotor system support stiffness

According to Figure 1(b), the form of micro-turbine bearing-rotor system fulcrum parts is in series, therefore, the stiffness of fulcrum and each part can be equivalently calculated, respectively.

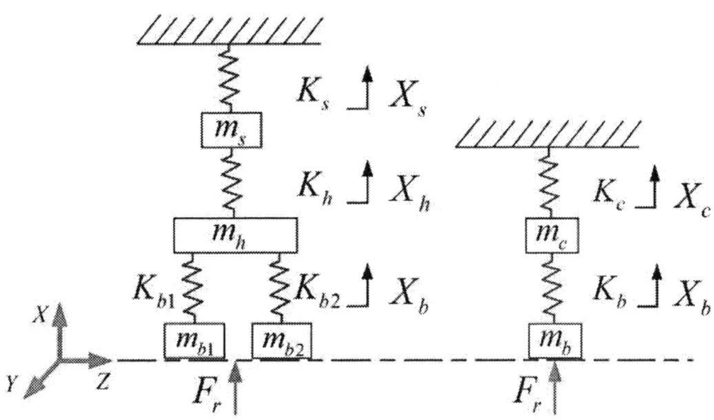

(a) (b)

Figure 2. Fulcrum Radial Stiffness Model of Test Rig.

Ignoring damping, the radial stiffness model of fulcrum structure is shown in Figure 2(a). K_{bi} ($i = 1,2$), K_h and K_s is the radial stiffness of bearing, house and pedestal, respectively.

m_{bi} ($i=1,2$), m_h and m_s is the mass of bearing, house and pedestal, respectively. F_r is the fulcrum radial load. The model illustrated in Figure 2(a) is further simplified as shown in Figure 2(b) by mixing the two bearings together and combining house with the pedestal, whose stiffness and mass are K_b, m_b and K_c, m_c, respectively. Therefore, the motion differential equations of mass-spring system model demonstrated in Figure 2(b) is presented as follows:

$$m_b\ddot{X}_b + K_b(X_b - X_c) = 0 \tag{1}$$

$$m_c\ddot{X}_c + K_b(X_c - X_b) + K_cX_c = 0 \tag{2}$$

Owing to the stiffness of house and pedestal is so strong that their deformation is equal to zero, that is $X_c=0$. Hence, the motion differential equations of mass-spring system model shown in Figure 2(b) becomes as follows:

$$m_b\ddot{X}_b + K_bX_b = 0 \tag{3}$$

Assuming that the fulcrum radial stiffness is K_t, the integral motion differential equations of mass-spring system model is shown as follows:

$$m_b\ddot{X}_b + K_tX_b = 0 \tag{4}$$

Comparing the Equation (3) and (4), we can conclude that the fulcrum radial stiffness is equal to the bearing radial stiffness, that is:

$$K_t = K_b = K_{b1} + K_{b2} \tag{5}$$

According to Equation (5), the bearing stiffness is the main contributor of micro-turbine bearing-rotor system support stiffness.

2.3. *Identification method for bearing stiffness*

According to Sect. 2.2, support stiffness is mainly affected by bearing stiffness in micro-turbine bearing-rotor system, hence, the bearing stiffness can be extracted from bearing-rotor system. In order to identify the bearing stiffness, the dynamics equation of micro-turbine bearing-rotor system is built. Steps as follows:

The rotor system mechanics schematic diagram of micro-turbine test rig is illustrated in Figure 3, the hypotheses is adopted as follows during modeling process:

1) The discs are rigid;

2) The shaft is rigid without mass, whose mass is lumped to discs;

3) Ignoring the torsional vibration and axial vibration, there is bending vibration in bearing-rotor system;

4) The bearing is elastic, which is modeled as the stiffness model with 4 degrees of freedom, but the house and pedestal are rigid;

5) The speed is constant.

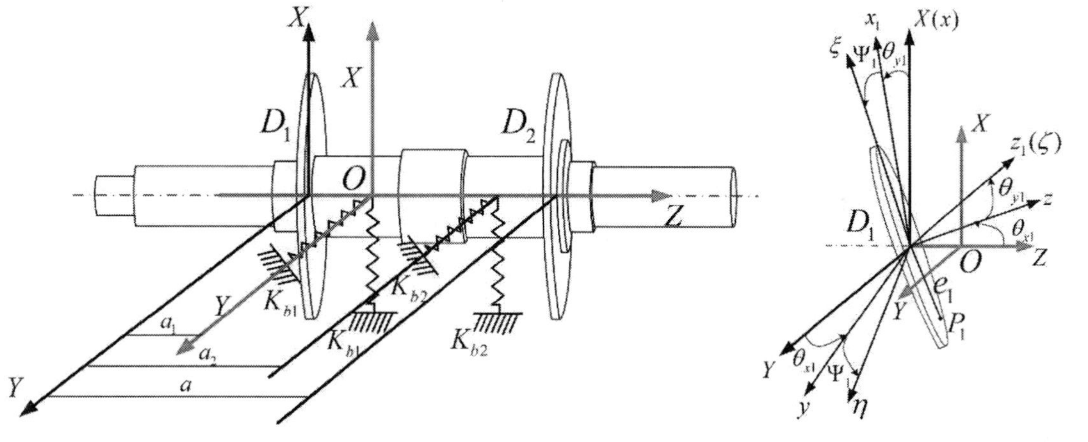

(a) (b)

Figure 3. Bearing-rotor System Mechanics Schematic Diagram of Micro-turbine Test Rig: (b) (In which P_1 and D_1 denote the mass center and geometric center of the disc 1, respectively).

The bearing-rotor system motion differential equation with 4 degrees of freedom established by employing the Lagrange energy method is shown as follows:

$$\begin{aligned}
&(m_1+m_2)\ddot{x}_1+(K_{b1}+K_{b2})x_1+m_2a\ddot{\theta}_{y_1}+(K_{b1}a_1+K_{b2}a_2)\theta_{y_1}\\
&=(m_1+m_2)e_1\Omega_1^2\cos(\Omega_1t+\varphi_1)\\
&(m_1+m_2)\ddot{y}_1+(K_{b1}+K_{b2})y_1-m_2a\ddot{\theta}_{x_1}-(K_{b1}a_1+K_{b2}a_2)\theta_{x_1}\\
&=(m_1+m_2)e_1\Omega_1^2\sin(\Omega_1t+\varphi_1)\\
&m_2a\ddot{x}_1+(K_{b1}a_1+K_{b2}a_2)x_1-(m_2a^2+J_{d1}+J_{d2})\ddot{\theta}_{y_1}+(J_{p1}+J_{p2})\Omega_1\dot{\theta}_{x_1}\\
&+(K_{b1}a_1^2+K_{b2}a_2^2+K_{1\theta_y}+K_{2\theta_y})\theta_{y_1}=m_2e_1\Omega_1^2a\cos(\Omega_1t+\varphi_1)\\
&-m_2a\ddot{y}_1-(K_{b1}a_1+K_{b2}a_2)y_1+(m_2a^2+J_{d1}+J_{d2})\ddot{\theta}_{x_1}-(J_{p1}+J_{p2})\Omega_1\dot{\theta}_{y_1}\\
&+(K_{b1}a_1^2+K_{b2}a_2^2+K_{1\theta_x}+K_{2\theta_x})\theta_{x_1}=m_2e_1\Omega_1^2a\sin(\Omega_1t+\varphi_1)
\end{aligned} \tag{6}$$

where m_i, J_{jpi}, J_{jdi} ($i=1,2$) is the mass, polar inertia moment and radial inertia moment of rotor system lumped to the disk D_i ($i=1,2$), respectively. x_i, y_i ($i=1,2$) is the displacement of disc D_i ($i=1,2$) along X and Y direction, respectively. θ_{x_i}, θ_{y_i} ($i=1,2$) is the angular displacement of disc D_i ($i=1,2$) along X and Y direction, respectively. Moreover, $(m_1+m_2)e_1\Omega_1^2$ is time-varying amplitude of force induced by rotor unbalance, $m_2e_1\Omega_1^2a$ is time-varying moment amplitude of D_2 acting on the disc D_1, which is originated from rotor unbalance.

Ignoring θ_{x_i}, θ_{y_i} ($i=1,2$), the Equation (6) becomes as follows:

$$\begin{aligned}
&(m_1+m_2)\ddot{x}_1+(K_{b1}+K_{b2})x_1=(m_1+m_2)e_1\Omega_1^2\cos(\Omega_1 t+\varphi_1)\\
&(m_1+m_2)\ddot{y}_1+(K_{b1}+K_{b2})y_1=(m_1+m_2)e_1\Omega_1^2\sin(\Omega_1 t+\varphi_1)\\
&m_2a\ddot{x}_1+(K_{b1}a_1+K_{b2}a_2)x_1=m_2e_1\Omega_1^2 a\cos(\Omega_1 t+\varphi_1)\\
&-m_2a\ddot{y}_1-(K_{b1}a_1+K_{b2}a_2)y_1=m_2e_1\Omega_1^2 a\sin(\Omega_1 t+\varphi_1)
\end{aligned} \tag{7}$$

According to modal analysis theory, the Equation (7) is further simplified as follows:

$$\begin{aligned}
m\ddot{x}+K_t x&=me_1\Omega^2\cos\Omega t\\
m\ddot{y}+K_t y&=me_1\Omega^2\sin\Omega t
\end{aligned} \tag{8}$$

where $m=m_1+m_2$ is the mass of system, $K_t=K_{b1}+K_{b2}$ is the support stiffness, e_1 is the geometric eccentricity and Ω is the rotational speed.

The equation becomes as follow:

$$\begin{aligned}
\ddot{x}+\omega_n^2 x&=e_1\Omega^2\cos\Omega t\\
\ddot{y}+\omega_n^2 y&=e_1\Omega^2\sin\Omega t
\end{aligned} \tag{9}$$

where $\omega_n=\left(\frac{K_t}{m}\right)^{1/2}$, is the system natural frequency.

Marking z as $z=x+iy$, the above equation can be written as follows:

$$\ddot{z}+\omega_n^2 z=e_1\Omega^2 \mathrm{e}^{i\Omega t} \tag{10}$$

where the particular solution is $z=A\mathrm{e}^{i\Omega t}$, then, the solution of equation (10) is obtained as follows:

$$z=\frac{e_1(\Omega/\omega_n)^2}{1-(\Omega/\omega_n)^2}\mathrm{e}^{i\Omega t} \tag{11}$$

Therefore, the unbalance response of rotor system is shown as follows:

$$\begin{aligned}
x&=\frac{e_1(\Omega/\omega_n)^2}{1-(\Omega/\omega_n)^2}\cos(\Omega t)\\
y&=\frac{e_1(\Omega/\omega_n)^2}{1-(\Omega/\omega_n)^2}\sin(\Omega t)
\end{aligned} \tag{12}$$

The corresponding vibration amplitude is written as follows:

$$|A|=\left|\frac{e_1\Omega^2}{\omega_n^2-\Omega^2}\right|=\left|\frac{e_1(\Omega/\omega_n)^2}{1-(\Omega/\omega_n)^2}\right| \tag{13}$$

According to Equation (13), the amplitude of vibration A is greater than zero, the expression of support stiffness along x or y direction is calculated as follows:

$$K_t=\frac{m\Omega^2*(e_1+A)}{A} \tag{14}$$

As shown in Equation (14), the support stiffness of micro-turbine bearing-rotor system is dependent on the unbalance, geometry eccentricity and vibration amplitude. Comparing Equation (5) and Equation (14), the relationship of each bearing stiffness and displacement is presented as follows:

$$K_{b1}=K_{b2}=\frac{m\Omega^2*(e_1+A)}{2A} \tag{15}$$

where geometry eccentricity e_1 is determined by modal test.

IOP Conf. Series: Materials Science and Engineering 1081 (2021) 012012 doi:10.1088/1757-899X/1081/1/012012

In a word, the stiffness revolution rule of bearing under excessive preload is achieved by adopting Equation (15). Firstly, the eccentricity e_1 of bearing-rotor system is obtained through modal test. Then, the micro-turbine bearing-rotor system test rig is employed to perform bearing run-to-failure test, at the same time, the shaft vibration of the whole life is recorded. Finally, the bearing stiffness identification and evolution rule are implemented based on Equation (15). Specific framework is explained in Figure 4.

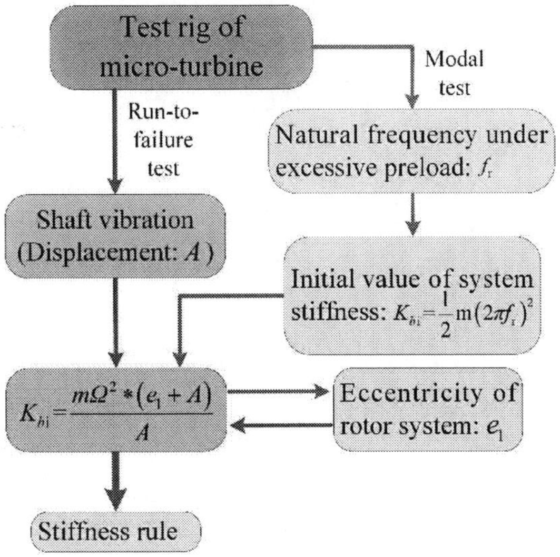

Figure 4. Framework of Stiffness Identification.

2.4. *Method for vibration test*

Micro-turbine bearing-rotor system test rig is illustrated in Figure 1(a), the collected vibration data is engaged to bearing stiffness identification over the service process. The rotor is supported by two angular contact ball bearings, which is coupled to AC motor for different revolutions. The preload mechanism is rigid preload as shown in Figure 1(b), the preload grade is realized with the help of modifying the spin depth of bolt to acquire different axial relative displacement e . The bearings are located in E, F of test rig, thereinto, the bearing in position E is tested bearing, the one in position F is contrast bearing. The run-to-failure test is last until the occurrence of fault or significant increment of vibration, which means bearing becomes invalid. In the meantime, test is completed. The total failure time (the whole life period) is calculated from bearing is running after installed.

The vibration of bearing and shaft is collected by acceleration sensors CAYD115V-100A mounted on house and eddy current sensors RP6606XL fixed near the shaft, respectively. As shown in Figure 5, sensor 1 is utilized to record vertical vibration of shaft, whose horizontal vibration is collected by sensor 2. Sensor 3 and 4 is employed to obtain vertical vibration of tested bearing and contrast bearing, respectively. Moreover, data acquisition equipment is composed of hardware NI Cdaq-9174 and software programmed based on NI LabVIEW, which can cover the command of real-time acquisition and demonstration. Speed is 2400rpm, sample frequency is 25.6Hz, sample duration is 20s, the data is recorded every 1 min. Tested bearing is angular contact ball bearing 7208AC.

IOP Conf. Series: Materials Science and Engineering 1081 (2021) 012012 doi:10.1088/1757-899X/1081/1/012012

Figure 5. Location of Sensors.

3. Methodology of stiffness identification and vibration test

3.1. Comparison of braces stiffness under different preload

Micro-turbine bearing-rotor system test rig consists of angular contact ball bearing, house and pedestal, whose stiffness is arranged in tandem. In order to prove that the main contributor of support stiffness is bearing stiffness, the stiffness of braces under different preload is achieved adopting LMS to perform modal test of braces. Preload grade is modified by swiveling bolt to changing axial relative displacement. As depicted in Figure 3, three eddy current sensors are arranged evenly along circumference of end cover near tested bearing. The mean of displacements of three points is treated as e , which is shown in Table 1, preload grades are divided into loose, tight, normal and overtight according to e .

Figure 6. The Relative Distance Test.

Table 1. Grade of Preload

Grade	e /mm	Tightness	e /mm
Loose	0.818	Normal	0.764
Tight	0.772	Overtight	0.682

According to literature[3], the stiffness of braces under different preload is achieved by modal test. From the results are illustrated in Table 2, there differs 1-2 order of magnitude between stiffness of bearing and that of house or pedestal. Consequently, bearing stiffness contributes mostly to support stiffness in micro-turbine bearing-rotor system test rig, namely, X_c=0 .

Table 2. Comparison of Braces under Different Preload

Braces \ Grade	Loose	Tight	Normal	Overtight
Pedestal	1.25e8N/m	1.25e8N/m	1.25e8N/m	1.25e8N/m
House	3.74e7N/m	3.74e7N/m	3.74e7N/m	3.74e7N/m
Bearing	2.02e6N/m	2.58e6N/m	4.96e6N/m	9.08e6N/m

The curves of bearing stiffness under different preload are described in Figure 7. The bearing stiffness along vertical radial direction is equal to that of horizontal radial direction. They are increasing sharply with the increment of preload. However, the stiffness along horizontal radial direction raises faster than that of vertical radial direction, the maximum difference is 15.65%.

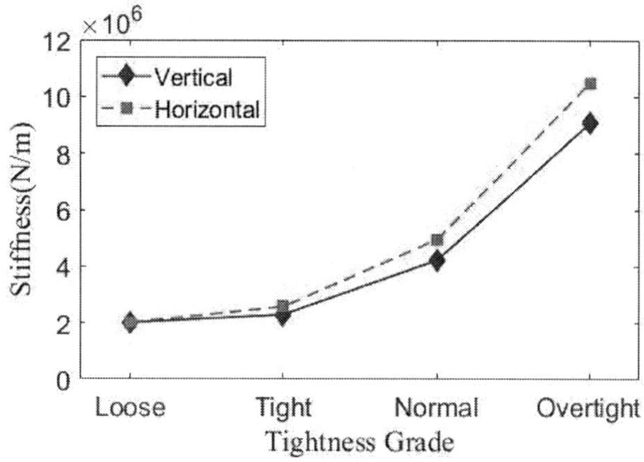

Figure 7. Bearing Stiffness of Different Preload.

3.2. Bearing stiffness evolution rule under excessive preload

The run-to-failure test is invalid under loose preload owing to roller has tendency to skid earlier, which results in instability between roller and cage. Nevertheless, it is time-consuming and costly to perform run-to-failure test due to the actual service life is too long under normal preload. Hence, the excessive preload is adopted in this test, which is corresponding to the grade of overtight shown in Table 1.

The test lasts 32 min, bearing loses efficacy due to seizure. As stiffness is the natural property of bearing-rotor system, in other words, the support stiffness is constant when bearing-rotor system is changeless. Therefore, the eccentricity e_1 induced by machining and installation is abstracted employing initial stiffness collected in Sect.3.1 during the process of bearing stiffness identification over the whole life. Next, In order to acquire bearing stiffness evolvement curve along vertical and horizontal direction during the whole life, the displacement every 2 min is introduced into the identification method of bearing stiffness explained in Sect.2.3.The result is illustrated in Figure 8. The development tendency of stiffness between vertical radial direction and horizontal radial direction is similar, which is approximately divided into three stages: first fluctuation (0-6min), stabilization (6-16min) and second fluctuation (16-32min). In stage one, the curve increases firstly and then decreases. The stiffness in stage two is almost stable. In stage three, the stiffness curve oscillates frequently and notably, after one sharp increment (16-18min), the curve keeps steady for a moment, then comes into volatility period (22-32min), finally, the stiffness drops down owing to bearing seizure. Comparing stiffness evolution between 0-4min and 28-32min, the trend of them is homology, which indicates that bearing is inclined to seizure in primary stage, the reason leading to this phenomenon is bearing friction moment enlarges due to bearing parts expand induced by violent rise of temperature, however, the amount of heat is not enough to produce abundant expansion at present, thus, the seizure is absent. The missing legend of Figure 9, Figure10 and Figure 11 is same to that of Figure 8.

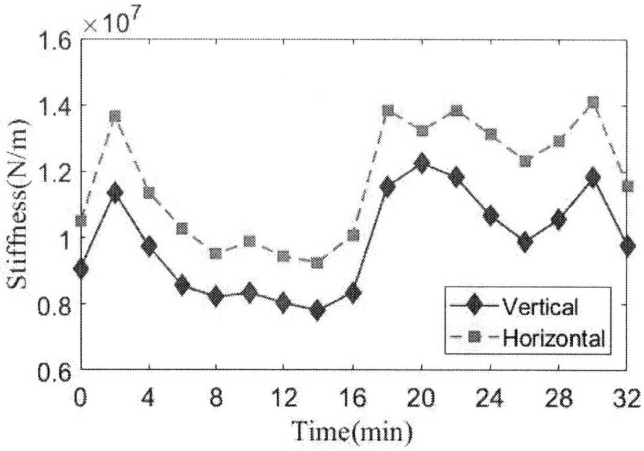

Figure 8. Bearing Stiffness Evolution Curve.

3.3. Rotor vibration characteristic during the whole life under excessive preload

During the running process of bearing, the variation of its stiffness has directly effect on rotor vibration and stability. As shown in Figure 9, there are system vibration responses in time domain of main infection moments during the whole bearing life. The corresponding responses in frequency domain are introduced in Figure 10. It is obvious that the rotor displacements are same along horizontal and vertical direction, the rotational frequency f_r evoked by unbalance of rotor system is dominant. In first fluctuation stage (0-6min), the amplitude of rotor displacement firstly decreases and then increases, the chief frequency is rotational frequency f_r, whose amplitude has the same tendency with that in time domain, however, the amplitude of fundamental train frequency f_c slightly go up in 2min. In stable running stage (6-16min), the amplitude of rotor displacement remains unchanged, the dominant frequency component is rotational frequency f_r, whose amplitude raises lightly. In second fluctuation stage (16-32min), the amplitude of displacement cuts down initially, subsequently, it keeps stable for a period until the occurrence of impulse signal, only in seizure stage, fundamental train frequency f_c, 2-times f_c and f_r are governing, whose amplitudes are abruptly increasing.

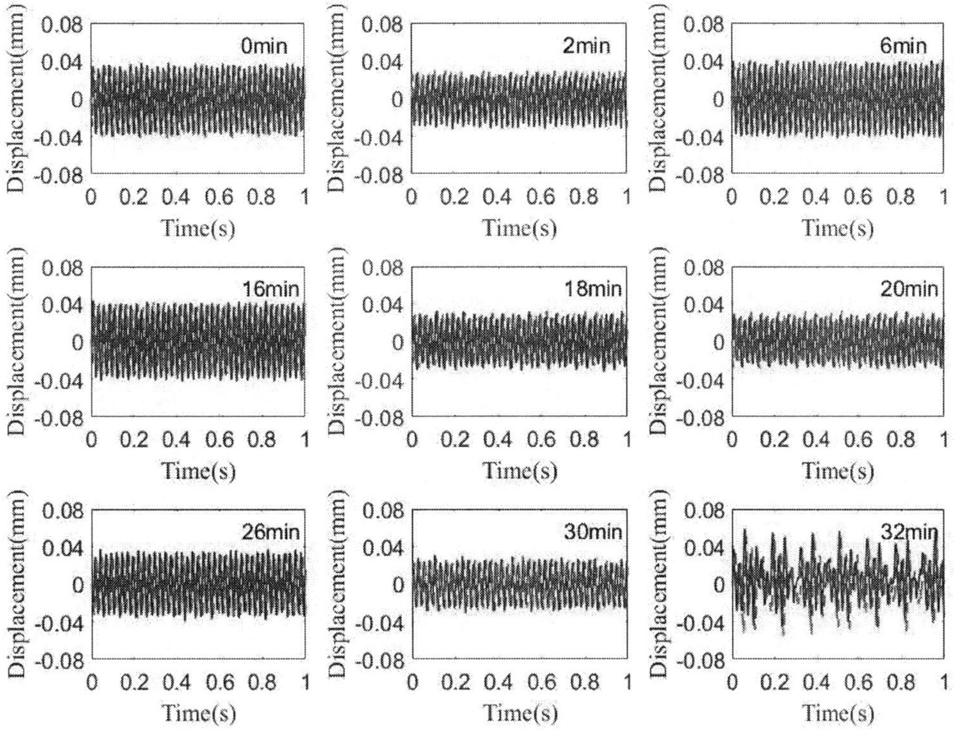

Figure 9. Time Responses of Various Moments.

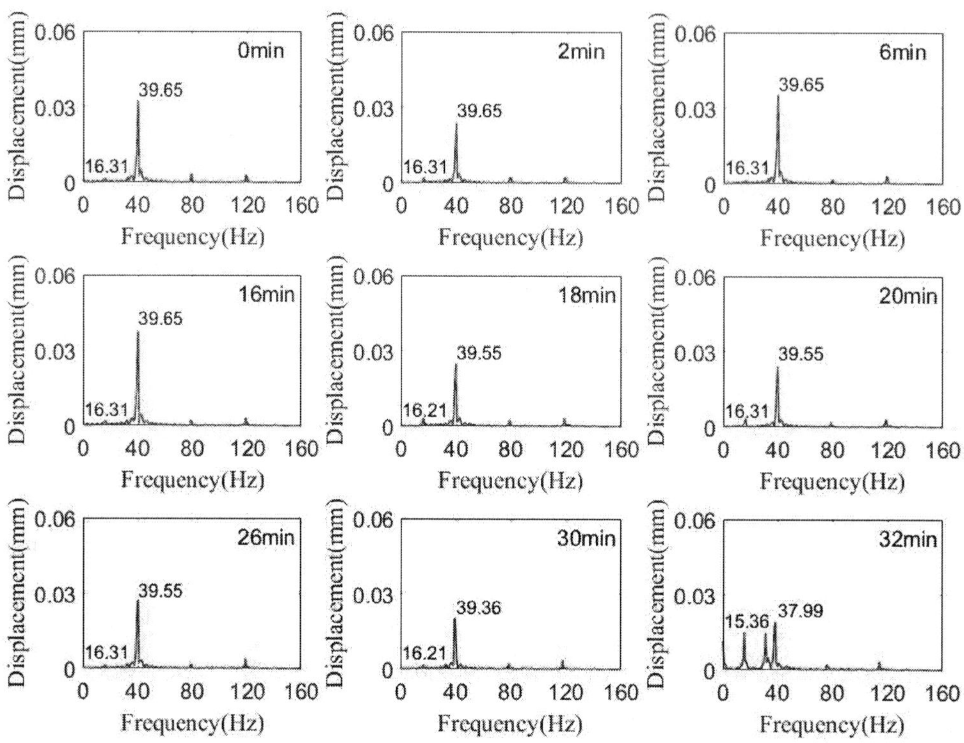

Figure 10. Frequency Responses of Various Moments.

As shown in Figure 11, the rotor axial orbit has the same variation tendency with displacement, which performs regular, stable and variable size “rhombus” whirling motion. Rotor radial runout changes in different degrees along different directions: the horizontal radial runout of rotor increases lightly in first increment stage (0-2min), the vertical radial runout of rotor enlarges in second addition stage (16-18min), besides, the radial runout has the notable expansion in third increment stage (26-30min) in both directions. In the moment of seizure, the axial orbit of rotor is complicated, which is caused by fundamental train frequency f_c and 2-times f_c.

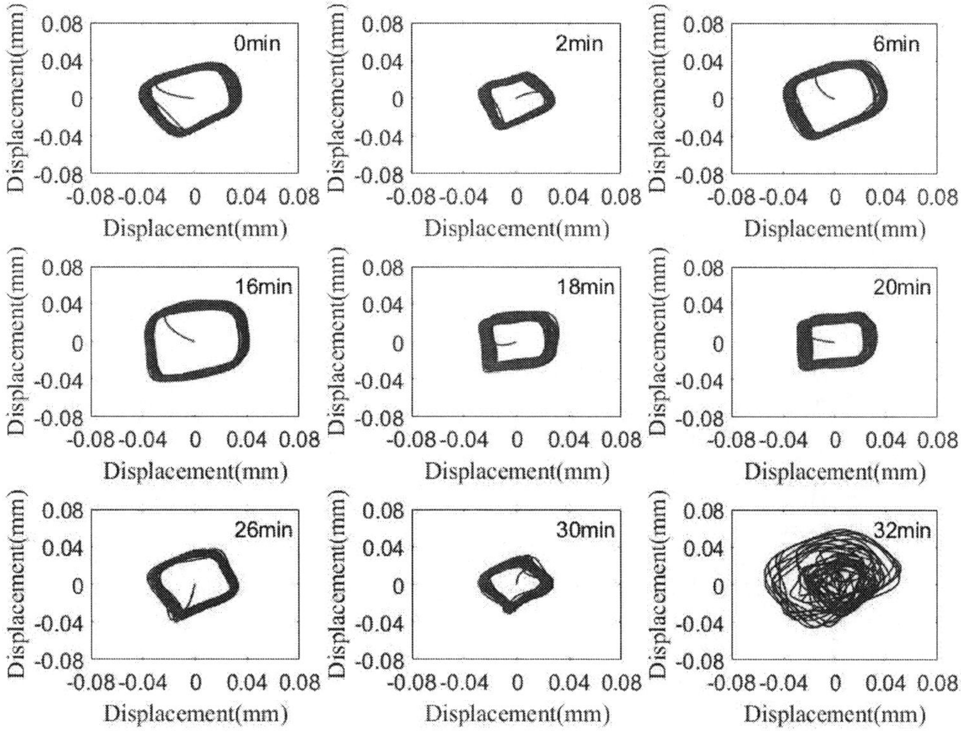

Figure 11. Rotor Orbits of Various Moments.

4. Conclusions

In this paper, the bearing stiffness identification method is promoted based on bearing-rotor system dynamics, the study on bearing stiffness under different preload and its evolution rule under excessive preload is carried out, the results are explained as follows:

(1) Preload has obvious effect on bearing stiffness of micro-turbine. With the increment of preload, bearing stiffness goes up rapidly.

(2) The bearing stiffness curve performs the inclination of "fluctuation-stability-significant fluctuation", on the basis of the curve, the bearing life is divided into three stages.

(3) With the development of bearing stiffness, the corresponding responses in time and frequency domain are varying in different extent. In first fluctuation stage, the amplitude of rotor displacement firstly cuts down and then increases, the chief frequency is rotational frequency f_r . In stabilization stage, the amplitude of rotor displacement remains steady, the dominant frequency component is rotational frequency f_r . In second fluctuation stage, the amplitude of displacement reduces initially, subsequently, it keeps unchanged for a period until impulse signal appears, only in seizure phase, fundamental train frequency f_c and 2-times f_c are dominant.

Acknowledgement

This research work was supported by the National Key R&D Program of China (No.: 2018YFB2000300) and the Natural Science Foundation of Liaoning Province (No.: 20180550792) and the High-level Talents of Dalian City (No.: 2018RQ18)

References

[1] A reference
This reference has two entries but the second one is not numbered (it uses the 'Reference (no number)' style.
[2] Another reference
[3] More references

[1] Tandon N and Choudhury A 1999 A review of vibration and acoustic measurement methods for the detection of defects in rolling element bearings *Tribol Int* **32** 469-80
[2] Aktürk N, Uneeb M and Gohar R 1997 The Effects of Number of Balls and Preload on Vibrations
Associated With Ball Bearings *Journal of Tribology* **119** 747-53
[3] Tiwari M, Gupta K and Prakash O 2000 Effect of Radial Internal Clearance of a Ball Bearing on the Dynamics of a Balanced Horizontal Rotor *Journal of Sound and Vibration* **238** 723-56
[4] Alfares M A and Elsharkawy A A 2003 Effects of axial preloading of angular contact ball bearings on the dynamics of a grinding machine spindle system Journal of Materials Processing Technology **136** 48-59
[5] Harsha S P, Sandeep K and Prakash R 2003 Effects of preload and number of balls on nonlinear dynamic behavior of ball bearing system *International Journal of Nonlinear Sciences and Numerical Simulation* **4** 265-78
[6] Cao H, Li Y, He Z and Zhu Y 2014 Time Varying Bearing Stiffness and Vibration Response Analysis of High Speed Rolling Bearing-rotor Systems Journal of Mechanical Engineering **50** 73-81
[7] Jin Y, Yang R, Hou L, Chen Y and Zhang Z 2017 Experiments and Numerical Results for Varying Compliance Contact Resonance in a Rigid Rotor–Ball Bearing System Journal of Tribology **139** 1-17
[8] Wardle F and Poon S Y 1983 Rolling bearing noise-cause and cure Chartered mechanical engineer
30 36-40
[9] Than V and Huang J 2016 Nonlinear thermal effects on high-speed spindle bearings subjected to
preload *Tribol Int* **96** 361-72
[10] Hwang Y and Lee C 2010 A review on the preload technology of the rolling bearing for the spindle of machine tools *International Journal of Precision Engineering and Manufacturing* **11** 491-8
[11] Harris T A 1965 How to compute the effects of preloaded bearings *Production Engineering* **19** 84-93
[12] James G, Katter J and Tu J F 1996 Bearing Condition Monitoring for Preventive Maintenance in a Production Environment *Tribol T* **39** 936-42
[13] Gunduz A, Dreyer J T and Singh R 2012 Effect of bearing preloads on the modal characteristics
of a shaft-bearing assembly: Experiments on double row angular contact ball bearings *Mechanical Systems and Signal Processing* **31** 176-95
[14] Cao H, Holkup T and Altintas Y 2011 A comparative study on the dynamics ospindles with respect to different preload mechanisms *The International Journal of Advanced Manufacturing Technology* **57** 871-83
[15] Zhang J, Fang B, Zhu Y and Hong J 2017 A comparative study and stiffness analysis of angular
contact ball bearings under different preload mechanisms *Mech Mach Theory* **115** 1-17
[16] Lundberg G and Palmgren A 1944 Dynamic capacity of rolling bearings *Acta Polytechnica Scandinavica.* **1** 7

[17] Hagiu G D and Gafitanu M D 1994 Preload-service life correlation for ball bearings on machine
tool main spindles *Wear* **172** 79-83
[18] G.Hagiu 2003 Reliable high speed spindles by optimum bearings preload *International Journal*
of Applied Mechanics and Engineerings **8** 57-70
[19] Kaczor J and Raczynski A 2014 The effect of preload of angular contact ball bearings on durability of bearing system *Proceedings of the Institution of Mechanical Engineers, Part J: Journal of Engineering Tribology* **229** 723-32
[20] Zhang J, Fang B, Hong J and Zhu Y 2017 Effect of preload on ball-raceway contact state and fatigue life of angular contact ball bearing *Tribol Int* **114** 365-72
[21] Wu P and Zhao X 2019 Incipient Fault Detection of Rolling Bearing Based on Fuzzy Entropy and Fractal Dimension *Chinese Journal of Turbomachinery* **61** 71-9
[22] Zhang C 2020 The Life Prediction of Bearing for Metro Fan Based on EMD Denoise and LSTM
Network *Chinese Journal of Turbomachinery* **62** 77-82
[23] Harsha S P 2005 Nonlinear dynamic analysis of an unbalanced rotor supported by roller bearing
Chaos, Solitons & Fractals **26** 47-66
[24] Harsha S P 2005 Non-linear dynamic response of a balanced rotor supported on rolling element
bearings *Mechanical Systems and Signal Processing* **19** 551-78
[25] Bai C and Xu Q 2006 Dynamic model of ball bearings with internal clearance and waviness *Journal of Sound & Vibration* **294** 23-48
[26] Bai C, Xu Q and Zhang X 2006 Nonlinear stability of balanced rotor due to effect of ball bearing internal clearance *Applied Mathematics and Mechanics (English Edition)* **27** 175–86
[27] Bai C, Zhang H and Xu Q 2007 Effects of axial preload of ball bearing on the nonlinear dynamic
characteristics of a rotor-bearing system *Nonlinear Dynamics* **53** 173-90
[28] Bai C, Zhang H and Xu Q 2013 Subharmonic resonance of a symmetric ball bearing–rotor system *International Journal of Non-Linear Mechanics* **50** 1-10
[29] Zhang X, Han Q, Peng Z and Chu F 2015 A new nonlinear dynamic model of the rotor-bearing system considering preload and varying contact angle of the bearing *Commun Nonlinear Sci* **22** 821-41
[30] Ashtekar A and Sadeghi F 2011 Experimental and Analytical Investigation of High Speed Turbocharger Ball Bearings *Journal of Engineering for Gas Turbines and Power* **133** 1-14
[31] Allaire P, Dimond T and Cao J 2019 Bearing Fix for Large Sub-synchronous Vibration in Steam
Turbine-gearbox-generator System *Chinese Journal of Turbomachinery* **61** 34-38
[32] Xiao P, Chen Y, Li Q and Wang W 2015 Effects of Tilting Pad Bearing Preload on the Rotor Stability *Chinese Journal of Turbomachinery* **57** 31-36
[33] Long L, Lv B and Yuan W 2018 Influence of the Support Dynamic Stiffness on Critical Speeds
of a Gas Generator Rotor of a Turbine Engine *Chinese Journal of Turbomachinery* **60** 66-70
[34] Li H and Shin Y C 2004 Analysis of bearing configuration effects on high speed spindles using
an integrated dynamic thermo-mechanical spindle model *International Journal of Machine Tools and Manufacture* **44** 347-64

Analysis of cage slip in angular contact ball bearing considering non-Newtonian behavior of elastohydrodynamic lubrication

Baogang Wen[1], Meiling Wang[2*], Xianghe Yun[3],Xu Zhang[1], Qingkai Han[4]

1. School of Mechanical Engineering and Automation, Dalian Polytechnic University, Dalian,116034, PR China,wbg_dlut@163.com

2.College of Locomotive and Rolling stock Engineering, Dalian Jiaotong University, Dalian, 116028, PR China, meilingcc@163.com
3. School of Mechanical Engineering, Dalian University of Technology, Dalian,116024, PR China, qk.han@hotmail.com

4. School of Mechanical Engineering and Automation, Northeastern University, Shenyang,116024, PR China, qk.han@hotmail.com

* Corresponding author: Tel.:+86-411-84106065; E-mail address: meilingcc@163.com

Abstract. A dynamic model of angular contact ball bearing considering non-Newtonian behavior and interaction between balls and cage is deduced to predict cage slip. And then the cage slip are investigated and compared with measured results under different rotational speeds and axial loads to verify the proposed dynamic model of bearing. It is concluded that the results based on the non-Newtonian behavior of elastohydrodynamic lubrication is more agreement in trend with the tests than that based on the Newtonian assumption, and so the non-Newtonian behavior of elastohydrodynamic lubrication are needed in the cage slip analysis.

Keywords.Angular contact ball bearing, cage slip, dynamic model, non-Newtonian

1. Introduction

Bearing is one of the important basic parts in the major equipment, such as aero-engine, turbo-machinery and so on [1,2].Its performance will directly affect the rotor dynamics, even the whole equipment[3,5].The dynamic characteristics of cage, considered as one of the most critical components in rolling bearings, greatly affect the bearing performance[6]. For example, the cage slip, i.e. the linear velocity of rolling elements and raceway contact differences, often happens at high speed and light load condition, which causes frequent collision between rollers and cage, and abnormal vibration and temperature of bearing abnormal changes, and even cause cage wear and fracture failures [7,8].

The sliding behavior of the cage cannot be obtained by the conventional quasi-static analysis, in which the forces between the rolling bodies and cage and the inertial angular acceleration of the cage are ignored. It is also be faced numerous difficulties by use of the complete dynamic model to investigate cage slip behavior, simply because the model involves many degree of freedoms and the complex relationship among parameters needed to determine, and between them. Kannel[9], Boesiger[10], Gupta[11,12], Meeks [13,14], Ghaisas [6]make an outstanding contribution.

Tu et al. [15] established a dynamic model of the roller bearing with a single degree of freedom considering the contact and friction between the rolling body and the cage, studied the sliding phenomenon of the cage during the acceleration of the bearing cage, and calculated the friction by using Coulomb's friction formula. Jain S[16] based on the multi-body dynamics theory and considering the elastohydrodynamic lubrication Newtonian fluid oil film force and the interaction between cage and rolling body, established the dynamics model of rolling bearing, and carried out under the condition of axial load skid characteristics research. But the contact angle in the model as a fixed value, without considering the change of the load, speed under the condition of contact angle, so the model is only applicable to low speed bearings. Han Qinkai et al. [17],on the basis of Jain S, considered the change of contact angle to establish the angular contact bearing sliding dynamics model, and carried out the influence of fixed radial load and variable radial load on the bearing sliding velocity, etc. However, the model should be based on the raceway control theory hypothesis. Wang Yulong[18] considered the contact and lubrication between the rolling body and the inner and outer rings, as well as the cage, and established the rolling bearing dynamic model with 6 degrees of freedom, and studied the sliding and sliding velocity changes under different working conditions.

The above dynamic models of bearing are based on Newtonian fluid model and Newtonian oil film thickness calculation, without considering the non-Newtonian fluid behavior in elastohydrodynamic force, the calculation of elastohydrodynamic friction tends to produce a large error. The shear stress of traditional Newtonian fluid has a linear relationship with shear strain rate, while the shear stress of non-Newtonian fluid has a nonlinear relationship with shear strain rate, which is more approximate with the actual. In order to predict the bearing performance more accurately, the non-Newtonian rheological properties of lubricating oil should be considered.

In this paper, the dynamic model of the rolling bearing is established by introducing the modified Carreau model and modified oil film thickness formula based on the non-Newtonian fluid theory, considering the interaction between the cage and the rolling bodies and the fluid resistance of the cage. And the cage slip measurements are also carried out to verify the proposed model.

2. Dynamic model of angular contact ball bearing

2.1. *Differential equations of ball motion*

The external load of the bearing is $\boldsymbol{F} = \{F_X \quad F_Z \quad F_Z \quad M_Y \quad M_Z\}^T$ .In order to describe the motion of the ball, three coordinate systems are developed as shown in Figure 1. The first coordinate system ($OXYZ$) is fixed at the bearing centre with X axis coinciding with the bearing axis; the second coordinate system ($o'x'y'z'$) is a moving coordinate system axis x' axis parallel with bearing axis and the coordinate origin attached to the centre of ball, rotating around x axis with speed ω_m .In this coordinate, the ball has three angular displacement components $\theta_{x'},\theta_{y'},\theta_{z'}$ around $x'y'z'$ axis, respectively.

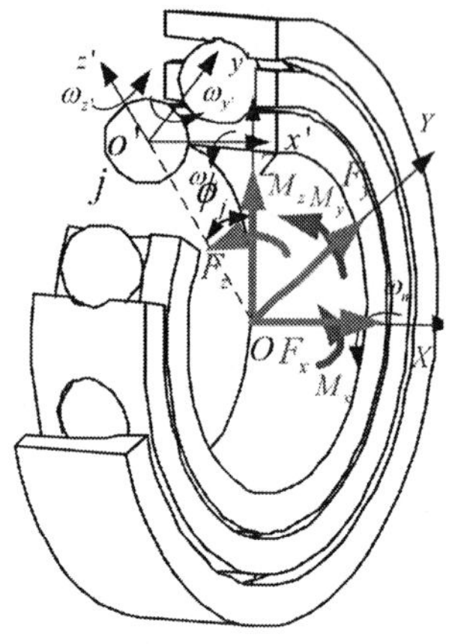

Figure 1. Coordinate system and forces of rolling bearing

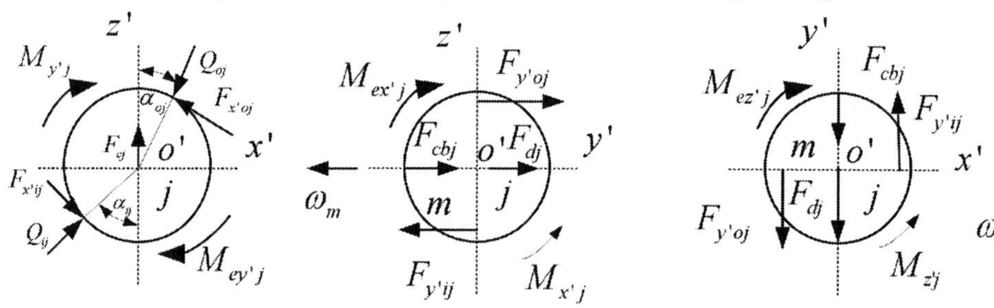

(a)$x'o'z'$ (b) $y'o'z'$(c)$x'o'y'$

Figure 2. Forces on the ball

Subscript *i*, *o* and *j* represent the parameter of the inner ring, outer ring and *j*th ball, respectively. Q is the contact force between ball and inner or outer raceway, which can be calculated based on the Hertz in Ref[19]. F_{cbj} is the interaction force acted on the cage by ball. F_{cj} the centrifugal force, which can be expressed $F_{cj}=\frac{1}{2}md_m\omega_{mj}$. M is the ball inertia moment can be decomposed to by $M_{x'j}$ 、 $M_{y'j}$ 、 $M_{z'j}$ in x, y and *z* directions respectively. F_{dj} and M_e is the viscous drag force and torque by lubricant oil on the ball which can be decomposed to by $M_{ex'j}$, $M_{ey'j}$, $M_{ez'j}$ in *x, y* and *z* directions respectively[12,16].

The dynamic equations of the *j*th ball can be expressed as follows

$$\begin{aligned} &Q_{ij}\sin\alpha_{ij}-Q_{oj}\sin\alpha_{oj}+F_{x'ij}\cos\alpha_{ij}-F_{x'oj}\cos\alpha_{oj}=0 \\ &Q_{ij}\cos\alpha_{ij}-Q_{oj}\cos\alpha_{oj}-F_{x'ij}\sin\alpha_{ij}+F_{x'oj}\sin\alpha_{oj}+F_{cj}=0 \end{aligned} \quad (j=1\cdots N) \tag{1a}$$

$$\begin{aligned} &I\dot{\omega}_{x'}=r(F_{y'o}\cos\alpha_o-F_{y'i}\cos\alpha_i)+\boldsymbol{M}_{so}\cos\alpha_o+\boldsymbol{M}_{si}\cos\alpha_i-M_{ex'} \\ &I\dot{\omega}_{y'}=r(F_{x'o}-F_{x'i})-M_{ey'} \\ &I\dot{\omega}_{z'}=r(F_{y'o}\sin\alpha_o-F_{y'i}\sin\alpha_i)-\boldsymbol{M}_{so}\sin\alpha_o-\boldsymbol{M}_{si}\sin\alpha_i-M_{ez'} \\ &I_m\dot{\omega}_{mj}=\frac{D_i}{2}F_{y'ij}+\frac{D_o}{2}F_{y'oj}-0.5F_{cbj}d_m-0.5F_{dj}d_m \end{aligned} \tag{1b}$$

where α is the contact angle of the bearing. $I=\frac{2}{5}mr^2$ is the rotational inertia of its own axis and $I_m=\frac{mD^2}{10}+\frac{d_m^{\ 2}}{4}m$ in which m is the mass of ball. Besides, $\omega_{x'j}$, $\omega_{y'j}$, $\omega_{z'j}$ denote the *j*th ball rotating angular velocities around its own axis. ω_{mj} is *j*th ball orbital revolution speed, $\omega_{mj}=\dot{\theta}_{mj}$. In addition, Q is the contact force and traction force between the ball and the inner or outer raceway. $F_{x'}$ 、 $F_{y'}$ $\boldsymbol{M}_s$ 、 $\boldsymbol{M}_s$ is traction force and traction moment between the ball and the inner or outer raceway can be obtained as follows. The traction force between the ball and the inner or outer raceway in Equation (6) can be calculated based on the elastohydrodynamic (EHD) lubrication model referring to Ref[19].

$$\boldsymbol{F}_{i/o}=-a_{i/o}b_{i/o}\int_{-1}^{1}\int_{-\sqrt{1-t^2}}^{\sqrt{1-t^2}}\boldsymbol{\tau}_{i/o}dtds \tag{2}$$

The traction moment is obtained as follows

$$\boldsymbol{M}_{si/o}=-\frac{1}{h_{i/o}}\int_{-a_{i/o}}^{a_{i/o}}\int_{-b_{i/o}}^{b_{i/o}}\boldsymbol{\tau}_{i/o}(x''\boldsymbol{i}+y''\boldsymbol{j})dx''dy'' \tag{3}$$

where *a* and *b* are the long and short axes of the contact ellipse, which can be obtained by Hertz contact. τ is the shear stress between the ball and raceway.
The shear stress considering non-Newtonian behaviour of elastohydrodynamic lubrication can be obtained as Carreau[20]

$$\begin{cases}\boldsymbol{\tau}=\eta\dot{\gamma}[1+(\eta\dot{\gamma}/G_r)^2]^{(n-1)/2} & \boldsymbol{\tau}<\lambda p\\ \boldsymbol{\tau}=\lambda(T)p & \boldsymbol{\tau}\geq\lambda p\end{cases} \tag{4}$$

where η is the oil viscosity at any temperature and pressure. G_r is the shear modulus of the lubricant and n is power-law exponent, λ is the ultimate shear. $\dot{\gamma}=\frac{\Delta \boldsymbol{v}}{h}$, in which Δv is the sliding velocity referring to[19], h is the oil film thickness between each ball and raceway. p is the contact pressure between each ball and raceway according to the Hertz contact theory. η is the oil viscosity at any temperature and pressure, which can be defined as

$$\eta=\eta_0(T)e^{\alpha p} \tag{5}$$

The oil viscosity at constant temperature is $\eta_0(T)=\eta_{40}e^{-\beta(T-T_0)}$, in which β temperature- viscosity coefficient, and T is the lubricant temperature.
The oil film thickness between each ball and raceway can be calculated considering the influence of non-Newtonian shear-thinning and thermal as follows, [21]

$$h=h_{nc}\varphi_t/\varphi_{nc} \tag{6}$$

where h_{nc} represents the classical evaluated film thickness, i.e., the Newtonian central film thickness. ϕ_t and ϕ_{nc} denote the adjustment factors for the thermal and non-Newtonian shear-thinning, respectively.
The Newtonian central film thickness h_{nc} in Equation (6) can be obtained by classical Hamrock and Dowson [22]

$$h_{iso}=2.69U^{0.67}G^{0.53}W^{-0.067}(1-0.61e^{-0.73\varsigma})R_x \tag{7}$$

IOP Conf. Series: Materials Science and Engineering 1081 (2021) 012013 doi:10.1088/1757-899X/1081/1/012013

where thee equivalents radius of curvature is defined as $R_x = \frac{1}{2}D(1 \mp \gamma)$, in which the dimensionlessparameter is $\gamma = D\cos\alpha / d_m$ and the "-"and"+" are chosen for the inner and outer raceways respectively. The dimensionless elastic modulus G, load W and rotating speed U are

$$\begin{aligned} G &= \alpha E' \\ W &= Q/(E' R_x^2) \\ U &= \eta u/(E' R_x) \end{aligned} \tag{8}$$

Where α denotes the pressure-viscosity coefficient and E' denotes the effective elastic modulus. u is the rolling velocity between ball and raceway.

The adjustment factors for thermal and non-Newtonian shear-thinning in Equation (6) can be evaluated by Gupta [23] and Bair [20] as follows

$$\phi_t = \frac{1 - 13.2(\frac{p}{E'})L^{0.42}}{1 + 0.213(1 + 2.23S^{0.83})L^{0.64}} \tag{9a}$$

$$\phi_t = [1 + 4.44(\frac{u\eta}{h_{iso}G_r})^{1.69}]^{1.26(1-n)^{1.78}} \tag{9b}$$

where the thermal loading parameter is defined as $L = \eta\beta\frac{u^2}{4k_f}$, in which k_f is the thermal conductivity. The slide to roll ratio is defined as $S = \frac{\Delta v}{u}$.

2.2. *Equilibrium equations of the inner ring*

The equilibrium equations of the inner ring under external loads must be established and solved iteratively combining with the Equation (1a) for unknown the contact parameters including the contact forces, contact areas of the Hertz and contact angles between the balls and the inner or outer raceway. The equilibrium equations of the inner ring are calculated as follows.

$$\begin{aligned} &F_X - \sum_{j=1}^{N}(Q_{ij}\sin\alpha_{ij} + F_{x'ij}\cos\alpha_{ij}) = 0 \\ &F_Y - \sum_{j=1}^{N}(Q_{ij}\cos\alpha_{ij} - F_{x'ij}\sin\alpha_{ij})\sin\phi_j = 0 \\ &F_Z - \sum_{j=1}^{N}(Q_{ij}\cos\alpha_{ij} - F_{x'ij}\sin\alpha_{ij})\cos\phi_j = 0 \\ &M_Y - \sum_{j=1}^{N} r_i(Q_{ij}\sin\alpha_{ij} + F_{x'ij})\cos\phi_j = 0 \\ &M_Z - \sum_{j=1}^{N} r_i(Q_{ij}\sin\alpha_{ij} + F_{x'ij})\sin\phi_j = 0 \end{aligned} \tag{10}$$

2.3. *Interaction between balls and cage*

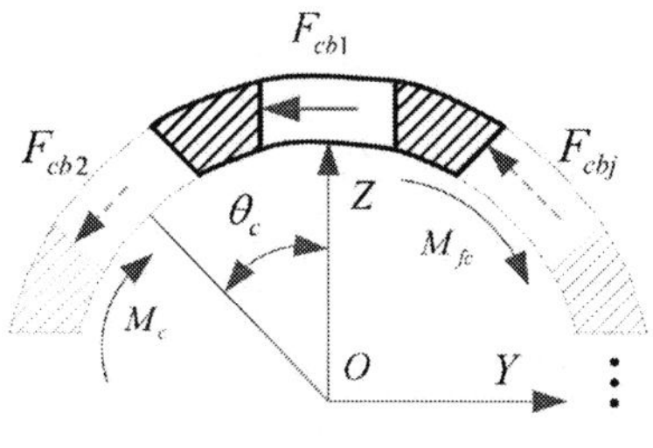

Figure 3. Forces on the cage

The differential equation governing the rotational motion of the cage is as follow:

$$I_c\dot{\omega}_c = \frac{d_m}{2}\sum_{i=1}^{z} F_{cbj} - M_{fc} \tag{11}$$

where I_c is the inertia moment of cage about X axis. M_{fc} is viscous drag moment by lubricant oil on cage. The interaction force acted on the cage by ball can be expressed as [16]

$$F_{cbj} = \frac{d_m}{2}(k_c\{\theta_{mj} - \theta_c\} + c_c\{\omega_{mj} - \omega_c\}) \tag{12}$$

where k_c and c_c are the interaction stiffness and damping between cage and ball, respectively。 θ_{mj} and ω_{mj} is the position angle and angular velocity of the *j*th ball at time *t*, and θ_c is the position angle and angular velocity of cage at time *t*.

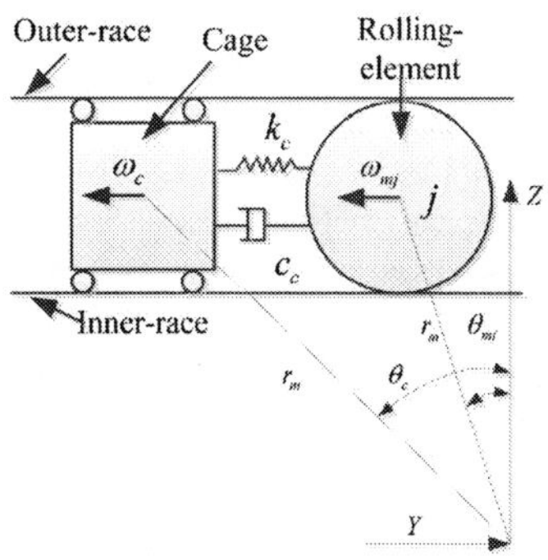

Figure 4. Interaction between balls and cage

2.4. Computation method and its flowchart

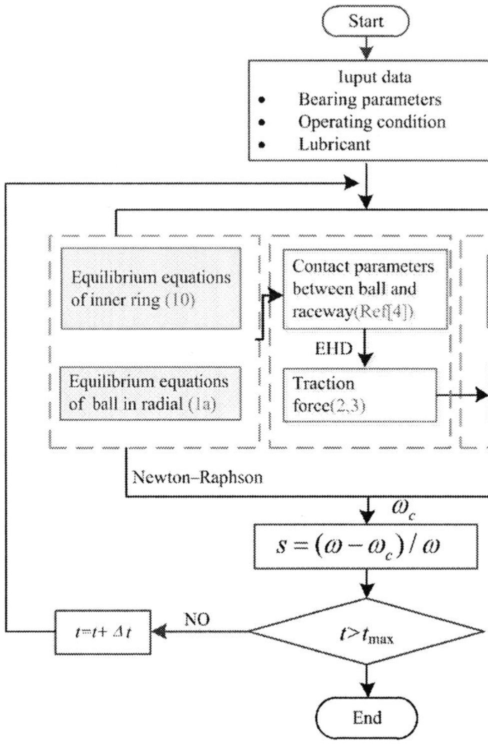

Figure 5. Flowchart for calculation of the dynamic model of bearing

Figure 5.shows the brief flowchart for the calculation. In the dynamic model, the input parameters include the geometric, lubrication parameters and operating conditions of the bearing for. The contact parameters between the ball and the raceway with the Newton–Raphson method are calculated based on the combined equilibrium equations of balls and ring. Based on the obtained results, the traction forces and the interactions of balls-cage are introduced in the dynamic model of balls. The dynamic equations of the cage and balls are solved numerically by using the Newmark- β method. Meanwhile, the computed results at the former step apply for the next initial values for the equations. The calculating process will stop when the time exceeds the setting time.

3. Model validation

3.1. Test cage slip of bearing

A test rig is developed to measure the cage motions of bearing and more details for test rig can be described in Ref. [24]. An eddy current sensor is fixed on a subpanel to measure the axial cage motion, as shown in Figure 6. The test results of axial cage motion are illustrated in Figure 7. The measured cage motions are analyzed by FFT and the cage rotational frequency f_c is obtained to calculate cage slip.

(a) Photo of test rig (b) Sensor for axial cage motion

Figure 6. Bearing test rig

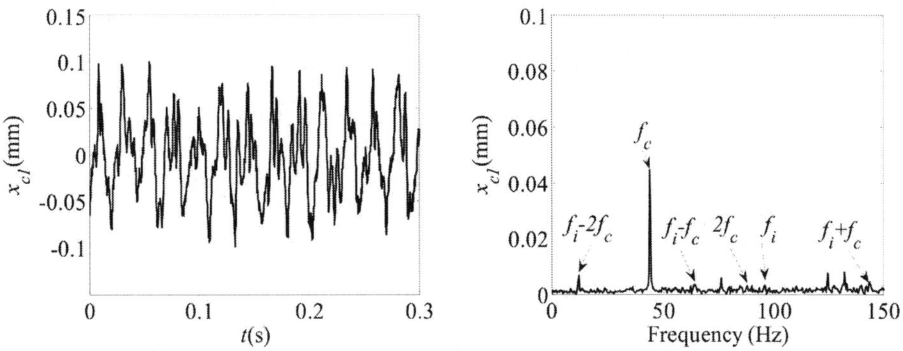

(a) Time-domain (b) Frequency- domain

Figure 7. Axial cage motion

3.2. Model validation

Firstly, the oil film thickness of the bearing taking account of the influences of non-Newtonian shear-thinning have been validated in Ref[25].Next, the skidding behaviors are measured and calculated under different rotational speeds and axial loads to validate the proposed model.

Table1. Geometrical and calculated parameters for an angular contact ball bearing of 7013AC

Parameter	Symbols	Value	Unit
Inner curvature factor	r_i	0.515	mm
Outer curvature factor	r_o	0.515	mm
all diameter	D	11.105	mm
Pitch diameter	d_m	85.25	mm
Initial contact angle	α^o	25	deg
Number of balls	Z	18	-
Cage inertia moment	I_c	3.4e-4	Kg.m^2
Interaction stiffness	k_c	1e8	N/m
Interaction damping	c_c	1e6	N/(m/s)

An angular contact ball bearing of 7013AC is chosen here, and its cage is made by 45# steel. Additionally, its geometrical and calculated parameters are listed in Table 1. Furthermore, the lubricant oil (46#) is used and its parameters are provided in Table 2.

Table 2. Parameters of oil lubrication

Parameters	Symbols	Value	Unit
Lubricant viscosity at 40℃	η	0.03956	Pa.s
Temperature-viscosity Coefficient	β	0.0368	K^{-1}
Thermal conductivity	k_f	0.14	W/m.K
Tested lubricant temperature	T	36	℃
Dielectric constant	ε_0	2.2	-
Relative permittivity	ε_r	8.85	-
Density of the lubricant	ρ_0	860	Kg/m^3
Pressure-viscosity coefficient	α	1.57e-8	1/Pa
Shear modulus of the Lubricant	G_r	5.4e4	Pa
Power-law exponent	n	0.54	

Figure 8 shows the comparison of cage slip under different axial loads between numerical results and the experimental results at the bearing speed of 2400 r/min and 3000 r/min, respectively.

As shown in Figure8, it obviously indicates that cage slip gradually increase with the increment of the rotating speeds. The behavior of lubrication have effect on the analysis of cage slip in angularcontactballbearing, and the calculated results obtained from present model is more agreement in trend with the test findings than the using the Newtonian of elastohydrodynamic, which verifies the developed model. It needs to consider the non-Newtonian behavior of elastohydrodynamic lubrication in cage slip analysis for angular contact ball bearing.

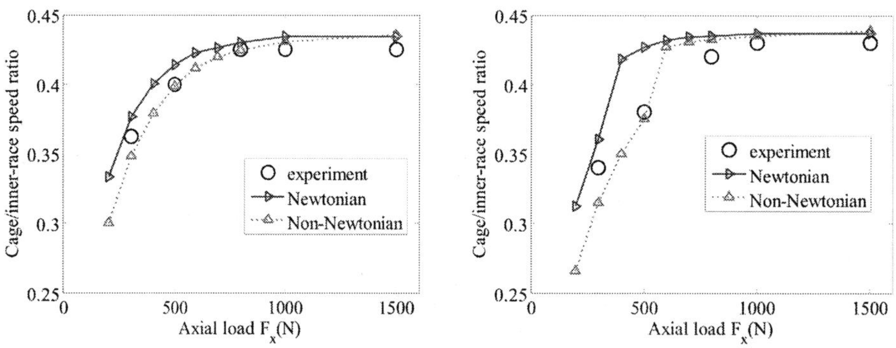

(a) 2400r/min (b) 3000r/min

Figure 8. Comparison of simulation results with experimental data

4. Conclusions

A dynamic model of angular contact ball bearing taking account of the influences of non-Newtonian shear-thinning and interaction between balls and cage is deduced to predict cage slip. And then the cage slip is calculated and the measurements are compared to verify the dynamic model of bearing.

The behavior of elastohydrodynamic lubrication has effect on the cage slip in angular contact ball bearing. The results based on the non-Newtonian theory of elastohydrodynamic lubrication are more agreement in trend with the tests than that based on the Newtonian assumption, and so the non-Newtonian behavior of elastohydrodynamic lubrication is needed in the cage slip analysis.

Acknowledgments

This work was financially supported by the National Natural Science Foundation of China (Grant No: 51905069) and the Natural Science Foundation of Liaoning Province(No.:20180550792) and the High-level Talents of Dalian City (No.: 2018RQ18)

Conflict of interest
The Authors declare that there is no conflict of interest.

References
[1] Zhang C 2020 The Life Prediction of Bearing for Metro Fan Based on EMD Denoise and LSTM Network Chinese *Journal of Turbomachinery* **62** 77-82
[2] Wang F and Liu Y 2016 Solution and Analysis of Bearing Overburning in the Side of Cantilever Centrifugal Fan Impellers *Chinese Journal of Turbomachinery* **58** 91-96
[3] Wu P and Zhao X 2019 Incipient Fault Detection of Rolling Bearing Based on Fuzzy Entropy and Fractal Dimension *Chinese Journal of Turbomachinery* **S1** 71-79
[4] Allaire P, Dimond T and Cao J 2019 Bearing Fix for Large Sub-synchronous Vibration in Steam Turbine-gearbox-generator System *Chinese Journal of Turbomachinery* **59** 39-44
[5] Du G, Wang Y and Yang G 2019 Rotor Dynamics Research of a 165kW Helium Blower with AMBs *Chinese Journal of Turbomachinery* **61** 34-8
[6] Ghaisas N, Wassgren C R and Sadeghi F 2004 Cage instabilities in cylindrical roller bearings *Journal of Tribology* **126** 681-689
[7] Kingsbury E and Walker R 1994 Motions of an unstable retainer in an instrument ball bearing *Journal of tribology* **116** 202-208
[8] Gupta P 1979 Dynamics of rolling-element bearings—part iii: Ball bearing analysis Journal of Tribology **101** 312-318
[9] Kannel J and Bupara S 1978 A simplified model of cage motion in angular contact bearings operating in the EHD lubrication regime *Journal of Tribology* **100** 395-403
[10] Boesiger E, Donley A and Loewenthal S 1992 An analytical and experimental investigation of ball bearing retainer instabilities *Journal of tribology* **114** 530-538
[11] Gupta P K 1991 Modeling of instabilities induced by cage clearances in ball bearings *Tribology Transactions* **34** 93-99
[12] Gupta P K 2012 Advanced dynamics of rolling elements: *New York: Springer Science & Business Media)*
[13] Meeks C R and Ng K O 1985 The dynamics of ball separators in ball bearings—part I: analysis *ASLE transactions* **28** 277-287
[14] Meeks C R 1985 The dynamics of ball separators in ball bearings—Part II: results of optimization study *ASLE transactions* **28** 288-295
[15] Tu W, Shao Y and Mechefske C K 2012 An analytical model to investigate skidding in rolling element bearings during acceleration *Journal of Mechanical Science and Technology* **26** 2451-2458
[16] Jain S and Hunt H 2011 A dynamic model to predict the occurrence of skidding in wind-turbine bearings. In: *Journal of Physics: Conference Series*, p 012027
[17] Han Q and Chu F 2015 Nonlinear dynamic model for skidding behavior of angular contact ball bearings *Journal of Sound and Vibration* **354** 219-235
[18] Wang Y, Wang W, Zhang S and Zhao Z 2015 Investigation of skidding in angular contact ball bearings under high speed *Tribology International* **92** 404-417
[19] Harris T A and Kotzalas M N 2006 Rolling bearing analysis (New York: CRC/Taylor & Francis)
[20] Bair S, Vergne P and Querry M 2005 A unified shear-thinning treatment of both film thickness and traction in EHD *Tribology Letters* **18** 145-152
[21] Echávarri Otero J, Guerra Ochoa E, Chacón Tanarro E, Díaz Lantada A and Munoz - Guijosa J M 2015 Analytical model for predicting friction in line contacts Lubrication Science
[22] Hamrock B J and Dowson D 1977 Isothermal elastohydrodynamic lubrication of point contacts: Part III—fully flooded results *Journal of Tribology* **99** 264-275

[23] Gupta P, Cheng H, Zhu D, Forster N and Schrand J 1992 Viscoelastic effects in MIL-L-7808-Type lubricant, Part I: Analytical formulation *Tribology transactions* **35** 269-274
[24] Wen B, Ren H, Zhang H and Han Q 2017 Experimental investigation of cage motions in an angular contact ball bearing Proceedings of the Institution of Mechanical Engineers, Part J: Journal of *Engineering Tribology* **231** 1041-1055
[25] Wen B, Ren H, Dang P, Hao X and Han Q 2018 Measurement and calculation of oil film thickness in a ball bearing *Industrial Lubrication and Tribology*. **70** 1500-1508

Effect of ball bearing misalignment on dynamic characteristics of rotor system

Baogang Wen[1], Meiling Wang[2,4*], Qingkai Han[3], Changxin Yu[4]

1. School of Mechanical Engineering and Automation, Dalian Polytechnic University, Dalian,116034, PR China,wbg_dlut@163.com
2.College of Locomotive and Rolling stock Engineering, Dalian Jiaotong University, Dalian, 116028, PR China, meilingcc@163.com
3. School of Mechanical Engineering and Automation, Northeastern University, Shenyang,116024, PR China, qk.han@hotmail.com
4.Wafangdian Bearing Group Corp, Wafangdian, 116300, China,yucx@zwz-bearing.com
* Corresponding author: Tel.:+86-411-84106065; E-mail address: meilingcc@163.com

Abstract. A theoretical model with 5 degrees of freedom (DOF) is derived for a rigid rotor system supported by two ball bearings considering the changes of the bearing stiffness due to misalignment. And then radial and axial vibrations of the rotor system and its bearing reactions under different misaligned offsets are investigated with theoretical and numerical analysis. Then, the effects of misalignment on the vibrations of rotor and the bearing, the variation of stiffness and reaction characteristics of the misaligned ball bearing are predicted theoretically. It is concluded that angular misalignment may cause axial vibration and parallel misalignment can bring out extra moment loads for misaligned bearing.

Keywords. Rotor system; bearing misalignment; bearing stiffness; theoretical model

1. Introduction

Bearing is one of the important basic parts in the major equipment, such as aero-engine, turbo-machinery and so on[1, 2]. Its performance will directly affect the rotor dynamics, even the security of the whole equipment[3-5].Misalignment of bearing is present because of improper assembly and thermal distortion, resulting in abnormal vibration and excessive preload[6]. However, perfect alignment cannot be obtained. The need for understanding the phenomena is important to practical engineers for the purpose of trouble shooting. There are two types of misalignment: coupling misalignment and bearing misalignment. Recently, the majority of the misalignment studies is about coupling misalignment, discussed the effects of its structure[7, 8] and type on the vibration of connected rotor system[9, 10], and evaluated the 2× vibration response as to be the characteristic feature of misalignment[11-15]. Hussain and Redmond[16, 17] pointed coupling misalignment is a source of both torsional and lateral excitations, of which angular misalignment may generate purely static forces and displacements. References[18, 19] showed that super-harmonic components are the most remarkable in the vibration spectrum for a hyper-static shaft-line with rigid coupling misalignment.

As to the bearing misalignment, no much work has been reported, and the corresponding dynamic mechanism is still unclear and insufficient. In this paper, a dynamic model of a misaligned rigid rotor system with 5 DOF is deduced based on Lagrangian theory, in which the variation of bearing stiffness due to the un-concentricity of supports is taken into account. The effects of misalignment are analyzed theoretically on the vibration of the rotor and bearing, the stiffness and reaction force of the misaligned bearing.

2. Motion equations of rigid rotor system with misaligned ball bearing

As the rigid rotor system shown in figure1(a) with misaligned ball bearing, an angular tilt is present for the shaft and offset are available for its right bearing B2, which refer to the angular and parallel misalignment, respectively. In addition, the upward point of bearing B1 may yaw around its initial position (Letting the deflection angle). A moving coordinate system O2X2Y2Z2 is introduced to illustrate the misaligned rotor system. At any time t, the relationship between OXYZ, and O2X2Y2Z2 is shown in figure 1(b).

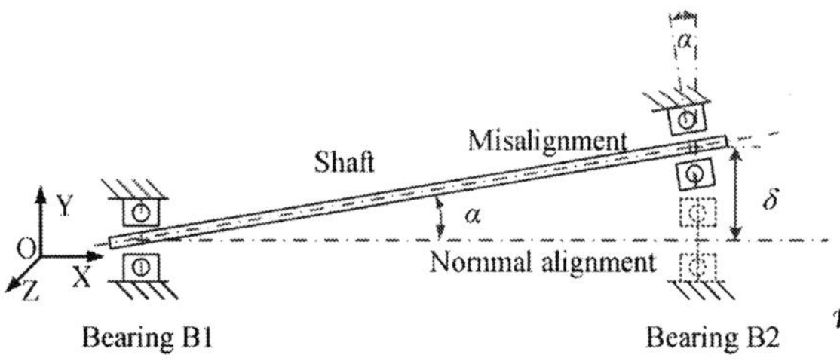

(a) Sketch

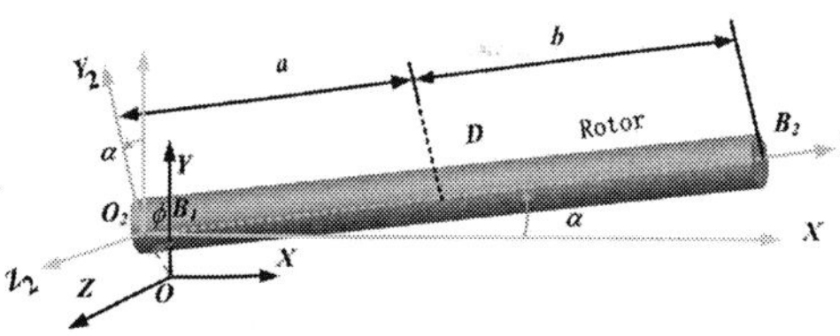

(b) Coordinates

Figure1. Misaligned rotor system

The equations of motion of misaligned rotor system are obtained following the Lagrangian theory, in which assumptions adopted are as follows: (1) the rotor is rigid and represented by one concentrated mass point; (2) bending and axial vibrations without torsional vibration of the rotor system are considered; (3) variable stiffness of misaligned bearing are considered; (4) shaft rotating speed is constant.

The vector $\mathbf{q}=\{x \quad y \quad z \quad \theta_y \quad \theta_z\}^T$ of concentrated mass point D in the fixed coordinate system $OXYZ$ is chosen as the generalized displacement vector, including three translational DOF and two rotational DOF around Y- and Z- axes. From the perspective of the rigidity assumption, rotational displacements of points on rotor are all θ_y, θ_z.

2.1 Kinetic and potential energy

The kinetic energy of the rotor system shown in figure1 is easily computed as a sum of the translational kinetic energy T_t and rotational kinetic energy T_r,

$$\begin{aligned}T_t &= \frac{1}{2}m(\dot{x}_P^2+\dot{y}_P^2+\dot{z}_P^2)\\ &= \frac{1}{2}m\left\{\left[\dot{x}+\alpha e\Omega\sin(\Omega t+\Psi_0)\right]^2+\left[\dot{y}-e\Omega\sin(\Omega t+\Psi_0)\right]^2+\left[\dot{z}+e\Omega\cos(\Omega t+\Psi_0)\right]^2\right\}\end{aligned} \tag{1}$$

$$T_r = \frac{1}{2}\left\{J_p(\Omega+\dot{\theta}_z\theta_y)^2+J_d[(-\dot{\theta}_z\sin\Omega t+\dot{\theta}_y\cos\Omega t)^2+(\dot{\theta}_z\cos\Omega t+\dot{\theta}_y\sin\Omega t)^2]\right\} \tag{2}$$

where m, J_p, J_d are the mass, polar moment of inertia and diameter moment of inertia respectively, Ω is the rotating speed of rotor, e and Ψ_0 are the eccentricity and initial phase angle.

The total potential energy of the rotor system is a sum of potential energy of the two supports U_1andU_2, both of which are

$$U_1 = \frac{1}{2}\mathbf{q}_{b1}^T\mathbf{K}\mathbf{q}_{b1}, \quad U_2 = \frac{1}{2}\mathbf{q}_{b2}^T\mathbf{K}\mathbf{q}_{b2} \tag{3}$$

Where K is a 5×5 stiffness matrix of the bearing, and its diagonal elements can be represented by K1-*K5* if ignoring the cross stiffness coefficients. $q_{b1}$and q_{b2} are the generalized displacement vectors at the two support points, respectively. In the misaligned condition, the relationshipsamongq_{b1}, q_{b2} and q can be expressed as

$$\mathbf{q}_{b1} = \mathbf{R}\mathbf{B}_1\mathbf{R}^T\mathbf{q} + \mathbf{S}_{b1}, \quad \mathbf{q}_{b2} = \mathbf{R}\mathbf{B}_2\mathbf{R}^T\mathbf{q} + \mathbf{S}_{b2} \tag{4}$$

Where $\mathbf{R} = \begin{bmatrix} 1 & -\alpha & 0 & 0 & 0 \\ \alpha & 1 & 0 & 0 & 0 \\ 0 & 0 & 1 & 0 & 0 \\ 0 & 0 & 0 & 1 & 0 \\ 0 & 0 & 0 & 0 & 1 \end{bmatrix}$, RT is the transpose matrix of R, $\mathbf{B}_1 = \begin{bmatrix} 1 & 0 & 0 & 0 & 0 \\ 0 & 1 & 0 & 0 & -a \\ 0 & 0 & 1 & a & 0 \\ 0 & 0 & 0 & 1 & 0 \\ 0 & 0 & 0 & 0 & 1 \end{bmatrix}$,

$\mathbf{B}_2 = \begin{bmatrix} 1 & 0 & 0 & 0 & 0 \\ 0 & 1 & 0 & 0 & b \\ 0 & 0 & 1 & -b & 0 \\ 0 & 0 & 0 & 1 & 0 \\ 0 & 0 & 0 & 0 & 1 \end{bmatrix}$, $\mathbf{S}_{b1} = \begin{Bmatrix} 0 \\ -\alpha^2\delta\cos\phi \\ 0 \\ -a\delta\sin\phi \\ a\delta\cos\phi \end{Bmatrix}$, $\mathbf{S}_{b2} = \begin{Bmatrix} 0 \\ -\alpha^2\delta\cos\phi \\ 0 \\ b\delta\sin\phi \\ -b\delta\cos\phi \end{Bmatrix}$, a and b are the distances between D and B1, B2 respectively.

2.2Stiffness matrix of the misaligned bearing

As to the rotor system shown in figure1, an angular tilt α and offset δ are present for the misaligned bearing bringing out extra preloads. So the displacements of the misaligned bearing (in figure2) can be represented by the axial and radial displacements δ_x^*, δ and angular displacement α respectively.

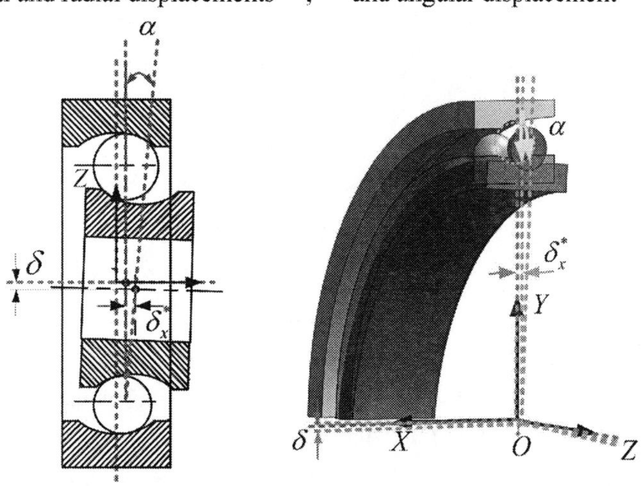

Figure2. Sketch of misaligned ball bearing

Under combined loads condition, the rolling bearing stiffness matrix(ignoring cross stiffness) $\mathbf{K}_{\text{lim}}$ with 5 degrees of freedom [20, 21] is

$$\mathbf{K}_{\text{lim}} = \begin{bmatrix} K_{11} & 0 & 0 & 0 & 0 \\ 0 & K_{22} & 0 & 0 & 0 \\ 0 & 0 & K_{33} & 0 & 0 \\ 0 & 0 & 0 & K_{44} & 0 \\ 0 & 0 & 0 & 0 & K_{55} \end{bmatrix} \tag{5}$$

Where $K_{11} \sim K_{55}$ are bearing translational stiffness coefficients in three directions and rotational coefficients around *Y*- and *Z*- axes, respectively.

2.3 Motion equations

The total kinetic energy of the rotor system is a sum of translational kinetic energy and rotational kinetic energy, namely, $T=T_t+T_r$. Total potential energy is a sum of elastic potential energy at two elastic supports, namely, $U=U_1+U_2$. By introducing these into Lagrangian equation and performing the relevant derivatives, the equations of motion of misaligned rotor system with 5-DOF are readily obtained

$$\mathbf{M}\ddot{\mathbf{q}} + \mathbf{G}\dot{\mathbf{q}} + \mathbf{K}_{\text{cg}}\mathbf{q} = \mathbf{Q} \tag{6}$$

Where, $\mathbf{M}$, $\mathbf{G}$, $\mathbf{K}_{\text{cg}}$, $\mathbf{Q}$ are mass matrix, gyroscopic matrix, stiffness matrix, generalized excitation force vector respectively.

$$\mathbf{M} = \begin{bmatrix} m & 0 & 0 & 0 & 0 \\ & m & 0 & 0 & 0 \\ & & m & 0 & 0 \\ & Sym. & & J_d & 0 \\ & & & & J_d \end{bmatrix}, \quad \mathbf{G} = \Omega \begin{bmatrix} 0 & 0 & 0 & 0 & 0 \\ 0 & 0 & 0 & 0 & 0 \\ 0 & 0 & 0 & 0 & 0 \\ 0 & 0 & 0 & 0 & J_p \\ 0 & 0 & 0 & -J_p & 0 \end{bmatrix} \tag{7}$$

$$\mathbf{K}_{\text{cg}} = \mathbf{R}\mathbf{B}_1^{\text{T}}\mathbf{R}^{\text{T}}\mathbf{K}\mathbf{R}\mathbf{B}_1\mathbf{R}^{\text{T}} + \mathbf{R}\mathbf{B}_2^{\text{T}}\mathbf{R}^{\text{T}}\mathbf{K}\mathbf{R}\mathbf{B}_2\mathbf{R}^{\text{T}}$$

$$= \begin{bmatrix} 2(1+2\alpha^2)K_1 & 0 & 0 & 0 & (a-b)\alpha(1+\alpha^2)K_1 \\ 0 & 2(1+2\alpha^2)K_2 & 0 & 0 & (-a+b)(1+\alpha^2)K_2 \\ 0 & 0 & 2K_3 & (a-b)K_3 & 0 \\ 0 & 0 & (a-b)K_3 & 2K_4+(a^2+b^2)K_3 & 0 \\ (a-b)\alpha(1+\alpha^2)K_1 & (-a+b)(1+\alpha^2)K_2 & 0 & 0 & 2K_5+\alpha^2(a^2+b^2)K_1+(a^2+b^2)K_2 \end{bmatrix}$$

$$\mathbf{Q} = \begin{Bmatrix} -\alpha m e \Omega^2 \cos(\Psi + \Psi_0) \\ m e \Omega^2 \cos(\Psi + \Psi_0) + 2K_2\alpha^2\delta\cos\phi \\ m e \Omega^2 \sin(\Psi + \Psi_0) \\ (a-b)K_4\delta\sin\phi \\ -(a-b)K_2\alpha^2\delta\cos\phi - (a-b)K_5\delta\cos\phi \end{Bmatrix} \tag{8}$$

In addition, by differentiating Lagrangian equation on, an algebraic equation is obtained to determine the angular position of the support point, which is related to y, θ_y, θ_z, namely

$$\begin{aligned}&2\alpha^2 K_2 y\sin\phi+(b-a)(\alpha^2 K_2+K_5)\theta_z\sin\phi+(b-a)K_4\theta_y\cos\phi+\cdots\\&+(a^2+b^2)\delta(K_4-K_5)\sin\phi\cos\phi=0\end{aligned} \tag{9}$$

3. Results and discussions

In order to investigate the influences of misalignment on the stiffness, vibration and reaction forces of bearings, numerical simulations are conducted with the fourth-order variable step Runge-Kutta method, based on the dynamic model of misaligned rotor system (Eqs. (6), (9)) in the above section.

The initial calculation conditions in simulations are as follows: the rotating speed is constant as 2400r/min (40Hz). The basic parameters of rotor system and the angular ball bearing of type 71906 are shown in Tables 1- 2.

Table 1. Parameters of the rotor system

Parameters	Value
Length of rotor	0.480m
Diameter of rotor	0.036m
Young modulus of rotor	2.06×10^{11}Pa
Poisson ratio of rotor	0.3
Mass of rotor	6Kg
Imbalance of rotor	6×10^{-3}kg.m
Density of rotor	7.85×10^{3}kg/m^3

Table 2 . Parameters of an angular ball bearing of type 71906

Parameters	Value
Diameter of inner ring D_i	0.030 m
Diameter of outer ring D_o	0.047m
Width	0.009m
Diameter of roller	0.004 m
Contact angle α_0	15°
No. of roller m	16
Load-deformation index n	3/2
Load deformation coefficient K_n	N/m^n
Axial preload	500N

As to the rotor with a length of 480mm and maximum elevation adjustment of 3mm on one side, the maximum angle misalignment amount obtained is about 0.00625rad (0.35 °). Thus, the δ =0mm, α =0~1° angular misalignment varying case is selected for the further simulation analysis. And also, the α =0°, δ =0~1mm parallel misalignment varying condition is adopted to grasp the effects of misalignment.

3.1 Changes of the stiffness of the misaligned bearing

An angular tilt and offset are available for misaligned bearing, resulting uneven and additional displacement loads, which will bring out the variation of stiffness of the misaligned bearing. An error ratio of bearing stiffness is defined to illustrate the variation,

$$err_K=\frac{\Delta K}{K_{norm}}\times100\% \tag{10}$$

Where ΔK is additional stiffness caused by misalignment, K_{norm} is the stiffness under normal state. Based on the proposed misaligned bearing model (Eq. (5)), variation of bearing stiffness is discussed under two cases mentioned above. The proposed error ratios of three translational stiffness coefficients of the bearing are altering along misalignment, as shown in Fig.3.

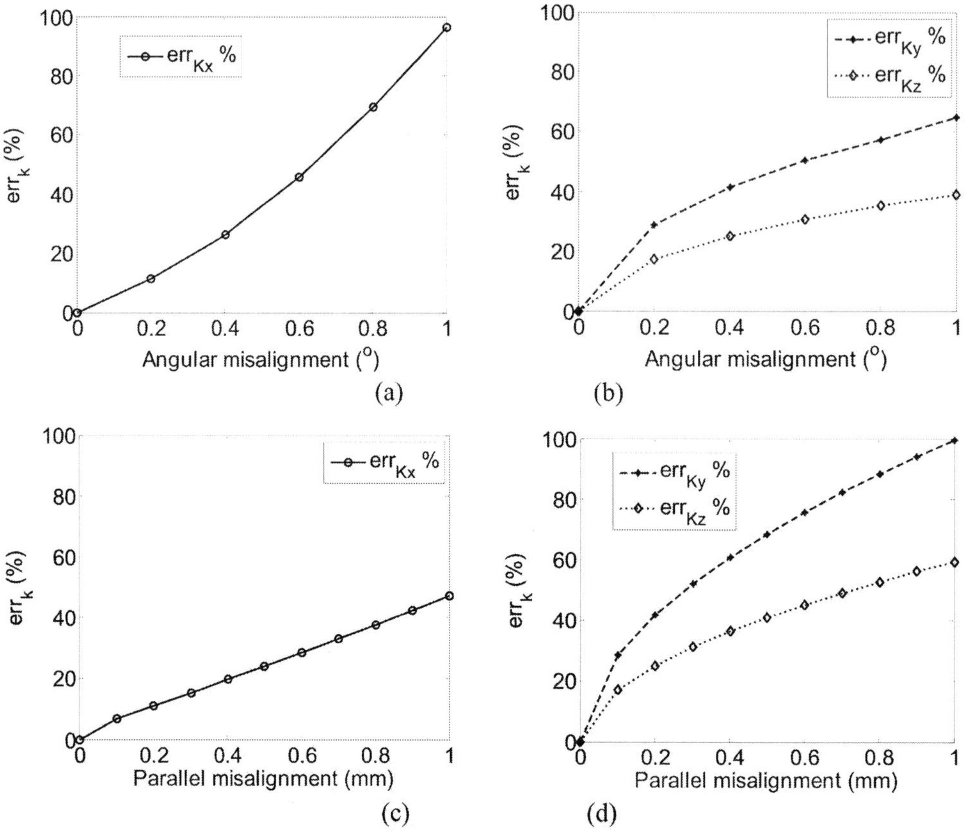

Figure3. Variation in error ratio of support's stiffness coefficients with (a-b) angular misalignment and (c-d) parallel misalignment

As seen from figure3, misalignment has a strong influence on the stiffness coefficients of bearing. Because the misalignment can directly result in uneven and additional displacement loads for misaligned bearing.

3.2 Vibration analysis of the misaligned rotor system

With the angular misalignment varying from 0° to 1° and the parallel misalignment varying from 0 mm to 1mm, the radial vibrations of the rotor and misaligned bearing are analyzed. The vibration response amplitudes of the rotor and bearing with angular and parallel misalignment are obtained, as shown in figure4.

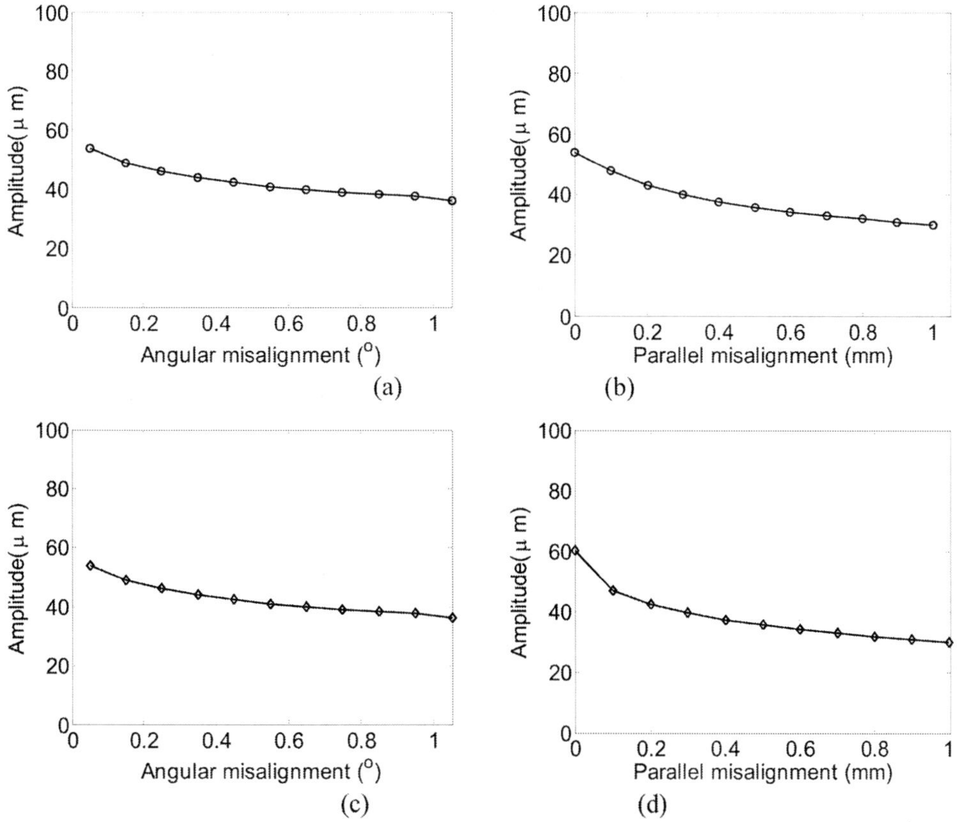

Figure4. Variation in amplitudes of vibration responses for the rotor (a-b) and misaligned bearing (c-d) with angular and parallel misalignment

For the both two cases, the radial vibration amplitudes of the rotor and misaligned bearing decrease slightly with increasing misalignment in the range of variation of α =0~1 °, δ =0~1mm, but the variation is not very evident.

3.3 Bearing reaction forces due to misalignment

The generalized reaction load F about the misaligned bearing can be easily obtained

$$\mathbf{F}=\mathbf{K}\,\mathbf{q}_b \tag{11}$$

Where K is 5×5 stiffness matrix of the bearing, as mentioned above. q_b denotes the generalized displacement vectors at the support point with 5 DOF.

Accordingly, the generalized reaction load F can be written as $\{F_x, F_y, F_z, M_y, M_z\}$, in which F_x, denotes the axial force. And the force F_y and the moment M_z are in *Y*-direction, the force F_z and M_y are in Z-direction. The changes of the generalized reaction forces of the misaligned bearing along with the misalignment increments are shown in Fig.5.

IOP Conf. Series: Materials Science and Engineering 1081 (2021) 012014 doi:10.1088/1757-899X/1081/1/012014

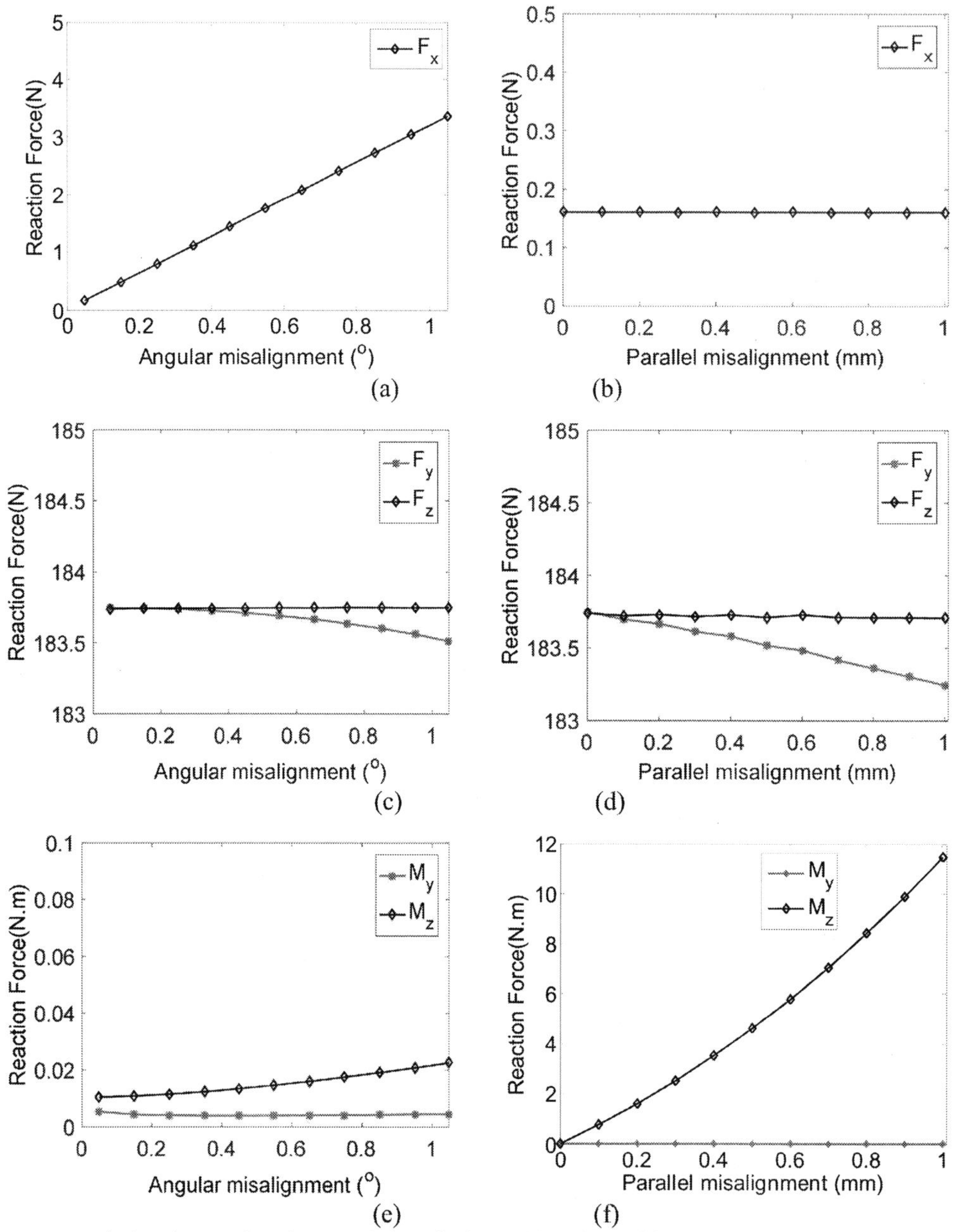

Figure5. Variation in reaction forces of the misaligned bearing with (a, c, e) angular misalignment and (b, d, f) parallel misalignment

As shown in figure5, the radial reaction forces F_y, F_z and the moments M_y, M_z of the misaligned rotor system do not change significantly, but the axial force F_x greatly increases, as the angular misalignment increasing in the range of variation of α =0~1 °. However, when the parallel misalignment varying from 0mm to 1mm, the translational forces F_x, F_y, F_z and the moment M_y, decrease slightly, and the moment M_z around Z-axis increase sharply. It is, therefore, deemed angular misalignment as the cause of axial vibration and parallel misalignment as extra moment loads for misaligned bearing.

4. Conclusions

The dynamic model of a misaligned rigid rotor system with 5 DOF is deduced by means of Lagrangian approach. The effects of misalignment are investigated numerically on the vibration responses of rotor and bearing, the stiffness and reaction force of the misaligned bearing.

The obtained results on such a rigid rotor system with misaligned ball bearing reveal that the vibration of the rotor system seems not to be sensitive to the misalignment, which has a strong influence on the stiffness coefficients of bearing. The transverse vibration amplitudes of rotor and bearing will decrease along the misaligned value, but the variations are not evident. And the axial and radial stiffness coefficient increases greatly with the value of angular and parallel misalignment. But the reaction loads change seriously attributed to the misaligned bearing. The angular misalignment can bring out extra axial force, and the parallel misalignment can give rise to additional moment loads.

Acknowledgements

This work was financially supported by the National Natural Science Foundation of China (Grant No: 51905069) and the Natural Science Foundation of Liaoning Province (No.: 20180550792 and 2019-ZD-0088) and the High-level Talents of Dalian City (No.: 2018RQ18)

Conflict of interest

The Authors declare that there is no conflict of interest.

References

[1] Zhang C 2020 The Life Prediction of Bearing for Metro Fan Based on EMD Denoise and LSTM Network *Chinese Journal of Turbomachinery* **62** 77-82
[2] Wang F and Liu Y 2016 Solution and Analysis of Bearing Overburning in the Side of Cantilever Centrifugal Fan Impellers *Chinese Journal of Turbomachinery* **58** 91-6
[3] Wu P and Zhao X 2019 Incipient Fault Detection of Rolling Bearing Based on Fuzzy Entropy and Fractal Dimension *Chinese Journal of Turbomachinery* **S1** 71-9
[4] Allaire P, Dimond T and Cao J 2019 Bearing Fix for Large Sub-synchronous Vibration in Steam Turbine-gearbox-generator System *Chinese Journal of Turbomachinery* **59** 39-44
[5] Du G, Wang Y and Yang G 2019 Rotor Dynamics Research of a 165kW Helium Blower with AMBs *Chinese Journal of Turbomachinery* **61** 34-8
[6] Piotrowski J 2007 *Shaft Alignment Handbook (3rd edition)*: Marcel Dekker Inc.)
[7] Gibbons C 1976 Coupling Misalignment Forces. In: *Proceedings of 5th Turbomachinery Symposium,* pp 111-6
[8] Palazzolo A, Locke S and Calistrat M 1992 Gear Coupling Misalignment Induced Forces and Their Effects on Machinery Vibration. In: *Proceedings of the 21st Turbomachinery Symposium,* pp 83-96
[9] Xu M and Marangoni R 1994 Vibration analysis of a motor-flexible coupling-rotor system subject to misalignment and unbalance, part I: theoretical model and analysis *Journal of Sound and Vibration* **176** 663-79
[10] Xu M and Marangoni R 1994 Vibration analysis of a motor-flexible coupling-rotor system subject to misalignment and unbalance, part II: experimental validation *Journal of Sound and Vibration* **176** 681-91
[11] Sekhar A and Prabhu B 1995 Effects of coupling misalignment on vibrations of rotating machinery *Journal of Sound and Vibration* **185** 655-71
[12] Ganeriwala S, Patel S and Hartung H 1999 The truth behind misalignment vibration spectra of rotating machinery. In: *Proceedings of Intentional Modal Analysis Conference,* pp 2078-205
[13] Lee Y and Lee C 1999 Modelling and vibration analysis of misaligned rotor-ball bearing systems *Journal of Sound and Vibration* **224** 17-32
[14] Lees A 2007 Misalignment in rigidly coupled rotors *Journal of Sound and Vibration* **305**

[15] Hu W, Miah H and Feng N 2000 A rig for testing lateral misalignment effects in a flexible rotor supported on three or more hydrodynamic journal bearings *Tribol Int* **33** 197-204
[16] Hussain K and Redmond I 2002 Dynamic response of two rotors connected by rigid type mechanical coupling with parallel misalignment *Journal of Sound and Vibration* **249** 483-98
[17] Redmond I 2010 Study of a misaligned flexibly coupled shaft system having nonlinear bearings and cyclic coupling stiffness-Theoretical model and analysis *Journal of Sound and Vibration* **329** 700-20
[18] Hariharan V and Srinivasan P 2011 Vibration analysis of parallel misaligned shaft with ball bearing system *Journal of Science & Technology* **33** 61-8
[19] Pennacchi P, Vania A and Chatterton S 2012 Nonlinear effects caused by coupling misalignment in rotors equipped with journal bearings *Mechanical Systems and Signal Processing* **30** 306-22
[20] Lim T and Singh R 1990 Vibration transmission through rolling element bearings, Part I: bearing stiffness formulation *Journal of Sound and Vibration* **139** 179-99
[21] Lim T and Singh R 1990 Vibration transmission through rolling element bearings, Part II: system studies *Journal of Sound and Vibration* **139** 201-25

Research on the Adjustable Guide Vane in a Compressor Sector Cascade

W F Xu[1], P Sun[2] and G G Yang[1]

[1] Dalian Maritime University, Dalian, China
[2] Civil Aviation University of China, Tianjin, China

E-mail: xuwenfeng@dlmu.edu.cn

Abstract. In order to simulate the inlet flow angle and Mach number of the tested blade, a row of guide vanes is set before the tested blade during the sector cascade experiment. In this paper, a kind of adjustable guide vane (AGV) is designed, which can rotate continuously and change the inlet flow angle of the tested blade conveniently and quickly. The variation rule of the clearance between the AGV and the end wall is introduced. The influence of the clearance on the flow field structure under different Mach number and incidence angle is studied by numerical and experimental methods. It is found that the clearance of the leading edge is small at the top and the root, and the leakage vortex is mixed with the passage vortex. The interaction between the leakage vortex and the passage vortex in the trailing edge results in the decrease of Mach number and the increase of flow angle at the AGV outlet. The clearance has little effect on the span of 20% - 80%. At the same incidence angle condition, with the increase of Mach number, the influence range of leakage vortex is expanded, and the variation magnitude of flow angle and Mach number at the top and root position is increased; At the same Mach number condition, with the decrease of the AGV rotation angle from positive to negative, the top clearance is increased gradually, and the root clearance is decreased gradually. The separation range at the top is expanded gradually, which results in the variation magnitude of flow angle and Mach number is increased, but the effect is opposite at the root of the AGV.

1. Introduction

With the high-speed development of the aerospace industry, there are more and more aerodynamic research methods for aero-engine, which are mainly divided into experimental research, theoretical analysis and numerical research. Wind tunnel experiment is one of the important research methods used to verify the correctness and reliability of the theoretical analysis and numerical calculation results, which has been widely used by researchers around the world [1]. The linear cascade can conveniently and quickly obtain some basic phenomena in supersonic and transonic cascades [2-3].

With the development of research on the three-dimensional flow field, the strong three-dimensional flow characteristics of the flow field in the linear cascade cannot be fully displayed [4-6]. In addition, the influence of inlet distortion on the performance of compressors needs to be studied through full-scale stage experiments [7]. However, the cost of the full-scale stage experiments is high, and the test method is difficult. Therefore, some researchers have made improvements on the basis of the linear wind tunnel to obtain the sector cascade wind tunnel, which made the measurement of the full three-dimensional flow field convenient and cheap [8-9].

In the process of wind tunnel cascade experiment, it is necessary to change the inlet Mach number and the flow angle of the cascade. The flow angle can be adjusted by changing the turntable of the tested rig in the linear cascade experiment [10-11]. In the sector cascade experiment, if the flow angle needs

IOP Conf. Series: Materials Science and Engineering 1081 (2021) 012015 doi:10.1088/1757-899X/1081/1/012015

to be changed, it is necessary to machine the tested blade with different stagger angles [12]. Liu J M et al [13] used the rectifier cascade to achieve the change of inlet flow angle during the study of turbine guide vane on the sector cascade. Li L L [14] and Sun P [15] et al. first proposed a sector cascade test method, in which a guide vane was installed before the stator to simulate the inlet flow conditions of the stator in the compressor stage. Sun P et al. [16] found that the clearance between the guide vane and the end wall would affect the outlet flow field and the test results in the sector cascade experiment with adjustable guide vanes. Tian C R et al [17] found that the leakage flow interfered with the main flow during the study of the effect of top clearance on the performance of the axial-flow fan, which affected the performance of the fan. Wu Y H et al [18] found that reducing the mixing of top leakage flow and secondary flow in the passage can increase the flow rate and reduce the flow loss of the compressor. In addition, it was found in the previous studies that the leakage vortex generated by the clearance can make the vortex structure change significantly at the top of the compressor stator, affecting the aerodynamic parameters of the outlet [19-20].

Therefore, an adjustable guide vane is designed in this paper, which can rotate around the rotation axis. The contact surface of the guide vane fixed block and the end wall is designed as spherical surface to reduce the clearance. Experimental and numerical methods were used to study the effect of leakage vortex on the flow field structure of the top and root under different Mach numbers and incidence angles.

2. Experimental facility and the guide vane structure

2.1. Sector cascade tunnel

The experimental investigation was carried out in a transonic wind tunnel, as shown in Fig.1. The wind tunnel is a continuous open jet type wind tunnel. After the compressed air passes through the steady flow section, it is accelerated through the contraction section and finally enters the test section.

Figure 1. Sketch of wind tunnel

The physical picture is shown in Fig. 2.The sector cascade consists of 15 adjustable guide vanes with a circumferential angle of 110° . The flow field was measured within the range of 8° circumferential angle of the middle outlet. The distribution of measurement points is shown in Fig. 2, there are 25 positions in the spanwise and 17 positions in the pitchwise, with a total of 425 measurement points.

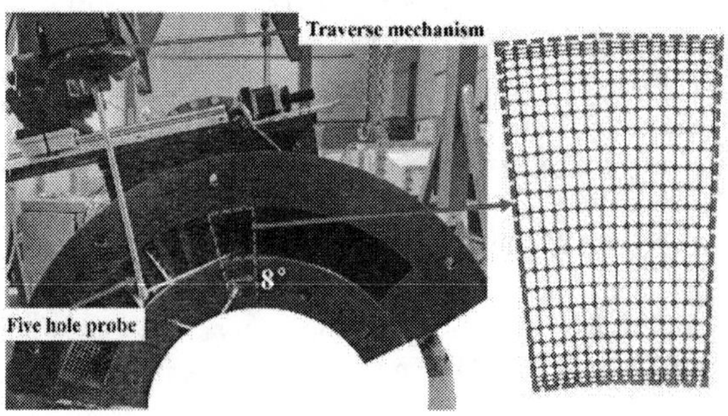

Figure 2. Tested cascade and range

2.2. The structure of the AGV

The schematic diagram of the AGV is shown in Fig.3. The stator represents the tested blades. The AGV is fixed on the end wall through the fixed block, and the AGV can rotate with the central axis of the fixed block. The inlet flow angle of the tested blade can be changed by means of the adjustment of the AGV stagger angle. The adjustment direction of the stagger angle is clockwise seen from the top, the incidence angle of the stator is defined as positive, otherwise, it is negative.

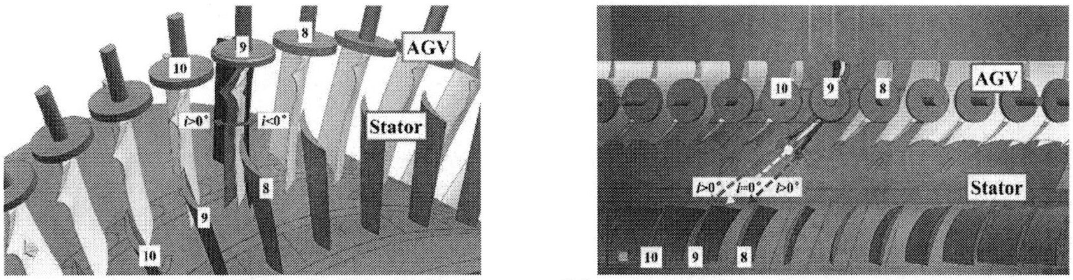

Figure 3. Schematic diagram of the agv stagger angle

As shown in Fig. 4, the meridian of the passage is straight, resulting in bumps and clearances between the fixed block and the end wall during the rotation of the AGV. Therefore, the meridian of the hub and casing is modified to a circular arc. Similarly, the meridian of the fixed block is also replaced by an arc, as shown in Fig. 4. ORI and OPT respectively represent the prototype and the modified structure. These two arcs take R1 and R2 as radii and take point e as the center, respectively, passing through points a (c) and b (d). The modified structure can make the fixed block fit the passage and eliminate the clearance, as shown in Fig. 5.

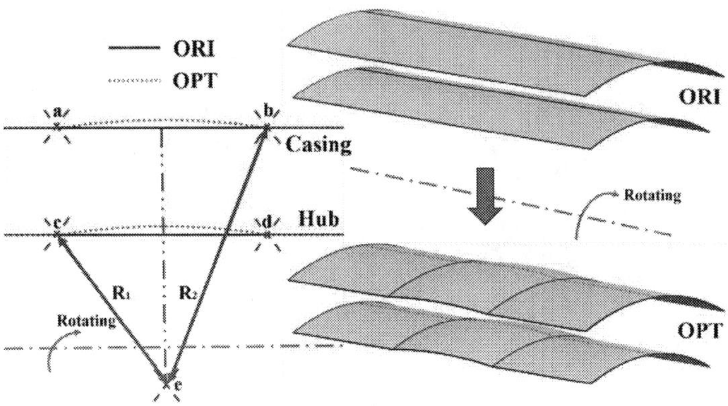

Figure 4. Structure diagram of the casing and hub

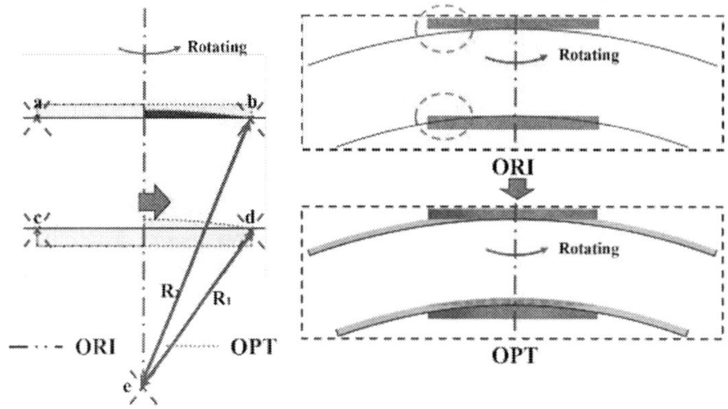

Figure 5. Structure diagram of the fixed block

However, limited by the solidity of the cascade, the fixed block cannot completely connect the top and root of the AGV, as shown in Fig. 6. There is still a clearance between the top and root of the AGV and the end wall. As the AGV is rotated, the clearance size also is changed, as shown in Fig.7. As the incidence angle of the AGV decreases from positive to negative, the top clearance is increased gradually, while the root clearance is decreased gradually.

After measuring the AGV clearance, it is found that the top clearance height of the leading edge is 0.5mm, the root clearance height is 0.3mm. There is no obvious change in the height of the clearance along the chordwise. The top clearance height of the trailing edge gradually increased from 0.5 to 1.4mm along the chordwise, and the root clearance height gradually increased from 0.4 to 1.3mm. According to the actual clearance height, the calculation model is modified to ensure that the structure of the actual model is approximately the same as the calculation model.

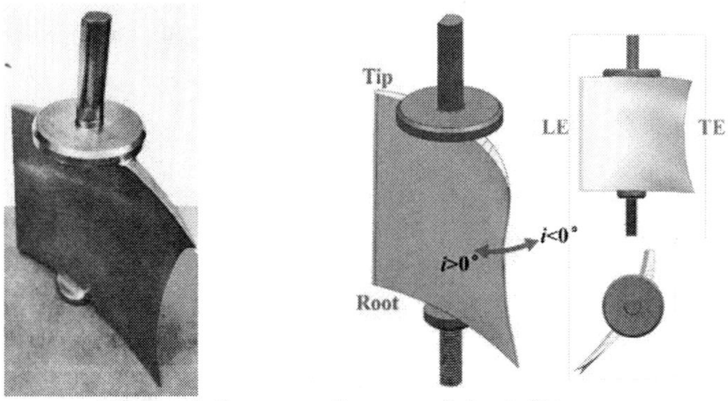

Figure 6. Structure diagram of the AGV

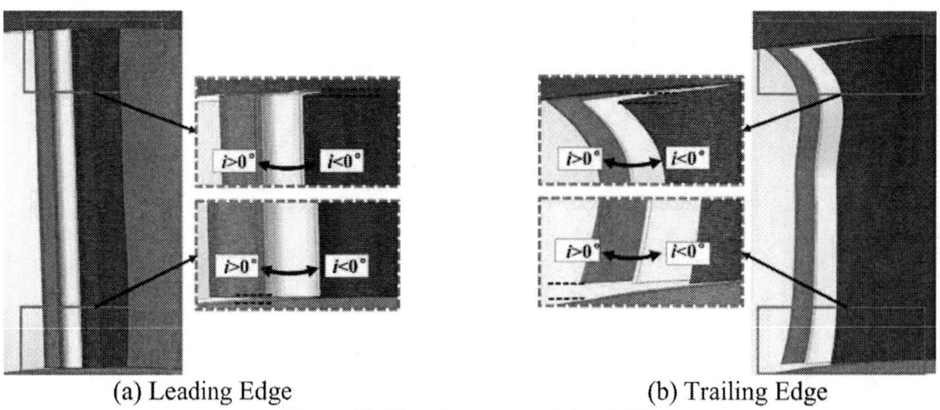

(a) Leading Edge (b) Trailing Edge

Figure 7. The clearance of the AGV

IOP Conf. Series: Materials Science and Engineering 1081 (2021) 012015 doi:10.1088/1757-899X/1081/1/012015

3. Numerical procedures

3.1. Numerical simulation methods

All of the numerical results presented in this paper were obtained by ANSYS CFX software. The meshes used in the investigation were generated by the grid generator ANSYS ICEM. The hexahedral structural meshes were generated in O-type topology. As shown in Fig.8, the calculation domain mesh corresponding to the physical structure of the piece is given. All of them had a similar structure, the total number of nodes was between 16 and 18 million. To ensure the accuracy of calculation, the y+ adjacent to the wall was approximately 10 to meet the need of the k-ε turbulence model. No-slip adiabatic wall condition was defined on the blade and end wall. The inlet pressure profile obtained from the cascade experiment was specified at the domain inlet. The static pressure (101325Pa) was used to set the outlet boundary condition.

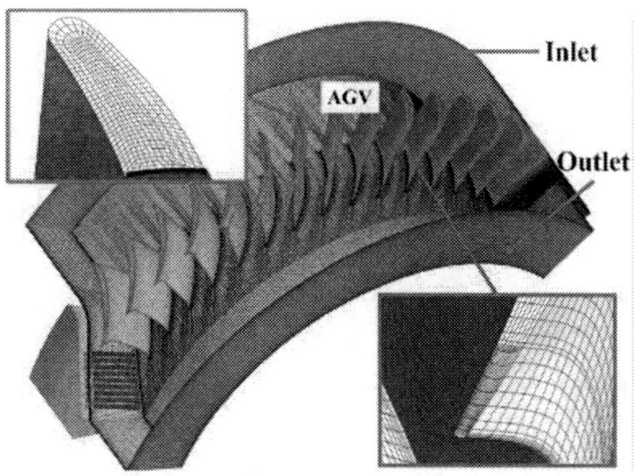

Figure 8. Computational mesh

3.2. Periodicity of the sector cascade

In this paper, the No.8 AGV in the middle region of the sector cascade is selected to analyze the flow field. In order to explore the repeatability of the flow field structure between the No.8 AGV and the adjacent blades, the aerodynamic parameter distribution of the cascade outlet is shown in Fig. 9. From Fig. 9(a), it can be found that there are three complete flow passages in the analyzed section, corresponding to No. 7, No. 8, and No. 9 AGV respectively. As shown in the contour of the Mach number and flow angle, there is no significant difference in the range and distribution of the corner region and wake area of each blade outlet.

The circumferential distribution curves of Mach number and flow angle at different spans are shown in Fig. 9(c). The outlet parameters corresponding to the three blades change periodically, which has good repeatability at different spans. It is verified that the flow field structure of the analyzed blades and the adjacent blades is approximately the same, and the aerodynamic parameters in the central region of the sector cascade are periodically distributed.

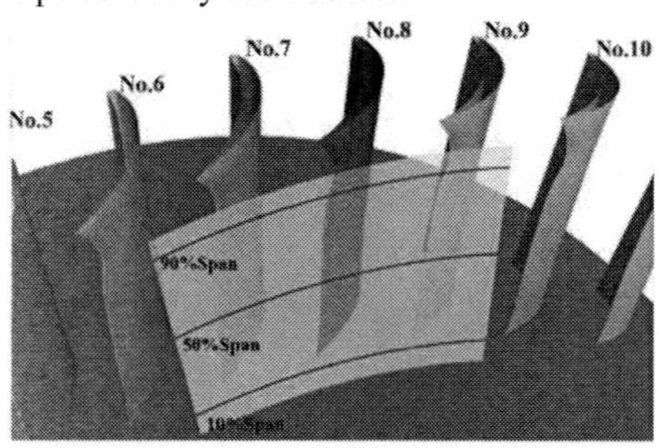

(a) The location of the analyzed section

IOP Conf. Series: Materials Science and Engineering 1081 (2021) 012015 doi:10.1088/1757-899X/1081/1/012015

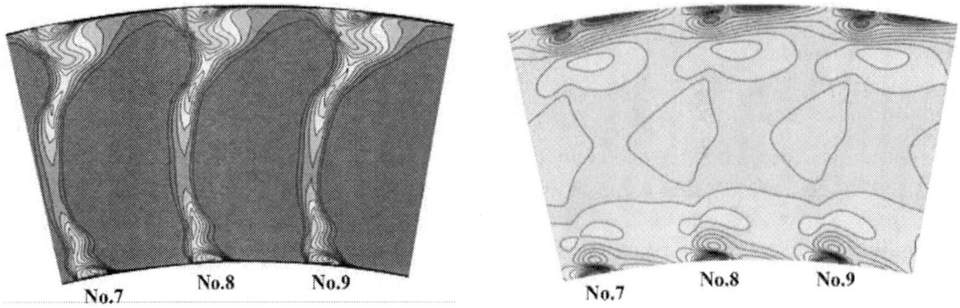

Mach number Flow angle

(b) Contours of aerodynamic parameters

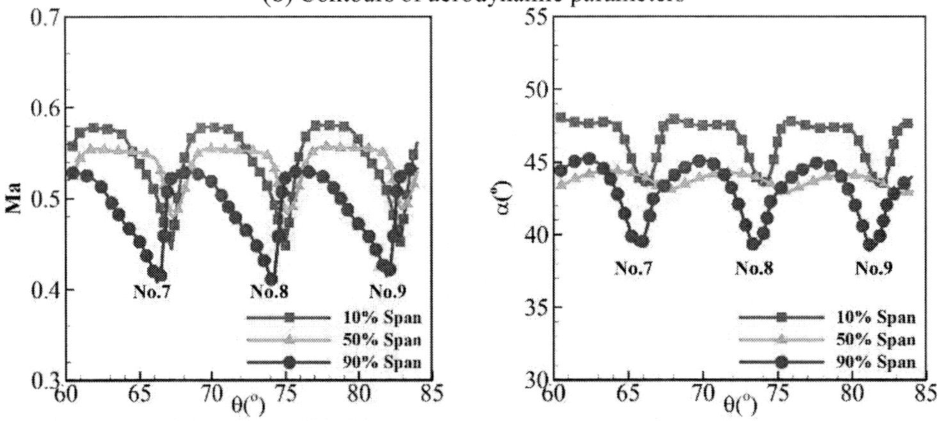

Mach number Flow angle

(c) Distribution of aerodynamic parameters along the circumference at different spans

Figure.9. Flow field structure at the AGV outlet

3.3. Validation of numerical simulation method

In order to validate the accuracy and reliability of the simulation method, the experimental and computational results were compared. The distribution of the flow angle and Mach number at the AGV outlet is shown in Fig.10. Due to the limitation of the test equipment, only the flow field structure in the span of 5-92% was measured during the experiment. It can be seen from the contour that the two results are almost the same, and there is a low-speed region with a large flow angle near the end wall. Compared with the experimental and numerical results, the maximum difference of flow angle is larger than that of Mach number, and the maximum difference is 12% near the tip of the AGV. It can be found from the pitch average curve that spanwise distributions of the experiment are similar to the numerical. The Mach number curve has the same change trend in the range above 65% span, but the difference in flow angle is large. In general, the numerical methods are reliable and can be used for the prediction of the aerodynamic performance in this study.

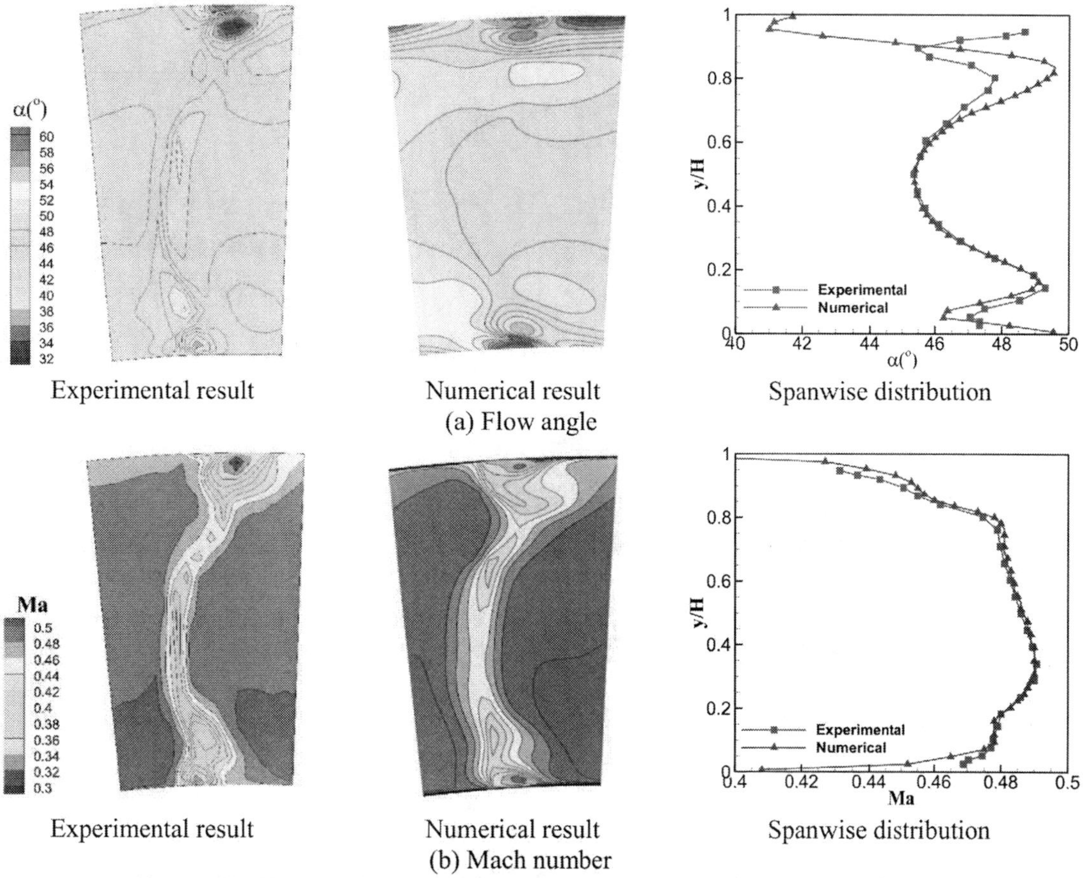

Figure 10. The experimental and computational results of the AGV outlet

4. Results and discussion

4.1. The effect of the AGV clearance on the flow field

Fig. 11 shows the contour and the pitch average curve of Mach number and flow angle at the outlet of guide vane when the incidence angle is 0°. The clearance leads to a low Mach number region at the top and root of the AGV, and the flow angle increases in the corresponding range. From the pitch average curve of the flow angle, it can be found that the clearance has little effect on the range of 17% -92% span. The flow angle increases significantly in the range above 92% span and below 7% span, while the flow angle value is slightly larger than that without clearance in the span of 7% -17%. From the pitch average curve of Mach number, the leakage flow results in the reduction of Mach number in the range above 79% and below 20%. Moreover, the influence range of clearance on the Mach number along the span is larger than the flow angle.

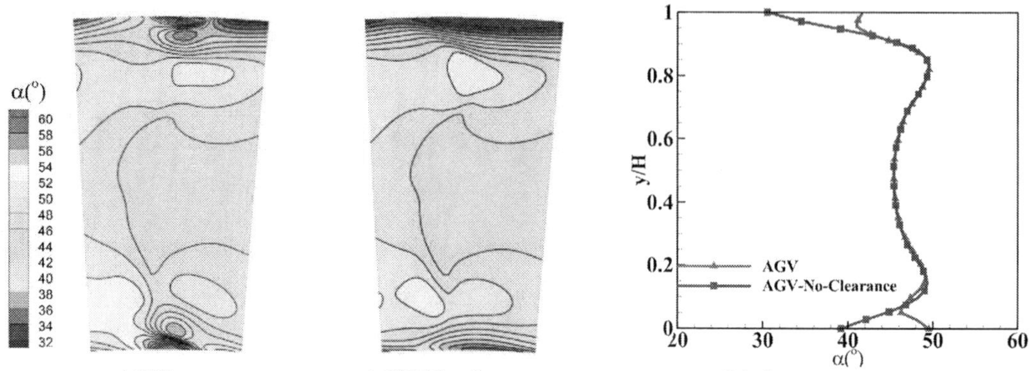

AGV AGV-No clearance Pitch average curve

(a) Flow angle

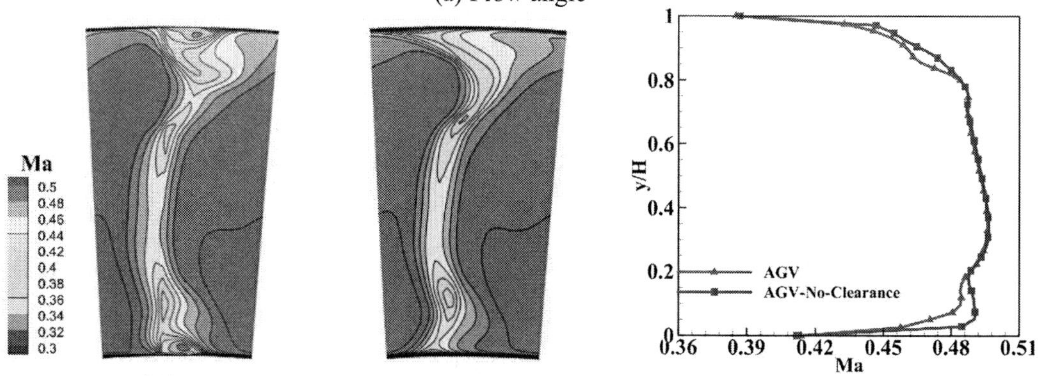

AGV AGV-No clearance Pitch average curve

(b) Mach number

Figure 11. Aerodynamic parameter of the AGV outlet

In order to study the influence of guide vane clearance on flow field structure in detail, Fig. 12 displays the 3D vortex structure in the AGV passage visualized by the iso-surface of Q and renders the iso-surface with the Mach number. It can be found that the clearance leads to the creation of a leakage vortex at the top (TLV) and root (RLV) of the AGV. The leakage flow between the top and the root of the leading edge is mixed with the suction branches of the horseshoe vortex to form the passage vortex. The leakage vortex squeezes out the passage vortex (PV) and narrows its influence range at the top and root of the trailing edge of the AGV. Therefore, it can be found that the leakage vortex has a significant effect on the vortex of the passage vortex. But it has little effect on the location and range of the concentrated shedding vortex (CSV) at the outlet. The outlet flow angle deviates from the main flow direction under the action of the cross secondary flow of the leakage vortex, resulting in the increase of the flow angle and the decrease of Mach number near the top and root.

IOP Conf. Series: Materials Science and Engineering 1081 (2021) 012015 doi:10.1088/1757-899X/1081/1/012015

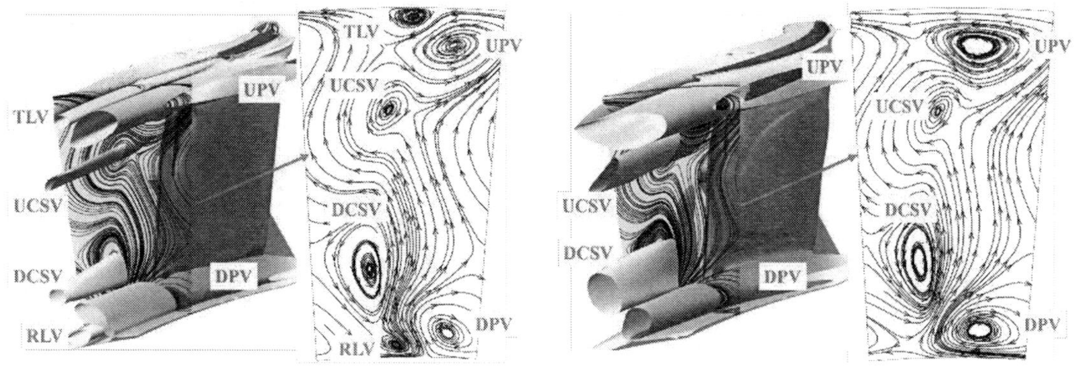

AGV AGV-No clearance

Figure 12. 3D vortex structure in the passage($Q=6.4*10^5$)

4.2. The effect of the AGV clearance at different Mach number on the flow field

The pitch average curve of flow angle and Mach number at different Mach numbers conditions is shown in Fig. 13. The Ma = 0.4 and no Ma = 0.4 respectively represent the results of guide vanes with and without clearance when Mach number is equal to 0.4. As shown in Fig. 13 (a), with the increase of Mach number, the flow angle difference of the guide vane with or without clearance increases gradually in the span above 92%. However, the clearance causes a small change in flow angle with the Mach number under the 17% span. In the span of 17% - 92%, the flow angle of the guide vane with or without clearance is almost the same under different Mach number conditions, and the clearance has no obvious influence on the range.

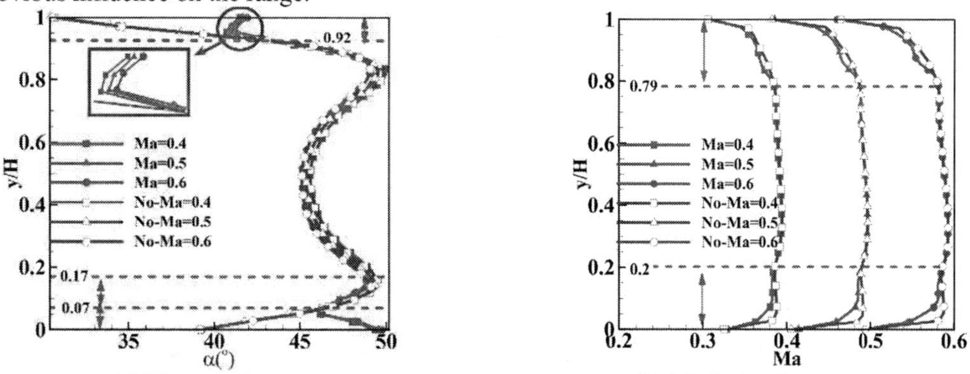

(a) Flow angle (b) Mach number

Figure 13. Aerodynamic parameter of the AGV outlet at different Mach number conditions

It can be seen from Fig. 13(b) that the clearance has little effect on the Mach number in the span of 20% - 79% under different Mach number conditions. The effect mainly exists at the hub and casing. With the increase of Mach number, the difference of Mach number of the same span increases gradually in the span of above 79% and below 20%. The maximum difference is 5.31% when the Mach number is equal to 0.4, and it increases to 5.85% when Mach number is equal to 0.6.

Figure 14 shows the 3D streamlines, the total pressure recovery coefficient contour, and secondary streamlines of the guide vanes at different Mach number. The color of the 3D streamline represents the Mach number. With the increase of Mach number, the speed of leakage flow increases, the strength of leakage vortex increases slightly, and the loss caused by leakage increases. The influence range of leakage vortex increases along spanwise, which leads to the decrease of Mach number around the end wall. The effect of clearance on the concentrated vortex is small, so it has little effect on the aerodynamic parameters in the span of 20%-79%.

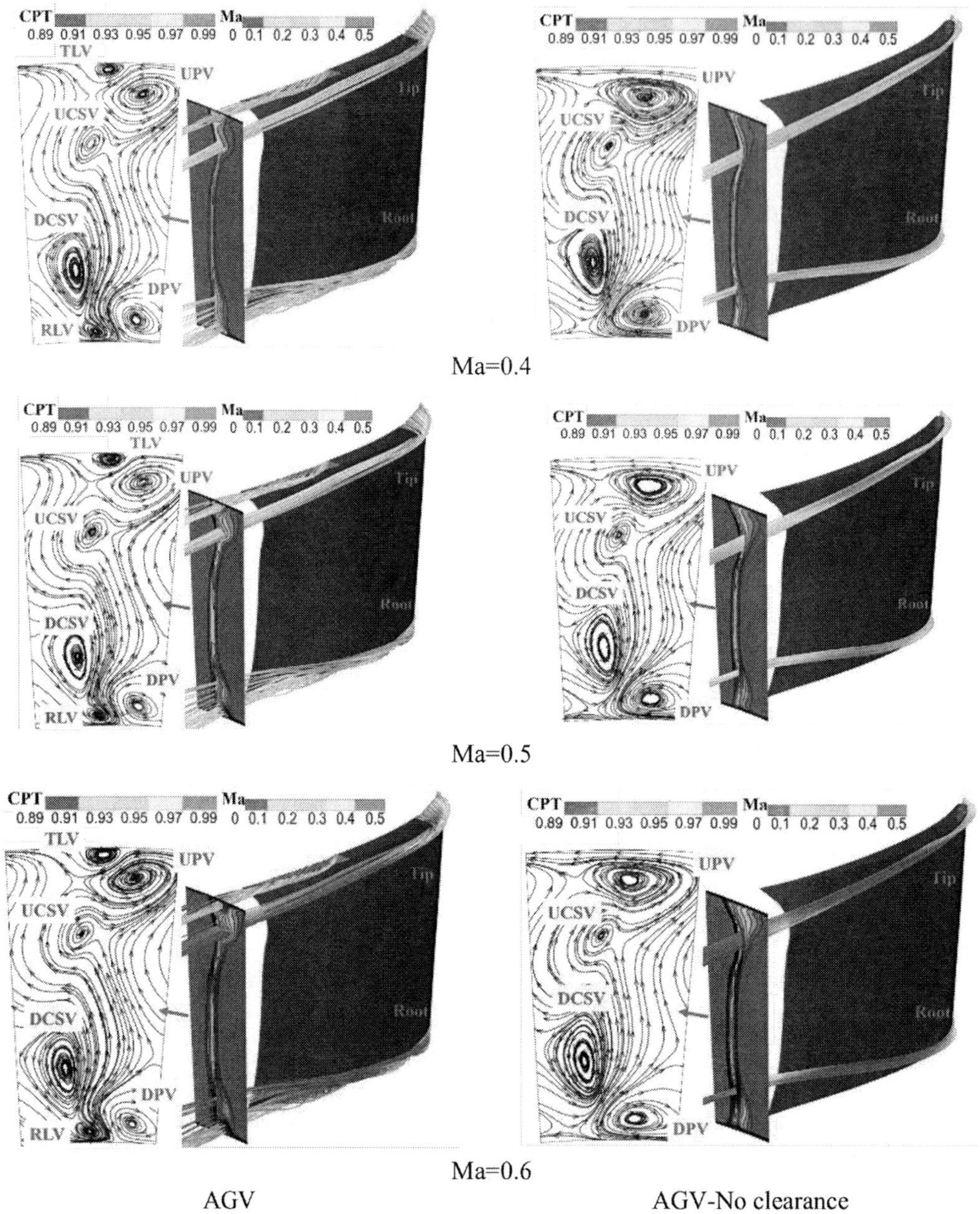

Figure 14. The 3D Streamlines, the total pressure recovery coefficient contour and secondary streamlines of the guide vanes at different Mach number conditions

4.3. The effect of the AGV clearance at different incidence angle on the flow field

As the AGV incidence angle decreases from positive to negative, the top clearance gradually increases and the root clearance gradually decreases. Fig. 15 shows the pitch average curve of the outlet Mach number and flow angle when the incidence angle is +6°, 0 °, and -6 °. The P6, P0, and N6 shown in the figure respectively represent the results of guide vane when the incidence angle is +6 °, 0 ° and - 6 °. It can be seen that clearance has no effect on the flow angle in the span of 18% - 90%. Above 90% span, with the decrease of incidence angle, the increasing variation magnitude of flow angle increases gradually, and the change is opposite under 18% span. From Fig. 15 (b), it can be found that the clearance has no significant effect on the Mach number in the span of 19% - 80%. The influence of clearance on Mach number under 19% span is larger than that above 80% span. With the decrease of incidence angle, the variation magnitude of the Mach number decrease gradually decreases near the root.

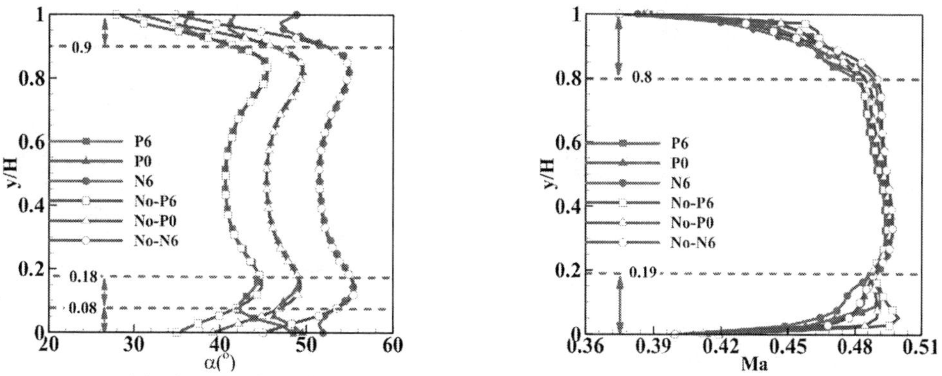

(a) Flow angle (b) Mach number

Figure 15. Aerodynamic parameter of the AGV outlet at different incidence angles conditions

It can be seen from Fig. 16 that as the incidence angle decreases, the separation range of the leakage vortex increases gradually at the top of the AGV outlet, while that decreases gradually at the root. The leakage vortex causes the airflow to deviate from the main flow direction near the end wall, which results in the increase of the flow angle compared with that of the guide vane without clearance. Under different incidence angle conditions, the leakage vortex has a significant inhibitory effect on the passage vortex. Although the separation range of the passage vortex is reduced, it can be found from the 3D streamline that the leakage flow causes the airflow mixing resulting in a reduction in airflow velocity near the end wall. In addition, the clearance has little effect on the middle region of the flow passage, and there is no obvious change in the position and range of the wake and the concentrated shedding vortex.

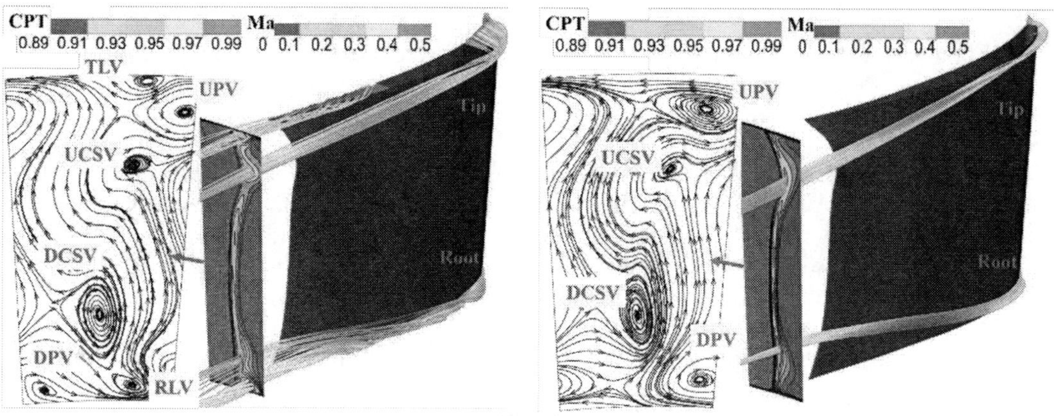

(1) P6

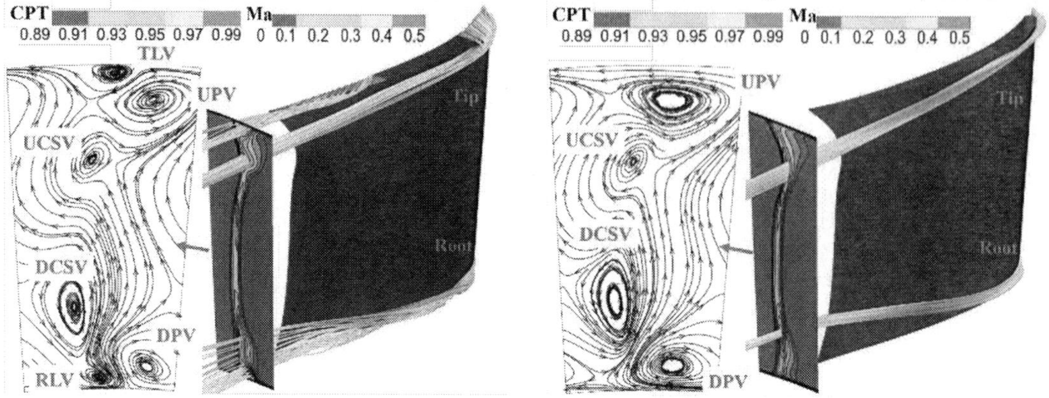

(2) P0

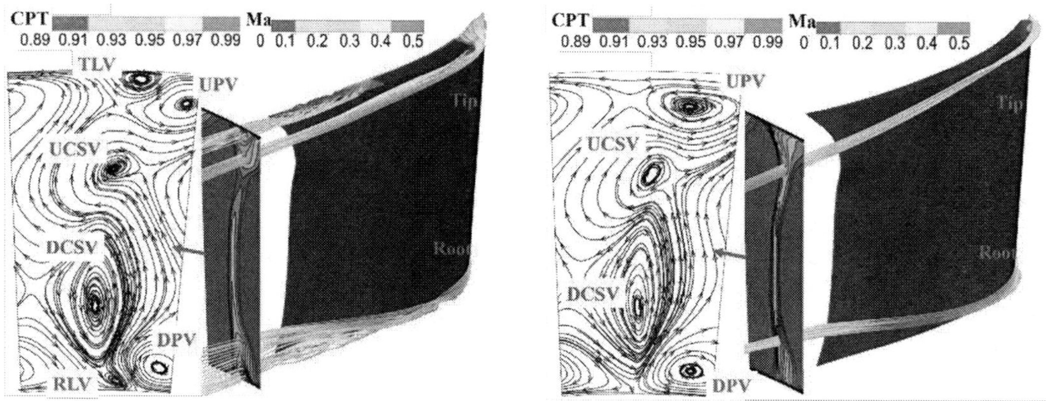

(3) N6

AGV AGV-No clearance

Figure 16. The 3D streamlines, the total pressure recovery coefficient contour and secondary streamlines of the guide vanes at different incidence angle conditions

5. Conclusions

The adjustable guide vane (AGV) is designed based on the transonic sector cascade wind tunnel. The variation rule of the AGV clearance is introduced. And the influence of the clearance on the flow field structure is studied by numerical and experimental methods. The results are summarized as follows:

1) The clearance between the AGV and the end wall is inevitable. The contact surface between the fixed block and the end wall is designed as a spherical surface to reduce the clearance.

2) Limited by the solidity of the cascade, there is still a clearance between the top and root of the AGV and the end wall. As the incidence angle of the AGV decreases, the top clearance gradually increases and the root clearance gradually decreases.

3) The clearance has no significant effect on the flow field structure in the span of 20% -79%. The leakage vortex leads to the increase of the flow angle and the decrease of the Mach number near the top and root of the AGV. The influence of the clearance on the Mach number is wider than the flow angle. With the increase of Mach number, the variation magnitude of the flow angle and Mach number gradually increases near the casing and hub.

4) The clearance has no obvious effect on the flow field structure in the span of 19%-80% under the same Mach number and different incidence angle conditions. With the decrease of the incidence angle, the top clearance gradually increases, which leads to an increase of the leakage vortex and the variation magnitude of the flow angle and Mach number. The rule is just the opposite at the root.

Acknowledgments

The author(s) would like to thank the members of the Small Marine Gas Turbine Technology Key Laboratory of Dalian Maritime University. This paper was supported by the National Natural Science Foundation of China (No. 51576024, 51436002), the Double First Class Construction Special Innovation Project of Dalian Maritime University (No. BSCXXM008).

References

[1] Lorenz M, Schulz A and Bauer H J 2010 Experimental study of surface roughness effects on a turbine airfoil in a linear cascade: Part II-aerodynamic losses *Int. ASME Turbo Expo: Power for Land, Sea, and Air* GT2010-23801

[2] OTTAVY, Xavier and Li P 2011 Experimental study of corner stall in a linear compressor cascade *J. Journal of Aeronautics and Astronautics* **24(3)** 235-242

[3] Chen S, Zhou Z and Meng Q 2017 Experiment study of the winglet-cavity top on the aerodynamic performance in a turbine cascade *J. Proceedings of the Institution of Mechanical Engineers* **231(7)** 676-685

[4] Li T H and Ye D J 1992 Experimental investigation of the three-dimensional flow in an annular compressor cascade at large incidence *J. Journal of Thermal Science* **1** 3-10

[5] Qu X L, Yu Y S, Lu J L and Zhang W Q 2019 The axial flow fan design for an acoustic wind tunnel *J. Chinese Journal of Turbomachinery* **61(6)** 56-60

[6] Liu B J, Zhang S and Yu X J 2018 Optimization design of a single stage low speed simulation axial compressor test facility *J. Journal of Aerospace Power* **33(7)** 1665-1675

[7] Song G X, Li J., Zhou Y T and Nie Y Z 2017 The experiment and simulation swirl distortion for axial compressor *J. Chinese Journal of Turbomachinery* **59(5)** 1-12

[8] Jiang Z L, Ling D J and Wang H 2006 Experimental investigation on the sector cascade of high pressure turbine guide *J. Gas Turbine Experimental and Research* **19(1)** 17-20

[9] Zhu L, Zhang J and Qing X J 2014 Experimental and improved design verification on the sector cascade of high pressure turbine nozzle *J. Gas Turbine Experimental and Research* **27(5)** 19-24

[10] Keerthi, M C, Rajeshwaran M S, Kushari A and De A 2016 Effect of leading edge tubercles on compressor cascade performance *J. AIAA Journal* **54(3)** 1-12

[11] Gao L M and Cai M 2018 Experimental investigations of compressor airfoil on cascade wind tunnel *J. Chinese Journal of Turbomachinery* **60(4)** 9-15

[12] Wei L, Fu J G and Chen L 2016 Experimental investigation on the annular sector cascade of a high Endwall angle turbine *Int. ASME Turbo Expo: Turbomachinery Technical Conference and Exposition* GT2016-57392

[13] Liu J M, Wang D and Ma Y F 2012 Experimental and numerical simulation of low-pressure turbine guide blade sector cascade *J. Aeroengine* **38(6)** 20-23

[14] Li L L, Pan R C, Li T and Sun P 2014 Investigation on the sector experimental method and numerical study on the aerodynamic characteristic of a fan stator blade *J. Journal of Dalian Maritime University* **40(4)** 97-102

[15] Teng L Z, Sun P and Fu W G 2017 Simulation research on the influence of baffle structure on flow periodicity in sectorial cascade *J. Thermal power engineering* **32(6)** 44-51

[16] Fu W G, Sun P and Zhong J J 2018 Aerodynamic characteristic research on final stage stator of a highly loaded fan *J. Journal of Thermal Science* **27(6)** 563-571

[17] Tian C R, and Li K B 2018 Large-Eddy simulation study on effect of blade top clearance on performance of the axial flow fan *J. Chinese Journal of Turbomachinery* **60(1)** 33-38

[18] Gan W J, Wu Y H and Chen Z Y 2019 Numerical investigation of centrifugal impeller stability enhancement and flow field using a self-recirculation casing treatment *J. Chinese Journal of Turbomachinery* **61(6)** 1-9

[19] Xu W F, Sun P and Zhong J J 2017 Numerical study for the influence of stator top clearance height on compressor performance *J. Journal of Dalian Maritime University* **43(3)** 63-70

[20] Xu W F, Sun P and Huang L L 2018 Influence of stator top clearance on compressor flow structure under near stall condition *J. Journal of Dalian Maritime University* **44(4)** 74-82

IOP Conf. Series: Materials Science and Engineering 1081 (2021) 012016 doi:10.1088/1757-899X/1081/1/012016

Effect of Guide Vane Openings and Different Flow Rates on Characteristics in Pump Mode of Pump-turbine with Splitter Blades

Jinfeng Zhang[1], Yajing Xiao[1], Shibin Jin[2], Haikun Cai[1], Haiqing Song[1]

[1]Research Center of Fluid Machinery Engineering and Technology, Jiangsu University, Zhenjiang, China
[2]Institute of Quality Supervision and Inspection, Wenling, China

E-mail:zhangjinfeng@ujs.edu.cn

Abstract. With the development of smart power grid, pumped storage has been developing on a large scale, and the pump-turbine with splitter blades is widely used. A combination of experimental and CFX numerical simulation was carried out to investigate the influence of guide vane openings on the flow characteristics in pump-turbine with splitter blades at different flow rates. The results show that change in the guide vane openings and flow rates have significant impacts on the internal flow characteristics in pump mode. The findings also revealed a poor flow stability in the condition of a small opening at a high flow rate and the condition of large opening at a low flow rate. Under the influence of pressure fluctuation, there exists local high-speed flow at the outlet of the draft tube and mid-span between the runner and the guide vane.

1. Introduction

Pumped storage power station uses the surplus power at the trough to pump water to the upstream reservoir for storage, and then uses it to supplement power at the peak load. It can also cooperate with other grid systems to carry out peak load regulation and valley filling [1-3]. It is recognized as a reliable peak load regulation power source in the world, so the stability of unit operation is crucial. The flow characteristic of conventional pump-turbine usually produces pressure disturbance and kinetic energy loss when it is in the working condition of the pump. However, its operational performance is characterized by challenges such as vibration, noise, mechanical fatigue, and so on, which are due to the unsteady flow state, pressure fluctuation, and energy loss during the operation. The interaction effect due to the guide vane opening and flow rate contributes to the pump external performance and flow characteristic [4-8].

Nicolet [9] carried out a one-dimensional acoustic simulation of the rotor-stator interaction of a mixed-flow pump-turbine under the resonant condition, and revealed the acoustic resonance mode and frequency characteristics generated by the rotor-stator interaction in the volute region, which laid a foundation for the simulation of the full acoustic model. Egusquiza [10] monitored the vibration of water turbine through the remote monitoring and control system and found that the vibration of the pump-turbine is much larger than the conventional hydraulic turbine, which is mainly because of rotor-stator interaction in the pump-turbine, especially in the working conditions of the turbine with a heavy load. Rodriguez[11] adopted the measurement method of sensor rotating along the shaft to monitor the rotor-stator interaction of pump-turbine, which avoids the measurement influence brought by bearing response and provides a more reliable measurement method for monitoring the rotor-stator interaction characteristics. Casartelli [12] investigated a CFD simulation of the flow field and pressure fluctuation change combined with FSI technology during the guide vane closing process, which provided a convenient and accurate CFD strategy for analyzing the transient characteristics of the unstable conditions of the pump-turbine as well as developing the guide vane control rules. Yefang Wang [13] resolve the vortical flows in a typical single-stage side channel pump by using the unsteady Reynolds-averaged Navier–Stokes (URANS) and three hybrid Reynolds-averaged Navier–Stokes-large eddy simulation (RANS-LES) models, to evaluate the suitability of these advanced turbulence models in predicting the pump hydraulic performance and unstable swirling flows. Yonezawa [14] studied the phase resonance phenomenon in centrifugal rotating machinery. By comparing and analyzing the phase resonance in the working condition of the pump and turbine respectively, they revealed the important influence of the direction of rotor-stator interaction and the radial velocity pulsation on the phase resonance. Pochylý [15] studied experimentally the influence of inserting shunt blades on the pressure fluctuation of pump-turbine passage. It is observed that the shunt blade could enlarge the high efficiency area of the pump-turbine, and the pressure pulsation could also disperse to a lower frequency range in the working condition of the turbine. Meanwhile, the pressure fluctuation could be significantly reduced in the working condition of the pump, especially in the unstable area. Shigemitsu [16] studied the performance and internal flow characteristics of a small centrifugal pump with a large blade outlet angle by using a shunt blade using experimental and numerical works. Namazizadeh [17] discussed the influence of diverging blades with different lengths, leading edge positions and distances on centrifugal pump head and efficiency, and adopted experimental design technology (DOE) to establish optimal design space, and used response surface method to calculate the optimal geometric size. Fan Zhang [18] reveals the exact regions of high hydraulic losses for side channel pump models using the entropy loss production method and contributes to the fundamental knowledge of side channel pumps by providing the regions of high-energy losses for further optimization of the main geometrical parts especially the impeller.

Several studies have demonstrated that numerical simulation had a reasonable accuracy in calculating the flow characteristic of the pump-turbine. Therefore, unsteady numerical simulation, performance test, and CFX test for pump-turbine with different splitter blades are carried out at different guide vane opening at different operating flow rates in the present study. The test results are compared with the numerical works to illustrate the reliability of the calculation. Furthermore, using analyzing the pressure fluctuations, streamlines, and turbulence kinetic energy in the different parts of the pump-turbine, the impacts of the guide vane opening on the flow characteristic at different flow rates are investigated to provide some references in the design of guide vane opening.

2. Numerical simulation and tests

2.1. Description of geometry model

The model of pump-turbine with splitter blades in this study is scaled by 1:7.47 for a pump-turbine in a domestic pumped storage power station. The basic parameters of the pump-turbine model are shown in Table 1.

Table 1. Basic parameters of the pump-turbine model.

Parameter	Data
Number of stay vanes	16
Number of guide vanes	16
Number of impeller vane	10
impeller outlet minimum diameter D_1(mm)	300
Impeller outlet maximum diameter D_2(mm)	584

runner outlet diameter(mm)	55(working condition)
volute outlet diameter (mm)	320(working condition)
Draft tube outlet diameter (mm)	600(working condition)

As pump-turbine is a reversible unit, two working modes should be taken into account. The pump mode is that the runner rotates anticlockwise, and the flow firstly enters from the draft tube. After the work and activity of the runner, it is then guided into the volute by the double-row cascade with a fixed guide blade. At that time, the casing converts the kinetic energy of the fluid to static pressure energy, to pump the water to the upstream reservoir. Similarly, the turbine mode is that the flow enters via the inlet nozzle. Flow rate of water will get reduced along the length of the volute, thus decreasing the area of volute will ensure a uniform velocity of the flow as it enters. Stay vanes and guide vanes are fitted at the entrance of the runner to convert one part of pressure energy to kinetic energy. Stay vanes steers the flow towards the runner section thus reducing the swirl at the inlet flow. In the runner, water enters radially and leaves axially, while both impulse force and lift force make the runner rotate by clockwise. During the flow, water glides over runner blades. The blades have a thin airfoil cross-section and when water flows over it, a low pressure is produced on one side and high pressure on the other side which subsequently results in a lift force which will hit and produce an impulse before leaving the runner and ends up flowing out from the draft tube. Lastly, the runner is connected to a generator via a shaft for electricity production. In this paper, Creo software is used to carry out 3D modeling and the full flow geometric model is shown in Figure 1.

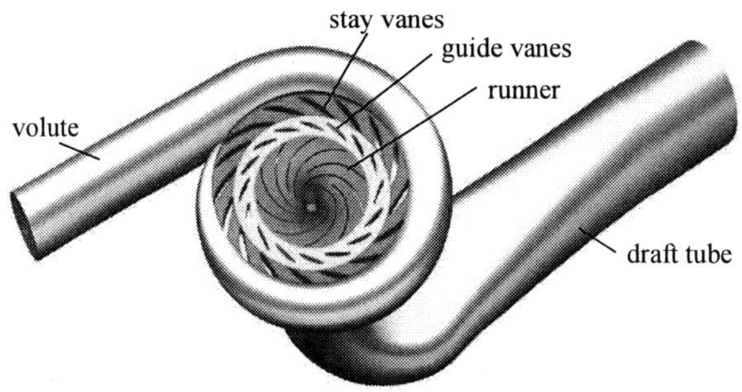

Figure 1. Full flow geometric model.

2.2. Transient computation procedure

The characteristics of structured mesh is that the nodes can be named orderly with fixed rules, which means that for small computations, the grid quality becomes easy to control, the boundary layer grid is guaranteed, and the convergence of computation is enhanced. The whole computational domain was split into five parts for modeling which are the volute, stay vanes, guide vanes, draft tube, and runner. The hybrid structured meshes were generated using mesh generation tool ANSYS-ICEM to the grid quantity, quality, and boundary layer to meet the requirements. Three groups of meshes were categorized into three groups of guide vane passages at the opening of 9.8, 17.5°and 24.8° selected in this paper. The mesh quality was higher than 0.5, and the angles were all greater than 25°.

Figure 2 shows the boundary layer details of the volute section, the boundary layer details of the runner blades, and the single channel grid of the guide blade at different guide vane opening 9.8°, 17.5° and 24.8° respectively.

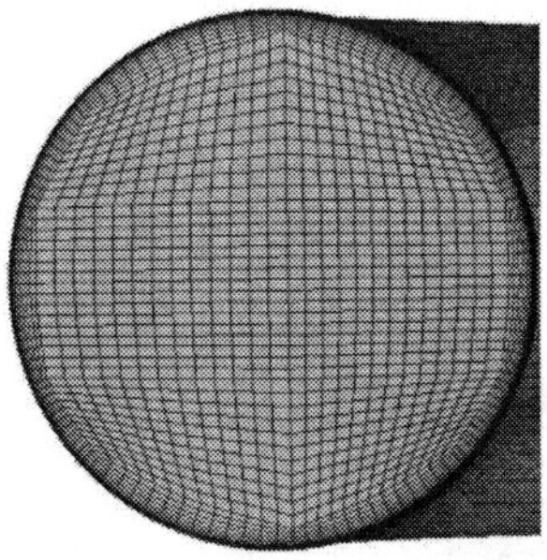

(a) Boundary layer of the volute section (b) Boundary layer of runner blades

9.8° 17.5° 24.8°

(c) Single channel mesh of the guide blade

Figure 2. Mesh.

The numerical calculations for different schemes were performed in the fluid simulation code ANSYS CFX by using the unsteady Reynolds-Averaged Navier-Stokes which is based on the finite volume method. [19.20] The boundary conditions in the calculation are shown in the Table 2.

Table 2. Basic parameters of the pump-turbine model.

Boundary Conditions	Settings
Turbulence model	Shear Stress Transport (SST)
Reference pressure	101,325Pa
Inlet	Total pressure
Outlet	Mass flow rate
Interface	Frozen-rator
Wall roughness	Smooth
Mass and momentum	No Slip
Advection Scheme	Upwind 、 High Resolution
Convergence criterion	10^{-5}

2.3. Mesh independence analysis

Since the number of mesh has a great impact on the calculation results, the mesh number independence analysis was performed for 5 sets of grids. According to the rated operating condition of 17.5° opening of the guide vanes, 8 groups of mesh number schemes were set at uniform intervals between 4 million and 10 million as shown in Table 3.

With the increase of mesh number, the relative head error tends to 1 as shown in Figure 3 which validates the reliability and accuracy of the simulation results. When the total number of grids in each computational domain is more than 7 million, the results of external characteristics remain stable and the number of grids tends to be independent. Therefore, Case 4 is selected, and the total number of grids is 7213,000.

Table 3. Mesh quantity statistics of each computational domain

	Total(10^6)	$\triangle H$(%)		Total(10^6)	$\triangle H$(%)
Case1	4.266	2.732	Case5	8.181	0.826
Case2	5.184	2.093	Case6	9.202	0.825
Case3	6.153	1.285	Case7	10.170	0,824
Case4	7.213	0.828			

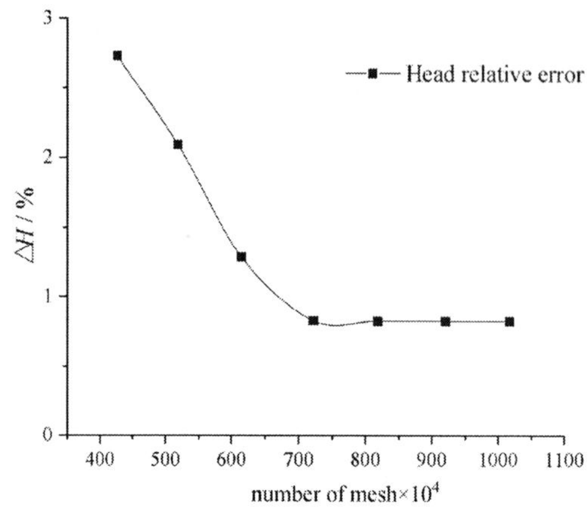

Figure 3. Mesh independence test.

3. Experimental validations

To study the external characteristics of the pump-turbine in different opening conditions, opening 9.8°, 17.5°and 24.8°are selected for analysis in this paper. The flow rate and head are normalized as flow rate coefficient φ and head coefficient by the following equations:

$$\psi = \frac{2gH}{u_2^2}$$

$$\varphi = \frac{Q}{u_2 R_2^2}$$

H—head; u_2—outlet circumferential velocity of runner; g—acceleration of gravity; Q—flow rate; R_2—radius of high pressure side of runner

As showed in Figure 4, the simulation head and efficiency both revealed a satisfactory agreement with the test. The maximum error of the lift coefficient in each condition is less than 4.5%, and the maximum error of the efficiency is less than 3.2%. Therefore, the prediction correctness of CFD calculation strategy in this paper can be evidenced, and the calculation results can be used as a benchmark for analysis. Furthermore, it is apparent from the flow-head curve that the head is large when the opening is small and the flow is small, the head is small when the opening is large and the flow is large; hump phenomenon exists in the vicinity of $0.5Q_d$-$0.75Q_d$, especially when the opening is large. It can be seen from the flow-efficiency curve that the small opening is more efficient at a low flow rate while a large opening is more efficient at a high flow rate, and the small opening is more sensitive to the change of flow rate. With the increase of the opening, the rated flow point 1.0 Q_d is shifted towards the large flow, and the high efficiency interval becomes wider. When the guide vane opening is 17.5°, the rated flow point had the highest efficiency, signifying that different flows have the corresponding optimal opening and too large or too small guide vane opening will lead to additional energy loss.

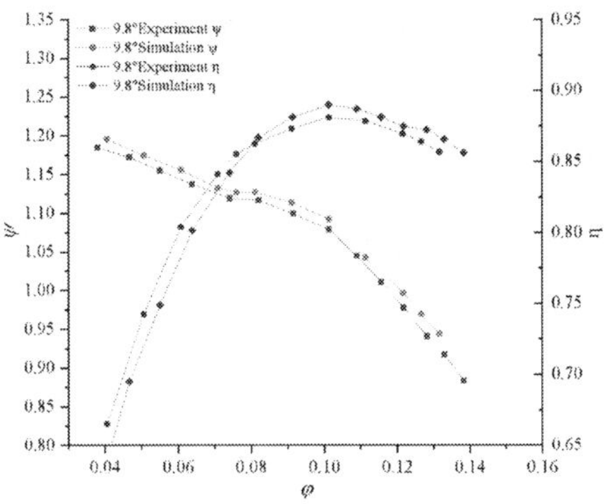

(a) 9.8°

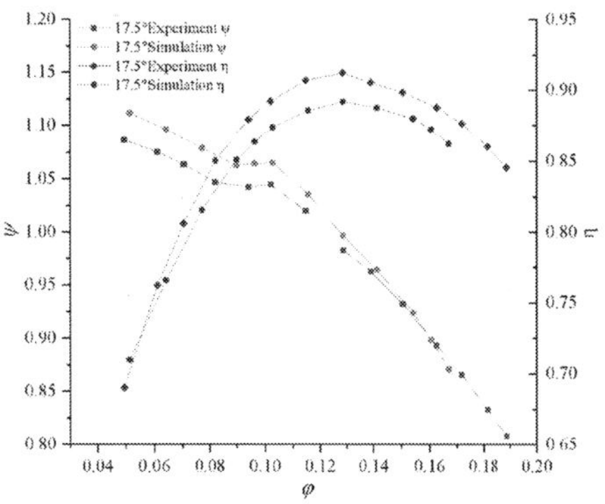

(b) 17.5°

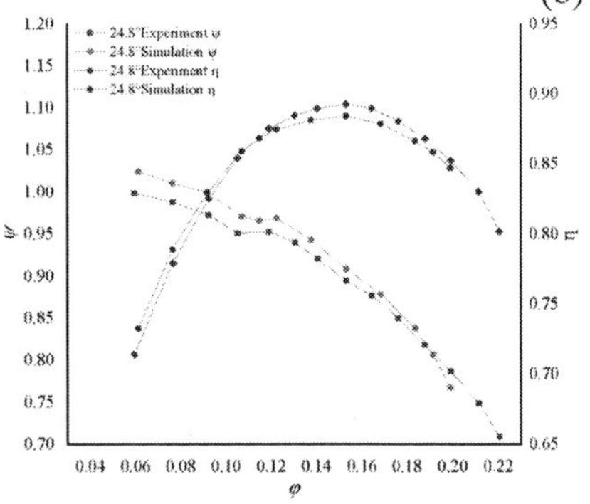

(c) 24.8°

Figure 4. Comparison of performance between simulation and experiment.

4. Results and discussion

Three operating conditions corresponding to the above three different openings, namely 0.75Q_d for small flow, 1.0Q_d for rated flow and 1.25Q_d for large flow, were selected respectively to analyze the internal flow characteristics of each channel and passage in different operating conditions of the pump with different guide vane openings and flows rates. Specific operating condition point parameters are shown in Table 4.

Table 4. Parameters of working conditions

Opening	0.75 Q_d	1.0 Q_d	1.25 Q_d
9.8°	φ=0.0760	φ=0.1013	φ=0.1266
17.5°	φ=0.0964	φ=0.1285	φ=0.1606
24.8°	φ=0.1149	φ=0.1532	φ=0.1915

4.1. Analysis on the mid-span between the runner and stay-guide vanes

Two different working conditions of the pump-turbine were taken into account, the double row cascade adopts the combination form of guide vanes and stay vanes in the turbine. The cascade plays the role of regulating diversion flow to ensure that the inflow of the runner inlet has an appropriate ring volume. While in the working condition of the pump, the double row cascade and volute transform the dynamic pressure of the runner's high-speed flow into the static pressure energy. The rotor-stator interaction in the bladeless region is stronger than that in the working condition of the turbine. Flow instability is more likely to occur at the mid-span between the runner and stay-guide vanes.

IOP Conf. Series: Materials Science and Engineering 1081 (2021) 012016 doi:10.1088/1757-899X/1081/1/012016

Figure 5 displays the streamline distribution of the flow surface streamlines at the mid-span between the runner and stay-guide vanes. The flow surface in the middle of the runner is set as span=0.5 by 'Blade to Blade'. It is apparent that the streamlines distribution in the runner is relatively uniform, and no vortex flow occurred at the inlet of the conventional blades, indicating that the design of splitter blades can improve the inflow of the runner and reduce the flow loss in the passage of runner. It was also revealed that at a low flow rate of $0.75Q_d$, separation vortex congestion occurs in the flow passages of stay vanes of double-row cascade at different openings. When the guide vane opening is large, the impact angle between the guide vanes and the inlet flow tends to be larger, leading to the separation of the vortex generated in advance to the flow passage of the guide vanes. Consequently, the flow velocity and hydraulic loss of adjacent guide vane passages are greatly increased. At rated flow rate $1.0Q_d$, the flow state in the condition of 17.5°exhibited the most stable small energy loss and higher efficiency. However, the vortex separation phenomenon still exists in the stay vane passage with a small opening and the guide vane passage with a large opening leading to large energy loss and reduced efficiency.

Therefore, it indicates that 17.5°is the optimal opening in the working condition of the pump. When the flow rate is$1.25Q_d$, the overflow capacity of both the runner and the double-row cascade is improved. The flow distribution in all channels with a large opening is similar. Under the influence of rotor-stator interaction, high-speed flows are mainly concentrated in the bladeless region, while the small opening will block the runner's outflow resulting in obvious flow separation in the stay vane channel near the tongue.

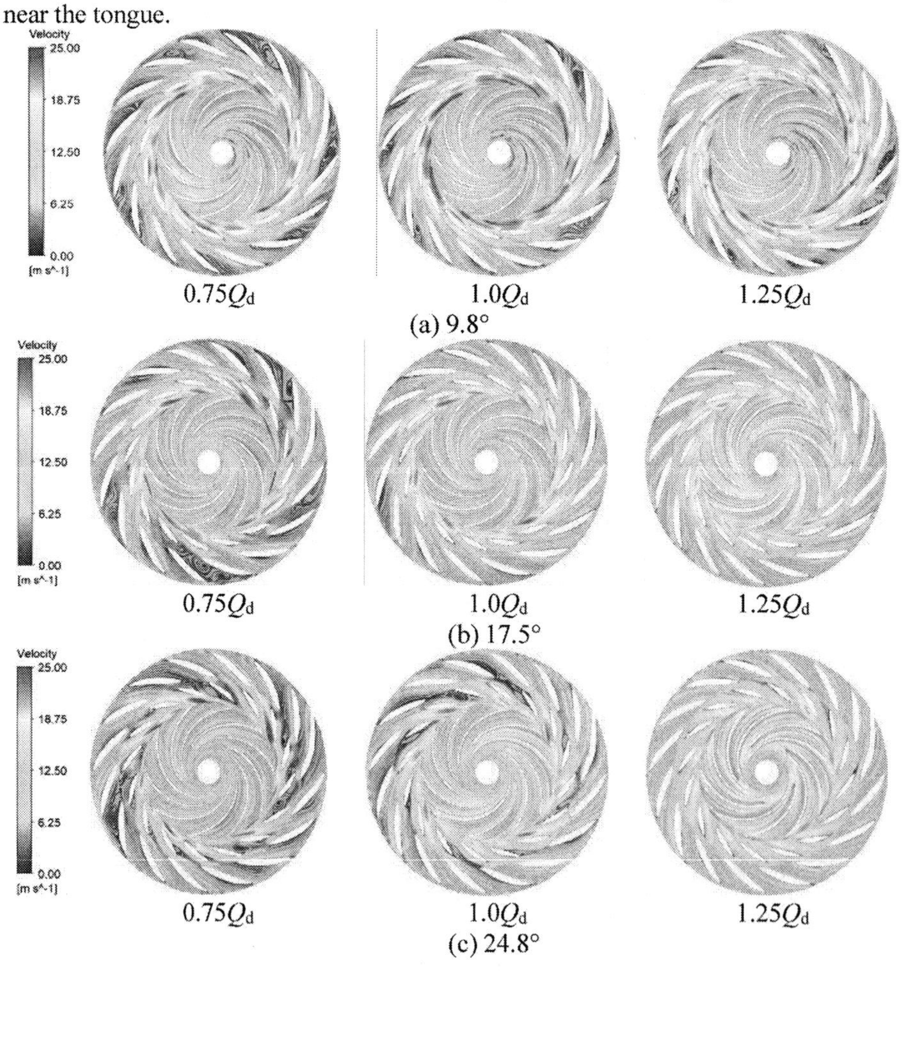

Figure 5. Streamlines at the mid-span in the runner and stay-guide vanes.

To visually reflect the energy loss caused by complex flows in the runner and double-row cascade, Figure 6 shows the turbulent kinetic energy distribution at the mid-span between the runner and stay-guide vanes. It can be seen that: at a low flow rate $0.75Q_d$ with a small opening, the high turbulent kinetic energy region is mainly bladeless region and stay-guide vane passage. However, the flow separation becomes problematic when the opening becomes large, which leads to greater turbulent kinetic energy and extremely uneven distribution at the mid-span between the runner and stay-guide vanes. In the cascade channel near the tongue, the energy loss is relatively apparent, indicating that it should try to avoid operating with the large opening at a low flow rate. At rated flow rate $1.0Q_d$, the flow state becomes stable with 17.5° opening, resulting in smaller turbulent kinetic energy, whilst higher turbulent kinetic energy region at a small opening is transferred to the stay vane channel. However, in the condition of large opening affected by the interaction of flow separation and rotor-stator interaction due to the increased impact angle, there exist 5 turbulent kinetic energies in the guide vane passage accompanied by intense energy exchange. At high flow rate $1.25Q_d$, the turbulent kinetic energy of the stay vane passage with a small opening is further strengthened, the turbulent kinetic energy is small with a large opening and the overall turbulent kinetic energy of each channel is evenly distributed, and the energy loss is reduced at the same time. In general, with the increase of the flow rate, the turbulent kinetic energy of the stay vane passage at the small opening is gradually intensified, while the turbulent kinetic energy at the large opening is gradually decreased, indicating that the flows at small and large openings are the most disordered while the energy losses are the most severe.

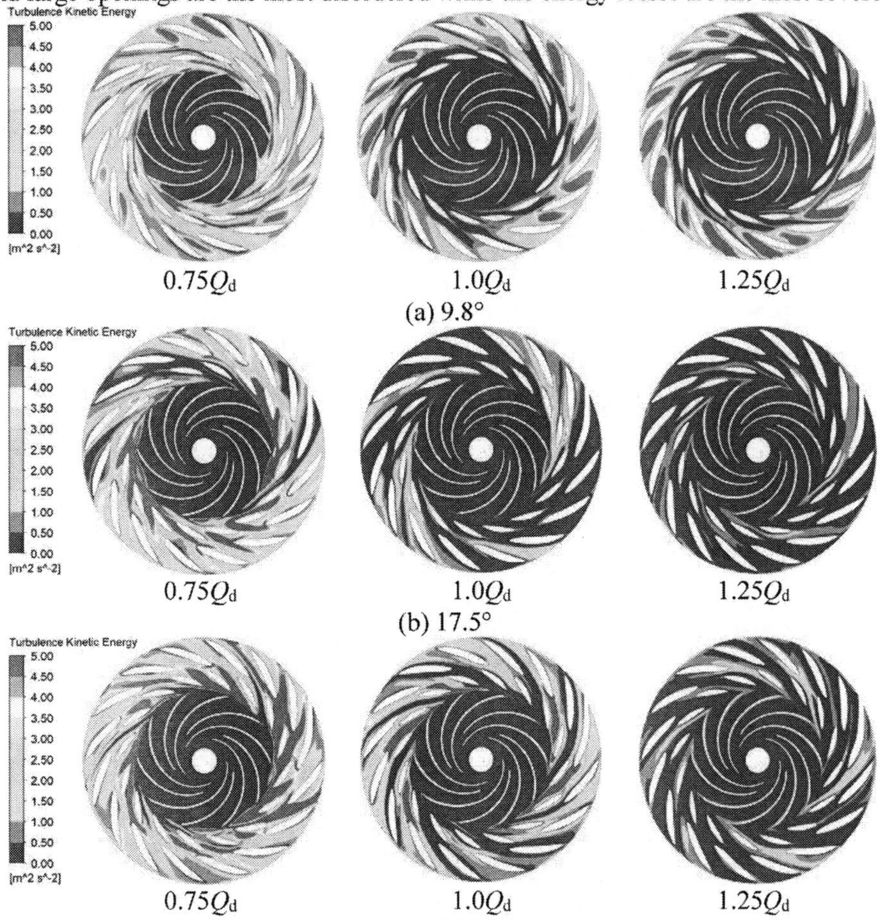

Figure 6. Turbulence kinetic energy at the mid-span between the runner and stay-guide vanes.

4.2. Analysis on the draft tube

Figure 7 shows the streamline distribution of the draft tube and different sections at the rated flow rate at 17.5°. Section A is the outlet of the draft tube, section B is the inlet of the straight taper tube (500mm away from section A), and section C is the normal section of the elbow tube along the profile. As shown in Figure 7, the flow is relatively regular in the working condition of the pump. The flow through the elbow can change the flow direction more smoothly, which is greatly affected by the impact on the outer wall of the elbow, and the flow velocity near the inside of the elbow which is relatively large.

From the streamline distribution of section C, it is also observed that, due to the impact of the outer wall of the elbow, the water collects from the outer wall to the inner wall, resulting in more dense streamlines at the middle position of the inner wall. From the streamline distribution of section B, it was also revealed that the inlet of the cone section is influenced by the centrifugal inertia force of the upstream cubital tube and the viscosity of the tube wall, hence leading to the formation of eddy currents in a small area near the outside of the tube wall. From the streamline distribution of section A, it was also revealed that, affected by the rotor-stator interaction, the coarse flow line of the draft tube presents 5 similar eddy current regions.

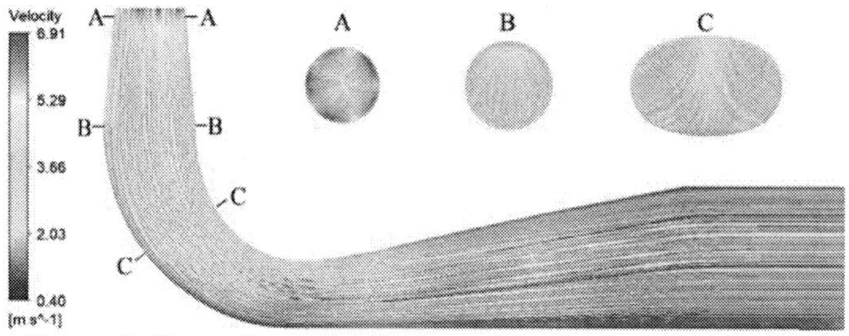

Figure 7. Streamlines distribution of draft tube at $1.0Q_d$ (17.5°).

To analyze the influence of rotor-stator interaction on the outlet of the draft tube in different working conditions, Figure 8 shows the streamline and pressure distribution of the draft tube outlet of section A in different working conditions. It can be seen that, when the flow rate is 0.75Qd and the draft tube with small guide vane opening, the pressure distribution becomes relatively regular. Affected by the rotor-stator interaction especially long blades, there exist 5 high-speed watersheds formed at the tube wall. Therefore, low-pressure areas and high-pressure areas are formed, while the flow state at the outlet of the draft tube with large guide vane opening becomes unstable, resulting in uneven and irregular distribution of the eddy current. With the increase of the opening as well as the flow rate, the flow distribution becomes more regular and the pressure changes tend to be more evenly. The high-pressure area gradually shifts to the center, demonstrating that the flow state at the outlet of the draft tube is more stable with the large opening at a high flow rate. In general, the outlet of the draft tube is most significantly affected by the inlet edge of the long blades with a small opening at a low flow rate. The flow distribution is more uniform with a large opening and a large flow rate, while the flow distribution is the most unstable with a large opening at a low flow rate, hence resulting in irregular flow line and pressure distribution.

IOP Conf. Series: Materials Science and Engineering 1081 (2021) 012016 doi:10.1088/1757-899X/1081/1/012016

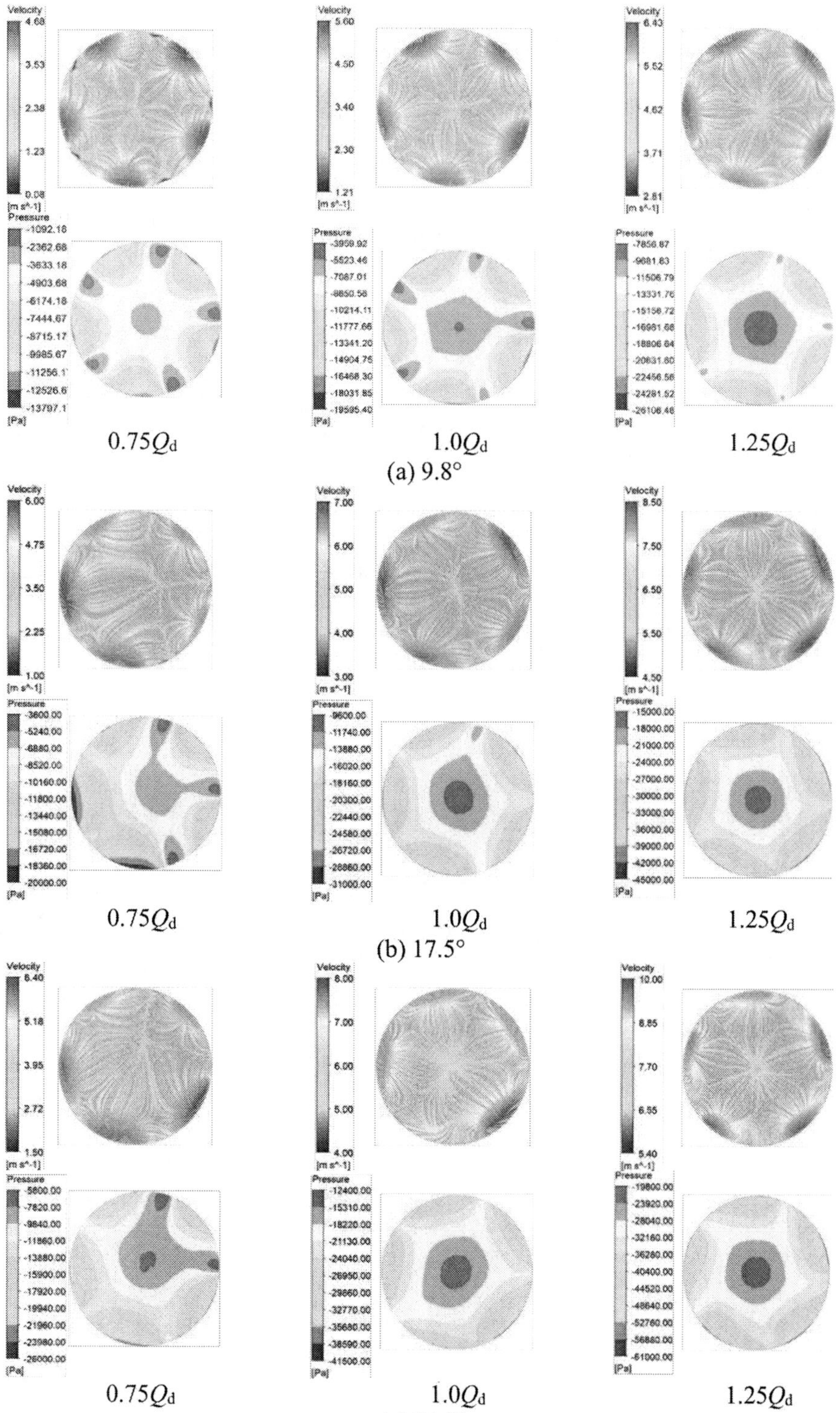

Figure 8. Streamline and Pressure distribution of draft tube outlet.

4.3. Analysis on the volute

In the working condition of the pump, the large section may not match the diffusion degree of the incoming flow, and the outflow of double-row cascade becomes disordered, leading to poor flow in the volute runner than in the working condition of turbine.

To analyze the flow characteristics of the volute, four volute flow passage sections were intercepted at a uniform angle in the circumferential direction, as shown in Figure 9.

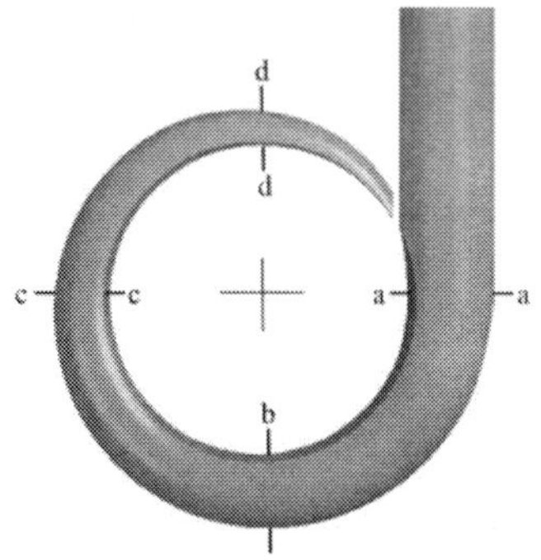

Figure 9. Diagram of volute section location.

Figure 10 shows streamlines in volute sections at 1.0 Q_d. It is observed that the inlet of the volute is perpendicular to all sections in the working condition of the pump. Water then spreads in an axial direction after entering the volute, which results in vortex rings at different sections. At the rated flow rate 1.0 Q_d, both sections a-a and b-b are evenly distributed on both sides, while vortex appears in the middle line of sections c-c and d-d. It shows that the flow distribution in the large section is more uniform at the rated flow rate, but at the same time, the backflow blockage in small section becomes worse. The flow differences between different sections are the most significant when the guide vane opening is small. With the increase of the openings, the flow in the volute section tends to a more roughly symmetrical distribution of backflow vortex on both sides.

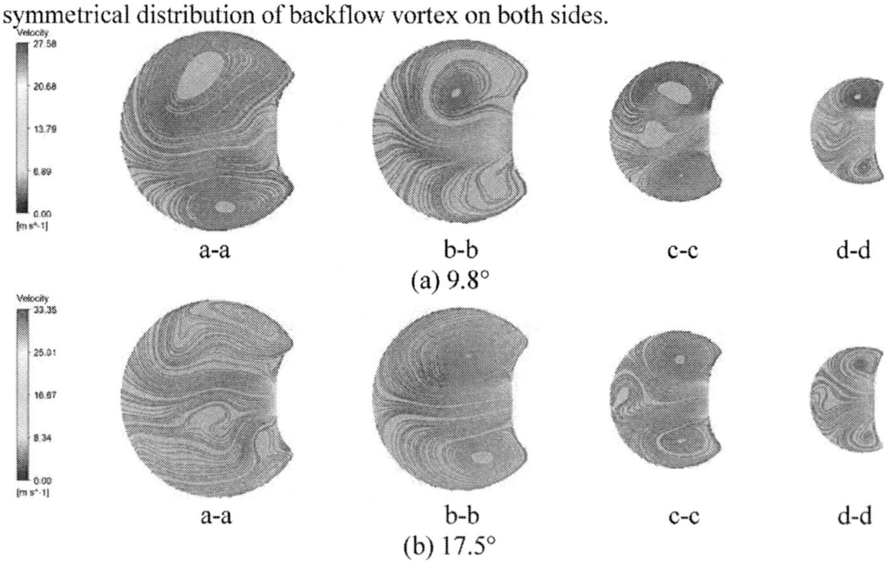

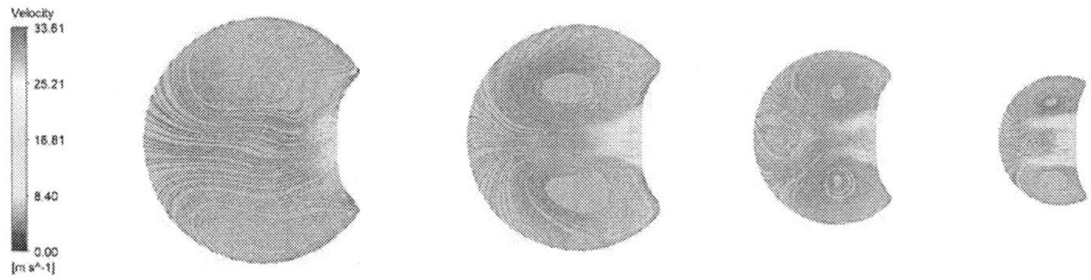

a-a b-b c-c d-d

(c) 24.8°

Figure 10. Streamlines in volute sections at 1.0Q_d.

5. Conclusion

In this paper, the influence of different guide vane opening on the flow characteristic in a pump-turbine with splitter blades has been investigated at different flow rates 0.75Q_d, 1.0 Q_d, and 1.25 Q_d. Performance and experimental data records both revealed a satisfactory agreement with the numerical simulation under different guide vane opening 9.8°,17.5°and 24.8°. Several monitor cross-sections were installed in the draft tube and mid-span between the runner and stay-guide vanes to capture the streamline and pressure signals during the transient numerical simulations. By analyzing the streamline and pressure distribution of draft tube and turbulence kinetic energy at the mid-span between the runner and stay-guide vanes at different rates and different openings, some conclusions are made as follows:

(1) Compared with different openings at different flow rates from 0.75 Q_d to 1.25 Q_d, the phenomenon of hump becomes more obvious when the opening is large and efficiency is more sensitive to changes in flow rates, specifically in a small guide vane opening. However, with the increase of the guide vane opening, the rated flow point 1.0Q_d shifts towards the large flow rate and at the same time, the high efficiency interval becomes wider, indicating that too large or too small opening will lead to additional energy loss. In this paper, the efficiency of the pump-turbine rated flow point is the highest at the opening 17.5°, which indicates that 17.5°is the optimal opening of the pump working condition.

(2) Compared with streamline, pressure distribution, and turbulence kinetic energy at different openings at different flow rates from 0.75Q_d to 1.25Q_d, changes in the guide vane opening and flow rate had a significant impact on the internal flow in the working conditions of the pump. The flow stability under the small opening and large flow tends to be poor. Especially at the small opening, the hydraulic loss caused by flow separation in the double-row cascade becomes very severe. Finally, under the influence of rotor-stator interaction, there exists a local high-speed flow at the outlet of the draft tube and bladeless region between the runner and stay-guide vanes.

References

[1] Haban V, Koutnik J, Pochyly F. 1-D Mathematical model of high-frequency pressure oscillations induced by RSI including an influence of fluid second viscosity[C]. Proceedings of the 21st IAHR Symposium on Hydraulic Machinery and Systems, 2002.

[2] Rodriguez C, Egusquiza E, Santos I. Frequencies in the Vibration Induced by the Rotor Stator Interaction in a Centrifugal Pump Turbine[J]. Journal of Fluids Engineering, 2007, 129(11): 1428-1435.

[3] Nicolet C, Zobeiri A, Maruzewski P, et al. On the upper part load vortex rope in Francis turbine: Experimental investigation[C]. IOP Conference Series Earth and Environmental Science. 2010: 793-798.

[4] Keller J, Blanco E, Barrio R, et al. PIV measurements of the unsteady flow structures in a volute centrifugal pump at a high flow rate[J]. Experiments in Fluids: Experimental Methods and Their Applications to Fluid Flow, 2014, 55(10): 1820-1-1820-14.

[5] Chitrakar S. Fully coupled FSI analysis of Francis turbines exposed to sediment erosion[J]. International Journal of Fluid Machinery and Systems, 2014, 7(3):101-109.

[6] Ali Rezghi,Alireza Riasi. The interaction effect of hydraulic transient conditions of two parallel pump-turbine units in a pumped-storage power plant with considering "S-shaped" instability region: Numerical simulation[J]. Renewable Energy,2018,118.

[7] Yabin Liu,Yadong Han,Lei Tan,Yuming Wang. Blade rotation angle on energy performance and tip leakage vortex in a mixed flow pump as turbine at pump mode[J]. Energy,2020,206.

[8] Zobeiri A, Kueny J L, Farhat M, et al. Pump-Turbine Rotor-Stator Interactions in Generating Mode: Pressure Fluctuation in Distributor Channel[C]. Proceecdings of 23th IAHR Symposium on Hydraulic Machinery and Systems, Yokohama, Japan, 2006.

[9] Nicolet C, Zobeiri A, Maruzewski P, et al. On the upper part load vortex rope in Francis turbine: Experimental investigation[C]. IOP Conference Series Earth and Environmental Science. 2010: 793-798.

[10] Rodriguez C, Egusquiza E, Santos I. Frequencies in the Vibration Induced by the Rotor Stator Interaction in a Centrifugal pump-turbine[J]. Journal of Fluids Engineering, 2007, 129(11): 1428-1435.

[11] Rodriguez C G, Mateos-Prieto B, Egusquiza E. Monitoring of Rotor-Stator Interaction in Pump-Turbine Using Vibrations Measured with On-Board Sensors Rotating with Shaft[J]. Shock and Vibration, 2014, 2014(pt.1): 276796.1-276796.8.

[12] Casartelli E, Mangani L, Ryan O, et al. Application of transient CFD-procedures for S-shape computation in pump-turbines with and without FSI[J]. Iop Conference, 2016, 49(4): 042008.

[13] Yefang Wang, Fan Zhang, Shouqi Yuan, et al. Effect of Unrans and Hybrid Rans-Les Turbulence Models on Unsteady Turbulent Flows Inside a Side Channel Pump, ASME Journal of Fluids Engineering, 2020, 142(6): 061503, doi:10.1115/1.4045995

[14] Yonezawa K, Toyahara S, Motoki S, et al. Phase Resonance in Centrifugal Fluid Machinery -A Comparison between Pump Mode and Turbine Mode Operations and a Discussion of Mechanisms of Flow Rate Fluctuation through a Stator-[J]. International Journal of Fluid Machinery and Systems, 2014, 7(2): 42-53.

[15] Pochylý F, Haluza M, Veselý J. The Francis Pump Turbine With Stochastic Blades[J]. Procedia Engineering, 2012, 39: 68-75.

[16] Shigemitsu T, Fukutomi J, Wada T, et al. Performance Analysis of Mini Centrifugal Pump with Splitter Blades[J]. Journal of Thermal Science, 2013, 22(6): 573-579.

[17] Namazizadeh M, Talebian Gevari M, Mojaddam M, et al. Optimization of the Splitter Blade Configuration and Geometry of a Centrifugal Pump Impeller using Design of Experiment[J]. Journal of Applied Fluid Mechanics, 2020, 13(1): 89-101.

[18] Fan Zhang, Desmond Appiah, Feng Hong, et al. Energy loss evaluation in a side channel pump

under different wrapping angles using entropy production method, International Communications in Heat and Mass Transfer, 2020 (113),

[19] Menter F R. Two-equation eddy-viscosity turbulence models for engineering applications[J]. AIAA journal, 1994, 32(8): 1598- 1605.

[20] Hinze J O. Turbulence[M]. New York:McG raw-Hill Publishing Co., 1975.

Automatic Optimization of Vertical Long-shaft Fire Pump Overload Based on Particle Swarm Optimization Algorithm

Jinfeng Zhang [*], Zhijun Yang [1], Liangqing Lai [1], Haiqing Song [1], Chengming Jing [1]

[1]Research Center of Fluid Machinery Engineering and Technology, Jiangsu University, Zhenjiang, China

E-mail: zhanajinfeng@ujs.edu.cn

Abstract. As a typical large-flow high-head pump, the vertical long-axis fire pumps have the advantages of rapid startup, superior structure, and strong applicability in multiple occasions. In order to shorten the non-overload optimization cycle, CFturbo, ICEM and CFX are integrated by writing batch processing commands to achieve automatic multi-load and no-load optimization of vertical long-axis fire pumps based on Isight's multi-disciplinary optimization platform. With the objective of achieving the non-overload characteristics of vertical long-axis fire pumps, the optimal Latin hypercube design is used to spatially sample the design variables of the impeller. After sampling, the particle swarm optimization (PSO) algorithm is used to optimize the design. The results show that, vertical long-axis fire pumps achieve non-overload characteristics within $1.5Q_0$ after optimization, while the head and efficiency at the rated operating point are almost unchanged.

1. Introduction

As a typical large-flow high-head pump, vertical long-axis fire pumps have the advantages of rapid startup, superior structure, and strong applicability in multiple occasions [1]. Therefore, they are widely used in Offshore platforms, flood control and drainage, urban water supply, boiler water supply, mines and other fields. With the rise in offshore platform oil development, large-scale terminals and other projects, the demand for vertical long-axis fire pumps continue to increase, which also places higher requirements on their performance. Affected by tides, waves, typhoons, etc., the power of vertical long-axis fire pumps increase extremely with the increase of flow, and it is easy to overload and burn the motor under large flow conditions. Therefore, the development of a vertical long-axis fire pump with no overload performance and high efficiency is the current research focus of the fire industry.

Presently, the domestic and foreign researches on the no-overload performance of centrifugal pumps are mainly concentrated on low specific speed centrifugal pumps [2]. There are few related studies on vertical long-axis fire pumps with medium specific revolutions, and the research on their no-overload characteristics is also non-existent. In recent years, intelligent optimization design has been widely used in all walks of life, and has been proved to be effective by actual engineering tests [3]. Fortunately, there are many cases of applying intelligent optimization algorithms to the optimization design of fluid machinery [4-13], which provides a reference for multi-parameter optimization of fluid machinery.

There are many impeller hydraulic parameters that affect the no-overload performance of vertical long-axis fire pumps. It is difficult and time-consuming to modify one by one. In order to achieve fast and efficient hydraulic optimal design of vertical long-axis fire pumps, this paper uses Isight optimization software to integrate

CFturbo, ICEM with CFX to establish a CFD-based no-overload automatic optimization platform for the vertical long-axis fire pump platform. Combined with the optimal Latin hypercube test design and particle swarm optimization algorithm, a vertical long-axis fire pump impeller is optimized for no overload.

2. Computational models and numerical methods

2.1 Computational models

The research object is a vertical single-suction, sin00gle-stage guide vane vertical long-axis fire pump with a specific revolution of 157.3. The main parameters are shown in Table 1. The pump structure is mainly composed of a water suction chamber, an impeller and a guide vane.

Table 1. Main parameters of vertical long-axis fire pump

Parameter	Parameter name	Value
Q_0	Design flow /$m^3 \cdot h^{-1}$	864
H	Head /m	43.3
n	Impeller speed/$r \cdot min^{-1}$	1485
N	Number of blades	6
D_1	Impeller inlet diameter/m	0.209
D_2	Impeller outlet diameter/m	0.383
b_2	Impeller outlet width /m	0.0587

Figure 1. Vertical long-axis fire pump

2.2. Numerical Methods

When the numerical simulation of the model pump is carried out, Creo Parametric 5.0 three-dimensional modeling software is used to perform three-position modeling on the pump inlet and outlet extension, the impeller water body and the space guide vane water body. CFX17.0 is used for steady calculation of the model pump in this research. Considering that this research focuses on the non-overload characteristics of vertical long-axis fire pumps, the RNG k-ε turbulence model is adopted, and the boundary conditions are pressure-inlet and massflow-outlet; the discrete solution is set to the second-order upwind style, and the convergence residual is set to10^{-5}.

Table 2. Analysis of grid independence

Scheme	Grid numbers	H/m	r_H/%	η/%	r_η/%
A	2.634×10^6	40.90	—	81.56	—
B	4.160×10^6	41.95	2.56	82.59	1.26
C	5.350×10^6	42.90	2.26	84.58	2.41
D	6.782×10^6	43.78	2.05	84.55	0.04
E	9.219×10^6	43.83	0.11	84.52	0.04
F	1.232×10^7	43.83	0	84.71	0.22

It can be seen from Table 2 that the error of the number of grids of scheme D for the simulation result is less than 1%. Therefore, the scheme of the number of grids of scheme D of 6.782 million was finally selected to start numerical simulation calculation in Figure 3. After analyzing the external characteristic curves of the experiments and simulations, it is found that the two trends are pretty consistent, so the simulation results can be considered reliable.

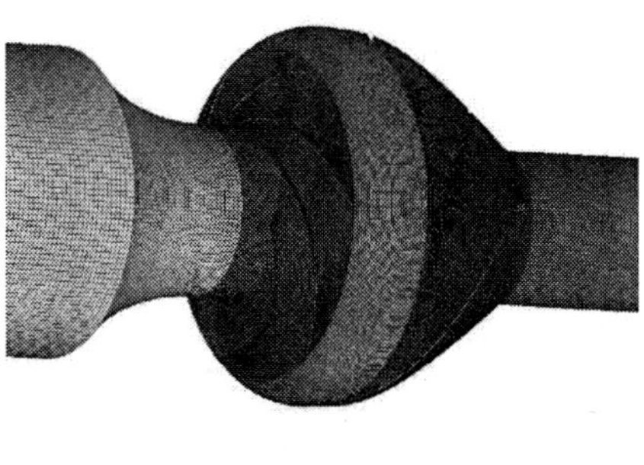

Figure 2. Structural grid of water body computing domain

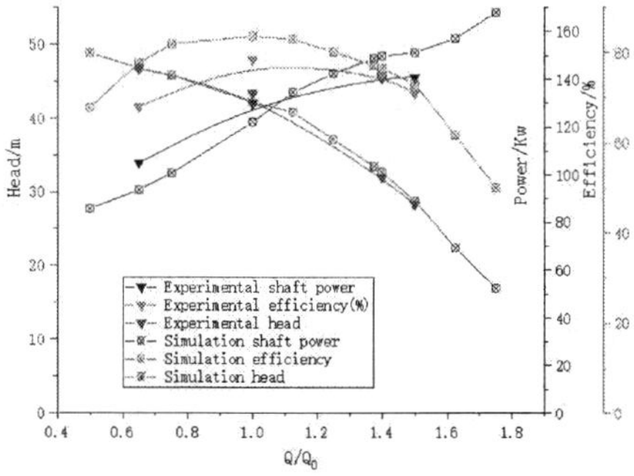

Figure 3. Experimental and simulated externa characteristic curves

3. Non-overload theory and optimization method

The theory of non-overload centrifugal pump means that when the centrifugal pump works in the full-flow working range, the pump's flow-axis power curve is saturated: the shaft power has a maximum change with the flow. However, the vertical long-axis fire pump requires no overload within $1.5Q_0$ (the maximum shaft power appears within $1.5Q_0$).

According to the traditional optimization method, the designer repeatedly changes the relevant parameters of the blade and performs numerical simulation calculations based on the calculated performance results in the blade optimization process. This repeated process not only has a long cycle but also has poor economic benefits. Practice has proved that most design schemes made in accordance with traditional design methods have room for improvement, which is not the best design.

This research combines calculation models with modern intelligent optimization algorithms, and applies mathematical methods and computer techniques to the optimization design process. First, design the vertical long-axis fire pump design variables and their value ranges. Secondly, conduct spatial sampling through experimental design to reduce the scope of optimization variables. Finally, use the optimization algorithm to optimize the calculation model. The research uses the Isight optimization platform to fully automate the design process to shorten the optimization cycle and reduce the optimization cost.

3.1. Non-overload constraint equation and optimization parameters

According to the theoretical conditions of the saturated shaft power characteristics of the centrifugal pump, Yuan Shouqi et al. deduced the constraint equations for the design of no overload pumps. They comprehensively considered the influence of geometric parameters on the performance of the pump, then modified the existing centrifugal pump calculation formulas and related coefficients, and finally proposed a more accurate set of constraint equations. According to Yuan Shouqi's research on the area ratio theory [13], the constraint equations for the non-overload centrifugal pumps (suitable for specific speed n_s=80-250) are followed:

$$\varphi_{\max} = \frac{1}{2} h_0 tg\beta_2, \frac{b_2}{D_2} = 0.000147 n_s^{4/3}$$
$$tg\beta_2 = \frac{n_s^{2/3}}{80 h_0 K_u^3}, 1.0 \le Y = \frac{\pi D_2 b_2 \psi_2 \sin\beta_2}{F_t} \tag{1}$$
$$P_{\max} = \frac{\rho}{4\eta_m} u_2^3 \pi D_2 b_2 \psi_2 h_0^2 tg\beta_2$$
$$Q_{\max} = \frac{1}{2} h_0 tg\beta_2 \eta_v \pi D_2 b_2 \psi_2 u_2$$

In order to ensure that the rated working head efficiency of the original model pump meets the requirements to the greatest extent. The parameters such as the outer diameter of the impeller and the width of the impeller outlet are kept. From the above non-overload constraint equations, it can be deduced that, the overload-free characteristic of the vertical long-axis fire pump is realized by changing the blade inlet placement angle β_1, reducing the blade outlet placement angle β_2 and increasing the blade wrap angle φ [14] .

Table 3 shows the range of design variables. In CFturbo's hydraulic design of vertical long-axis fire pump impellers, three streamlines are used to control the blades. β_{1-a}, β_{1-b} and β_{1-c} is the inlet placement angle on the three streamlines; β_{2-a}, β_{2-b} and β_{2-c} is the outlet placement angle corresponding to the three flow lines; φ_1, φ_2 and φ_3 is the blade wrap angle corresponding to the three flow lines; in order to reduce the optimization variables, let

$$\begin{aligned} &\beta_{1-b} = \beta_{1-a} - 15.4, \\ &\beta_{1-c} = \beta_{1-a} - 21, \\ &\beta_{2-c} = \beta_{2-b} = \beta_{2-a}, \\ &\varphi_3 = \varphi_2 = \varphi_1 \end{aligned} \tag{2}$$

Table 3. Design variable range

Variables	Min	Max	Initial value
β_{1-a} / (°)	25.8	51.6	40.4
β_{1-b} / (°)	10.4	36.2	25
β_{1-c} / (°)	4.8	30.6	19.4
β_{2-a} / (°)	21	33	27
β_{2-b} / (°)	21	33	27
β_{2-c} / (°)	21	33	27
φ_1 / (°)	110	155	120
φ_2 / (°)	110	155	120
φ_3 / (°)	110	155	120

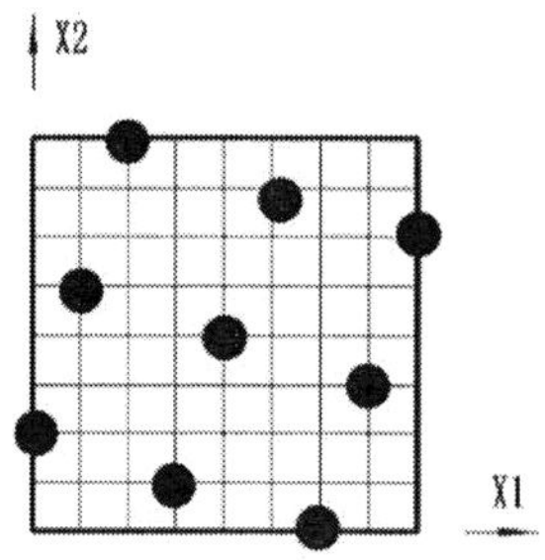

Figure 4. Optimal Latin hypercube test

3.2 Automatic optimization without overload

3.2.1. Optimal Latin Hypercube Test Design

Figure 4 is a schematic diagram of the optimal Latin hypercube test. The test design provides data samples for the construction of the calculation model. For the test design, the number of tests is required to be as small as possible, and the design space characteristics are fully reflected as much as possible. The optimal Latin hypercube test design method used in this research improves the uniformity of the random Latin hypercube design, so that all test points are distributed as evenly as possible in the design space, and have the ability to fit second-order or non-linear relationships[15].

3.2.2. Automatic Optimization Platform Construction

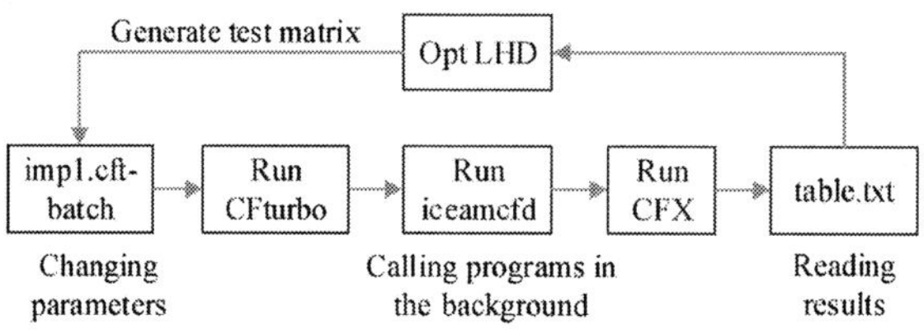

Figure 5. Software integration block diagram

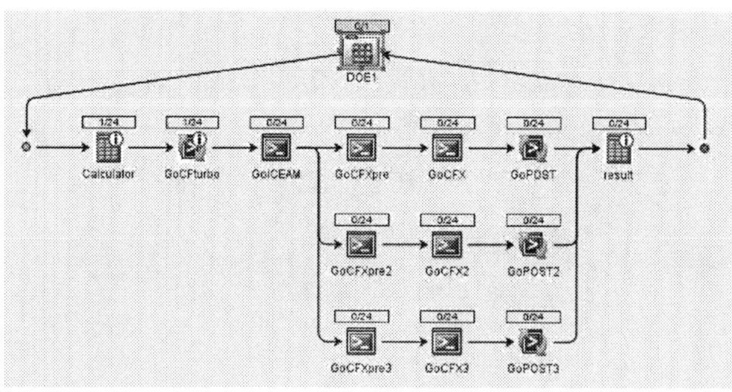

Figure 6. Software operation flowchart

Parametric modeling is the key to optimal design. In this research, the rotating machine design software CFturbo is used to hydraulically design the impeller. Based on the Isight optimization platform, this research integrates the three-dimensional modeling software Cfturbo, meshing software ICEM and numerical simulation software CFX to optimize the corresponding parameters of the variable blades, and perform multi-parameter optimization with non-overload characteristics as the optimization goal. The "design-simulation" integrated automatic optimization platform of the vertical long-axis fire pump is set up, and the integration process is shown in Figure 5.

- Export the file that records the CFturbo design process: impl.cft-batch, write the parameters to be optimized, and edit the equation (2) to modify the corresponding parameters.
- Establish a batch command for software operation as follows:

Table 4. A batch command for software operation

Softwares	Btch commands
CFturbo	"E:\Program Files\CFturbo 10\CFturbo.exe" -batch impl.cft-batch
icemcfd	"D:\Program Files (x86) \ANSYS Inc\v170\icemcfd\win64_amd\bin\icemcfd.bat" -batch -script yl xjt.rpl
cfx5pre	"D:\Program Files (x86) \ANSYS Inc\v170\CFX\bin\cfx5pre.exe" -session impeller_h50.pre-
cfxsolve	"D:\Program Files (x86) \ANSYS Inc\v170\CFX\bin\cfx5pre.exe" -session impeller_h50.pre cfx5solve.exe"-def C:\Users\Administrator\Desktop\isight_leon\impeller_h50.def -par-local -partition 2
cfxpost	"D:\Program Files (x86) \ANSYS Inc\v170\CFX\bin\cfx5post.exe"-session impeller_h50.cse

- Read the efficiency, head and shaft power from the table.txt of the CFX calculation results.
- The running process is shown in Figure 6. Set up and run the test design component to drive the platform automatically, then obtain the numerical simulation results. In order to provide enough sample points for the calculation model, 24 cases are designed using the optimal Latin hypercube, and the specific scheme is automatically generated in the Isight test component.

3.2.3. Particle Swarm Optimization (PSO) Algorithm optimization

Particle swarm optimization is a stochastic optimization algorithm proposed by Kennedy and Eberhart in 1995. It is a bionic algorithm inspired by group behaviors such as bird foraging. The optimization process is based on a particle group, and through iterative calculation, these particles are constantly moved from the decision space to search the space. Information is shared between particles, and speed and position updates are affected by their own experience and group experience, so that particles can move to a better position with the greatest probability. The basic speed and position update formulas are shown in equations (3) and (4):

$$v_{t+1} = v_t + c_1 \cdot r_1 \cdot \left(x_{pest,t} - x_t\right) + c_2 \cdot r_2 \cdot \left(x_{pest,t} - x_t\right) \tag{3}$$

$$x_{t+1} = x_t + v_{t+1} \tag{4}$$

In the formula, x is the particle position; v is the particle velocity; c is the individual learning factor and group learning factor, generally take 2; r is a random number between [0,1]; t represents the number of iterations, *gbest* represents. The optimal position of the group, *pbest* represents the optimal position of the individual.

The characteristics of the PSO algorithm are as follows:

- The PSO algorithm does not have cross operations and mutation operations like genetic algorithms, but relies on particle speed to complete the search, and in the iterative evolution, only the information of the optimal particle is passed to other particles, and the search speed is fast;
- The PSO algorithm is memorable and can remember the historical optimal position of the particle group and pass it to other particles:

- The structure is simple, there are few parameters to be adjusted, and it is easy to implement in engineering
- Using real number coding, the variable number of the problem is directly used as the dimension of the particle, which directly points to the problem.

The PSO optimization process is shown in Figure 7. The PSO algorithm is used to optimize the Latin hypercube design after sampling, to find the optimal solution that meets the requirements of no overload as quickly as possible.

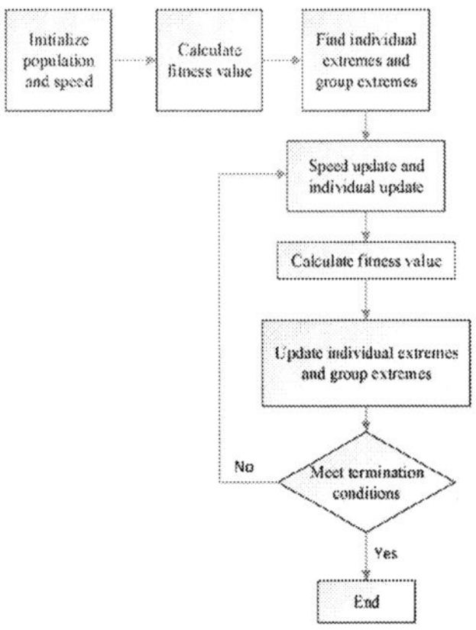

Figure 7. PSO optimization process

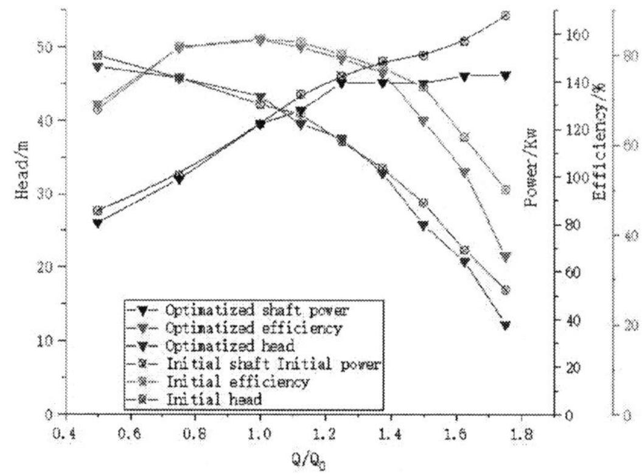

Figure 8. Comparison of optimization

Table 5 is the geometric parameter table. After the optimization, the shape of the impeller changes greatly, in which the exit edge of the impeller is appropriately inclined, the blade wrap angle increases substantially, the blade inlet placement angle slightly decreases, and the blade outlet placement angle becomes larger, but the head, efficiency and shaft power of the flow design point almost unchanged.

Figure 8 shows the efficiency, head and shaft power curve of the vertical long-axis fire pump before and after optimization. It can be found that after optimization, the vertical long-axis fire pump achieves no overload characteristics within $1.5Q_0$, and the head and efficiency at the rated operating point are almost constant. However, it is worth noting that by increasing the blade wrap angle and reducing the blade exit placement angle to achieve no overload performance, the efficiency and head change of the vertical long-axis fire pump under low flow conditions is relatively small. However, under high flow conditions, efficiency and head are reduced to a certain extent.

Table 5. Results of optimization

Variables	Initial values	Optimized values	Variables	Initial values	Optimized values
$\beta_{1-a}/(^{o})$	40.4	39.90	$\varphi_1/(^{o})$	120	127.96
$\beta_{1-b}/(^{o})$	25	24.6	$\varphi_2/(^{o})$	120	127.96
$\beta_{1-c}/(^{o})$	19.6	19.2	$\varphi_3/(^{o})$	120	127.96
$\beta_{2-a}/(^{o})$	27	29.79	$\eta/(\%)$	83.56	83.38
$\beta_{2-b}/(^{o})$	27	29.79	$H/(\mathrm{m})$	43.319	43.24
$\beta_{2-c}/(^{o})$	27	29.79	$P/(\mathrm{kW})$	122.02	122.06

3.3. Pareto graph analysis

In order to understand the contribution of design variables to the efficiency of the vertical long-axis fire pump, a Pareto diagram analysis of the experimental design results is shown in Figure 9. Figure 9 is Analysis graphs of optimization variables in Q_0、 $1.25Q_0$and $1.5Q_0$, including Pareto graph and Main effect diagram. The Pareto graph reflects the percentage of the contribution of all inputs to each target in the fitted quadratic regression model of the samples. Its absolute values indicate the significance of the impact on the target, and positive and negative indicate the positive and negative effects on the target. The order of significance of the influence of each design variable on efficiency is as follows: φ_1 , φ_1^2 , β_{2-a}^2 , β_{1-a}^2 , $\beta_{2-a}{\bullet}\beta_{1-a}$, $\beta_{2-a}{\bullet}\varphi_1$, $\beta_{1-a}{\bullet}\varphi_1$, β_{1-a} , β_{2-a} . It can be seen that the non-overload performance of the vertical long-axis fire pump is greatly affected by the blade wrap angle. The blade wrap angle has a large negative effect on the non-overload performance, indicating that the non-overload design of the vertical long-axis fire pump can be appropriately increased; while the blade inlet placement angle has a slight negative effect on the non-overload performance, indicating that the non-overload design of the vertical long-axis fire pump can appropriately increase the blade outlet placement angle; and the blade outlet placement angle has a non-overload performance. The small positive effect shows that appropriately reducing the blade inlet placement angle can achieve the non-overload design of the vertical long-axis fire pump. This conclusion is consistent with the results of the Main effect diagram.

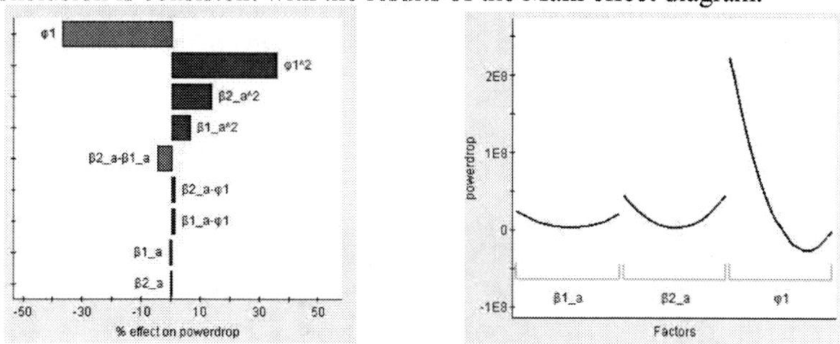

(a)Pareto graph (b)Main effect diagram

Figure 9. Analysis graphs of optimization variable in Q_0、 $1.25Q_0$ and$1.5Q_0$

4.Inflow Characteristics Analysis

(a)Initial impeller (b)Optimized impeller

Figure 10. Comparison of the static pressure distribution in the middle section of the front and rear impellers in Q_0

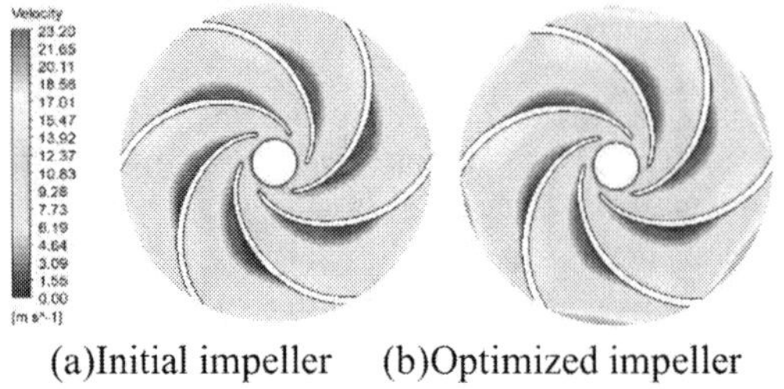

(a)Initial impeller (b)Optimized impeller

Figure 11. Comparison of the relative velocity distribution in the middle section of the front and rear impellers in $1.5Q_0$

Figure10 is a comparison of the static pressure distribution of design flow in the middle section of the impeller. Compared with the initial impeller, the optimized impeller wrap angle increases, the exit placement angle decreases, the blade length increases, and the constraint on the flow field is enhanced [16]. Compared to initial impeller, pressure gradient in the impeller inlet and outlet, blade working surface and back of optimized impeller is almost the same.

Figure11 is a comparison of the relative speed distribution of design flow in the middle section of the impeller. Compared with the initial impeller, the relative velocity distribution at the impeller inlet, blade working surface and back of optimized impeller is almost the same, and only the speed at the outlet is significantly increased. This indicates that the water velocity at the exit of the impeller is increased, and the ability of the impeller to convert mechanical energy into kinetic energy of water is improved.

Figure 12. Pressure monitoring points

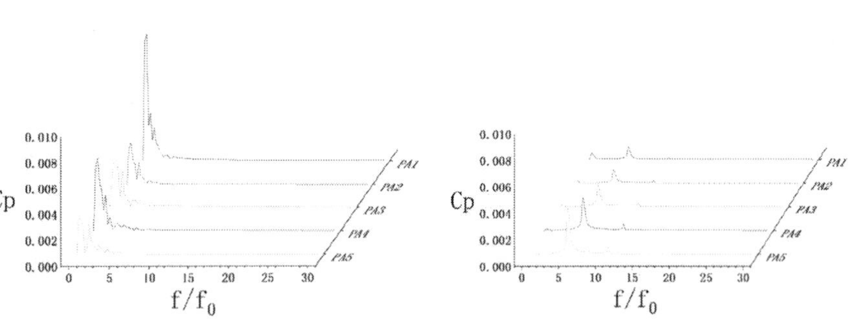

(a) Initial plan in Q_0 (b) Initial plan in $1.5Q_0$

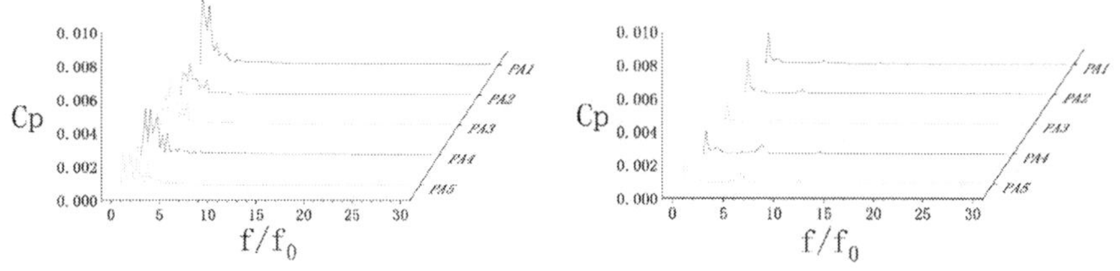

(c) Optimized plan in Q_0 (d) Optimized plan in $1.5Q_0$

Figure 13. Pressure pulsation diagram of blade pressure surface

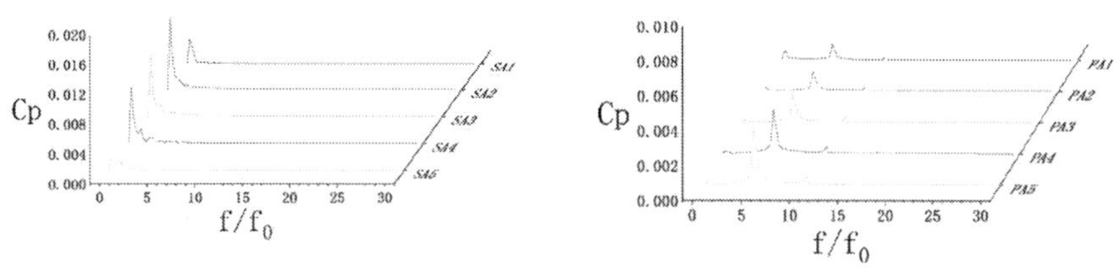

(a) Initial plan in Q_0 (b) Initial plan in $1.5Q_0$

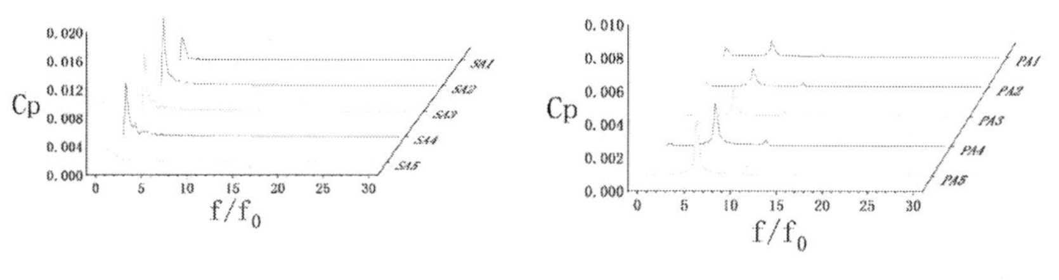

(c) Optimized plan in Q_0 (d) Optimized plan in $1.5Q_0$

Figure 14. Pressure pulsation diagram of blade suction surface

For unsteady calculation, the time step is 1/180T and the total time is 10T. The last 4T data is selected to do the pressure analysis. The distribution of pressure monitoring points is shown in Figure 12. The red balls are the pressure surface monitoring points, from the blade inlet to the outlet direction, respectively PA1, PA2, PA3, PA4, PA5. The yellow balls are the suction surface monitoring point, from the blade inlet to the outlet direction, respectively SA1, SA2, SA3, SA4, SA5.

Since the pressure fluctuation at the position of the monitoring point in the pump is very small relative to the pressure value itself at this point, the dimensionless pressure coefficient Cp is introduced in this paper, which can reduce the effect of static pressure to a certain extent and at the same time make the pressure at each monitoring point. The observation of the value is more intuitive and convenient. The calculation formula of *Cp* is as formula (5):

$$C_p = (p_i - \overline{p})/(0.5\rho u_2^2) \quad (5)$$

Among them: *pi* represents the instantaneous pressure of the monitoring point; represents the average value of all pressure data obtained at the monitoring point; u_2 represents the outlet circumferential speed of the impeller; ρ represents the fluid density.

Fig. 13 is the pressure pulsation diagram of the initial model and optimized model at the blade pressure surface of Q_0 and $1.5Q_0$, and Fig. 14 is the pulsation diagram of the initial model and optimized model at the blade suction surface in Q_0 and $1.5Q_0$. The pressure pulsation is very stable after $15f_0$, so we focus on analyzing the pressure pulsation before $15f_0$. The optimization model shows that the pressure pulsation on the pressure surface and the suction surface decreases significantly in Q_0, while the pressure pulsation amplitude on the pressure surface and suction surface does not change significantly in 1.5 Q_0, which also verifies that the optimization model achieves no overload characteristics in 1.5 Q_0. It sacrifices part of the efficiency of large flow.

5.Conclusion

The self-editing batch processing command integrates CFturbo, ICEM and CFX to establish a CFD automatic optimization platform for vertical long-axis fire pumps based on Isight. The platform is used to optimize the impeller of the vertical long-axis fire pump with a specific speed of 157, which can realize the non-overload optimization. The conclusions are as followed:

- Use the optimal Latin hypercube test design to uniformly sample and reduce the range of optimized variables, and then optimize the calculation model by PSO algorithm to obtain a set of optimal geometric parameters. The vertical long-axis fire pump achieves non-overload characteristics within 1.5 Q_0, and the head and efficiency of the rated point is almost unchanged.
- Based on the Pareto diagram analysis, the primary and secondary sequence of the design variable's impact on the objective function is obtained. The impeller wrap angle has a greater impact on the non-overload performance of the vertical long-axis fire pump. Researchers should focus on the impeller wrap angle when designing the vertical long-axis fire pump without overload.
- By comparing the internal flow characteristics of the original model and the optimized model, the impeller wrap angle of the optimized model is increased, the outlet placement angle is reduced, the blade length is increased, the binding force on the flow field is enhanced, and the flow channel diffusion capacity is reduced, while achieving the non-overload characteristics, a part of the efficiency of large flow is sacrificed.

Reference

[1] Kui Yu. Analysis of the characteristics of vertical long-axis fire pumps, the First National Conference on Fire Protection of Supertall Buildings, 2014

[2] Si Qaiorui, Yuan Shouqi, Wang Chuan, et al. Optimal design of submersible multistage pumps with low specific speed[J]. Transactions of the Chinese Society of Agricultural Engineering (Transactions of the CSAE), 2012, 28(8): 122—127. (in Chinesewith English abstract).

[3] Si Qaiorui, Yuan Shouqi, Wang Chuan, et al(2012). Optimal design of submersible multistage pumps with low specific speed[J]. Transactions of the Chinese Society of Agricultural Engineering (Transactions of the CSAE), 2012, 28(8): 122—127. (in Chinesewith English abstract).

[4] Si Qaiorui, Yuan Shouqi, Wang Chuan, et al(2012). Optimal design of submersible multistage pumps with low specific speed[J]. Transactions of the Chinese Society of Agricultural Engineering (Transactions of the CSAE), 2012, 28(8): 122—127. (in Chinesewith English abstract).

[5] Mounier Violette, Mendoza Luis Carlos, Schiffmann(2017). JurgThermo-economic optimization of an ORC driven heat pump based on small scale turbomachinery and comparison with absorption heat pumps[J]. International Journal of Refrigeration, 2017, 81: 96-110.

[6] Zhounian, Lai,Qian Li,An Zhao(2020). Improving Reliability of Pumps in Parallel Pump Systems Using Particle Swam Optimization Approach [J]. IEEE Access, 2020, 8 : 58427-58434.

[7] LU Rong, YUAN Jianping, LI Yanjun(2017). Automatic optimization of axial flow pump based on radialbasis functions neural network and CFD [J]. Journal of Drainage and Irrigation Machinery Engineering, 2017, 35(6): 481—487.

[8] Liu Yi, Tan Lei, Cao Shuliang(2012). Micro-genetic algorithm and its application in optimization design of mixed flow pump impeller[J]. Machinery Design & Manufacture, 2012(9): 1—3.

[9] Ji-lei,Wang;Jin-guang,Yang;Yan Liu(2020). Multi-objective Optimization Design of a Centrifugal Impeller [J]. Chinese Journal of Turbomachinery, 2020, 62(1): 1006-8155.

[10] Chen Cheng,Jun Liu,Hao Xu(2020). Design of Performance Prediction Software for Multistage Centrifugal Compressor [J]. Chinese Journal of Turbomachinery, 2020, 62(1): 1006-8155.]

[11] Xu-jia Chao,Jun Xu(2020). Digital Modeling Algorithm and Structure Visualization of Shrouded Centrifugal Impeller [J]. Chinese Journal of Turbomachinery, 2020, 62(1): 1006-8155.

[12] Jin-xin,Yao,Shu-Hua Yang(2020). Optimal Design of Centrifugal Compressor Impeller Structure [J]. Chinese Journal of Turbomachinery, 2020, 62(1): 1006-8155.

[13] Jin-xin Yao,Shu-Hua Yang (2020). Rotating Stall State Identification of Axial Compressor Based on MPE-CEEMD Algorithm [J]. Chinese Journal of Turbomachinery, 2020, 62(1): 1006-8155.

[14] Shouqi Yuan(1996).Theory and Design of Low Specific Speed Centrifugal Pump [M]. Beijing: Machinery Industry Press, 1996.

[15] HAN Ding,Heng Jianrong(2012).A survey of metamodelingtechniques in engineering optimization[J].Journal of East China University of Science and Technology(natural cience edition)，2012，38 (6) : 762—768.

[16] Chunlin Wang, Beibei Hu(2019).Multi-objective optimization of double vane pumps based on radial basis neural network and particle swarm optimization, of the Chinese Society of Agricultural Engineering [A], 2019, 1002-6819 (2019) -02-0025-08.

IOP Conf. Series: Materials Science and Engineering 1081 (2021) 012018 doi:10.1088/1757-899X/1081/1/012018

Analysis of jet suppressing root leakage in plane cascade

Meng Tang[1] **Xin Fu**[1] **Ying Y Xu**[1]
Nanjing University of Aeronautics and Astronautics,Nan Jing, CHN
E_mail:fu_xin@nuaa.edu.cn

Abstract: The three-dimensional numerical simulation of the sealing structure of compressor cascade with jet structure is carried out to study the flow structure of the tooth cavity under the influence of jet. Based on this, the influence of jet on the flow performance in the tooth cavity was analyzed from four aspects of vorticity distribution, turbulent kinetic energy distribution, Mach number distribution and static pressure distribution. The results show that the interaction of the jet flow and the leakage flow makes the internal flow structure of the tooth cavity complicated and changeable. The two fluids are connected and mixed with each other and keep separating and converging, which makes the flow loss increase sharply and the leakage coefficient is greatly reduced. The jet flow drastically changes vorticity distribution and static pressure distribution in the tooth cavity. In the jet-affected section, the jet-affected range is approximately 2 times the circumferential width of the jet port.

1. Introduction

With the development of aero-engine technology, the pressure ratio of high-pressure compressors continues to increase. This means a further increase in clearance leakage, which will affect the performance of the entire engine. Sealing labyrinth is a non-contact sealing structure whose sealing efficiency is mainly determined by the radial clearance between rotor parts and stator parts and the number of labyrinth seals. Because of its simple structure, high sealing efficiency, suitable for harsh working environment and other characteristics, it is widely used in aircraft engine. After adopting new technology and new structure, it has broader application prospects.

Stoff[1] first used the finite difference method and the standard turbulence model to numerically simulate the incompressible flow field inside the sealing labyrinth. The results show that the numerical calculation can predict the flow phenomenon in the cavity to a certain extent. Rhode[2] studied the effects of throttling gap width, step height, and tip diameter of sealing labyrinth on the leakage of the sealing labyrinth on the basis of studying the sealing performance of the sealing labyrinth. In addition, he also studied the leakage characteristics of the special cavity shape of the stepped sealing labyrinth, and found out the cavity geometry with the greatest leakage resistance. The results of research by

Romulo et al.[3-4] show that the flow coefficient can be reduced by reducing the tip width of sealing labyrinth, rounding the cavity or adding grooves on the surface of stator parts, and the sealing requirements can be basically satisfied by the number of sealing teeth reaching three. Research by Doukelis et al. [5] showed that the vortex structure of the leakage flow is more sensitive to the gap size when the hub rotates. Dento et al. [6] studied the influence of inlet and outlet geometry of stator cavity on stator leakage flow characteristics and analyzed the mechanism of leakage flow affecting the main flow.

Studies by Meng Dejun[8] and others show that the seal clearance of sealing labyrinth is linearly proportional to the leakage, but the influence of the sealing clearance on the total pressure loss is non-linear. Luo Yan et al. [9] Carried out an experimental test on a two-stage low-speed axial-flow compressor and found that the flow loss inside the compressor was mainly caused by the leakage flow at the root of the blade and the separation of the angular region at the blade tip. The leakage flow at the root of the blade and the separation flow at the corner of the blade are constantly mixed with the surrounding mainstream gas, causing total pressure loss. Wang Pengfei[10], Wang Suofang[11-12], Ji Guojian[13], Wang Daijun[14], etc. respectively conducted relevant researches on the effects of rotation speed and pressure ratio on the sealing characteristics of the sealing labyrinth. The study of ma yaru et al. [15] shows that the anti-friction structure design of the tooth-tip groove can not only adapt to the radial vibration and axial movement of the rotor, but also reduce the amount of sealing leakage and improve the sealing efficiency of the sealing labyrinth. Cao Yonghua[16] and others conducted a two-dimensional reverse jet test and found that when the jet angle was 45 °and the jet position was in the middle of the first tooth cavity, the leakage coefficient was reduced by 11.5% compared with that without the jet.

2. Computational models and numerical methods

2.1. Introduction to computational models

A simplified compressor cascade sealing labyrinth model (figure 1) which is composed of three labyrinth seals, platform bushing, wheel hub and reverse jet port is adopted in this study. The three-tooth sealing structure is located between the platform bushing and the wheel hub, and the reverse jet port is located in the axial middle of the first tooth cavity. Due to the periodicity of the flow structure, a simplified model of 10°is taken here for simulation. The sealing labyrinth model is divided into a rotational upstream segment, a jet-affected segment and a rotational downstream segment along the circumferential direction (Figure 2). The jet-affected segment is the area where the jet flow has a significant effect on the structure and performance of leakage flow in the tooth cavity. The axial width of the jet port is defined as *l*, the circumferential width of the jet port is defined as *d*, and the height of the jet port (H) is uniformly set as 1mm. The parameters of sealing labyrinth are shown in Figure 3.

The leakage clearance (*e*) is 0.39mm, the thickness of the tooth tip (*d*) is 0.2mm, the depth of the tooth cavity (h) is 4.3mm, and the width of the tooth cavity (D) is 4mm. The total length of the leakage flow channel is 18 times the width of the tooth cavity, the inlet of the channel (a) is 7 times the width of the tooth cavity, and the outlet of the channel (b) is 8 times the width of the tooth cavity.

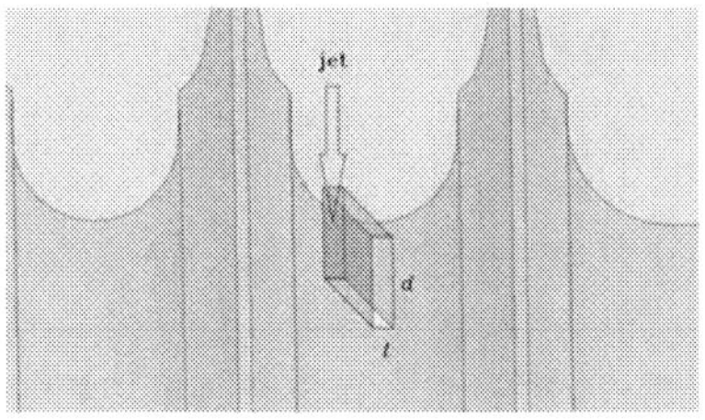

Figure 1.A simplified compressor cascade sealing labyrinth model

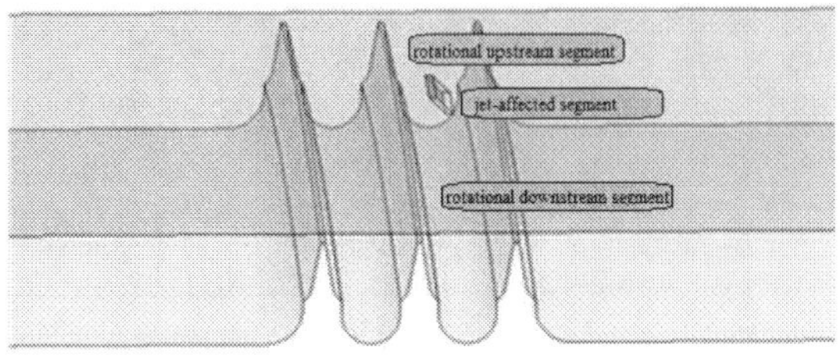

Figure 2. Three parts along the circumferential direction

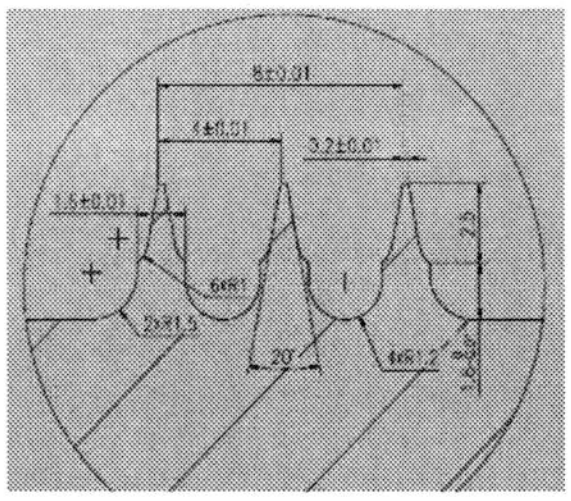

Figure 3. The parameters of sealing labyrinth

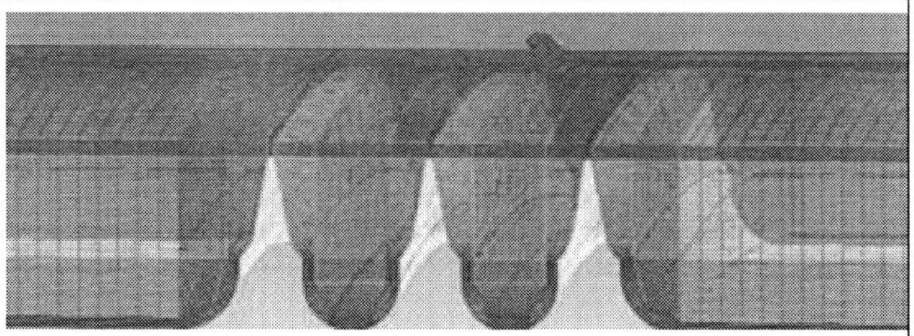

Figure 4. The grids of the computational domain

2.2. Computational grids and boundary

In this paper ,the grids of the computational domain are manually divided into blocks using ANSYS ICEM CFD software. The computational domain is divided into four parts: upstream cavity, downstream cavity, tooth cavity and jet port. As shown in Figure 4, the grids are all structured grids. Grids are encrypted along the wheel hub and the platform bushing and the height of the first layer of the grids between the platform bushing and the wheel hub is taken to be 0.008mm. The number of near-wall surfaces y + is controlled by adjusting the number of grid nodes and the extension ratio of the grid to meet the accuracy requirements of the calculation.

The grid-independent analysis is performed here. It can be seen from Figure 5 showing the changes of channel flow under six different numbers of grids that when the number of grids is around 2 million, the flow in the channel remains around 1.996×10-2kg/s, As the number of grids continue to increase, the flow remains almost unchanged. Therefore, the number of computational grids in this study is set at about 2 million.

Calculation boundary conditions: the total inlet pressure is set to standard atmospheric pressure of 101325pa, the total inlet pressure of jet is set to 102000Pa, the total temperature is set to 300k, the solid wall is set to adiabatic and non-slip boundary, the periodic boundary is given along the circumference, and the speed and outlet pressure are set according to different working conditions.

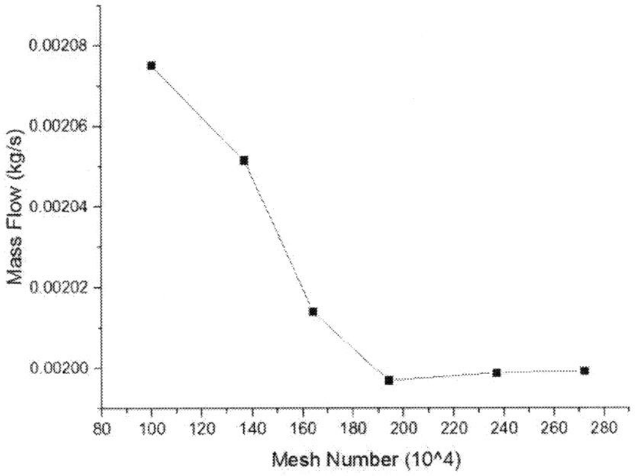

Figure 5. The changes of channel flow under six different numbers of grids

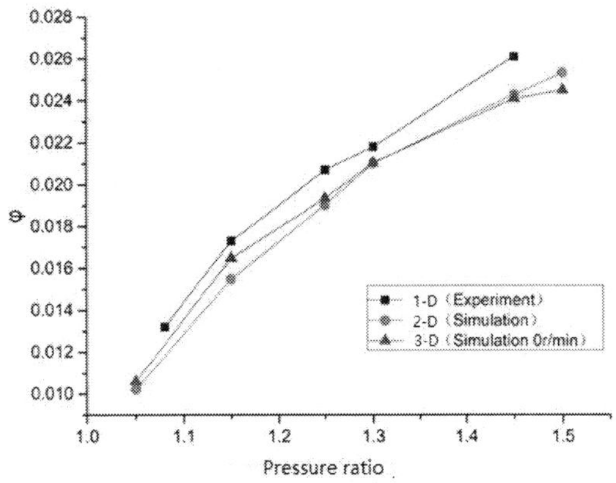

Figure 6. Verifying the accuracy of the calculation model

2.3. Calculation accuracy verification

In order to verify the accuracy of the calculation model and compare the aerodynamic losses caused by the sealing structure, this study sets the rotating speed at 0r/min, the total temperature at 300K, the total inlet pressure of sealing labyrinth at 101325Pa and the total inlet pressure of the jet at 102000Pa. The curve of the leakage coefficient of sealing labyrinth under different pressure ratios is obtained and compares with the experimental results under the same experimental conditions in literature [16].It can be seen from Figure 6 that the 3D numerical simulation results are slightly lower than the experimental results, but the overall agreement is good. Therefore, the numerical simulation results are reliable.

3. Calculation results and analysis

3.1. Fluid flow structure in the tooth cavity of labyrinth seals under the influence of jet

In this paper, the wheel speed is set as 3000r / min and the inlet and outlet pressure ratios are both set as 1.5. In order to analyze the flow structure of the fluid in the tooth cavity under the influence of the jet,the axial width of the jet port is set as 0.3mm, the circumferential width is set as 1.0° and the incidence Angle of the jet (α) is set as 45°.

Figure 7 is the fluid flow diagram in tooth cavity 1. It can be found from the Figure (a) showing the fluid flow structure in the tooth cavity 1 that the leakage flow and the jet flow are violently mixed near the jet port, which results in a very complicated fluid flow structure in the tooth cavity.

Figure (b) shows the flow of jet fluid after it enters the tooth cavity, where the black streamline represents the breathable airflow. It can be seen from the figure that the flow direction of the black fluid is deflected toward the middle of the tooth cavity near the jet port under the influence of the jet flow. After the jet flow enters the tooth cavity, a part of the fluid (cyan streamline) forms a small clockwise jet vortex downstream of the jet port whose circumferential width is the same as the circumferential width of the jet port and the other part of the fluid (purplish red streamline) is impacted by the fluid in the breathable area of the upper wall. Under the influence of the black fluid, the purplish red fluid gradually changes its flow direction and wraps the cyan vortex in it. Finally, both flow to the next tooth cavity with the black fluid.

Figure (c) shows the flow structure of the leakage fluid entering the tooth cavity near the jet port. It can be seen that the green fluid forms a large-size vortex together with the jet flow in the tooth cavity. The large-size vortex develops towards the rotation direction of labyrinth and eventually forms a small-sized vortex opposite to the direction of its own vortex rotation at the bottom of the tooth cavity. After entering the tooth cavity, the pink fluid also forms a large-size vortex which points to the opposite rotation direction of the labyrinth under the influence of the jet flow. Finally, it flows directly into the next tooth cavity near the jet port.

Figure (d) shows the development of the main vortex line in the first tooth cavity. The development of the blue fluid in the tooth cavity is extremely irregular. In the rotational upstream segment, the closer the blue fluid is to the jet-affected segment, the smaller its size will be. After passing through the jet-affected segment, there is a certain accumulation of blue fluid. Then the size of the vortex gradually increases and flows into the next cycle. In the rotational upstream segment, the light blue fluid develops in the form of small-sized vortices in a counterclockwise direction. When the light blue fluid flows near the jet port, it is rolled up by the pink fluid to change the flow direction. Then the two fluid flow together toward rotational upstream segment. However, when the pink fluid enters the next tooth cavity, it is mixed with the blue fluid and eventually flows toward rotational downstream segment in the form of large-size vortex.

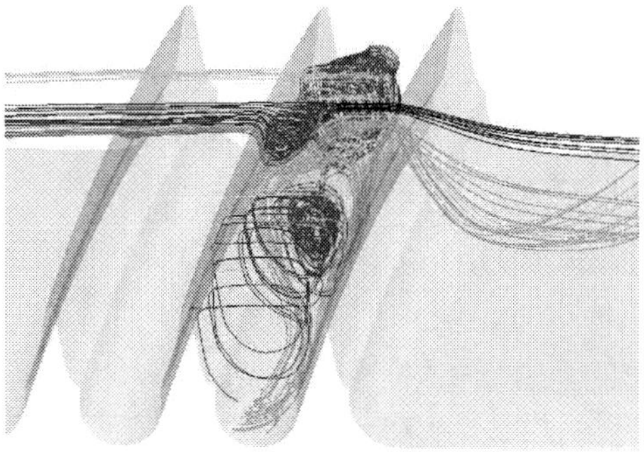

Figure (a).

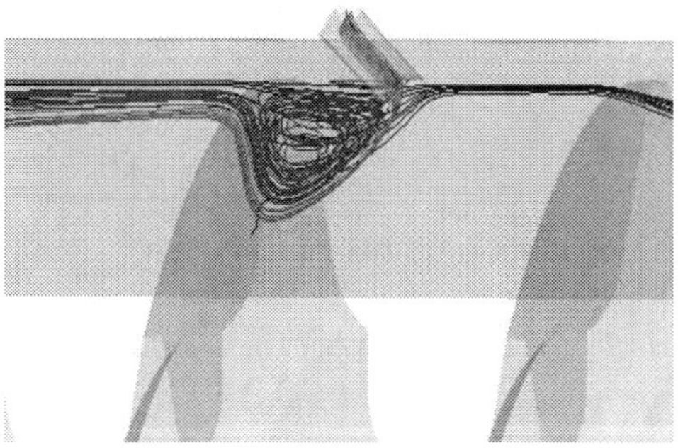

Figure (b).

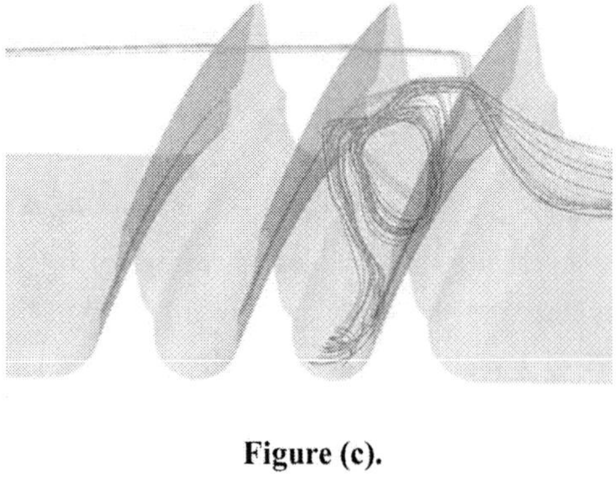

Figure (c).

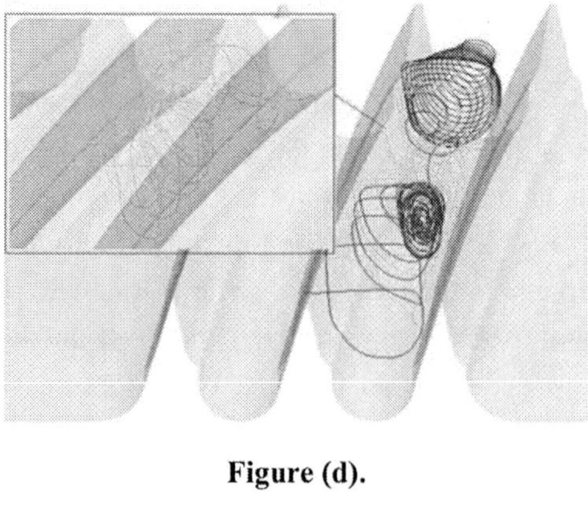

Figure (d).

Figure 7. The fluid flow diagram in tooth cavity 1

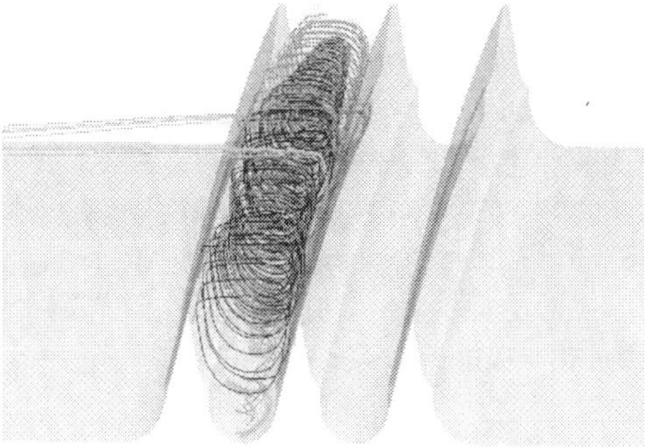

Figure (a).

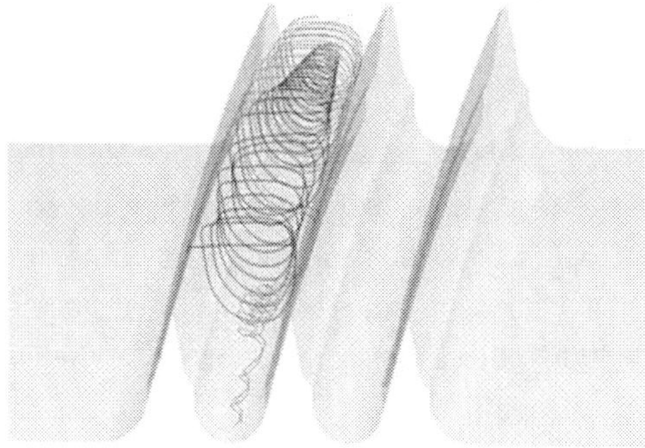

Figure (b).

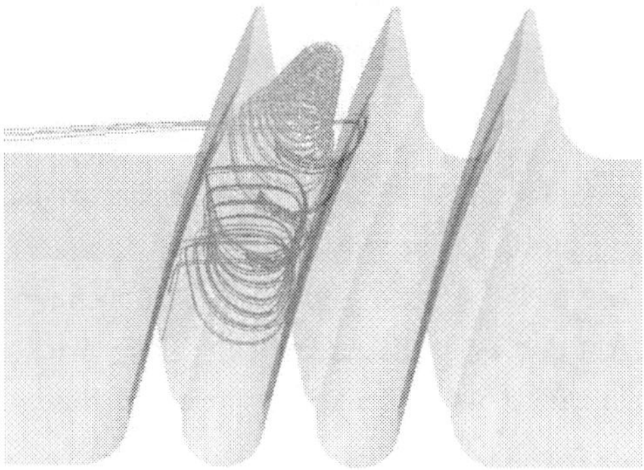

Figure (c).

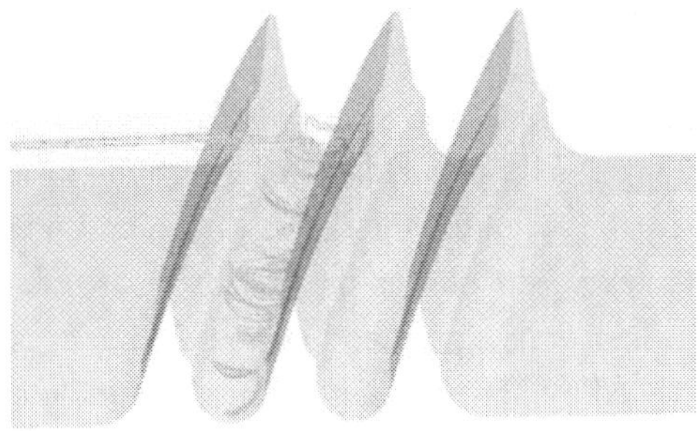

Figure (d).

Figure8. The fluid streamline diagram in the tooth cavity 2

Figure 8 is the fluid streamline diagram in the tooth cavity 2. It can be seen from Figuer (a) which shows the overall flow structure of the fluid in the tooth cavity 2 that the fluid flow structure in the tooth cavity 2 is relatively simple and regular compared to the flow structure in the tooth cavity 1.

Figures (b) and (c) show the development of the main vortex fluid in the tooth cavity 2. By comparing the flow structure of the fluid in cavity 2 with that in cavity 1, it can be found that the flow structure of a part of the main vortex fluid in cavity 2 is not damaged. The blue fluid in the Figure (b) is a small-size vortex in the ratational upstream segment but still develops as a large-size vortex in the entire tooth cavity. The green fluid in the Figure (b) develops in the form of a main vortex in the ratational upstream segment and changes in the flow direction near the jet-affected segment. Finally, it flows downstream of the rotation in the form of a small-sized vortex in a counterclockwise direction. It can be seen from the Figure (c) that a part of the main vortex fluid represented by light pink streamlines is mixed with the jet flow near the jet port and directly enters the upstream cavity through the next tooth tip. We can alse find that the orange fluid in the rotational upstream segment develops in the form of a small-size vortex and the orange fluid in the vicinity of the jet-affected segment is lifted up by the jet flow. Finally, the orange fluid forms large-scale vortex in the tooth cavity.

Figure (d) shows the development of the small-size vortex represented by a pink fluid at the bottom of tooth cavity 2. The small-size vortex develops irregularly along the circumferential direction and a part of the small-size vortex is sucked by the jet flow in the jet-affected segment and mixed with the jet flow to enter the upstream cavity through the third tooth.

In general, the jet flow dose not make the flow structure in the tooth cavity develop into regular main vortex and a small-size vortex that rotates in the opposite direction. On the contrary, the interaction between jet flow and leakage flow makes the flow structure in the tooth cavity complex and changeable. Although the leakage flow still forms a main vortex and a reverse small-size vortex in the tooth cavity, the two fluids mix with each other and permeate each other in the jet-affected section, which causes the fluids to constantly change direction ,separate and merge. In the end, the flow loss of the fluid increases sharply, thereby achieving the overall sealing effect and greatly reducing the leakage coefficient.

3.2. *Analysis of the influence of jet on the flow performance in the tooth cavity*

In order to study the influence of the jet flow structure on the performance of the sealing labyrinth, the changes of vorticity and static pressure in the tooth cavity caused by the jet flow are analyzed.

3.2.1. *Analysis of vorticity in the tooth cavity*

Figure 9 is the cloud diagram of vorticity distribution in the tooth cavity under the influence of jet flow. It can be seen from the Figure (a) which is the axial distribution of vorticity at the jet position (5 °)that the low vortex fluid in the breathable area develops close to the surface of the platform bushing from the first tooth tip to the location of the jet port. When it comes to the jet port, the low vortex fluid affected by jet vortex deflects to the middle of the tooth cavity, which makes a part of the fluid in the breathable area enter the tooth cavity to participate in the dissipation of the vortex. It can also be found in the figure that the jet flow accumulates between the jet port and the second tooth tip to generate a region of high vortex which can greatly block the leakage flow into the next tooth cavity. There is an obvious high vortex area caused by the fluid forming a small-size vortex in the tooth cavity opposite to the rotation direction of the main vortex in the convex platform area to the left of tooth cavity 2. This is consistent with the flow structure in tooth cavity 2 shown in figure 8.

Figure (b) shows the vorticity distribution at seven different positions along the circumferential direction: 0°, 2.5°, 4°, 5°, 6°, 8°, and 10°. The high vortex area formed by the jet flow downstream of the jet port is very small at 4°and 6°circumferential positions, which indicates that the range of the jet flow affecting the breathable fluid is about twice the circumferential width of the jet port. From the position of the jet port to the rotation surfaces on both sides, the area of the high vorticity region inside the tooth cavity 1 gradually decreases and the high vorticity region at the convex table on the left side of the tooth cavity gradually becomes obvious. This indicates that, outside the influence range of the jet, a small scale vortex is formed at the bottom of the tooth cavity again. The vorticity in cavity 2 does not change significantly along the circumferential distribution, but on the 4°and 6°cross sections, a clear high vortex area appears at the center of the backflow region of tooth cavity 2. This high vorticity region is not connected to the main vortex region, mainly because the light pink fluid and orange fluid in Figure 8 (c) are separated from the main vortex on the boundary of jet-affected segment and directly enter the next tooth cavity.

IOP Conf. Series: Materials Science and Engineering 1081 (2021) 012018 doi:10.1088/1757-899X/1081/1/012018

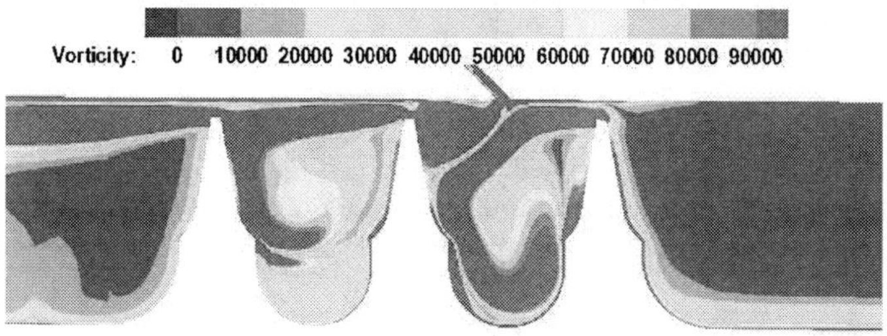

Figure(a).the axial distribution of vorticity at the jet position (5 °)

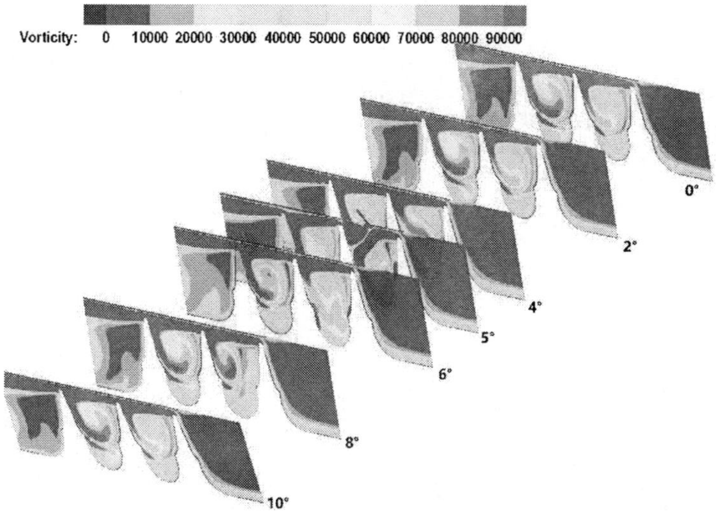

Figure(b).the vorticity distribution at seven different positions along the circumferential direction: 0°, 2.5°, 4°, 5°, 6°, 8°, and 10°

Figure9.the cloud diagram of vorticity distribution in the tooth cavity under the influence of jet flow

3.2.2. Analysis of turbulent kinetic energy in the tooth cavity

It can be found from the Figure 10 (a) shwing the axial distribution cloud diagram of the turbulent kinetic energy at the jet position (5 °). that the turbulent kinetic energy close to the platform bushing is small and the gradient of the turbulent kinetic energy in the middle of the tooth cavity is small. The turbulent kinetic energy is very low in most areas of the entire tooth cavity, but at the tip of the tooth, the turbulent kinetic energy changes violently due to the large disturbance of the fluid here. At the left side of the jet port, the turbulence intensity increases due to the mixing and collision of the leakage flow and the jet fluid. Therefore, the turbulent kinetic energy increases sharply. With the influence of labyrinth configuration, the turbulent kinetic energy reaches the maximum value at the second tooth tip. The turbulent kinetic energy on the left side of the tooth cavity 1 increases, forming a local high turbulent kinetic energy region, which is consistent with the flow structure of the tooth cavity described in FIG. 7, where a local small-scale vortex is formed.

Figure 10 (b) shows the distribution of the turbulent kinetic energy of the tooth cavity at different circumferential positions. It can be seen from the figure that the distribution of the turbulent kinetic energy contour at the jet position in the tooth cavity 1 is the most complicated, and the gradient of the turbulent kinetic energy at the tip of the second tooth is the largest. From the position of jet flow along the circumference to the two sides, the turbulent kinetic energy changes gently in the cavity 1, and the region of low turbulent kinetic energy gradually expands to the bottom of the cavity 1. As can be seen from the cloud diagram of the sections of 0°, 2°, 8° and 10°, the contour line k=400J of turbulent kinetic energy in the tooth cavity 2 splits the low-turbulent kinetic energy region into two from the convex stage of the Sealing labyrinth, and forms a crescent-shaped low-turbulent kinetic energy region at the bottom of the tooth cavity. It shows that the influence of the jet fluid on the energy of the leakage flow is mainly in the area near the jet port in the tooth cavity 1, while the main influence area in the tooth cavity 2 is far away from the jet port.

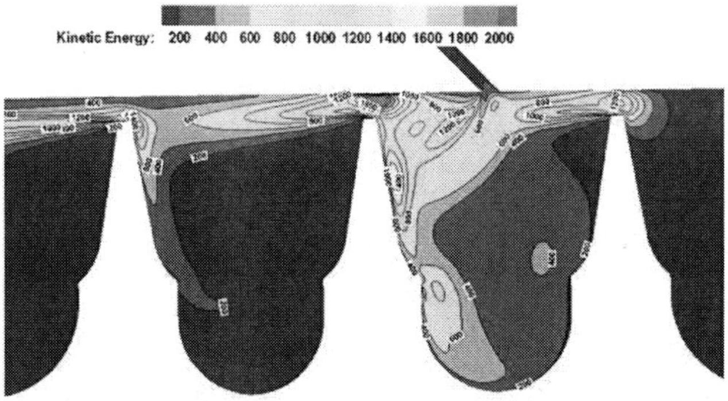

Figure(a).the axial distribution of the turbulent kinetic energy at the jet position (5 °)

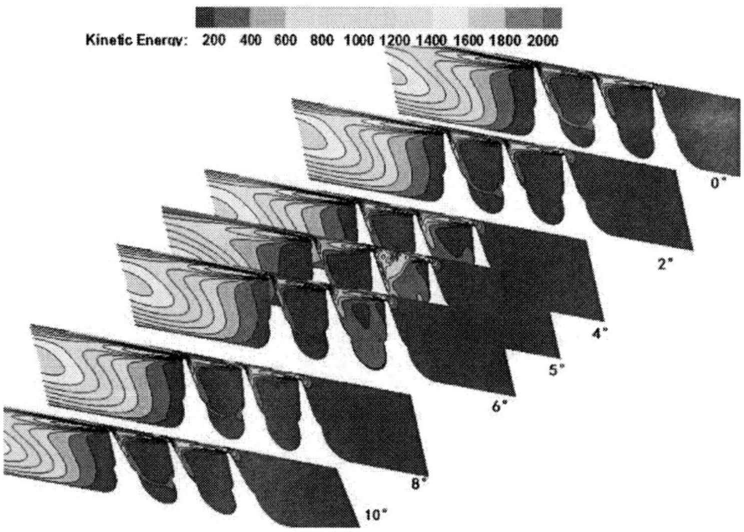

Figure(b).the distribution of the turbulent kinetic energy of the tooth cavity at different circumferential positions

Figure10.the cloud diagram of the turbulent kinetic energy distribution in the tooth cavity under the influence of jet flow

3.2.3. Analysis of the mach number distribution in the tooth cavity

Figure 11 is an axial distribution cloud diagram of the Mach number of the tooth cavity at the jet position (5°). It can be seen that the high Mach region in the upper part of the tooth cavity 1 is blown into two parts by the jet fluid, one part is on the left side of the first tooth tip, and the other part is on the left of the jet position mouth and deflected into the tooth cavity at the same time. On the left side of the jet port close to the platform bushing, the Mach number is reduced to the minimum Mach,

IOP Conf. Series: Materials Science and Engineering 1081 (2021) 012018 doi:10.1088/1757-899X/1081/1/012018

where the fluid forms a vortex. There is a low Mach region at the bottom of tooth cavity 2 that is not connected with the low Mach region at the main vortex core, which is similar to the streamline distribution in tooth cavity 2 in Figure 8.

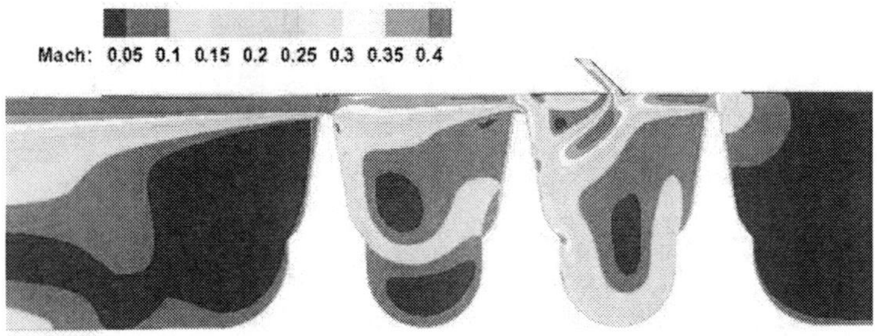

Figure11. The axial distribution of the Mach number of the tooth cavity at the jet position (5°)

3.2.4. Analysis of static pressure distribution in the tooth cavity

Figure 12 shows the static pressure distribution of the tooth cavity under the influence of the jet. It can be found from the Figure12 (a) showing the axial distribution of static pressure in the tooth cavity at the jet position (5°) that the static pressure of the fluid is significantly reduced after the fluid passes through the tooth tip throttling. At the jet port, the jet flow and the breathable flow collide with each other, resulting in the decrease of kinetic energy and the increase of static pressure, so a small area of high pressure appears in this area. In the downstream of the jet port, the static pressure decreases due to the dissipation of the vortex. The existence of the vortex generated by the jet flow causes the breathable flow to deflect into the tooth cavity and hit the left wall of the tooth cavity at a high speed, which causes the kinetic energy of the fluid to drop sharply and the static pressure to increase. Therefore, it can be found from the figure that a small area of high pressure appears on the left wall of the tooth cavity 1.

Figure12 (b) shows the static pressure distribution at seven different positions along the circumferential direction: 0°, 2°, 4°, 5°, 6°, 8°, and 10°. It can be seen that the static pressure distribution in the tooth cavity 2 changes little along the circumference. Comparing the static pressure distribution cloud diagram in tooth cavity 1, it can be found that the pressure gradient formed by the jet fluid at the jet port is not obvious on the 4°section and the 6°section. At the same time, the low-pressure region downstream of the jet port is also weak on these two sections, indicating that the jet fluid has a small effect on the static pressure here. This is similar to the vorticity distribution under the influence of the jet, which shows that the boundary of the influence of the jet fluid on the performance of the cavity fluid is indeed in the range of 4°and 6°.

IOP Conf. Series: Materials Science and Engineering 1081 (2021) 012018 doi:10.1088/1757-899X/1081/1/012018

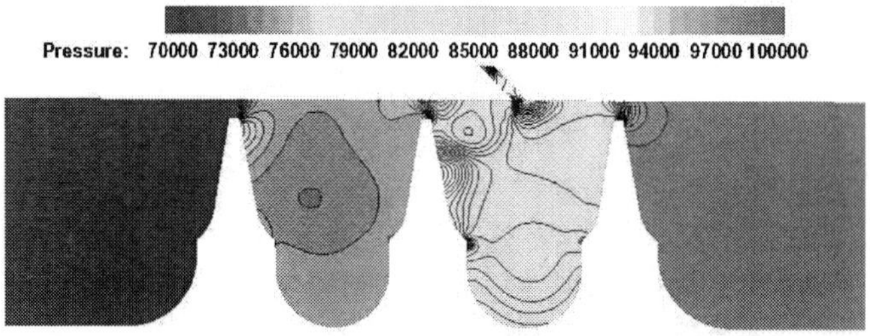

Figure (a). The axial distribution of static pressure in the tooth cavity at the jet position (5°)

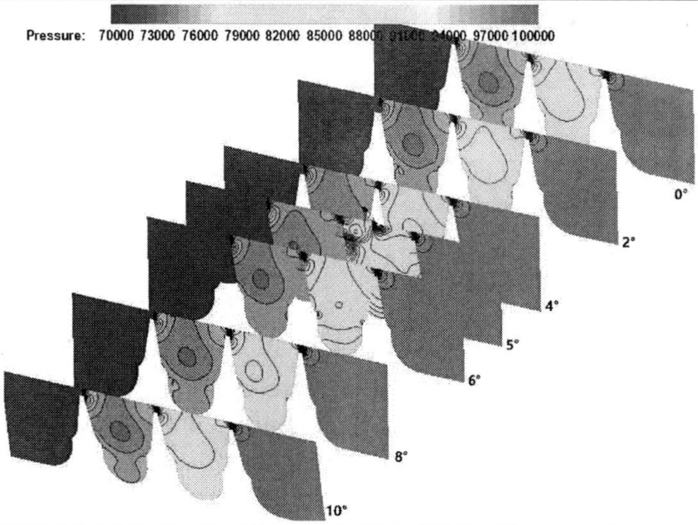

Figure (b). The static pressure distribution at seven different positions along the circumferential direction: 0°, 2°, 4°, 5°, 6°, 8°, and 10°.

Figure 12. the cloud diagram of the static pressure distribution in the tooth cavity under the influence of jet flow

4. Conclusions

In this paper, through the numerical simulation of the three-tooth sealing structure with jet structure, the flow structure of sealing labyrinth and the flow performance of sealing labyrinth under the influence of jet are recognized and analyzed. The conclusions are as follows:

(1) Within the jet-affected section, the jet fluid and the leaked fluid interact with each other, which makes the flow structure in the tooth cavity complicated and changeable, resulting in the failure to form a small-scale vortex at the bottom of the tooth cavity 1. Downstream of the jet, the jet fluid forms a vortex to achieve the purpose of blocking the leakage flow.

(2) A small-size vortex appears in the tooth cavity 2, but near the jet section, a part of the small-scale vortex fluid entrained by the main vortex fluid flows together with the main vortex fluid. The other part climbs upward along the wall of the tooth cavity. Finally, it is sucked by the breathable fluid and flows directly into the upstream cavity through the third tooth.

(3) The jet flow makes the vorticity distribution, turbulent kinetic energy distribution, Mach number distribution and static pressure distribution of the cavity change dramatically and the range influenced by the jet flow is approximately 2 times the circumferential width of the jet port.

(4) Within the jet-affected section, the low vortex of the breathable area is deflected into the

tooth cavity to participate in the dissipation of vortex.

Reference

[1] Stoff H 1980 Incompressible Flow in a Labyrinth Seals *Journal of Fluid Mechanics J* (Translated of ASME) 110:P817-829

[2] Romulo B.Freitas 2012 Computational Evaluation of Labyrinth Seal Congurations *J* AIAA-2012-04 76

[3] Rhode D L and Younger J S 1997 Flow Visualization and Leakage Measurements of Stepped Labyrinth Seals: Part 2 Sloping Surfaces *J* (Transactions of the ASME) 119:884

[4] Rhode D L 2002 Flow Visualization and Leakage Measurement of Stepped Labyrinth Seal Part 1-Annular Groove *J*

[5] Doukelis A, Mathioudakis K and Papailiou K 1998 The Effect of Tip Clearance Gap Size and Wall Rotation on the Performance of a High-Speed Annular Compressor Cascade *C Asme International Gas Turbine & Aeroengine Congress & Exhibition*

[6] John D.Denton and Eric M.Crutis 2008 The influence of shroud and cavity geometry on turbine performance:An experimental and computational study Part1:Shroud geometry *J Journal of Turbomachinery* 130:041001-1

[7] SHEN Hong, ZHEN Tian-hui, and CHEN Yu-jie 2011 Improvement of Aero-Engine Sealing Technology *J Gas Turbine Experiment and Research* 24(4)： P51-55

[8] MENG De-jun, XING Lei and LI Jian 2013 Influence of Shrouded Stator Cavity Flow on Compressor Stator Performance *J Aeroengine* 39(6):P31-35.

[9] Luo Ju, Hu Jun, Yan Wei, Zhang Chenkai,and Yin Chao 2013 Measurement of Flow Field in Stator Passages of a Compressor with Hub Gap *J China Mechanical Engineering* 24(18):P2458-2468

[10] WANG Peng-fei, LIU Yu-fang, GUO Wen, LI Jiang-hai and LIU Bo 2007 Influence of High Rotational Speeds on the Labyrinth Sealing Char acteristics *J Gas Turbine Experiment and Research* 120(2):P45-48

[11] WANG Suo-fang, LU Hai-feng and XIA Pan-yong 2004 Numerical Analyses and Experimental Investigations of Seal Labyrinth Characteristics *J Journal of Nanjing Univestics of Aeronautics&Astronautics* 36(6):P732-735

[12] WANG Suo fang and LU hai-feng 2005 Numerical Analysis of Flow and Experimental Investigation of Labyrjnth Sealjng Characteristics at Rotational and Static States *J Journal of Aerospace Power* 20(3):P444-449

[13] Ji Guojian, Ji Honghu and Cao Guangzhou 2008 Effect of Rotating Speed on Leakage Characteristics of Labyrinth-Honeycomb Seal with Small Clearance [J]. Journal of Nanjing University of Aeronautics & Astronautics 40(2):142-145

[14] WANG Dai-jun, SU Yun-liang and LUO Xiang 2007 Experimental Investigation on Leakage Flow Char acteristics in Rotating Labyrinth Seals *J Gas Turbine Experiment and Research* 20(2)

[15] MA Ya-ru, HUO Wen-hao, LIU Jing， JIANG Li-jun, FANG Zhi and LI Zhi-gang 2019 Investigations on the Leakage Perfoemance of Wear Failure and Anti-rubbing Structure Design for a Labyrinth Seal *J Journal of Turbomachinery* (05)P64-71

[16] CAO Yong-hua 2017 A Study Based on Using Jet to Suppress Leakage Flow in the Labyrinth Seal *D Nanjing University of Aeronautics & Astronautics*
[17] Zhang Yibin 2015 Experiment Research of Compressor Labyrinth Seal and Ensemble Cupling Numerical Analysis *D Nanjing University of Aeronautics & Astronautics*
[18] CHEN Zhen-yi, PENG Tao, JIANG Bin and ZHENG Qun 2018 Effect of Axial Clearance on Radial Flow in Axial Compressor Near to Stall Condition *J Journal of Turbomachinery* (04)P16-23
[19] Jinjing Sun, Yangwei Liu, Lipeng Lu and Xavier OTTAVY 2016 Control of corner Separation By Optimized Slot Configuration in a Liner Compressor Cascade *J Journal of Turbomachinery* (06)P9-18
[20] FU Xin, CAO Yong-hua, ZHANG Yi-bin and ZHU Chao 2017 Study of Leakage Flow Boundary Model-based Sealing Labyrinth Analysis Method *J ACTA ARMAMENTARII* (04)P824-832
[21] ZHOU Bin-bin, YANG Bo and GU Chuan-gang 2012 Numerical Simulation of the Influence of Changeable Axial Gap on Compressor Clocking Effect *J Journal of Turbomachinery* (03)P9-15
[22] CHEN Zhe and LIU Ying-zheng 2012 Numerical Investigation on the Effect of Bleeding to the Flow-field of the Downstream Row in the Axial Compressor *J Journal of Turbomachinery* (03)P22-25

Design and Research of New Type High-speed Hydrostatic Spindle Test Rig

Yunfeng Xiao[1], Di Song[2],Bo Meng[1], Lingyu Wang[1], Hao Zhang[1], Mingyue Wang[2]

(1.Beijing University of Petrochemical Technology, Beijing 102617, China;
2.Beijing University of Technology,, Beijing 100124, China)

E-mail: mengbo_ty@bipt.edu.cn

Abstract. Aiming at the air dynamic pressure foil bearing performance test rig, design and research a new high-speed static pressure spindle test rig. Carry out aerodynamic design of the driving turbine and compressor, obtain the three-dimensional impeller shape and carry out the strength calculation, design a new high-speed static pressure spindle test rig; use the finite element method to model the test bed rotor system, and use the transfer matrix method to analyse the shaft system structure critical speed and mode, so that the test rig can effectively avoid the critical speed and run stably. Analyse the unbalanced response and stability of the rotor system, and reasonably arrange the test rig device to achieve the purpose of measuring the performance of the dynamic pressure foil bearing, and ensure that the turbine static pressure spindle test rig can provide for the measurement of the air dynamic pressure foil bearing performance stable high speed.

1. Introduction

The Air foil bearing is a kind of self acting dynamic pressure gas bearing. It uses surrounding air as lubricant[1], elastic foil as supporting element, and relies on the dynamic pressure effect between the flexible foil surface and the shaft to support the rotor system[2]. With the characteristics of high speed, low friction, no pollution and long service life[3], air foil bearing is widely used in rotating machinery such as air dynamic pressure suspension turbine cooler, oil-free lubricating turbine expander, centrifugal blowers with foil bearings, etc[4].

Since the 1990s, some universities and research institutions in China have only begun theoretical and experimental research on air foil bearings. The team of Professor Hou Yu[5]from the Institute of Refrigeration and Cryogenics of Xi'an Jiaotong University has carried out theoretical and experimental research on air foil bearings in turboexpanders. Yang Lihua et al. [6] built a radial bearing experimental platform, and used the friction moment method and radial displacement response spectrum method to test the bearing lift off speed. The Professor Feng Kai [7]of Hunan University has recently research on foil bearing and rotordynamic. Meanwhile, the member of his research team, Lv Peng and Liu Yuman[8,9] carried out the theoretical and experimental research on flexibility aerodynamic pressured bearing, and experimentally studied the static characteristics of air foil bearings. Xie Shilong, Nanjing University of Aeronautics and Astronautics[10], discussed the influence of the fixing method on the structural rigidity of the wave foil gas bearing foil, and an experimental bench was set up. The experimental bench consisted of the experimental bench body, the lubrication cooling system, and the measurement system.

In 2011, Tae Ho Kim and Luis San Andresde et al.[11] Proposed a new experimental method to identify the structural stiffness and damping of bearings. However, The second generation wave foil bearing has not been tested. In 2015, Franck Balducchi et al.[12]built a pneumatic rotor test rig. The nonlinear response of a rigid rotor supported by gas foil bearing to different unbalance is studied. However, due to the heating problem of the motor, it is impossible to test the performance of the hydrodynamic foil bearing at high speed for a long time, which leads to the inaccuracy of the relevant theoretical model. Therefore, a new high-speed hydrostatic spindle test-bed using aerostatic bearing to carry the spindle is proposed, which can maintain high-speed and stable operation for a long time, which is important for the performance test of aerodynamic foil bearing research significance.

2.Aerodynamic design and strength calculation of turbine and compressor

2.1 Aerodynamic design

Referring to the parameters of the air turbine expander and the existing domestic and foreign turbine expanders, and according to the specific characteristics and requirements, the basic thermal parameters and structural parameters of the driving turbine and compressor are selected to optimize the reaction degree, wheel diameter ratio and other parameters. In addition to its thermal performance, the stability of mechanical performance is also an important factor, which determines the feasibility and operation reliability of mechanical design[8]. Therefore, it is necessary to carry out multi-objective optimization design for its thermal performance and mechanical performance. The constraints are considered comprehensively, and other thermal and structural parameters are selected from the database, and the experience of experts is added to improve the search efficiency. Main design results: The operating speed is 70000rpm, the air mass flow is 0.276kg/s, the compressor diameter is D = 92mm, the driving turbine D = 85mm, ρ = 0.53, U1 =0.624,and the adiabatic efficiency is greater than 0.65. Three-dimensional modeling of the final turbine and compressor are shown in Figures 1(a) and (b).

(a) (b)

Figure 1. (a)Three-dimensional modeling of turbine;(b) Three-dimensional modeling of compressor.

The calculation and simulation results of the turbine and compressor are as follows: the turbine inlet flow is 0.27kg/s, the outlet flow is 0.22kg/s, the overall efficiency is 92.5%, the inlet/outlet pressure ratio is 1.5, the compressor inlet flow is 0.27kg/s, the outlet flow is 0.22kg/s, the overall efficiency is 74.1%, the inlet / outlet pressure ratio is 2.04, and the error of the simulation results is within the allowable range.

2.2 Strength calculation

IOP Conf. Series: Materials Science and Engineering 1081 (2021) 012019 doi:10.1088/1757-899X/1081/1/012019

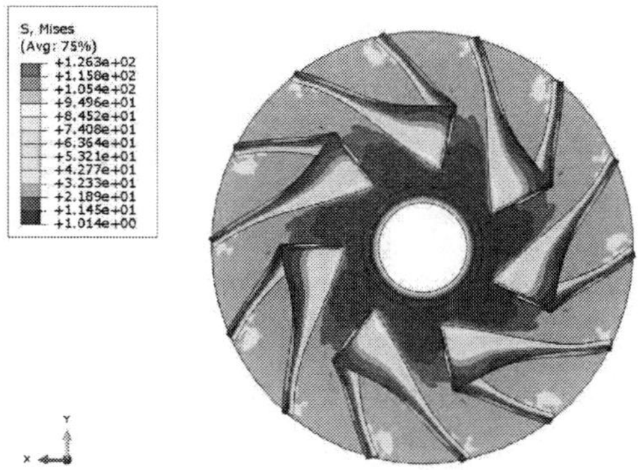

Figure 2. Stress contour

The computer-solved Mises stress value is obtained. The resulting stress distribution of the compressor impeller is shown in Figure 2. According to the cloud diagram display, the maximum stress value of the impeller occurs at the root of the leading edge of the blade is 126 MPa, and the yield strength of the material is 300MPa, which is less prone to yield failure than the allowable stress of the material, and is in a safer range. After the aerodynamic strength calculation, the results are almost the same.

3.Rotor dynamics calculation of hydrostatic spindle system

Using the transfer matrix method, the state vector of the ith section of the rotor is Zi, which is composed of the radial displacement Xi of the section, the deflection angle αi, the bending moment Mi and the amplitude of the shear Q_i

$$Z_i = [X, A, M, Q]_i^T \tag{1}$$

It has a certain relationship with the state vector Zi+1of section i+1, that is

$$Z_{i+1} = T_i Z_i \tag{2}$$

Where Ti is called the transfer matrix of a component between two sections. When there are r elements in the state vector, Ti is a square matrix of order r×r whose elements can be obtained by analyzing the force and deformation relations on the stressed components.

$$T_i = B_i D_i \begin{bmatrix} 1 + \frac{l^3(1+\nu)(m\omega^2 - K_{sj})}{6EJ} & 1 + \frac{l^2(I_p - I_d)\omega^2}{2EJ} & \frac{l^2}{2EJ} & \frac{(1+\nu)}{6EJ} \\ \frac{l^2(m\omega^2 - K_{sj})}{6EJ} & 1 + \frac{l(I_p - I_d)\omega^2}{EJ} & \frac{l}{EJ} & \frac{l^2}{2EJ} \\ l(m\omega^2 - K_{sj}) & (I_p - I_d)\omega^2 & 1 & l \\ (m\omega^2 - K_{sj}) & 0 & 0 & 1 \end{bmatrix} \tag{3}$$

Establish the rotor system model, use the small disturbance method to calculate the stiffness and damping value of hydrostatic bearing, select 25th node and 35th node to apply the bearing position, the model is shown in the figure 3.

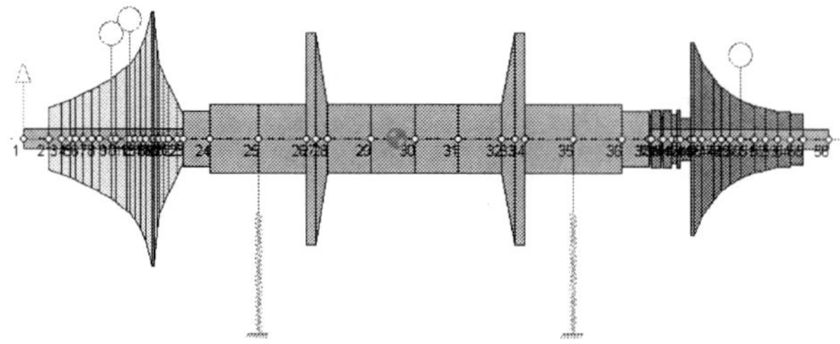

Figure 3. Dynamic model of rotor system

3.1 Calculation of critical speed under rigid support

The natural frequency of the rotor is determined by the stiffness and mass of the rotor and their distribution, independent of the initial conditions. The former three steps natural frequency of the rotor is the first three critical speed under the rigid support. It reflects whether the design of the rotor itself is reasonable.

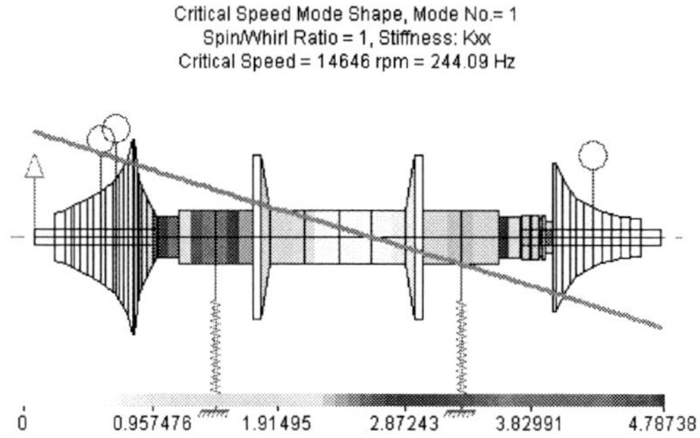

Figure 4. The first order modes and modes of the rotor

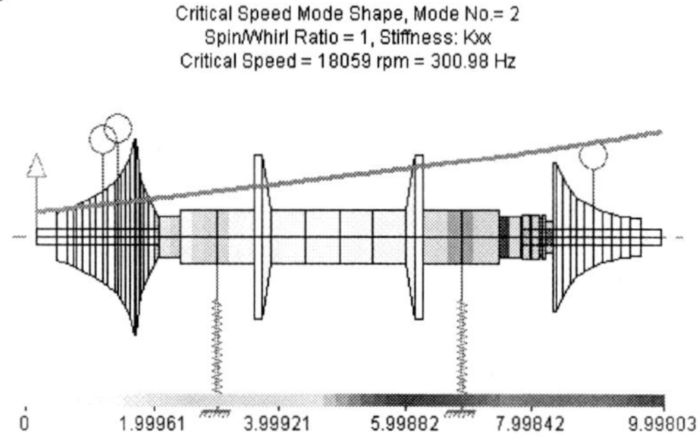

Figure 5. Second order modes and modes of vibration

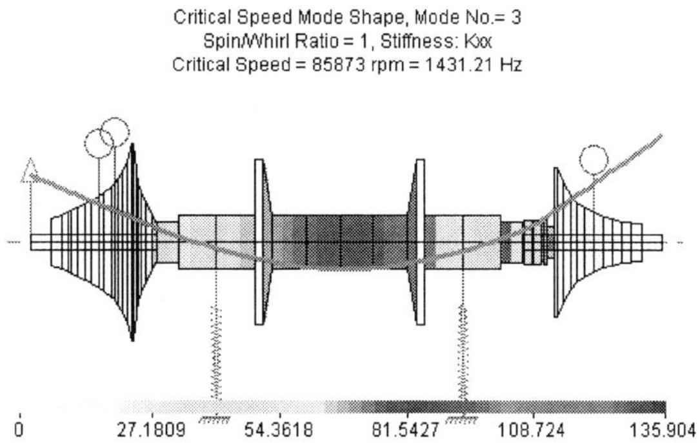

Figure 6. Third mode and mode shape of rotor

3.2 Unbalance response calculation of the rotor system of the experimental platform

The unbalance response is used to study the sensitivity of the rotor to the unbalance on the position. The unbalance response of the rotor is analyzed by the transfer matrix method, which is similar to the calculation of the critical speed, The recurrence of the state vector starts from the left most section of the rotor. The state vector of section i(i=2,3,…,N,N+1) is

$$\begin{Bmatrix} X \\ A \\ M \\ Q \end{Bmatrix} = \begin{bmatrix} a_{11} & a_{12} & a_{13} & a_{14} \\ a_{21} & a_{22} & a_{23} & a_{24} \\ a_{31} & a_{32} & a_{33} & a_{34} \\ a_{41} & a_{42} & a_{43} & a_{44} \end{bmatrix}_{i-1} \begin{Bmatrix} X \\ A \\ 0 \\ 0 \end{Bmatrix}_1 + \begin{Bmatrix} b_1 \\ b_2 \\ b_3 \\ b_4 \end{Bmatrix}_{i-1} = \begin{bmatrix} a_{11} & a_{12} \\ a_{21} & a_{22} \\ a_{31} & a_{32} \\ a_{41} & a_{42} \end{bmatrix}_{i-1} \begin{Bmatrix} X \\ A \end{Bmatrix}_1 + \begin{Bmatrix} b_1 \\ b_2 \\ b_3 \\ b_4 \end{Bmatrix}_{i-1} \tag{4}$$

For the right most N+1 section, and considering its boundary conditions are

$$\begin{Bmatrix} M \\ Q \end{Bmatrix}_{N+1} = \begin{bmatrix} a_{11} & a_{12} \\ a_{21} & a_{22} \end{bmatrix}_N \begin{Bmatrix} X \\ A \end{Bmatrix}_1 + \begin{Bmatrix} b_3 \\ b_4 \end{Bmatrix}_N = \{0\} \tag{5}$$

For the elastic undamped critical speed in the speed range of 0% - 125% of the trip speed, a separate damping unbalance response analysis is carried out. The unbalance shall be applied according to the specific shape of undamped vibration mode. The applied unbalance is calculated based on the total mass of the cantilever end and applied at the maximum displacement of the shafting.

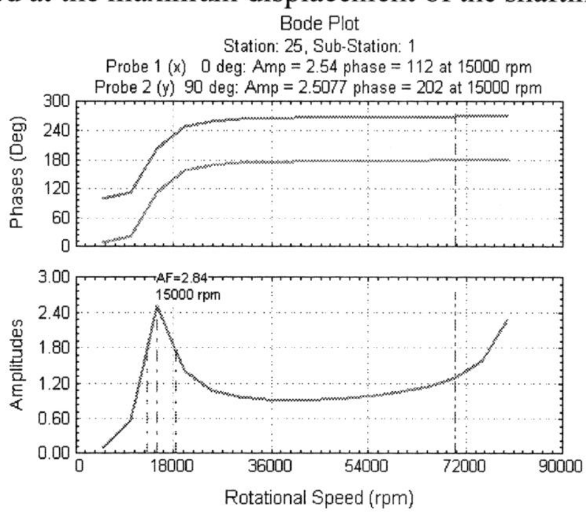

Figure 7. Response at the left bearing of the first-order unbalance response analysis of the shaft system

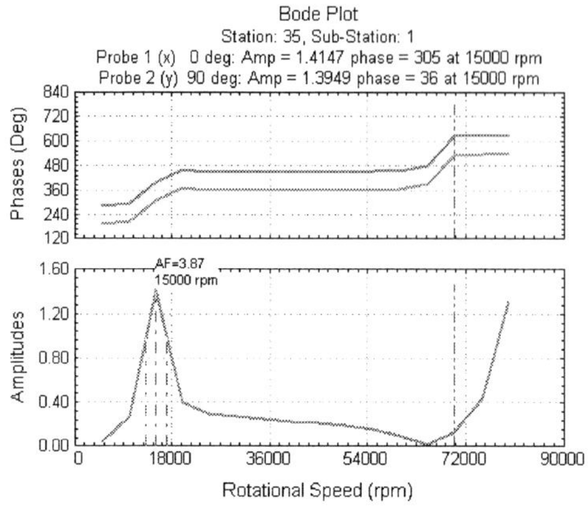

Figure 8. Response at the right bearing of the first-order unbalance response analysis of the shaft system

When calculating the unbalanced response, the AF value corresponding to the peak value of the left and right bearings is 2.84 and 3.87 respectively, and the calculation of the isolation margin is as follows:

14400rpm:

$$SM_{左}=17\left(1-\frac{1}{AF-1.5}\right)=17\left(1-\frac{1}{2.84-1.5}\right)=4.31 \tag{6}$$

15000rpm:

$$SM_{左}=17\left(1-\frac{1}{AF-1.5}\right)=17\left(1-\frac{1}{3.87-1.5}\right)=9.82 \tag{7}$$

The vibration peak value of the left bearing is large at 14400rpm, and its isolation margin is 4.31%. The working speed of 70000rpm falls outside the isolation margin, and the unbalanced response of the left bearing is qualified.

The isolation margin SM of the right bearing is the minimum value between the calculated value and 16, so it is 9.82.

The vibration peak value of the right bearing is large at 150000 rpm, its isolation margin is 9.82%, and the working speed of 70000 RPM falls outside the isolation margin. The unbalanced response at the right bearing is qualified.

According to rule 2.1.3 of API617[13], the unbalanced response at the left and right bearings is qualified.

3.3 Stability analysis

During the operation of the rotor, large-scale vibration may be caused by some reasons. These reasons are caused by the oil film force in the bearing, the fluid force in the seal, the friction between the rotor and the stator, and the mass and stiffness of the rotor are not equal in each radial direction And many factors that have yet to be recognized. Rotation of the rotor will cause the rotor to lose stability, referred to as instability. Instability is extremely dangerous for the safe operation of the rotor. The stability analysis is shown in Figure 11.

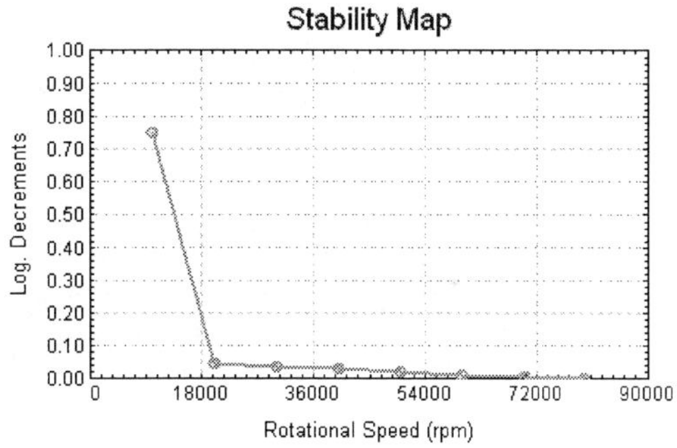

Figure 9. Stability map

For the elastic undamped critical speed in the speed range of 0% - 125% of the trip speed, The damping unbalance response should be analyzed. The unbalance shall be applied according to the specific shape of undamped vibration mode. The applied unbalance is calculated based on the total mass of the cantilever end and applied at the maximum displacement of the shafting.

4.Turbine static pressure spindle experimental bench construction

According to the design and calculation of the experiment table, the turbine static pressure spindle experiment table was constructed. The layout of the experiment table is shown in figure 9. Divided into drive module, loading module, measurement module and visualization system. The driving module uses an air compressor to input the high-pressure air source. The rotor system has been dynamically detected in the previous work, and the speed can reach the working speed. The measurement module consists of a series of sensors. The visualization system is output and processed by Labview software written by the team.

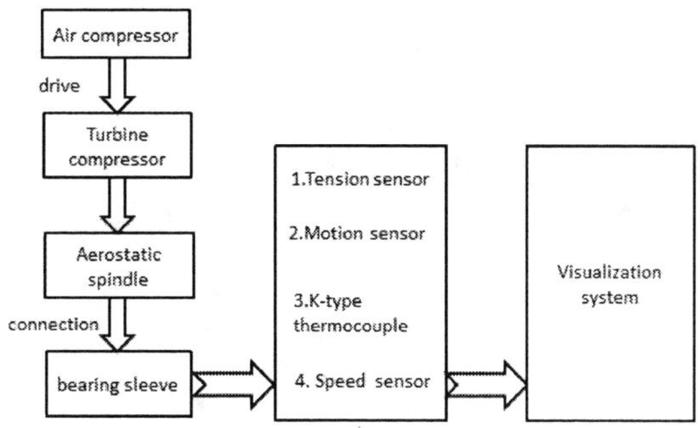

Figure 10. Layout of experiment table

Figure 11. Site map of aerodynamic spindle experiment table

4.1 Sensor arrangement

After the experiment table is set up, various sensors need to be reasonably arranged, mainly including speed sensors, displacement sensors, temperature sensors, and force sensors, to measure the speed, displacement, temperature, and friction torque of foil dynamic pressure air bearings. The performance test of the bearing capacity, start-stop times, take-off speed, and stability of the foil dynamic pressure air bearing is realized.

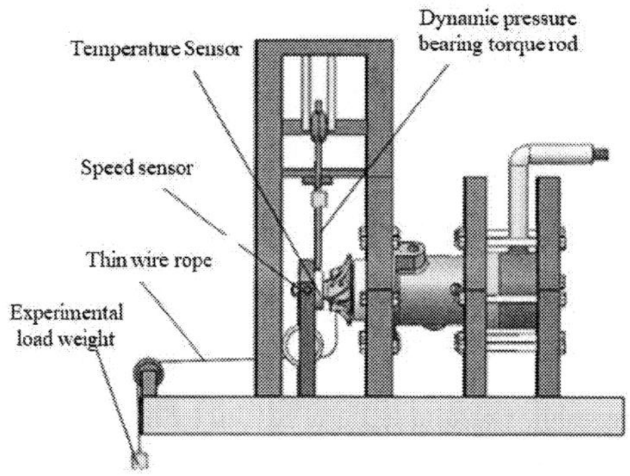

Figure 12. Sensor layout of aerodynamic spindle experiment table

5.Air foil bearing lift off speed and bearing capacity test

Lift off speed test of aerodynamic bearing is carried out by using the built experimental platform. Figure 11 is the actual picture of foil bearing, and the specific parameters of bearing are shown in the table.

Figure 13. Experimental bearing

Table 1.Geometrical parameters of foil bearing

Dimension parameters	Value
Foil thickness	0.07[mm]
Foil height	0.22[mm]
Bump pitch	3.3[mm]
Diameter	28[mm]
Foil widch	25[mm]

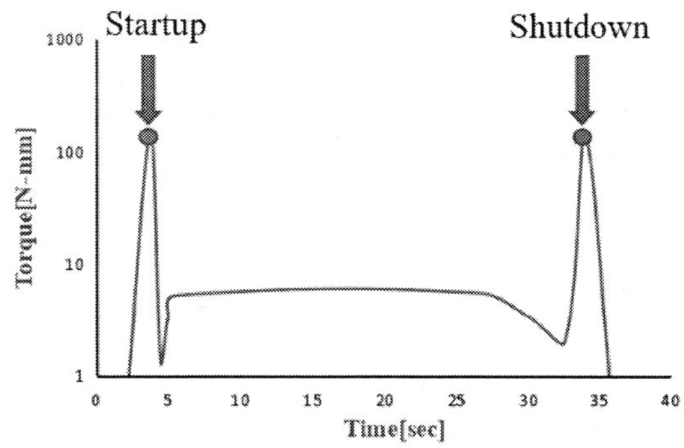

Figure 14. friction torque diagram

At the beginning of the experiment, the friction torque of the foil bearing is transmitted to the tension sensor through the torque rod, and then it is obtained through the data analysis platform. When the friction torque reaches the maximum value, it indicates that the hydrodynamic foil bearing is in the lift off stage, when the friction torque reaches a stable value, the rotational speed is the lift off speed of the bearing,the lift off speed is 4000rpm. when test bench runs for about 28 s, the friction torque decreases slightly and then increases rapidly when the driving device is turned off. The lift off speed measured in the experiment is 4000rpm.

6.Conclusion

(1)At a safe working speed, the aerodynamic design and strength check of the drive mechanism of the static pressure spindle experiment table were performed. The Mises stress was displayed as MPa, which met the requirements of the experiment table working speed.

(2)Based on the finite element method, the rotor system of the experiment table was modeled. The first three critical speeds and modes of the rotor system were studied, and the working speed of the experiment table was effectively avoided. Calculation, according to API617 Rule 2.1.3, the unbalance response verification is qualified; the stability analysis also meets the requirements.

(3)Based on the previous design and theoretical calculation of the experimental bench, a turbine static pressure spindle experimental bench was set up to rationally arrange the various sensors. Prepare for subsequent experiments with air foil bearings.

(4) The lift off speed of the five foil aerodynamic bearing was measured on the experimental platform, and the lift off speed of the bearing was 4000 rpm.

IOP Conf. Series: Materials Science and Engineering 1081 (2021) 012019 doi:10.1088/1757-899X/1081/1/012019

References

[1] NALEPA K,PIETKIEWICZ P， ZYWICA G. Development of the foil bearing technolog ［J］.Technical Sciences，2009（12）：229—240.

[2] AGRAWAL G L. Foil air/gas bearing technology: an overview:97-GT-347[R]. USA:ASME,1997:1-11.

[3] DELLACORTE C.Oil-free shaft support system rotor dynamics：past，present and future challenges and opportunities ［J］.Mechani－cal Systems and Signal Processing,2012,29 67—76.

[4] Shu Xingjun,Xu Gang,Zheng Yueqing,et al.Key technologies for centrifugal blowers with foil bearings and high speed magnet motor [J]. Chinese Journal of Turbomachinery,2017,59(02):36-42.

[5] Zhu Zhaohui,Yao Yanxia,Hou Yu,et al.Study of new aerodynamic foil journal bearing for cryogenic turbo expander [J]. Low Temperature and Specialty Gase,2003(02):28-31.

[6] Yang Lihua,Shi Jianhua,Liu Heng,et al. The experimental study on the lift of speed and load capacity of cpmpliant aerodynamic foil journal air bearings[J]. Lubrication Engineering,2006(03):15-18.

[7] Feng Kai,Huang Ming,Li Chengqin,et al.Test of high speed and heavy load gas foil bearing and set up of test rig[J]. Journal of Hunan University（Natural Sciences）,2020,47(02):1-5.

[8] Lv Peng. Design and experimental study on air cycle machine rotor system supported by gas foil bearing[D]. Hunan University,2017.

[9] Liu Yuman.The design of test rig and experimental research for high-speed gas bearings[D].Hunan University,2016.

[10] Xie Shilong Ma Xizhi Liu Xiangcheng. Influence of fixed types to structural stiffness of foil in bump-type gas bearing[J]. Lubrication Engineering. 2015,40(06):40-45+69.

[11] San Andrds L，Ryu K，Kim T H. Identification of Structural Stiffness and Energy Dissipation Parameters in a Second Generation Foil Bearing：Effect of Shaft Temperature. Journal of engineering for gas turbines and power，201 1，133(3)：032501

[12] Balducchi F, Arghir M, Gauthier R. Experimental analysis of the unbalance response of rigid rotors supported on aerodynamic foil bearings. Journal of Vibration and Acoustics, 2015, 137(6): 061014

[13] Zhang Xiaolong,Xie Yonghui. Study of the rotordynamic characteristics of a centrifugal compressor based on the API 617 standard[J]. Chinese Journal of Turbomachinery. 2017,59(01):32-37.

IOP Conf. Series: Materials Science and Engineering 1081 (2021) 012020 doi:10.1088/1757-899X/1081/1/012020

Effect of Gradient Layer Number on Microstructure and Property of Laser Deposition TA15/Ti_2AlNb Transition Zone

Bo He[1] Ying-tong Lin[1] Guang Yang[1*] Qing-jun Zhou[2] Xiang-ming Wang[3]

[1] School of Mechanical and Electrical Engineering, Shenyang Aerospace University, Shenyang 11036, China;

[2] Capital Aerospace Machinery Corporation Limited, Beijing 100076, China;

[3] Shenyang Aircraft Design Institute, Shenyang 110035, China;

*Corresponding author: yangguang@sau.edu.cn

Abstract. 3 numerical models of laser deposited TA15/Ti_2AlNb gradient composite structure were established to simulate the temperature variation of gradient composite structures with different gradient layers in different regions. Then TA15/Ti_2AlNb gradient composite structures with different gradient layers were prepared by laser deposition technology. The microstructure, room-temperature tensile properties and microhardness of the deposited samples were observed by optical microscope, tensile testing machine, scanning electron microscope and microhardness tester. The results showed that as the number of gradient layers increased, the temperature in the deposition process of the transition zone increased, while the microstructure difference in the transition zone decreased gradually. As the number of transition layers increased,the tensile strength decreased, but the elongation after fracture increased. All the fracture locations were in the transition zone, and with the increasing of gradient layers, the fracture location gradually shifted to the Ti_2AlNb side. The microhardness of the transition zone increased with the increasing of gradient layers. This study will provide reference for the optimization design of TA15/Ti_2AlNb gradient composite structure.

1. Introduction

Laser deposition manufacturing technology is an integrated manufacturing technology that can meet the requirements of forming on performance and precision. It is the most practical and reliable method in metal additive manufacturing[1]. Before the laser deposition starts, the deposition path of the laser head is planned in advance on a computer. The laser beam is used as high-temperature heat source to melt the surface of the substrate and form a melting pool. The metal powder is transported to the melting pool synchronously by the powder feeder. After rapid cooling, the powder solidifies and forms a metallurgical bond with the matrix material, and then solid components are manufactured by stacking layer by layer. This technology has been used more and more widely in the manufacturing of metal parts in aerospace field. With the rapid development of advanced manufacturing technology, its design concept has also promoted the liberation and improvement of structural design ideas[2~3].

Gradient composite structure means that the chemical composition of the structure changes continuously in one, two or three-dimensional space, so that the performance of the structure also presents a continuous gradient change. Compared with traditional composite structure, gradient composite structure has obvious advantages and unique characteristics. Firstly, gradient composite structure can be designed according to specific performance requirements under different service environments. By choosing different composition gradient, the materials can give full play to their own characteristics and performance advantages. Secondly, when two different materials intersect at the transition interface, the sudden change of physical properties such as thermal conductivity can be reduced by replacing the sudden change of interface between the two materials and the continuous change of component gradient, thus reducing the thermal stress caused by the difference of physical properties. Gradient composite structure has been widely used in aerospace, energy engineering, optical engineering and other fields because of its good performance[4].

TA15 titanium alloy is a near α titanium alloy rich in aluminum. It has the advantages of both α and α+β titanium alloys. It is characterized by high strength at medium and high room temperature, excellent thermal stability and outstanding welding performance, and it can work continuously at 500℃. Therefore, it is mainly used for aero-engines and various aircraft structural parts[5~9]. With the advantages of low density, high specific strength and fracture toughness, Ti_2AlNb alloy can work at high temperature of 600~850℃[10~12]. It is a high-temperature structural material commonly used in aerospace engines. TA15/Ti_2AlNb gradient composite structure was prepared by laser deposition technology, which could meet the performance requirements of different parts at different temperatures.

Zhong-liang Qi et al[13]. prepared TA15-Ti_2AlNb gradient material with different mass fractions of Ta15-Ti_2AlNb by using laser stereoscopic forming technology, and studied the relationship between the content, microstructure and hardness of Ti_2AlNb. Yan-tao Liu et al[14]. mixed TA15 and Ti_2AlNb powders in different proportions, prepared the thin walls of alloy with three different proportions by laser melting deposition process, and then analyzed the composition, microstructure and mechanical properties of the thin wall. Xian-zheng Bu et al[15]. connected TA15 and Ti_2AlNb alloys by laser welding process, and analyzed the microstructure, hardness and tensile properties of the welded joints. However, due to the characteristics of "rapid heating and rapid cooling" of laser deposition manufacturing technology and the different chemical compositions of the alloy, the temperature difference often occurs at the deposition transition interface, leading to the change of microstructure and performance. The performance at the transition interface of the composite structure composed of two materials is relatively weak[16~19]. Therefore, it is one of the current research hotspots to explore the formation law of the transition interface and find the regulation method to improve the performance of transition interface. By optimizing the transition mode, that is, using the gradient transition method, the temperature difference of the transition interface can be alleviated, and the structure and performance of the transition interface can be effectively controlled. In this paper, a temperature field finite element model of laser deposition TA15/Ti_2AlNb gradient composite structure was established by APDL language[20~23]. Then the temperature variation with time during the deposition of TA15/Ti_2AlNb gradient composite structure with different gradient layers was simulated. The TA15/Ti_2AlNb gradient composite structure with different layers was prepared by laser deposition manufacturing technology, and their microstructure and mechanical properties were tested. The effects of gradient layers on deposition temperature, microstructure and mechanical properties in the laser deposition manufacturing process were explored, so as to provide reference for further optimization of laser deposition manufacturing to manufacture gradient composite structure.

2. Modelling and Experiment

2.1. The establishment of temperature field model

Before the experiment, a finite element model of TA15/Ti_2AlNb gradient composite structure with three different gradient layers was established in the finite element analysis software, as shown in Figure 1. A hexahedral mesh was used. The size of the substrate was 70 mm×30 mm×10 mm and that of the deposition block was 50 mm×10 mm×30 mm, the thickness of each gradient layer was 3 mm, the laser power was 2 KW, the scanning rate was 8 mm/s, and the diameter of laser spot was 4 mm. Using the "life and death unit" method, the corresponding unit of the position that was not scanned by the laser shall be "killed", which shall be activated again when the corresponding unit was reached by the laser next time. The moving Gauss heat source model was selected to more accurately describe the temperature field. At the same time, the influence of air convection and total heat transfer coefficient was considered to ensure the accuracy of the calculation and the validity of the temperature field simulation results.

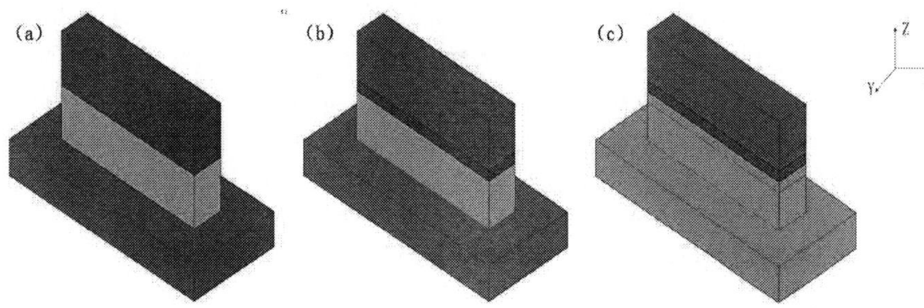

Figure 1 Finite element model of laser deposited Ta15/Ti_2AlNb gradient composite structure (a)direct transformation; (b) 1 layer; (c) 3 layers

2.2. Experimental materials and methods

The TA15/Ti_2AlNb gradient composite structure was prepared by the laser deposition manufacturing system, which was mainly composed of laser, powder feeder, water chiller, inter atmosphere protection box, oxygen analyzer, etc. The parameters of laser deposition process were: laser power was 2 KW, scanning speed was 8 mm/s, powder feeding speed was 0.8 r/min, spot diameter was 4 mm, lap rate was 50%, and laser head rising distance was 1 mm. The whole experiment was conducted under the protection of argon inert gas. The particle size of TA15 and Ti_2AlNb powder was about 60 ~ 160 m. The chemical composition of TA15 and Ti_2AlNb powder is shown in Table 1. Before the experiment, TA15 and Ti_2AlNb powder were dried in a vacuum at 100 ℃ for 120 min to remove the moisture adsorbed on the surface of the powder. The experimental substrate was forged TA15 plate with a size of 70 mm×30 mm×10 mm. Before the experiment, the surface was polished with sandpaper and wiped with acetone to prevent surface oxidation film and oil stains from affecting the experimental results.

Table l. Chemical compositions of TA15 and Ti_2AlNb powders (ω/%)

Powder	Composition						
TA15	Al	Mo	Zr	V	Fe	Si	Ti
	6.5	1.78	1.92	2.28	0.25	0.15	Bal.
Ti_2AlNb	Al	Nb	V	O	N	H	Ti
	9.68	41.37	<0.05	0.053	0.0062	0.0024	Bal.

The overall size of the specimen prepared in this experiment was 50mm×10mm×30mm, and multi-channel and multi-layer laser deposition manufacturing technology was adopted. In the preparation of the specimen with a gradient layer of 0 (direct transition), 15mm TA15 titanium alloy powder and 15mm Ti_2AlNb alloy powder were successively deposited. In the preparation of the specimen with a gradient layer of 1, 13.5 mm of TA15 titanium alloy powder was deposited, followed by 3 mm 50% TA15+ 50% Ti_2AlNb mixed powder and 13.5 mm of Ti_2AlNb alloy powder. In the preparation of the specimen with a gradient layer of 3, 10.5 mm of TA15 titanium alloy powder was deposited, followed by mixed powder of 75%TA15+25%Ti_2AlNb, 50%TA15+50%Ti_2AlNb and 25%TA15+75%Ti_2AlNb from bottom to top. Each kind of powder was deposited for 3mm, and the remaining 10.5mm was deposited by Ti_2AlNb alloy powder.

The deposited specimens were ground and polished with sandpaper, and the surface of the specimens was corroded with an etchant (VHF:VHNO3:VH2O2:VH2O=2:1:7:20) for 30~50 s. The microstructure of the three specimens prepared was observed by GX51 OLYMPUS optical microscope. The stretched rods were prepared, as shown in Figure 2. Then the stretched rods prepared from the deposited specimens were stretched at room temperature using the INSTRON5982 electronic universal testing machine. The tensile fracture was observed by ZEISS-SIGMA scanning electron microscope. Finally, the microhardness of TA15 side, Ti_2AlNb side and each gradient layer was tested using the HVS-1000A microhardness tester. The load was 200 g and the duration was 10 s.

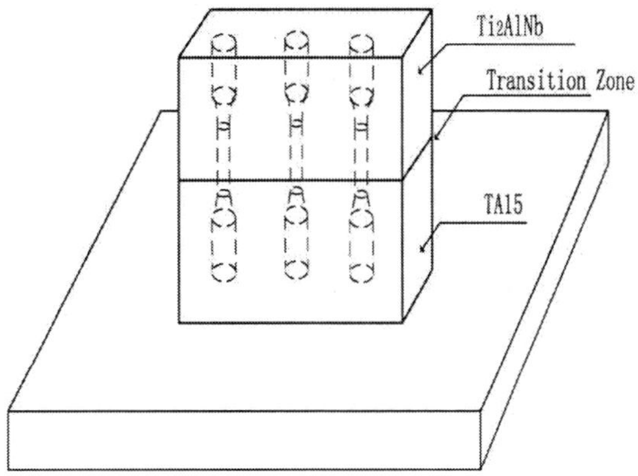

Figure 2 Schematic diagram of tensile rod sampling

3. Result and analysis

3.1. *Simulation and analysis of temperature field*

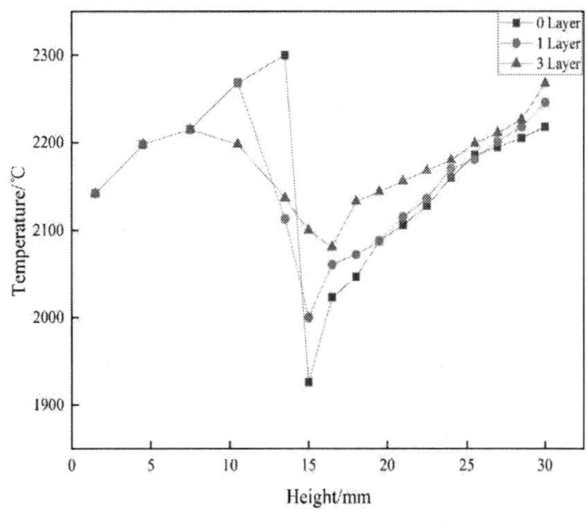

Figure 3 Simulation curve of maximum temperature change in each region

The temperature variation curves of TA15/Ti_2AlNb gradient composite structure with different gradient layers along the direction of deposition height are shown in Figure 3. The deposition temperature on both TA15 side and Ti_2AlNb side increased with the increase of the deposition height, which was mainly due to the fact that the temperature gradient inside the molten pool was higher during the deposition process, and the temperature gradient of the component along the Z axis was the largest, so the heat was mainly transferred along the Z axis. As the deposition height increased, heat was accumulated at the bottom and the temperature of the new deposition layer gradually increased. When the number of gradient layer was 0, the deposition materials directly transited from TA15 to Ti_2AlNb, the high temperature thermal conductivity of TA15 and Ti_2AlNb were 18 K/(W • m^{-1} • K^{-1}) and 22 K/(W •m^{-1} •K^{-1}), respectively. The thermal conductivity of the material was enhanced, the heat was quickly transferred externally and the heat accumulation was reduced, resulting in a decrease in the maximum temperature at the transition interface compared with TA15 side. When the gradient layer number was 1, the high temperature thermal conductivity of 50%TA15+50%Ti_2AlNb mixed powder is between that of TA15 and Ti2AlNb, leading to a less significant temperature drop than that when the number of gradient layer was 0. When the number of gradient layers was 3, the difference of high temperature thermal conductivity between adjacent gradient layers decreased, resulting in a slow decrease in the deposition temperature of each gradient layer and a gradual decrease in the temperature gradient during deposition. As a result, the thermal accumulation in the transition region increased, resulting in the deposition temperature of Ti_2AlNb powder at each height in the subsequent deposition process higher than that when the number of gradient layer was 0 and 1.

3.2. Analysis of strcuture

Figure 4(a) shows the microstructure of the transition interface of TA15/Ti_2AlNb composite structure without gradient layers. The microstructure of TA15 side is a typical basket-like near α titanium alloy structure. The Ti_2AlNb side consisted of B2, O and α_2 phases. Since a small amount of deposited parts were melted by high-energy laser during deposition, the powder feeder sent the powder into the molten pool, so the transition interface contained both TA15 and Ti_2AlNb alloy powder. Therefore, the microstructure of the transition interface consisted of α, β, B2, O and α_2 phases. It was observed that the microstructure on both sides of the transition interface maintained its own characteristics.

The microstructure of TA15 and 50%TA15+50%Ti_2AlNb when the number of gradient layer is 1 is shown in Figure 4(b). It can be seen from the figure that the microstructure morphology of the TA15 side significantly changed. Laser deposition is a cyclic process in which the formed cladding undergoes periodic "heating and cooling" cycles, making the formed part undergo multiple high temperature cycles. When the number of gradient layer was 1, the deposition temperature of Ti_2AlNb side was higher than that of 1, which promoted the cross-growth of laminates at TA15 side in all directions, leading to the truncation of each other and the occurrence of a large number of short rod sheets. The microstructure of the 50%TA15+50%Ti2AlNb side consisted of α, α_2 and β phases. The increase of Al content resulted in a small amount of α_2 phase precipitation. Figure 4(c) shows the transition interface between 50%TA15+50%Ti_2AlNb and Ti_2AlNb. Since the laser deposition melted a small amount of the deposited 50%TA15+50%Ti_2AlNb part, there was still a small amount of α phase. It was mainly composed of B2, O and α_2 phases. The difference of the transition interface was smaller than that when there was no gradient layer.

The microstructure of the transition interface between TA15 and 75%TA15+25%Ti_2AlNb when the number of gradient layers is 3 is shown in Figure 4(d). The temperature of the transition region and Ti_2AlNb side was higher than that when the number of gradient layer was 0 and 1. The α clusters on the TA15 side grew more fully and the α sheets were cut off each other to generate more fine α sheets. On the 75%TA15+25%Ti_2AlNb side, due to the increase of temperature in the transition region, part of α phase precipitated into α_2 phase, but the internal phase was still dominated byα and β phases. Figure 4(e) shows the transition interface between 75%TA15+25%Ti_2AlNb and 50%TA15+50%Ti_2AlNb. The content of Ti_2AlNb at this position continued to increase, resulting in a further increase in the content of Al. Thus, more α_2 phase began to precipitate. The content of Nb continued to increase, and O phase began to precipitate, the microstructure at this position mainly consisted of α_2 and B2 phases, with only a small amount of O phase. Figure 4(f) shows the transition interface between 50%TA15+50%Ti_2AlNb and 25%TA15+75%Ti_2AlNb, where the content of Al increased and a large number of α_2 phase precipitated. The content of Nb continued to increase, so did the contents of B2 and O phases. The microstructure of this region was mainly composed of α_2, B2 and O phases. Figure 4(g) shows the transition interface between 25%TA15+75%Ti_2AlNb and Ti_2AlNb. The content of both Al and Nb at this position increased. As a result, the microstructure of the transition interface was almost the same as that of Ti_2AlNb side, and the transition interface almost disappeared. Similarly, with the increase of deposition height, the temperature increased gradually during the deposition process. The higher the temperature, the greater the diffusion speed, the faster the grain growth and the finer the grain. Therefore, the microstructure of Ti_2AlNb side in Figure 4(g) is obviously finer than that in Figure 4(a) and Figure 4(c).

Comparing the microstructure of the transition interface of TA15/Ti2AlNb gradient composite structure with different gradient layers, it can be found that as the number of gradient layers increased, the difference of the transition interface gradually decreased and the transition gradually became smooth.

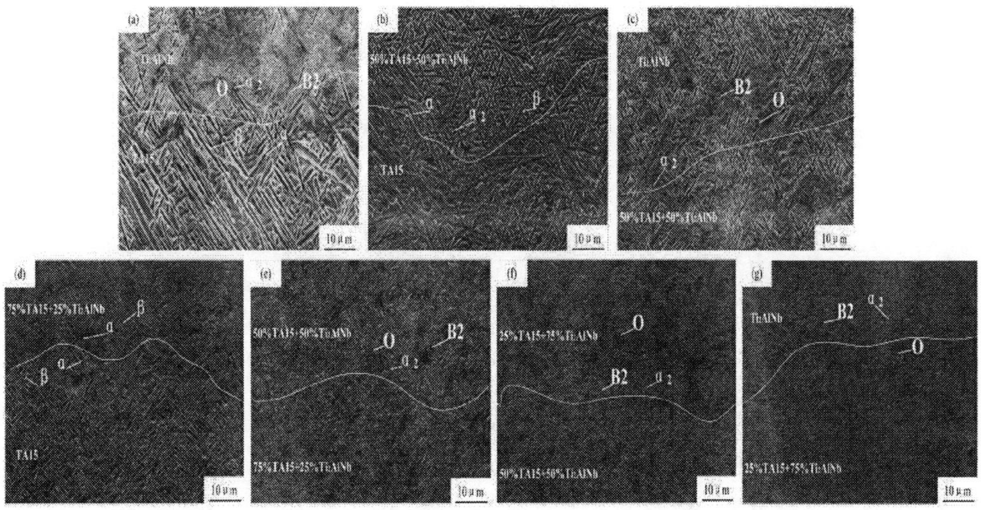

Figure 4 Morphology of microstructure at different interface

(a) transition interface between TA15 and Ti_2AlNb; (b) transition interface between TA15 and 50%TA15+50%Ti_2AlNb; (c) transition interface between 50%TA15+50%Ti_2AlNb and Ti_2AlNb; (d) transition interface between TA15 and 75%TA15+25%Ti_2AlNb; (e) transition interface between 75%TA15+25%Ti_2AlNb and 50%TA15+50%Ti_2AlNb; (f) transition interface between 50%TA15+50%Ti_2AlNb and 25%TA15+75%Ti_2AlNb; (g) transition interface between 25%TA15+75%Ti_2AlNb and Ti_2AlNb

3.3. Analysis of tensile property

The tensile properties of TA15/Ti_2AlNb gradient composite structures with different gradient layers at room temperature are shown in Table 2. According to tensile property data, the tensile strength of the composite structure without gradient layer was the largest, and the tensile strength of the composite structure was close to 0 when the number of gradient layers was 1, and the tensile strength of the composite structure with 3 gradient layers was the lowest. However, the elongation after fracture of the composite structure with 3 gradient layers was significantly higher than that of the other two transition modes. It can be seen that the plasticity of TA15/Ti_2AlNb gradient composite structure increased with the increase of the gradient layers.

Table 2 Room temperature tensile properties of TA15/Ti2AlNb Gradient composite structure

Number of layers	σ_b/MPa	δ/%
0	928.0	5.27
1	926.45	5.5
3	892.5	8.76

Figure 5 shows the EDS analysis results of the tensile fracture of TA15/Ti_2AlNb gradient composite structure with different gradient layers. When the number of gradient layers was 0, the fracture position contained both the unique Zr element of TA15 and the unique Nb element of Ti_2AlNb, indicating that the fracture occurred at the transition interface. When the gradient layer number was 1, the Nb element at the fracture surface increased, indicating that the fracture occurred on the side close to Ti2AlNb at the transition interface. When the number of gradient layers was 3, the Nb element at the fracture continued to increase, indicating that the fracture is continuously approaching the Ti_2AlNb side. Combined with the microstructure analysis, as the number of gradient layers increased, the proportion of B2 phase in the transition zone gradually increased, while B2 phase had a volume-centered cubic structure, which belonged to the plastic phase, and its resistance to deformation was relatively poor, so the tensile strength gradually decreased. The B2 phase on the Ti_2AlNb side was refined, enhancing its ability to resist deformation, so the fracture occurred in the transition region. With the increase of the number of gradient layers, the distribution of B2 phase in the transition region gradually moved to the Ti_2AlNb side, which is the main reason why the fracture occurred near the Ti_2AlNb side.

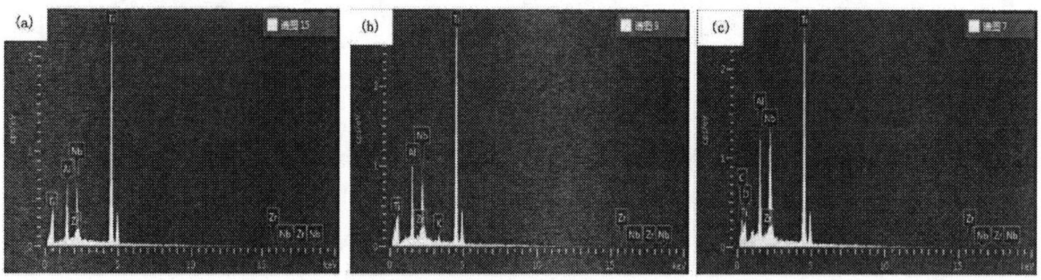

Figure 5 EDS analysis of tensile fracture of TA15/Ti_2AlNb gradient composite structure with different transition layers

(a) direct transformation; (b) 1 layer; (c) 3 layers

Figure 6(a)~(c) shows the tensile fracture morphology of TA15/Ti_2AlNb gradient composite structure specimen at room temperature. It can be seen from the figure that the fracture surface of specimen with different gradient layers consisted of fiber region, radiation region and shear lip. The

IOP Conf. Series: Materials Science and Engineering 1081 (2021) 012020 doi:10.1088/1757-899X/1081/1/012020

fracture morphology was relatively flat, exhibiting typical fluvial pattern and cleavage stage, which was determined to be cleavage fracture. It can be seen from Figure 6(d)~(f) that the fracture dimples of the specimen with 3 gradient layers were large and deep, while those of other specimen were small and sparse, indicating that the plasticity of the composite structure with 0, 1 and 3 gradient layers increased, which is consistent with the test results of mechanical properties.

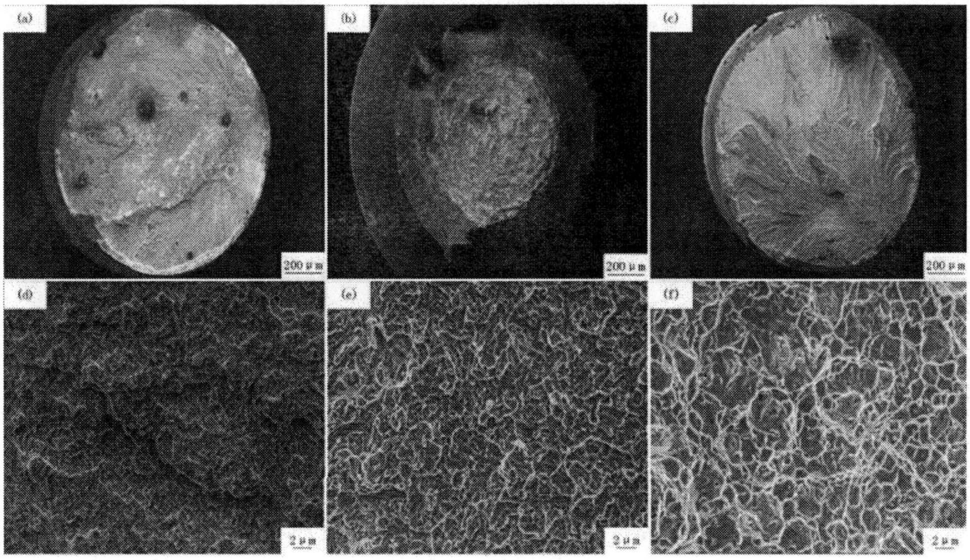

Figure 6 Tensile fracture morphology of TA15/Ti_2AlNb gradient composite structure with different transition layers

(a) direct transformation; (b) 1 layer; (c) 3 layers; (d)direct transformation; (e) 1 layer; (f) 3 layers

3.4. Analysis of microhardness

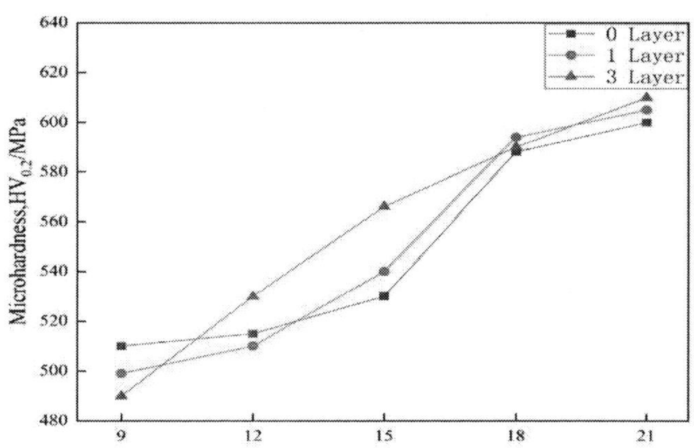

Figure 7 Microhardness of Transition zone of TA15/Ti_2AlNb gradient composite structure with different transition layers

The microhardness of the central region of three composite structures was measured from the TA15 side to the Ti2AlNb side along the direction of deposition height. Five test points were selected from each height position. The average value was obtained after removing the maximum and minimum values. The results are shown in Figure 7. It can be seen that the microhardness significantly changed with the increase of gradient layers. With the increase of the number of gradient layers, the deposition temperature of the transition region gradually increased, the microstructure was refined, and the alloy was more difficult to deform, so the microhardness of the transition region increased gradually. With the increase of the number of gradient layers, the fluctuation of microhardness in the transition region decreased gradually because as the number of gradient layers increased, the composition gradient of the transition region gradually decreased, so did the microstructure difference of each layer in the transition region.

4. Conclusion

1) The number of gradient layers has a great influence on the temperature of the transition region during the deposition of TA15/Ti_2AlNb gradient composite structure. With the increase of the gradient layers, the temperature in the transition region of TA15/Ti_2AlNb gradient composite structure gradually increased and the temperature gradient gradually decreased.

2) The number of gradient layers has a great influence on the microstructure of TA15/Ti_2AlNb gradient composite structure. With the increase of the number of gradient layers, the microstructure difference of each transition interface gradually decreased, and the microstructure of both sides of the transition region was refined.

3) As the number of gradient layers increased, the tensile strength of TA15/Ti_2AlNb gradient composite structure at room temperature decreased, but the elongation increased after fracture. The fracture occurred in the transition region when the number of gradient layers was 0, 1, and 3, which was determined to be cleavage fracture. However, as the number of gradient layers increased, the B2 phase in the transition region gradually shifted to the Ti_2AlNb side, resulting in a gradual shift of the fracture position to the Ti_2AlNb side and an increase of the dimples at the fracture.

4) As the number of gradient layers increased, the temperature of the transition region increased, the microstructure was refined, and the alloy was more difficult to deform, resulting in an upward trend in the microhardness of the transition region. In addition, the difference of the transition interface was reduced, so was the fluctuation of the microhardness in the transition region.

Acknowledgement

National Key Research and Development Programs(2018YFB1105809)
National study abroad fund(201908210093)
National Key Research and Development Programs(2016YFB1100504)

References

[1] Hu M J, Ji L K, Ma Q R, et al. Overview of Laser Additive Manufacturing Technology and Status[J]. Petroleum Tubular Goods & Instruments, 2019, 5(05): 1-6.

[2] Yang J X, Ke H, Cui Z, et al. Research and Application Progress of Laser Metal Deposition[J]. Aeronautical Manufacturing Technology, 2020, 63(10): 14-22.

[3] Gong S L, Suo H B, Li X H, Development and Application of Manufacturing Technology[J], Aeronautical Manufacturing Technology, 2013(13) :66-71.

[4] Liu S, Liu C S, Li J G, et al. Research Progress of Laser Rapid Forming Technique in The Preparation of Functionally Gradient Materials[J].Hebei Metallurgy, 2020(04) :1-6.

[5] Xue K M, Guo W W, Shi Y B, et al. Microstructure and Tensile Properties of Multi-Directional Forging of TA15 Titanium Alloy[J]. Rare Metal Materials and Engineering, 2019, 48(10) :3340-3345.

[6] Hongchao Ji, Zhanshuo Peng, Weichi Pei, et al. Constitutive equation and hot processing map of TA15 titanium alloy. 2020, 7(4).

[7] Sun C J, Zhao J B, Zhao Y H, et al. Effects of Laser Melting Deposition Process Parameters on Ultrasonic Testing Accuracy of TA15 Titanium Alloy[J]. Acta Optica Sinica, 2019, 39(10): 210-217.

[8] XU Wei-feng, MA Jun, LUO Yu-xuan, et al. Microstructure and high-temperature mechanical properties of laser beam welded TC4/TA15 dissimilar titanium alloy joints. 2020, 30(1):160-170.

[9] Zhao Y, Guo H Z, Wang T, et al. Effect of Channel Angular Pressing on Microhardness of TA15 Alloy[J]. Rare Metal Materials and Engineering, 2011, 40(05): 885-888.

[10] Xu L, Wu J, Cui X X, et al. Cracking Analysis of Powder Metallurgy Ti2AlNb Component During Heat Treatment[J]. Aeronautical Manufacturing Technology, 2020,63(16): 14-20.

[11] Yang J J, Liu L W, Zhao Y, et al. Thermodynamic analysis of Ti2AlNb-based Alloy in the Initial Stage of High Temperature Oxidation[J]. Materials Science and Engineering of Powder Metallurgy[J]. 2020, 25(02): 98-103.

[12] Chen Y X, Gao Y K. Simulation of the Residual Stress and Fatigue Prediction of Ti2AlNb Intermetallic Compound under Shot Peening[J]. Surface Technology, 2019, 48(06): 167-172+188.

[13] Qi Z L, Chen J, Zhou Q J, et al. Study on the Microstructure and Mechanical Property of Laser Solid Formed TA15-Ti2AlNb Compositional Graded Material[J]. Applied Laser, 2020 40(02): 214-220.

[14] Liu Y T, Zhang Y Z, Chen Y Q, et al. Microstructure and Mechanical Properties of Laser Melting Deposited TA15+Ti2AlNb Alloys[J]. Journal of Aeronautical Materials, 2017, 37(03): 61-67.

[15] Bu X Z, Li H W, et al. Wang Z M, et al. Microstructure and mechanical properties of titanium alloy TA15 and Ti2AlNb laser welding[J]. Welding Technology, 2016, 45(03): 9-11.

[16] He B, Liu J, Yang G, Microstructure and Properties of Direct Transition Interface of TC4/TC11 Fabricated By Laser Deposition Manufacturing[J]. Rare Metal Materials and Engineering, 2019, 48(01): 293-299.

[17] Yang S Z. Study on the Erosion Resistance of Cr3C2/NiCr, Stellite6, Inconel625 and Inconel718 Coatings Sprayed by HVOF[J]. Chinese Journal of Turbomachinery, 2017,59(02):66-70.

[18] Zhou J Z, He J L, Tan W S, et al. Strengthening Mechanism and Wear Resistance of Al2O3/Fe901 composite Coating Prepared by Laser Cladding[J]. Acta Optica Sinica, 2019, 39(05): 219-227.

[19] Yin X C,Liu J R, Wang Q J, et al. Microstructure and Tensile Properties of TC17/TC11 Dual Alloy Fabricated with Laser Melting Deposition Method[J]. Rare Metal Materials and Engineering, 2020, 49(03): 1024-1030.

[20] Zuo S G, Liu Y M, Lu W. FEA of High Temperature Centrifugal Fan Impeller Based on Ansys-Workbench[J]. Chinese Journal of Turbomachinery, 2019, 61(S1): 80-86.

[21] Li X, Qu Y J, Song L L, A Numerical Investigation on Cooling Effectiveness for Advanced Fan-shaped Film Cooling[J]. Chinese Journal of Turbomachinery, 2019, 61(03): 72-77+5.

[22] Wang R, Jing L L. Finite Element Analysis of Centrifugal Fan Impeller with Different Blade Profile Based on ANSYS[J]. Chinese Journal of Turbomachinery, 2018, 60(S1): 14-18.

[23] Ji C J, Gao Y W, Sun Q, et al. Structure Optimization Study of Centrifugal Fan Impeller Based on ANSYS[J]. Chinese Journal of Turbomachinery, 2016, 58(06): 49-52.

Numerical Investigation of Leaned Strake in the Ram-rotor

Jian Guan[1], Ji-ang Han[1], Kexin Yu[1], Jingjun Zhong[2]

[1] Naval Architecture and Ocean Engineering College, Dalian Maritime University, Dalian, PR China
[2] Merchant Marine College, Shanghai Maritime University, Shanghai, PR China

E-mail: hja6065@126.com

Abstract. The concept of lean commonly adopted in conventional compressors is introduced into the design of the ram-rotor strake, and numerical simulation of the effect of lean on the aerodynamic performance of the ram-rotors has been conducted in this paper. Three types of leaned]0strake schemes are modeled and analyzed first to investigate the effect of lean on pressure rise and loss characteristics, and then based on C-mid scheme, different amounts of leaning angle are chosen for further research so as to improve the performance of leaned ram-rotor. Results show that the application of lean can affect the intensity and position of shock system. Radial transport and accumulation of low momentum fluid in the boundary layer of suction surface is suppressed. Compared to baseline, C-cas scheme increases total pressure ratio by 2.338 %, however, C-hub scheme improve the adiabatic efficiency by 0.462%. For C-mid scheme, a slight reduction of performance is observed.

1. Introduction

Axial flow compressor is an important power equipment in industrial production, but also an important energy consumption equipment. In the current situation of worldwide energy shortage, it is an important problem to save energy and reduce the consumption of axial flow compressor with high energy consumption[1].In modern aeroengine and land gas turbine, compressors stages are required to operate with high compression ratio and efficiency, which helps to minimize the fuel consumption and make a compact component configuration. If the Mach number of the flow passage is high, a slight change in flow will cause congestion[2].Transonic and supersonic compressors have been gradually developed to achieve this goal. Conventional supersonic axial compressor rotors have many blades and the blade surface produces a shock system with low efficiency, resulting in a large loss in the compression process[3].Transonic compressor with good performance is one of the main components of high performance turboshaft engine.In order to reduce the compressor class, reduce the overall quality and improve the thrust-to-weight ratio of the engine, it is necessary to carry out the design and experimental research of the compressor with high pressurization ratio, high load and wide stability range [4].Kantrowitz, et al [5]and Klapproth, et al [6] proposed two types of supersonic compressor rotors which are named the shock-in-type rotor and the impulse-type rotor, and demonstrated that supersonic compressor has a more preferable aerodynamic performance. With the development of transonic and supersonic compressors, a new kind of supersonic compressor rotor named ram-rotor or ram-pressor was presented recently, which has high hub-to-tip ratio and shallow stagger angle. As pointed out by Lawlor, et al [7], initial proof-of-concept of the ram-rotor was designed and tested to demonstrate the basic operational characteristics. Numerically predicted results and experimental study are reported by Grosvenor, et al [8][9], which obtained the measured pressure ratio of 7.8:1. Steele, et al [10] and Lawlor, et al [11] investigated the potential prospect of shock compression technology into micro turbines and CO_2 compressors. Many useful results also can be found by Han, et al [12-15] and Yang, et al [16] whose work focuses on the ram-rotor flow field and performance under the environment of whole stage system and single strake row respectively.

In order to improve the flow field in ram-rotors, many blades design means of conventional turbomachinery can be adopted into the strake of ram-rotor. Some successful efforts about the use of sweep on the strake and the consideration of the strake section shape in the ram-rotor were performed by Han, et al [17] and Yang, et al [18]. It is also known that blade curve or lean design has been demonstrated an effective flow control method to improve compressor performances, and many research efforts were spent on lean for compressor performance improvement. As pointed out by Bruges man, et al [19] that dihedral(lean) influenced strongly the development of the secondary flow. The overall loss and loss distributions in the spanwise direction were also modified by the lean effect. Bergner, et al [20] reported an investigation of the impact of various amounts of lean on the aerodynamic behavior and found that lean controlled the radial distribution of the mass flow and changed the shock pattern. The work by Benini, et al [21] showed that a substantial increment in the overall efficiency was obtained by using forward lean. Neshat, et al [22] drew his conclusion that blade lean was clearly more effective than blade sweep and it increased both the stage efficiency and the chocking flow rate.

From the published literatures, one of the main drivers for using lean in axial compressors is to add a radial blade force to the fluid cell, so that the original radial pressure field is modified for good control of secondary flow. Moreover, it has been found by the authors that in ram-rotors there exists massive flow separation near suction surface of strake aft part, which is one of the dominant sources of loss and detrimental to get a high aerodynamic performance. The challenge is to limit the loss and blockage produced by the flow separation. Hence, in view of the beneficial effect of lean in improving secondary flow and flow separation, the lean design is introduced into the strake of ram-rotors, with the purpose of controlling the flow separation and blockage near the strake aft part. Meantime, the impact of lean design on the aerodynamic behavior of the ram-rotor is obtained for further flow field and performance improvement.

2. Model and numerical method

2.1 Basic Ram-rotor Model

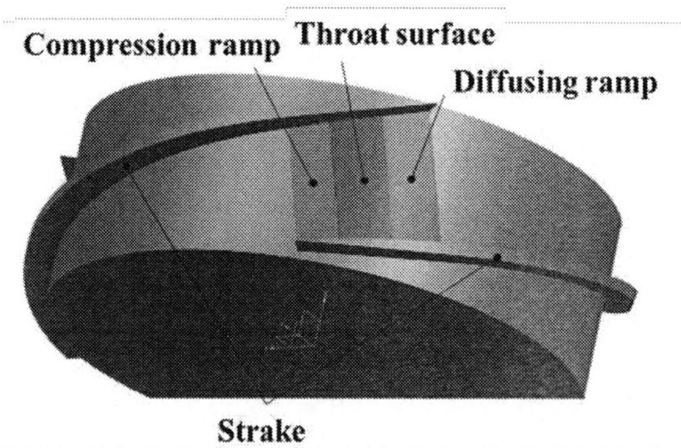

Figure 1. Schematic Diagram of a Ram-rotor

The basic three-dimensional ram-rotor model in present work is shown in Figure 1. On the ram-rotor rim configure three flow paths, and each path consists of two neighboring strake walls, hub wall and casing wall. The hub wall which consists mainly of the compression ramp, the throat surface and the diffusing ramp is the core part of the ram-rotor. Significant pressure rise is achieved across the shock system which is produced, reflected and hold between the casing wall and hub wall. The specific geometric parameters of the ram-rotor model are listed in Table 1.

Table 1. Geometric Parameters of Ram-rotor

Parameters	Values
Outer diameter/mm	350
Hub to casing ratio	0.9
Throat to strake ratio	0.857
Solidity at hub	0.213
Compression ramp angle/deg	6
Diffusing ramp angle/deg	5
Stagger angle/deg	15
Radius of LE and TE/mm	0.15
Tip clearance height/mm	0.35
Strake maximum thickness/mm	7.115

2.2 Geometric Leaning Definition

For baseline ram-rotor, the strake is radial stacked, and with respect to leaning model, there are different methods to create leaned blades such as shifting stacking line, leading edge line or trailing line in compressors. Shifting trailing edge is chosen in present work to make geometric modification in the ram-rotor, and the leaned strake is obtained by varying the strake profile perpendicular to the chord line direction, which is Y direction in Figure 2. The two points of O1 and O2 present the leading edge and trailing edge of strake, respectively. Point A and Point B separate the whole strake into such three parts as fore, middle and aft parts. In order to control the flow radial transportation near strake aft part and do not affect the incoming supersonic flow, the lean design is only applied into the design of the strake aft part. With keeping tip section unchanged to the baseline scheme, C-cas scheme is formed through moving the rest spanwise sections (strake aft part) from root to tip linearly perpendicular to the chord line (-Y direction in Figure 2). That is to say that the amount of movement is zero at tip section and highest at root section. Likewise, C-mid and C-hub schemes are obtained with keeping middle section and root section intact, respectively, to shift the rest strake sections. For C-mid scheme, strake sections from middle to tip are moved along +Y direction, with sections form middle to root sections along –Y direction. C-hub scheme moves the section from root to tip linearly along +Y direction.

Additionally, for the sake of smooth connection between middle part and aft part, the amount of section movement is zero at location B and highest at the trailing edge (location O2), which indicates no lean at the initial position of strake aft part and maximum lean at trailing edge position. Leaning angle in present work is defined as the acute angle between trailing edges of leaned ram-rotor and baseline ram-rotor.

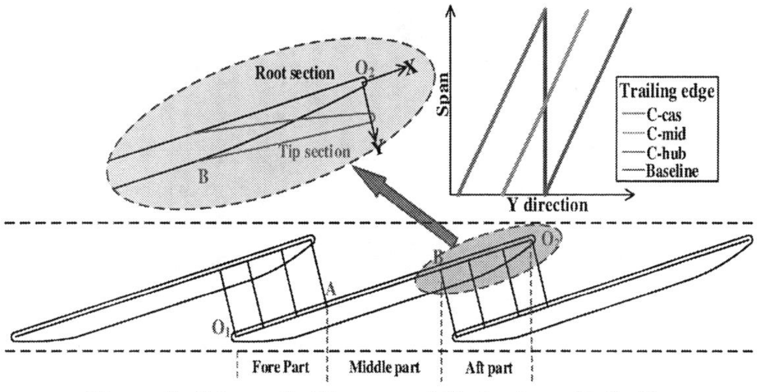

Figure 2. Schematic Diagram of Strake Lean Definition

It is worth noting that the strake profile from leading edge (location O1) to location B of all three leaned ram-rotors maintain the same as that of baseline ram-rotor. Moreover, in order to investigate the impact of leaning angle on the flow characteristics in the leaned ram-rotors, six leaning angles are applied in C-mid scheme. The leaning angle varies in the range from 5 to 30 degrees, with the intervals of 5 degrees.

2.3 Grid and Numerical Method

The commercial CFD code FLUENT 6.3[23] was used to predict accurately the aerodynamic behavior of ram-rotors. Computations are performed on a single ram-rotor path for the purpose of reducing computational cost, and all simulations are steady-state solutions. Figure 3 shows the computational grid used for the analysis. Hexahedral grid of about 1.6 million cells are employed with fine mesh near the walls and the region where shocks occur. Grid independence study has been carried out to check the effect of the grid number on the convergence and calculation solutions in the previous research[13], and it is found that the solutions in terms of mass flow rate, total pressure ratio and adiabatic efficiency don't show any appreciable variation with different grid precisions. The solutions show no correlation with grid size when grid number is above 1.41 million. Thus, the grid size of ram-rotor model employed in the current study is adequate.

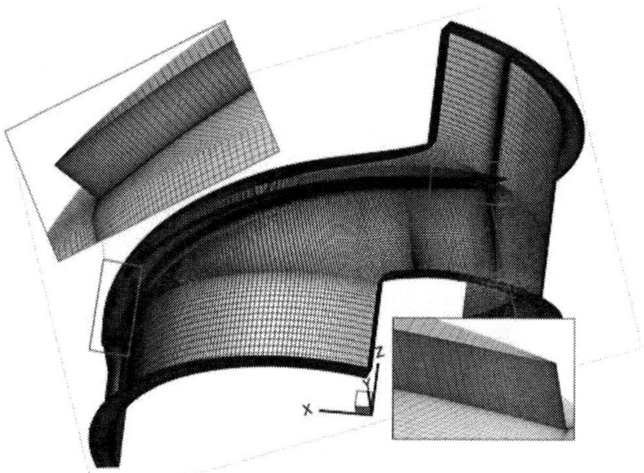

Figure 3. Grid of Computational

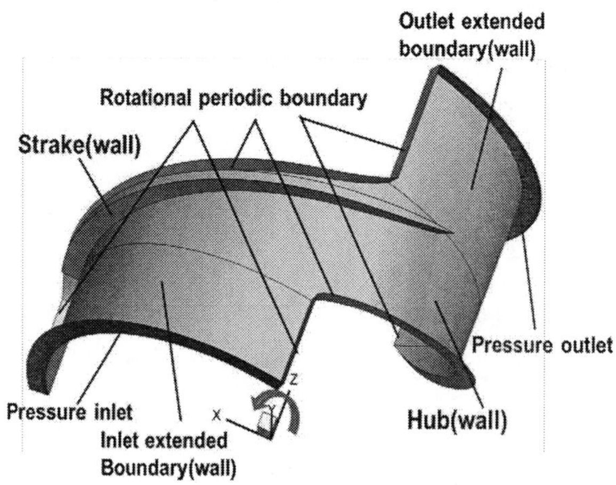

Figure 4. Boundary Conditions Setting of Ram-rotor

The numerical method used for the current work is from reference [17]. The stability of Spalart-Allmaras turbulence model and the convergence of residual are better[24].Density-based solver is chosen to solve Navier-Stokes (RANS) equations, and turbulence closure is achieved through S-A turbulence model.The second order unwind scheme is employed to spatially discretize convective term. The fluid is set as ideal gas and the Sutherland law is chosen for viscosity. The computational domain with given boundary conditions is shown in Figure 4. Total pressure, total temperature and flow direction are specified at the inlet of the computational domain, whereas static pressure is specified at the outlet of the computational domain. Hub and strake walls are defined as rotating wall with design rotating speed, and the casing wall is set as absolutely stationary wall. No-slip and adiabatic conditions are applied at all solid boundaries. The design rotational speed of the ram-rotor is 26728 RPM. Rotational periodicity is imposed on the circumferential boundaries. In the previous research [17], the case of NASA rotor 37 was used to validate the reliability of numerical method used in this paper. In terms of the relative Mach number distribution, the shock structure, shock strength and shock position obtained by numerical simulation agree well with the experimental data [25-27]. The numerical results of spanwise variation of adiabatic efficiency and total pressure ratio at outlet measurement station also match well with those of the experiment data. Hence, the numerical method in current work predicts overall flow pattern, including the shocks, reasonably well, and can be used to reliably capture shock and predict ram-rotor performance.

3. Results and discussion

3.1 Comparison of Ram-rotor Performance

The total pressure ratio and adiabatic efficiency of baseline and three types of leaned ram-rotors are plotted against mass flow rate in Figure 5. It can be seen that the performance curves produce distinct changes compared to the baseline. As a whole, C-cas scheme shows an increment in pressure rise, while C-hub scheme is beneficial in improving adiabatic efficiency of the ram-rotor. With respect to C-mid scheme, although this kind of lean exerts little impact on the improvement of pressure ratio and efficiency, the performance curve shows a little left displacement, which indicates that this scheme can work on lower mass flow condition and is believed to have a wider working range than the baseline.

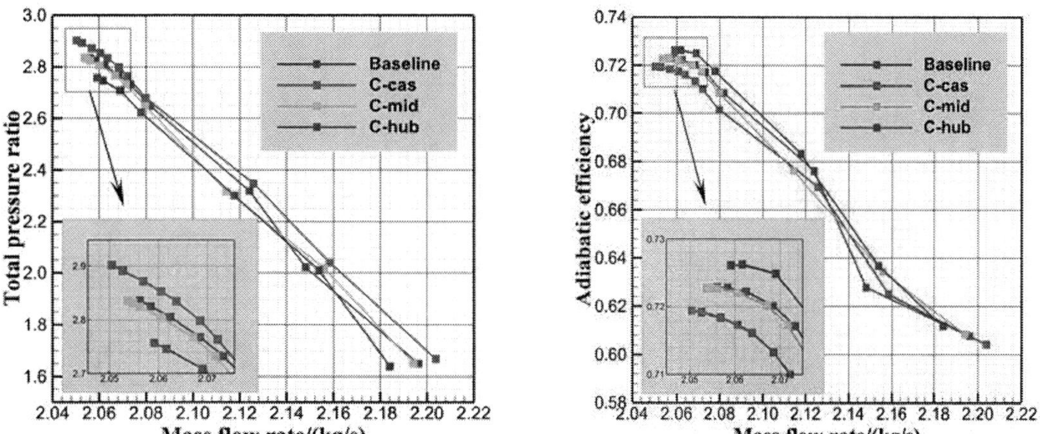

a. Total Pressure Ratio v.s. Mass Flow Rate b. Adiabatic Efficiency v.s. Mass Flow Rate

Figure 5. Performance Maps of Four Ram-rotors

For quantification analysis of aerodynamic performance of ram-rotors, the performance and corresponding increment of four ram-rotors at their peak efficiency conditions are listed in Table 2. Different levels of improvement and deterioration are conducted by the use of different lean schemes. It is shown that the C-cas scheme generates a significantly higher total pressure ratio that is 2.338 %

IOP Conf. Series: Materials Science and Engineering 1081 (2021) 012021 doi:10.1088/1757-899X/1081/1/012021

larger than the baseline, while the adiabatic efficiency is decreased by 0.483%. For C-mid scheme, the pressure ratio and efficiency appear to be similar to that of baseline, with only a slight reduction of 0.169% and 0.025%, respectively. When it comes to C-hub scheme, the pressure rise capacity is worsened, with a total pressure ratio decrease of 3.163%, however, the adiabatic efficiency of this scheme is improved by 0.462% compared to baseline. Overall, different lean design schemes are appreciated for different application intention. For the purpose of higher pressure rise capacity, such lean design as C-cas scheme is preferable, and the C-hub scheme is more favorable for increasing adiabatic efficiency. As for C-mid scheme, the detailed analysis needs to be further studied.

Table 2. Ram-rotor Performance At Peak Efficiency Conditions

Schemes	Total pressure ratio and variation		Adiabatic efficiency and variation	
Baseline	2.835	0	0.7229	0
C-cas	2.902	2.338%	0.7194	-0.483%
C-mid	2.831	-0.169%	0.7227	-0.025%
C-hub	2.746	-3.163%	0.7262	0.462%

3.2 Pressure Rising and Flow Loss Characteristics Inside Ram-rotor

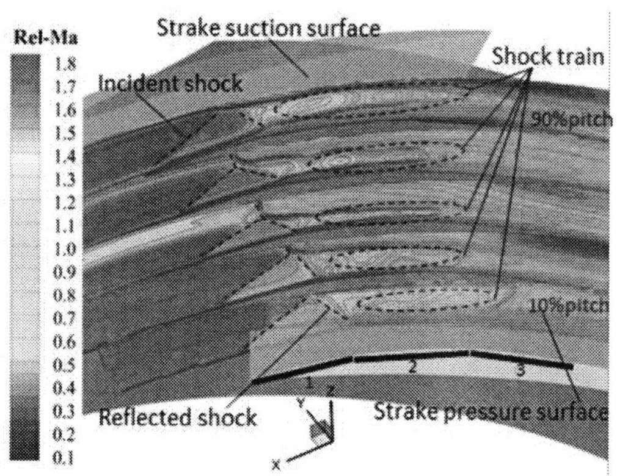

a. S2 Stream Surface at Different Pitch Position

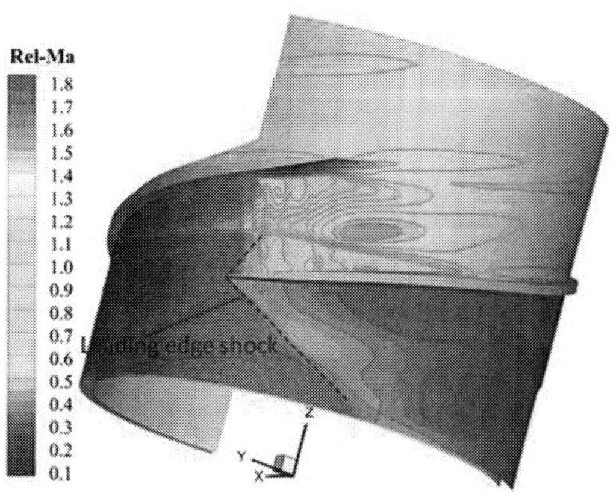

b. S1 Stream Surface At 50% Throat Height

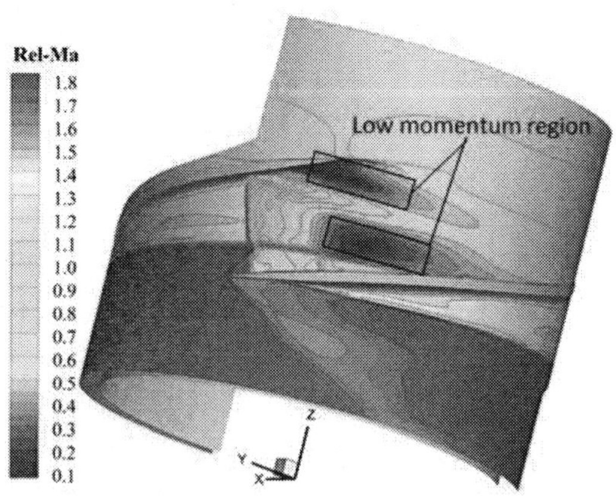

c. S1 Stream Surface At 90% Throat Height

Figure 6. Relative Mach Number Contours in The Baseline

The numerical results reveal that the shock pattern in the ram-rotor is similar, and the use of lean can give rise to little change in shock system structure. Figure 6 illustrates the relative Mach number distribution at peak efficiency point for the basic knowledge of shock system and flow feature in the baseline. There are two kinds of shock system that can be observed. One is the shock system featured by incident shock, reflected shock, the Mach stem(arising from low pre-shock Mach number under the effect of the tip leakage flow) and shock train on S2 stream surface, which is produced and held by hub surface(see Figure 6a), and most of the pressure rise occurs across this shock system. The other one is the shock system on S1 stream surface featured by leading edge shock which is induced by strake leading edge (see Figure 6b), and the pressure side leg of the leading edge shock spreads into the flow path which interacts with the shock system on S2 stream surface, while the suction side leg of this shock extends towards inlet which exerts an influence on the incoming flow. It is worth noting that these two kinds of shock system can both been viewed on the S1 or S2 stream surface. Moreover, in Figure 6c, two evident low momentum regions are formed near suction surface of strake aft part and near the middle region of flow path, which limit the aerodynamic performance of ram-rotors.

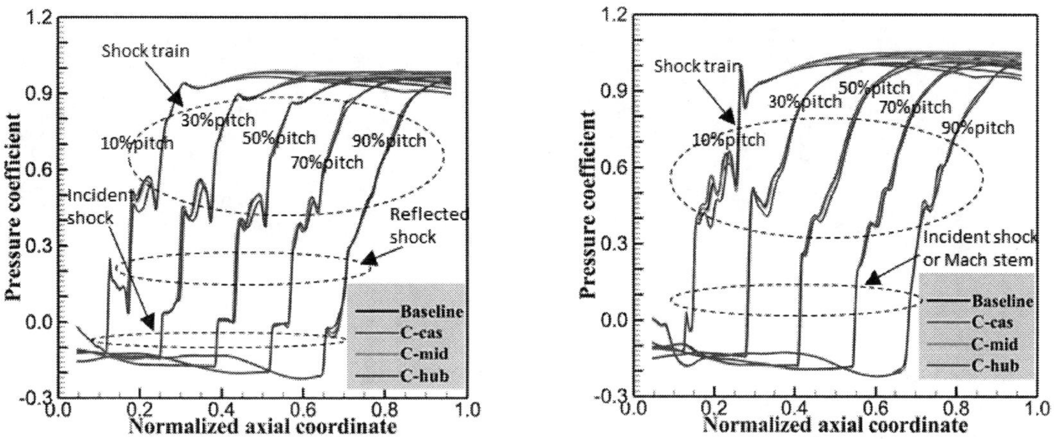

a. At 10% Throat Height b. At 99% Throat Height

Figure 7. Axial Distributions of Pressure Coefficients of Different Pitch Positions

Despite the fact that the flow feature of the ram-rotors is not altered in essence, the application of lean does affect the intensity and position of shocks. Figure 7 shows the axial distributions of pressure coefficients in the ram-rotor path. Every sharp pressure rise corresponds to the occurrence of shock or shock train. It can be observed that the shock system location is moved at the whole span height of strake aft part when applying lean. The principal location change at 10% throat height is the reflected shock and the initial position of shock train (see Figure 7a). C-mid scheme shows little displacement of shock, which is characterized by the same shock or shock train position, and has almost the same pressure rise characteristics of the baseline in the flow path. As for the C-cas scheme, forward motion of reflected shock is observed with a higher pressure rise, compared to the baseline. It is also the same case for the shock train. The pressure rise after shock train is also higher for C-cas scheme than baseline. This effect is more evident near the pressure surface region of flow path, which corresponds to the position range between 10% to 50% pitch. The forward motion of reflected shock and shock train in C-cas scheme makes a more compact distance between shocks on S2 stream surface, and the scale and intensity of shock interaction is enhanced, which produces an adverse effect on the ram-rotor efficiency in spite of high pressure-ratio. However, different effect is obtained for the C-hub scheme, which has a more backward displacement of reflected shock and shock train. Correspondingly, the pressure rise through the C-hub scheme is also reduced more, just as shown in Table 2. When at 99% throat height (see Figure 7b), the effect is similar but with a slight difference. Due to close distance from the casing wall, the shock system on S2 stream surface at the front of shock train presents only one single shock, and once pressure rise is realized by the incident shock or the Mach stem of lambda-

IOP Conf. Series: Materials Science and Engineering 1081 (2021) 012021 doi:10.1088/1757-899X/1081/1/012021

shock. This pressure rise is hardly influenced by the use of lean. The variation of shock train at this throat height is just as that at the 10% throat height.

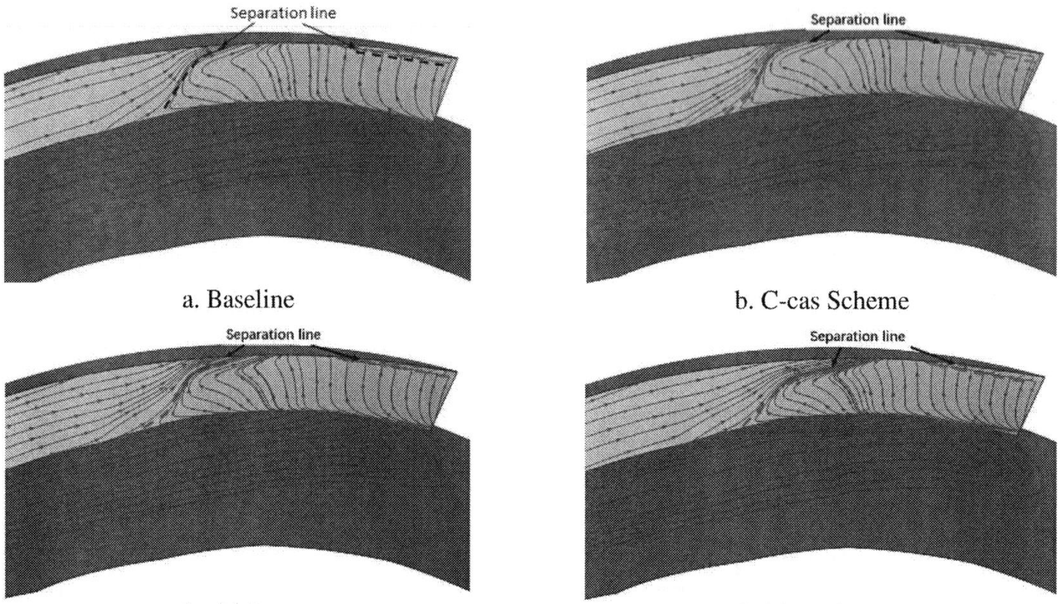

a. Baseline b. C-cas Scheme

c. C-mid Scheme d. C-hub Scheme

Figure 8. Limited Streamlines on The Suction Surface

Limited streamlines on the suction surface of strake wall and on the hub are given in Figure 8. Two main features are seen in the baseline, and one is that the flow in the boundary layer separates from strake suction surface at the region near compression ramp. This flow phenomenon is attributed to the occurrence of incident shock. The fluid in the region of strake suction surface can't undergo the high pressure rise of incident shock so that they fail to maintain the attachment to the wall. The other feature is that when the fluid near the hub region flow into the region near the diffusing ramp, they are deflected towards the corner between suction surface and hub wall, and then climb up along the spanwise direction. Thus, another separation line is formed when these upward-fluid encounter the strake tip leakage flow. This separation line is a result of the great pressure gradient in the rear of the ram-rotor path. Because the large pressure rise is achieved after diffusing ramp, the fluid at the corner is influenced by the great pressure gradient, which explains the reason of streamlines behavior. For the three leaned ram-rotor schemes, an important variation is scope and scale of the first separation region. Due to forward motion and intensified strength of shock system on S2 stream surface, flow separation is shifted slightly forward in C-cas scheme and less streamlines run downstream, which means that the flow blockage is aggravated in the first separation region. In spite of the fact that there is a slight reduction in terms of the ram-rotor performance, C-mid scheme still has a positive effect to some extent on the streamlines behavior on the suction surface compared to the baseline. The first separation line on the suction surface extends more downstream, which is an indication of delaying separation. As for C-hub scheme, the beneficial effect is more obvious at the first separation region with a more backward separation line, and the flow capacity at suction surface is the strongest among the four schemes.

Figure 9 gives the low momentum fluid denoted by iso-surface of relative Mach number=0.2, which is colored with static entropy. There are two main high loss regions or low momentum regions at the corners between strake aft part suction surface and end-wall. The development of secondary flow is governed by the balance of forces acting on the fluid, and the introduction of lean may produce an additional blade radial force component, also the spanwise pressure gradient, on the fluid so that the secondary flow evolution is under control to some extent, which weakens the flow migration towards strake tip.

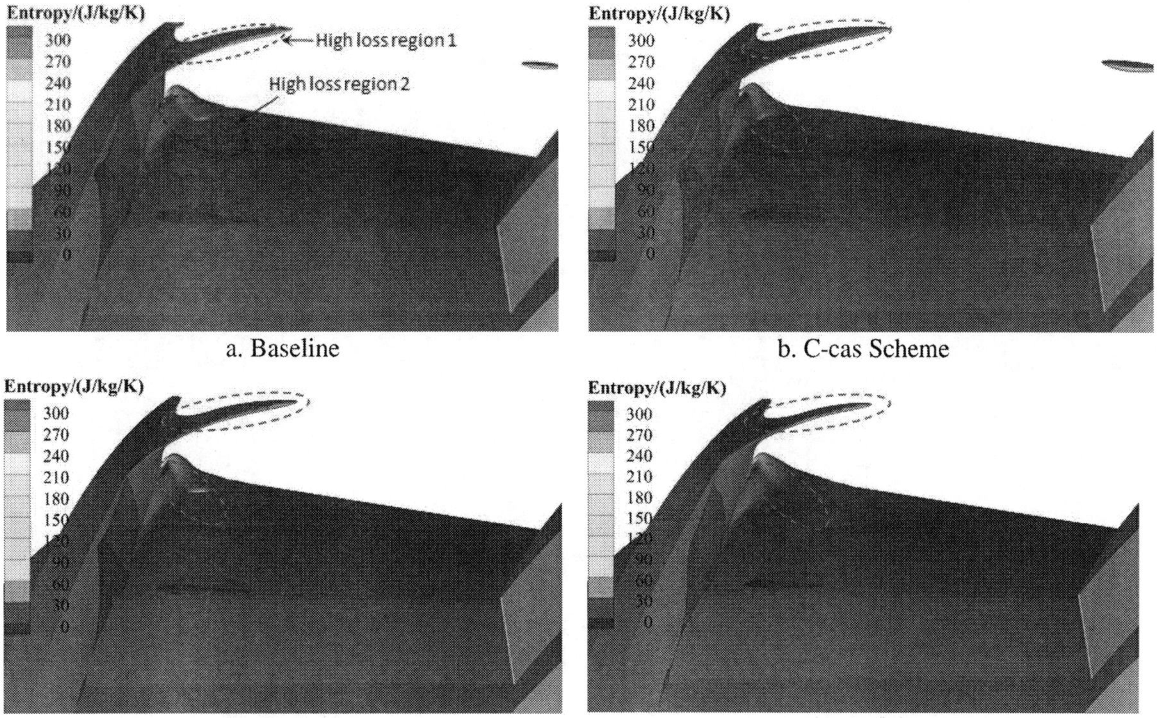

a. Baseline b. C-cas Scheme

c. C-mid Scheme d. C-hub Scheme

Figure 9. The Low Momentum Fluid Colored by Static Entropy (Iso-surface Of Rel-Mach=0.2)

Under the beneficial effect of lean, C-mid scheme shows a slight reduction of extension of high loss region 1, in spite of the fact that the beneficial effect is not so obvious. Due to high pressure rise obtained by C-cas scheme, the flow loss is increased at region 1. Because the diverging degree of C-cas scheme is greatest at lower flow path near the hub region, the supersonic flow blows out some low momentum fluid to some extent. Hence, the flow loss is somewhat reduced at high loss region 2. As for C-hub scheme, the flow migration to the tip on the suction surface of strake is suppressed more to alleviate the low momentum fluid accumulating near the corner between suction surface and casing wall. The high loss region 1 is decreased much more compared to the baseline. However, a disadvantage is that relatively more low momentum fluid is accumulated at the corner between hub and suction surface, but the overall effect is positive according to Table 2.

Fifteen slices which are cut and numbered along streamwise direction of the ram-rotor are extracted in Figure 10. Slice positions numbered from 5 to 10 represent the shock system compression region near compression ramp, throat surface and diffusing ramp. The total pressure ratio and total pressure loss coefficient are plotted against stream position in Figure 11. Flow loss is increased remarkably at shock region with abrupt pressure rise in four schemes. It is shown that the use of lean affects the pressure ratio and loss characteristics along stream direction. C-cas scheme performs better in terms of total pressure ratio at the expense of high flow loss. C-mid scheme indicates little improvement at shock region and in front of shock region, however, there is a loss reduction in the region after shock train. Total pressure ratio maintains the same as baseline. As discussed above, the use of lean alleviates the radial transportation and shows an effective loss reduction along stream direction. It is also known that the pressure ratio of this scheme is decreased.

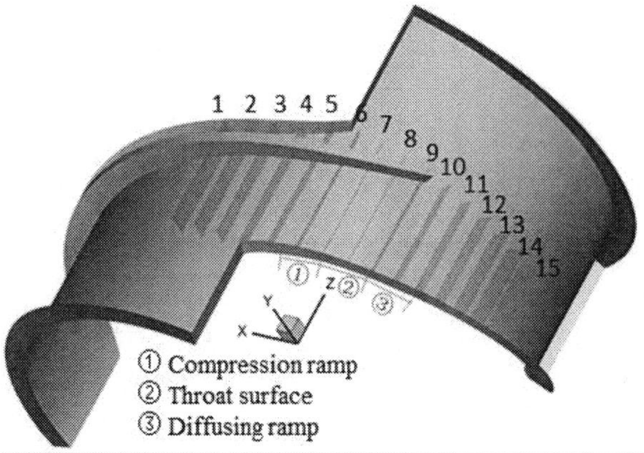

Figure 10. Schematic Diagram of Slices Along Streamwise Direction

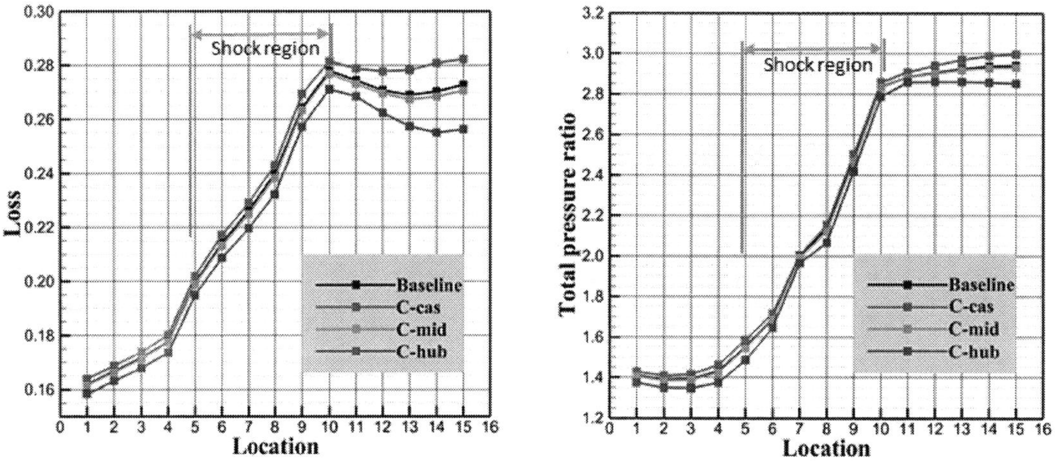

a. Total Pressure Loss Coefficient b. Total Pressure Ratio

Figure 11. Flow Characteristic Distribution Along Streamwise Direction

3.3 Flow Characteristics at Ram-rotor Outlet

The exit flow is important to the flow field of possible static blade row, and Figure 12 plots the radial distribution of flow parameters at outlet. The radial migration trend can be observed from Figure 12a. Compared to baseline, the radial velocity of three leaned schemes is reduced in the mid-span region, which is a beneficial effect on the suppression of low momentum fluid accumulation. The trend of streamlines towards the tip corner is reduced in three leaned ram-rotors. C-hub scheme shows a greater improvement in the bottom half of the strake height, which corresponds to the high efficiency of this scheme in Table 2. Different with C-hub scheme, great reduction of radial velocity occurs in the upper half of strake height in C-cas scheme, and the radial velocity below 30% strake height is even raised. The beneficial effect of suppressing radial migration can't compensate the detrimental impact due to large pressure rise. As seen in Figure 12b, C-cas scheme gives rise to a significant increase of axial velocity density below 40% strake height, so more mass flow goes through near hub region. The C-mid scheme reveals almost the same distribution with baseline above 40% strake height, with a little increase below this height. For C-hub scheme, the redistribution of flow makes a reduction of mass flow below 40% strake height, and the opposite variation is seen above this height. Absolute flow angle at outlet is plotted in Figure 12c, and the flow angle in present paper is defined as the acute angle between flow direction and circumferential direction. It is found that the flow angle is increased above 20% strake height for C-hub scheme. It can be explained by the fact that the high loss region located at the corner between suction surface and casing wall is suppressed compared to baseline (Figure 9d). The reduction of blockage in this region makes more amount of flow run through the upper part of flow path, with the result of larger flow angle. When it comes to C-cas scheme, an opposite trend is observed above 20% strake height. The flow blockage caused by high loss region 1(Figure 9a) makes a more non-uniformed distribution of flow angle, which is more unfavorable. From the distribution of pressure ratio, C-hub scheme performs a reduction of pressure rise from hub to tip. C-mid scheme indicates an obvious decrease at the tip region and a little improvement below 40% strake height. Under the influence of flow redistribution, the pressure rise of this scheme shows almost no change. As for C-cas scheme, the pressure ratio presents an increase almost covering the whole span except for the spanwise position above about 95% strake height. It is attributed to the severe tip blockage, which decreases the pressure rise capacity.

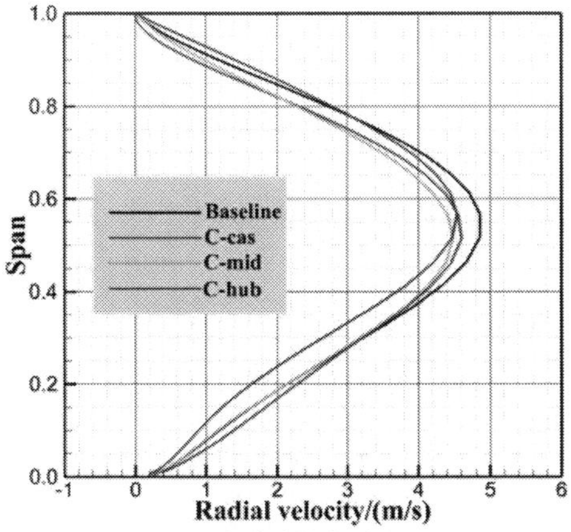

a. Radial Velocity

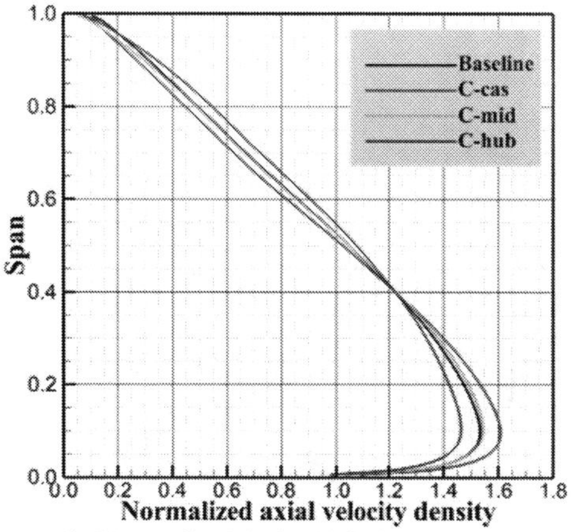

b. Normalized Axial Velocity Density

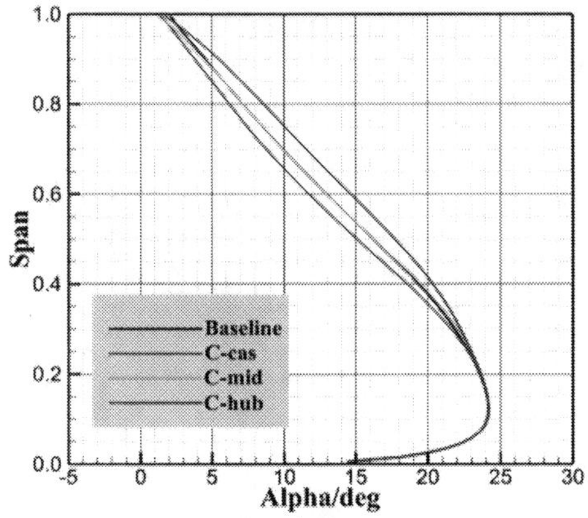

c. Absolute Flow Angle

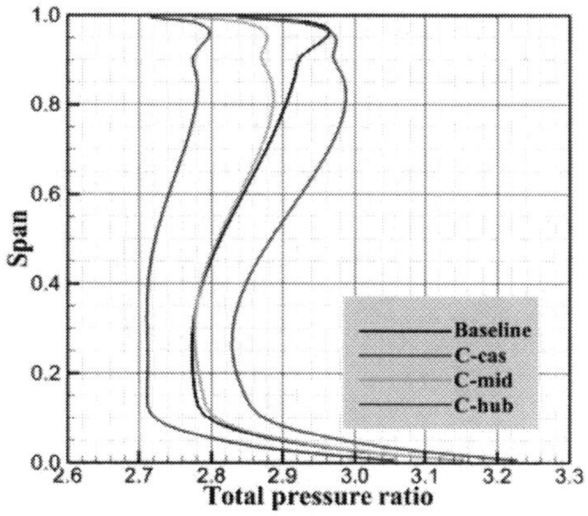

d. Total Pressure Ratio

Figure 12. Radial Distribution of Flow Parameters at Ram-rotor Outlet

3.4 Strake Tip Leakage Flow

The strake tip clearance leakage is one of the main flow features in ram-rotors, and Figure 13 provides the pressure coefficient distributions at 95% throat height of four ram-rotors. It is seen in Figure 13 that in C-cas scheme, the forward motion and intensified strength of shock train lead to an increase of static pressure on pressure surface at axial position from about 10% to 20%, which increases the aerodynamic load of strake fore part. Hence, the strake tip leakage is intensified at strake fore part. Great variation occurs at strake aft part (from about 80% axial position to trailing edge). Compared to baseline, C-mid and C-hub schemes both yield a decrease of strake load due to reducing pressure difference between both sides of strake, and the resulting tip leakage extent is smaller. The improvement of tip leakage in C-hub scheme is more preferable. There is an opposite effect for C-cas scheme which causes more tip leakage.

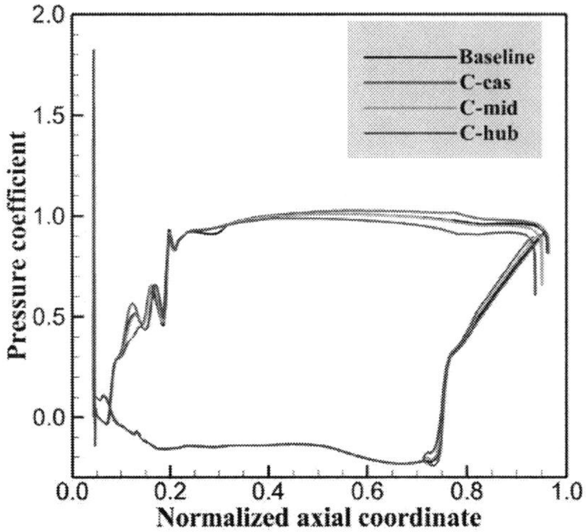

Figure 13. Pressure Coefficient Distributions At 95% Throat Height

Figure 14 illustrates three dimensional streamlines and entropy distribution with a color bar range from 150 to 300. It is known from these figures that the tip leakage initiates near the leading edge, and the leakage vortex is formed and develops downstream. Strake tip loss arises from the mixing between leakage flow and main flow. The leakage is enhanced and more significant loss is generated when the low-momentum leakage flow encounters the pressure surface of adjacent strake (The encountering location is denoted with blue filled arrow in Figure 14a). C-cas scheme gives a more severe leakage loss compared to baseline due to higher pressure rise capacity, whereas there is a low leakage loss in C-hub scheme resulting from smaller strake load and pressure difference. As for C-mid scheme, the effect is a compromise between C-cas and C-hub schemes, which shows a similar loss to baseline.

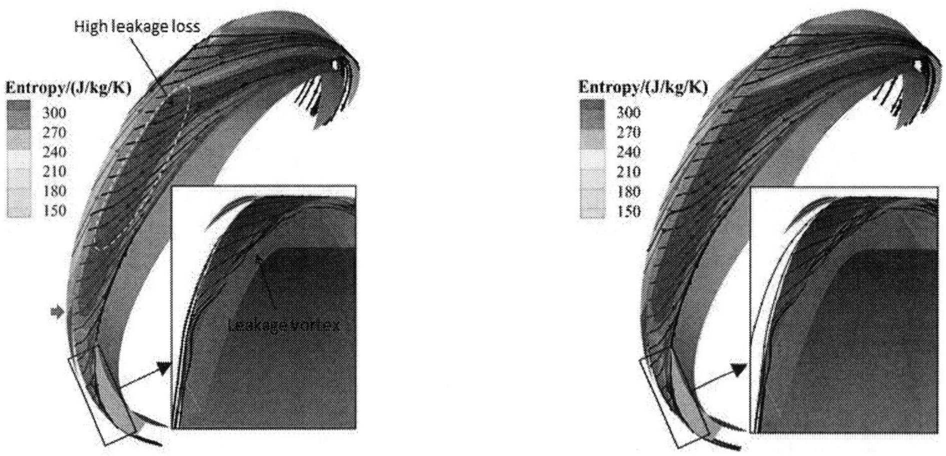

a. Baseline b. C-cas Scheme

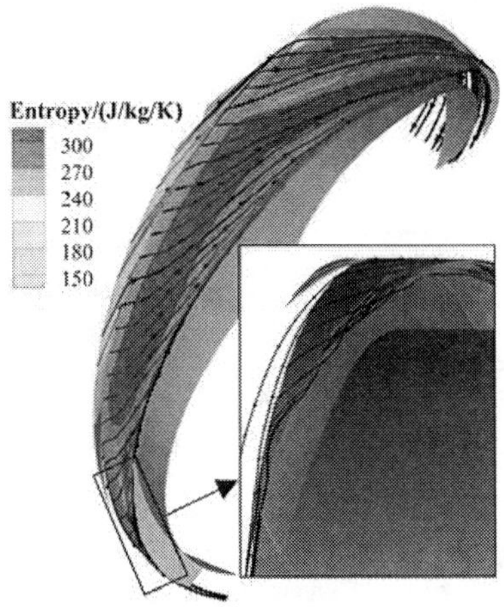

c. C-mid Scheme

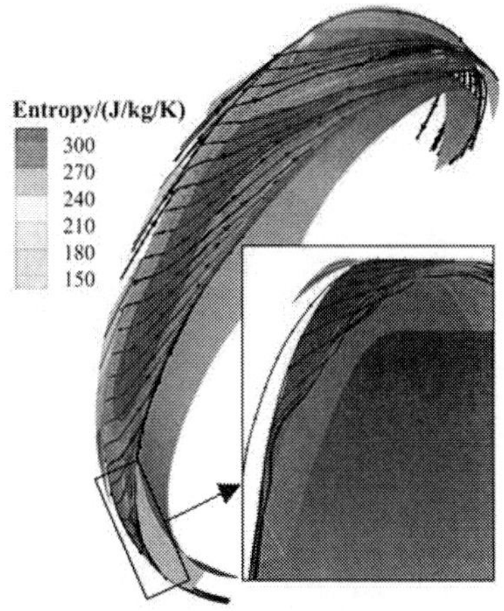

d. C-hub Scheme

Figure 14. Streamlines and Entropy Distribution

3.5 Effect of Leaning Angle

As mentioned above, it is true that C-mid scheme performs better in terms of flow capacity near the suction surface of strake (see figure 8c) and the radial migration tendency of boundary layer fluid (see figure 12a), despite the fact that little positive effect is gained. Maybe the amount of lean is not enough to get a more optimized ram-rotor by using such leaned ram-rotor as C-mid scheme. Hence, in this section the impact of leaning angle on the aerodynamic characteristics of C-mid scheme is investigated. It can be observed in figure 15 that pressure ratio and efficiency both rise first and then drop with the increase of leaning angle, while the optimum angles are different. The total pressure ratio reaches maximum value of 2.831 at the leaning angle of 10 degrees, and 20 degrees of leaning angle can gain a more preferable adiabatic efficiency of 0.7235. For a better ram-rotor performance, there need to be a compromise selection of leaning angle.

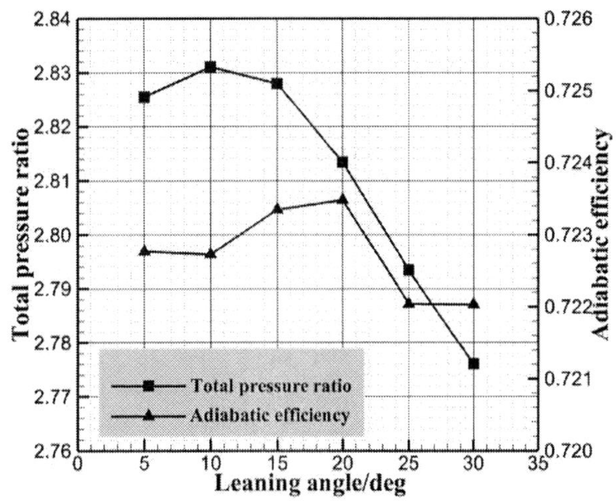

Figure 15. Ram-rotor Characteristic Parameters Variation with Sweeping Angles

Figure 16 gives the pressure coefficient distribution through the ram-rotor path at different leaning angles. The pressure rise characteristics is similar when using different leaning angles. The shock system ahead of throat surface (incident shock and reflected shock) is almost the same, and the difference is mainly expressed by the strength and position of shock train. As a whole, large leaning angle makes a more downstream initial position of shock train, except that 10 degrees lean makes its position a little forward. Because the backward motion of shock system produces no benefit to the strength of shock train, the pressure rise resulting from shock train is reduced when applying overlarge leaning angle.

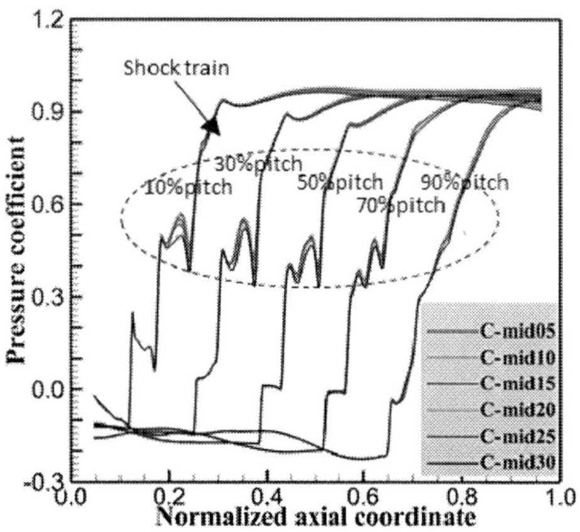

a. At 10% Throat Height

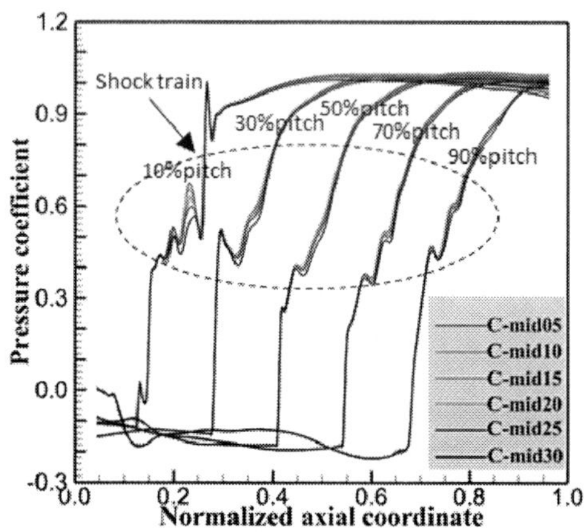

b. At 99% Throat Height

Figure 16. The pressure coefficient distribution through the ram-rotor path at different leaning angles

The pressure coefficient distributions at 95% throat height is plotted in figure 17, and figure 18 gives the axial distribution of normalized velocity density at mid-gap height of tip clearance region. Due to the weakened shock train and decreased pressure rise capacity at the strake aft part, the aerodynamic load promotes an almost descending trend. Correspondingly, there is a reduction of leakage mass flow especially in the shock train region near pressure surface (dotted oval box) and in the tip low momentum corner near the suction surface (dotted rectangle box). Overall, the beneficial effect is more positive when increasing leaning angle.

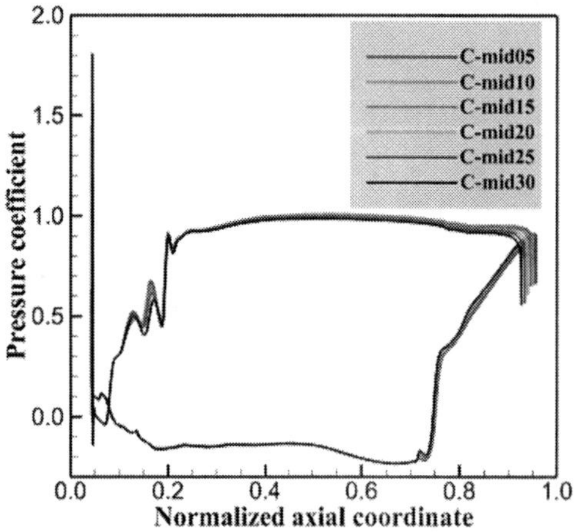

Figure 17. Pressure Coefficient Distributions at 95% Throat Height

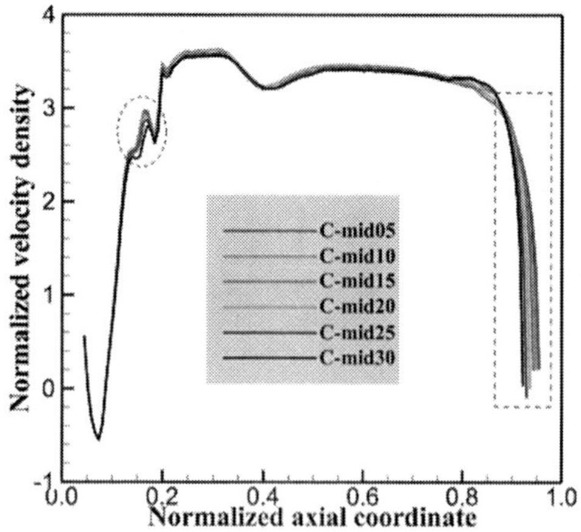

Figure 18. Axial Distributions of Normalized Velocity Density at Mid-gap Height of Tip Clearance

IOP Conf. Series: Materials Science and Engineering 1081 (2021) 012021 doi:10.1088/1757-899X/1081/1/012021

Figure 19 gives radial distribution of radial velocity and loss coefficient with different leaning angle at ram-rotor outlet. From Figure 19a, it can be seen that there is a significant reduction of radial velocity at midspan height when initially increasing angle from 5 to 15 degrees. The radial transport of high loss fluid is gradually inhibited, so is the separation area of reflux tip. Just as shown in Figure 19b, the total pressure loss coefficient presents a reducing trend above about 40% strake height. It is beneficial to reduce the flow loss and to further increase efficiency of ram-rotor. As the leaning angle varies from 20 degrees to 30 degrees the reduction of radial velocity is mainly realized in the region below 40% strake height, however the radial velocity above 85% height is even increased, which can reduce the beneficial effect of suppressing the accumulation of low momentum in the tip corner. It is necessary to point that large leaning angle leads to a large diverging S1 stream surface at hub section, and it is easy to induce more low momentum fluid accumulation at hub corner and raise the flow loss below about 30% strake height (see figure 19b). When at 25 or 30 degrees, the improvement of radial transport can't compensate the detrimental impact caused by the increase of low momentum in the hub corner, which leads to a decrease of adiabatic efficiency in figure 15.

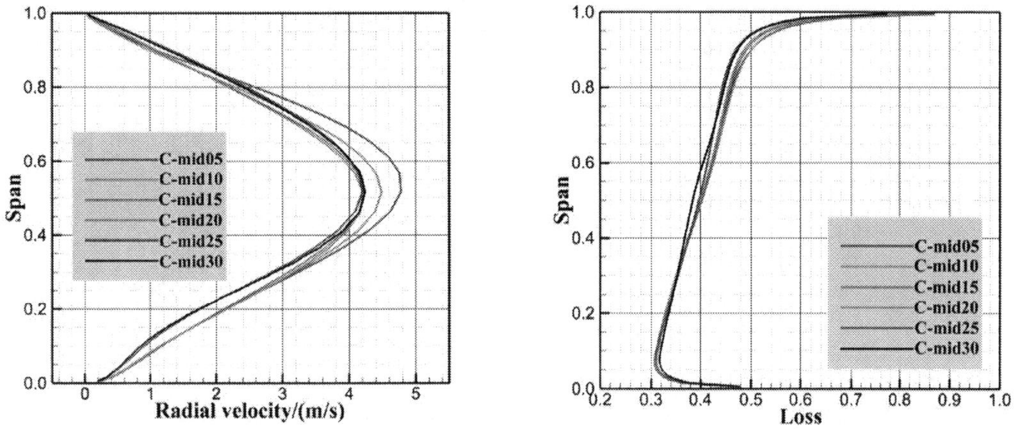

a. Radial Velocity b. Total Pressure Loss Coefficient

Figure 19. Radial Distribution of Flow Parameters at Ram-rotor Outlet

4. Conclusions

A detailed numerically investigation has been conducted to better understand the strake lean effect on the aerodynamic behavior of ram-rotors. Three types of leaned ram-rotors and one leaned scheme with six leaning angles are modeled and analyzed, respectively. It is found that the application of lean can affect the intensity and position of shock system, radial transport and accumulation of low momentum fluid in the boundary layer of suction surface.

The leaned ram-rotor which keeps tip section unchanged (C-cas scheme) performs better in pressure rise, with an increase of 2.338 % in terms of total pressure ratio compared to baseline. Reflected shock and shock train shift upstream with a high pressure rise, which worsens flow blockage and separation near suction surface of strake. The beneficial effect of lean can't compensate the detrimental impact due to large pressure rise. The increase of strake load aggravates the tip leakage flow.

While the leaned ram-rotor which maintains identified sections of root (C-hub scheme) is more preferable in reducing loss, with an improvement of 0.462% in terms of adiabatic efficiency compared to baseline. It shows a more backward displacement of shock system and delays separation on the suction surface. The blockage at the tip corner of strake aft part is relieved. Loss is also reduced due to low strake load.

With respect to the leaned ram-rotor which has the same middle section with baseline(C-mid scheme), a slight reduction of 0.169% and 0.025% in terms of total pressure ratio and adiabatic efficiency is obtained, but it still has a positive effect to some extent on the streamlines behavior on the suction surface compared to the baseline. Different leaning angles are applied in C-mid scheme, and it is concluded that there exist different optimal leaning angle values in terms of total pressure ratio and adiabatic efficiency as leaning angle rises. Large leaning angle makes the shock train more backward and reduces its strength. A gradually descending trend of tip leakage occurs when leaning angle rises. However, if overlarge leaning angle, such as 25 or 30 degrees, is introduced, the improvement of radial transport is not enough to compensate the detrimental impact of low momentum accumulation in the hub corner.

5. References

[1] Zhang J Z, Chen J H, Li K L.Preliminary Study on the Application of Flow Control Technology in Axial flow Compressor [J]. Fan Technology,2013(03):15-20.
[2] Lv Yaguo, Huang Jian.Numerical Simulation of combined Compressor Performance and Flow Field [J]. Fan Technology,2011(03):11-14+22.
[3] Hu Yang, WANG Zhiheng, XI Guang.Research progress of CARBON dioxide compressor in carbon capture and Storage (CCS) system [J].Fan Technology,2013(04):70-76+86.
[4] Gao Limin, GAO Lei, XIE Jian.Research on design of high pressure specific transonic centrifugal compressor [J]. Fan technology,2016,58(03):42-49.
[5] Kantrowitz A, (1954). The Supersonic Axial-flow Compressor. NACA-Report 974
[6] Klapproth J F, Ullman G N and Tysl E R, (1952). Performance of an Impulse-type Supersonic Compressor with Stators. NACA RM52B22
[7] Lawlor S P, Hinkey J B, Mackin S G, et al, (2004). Supersonic Compression Stage Design and Test Results. ASME Paper 2004-IMECE-59914
[8] Grosvenor A D, Brown P M, Lawlor S P, (2006). Design Methodology and Predicted Performance for a Supersonic Compressor Stage. ASME Paper 2006-GT-90409
[9] Grosvenor A D, Taylor D A, Bucher J R, et al, (2008). Measured and Predicted Performance of a High Pressure ratio Supersonic Compressor Rotor. ASME Paper 2008-GT-50150
[10] Steele R, Baldwin P, Kesseli J, et al, (2005). Insertion of Shock Wave Compression Technology into Micro Turbines for Increased Efficiency and Reduced Costs. ASME Paper 2005-GT-68203
[11] Lawlor S P, Baldwin P, (2005). Conceptual Design of A Supersonic CO_2 Compressor. ASME Paper 2005-GT-68349
[12] Han J A, Zhong J J, Pan Z S, et al, (2014). Effect of Tip Clearance Size on the Ram-rotor Performance. ASME Paper 2014-GT-26084
[13] Guan J, Han J A, Zhong J J, et al, (2017). Numerical Research of Shock Wave/Boundary Layer/Leakage Flow Interaction in the Ram-rotor. Journal of Engineering Thermophysics, Vol.38, No.5, 950-956(in chinese)
[14] Han J A, Zhong J J, Yan H M, et al, (2009). Numerical Research of Three Dimensional Flow-path in a Ram-rotor. Journal of Aerospace Power,Vol.24, No.5, 1079-1088(in chinese)
[15] Han J A, Guan J, Zhong J J, et al, (2016). Effect of Vane Opening on Aerodynamic Performance of the Ram-rotor Test System. Journal of Thermal Science, Vol.25, No.3, 207-215
[16] Yang L, Zhong J J, Han J A, et al, (2012). Numerical research on the flow field and performance of a ram-rotor and a scrampressor. ASME Paper, 2012-GT-69358
[17] Guan J, Han J A, Zhong J J, et al, (2017). Numerical Research of Aerodynamic Sweep on Leading Edge in the Ram-rotor. ASME Paper 2017-GT-64202
[18] Yang L, Zhong J J, Han J A, (2011). Numerical Research of the Ram-rotor with Different Geometric Parameters. ASME Paper 2011-GT-46051
[19] Breugelmans F A H, Carels Y, Demuth M, (1984). Influence of Dihedral on the Secondary Flow in a Two-dimensional Compressor Cascade. Journal of Engineering for Gas Turbines and Power, Vol.6, No.3, 578-584
[20] Bergner J, Hennecke D K, Hoeger M, et al, (2003). Darmstadt Rotor NO.2, II: Design of Leaning Rotor Blades. International Journal of Rotating Machinery, Vol.9, No.6, 385-391

[21] Benini E, Biollo R, (2007). Aerodynamics of Swept and Leaned Transonic Compressor-Rotors. Applied Energy, Vol.84, No.10, 1012-1027
[22] Neshat M A, Akhlaghi M, Fathi A, et al, (2015). Investigating the Effect of Blade Sweep and Lean in One Stage of An Industrial Gas Turbine's Transonic Compressor. Propulsion and Power Research, Vol.4, No.4, 221-229
[23] Fluent Inc, (2006). Fluent 6.3 user's guide.
[24] Zhong Yikai, Yang Jinguang, ZHANG Min, Yang Shuai, Liu Yan, WANG Chunxue. Comparative study on complex turbulence model in numerical simulation of axial flow compressor [J]. Fan technology,2020,62(02):9-18.
[25] Suder K L, Celestina M L, (1994). Experimental and Computational Investigation of the Tip Clearance Flow in a Transonic Axial Compressor Rotor. ASME Paper 94-GT-365
[26] Denton J D, (1997). Lessons from Rotor 37. Journal of Thermal Science, Vol.6, No.1, 1-13
[27] Chima R V, (1998). Calculation of Tip Clearance Effects in a Transonic Compressor Rotor. Journal of Turbomachinery, Vol.120, No.1, 131-140

Acknowledgments
The author(s) disclosed receipt of the following financial support for the research, authorship, and/or publication of this article: This work was supported by the National Natural Science Foundation of China (No.51406022, No.51436002), Program for Liaoning Innovative Research Team in University (LT2015004) and State Foundation for Studying Abroad.

Investigation of Time-series Prediction for Turbine Machinery Condition Monitoring

Chang-Jun Liu[1], Hai-Tao Liu[1], Chao Bian[1], Xu-Dong Chen[1], Shu-Hua Yang[1,2], Xiao-Fang Wang[1*]

[1] School of Energy and Power Engineering, Dalian University of Technology, Dalian 116024, China
[2] Shenyang Blower Works Group Corporation, 110869 Shenyang, China

*Corresponding Author, Email: dlwxf@dlut.edu.cn

Abstract. A running compressor generates a huge amount of data every day. Some of the data contains useful information that allows the user to determine in advance whether the machine will fail or not. Forecasting in advance cannot only avoid the huge loss caused by unnecessary downtime but also ensure the safety of staff. The main tasks of compressor condition monitoring include data acquisition, data preprocessing, data analysis, alarm module, response strategy analysis, etc. Prediction is one of the tasks of condition monitoring. This paper mainly studies the performance of different time prediction models on the running data produced by a centrifugal compressor. In the data preprocessing part, this paper uses Bayesian wavelet denoising and probabilistic principal component analysis methods proposed by the research group to remove the useless parts in the data. In the part of time series model, the prediction performance of LSTM, GRU and ARIMA on the same dataset is compared. The comparative results indicate that since the dataset is non-stationary, the LSTM and GRU outperform the statistical ARIMA model.

1. Introduction

The diagnosis of rotating machinery includes the diagnosis of equipment performance and fault diagnosis of equipment [1]. Complex rotating machinery, such as fans, compressors, steam turbines and gas turbines, is a key production tool in important modern production departments such as petroleum, chemical industry, metallurgy, aerospace and electric power, etc. It is of great significance to carry out performance monitoring and fault diagnosis for the equipment.

With the increasing complexity and reliability requirements of rotating machinery, the traditional o peration and maintenance mode has changed from traditional post maintenance and regular maintenance to more efficient condition-based maintenance (CBM) and time-based maintenance(TBM). Rosmai n-i Ahmad et al. summarized the development of CBM [2].

One task of CBM is condition-based monitoring. According to the monitoring principle, it can be divided into non-performance based CBM, which contains oil sample, thermography, strain analysis, vibration, metal temperature, load analysis, acoustics; and performance-based CBM which usually refers to gas analysis. In addition, the monitoring method can be divided into online monitoring and offline monitoring [3].

An important purpose of condition monitoring is to reduce the cost of equipment operation [4]. And the loss caused by the unexpected shutdown is very heavy for an enterprise relying on centrifugal compressor for production. It will not only cause losses, but also cause casualties [5].

If the trend of turbine performance can be predicted in advance, the corresponding strategy can be formulated in advance to avoid unexpected shutdown through the analysis results. Therefore, researchers need a system that can predict the performance of centrifugal compressor units in advance.

With the advent of the era of Engineering 4.0, the storage technology of data is more and more advanced. These massive data bring unprecedented opportunities to the technology based on digital drive. The problem of using deep learning and other methods to predict data has been widely studied. The establishment of deep network is helpful for mining high-order and abstract information in data. When the effective feature representation of data is relatively complete, it can get better results whether it is used for damage classification or prediction regression [6].

Before developing a time series prediction model, we need to pre-process the data due to the possible noise. Si et al. [7] used wavelet packet technology for noise elimination, and took out the characteristic values of signal faults as the sample data for network training and testing. Xu and Liu et al. [8] used Hilbert-Huang transformation to obtain the instantaneous frequency of the decomposed signal, and calculated the statistical feature set of the basic mode component and the instantaneous frequency. Xu and Jiang et al. [9] proposed a probabilistic signal processing method consisting of discrete wavelet packet transform, Bayesian hypothesis test and probabilistic principal component analysis.

After data pre-processing, we turn to build up the time series prediction model. In recent years, Bottieau et al. [10] used LSTM to estimate the impact of prediction accuracy on the ability of wind turbines, Yin et al. [11] proposed a fault diagnosis method for wind turbine gearboxes based on optimized LSTM neural networks with cosine loss. The above researches mainly focus on the analysis of vibration signals of wind turbines by using neural network models, while the data generated by compressors and other forms of data other than vibration are rarely studied. Zhao et al. [12] used the LSTM neural network to predict the dynamic response of the rotating machinery under external excitation, and obtained the health probability of the machinery by using residual data between sensors and model prediction through Bayesian method. However, this study did not compare the performance of LSTM model with other models. In view of the lack of horizontal comparison between models and the lack of experiments on compressor data, this paper makes a comparative analysis of three different models on the same compressor data set.

In this paper, the structure is as follows: section 1 briefly introduces the basic methods including LSTM (long-short term memory), GRU (gated recurrent unit) and ARIMA (autoregressive integrated moving average model). Thereafter, sections 2 and 3 conduct comparative study of the three time series prediction models on a real centrifugal compressor dataset to comprehensively investigate their characteristics and performance. Finally, section 4 gives the conclusion.

2. Basic methods

2.1. Time-series prediction models

2.1.1. LSTM neural network prediction model. Before we introduce the model, let's determine the form of the input data. There are many input forms of LSTM neural network. Here we introduce the single variable time series data. Let's assume that there is a time axis with n time points, which are evenly distributed on the time axis in a discrete form, X_t ($t = 1, 2, 3 \ldots n$) is a multi-dimensional time variable at each time point. These random variables constitute the input of LSTM neural network [13][13].

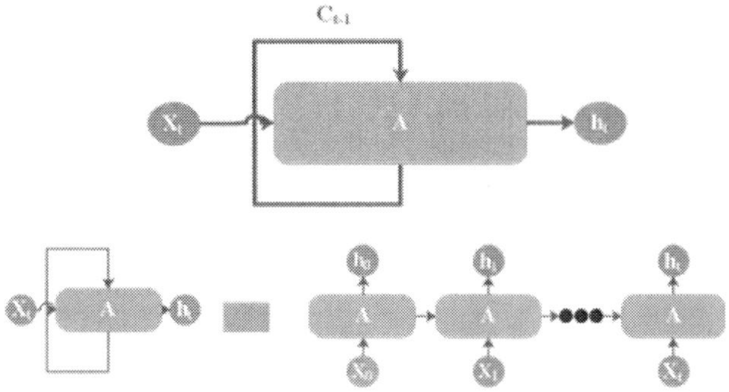

Figure 1. Recurrent neural networks and its unrolled form.

The LSTM neural network contains a lot of parameters [14]. Specifically, C_t is called cell state, it contains part of the information summarized from the historical data of neural network, and runs through the whole unit. h_t is the output of each cell. X_t is the input for each cell. There are four gate-like structures in LSTM. Their function is to filter through the information [15].

The first is the forgetting gate formulated as:

$$f_t=\sigma\left(W_f[h_{t-1},x_t]+b_f\right), \tag{1}$$

where $\sigma(\cdot)$ is the sigmoid activation function used for the nonlinear transformation of input, W_f is the weight of the neural network, and b_f is the bias.

Secondly, the update gate, also known as the input gate, is used to update the state of the cell. The specific formula is as follows:

$$i_t=\sigma(W_i[h_{t-1},x_t]+b_i), \tag{2}$$

$$\widetilde{C}_t= tanh(W_C[h_{t-1},x_t]+b_C), \tag{3}$$

$$C_t=f_t{*}C_{t-1}+i_t{*}\widetilde{C}_t, \tag{4}$$

where C_t represents the state of the unit, which is used to store the historical information, and is the degree of updating information.

Finally, the output gate responsible for the output results and is formulated as follows:

$$o_t=\sigma(W_o[h_{t-1},x_t]+b_t), \tag{5}$$

$$h_t=o_t{*}\,tanh(C_t) \tag{6}$$

2.1.2. GRU neural network prediction model. GRU is a special form of LSTM. It also has a gate-like structure in LSTM. However, the difference is that the GRU neural network simplifies four gates into two gates, namely reset gating and update gating.

The first is to reset the gating r_t. Its main function is similar to the hidden state in LSTM. It integrates the input information into $h_{(t-1)}$, and controls how much information is written to the current candidate set $h_{(t-1)}$ at the previous moment. The formula is as follows:

$$r_t=\sigma(W_r\cdot[h_{t-1},x_t]), \tag{7}$$

$$\tilde{h}_t= tanh(W\cdot[r_t{*}h_{t-1},x_t]) \tag{8}$$

The second is update gate z_t. The larger the update gate is, the more state information of the previous moment is brought in. The second is the update gate z_t, which is expressed as:

$$z_t=\sigma(W_z\cdot[h_{t-1},x_t]), \tag{9}$$

$$z_t=\sigma(W_z\cdot[h_{t-1},x_t]), \tag{10}$$

$$h_t=(1-z_t)*h_{t-1}+z_t*\tilde{h}_t \tag{11}$$

The larger the update gate is, the more state information of the previous moment is brought in.

2.1.3. ARIMA time series prediction model. The ARIMA model is one of the commonly used statistical models for time prediction. It is based on the integration of autoregressive model (AR) and moving average model (MA). Before introducing the model, we need to explain the conditions they used.

2.1.3.1. Stability requirement. Stationarity means that the curve fitted based on time series samples has the inertia which continues according to the existing form. It can be divided into strict stability and broad stability. Strict stationarity means that no matter how much window size, as long as the same number of consecutive data, they constitute the same joint distribution. But usually, the strict stability condition is difficult to satisfy. Hence, the strict stability condition is weakened to the wide stability condition. At this time, only the second moment is required to exist at any time. The mathematical expectation of random variables does not change with time. The autocorrelation coefficient of random variables at two different time points is only related to the time difference and does not change with time.

2.1.3.2. Basic theory of ARIMA. Autoregressive model is a time series model based on its own data to predict the future. It needs to meet the stationary condition, and the data must have correlation. If the autocorrelation coefficient is less than 0.5, it should not be used. The p-order autoregressive process formula is as follows:

$$y_t=\mu+\sum_{i=1}^{p}\gamma_i y_{t-i}+\varepsilon_t, \tag{12}$$

where μ is a constant term, p is the order, and γ_i is the autocorrelation coefficient, ε_t is the error.

The moving average model focuses on the accumulation of errors in the autoregressive model. The formula is as follows:

$$y_t=\mu+\varepsilon_t+\sum_{i=1}^{q}\theta_i\varepsilon_{t-1} \tag{13}$$

The moving average method can effectively eliminate the random fluctuation in the prediction.

The ARMA method is a combination of autoregressive model and moving average model as:

$$y_t=\mu+\sum_{i=1}^{p}\gamma_i y_{t-1}+\varepsilon_t+\sum_{i=1}^{q}\theta_i\varepsilon_{t-1} \tag{14}$$

The ARIMA model is based on the ARMA model by adding difference steps to ensure that the data meets the stability conditions. The ARIMA model contains three parameters p, q, and d. The parameter p is the order of autoregression, which is also the lag number of time series data itself, called the AR term. The parameter q is the lag number (lags) of the prediction model using random error, which is called the MA term. The parameter d means that the data has experienced several orders of difference before obtaining wide stationarity, also known as the integrated term.

3. Comparative study

3.1. Dataset

The data comes from a group of centrifugal compressors with the shaft speed of 5556 RPM. The data contains 11 variables, and the specific variables are shown in the figure 2. One of the variables (inlet air temperature) selected from the raw data without any processing is shown in figure 3. From March 1, 2013 to October 16, 2013, the data was obtained hourly, including a total of 3520 data points. Note that the data points during the maintenance outage are excluded in this study. There is one outage

IOP Conf. Series: Materials Science and Engineering 1081 (2021) 012022 doi:10.1088/1757-899X/1081/1/012022

happened because of the failure of multiple impellers. The run data is recorded manually, so the data inevitably contains outliers. human error, outliers, missing values, and noise.

3.2. Data pre-processing
In this section, the data used in the data analysis model is processed by the Bayesian wavelet denoising and the probabilistic principal component analysis (PPCA), see Appendix. That is, we first use the Bayesian wavelet to clean the data because there are a large number of noises which contains useless information. By this way we can get the denoised data, see figure 3. Thereafter, we use PPCA to reduce the dimension of the data to make the prediction easier. Consequently, the resulting data contains six principal components that contribute the most to the original information. In the experiments below, we take the first principle component (PPCA1) for analysis.

The purpose of this example is to compare the performance of the aforementioned three time series models on the real data. We run the experiments on a personal laptop with intel 2.40 GHz CPU and 16 G RAM.

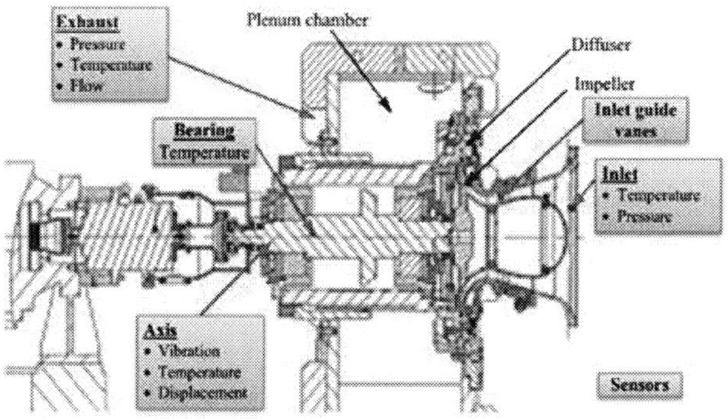

Figure 2. Monitor variables of the centrifugal compressor [9].

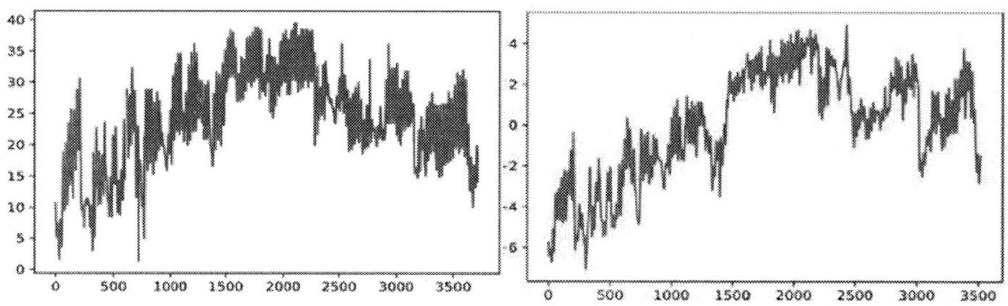

(a) (b)

Figure 3. (a) The raw data and (b) the denoised data regarding the inlet air temperature.

4. Results and discussions

4.1. The predictions of LSTM
The process predicted by the LSTM model is shown in figure 4.

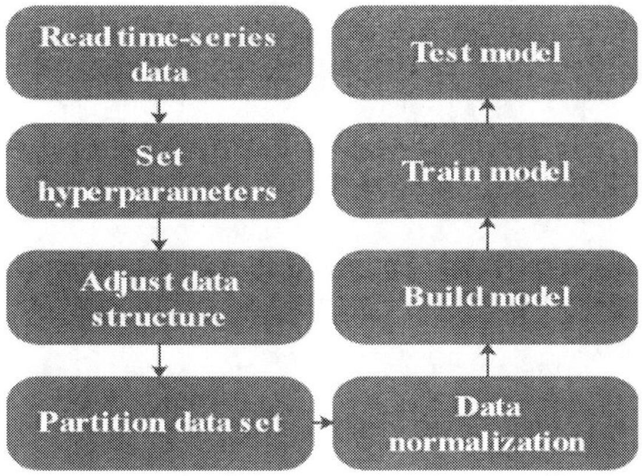

Figure 4. Flow chart of LSTM neural network prediction.

First of all, we read the data file and change the form of the data. The random variables corresponding to each time point ($t= t_1,t_2,t_3...t_{3520}$) are defined as x_1, x_2, ... x_{3520}. A window of size 2 is specified. After traversing the entire data set by this window, the format of the data set is changed to (x_1, x_2), $(x_2, x_3)...(x_{3519}, x_{3520})$, which ensures that the input meets the input form required by the LSTM neural network.

The parameters of the model should be set before training. In machine learning, these parameters are called hyperparameters. We set the training epoch (training rounds) as 50, the input size (input dimension) as 2, the number of LSTM hidden layer as 1, the number of hidden units (hidden layer neurons) as 50, and the learning rate as 0.01.

After setting the parameters, the dataset should be divided into the training set and the test set. We choose 70% of the data as the training set for model training and the remaining 30% as the test set to test the model performance.

After dividing dataset, in order to optimize the prediction ability of the model, the Z-score criterion is used to normalize the data. This method changes the overall distribution of data into normal distribution along each dimension, and can eliminate the gradient dispersion or gradient disappearance caused by the outliers in the error back propagation.

Next step is the final step before training model, we should make the data normalization. Because when we calculate the minimum value of the loss function, we used the method based on gradient descent. And the data normalization can make the gradient descent distance shorter, so the time of calculating will be cut.

Then we start to train the model. The prediction experiment is designed to predict the next time step with double time steps. Considering the low complexity of the task, the single-layer one-way LSTM neural network was used for prediction.

The training process is as follows the data from the LSTM network output is fed into a fully connected layer, and the output of the full connection layer should input into the MSE loss function together with the truth value. Then, the Adam [16] optimization method based on gradient descent mentioned above is used to optimize the function value, and the model training is completed after 50 rounds of repetition.

After training, we begin to test model on the test set. There are 1056 data points in the test set. The test results are shown in the figure 5.

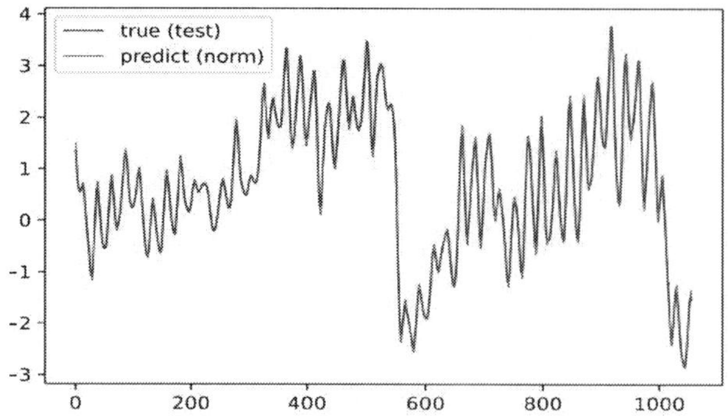

Figure 5. Performance of LSTM on test set.

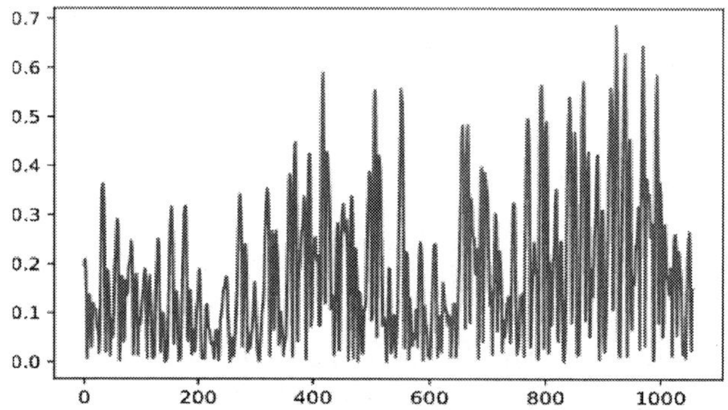

Figure.6 Residual error between the predicted value and the true value.

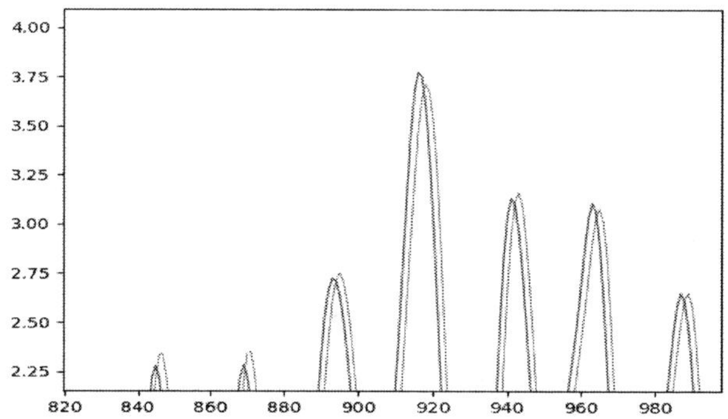

Figure 7. The local hysteresis of LSTM predictions.

The root-mean-square error of the model on the training set is 0.00857. In order to exclude the possibility of overfitting, the performance of the model on the test set should be evaluated. In figure 4, the green curve is the true value on the test set, and the red curve is the predicted value on the test set after inversed normalization. Additionally, figure 6 shows the residual curve between the predicted value and the true value. It can be observed that the maximum absolute residual error is 0.687. It can be found that, although the fluctuation range of the true data is large, the trained LSTM network well captures the trend of data fluctuation on the train set, and generalizes well on the data never seen by the model, which may be related to the ability of the LSTM model itself to remember the features of historical data. However, based on experience, the model cannot reach such a degree of accuracy in practical engineering applications. After analysis, the most likely reason is that the task of single-step prediction is too simple for the model. The LSTM model can easily predict the trend of the next time step by summarizing the data characteristics of the first two time steps by the ability to remember historical data characteristics. It can be seen from the figure 7 that the predicted value lags behind the real value to a certain extent. The hysteresis of the predicted value is also related to the characteristics of the model itself. The reason for "hysteresis" is that LSTM pays too much attention to the data at the last *t-1* moment in the time step, resulting in the prediction result being the repetition of the data at the *t-1* moment. However, from the overall prediction results, the LSTM model performs very well in this group of data.

4.2. The predictions of GRU

Since the GRU is a variant of the LSTM, it is the same as the previous section in data input. It can be seen from the figure 8 and figure 9 that the prediction of GRU is slightly better than that of LSTM, and the maximum error on the test set is 0.39. Similarly, after 50 rounds of data training, the GRU converges slightly faster than the LSTM, which can be explained by the structure of the model. The GRU has only two doors (update and reset gates) and the LSTM has three doors (forget, input, and output gates). The GRU directly passes hidden state to the next unit, while the LSTM uses the memory cell to wrap the hidden State. Hence, the GRU is easier to converge because of fewer parameters. However, in the case of large data sets, the LSTM representation performance is better.

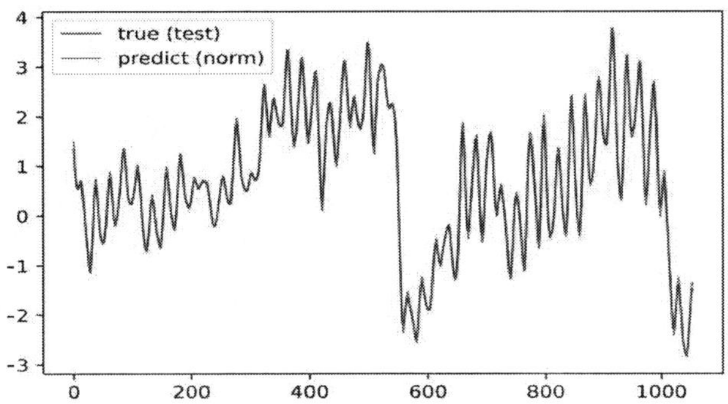

Figure 8. Performance of GRU on test set.

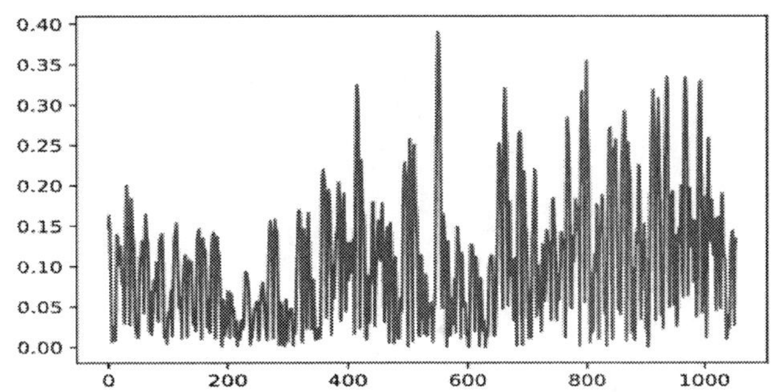

Figure 9. Residual error between the predicted value and the true value.

4.3. The predictions of ARIMA

According to the requirements, the Autocorrelation coefficient (ACF) stability test was first conducted on the original data.

From the ACF stability test, we can find that $P > 0.05$, and the t statistic is outside the 5% confidence interval. Hence, it can be determined that the time series is not a stationary time series, and further difference processing is needed for the data.

After the first-order difference, the data meets the demand of the Dickey fuller detection, but the mean and standard deviation still fluctuate.

We thereafter use the Bayesian information criterion to find p and q for the ARIMA model. This method is based on the principle of making the model as simple as possible to select parameters. The smaller the BIC value is, the simpler the model is. As shown in figure 10, we arrive at p = 20, q = 4.

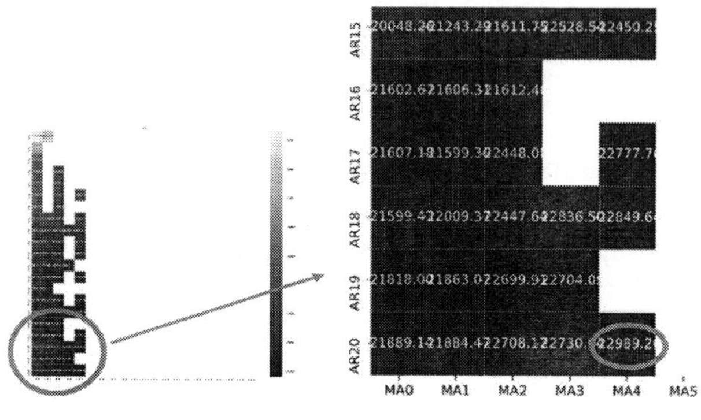

Figure 10. Bayesian Information Criterion diagram.

However, the ARIMA (20,1,4) model does not perform well in the prediction set as shown in figure 11, since it only catching part of the trend of the data. It is completely distorted after 200 steps of prediction. The reason for the poor predictive performance of ARIMA is that the ARIMA model is limited. The reason why it requires the data to be stationary is that the model relies on the continuation of historical data when making predictions. If the existing underlying system is too complex and the data lacks "inertia", it is impossible to find an ARIMA model with the right parameters to predict. Nonstationary sequences such as random walk sequences are not suitable for prediction with this model. However, the ARIMA also has its advantages in data prediction. If the basic model is simple enough, it is much more efficient than other deep learning methods and saves a lot of computing costs [17].

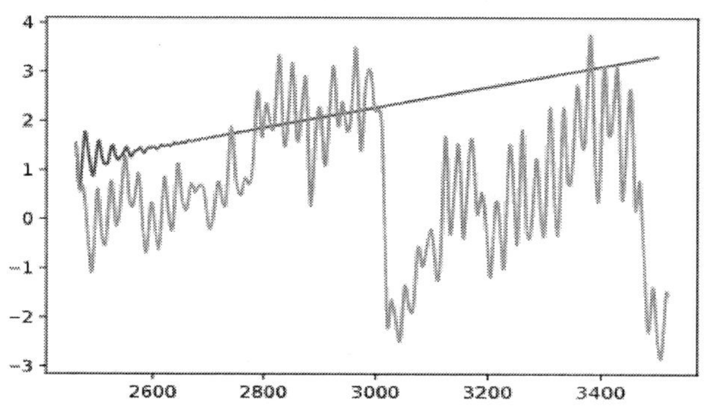

Figure 11. Performance of ARIMA on test set.

5. Conclusion

This paper compares the performance of three time-series prediction models. In this group of compressor operation data, two time series models based on recurrent neural network (LSTM and GRU) perform very well by accurately predicting the trend of the data. Considering that the GRU is not only slightly better than LSTM in terms of test results, but also simpler in structure than LSTM, its capability in time series prediction is worth further exploration.

The ARIMA model based on statistical method does not perform well in this dataset. But the problem can be attributed to the following reasons:

- The data itself is non-stationary, the autocorrelation of data is not strong, and the difference of data cannot meet the requirements of the model.
- Unlike the SARIMA model [19], the ARIMA model does not consider some factors, for example, the seasonality, which could further improve the prediction.

Appendix A

Bayesian wavelet denoising

Discrete wavelet packet decomposition, also known as the optimal sub-band tree structure, is the optimization of wavelet transform [19].

In wavelet transform, if the scale parameter is 2^i , that is, the scale is discretized by power series. Then, this kind of wavelet transform is called dyadic wavelet transform.

In dyadic wavelet transform, the scale function and wavelet function of adjacent series also have a recursive relationship when the wavelet packet is decomposed at all levels.

Note that the father wavelet $\Phi(t)$ in the wavelet packet transform is $\mu_0^0(t)$, the mother wavelet $\Psi(t)$ is $\mu_1^0(t)$,where the superscript represents the decomposition order of the wavelet packet, and the subscript represents the position of the wavelet packet in its level. The recurrence relation is expressed as follows:

$$\begin{cases} \mu_{2n}(t)=\sum_k s_k\mu_n(2t\text{-}k) \\ \mu_{2n+1}(t)=\sum_k w_k\mu_n(2t\text{-}k) \end{cases} \tag{A.1}$$

where s_k represents the scale coefficient and w_k represents the wavelet coefficient.

Because the noise should account for a small part of the whole data, we carry out Bayesian hypothesis test on each group of coefficients on the decomposition time series and set the coefficients outside the confidence interval to zero. Then we reorganize the denoised parts to obtain the following formula:

$$f(t_i)=\sum_{k\in Z}\sum_{J\in Z}\left[S_J(k)\varphi_{j,k}(t_i)\right]+w_j(k)\omega_{j,k}(t_i), \tag{A. 2}$$

where $f(t_i)$ is the time series after a series of decomposition and reorganization. The double summation symbol decomposes the scale and wavelet subspace to the second level subspace at the same time, so as to decompose the signal in time domain and frequency domain.

Probabilistic principal components analysis

Probabilistic PCA model [20] is a special form of linear factor model. The data generation process of the linear factor model is as follows:

$$h\sim p(h), \tag{A.3}$$

where $p(h)$ is the distribution of h satisfying $p(h)=\prod_i p(h_i)$, which is easy to take sample from the distribution. Next, under the given h distribution, the observable variables of true value are sampled as:

$$x=Wh+b+noise \tag{A.4}$$

In factor analysis, the priori distribution of h is a Gaussian distribution.

$$h \sim N(h;0,I) \tag{A.5}$$

It is assumed that the noise is extracted from the Gaussian distribution of the diagonal covariance matrix. The covariance matrix is $\psi=\text{diag}(\sigma^2)$, where $\sigma^2=\left[\sigma_1^2,\sigma_2^2,\sigma_3^2...,\sigma_n^2\right]^T$ denotes a vector, and each variable represents the variance of a variable.

Consequently, x obeys multidimensional normal distribution and satisfies the following conditions:

$$x \sim N\left(x;b,WW^T+\psi\right) \tag{A.6}$$

For the PPCA method, we need to modify the factor analysis model to make the conditional variance σ_i^2 being equal to the same value. Then the conditional distribution of x becomes:

$$x \sim N\left(x; b, WW^T + \sigma^2 I\right) \tag{A.7}$$

Or, it is equivalent to:

$$x = Wh + b + \sigma z, \tag{A.8}$$

where $z \sim N(z;\ 0,\ I)$ is Gaussian noise. The probabilistic principal component analysis model takes advantage of the observation that most of the changes in the data can be described by the latent variable h, except for some small residual reconstruction errors. When $\sigma \rightarrow 0$, the probability PCA degenerates to PCA.

References

[1] He Y, An Y and W H 2007 *Chinese Journal of Turbomachinery* **6** 68-72
[2] Ahmad R and Kamaruddin S 2012 *Comput. Ind. Eng.* **63** 135-49
[3] Tahan M, Tsoutsanis E and Muhammad M 2017 *Appl. Energy* **198** 122-44
[4] Bernalte Sanchez P J and Garcia Marquez F P 2020 *Fourteenth Int. Conf. on Management Science and Engineering Management (Chisinau)* vol 2 ed Xu J *et al* (Cham: Springer Internatinal Publishing) pp 791-800
[5] Yang S, Jiang X, Xu S and Wang X 2019 IEEE *Trans. Ind. Electron.* **67** 10887-96
[6] Wen C, Lv F, Bao Z and Liu M 2020 *J. Electron. Inf. Technol.* **42** 234-48
[7] Si J, Ma J, Niu J and Wang E 2017 *J. Vib. Shock* **36** 164-71
[8] Xu Z, Liu K, Zhang H, Wang D, Zhang M and Wu X 2014 *J. Vib. Shock* **23** 133-9
[9] Xu S, Jiang X, Huang J, Yang S, Wang X 2016 *Mech. Syst. Signal Process.* **80** 1-18
[10] Bottieau J, Vallee F, De Greve Z and Toubeau J F 2018 *IEEE Int. Energy Conf. (Limassol)* (Piscataway: IEEE) p 6
[11] Yin A, Yan Y, Zhang Z, Li C and Sanchez R 2020 *Sensors* **20** 1-13
[12] Zhao H, Jiang X, Xu S, Lin L and Gong Y 2020 *Chinese Journal of Turbomachinery* **62** 69-76
[13] Zhang C 2020 *Chinese Journal of Turbomachinery* **62** 77-82
[14] Hochreiter S and Schmidhuber J 1997 *Neural Comput.* **9** 1735-80
[15] Liu X, Zhou Q, Zhao J, Shen H and Xiong X 2019 *Sensors* **19** 1-20
[16] Li C, Wu X, Zhu P, Song L, Li J and Feng Z 2018 *Chinese Journal of Turbomachinery* **60** 1-11
[17] Contreras J, Espinola R, Nogales F J and Conejo A J 2003 *IEEE Trans. Power Syst.* **18** 1014-20
[18] Dimri T, Ahmad S and Sharif M 2020 *J. Earth Syst. Ence* **129** 1-16
[19] Dong L, Wei G, Hui H, Li H and Wang J 2011 *Chinese Journal of Turbomachinery* **5** 37-41
[20] Zhou Q, Liu C, Zhao J, Shen H, Xiong X 2018 *J. Vib. Shock* **37** 31-7

Acknowledgments

This work was supported by the National Key Research and Development Project (2020YFA0714403), the MIIT Marine Welfare Project (Z135060009002), the research and Innovation in Science and Technology Major Project of Liaoning Province (2019JH1-10100024), the National Natural Science Foundation of China (52005074), and the Fundamental Research Funds for the Central Universities (DUT19RC(3)070) at Dalian University of Technology.

Impact of exploration-exploitation trade-off on UCB-based Bayesian Optimization

Chao Bian, Xiaofang Wang, Changjun Liu, Xinyu Xie, Haitao Liu*

School of Energy and Power Engineering, Dalian University of Technology, Dalian 116024, China

*Corresponding Author, Email: htliu@dlut.edu.cn

Abstract. As an efficient global optimization method to deal with expensive black-box functions, Bayesian optimization (BO) has great potential in the field of turbomachinery component design. It can relieve the complexity, long cycle, and excessive dependence on experience of the design process. A core of Bayesian optimization is acquisition functions which aim to guide the search to efficiently find the optimum. An excellent acquisition function can greatly improve the performance of the algorithm. The well-known acquisition function, upper confidence bound (UCB), in Bayesian optimization achieves the balance between local exploitation and global exploration through an explicit trade-off coefficient. The trade-off coefficient is a key to the quality of UCB-based Bayesian optimization. Currently, this coefficient is either manually specified according to some criterion or sampled from a specific distribution. Due to the lack of a comparative study on existing trade-off strategies, this paper attempts to comprehensively analyse and highlight the impact of various trade-off strategies on the performance of UCB-based BO algorithm through eleven numerical examples and two turbomachinery design cases. The comparative results indicate that the proper trade-off is key for the optimization quality, and the strategy of looping through a trade-off set shows superiority in our numerical experiments.

1. Introduction

In the field of turbomachinery design, such as the design of compressor, steam turbine, and turbine, there is often no explicit expression between design variables and output results. That is, the simulation model is a black-box function with only input and output. Moreover, the cost of single simulation calculation may be very expensive. In view of these problems, traditional optimization methods, such as gradient based optimization method, cannot meet the needs of efficient design optimization.

To this end, surrogate based optimization has been proposed and studied for simulation-based optimization. For example, based on the combination of artificial neural network and genetic algorithm, the optimization of compressor return channel blades and low flowrate coefficient centrifugal compressor impellers have significantly improved the aerodynamic performance of centrifugal compressors [1]-[2]. The impeller disk is optimized based on the kriging model [3]. After optimization, the mass of the impeller is reduced by 15%, and the axial deformation of the friction end surface is reduced by 29%. Based on the CORS-BRF constrained optimization algorithm [4], the impeller geometry is optimized and the stress on the impeller is reduced. Using response surface methodology to optimize the centrifugal fan volute, the fan efficiency increased by 4.6% [5].

Particularly, this paper focuses on Bayesian optimization for simulation-based optimization. Bayesian optimization is an efficient surrogated-based global optimization algorithm, which is composed of two core parts: probabilistic surrogate model and acquisition function [6]. The probabilistic surrogate model can be used to approximate the complex problems, and the acquisition function plays the role of selecting the next most potential evaluation point. When the objective function of optimization is expensive black-box function or lack of gradient information, Bayesian optimization has obvious advantages in comparison to other optimization algorithms. It can use a small amount of data from previous samples to generate a statistical model of the system. This model is then used to suggest the next input through an acquisition function. By sequentially selecting new sample points to update the model, the optimum of the complex problem is finally obtained. Bayesian optimization has been successfully applied in machine learning [7] and design optimization [8]. It also has been applied and studied in the field of turbomachinery. For example, the parallel constrained Bayesian optimization [9] has been proposed to optimize the geometric parameters of the centrifugal slurry pump impeller and effectively reduce the wear rate of the impeller. Besides, to obtain maximum efficiency of the pin fin array in the developing region of the flow, Bayesian optimization is used to change the shape of the cooling channel in the trailing edge of the blade [10].

As a core component of Bayesian optimization, the acquisition function usually determines the optimization performance of the algorithm. The acquisition function suggests the next input point to update the surrogate model. A reasonable acquisition function needs to fully consider the trade-off of local exploitation and global exploration, in order to ensure that the algorithm has a fast convergence rate and the ability to jump out of the local optimum. Along this line, there are many species of acquisition functions, which consider the balance between exploitation and exploration from different perspectives. This paper only focuses on Bayesian optimization based on upper confidence bound (UCB) method. The UCB usually consists of posterior mean and posterior standard deviation provided by the probabilistic surrogate model. The exploitation and exploration can be balanced by multiplying the posterior standard deviation by a trade-off coefficient.

The trade-off coefficient, which often has a great impact on the performance of the Bayesian optimization algorithm, could be specified by the user. In practical application, people usually set this parameter as a fixed constant for convenience [11]. Subsequently, the Gaussian process upper confidence bound (GP-UCB) was proposed, which provides strong theoretical guarantees on the overall convergence rate while balancing local and global [12]. As an improvement of GP-UCB, the randomized Gaussian Process Upper Confidence Bound (RGP-UCB) is proposed to retain the theoretical guarantee of convergence speed and improve the local and global trade-off effect through sampling from gamma distribution [13].

Due to the lack of a comparative study on existing trade-off strategies, this paper attempts to comprehensively analyse and highlight the impact of various trade-off strategies on the performance of UCB-based BO algorithm through eleven numerical examples and two turbomachinery design cases.

The rest of the paper is organized as follow. Section 2 first gives a brief introduction of the UCB-based Bayesian optimization as well as Gaussian process. Thereafter, section 3 reviews existing trade-off strategies for Bayesian optimization, followed by the numerical experiments on benchmarks and turbomachinery cases in sections 4 and 5, respectively. Finally, section 6 offers the conclusions.

2. Bayesian optimization revisited

In this section, we provide a brief review of Bayesian optimization and Gaussian process, with an emphasis on the selection of reasonable trade-off coefficients among various UCB-based Bayesian optimization methods.

2.1. Bayesian Optimization

Bayesian optimization is an efficient global optimization algorithm for solving expensive black-box functions. The goal of the algorithm is to find the global optimal solution of (1) in a bounded search space.

$$\boldsymbol{x}^* = \arg\max_{\boldsymbol{x}\in\chi} f(\boldsymbol{x}) \tag{1}$$

where χ is often a compact subset of $\mathbb{R}^d$ and represents the design space of interest, $\boldsymbol{x}$ is a d-dimensional decision vector, and f is the objective function, which is usually an unknown black box function without explicit expression, but the corresponding response value can be obtained at input $\boldsymbol{x}$ in the domain. This corresponding response value is a noise-corrupted output $y = f(\boldsymbol{x}) + \epsilon_{noise}$, where $\epsilon_{noise} \sim \mathcal{N}(0, \sigma^2)$. This type of problem is difficult to solve using the classic gradient ascent methods. Generally speaking, Bayesian optimization is a sequence model-based method to solve such problems. We specify a prior belief for the possible objective function, and then combine the data we already have to build a probabilistic surrogate model. Thereafter, we use the Bayesian posterior information of the model, which represents our updated beliefs about the possible objective function being optimized, to establish an appropriate acquisition function, $\alpha(\boldsymbol{x})$. The acquisition function actively selects the most "potential" evaluation points for evaluation, avoiding unnecessary sampling. The most commonly used probabilistic surrogate model in Bayesian optimization is Gaussian process (GP), which is the model used in our work. It is notable that other models such as random forest and deep neural network can be used, once they could offer probabilistic predictions. The most popular acquisition functions in BO are improved probability (PI) [14], expected improvement (EI) [15], and upper bound confidence bound (UCB) [11].

2.2. Gaussian Process

Gaussian process (GP) is a nonparametric model which is the most common statistical model used in Bayesian optimization. It consists of a mean function, $m: \chi \to \mathbb{R}$ and a semidefinite covariance function, $k: \chi \times \chi \to \mathbb{R}$ (an effective covariance function must be semi positive definite [16]).

$$f(\boldsymbol{x}) \sim GP\big(m(\boldsymbol{x}), k(\boldsymbol{x}, \boldsymbol{x}')\big) \tag{2}$$

where the mean function $m(\boldsymbol{x}) = \mathbb{E}[f(\boldsymbol{x})]$,for simplicity, is usually set to $m(\boldsymbol{x}) = 0$, the covariance function $k(\boldsymbol{x}, \boldsymbol{x}') = \mathbb{E}[f(\boldsymbol{x}) - m(\boldsymbol{x})(f(\boldsymbol{x}') - m(\boldsymbol{x}'))]$ quantifies the correlation between two data points. In practical application, the most popular covariance function is the squared exponential kernel [17].

$$k_{SE}(\boldsymbol{x}_i, \boldsymbol{x}_j) = \sigma_f^2 \exp\,(-\frac{1}{2}\sum_{i=1}^{t}\frac{(\boldsymbol{x}_i - \boldsymbol{x}_j)^2}{l_i^2}) \tag{3}$$

where σ_f^2 is an output scale, and l_i is an input scale along the ith dimension.

Gaussian process is a set of random variables and any finite random variables satisfy a joint Gaussian distribution. Suppose that the latent function values $\boldsymbol{f} = \{f(\boldsymbol{x}_1), f(\boldsymbol{x}_2), \cdots, f(\boldsymbol{x}_t)\}$ follow a zero-mean GP prior:

$$p(\boldsymbol{f}|\boldsymbol{X}, \boldsymbol{\theta}) = \mathcal{N}(\boldsymbol{0}, \boldsymbol{\Sigma}) \tag{4}$$

where $\boldsymbol{X} = \{\boldsymbol{x}_1, \boldsymbol{x}_2, \cdots, \boldsymbol{x}_t\}$ is the training set , $\boldsymbol{\Sigma}$ is the covariance matrix composed of $k(\boldsymbol{x}, \boldsymbol{x}')$ ($\boldsymbol{\Sigma}_{i,j} = k(\boldsymbol{x}_i, \boldsymbol{x}_j)$), and finally $\boldsymbol{\theta}$ comprises the hyperparameters.

When there is observation noise, i.e., $y = f(\boldsymbol{x}) + \epsilon_{noise}$, where $\epsilon_{noise} \sim \mathcal{N}(0, \sigma^2)$, the likelihood distribution is obtained:

$$p(\boldsymbol{y}|\boldsymbol{f}) = \mathcal{N}(\boldsymbol{f}, \sigma^2\boldsymbol{I}) \tag{5}$$

The corresponding marginal likelihood distribution of observations $\boldsymbol{y} = \{y_1, y_2, \cdots, y_t\}$ is:

$$p(\boldsymbol{y}|\boldsymbol{X}, \boldsymbol{\theta}) = \int p(\boldsymbol{y}|\boldsymbol{f})p(\boldsymbol{f}|\boldsymbol{X}, \boldsymbol{\theta})\mathrm{d}\boldsymbol{f} = \mathcal{N}(\boldsymbol{0}, \boldsymbol{\Sigma} + \sigma^2\boldsymbol{I}) \tag{6}$$

Usually, we optimize the hyperparameters $\boldsymbol{\theta}$ by maximizing the marginal likelihood distribution. According to the properties of Gaussian process, the following joint distribution can be obtained:

$$\begin{bmatrix} \boldsymbol{y} \\ f_* \end{bmatrix} \sim \mathcal{N}\left(\boldsymbol{0}, \begin{bmatrix} \boldsymbol{\Sigma} + \sigma^2\boldsymbol{I} & K_*^T \\ K_* & k_{**} \end{bmatrix}\right) \tag{7}$$

where f_* is the prediction function values, $\boldsymbol{x}_*$ is the prediction input, $\boldsymbol{k}_* = \{k(\boldsymbol{x}_1, \boldsymbol{x}_*), k(\boldsymbol{x}_2, \boldsymbol{x}_*), \cdots, k(\boldsymbol{x}_t, \boldsymbol{x}_*)\}$, $k_{**} = k(\boldsymbol{x}_*, \boldsymbol{x}_*)$. From this joint distribution, we can get the following prediction distribution:

$$p(f_*|\boldsymbol{X}, \boldsymbol{y}, \boldsymbol{\theta}) = \mathcal{N}\left(\bar{f}_*, \text{cov}(f_*)\right) \quad (8)$$

$$\bar{f}_* = \boldsymbol{k}_*(\boldsymbol{\Sigma} + \sigma^2\boldsymbol{I})^{-1}\boldsymbol{y} \quad (9)$$

$$\text{cov}(f_*) = k_{**} - \boldsymbol{k}_*(\boldsymbol{\Sigma} + \sigma^2\boldsymbol{I})^{-1}\boldsymbol{k}_*^T \quad (10)$$

where $\bar{f}_*$ is the prediction mean, $\text{cov}(f_*)$ is the prediction covariance.

2.3. Acquisition Function

In the framework of BO, we first use Gaussian process to construct the probability surrogate model; next, we combine the existing data set $\boldsymbol{D}$ and the surrogate model to create an acquisition function, $\alpha(\boldsymbol{x})$; finally, we select the next evaluation point $\boldsymbol{x}_{t+1}$ by maximizing the acquisition function.

$$\boldsymbol{x}_{t+1} = \arg\max_{x \in \chi} \alpha(\boldsymbol{x}_t; \boldsymbol{D}) \quad (11)$$

In the design of acquisition function, we need to balance local exploitation and global exploration reasonably. Local exploitation means that the sampling process pays attention to the improved locations, which have high probability of containing the global optimum. On the contrary, global exploration means that the sampling process tends to make the samples far away from each other in order to explore the unknown region which may contain the optimum. Excessive exploitation will cause the acquisition function to fall into the local optimum, while excessive exploration will slow the convergence speed. Hence, it is important to make a reasonable trade-off between exploitation and exploration in order to improve the optimization quality. Our work only focuses on Bayesian Optimization based on UCB. In the UCB method, a trade-off coefficient is used to balance the local and global. We are concerned about the influence of the trade-off coefficient set by different methods on Bayesian Optimization. Next, we will introduce different trade-off strategies for UCB.

3. Trade-off strategies for UCB-based BO

3.1. Bayesian Optimization

In the Sequential Design for Optimization (SDO) algorithm proposed by Cox and John [11], they select the prediction site which has the smallest value of lower confidence bound (LCB) as the next evaluation point:

$$\alpha_{LCB}(\boldsymbol{x}; \boldsymbol{D}) = -\left(\mu_t(\boldsymbol{x}) - k\sigma_t(\boldsymbol{x})\right) \quad (12)$$

where $k \geq 0$. When we are concerned with the maximum value of the objective function, we usually use the upper confidence bound (UCB) strategy:

$$\alpha_{UCB}(\boldsymbol{x}; \boldsymbol{D}) = \mu_t(\boldsymbol{x}) + k\sigma_t(\boldsymbol{x}) \quad (13)$$

which is the acquisition function commonly used in Bayesian optimization algorithm. For any input $\boldsymbol{x}$, it has a mean $\mu_t(\boldsymbol{x})$ and a standard deviation $\sigma_t(\boldsymbol{x})$ (indicating uncertainty). We believe that $k\sigma_t(\boldsymbol{x})$ is a relatively reliable bound. Hence, $f(\boldsymbol{x})$ has a relatively high probability to reach $\mu_t(\boldsymbol{x}) + k\sigma_t(\boldsymbol{x})$. The process of maximizing the acquisition function is to maximize our belief. In the process of maximizing the acquisition function, $\mu_t(\boldsymbol{x})$ makes the collection function tend to exploit the improved area, which belongs to exploitation item. $\sigma_t(\boldsymbol{x})$ makes the collection function choose the new point that is that is usually indicated with large standard deviation, which belongs to the exploration item. As a local and global trade-off coefficient, k is usually set as a constant. In [11], they set $k = 1.96$ to indicate the meaning of the "confidence factor". In our work, we set $k = 1.96$. As a comparison, we also set $k = 10$, which is a trade-off coefficient value favorable for exploration.

3.2. Gaussian process upper confidence bound

Srinivas et al. [12] proposed the Gaussian process upper confidence bound (GP-UCB) method based on cumulative regret boundary. The convergence of the algorithm is proved by strict theory.

$$\alpha_{GP-UCB}(\boldsymbol{x};\boldsymbol{D}) = \mu_t(\boldsymbol{x}) + \sqrt{\beta_t}\sigma_t(\boldsymbol{x}) \tag{14}$$

GP-UCB regards the Bayesian optimization as a multi-armed bandit and obtains instantaneous regret function:

$$r(\boldsymbol{x}) = f(\boldsymbol{x}_*) - f(\boldsymbol{x}) \tag{15}$$

The goal of optimization in the framework is equivalent to:

$$\min \sum_t^T r(\boldsymbol{x}_t) = \max \sum_t^T f(\boldsymbol{x}_t) \tag{16}$$

where T is the number of optimization iterations. It is proved that if the kernel satisfies the following formula:

$$P\{\sup_{x\epsilon\chi} |\partial f/\partial \boldsymbol{x}_i| > L\} \leq ae^{-(L/b)^2} \tag{17}$$

where $i = 1, \cdots, t$, the constants $a, b > 0$, $\chi \subset [0, m]^d$, $d \in \mathbb{N}$, and $m > 0$. Given $\delta \in (0,1)$ and

$$\beta_t = 2\log\left(\frac{t^2 2\pi^2}{3\delta}\right) + 2d\log\left(t^2\, dbm\sqrt{\log(4da/\delta)}\right) \tag{18}$$

Then with high probability of $1 - \delta$, the algorithm is no regret, i.e. $\lim_{T\to\infty} R_T/T = 0$, where R_T is the cumulative regret:

$$R_T = \sum_{t=1}^T f(\boldsymbol{x}_*) - f(\boldsymbol{x}_t) \tag{19}$$

This implies a lower-bound on the convergence rate for the optimization problem, which guarantees the convergence of the algorithm.

3.3. Randomized Gaussian process upper confidence bound

Although the selection of β_t in GP-UCB provides a regret bound, it is considered that the value of β_t is often greater than the required value, which will lead to the algorithm taking too much consideration of global exploration [13]. Hence, a method called Randomized Gaussian Process Upper Confidence Bound (RGP-UCB) has been proposed. It can keep the regret bound and select a smaller trade-off coefficient. Consequently, the algorithm can better balance local exploitation and global exploration as:

$$\alpha_{RGP-UCB}(\boldsymbol{x};\boldsymbol{D}) = \mu_t(\boldsymbol{x}) + \sqrt{\breve{\beta}_t}\sigma_t(\boldsymbol{x}) \tag{20}$$

where $\breve{\beta}_t$ is a random number following the gamma distribution $\Gamma(\gamma_t, \theta)$, with $\theta > 0$ being a constant used to adjust the size of $\breve{\beta}_t$, but setting a proper θ requires certain prior knowledge. In order to ensure that the algorithm maintain a regret bound, γ_t has been proven to follow:

$$\gamma_t = \frac{\log\left(\frac{1}{\sqrt{2\pi}}(t^2+1)\right)}{\log(1+\theta/2)} \tag{21}$$

Through the use of $\Gamma(\gamma_t, \theta)$, the trade-off coefficient $\breve{\beta}_t$ selected in RGP-UCB is smaller than β_t in the original GP-UCB while still maintaining convergence guarantees.

IOP Conf. Series: Materials Science and Engineering 1081 (2021) 012023 doi:10.1088/1757-899X/1081/1/012023

4. Numerical experiments

In this section, numerical examples and engineering cases are used to investigate the performance of UCB-based Bayesian optimization algorithms using various trade-off strategies.

In each case, the experiment was repeated 10 times with $10d$ initial points [18] generated by the TPLHD (translational propagation Latin hypercube design) sample scheme [19]. The number of iterations required for each algorithm to reach a relative error of 0.01% (the absolute error is used when the optimal value is 0) is recorded. And the allowed maximal number of function evaluations is set to 400 in our experiments. The methods involved in the comparison are:

- Upper Confidence Bound with $k = 1.96$.
- Upper Confidence Bound with $k = 10$.
- Upper Confidence Bound with $k \in [0, 1, 100]$.
- Standard Gaussian process upper confidence bound (GP-UCB).
- Standard randomized Gaussian process upper confidence bound (RGP-UCB).

In the above competitors, the UCB using a small, fixed trade-off coefficient $k = 1.96$ favors local exploitation; while the UCB using a large $k = 10$ prefers global exploration; the UCB using a trade-off set $k \in [0, 1, 100]$ performs a search loop using various k values, thus resulting in a comprehensive consideration of local exploitation and global exploration; finally, the last two methods adopt an adaptive k during the optimization.

4.1. Test functions

In this section, the performance of the algorithms is verified numerically on 11 benchmarks with different characteristics. The problem dimensions, the abbreviation used, the bounds of design space and the known global optimum of each test function are summarized in Table 1. Additionally, Figure 1 depicts the convergence histories of these algorithms on eleven benchmarks, respectively. The mathematical formulas of these benchmark functions can be found on the web page[1].

Table 1. Summary of the test function set.

Test functions	Abbreviation	Dimension	Domain	Global minimum
Branin	BA	2	$[-5,10]\times[0,15]$	0.3978
Bohachevsky	BO	2	$[-2,2]^2$	0.0000
easom	EA	2	$[-10,10]^2$	-1.0000
Schwef	SCH	2	$[-400,500]^2$	0.0000
Shubert	SHU	2	$[-2,2]^2$	-186.7309
Six-hump camel	SC6	2	$[-3,2]^2$	-1.0316
Hartman3	H3	3	$[0,1]^3$	-3.8627
Rosenbrock	RO	3	$[-2,2]^3$	0.0000
Levy	LE	3	$[-8,8]^3$	0.0000
Shekel 5	SH5	4	$[0,5]^4$	-10.1532
Sphere	SP	5	$[-3,5]^5$	0.0000

4.2. Results and discussions

Table 2 showcases the number of functional evaluations [mean and standard deviation (SD)] required to achieve the specified relative error (0.01%) for the UCB-based BO using different trade-off strategies, with the best results marked in bold.

[1] http://www.sfu.ca/~ssurjano/optimization.html

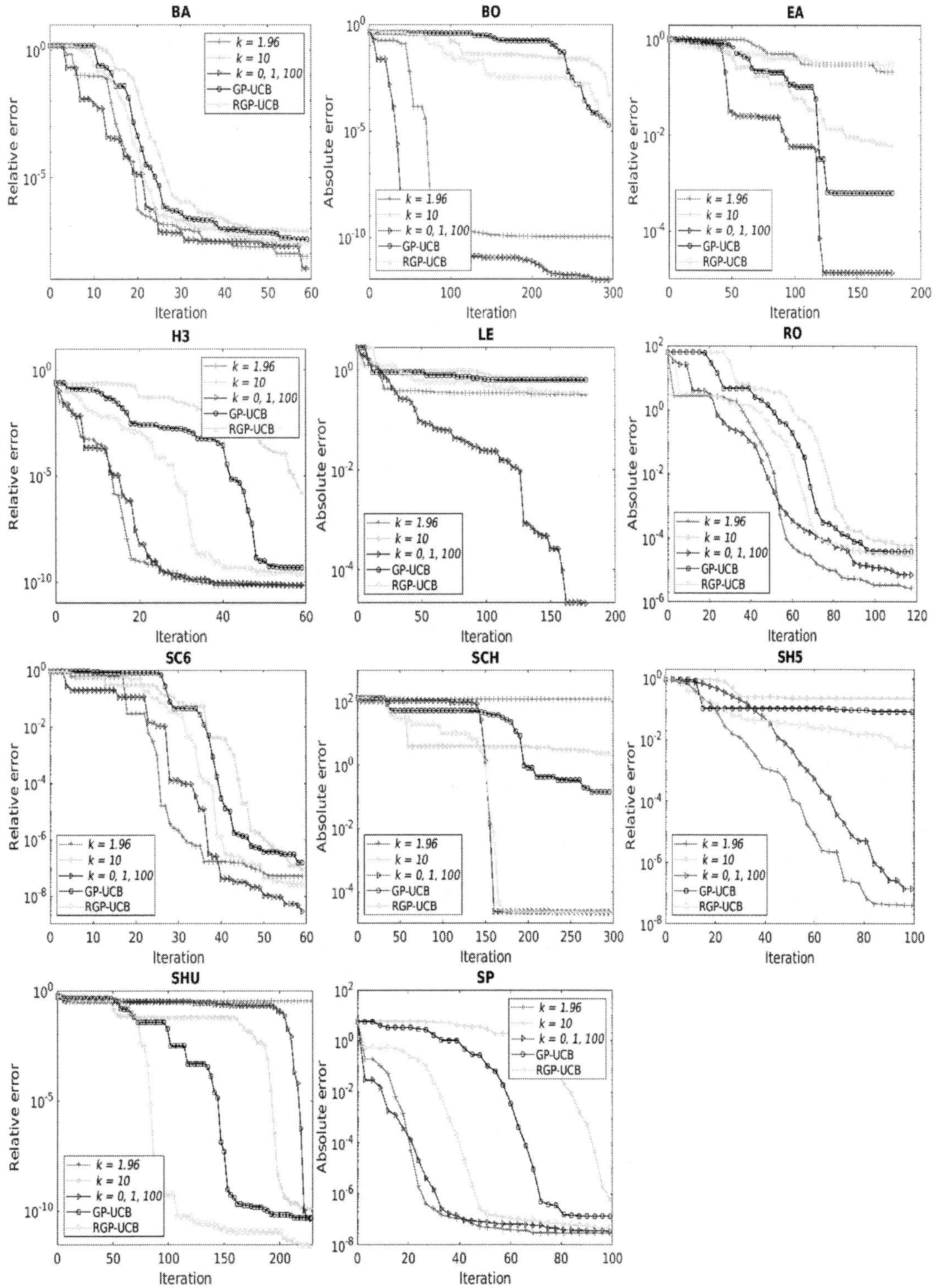

Figure 1. Performance comparison of UCB-based Bayesian optimization using different trade-off strategies on eleven benchmarks.

Table 2. Number of function evaluations (mean and SD) required to achieve a specified relative error (0.01%).

Test functions	UCB (k = 1.96)	UCB (k = 10)	UCB ($k \in [0,1,10]$)	GP-UCB	RGP-UCB
BA	38.3(1.3)	44.2(1.4)	**35.8(2.7)**	41.4(0.8)	39.6(1.5)
BO	71.3(9.8)	318(19.1)	**51.9(1.5)**	292.4(11.2)	190.3(70.4)
EA	125.1(57.3)[a]	227.0(46.7)[b]	**113.2(73.6)**	179.6(47.1)[c]	142.1(78.7)[d]
SCH	>400	>400	**172.8(13.4)**	258.6(25.4)[c]	178.9(4.5)
SHU	>400	210.5(1.2)	195.5(52.2)	154.1(12.4)	**94.2(6.2)**
SC6	**44.0(0.8)**	63.7(1.9)	49.0(2.3)	58.9(1.0)	55(2.1)
H3	**42.6(1.6)**	81.5(3.4)	43(0.0)	64.6(4.0)	54.5(3.2)
RO	**86.9(6.1)**	124.9(14.2)	98.9(9.8)	120.2(13.3)	109.3(16.5)
LE	367.0(41.0)[a]	>400	**144.4(34.3)**	>400	>400
SH5	**86.8(8.0)**	>400	103.7(5.4)	>400	183.1(21.3)[d]
SP	70.8(0.6)	141.9(1.1)	**69.9(2.5)**	115.4(1.6)	89.9(1.4)

[a] The algorithm only succeeded to achieve the required relative error within the limited computing resource in eight out of ten runs for EA, and in two out of ten runs for LE. Hence, the data is the average of successful runs.
[b] The algorithm only succeeded in two out of ten runs for EA.
[c] The algorithm only succeeded in five out of ten runs for EA, and in three out of ten runs for SCH.
[d] The algorithm only succeeded in seven out of ten runs for EA, and in nine out of ten runs for SH5.

The comparative results in Table 2 imply that for the UCB-based BO using a trade-off set $k \in [0, 1, 100]$, the global optimum can be quickly located in the most test functions with multiple local optimal solution or flat interested area, see for example, the BA, BO, EA, SCH, LE, and SP cases. Taking a set of trade-off coefficient values which are biased towards local exploitation can speed up the convergence while having the ability to jump out of the local optimal solution.

When the trade-off coefficient value $k = 1.96$ is employed, the performance of BO show superiority on the SC6, H3, RO, and SH5 functions. This is because these test functions are relatively simple, and pure local exploitation is conducive to rapid convergence to global optimum. Contrarily, for the SHU comprising multiple global optimal solutions with large gradients, the small trade-off coefficient $k = 1.96$ makes the algorithm fall into the local optimum. And the set [0, 1, 100] tends to local exploitation, which wastes too much resources on local solutions. RGP-UCB can just jump out of the local optimal solution without being too local.

Differently, we can intuitively observe that when $k = 10$, the acquisition function is biased towards global exploration, and excessive resources are placed on the exploration of the design space. Even after locating the interested area, it cannot quickly converge to the required accuracy. Hence, the UCB-based BO using $k = 10$ yields the poorest results in general.

For the GP-UCB algorithm, although it provides a regret bound to ensure convergence, is the algorithm prefers a large trade-off coefficient, which is biased towards global exploration. On the contrary, RGP-UCB uses the gamma distribution to retain the regret bound while generate relatively small trade-off coefficient, which hence improves the performance of the algorithm. However, as shown in Figure 1, the performance of RGP-UCB is poor on the simple functions (for example BO, EA, SH5, and SP), because the value of the trade-off coefficient is still large.

5. Engineering applications

This section investigates the characteristics and performance of the UCB-based BOs using different trade-off strategies on two turbomachinery design cases.

5.1. Weight optimization of aircraft wing

The first engineering case is the task of reducing the wing weight of the Cessna C172 Skyhawk aircraft. According to [20], we use the following analytical expression to approximate the weight of the aircraft wing:

$$f(x) = 0.036 S_w^{0.758} W_{fw}^{0.0035} \left(\frac{A}{\cos^2(\Lambda)}\right)^{0.6} q^{0.006} \lambda^{0.04} \left(\frac{100 t_c}{\cos(\Lambda)}\right)^{-0.3} \left(N_z W_{dg}\right)^{0.49} + S_w W_p \quad (22)$$

Table 3 contains the value range and physical meaning of each variable corresponding to the analytical expression, which roughly represents the design variable and corresponding value range of the Cessna C172 Skyhawk aircraft. The algorithm configurations keep the same as the benchmarks in section 2.

Table 3. The input variables of a light aircraft wing and their input ranges.

Symbol and Domain	Parameter
$S_W \in [150, 200]$	Wing area (ft^2)
$W_{fw} \in [220, 300]$	Weight of fuel in the wing (lb)
$A \in [6, 10]$	Aspect ratio
$\Lambda \in [-10, 10]$	Quarter-chord sweep (degrees)
$q \in [16, 45]$	Dynamic pressure at cruise (lb/ft^2)
$\lambda \in [0.5, 1]$	Taper ratio
$t_c \in [0.08, 0.18]$	Aerofoil thickness to chord ratio
$N_z \in [2.5, 6]$	Ultimate load factor
$W_{dg} \in [1700, 2500]$	Flight design gross weight (lb)
$W_p \in [0.025, 0.08]$	Paint weight (lb/ft^2)

Figure 2 shows the performance of different UCB-based BOs for optimizing the weight of aircraft wings. It can be observed that for this complex engineering case, in the initial stage of optimization, fully exploring the design space will bring benefits. As a result, $k = 10$ and GP-UCB have a better convergence rate in the initial iteration. However, as the number of iteration steps increases, the algorithms that are biased towards local exploitation can finally achieve a better convergence accuracy. Therefore, after enough iterations, the acquisition function with a trade-off set $k \in [0, 1, 100]$, which fully considers the trade-off between local exploitation and global exploration, has a better performance over the others.

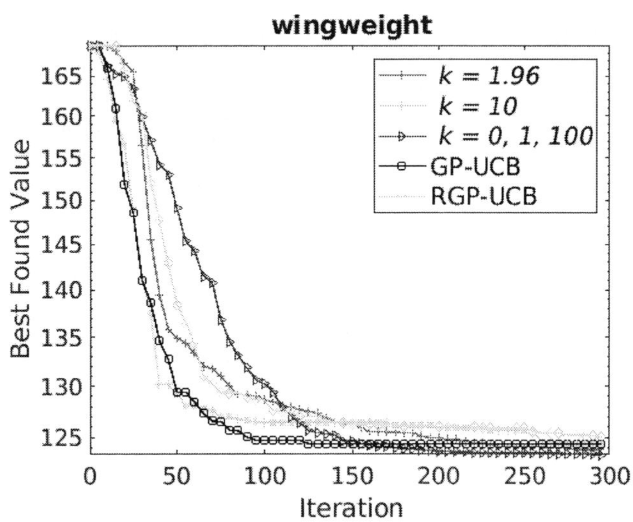

Figure 2. Performance comparison of UCB-based Bayesian optimization using different trade-off strategies on reducing the weight of light wing.

5.2. Drag optimization of turbofan engine

The second engineering case focuses on optimizing the ram drag of the turbofan engine, which can be solved by the EngineSim[2] simulator developed by NASA Glenn Research Center. This simulator can model the design and testing of jet engines and explore the influence of different design variables on engine performance. Figure 3 shows the geometry of the turbofan engine. Table 4 shows the selected design variables and the corresponding value ranges in our optimization case.

Figure 3. Geometry of turbofan engine[2].

Table 4. The input variables of a light aircraft wing and their input ranges.

Symbol and Domain	Parameter
$FanBPR \in [3.0, 3.6]$	Fan bypass pressure ratio
$FanPR \in [1.7, 1.8]$	Fan pressure ratio
$CompPR \in [18, 24]$	Compressor pressure ratio
$CombTMax \in [2200, 2800]$	Combustion temperature maximum (K)
$Mach \in [0.6, 0.9]$	Flight Mach number
$Altitude \in [5000, 10000]$	Flight altitude (m)

Figure 4 shows the performance of different algorithms to optimize the ram resistance of the turbofan engine. For this extreme problem that is biased towards global exploration, it can be observed that the larger trade-off coefficient yields better performance., see for example the results of $k = 10$ and GP-UCB. As for the BO using a set of trade-off coefficient values, since there is a case of $k = 100$, it converges quickly at the beginning of the iteration. However, with the optimization proceed, the cases of $k = 0{,}1$ makes the BO using a set of trade-off coefficient values more concerned about local exploitation, and the final result is poor. Therefore, it can be concluded that the proper trade-off set is problem-dependent, and how to achieve an adaptive and dynamic trade-off set is left for our future work.

[2] https://www.grc.nasa.gov/WWW/K-12/airplane/ngnsim.html .

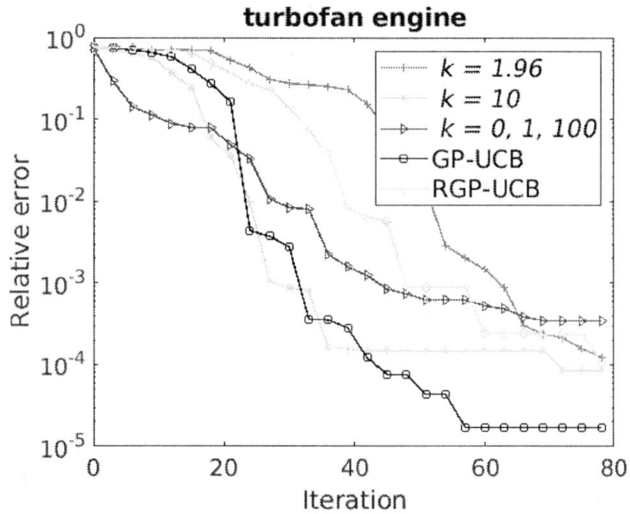

Figure 4. Performance comparison of UCB-based Bayesian optimization using different trade-off strategies on reducing the ram drag of turbofan engine.

6. Conclusion

Through numerical benchmarks and engineering applications, the impact of trade-off coefficients of local exploitation and global exploration on the performance of UCB-based Bayesian optimization is compared and studied. The results show that the performance of the algorithm varies greatly when different trade-off strategies are used. When a fixed value $k = 1.96$ is used, the algorithm favors local exploitation and has a faster convergence speed on simple benchmarks, which however is not the case for complex benchmarks. When $k = 10$, the algorithm is biased towards global exploration. Consequently, most of the resources might be wasted in exploring the design space, hence slowing the convergence speed. When GP-UCB is used, although the algorithm provides a regret bound and ensures convergence, it usually provides a relatively large value for the trade-off coefficient, which cannot well balance local and global well, thus resulting in poor performance. Besides, the RGP-UCB uses a gamma distribution to sample small trade-off coefficient with high probability while ensuring convergence; but it requires a certain prior knowledge to set a reasonable θ, which will adjust the trade-off coefficient. This is difficult for complex problems. The trade-off set $k \in [0, 1, 100]$ is generally biased towards local exploitation, which improves the convergence speed while ensuring the ability to jump out of the local optimum, and has a better performance in general. But this specific trade-off set has also been found to converge slowly for several cases. Therefore, it can be concluded that the proper trade-off set is problem-dependent, and the issue of how to achieve an adaptive and dynamic trade-off set is left for our future work.

7. References

[1] Chen H, Wen S, Xu Z and Xue P 2013 *Chinese Journal of Turbomachinery* 43-46
[2] Yin X, Wang J, Fan Y, Wen S and Zhang P 2018 *Chinese Journal of Turbomachinery* **60** 16-22
[3] Ji C, Xu H, Sun Q and Sun M 2017 *Chinese Journal of Turbomachinery* **59** 44-48
[4] Xu S, Zhong S, Guo Z, Yang S and Meng J 2018 *Chinese Journal of Turbomachinery* **60** 64-69
[5] Jiang W, Zhang P, Yang Z and Pei Q 2019 *Chinese Journal of Turbomachinery* **61** 23-28
[6] Cui J and Yang B 2018 *Journal of Software* **29** 3068-3090
[7] Rasmussen C E 2003 *Machine Learning Summer School Conf. (Tubingen)* vol 3176 (Berlin: Springer-verlag Berlin) pp 63-71

[8] Beyer H G and Sendhoff B 2001 *Comput. Meth. Appl. Mech. Eng.* 196 3190-3218
[9] Tran A, Sun J, Furlan J M, Pagalthivarthi K V, Visintainer R J and Wang Y 2019 *Comput. Meth. Appl. Mech. Eng.* 347 827-852
[10] Ghosh S, Mondal S, Jayanta S and Asok R 2019 *AIAA Propulsion and Energy Forum (Indianapolis)*
[11] Cox D D and John S 1992 *IEEE Int. Conf. on Systems (Chicago)* (New York: IEEE) pp 1241-1246
[12] Srinivas N, Krause A, Kakade S M and Seeger M 2010 *Proceedings of the 27th Int. Conf. on Machine Learning (June 21-24 2010 Haifa)*
[13] Berk J, Gupta S, Rana S and Venkatesh S 2020 *Int. Joint Conf. on Artificial Intelligence (Yokohama)* ed Christian B (California: IJCAI) pp 2284-2290
[14] Kushner H J 1964 *J. Fluids Eng.* **86** 97-106
[15] Mockus J, Tiesis V and Zilinskas A 1978 *Towards Global Optimization 2* Dixon L and Szego G (New York: North-Holland Press) pp 117-129
[16] Rasmussen C E and Williams C K I 2006 *Gaussian Processes for Machine Learning* (Cambridge: MIT Press) pp 7-31
[17] Schölkopf B, Tsuda K and Vert J 2004 *Kernel Methods in Computational Biology* (Cambridge: MIT Press)
[18] Welch W, Sacks J and Loeppky J 2009 *Technometrics* **51** 366-376
[19] Venter G, Balabanov V and Viana F A C 2010 *Int. J. Numer. Methods Eng.* **82** 135-156
[20] Forrester A I J, Sobester A and Keane A J 2008 *Engineering Design Via Surrogate Modelling: A Practical Guide* (New Jersey: John Wiley & Sons Inc) pp 10-13

Acknowledgments
This work was supported by the National Key Research and Development Project (2016YFB0600104), the National Natural Science Foundation of China (52005074), the research and Innovation in Science and Technology Major Project of Liaoning Province (2019JH1-10100024), and the Fundamental Research Funds for the Central Universities (DUT19RC(3)070) at Dalian University of Technology.

IOP Conf. Series: Materials Science and Engineering 1081 (2021) 012024 doi:10.1088/1757-899X/1081/1/012024

Influence of inlet flow field uniformity on the performance of the nuclear main pump

Long Sha, Shuhua Yang, Xiangdong Kong, Hui Sun and Lin Pei

Shenyang Blower Works Group Corporation

E-mail: www.salon.com@163.com

Abstract. A test loop system with adjustable spoiler function and visual structure was built in this paper. A PIV test system was adopted to investigate the influence of the uniformity of the inlet flow field on the performance of a nuclear main pump scale model, and the NUMECA software was used to numerically simulate the test cases. The research results show that when the rotation direction of the spoiler blade is opposite to that of the impeller, the greater the deviation of the spoiler blade from the axial angle, the more the pump efficiency and head drop; proper positive pre-rotation will not reduce the pump efficiency, but will increase the pump head; for the same spoiler blade deflection angle, when its structure produces irregular flow field disturbances under non-centrosymmetric conditions, the flow field downstream of it will not produce clear swirling flow, and it will not have more impact on the performance of the pump; the axial velocity distribution uniformity δ and the velocity weighted average angle θ, θ' have obvious direct effects on efficiency, and in the δ interval of each working condition in this paper (the difference between the maximum value and the minimum value of δ is 0.093), the speed weighted average angle has a greater impact on efficiency.

Nomenclature.

Q	Rate of flow
δ	Axial velocity distribution uniformity
θ	Weighted average angle parameter of velocity
θ_i	Absolute axial velocity angle
θ'	Weighted average angle parameter of velocity
η	Efficiency
H	Head

1. Introduction

With the rapid development and construction of nuclear power in China, the market demand of nuclear power pump, which is the transport heart of hundreds of large and small systems, is also increasing day by day. Different from pumps used in petroleum, chemical, thermal power and other industrial fields, nuclear power pumps should not only meet the normal operation of the whole main and auxiliary systems, but also strictly ensure the safety, reliability, seismic behavior and durability of the equipment, in particular, as the heart of nuclear safety and related systems, nuclear safety grade pumps must meet stringent safety and quality requirements.

Some scholars have done some research on the influence of inlet flow field uniformity on the flow field and performance in the pump. By directly solving the three dimensional Reynolds average N-S equation and the standard k-ε turbulence model equations, Lu Guanglin simulated the flow field in the inlet channel of large and medium-sized pumping stations, and then performed hydraulic optimization calculation for the inlet channel with complex geometry [1]. Tang Fangping introduced the research progress of the inlet flow field in China, discussed the research methods of the inlet flow field, and analyzed the advantages and disadvantages of various methods [2]. Liu Haijian designed three kinds of flow field uniformity devices for the inhomogeneity of water pump inlet flow field. Numerical simulation and experimental study show that the resistance coefficient of rectangular grid type is more than 3 times that of the diversion type and the concentric circle type uniform device, and the axial velocity angle of the section is the most important factor affecting the uniformity of the flow field and all the three flow field devices can significantly increase the axial velocity angle of the inlet section [3]. In view of the influence of elbow inlet channel on cavitation characteristics of large axial flow pump unit, Yan Hao derived the relationship between pump cavitation allowance and inlet velocity uniformity based on inlet velocity triangle theory, and simulated the external characteristic curve of large axial flow pump with CFD technology, and compared with the test results [4]. Yang Minguan designed 5 kinds of water guide cones, and used Fluent software to calculate the three-dimensional flow field of each type of water guide cones, analyzed the flow characteristics of the water guide cones, and summarized the three flow parts of the water guide con flow field and the velocity distribution law of the axial surface of the flow field. The variation of axial velocity distribution uniformity and weighted average deflector angle with the deflector of water conducting cone are summarized [5].

The interactive combination of numerical simulation and experimental research is an important means to study the flow law in pump and guide the design of pump hydraulic model [6-13]. However, the accuracy of the numerical simulation results must be verified by a large amount of objective and reliable data, and the appropriate testing method is the key to ensure the reliability of the test results [14-20]. PIV particle velocimetry system can record the relevant information of the entire smooth surface at the same time to obtain the instantaneous velocity field of flow, which has become an advanced measurement method for studying the flow field inside turbomachinery [21-28].

At present, there is a lack of visual experimental research on the uniformity of flow field at pump inlet. In this paper, a visualization test system of flow field was established, and PIV produced by TSI was used as a test method to investigate the influence of flow field uniformity at the inlet of a nuclear main pump model on pump performance.

2. Research methods

2.1. Test system

In this paper, the flow field visualization experiment and numerical simulation are combined to study the shrinkage model of a nuclear main pump. Build the test system with adjustable spoiler function and visual structure, hollow polystyrene glass beads with average particle size of 20-30 μm were used as tracer particles for flow field measurement and PIV system was used for flow field measurement. At the same time, NUMECA software is used for numerical calculation of corresponding operating cases. The computational domain includes inlet extension section, spoiler section, impeller, guide blade, volute and outlet extension section, and the calculated number of grid nodes is 42.096 million.

In order to achieve adjustable flow uniformity at the pump inlet, a spoiler section is set upstream of the inlet. Four spoiler blades are evenly arranged in the inlet spoiler section, and each one can swing freely independently. In order to carry out PIV visualization experiment, the pump inlet was designed as a visual structure. The structure consists of plexiglass tubes and a transparent square water tank on both sides to reduce the distortion of light due to different refractive indices [29-31]. At the same time, only the two sides of the water tank that need to project laser and shoot are made into transparent structure, while the other four sides are all opaque, which can effectively reduce the interference of external light. The test system is shown in figure 1.

Figure 1. Test system.

The test speed was selected as 896rpm, and six inlet flow cases were designed for the spoiler angle of 4 blades 0°, 4 blades +15°, 4 blades +30°, 2 blades +30°&2 blades -30°, 4 blades -15°, and no spoiler section. In order to investigate the influence of flow rate on the unevenness of velocity distribution, three flows (0.8 design flow rate, design flow rate and 1.2 design flow rate) were selected for each inlet. The arrangement of test cases is shown in table 1.

Table 1. Test cases.

(a)

Cases	1	2	3	4	5	6
inlet condition (blade angle × number of blades)	0°×4	0°×4	0°×4	15°×4	15°×4	15°×4
Q [m^3/h]	657	821	985	657	821	985

(b)

Cases	7	8	9	10	11	12
inlet condition (blade angle × number of blades)	30°×4	30°×4	30°×4	30°×2 -30°×2	30°×2 -30°×2	30°×2 -30°×2
Q [m^3/h]	657	821	985	657	821	985

(c)

Cases	13	14	15	16	17	18
inlet condition (blade angle × number of blades)	-15°×4	-15°×4	-15°×4	no spoiler	no spoiler	no spoiler
Q [m^3/h]	657	821	985	657	821	985

2.2. Uniformity parameters

In order to quantitatively measure the uniformity of velocity distribution of each fluid element in the flow field at the pump inlet, the axial velocity distribution uniformity parameter of the flow field is introduced and defined as:

$$\delta = 1 - \frac{1}{\overline{u_a}}\left(m^{-1} \times \sum_{i=1}^{m}\left(u_{ai} - \overline{u_a}\right)^2\right)^{1/2} \times 100\% \tag{1}$$

Here, $\overline{u_a}$ is the average axial velocity of the inlet section, u_{ai} is the axial velocity of each element in the inlet section, m is the number of measuring points on the inlet section. The axial velocity distribution uniformity of section flow field δ reflects the design quality of the inlet flow passage of the pump, the closer to 100%, the more uniform the axial velocity distribution of the inlet flow passage section is.

In addition, in order to quantitatively measure the uniformity of velocity angle distribution of each fluid element in the flow field at the pump inlet, a weighted average angle parameter of velocity in the flow field at the section θ is introduced, which is defined as:

$$\theta = \left(\sum_{i=1}^{m} u_{ai}\right)^{-1} \times \sum_{i=1}^{m} u_{ai}\left[90° - \arctan\left|\frac{u_{ti}}{u_{ai}}\right|\right] \tag{2}$$

Here, u_{ti} is the radial velocity of each element in the inlet section. The weighted average angle of cross section flow field velocity θ reflects the radial velocity of each element in the cross section of the flow field at the pump inlet. The closer the θ value is to 90°, the smaller the radial velocity is, and the higher the perpendicularity between the flow direction at the pump inlet and the cross section of the pump inlet is.

According to the definition of θ, it can be seen that this coefficient cannot reflect the direction of radial velocity, that is, it cannot describe the positive or negative character of the included angle between radial velocity and axial velocity. Therefore, we define the absolute axial velocity angle of each unit on the pump inlet cross section as follows:

$$\theta_i = \arctan\left|u_{ti}/u_{ai}\right| \tag{3}$$

And another velocity weighted average angle parameter θ' is proposed, which is defined as:

$$\theta' = \left(\sum_{i=1}^{m} u_{ai}\right)^{-1} \times \sum_{i=1}^{m} u_{ai}\left[90° - \arctan\left|\frac{u_{ti}}{u_{ai}}\right|\right] \tag{4}$$

3. Results and analysis

3.1. Analysis of the influence of turbulence conditions on pump performance

Comparing the flow field velocity distribution of each case obtained by the test and CFD, it can be seen that the two are roughly similar. For the performance test, if the inlet pressure measuring device is installed downstream of the turbulent section, it is impossible to obtain a stable and accurate pressure value because the flow field here will be in a turbulent state. If the inlet pressure measuring device is installed upstream of the turbulence section, it is also impossible to obtain accurate pump performance because the measured efficiency will include the turbulence section piping. Therefore, this test only performed performance tests on 9 cases under the condition of no turbulence, 4 blades 0° and 4 blades -15° turbulent flow conditions with a relatively uniform inlet flow field velocity distribution. The test results are shown in table 2. Here we make case 17 the base case (η and H of which are 1), η and H values of other cases are ratios to this. The numerical calculation objects of pump performance include all cases, and the calculation results are shown in table 3.

Table 2. Results of pump performance tests.

Cases	1	2	3	13	14	15	16	17	18
inlet condition (blade angle × number of blades)	0°×4			-15°×4			no spoiler		
η [%]	1.04	1.01	0.76	1.07	1.00	0.74	1.06	1.00	0.71
H [m]	1.40	1.06	0.60	1.47	1.09	0.64	1.38	1.00	0.57

Table 3. Results of pump performance numerical calculation.

(a)

Cases	1	2	3	4	5	6	7	8	9
inlet condition (blade angle × number of blades)	0°×4			15°×4			30°×4		
η [%]	0.99	0.99	0.80	0.98	0.96	0.73	0.95	0.93	0.61
H [m]	1.25	1.00	0.61	1.18	0.90	0.51	1.10	0.80	0.36

(b)

Cases	10	11	12	13	14	15	16	17	18
inlet condition (blade angle × number of blades)	30°×2, -30°×2			-15°×4			no spoiler		
η [%]	0.95	0.93	0.71	0.99	0.98	0.80	0.99	1.00	0.81
H [m]	1.21	0.97	0.54	1.30	1.07	0.69	1.26	1.00	0.62

Figure 2 and figure 3 show the comparison of pump efficiency and head calculation results under various turbulent flow conditions. It can be seen from the figure that the pump efficiency under the condition of no turbulence section, 4 blades 0°, and 4 blades -15° turbulence are similar at each flow point. The efficiency of 4-blade 0° and 4-blade -15° cases are not decrease, on the contrary, the head of 4-blade -15° case is slightly higher than other cases. Because under the condition of 4 blades -15°, the fluid passing through the turbulence section will produce the same rotating flow as that of the impeller, which is equivalent to applying a positive pre-rotation at the pump inlet, so it will not cause disturbance to the flow field in the pump. At the same time, for uniform flow, the spoiler blade added in this experiment will not have a significant impact on the flow field when it is at the 0° position.

For the case of 4 blades at 15°, the efficiency and head are significantly reduced, and for the case of 4 blades at 30°, the efficiency and head are further reduced significantly. This shows that the disturbance generated by the spoiler blades that is opposite to the impeller rotation will have a non-negligible impact on the pump performance, and the greater the disturbance, the more performance degradation. For the cases of 2 blades 30° & 2 blades -30°, the efficiency in medium and small flow is similar to that of 4 blades 30°. In the case where the efficiency of large flow is higher than that of 4 blades 30°, the head at each flow point is higher than 4 blades 15°. This shows that the flow field disturbance generated by the spoiler blades under non-centrosymmetric conditions is not as regular as symmetrical, and will not produce a swirling flow with a clear direction, but it will not bring greater adverse effects on the pump performance.

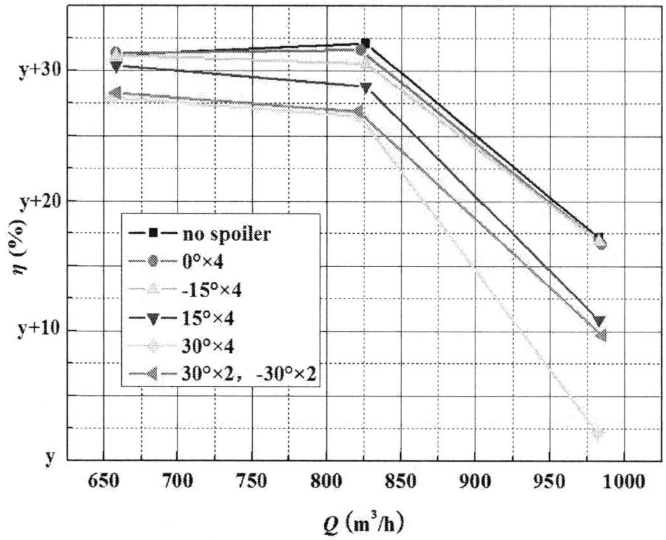

Figure 2. Comparison of pump efficiency under various turbulent flow conditions calculated.

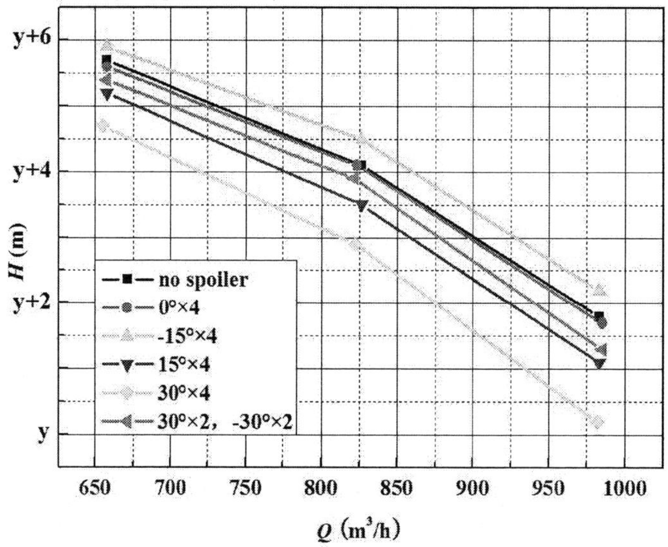

Figure 3. Comparison of pump heads under various turbulent flow conditions calculated.

Figure 4 and figure 5 show the test and calculation results of the efficiency and head under the case of no turbulence section, 4 blades 0°, and 4 blades -15°. It can be seen that the test efficiency is slightly lower than the calculated value under small flow conditions. With the increase of the flow rate, the test efficiency decreases rapidly and is significantly lower than the calculated value. This shows that the disturbance in the flow field except for the spoiler blades (there is a three-way structure upstream of the pump inlet) has a more obvious impact on the performance of the pump, and the impact is strengthened with the increase of the flow rate. Overall, the test head is lower than the calculated head. The head of the 4-blade -15° case is the highest, which is consistent with the calculation trend.

For the test results, the efficiency values under the three turbulent conditions are similar, and the 4-blade 0° and 4-blade -15° cases are even higher than the case without turbulent blades at large flow.

Similarly, the head without turbulence blades is the smallest under the three turbulence conditions. This is because the 0° and -15° spoiler blades not only have no adverse factors on the flow field, but also weaken and rectify the inherent external turbulence of the system.

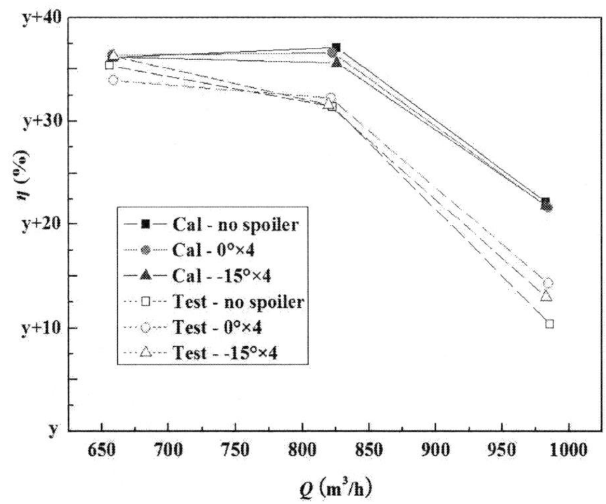

Figure 4. Comparison of pump efficiency obtained by experiment and calculation.

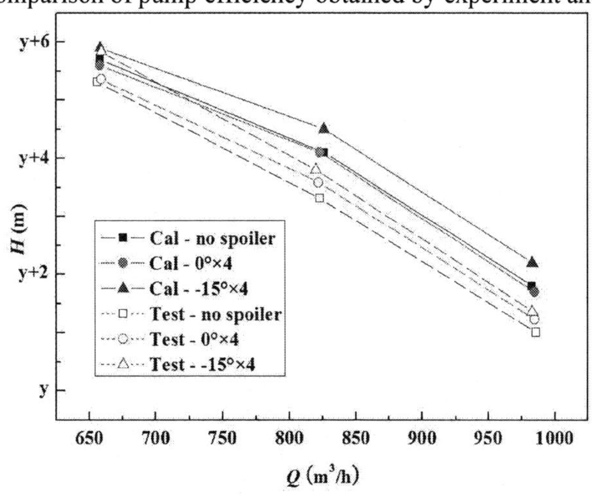

Figure 5. Comparison of pump heads obtained by experiment and calculation.

3.2. Analysis of uniformity of velocity distribution

In order to study the flow field under various turbulent flow conditions more intuitively and in-depth, the cross section of the pipeline flow field was taken as the research object in the experimental observation section. The diameter of the cross-section is 1.18 times that of the inlet pipe and is 414 mm from the pump inlet. Figure 6 to figure 8 show the axial velocity distribution on the inlet section of each flow rate and each turbulent flow condition. The Y coordinate value from small to large is the bottom-up direction of the pipe in the figure.

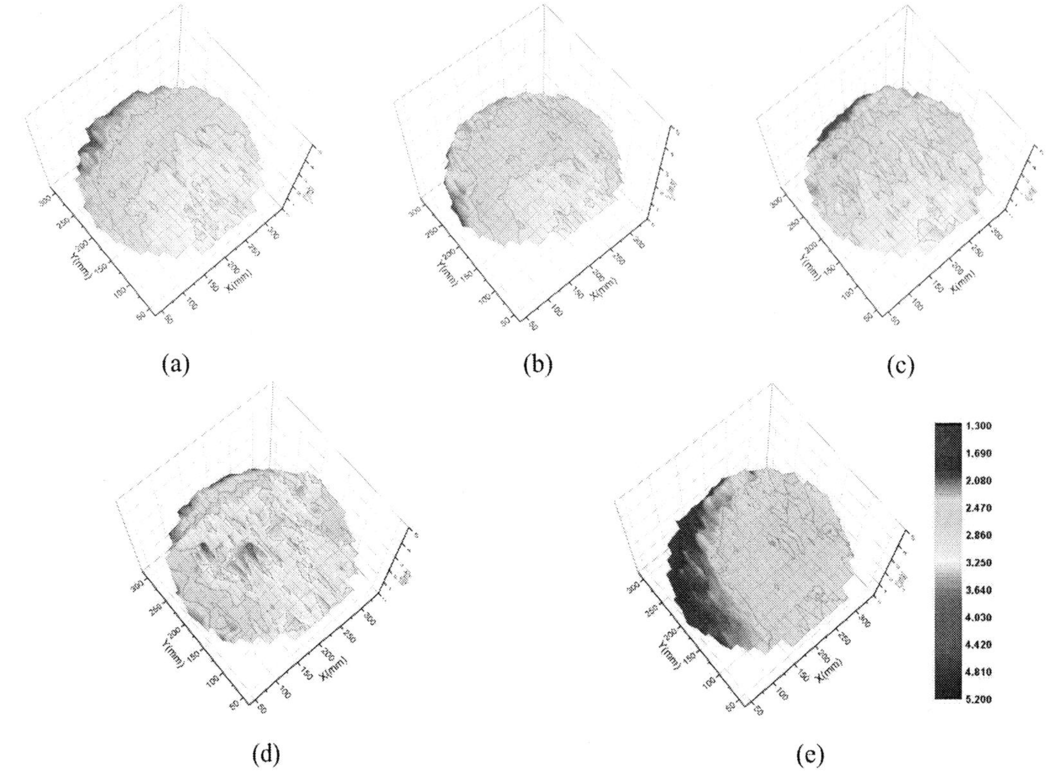

Figure 6. Axial velocity distributions on the inlet sections of each turbulent flow case under 0.8 times design flow: (a) 4 blades 0°, (b) 4 blades -15°, (c) 4 blades +15°, (d) 4 blades +30°, (e) 2 blades +30°&2 blades -30°.

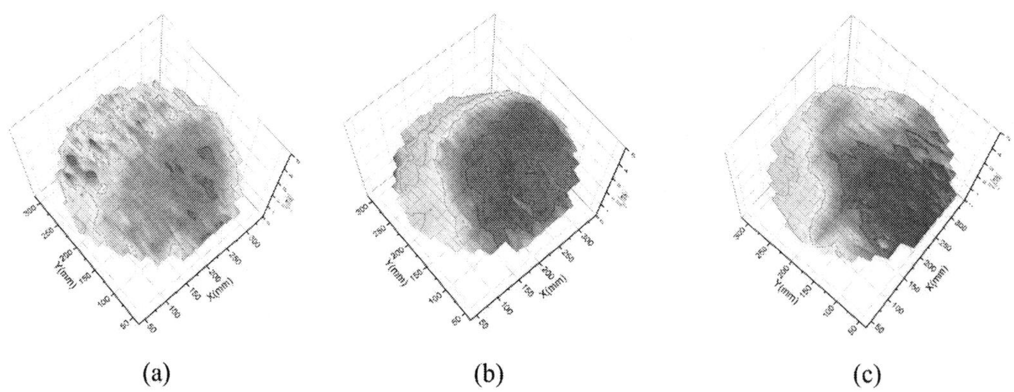

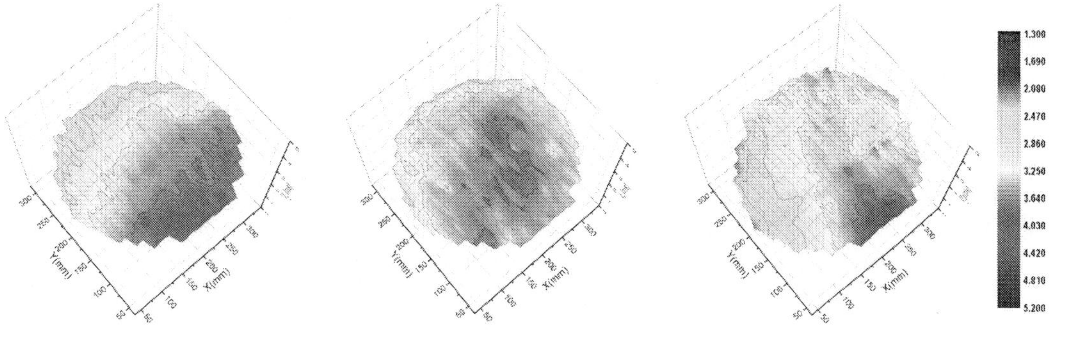

(d) (e) (f)

Figure 7. Axial velocity distributions on the inlet sections of each turbulent flow case under design flow: (a) no spoiler section, (b) 4 blades 0°, (c) 4 blades -15°, (d) 4 blades +15°, (e) 4 blades +30°, (f) 2 blades +30°&2 blades -30°.

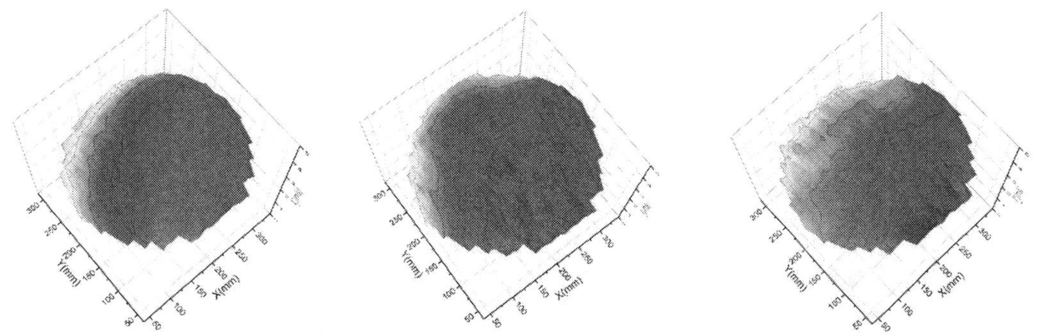

(a) (b) (c)

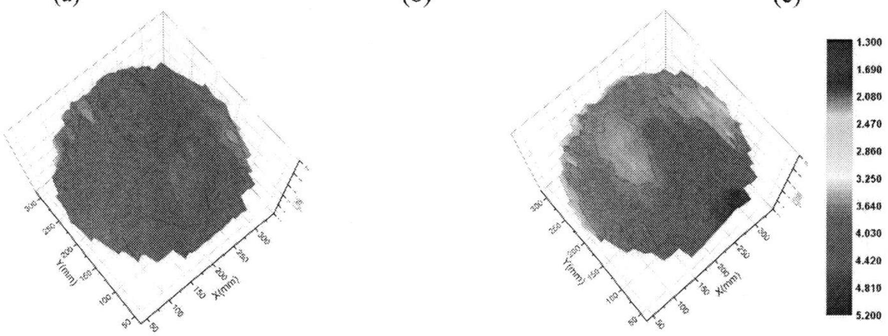

(d) (e)

Figure 8. Axial velocity distributions on the inlet sections of each turbulent flow case under 1.2 times the design flow: (a) 4 blades 0°, (b) 4 blades -15°, (c) 4 blades +15°, (d) 4 blades +30°, (e) 2 blades +30°&2 blades -30°.

By comparison, it can be seen that, in general, for all cases except 2 blades +30° & 2 blades -30°, the axial velocity of the upper part of the cross-sectional flow field is small, and the axial velocity of the lower part is large. This shows that the disturbance generated by the upstream elbow presents the radial velocity distribution characteristics here. Among them, the distribution characteristics of the three cases of 4 blades 0°, 4 blades -15°, and 4 blades 15° are relatively similar. However, due to the different angles of the spoiler blades, the effect on the elbow disturbance is not exactly the same, so the characteristics of the high-speed zone are slightly different. The radial gradient of the cross-sectional velocity of the undisturbed section and the 4-blade 30° case is smaller than other cases, that is, the axial velocity distribution of these two cases is more uniform. For the cases of 2 blades +30° & 2 blades -30°, the axial velocity distribution not only has a gradient in the longitudinal direction of the pipeline, but also has an obvious velocity gradient in the transverse direction. As the flow rate increases, the axial velocity of the cross section will increase correspondingly, but the distribution trend is consistent, and the radial velocity gradient has no obvious regular changes.

Figure 9 and figure 10 show the comparison of the axial velocity value and the absolute axial angular distribution at the inlet section of the 4-blade 0° and the 4-blade 30° at the design flow. Through comparison, it can be found that for these two cases, both have the same characteristics that the position with a high axial velocity value has a small absolute axial angle, and the position with a low velocity value has a large absolute axial angle. But there are obvious differences under the same characteristics. Due to the guiding effect of the spoiler blades, the absolute axial angles of the speeds at various locations within the case of the 4 blades 30° are quite different. For the case of 4 blades 0°, the total velocity values are similar everywhere in the pipeline. The fluid energy in each micro-element flow tube remains basically constant, indicating that the attenuation of the flow field energy by the spoiler blades is small. For the case of 4 blades 30°, it can be seen from the comparison of axial velocity and angle that the spoiler blade obviously has a certain influence on the momentum of each micro-element flow tube, so the flow field energy loss is greater.

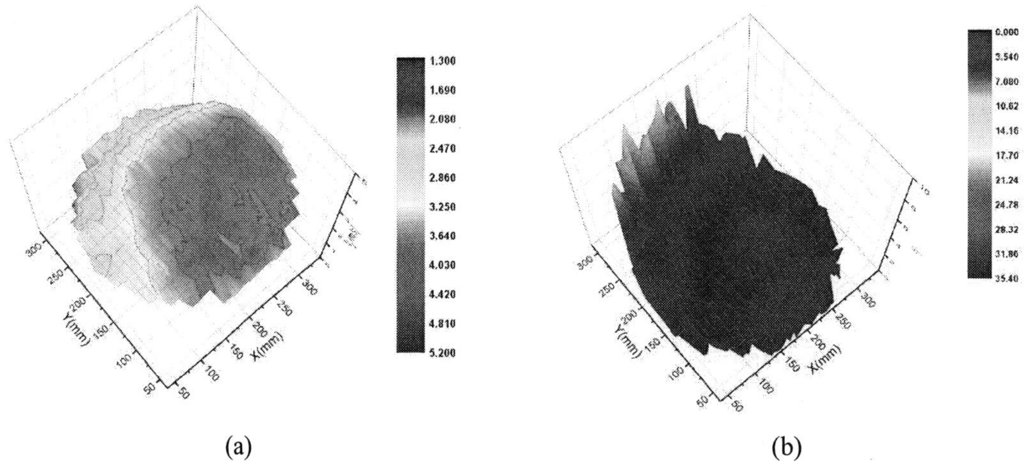

Figure 9. Axial velocity value and angle distributions of inlet section of 4 blades 0° under design flow: (a) axial velocity value, (b) absolute axial angular of velocity.

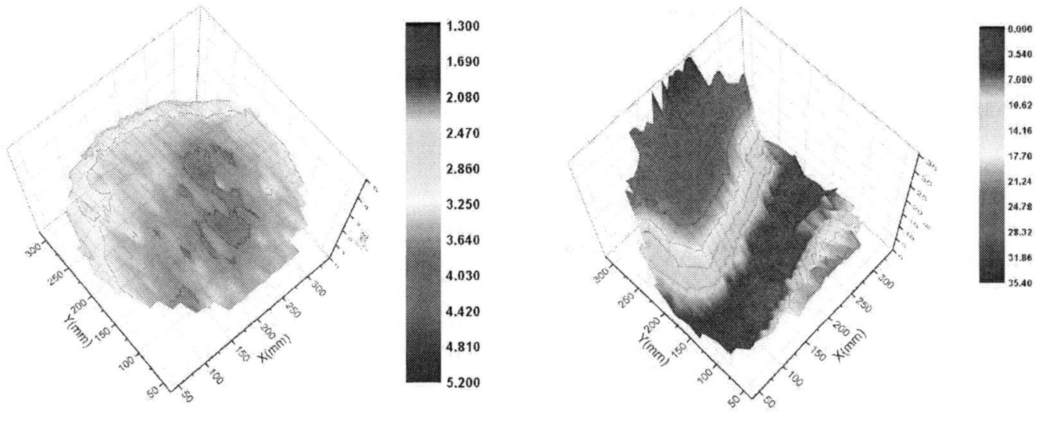

(a) (b)

Figure 10. Axial velocity value and angle distributions of inlet section of 4 blades +30° under design flow: (a) axial velocity value, (b) absolute axial angular of velocity.

Table 4 is a summary of the axial velocity distribution uniformity δ and the velocity-weighted average angle θ, θ' for each case obtained from the test. Figure 11 shows the comparison of the uniformity δ of axial velocity distribution for each cases. Figure 12 and Figure 13 show the comparison of the weighted average angles θ and θ' of speeds for each cases. For the uniformity δ, the closer its value is to 1, the better the uniformity. It can be seen that the undisturbed section is the highest, followed by the 4 blades 30° case, and then the 4 blades -15° case. The cases of 4 blades 0°, 4 blades 15°, 2 blades +30° & 2 blades -30° are relatively low. There is no obvious trend law between uniformity δ and flow.

For the speed-weighted average angles θ and θ', the closer their values are to 90°, the better the uniformity. It can be seen that the cases of 4 blades 0° and 4 blades 15° are the closest to 90°, followed by no turbulence and 2 blades +30°&2 blades -30°. The θ value of the 4 blades 15° case is lower, and the 4 blades 15° case is the lowest. The distribution of the weighted average angle and the flow rate also have no obvious trend law, and there is no direct relationship with the uniformity δ, it only depends on the incoming flow conditions and the turbulent flow conditions.

Table 4. δ, θ and θ' on the entrance section of each case in the test.

(a)

Cases	1	2	3	4	5	6	7	8	9
inlet condition (blade angle × number of blades)		0°×4			15°×4			30°×4	
δ	0.876	0.866	0.899	0.868	0.884	0.888	0.863	0.921	0.942
θ [°]	88.538	89.032	90.283	83.828	85.980	85.938	75.321	83.627	80.533
θ' [°]	87.905	88.170	88.178	82.031	82.661	84.352	74.298	76.419	75.161

(b)

Cases	10	11	12	13	14	15	16	17	18
inlet condition (blade angle × number of blades)		30°×2, -30°×2			-15°×4			no spoiler	
δ	0.849	0.881	0.909	0.890	0.903	0.902	-	0.935	-
θ [°]	81.672	87.135	92.331	90.233	90.657	90.690	-	88.050	-
θ' [°]	81.129	85.962	86.425	87.410	87.092	85.331	-	83.405	-

Combine the previous comparison of pump performance with the three figures of speed uniformity, we could get some results: the axial velocity distribution uniformity δ is the highest under the no spoiler case, the weighted average angles θ and θ' are not low either, therefore, its efficiency is high; δ of the 4 blades 0° case is low, while θ and θ' are closest to 90° in this case, therefore, its efficiency is also high; δ of the 4 blades -15° case is in a medium level, while θ and θ' are very close to 90° which are just after the 4 blades 0° case, therefore, its efficiency is high too; δ of the 4 blades 15° case is low, and θ and θ' are also low in this case, therefore, its efficiency is obviously lower than the previous cases; δ of the 4 blades 30° case is high which is just after the no spoiler case, while θ and θ' of this case are the lowest, so its efficiency is the lowest; δ of the 2 blades +30° & 2 blades -30° case is low, and θ and θ' are in a medium level, therefore, its efficiency is low too which is just higher than δ of the 4 blades 30° case. In conclusion, δ, θ and θ' all have obvious direct effects on the efficiency of the pump. In the range of δ of cases in this paper (the difference between the maximum value and the minimum value is 0.093), the effects of θ and θ' on efficiency are higher than δ obviously.

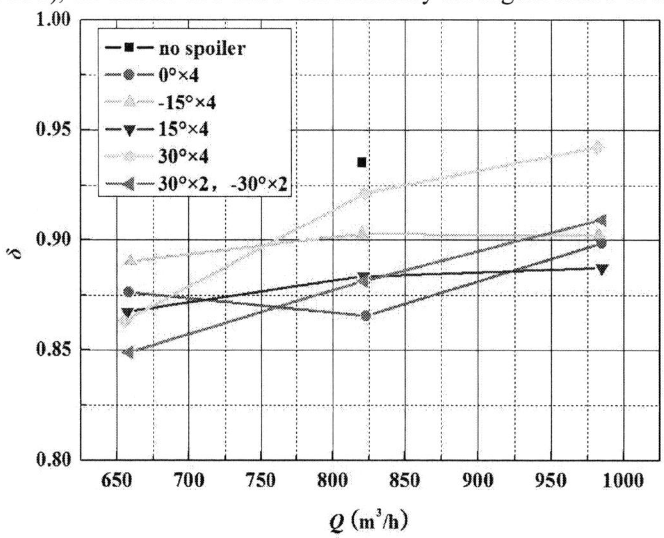

Figure 11. Comparison of δ under various cases.

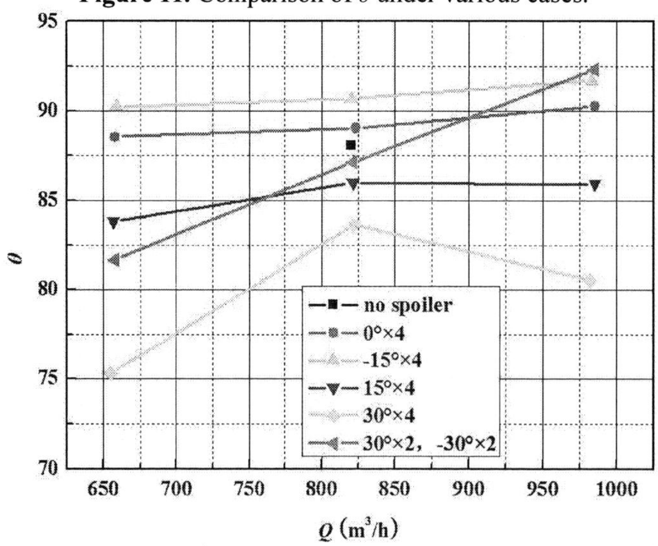

Figure 12. Comparison of θ under various cases.

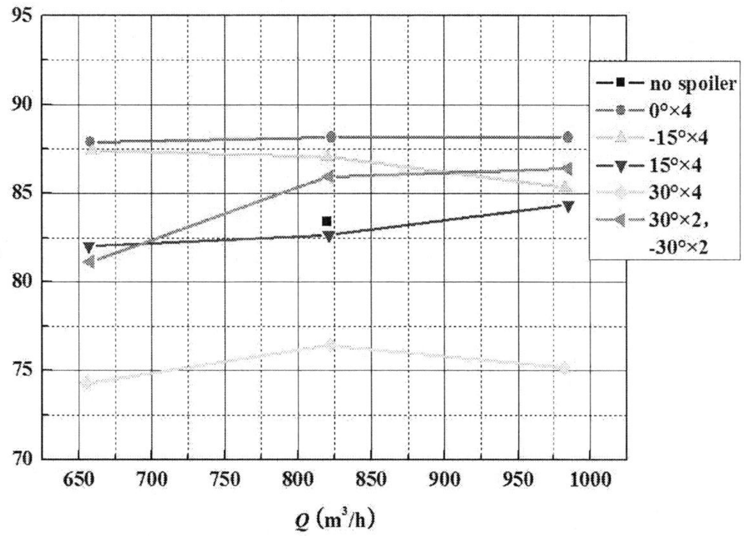

Figure 13. Comparison of θ' under various cases.

4. Conclusions

A PIV test system was adopted to investigate the influence of the uniformity of the inlet flow field on the performance of a nuclear main pump scale model, and the NUMECA software was used to numerically simulate the test cases. The conclusions can be drawn as follows:

1) When the rotation direction of the spoiler blade is opposite to that of the impeller, the greater the deviation of the spoiler blade from the axial angle, the more the pump efficiency and head drop.

2) Proper positive pre-rotation will not reduce the pump efficiency, but will increase the pump head.

3) For the same spoiler blade deflection angle, when its structure produces irregular flow field disturbances under non-centrosymmetric conditions, the flow field downstream of it will not produce clear swirling flow, and it will not have more impact on the performance of the pump.

4) The axial velocity distribution uniformity δ and the velocity weighted average angle θ, θ' have obvious direct effects on efficiency, and in the δ interval of each working condition in this paper (the difference between the maximum value and the minimum value of δ is 0.093), the speed weighted average angle has a greater impact on efficiency.

References

[1] Lu L G, Zhou J R 1995 Three-dimensional numerical simulation of turbulent flow and hydraulic optimal design for inlet channel of pumping station *Journal of Hydraulic Engineering* **12** 67-75.

[2] Tang F P, Geng W M, Yang G P 2004 Influence of the flow pattern in the sump on the flow field of pump inlet *Journal of Drainage and Irrigation Machinery Engineering* **22** 12-14.

[3] Liu H J, Xu J, Zhou F C, Su S L, Yao W P 2016 The analysis of the flow uniformity at the inlet of pump *Ship Science and Technology* **15** 79-85.

[4] Yan H, Liu M Q, Liang X, Lin P 2014 Influence on cavitation characteristics of large axial-flow pumping unit by uniformity of conduit inflow *Journal of Huazhong University of Science and Technology: Natural Science Edition* **10** 108-12.

[5] Yang M G, Meng Y, Li Z, Gao B, Ni D 2015 Design of axial-flow impeller guide cone and simulation on hydraulic performance of its passage *Transactions of the Chinese Society of Agricultural Engineering* **31** 81-88.

[6] Cavazzini G, Dupont P, Pavesi G, Dazin A, Bois G, Atif A 2011 Analysis of unsteady flow velocity fields inside the impeller of a radial flow pump: PIV measurements and numerical

calculation comparisons *Proc. of ASME-JSME-HSME Joint fluids engineering conference* **2011** 24-29.

[7] He Z N, Chen S S, Pan G X, Zhou Z F, Xue F J, Li S H 2013 CFD and PIV test of centrifugal pump with long and short blades *Journal of Yangzhou University: Natural Science Edition* **16** 40-44.

[8] Huang H M, Gao H, Shen F, Du C H 2008 Numerical simulation and experimental validation of the flow field in axial flow pump *Transactions of The Chinese Society of Agricultural Machinery* **39** 66-69.

[9] Su W T 2014 Study on the inner flow stabilities of large-scale francis hydro-turbin model *Harbin: Harbin Institute of Technology* **6** 122.

[10] Yuan J P 2008 Inner flow PIV measurement and its unsteady turbulent flow numerical simulation for a centrifugal pump with multi-programs *Zhenjiang: Jiangsu University* **5** 18-19.

[11] Westra R W, Broersma L, van Andel K, Kruyt N P 2010 PIV measurements and CFD computations of secondary flow in a centrifugal pump impeller *Journal of Fluids Engineering* **132** 061-104.

[12] Du C S 2012 Experimental study of inner flow with structural parameters in cyclone gasifier *Harbin: Harbin Institute of Technology.*

[13] Westra R W, Broersma L, van Andel K, Kruyt N P 2009 Secondary flows in centrifugal pump impellers: PIV measurements and CFD computations *ASME 2009 Fluids Engineering Division Summer Meeting. American Society of Mechanical Engineers*.305-14.

[14] Shao C L, Gu B Q, Huang X L, Zhou J F 2010 Experimental study on flow in the impeller of low specific speed pump using particle image velocimetry *Journal of Aerospace Power* **9** 2091-96.

[15] Li X M 2002 Experimental research on flow in the closed pump sump with PIV *Beijing: Tsinghua University.*

[16] Yang H, Tang F P, Zhou J R, Liu C, Yu L H 2007 A method for evaluating the PIV measurement error in a centrifugal pump *ASME/JSME 2007 5th Joint Fluids Engineering Conference. American Society of Mechanical Engineers* **9** 33-37.

[17] Wang X J, You B 2004 PIV experimental comparison of internal flow field in axial fan with and without external duct Compressor, *Blower & Fan Technology* **2** 20,21,38.

[18] Paone N, Riethmuller M L, Van den Braembussche R A 1989 Experimental investigation of the flow in the vaneless diffuser of a centrifugal pump by particle image displacement velocimetry *Experiments in Fluids* **7** 371-78.

[19] Wu Y L, Liu S H, Yuan H J, Shao J 2011 PIV measurement on internal instantaneous flows of a centrifugal pump *Science China Technological Sciences* **54** 270-76.

[20] Wen S P, Gong W Q, Zhang Y Y, Wang X J, Sun Y S, Ma X M 2004 PIV measurement of unsteady flow field vaneless diffuser *Compressor, Blower & Fan Technology* **3** 10-13.

[21] Benra F K, Feng J, Dohmen H J 2007 Measurement of the periodic flow field in a radial diffuser pump by the PIV method *ASME-Paper FEDSM2007-37400 [CD-ROM].*

[22] Yang G A, Yin X, Song Z, Huang C 2010 Particle image velocimetry measurement of the flow field in the play of the drilling pump valve *Chinese Journal of Mechanical Engineering* **23** 1.

[23] Wen S P, Zhang Y Y, Zhang C H, Wang X J, Sun Y S, Zhang Y, Ma X M 2004 PIV measurement of unsteady flow within blade diffusor. *Compressor, Blower & Fan Technology* **4** 1-4.

[24] Zhang L, Wang Q J,Chen H P 2004 PIV measurement of flow field within vaneless diffuser of centrifugal compressor *Compressor, Blower & Fan Technology* **6** 18-21,41.

[25] Shao C L, Gu B Q, Chen Y 2010 Measurement of steady and unsteady flow in centrifugal pumps spiral casing using PIV *Transactions of the Chinese Society of Agricultural Engineering* **7** 128-33.

[26] Di Cristo C 2011 Particle imaging velocimetry and its applications in hydraulics: A state-of-

The-art review *Experimental Methods in Hydraulic Research. Springer Berlin Heidelberg* 49-66.

[27] Zhi M, Li W, Wang T 2017 PIV measurements of the flow field characteristics in an inertial particle separator *Compressor, Blower & Fan Technology* **59** 56-60,70.

[28] Charoenngam P, Subramanian A, Kadambi J R, Wernet M, Courtwright R 2002 PIV investigations of particle velocities in the tongue region of a slurry pump *ASME 2002 Joint US-European Fluids Engineering Division Conference. American Society of Mechanical Engineers* 211-22.

[29] Geng W M, Liu C, Tang P F 2010 3D-PIV measurements of flow fields at exit to impeller of an axial flow pump *Journal of Hohai University: Natural Sciences* **38** 516-21.

[30] Che C C, Tian M C 2013 PIV experiment on flow disturbance characteristice of embedded trapezoid winglets in tube *CIESC Journal* **64** 3976-84.

[31] Wang C G 2009 Numerical simulation and PIV experimental rsearch of two-dimensional on power-law fluid in circular pipe *Daqing: Daqing Petroleum Institute.*

Modeling and analysis of the compressor for the closed cycle system on anti-surge process

Y R Sun, L X Yao, X Jin, Y Y Sun, C Tao
(Shenyang Blower Works Group Corporation, No.16A Development Road, Economic & Technical Development Zone, Shenyang, China.)

Email: sunyongrui@shengu.com.cn

Abstract. In order to study the interaction mechanism of the pipe network, the compressor and the control system in the process of the anti-surge regulation, the simulation model of the overall Closed Cycle System (CCS) is established in this paper. By comparing the simulation results with the target values at the Design Operation and the Maximum Mach Operation, the calculation deviation is less than 1%, which verifies the accuracy of the model. At the Maximum Mach Operation, the analysis of the CCS during the quasi steady process and the dynamic process is carried out respectively. According to the result of quasi steady state analysis, the effectiveness of the designed anti-surge circuit during the normal operating process of the CCS is verified. By opening the Anti-Surge Valve, additional 38.5% surge margin is provided. And the surge margin at the simulation process would be slightly higher than the design value for the variation of the inlet condition from the different operations. Through the dynamic analysis, the changes of the different performance parameters are studied in the process of sudden blockage in the pipe network, and during this process the compressor surge could be induced easily due to the sharp decline of the inlet pressure. To ensure the safety of the compressor during the acceleration and deceleration process, the moving trajectory of the operation point on the Compressor Characteristic Map should be studied by considering the effect of the resistance characteristics of the CCS pipe network. In addition, the anti-surge system should have a high response frequency to deal with the sudden blockage.

1. Introduction

The test bed of Closed Cycle System (CCS) is very important for the compressor manufacturer such as Shenyang Blower Works Group Corporation. This test bed could be used for the special gas experiments (like Co2, N2 and He) and the experiments of the sensor calibration. It could save a lot of time and fund. The compressor is usually used to drive the air in the wind tunnel circuit, and to provide the required pressure ratio for the test section. Since the wide operating range and abundant regulation means of continuous wind tunnel, the higher safety requirements of the CCS tunnel compressor is put forward. However the relevant researches are very rare. Due to similarity of the wind tunnel and the closed cycle compressor system, the researches on the anti-surge of the continuous wind tunnel can provide the effective references.

The Continuous wind tunnel is a closed cycle system, and has the advantages of the excellent flow field quality, wide operation range and continuous operation, so requirements of the anti-surge is very strict [1]. The flow uniformity and the mass flow at compressor inlet would affect the blade inlet flow angle, and once the air flow angle increases to a certain value, the non-working face of the blade will stall [2]. Coupling with the wind tunnel pipe network system, the strong abrupt stall could cause the compressor surge. Parameters in wind tunnel such as pressure and flow would present low-frequency and large amplitude longitudinal pulsation oscillation, which could result in the interference between rotor and stator, leading to major accidents [3-4].

The mechanism of compressor surge has been studied deeply. By analyzing a transonic compressor, James gives the evaluation criteria of the compressor performance and stall line [5]. Xu Xiaoqiu puts forward the surge mechanism and anti-surge control method for an industrial compressor [6]. In references [7] and [8], the feasibility of two anti-surge measures, namely direct venting regulation and bypass regulation, of wind tunnel compressor is analyzed. The compressor surge boundary of two wind tunnels in china (namely NF-6 and av90-3) has been tested by Gao Chao and Zhou Enmin [9-10]. Zhang Wen achieves the real-time judgment and automatic control of surge by the use of PLC process control and WinCC platform [11]. In the existing literatures about the anti-surge control of wind tunnel compressor, the researchers are still mainly focused on experimental research [6-11]. Although the experimental study is more reliable, it needs to establish a complete wind tunnel test system before the research, which has significant lag. And considering the safety and cost, it is impossible to carry out destructive test to obtain the wind tunnel system operating parameters under the limit conditions. Through establishing the simulation model, the surge process is analyzed in the design stage, which can be used to guide the design of anti-surge system and formulate anti surge control strategy, and can simulate the operation process of extreme conditions. A series of the related researches have been carried out for the industrial compressor[12-14], but there is no relevant research report about wind tunnel compressor.

Ref. [11] points out the factors such as the Mach number, the opening aero of Two-Throat, the attack angle and the pipe network characteristics are available to cause compressor surge. However, the above situation are caused by personnel adjustment at normal operation point, and the occurrence of surge is gradual and predictable. However, due to the sudden falling off of the components or test pieces, the throat of the pipe network is blocked and could cause surge. This process is unpredictable and difficult to carry on an experimental study for its destructiveness. The dynamic simulation is an effective research method to solve this kind of problems. In this paper, the quasi steady model and the dynamic model of the Closed Cycle System (CCS) are established to study the surge process, especially in the extreme conditions, to obtain the influencing mechanism. By this way, it's helpful to further understand the surge process and formulate reasonable anti surge strategy.

2. Analysis of the quasi steady state process

2.1 Analysis area of the CCS system

The analysis area includes the overall CCS, as shown in Figure 1, including the motor, the compressor, the heat exchanger, the silencing section, the main valve, the anti-surge circuit, the anti-surge valve and the pipeline system. The input parameters are listed as follows: (1) the compressor characteristic data, (2) the size, friction coefficient and heat transfer coefficient of pipeline system, (3) the wall temperature and the heat transfer coefficient of heat exchanger, (4) the corresponding relationship between the flow coefficient and the valve opening angle. Valve A is the main valve, and the impact of pipeline local blockage on the CCS is simulated by adjusting the angle of this valve. Valve B is anti -surge valve, which controls the open/close status of the anti-surge circuit.

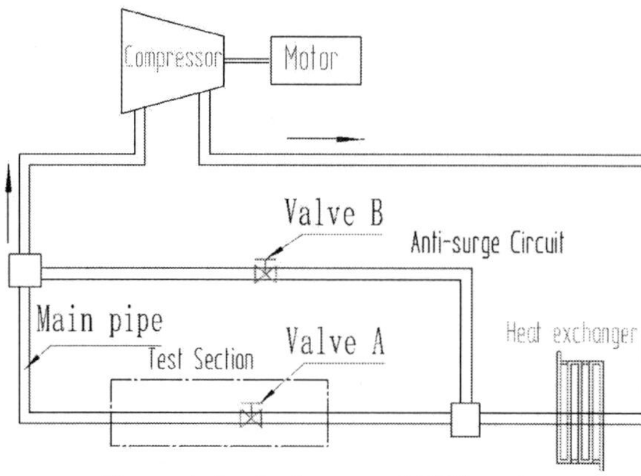

Figure 1 Analysis area of CCS system

2.2 Mathematical model

2.2.1 Governing equations

The flow process in pipeline is described by the gas state equation, the mass conservation equation, the energy conservation equation and the momentum equation, the equations are shown as follows:

$$pv = R_g T \tag{1}$$

$$\frac{dm}{dt} = \sum_{boundaries} \dot{m} \tag{2}$$

$$\frac{d(me)}{dt} = -p\frac{dV}{dt} + \sum_{boundaries} (\dot{m}H) - hA_s (T_{fluid} - T_{wall}) \tag{3}$$

$$\frac{d\dot{m}}{dt} = \left[dpA + \sum_{boundaries} (\dot{m}u) - 4C_f \frac{\rho u|u|}{2} \frac{dxA}{D} - Cp\frac{\rho u|u|}{2} A \right] / dx \tag{4}$$

Where As is the heat transfer area, Twall is the wall temperature, Cf is the friction coefficient of the pipeline, and CP is the local pressure loss coefficient.

2.2.2 Compressor Model

The compressor model is established by the compressor characteristic data. The compressor performance at different inlet conditions is accurately calculated by using reduced parameters, the equations of the reduce parameters are shown as follows:

$$RPM_{re} = \frac{RPM_{act}}{\sqrt{T_{t,in}}} \tag{5}$$

$$\dot{m}_{re} = \frac{\dot{m}_{act}\sqrt{T_{t,in}}}{P_{t,in}} \tag{6}$$

Where: the subscript *re* represents the reduced parameter according to the entrance conditions, and the subscript *act* represents the actual parameter.

The enthalpy, power and enthalpy are shown as the following equations.

$$h_{out} = h_{in} + \Delta h_s \, {}^{1}\!/_{\eta_s} \tag{7}$$

$$P = \dot{m}(h_{in} - h_{out}) \tag{8}$$

$$\Delta h_s = c_p T_{t,in} \left(PR^{\frac{\kappa-1}{\kappa}} \text{-}1 \right) \tag{9}$$

Where: $\dot{m}$ is the mass flow, η_s is the isentropic efficiency, *PR* is the total pressure ratio, the above parameters are determined by interpolation according to the compressor characteristic data.

2.3 Validation of the steady state model

The CCS Design Operation (Oper_D) and the Maximum Mach Operation (Oper_M) are selected as the verification operations, the calculation accuracy of the model is determined by comparing the calculated and the target values of the heat exchanger outlet temperature, the Mach number at the test section, the compressor inlet parameters (total temperature, total pressure and mass flow), and the compressor performance parameters (pressure ratio, efficiency and power). The calculation accuracy is shown in Table 1. In Table 1, T_Cooler is the outlet temperature of heat exchanger, Tt is the total temperature, Pt is the total pressure, PR is the pressure ratio, Eff is the isentropic efficiency, and Power is the compressor power. The calculation deviation of each parameter in Table. 1 is less than 1%, the calculation accuracy of the model is accurate.

Table 1 Calculation deviation of steady state model (Unit, %)

Parameter	T_cooler	Compressor inlet			Compressor performance		
		Tt	Pt	Mass flow	PR	Eff	Power
Oper_D	-0.03	-0.03	0.11	-0.02	0.00	-0.24	0.39
Oper_M	0.00	0.00	0.16	0.03	-0.13	0.00	-0.10

Note: $Calculation\ Deviation = (Calculated\ Values - Target\ Values)/Target\ Values \times 100\%$

2.4 Anti-surge control system

In order to ensure the safety of the compressor, three protection lines are set, namely Alarm Line, Valve Opening Line and Deceleration Line, as shown in Figure 2. When the combination operation point of the compressor and the pipe network system moves to the Alarm Line, the sound-light alarm would be sent out; when the combination operation point moves to the Valve Opening Line, the anti-surge valve would be turned on, and the response time of the anti-surge valve is 3s; when the combination operation point moves to the Deceleration Line, the compressor speed would reduce by 20%, and the response time of speed reduction is 3s. The anti-surge control system is established according to the above three control lines. When the combination operation point moves to different areas, the corresponding protection measures would be triggered such as alarming, valve opening and decelerating in the dynamic model.

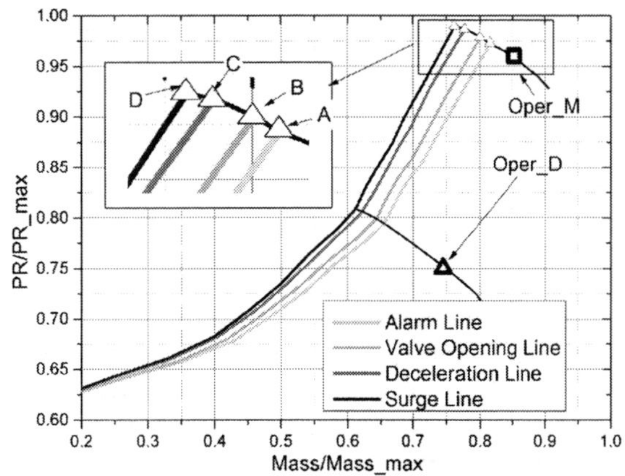

Figure 2 Protection lines of the anti-surge control system

2.5 Establishment of Dynamic model

Macro dynamic processes discussed in this paper include the quasi steady state process and the dynamic process. During the quasi steady state process, the parameters of compressor change slowly and continuously, which usually occurs in the normal operation process of the CCS compressor. The dynamic process simulation is used to study the emergency situations, such as the sudden falling off of the components or test pieces, which could cause local sudden blockage, make the parameters change sharply and induce surge. The speed change in the above process is necessary to simulate by considering the power difference between the motor and the compressor, as well as the influence of inertia. In this model, PID algorithm is used to simulate the motor power control system in real time to ensure that the compressor speed can converge to the target speed. The control process is shown in Figure 3. When the actual speed of the compressor deviates from the target speed, the power correction is calculated by Proportion (P), Integral (I) and Derivative (D), and the motor output power is adjusted. The compressor speed could converge to the target speed. PID control parameters are determined by debugging.

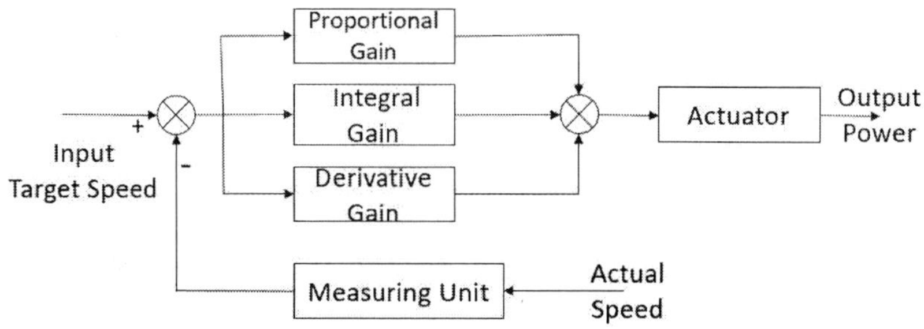

Figure 3 Algorithm diagram of PID

3 Analysis of the quasi steady state process

As shown in Figure 2, the surge margin of Oper_M operation is minimum, more attention should be paid to this operation to avoid the compressor surge. So Oper_M is selected to study the influence mechanism. The position of the combination operation point on the Compressor Characteristic Map is controlled by adjusting opening of Valve A on the main pipe. By reducing the opening of Valve A gradually, the combination operation point moves from Oper_M to the small flow area along the constant speed line. By this way, the slow and conventional adjustment process of the CCS is simulated, so namely quasi steady state process.

Seven operating points are selected for further analysis on the constant speed operation line. The first five operating points are Oper_M, Alarm Point (Point A in Figure 2), Valve Opening Point (Point B), Deceleration Point (Point C) and Surge Point (Point D). The last two operating points are Valve-Opening-Deceleration Point (point E, coinciding with point C) and Valve-Opening-Surge Point (Point F, coinciding with Point D), the last two points separately operate at the Deceleration line and Surge line with the Anti-Surge Valve opening. The last two points are used to analyze the ultimate anti-surge capacity of the designed anti-surge circuit. During all the seven quasi steady state processes (from Oper_M to A, B…F), the anti-surge control system is effective, which ensures compressor operating safely. The results of the seven operation points are shown in Table 2. The design value of surge margin is obtained in the design process when the inlet conditions are constant. The negative value for the calculated surge margin means the additional surge margin provided by anti-surge circuit.

According to Table 2, from Oper_M to Point D, the opening of the Valve A gradually decrease. With the operation moving to the small flow area, the outlet temperature of the compressor increases, while the capacity of the heat exchanger remains constant, so the total temperature at the outlet of the heat exchanger increases. By reducing the opening of Valve A, and the total pressure loss in the main pipe increases and the compressor inlet pressure decreases. According to the compressor similarity criterion, the reduced flow rate of compressor surge point remains constant under different compressor inlet conditions. When the operating point moves to the small flow area, the total temperature increases and the total pressure decreases, according to Eq.(6), which leads to the relative increase of surge margin, from 10% to 13.1%. At the Point F, the flow rate of the main pipeline in the main pipe is ultimate minimum value of the CCS, and the anti-surge circuit can provide an additional 38.5% surge margin.

Table 2 Compressor inlet parameters and surge margin for each operating point

Operation	Inlet Temperature /K	Inlet pressure /Pa	Calculated surge margin /%	Design surge margin /%
Oper_M	300	70857	13.1	10
A	300.31	69757	8.5	7
B	300.62	69413	5.9	5
C	301.0	69000	2.2	2
D	301.17	68810	0.0	0
E	301.02	68964	-36.3	-
F	301.18	68791	-38.5	-

4 Analysis of dynamic process

In the dynamic process, the influence of the local blockage on the CCS compressor is simulated by suddenly closing the flow area of the Valve A. Three cases are listed as follow. (1) The angle of the Valve A reduces from the open angle of Oper_M to the open angle of Point A within 0.1 s, namely Oper_M to Point A, and the Valve B keeps close during this process. (2) The angle of the Valve A reduces from the open angle of Oper_M to the open angle of Point D within 0.1 s, namely Oper_M to Point D, according to the anti-surge control strategy, Valve B is open and the compressor speed is reduced by 20%. (3) The angle of the Valve A reduces from the open angle of Oper_M to the open angle of Point E within 0.1 s, namely Oper_M to Point E, Valve B is open and the compressor speed is reduced by 20%.

4.1 Dynamic process from Oper_M to Point A

The changes of different parameters in the dynamic process are shown in Figure 4. According to the changes of the parameters in the Figure 4， the open angle of the Valve A suddenly is reduced to the open angle of Point A at t=150s, which simulates the effect of the sudden local blockage in the CCS on the compressor. The mass flow through the Valve A suddenly decreases (reduced by 17.2%), meanwhile the compressor inlet mass flow does not change , so that the air in the pipeline between Valve A and compressor is pumped out, the air temperature and pressure drop. While the gas pressure at compressor outlet does not change significantly, the compressor pressure ratio suddenly increase (increased by 4.2%). According to the compressor characteristics, the operating point quickly moves to the small flow area. The compressor surge occurs almost at the same time as the local blockage (t=150s). Due to the reduction of mass flow, the power consumption used for compressing the air decreases, and the output power of the motor keeps constant, so the compressor speed increases.

When the open of the Valve A suddenly decrease, a part of air is blocked in front of the Valve A, the subsequent air would compress this part air and make its pressure and temperature increase. When this part air with the higher temperature and pressure flows into the compressor, the pressure ratio decreases rapidly, and the flow rate begins to increase, the compressor gradually exits the surge state, finally the operating point coincides with Point A. The moving trajectory of the compressor operating point is shown in Figure 5 and the trajectory is consistent with the previous analysis. When the Valve A is turned down, the compressor enters the surge state, and then exits the surge after 2s. Generally, the anti-surge system needs to monitor the temperature and pressure to determine the position of the compressor operating point and the corresponding protection plan, which has a certain hysteresis (delay time set as 3s). The designed protection system exists certain shortcomings in dealing with this kind of dynamic processes. The response frequency should be increased to deal with the sudden blockage.

Comparing the results in Table 2, the actual surge margin of the Point A is 8.5%, but the dynamic simulation results show that the compressor enters surge state during the process. According to the above analysis, local blockage leads to the pressure at the compressor inlet reduce sharply, and so the pressure ratio increase significantly, and could induce surge easily. The existing experimental and simulation researches on processes of the steady-state or quasi steady state can't present actual operation of compressor for these sudden events.

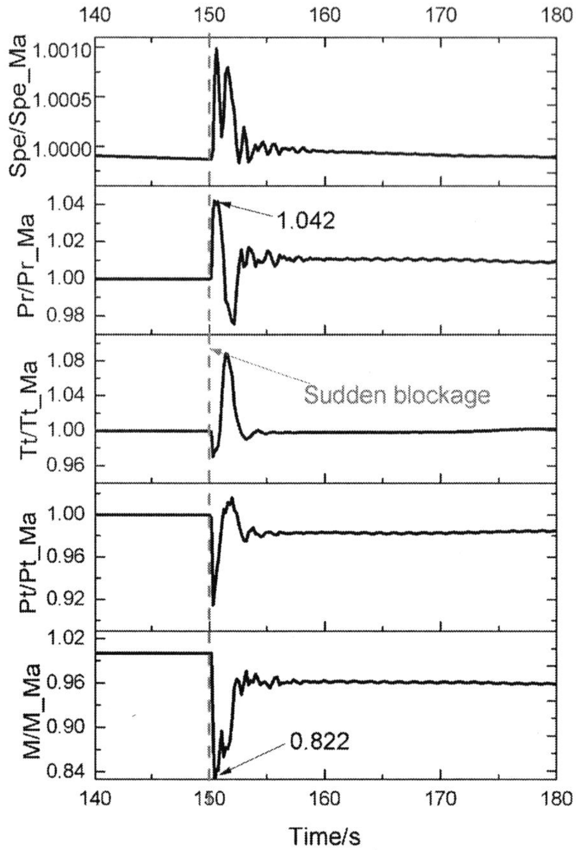

Figure 4 Changes of the different parameters in the dynamic process from Oper_M to Point A

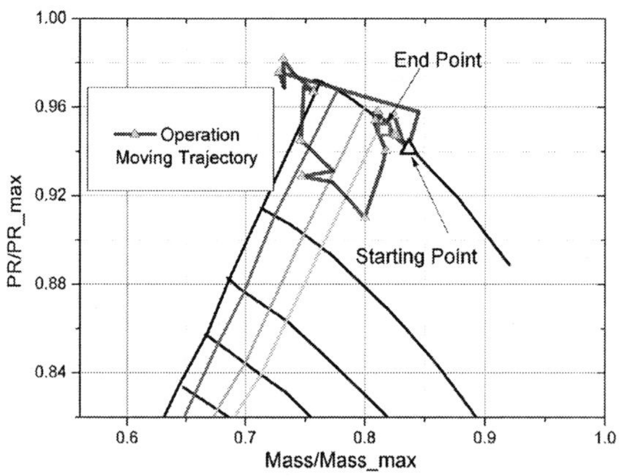

Figure 5 Moving trajectory of the compressor operating from Oper_M to Point A

4.2 Dynamic process from Oper_M to Point D

The changes of different parameters in this process are similar to that shown in Figure 4. The differences are that the anti-surge valve is open and the compressor speed is reduced by 20% at t=153s. The moving trajectory of operation point is shown in Figure 6. The blockage first causes a significant increase of the pressure ratio, and compressor enters the surge state. With the Valve A opening, the pressure ratio decreasing and mass flow increasing, the compressor is gradually exit surge state. The compressor speed is reduced safely under PID control.

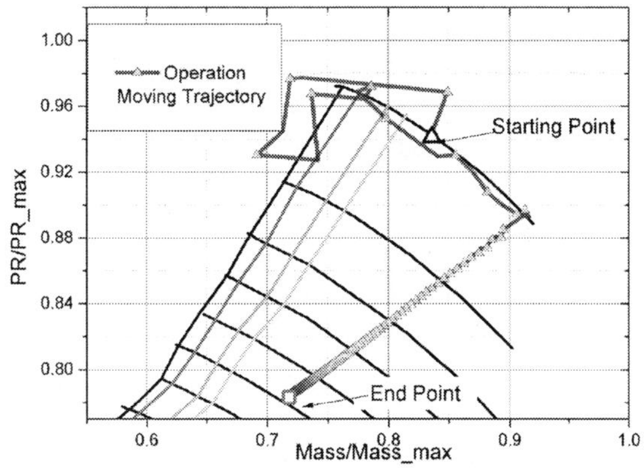

Figure6 Moving trajectory of the compressor operating from Oper_M to Point D

4.3 Dynamic process from Oper_M to Point E

The changes of different parameters in the dynamic process are similar to that in Section 3.2. With the Valve A opening, the compressor will exit the surge state, but during the process of deceleration, the compressor will enter the surge state again. According to Table 2, the surge margin of Point E is about 2%, but as shown in Figure7 the speed can't decrease safely. The reason is that the CCS has its special pipe network resistance characteristics, and result in a specific moving trajectory of the combination operation point on the compressor Characteristic Map. Due to the impossibility to decelerate safely for Point E without any adjustment means, the mass flow of this operation in main pipe may be unreachable, the additional surge margin of the anti-surge circuit calculated by during the quasi steady state process is not surely to be reasonable. So the simulation analysis of the dynamic process is an effective way to test the safety of the compressor during the dynamic process, especially for the extreme operations.

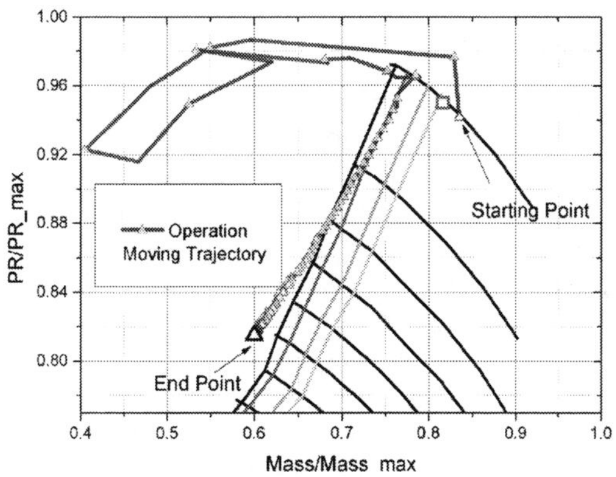

Figure7 Moving trajectory of the compressor operating from Oper_M to Point E

5 Conclusion

In this paper, the overall model of the CCS is established, including the pipeline model, the compressor model, the valve model, the anti-surge control model and PID control model. The accuracy of the model is verified by comparing the steady-state calculation results with the target data. Then, the analysis of the quasi steady state process and dynamic process are carried out respectively, and the following conclusions are obtained.

(1) According to the results of the quasi steady state analysis, the design anti-surge circuit is effective during the normal operating process of the CCS, the design anti-surge circuit could provide additional 38.5% surge margin. And the surge margin at the actual operation process would be slightly higher

than the design value for the variation of the inlet condition from the different operations, the surge margin of Maximum Mach Operation increases from design value of 10%to 13.1%.
(2) Through the dynamic analysis, the changes of the different performance parameters are studied in the process of the sudden blockage in the pipe network, the sharp decline of the inlet pressure is the main reason to induce the compressor surge. As the blockage could cause surge in a short time, the anti-surge system should have a high response frequency to deal with the sudden blockage.
(3) In the deceleration process of the compressor, the operation point would enter the surge state again from the safe state due to the effect of the special resistance characteristics for the pipe network. So it's an effective method to accurately study the interaction mechanism of the pipe network, the compressor and the control system in the process of the anti-surge regulation, and analyze the safety of the compressor during the dynamic process, especially for the extreme operations.

References
[1]Cai Tao, Shi lun 2012 *J* An Optimized Design and Discussion about the Divergence Angle of Diffuser in Low-speed Wing Tunnel *Chinese Journal of Turbomachinery*, **01**:7-9.
[2]Zhang Wen and Zhou Enmin, 2015 *J.* Safety Characteristics Reshearch and Operational Protection of an Axial Compressor in the Wind Tunnel. *Chinese Journal of Turbomachinery*, **57(01)**:78-9.
[3] Nhan Nguyen, Roy Guist and Doug Muzzio. 2013 *C.* Experimental investigation of the rotor blade vibration in the three-stage compressor of the 11- by 11-Foot Transonic Wind Tunnel. *Joint Propulsion Conference & Exhibit.*
[4]Ge Yongfu 1999 *J.* The Preliminary Study on Anti-surge Curve of Axial Air Compressor. *AUTOMATION IN PETRO-CHEMICAL INDUSTRY*, **002**:2-5.
[5]Gunn J, Martindale W and Wagner D 1992 *R.* Performance evaluation of a transonic wind tunnel compressor. *AIAA*. **92-3927**.
[6]Xu Xiaoqiu, Mei Xiaoqin and Chen Yongjian 2002 *J.* Anti-surge Control System of the Turbocompressors. *Journal of Sichuan University (Engineering Science Edition)*, **34(1)**:85-88.
[7]Li Yibin and Qiao Zhide 2001 *J.* The study of application of by-pass to prevent surging in pressurized continuous wind tunnel. *Experiments & Measurements in Fluid Mechanics*, **15(4)**:53-58.
[8]Li Yibin and Qiao Zhide. 2001 *J.* The study of the application of exhausting to prevent surging in pressurized continuous transonic wind tunnel. *Experiments & Measurements in Fluid Mechanics*, **015(003)**:24-29.
[9]Gao Chao, Wang Na and Yuan Xianshi 2013 *J.* Determination of the axial-flow compressor surge boundary in NF-6 pressurized continuously-operating high-speed wind tunnel. *Experiments & Measurements in Fluid Mechanics*, **27(005)**:61-66.
[10]Zhou Enmin, Cheng Song and Xu Jing 2014 *J.* Surge margin test and research AV 90-3 axial compressor in 0.6m continuous trainsonic wind tunnel. *Journal of Experiments in Fluid Mechanics*, **28(5)**: 81-85.
[11]Zhang Wen, Zhou Enmin and Cheng Song 2017 *J.* Anti-surge model and control strategy of fan in wind tunnel. *Journal of Aerospace Power*. **32(6)**:1434-1440.
[12]Fei Qizhi, Lin Fengbao and Zhang Jiarui 2016 *J.* Research on the Starting of Compressor Unit with Pressure Based on UNISIM. *Chinese Journal of Turbomachinery*. **58(03)**:89-93.
[13]Ma Yiwei and Ma Liangyu 2008 *J.* Real-time Dynamic Simulation Model of Axial-flow Fan based on BP Neural Network. *Chinese Journal of Turbomachinery*, **2008(06)**:48-50.
[14]Meng Lei, Wang Xuejun and Wang Ling 2008 *J.* Reseach on the Model of Dynamic Integrated System for Computer Aided Based on MAS. *Chinese Journal of Turbomachinery*, **2008(03)**:60-62.

Acknowledgment
The work is carried out with the support by Natural Science Foundation of Liaoning Province of China (No.2020-BS-274). The author gratefully acknowledges President Yang Shuhua and Chief Engineer of Liu Changsheng for the Research Institute of Shenyang Blower Works Group Corporation.

IOP Conf. Series: Materials Science and Engineering 1081 (2021) 012026 doi:10.1088/1757-899X/1081/1/012026

Aerodynamic optimization of the last stage turbine blade for an industrial gas turbine

Yongfeng Sui*,Jinquan Chen, Peng Chu, Jibing Lan, Sheng Ren, Hongchao Wu, Yunbin Xu, Jian Xie and Jun Ding

Hangzhou Steam Turbine & Power Group Co., Ltd., Hangzhou, 310022, China

* Corresponding author: Suiyf@htc.cn

Abstract. A multi-objective, aerodynamic optimization of the last stage turbine blade of an 100Hz industrial gas turbine is presented. We aim at maximizing the stage's efficiency, while at the same time meeting mechanical integrity. The optimization procedure was build, during the optimization process the mass flow rate and stage meridian were kept constant to obtain the required stage loading without changing the operating point. The optimized geometry are chosen taking into account aerodynamic constraints as well as mechanical constrains, such as centrifugal stress and bending stress. The SST turbulence model was used to calculate the flow field of the original and optimized stage, respectively. The geometry of the original and optimized blade was compared. The stage efficiency of the design point and under different operate conditions were compared. Stage and blade efficiency along span was analyzed in details. The blade loading at root mid top span of the original and optimized blade were discussed. Detailed flow field analysis shows that, after optimization, the peak Mach number of the root span profile was decreased from 1.22 to 1.09, which improves the flow field. Profile loss and end wall loss were also reduced. After optimization, maximum centrifugal stress and the maximum air bending stress were decreased which leads to improved blade life and safety. The number of blades and b/h were suggested for the last stage blade.

1. Introduction

Industrial gas turbine are used for power generation and mechanical drive. It requires high efficiency and low emissions to meet economic requirement and environmental friendliness. The last stage turbine blade has complex three- dimensional flow in the flow passage and large centrifugal force. Novak[1] developed streamline curvature computing procedures. Dorman[2] developed a technique for improving turbine aerodynamic performance with application of controlled-vortex aerodynamics to axial flow turbines, which has been made practical by advanced computer technology. It is still widely used in turbine design process nowadays.

The large steam turbine last stage blade flow passage occurs supersonic flow up to Mach number 1.6, which is difficult to deal with. McBean [3]designed three different length of steam turbine last stage blade of 41in,45in and 49in, the design process include aerodynamic design, mechanical design. Xin[4] developed new low pressure stages for 1000MW boiler feed water pump steam turbine which demands good strength and aerodynamic performance. Tip section of the steam turbine last stage blade was analyzed by Sreedharan[5] using 2D simulations, major airfoil design parameters such as un-guided turning and trailing edge angle were studied, a new design feature on the suction side of the LSB which has higher performance benefit was proposed. Static strength criteria based on allowable stress and dynamic strength criteria based on safety factor are currently the main assessment criteria for the blade design and reliability consideration. Blade life evaluation studies such as static stress analysis, dynamic stress analysis, high cycle and low cycle fatigue life analysis, the ideas and methods for further studies were summarized by XIE[6][7].

In gas turbine blade design and research area, aerodynamic redesign of a four-stage heavy-duty gas turbine was done by Filippo Rubechini[8], through-flow methods, three-dimensional RANS computations, were applied according to a given hierarchical criterion, the overall redesign procedure relies on a neural-network-based approach aimed at maximizing the turbine's power output while satisfying geometrical and mechanical constraints. Two dimensional through-flow using an automatic optimization strategy of multistage turbomachinery[9], blade-to-blade viscous solutions for aerodynamic profile design[10][11][12], three dimensional Navier-Stokes with a number of heuristic and gradient based optimisers for a single row simulation [13][14] have been developed in the early years. Bellucci[15] and Cotroneo[16] developed a multi-objective, aerodynamic optimization method for a high-pressure steam turbine stage which contains prismatic vane and blade. The aerodynamic constraints, such as limitation of the pressure recovery in the uncovered part of the suction side, as well as mechanical constraints, such as root tensile stress and dynamic behavior are chosen taking into account. Redesign procedure for a 17-stage steam turbine which contains optimization of stator lean and rotor twist single for a single stage, geometrical transformation between the original reference stage and the optimized one, neural-network-based refinement of the stator and rotor twist of each stage, is presented by Rubechini [17], Hasenjäger [18] and Öksüz [19] developed a multiploid distance based MOGA aerodynamic shape design optimization tool capable of handling surrogate models for a turbine blade. The multi-objective aerodynamic design objectives are maximizing the adiabatic efficiency and torque so as to reduce the weight, size and cost of the gas turbine engine. An automated multi-objective and multidisciplinary design optimization (MDO) of a transonic turbine stage to maximize the isentropic efficiency and minimize the maximum stress of the rotor with constraints on mass flowrate and dynamic frequencies is presented by Song[20][21]. the isentropic efficiency decreased by 0.02%, but obtained a decrease in the profit of the maximum stress by 15.4%.

Supersonic flow in the gas turbine last stage blade passage is not as strong as the large steam turbine last stage blade. Power generated by the last stage accounts for 20% of a four stage turbine, transonic flow always occurs in the turbine last stage, it is a challenge for designer to acquire high blade efficiency. Another important aspect of last stage blade design is the need for a relatively long operational life. The requirement for a reliable blade design, that must operate under difficult conditions in terms of the high temperature (over 850K) without cooling and also under high centrifugal load, poses a great challenge to blade designers. The blade should also be manufactured in a cost effective manner, and be relatively simple to assemble. So that the final product meets all of the requirements in terms of aerodynamic performance and mechanical integrity.

A lot of research has been done for large steam turbine last stage blade. There are relative few research focus on improving the gas turbine last stage efficiency while meeting mechanical integrity. To consider aero thermal performance and mechanical integrity at the same time in the design process is highly demanded due to the fact that an aerodynamic optimum may not satisfy mechanical and aeromechanical criteria. Thus, an aerodynamics only design optimization might lead to considerable time delay and extra costs. The aim of the present study is to maximize the stage's efficiency, while meeting the mechanical integrity. Efforts have been made to make the optimization process time cost effectively and exactly. Another purpose of this paper is to find proper number of blades and b/h which can contribute to get high efficiency and good strength for gas turbine last stage blade.

2. Numerical tool

2.1. Optimization procedure

The present optimization is aimed at maximizing the last stage's total to total efficiency which was selected as the objective function. The optimization of the blade is an iterative complex analysis process to satisfy all the design constraints, related to the different disciplines, such as aerodynamics, structural mechanics.

By previous analysis, the original last stage has good mean line design and 2D(through flow) design, during optimization, 1D and 2D aerodynamic design results which contains velocity triangles were kept constant. So during the optimization process several constraints were imposed in order to meet the design rules. The mass flow rate was kept constant, to obtain the required stage loading without changing the operating point when replacing the original blade with the optimized one. The stage meridian was also kept constant.

The optimization procedure is shown in figure 1. The blade optimization procedure starts with choosing number of blades which can effect both the efficiency and strength. The following results show that blade with fewer number and longer chord may achieve higher efficiency while improve the blade strength. Seven blade profile was chosen to be optimized, which can represent the blade aerodynamic performance while saving optimization time. In this step reasonable t/b was chosen to reduce the number of iterations. For the blade loading distribution of the airfoil surface especially on the rear part of the suction side, limitation of the diffusion rate was imposed to avoid boundary layer separation. In the root span, the blade airfoil was optimized as aft loaded to reduce secondary loss. 3D blade was optimized to stack 2D profiles and smooth the surface. 3D CFD was used to simulate the aerodynamic performance of the optimized stage to determine if the expected goal were achieved. The main complex problem of such a procedure is that a good aerodynamics design may not satisfy mechanical requirement. In the present optimization, the mechanical integrity checks relied on simplified design rules based on the evaluation of centrifugal stress and bending stress, ensuring the same or better mechanical capability for the optimized geometry as the original one. Simplified design rules play an important role in reducing optimization time.

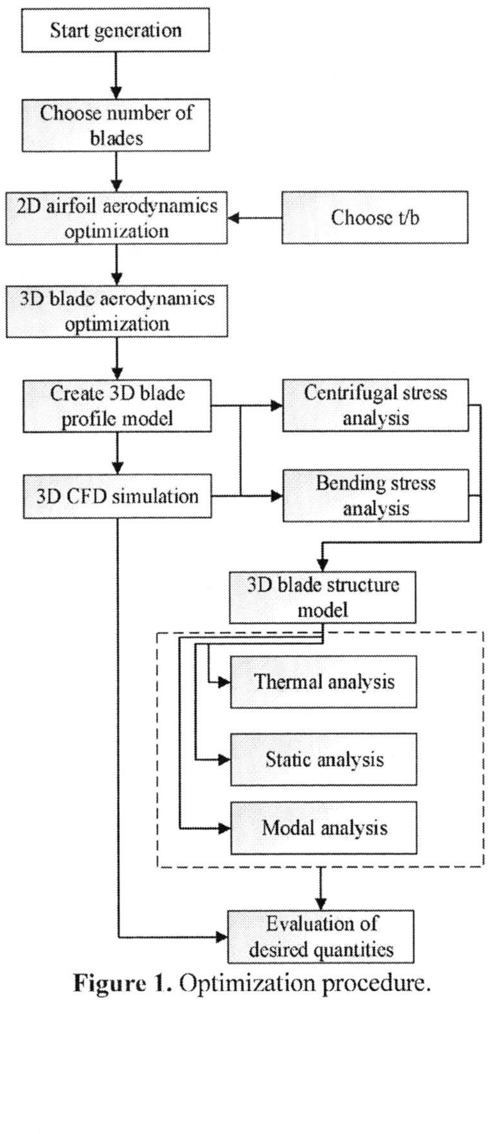

Figure 1. Optimization procedure.

2.2. Computational method

The computational domain shown in figure 2 consists of the last vane and the last blade for an 100Hz industrial gas turbine which has four stages turbine. The structured grid is generated using Turbogrid. The total nodes of the computational domain is 1.4million, which contains 6800,000 nodes for the vane, 720,000 nodes for the blade. The height of the first layer was kept to ensure $y+<1$, which satisfies the selected SST turbulence model.

The numerical simulations in the present work were conducted by using the commercial CFD software ANSYS CFX. The steady compressible RANS equations were solved with the SST turbulence model, which combines the advantages of the k-ω model and the k-ε model. The SST turbulence model has good precision when solving the boundary layer flow with the inverse pressure gradient [22]. The g-θ model which is proposed by Menter[23]and Langtry[24] was used as transition model. The g-θ model which has high precision judges the transition process based on local variables, and has strong adaptability to the grid. The convection term is discretized using a high precision differential format.

Boundary conditions are given in table 1. Total pressure and total temperature with radial distribution as shown in figure 3 was given for the inlet which comes from the simulation of the whole four stage turbine. The outlet boundary condition was given by a uniform static pressure distribution. The rotation speed is 6000r/min. The working fluid is real gas which contains CO_2, H_2O, Ar, N_2 and O_2. The walls were simulated as no-slip, adiabatic boundary conditions. Steady simulations were conducted with mixing planes between the stator and rotor.

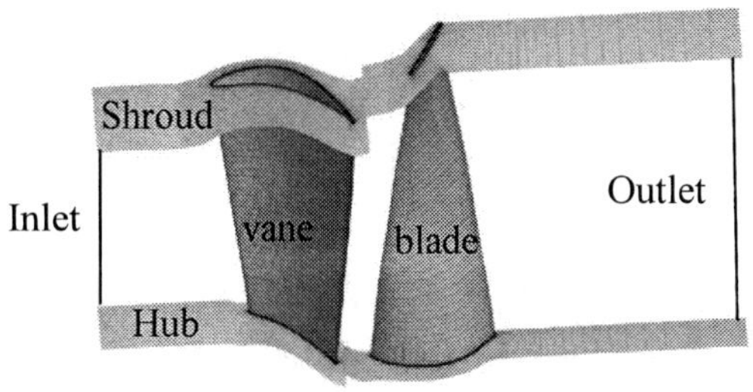

Figure 2. Computational domain and grid.

Table 1. Boundary conditions.

Parameters	Value
p_{t0}	2.2（radial distribution）
T_{t0}	980（radial distribution）
p_2	0.9（uniform）
ω	6000
Material	real gas

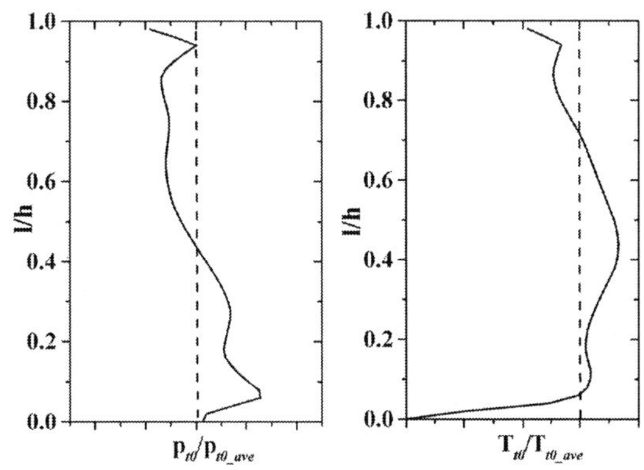

Figure 3. Total pressure and total temperature distribution at inlet.

3. Aerodynamic results and discussion

3.1. Geometry

The original and optimized blade section geometric features are compared in table 2. The biggest difference between the original and optimized blade is the reduction of the number of blades which are decreased from 101 to 81, followed by the C increased along span and aspect ratio b/h changed from 0.2 to 0.24. But the pitch/chord along blade span remains almost unchanged.

In figure 4 cross section of the last stage is compared between the original and optimized case. Hub and shroud remains unchanged, the vane and distance between the vane and the blade were also kept constant. The optimized blade has longer axial chord.

Table 2. Blade geometric, original vs optimized.

Parameters	Original	Optimized
Number of blades	101	81
t/b at root span	0.34	0.34
C at root span	84.6	106.5
t/b at mid span	0.57	0.57
C at mid span	60.6	72.2
t/b at top span	1	0.98
C at top span	25.2	30.6
b/h	0.2	0.24

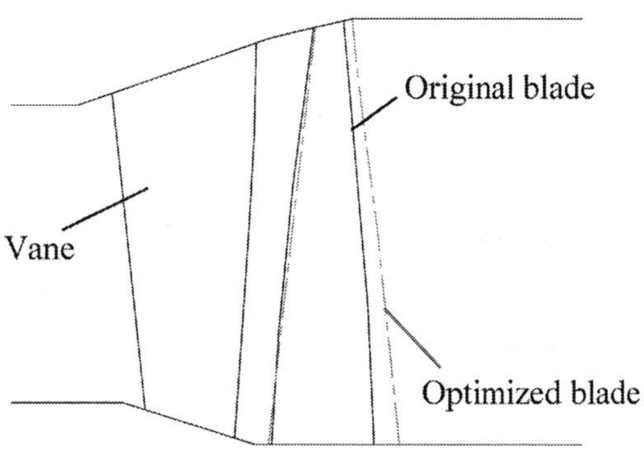

Figure 4. Cross section of the last stage, original vs optimized.

3.2. Turbine efficiency

The total isentropic stage efficiency $\eta_{stage,is,tt}$ and the total isentropic blade efficiency $\eta_{blade,is,tt}$ are defined:

$$\eta_{stage,is,tt} = \frac{m \cdot \Delta h}{m \cdot \Delta h_{is,t}} = \frac{h_{0,t,abs.} - h_{2,t,abs.}}{h_{0,is,t,abs.} - h_{2,is,t,abs.}}$$

$$\eta_{blade,is,tt} = \frac{m \cdot \Delta h}{m \cdot \Delta h_{is,t}} = \frac{h_{1,t,rel.} - h_{2,t,rel.}}{h_{1,is,t,rel.} - h_{2,is,t,rel.}}$$

Where $h_{0,t,abs.}$ is the vane inlet absolute total enthalpy, $h_{2,t,abs.}$ is the blade outlet absolute total enthalpy, $h_{1,t,rel.}$ is the blade inlet relative total enthalpy, $h_{2,t,rel.}$ is the blade outlet relative total enthalpy.

In table 3 the significant improvement of isentropic stage efficiency is gained by optimization.

Table 3. Aerodynamics results.

Parameters	Original	Optimized
Stage efficiency	93.4%	94.3%
Stage reaction	0.5	0.51
Blade inlet rel. Ma	0.39	0.4
Blade outlet rel. Ma	0.93	0.92
Blade outlet abs. Ma	0.52	0.52
$\Delta h/u^2$	1.37	1.37
m	163	163

In figure 5 the stage efficiency along span is compared between the original and optimized case. The optimized blade has much higher stage efficiency in the first 50% span. From 50% to 100%, with small exception around 85% l/h, the optimized case results in a little higher stage efficiency (almost the same as the original case). In figure 6, the blade efficiency is compared between the original and optimized case. The optimized case has much higher stage efficiency in the first 50% span. From 50% to 100%, the optimized case results in little higher(0~0.5%) stage efficiency. The results indicate that optimized stage efficiency gains are attributed to 0~50% l/h blade profile optimization. The reason for the optimized case has little higher stage efficiency at 50%~100% l/h compared with the original case is that the original case already has 95%~97.5% blade efficiency at 50%~100% l/h, which has small potential for optimization.

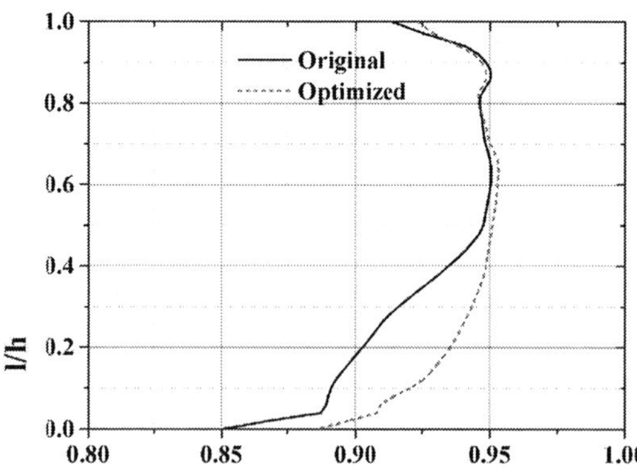

Figure 5. Stage efficiency along span, original vs optimized.

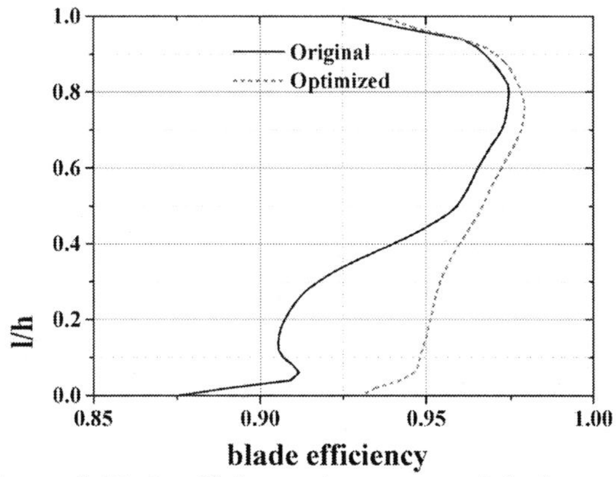

Figure 6. Blade efficiency along span, original vs optimized.

Streamwise distribution of the mass averaged entropy on the blade cross section from inlet to outlet is shown in figure 7. When gas flows in the blade passage, flow loss consists of two main parts: profile loss and secondary flow loss[25]. Flow loss grows very slow before 0.5 axial chord in the blade passage for the original and optimized case, the difference between them is also very small. The reason is that in the suction side and pressure side, flow are under the influence of accelerating pressure gradient, profile loss is very small, and the secondary loss is small because the area of the passage vortex and horse shore vortex is relative small for the long blade. Incidence loss is also small. Based on the above reasons, flow loss grows very slow for the original and optimized blade before 0.5 axial chord.

From 0.5 to 1.0 axial chord, as the fluid velocity increases, the boundary layer thickens and supersonic speed appears in some areas， the loss increases quickly from 0.7 to 1.0 axial chord, adverse pressure gradient occurs at the blade suction surface, boundary layer thickens more, even separation and transition occur. Entropy increases much faster at the end of the blade because of the wake loss. For the optimized case, the lower pressure point position of suction surface moves downstream, adverse pressure gradient decreases, adverse pressure gradient segments shorten, so flow loss is less than that of the original.

Figure 8 provides a comparison of the stage efficiency between the original and optimized blade over the pressure ratio ranging from 1.6 to 2.5 which can cover 20%~110% load of the gas turbine under different back pressure. With the pressure ratio increasing, the increases of stage efficiency become higher and then become lower, resulting from 0.2% to 1.1%, and from 1.1% to 0.8%, which indicates after optimization the stage have better aerodynamic performance over wide range of pressure ratio. Compared with the original blade, the optimized case has much higher efficiency near the design point, away from the design point the increases of the stage efficiency becomes lower which achieves the goal of optimization.

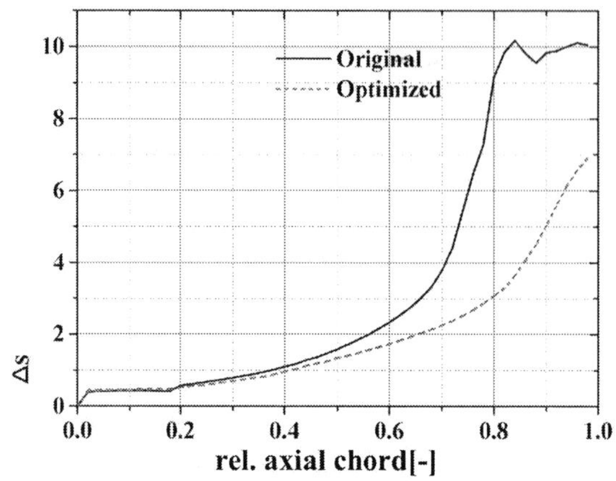

Figure 7. Blade entropy distribution from inlet to outlet, original vs optimized.

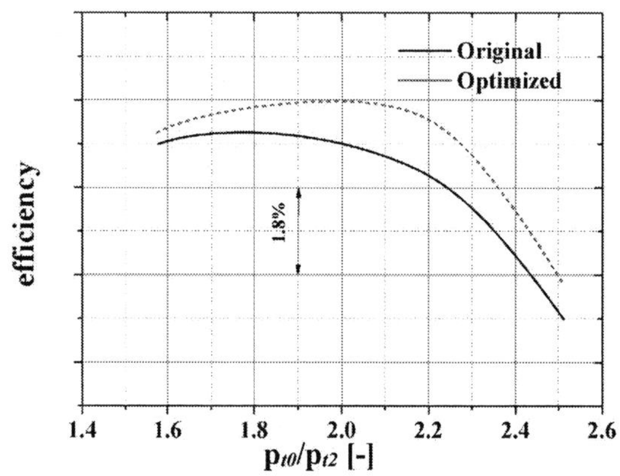

Figure 8. Stage efficiency under different pressure ratio original vs optimized.

3.3. Blade Loading

The last stage blade aerodynamic optimization involves a large number of iterations to reach a design satisfying mechanical integrity and assembly constraints. The original and optimized blade loading and the separation character of the flow are studied.

The hub section of last stage blade typically has to cope with high flow turning in the order of 90° and entails transonic inlet and outlet flow conditions. The challenge in the optimization of such sections is to accommodate for the required flow turning whilst minimizing leading, passage, and trailing shock losses. Such high turning sections feature relatively low level of pitch/chord ratio (t/b=0.34), and by nature, approaches aerodynamic limit load quite early (M2 =0.93). The hub section must have appropriate area to meet mechanical integrity. In figure 9 airfoil pressure distribution at root span is compared between the original and optimized case. The original balde is almost uniform loaded, which leads to higher peak Mach number. The optimized blade is highly-aft-loaded, in the first 80% of the axial chord, the pressure on the pressure surface remains almost constant, in the first 60% of the axial chord, the pressure on the suction surface decreases slowly, and then the pressure on the suction surface decreases fast, the blade loading increases rapidly. Compared with the original blade, in the first 50% of the axial chord, the flow in the optimized blade is decelerated resulting in a diminished cross passage pressure gradient. After a short acceleration period, the negative pressure gradient at the rear part of the passage is decreased compared to the original design. This optimization leads to an increased cross passage pressure gradient in the after part and a homogenized static pressure distribution at the throat of the passage.

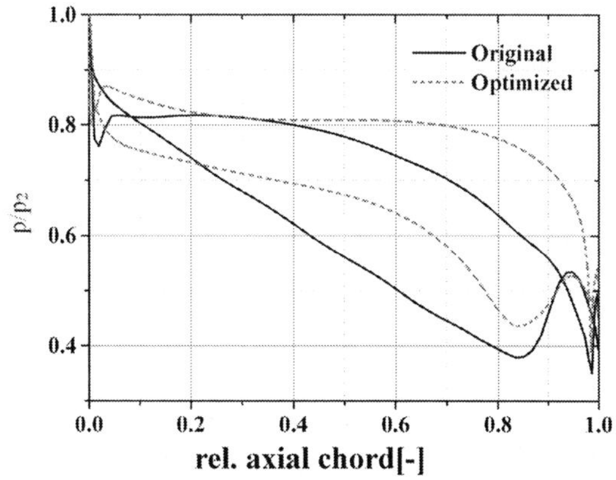

Figure 9. Airfoil pressure distribution at root span, original vs optimized.

Airfoil pressure distribution at mid span is shown in figure 10. Compared with the original case the optimized blade seems more aft-loaded. From the 50% axial chord to 80% axial chord, pressure in the optimized blade decreases very fast which means the blade loading increases fast. Like the root span section, the optimized mid span section leads to an increased cross passage pressure gradient in the after part and a homogenized static pressure distribution at the throat of the passage.

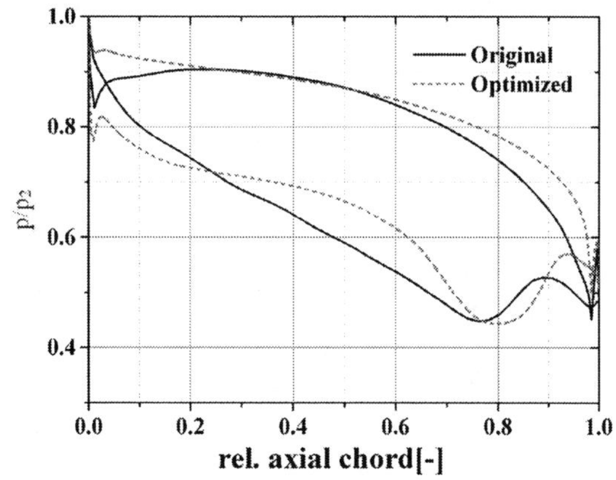

Figure 10. Airfoil pressure distribution at mid span, original vs optimized.

Last stage blade tip section design is one of the most challenging tasks a gas turbine aero designer faces due to the transonic/supersonic flow. Considerable amount of effort is required during the optimization process to ensure a mechanically robust tip section with low aerodynamic losses. In figure 11 airfoil pressure distribution at top span is compared between the original and optimized case. The blade loading of the optimized blade reaches maximum earlier than the original case. Flow in the optimized blade accelerating fast when goes into the passage, that is different from the root and medium section. Maybe this is contributed to structural integrity considerations. figure 5 and figure 6 show that the original and optimized blade has similar high efficiency at the top span. So the optimized section profile at top span is accepted.

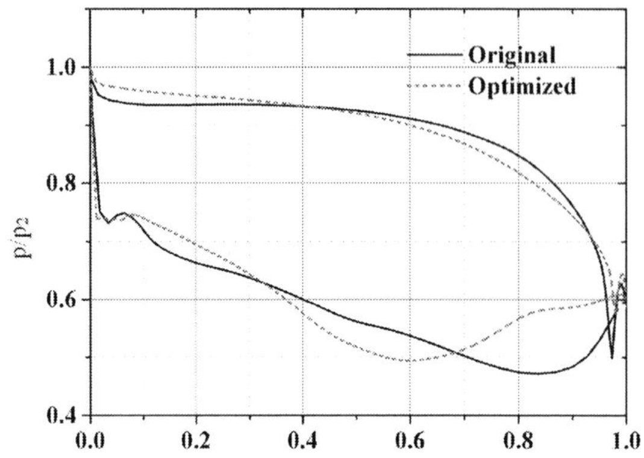

Figure 11. Airfoil pressure distribution at top span original vs optimized.

3.4. Blade flow field analysis

Controlled-vortex technique was used to design the last stage. During optimization the blade outlet flow angel a is kept constant. In figure 12 a along blade span is compared between the original and optimized case. The blade outlet flow angle increases from 78° in the hub to 110° in the shroud. The flow angle increasing form hub to shroud is almost linear. But the original case seems more fluctuation than the optimized blade. Due to the application of controlled-vortex design, the streamline in blade outlet is upturned which leads to uniform flow angle, uniform static pressure, uniform velocity. It should be careful to design the exhaust hood to increase the static pressure recovery in order to gain high turbine efficiency, because absolute velocity along the blade span outlet is not axial.

The stage reaction is based on the isentropic enthalpy drop over the blade row and the turbine stage, which can be defined as follows:

$$\Omega = \frac{h_{s,1} - h_{s,2}}{h_{t,0} - h_{s,2}}$$

Where the stage inlet total enthalpy $h_{t,0} = f(p_{t,0}, T_{t,0})$, the blade inlet static isotropic enthalpy $h_{s,1} = f(p_1, s_0)$, the blade outlet static isotropic enthalpy $h_{s,2} = f(p_2, s_0)$.

In figure 13 the stage reaction distribution of the original and optimized case is compared. The stage reaction along the span was kept very well, 0.46~0.57 from hub to shroud. The stage reaction of the optimized case is similar to the original, except in the hub aera where the optimized case has a little higher stage reaction than the original. The optimized stage has more uniform stage reaction along blade which can ensure the stage have high efficiency during variable working condition.

Because of the better stage reaction control, in figure 14 relative Mach number along the blade span outlet is more uniform for the optimized blade. Although during optimization the 2D design results was kept constant, only seven section of profile was chosen for optimization, Mach number along the blade acquired from 3D analysis occurs some difference. If free-vortex aerodynamics method was used, the range of the Mach number would have greater difference between the hub and shroud.

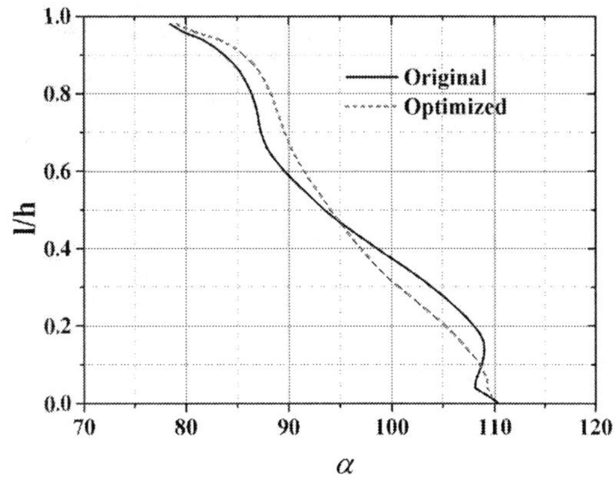

Figure 12. α along blade span, original vs optimized.

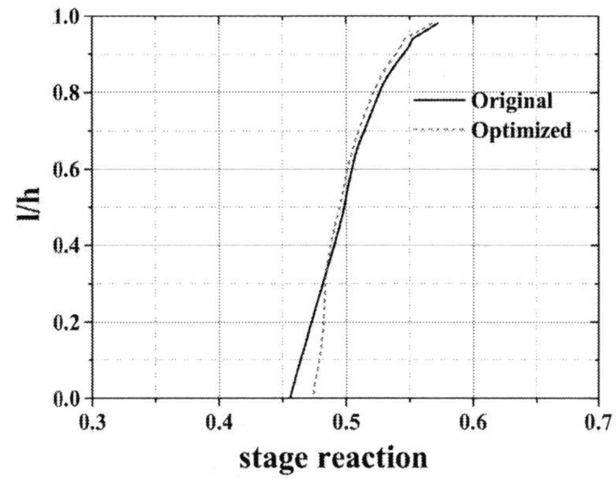

Figure 13. Stage reaction distribution, original vs optimized.

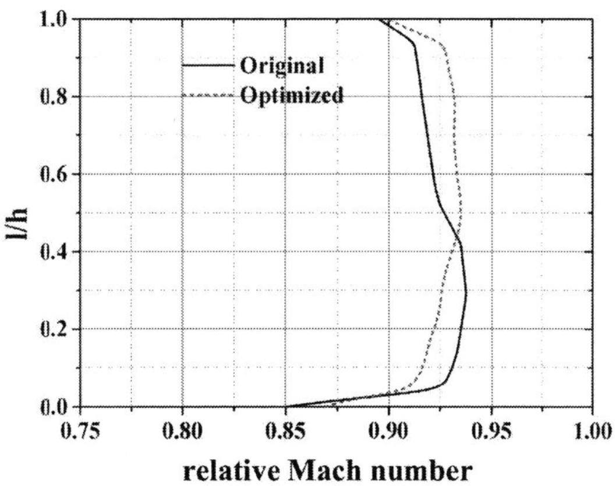

Figure 14. Relative Mach number distribution along blade span, original vs optimized.

To examine further the loss reduction mechanism, the normalized entropy contours at the exit plane are shown in figure 15. The pictures contain two blade flow passage, the color levels of entropy are the same in both pictures (red is high entropy, blue is low entropy). It is clear that the optimized blade has much less entropy generation in the first 50% span compared with the original case.

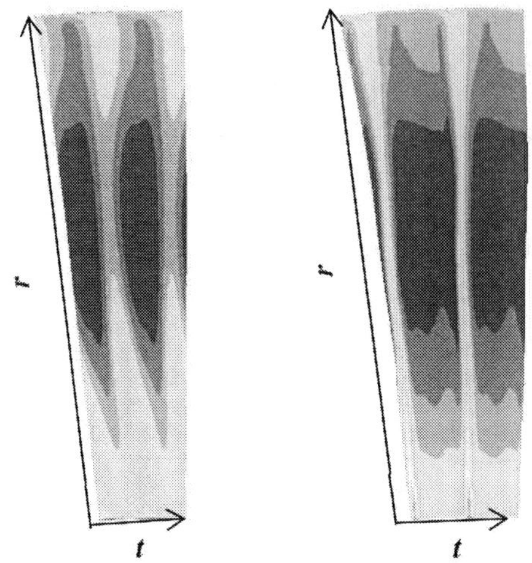

Figure 15. Entropy generation contour at the rotor exit.
Left: original, right: optimized.

The results of the simulations of the original and optimized blade is compared in figure 16 and figure 17. The span position of 5%~20% l/h, where the data from the line plots is extracted. The color levels of static pressure are the same in both pictures (red is high pressure, blue is low pressure). The peak suction is reduced in the optimized case resulting in weaker diffusion.

The weaker diffusion and the reduced S-shape of the static pressure distribution suggest later reattachment and a smaller separation bubble for the optimized airfoil. Therefore separation likely occur further downstream than in the original.

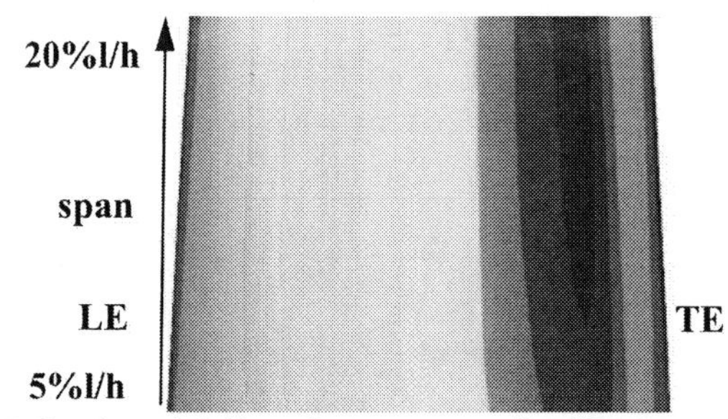

Figure 16. Suction airfoil surface pressure distribution at root span of original.

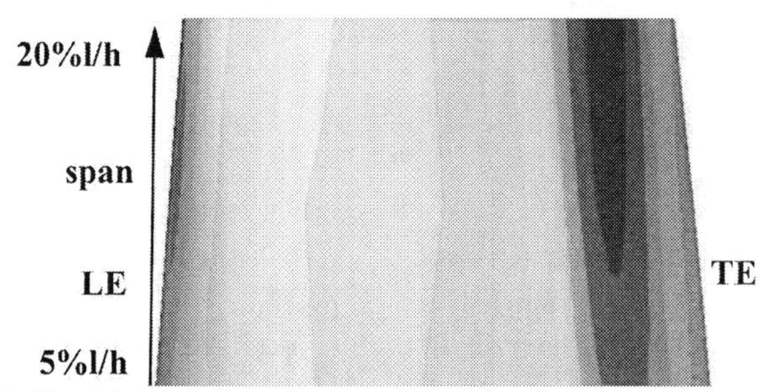

Figure 17. Suction airfoil surface pressure distribution at root span of optimization.

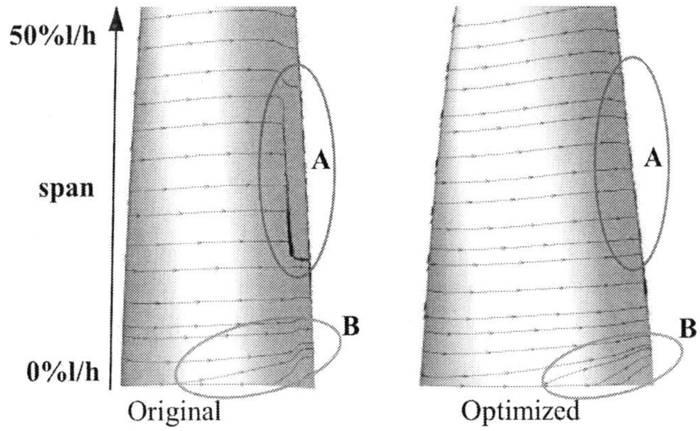

Figure 18. Streamline on blade suction surface.

To study the state of the suction side flow, the streamline are compared in original and optimized case. In figure 18 on the original airfoil a small separation bubble with downstream reattachment is visible between 15%~35% span (area A). In contrast there is no flow separation on the optimized blade. The feature is located much further downstream than the original one which supports the earlier interpretations of airfoil pressure distributions. Longer distance in laminar flow on the airfoil upstream of the separation and shorter turbulent boundary layer downstream result in lower profile losses, see Denton[25]. This is seen as one indication for the increase of the overall stage efficiency. End wall secondary flows may be responsible for as much as a third of the losses found in a turbine row depending on factors such as aspect ratio [25]. In the last stage long blade that the paper described, the secondary flow effect area B which is below 5% l/h. Long blade secondary flow is not as strong as short blade because of the aspect ratio. After the horseshoe vortex is formed, the pressure side leg of the horseshoe vortex collide and combine with or wrap around the suction side leg of the horseshoe vortex and together climb the suction surface. Compared with the original blade, horseshoe vortex climb the suction surface later, which indicates highly-aft-loaded profile at the optimized blade end wall delayed flow separation. Compared with the original blade, area effected by secondary flow is smaller which means the secondary flow strength is weakened.

Iso-surface of vorticity is compared in figure 19. On the original airfoil suction side, high vorticity is observed between 0%~40% span, while the optimized blade does not have it, which leads to higher efficiency. Compared with the optimized blade, the original blade has much stronger transverse flow at hub which leads to higher vorticity, while the optimized has weak vorticity distribution. High vorticity which represent corner vortex is also found at blade trailing edge outlet, while the optimized blade seems a little weaker.

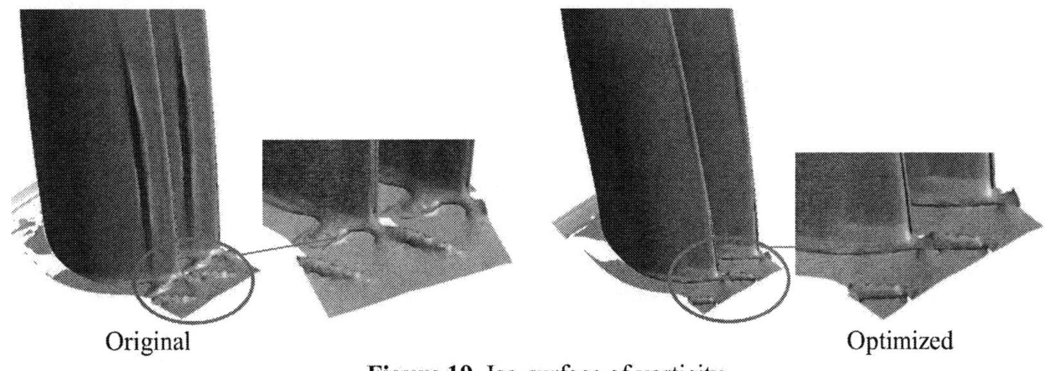

Figure 19. Iso-surface of vorticity.

To study the detailed flow field, Mach number distribution at different span is compared in figure 20 and figure 21. The color levels of Mach number are the same in the same blade height.

At root span, the optimized blade has an increase in flow acceleration ahead of the exit throat, thereby reducing the peak Mach number from 1.22 to 1.09, which indicates the profile loss is reduced. Compared with the original blade at root span section, the optimized blade minimizes the trailing shock strength and can avoid shock induced boundary layer separation. At mid span, compared with the original blade, the optimized blade has a homogenized Mach number distribution after the throat of the passage, which can reduce trailing shock losses. At top span, both the original and the optimized blade have relative low peak Mach number(less than root and mid span). At top span, the optimized blade reaches the peak Mach number earlier than the original case, but the boundary layer of the optimized blade gets reattached before leaving the trailing edge, so the optimized blade has high efficiency as the original case.

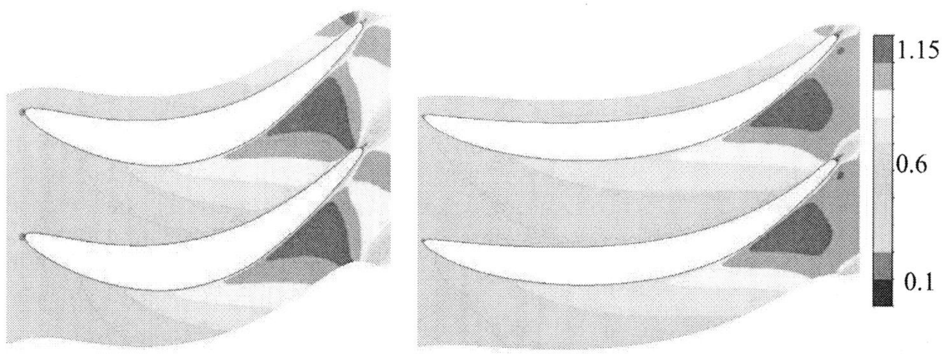

Figure 20. Mach number distribution at root span.

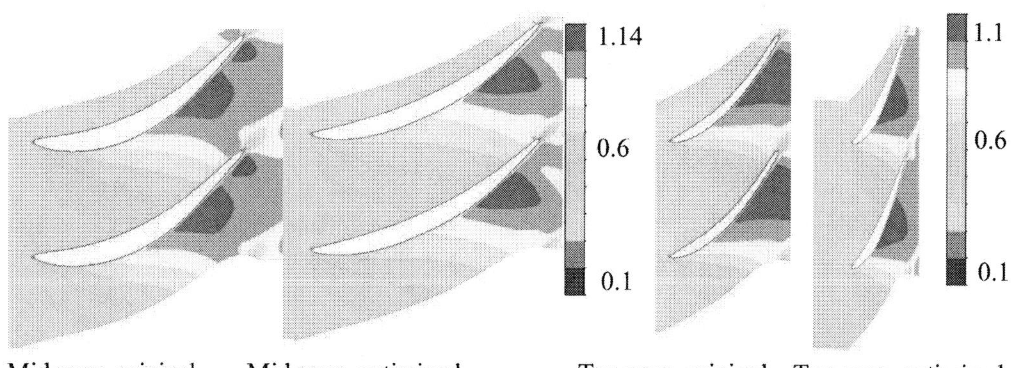

Figure 21. Mach number distribution at mid span and top span.

4. Mechanical design and analysis

Mechanical design is achieved in figure 22. Tip shroud was redesigned for the optimized blade which can prevent the migration of the flow from pressure to suction side resulting in higher loading of the blade tip section and less leakage. Tip shroud can also add stiffness to the blade which can reduce the risk of buffeting stresses. The same five teeth fir-tree blade root which can support big centrifugal force was applied for the original and optimized case.

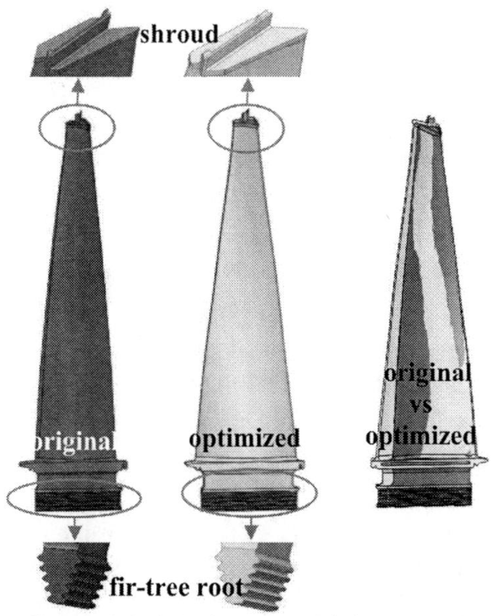

Figure 22. 3D solid model of the last stage blade, original vs optimized.

Average centrifugal stress along span is shown in figure 23, from 0 to 10% l/h where the blade has maximum average centrifugal stress, the averaged centrifugal stress of the optimized blade is a little smaller than the original case. From 10% to 70% l/h, averaged centrifugal stress of the optimized blade is a little bigger than the original case, the maximum difference is less than 20MPa. Generally, the optimized blade has similar average centrifugal stress distribution to the original case.

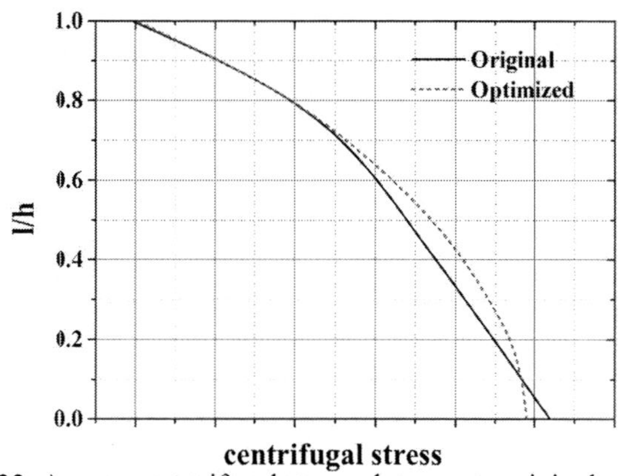

Figure 23. Average centrifugal stress along span, original vs optimized.

In figure 24 bending stress along span of the two blades is compared. With l/h increasing from 0% to 50% the bending stress increases, and then it decreases. The maximum bending stress of the optimized blade is decreased by 25% compared with the original case. The position of the maximum bending stress of the optimized blade is located at 50% span, which is similar to the original case. After optimization, the blade has lower bending stress which can contribute to extend the blade life.

Detailed finite element mechanical analysis considering temperature distribution and material properties was carried out to predict the LCF life of the blade. The results show that after optimization the blade has more than 100000 equivalent operating hours.

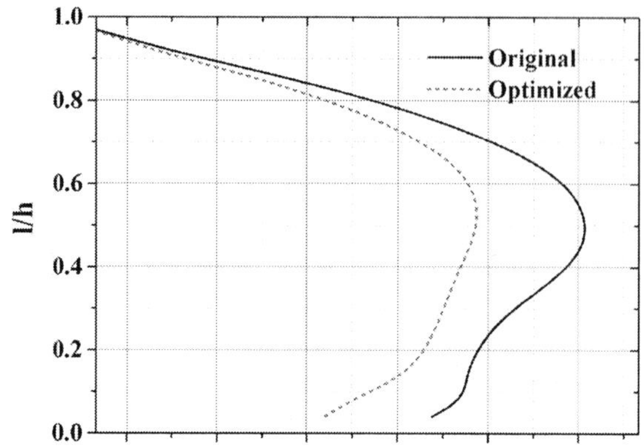

Figure 24. Bending stress along span, original vs optimized.

5. Conclusions

The present paper describes the aerodynamic optimization of the last stage blade of an industrial gas turbine. The optimization strategy aimed at maximizing the stage's efficiency, while meeting the mechanical integrity. During optimization, the original 1D and 2D aerodynamic design results were kept constant, so the mass flow rate and the stage meridian were kept constant. The SST turbulence model was used to calculate the flow field of the original and optimized stage, coupled with mechanical integrity checks by means of simplified design rules based on the evaluation of centrifugal stress and bending stress in order to ensure the optimized blade has same or better mechanical capability of the original one.

1. Numerical results show that the optimized stage efficiency was increased from 93.4% to 94.3% at the design point, while the optimized blade has much higher stage efficiency in the first 50% span. The stage efficiency was increased by 0.3%~1.1% with the stage pressure ratio ranging from 1.6~2.5.

2. Detailed flow field analysis shows that, after optimization, the root span section profile of the optimized blade is highly-aft-loaded, which delays flow separation and reduce the profile loss and secondary loss. The peak Mach number of the root span profile was decreased from 1.22 to 1.09, which improves the flow field. The optimized blade eliminates the separation on the suction surface which occurred on the original blade.

3. The 3D solid model of the optimized blade was built with shroud and fir-tree root. Simplified structural analysis shows that the averaged centrifugal tress of the optimized blade is a little smaller than the original case, the maximum bending stress of the optimized blade was decreased by 25%. The strength performance of the blade is improved.

4. We suggest choosing fewer number of blades with longer chord when designing the last stage blade, which can achieve higher efficiency and good strength.

Acknowledgements

The authors disclosed receipt of the following financial support for the research, authorship, and/or publication of this article: The present study was supported by China Postdoctoral Science Foundation No. 2020M671672.

References

[1] Novak R 1967 *J. Eng. Gas. Turb. Power.* **89** 478-90
[2] Dorman T 1968 *J. Eng Gas. Turb. Power.* **90** 245-50
[3] McBean I, Havakechian S and Masserey P 2010 *ASME GT*2010-22747
[4] Xin X, Sui Y, Mao H, Chu P, Kong J and Ye Z 2015 *ASME GT*2015-42135
[5] Sreedharan S, Mistry H, Ostrovskiy V and Guo T 2015 *ASME GT*2015-44084
[6] Xie Y and Zhang D 2007 *Turb. Technol.* **49** 401-22

[7] Yuan R, Zhou Q, Zhang Q and Xie Y 2014 *Adv. Math. Phys.* **549430** 1-7
[8] Rubechini F, Schneider A, Arnone A, Daccà F, Canelli C and Garibaldi P 2011 *ASME GT*2011-46258
[9] Cravero C and Dawes W 1997 *ASME 97-GT*-294
[10] Trigg M, Tubby G and Sheard A 1999 *J. Turbomach.* **121** 11-17
[11] Cravero C and Satta A 2001 *AIAA Paper* **2001** 3044
[12] Cravero C and Satta A 2001 *ASME 2001-GT*-0508
[13] Shahpar S 2000 *ASME 2000-GT*-53
[14] Dennis B, Egorov I, Han Z, Dulikravich G and Poloni C 2000 *AIAA paper* **2000** 4876
[15] Bellucci J, Rubechini F, Arnone A, Arcangeli L, Maceli N and Dossena V 2012 *ASME GT*2012-69529
[16] Cotroneo J, Cole T and Hofer D 2007 *ASME GT*2007-27315
[17] Rubechini F, Schneider A, Arnone A, Cecchi S and Malavasi F 2012 *ASME GT*2009-60083
[18] Hasenjäger M, Sendhoff B, Sonoda T and Arima T 2005 *ASME Paper GT*2005-68680
[19] Öksüz O and Akmandor I 2010 *J. Turbomach.* **132** 041009
[20] Song L, Luo C, Li J and Feng Z 2012 *P. I. Mech. Eng., A. – J. Pow.* **226** 262-76
[21] Li H, Song L, Li Y and Feng Z 2009 *J. Turbomach.* **133** 031014
[22] Menter F, Kuntz M and Langtry R 2003 *Proceedings of the 4th International Symposium on Turbulence, Heat and Mass Transfer* **4** 625-32
[23] Menter F, Langtry R, Likki S, Suzen Y, Huang P and Völker S 2004 *J. Turbomach.* **128** 413-22
[24] Langtry R and Menter F 2005 *AIAA Paper* **2005** 522
[25] Denton J 1993 *J. Turbomach.* **115** 621–56

Study on the Aerodynamic Performance and Internal-External Flow Interaction Effects of Distributed Ducted Fan in the Climbing-Condition

Tian Yu Gong[1], Peng Xiang Liang[1], Zhuo Qi Wang[2] and Wei Yuan[1, *1]

[1]National Key Laboratory of Science and Technology on Aero-Engine Aero-thermodynamics, School of Energy and Power Engineering, Beihang University, Beijing 100191, China

[2] Unmanned System Institute, Beihang University, Beijing, 100191, PR China

E-mail: yuanwei@buaa.edu.cn

Abstract. Distributed electric propulsion ducted fans play an important role in saving energy and reducing environmental pollution of the airplanes. In this research, unsteady RANS CFD numerical simulations were used to analyse the possible impact of aerodynamic performance and interaction effects of distributed ducted fan in the climbing condition, which is verified with experimental method. Firstly, this paper will explore the effect of internal/external flow interaction on the flow field of the propulsion system at the same time. Secondly, from the perspective of propulsion efficiency and energy consumption, this paper will find the thrust relationship between the maximum continuous working state and the maximum working state and study the performance of the ducted fan. This paper will aim to reduce energy consumption during the propulsion of ducted fans and obtain longer flight time, and find a balance point between power and propulsion efficiency. The results show that the maximum continuous working state is 85% of the maximum thrust state approximately. By combining the aerodynamic interaction phenomena with the aerodynamic performance of the distributed ducted fan, this research provides an important basis and advice for enhancing the flight time of distributed ducted fans and reducing energy consumption during the process of engineering and flow mechanisms.

1. Introduction

Social development promotes requirements on the safety of civil aviation propulsion systems and environmental protection and prompts scholars to carry out more research [1-2]. Some researchers have conducted internal flow field analysis in the field of axial flow propulsion systems. Bryce's research results suggest that the increase in blade load will promote the separation in blades' end wall and boundary layer, and the performance will deteriorate [3]. Z Q Wang et al conducted experiments in a wind tunnel, the results showed that curved blades can effectively control the formation of flow channel vortices [4]. Alexej studied the effect of the tip clearance of the axial flow fan on the tip clearance flow through large eddy simulation method. The results showed that the size of the blade tip will affect the

*[1] Corresponding Author

size and shape of the blade tip leakage vortex [5]. Zhen studied the effect of axial clearance on radial flow. The results showed that the radial flow in the moving blade junction area gradually develops toward the blade tip as the flow rate decreases [6]. De-lei Chen studied the effect of setting angle on the performance of axial fans. The results show that at low flow rates, blade setting angle of the fan can be adjusted properly and the efficiency of the fan can be improved, and the appropriate blade setting angle can effectively improve the flow distribution in the fan, reduce flow loss, and delay the fan stall [7]. Hai Xin Chen. et al use genetic algorithm to perform multi-objective optimization designing on rotor blades. The research results show that the optimized propulsion performance can be improved very well [8]. Yang Wei Liu and Li Peng Lu et al. used the DDES method to study the flow structure of the blade tip leakage vortex at different blade tip sizes. The results of the study show that the flow field of small tip clearance is more stable and the anisotropy of the flow vortex is weaker [9]. However, the traditional flight / airplane layout has been limited in reducing fuel consumption and reducing pollutant emissions. There is an urgent need to find an alternative propulsion system to meet the increasingly stringent development requirements [14]. The advantages of circle ducted fan performance at lower incoming Mach number are more obvious[10-11]. Hubbard pointed out in early research that circle ducted fans can increase the safety of the propulsion system and reduce the noise impact of the propulsion system, more and more ducted fans are used in aircraft propulsion systems [12].

In recent years, the proposed 'Green Aviation' initiative promotes the motivation to reduce the environmental impact of pollutants during aviation flight, which has enhanced researchers in-depth research on distributed electric propulsion systems. Distributed ducted fans are gradually paid attention by researchers because of their advantages of reducing energy consumption and pollutant emissions [14]. A number of small or miniature engines installed on the wing or fuselage replace the current 2 to 4 traditional engines. Fuel consumption and polluting emissions can be reduced effectively with reasonable distribution of thrust, more advantages in aerodynamic stability can also be achieved [13-14].In order to further explore the effect of internal and external flow coupling effects on the aerodynamic performance of distributed axial ducted fans, some scholars have begun to study the aerodynamic performance of distributed ducted fan power systems. At present, in order to avoid the huge amount of calculation in the integrated numerical calculation process of internal and external flow coupling, the US GLENN research center has developed Chima volume force model for integrated numerical calculation, which is based on a fan with a diameter of 300 mm [15]. Mantic.lugo simulates the coupling of internal and external flow by setting the inlet and outlet boundary conditions to explore the interaction effect [16]. Kim. H. and Gong .Y. replaced the rotor blades and stator vanes by a circumferential volume force model [17-18]. Cui Rong[19]Calculated and analysed a distributed propulsion system using the integrated flight / engine numerical simulation method based on the volume force model along the streamline. Although using volume force model saves a lot of computing resources, it does not reflect the effect of internal and external coupling on ducted fans. Due to the coupling effect of internal and external flow, there is a certain degree of radial inlet total pressure distortion in the distributed ducted fan studied in this paper. Hamed. A conducted experiment to study the effect of distortion location on the ground test of the propulsion system and discussed the effects of total pressure distortion, vortex and turbulence on the flow [20]. Rodriguez J. designed a new type of compressor lip to improve the flow stability. The study also showed that the flow distortion on the surface of the propulsion system mainly originates from the flow separation on the outer wall of the intake system, the internal intake of the propulsion system and the effect on the rotating vortex [21].

In general, most of the previous research works have focused on exploring the internal flow field of the axial flow propulsion system and using the volume force model to study the internal and external flow coupling between the aircraft and propulsion system. Few comprehensive considerations have been given to the coupled internal and external flows of distributed ducted fans. Little attention has been paid on the distributed ducted fan performance under the coupling effect of internal and external flow and the details of the flow field. For the distributed electric ducted fans studied in this paper, on the one hand it is required to reduce energy consumption and get longer flight time, on the other hand, it is expected to work with high propulsion efficiency. The objective of this research is to analyse aerodynamic performance of the maximum propulsion efficiency and maximum working state, and then understand the aerodynamic performance difference between them. The unsteady RANS computational fluid dynamics numerical simulation were conducted in order to acquire the internal and external flow interaction field. Firstly, by contrasting the numerical results with experimental results, the unsteady RANS numerical simulation results were evaluated preliminary, which can provide a reference for the unsteady RANS numerical simulation results. Secondly, the unsteady flow field was analysed to provide a clear understand of the internal and external flow interaction aerodynamic performance. Finally, the full passage distributed ducted fan is considered in specific working conditions, numerical simulations were used to research the performance.

2. Computational setup

In this section, the geometry characteristic of the distributed ducted fan will be introduced and the numerical simulation methods will be described. Section A will present the geometries that is employed in this study, section B will describe the solver setup and discuss the uncertainty quantification and experimental validation, section C will present the domain and boundary conditions that used in this study.

2.1. Geometry

The current research object is a distributed ducted fan similar to that shown in 'Figure 1[22]'. The ducted fan is not embedded in the upper surface of the wing. it is mounted above the wing near the trailing edge.

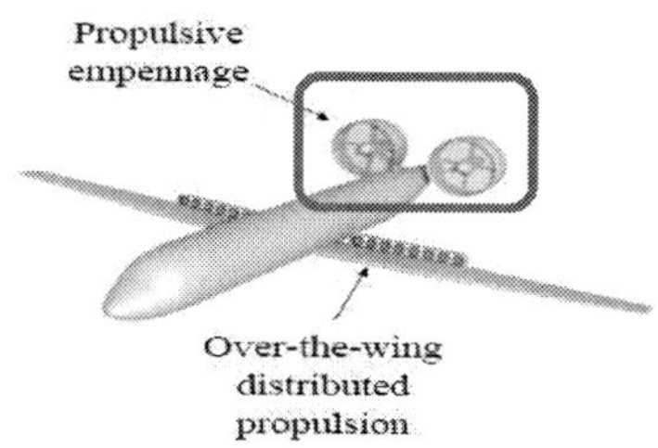

Figure 1: Schematic of Distributed Ducted Fan[22].

The diameter of the ducted fan is 240mm, which is expressed by D. The number of vane, rotor blade and stator vane is three, eight and seven respectively. The distributed ducted fan rotates in counter-clockwise direction. The details of the distributed ducted fan are given in table 1.

Table 1 Parameter of ducted fans.

Parameter	Value
Blade tip speed (m/s)	115
Flight speed(m /s)	10-30
Rotating speed(rpm)	0-9100
Thrust range(N)	0-80

2.2 Solver Setup and Discuss the Uncertainty Quantification

The unsteady RANS equations were solved by using ANSYS, namely, CFD solver CFX. The compressibility effects have not taken into account, for the inlet wind velocity was 0.1Ma. The fluid was ideal gas. In the calculation and solution process, second order backward transient scheme is used, and the time is set to automatic.

2.3 Domain and Boundary Conditions

The domain for the numerical simulation of the distributed ducted fan is presented in 'Figure 3'. The flow direction, total pressure and total temperature were specified at the inlet. In the far-field boundary, opening boundary was prescribed, apart from velocity, static temperature is applied in the opening boundary. At the outlet, the static pressure was prescribed. The unsteady simulation physical step is 50, the loop is 5. The grids were generated with *ANSYS ICEM*、*NUMECA* (Auto-grid and IGG) software. Both structured mesh and unstructured mesh were used in the domain. Furthermore, Turbulence has been modelled with the standard two equations k-ε model. With the full blade model, the unsteady numerical simulations were performed. Near the no slip walls, a scalable wall model has to be employed, the first layer height is specified in order to achieve in a maximum y+ lower than 50, more than 30. 'Figure 3' shows the schematic of grid computing domain for internal/external flow of ducted fans, the interface in the figure is set as the transient rotor stator interface. Transient rotor stator interface model should be used, which is important to account for transient interaction effects at a sliding interface and can predicts the true transient interaction effects of the flow between a stator and rotor passage. The interface between the internal and external flow field can be connected "General". According to the software CFX. A general connection can be applied to fully transient sliding interfaces between domains. Assuming that the diameter of the ducted fan is D, the upstream distance is 15D, the downstream distance is 15D, and the far field distance is 15D. 'Figure 2'is based on the grid-independent verification of the distributed ducted fan thrust coefficient. Based on the expected error on thrust coefficient when choosing the suitable value, and CPU computing power, the total number of grids is about 8.5 million, considering the number of calculation grids and calculation accuracy.

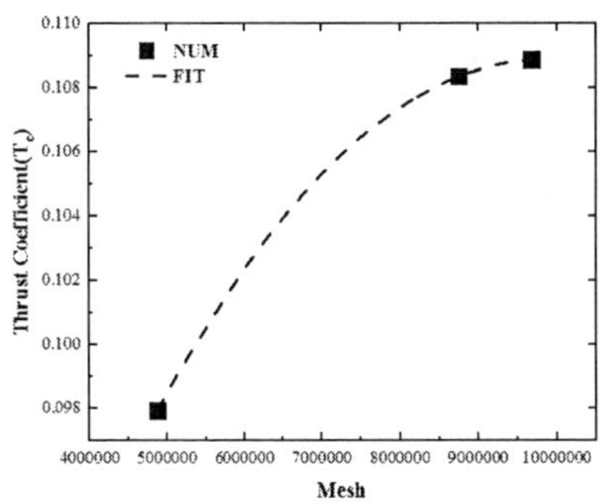

Figure 2. Grid independence.

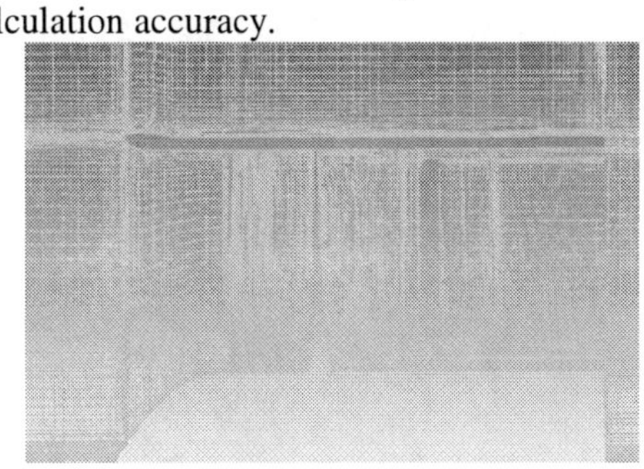

Figure 3. Schematic of grid computing domain.

3. Experimental methods

3.1. A Wind Tunnel Model and Data Acquisition Device

Experimental wind tunnel testing of the distributed ducted fan model was conducted in a 15m×6.85m subsonic speed, opened-environment, low-turbulence wind tunnel, which is located in the Beijing 35 Middle School, as shown in the 'Figure 4'. The wind tunnel is equipped with data acquisition device, including Pitot tubes, model angle of attack, sideslip angle mechanism control system, six-component strain balance and signal conditioning system, digital DC speed control system, etc. The maximum wind speed which is able to be measured is 55m/s. All experimental data recorded in this paper is based on blade diameter, the Reynolds number is 5E5 approximately.

The test wind tunnel contains the test section, diffusion section and other parts of the structure. This test is conducted in the test section. The test equipment used in the test process is shown in the 'Figure 5', which mainly includes the distributed full-scale test model and measuring elements of the distributed ducted fan. The distributed full-scale test model of the distributed ducted fan is installed on the six-component balance through the installation hole and the fixing belt. The full-scale test model of distributed ducted fan consists of fan, motor, electronic speed controller, cooling channel and photoelectric sensor. The motor drives the fan to provide power, and the electronic speed controller controls the fan rotating speed.

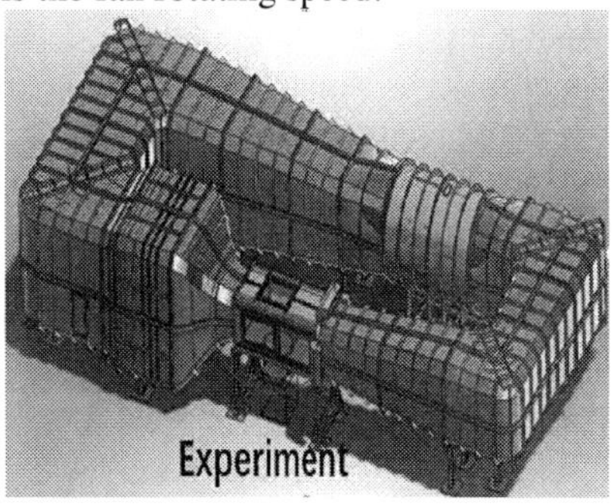

Figure 4. Schematic of Low-Speed Wind Tunnel. Figure 5. Schematic of Test Layout.

In order to study the aerodynamic performance of the distributed ducted fan propulsion system and the influence of design parameters on the aerodynamic performance through experiments, the ducted fan was tested. The experiment results are compared with the numerical simulation results to verify the reliability of the numerical simulation results, and analyze the numerical results of the flow field and explore the effect of internal and external flow coupling on the aerodynamic performance of distributed ducted fans and flow mechanism.

Table 2 Wind tunnel test conditions.

Parameter	Value
Inlet flow speed (m/s)	5-55
Rotating speed (rpm)	0-9100
Height (m)	0-3
Temperature (K)	292
Pressure (Pa)	100200

4. Analysis and discussion

4.1. A The Validation

The numerical simulation results were compared with the experimental data. That qualitative comparison is developed to increase the reliability in the simulation results in the ‘Figure 6’.

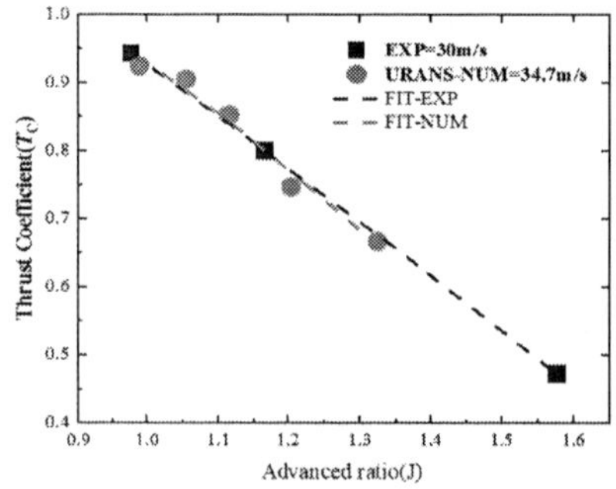

Figure 6. Advanced Ratio-Thrust Coefficient.

The atmospheric pressure measured by the device is 100200Pa, and the atmospheric temperature is 292K. The horizontal measurement device maintains the axial intake of the ducted fan at 0 ° angle of attack. During the test, different inflow velocity V_{in} and corresponding rotating speeds were obtained to obtain different advanced ratio J. In general,regarding of the ducted fan performance, there is an error of less than 5% between the calculated results shown in the figure and the results obtained from the experiment based on the thrust coefficient. The difference may be results from the number of grids, the selection of turbulence models, and the interface and calculation method and so on.The stability of the measurement during the experiment and the reading of the data of the sensor's measuring point, including post-processing, will lead to the appeal result.

4.2. Aerodynamic Coupling Performance and Interaction Phenomena

In this section, we firstly analyse the performance of the electric propulsion distributed ducted fan causes of different working conditions, afterwards, we analyse the internal and external flow coupling flow field in the maximum working state and the longest continuous working state. Finally, based on the above results, we further analyse the aerodynamic performance from the fluid flow contour. We have made a regulation that the electric propulsion distributed ducted fan is at its maximum working state when it works at Rated rotating speed.

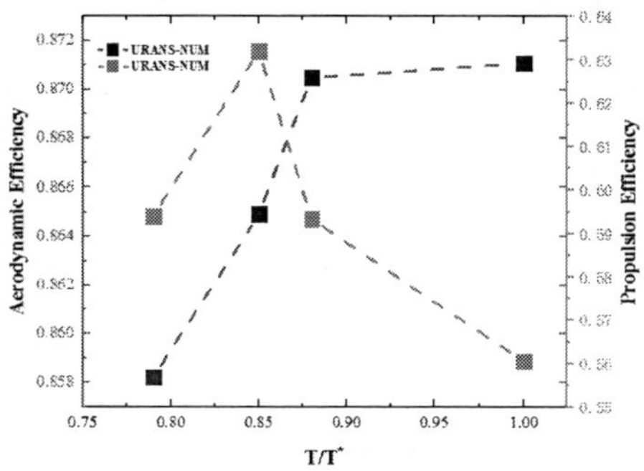

Figure 7 Ratio of Thrust- Aerodynamic Efficiency/ Propulsion Efficiency.

'Figure 7'and 'Figure 8' will analyse the effects of different rotating speed and internal/external flow coupling effects on the aerodynamic efficiency and propulsion efficiency of electric propulsion distributed ducted fans at the same flight velocity. This paper defines a new dimensionless parameter, which is expressed to the ratio of thrust between working state corresponding to random rotating speed and the maximum working state (T/T*) in flight.

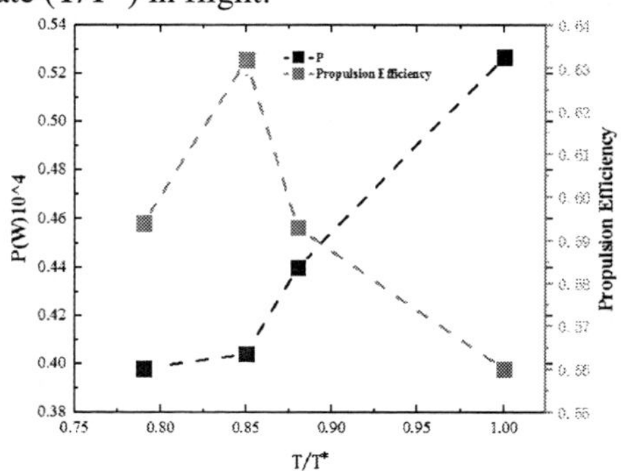

Figure 8 Ratio of Thrust-P /Propulsion Efficiency.

The ratio of thrust represents the magnitude of the thrust value when distributed ducted fan rotating speed changes. In this paper, firstly: as shown in 'Figure 7', as the ratio of thrust increases, aerodynamic efficiency increases. Secondly: as shown in 'Figure 8', as the ratio of thrust decreases, the electric propulsion distributed ducted fan rotating speed decreases, the propulsion efficiency first increases, and then decreases. For the research object of this paper, there is a ratio of thrust corresponding to the optimal propulsion efficiency. Regarding of the working state, the maximum aerodynamic efficiency is not consistent with the maximum propulsion efficiency. Finally, as shown in 'Figure 8', as the ratio of thrust increases, the power consumption also increases. For the distributed electric propulsion ducted fan studied in this paper: This paper chooses the longest continuous working state from following three reasons: low power consumption, high propulsion efficiency and high aerodynamic efficiency. Based on the above three reasons, for the electric propulsion distributed ducted fan that researched in this paper, the working state of the highest efficiency point is the longest continuous working state discussed in this paper, and the ratio of thrust of the longest continuous working state is 85%. At this time, the propulsion efficiency is the highest, and the energy consumption and aerodynamic efficiency meet the requirements that defined earlier in this paper.

4.3. An Aerodynamic Efficiency and Power Consumption

Distributed ducted fan works near the design point, the change influence of unsteady flow fields between the highest propulsion efficiency point and maximum energy consumption point is small, therefore, one of the same moments is selected for analysis and discussion. The internal flow of the rotor blade has a significant influence on the aerodynamic efficiency. Overall, the aerodynamic efficiency of the maximum energy consumption point in 'Figure 7'is higher than the aerodynamic efficiency of the longest continuous working state. In the following discussion, the flow mechanism of different working states on aerodynamic efficiency will be analysed in detail from the rotor blade flow field and angle of attack.

'Figure 9'is an entropy contour of rotor blades at 90% of the blade height about different working states. 'Figure 10'is an entropy contour of rotor blades at 10% of the blade height in different working states. At the rotor tip, the entropy in the longest continuous working state is much larger than the entropy area in the maximum working state; The entropy distribution of the blade root is not significantly different. Generally speaking, the high entropy of the longest continuous working state is greater than that of the maximum working state, hence, the aerodynamic efficiency of the maximum working state is higher than that of the longest continuous working state.

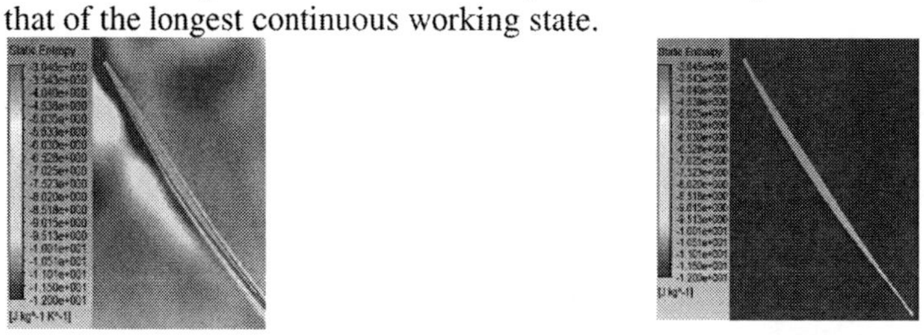

The Highest Propulsion Efficiency Point Maximum Energy Consumption Point

Figure 9 Rotor Tip Entropy Distribution.

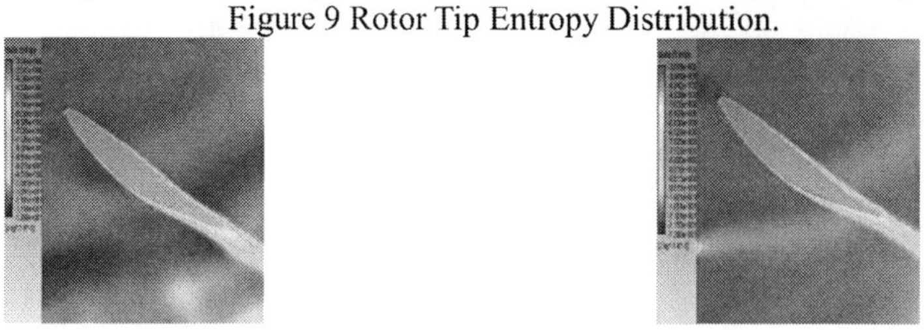

The Highest Propulsion Efficiency Point Maximum Energy Consumption Point

Figure 10 Rotor Root Entropy Distribution.

'Figure 11'and 'Figure 12' show the distribution of the inlet airflow angle and geometry inlet angle and outlet airflow angle along the blade span from the rotor blade root to the tip (the airflow is circumference averaged in the blade passage, and the relative inlet airflow angle is defined as arctan(Wz/Wt) *180/π, where Wz is the axial velocity of the inlet and Wt is the tangential component of the relative velocity). It can be found from 'Figure 11' that by comparing the inlet flow angle distribution of the rotor blade root to the rotor blade tip, the airflow angle in the longest continuous working state is smaller than that in the maximum working state. 'Figure 12'shows outlet flow angle at the rotor blade has little difference under two different working state. This shows that the radial distortion of the blade tip and blade root in this paper has the same degree of influence on the under different rotating speeds.

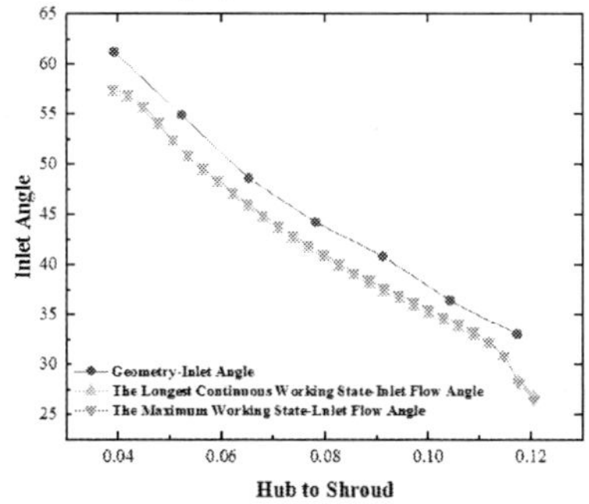

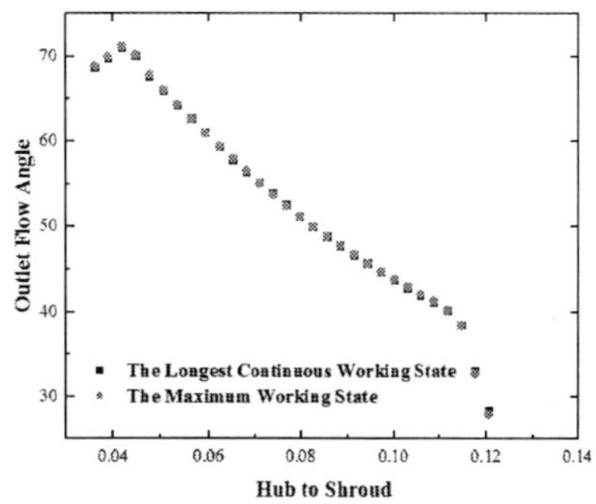

Figure 11:Rotor Inlet Angle. Figure 12:Rotor Outlet Flow Angle.

In the 'Figure 13', C_{1i} and W_{1i} (i=1,2) express the absolute and relative inlet velocity of the rotor blade in the longest continuous working state and the maximum working state, respectively, C_{2i} and W_{2i} (i=1,2) express the absolute and relative outlet velocity of the rotor blade in the longest continuous working state and the maximum working state, respectively, U_i (i=1,2) represents the longest continuous working state and the maximum working state rotor blade circumferential rotating speed, respectively. Make a regulation: The angle between C and U is α, and the angle between W and U is ß.

Combining 'Figure 11' and 'Figure 13' to analyze the mechanism of aerodynamic efficiency. When the electric propulsion distributed ducted fan works at different rotating speeds and the same flight inlet velocity, the ducted fan rotating speed increases, in the same time, the relative airflow angle increases. As the fan rotating speed increases, the rotor leading edge relative airflow angle is closer to the design geometric airflow angle, the angle of attack is smaller. In this paper, the electric propulsion distributed ducted fan works at different conditions, rotor inlet axial velocity increases faster than rotor inlet rotating speed. The angle of attack in the maximum working state is smaller than that in the longest continuous working state. And the aerodynamic efficiency in the maximum working state is higher than that in the longest continuous working state.

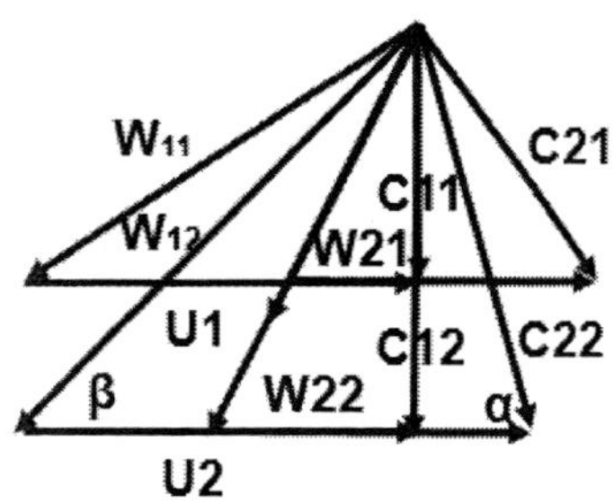

Figure 13 Velocity Triangle Distribution.

IOP Conf. Series: Materials Science and Engineering 1081 (2021) 012027 doi:10.1088/1757-899X/1081/1/012027

4.4. Propulsion Efficiency

The external resistance has a significant effect on the effective propulsion power. This paper will discuss the influencing factors of propulsion efficiency from two aspects: differential pressure resistance, friction resistance.

First, the streamline and the static pressure of the lip distribution are discussed to analysis the lip flow field of the electric propulsion distributed ducted fan. 'Figure 14'shows the streamline of the lip. As can be seen from 'Figure 14', the density of red streamlines at the highest efficiency point is fewer than that at the maximum working point. For the electric propulsion ducted fan studied in this paper, under the conditions of the same inlet wind velocity and different rotating speeds, as shown in the 'Figure 14', the separate position of the lip streamline is basically the same, located at the upper of the lip wall surface, and the fluid flows through the surface of the lip, the actual fluid flow into the ducted fan is more than the ideally defined fluid flow. When the rotating speed is higher, the fluid ingests into the ducted fan faster. By comparing 'Figure 15(a)' with 'Figure 15(b)', the airflow accelerates on the lip surface and the static pressure also decreases. The static pressure of the front of the lip on the highest efficiency point is greater than in the maximum working state, the maximum working state produces an obvious semi-circular low static pressure area on the lip.

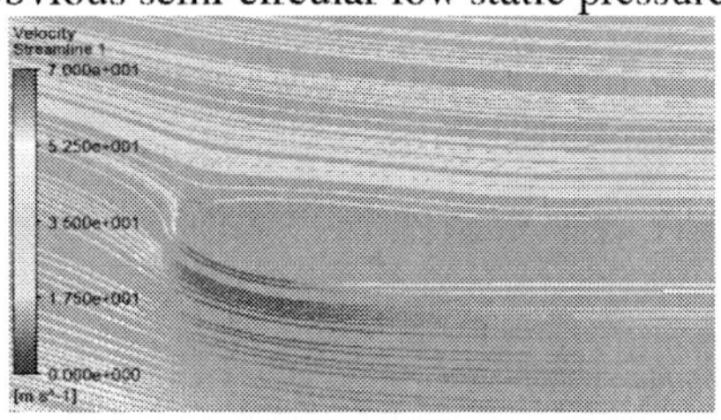

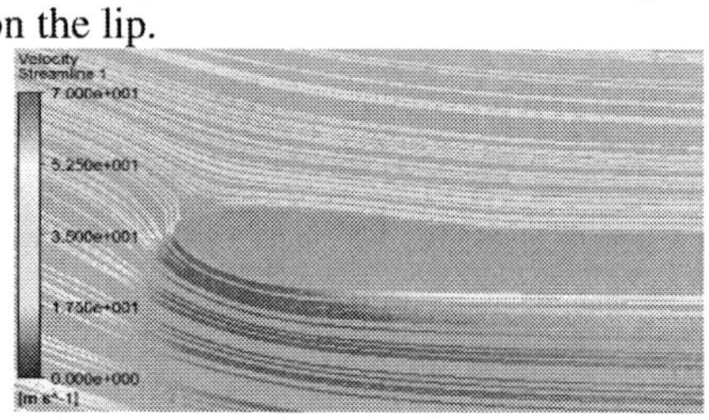

a. The Highest Propulsion Efficiency b. Maximum Energy Consumption

Figure 14. Axial Streamlines Distribution of Lip.

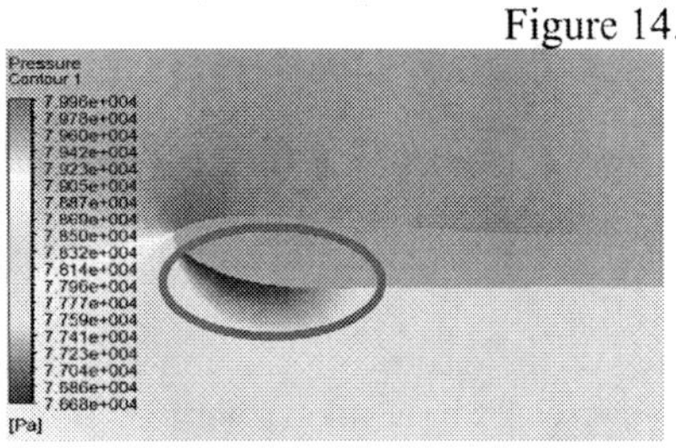

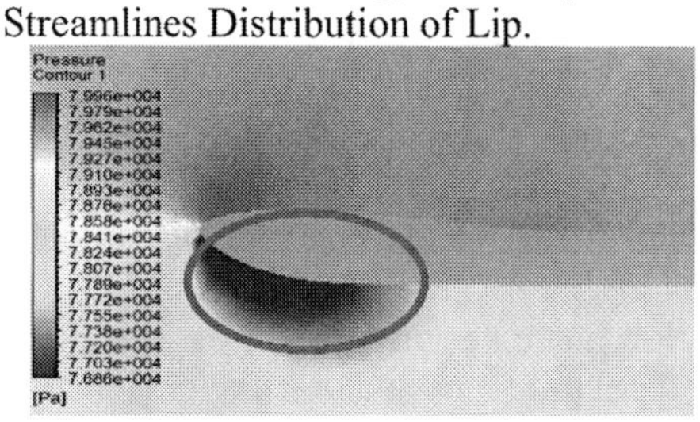

a. The Highest Propulsion Efficiency b. Maximum Energy Consumption

Figure 15. Pressure Distribution of Lip.

The following part will analyse the propulsion efficiency of the electric propulsion distributed ducted fan at the same flight velocity and different fan rotating speeds. For the research object in this paper, the location where the geometric curvature has obvious change is the lip wall of the ducted fan. We define the lip geometry as follows: as shown in 'Figure 16' Lip geometric BAC surface is the main part of lip, which is researched on the mechanism of influence on propulsion efficiency in the subsequent research and discussion.

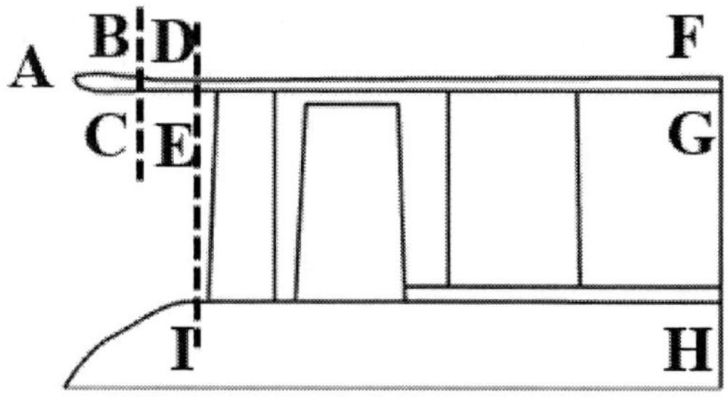

Figure 16 Geometric Schematic of Electric Propulsion Distributed Ducted Fan.

In the following discussion, two dimensionless parameters defined in this paper will be introduced, namely, the dimensionless static pressure coefficient, the axial velocity gradient coefficient. The reason for choosing the two dimensionless parameters is that for the electric propulsion distributed ducted fan studied in this paper, the inlet flow velocity is the same, the ducted fan works at different rotating speeds, in order to eliminate the interaction effects on different rotating speeds.

$$C_p = \frac{P}{0.5 * \rho * U * U} \tag{1}$$

$$C_{VG} = \frac{VG}{U} \tag{2}$$

'Figure 17(a)' shows the static pressure distribution along the axial direction of the lip BAC wall surface. In "Figure 17(b)" shows wall static pressure with dimensionless, 'as in equation (1)'. In the upper wall surface AB, regarding of the wall static pressure, there is no difference between the maximum working state and the longest continuous working state. In the lower wall AC, the static pressure of the wall in the maximum working state is lower than that of the longest continuous working state. For the lip wall BAC, the resistance of the wall pressure difference in the maximum working state is greater than that of the longest continuous working state. The 'Figure 17(b)'is the dimensionless wall static pressure coefficient distribution. Through numerical integration, the area surrounded by the black line area is 2.33, and the area of the red line area is 2.40, The pressure resistance coefficient of the longest continuous working state is smaller than that of the maximum working state.

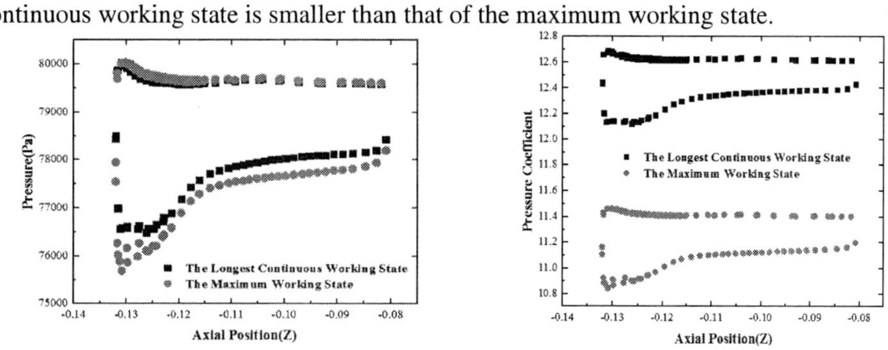

a: Static Pressure Distribution

b: Dimensionless Static Pressure Coefficient Distribution

Figure 17 Lip Axial Static Pressure.

'Figure 18(a)'is the velocity gradient distribution of the lip BAC wall surface in the axial direction. In "Figure 18(b)" shows wall velocity gradient with dimensionless, 'as in equation (2)'. In the upper

IOP Conf. Series: Materials Science and Engineering 1081 (2021) 012027 doi:10.1088/1757-899X/1081/1/012027

wall surface AB, the wall velocity gradient in the maximum working state is greater than that in the longest continuous working state. In the lower wall surface AC, the wall velocity gradient in the maximum working state is similar to the longest continuous working state.

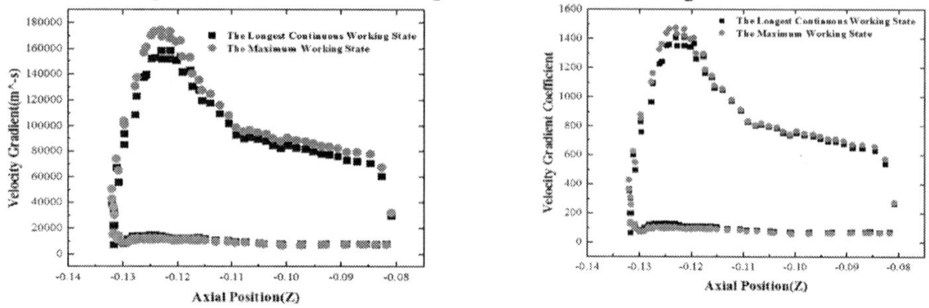

a: Velocity Gradient Distribution b: Dimensionless Velocity Gradient Coefficient Distribution

Figure 18 Lip Axial Velocity Gradient.

For the lip wall surface BAC, the wall surface friction resistance in the maximum working state is greater than the wall surface friction resistance in the longest continuous working state.

'Figure 19' is an entropy contour. The airflow does not produce an obvious high entropy area on the upper wall of the lip, while the airflow produces a relative high entropy area on the lower wall of the lip. The maximum entropy loss in the maximum working state is higher than in the longest continuous working state. The flight velocity remains unchanged, and the rotating speed of the electric propulsion distributed ducted fan changes. The fluid on the wall of the lip separates, resulting in frictional resistance. At the same time, the subsonic fluid decelerates and increase pressure on the concave surface of the lip, and the pressure on the convex surface accelerates to reduce pressure, resulting in differential pressure resistance.

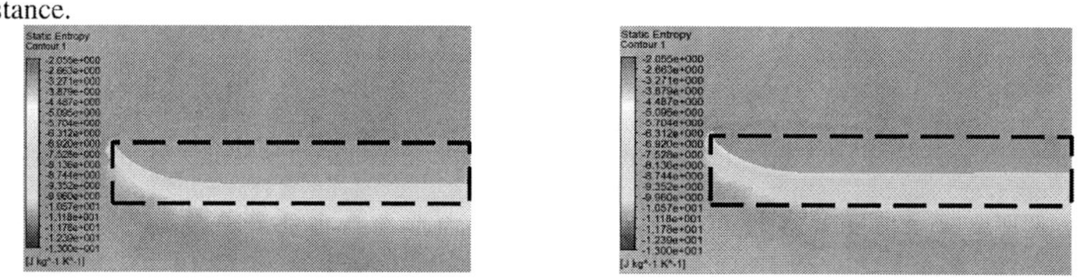

a. The Highest Propulsion Efficiency b. Maximum Energy Consumption

Figure 19 Static Entropy Distribution of Lip.

Differential pressure resistance and fluid friction resistance together affect the lip loss of the ducted fan. The greater the friction resistance and differential pressure resistance are, the lower the propulsion efficiency is.

As we all know: propulsion efficiency has a significant impact on the aircraft. This paper analyses the factors affecting the propulsion efficiency of electric propulsion distributed ducted fans to exclude the impact of different rotating speed. Including friction drag coefficient, differential pressure drag coefficient and have an impact on propulsion efficiency. In the future, from the perspective of reducing frictional resistance and differential pressure resistance, the researches should be taken measures to reduce the frictional resistance and differential pressure resistance of the electric propulsion distributed ducted fan, and in the same time, the researches should be taken measures to reduce power consumption, to expand the peak range of propulsion efficiency and increase the low-loss range of angle of attack. The researchers can optimize the lip profile, reducing frictional resistance and differential pressure resistance is an effective way to achieve the above objectives.

IOP Conf. Series: Materials Science and Engineering 1081 (2021) 012027 doi:10.1088/1757-899X/1081/1/012027

Conclusions and outlook

As the ratio of thrust increases, the aerodynamic efficiency increases, and the propulsion efficiency increases first and then decreases. For the research object of this paper, there is a best ratio of thrust corresponding to the maximum propulsion efficiency. As the ratio of thrust increases, the power consumption also increases. The ratio of thrust in the longest continuous working state is 85%, and the propulsion efficiency at this time is the highest.

The main factor affecting aerodynamic efficiency is the airflow angle of attack. When the rotating speed of the electric propulsion distributed ducted fan increases, the relative inlet angle of the airflow increases, which is closer to the design geometric inlet airflow angle, the airflow inlet angle of attack decreases, and the aerodynamic efficiency increases.

The determination of the propulsion efficiency are the values of frictional resistance, differential pressure resistance. The differential pressure resistance, friction resistance in the maximum working state are greater than in the longest continuous working state. Differential pressure resistance, fluid friction resistance together affects the ducted fan lip loss. The key step to improve propulsion efficiency is to reduce the influence of resistance. The researchers should be taken measures to expand the peak range of propulsion efficiency and increase the low-loss range of attack angle. The researchers can optimize the lip profile, reducing frictional resistance and differential pressure resistance is an effective way to achieve the above objectives.

References

[1] Greitzer E M et al 2010 N+3 Aircraft Concept Designs and Trade Studies *NASA/CR-2010-216794* Glenn Research Center
[2] Smith L H 1993 Wake Ingestion Propulsion Benefit *J. Propul. Power* **9** 74-82
[3] Bryce J D, Cherrett M A and Lyes P A 1995 Three-Dimensional Flow in a Highly Loaded Single-Stage Transonic Fan *J. Turbomach* **117** 22-28
[4] Shang E and Z Q Wang 1993 The Experimental Investigations on the Compressor Cascades with Leaned and Curved Blade *ASME International Gas Turbine and Aeroengine Congress and Exposition*
[5] Alexej P, Matthias M and Wolfgang S 2018 Large-Eddy Simulation of the Tip-Leakage Flow in an Axial Fan *Chinese J. Turbomach* **60**
[6] Zhen Y, Tao P, Bin J and Qun Z 2018 Effect of Axial Clearance on Radial Flow in Axial Compressor Near to Stall Condition *Chinese J. Turbomach*
[7] Delei C, Huashu D, Hantao M, Jiong Z, Zhenhai L 2019 Effect of Front Blade Mounting Angle on Performance of Counter-rotating Axial Flow Fan *Chinese J. Turbomach*
[8] Yincheang N and Haixin C 2019 Rapid Aerodynamic Design of Prop-rotor Blade with Optimization *Chinese J. Turbomach* **61**
[9] Luyang Z, Yangwei L and Lipeng L 2019 Unsteady Behavior of Tip Leakage Vortex in an Axial Compressor with Different Rotor Tip-gap Sizes Using DDES *Chinese J. Turbomach* **61**
[10] Black D M, Wainauski H S and Rohrbach C 1968 Shrouded Propellers A Comprehensive Performance Study *AIAA 5th Annual Meeting and Technical Display*
[11] Bogdanski K, Krusz W and Rodzwicz M 2014 Design and Optimization of Low Speed Ducted Fan for a New Generation of Joined Wing Aircraft *29th Congress of Interational Council of the Aeronautical Sciences (Saint Peterburg)*
[12] Hubbard H H 1950 Sound Measurements for Five Shrouded Propellers at Static Conditions *Tech NACA-TN-2024*
[13] Sehra A K and Whitlow W 2004 Propulsion and Power for 21st Century Aviation *Prog. Aerosp. Sci* **40** 199-235
[14] Amir S G, Georgios D and Riti S 2011 Challenges of Future Aircraft Propulsion: A Review of Propulsion Technology and Its Potential Application for the All-Electric Commercial Aircraft *Prog. Aerosp. Sci* **47** 369-391

[15] Chima R V, Arend D J and Castner R S 2010 CFD Models of a Serpentine Inlet, Fan and Nozzle *AIAA*
[16] Mantic L, Doulgeris G and Gohhardani A 2012 Computational Analysis of Aerodynamic Performance of a Jet-flap Airfoil in Transonic Flow *J. Aerosp. Eng* **226** 664-678.
[17] KIM H and LIOU M S 2015 Mail-slot Nacelle Shape Design for N3-X Hybrid Wing-body Configuration *AIAA*
[18] Gong Y and Greitzer E M 1999 A Computational Model for Short Wavelength Stall Inception and Development in Multistage Compressors *J. Turbomach*
[19] Cui R, Li Qiushi and PAN Tianyu 2016 Streamwise-body-force-model for Rapid Simulation Combining Internal and External Flow Field *Chinese J. Aeronaut* **29** 1205-12
[20] Hamed A and Numbers K 1997 Inlet Distortion Considerations for High Cycle Fatigue in Gas Turbine Engine *Joint Propulsion Conference and Exhibit*
[21] Rodriguez J, Klumpp S and Biesinger T 2013 A New Inlet Distortion and Pressure Loss Based Design of an Intake System for Stationary Gas Turbines *ASME Turbo Expo: Turbine Technical Conference and Exposition*
[22] Hugo F M, Bento, Reynard de V, Leo L M and Veldhuis 2020 Aerodynamic Performance and Interaction Effects of Circular and Square Ducted Propellers *AIAA SciTech Forum (Orlando)*

Acknowledgments
The authors are grateful to the Beijing 35 Middle School sincerely, which provides low-speed wind tunnel experimental equipment. The authors also thank Yang Dong for providing effective discussions in Beihang University.

IOP Conf. Series: Materials Science and Engineering 1081 (2021) 012028 doi:10.1088/1757-899X/1081/1/012028

Variation Law and Experimental Study on Multi-stage Critical Speed of an Aero-engine Power Turbine Rotor

Weijian Nie[1,2], Wangqun Deng[1,2] and Yanong Chen[1,2]

[1] AECC Hunan Aviation Powerplant Research Institute, Zhuzhou 412002, China
[2] Key Laboratory of Aero engine Vibration Technology, Aero Engine Corporation of China, Zhuzhou 412002, China

E-mail: nwjxj@126.com

Abstract. Finite element method was used to establish the calculation model of the dynamic characteristics of the simulated power turbine rotor, and the dynamic characteristics of the rotor were systematically calculated and analysed. The variation law of the first three stages critical speeds of the rotor with the supporting stiffness and the mass of each disk were revealed. The dynamic characteristics test of the simulated power turbine rotor in the full speed range was completed, and the calculation results and test results were compared and analysed. The results show that the calculation model reflects the dynamic characteristics of the rotor well, and the rationality of the rotor structure design, supporting layout and critical speed distribution were verified. The study provides a theoretical basis for the adjustment of critical speed based on the supporting stiffness and disk's mass, it also provides technical reference for the real rotor structure design and test.

1. Introduction

Rotor is an important rotating part of turbine machineries such as motors, pumps, compressors, gas turbines and aero-engines. It is of great significance to study rotor dynamics. Many scholars like Zhou [1], He[2], Zhang[3] , Tai[4], Du[5] and other researcher[6-9] have studied the dynamics of turbo machinery rotor and achieved notable results. Compared with the rotor of other turbo machinery, the rotor of aeroengine was characteristic of high speed, complex structure and bad working condition, there were great risks and difficulties in the dynamic research of the real rotor directly. Therefore, the dynamic analysis and research of the simulated rotor (connection structure, supporting layout, mass inertia, etc. were basically consistent with the real rotor) were usually carried out firstly. Deng [10,11] , Liu [12] and Wu [13], conducted dynamic analysis and experimental research on the simulated rotor of the aeroengine , which provided a reference for the structure design and test of the real rotor. Critical speed design is the core content of the rotor dynamics design of small and medium-sized aeroengine, which plays an important role in the development of small and medium-sized aeroengine. In order to meeting the margin requirements of the rotor critical speed design, it can be achieved by selecting appropriate supporting stiffness, changing the mass distribution, optimizing the rotor structure and other methods. Deng [14,20], Nie [15,16], Hong [17], Bai [18], Mei [19] and other scholars have carried out a systematic study on the influencing factors of the critical speed of the aeroengine, providing a theoretical basis for the adjustment of the critical speed. Whether the structure design, supporting layout and critical speed distribution of the rotor are reasonable or not needs to be verified by experiments.

IOP Conf. Series: Materials Science and Engineering 1081 (2021) 012028 doi:10.1088/1757-899X/1081/1/012028

In this paper, dynamic analysis and experimental research were carried out on the flexible rotor with complex structure of aeroengine. The dynamic characteristic calculation model of the simulated power turbine rotor was established by finite element method. The dynamic characteristic of the rotor was systematically calculated and analysed. The laws of the first three critical speeds of the rotor change with supporting stiffness and each disk's mass were revealed, and the dynamic characteristic in the full speed range was carried out. The rationality of rotor structure design and dynamic design was verified by experimental research, which laid a solid foundation for the dynamic research of real rotor.

2. Rotor Structure

The schematic diagram of the simulated power turbine rotor is shown in figure 1. The whole rotor was mainly composed of power turbine shaft, first stage simulated power turbine disk, second stage simulated power turbine disk and other parts. The two stages simulated power turbine disks were connected by end teeth, the first stage simulated power turbine disk and power turbine shaft were connected by splines for transmitting torque with cylindrical centering at both ends, and the second stage simulated power turbine disk and power turbine shaft were connected by circle cylinder interference fit. The length of the rotor's power turbine shaft was nearly 1.5 meters, and the hollow structure was adopted. The mass center position, mass and moment of inertia of the simulated power turbine disks were in good agreement with the real disks. There were four bearings in the rotor, among which the No.1 bearing was ball bearing, the No.2, No.6.5 and No.7 bearings were all roller bearing. The simulated power turbine rotor was a slender, hollow structure flexible rotor with large aspect ratio.

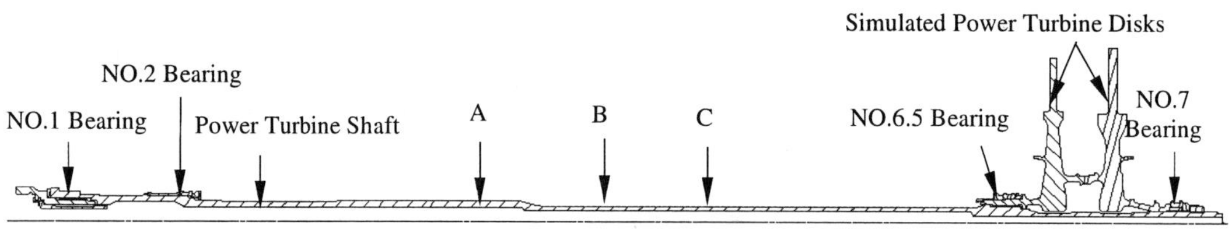

Figure 1. Structural Sketch of the Rotor.

3. Finite Element Analysis Model

3.1. Concentrated Mass

In order to facilitating the model, parts of the body of the two stages simulated power turbine disks were simulated with the concentrated mass, which characteristics are shown in table 1.

Table 1. Concentrated Mass Characteristics of Disks.

Parts	Mass (kg)	Polar Moment of Inertia (kg m^2)	Diameter Moment of Inertia (kg m^2)
Partial First Stage Simulated Power Turbine Disk	10.399	89169	177583
Partial Second Stage Simulated Power Turbine Disk	14.337	137028	272823

3.2. Supporting Stiffness

The supporting stiffness of the rotor is shown in table 2.

Table 2. Supporting Stiffness.

Supporting Number	No.1	No.2	No.6.5	No.7
Stiffness(E+7N/m)	5	1.78	3.5	0.0154

3.3. Finite Element Model

During the pre-processing of the model, the structure of the rotor has been simplified, and some small local structures such as chamfering and threaded holes have been neglected. The finite element software PATRAN was used to establish the model based on beam element, and then the analysis software SAMCEF/ROTOR was used to improve the calculation model (mainly including the establishment of concentrated mass elements and bearing elements). The model has 708 beam elements, 717 nodes, 2 concentrated mass elements and 4 bearing elements. The finite element calculation model of the rotor established is shown in figure 2.

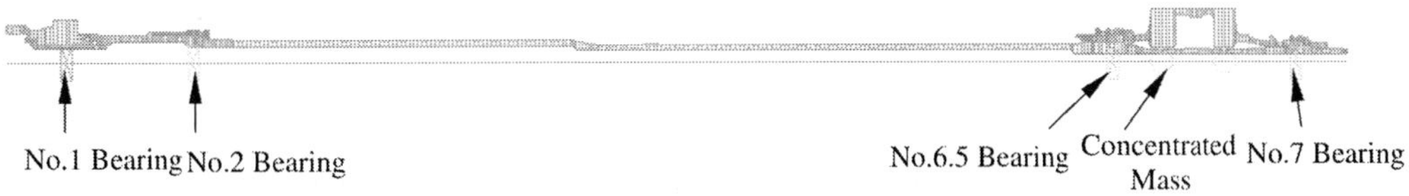

Figure 2. Finite Element Model of The Rotor.

4. Calculation Results of Dynamic Characteristics

4.1. Results of Critical Speeds and Mode Shapes

Under the condition of supporting stiffness in table 2, the critical speeds and mode shapes of the first three stages of the simulated power turbine rotor were calculated. The results of the first three stages critical speed and their margins are shown in table 3, and the first three stages mode shape are shown in figure 3.

Table 3. Calculation Results of the First Three Stages Critical Speed.

Stage	First	Second	Third
Critical Speed(rpm)	4730	9250	26533
Margin (%)	>35	>18	>50

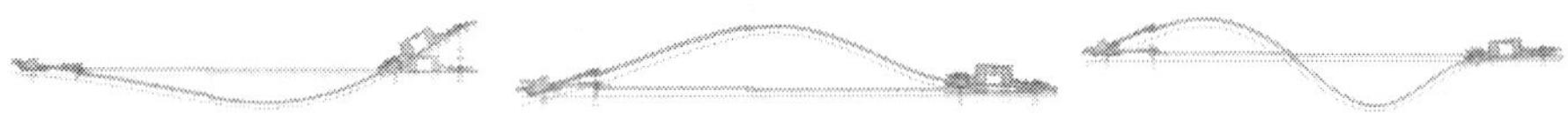

(a) First Stage. (b) Second Stage. (c) Third Stage .

Figure 3. The First Three Stages Mode Shape of The Rotor.

It can be seen that the simulated power turbine rotor worked above the two stages critical speed, the second stage critical speed margin was slightly less than 20%, and the other two stages critical speed meet the design requirement. The first three stages mode shapes of the simulated power turbine rotor were all bending mode shapes, and the slender power turbine shaft was the main reason for the bending deformation of the rotor.

4.2. Law of Critical Speeds Change with Supporting Stiffness

Without changing the structure and mass distribution of the simulated power turbine rotor, the stiffness of the rotor in the table 2 was taken as the reference stiffness, and the law of the first three stages critical speeds of the rotor change with the supporting stiffness was obtained through calculation, which provides a theoretical basis for the adjustment of critical speeds based on the supporting stiffness.

4.2.1. Law of Critical Speeds Change with No.1 Supporting Stiffness. When the stiffness of No.2, No.6.5 and No.7 supporting stayed unchanged, when No.1 supporting stiffness varied from 2E+6N/m to 20E+7N/m, curve of the first three stages critical speeds of the simulated power turbine rotor change with No.1 supporting stiffness is shown in figure 4, and the change rates of the first three stages critical speed are shown in table 4. It can be seen that the first, second and third stage critical speeds of the simulated power turbine rotor increased by 8.36%, 11.65% and 81.97% respectively when the No.1 supporting stiffness increased from 2E+6N/m to 5E+7N/m. With the continuous increase of the No.1 supporting stiffness, the first and third critical speeds of the rotor almost no longer changed.

4.2.2. Law of Critical Speeds Change with No.2 Supporting Stiffness. When the stiffness of No.1, No.6.5 and No.7 supporting stayed unchanged, when No.2 supporting stiffness varied from 2E+6N/m to 20E+7N/m, curve of the first three stages critical speeds of the simulated power turbine rotor change with No.2 supporting stiffness is shown in figure 5, and the change rates of the first three stages critical speed are shown in table 5. It can be seen that the first, second and third stage critical speeds of the simulated power turbine rotor increased by 21.22%, 37.20% and 72.49% respectively when the No.2 supporting stiffness increased from 2E+6N/m to 5E+7N/m. With the continuous increase to 20E+7N/m of the No.2 supporting stiffness, the first three stages critical speed increased by 1.91%, 8.12% and 7.72% respectively.

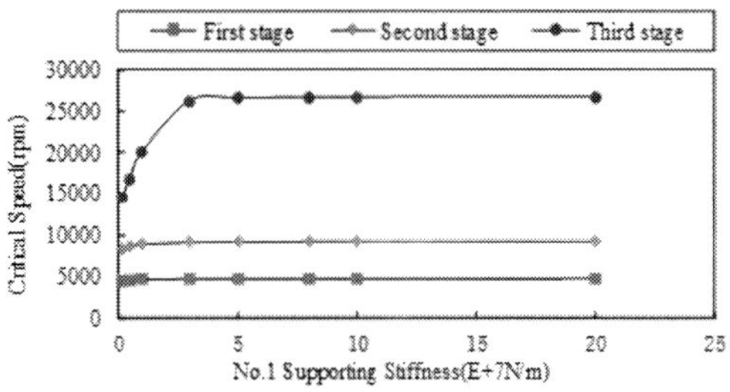

Figure 4. Curve of the First Three Stages Critical Speed Change with No.1 Supporting Stiffness.

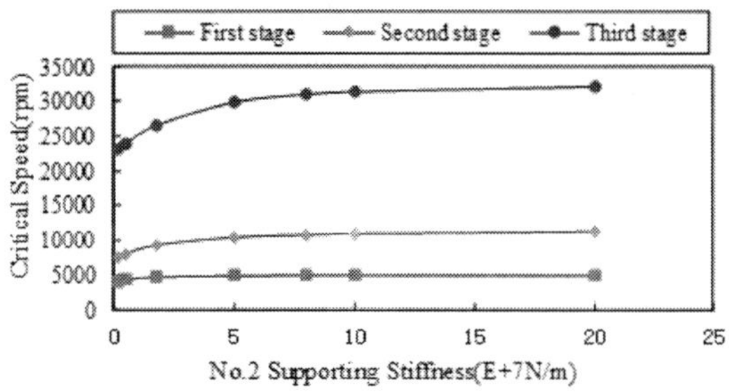

Figure 5. Curve of the First Three Stages Critical Speed Change with No.2 Supporting Stiffness.

Table 4. Change Rates of the First Three Stages Critical Speeds (No.1 Supporting Stiffness Changes).

Ranges of No.1 Supporting Stiffness(E+7N/m)	0.2-5	5-20
Change Rate of First Stage Critical Speed (%)	8.36	0.59
Change Rate of Second Stage Critical Speed (%)	11.65	0.80
Change Rate of Third Stage Critical Speed (%)	81.97	0.52

Table 5. Change Rates of the First Three Stages Critical Speeds (No.2 Supporting Stiffness Changes).

Ranges of No.2 Supporting Stiffness(E+7N/m)	0.2-5	5-20
Change Rate of First Stage Critical Speed (%)	21.22	1.91
Change Rate of Second Stage Critical Speed (%)	37.20	8.12
Change Rate of Third Stage Critical Speed (%)	72.49	7.72

4.2.3. Law of Critical Speeds Change with No.6.5 Supporting Stiffness. When the stiffness of No.1, No.2 and No.7 supporting stayed unchanged, when No.6.5 supporting stiffness varied from 2E+6N/m to 20E+7N/m, curve of the first three stages critical speeds of the simulated power turbine rotor change with No.6.5 supporting stiffness is shown in figure 6, and the change rates of the first three stages critical speed are shown in table 6.When the No.6.5 supporting stiffness increased from 2E+6N/m to 5E+7N/m, the first stage critical speeds of the rotor increased by 131.28%, with the continuous increase to 20E+7N/m of the No.6.5 supporting stiffness, the first stage critical speeds increased by 8.23%. The second stage critical speeds of the rotor increased by 35.40% when the No.6.5 supporting stiffness increased from 2E+6N/m to 20E+7N/m. There was no substantial change in the third stage critical speed of the rotor when the No.6.5 supporting stiffness increased from 2E+6N/m to 20E+7N/m.

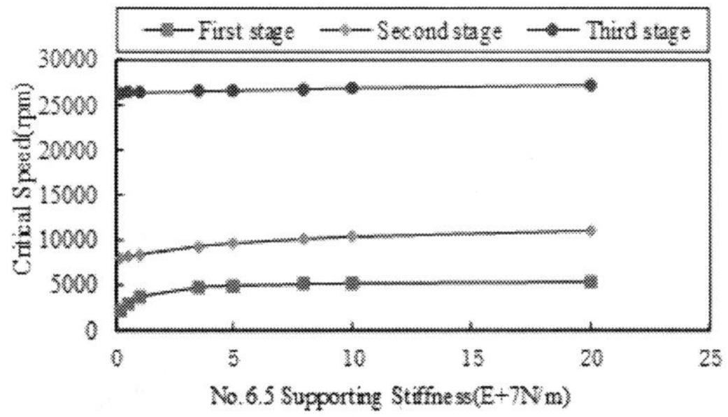

Figure 6. Curve of the First Three Stages Critical Speed Change with No.6.5 Supporting Stiffness.

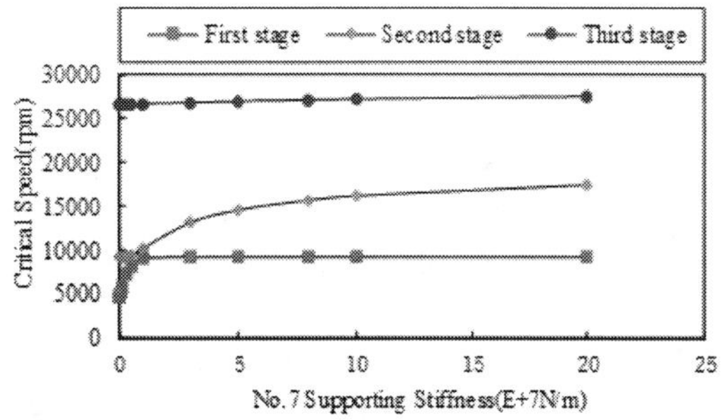

Figure 7. Curve of the First Three Stages Critical Speed Change with No.7 Supporting Stiffness.

Table 6. Change Rates of the First Three Stages Critical Speeds (No.6.5 Supporting Stiffness Changes).

Ranges of No.6.5 Supporting Stiffness (E+7N/m)	0.2-5	5-20
Change Rate of First Stage Critical Speed (%)	131.28	8.23
Change Rate of Second Stage Critical Speed (%)	20.62	14.78
Change Rate of Third Stage Critical Speed (%)	1.04	1.88

Table 7. Change Rates of the First Three Stages Critical Speeds (No.7 Supporting Stiffness Changes).

Ranges of No.7 Supporting Stiffness/(E+7N/m)	0.01-5	5-20
Change Rate of First Stage Critical Speed (%)	101.29	0
Change Rate of Second Stage Critical Speed (%)	57.19	20.05
Change Rate of Third Stage Critical Speed (%)	1.50	2.16

4.2.4. Law of Critical Speeds Change with No.7 Supporting Stiffness. When the stiffness of No.1, No.2 and No.6.5 supporting stayed unchanged, when No.7 supporting stiffness varied from 1E+5N/m to 20E+7N/m, curve of the first three stages critical speed of the simulated power turbine rotor changes with No.7 supporting stiffness is shown in figure 7, and the change rates of the first three stages critical speed are shown in table 7.When the No.7 supporting stiffness increased from 1E+5N/m to 5E+7N/m, the first stage critical speeds of the rotor increased by 101.29%, with the continuous increase to 20E+7N/m of the No.7 supporting stiffness, the first stage critical speed almost no longer changed. The second stage critical speeds of the rotor increased by 77.24% when the No.7supporting stiffness increased from 1E+5N/m to 20E+7N/m. There was no substantial change in the third stage critical speed of the rotor when the No.7 supporting stiffness increased from 1E+5N/m to 20E+7N/m.

4.3. Law of Critical Speeds Change with Simulated Power Turbine Disks

Without changing the supporting stiffness and the disks' structure, it was revealed that the first three stages critical speed of the rotor varied with the disks' mass, which only depends on the density of the disks. The supporting stiffness for calculation is shown in table 2, and the disk material and their properties are shown in table 8.

Table 8. The Disks' Properties.

Material	Density (kg m^{-3})	Modulus of Elasticity (GPa)	Poisson's Ratio
Magnesium Alloy	1780	40.2	0.34
Aluminium Alloy	2800	71	0.31
Titanium Alloy	4440	109	0.34
Stainless Steel	7800	196	0.3
Superalloy	8010	205	0.29
Superalloy	8400	191	0.3

4.3.1. Law of Critical Speeds Change with Mass of The First Stage Simulated Power Turbine Disk. When the density of first stage simulated power turbine disk varied from 1780 kg m^{-3} to 8400 kg m^{-3}, the curve of the first three stages critical speeds of the rotor change with density of the first stage simulated power turbine disk is shown in figure8, and the change rates of the first three stages critical speeds are shown in table 9. When the density of the first stage simulated disk increased from 1780 kg m^{-3} to 8400 kg m^{-3}, the change rates of the first three stages critical speeds of the simulated power turbine rotor were all less than 5%, indicating that the mass of the first stage disk has little effect on the first three stages critical speeds of the simulated power turbine rotor.

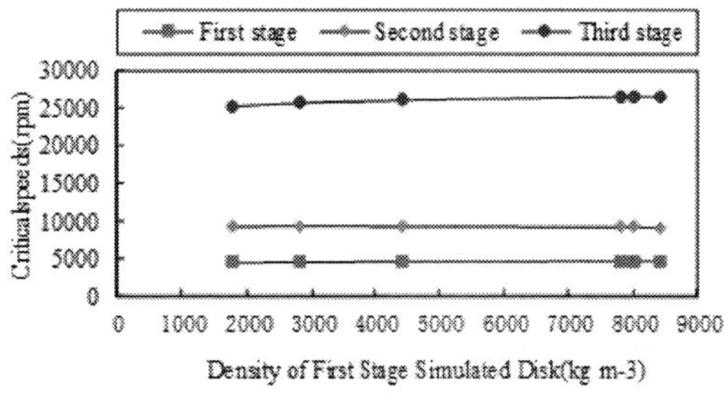

Figure 8. Curve of Critical Speeds Change with Density of The First Stage Disk.

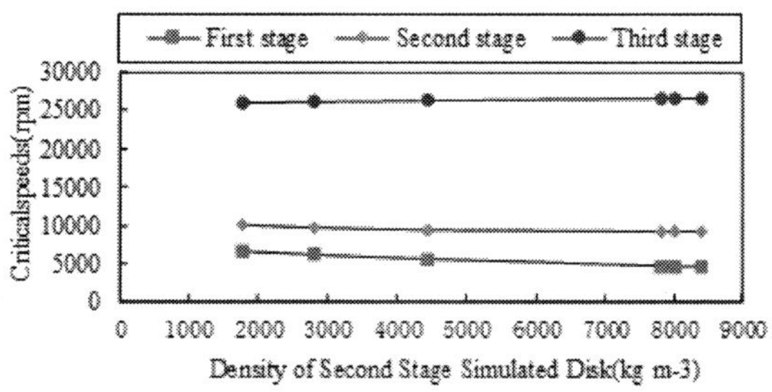

Figure 9. Curve of Critical Speeds Change with Density of The Second Stage Disk.

4.3.2. Law of Critical Speeds Change with Mass of The Second Stage Simulated Power Turbine Disk. When the density of second stage simulated power turbine disk varied from 1780kg m^{-3} to 8400 kg m^{-3}, the curve of the first three stages critical speeds of the rotor change with density of the second stage simulated power turbine disk is shown in figure9, and the change rates of the first three stages critical speeds are shown in table 10. When the density of the second stage simulated disk increased from 1780 kg m^{-3} to 8400 kg m^{-3}, the change rates of the first two stages critical speed of the rotor were reduced 30.67% and 8.57% respectively, and there was no substantial change in the third critical speed. It shown that the mass of the second stage simulated disk has a certain influence on the first two critical speeds of the rotor, but has no substantial influence on the third critical speed.

Table 9. Change Rates of the First Three Stages Critical Speeds (Density of The First Stage Disk Changes).

The First-stage Disk's Density (kg m^{-3})	1780-8400
Change Rate of First Stage Critical Speed (%)	3.25
Change Rate of Second Stage Critical Speed (%)	-0.89
Change Rate of Third Stage Critical Speed (%)	4.89

Table 10. Change Rates of the First Three Stages Critical Speeds (Density of The Second Stage Disk Changes).

The Second-stage Disk's Density (kg m^{-3})	1780-8400
Change Rate of First Stage Critical Speed (%)	-30.67
Change Rate of Second Stage Critical Speed (%)	-8.57
Change Rate of Third Stage Critical Speed (%)	1.85

5. Dynamic Characteristics Test and Result Analysis

Dynamic characteristics test of the rotor was carried out on the high-speed rotating rig. In figure 1, a displacement sensor D_1 was arranged in the vertical direction of position A, a displacement sensor D_2 was arranged in the vertical direction of position C, and two displacement sensors D_3 and D_4 were respectively arranged in the vertical and horizontal directions of position B. The rotor deflection at positions A, B and C were measured respectively.

The curves of rotor deflection change with relative speed the rotor in the full speed range measured by the for sensors are shown in figure 10. In figure 10, the relative speed is defined as: *relative speed = (actual speed / rated working speed) × 100%.*

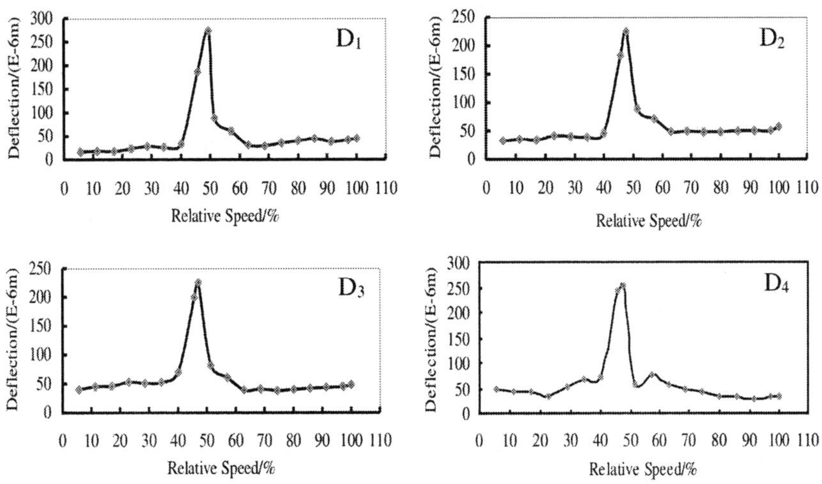

Figure 10. Curves of Rotor Deflection Change with Relive Speed Measured by Four Sensors.

The analysis of the critical speed calculation results and the experimental results were obtained, as shown in table 11. In table 11, calculation error was defined as: *Calculation Error = ((Test Result-Calculation Result)/Test Result) *100%.*

Table 11. Analysis of Calculation Result and Test Result of Critical Speeds.

Calculation Result(rpm)	First Stage	Second Stage
Test Result(rpm)	4730	9250
Error (%)	Not Obvious	8456~8638
Margin of Test Result (%)	—	7.08~9.38

The calculation error of the second critical speed was 7.08-9.38%, less than 10%. The difference between rotor calculation model and actual structure is the main cause of error, and the measurement error is another source of the error. In fact, if the actual error of oil film stiffness, the measurement error of supporting stiffness and the complex structure of the rotor were considered, the calculation results still have good consistency with the test results. The rotor has obvious resonance peak at the critical speed, indicating that the rotor has been bending change, it is also consistent with the calculated vibration modes. From the test results, the margin of the second critical speed of the rotor was 23.71~25.31%, greater than 20%, which meets the critical speed margin requirement [21].

6. Conclusion

In this paper, the finite element calculation model of the dynamic characteristics of the slender simulated power turbine rotor was established, and the dynamic characteristics of the rotor were systematically calculated and analysed. The laws of the first three critical speeds of the rotor change with the supporting stiffness and the mass of each disk were revealed, which provides the theoretical basis for the adjustment of the critical speeds of the rotor, and the dynamic characteristics test verification of the rotor in the full speed range was completed. The main conclusions are as follows.

1.No.1 supporting stiffness has obvious influence on the third critical speed, No.2 supporting stiffness has obvious influence on the first three critical speeds, No.6.5 supporting stiffness has obvious influence on the first and second critical speeds, and No.7 supporting stiffness has obvious influence on the first and second critical speeds.

2. In a certain range of supporting stiffness, the first three critical speeds of the simulated rotor can be adjusted by adjusting the supporting stiffness.

3. The mass of the first stage simulated power turbine disk has no substantial influence on the first three critical speeds of the rotor. The mass of the second stage simulated power turbine disk has a significant influence on the first stage critical speed of the rotor, a certain influence on the second stage critical speed, and no substantial influence on the third stage critical speed.

4. Mode shapes of the first three stages mode shapes of the simulated power turbine rotor are all flexural mode shapes. The test results verified the rationality of the rotor structure design, supporting layout and critical speed distribution.

5.The calculated error of the finite element model is less than 10%, which is in good agreement with the experimental results.

References

[1] Zhou C Y, Xue Z H, Liu X F and Wen J X, 2018, The rotor dynamic calculation analysis of pump rotor system, *Chinese Journal of Turbomachinery*, **60**(S1), pp.39–48.
[2] He C H, Xue Z H and Liu X F, 2019, The rotor dynamic analysis of centrifugal pump rotor system, *Chinese Journal of Turbomachinery*, **61**(Z1), pp.65–70.
[3] Zhang Y, Zhang Y C and Li R M, 2017, Rotor-dynamics system analysis of centrifugal blowers with high speed and efficiency, *Chinese Journal of Turbomachinery*, **59**(3), pp.40–43.
[4] Tai X Y, Li Y, Yang S H and Zhang Y, 2019, Analysis of rotor dynamics characteristic for marine floating centrifugal compressor, *Chinese Journal of Turbomachinery*, **61**(1), pp.63–70.

[5] Du G W, Wang Y Z and Yang G J, 2017, Rotor dynamic research of a 165kW helium blower with AMBs, *Chinese Journal of Turbomachinery*, **59**(4), pp.63–70.
[6] Li Z H, Ren H and Luo Z Y, 2013, Study on dynamic behaviors of a certain type aero-engine rotor system and based on co-simulation technology, *Machinery Design & Manufacture*, (12), pp.174–176,180.
[7] Deng W Q, Guo F Y and Gao D P, 2016, Dynamic characteristics calculation for a high speed flexible rotor of an aeroengine, *Journal of Vibration and Shock,* **25**(5), pp.130–133.
[8] Miao H, Wang K M and Ai S M, 2011, Finite element analysis on critical speeds of a dual-rotor system, *Journal of Shenyang Aerospace University*, **28**(5), pp.27–31.
[9] Mo Y Y, Li Q T and Zhang B, 2011, Dynamic characteristic calculation of a engine's double rotors system, *Machinery Design & Manufacture*, (7), pp.117–119.
[10] Deng W Q, Tang G abd Shu S R, 2010, Dynamic characteristics research on simulated low pressure rotor of contra rotating engine, *Aeroengine*, **36**(6), pp.6–8.
[11] Deng W Q, Fan P P and Xu Y L, 2017, Dynamic characteristic study of a simulated low-pressure rotor of a turboprop engine, *Aeronautical Science & Technology*, **28**(8), pp.72–77.
[12] Liu W K, Deng W Q and Lu B, 2019, Research on the dynamic of a cantilever power turbine rotor with flexible transition section, *Gas Turbine Experiment and Research*, **32**(2), pp.34–41.
[13] Wu G Fn, Chen G Z and Tu M P, 2006, Analysis and experimental study of the high speed flexible rotor dynamic behaviors, *Journal of Aerospace Power*, **26**(3), pp.563–568.
[14] Deng W Q, Nie W J and He P, 2015, Change laws of critical speeds with supporting stiffness of a high-speed flexible rotor, *Noise and Vibration Control*, **35**(3), pp.98–101.
[15] Nie W J, Deng W Q and He P, 2015, Influence of different draw-bars on dynamic characteristics of a high speed flexible rotor, *Noise and Vibration Control*, **35**(3), pp.135–139.
[16] Nie W J, Deng W Q and Xu Y Y, 2015, Laws analysis on critical speeds change with supporting stiffness and disk mass of a high-speed flexible rotor, *Gas Turbine Experiment and Research*, **28**(3), pp.19–24.
[17] Hong J, Wang H and Xiao D W, 2008, Effects of dynamic stiffness of rotor bearing on rotor dynamic characteristics, *Aeroengine*, **34**(1), pp.23–27.
[18] Bai Z X, Wu W L and Liu H Z, 2012, The effect of the stiffness of a rotor's supporting system on its dynamic characteristics, *Machine Design and Research*, **28**(4), pp.18–21.
[19] Mei Q and Ou Y X, 2004, Effects of supporting distribution on dynamic characteristics of a rotor with double bearing, *Journal of Vibration Engineering*, **17**(S), pp.156–158.
[20] Deng W Q, Wang Y and Nie W J, 2016, Influence analysis of supporting stiffness and supporting axial location on critical speeds of a low-pressure rotor of a counter rotating engine, *Aeroengine*, **42**(3), pp.7–11.
[21] Fu C G, 2000, *Rotor Dynamics and Whole Engine Vibration* vol.19, (Beijing: Aviation Industry Press).

IOP Conf. Series: Materials Science and Engineering 1081 (2021) 012029 doi:10.1088/1757-899X/1081/1/012029

Dynamic Analysis and Experimental Research on the Cantilever High Speed Rotor of Turbofan Engine

Weijian Nie[1,2], Wangqun Deng[1,2], Sheng Yuan and Yanong Chen[1,2]

[1] AECC Hunan Aviation Powerplant Research Institute, Zhuzhou 412002, China
[2] Key Laboratory of Aero engine Vibration Technology, Aero Engine Corporation of China, Zhuzhou 412002, China

E-mail: nwjxj@126.com

Abstract. Taking a simulated low-pressure rotor of a turbofan engine as the research object, the dynamic characteristic calculation model of the simulated low-pressure rotor with hollow shaft and solid shaft was established respectively by using the finite element method. The first three order critical speeds and mode shapes of the rotor were calculated systematically. The dynamic balance test of the simulated low-pressure rotor and the research on the damping effect of the squeeze film damper were carried out on the high-speed rotating test rig, and the dynamic characteristic test was completed in the full speed range. The results show that the finite element results agree well with the experimental results. The rationality of the rotor's hollow shaft structure and rotor dynamic design was verified. The research provides reference and technical support for structural design, dynamic design and experimental research of real low-pressure rotor in engine.

1. Introduction

Rotor is an important rotating part of turbine machineries such as motors, pumps, compressors, gas turbines and aero-engines. It is of great significance to study rotor dynamics. Many scholars like Zhou [1], He[2], Zhang[3] , Tai[4], Du[5] and other researcher[6-9] have studied the dynamics of turbo machinery rotor and achieved remarkable results. Compared with the rotor of other turbomachinery, the rotor of aeroengine was characteristic of high speed, complex structure and bad working condition. Establishing a calculation model which can reflect the actual situation of aero-engine rotor was the basis of rotor analysis. At present, the finite element method was widely used to establish the calculation model of rotor. Deng [6,7], Zhu [8], Liu [9], Li [10], Miao [11], Mo [12] and other scholars used the finite element method to carry out the dynamic analysis of the aeroengine rotor, and achieved considerable results. The dynamic characteristics of the rotor were affected by many factors, in the structural design stage of the rotor, it was necessary to comprehensively consider many factors, such as the supporting layout, supporting mode, mass distribution and critical speed. Therefore, it was often necessary to design the rotor with various structural forms, and calculate and analyze the dynamic characteristics of the rotor with different structural forms. Through comparative analysis, select one of the more reasonable scheme of rotor structure. Deng [13,14] studied the law of critical speed of high-speed flexible rotor changed with support stiffness and the influence of support layout on the critical speed of a low-pressure rotor, which provided a theoretical basis for the design and adjustment of the critical speed of the rotor. Nie [15,16] analyzed the influence of tie bars on the dynamic characteristics of high-speed flexible rotor, and studied the influence of support stiffness and disk mass on critical speeds of that rotor. However, whether the rotor structure was reasonable need to be verified by test. In the process of test, it was often unable to operate safely to the rated working speed due to excessive imbalance, and high-speed dynamic balance test was often required. Deng [17,18,19] conducted high-speed dynamic balance research on power turbine rotor and low-pressure rotor of aeroengine, which developed high-speed dynamic balance technology of flexible rotor.

In this paper, a simulated low-pressure rotor of a turbofan engine was taken as the research object to carry out dynamic analysis and experimental research. Through calculation analysis and test, the rationality of hollow structure and critical speed distribution were verified. Aiming at the actual complex structure of the simulated low-pressure rotor, high-speed dynamic balance test was carried out, and the effect was good. Besides, the vibration reduction effect of squeeze film damper was verified, and the comparative analysis of calculation results and test results were made. The results show that the finite element model well reflected the real dynamic characteristics of the rotor, and the test verified that the structure and dynamic design of the simulated low-pressure rotor were reasonable, which provided a reference for the dynamic design, structural design and test of the real low-pressure rotor in engine.

2. Rotor Structure

The structural sketches of the simulated low-pressure rotor with hollow shaft and solid shaft are shown in figure 1 and figure 2 respectively. Except for the different structure of the low-pressure shaft, the other structures of the rotors were coincident. The whole rotor was mainly composed of simulated fan disk, simulated turbine disk, low-pressure shaft and other main parts. The cantilever of simulated fan disk was adopted. No. 1 bearing was ball bearing, No. 2 and No. 5 bearing were roller bearing. The rotor is a typical high-speed flexible rotor.

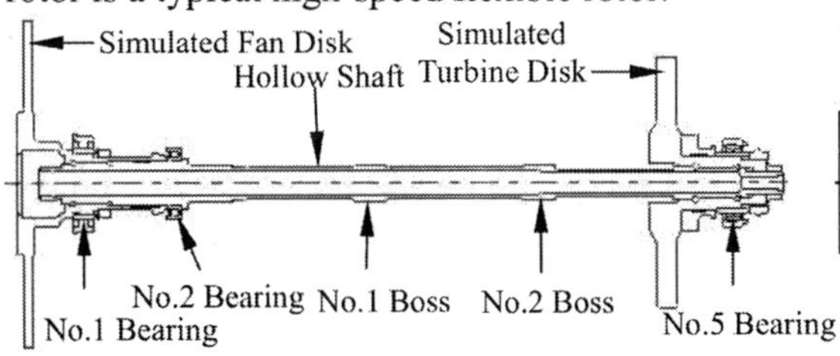

Figure 1. Structural Sketch of the Simulated Rotor with Hollow Shaft.

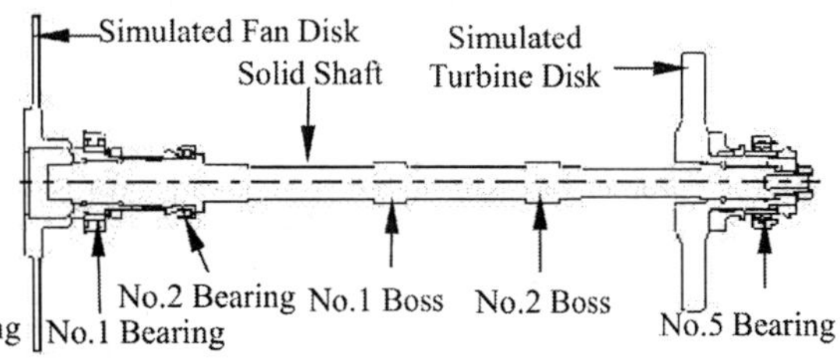

Figure 2. Structural Sketch of the Simulated Rotor with Solid Shaft.

3. Finite Element Analysis Model

The finite element calculation models of simulated low-pressure rotor with hollow shaft and solid shaft are shown in figure 3 and figure 4 respectively. In the model, beam element was used to simulate the main body of the rotor, bearing element was used to simulate the support of the rotor, five hundred and fifty-seven beam elements, three bearing elements and fourteen rigid connection elements were used to simulate the low-pressure rotor with hollow shaft, four hundred and eighty-five beam elements, three bearing elements and fourteen rigid connection elements were used to simulate the low-pressure rotor with solid shaft.

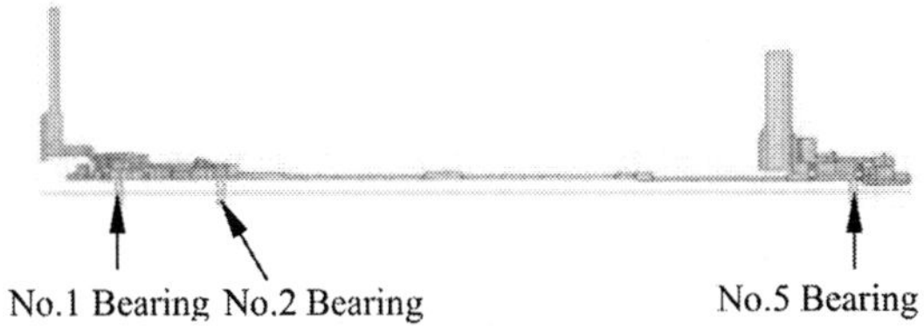

Figure 3. Finite Element Model of the Simulated Low-pressure Rotor with Hollow Shaft.

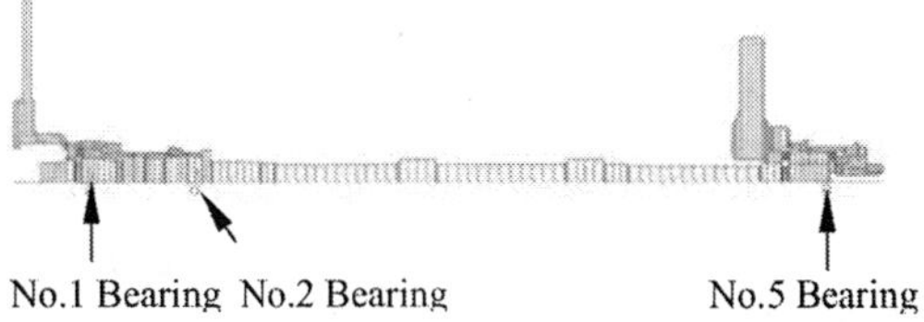

Figure 4. Finite Element Model of the Simulated Low-pressure Rotor with Solid Shaft.

4. Calculation Results of Dynamic Characteristics

4.1. Supporting Stiffness

The rotor is supported by squirrel cage elastic support structure, as shown in figure 5. The actual stiffness of each elastic support is shown in table 1. Under the stiffness of each elastic support, the first three order critical speeds and mode shapes of the simulated low-pressure rotor with hollow shaft and solid shaft were calculated respectively.

Figure 5. Structural Sketch of the Elastic Support.

Table 1. Actual Stiffness of Each Elastic Support.

Supporting Number	No.1	No.2	No.5
Stiffness(E+7N/m)	1.21	2.30	0.44

4.2. Comparative Analysis of Critical Speed

The calculation results and margins of the first three order critical speeds of the simulated low-pressure rotor with hollow shaft and solid shaft are shown in table2. The critical speed margin related to rated working speed was defined as: *Critical Speed Margin = ((Rated Working Speed- Critical Speed)/ Rated Working Speed) *100%.*

Table 2. The First Three Order Critical Speeds and Margins.

Stage		First	Second	Third
Rotor with Hollow Shaft	Critical Speed(rpm)	7831	15237	38322
	Margin(%)	>70	>40	>30
Rotor with Solid Shaft	Critical Speed(rpm)	7780	15649	28178
	Margin(%)	>70	>40	<5

The simulated low-pressure rotor with hollow shaft has two critical speeds within the rated working speed range, and the simulated low-pressure rotor with solid shaft has three critical speeds within the rated working speed range. The first two order critical speeds of the simulated low-pressure rotor with hollow shaft and the solid shaft were not very different. It is mainly because the other factors affecting the critical speed and the structure of the two types of rotors were the same except for the low-pressure shaft, However, the critical speed of the third stage was quite different, indicating that the structure of the low-pressure shaft has no effect on the first two order critical speed of the rotor, but it has a great influence on the third stage critical speed. The first three order critical speed margins of simulated low-pressure rotor with hollow shaft were all greater than 30%, which meet the requirements of critical speed design criteria. The first two order critical speed margins of simulated low-pressure rotor with solid shaft were more than 40%, but the third stage critical speed margin was less than 5%, which does not meet the requirements of critical speed design criteria. From the perspective of critical speed margin, the simulated low-pressure rotor with hollow shaft should be selected.

4.3. Comparative Analysis of Mode Shape

The calculation results of the first three order mode shapes of the simulated low-pressure rotor with hollow shaft and solid shaft are shown in table 3.It can be seen from table 3 that the first three order mode shapes of the simulated low-pressure rotor with hollow shaft and solid shaft were all bending mode shapes, and the simulated low-pressure rotor was a typical high-speed flexible rotor; the first three order mode shapes of the simulated low-pressure rotor with hollow shaft and solid shaft were basically the same, and the structural form of the low-pressure shaft has no effect on the first three order mode shapes of the rotor.

Table 3. The First Three Order Mode Shape of The Rotor.

	First	Second	Third
Rotor with Hollow Shaft		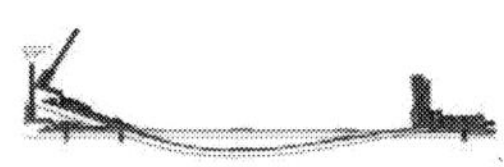	
Rotor with Solid Shaft			

The first three order mode shapes of the simulated low-pressure rotor with hollow shaft and solid shaft were all bending mode shapes, and the simulated low-pressure rotor was a typical high-speed flexible rotor; the first three order mode shapes of the simulated low-pressure rotor with hollow shaft and solid shaft were basically the same, and the structural form of the low-pressure shaft has no effect on the first three order mode shapes of the rotor.

5. Dynamic Characteristics Test and Result Analysis

The dynamic characteristic test of the simulated low-pressure rotor was carried out on a high-speed rotating test rig. The schematic diagram of the installation and test of the rotor is shown in figure 6. The displacement sensors were arranged at positions ①, ② and ③ as shown in figure 6 to measure the deflection of the rotor. The displacement sensors D_1 and D_2 were arranged at vertical direction of position ① and ③ respectively, and D_3 and D_4 were arranged at vertical and horizontal direction of position ② respectively. Four vibration acceleration sensors were arranged on two rigid supports to measure the vibration accelerations. In Figure 6, " ⊥ " indicates vertical direction, "=" indicates horizontal direction, A_1-A_4 indicates acceleration sensor, D_1-D_4 indicates vibration displacement sensor. The installation photo of the rotor on the tester is shown in figure 7.

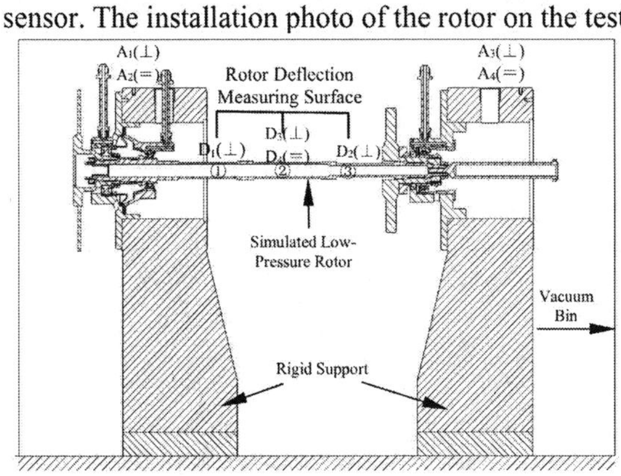

Figure 6. Schematic Diagram of the Installation and Test of the Simulated Low-pressure Rotor.

Figure 7. Photo of the Rotor on the Tester.

5.1. Test Result

According to the test results, the curves of rotor deflection change with rotation speed measured by four displacement sensors are shown in figure 8. The critical speeds measured by each sensor and the corresponding deflection value are also identified.

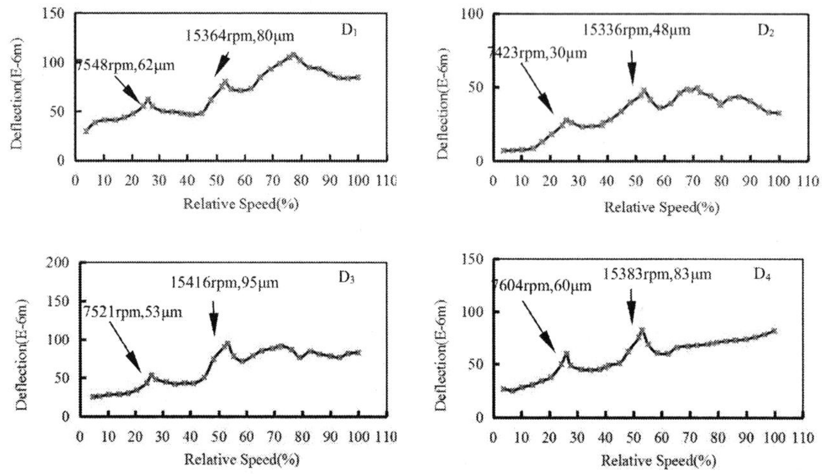

Figure 8. Curves of Rotor Deflection Change with Relive Speed Measured by Four Sensors.

5.2. Result Analysis

The analysis of the critical speed calculated results and the experimental results (average value) were obtained, as shown in table 4. The calculation error was defined as: *Calculation Error = ((Calculation Result- Test Result)/Test Result) *100%.*

Table 4. Analysis of Calculation Result and Test Result of Critical Speeds.

Stage	First	Second
Calculation Result(rpm)	7831	15237
Test Result(rpm)	7524	15375
Error (%)	4.08	0.90
Margin of Test Result (%)	74.06	46.98

The calculated errors of the first two order critical speeds are all less than 5%, the calculated results are in good agreement with the test results. The rotor has obvious resonance peak at the critical speed, indicating that the rotor has been bending deformation, it is also consistent with the calculated vibration mode shapes. From the test results, the margin of the first two order critical speed of the rotor are all greater than 20%, which meets the critical speed margin requirement [20].

6. Experimental Study on High Speed Dynamic Balance

It can be seen from figure 8 that the rotor can smoothly cross the first two order critical speeds and run to the rated working speed, but the deflection at the rated working speed is large, and the deflections measured by D_1, D_3 and D_4 sensors are not less than 82μm, high-speed dynamic balance for the rotor is required to reduce the rotor deflection, in order to running safely and stably at the rated working speed.

6.1. Balance Difficulties

The simulated low-pressure rotor working across the two-stage bending critical speed, it is a typical high-speed flexible rotor and it is very sensitive to the unbalance, so it is very difficult to carry out the high-speed dynamic balance test on it, which is mainly reflected in the following aspects.

(1) The ratio of length to diameter of low-pressure shaft is greater than 20. It is a high-speed flexible rotor with long diameter ratio slender flexible shaft, and the additional imbalance caused by vibration mode is very large.

(2) The low-pressure shaft adopts hollow thin-walled structure, and the coaxially of the inner and outer walls is difficult to guarantee, which inevitably leads to a certain initial imbalance of the rotor.

(3) The balance surface on the rotor is very limited, with only two balance surfaces and two balance bosses (No.1 boss and No.2 boss as shown in figure 1) on the low-pressure shaft.

(4) The length of the two balance bosses is only 20 mm and the height is only 1.5 mm. Not only the amount of material removed by grinding is very limited, but also the low-pressure shaft cannot be damaged in any form during grinding, so it is difficult to operate.

6.2. Balance Process

The influence coefficient method was used to balancing, and the rated working speed with a balancing speed of 89.66% was selected. The measuring surface was near the No.2 boss. The sensor was D_3, and the balance face was No.1 boss. The high-speed dynamic balance process is shown in table 5.

Table 5. High Speed Dynamic Balance Process.

Results Measured by D_3 Sensor		Process
Deflection(μm)	Phase(°)	
86	102	Initial State
103	104	After Adding the Fitting Weight at 180 ° Position of No. 1 boss
70.5	80	According to the Calculation of the Test Instrument, It is Necessary to Add 2.21 Times of the Fitting Weight at the 353 ° Position of No.1 boss, and Actually Move the Original Fitting Weight to the 353 ° Position
26.5	117	After Adding 2 Times of Original Weight at 353 ° Position of No.1 Boss
33.5	359	After Grinding the Material at the Corresponding Position of No.1 Boss

As shown in table5, the rotor deflection of the D_3 sensor was reduced from 86μm to 33.5μm after high speed dynamic balance, and a good balance effect was achieved.

6.3. Balance Effect

After high-speed dynamic balance the deflection of the rotor measured by four displacement sensors at rated working speed were reduced, and the effect of high-speed dynamic balance at rated working speed is shown in table 6.

Table 6. High Speed Dynamic Balance Effect at Rated Working Speed.

State	D_1	D_2	D_3	D_4
Before Balancing(E-6m)	84.5	30	82.5	83
After Balancing(E-6m)	23	26	10	16
Effect (%)	72.78	13.33	87.88	80.72

As can be seen from table 6, the dynamic deflection of the four deflection measuring points at the rated working speed has decreased by 72.78%, 13.33%, 87.88% and 80.72% respectively, and the balance effects are very satisfactory.

IOP Conf. Series: Materials Science and Engineering 1081 (2021) 012029 doi:10.1088/1757-899X/1081/1/012029

7. Experimental Study on High Speed Dynamic Balance

In order to verify the damping effect of squeeze oil film damper, two structures were set at No. 2 bearing: one with squeeze oil film and the other without squeeze oil film. And verification test on simulated low-pressure rotor with hollow shaft was carried out. The oil film parameters are shown in table 7, and the schematic diagram is shown in figure 6.

Table 7. Oil Film Parameter.

Parameter	Oil Film Length (mm)	Oil Film Thickness (mm)
Value	16	0.16

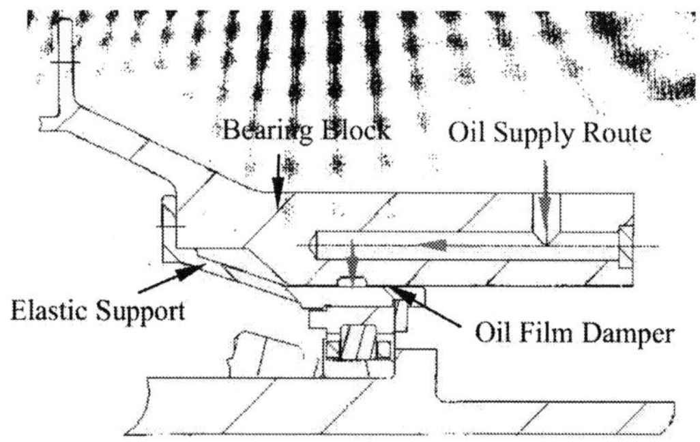

Figure 9. Schematic Diagram of Squeeze Film Damper.

The test results of the rotor with squeeze oil film damper are shown in Figure 10.

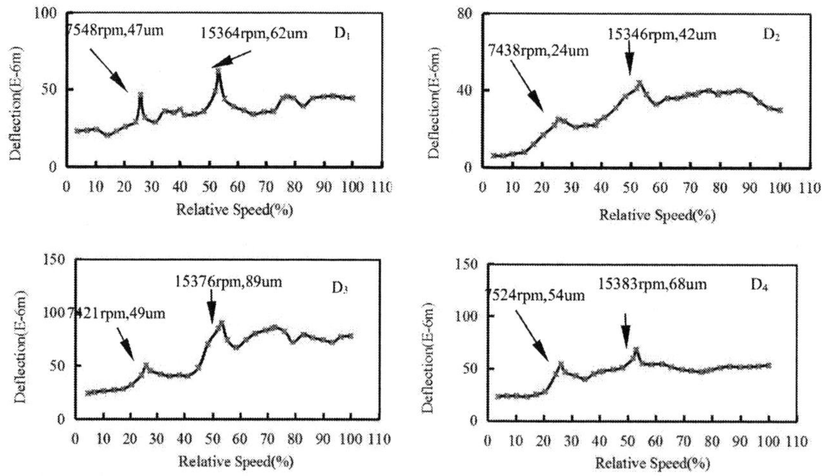

Figure 10. Curves of Rotor Deflection Change with Relive Speed Measured by Four Sensors.

The deflection measured by each vibration displacement sensor at critical speeds is shown in table 8.

Table 8. Comparison of Deflection Values at Critical Speed.

	First Order Critical Speed				Second Order Critical Speed			
Senor	D_1	D_2	D_3	D_4	D_1	D_2	D_3	D_4
Without Squeeze Film	62	30	53	60	80	48	95	83
With Squeeze Film	47	21	49	54	62	42	89	68
Decrement (%)	24.19	30.00	7.55	10.00	22.50	12.50	6.32	18.07

As shown in table 8, the deflection values of all measured points at critical speeds are significantly reduced, and the reduction ranges are not less than 6.32%, indicating that the squeeze film damper has a good effect of vibration reduction in the full speed range.

8. Conclusion

In this paper, the first three order critical speeds and mode shapes of simulated low-pressure rotor with solid shaft and hollow shaft were calculated and compared. The high-speed dynamic balance test of the rotor was carried out. The balance effect was good and the vibration reduction effect of the squeeze film damper was verified. The research results can be used directly to the real rotor in engine. The main conclusions are as follows.

1.The first three order mode shapes of the simulated low-pressure rotor with hollow shaft and solid shaft are basically the same, and they are all bending modes. The structure of the low-pressure shaft has no substantial influence on the first two critical speeds of the simulated low-pressure rotor, but has a great influence on the third critical speed, and has no substantial influence on the first three order mode shapes.

2. Low-pressure shaft should adopt hollow shaft structure to making the rotor critical speed meet the design criteria.

3. The critical speed calculation error is less than 5%, which reflects the real dynamic characteristics of the rotor very well.

4. The balance method can achieve high speed dynamic balance of cantilever flexible rotor, and the balance effect is good.

5.Squeeze film damper has significant effect on vibration reduction of slender shaft high-speed flexible rotor.

References

[1] Zhou C Y, Xue Z H, Liu X F and Wen J X, 2018, The rotor dynamic calculation analysis of pump rotor system, *Chinese Journal of Turbomachinery*, **60**(S1), pp.39–48.

[2] He C H, Xue Z H and Liu X F, 2019, The rotor dynamic analysis of centrifugal pump rotor system, *Chinese Journal of Turbomachinery*, **61**(Z1), pp.65–70.

[3] Zhang Y, Zhang Y C and Li R M, 2017, Rotor-dynamics system analysis of centrifugal blowers with high speed and efficiency, *Chinese Journal of Turbomachinery*, **59**(3), pp.40–43.

[4] Tai X Y, Li Y, Yang S H and Zhang Y, 2019, Analysis of rotor dynamics characteristic for marine floating centrifugal compressor, *Chinese Journal of Turbomachinery*, **61**(1), pp.63–70.

[5] Du G W, Wang Y Z and Yang G J, 2017, Rotor dynamic research of a 165kW helium blower with AMBs, *Chinese Journal of Turbomachinery*, **59**(4), pp.63–70.

[6] Deng W Q, Tang G abd Shu S R, 2010, Dynamic characteristics research on simulated low pressure rotor of contra rotating engine, *Aeroengine*, **36**(6), pp.6–8.

[7] Deng W Q, Fan P P and Xu Y L, 2017, Dynamic characteristic study of a simulated low-pressure rotor of a turboprop engine, *Aeronautical Science & Technology*, **28**(8), pp.72–77.

[8] Haimin Zhu, Weifang Chen, Rupeng Zhu, Jie Gao, Meijun Liao(2020). Study on the Dynamic Characteristics of a Rotor Bearing System with Damping Rings Subjected to Base Vibration. Journal of Vibration Engineering & Technologies, Vol.8, No. 1, pp.121-132.

[9] Liu W K, Deng W Q and Lu B, 2019, Research on the dynamic of a cantilever power turbine rotor with flexible transition section, *Gas Turbine Experiment and Research*, **32**(2), pp.34–41.

[10] Li Z H, Ren H and Luo Z Y, 2013, Study on dynamic behaviors of a certain type aero-engine rotor system and based on co-simulation technology, *Machinery Design & Manufacture*, (12), pp.174–176,180.

[11] Miao H, Wang K M and Ai S M, 2011, Finite element analysis on critical speeds of a dual-rotor system, *Journal of Shenyang Aerospace University*, **28**(5), pp.27–31.

[12] Mo Y Y, Li Q T and Zhang B, 2011, Dynamic characteristic calculation of a engine's double rotors system, *Machinery Design & Manufacture*, (7), pp.117–119.

[13] Deng W Q, Nie W J and He P, 2015, Change laws of critical speeds with supporting stiffness of a high-speed flexible rotor, *Noise and Vibration Control*, **35**(3), pp.98–101.
[14] Deng W Q, Wang Y and Nie W J, 2016, Influence analysis of supporting stiffness and supporting axial location on critical speeds of a low-pressure rotor of a counter rotating engine, *Aeroengine*, **42**(3), pp.7–11.
[15] Nie W J, Deng W Q and He P, 2015, Influence of different draw-bars on dynamic characteristics of a high speed flexible rotor, *Noise and Vibration Control*, **35**(3), pp.135–139.
[16] Nie W J, Deng W Q and Xu Y Y, 2015, Laws analysis on critical speeds change with supporting stiffness and disk mass of a high-speed flexible rotor, *Gas Turbine Experiment and Research*, **28**(3), pp.19–24.
[17] Deng W Q, Nie W J and Xu Y L, 2018, Vibration fault analysis and balance technique study of a high speed flexible rotor of a turbofan engine, *Gas Turbine Experiment and Research*, **31**(1), pp.24–30,58.
[18] Deng W Q, Wang Z and Shu S R, 2012, Dynamic Ccharacteristics and high speed dynamic balance technique for a power turbine rotor of a turbo-shaft engine, *Journal of Vibration and Shock*, **31**(7), pp.162–165,170.
[19] Deng W Q, Tang G and Shu S R, 2010, High speed dynamic balance experiment research on a simulated low-pressure rotor of a counter rotating engine, *Gas Turbine Experiment and Research*, **23**(4), pp.5-9.
[20] Fu C G, 2000, *Rotor Dynamics and Whole Engine Vibration* vol.19, (Beijing: Aviation Industry Press).

IOP Conf. Series: Materials Science and Engineering 1081 (2021) 012030 doi:10.1088/1757-899X/1081/1/012030

Numerical study on film cooling characteristics of BESA hole with water mist/air two-phase flow

Huang Kang, Shi Pei-jie*, Jiang An-lin, Ma Hu-sheng

China Aerodynamics Research and Development Center, Mianyang 621000,China

Shipeijie@cardc.com; huangkang@cardc.com

Abstract. In view of a new type of film cooling hole, the backward-expanding shoulder arm(BESA) hole proposed by the author, a numerical model of two-phase flow film cooling with water mist is established. The numerical simulation of water mist / air two-phase flow film cooling characteristics shows that the comprehensive performance of water mist / air two-phase flow film cooling is significantly improved compared with pure air film cooling. For the round hole plate model, when the blowing ratio is 0.5, the film cooling efficiency of two-phase flow film cooling with 2% water spray is 37.5% higher than that of pure air film cooling (20D from the film hole outlet), When the blowing ratio is 1.0, the film cooling efficiency of the flat two-phase flow with 2% water spray is 49.0% higher than that of the pure air film cooling (20D away from the film hole outlet).When the spray amount of water is the same, the film cooling efficiency of the BESA hole is higher than that of the round hole, the expanding hole and the shoulder arm hole. For the flat film cooling model of the BESA hole, When other factors remain unchanged, the cooling efficiency of the two-phase flow film increases monotonously from 2% to 5%.The cooling characteristics of the two-phase flow film on the pressure surface of the turbine blade of the BESA hole are similar to those of the flat model, and the film coverage effect is better than that of the round hole, the expanding hole and the shoulder arm hole.

1. Introduction

As one of the main cooling modes of gas turbine components, film cooling is widely used in turbine end wall, blade leading edge and blade trailing edge of gas turbine[1-3]. As the expected performance of gas turbines continues to improve, the contradiction between cooling demand and cooling gas volume becomes increasingly prominent: on the one hand, the amount of air used for cooling turbines in some advanced gas turbines has reached as high as 15%~20%, a large amount of high-pressure air used for cooling will inevitably lead to the loss of power plant performance; on the other hand, while increasing the air compression ratio, it will inevitably increase the temperature of cooling air (the cooling air temperature of advanced gas turbine has reached 900K), reduce its heat absorption capacity, and make cooling more difficult[4-9]. Therefore, how to further tap the potential of film cooling technology and improve the effect of film cooling is one of the technical difficulties faced by heat transfer researchers.

A lot of research shows that improving the structure parameters such as film cooling hole type, film incidence angle and hole spacing can improve the film cooling effect. In recent years, researchers in the United States, Russia and other countries have found that spraying a small amount of water mist into the cooling gas can significantly improve the film cooling efficiency, which provides another feasible solution to the current contradiction between cooling demand and cooling gas volume.

In this paper, the heat transfer problems in two-phase flow film cooling with different film cooling hole and water mist / air two-phase flow film cooling mode are studied by means of steady numerical simulation, especially the backward-expanding shoulder arm(BESA) hole proposed by the author earlier.

2. Physical model

In order to reduce the computational complexity and facilitate the study of single influencing factors, a simplified physical model of the main flow gas and cooling air passages from the compressor exhaust port to the cooling stave surface is established. Fig. 1 is a flat film cooling model similar to the small curvature section of a turbine blade and Fig. 2 is a single channel turbine blade model. The main shapes of the film holes involved in this study are round hole, expanding hole, shoulder arm hole and backward-expanding shoulder arm(BESA) hole. The flow path diagrams of the holes involved are shown in Fig. 3, and the entrance diameters of the holes are the same. The shoulder arm hole is composed of a main jet hole and two side holes with an angle of 30 deg to the mainstream[10]. They are all round holes. The diameter d of the two side holes is half of the diameter D (D=3mm) of the main jet hole, and the width of the two side holes is 3D. The structure of the backward-expanding shoulder arm(BESA) hole is that the central plane of the two sides of the hole is inclined backward, and the angle between the central plane of the main jet hole and the central plane of the main jet hole is 10 deg. The main jet hole adopts a progressively expanding hole with an expansion angle of 5 deg. The expansion angle of expanding hole is 5 deg[11].

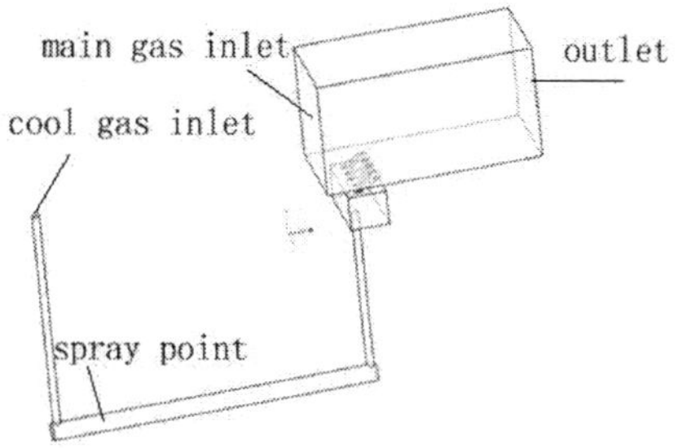

Figure 1. Computing domain of flat model.

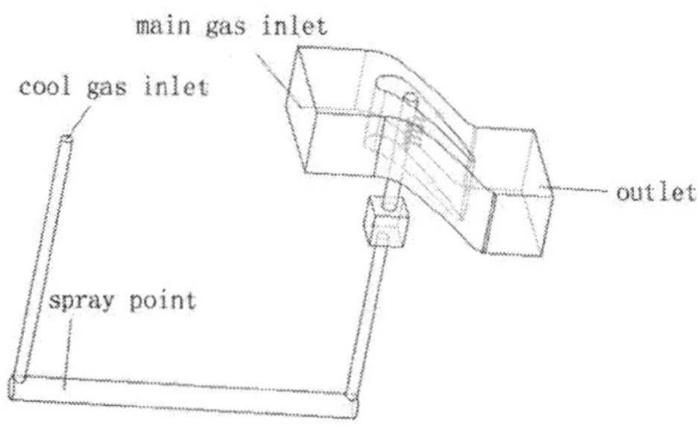

Figure 2. Computational domain of turbine blade model.

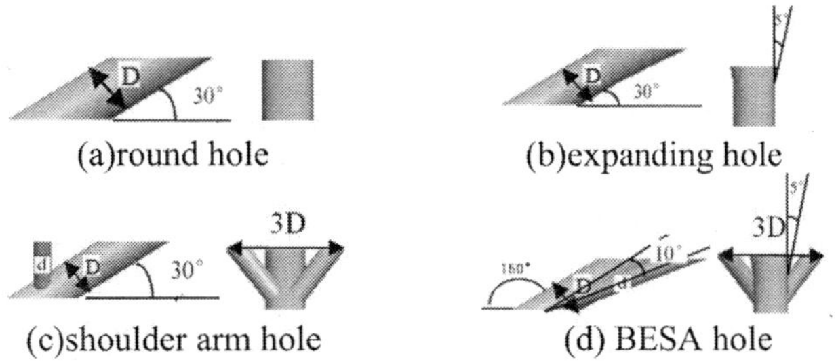

Figure 3. Geometry of different gas film hole channel.

3. Numerical model and verification

3.1. Computational model

The volume fraction of water mist particles involved in this paper is less than 10%, so Discrete Phase Model is used to calculate the discrete phase. Assuming that the second phase is very sparse, the interaction between particles and particles and the effect of volume fraction on the continuous phase can be neglected. When particles are treated as discrete particles, the continuous phase flow field is first calculated, and then the force on each particle is solved by combining the flow field variables to obtain the particle velocity, so as to track the trajectory of each particle. The discrete phase model can not only calculate the trajectories of these particles, but also calculate the heat and mass transfer caused by the particles. In the process of calculation, the coupling between phases and the influence of the coupling results on the trajectories of discrete phases and the flow of continuous phases can be taken into account.

The interaction between droplets and walls is caused by many complex factors, such as wall material, solid wall temperature, wall roughness, incident angle of particles, incident velocity of particles, wall liquid film and so on. Because the inner surface temperature of the pipeline is higher than the boiling temperature of water, wall-jet model is used to deal with the process of droplet impingement.

In this paper, SST turbulence model is selected in the calculation process. The inlet turbulence is 5%. The inlet velocity of cooling gas is determined by the corresponding blowing ratio. The mainstream temperature is 1600K and the cooling gas inlet temperature is 800K. In order to calculate the adiabatic film cooling efficiency conveniently, the adiabatic wall is adopted. The two sides of the main gas channel are set as periodic boundary conditions; the wall velocity does not slip, and the convergence residual is less than 10^{-5} in the whole calculation process. Mass flow rate is 2%, 3.5% and 5%; main gas temperature is 1600K; droplet diameter is 10 μ m.

3.2. Parameter definition and feasibility analysis of computing moel

a. Blow ratio M

$$M = \frac{\rho_c U_c}{\rho_m U_m} \tag{1}$$

In the formula, ρ_c is the density of the cooling gas and ρ_m is the density of mainstream gas, U_c is the velocity of the cooling gas and U_m is the velocity of mainstream gas.

b. Adiabatic film cooling efficiency η_{aw}

$$\eta_{aw} = \frac{T_m - T_{aw}}{T_m - T_c} \tag{2}$$

In the formula: T_m - mainstream temperature, T_c - cooling temperature, T_{aw} - adiabatic wall temperature. It can be seen that the adiabatic film cooling efficiency is actually a dimensionless temperature value, hereinafter referred to as cooling efficiency.

c. Directional average adiabatic film cooling efficiency $\overline{\eta}_{aw}$

At present, the spread average adiabatic film cooling efficiency is widely used, which reflects the comprehensive cooling effect of a row of holes on downstream coverage, and has more practical significance in engineering. Therefore, the development of cooling efficiency along the mainstream direction is assessed by the spread average adiabatic film cooling efficiency in this project. Assuming that the spreading direction is x and its width is p, the spreading average adiabatic wall temperature is defined as:

$$\overline{T}_{aw} = \frac{1}{p}\int_0^p T_x dx \tag{3}$$

The extended average cooling efficiency is defined as:

$$\overline{\eta}_{aw} = \frac{T_m - \overline{T}_{aw}}{T_m - T_c} \tag{4}$$

T_m - Mainstream temperature, T_c - cooling temperature and $\overline{T}_{aw}$ - average adiabatic wall temperature. In this paper, the spread width p is taken as three times of the inlet diameter D of cooling gas.

3.3. Computing model verification

In order to verify the numerical calculation model in this paper, a geometric model with the same experimental conditions as in reference[12] is established. The comparison between the numerical calculation results and the experimental results is shown in Fig. 4. Although the calculated values are different from the experimental values, the trend of the numerical calculation model in this paper is in good agreement. It shows that the numerical calculation model established in this paper can be used for the trend comparison and analysis of various schemes.

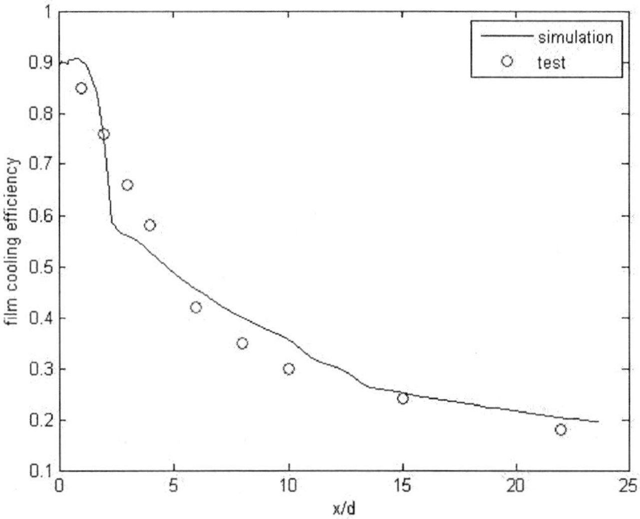

Figure 4. Film Cooling Efficiency of Circular Hole Center Line at M=0.5.

4. Analysis of calculation results

4.1. Effect of Spray Rate of Water Mist on Cooling Air Temperature and Physical Properties

Fig.5 shows the average temperature ratio and constant pressure specific heat ratio of the cooling air at the outlet of the film cooling hole of the circular-hole flat-plate model when the blowing ratio is 1.0 with different water spraying quantity. It can be seen from the figure that :the average temperature of cooling air at the outlet of film hole decreases significantly with the increase of water mist injection; the average specific heat capacity of the cooling medium at the outlet of the film hole increases gradually with the increase of water mist injection.

This is mainly because the latent heat of vaporization of water reduces the temperature of cooling medium, and the specific heat capacity of water at constant pressure is larger. When water and cooling gas are mixed, the specific heat capacity of cooling medium at constant pressure is larger than that of pure air.

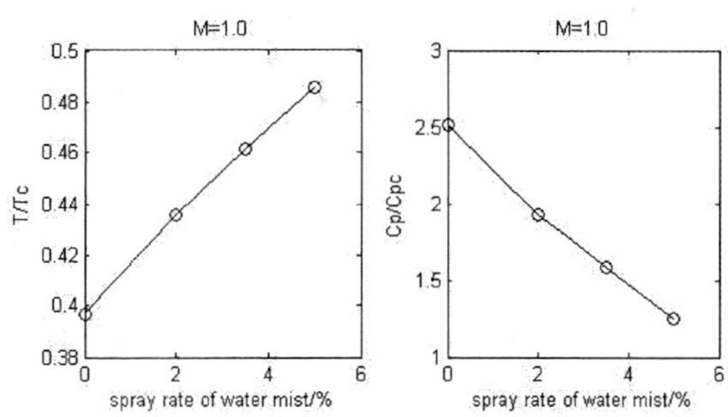

Figure 5. The influence of different water sprays on cooling air properties.

IOP Conf. Series: Materials Science and Engineering 1081 (2021) 012030 doi:10.1088/1757-899X/1081/1/012030

4.2. Effect of Spray Rate of Water Mist on Film Cooling in Two-Phase Flow of Flat Plate

Fig. 6 is a comparison of the extended average film cooling efficiency under different spray rates. It can be seen from the figure that the comprehensive performance of film cooling of water mist/air two-phase flow is significantly improved than that of traditional pure air film cooling. When the blowing ratio is 0.5, the film cooling efficiency of flat two-phase flow with 2% water injection is 37.5% higher than that of pure air film cooling at x/D=20. When the blowing ratio is 1.0, 2% water injection is added. Compared with pure air film cooling, the film cooling efficiency of flat plate two-phase flow at x/D=20 increases by 49.0%.

In order to further analyze the influence of water mist injection rate on film cooling effect of two-phase flow film cooling scheme with circular hole, the film cooling efficiency distribution on the cooling wall surface of two-phase flow film cooling with different water mist injection rate was analyzed in detail when blowing ratio was 0.5 and 1.0. Fig. 7 shows the distribution of adiabatic film cooling efficiency on the cooling stave of two-phase flow film cooling schemes with different water mist injection rates at blowing ratios of 0.5 and 1.0. After adding micro water mist into the cooling gas, the film cooling efficiency values along the downstream and forward direction of the film hole center line is significantly increased, and the value increases with the increase of water mist injection.

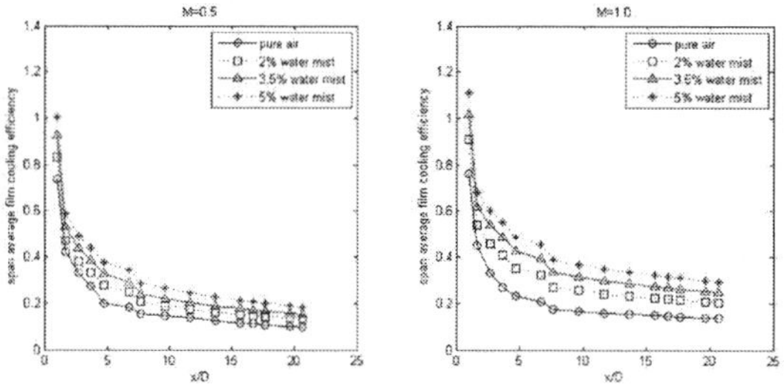

Figure 6. The influence of water injection rate of the circular hole on the spread average film cooling efficiency.

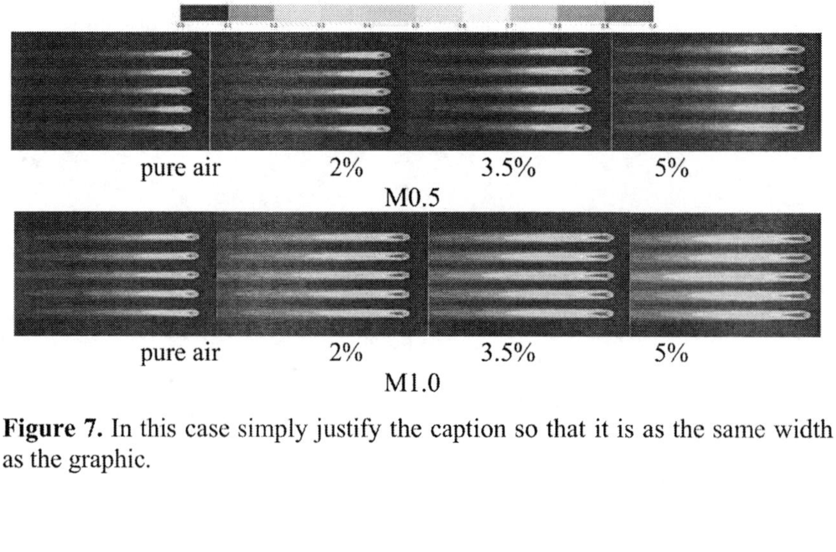

Figure 7. In this case simply justify the caption so that it is as the same width as the graphic.

IOP Conf. Series: Materials Science and Engineering 1081 (2021) 012030 doi:10.1088/1757-899X/1081/1/012030

4.3. Effect of hole geometry on Film Cooling of Flat Two-Phase Flow

Fig.8 and Fig.9 show the effects of different hole geometry on the film efficiency of two-phase flow under the conditions of blowing ratio 0.5 and blowing ratio 1.0, respectively, when the spray amount of water mist is 2%, 3.5% and 5%. It can be seen from the figure that the film cooling efficiency of two-phase flow in the scheme of the backward-expanding shoulder arm(BESA) hole is the highest, followed by expanding hole, and Shoulder-Arm hole is slightly higher than that of round hole, and the trend is more obvious with the increase of spray amount. It can be seen that the method of adding micro water mist to the cooling gas, combined with the selection of suitable hole geometry, can achieve better protection of the cooling stave surface under the same consumption of cooling gas. When other factors remain unchanged, the film cooling efficiency of BESA hole of two-phase flow increases monotonously with the increase of water mist injection from 2% to 5%.

In order to further explore the influence of hole geometry on film cooling efficiency of two-phase flow film cooling scheme, the film cooling efficiency distribution on the cooling stave surface of two-phase flow film cooling scheme with different hole geometry at blowing ratio of 1.0 was analyzed. Fig. 10 shows the distribution of adiabatic film cooling efficiency on the cooling stave of two-phase flow film cooling schemes with different hole geometry when blowing ratios at 1.0. It can be seen from the figure that the effective film coverage area of the four types of holes increases with the increase of water mist injection. Under the same water spray amount, the film cooling coverage of the backward-expanding shoulder arm(BESA) hole scheme is more uniform than that of the round hole, shoulder arm hole and progressively enlarged hole not only in the mainstream streamline direction, but also in the spreading direction.

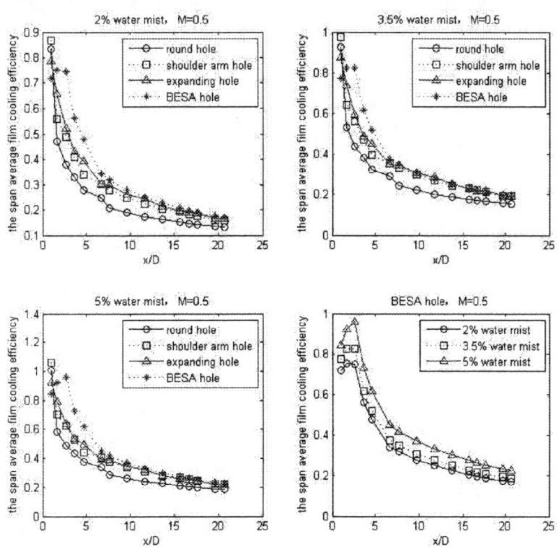

Figure 8. The effect of hole geometry on the spread average film cooling efficiency when M = 0.5 with different water mist.

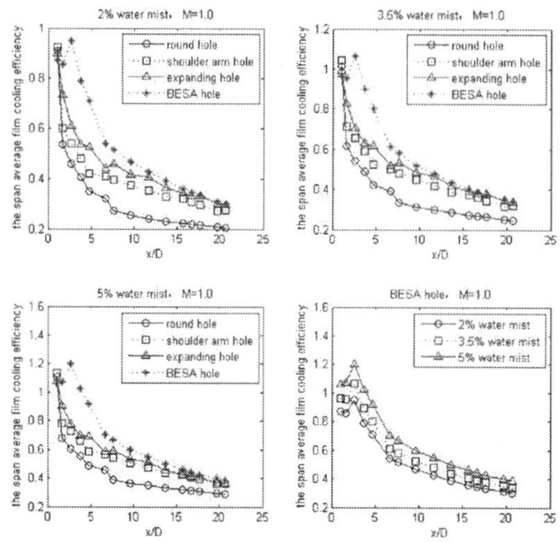

Figure 9. The effect of hole geometry on the spread average film cooling efficiency when M = 1.0 with different water mist.

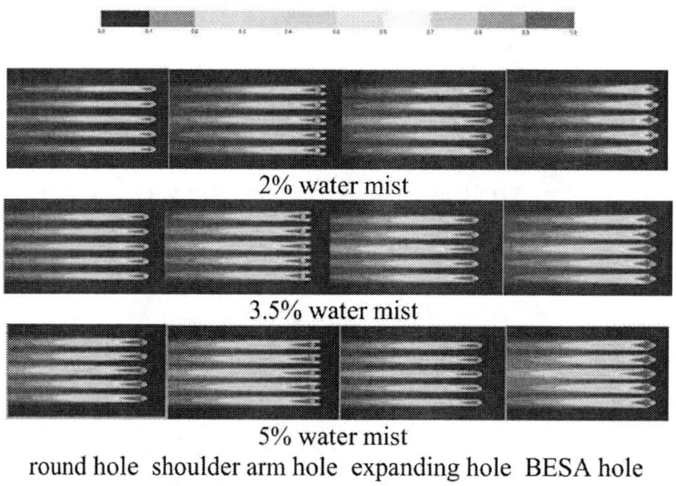

Figure 10. Effect of hole geometry on Film Cooling Efficiency Distribution with Different Water Mist Injection Rates when M = 1.0.

4.4. Effect of hole geometry on Two-Phase Flow Film Cooling of Turbine Blades

In order to study the effect of different hole geometry schemes on film cooling of two-phase flow in turbine blade, a row of film cooling holes were arranged on the pressure surface of turbine cascade with water mist injection rate of 2%.

Fig. 11 shows the film cooling efficiency distribution of different hole geometry schemes on the pressure surface of the turbine cascade. The blowing ratio of the cooling hole on the pressure surface is 1.0. From the figure, it can be seen that the air film distribution on the pressure surface is very uniform in scheme of BESA hole, and the blade surface can be well protected. The shoulder arm hole and the expanding hole form a good effective gas film covering along the main stream streamline direction on the pressure surface, but the coverage effect is slightly worse than that of the BESA hole scheme. The film cooling efficiency of the round hole scheme is lower than that of the other three hole schemes.

From the above results, it can be seen that the research results of different hole geometry schemes on the turbine cascade are basically the same as those obtained in the plate model. The film cooling effect of two-phase flow can be significantly improved by the use of BESA holes on flat plate model and turbine blade.

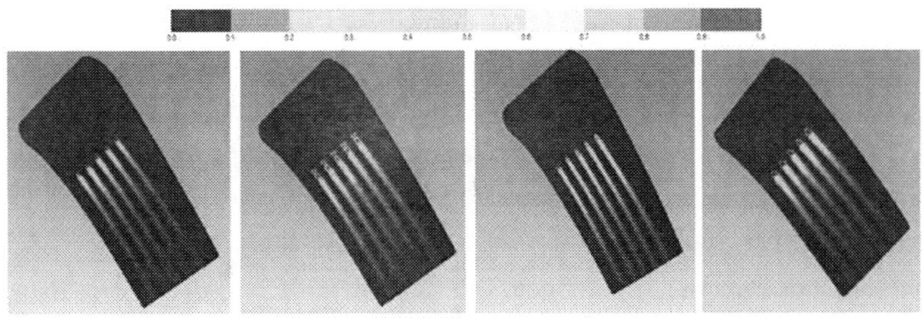

round hole shoulder arm hole expanding hole BESA hole

Figure 11. Distribution Chart of Film Cooling Efficiency on pressure Surface of Turbine Blade.

5. Conclusion

Through the study of film cooling characteristics of two-phase flow in flat plate model and turbine stator blade model with different water spray quantity, water spray location, blowing ratio, hole geometry, the following understandings are obtained:

(1) Compared with single-phase pure air film cooling, the adiabatic film cooling performance of mist/air two-phase flow is significantly improved. For the circular hole plate model, when the blowing ratio is 0.5, the film cooling efficiency of two-phase flow with 2% water injection is 37.5% higher than that of pure air film cooling. When the blowing ratio is 1.0, the film cooling efficiency of two-phase flow with 2% water injection is 49.0% higher than that of pure air film cooling.

(2) Under the same blowing ratio and other conditions, the spray volume of water mist increases from 0% to 5%, the dimensionless temperature of the mixture of cooling gas and water mist decreases from 1.0 to 0.498, and the dimensionless specific pressure heat capacity increases from 1.0 to 1.223.

(3) Under the same blowing ratio and other conditions, the spray amount of water mist increases from 2% to 5%, and the spreading average film cooling efficiency of two-phase flow increases gradually, showing a monotonic increasing trend.

(4) The effect of different hole geometry types on film cooling of two-phase flow was studied. It was found that, under the same other conditions, the film cooling efficiency of BESA hole was significantly higher than that of expansion hole, Shoulder-Arm hole and round hole. It was concluded that choosing appropriate pass types in film cooling of two-phase flow could further improve the cooling effect.

(5) The study of the effect of the hole geometry on the film cooling of two-phase flow in turbine blade shows that the effect of the hole geometry on the film cooling of turbine blade is similar to that

obtained from the study of plate film cooling. The use of the BESA hole in both flat plate model and turbine blade can significantly improve the two-phase flow Film cooling efficient.

References

[1] Zhu Uan-xin,Tan Xiao-ming,Guo Wen, et al.Numerical Simulation on Effects of Different Film Cooling Holes on Plat[J]Journal of Propulsion Technology,Apr. 2013,Vol. 34 No. 4

[2] Ge Shao-yan, Xu Jing-zhong. Film cooling[M].Bei Jing:Science Press,1985.

[3] LI Shao-hua, SONG Dong-hui, LIU Jian-hong, et al. Numerical Simulations of Flat Plate Film Cooling Using Respectively Different Shaped Jet Holes[J].Proceedings of the CSEE, Vol.26 No.17 Sep. 2006.

[4] Leylek J. H. , Zerkle R. D. . Discrete Jet Film-cooling: A Comparison of Computational Results with Experiments. ASME Journal of Transaction, 1994,116: 358-368.

[5] Goldstein R. J. , Eckert E. R., Burggraf F. . Effects of Hole Geometry and Density on Three-dimensional Film Cooling. International Journal of Heat Mass Transfer, 17: 595-607.

[6] Okita Y,Nishiura M. Film Effectiveness Performance of an Arrowhead-Shaped Film-Cooling Hole Geometry. Journal of Turbomachinery. 2007,129(2):331-339.

[7] Kusterer K, Bohn D, Sugimoto T, et al. Double-Jet Ejection of Cooling Air for Improved Film Cooling. Journal of Turbomachinery. 2007, 129(4); 809-815.

[8] Goldstein, R. J.Adv. Heat Transfer[M]. 1971, 7, pp. 321–379.

[9] Gustafson R., Mahmood G I, and Acharya S. Flowfield in a Film-Cooled Three-Dimensional Endwall Passage: Aerodynamic Measurements[C].ASME Paper 2007,No. GT2007-28154.

[10] James D.Heidmann , Srinath Ekkad. A Novel Antivortex Turbine Film-Cooling Hole Concept[J]. Journal of Turbomachinery, JULY 2008, Vol.130/031020-1.

[11] Huang Kang, Ma Hu-sheng. Numerical simulation on cooling characteristics of plate film from backward-expanding shoulder arm hole [J],ACTA AERODYNAMICA SINICA.2017,Vol.35,No.6 886–889.

[12] Sinha A K, Bogard D G. Film-cooling effectiveness downstream of a single row of holes with variable density ratio[J]. ASME Journal of Turbomachinery, 1991, 113(3): 441-449.

[13] Kapadia S, Roy S, and Heidmann J. First Hybrid Turbulence Modeling for Turbine Blade Cooling[J]. Thermophys Heat Transfer,2004, 181, pp.154–156.

[14] Bunker R. S.,Film Cooling Effectiveness Due to Discrete Holes Within a Transverse Surface Slot[C]. ASME Paper 2002,No. GT-2002-30178.

[15] Dhungel S, Phillips A, Ekkad S V, et al. Experimental Investigation of a Novel Anti-Vortex Film Cooling Hole Design[C],ASME Paper No. GT2007–27419.

[16] Bogard D G, and Thole K A. Gas Turbine Film Cooling,[J]. Propul Power, 2006, 222, pp. 249–270.

[17] Ekkad S V, Ou S, and Rivir R. B. Effect of Jet Pulsation and Duty Cycle on Film Cooling From a Single Jet on a Leading Edge Model[C], ASME J. Turbomach., 2005,128, pp. 564–571.

[18] Heng-jie Shan, Dao-rui Zhang,Rui-hua Li,et al.Analysis and Treatment of Vibration for an Adjustable Primary Axial Fan[J].Chinese Journal of Turbomachinery,2017(02).

[19] Ma Xin, Sun Lili,Chen Yan,et al.Design of Supported Thrust Bearing in Single Stage Recycle

Gas Compressor[J].Chinese Journal of Turbomachinery,2014(03).

[20] Chen Huaxin,Hou Bairong.Vibration Analysis and Solution of Rotor Blade Adjustable Axial Induced Draft Fan for 1000MW Thermal Power[J].Chinese Journal of Turbomachinery,2015(02).

[21] Zhang Dengchun, Chen Huanxin,Chen Yazhou,et al.Numerical Simulation of Internal Airflow Field in a Centrifugal Blower[J].Chinese Journal of Turbomachinery,2015(03).

[22] Jing Ling, Qian-kun Liu,Zhen-qin Ji,et al.Effect of the Cascade Solidity on the Performance of a Multi-blade Centrifugal Fan[J].Chinese Journal of Turbomachinery,2017(03).

Experimental and numerical investigation on the influence of the diffuser blade setting angle on the performance of centrifugal compressor

Junyi Fang[1], Chunjun Ji[1], Chunyang Li[1], Ruiwu Zhang2 and Shalyaev Denis[2]
[1] School of Power Engineering. Dalian University of Technology, Dalian, China,

[2] Liaoning Fu An gas turbine Co. Anshan, PR China

E-mail: 719783952@qq.com;

Abstract: Setting angle of the front blade row of tandem cascade diffuser was investigated experimentally to understand its influence on the performance of centrifugal compressor. Three diffusers with different setting angles were tested. The change of front blade setting angle mainly alters. The angle of attack (AOA) on the leading edge of front blades, which affects the surge point of the compressor. The throat area of the blade row is also modified, which affect the chock point of the compressor. With the increase of setting angle, performance curves of centrifugal compressor move to large flow rate side, the operation range is increased at the same time. In addition, numerical simulation was used to investigate the flow details on the passage of diffuser blades. Overall, the setting angle of front blade row has a significant influence of the performance of centrifugal compressor.

Nomenclature

AOA	Angle of attack, °
b	Blade height, mm
C_p	static pressure recovery coefficient
D	Setting diameter, mm
G_m	Mass flow rate, kg/s
Ma	Relative Mach number
k	Adiabatic exponent
P	Static pressure

P^*	Total pressure
R	Ideal gas constant
β	Setting angle
$\triangle K_y$	Surge margin
π^*	Total pressure ratio
η	Isentropic efficiency
ζ	Total pressure loss coefficient

Subscript and superscript

3	Leading edge of front blade row
4	Trailing edge of front blade row
5	Leading edge of rear blade row
6	Trailing edge of front blade row
*	"Total" state
design	Design point
surge	Surge point

1. Introduction

The design of diffuser is an important part of a compressor design, especially for the design of transonic compressors. Transonic compressor has a high rotating speed and large backward curved angle, it means the flow angle at the impeller outlet will be large and the flow speed will be fast. In this condition, the diffuser will bear a large load. A large flow angle will cause more friction loss[1] . Vane-diffuser is widely used in transonic compressor, many researches[2-4] showed that vane-diffuser could generally attain a higher efficiency and pressure ratio than vane less diffuser. The single row vane-diffuser has a good performance but its operation range is narrow. To compromise, low solidity vane-diffuser was put forward[5]. Another choice is the tandem cascade diffuser[6]. Tandem cascade diffuser can get a better performance than single row vane-diffuser, its flow field is more uniform[7].

The current research on cascade diffusers mainly focuses on the aspect ratio[8], blade bending angle[9], number of blades, and on the relative positions of the tandem cascades[10-12]. These studies have accumulated an extensive number of calculations and experimental data. Liu Lei used numerical simulations to reconstruct the diffuser of a centrifugal compressor, and quantified the different performances with the three chord length ratios of 0.7, 1.0, and 2.0[13]. The results indicate that the efficiency of the cascade diffuser is better than that of the wedge diffuser, and the compressor stability margin can be significantly widened by the cascade diffuser. The circumferential offset has considerable influences on the load distribution of the front and rear row blades. When the circumferential offset increases, the load of the front row will also increase. Minor changes of the inlet flow angle of the rear row are documented as a function of the inlet attack angle of the tandem cascade. However, when the front row is associated with a large attack angle, the rear row will have a small attack angle so that the tandem cascade can maintain a small loss within a large range of attack angles[14]. Zhou[15] used an experimental design method to improve the load of tandem cascades and noted that the inlet stagger angle of the rear row has a significant influence on the response value of the total pressure ratio and isentropic efficiency. Studies on the flow angle of diffuser blades are mostly aimed at the single blade diffuser. Tamaki[16] showed the effect of the diffuser throat area on the stable range of a centrifugal compressor and categorized the matching between an impeller and vaned diffuser into four types. The other experiment of Tamaki presented the effect of diffuser vane setting angle on the performance of a centrifugal compressor with volute. The experiment provided the evidence that the choke flow was independent of the van setting angle but determined by the throat area. The surge flow rate tended to decrease as the vane setting angle increase. Being closed to the impeller, vaned diffuser will be greatly affected by the unsteady flow form the impeller outlet, it is called impeller-diffuser interactions. Zhao[17] indicated that the generation and shedding of leading-edge vortex is closely related to the altering inlet flow angle in the circumferential direction. Anish[18] noted that the main factor influencing the diffuser instability is the circumferential variation of the flow angle at the leading edge of the diffuser vane. The aerodynamic loss in the diffuser is induced mainly by the generation and shedding of the vortexes which exist on the diffuser vane leading edge, on the suction side of the diffuser vane near the hub and on the diffuser vane trailing edge.

However, there are few researches on the setting angle of the front vanes of the tandem cascade diffuser. In this study, three different setting angles of the front blades of a tandem cascade diffuser are simulated and tested on a centrifugal compressor. This study proves that the setting angle plays an important row in the tandem cascade, it greatly affect the performance of the centrifugal compressor. What is more, the setting angle will play a same role as the relative positions with the tandem cascades. The results of this study have high reference value for the design of tandem cascade diffuser.

2. Geometric model and test procedures

The diffuser is from a centrifugal compressor of a 4 MW gas turbine. It is the second stage of the gas turbine compressor. At design point, its mass flow rate is 5.2 kg/s, pressure ratio is 2.56 and efficiency is 0.85. There are 13 main blades and 13 split blades on the impeller, and 23 blades in each row of the diffuser. Figure 1 presents the flow path and the geometric model of the diffuser.

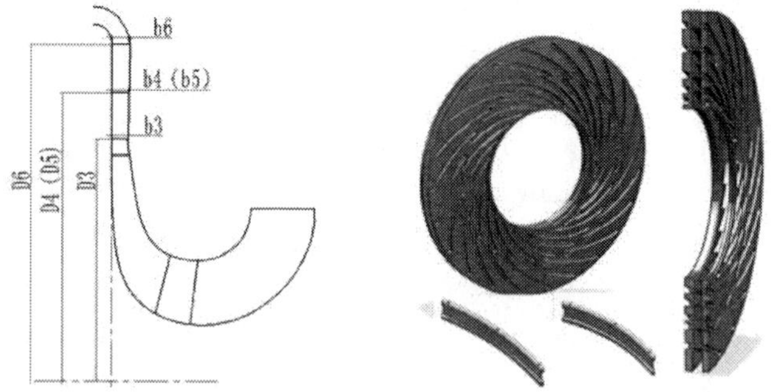

Fig 1. Meridional flow path and geometric model of diffuser.

Table 1 lists the geometric parameters of the diffuser. This study provides 3 different setting angles, named as case 1, 2 and 3 respectively. In the three cases only the setting angle of the front blade row is changed, the vane profile remains the same. Figure 2 shows the relative position of the front and rear rows and the test model. It is showed that the front row of blade is detachable, so that all the three cases could be tested on the compressor. When the setting angle changes, the chink between the front and rear blades change greatly the throat area of front blade is also changed.

Table 1. Geometric parameters of the diffuser

	Case 1	Case 2	Case 3
Inlet setting angle of first row, β_3(°)	67.3	69.3	71.3
Outlet setting angle of first row, β_4(°)	64.0	66.0	68.0
Inlet setting angle of second row, β_5(°)	66.0	66.0	66.0
Outlet setting angle of second row, β_6(°)	60.0	60.0	60.0
Blade Number of first row, Z_3	23	23	23
Blade Number of second row, Z_4	23	23	23
Throat area, mm^2	11232	10200	9167

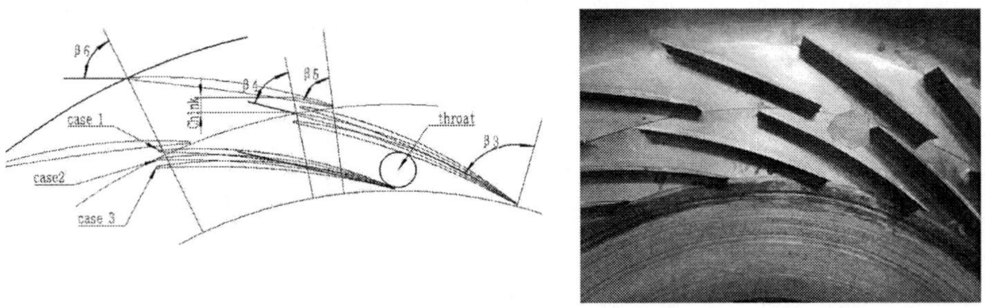

Fig 2. Relative position of the blades and the test model

Figures 3 and 4 show the compressor instrumentation and the diagram of test facility. There are six fixed measuring positions showed in Figure 4, which are used to measure the environmental data and the stage performance parameters of the compressor during the experiment. Figure 3 presents the additional measuring positions at the inlet of impeller, inlet of the diffuser and outlet of the diffuser, each position has probes for P^*, P and T^*. The systematic error of the probe is 0.025%. During the experiment, the data were collected after 30 minutes of stable operation. All the total pressure and total temperature were obtained as time and area averaged.

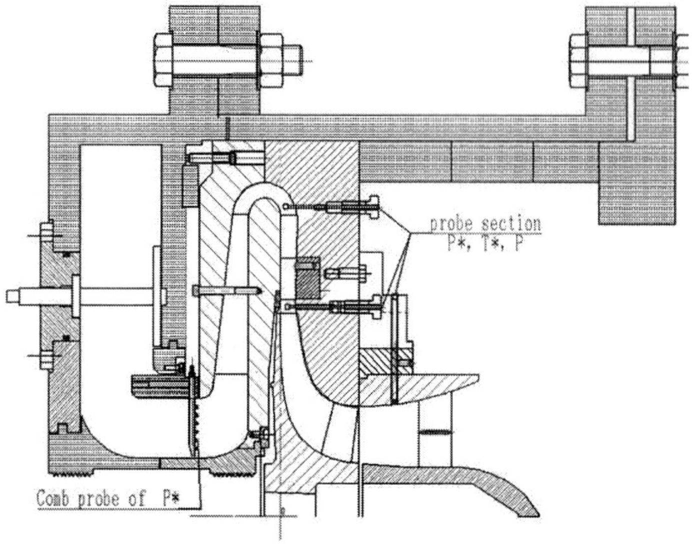

Fig 3. Instrumentation for diffuser

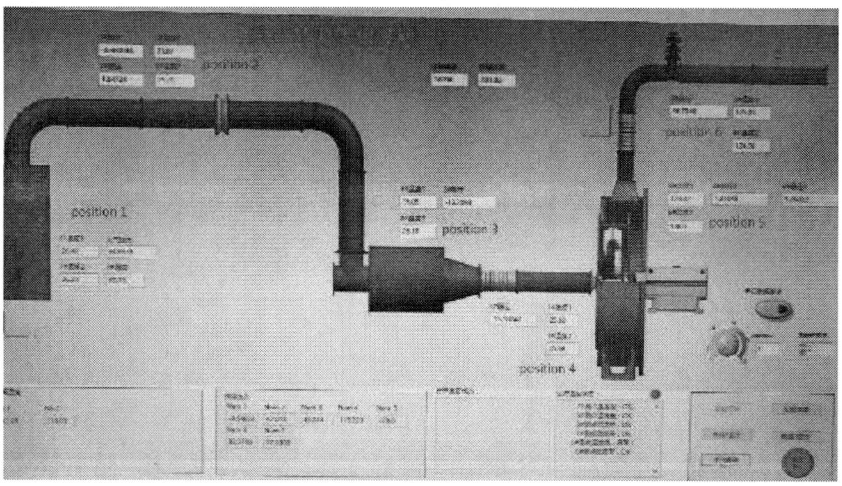

Fig 4. Test setup

3. Experimental result

The experiment for this study is a performance test, and the raw data of the experiment is P^*, P and T^* at each measuring position. The total pressure ratio (π^*) is define in formula 1, and isentropic efficiency (η) formula 2. The result is showed on Figure 5. It is obviously that the three cases make a great difference on the performance characteristics. With the increase of the setting angle, the total pressure ratio is slightly increased, and the efficiency is slightly reduced. The operation range is also different, case 3 has the minimum surge margin ($\triangle K_y$) and case 1 has a maximum surge margin. $\triangle K_y$ is define in formula 3. There is an abnormal point in case 1, on which point the compressor is already surged. Table 2 presents the surge point and $\triangle K_y$ of the experiment.

$$\pi^* = \frac{P_6^*}{P_1^*} \quad (1)$$

$$\eta = \frac{\pi^{*(k-1)\times k^{-1}}}{\frac{T_6^*}{T_1^*}-1} \quad (2)$$

$$\triangle K_y = \left[\frac{\pi^*_{surge}\times G_{m,design}}{\pi^*_{design}\times G_{m,surge}} - 1\right] \times 100\% \quad (3)$$

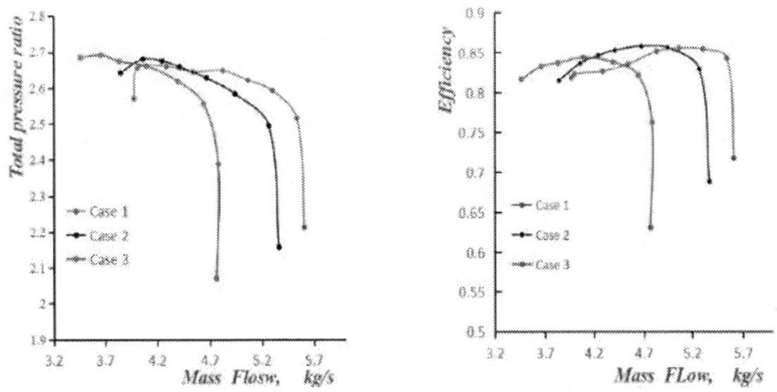

(a) total pressure (b) efficiency

Fig 5. Performance curves of the compressor

Table 2. Surge point and $\triangle K_y$

	$G_{m,surge}$ kg/s	π^*_{surge}	$G_{m,design}$ Kg/s	π^*_{design}	$\triangle K_y$ %
Case 1	4.011	2.65	5.30	2.59	35.53
Case 2	3.84	2.64	4.94	2.58	31.48
Case 3	3.46	2.68	4.39	2.61	30.14

On the flow rate of 5.2 kg/s, which is the target design point, case 1 has the best performance. While under the condition of small flow rates, case 2 is better than case 1. Case 3 fails to provide sufficient mass flow to meeting the requirement of design.

The total pressure loss coefficient (ζ) is widely used to evaluate the flow loss of diffuser. Figure 6 present the change of ζ with the flow rate of the three cases. Each case has a minimum ζ point, on the right side of this point, the loss increases quickly with the flow rate. Overall, ζ of case 1 is lower than the other cases. It is consistent with figure 5, on the condition of large flow rates, case 1 shows higher pressure ratios.

$$\zeta = \frac{P_3^* - P_6^*}{P_3^* - P_3} \quad (4)$$

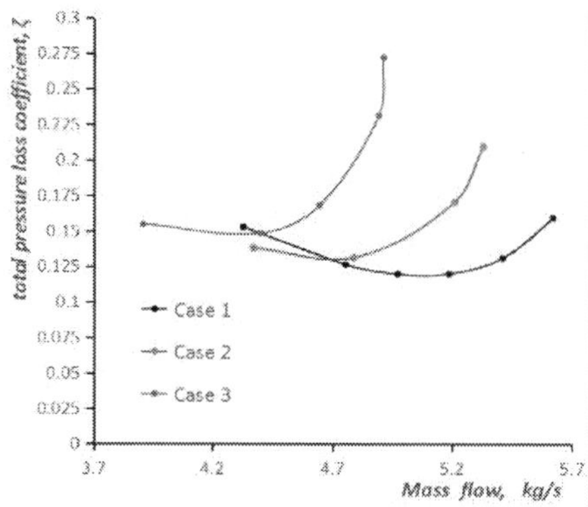

Fig 6. Total pressure loss coefficient (ζ)

The experiment could only provide an overall performance. To investigate the mechanism of the experimental result, a numerical simulation was made for this study.

4. Numerical method and analyses

In this study the software NUMECA/Fine turbo was use to perform the numerical computations. It contains a solver to 3-D compressible Reynolds-averaged Navier-Stokes equations, which employs an explicit fourth-order Runge–Kutta scheme for the temporal discretization and a second-order central-differenced scheme for the spatial discretization. The shear-stress transport (SST) model is chosen as the turbulence model, because it is more accurate for gap flows[19]. Another advantage for SST-extend wall function is that the Y^+ allowed is 20~50, it means that under the limit of grid aspect ratio, mesh for this turbulence model will allow a bigger wall grid size. In this way, the mesh number can be less than other models that without wall extend function.

The mesh is created using "Auto Grid 5", the mesh number for the compressor is 2.2 million, mesh quality totally meets the requirement of the software. Grid independence verification was carried out to ensure the mesh number is enough for the accuracy in computation. Figure 7 shows the mesh of the compressor, there is a filet at the root of the diffuser blade, so that the model is more consistent with the real diffuser. Periodic conditions are adopted in the grid. Periodicity of impeller is 13, the diffuser is 23. There are three rotor-stator (R/S) surface in the mesh. R/S surface between the impeller and diffuser is set as conservative coupling by pitch wise row, which means a strict conservation of mass, momentum and energy. R/S surface between the two blade rows of diffuser is set as full non matching frozen rotor, which acquire the periodicity of upstream and downstream must be equal and ignore the rotating speed in calculation.

IOP Conf. Series: Materials Science and Engineering 1081 (2021) 012031 doi:10.1088/1757-899X/1081/1/012031

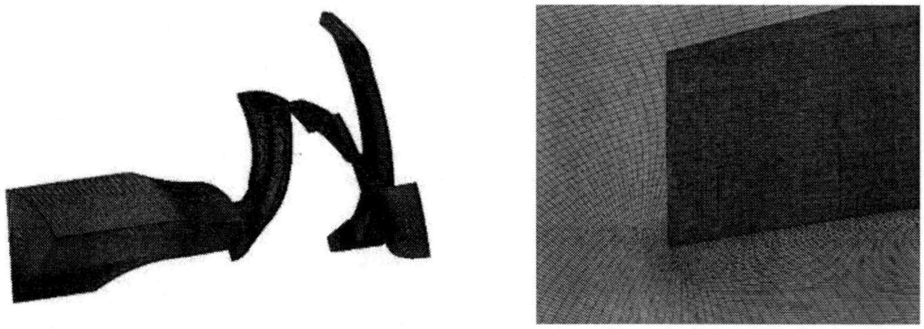

Fig 7. Mesh of the compressor and leading edge grid of the diffuser blade.

Figure 8 presents the comparison between the calculation and the experiment. With the speed of 87% and 100% to the design speed, the comparison shows little differences. The 105% speed curves show a slight larger error, that is because under the speed of 105%, the power of electric motor cannot provide enough power, so an inlet throttling have to be used to complete the performance curve near surge (the brown curve labelled 105% TEST RUN). In this study only the performance of 100% speed is analyzed, but it must be noted that the conclusion of this study is applicable to other speeds.

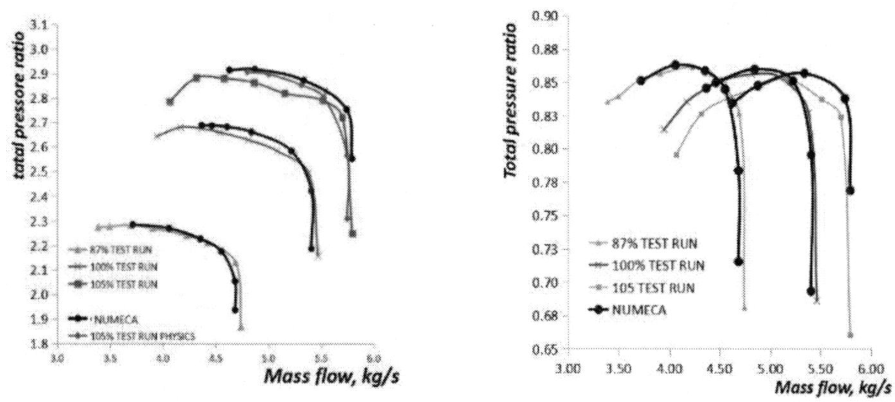

(a) Pressure ratio (b) Efficiency

Fig 8. Comparison of the compressor performance between the experiment and calculation

The static pressure coefficient (C_p) of the diffusers, which is defined in expression (5):

$$C_p = \frac{P_s - P_3}{P_3^* - P_3} \tag{5}$$

Figure 9 presents the C_p at design point of the three cases. In case 1, there is a stagger between the front and read blade rows, meaning that the outlet diameter of the front blade row is larger than the inlet diameter of the rear blade row. In case 2, the two diameters are equal, and in case 3 the outlet diameter of the front blades is smaller than the inlet diameter of the rear blades. As can be seen in the figure, in case 1, the front blades carry more load than the other cases, while in case 3 the opposite is true. Both case 1 and case 2 show good static pressure distribution at the front row, but case 3 shows a large negative loading near the leading edge of the front blades. It means that the leading edge of the diffuser has a positive attack angle at design point of case 3. In this condition, case 3 will get a smaller surge margin than others. Case 1 still keep a negative incidence angle at the leading edge of the front blades at the design mass flow. The load distribution of the rear blade is also affected, in case 3 there is a big pressure difference between the pressure surface and suction surface of the rear blade. Too much load on the rear blades will have a negative effect on compressor efficiency.

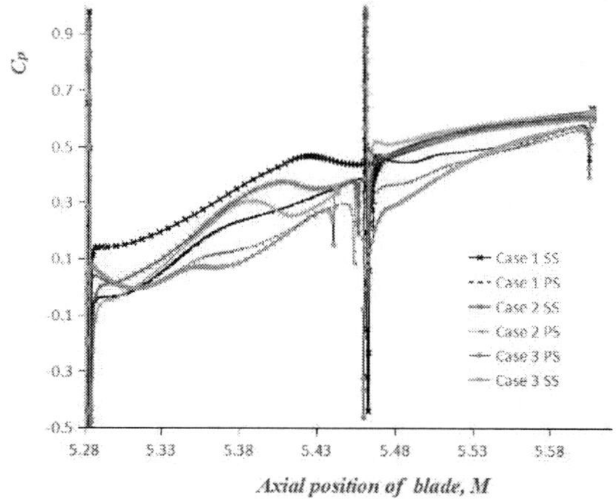

Fig 9. Cp distribution on 50% span of the diffuser blades at design point.

Figure 10 and 11 present the velocity vector and flow angle at design point. It is obviously that at the leading edge of front blade row, case 1 has a flow angle that about 2~3 degree smaller than case 3, while the setting angle of case 1 is 4 degree smaller than case 3. It can be seen from the streanline that case 1 has a very small negative angle of attack (AOA) on the leading edge of front blade. Flow angle of case 2 is slightly higher than case 1 on the leading edge of front blade, while it is obviously higher than case 1 on the leading edge of the rear blades. In all sections the case 3 has a highest flow angle. Compared with the setting angle, case 3 has a small positive AOA at design point. This result is consistent with that in figure 9. Tandem cascades are effective in suppressing the separate flow on the diffuser blades. Figure 10 shows the start point of separation is near the trailing edge of SS. In Case 1, affected by the narrow gap between the front and rear blade rows, the separation on front blades is weaker than other cases. Another obvious difference is that in case1 the wake of front blades is very weak in the passage of the rear blades. But in case 3, the influence area of wake occupies almost two-thirds of the passage. The influence of wake results in poor uniformity in the passage, and then the pressure difference between PS and SS of the rear blades is rising, the load is rising at the same time. It should be noted that in case 1, the rear blades are set as splitter blade of the front blade row in the mesh, because of the stagger. In this condition, there is no R/S surface between the two blade rows, and no break in the streamline.

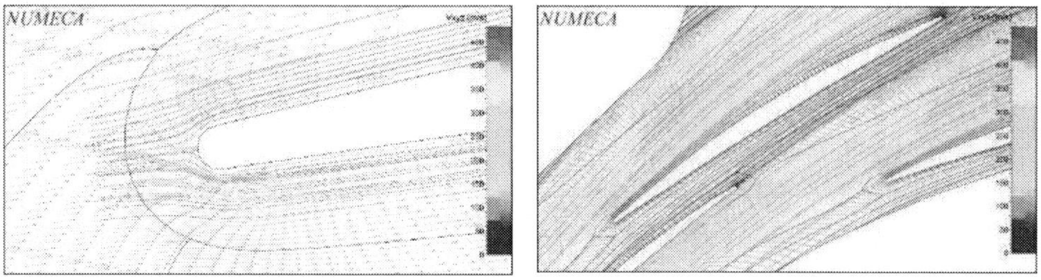

Case 1 (a) (b)

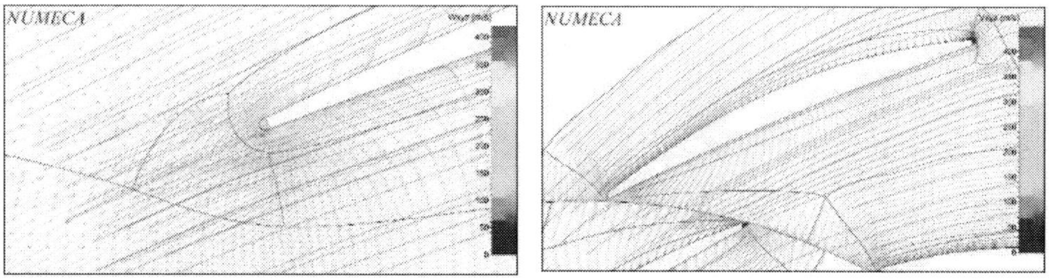

Case 2 (a) (b)

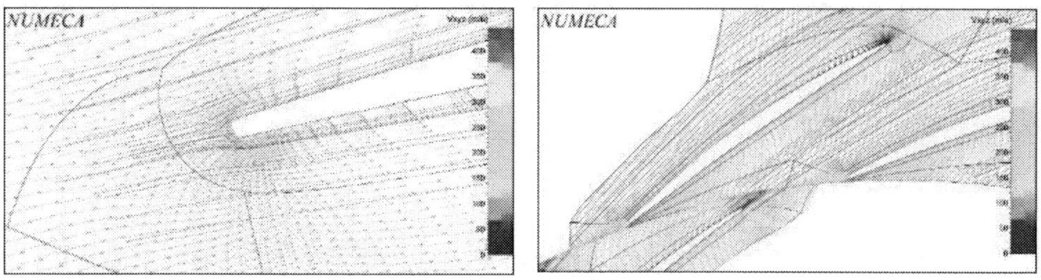

Case 3 (a) (b)

Fig10. Velocity vector, (a) leading edge of front blade, (b) rear blade.

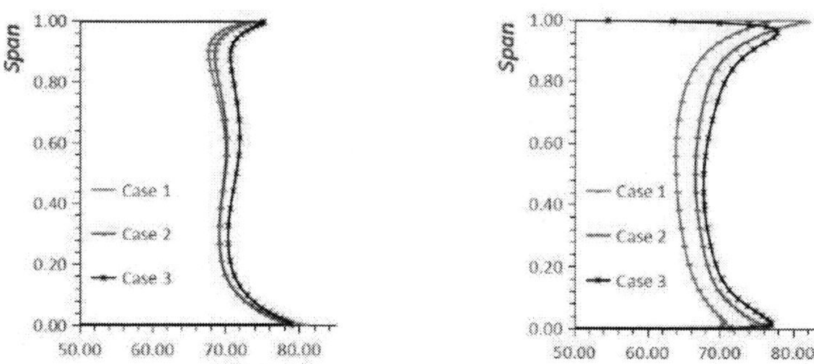

(a) (b)

Fig 11. Flow angle at the leading edge of blade, (a) front blade; (b) rear blade.

The relative Mach number (Ma) distribution on the 5%, 50% and 95% spans are showed in Fiugre 12. As a suoolenemt to figure 10, Ma can reflect the velocity variation in the diffuser passage more directly. Relatively, Case 1 and case 2 have a higher Ma on the outlet of diffuser, that is because they have a higher mass flow rate at design point. Case 3 has the most change of Ma form inlet to outlet, which means a larger static pressure rise and a higher ζ. Ma distribution of case 1 is more uniform, so the load changes more evenly. The narrow gap between the front and rear blades of case 1 results in a jet flow, which effectively depress the wake of the frond blade. The boundary later separation is a main reason for the unstable operation. A jet flow will also blew away the boundary layer on the pressure surface of the rear blade. In this way, case 1 presents a lower ζ.

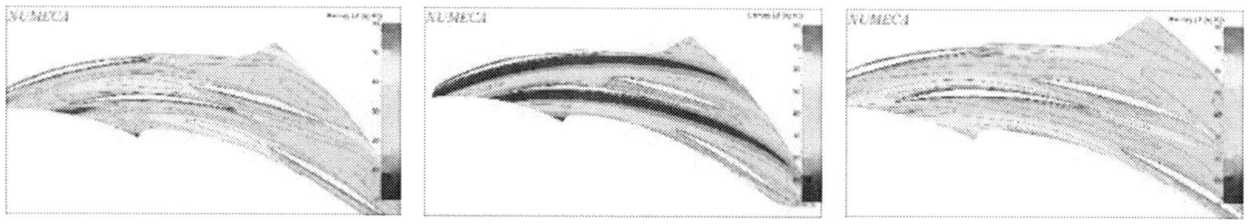

Ca

se 1 5% span 50% 95%

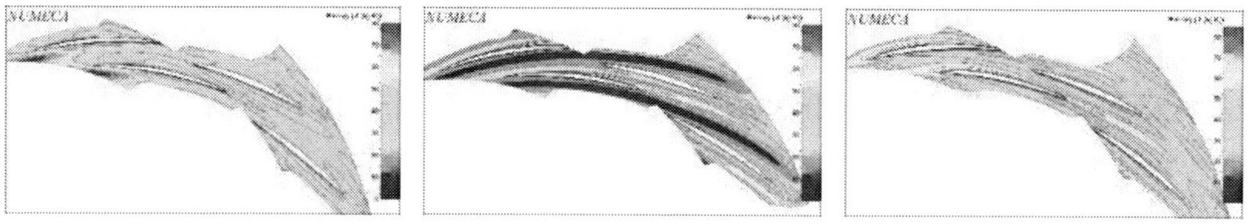

Ca

se 2 5% span 50% 95%

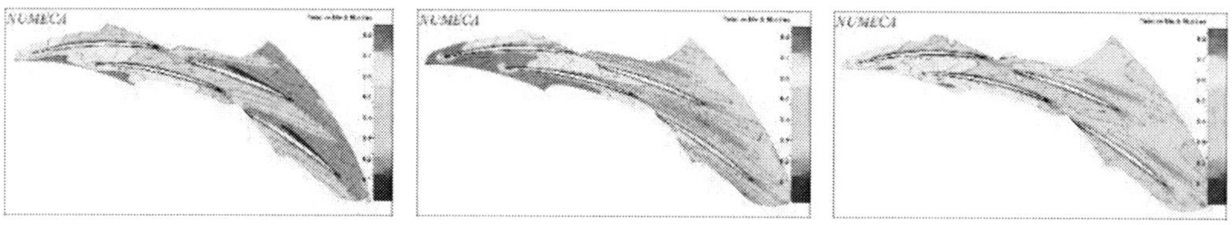

Case 3 5% span 50% 95%

Fig 12. Ma distribution at 5%, 50% and 95% spans at design point.

Figure 13 is the entropy distribution. The entropy distribution has little difference in three cases. In case 1 and case 2, the low entropy flow avoid the rear blades. The fluid with relatively high entropy passes through the rear blade, in this condition, the entropy will not increased too much by the rear blades. In case 3 the fluid with lowest entropy just passes the rear blades, that means the entropy increase is larger. That is , case 3 has more energy loss, and lower efficiency at design point. The highest entropy fluid is mainly on the wake of blades, while in case 1, the wake of front blade is obviously depressed by the gap between the front and rear blades. With the increase of setting angle, the uniformity of entropy distribution is getting worse, it provide that the loss is increased. The entropy distribution display a consistent conclusion with figure 5 and 6, case 1 has the lowest energy loss and highest efficiency.

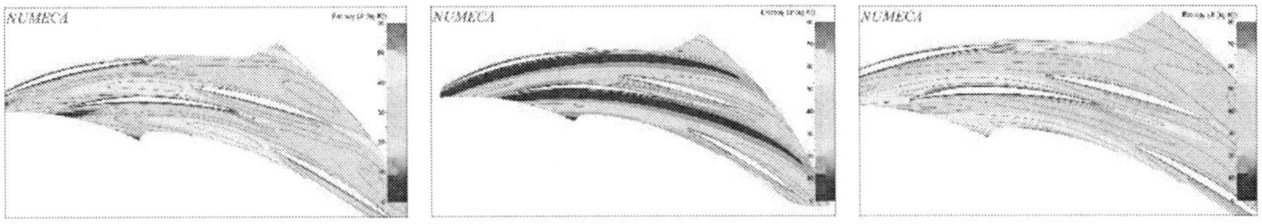

Case 1 5% span 50% 95%

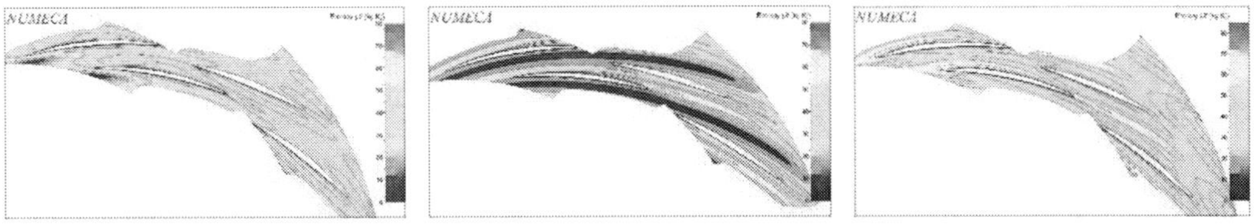

Case 2 5% span 50% 95%

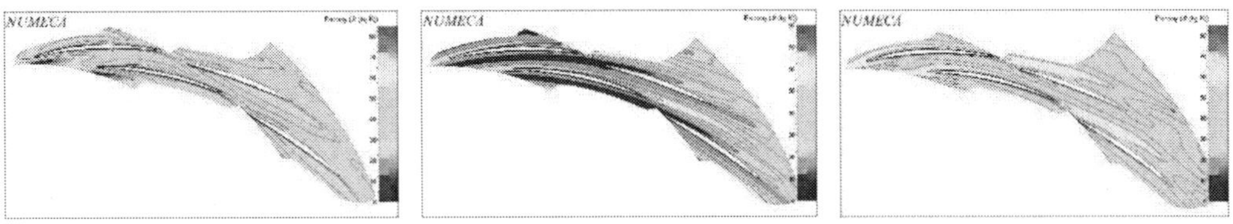

Case 3 5% span 50% 95%

Fig 13. Entropy distribution at 5%, 50% and 95% spans at design point

Figure 14 is the static and Ma disrtribution on the flow path of case 1. It is showed that on chock point, the sudden drop of static pressure and critical mach number are all located in the throat area of the front blade. This indicates that the chock mainly occurs in the throat of the front blade. With the change of setting angel, the throat area is also changed. Case 1 has the largest throat area, so it has the largest mass flow rate on chock point.

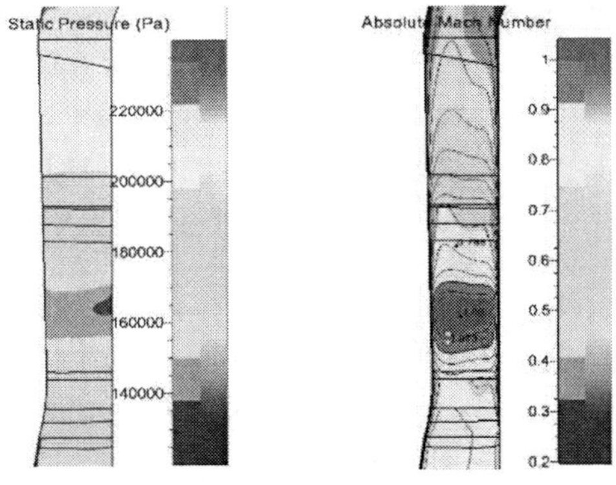

Fig 14. Static and Ma distribution on the flow path, on chock point.

5. Conclusions

In this study, the setting angle of front blade of tandem cascade diffuser influence on the performance of a centrifugal compressor is investigated experimentally. In addition, the numerical simulation is used to analysis the flow details. The main conclusions are obtained as following:

1) The setting angle of front blade of tandem cascade diffuser will greatly influence the performance of centrifugal compressor. A 2° change on the setting angle will change the surge margin by 1.5%~4.2%, the design point will also move to small flow rate side with the increase of setting angle. Moreover, lower setting angle will perform a higher efficiency.
2) Under the condition of constant vane profile, the change of setting angle will change the relative position of the front blades and rear blades. The narrow gap between the front and rear blade will inhibit the wake of the front blade, which is beneficial to reduce the load of the rear blades and energy loss.
3) Surge and chock all occur in the passage of front blade. Change of setting angle changes the AOA, which affect the surge point. The change of throat mainly affects the chock point.

Reference

[1] Zangeneh M, Schleer M and Ploger F, et al (2004).Investigation of an inversely designed centrifugal compressor stage part 1 : Design and numerical verification. Journal of Turbomachinery,126: 73-81 .

[2] Moustapha H (2003). Small Gas Turbine Technology: Evolution and Challenge. AIAA :2559

[3] Dehkharqani A, Boroomand M and Eshraghi, H (2014).A numerical investigation of loss coefficient variation in various incidence angles in tandem blades cascade. Volume 1: Advances in aerospace technology, doi:10.111539881

[4] Tan Gaoming, He xianghui and Liu Houlin, et al (2016), Design and analysis of a radial diffuser in a single-stage centrifugal pump. Engineering Applications of Computational Fluid Mechanics,10:1,500-511

[5] Senoo Y (1981). Low solidity circular cascade for flow range blower. Proceedings of advanced concepts in Turbomachinery. Hanover, New hampshire.

[6] Lars M, Dragan K and Detlev W, et al (2011). High Turning Compressor Tandem Cascade for High Subsonic Flows Part 2:Numerical and Experimental Investigations, AIAA, 5602

[7] Obukhov O, Smirnov and Gysak O (2015). Numerical and experimental investigation of efficiency of waned diffuser of centrifugal compressor. International communication in heat and mass transfer available online.649-658

[8] Sanger L (1971). Analytical study of the effects of geometric changes on the flow characters of tandem blade compressor states. NASA TN, d-6264.

[9] Hautr C (1975). Experimental study of tandem blades for rotor blade usage in a single stage axial flow compressor. Knoxville, US: University of Tennessee.

[10] Tan Gaoming, He xianghui and Liu Houlin, et al, Design and analysis of a radial diffuser in a single-stage centrifugal pump. Engineering Applications of Computational Fluid Mechanics,2016,10:1,500-511

[11] Oh J, Buckley C and Agrawa L, et al (2012). Numerical investigation of low solidity vaned diffuser performance in a high-pressure centrifugal compressor. Part III :Tandem vanes .Journal of Turbomachinery. 134 : 061025-1 .

[12] Xie R, Yang Y and Hai Y, et al (2010) . Research on centrifugal compressor staggered arrangement cascade diffusers . Gas Turbine Technology 2: 48-52 .

[13] Liu lei, Wu Shixun and Zhao Honglei (2019). Numerical Study of the Tandem Cascade Chord Length Ratio on the Performance of Centrifugal compressor. Gas Turbine Technology,3-32

[14] Shan Shujun, Hou Anping and Li Jianxiong (2018). Effects of Percent Pitch on Performance of High Load Tandem Cascade. Journal of Aerospace Power. 33(5).

[15] Zhou Li, Wang Zhanxue and Liu Zengwen (2014). Investigation on influence of design parameters for tandem cascades diffuser using DOE method. Engineering Applications of Computational Fluid Mechanics, 8(2), 240-251

[16] Tamaki H, Nakao H and Saito M (1999). The Experimental Study of Matching Between Centrifugal Compressor Impeller and Diffuser,ASME J. Turbomachinery. 121(1), pp. 113-118.

[17] Zhao Pengfei, Liu Yan and Li Hongkun,, ed al (2016). The effect of impeller – diffuser interactions on diffuser performance in a centrifugal compressor. Engineering Applications of Computational Fluid Mechanics. 10:1,565-577

[18] Anish S, Sitaram N and Kim H D. (2014). A numerical study of the unsteady interaction effects on diffuser performance in a centrifugal compressor. Journal of Turbomachinery, 136, 011012.

[19] Guo Ran, Jia Liping and Fan xiaoli, et al. NUMECA tutorials. China Machine press.2013

Acknowledgement

Firstly, I would like to appreciate the careful guidance form my tutor, professor Chunjun Ji. Then, I pleased to acknowledge Liaoning Fu-An gas turbine Co, who funded this study. At last many thanks for all the staff of the laboratory.

IOP Conf. Series: Materials Science and Engineering 1081 (2021) 012032 doi:10.1088/1757-899X/1081/1/012032

Research on flutter method for three dimensional wing rudder

X Zhou [1], L Liang [1], L Yang [1], J Xiao [1], Q Zhang [1]

[1]Aerospace System Engineering Institute, Beijing 100076, China

E-mail: Zhouxiaoya0826@163.com

Abstract. Aiming at providing high efficiency and accurate flutter analysis for high speed aircraft in transonic flow, flutter analysis in frequency domain based on three-dimensional wing rudder is employed for AGARD 445.6 wing. The simulation results are consistent with both experiment and reference, which validates the flutter mechanism in the subsonic and transonic regions. This simulation not only improves the accuracy of flutter analysis, but also saves calculation time and improves efficiency. Thus, it lays a foundation for high speed aircraft in transonic flow flutter prediction.

1. Introduction

Aerodynamic performance prediction is the primary consideration in many structure design [1-3].Flutter is a large-scale vibration caused by the coupling of unsteady aerodynamic, elastic and inertial forces of elastic structures subjected to uniform air flow. When the flutter occurs, if the structure's speed exceeds the critical flutter speed, a small disturbance will cause divergent vibrating, leading to catastrophic structural damage within a few seconds or even tens of milliseconds [4-5]. Therefore, accurate prediction of structure flutter analysis plays an important role in aircraft dynamic analysis and structural optimization design. Recently, using the fluid-structure interaction simulation to analyze the mechanism of structural flutter and obtain its results has received growing attention in the aeroelasticity area [6-7].

The traditional aircraft-oriented flutter simulation is the frequency-domain flutter analysis method. This method derives the aerodynamic force on the surface assuming that the structure is in the critical state of flutter, and finally obtains the flutter speed and frequency [8]. However, this method is based on a planar model without taking into account detailed information of the structure such as thickness and angle of attack, which affects the accuracy of the flutter analysis results. Thus, researchers proposed a time-domain flutter analysis method for three-dimensional structural models, which precisely simulates the motion of the structure in the flow field. Nevertheless, it requires plenty of time for flutter boundary calculation [9]. Therefore, the research on simulation method which provides efficient and accurate flutter analysis for aircraft structure is of great significance in aircraft structure design and engineering practice.

Based on Nastran, this paper extends a flutter method to three dimensional wing rudder. Test case of AGARD WING 445.6 is investigated to confirm the validity of the method. Through comparing results in the experiment and reference, it proves that this method can improve the accuracy of the flutter analysis while significantly save computing cost. In addition, this method can be expanded to transonic unsteady flutter of different airfoils, laying a foundation for a deep understanding of the flutter mechanism and the efficient and accurate flutter prediction.

2. AGARD WING 445.6 Standard Model

The AGARD WING 445.6 wing was first used in the 1961 NASA Langley Center TDT wind tunnel test. This benchmark model is made by mahogany and is covered with small holes. Its airfoil is NACA65A004 along the flow direction. Plane characteristic parameters are as follow: the aspect ratio is 1.644, tip root ratio is 0.6529 and quarter chord wing sweep angle is 45°. In the wind tunnel test, the No. 3 wing model with weak stiffness occurred flutter, which has become the internationally accepted assessment standard for the simulation method of transonic wings flutter characteristics [10-11].

The plane shape and geometric parameters of the model are shown in Figure 1. The material density is 351.98kg/m^3. The elastic modulus of x, y, and z directions of the wing are: 1GPa, 1.54GPa, 1Gpa. The shear moduli of the three directions of the wing are 2.6GPa. The Poisson's ratio is 0.31.

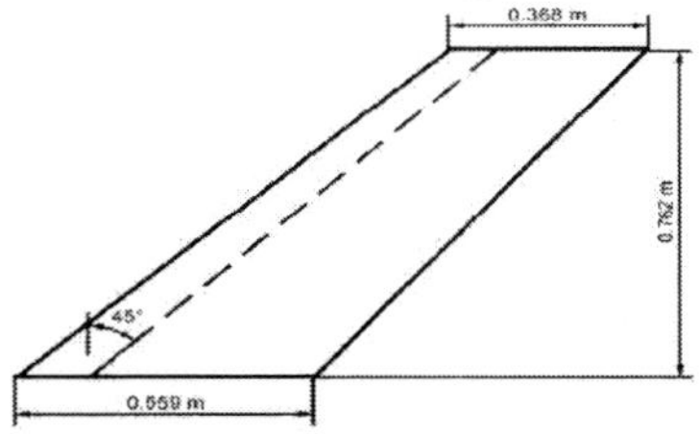

Figure 1. AGARD WING 445.6 geometric model

3. Three dimensional wing rudder flutter method

3.1. Modeling Method

According to the reference [11], a three-dimensional model of the wing is obtained using Pro/E Creo. The main steps are as follow:

a) Generate a two-dimensional airfoil section model based on NACA 65A004 airfoil profile data provided in the literature.

b) Around the main axis of the wing, superimpose the two-dimensional airfoil at a quarter-chord length along the 45-degree direction to generate a three-dimensional surface model.

c) Use the surface model to generate a three-dimensional rudder model first. Then smooth transition trimming is performed on the wing tip and wing root. Finally, the three-dimensional rudder model is smoothed.

Based on the geometric parameters of the wing standard model AGARD WING 445.6, the final 3D solid model is derived as shown in Figure 2.

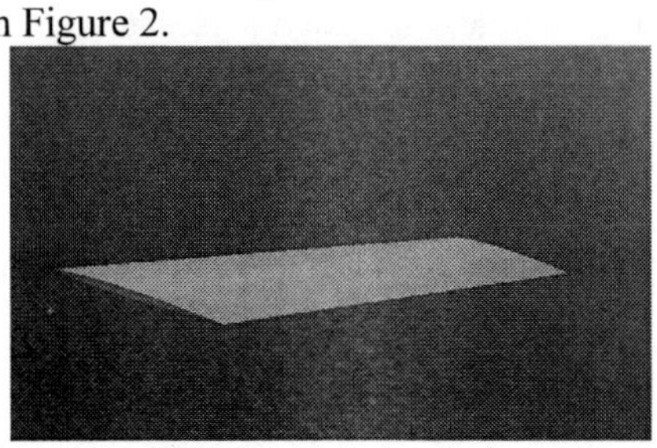

Figure 2. AGARD WING 445.6 simulation geometry model

3.2. Frequency domain flutter analysis method

The establishment of aeroelastic model directly determines whether the results of flutter analysis are correct. It includes the setting of aerodynamic parameters, the selection of reduced frequency, and the method of spline interpolation. There are two common methods for calculating flutter: KE method and PK method. The PK method derives the flutter results through obtaining eigenvalues of a real number matrix. This method also shows certain subcritical characteristics. Even the calculation is more complicated, PK method is more applicable than KE method [12].

IOP Conf. Series: Materials Science and Engineering 1081 (2021) 012032 doi:10.1088/1757-899X/1081/1/012032

According to the reference [13-14], considering characteristics of the three-dimensional rudder structure model of the wing, the PK method is selected as the flutter calculation method as the flutter calculation method and the flutter equation is:

$$\left[-M_{hh}p^2+\left(B_{hh}-\frac{1}{4}\rho\bar{c}VQ^{i}_{hh}/k\right)p+\left(K_{hh}-\frac{1}{2}\rho V^2Q^{R}_{hh}\right)\right]\{u_h\}=0 \tag{1}$$

Where Q^{i}_{hh} is modal aerodynamic damping matrix, Q^{R}_{hh} is modal aerodynamic stiffness matrix, p is characteristic value.

As shown in Equation (2), the range of reduced frequency k can be calculated according to the speed V and range of natural frequency ω, and the relevant values can be automatically derived using Nastran.

$$k=\frac{\omega c_{ref}}{2V} \tag{2}$$

Where c_{ref} is the reference chord length.

The main technical route is as follow:

a) Set the atmospheric density and Mach number; b) Assume the flight speed; c)For each structural mode ω_i, assuming the initial reduced frequency $k_i^{(0)}$ equals to $\frac{\omega_i c_{ref}}{2V}$,then obtaining the aerodynamic coefficient; d) Obtain two characteristic value p of the flutter equation (1), where $p_1=\gamma k_{i1}+ik_{i1}$, $p_2=\gamma k_{i2}+ik_{i2}$;e) Select the imaginary part of characteristic value which is close to the structural mode ω_i and used as the initial reduced frequency of the next iteration $k_i^{(1)}$; f) During the iteration process, when the difference between $k_i^{(j)}$ and $k_i^{(j\text{-}1)}$ is less than the preset index, stop the iteration and the reduced frequency value obtained at this time is the final solution.

MSC.Nastran is used to calculate AGARD WING 445.6 flutter results using PK method. Firstly, establish the aerodynamic grid of the wing, load various conditions, and modify the generated bdf file.

Secondly, select the spline interpolation method. Different from the traditional method, which uses the shell element to generate a two-dimensional model, this article uses a three-dimensional rudder model for flutter analysis. And hence the aerodynamic interpolation method is set to the TPS interpolation method (a spline method for high-level aerodynamic model interpolation in three-dimensional space).

Finally, different aerodynamic density and speed range under Ma = 0.499, 0.678, 0.901, 0.96, 1.072, 1.141 are obtained. Corresponding to different Mach number, the aerodynamic density and speed range are shown in Table 1.

Table 1. Aerodynamic density and speed range at different Mach numbers

Mach number	Aerodynamic density(kg/m^3)	Speed range (m/s)
0.499	0.4277	100~200
0.678	0.2082	100~300
0.901	0.0995	150~350
0.960	0.0634	200~400
1.070	0.0551	250~450

Using Equation (2), the Mach number-reduced frequency pair could be set for flutter analysis.

4. Agard wing 445.6 results and analysis

4.1. Agard wing 445.6 modal results and analysis

IOP Conf. Series: Materials Science and Engineering 1081 (2021) 012032 doi:10.1088/1757-899X/1081/1/012032

Use Nastran to obtain modal results. Based on the three-dimensional rudder model, select the Tetmesher mesh generator and the topology type, set the element size, the material properties, element properties and impose constraints. The natural modal distribution range of the AGARD wing 445.6 is relatively wide, and thus the results of the first four orders can be analyzed.

Table 2 shows the first four modal frequencies from simulation values, experiment and reference [11, 15]. The vibration mode comparison of Nastran simulation and NASA experiment is shown in Figure 3.

Table 2. AGARD WING 445.6 natural frequency Hz

Natural frequency	Experiment	Reference	Simulation
First order	9.5992	9.74	9.2205
Second order	38.165	38.4	37.905
Third order	48.3482	50.3	54.737
Fourth order	91.5448	91.8	94.713

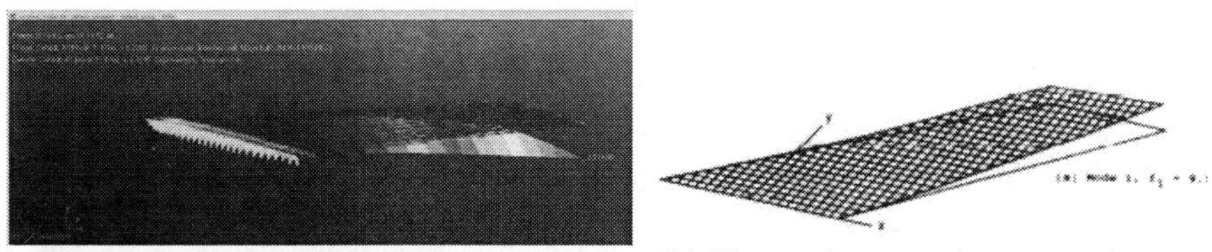

(a)First order simulation (b) First order experiment results

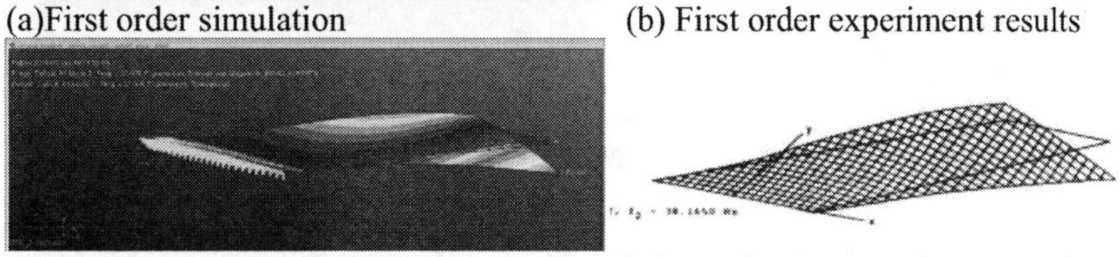

(c) Second order simulation (d) Second order experiment results

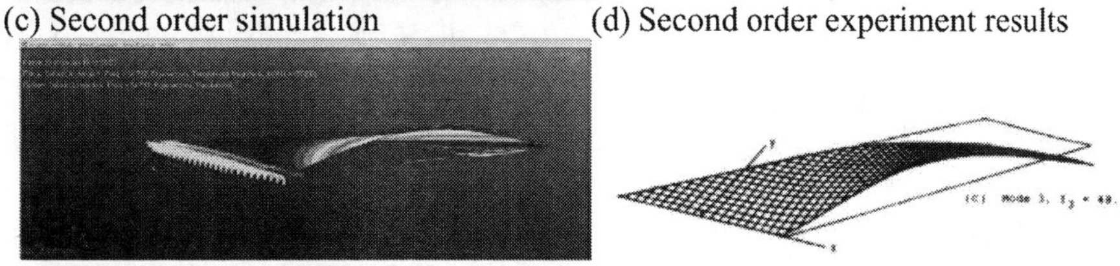

(e) Third order simulation (f) Third order experiment results

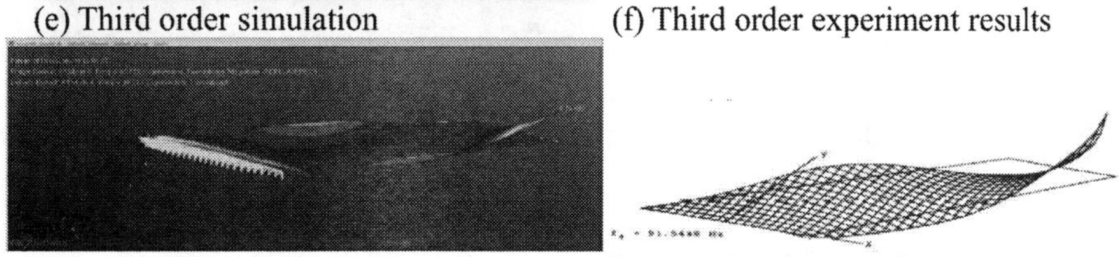

(g) Fourth order simulation (h) Fourth order experiment results

Figure 3. AGARD WING 445.6 vibration mode of simulation and experiment

It should be pointed out that the first mode of simulation shown in Figure 3 a) is downward, while the mode given in Figure 3 b) of the experiment is upward. This results is only caused by a phase difference of 180 °, the first modes of simulation and experiment are still consistent. There are no phase differences between the other analytical and experimental modes. Thus it concludes that simulation vibration modes are in good agreement with NASA experimental modes.

According to the natural frequency values in Table 3 and the vibration mode in Figure 3, the simulation results are basically consistent with the experimental and reference results and the error range is relatively small. The mode shape and frequency are in reasonable agreement with the experimental results. It indicates that the three-dimensional rudder modeling method is correct and feasible, and can accurately obtain the modal analysis results.

4.2. *Agard wing 445.6 flutter results and analysis*

The first four velocities and damping at different Mach numbers are showing in v-g diagrams, and the first four velocities and frequencies at different Mach numbers are depicted in v-f diagrams.

Taking the flutter results of Ma = 0.499 as an example, as shown in Figure 4a), Figure 4b) gives the relevant results in Reference [10].

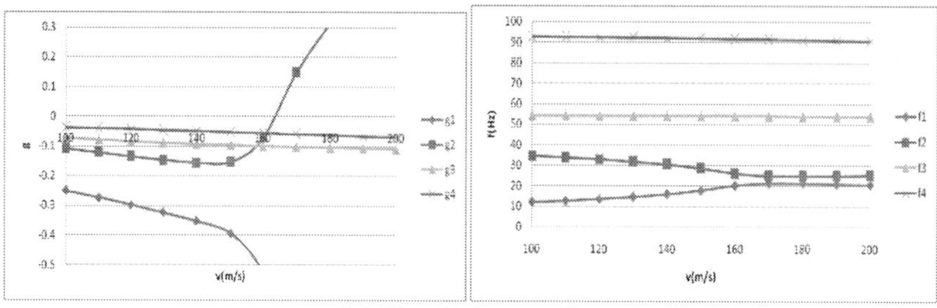

(a) Simulation results

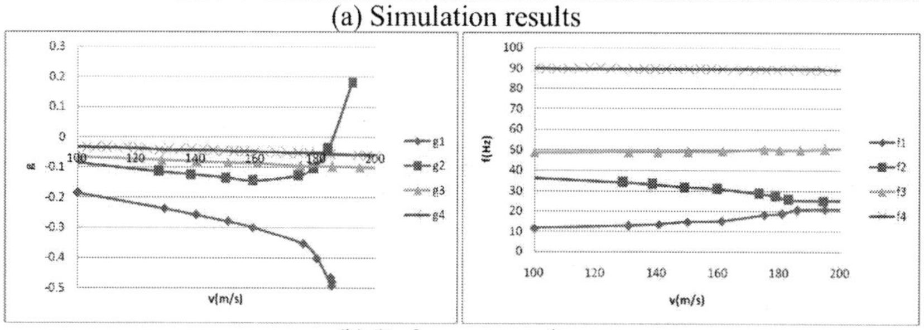

(b) Reference results

Figure 4. v-g and v-f diagram(Ma=0.499)

Figure 4 proves that the flutter analysis results at Ma = 0.499 are basically consistent with the v-g and v-f curve trends obtained by the reference [10], and the critical flutter velocity and flutter frequency gives satisfactory accuracy comparing with the reference results, which proves that the correctness of the flutter analysis method.

Similarly, flutter results at Ma = 0.678, 0.901, 0.96, 1.072, 1.141 can also be obtained.

Introducing flutter speed coefficient $\overline{V_F}$:

$$\overline{V_F} = \frac{V}{b\omega_\alpha\sqrt{\overline{\mu}}} \tag{3}$$

Where V is the inflow velocity, b is the reference length (generally defined as half of the root chord of the wing), ω_α is the reference frequency (generally defined as the first order torsion frequency of the wing), $\overline{\mu}$ is the mass ratio.

The flutter characteristics of wings in transonic region can be analyzed by using the flutter velocity coefficient. Depict the curve of flutter speed coefficient with Mach number, i.e. flutter speed characteristic curve, as shown in Figure 5:

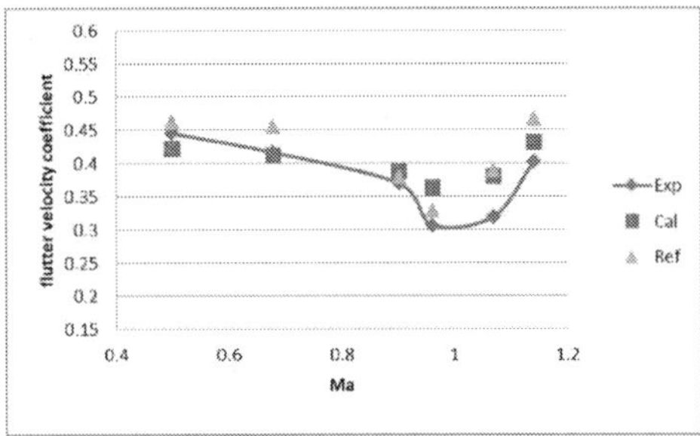

Figure 5. Flutter speed characteristic curve of transonic wing

According to Figure 5, the calculated and experimental values of transonic flutter speed characteristics are generally consistent and slightly deviated. Compared with the data in reference [10], the simulation results are closer to the experimental results. Thus the accuracy and correctness of the flutter analysis method used in this article are validated again. In addition, flutter results clearly reflect the physical characteristics of "pits" in the transonic region when the critical flutter speed of transonic wings changes with the Mach number [16-18]. It shows that this kind of wing is rather dangerous when flying in the transonic region.

Similarly, the curve of flutter frequency coefficient ($\frac{\omega}{\omega_\alpha}$) changing with Mach number can be depicted, i.e. flutter frequency characteristic curve, as shown in Figure 6:

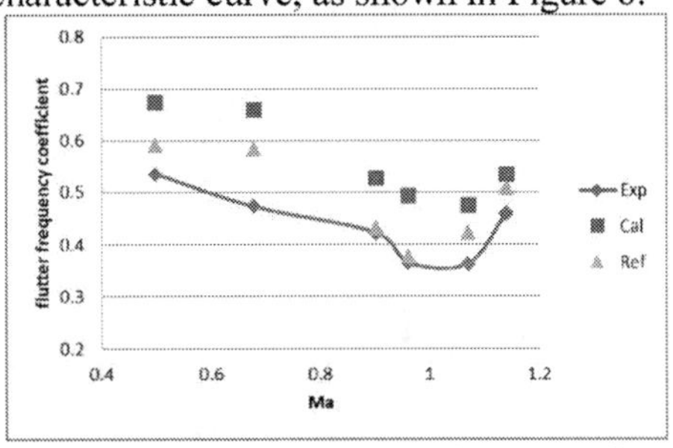

Figure 6. Flutter frequency characteristic curve of transonic wing

Figure 6 shows that the flutter frequency characteristic curve obtained in this article is consistent with the experimental results and reference data trends. However compared with the literature data, the error between the simulation and experiment is larger, which may be caused by the geometric model accuracy and modal parameters. In addition, the aerodynamic calculation in flutter analysis uses the panel method for a three-dimensional rudder model, which may also cause errors. It can be subsequently compared and analyzed through the ZAERO software aerodynamic calculation results for the further research. In addition, similar to the flutter velocity characteristic curve, the relationship between the flutter frequency and the Mach number also indicates the physical characteristics of the appearance of "pits" in the transonic flutter critical velocity as the Mach number changes.

5. Conclusion

Based on the three-dimensional wing rudder, this article uses the frequency domain flutter simulation method to provide a flutter analysis and simulation process. Taking the NASA wing standard model AGARD WING 445.6 as an example, it is obtained the following conclusions:

(1) According to the characteristics of high speed aircraft in transonic flow, an extended method for flutter analysis of wing rudder based on three-dimensional prototype model is explored.

(2) This extended method is different from the traditional method, which analyzes flutter after model simplification. It uses a three-dimensional prototype structure for analysis, taking into account the influence of the thickness and angle of attack of the structure on the flutter. Therefore, it simulates the wing in the flow field accurately. The flutter simulation is basically consistent with the reference

and experiment, and the error is relatively small, which proves the correctness of the flutter analysis method used in this paper. This method can be further generalized and applied to solve the nonlinear fluid-structure coupling problem of the aircraft structure.

(2) Using this extended method, the transonic flutter mechanism has been analyzed and studied, which are consistent with other theoretical analysis.

(3) Compared with the traditional time-domain and frequency-domain analysis methods, the extended method greatly improves the accuracy of flutter analysis results while greatly saving computation time and improving efficiency.

References

[1] Yang S, Liu Y, Zhang M, Yang J, Comparison of Aerodynamic Performance Prediction of Winglet Shroud Turbine Cascades Based on Source Terms and Body-Fitted Grid CFD (in Chinese) [J]. *Chinese Journal of Turbomachinery*, 2020, (1).

[2] Li D, Liu X, Aerodynamic Performance of Typical Bionic Owl-based Airfoil Applied on Design of Fan Blade (in Chinese) [J]. *Chinese Journal of Turbomachinery*, 2016, (1):28-34+55.

[3] Li S, Liu X, Qin Z, Numerical Study of the Effects of an Eccentric Impeller on the Aerodynamic Performance and Noise of a Multi-blade Centrifugal Fan (in Chinese) [J]. *Chinese Journal of Turbomachinery*, 2017,(1).

[4] Liu W, Huang X, Wang D, R. Liu, Investigation of Influence Coefficient Method for Flutter Analysis (in Chinese) [J]. *Chinese Journal of Turbomachinery*, 2019, **61**(1):5+59-67.

[5] Zhao F, Vahdati M, A Review of Computational Methods for Aeroelastic Instabilities of Multi-stage Compressor [J]. *Chinese Journal of Turbomachinery*, 2018,(2):1-10.

[6] Yang C, Wan Z, Chen G, Principle of aircraft aeroelasticity (in Chinese) [M]. Beijing: Beijing University of Aeronautics and Astronautics,2011.

[7] Qi X, Zhu X, Ji C, Online Monitoring for IRIG Standard Chatter Singa(in Chinese) [J],*Computer Measurement and Control*,2016,**24**(4):42-44.

[8] He S, Yang Z, Gu Y, Frequency domain analysis for wing transonic flutter (in Chinese) [J], *Sci Sin-Phys Mech Astron*, 2014, **44**(3): 285–292

[9] Zhang S, Zhao Y, Computing Transonic Flutters of a Wing with an External Store Using Time-Domain Method (in Chinese) . *Chinese Conference on Computational Mechanics* [C].2012

[10] Qian W, Bai Y, Wang D, Design, Manufacture and frquency experiment of a Novel flutter model for AGARD WING 445.6 with structure-similarity [A]. *30th Congress of the International Council of the Aeronautical Sciences*, ICAS 2016,2016

[11] *AGARD Standard Aeroelastic Configurations for Dynamic Response l-Wing 445.6*[R].ASA Technical Memorandum 100492, August 1987.

[12] Yuan P, *Flutter Analysis of Unsteady Transonic Wing* (in Chinese) [D]. Zhejiang: Zhejiang University, 2011

[13] Rodden W P , Johnson E H, *MSC. Nastran Version 68/Aeroelastic Analysis User's Guide*[M]. 1994.

[14] *Nastran user's guide-level 15*[R],NASA,April 1975

[15] Guo P, *Simulation of Fluid-Structure Interaction about Airfoil Vibration at Subsonic Speed* (in Chinese) [D].Zhejiang: Zhejiang University, 2013

[16] Lu X, Ye Z, Zhang C, A Coupled ANSYS/CFX Method for the AGARD 445.6 Wing Flutter Calculation (in Chinese) [J]. *Computer Simulation*.2010,**27**(9):88-91

[17] Liu N, *Investigation of Transonic Nonlinear Flutter and Efficient Analysis Approach* (in Chinese) [D]. Xi'an:Northwestern Polytechnical University, 2016

[18] Elizabeth M. Lee-Rausch. Calculation of AGARD WING 445.6 flutter using Navier-Stokes aerodynamics[C]. *Proceedings of AIAA 11th Applied Aerodynamics Conference*, California. 1993. AIAA paper No. 93-3476

Effect of Hub Gap on Expanding Stability of Centrifugal Compressor

Chunyang Li[1] , Chunjun Ji[1] and Junyi Fang[1,2]

[1] Dalian University of Technology, School of Energy and Power Engineering

[2] Liaoning Fu An gas turbine Co. Anshan, China

E-mail: 906549286@qq.com

Abstract. In this paper, the possibility of using hub gap (" HG "for short) to suppress gas stall and expand the range of stable working condition of compressor is verified by numerical method. The research shows that when designing flow rate, HG increases, the efficiency and pressure ratio of model level decrease. When there is a large flow, the efficiency and pressure ratio of HG model are higher than those without HG. The larger HG is, the higher the efficiency and pressure ratio is. From the analysis of the expanding stability capacity, the model-level flow margin of small HG and large HG is 5% and 15% higher than that of models without HG, and the comprehensive stall margin is 6.87% and 23.59% higher on average. When operating at unstable working conditions, the action of HG causes the airflow in the high pressure zone to flow into the gap and out of the low pressure zone, which improves the flow distribution in the blade passage and makes the flow field stable. However, when operating near the design condition, the leakage flow and its mixing with the momentum of the mainstream are the main sources of aerodynamic loss. The larger HG is, the greater the pressure loss caused by the leakage flow is. Generally speaking, HG can play the role of compressor expansion stability, but at the cost of efficiency; The bigger HG is, the stronger the role of expanding stability is and the greater the efficiency loss is.

1. Introduction

Centrifugal compressor has a certain range of work, the flow is too large or too small can affect its long-term, stable, safe operation, to production. In order to expand the stable working range of compressors, scholars have adopted a variety of measures, such as air injection to affect fluid flow. Researchers include Spakovszky[1] and Skoch[2] et al. Spakovszky's experiment shows that when the injection mass flow is 0.5% of the compressor mass flow, the surge margin increases by 25%. Marsan et al.[3] evaluated the possibility of using boundary layer suction technology to delay the surge of centrifugal compressor, and designed a multislot control strategy according to the characteristics of unstable pressure field. Casing treatment is also a mature method of expanding stability, which is widely used. He and Zheng[4] pointed that, casing treatment device can effectively eliminate the rotation instability and stall of the impeller inlet at low speed and suppress the surge of the compression system at medium and high speed. Sun et al.[5] explained the mechanism of expanding stability of casing treatment structure and proved its effectiveness.

In the bend of multi-stage centrifugal compressor, influenced by the turning centrifugal force, the air flow angle changes sharply, and the air flow speed changes greatly along the spanwise. Especially in the hub side, there is usually a low-speed airflow region. This part of low-speed gas is easy to break away from the main stream and move along the flow path, which affects the inlet of the downstream reflux, causing the instability of the compressor flow field and the decrease of efficiency. In this paper, HG is used to suppress gas stall, which is verified by numerical method. From the existing literature, the research of HG is basically based on axial compressor. In the axial compressor, in order to manufacture a variable stator vane, there is inevitably a radial gap between the blade and the endwall. The leakage flow of the gap interferes and mixes with the channel vortex and the boundary layer of the suction surface, which increases the aerodynamic loss. Si et al.[6] showed that the larger the hub clearance is, the greater the total pressure loss is. Although the efficiency decreases with the increase of the clearance, it enlarges the working range of the compressor[7]. Mao et al.[8] pointed the loss near the hub increases due to the leakage flow, and the improvement of stall margin is due to the redistribution of mass flow along the spanwise direction. Proper hub clearance can improve the flow condition near the hub of high load stator[9][10]. Lu and Li[11] studied the influence of HG shape (uniform, expansion and contraction gap) on the flow field, pointed out that the expansion gap can weaken the leakage flow near the leading edge, restrain the leakage vortex, and obtain good aerodynamic performance in the hub area.

The research in this paper has two meanings: first, although there are many researches on HG in axial compressor, the influence of HG on compressor's overall performance, stability expansion ability and flow field is still to be studied because there is no gap in the blade root of centrifugal compressor's stationary blade usually; second, if HG can be used to control gas flow instability near the bend, it can be extended to other parts and flow channels with larger bending degree in turbomachinery.

2. Numerical Setup

In this paper, the evaluation of HG expansion stability method is based on the structural form of "impeller + integrated static blade", which ensures the existence of blade structure on the wall with large bending degree. The impeller consists 13 blades, connected with 10 integrated static blades. The speed of impeller is 8129 rpm, outlet diameter is 600 mm. The efficiency and pressure ratio at the design point (7.6kg/s) are 82.82%, 1.483. HG mesh was created in Numeca software AutoGrid5 module. When meshing the blade, the "Row Wizard" function of the module is used to automatically form the blade with HG and generate a structured grid. Model HG was 0 by default. Four simulation models of different gaps (HG = 0.2mm, 0.5mm, 1mm and 3mm) were established by changing the "Hub Gap", and HG was evenly distributed at the blade root. The impeller grid adopts the topology structure of H&I, and the static blade grid adopts the topology structure of O4H. The grid points distribution of the model is shown in Figure 1. The model and grid with HG (as arrow) are shown in Figure 2.

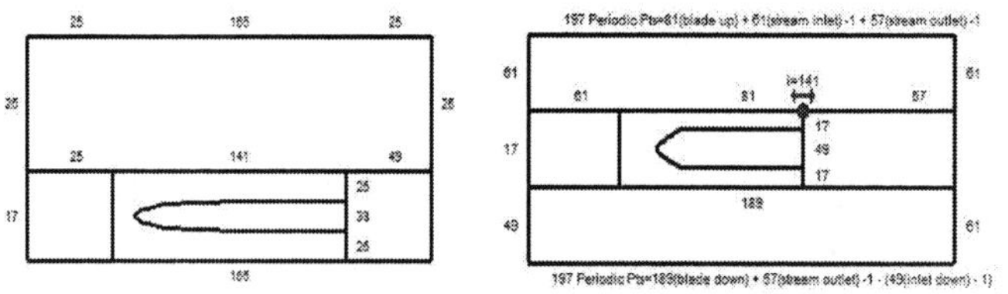

Figure 1. The grid points of blades.

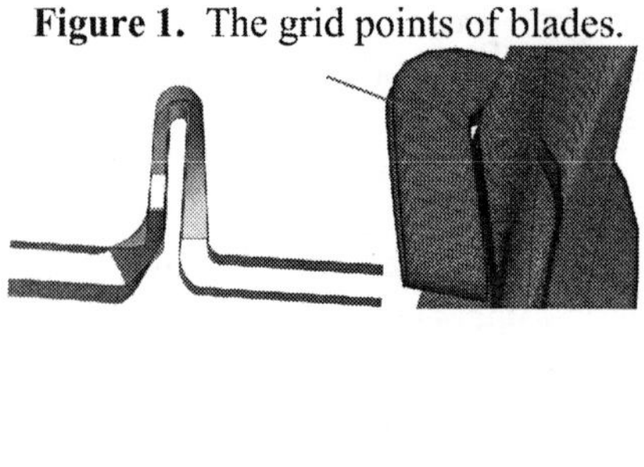

Figure 2. The geometric model.

Model mesh satisfies the min. Orthogonal Angle >10°, the max. Asp. Ratio <2000, and the max. Exp. Ratio <5. Ji et al. [12] has carried out the independence verification of mesh number. The Fine Turbo module was used to solve the three-dimensional Reynolds Navier-Stokes equations, and the turbulence model was Sparlart-Allmaras. The mixed plane method was used to connect rotor to stator, the mesh was encrypted near the wall, and the thickness of the first wall surface was 0.01mm, which met the requirements of the turbulence model y+ value 1-10.

The working medium is air, which is a viscous, compressible gas. The inlet condition is given the flow direction and total temperature and pressure, the outlet condition is given the mass flow rate, and the wall condition is the adiabatic wall with constant speed. The convergence criterion is global residuals and residuals in each block decrease by more than three orders of magnitude. The relative error of import and export mass flow is less than 0.5%, and the flow is no longer changing.

3. Overall performance

Figure 3 shows the efficiency and pressure ratio characteristic curves of different HG models. It can be seen that at the design flow rate (7.6kg/s), with the existence and increase of HG, the efficiency and pressure ratio of model level both show a downward trend. However, when the flow rate increased to the near blocking point, the efficiency and pressure ratio of the model with Gap=0.2mm was the highest, and the efficiency and pressure ratio of the other models decreased with the increase of HG. If the flow continues to increase, the efficiency and pressure ratio of the model level will increase with HG. At this point, the efficiency and pressure ratio of all models decrease substantially compared to design points.

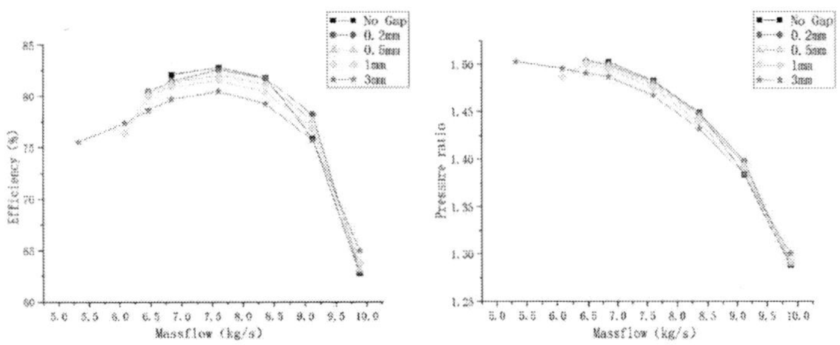

Figure 3. Efficiency and pressure ratio curve of models.

Figure 4 shows the efficiency and pressure ratio attenuation of models with HG compared with those without HG. From the flow rate (6.84~9.12) kg/s, the efficiency and pressure ratio of HG model were lower than those without HG, and HG increased, while the efficiency and pressure ratio decreased. At small HG, the flow increased and the attenuation degree decreased. At large HG, the flow rate increases and the attenuation degree increases. This is because, in the small HG model, the loss of gas in the boundary layer is weakened by the increase of leakage flow, while in the large HG model, leakage loss occupies a major position. At flow rate (9.12~9.88) kg/s, the model showed the opposite characteristics. The model efficiency and pressure ratio with HG were higher than those without HG, and the higher the HG was, the higher the efficiency and pressure ratio was. It indicates that HG has an adverse effect on the flow field when the compressor is running near the design point. When not running near the design point, HG is beneficial to the flow field.

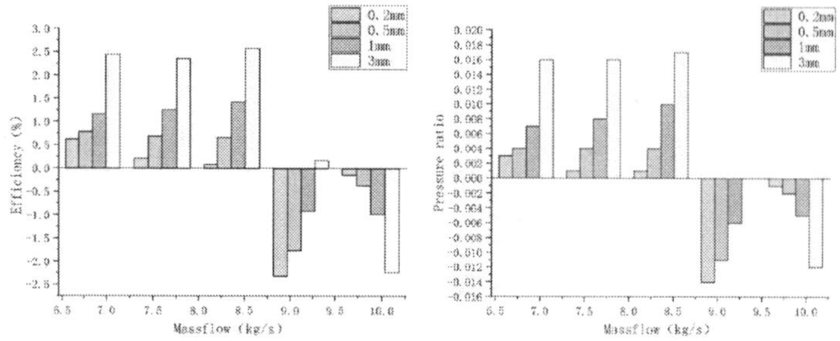

Figure 4. Efficiency and pressure ratio attenuation histogram.

Two aerodynamic performance parameters are defined as follows:

Comprehensive stall margin (SM):

$$SM = \left(\frac{\varepsilon_S/m_s}{\varepsilon_d/m_d} - 1\right) \times 100\% \tag{1}$$

Flow margin (SM'):

$$SM' = \frac{q_{md} - q_{ms}}{q_{md}} \times 100\% \tag{2}$$

Where, ε_s and ms are the total pressure ratio and mass flow at the stall point of the compressor; ε_d and md are the total pressure ratio and mass flow at the design point at the same speed. q_{md} and qms are the mass flow at the design point and stall point.

After calculation, it is found that the model SM with HG is higher than that without HG, and the model with large HG is higher than that with small HG. Compared with the model without HG, SM'of small HG and large HG increased by 5% and 15%, and SM increased by 6.87% and 23.59% on average. Generally speaking, HG can play a role in compressor stability expansion, but at the same time at the expense of efficiency; the larger the HG, the stronger the stability expansion, the greater the efficiency loss.

4. Flow Field

4.1. *Flow Field*

Figure 5 shows the total pressure distribution of the model in the meridional plane at the design point, and the models without HG, 0.2mm and 3mm gap from left to right.

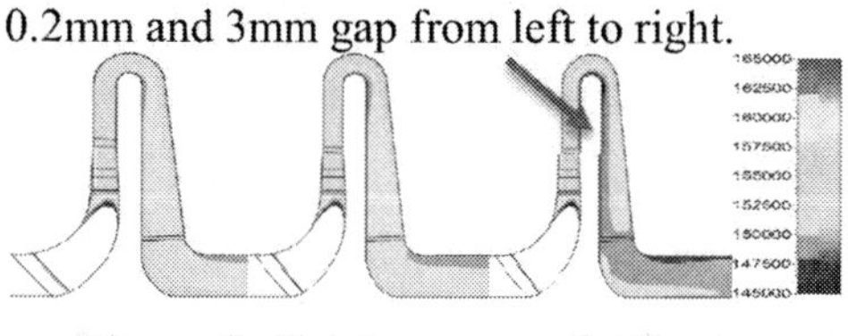

Figure 5. Total pressure distribution.

As can be seen from Figure 5, with the presence and increase of HG, significant low pressure loss occurs on the hub side of the stationary blade, especially from the bend outlet to the stationary blade outlet. In the middle and downstream of the stator flow passage, the area of low pressure formed by the mixing of leakage flow and main flow expands and the loss increases. Before the turn, the difference of gas pressure distribution in the flow channel was not obvious, especially for the model with small HG. After the turn, the difference of total pressure increased significantly. At the same time, there is also a low pressure area at the exit of the compressor stage, resulting in flow loss. At the design point, leakage flow is the main factor of aerodynamic loss. The larger HG is, the greater the pressure loss is.

Below, the velocity and entropy distribution of 5% blade height section in the No HG, 0.2mm and 3mm clearance models are analyzed at the blocking condition (flow rate was 9.88kg/s) The above row charts are the inlet channel of the model, and the below are the outlet channel of the model, from left to right are No HG, 0.2mm and 3mm HG (Figure 6 and Figure 7).

Figure 6. Velocity distribution.

Figure 7. Entropy distribution.

Because the three models are all at the blocking condition, a large range of low-speed zone appears in the blade root region of the flow field, which causes the flow loss. With the presence and increase of HG, the low-velocity gas near the stationary blade inlet gradually concentrated in the middle of the flow passage. According to the velocity distribution diagram at the static blade outlet, in models without HG and with small HG, low-speed gas is concentrated in the wake area at the downstream of the stationary blade. With the increase of HG, the low-speed area is separated from the blade, and the suction surface of the blade tends to be in the middle of the flow path. The effect of this phenomenon is that the models without HG and with small HG have a significant increase in entropy at the entrance of the stationary blade, while the models with large HG have a high entropy zone near the bend. When the compressor flow is too large, due to the pressure difference, the gas flows from the pressure surface into the suction surface, resulting in leakage flow. The leakage flow and the gas in the main flow area flow together along the flow direction. As the gas is affected by centrifugal force at the bend, the direction of movement is from radial to circumferential, and the two gases mix and mingle at the bend. The leakage gas at the inlet of the static blade can reduce the blockage of the flow field, correct the flow rate of the compressor, reduce the flow loss within a certain range, and make the high-entropy area change to the bend.

4.2. Stability Extension Mechanism

According to the results of numerical calculation, when the flow rate is 6.46kg/s, the flow field of the model with HG is stable, while the model without HG has stall phenomenon in the model. The following is the analysis of the total pressure and streamline distribution in the meridional plane of the two models (the left side is No HG, and the right side is HG=0.2mm model).

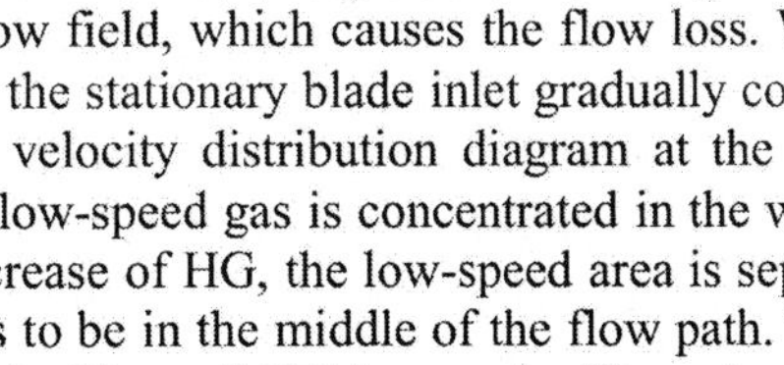

Figure 8. Total pressure distribution.

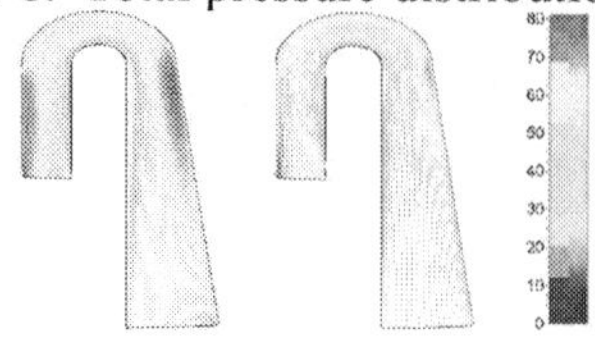

Figure 9. Streamline distribution.

From the total pressure and streamline distribution of the two models, it can be seen that in the model without HG, there is an obvious area of low pressure on the hub side, which will continue to the exit of the bend, and there is a gas separation area. These low pressure zones and flow stall separation zones contribute to the instability of flow in the model stage. With HG, HG leakage flow will reduce secondary flow and vortex movement at the hub, reduce flow loss, stabilize the flow field, and improve compressor surge margin. Figure 10 shows the entropy distribution of the two models (95%, 50% and 5% height sections from top to bottom).

(a) No HG (b) HG=0.2mm

Figure 10. Entropy distribution.

According to Figure 10, for the three height sections along the spanwise direction, the high-entropy regions of the model without HG are all larger than those of the model with HG. From the analysis of the distribution position of the high-entropy region, in the tip region, the high-entropy region of the model without HG is concentrated in the middle and downstream of the flow channel. This phenomenon can be explained by Figure 11. In the middle and downstream the flow channel of the model without HG, the low-velocity region of the fluid appears, resulting in separation and backflow. In the blade root region, the flow is very complex and harsh, and both models have high loss areas at the inlet of the stationary blade and the upstream of the flow passage. In the middle-leaf area, the high-loss area of the without HG model is concentrated in the upstream and the bend part of the flow passage. As can be seen from Figure 12, the flow angle of the gas at the flow passage varies greatly, and the gas has backflow phenomenon. At the entrance of the stationary blade of the model with HG (as shown by the arrow), the flow is separated. This is because the flow direction of the gas at the entrance is changed due to the presence of HG, but it does not spread, and the gas flows steadily in the entire flow field.

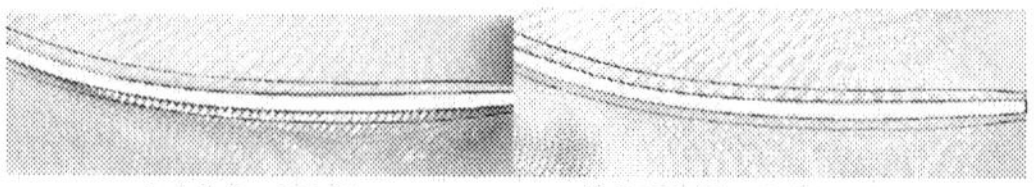

(a) No HG (b) HG=0.2mm

Figure 11. Velocity distribution.

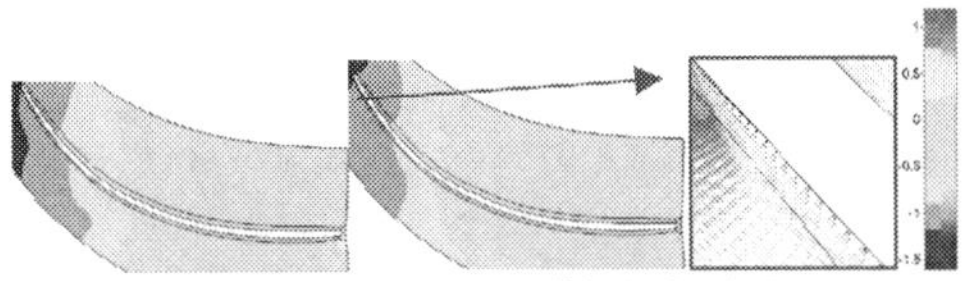

(a) No HG (b) HG=0.2mm

Figure 12. Flow angle distribution.

At the condition of small flow rate, the inlet angle of the compressor increases, the inlet flow field becomes uneven, the leading edge of the blade is separated locally, the flow passage on one side of the blade is blocked, and the gas is deflected. These low-energy fluids expand in the flow channel, resulting in a larger separation zone, which promotes the unstable development of the flow field and the loss of stage efficiency. In the model with HG, due to the pressure difference between the two sides of the blade, there is a strong leakage flow, and the leakage gas flows from the pressure surface to the suction surface, which alleviates the unevenness of the inlet passage, inhibits the development of stall disturbance of the flow to a certain extent, and makes the system more stable.

5. Conclusion

In this paper, HG is used to suppress gas stall and expand the range of stable working condition of compressor. This paper mainly draws the following conclusions:

(1) At the design flow rate, the model level efficiency and pressure ratio both showed a downward trend as HG grew from zero to large. At the large flow rate, the model showed the opposite characteristics. The model efficiency and pressure ratio with HG were higher than those without HG, and the larger HG was, the higher the efficiency and pressure ratio was. It indicates that HG has an adverse effect on the flow field when running near the design point of the compressor. When not running near the design point, HG is beneficial to the flow field. The flow margin of the two groups of models with small HG and large HG was increased by 5% and 15% compared with the model without HG, and the comprehensive stall margin increased by 6.87% and 23.59% on average. Generally speaking, HG can play a role of compressor expansion stability, but at the same time at the cost of efficiency; The bigger HG is, the stronger the role of expanding stability is and the greater the efficiency loss is.

(2) When operating at the unstable working conditions, the role of HG is to make the airflow in the high pressure zone flow into HG and then flow out of the low pressure zone, which improves the flow distribution in the blade passage, reduces secondary flows and vortices, and makes the flow field stable. However, when operating near the design point, the gas leakage flow in HG and its mixing with the momentum of the main flow become the main source of compressor loss, and the greater HG is, the greater the pressure loss caused by the leakage flow.

The study in this paper reveals the influence of static blade HG on the aerodynamic performance, stability expansion and flow field of centrifugal compressor, and it is found that HG can control the flow instability of gas at the bend, and it can be extended to other parts of turbine machinery with a large degree of bending (such as the L-shaped channel of the return channel between two stages of centrifugal compressor). Although HG plays an expanding stability role, it reduces the efficiency of the whole stage. The next design goal is to avoid the loss of efficiency while expanding the stability, and to apply it to the multi-stage compressor to ensure the reasonable matching of flow field between the stages. At the same time, the experiment will be used to verify the simulation.

References

[1] Z.S. Spakovsky. Backward traveling rotating stall waves in centrifugal compressors[J]. Journal of Turbomachinery-Transactions of the ASME, 2004,126(1): 1-12.

[2] Gary J. Skoch. Experimental investigation of centrifugal compressor stabilization techniques [J]. Journal of Turbomachinery-Transactions of the ASME, 2003,125(4): 704-713.

[3] Aurélien Marsan, Isabelle Trébinjac, Stéphane Moreau, Sylvain Coste. Active Flow Control in a

Radial Vaned Diffuser for Surge Margin Improvement: A Multislot Suction Strategy[J]. International Journal of Rotating Machinery, 2017, 2017:1-12.
[4] Xiao He, Xingqian Zheng. Roles and mechanisms of casing treatment on diferent scales of fow instability in high pressure ratio centrifugal compressors.Aerospace Science and Technology, 2019, 84: 734-746.
[5] Xiaofeng Sun, Xu Dong, Dakun Sun. Recent development of casing treatments for aeroengine compressors [J]. Chinese Journal of Aeronautics, 2019, 32(1): 1-36.
[6] Xiayi Si, Jinfang Teng, Xiaoqing Qiang, Jinzhang Feng. Different effects of cantilevered and shrouded stators on axial compressor performance, ASME paper GT2017-63261, 2017.
[7] Chengwu Yang, Xingen Lu, Yanfeng Zhang, Shengfeng Zhao,Junqiang Zhu. Numerical investigation of a cantilevered compressor stator at varying clearance sizes, ASME paper GT2015-42124, 2015.
[8] Xiaochen Mao, Bo Liu, Botao Zhang. Hub clearance effects of a cantilevered tandem stator on the performance and flow behaviors in a small-scale axial flow compressor[J]. Aerospace Science and Technology, 2019,91:219-230.
[9] Kurian K. George, S. N. Agnimitra Sunkara, Jubin Tom George, Melvin Joseph, A.M. Pradeep, Bhaskar Roy. Investigations on stator hub end losses and its control in an axial flow compressor. ASME paper GT2014-26845, 2014.
[10] Baojie Liu, Ying Qiu, Guangfeng An, Xianjun Yu. Experimental investigation of the flow mechanisms and the performance change of a highly loaded axial compressor stage with/without stator hub clearance[J]. Applied Sciences,2019, 9(23):1-19.
[11] Hanan Lu, Qiushi Li. Cantilevered stator hub leakage flow control and loss reduction using non-uniform clearances[J]. Aerospace Science and Technology, 2016,51:1-10.
[12] Chunjun Ji, Chunyang Li, Junyi Fang, Qi Sun. Loss Mechanism of Static Interstage Components of Multistage Centrifugal Compressors for Integrated Blade Design[J]. Mathematical Problems in Engineering, 2018(4):1-16.

Numerical Study on Hydrodynamic Characteristics in a Centrifugal Pump at off Design Conditions Based on a Cycling-System Model

Wen-Hui Yin[1], Xi Wang[1], Tao Yu[1], Jie Jian[1], Xiang-Yuan Zhang[1], Zhi-Jun Shuai[1]*, Wan-You Li[1] and Chen-Xing Jiang[2]*

[1] College of Power and Energy Engineering, Harbin Engineering University, Harbin 150001, China
[2] College of Energy, Xiamen University, Xiamen 361005, China

E-mail: [1]shuaizhijun98@163.com; [2]jiangchx@xmu.edu.cn

Abstract. In this paper, the unsteady hydrodynamic characteristics of a centrifugal pump handled at low-flow-rate conditions are numerically investigated. A cycling-system model is established to capture the influence of off-design operation on the internal flow field and the pressure pulsation. The preliminary results show that the cycling-system model can predict the flow field of the pump more conforming to the practical situation than that of the pump alone; besides, the generation of stall clusters in the impeller will induce low-frequency pressure pulsation.

1. Introduction

The performance of a centrifugal pump depends on its internal flow characteristics. The smoother the internal flow is, the more regularly the velocity and pressure distribute, and the more stable operating state is [1]. With the development of computational fluid dynamics (CFD) techniques, many numerical studies have been performed to investigate the internal flow characteristics inside the pumps, and most researchers consider pumps as a single component, with fixed boundary conditions in their numerical simulations [2-9]. However, when the pump is handled at off-design conditions, the internal flow field changes greatly and subsequently causes unstable operation, such as backflow, vortex, stall, and flow separation [10-17]. When stall clusters are generated inside the pump, the uniformity of the internal flow field will be destroyed. In special, when the rotation frequency of the stall cluster is the same as or close to the natural frequency of the pump system, resonance occurs [18-21], which will seriously affect the efficiency and stability. In the circumstances, fixed boundary conditions are no longer applicable. In this paper, to capture the influence of the off-design operation on internal flow field and pressure pulsation, a simplified version of the cycling pipeline model is established. It does not need to fix the inlet and outlet boundary conditions. Based on the cycling-system model, the unsteady simulation of the centrifugal pump at rated and off design conditions are performed.

2. Physical Model

IOP Conf. Series: Materials Science and Engineering 1081 (2021) 012034 doi:10.1088/1757-899X/1081/1/012034

The whole cycling system is composed of three parts: a centrifugal pump, a valve, and a water tank. Since the valve is mainly used to control the flow rate, it is simplified to a converging-diverging pipeline. The cycling-system model is shown in Figure 1.

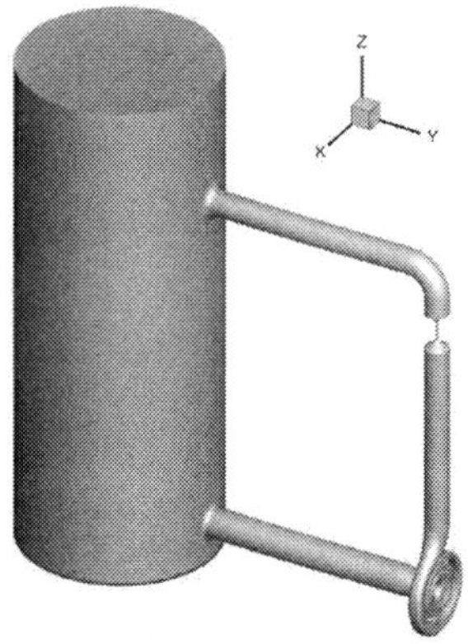

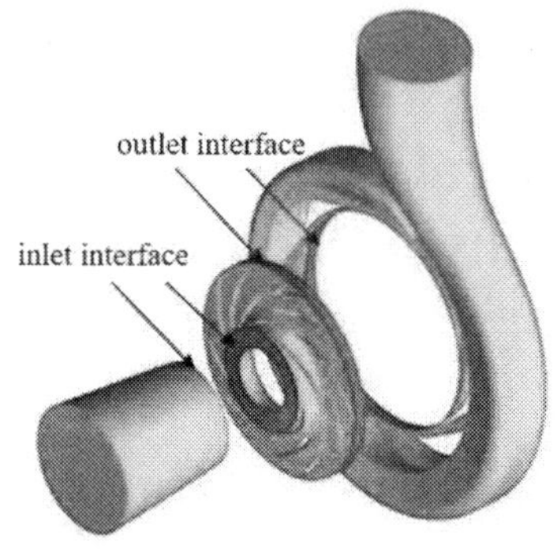

Figure 1. 3D Perspective of the Cycling-system Model. **Figure 2.** Flow Field of the Model Pump.

The main structural parameters of the research pump are shown in Table 1, and the whole flow field of the model pump is shown in Figure 2.

Table 1. Main Parameters of the Centrifugal Pump.

Parameter	Centrifugal pump
head(m)	325
inlet diameter (mm)	150
outlet diameter (mm)	125
number of the long impeller blade	5
number of the short impeller blade	5
number of volute guide vanes	7
number of inlet guide vanes	3

Considering the complex flow characteristics inside the pump, an unstructured tetrahedral grid is selected to generate the mesh system. After grid-independence analysis, the total cell number of the mesh is 4 million shown in Figure 3. SST is used as the turbulent model.

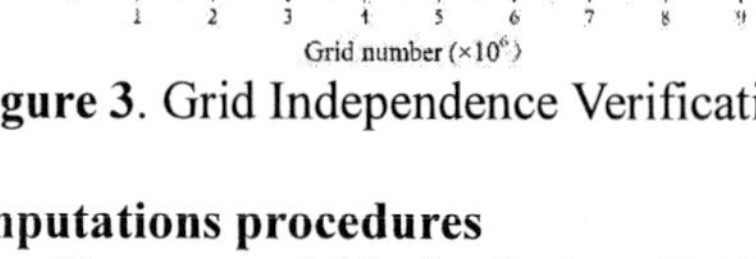

Figure 3. Grid Independence Verification.

3. Governing equations and computations procedures

The unsteady incompressible Reynolds-averaged Navier-Stokes (RANS) equations are solved for the flow passing through the centrifugal-flow pump. The basic equations in a manner of RANS are the following continuity and momentum equations, respectively.

IOP Conf. Series: Materials Science and Engineering 1081 (2021) 012034 doi:10.1088/1757-899X/1081/1/012034

3.1. Mass conservation equation

$$\frac{\partial}{\partial t}(\rho)+\frac{\partial}{\partial x_i}(\rho u_i)=0 \tag{1}$$

3.2. Navier-Stokes Equations

$$\frac{\partial}{\partial t}(\rho u_j)+\frac{\partial}{\partial x_i}(\rho u_i u_j)=-\frac{\partial P}{\partial x_j}+\frac{\partial}{\partial x_i}[\mu(\frac{\partial u_i}{\partial x_j}+\frac{\partial u_j}{\partial x_i})]+\rho g+F_i \tag{2}$$

3.3. Simulation set-up

The relevant settings of the two models, a single pump and a cycling-system, are shown in Table 2.

Table 2. Calculation Set-up of the Two Models.

Boundary Conditions	Single Pump	Cycling-System
Pressure-inlet	2×10^5 Pa	/
Rated condition	44.36 kg/s	/
Low-flow condition	7.5 kg/s	/

3.4. Monitoring points set-up

The layout of the main monitoring points is shown in Figure 4.

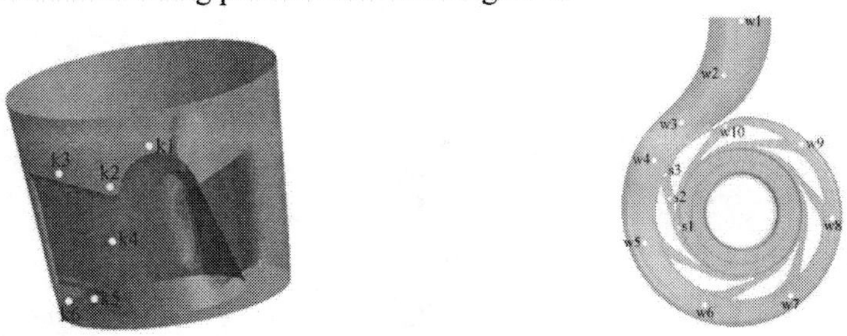

(a) (b)

Figure 4. monitoring points set-up. (a) Monitoring Points in Inducer. (b) Monitoring Points in Diffuser and Volute.

4. Results and Discussions

4.1. External Characteristics

In unsteady simulations, the pump head and efficiency also fluctuate within a certain range. The average value is taken in the characteristic analysis (seeing Figure 5).

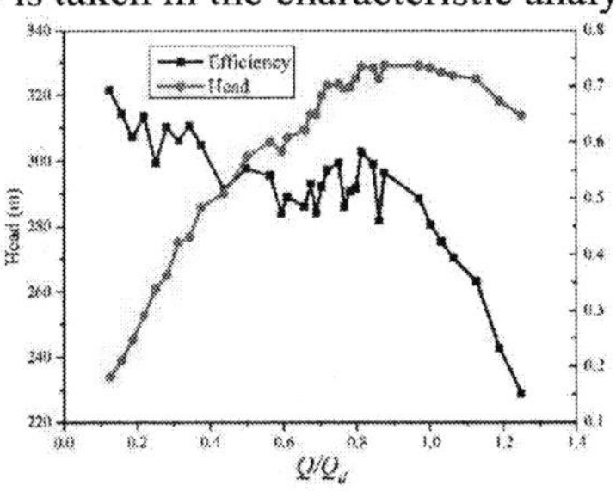

Figure 5. Pump Characteristics Curves.

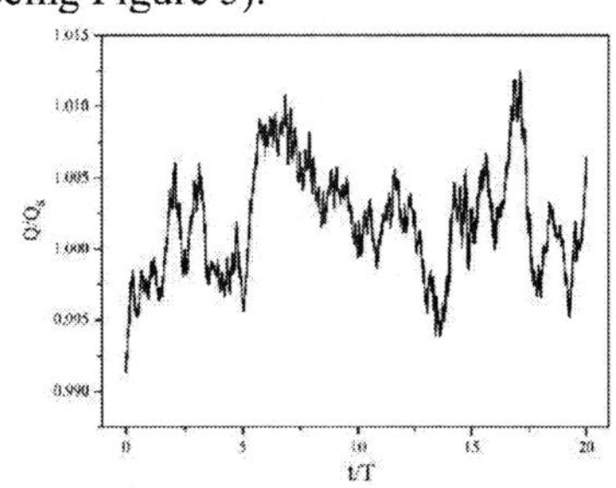

Figure 6. Flow Rate in Time Domain.

For a centrifugal pump with guide vanes, the flow range is mainly divided into partial load swirl and full load instability due to the rotating stall [22]. It can be seen from the characteristics curves that the head has a partial depression at a low flow rate (less than 0.6 Q_d), and this unstable form is called a partial load swirl; as the flow rate increases, around 0.8 Q_d , the depression appears in both head and efficiency curves, that unstable form is called full-load instability.

The generation of the rotating stall in impeller of diffuser will induce the flow instability and fluctuation, where a fixed boundary value does not match the physical truth. However, the cycling-system model does not need to impose the inlet and outlet boundary conditions and is self-coupled to solve, and the parameters of the flow field inside the pump are obtained through internal calculation. The flow rate simulated by the cycling-system model is monitored at 0.17 Q_d condition, and the flow rate changes at time domain is shown in Figure 6. It is clear that the outlet flow rate is not a fixed value but fluctuates with time.

4.2. Analysis of internal flow field

The centrifugal pump pressure contours are shown in Figures 7 and 8. The overall pressure at low-flow-rate condition is higher than that at rated condition, which is caused by the reduced velocity. In the middle section of the impeller and volute, there are a low-pressure area at the root of the blade and a high-pressure area around at both conditions due to the increase of the flow velocity. At rated condition, the isobaric line is vertical to the blade surface, indicating that the pressure gradient is small at the pressure surface, which is not easy to cause flow separation; while at low-flow-rate condition, the isobaric line is no longer vertical to the blade surface. The increased surface pressure gradient will cause unstable flow structures such as flow separation.

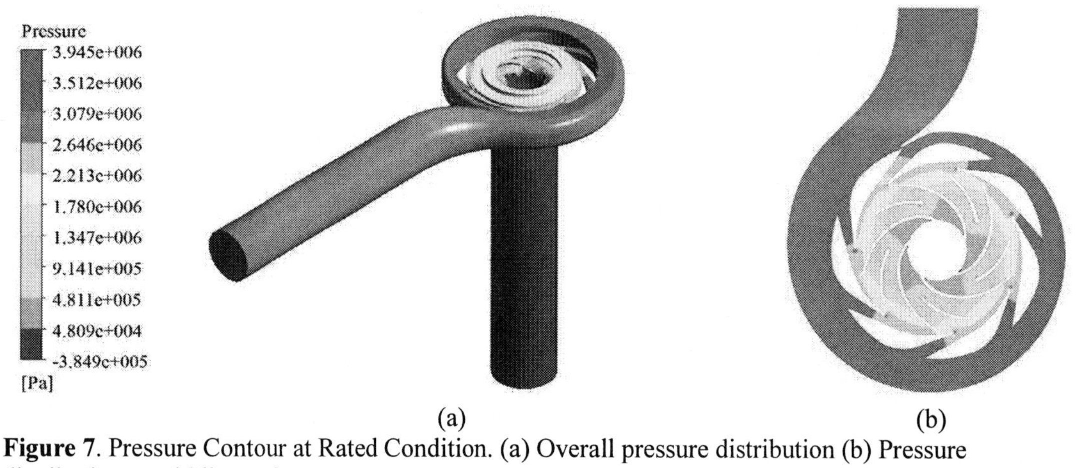

Figure 7. Pressure Contour at Rated Condition. (a) Overall pressure distribution (b) Pressure distribution at middle section.

Figure 8. Pressure Contour at 0.17 Q_d Condition Condition. (a) Overall pressure distribution (b) Pressure distribution at middle section.

Figure 9 illustrates the streamline diagrams of the impeller-volute section at rated and 0.17 Q_d conditions. At the junction of the diffuser guide vane and the volute, flow separation happens. In the flow channel of the volute, there is a vortex caused by unreasonable structural design. At low-flow rate condition, a vortex, in the opposite direction of the impeller rotation, is generated on the suction side of the impeller short blade and block the flow channel. Besides, the unstable structures in the guide vanes of the diffuser causes backflow in the flow channel, and some flow channels are completely blocked by the vortex, which seriously affects the performance of the pump. The streamlines in the flow path of the entire volute are no longer smooth, which are caused by the off-design operation.

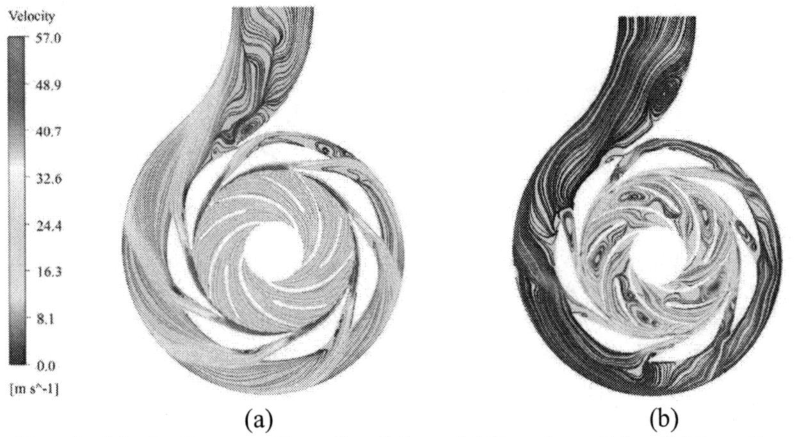

Figure 9. Streamlines inside the Pump at Two Conditions. (a) Rated condition (b) Low-flow-rate condition.

Figure 10 shows the velocity streamline of the volute section at two conditions. It can be seen that at rated condition, there is a pair of reverse vortices in the volute section and they maintain symmetrical distribution proximately; while at low-flow-rate condition, on one side of the volute cross-section, there aren't any secondary flow structures, while the other side, two vortices of the secondary flow structure are generated with asymmetrical distribution. The dominant vortex almost fills the entire volute flow path. The off-design operation influences the flow field inside the volute.

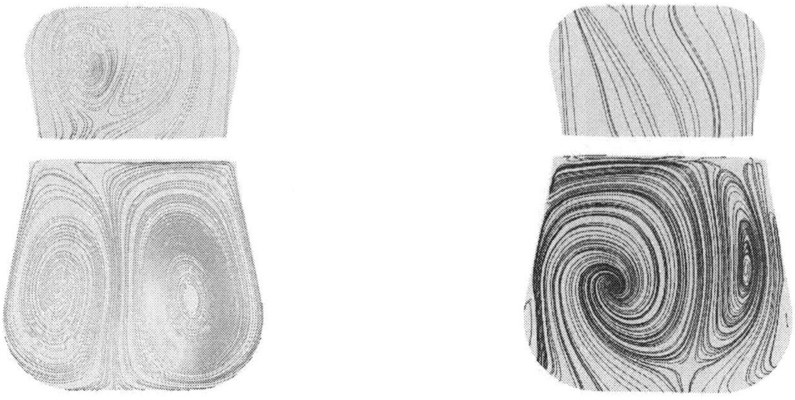

(a) (b)

Figure 10. Volute Secondary Flow at Two Conditions. (a) Rated condition (b) Low-flow-rate condition.

4.3. Pressure fluctuation characteristics

Figure 11 shows the pressure fluctuation in the region of inlet guide vane in a cycle. At rated condition, the pressure near impeller (seeing locations Points 5 and 6 presents obvious regular fluctuation with five peaks within a cycle, which is related to the rotor-stator interaction between impeller and inlet guide vane. The other locations have no obvious regularity. In the case of low-flow-rate condition, only the location Point 6, closest to the impeller, has obvious regularity.

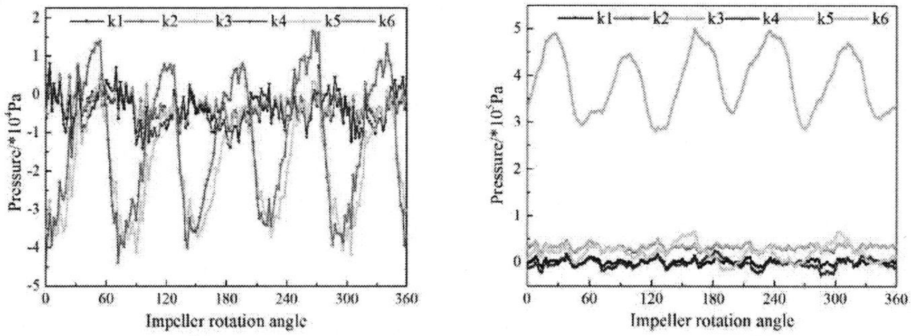

(a) (b)

Figure 11. Pressure Fluctuation in Time Domain at Inducer. (a) Rated condition (b) Low-flow-rate condition.

For further processing, Fast Fourier transform (FFT) is employed to obtain the frequency domain information (Figure 12). At rated condition, the locations Points 5 and 6 have higher pressure amplitude at 5 times the shaft frequency, corresponding to the blade passing frequency (BPF). While at low-flow-rate condition, only the location Point 6 has a higher peak at BPF. Besides, compared the two working conditions, the pressure amplitude at BPF is higher at low-flow-rate condition near the area of rotor-stator interaction.

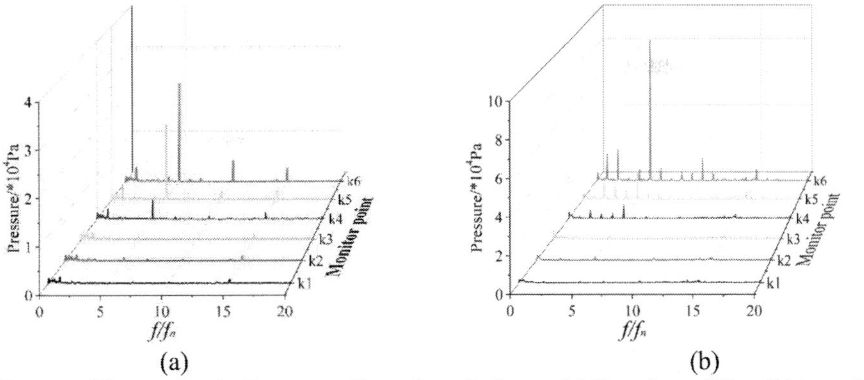

Figure 12. Pressure Fluctuation in Frequency Domain at Inducer. (a) Rated condition (b) Low-flow-rate condition.

Considering the similar results of 7 diffuser guide vanes, only one group data is shown to explain the pressure fluctuation characteristics of the two working conditions. Figure 13 shows the pressure fluctuation of the diffuser in time domain. Point S1 near the impeller has 10 peaks within a cycle, corresponding to the blade number of the impeller. Besides, the sudden increase and decrease of the pressure are sharper at location Point S1 due to 5 long blades and 5 short blades swept. The difference is that at low-flow-rate condition, the average pressure amplitude and fluctuation amplitude are higher.

The frequency domain information at diffuser is shown in Figure 14. Both working conditions have obvious peaks at 5, 10, and 15 times the shaft frequency. While at low-flow-rate condition, many random pressure pulsations are generated, which is due to the stall clusters generation at low-flow-rate condition.

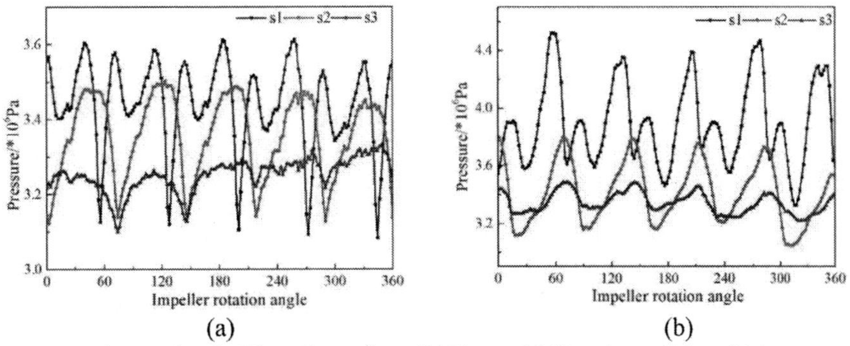

Figure 13. Pressure Fluctuation in Time Domain at Diffuser. (a) Rated condition (b) Low-flow-rate condition.

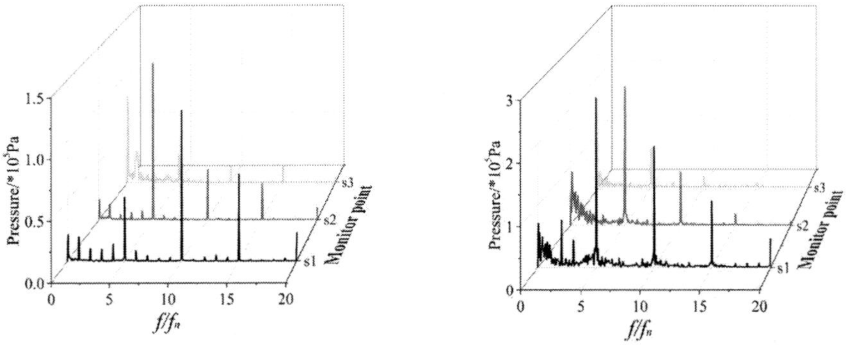

Figure 14. Pressure Fluctuation in Frequency Domain at Diffuser. (a) Rated condition (b) Low-flow-rate condition.

IOP Conf. Series: Materials Science and Engineering 1081 (2021) 012034 doi:10.1088/1757-899X/1081/1/012034

Figure 15 shows the pressure fluctuation at the volute in time domain. The pressure fluctuations have no obvious regularity at two working conditions, which is caused by asymmetry of volute structure. The difference is that the amplitude at low-flow-rate condition is higher than that at rated condition. According to results in frequency domain (Figure 16), both two conditions have obvious pressure amplitude at 5,10,15 times the shaft frequency. But the amplitude is higher at low-flow-rate condition than that at rated condition. Besides, 7 and 14 times the shaft frequency are generated under low-flow-rate condition corresponding to the blade number of the diffuser.

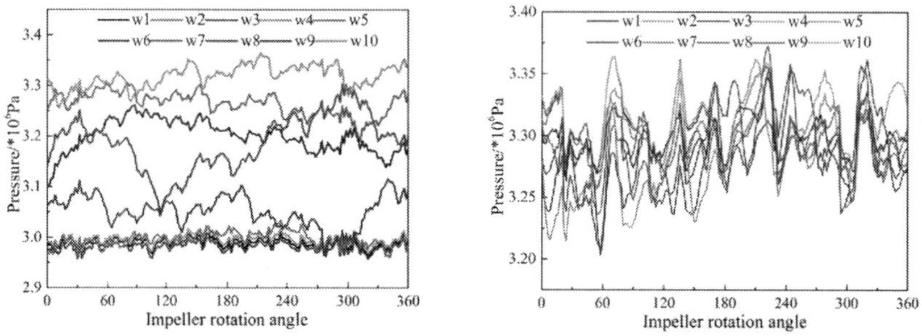

Figure 15. Pressure Fluctuation in Time Domain at Volute. (a) Rated condition (b) Low-flow-rate condition.

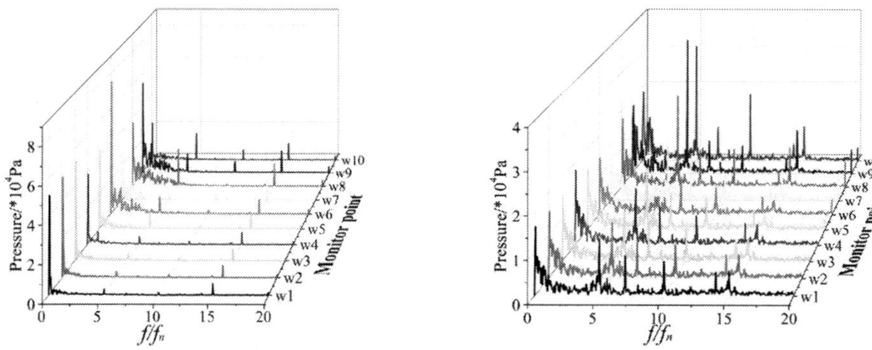

Figure 16. Pressure Fluctuation in Frequency Domain at Volute. (a) Rated condition (b) Low-flow-rate condition.

5. Conclusions

This paper establishes a cycling-system model to numerically investigate the unsteady hydrodynamic characteristics and the pressure pulsation of a centrifugal pump handled at low-flow-rate conditions. The conclusions are as follows:

(1) The numerical method of the cycling-system model does not impose fixed inlet and outlet boundary conditions, and the parameters are obtained through internal calculations, which are more conforming to practical flow especially at off design conditions.

(2) At low-flow-rate condition, the generation of stall clusters in impeller and diffuser domains will influence the pressure distribution and flow pattern inside the pump; the existence of the diffuser worsen the pulsation, which needs further investigation.

(3) Low-frequency random pressure pulsations are generated at low-flow-rate condition, which is dangerous if the rotation frequency of the stall cluster is the same as or close to the natural frequency of the pump system. In other words, if a pump must be handled at some off design conditions, we have to consider the possibility due the low-frequency excitation in advance.

Acknowledgment

This research work was supported by the National Science Foundation of China (Grant number 51806042)

Nomenclature

u、v、w three Eulerian velocity components

Q_d rated flow rate

Q_s low flow rate

T one cycle

f frequency

f_n shaft frequency

References

[1] Zhang L, Zhang W and Wang C 2018 Effect of impeller blade leading-edge chamfer on internal flow field of centrifugal pump *Power System Engineering* **34** 1-5

[2] Li X J, Yuan S Q, Pan Z Y, Yuan J P and Si Q R 2013 Numerical simulation of the whole flow field for centrifugal pump with structured grid *Transactions of the Chinese Society for Agricultural Machinery* **44** 50-54

[3] Li Z F 2009 Numerical simulation and experimental study on the transient flow in centrifugal pump during starting period Zhejiang University

[4] Tang H and He F 2002 Numerical simulation of flow field in centrifugal pumps *Pump Technology* 3-8+14

[5] Wang X, Rao J, Li Z S, Zhu X N and Chang C 2019 Numerical study of the centrifugal compressor impeller with gurney flaps *Chinese Journal of Turbomachinery* **61** 37-40

[6] Wang J L, Yang J G, Liu Y and Zhang T C 2019 Multi-objective optimization design of a centrifugal impeller *Chinese Journal of Turbomachinery* **61** 26-31+4

[7] Liu Y Y, Wang W J and Li T L 2019 Reach on cavitation flow of centrifugal pump based on slotting jet flow *Chinese Journal of Turbomachinery* **61** 18-22

[8] Yao H Y, Wu D Z, Cao L L and Wu P 2019 Numerical investigation of internal flow in cavitation resistant centrifugal pump *Chinese Journal of Turbomachinery* **61** 10-17+4

[9] Fang Y J, Ji C J, Sun Q, Liu J N and Zhang X W 2020 Validation of numerical simulation for centrifugal compressor under different inlet directions *Chinese Journal of Turbomachinery* **62** 15-21

[10] Yao Z F, Wang F J , Qu L X, Xiao R F, He C L and Wang M 2011 Experimental investigation of time-frequency characteristics of pressure fluctuations in a double-suction centrifugal pump *Journal of Fluids Engineering* **133**

[11] Kaupert K A and Staubli T 1999 The unsteady pressure field in a high specific speed centrifugal pump impeller-part 1:influence of the volute *Journal of Fluids Engeering* **121**

[12] Lu Y J 1992 Rotating stall and surge phenomenon and separation vortex flow in fans and blade machines *Water Conservancy & Electric Power Machinery* **5** 20-44

[13] Jia X Q 2017 Investigation on unstable flow and vibration characteristics of a centrifugal pump Zhejiang University

[14] Zhou P J 2015 Investigation of characteristics in centrifugal pumps China Agricultural University

[15] Qu L X, Wang F J, Cong G H and Yao Z F 2011 Pressure Fluctuations of the impeller in a double suction centrifugal pump *Transactions of the Chinese Society for Agricultural Machinery* **42** 79-84+78

[16] Zhou P J, Wang F J and Yao Z F 2015 Investigation of pressure fluctuation in centrifugal pump impeller under rotating stall conditions *Transactions of the Chinese Society for Agricultural Machinery* **46** 56-61

[17] Cong G H and Wang F J 2008 Numerical investigation of unsteady pressure fluctuations near volute tongue in a double-suction centrifugal pump *Transactions of the Chinese Society for Agricultural Machinery* 60-63+67
[18] Zhou P J, Wang F J and Yao Z F 2016 Study on effects of blade number on stall characteristics for centrifugal pump impeller *Journal of Mechanical Engineering* **52** 207-212
[19] Wang Z B 2020 Analysis of pressure pulsation under rotating stall of pump-turbine in pump mode Xi'an University of Technology
[20] Huang K L 2003 Internal flow and induced vibration and noise characteristics of cooling pump in internal combustion engine Jiangsu University
[21] Wang H and Tsukamoto Hiroshi 2003 Experimental and numerical study of unsteady flow in a diffuser pump at off-design conditions *Journal of Fluids Engineering* **125**
[22] Yang J, Yuan S Q, Pei J and Zhang J F 2015 Overview of rotating stall in centrifugal pump with vaned diffuser *Journal of Drainage and Irrigation Machinery Engineering* **33** 369-373

IOP Conf. Series: Materials Science and Engineering 1081 (2021) 012035 doi:10.1088/1757-899X/1081/1/012035

Simulation study on the relationship between catalyst particle concentration and flow field of a gas turbine

Guangkui Liu[1*], Sicong Sun[1], Long Sha[2], Kui Liang[1], Xisheng Yang[1], Dong An[1] and Qi Wen[1]
[1] Chinese Special Equipment Inspection & Research Institute
[2] Shenyang Blower Works Group Corporation

E-mail: liuguangkui9527@163.com

Abstract. In this paper, the characteristics of the flow field in the gas turbine under different catalyst particle concentrations are simulated and analyzed by FLUENT software. The influences of catalyst particle concentration on the high-temperature smoke flow field, the movement track of catalyst particles and the erosion of the moving blade are studied in detail. The results show that the change of particle concentration has no obvious influence on the high temperature smoke flow field. The erosion wear of moving blade mainly occurs near the leading edge and trailing edge of the blade. The possibility of erosion wear at the leading edge is obviously greater than that at the trailing edge. The concentration of catalyst particles has a great influence on the erosion wear of moving blade. The erosion wear will be aggravated with the increase of catalyst particle concentration.

1. Introduction

The flue gas turbine is the main energy recovery equipment in the heavy oil catalytic cracking unit[1, 2]. The flue gas turbine generally works long hours under a high speed of 6000-7000r/min, 600~700℃ and 0.2Mpa[3, 4]. When the catalyst separation effect is poor, a considerable number of catalyst particles are attached into the flue gas, which will cause serious erosion to the moving blade. The blade erosion wear of the smoke turbine is a typical gas-solid erosion wear[5, 6]. The high-temperature flue gas with solid catalyst particles enters into the smoke turbine and moves rapidly in the flow passage between the stationary blade and moving blade. They collide and impact the smoke turbine blade, which results in blade wear[7].

Many scholars have carried out numerical simulation research on gas turbine flow[8-10]. Du et al.[11] studied the distribution of the distribution of gaseous phase flow field in flue gas turbines by using a CFD-based numerical simulation method. The results showed that the fine catalyst particles entrained by exhaust gases more readily accumulate and melt on the pressure surface of blades. Tan et al.[12] studied the numerical simulations of gas-solid two-phase flow in the flue gas turbine stage cascade for FCC based on Euler-Lagrange model. The results showed that the impact erosion at top of blade trailing edge is caused by the particles no smaller than 10μm. Dong et al.[13] used the finite element software ANSYS/LS-DYNA to study the fatigue fracture of flue gas turbine blade caused by the erosion and abrasion of the solid catalyst particles. The erosion rule on the target material with the same particle size and different impact velocity as well as the erosion effect on the different size particles were gotten. It can be seen that larger particles will have erosion wear on the turbine blades. The influence of particle concentration on the flow field and dynamic blade erosion in the turbine needs to be further investigated.

Therefore, in order to study the influences of catalyst particle concentration on the flow field in the gas turbine, FLUENT software was used in this paper to simulate and analyze the characteristics of the flow field in the gas turbine under different catalyst particle concentrations. The influences of catalyst particle concentration on the high-temperature smoke flow field, the movement track of the catalyst and the erosion of the moving blade were studied in detail. The simulation method and results are described carefully in this paper.

2. Simulation method

2.1. Calculation model

In this paper, the numerical simulation of the internal flow field at the gas turbine was conducted by FLUENT software. The species transport model was selected. The turbulence model was k-ε model. DPM model was used for catalyst simulation. Based on the flow field characteristics in the gas turbine, governing equations were selected as follows:

Conservation of mass equation:

$$\frac{\partial \rho}{\partial t} + div(\rho \vec{u}) = 0 \tag{1}$$

Momentum conservation equation:

$$\frac{\partial (\rho u)}{\partial t} + div(\rho u \vec{u}) = div\,(\mu\; grad\, u) - \frac{\partial p}{\partial x} + S_u \tag{2}$$

$$\frac{\partial (\rho v)}{\partial t} + div(\rho v \vec{u}) = div\,(\mu\; grad\; v) - \frac{\partial p}{\partial y} + S_v \tag{3}$$

$$\frac{\partial (\rho w)}{\partial t} + div(\rho w \vec{u}) = div\,(\mu\; grad\, w) - \frac{\partial p}{\partial z} + S_w \tag{4}$$

Energy conservation equation:

$$\frac{\partial (\rho T)}{\partial t} + div(\rho \vec{u} T) = div(\frac{k}{c_p}\; grad\; T) + S_T \tag{5}$$

2.2. Geometric model and mesh generation

According to the actual structure of the flue gas turbine, fluid domain geometry model of the flue gas turbine is established. The fluid domain of the flue gas turbine consists of inlet section, stator blade, moving blade and outlet section, which contained 38 stator blades and 57 moving blades. The actual structure of the flue gas turbine was shown in Figure 1. The geometric model of numerical simulation was shown in Figure 2.

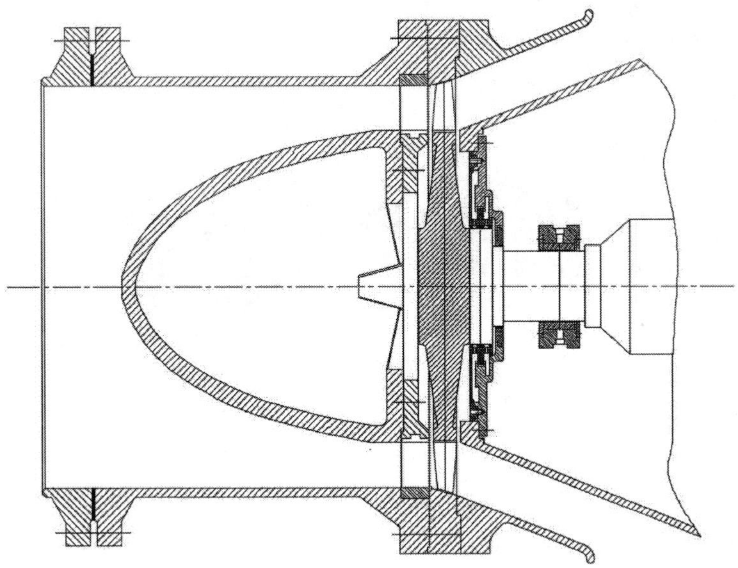

Figure 1. Structure diagram of the smoke turbine

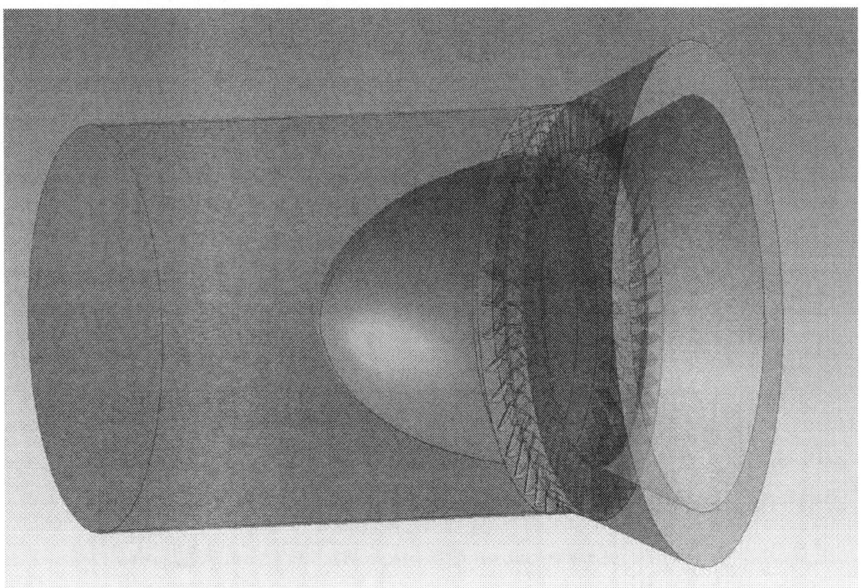

Figure 2. Fluid domain model of the flue gas turbine

The Mesh module in ANSYS was used to divide the geometric model. In this paper, tetrahedral unstructured grids were used and the number of grid nodes in the computational domain was 15.7 million. The Mesh division result of the geometric model was shown in Figure 3.

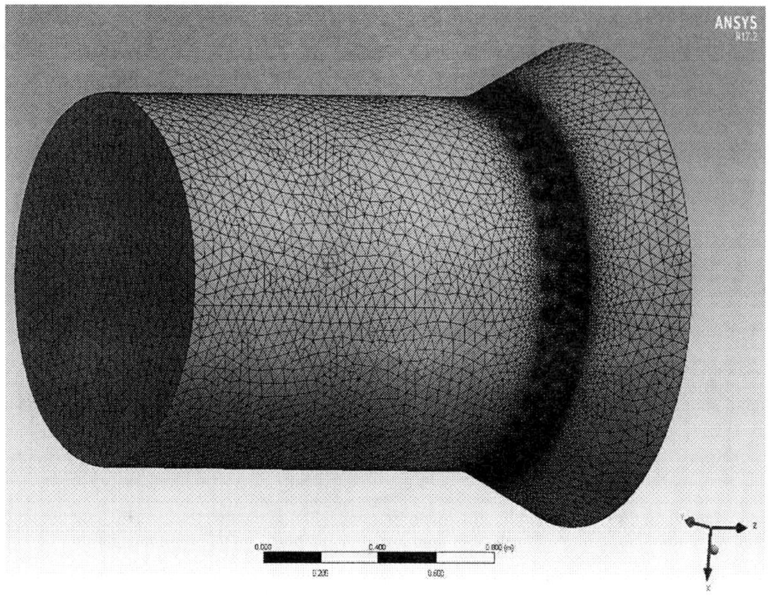

Figure 3. Mesh generation results of the geometric model

2.3. Boundary conditions and working condition arrangement

This study used a Density-Based Solver (a dense-based Solver that can achieve high computational accuracy) for numerical simulation. The solution method was implicit (implicit). The working medium is high-temperature flue gas. The inlet temperature of the flue gas is 670 °C. The rotor speed of the flue gas turbine is 6263 r/min. The flue gas flow is 1796.0 Nm^3/min under rated conditions. The specific simulation parameters and conditions were shown in Table 1.

Table 1. Numerical simulation parameters of catalyst particle concentration study

	High-temperature flue gas traffic [Nm^3/min]	High-temperature flue gas inlet temperature [K]	Gas turbine speed [r/min]	Catalyst particle concentration [g/m^3]	Catalyst particle size [m]
M_1	1796.0	943.0	6263	0.2	1.5×10^{-6}
M_2	1796.0	943.0	6263	0.17	1.5×10^{-6}
M_3	1796.0	943.0	6263	0.13	1.5×10^{-6}
M_4	1796.0	943.0	6263	0.09	1.5×10^{-6}

3. Simulation results and analysis

3.1. Influence of catalyst particle concentration on smoke flow field at high temperature

Figures 4(a)~(d) show the pressure distribution cloud map of blade median diameter (50% blade height) at different particle concentrations. It can be seen that the pressure of high-temperature flue gas decreases gradually along the flow direction. There is no obvious difference between the calculated pressure value and the change of pressure gradient when the catalyst particle concentration changes. Simultaneously, at the moving blade work and non-work face, the change trend and the velocity distribution pattern of the pressure distribution along the blade shape, each moving blade high pressure distribution curve of the line, temperature of the flow field and temperature gradient did not change obviously with the concentration of the catalyst particles. The reason is that the flow field particle concentration level was low, which can't have obvious influence on the momentum and thermodynamic characteristics of whole flow field. Therefore, the numerical simulation results show that the pressure distribution of flow field in the gas turbine has no obvious change with changing the concentration of catalyst particles.

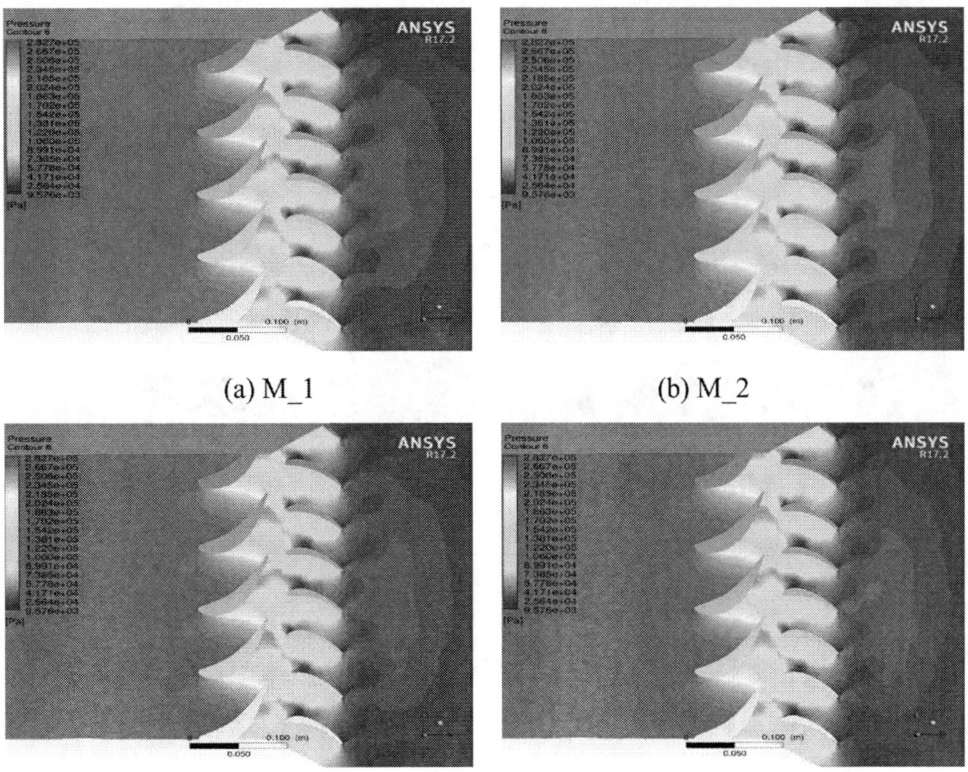

(a) M_1 (b) M_2

(c) M_3 (d) M_4

Figure 4. Cloud diagram of middle diameter section pressure in the fluid domain of the gas turbine with different particle concentrations

3.2. Influence of the concentration of catalyst particles on the trajectory of the catalyst

Figures 5(a)~(d) show the particle movement trajectory at the rotor blade under different catalyst particle concentrations. As can be seen from the figure, there is no obvious difference in the movement trajectory of catalyst particles with the change of particle concentration. The catalyst particles impinged on the moving blade under the

IOP Conf. Series: Materials Science and Engineering 1081 (2021) 012035 doi:10.1088/1757-899X/1081/1/012035

action of different factors such as the turbulent action of high temperature flue gas, inertia force and thermal swimming force. When the catalyst particles are accelerated through static blades, they impact the rotor blades at a high speed. When they collide with the rotor blades, their velocity decreases. After collision and reflection, the catalyst particles accelerate away from the rotor blades with the high-temperature flue gas. Numerical simulation shows that the erosion positions of rotor blades are different at different catalyst concentrations, but most of the erosion occurs near the leading edge and trailing edge of the blades.

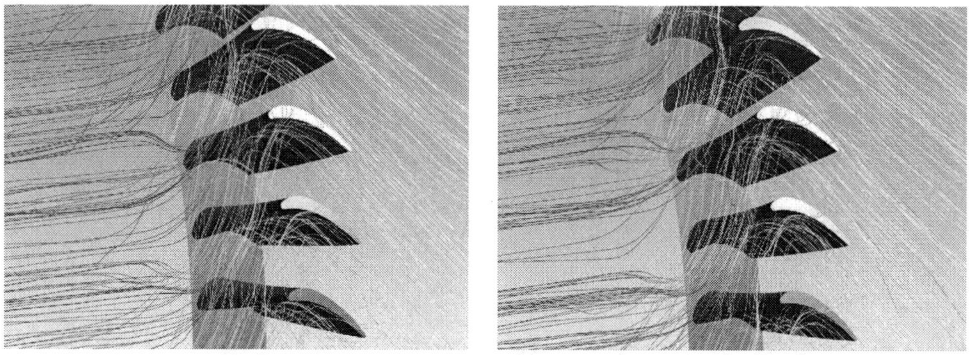

(a) M_1 (b) M_2

(c) M_3 (d) M_4

Figure 5. Influence of different catalyst particle concentrations on particle trajectory

3.3. Influence of catalyst particle concentration on dynamic blade erosion

Figures 6(a)~(d) show the axial distribution of particle erosion wear on moving blade when the catalyst particle concentration is changed. Numerical simulation results show that the change of catalyst concentration has a great influence on the erosion and wear of rotor blades. With the decrease of catalyst concentration, the axial erosion rate of the moving blade decreased to different degrees. The erosion wear mainly occurred in the leading edge and trailing edge of the moving blade as well. The erosion wear of the leading edge of the blade was much greater than that of the trailing edge. With the increase of catalyst particle concentration, the erosion and wear in the middle of the blade also increased to different degrees.

IOP Conf. Series: Materials Science and Engineering 1081 (2021) 012035 doi:10.1088/1757-899X/1081/1/012035

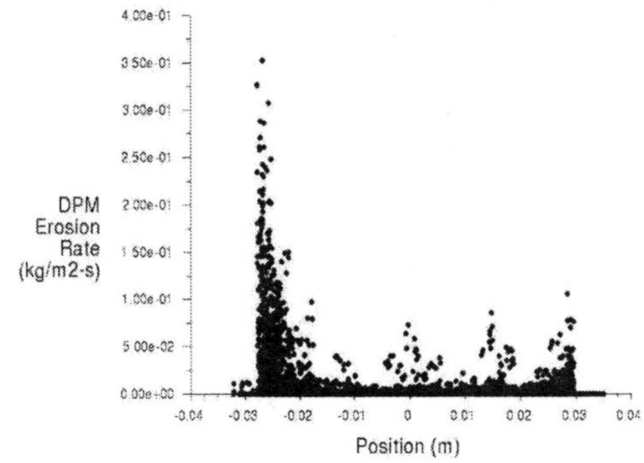

(a) M_1

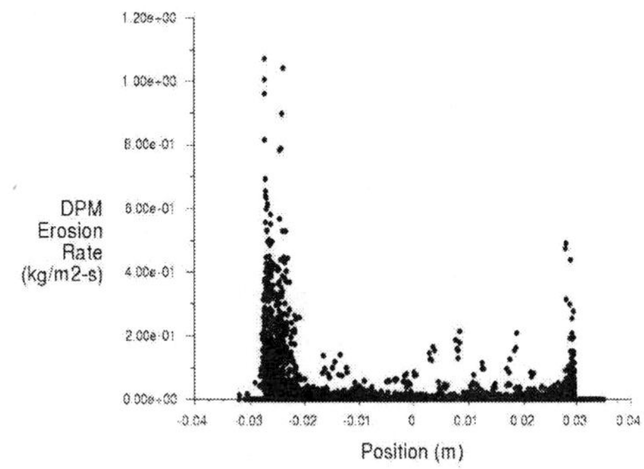

(b) M_2

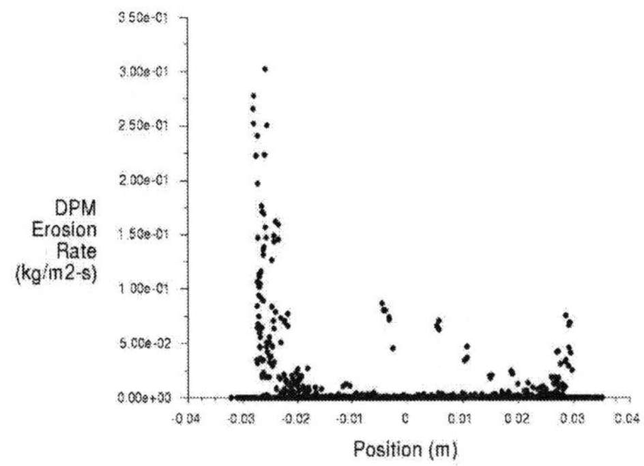

(c) M_3

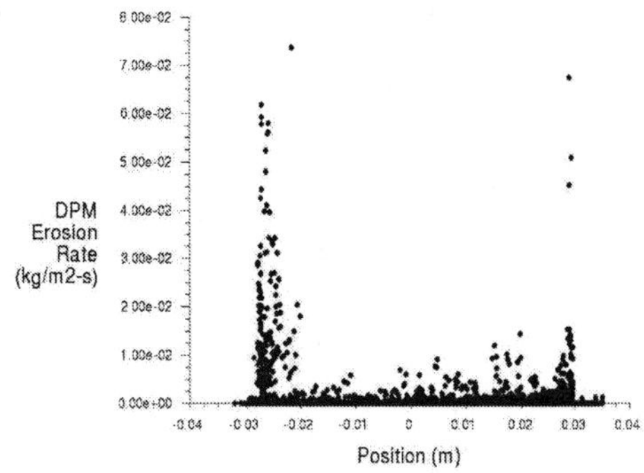

(d) M_4

Figure 6. Influence of different catalyst particle concentrations on the axial distribution of corrosion wear

4. Conclusion

The influence of catalyst particle concentration on the flow field in the gas turbine was investigated by numerical simulation of the gas-solid two-phase flow field in the gas turbine. The influence of the particle movement trajectory and the erosion of the moving blade were also investigated. Based on the numerical research, theoretical support and guidance are provided for the failure analysis and optimization direction of gas turbine. The specific conclusions of this paper are as follows:

(1) Within the range of catalyst particle concentration studied in this paper, the change of particle concentration will not have a significant impact on the high-temperature smoke flow field.

(2) Moving blade erosion wear mainly occurs near the leading edge and trailing edge of the blade. The possibility of erosion wear at the leading edge is obviously greater than that at the trailing edge.

(3) The catalyst particle concentration has a great influence on the erosion wear of moving blade. With the increase of particle concentration, the erosion wear of moving blade will be intensified to different degrees. Meanwhile, the location of erosion wear of moving blade is also different, but it still mainly occurs near the leading edge and trailing edge of blade.

References

[1] H. Sun, M. Zhu, Z. Sun, COMMERCIAL APPLICATION OF CATALYST COKC-1, QILU PETROCHEMICAL TECHNOLOGY, 36 (2008) 17-20.

[2] H. Liu, Commercial application of COKC-1 catalyst in RFCC unit, INDUSTRIAL CATALYSIS, 25 (2017) 60-64.

[3] Z. Wang, Solving the vibration problem of axial flow compressor unit of smoke turbine, Chinese Journal of Turbomachinery, (1995) 36-39.

[4] F. Yu, B. Huang, Transformation of Domestic Hydrauli Control System in the Flue Gas Fan Un, Chinese Journal of Turbomachinery, 57 (2015) 90-94+98.

[5] B. Yang, F. Zhong, H. Yuan, Experimental Study on internal Flow of cross-flow Fans, CHINESE JOUNERAL OF MECHANICAL ENGINEERING, 36 (2000) 68-70.

[6] W. Shangguan, W. Mo, H. Shu, N. Yu, P. He, F. Zhu, Study on the Effect of Structure and Parameters of Engine Annular Cooling Fans on Aerodynamic Performances, Chinese Internal Combustion Engine Engineering, 038 (2017) 56-62.

[7] Z. Lv, Numerical Simulation and Analysis of the impact by Non-spherical Catalyst Particles on the Surface of Component, Zhejiang University of Technology, 2015.

[8] P.R. Childs, Gas Turbine Engine Internal and Secondary Flow System, Chinese Journal of Turbomachinery, 60 (2018) 53-58.

[9] Z. Yang, Progress and Challenges in Large Eddy Simulation of Gas Turbine Flows, Chinese Journal of Turbomachinery, 60 (2018) 11-24.

[10] Y. Wang, W. Zhao, L. Liang, Q. Wu, H. Jin, C. Wang, A Review on Optimal Design of Fluid Machinery Based on Machine Learning Method, Chinese Journal of Turbomachinery, 62 (2020) 77-90.

[11] Y. Du, H. Zhao, C. Yang, R. Hu, Y. Zhang, Catalyst fines behavior among FCC flue gas turbine blade rows—Effect of gaseous phase flow field distribution, CHEMICAL ENGINEERING(CHINA), (2012) 63-66.

[12] H. Tan, J. Wang, Y. Jin, Numerical Simulation of Gas-solid Two-phase Flows in

Stage Cascade of Flue Gas Turbine Used for Fcc Unit, TURBINE TECHNOLOGY, 54 (2012) 437-441.

[13] X. Dong, H. Zhang, X. Cai, L. Wang, Numerical study on the erosion and abrasion behavior of the high temperature alloy material of flue gas turbine blade, JOURNAL OF PLASTICITY ENGINEERING, (2014) 122-127.

IOP Conf. Series: Materials Science and Engineering 1081 (2021) 012036 doi:10.1088/1757-899X/1081/1/012036

An Experimental Platform with Rotating Airflow Excitation and Stationary Blisk Vibration Measurement

Luohui Ouyang [1], Qingzhen Bi [1*], Hai Shang[1], Limin Zhu [1] and Jianhua Yu [2]

[1] School of Mechanical Engineering, Shanghai Jiao Tong University, Shanghai, PR China
[2] AECC Commercial Aircraft Engine Co., LTD, Shanghai, PR China

E-mail: biqz@sjtu.edu.cn

Abstract. With the question that the vibration of the realistic blisk in the engine is hard to be predicted, researchers need an experimental test platform to study the realistic dynamic properties and monitor the vibration of the designed blisk. For keeping the relative motion between airflow and blisk in a real engine, and facilitating the measurement of the blisk vibration, a new experimental platform with rotating air excitation and stationary blisk vibration measurement is developed in this paper. Firstly, the schematic diagram of the test platform is shown and all the components are illustrated clearly, including the stationary airflow excitation system, the test rig, and the vibration measurement system. Then two typical experiments are conducted with the designed academic blisk. A stationary excitation and a transient acceleration excitation are employed to the stationary blisk. Through the time-frequency analysis of the vibration signal measured by the Laser Doppler Vibrometer, the source of each component of the vibration signal is obtained. The experiment results show that the stationary and transient vibration of the blisk can be measured and identified correctly.

1. Introduction

The safe operation of the blisk is an important prerequisite for the normal working of the engine. While the vibration of the realistic blisk in the engine is hard to be predicted due to the actual intricate working conditions. Even some researchers have proposed some predictions of the maximum amplitude of blisk in the stationary [1,2] and transient [3,4] excitations, experimental investigations of the blisk vibration are still needed for validating the predictions. Moreover, for studying the dynamic properties of the designed blisk, an experimental platform with excitation and measurement is demanded.

Experimental studies published in the field of platforms with blisk excitation and vibration measurement can be divided into two categories, depending on whether the blisk is rotating or stationary. (i) The blisk is rotating by the airflow. The vibration of the blisk can be measured by strain gauges with a slip ring or telemetry system [5], Laser Doppler Vibrometer (LDV) with a self-tracking mirror [6,7], and tip timing [8-11]. These techniques can obtain blisk vibration in real operation status. However, they contain additional centrifugal effect that is usually not considered in the analysis and prediction of the vibration. And the measurement techniques are too expensive. (ii) The blisk is stationary. The blades are excited by the electromagnets [12,13], piezoelectric actuators [14,15], or acoustic excitations [16,17] to obtain any deterministic excitation. The vibration of the stationary blisk can be conveniently measured by a variety of techniques, such as strain gauge[18,19], accelerometer [20], proximity probe [21], LDV

[22], or holographic equipment [23]. Even these tests can obtain an accurate response under the deterministic excitation, they cannot form a real airflow excitation in engine operation.

To conveniently measure the blisk vibration in the real airflow excitation, a new experimental test platform is needed to be designed. The blisk is set to be stationary for avoiding the centrifugal effect and employing convenient measurement techniques. The contact measurement techniques, such as strain gauge and accelerometer, may bring additional mass effect [24]. As for the non-contact measurement, the holographic equipment is too expensive to afford. Then LDV is employed to measure the vibration of blisk for its non-intrusive nature and mature commercial use. The nozzle is driven to be rotating for keeping the relative motion between airflow and blisk, also this setting can ensure that the airflow excitation is similar to the real operation condition.

In this paper, a new experimental test platform with rotating nozzle and stationary blisk is developed. The vibrations of the blisk in stationary and transient excitation are measured and analyzed. In Section 2, the experiment test platform is introduced in detail. The schematic diagram of the test platform is shown and explain all the components clearly. The detailed structure of the test rig and the actual test arrangement for the experiment are demonstrated. Then a simplified academic blisk is designed and the mode characteristic is analyzed. In Section 3, Two experiments are conducted. A stationary excitation and a transient excitation are employed to the blisk and the vibrations are measured and analyzed. Moreover, an analysis of the truncated excitation is given. The last is the conclusion.

2. Experiment test platform

2.1. Test platform design

For measuring the vibration of the stationary blisk under the rotating airflow excitation in a non-contact way, the experimental test platform is designed with three parts: the airflow excitation system, the test rig, and the vibration measurement system. The schematic diagram of the platform is shown in Figure 1.

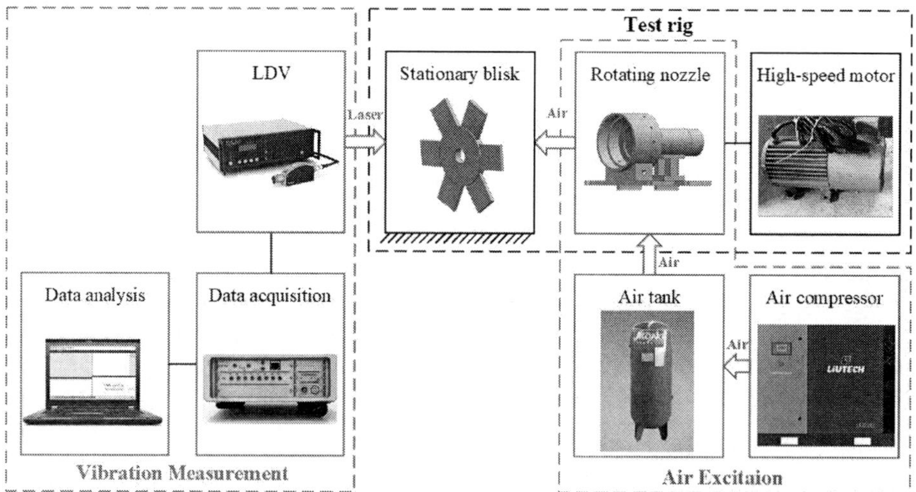

Figure 1. Schematic diagram of the experimental test platform.

The airflow excitation system provides the rotating excitation which is similar to the real air jet in the engine to the blisk. The air is compressed in the air compressor (LiuTECH LU22G-8) and transferred to the air tank (Haikong C1/8). The air tank can provide stable air pressure. Then the air is imported to the test rig. The nozzle number in the nozzle plate indicates the engine order and can be replaced according to the requested engine order.

The test rig combines the high-speed rotating mechanical structure and high-pressure airflow, which contains the high-speed motor, rotating nozzle system, and stationary blisk. The detailed design of the test rig is shown in Figure 2. The high-speed motor (Hualian TYG-7.5kW) drives the nozzle system to rotate, and the airflow spurted from the rotating nozzle plate to form a travelling wave excitation. the tested blisk is mounted to be stationary for avoiding the centrifugal effect.

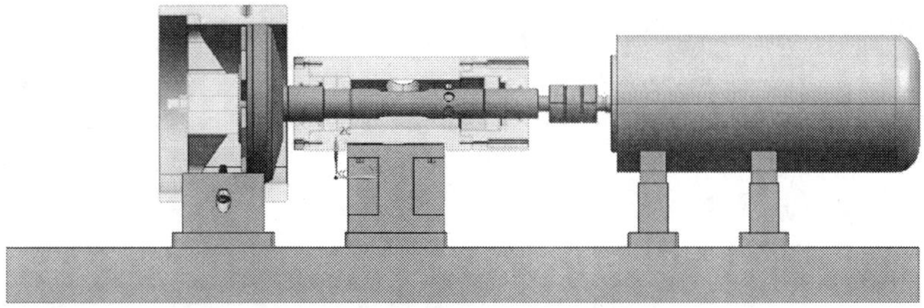

Figure 2. The detailed structure of the test rig.

Under the excitation of the travelling airflow excitation, the blade will vibrate. To measure the vibration of the blade tip in a non-contact technology, LDV (Polytec CLV-2534) is employed based on the Doppler principle. The vibration data is collected by the data acquisition (SIEMENS LMS-SCADAS-mobile-202) and is analyzed by the vibration analysis software (SIEMENS Test.Lab) on the computer.

Figure 3 shows the test arrangement for the experiment. Figure 3(a) demonstrates the airflow excitation setups and test rig. Moreover, an air dryer is added in the air excitation setups for drying the airflow in a humid environment. The air pressure to excite and the speed of the motor are controlled in the software compiled by NI LabVIEW. The side view of the test rig in Figure 3(b) shows clearly the rotating nozzle and stationary blisk. The nozzle and blisk can be replaced quickly and easily for any experiment needs. Figure 3(c) shows the vibration measurement and controllers. Besides, a pressure sensor is equipped in the inlet of the test rig and a PID closed-loop control is employed for ensuring that the input pressure is stable.

2.2. Blisk design

The blisk is the test piece of the experiment and can be designed to for any shapes and sizes. For simplicity and convenience of simulation, the blisk in the experiment is designed to be flat, and the sector number is 6. The three-dimensional model of the blisk is shown in Figure 4(a). The key geometric parameters of the blisk are listed as follows. The disk is an annulus with 80 mm outer diameter, 20 mm inner diameter, and 10 mm thickness. The blade tip is on the 160 mm diameter circle. The width and thickness of the blade are 34 mm and 5 mm respectively.

For prior knowledge of the dynamic characteristic of the blisk, a simulative modal analysis of the blisk is given based on the ANSYS software. The cyclic symmetry modelling method is employed that only one sector of the blisk model is needed. The mesh grid is controlled in the software with the 1 mm element size by adaptive size control function. The number of elements in the meshed model is 11417. The material of the blisk is aluminium alloy with Young's modulus = 80 GPa and density = 2600 kg/m3.

With these setting parameters, the first three family natural frequencies in different nodal diameters can be obtained and plotted in Figure 4(b). It can be inferred that the first family natural frequencies are the blade dominated frequencies, and are chosen as the excited frequencies in the latter experiments. Figure 4(b) also shows parts of the mode shapes in zero and one nodal diameters. In the first mode with zero nodal diameter, every blade arrives to the maximum deformation at the same time and can be conveniently measured. While in the second mode with zero nodal diameter which the disk mode dominates, the blade mode follows the disk mode. Then blade vibrates in the circumferential direction.

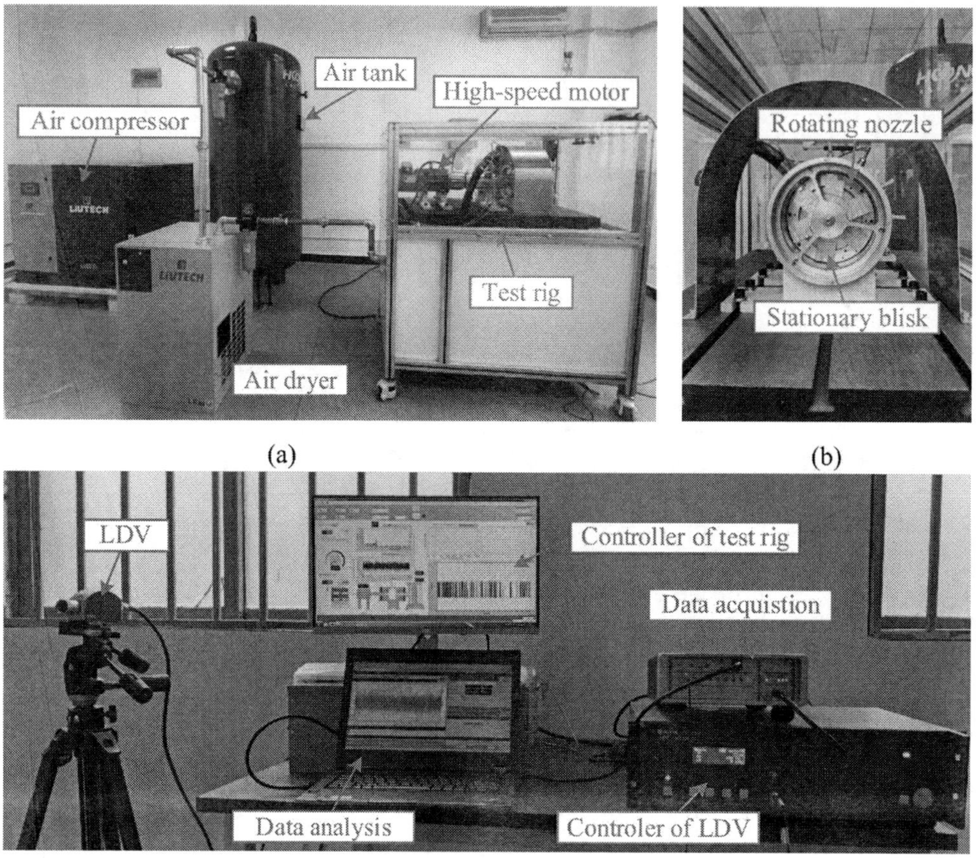

(a) (b)

(c)

Figure 3. Test arrangement for the experiment: (a) air excitation system and test rig (b) side view of the test rig (c) vibration measurement and controllers.

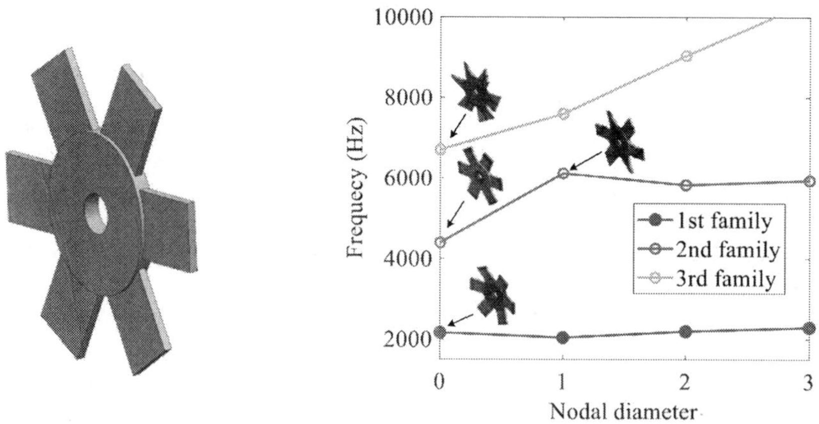

(a) (b)

Figure 4. Basic characteristic of the blisk model: (a) geometry of the model (b) the first three family natural frequencies in different nodal diameters with part of the mode shapes.

Because LDV measures the vibration in the normal direction of the blisk, then in this mode, the blade vibration cannot be measured. And in the third mode with zero nodal diameter, the torsional vibration of the blade is obvious and is the same as the blade mode in the second mode with one nodal diameter except for the different nodal diameter.

3. Experiments on the stationary and transient excitation

3.1. Response measurement and analysis

To testify the experiment ability of the test platform developed in Section 2, a stationary excitation and a transient excitation are employed in the experiments. The air pressure of the experiments is set to be 0.1 MPa that the blade will be excited at an observable level. The number of exhaust ports of the rotating nozzle is 12 so that the engine order of the excitation is 12. Because there are small differences in natural frequency between the real blisk and the designed model, the natural frequency of the experimental blisk is pre-tested and is 2332 Hz. As for the setting for measurement, the sampling frequency is 51200 Hz. Then even in the resonance, the sampling number in a vibration period is 22. The measurement would not lose the peak information of the vibration. Moreover, because of the LDV equipment, the velocity of the vibration is measured and is regarded as a description of the vibration. And the measured blades in the experiments are all the same. After these settings are completed, two experiments are conducted.

Firstly, a stationary excitation is employed to the blisk. The rotating speed of the rotating nozzle is constant with 11660 rpm (revolutions per minute). Combined with this rotating speed and engine order, the blisk will be excited exactly in the resonance. The stationary response is measured and shown in Figure 5(a). From the partial enlarged drawing, the vibration signal is fully sampled. After the time-frequency analysis, the autopower spectrum of the stationary vibration is shown in Figure 5(b). It can be indicated that the natural frequency of the blisk is truly excited and dominates among all frequency bandwidth. Moreover, because of the interval of the exhaust ports are not ideally evenly spaced, the lower eleven orders of the excitations will be shown in the figure. And the high order harmonic is shown because of the truncated excitation and will be discussed latter.

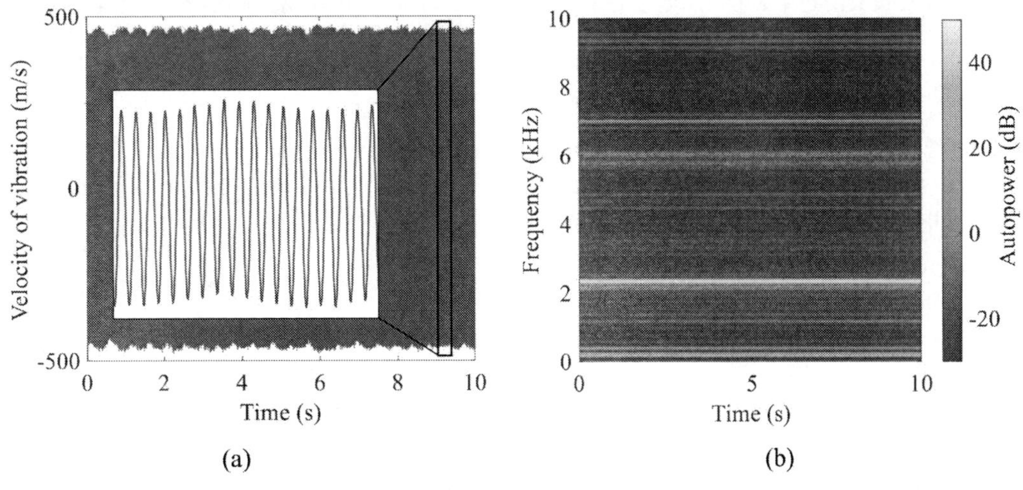

Figure 5. Stationary response and analysis in 11660 rpm: (a) stationary response in the time domain (b) time-frequency analysis of the stationary response.

Secondly, a transient excitation is employed to the blisk. The transient excitation is constant acceleration with the acceleration rate is 600 rpm/s, so that in a time duration of 30 seconds, the rotating speed will linearly increase to 18000 rpm, and will pass through the resonance of the blisk with the 12-engine order. The transient response in the time domain is measured and shown in Figure 6(a).

Combined with the time-frequency analysis of the transient response in Figure 6(b), the multiple resonances in the time domain can be explained.

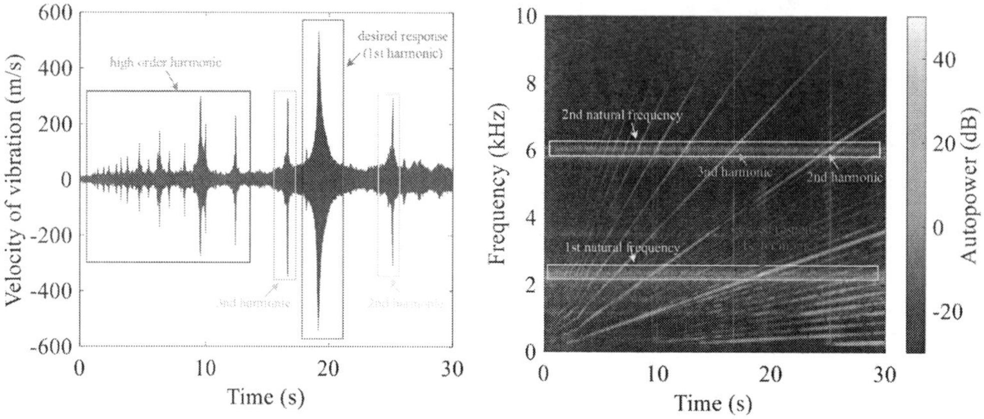

(a) (b)

Figure 6. Transient response and analysis in 10 Hz/s acceleration: (a) transient response in the time domain (b) time-frequency analysis of the transient response.

The resonance in the red frame is the desired response because of the time and frequency matches. The blisk is passing through its first natural frequency. The first and second natural frequency of the real blade is nearly the same with the simulated model in Figure 4(b). Because of the truncated excitation, there exist multiple high harmonic excitations. The second and third harmonic excitations are passing through the second natural frequency of the blisk, therefore, two resonance are shown in cyan frames and yellow frames respectively. Moreover, other high order harmonic excitations will excite the first two natural frequencies in the beginning of time period in green frame. In addition, the responses due to lower eleven orders of the excitations can be shown because the exhaust ports are not ideally evenly spaced. Besides, when the second and third harmonic excitations are passing through the second natural frequency of the blisk, the drastic vibration of the blisk may excite other test setups and bring white noise in the whole spectrum at that moment.

3.2. Effect of excitation truncation

Theoretically, the acceleration excitations in the industrial jet are sinusoidal waves with frequency linearly increased, which is shown in the blue dashed curve in Figure 7 (a). However, because of the structural design of the blisk and the air passage, the rotating nozzle only can provide positive pressure to the blisk, and cannot provide the negative pressure to the blisk. Therefore, the excitation signal is truncated in the negative parts. Besides, when the airflow sweeps across the blade surface, the pressure will keep in constant in a short time, and it can be regarded as the truncated in the top parts of the excitation signal. Figure 7 (b) shows the time-frequency spectrum of the truncated excitation. It can be inferred that except for the first harmonic excitation, which is corresponding to the theoretical excitation, there exist higher multiple harmonic excitations. Therefore, the truncated excitation is the reason for multiple harmonic responses in transient vibration shown in Figure 6 (a).

4. Conclusions

In this work, a new experimental platform with rotating airflow excitation and stationary blisk vibration measurement is developed. The experimental platform contains the rotating airflow excitation system, test rig, and non-contact vibration measurement system. The test rig unites the air passage and rotor to generate a rotating airflow excitation. Compare to the realistic engine with stationary nozzle and rotating blisk, the new experimental test platform can keep the relative motion between airflow and blisk, and

IOP Conf. Series: Materials Science and Engineering 1081 (2021) 012036 doi:10.1088/1757-899X/1081/1/012036

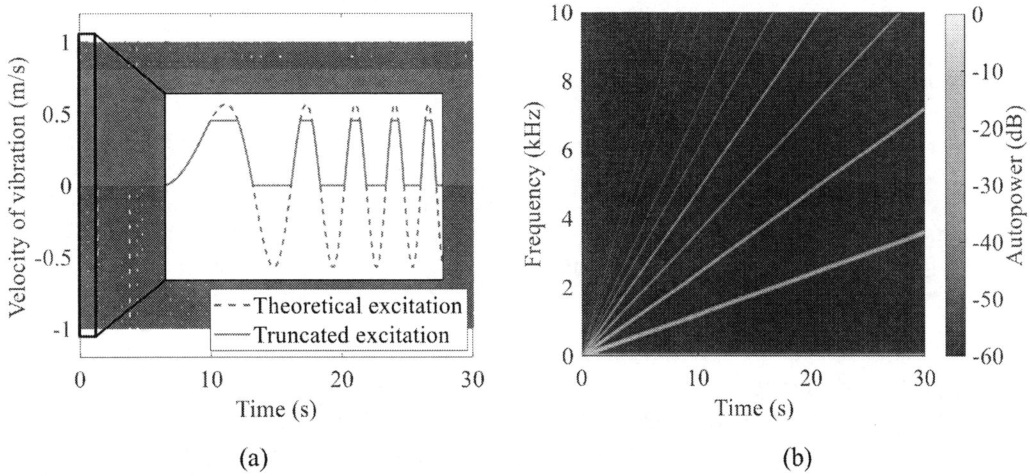

Figure 7. Analysis of the truncated excitation in the experiment: (a) comparison the theoretical and truncated excitation in the time domain (b) time-frequency analysis of the truncated excitation.

can avoid the centrifugal effect of the rotating blisk. Also, the vibration of blisk is conveniently to be measured by LDV in a non-contact way because the blisk is stationary. Besides, the blisk and the nozzle is replaceable, which satisfies any shaped blisk in any engine order.

The vibrations of the blisk in stationary and transient excitations in the experiment platform are measured and analyzed. For stationary excitation, the frequency of the vibration is matched with the product of the excitation frequency and engine order. For transient acceleration excitation, the blisk is passing through its first natural frequency band. Besides, the lower engine order is excited for the non-strictly distributed intervals of the exhaust ports, and higher harmonic excitation is brought in the blisk for the design induced excitation truncation. Also, the white noise caused by high harmonic induced resonance is introduced in the vibration measurement. Even so, the first harmonic resonance which is the desired one is recognized and dominates among all other vibrations.

For further research, combined with the vibration amplitude prediction proposed by the authors before, several validations will be conducted. And the mistuning will be intentionally added in the tested blisk in the later investigation.

Acknowledgement
The authors gratefully acknowledge the financial support of the National Science and Technology Major Project (No. 2018ZX04005001).

References
[1] Petrov E P, Sanliturk K Y and Ewins D J 2002 A new method for dynamic analysis of mistuned bladed disks based on the exact relationship between tuned and mistuned systems *Journal of Engineering for Gas Turbines and Power* **124** 586-597.
[2] Feiner D M and Griffin J H 2002 A fundamental model of mistuning for a single family of modes *Journal of Turbomachinery* **124** 597-605.
[3] Bonhage M, Pohle L, Scheidt P L and Wallaschek J 2015 Transient amplitude amplification of mistuned blisks *Journal of Engineering for Gas Turbines and Power* **137** 112502.
[4] Ouyang L, Shang H, Chen H, Bi Q and Zhu L 2020 The mechanism on prediction of transient maximum amplitude for tuned and mistuned blisks *Journal of Engineering for Gas Turbines and Power* **142** 051013.
[5] Li Z and Wen Y 2004 Vibration monitor system for fan *Chinese Journal of Turbomachinery* **06** 56-58.

[6] Stanbridge A B, Sever I A and Ewins D J 2002 Vibration measurements in a rotating blisk test rig using an LDV *Fifth International Conference on Vibration Measurements by Laser Techniques: Advances and Applications* **4827** 1-8.

[7] Sever I A, Petrov E P and Ewins D J 2008 Experimental and numerical investigation of rotating bladed disk forced response using underplatform friction dampers *Journal of Engineering for Gas Turbines and Power* **130** 042503.

[8] He C, Li H and Zhang X 2015 Review of blade fatigue failure mechanism and detection method based on flow induced vibration *Chinese Journal of Turbomachinery* **57** 69-77.

[9] Zhai G, Wang W, Chen Z and Shao X 2019 Research on monitoring method and system of rotor Torsional vibration based on tip timing *Chinese Journal of Turbomachinery* **61** 40-47.

[10] Heath S and Imregun M 1998 A survey of blade tip-timing measurement techniques for turbomachinery vibration *Journal of Engineering for Gas Turbines and Power* **120** 784–791.

[11] Battiato G, Firrone C M and Berruti T M 2017 Forced response of rotating bladed disks: Blade Tip-Timing measurements *Mechanical Systems and Signal Processing* **85** 912-926.

[12] Kruse M J and Pierre C 1997 An experimental investigation of vibration localization in bladed disks: part I—free response *Turbo Expo: Power for Land, Sea, and Air* **78712** V004T14A068.

[13] Kruse M J and Pierre C 1997 An experimental investigation of vibration localization in bladed disks: part II—forced response *Turbo Expo: Power for Land, Sea, and Air* **78712** V004T14A069.

[14] Fabunmi J A 1980 Forced vibrations of a single stage axial compressor rotor *Journal of Engineering for Gas Turbines and Power* **102** 322–328.

[15] Judge J, Pierre C and Mehmed O 2001 Experimental investigation of mode localization and forced response amplitude magnification for a mistuned bladed disk *Journal of Engineering for Gas Turbines and Power* **123** 940-950.

[16] Jones K W and Cross C J 2003 Traveling wave excitation system for bladed disks *Journal of Propulsion and Power* **19** 135-141.

[17] Judge J A, Ceccio S L and Pierre C 2003 Traveling-wave excitation and optical measurement techniques for non-contacting investigation of bladed disk dynamics *The Shock and Vibration Digest* **35** 183-190.

[18] Russhard P 2015 The rise and fall of the rotor blade strain gauge *Vibration Engineering and Technology of Machinery* ed Jyoti K. Sinha (Switzerland: Springer) p 27-37.

[19] Lerche A H, Moore J J and Allison T C 2011 Experimental study of blade vibration in centrifugal compressors *Turbo Expo: Power for Land, Sea, and Air* **54662** 1057-1066.

[20] Yue F, An J and Duan F 2005 Application and analysis of vibration measured parameters on trouble diagnosis *Chinese Journal of Turbomachinery* **03** 63-64.

[21] Diamond D H, Heyns P S, Oberholster A J 2015 A comparison between three blade tip Timing algorithms for estimating synchronous turbomachine blade vibration *9th WCEAM Research Papers* ed J. Amadi-Echendu (Switzerland: Springer) p 215-225.

[22] Ding Q, Shen X and Liu S 2003 Measuring system of flow velocity & flowrate of fan measured by laser doppler *Chinese Journal of Turbomachinery* **03** 8-12.

[23] Haupt U and Rautenberg M 1984 Blade vibration measurements on centrifugal compressors by means of telemetry and holographic interferometry *Journal of Engineering for Gas Turbines and Power* **106** 70–78.

[24] Beirow B, Kuhhorn A and Nipkau J 2009 On the influence of strain gauge instrumentation on blade vibrations of integral blisk compressor rotors applying a discrete model *Turbo Expo: Power for Land, Sea, and Air* **48876** 245-253.

A comparative study on the numerical solution algorithms of gas Reynolds equation

Yongliang Wang, Yi Wang, Longkai Zheng

Naval Architecture and Ocean Engineering College, Dalian Maritime University, Dalian 116026, China

E-mail:wangyl@dlmu.edu.cn

Abstract. The improved P* algorithm is studied for solving the Reynolds equation by finite difference method. P* algorithm has the advantages of distinct mathematical concepts and simple programming. Firstly, the P* algorithm is combined with the Successive Over-Relaxation (SOR) method, considering the influence of the selection of relaxation factor on the calculation results. The optimal relaxation factor is determined to be 0.97, in this case, the minimum film thickness of flexible gas foil bearing can be calculated to 3.8 μm. Then, by changing the position of P*, the algorithm accuracy, sensitivity to the initial value, and computational efficiency of different P* algorithms are compared and studied. In the end, the P* algorithm is compared with the Newton-Raphson algorithm, it is found that although the Newton-Simpson algorithm computations the matrix dimension is large, but it has the advantages of fast computing speed and high efficiency.

Nomenclature

R	radius of bearing (m)
L	length of bearing (m)
C	nominal radial clearance of the bearing (m)
P_a	ambient pressure (Pa)
h	film thickness (m)
$\bar{h}$	dimensionless film thickness
p	film pressure (Pa)
P_a	ambient pressure (Pa)
$\bar{P}$	dimensionless film pressure
z	axial coordinate
ξ	dimensionless axial coordinate
θ	circumferential coordinate
T	temperature of lubrication (K)
T_0	temperature of ambient gas (K)
$\bar{T}$	dimensionless temperature
μ_0	viscosity of ambient gas (N·s/m^2)
μ_g	viscosity of lubrication (N·s/m^2)

$\bar{\mu}_g$ dimensionless air viscosity
ω angular speed of shaft (rpm)
ω_s whirl speed (rpm)
t time (s)
$\bar{t}$ dimensionless time
γ excitation frequency ratio (ω_s / ω)
Λ_x bearing number
s bump pitch (m)
l half of bump length (m)
t_B bump foil thickness (m)
E_b young's modulus of bump foil
v_b nominal radial clearance of the bearing
A_0 effective area that one bump covers
w_d foil deformation (m)
$\bar{w}_d$ dimensionless foil deformation
k_{b0} bump stiffness (N·s/m^3)
α bump foil compliance number
ε eccentricity ratio of bearing
φ attitude angle (degree)

1. Introduction

The traditional blower and air compressor are driven by gears, and the mechanical friction in each gear drive produces a large energy consumption [1], moreover, the noise problem is prominent and the stability performance needs to be strengthened [2]. At the same time, whether the design of the rotor bearing system is reasonable is related to the safety and stability of the whole unit [3]. At this point, the choice of bearing is very important. Gas foil bearing is a kind of self-acting dynamic pressure gas bearing with foil as elastic surface. Compared with traditional gas bearing, foil bearing has the advantages of high speed, long working life, low temperature resistance, high temperature resistance, low energy loss, no pollution to the environment and excellent dynamic characteristics. It can be used in air compressor [4]、 blower [5] and aerospace and so on [6].

In the process of fluid lubrication calculation by numerical analysis method [7], the solution of gas Reynolds equation [8] can be said to be always in a core position. For a long time, people have done a lot of exploration and research on how to solve the gas Reynolds equation efficiently, quickly and accurately. In the process of numerical iteration, the Reynolds equation will be solved once for each iteration, so the efficiency of solving the Reynolds equation algorithm directly determines the speed of the simulation platform. Among them, the finite difference method is a common algorithm for solving differential equations. The original differential equation is discretized into difference equations and solved. Many scholars have done a lot of research on the basis of the finite difference method.

Huapen Wu et al. [9] studied GMRES algorithm and the traditional SOR algorithm to solve the Reynolds equation in terms of speed and accuracy. Through comparison, they found that the pre-processed GMRES algorithm had obvious advantages. Zhang Dexuan et al. [10] proposed an iterative algorithm combining the golden section method to determine the relaxation factor with the improved iterative scheme of symmetric successive super-relaxation method, which has the advantage of high precision and fast convergence. Xu Shengli [11] used the multi-grid method introduced in this paper is an accelerated over-relaxation iterative method with shorter computation time, higher precision and faster convergence rate than finite difference method. Zhang Xi et al. [12] proposed a method for grid optimization based on the Reynold equation difference operator based on the idea of grid hierarchy, and on this basis developed an adaptive regional refinement algorithm based on pressure gradient change and regional integral value. Wang Qun [13] studied the effect of the number of grids on the accuracy of the algorithm and the effect of the number of grids on the efficiency of the algorithm.

Although the solution method has been continuously improved and the convergence speed has been continuously improved, some algorithms have some difficulties in programming, especially in the process of modifying after changing boundary conditions. Based on the simplicity of programming, the P* algorithm is studied. The algorithm has the advantages of clear physical concept, easy programming and accurate calculation results. At the same time, the P* algorithm is combined with the dividing interactive method, which has a good convergence rate.

2. Theoretical analysis

2.1. General structure of gas foil journal bearing

Figure 1 shows the configuration of a first-generation bump type Gas Foil Bearing (GFB). Figure 2 shows the Cartesian coordinate system. The GFB consists of a top foil and a series of corrugated bumps (bump foil) to support the top foil. The leading edge of the top foil is free, and the foil trailing edge is fixed to the bearing sleeve.

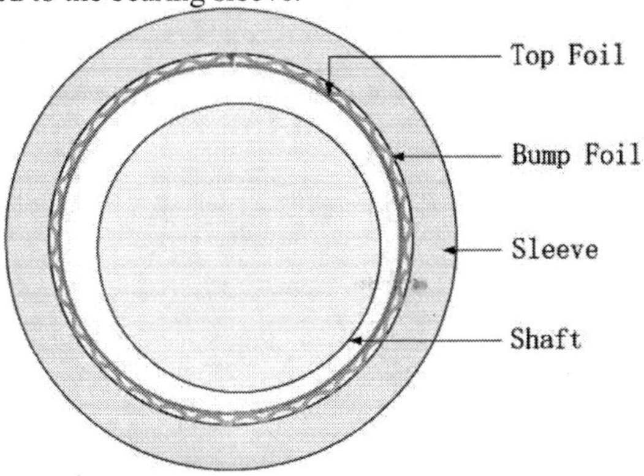

Figure 1. Schematic view of gas bump foil journal bearing.

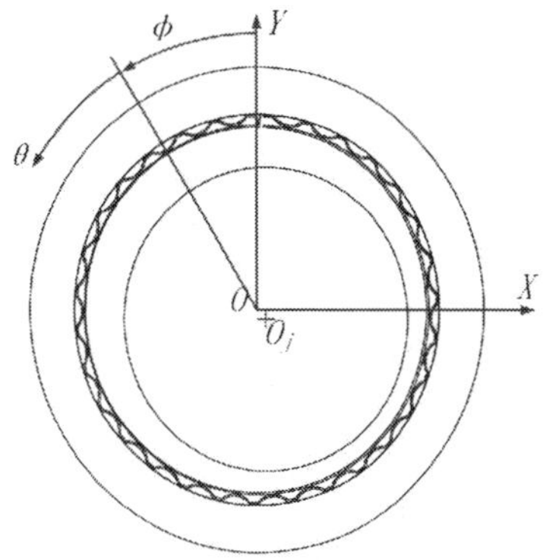

Figure 2. Cartesian coordinate system.

According to the bearing geometry shown in figure 2, assuming ideal gas for the working fluid, the dimensionless form of isothermal laminar Reynolds equation can be written as:

$$\frac{\partial}{\partial\theta}\left(\overline{P}\overline{h}^3\frac{\partial\overline{P}}{\partial\theta}\right)+\left(\frac{2R}{L}\right)^2\frac{\partial}{\partial\xi}\left(\overline{P}\overline{h}^3\frac{\partial\overline{P}}{\partial\xi}\right)=\Lambda_x\frac{\partial\left(\overline{P}\overline{h}\right)}{\partial\theta}+2\gamma\Lambda_x\frac{\partial\left(\overline{P}\overline{h}\right)}{\partial\overline{t}} \quad (1)$$

Where

$$\overline{\mu}_g=\frac{\mu_g}{\mu_0}\quad \overline{T}=\frac{T}{T_0}\quad \overline{P}=\frac{p}{p_a}\quad \overline{h}=\frac{h}{C}\quad \overline{t}=\upsilon t\quad \xi=\frac{z}{0.5L}\quad \gamma=\frac{\omega_s}{\omega}\quad \Lambda_x=\frac{6\mu_0\omega R^2}{C^2p_a}$$

2.2. Film thickness equation

According to references [14], the thickness of the air film in a gas bearing can be formulated as:

$$\overline{h}=1+\varepsilon\cos(\theta-\varphi) \quad (2)$$

As for the foil bearing, the amount of foil deformation should be considered. So the foil bearing thickness should be written as:

$$\overline{h}=1+\varepsilon\cos(\theta-\varphi)+\frac{w_d}{C} \quad (3)$$

The bump foils are equivalent to linear spring supports, without considering the stiffness characteristics of top foils and the damping effect between top and bump foils. It also ignores the damping effect between bump foils and the inner wall of the bearing. The action relation between lubrication film pressure and foil deformation is given by $(p-p_a)A_0 = k_{b0}Lw_d$. Here, A_0 is an effective area that the bump covers, and the k_{b0} is the bump stiffness, which is given by $k_{b0} = E_b \cdot t_b^3 / \left[2l^3\left(1-v_b^2\right)\right]$. The specific symbols are shown in figure 3 [15].

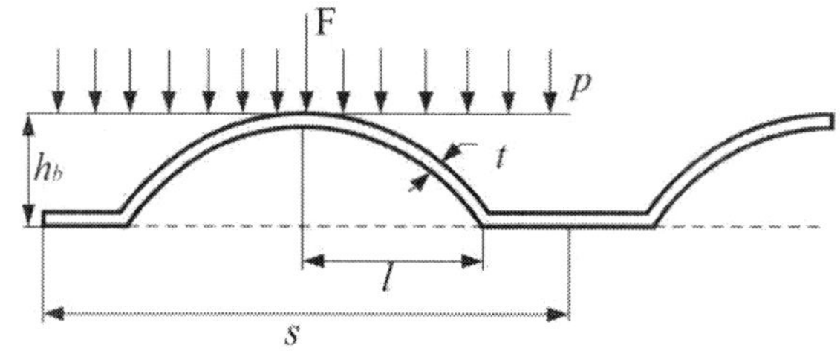

Figure 3. Schematic diagram of bump foil and top foil structure [15].

Converting it to dimensionless form can be written as $\overline{w}_d = \alpha\left(\overline{P}-1\right)$, α is the bump foil compliance number given by $\alpha = p_a \cdot s/\left(k_{b0} \cdot C\right)$. And dimensionless foil deformation is defined as $\overline{w}_d = w_d/C$. The corresponding dimensionless film thickness expression is as follows [16]:

$$\overline{h}=1+\varepsilon\cos(\theta-\varphi)+\alpha\left(\overline{P}-1\right) \tag{4}$$

2.3. Grid generation

The meshing of the inner surface of the bearing is shown in figure 4 [17]. Since the bearing structure is symmetrical about the cross-section of the bearing, only half of the bearing is taken here for calculation. In figure 4, n_θ is equally divided in the circumferential direction of the bearing and n_z is divided into the length of the axial half bearing. Therefore, there are $n_\theta+1$ nodes in the circumferential direction and $nz+1$ nodes in the length direction of the half bearing. The solution domain of bearing is $(1:n_\theta+1, 1:n_z+1)$.

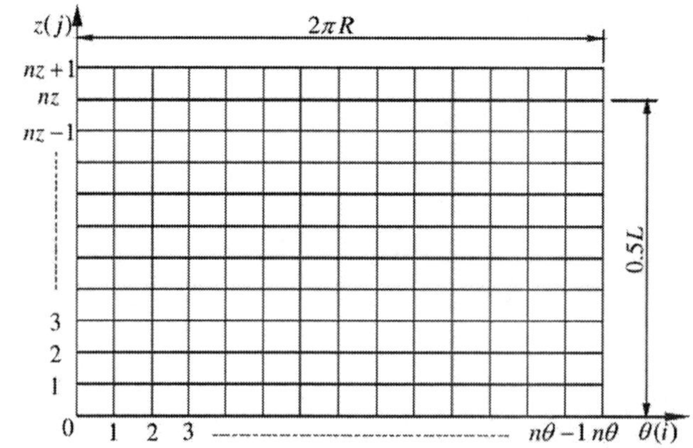

Figure 4. Diagram of grid generation in computing domain [17].

2.4. Boundary conditions

From the structure of figure 1, we can see that after the top foil is rolled into a cylinder, its head and tail are not connected, which means that the leading edge and trailing edge are not overlapped. At the same time, the top foil is supported by bump foil. Therefore, both the head and tail of the foil two sides of the bearing edge are connected with the external environment, the film pressure of these parts is equal to the atmospheric pressure.

The boundary conditions of environmental pressure are as follows:

$$\overline{P}\left(1:n_\theta+1,0\right)=1 \tag{5}$$

$$\overline{P}\left(1,0:n_z+1\right)=\overline{P}\left(n_\theta+1,0:n_z+1\right)=1 \tag{6}$$

As mentioned above, only the length of the half bearing is considered here. The symmetric boundary conditions are as follows:

$$\overline{P}\left(1:n_\theta+1,n_z-1\right)=\overline{P}\left(1:n_\theta+1,n_z+1\right) \tag{7}$$

$$\frac{\partial \overline{P}_{i,j}}{\partial \xi}=0\left(i=1:n_\theta+1,j=n_z\right) \tag{8}$$

2.5. Convergence criteria

When the result of the iteration reaches a certain precision, the iteration will terminate. The convergence conditions commonly used are as follow:

$$\sqrt{\sum_{i=1}^{n\theta}\sum_{j=1}^{nz}\left(\frac{\Delta\overline{p}_{i,j}}{\overline{p}_{i,j}}\right)^2}\le 10^{-6} \tag{9}$$

In this paper, finite element method and finite difference method are used to solve Reynolds equation and gas film thickness equation. And then we can get the static characteristic parameters such as film pressure distribution and film thickness distribution of gas foil bearing. In the following content, P* algorithm will be explained in detail, and different algorithms will be compared to it.

3. Algorithm P*

The method used in the reference [18] to solve The Reynolds equation is adopted by most researchers, namely the Newton-Raphson iterative method. This method is mainly to solve the linear equation $Ax=B$. In many studies, the method is considered to be improved. For example, the SOR iteration method is adopted in the reference [10], and the multi-grid computing method is adopted in the reference [11]. Although the solution speed is improved with continuous improvement, it is more complex in programming, especially after the boundary conditions are changed, the corresponding positions in the sparse matrix need to be revised again. Based on the simplicity of programming, the P* algorithm is studied in this paper.

3.1. Algorithm P introductions*

By expanding the partial differential items in equation (1), we can get the following results:

$$\begin{aligned}&\overline{h}^3\frac{\partial\overline{P}}{\partial\theta}\frac{\partial\overline{P}}{\partial\theta}+3\overline{h}^2\overline{P}\frac{\partial\overline{h}}{\partial\theta}\frac{\partial\overline{P}}{\partial\theta}+\overline{P}\overline{h}^3\frac{\partial^2\overline{P}}{\partial\theta^2}+\left(\frac{2R}{L}\right)^2\left(\overline{h}^3\frac{\partial\overline{P}}{\partial\xi}\frac{\partial\overline{P}}{\partial\xi}+3\overline{h}^2\overline{P}\frac{\partial\overline{h}}{\partial\xi}\frac{\partial\overline{P}}{\partial\xi}+\overline{P}\overline{h}^3\frac{\partial^2\overline{P}}{\partial\xi^2}\right)\\&=\Lambda_x\cdot\overline{h}\frac{\partial\overline{P}}{\partial\theta}+\Lambda_x\cdot\overline{P}\frac{\partial\overline{h}}{\partial\theta}\end{aligned} \tag{10}$$

Considering the simple programming, the P* algorithm is used for finite difference iteration. In equation (10), find outside the partial derivative on the left side of the equal, change it to P^* and define it as the initial pressure. Every time the calculation is completed, the value of P^* is updated, and then continues to participate in the next step of calculation in the form of a given value. After repeated iterative calculation, the convergence condition is finally satisfied and the convergence is realized. When P^* is set, the equation (10) becomes:

$$\bar{h}^3\frac{\partial\bar{P}}{\partial\theta}\frac{\partial\bar{P}}{\partial\theta}+\bar{P}^*\left(3\bar{h}^2\frac{\partial\bar{h}}{\partial\theta}\frac{\partial\bar{P}}{\partial\theta}+\bar{h}^3\frac{\partial^2\bar{P}}{\partial\theta^2}\right)+\left(\frac{2R}{L}\right)^2\left[\bar{h}^3\frac{\partial\bar{P}}{\partial\xi}\frac{\partial\bar{P}}{\partial\xi}+\bar{P}^*\left(3\bar{h}^2\frac{\partial\bar{h}}{\partial\xi}\frac{\partial\bar{P}}{\partial\xi}+\bar{h}^3\frac{\partial^2\bar{P}}{\partial\xi^2}\right)\right]$$
$$=\Lambda_x\cdot\bar{h}\frac{\partial\bar{P}}{\partial\theta}+\Lambda_x\cdot\bar{P}\frac{\partial\bar{h}}{\partial\theta} \tag{11}$$

The equation (11) is reformulated with the central difference scheme as follow:

$$\bar{P}^*_{i,j}\left[\bar{h}^3_{i,j}\frac{\bar{P}_{i+1,j}-2\bar{P}_{i,j}+\bar{P}_{i-1,j}}{\Delta\theta^2}+\left(\frac{2R}{L}\right)^2\bar{h}^3_{i,j}\frac{\bar{P}_{i,j+1}-2\bar{P}_{i,j}+\bar{P}_{i,j-1}}{\Delta\xi^2}\right]$$
$$+\bar{P}^*_{i,j}\left(3\bar{h}^2_{i,j}\frac{\bar{h}_{i+1,j}-\bar{h}_{i-1,j}}{2\Delta\theta}\frac{\bar{P}_{i+1,j}-\bar{P}_{i-1,j}}{2\Delta\theta}+3\bar{h}^2_{i,j}\frac{\bar{h}_{i,j+1}-\bar{h}_{i,j-1}}{2\Delta\xi}\frac{\bar{P}_{i,j+1}-\bar{P}_{i,j-1}}{2\Delta\xi}\right) \tag{12}$$
$$+\bar{h}^3_{i,j}\frac{\left(\bar{P}_{i+1,j}-\bar{P}_{i-1,j}\right)^2}{4\Delta\theta^2}+\left(\frac{2R}{L}\right)^2\bar{h}^3_{i,j}\frac{\left(\bar{P}_{i,j+1}-\bar{P}_{i,j-1}\right)^2}{4\Delta\xi^2}=\Lambda_x\bar{h}_{i,j}\frac{\bar{P}_{i+1,j}-\bar{P}_{i-1,j}}{2\Delta\theta}+\Lambda_x\bar{P}_{i,j}\frac{\bar{h}_{i+1,j}-\bar{h}_{i-1,j}}{2\Delta\theta}$$

Where i,j is the node number. After simplifying equation (12), we can get:

$$a_{i,j}\bar{P}^{(k)}_{i-1,j}+b_{i,j}\bar{P}^{(k)}_{i+1,j}+c_{i,j}\bar{P}^{(k+1)}_{i,j}+d_{i,j}\bar{P}^{(k)}_{i,j-1}+e_{i,j}\bar{P}^{(k)}_{i,j+1}+\left(\bar{P}^{(k)}_{i+1,j}-\bar{P}^{(k)}_{i-1,j}\right)^2m_{i,j}+\left(\bar{P}^{(k)}_{i,j+1}-\bar{P}^{(k)}_{i,j-1}\right)^2n_{i,j}=0 \tag{13}$$

Where $a_{i,j}$、$b_{i,j}$、$c_{i,j}$、$d_{i,j}$、$e_{i,j}$、$m_{i,j}$、$n_{i,j}$ are the coefficient matrixs as follows:

$$a_{i,j}=\bar{P}^*_{i,j}\cdot\left(\frac{\bar{h}^3_{i,j}}{\Delta\theta^2}-\frac{3\bar{h}^2_{i,j}}{2\Delta\theta}\frac{\bar{h}_{i+1,j}-\bar{h}_{i-1,j}}{2\Delta\theta}\right)+\frac{\Lambda_x\cdot\bar{h}_{i,j}}{2\Delta\theta} \tag{14}$$

$$b_{i,j}=\bar{P}^*_{i,j}\cdot\left(\frac{\bar{h}^3_{i,j}}{\Delta\theta^2}+\frac{\bar{h}_{i+1,j}-\bar{h}_{i-1,j}}{2\Delta\theta}\frac{3\bar{h}^2_{i,j}}{2\Delta\theta}\right)-\frac{\Lambda_x\cdot\bar{h}_{i,j}}{2\Delta\theta} \tag{15}$$

$$c_{i,j}=-2\bar{P}^*_{i,j}\cdot\left[\frac{\bar{h}^3_{i,j}}{\Delta\theta^2}+\left(\frac{2R}{L}\right)^2\frac{\bar{h}^3_{i,j}}{\Delta\xi^2}\right]-\Lambda_x\frac{\bar{h}_{i+1,j}-\bar{h}_{i-1,j}}{2\Delta\theta} \tag{16}$$

$$d_{i,j}=\bar{P}^*_{i,j}\cdot\left(\frac{2R}{L}\right)^2\left(\frac{\bar{h}^3_{i,j}}{\Delta\xi^2}-\frac{\bar{h}_{i,j+1}-\bar{h}_{i,j-1}}{2\Delta\xi}\frac{3\bar{h}^2_{i,j}}{2\Delta\xi}\right) \tag{17}$$

$$e_{i,j}=\bar{P}^*_{i,j}\cdot\left(\frac{2R}{L}\right)^2\left(\frac{\bar{h}^3_{i,j}}{\Delta\xi^2}+\frac{\bar{h}_{i,j+1}-\bar{h}_{i,j-1}}{2\Delta\xi}\frac{3\bar{h}^2_{i,j}}{2\Delta\xi}\right) \tag{18}$$

$$m_{i,j}=\frac{\bar{h}^3_{i,j}}{4\Delta\theta^2} \tag{19}$$

$$n_{i,j}=\left(\frac{2R}{L}\right)^2\frac{\bar{h}^3_{i,j}}{4\Delta\xi^2} \tag{20}$$

3.2. The accuracy of P algorithm*

Firstly, we need to compare the results of the P* algorithm and the Newton-Raphson algorithm. The parameters in table 1 are used for calculation, and ε=0.8. Shown in figure 5 is the dimensionless pressure for the foil bearing running at different speeds. The corresponding film thickness for a foil bearing is also plotted for comparison as figure 6. From the comparison, it can be found that the calculation results of the two algorithms are consistent, thus proving the accuracy of the P* algorithm.

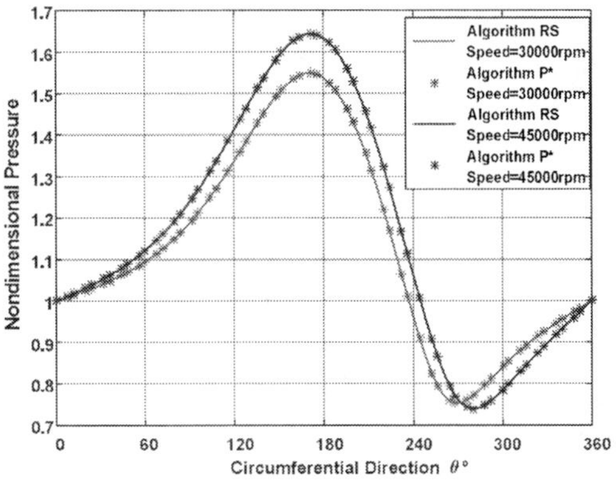

Figure 5. Dimensionless pressure at the axial midsection of foil bearing compared with Newton-Raphson algorithm.

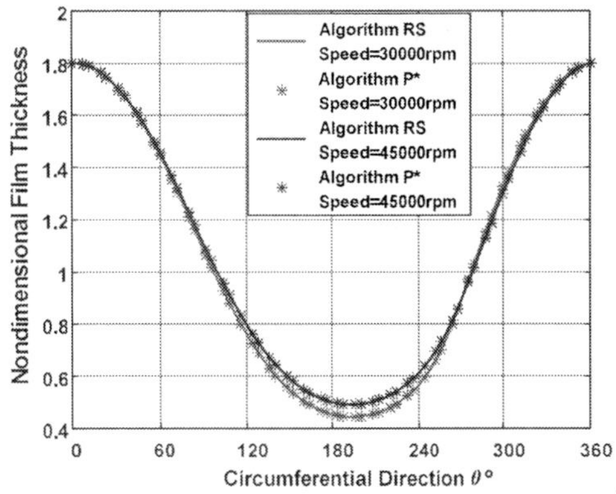

Figure 6. Dimensionless film thickness at the axial midsection of foil bearing compared with Newton-Raphson algorithm.

The parameters in reference [19] are selected for calculation. Shown in figure 7 is the dimensionless pressure for the rigid bearing running at 30 krpm with an assumed minimum film thickness of about 16 μm. The corresponding film thickness for a rigid bearing is also plotted for comparison as figure 8. The numerical analysis results calculated by the P* algorithm is compared with the curve data in reference [19]. For the gas film pressure, the calculated results of this model are basically consistent with the data obtained in references. The maximum error occurs at the maximum film pressure, and the error value is less than 5%. As can be seen from figure 8 that the film thickness distribution of gas bearing with rigid surface coincides with the reference data completely. This is because the theoretical model of rigid surface bearing is simple and does not consider the influence of deformation and other factors. First of all, we need to compare the results of the P* algorithm and the Newton-Raphson algorithm.

Table 1. The bearing parameters [18]

Foil bearing parameters	Values
Shaft Radius, R (m)	19.05×10^{-3}
Bearing Length, L (m)	38.10×10^{-3}
Nominal clearance, C (m)	3.18×10^{-5}
Top foil thickness, t_p (m)	76.2×10^{-6}
Bump foil thickness, t_b (m)	101.6×10^{-6}
Bump Length, h_b (m)	0.63×10^{-3}
Bump pitch, s (m)	4.572×10^{-3}
Bump length, $2l$ (m)	3.556×10^{-6}
Young's Modulus of bump foil, E (Pa)	214×10^{9}
Poisson's ratio, v	0.29
Viscosity, μ (pa·s)	1.932×10^{-5}
Density, ρ (kg/m^3)	1.1614

IOP Conf. Series: Materials Science and Engineering 1081 (2021) 012037 doi:10.1088/1757-899X/1081/1/012037

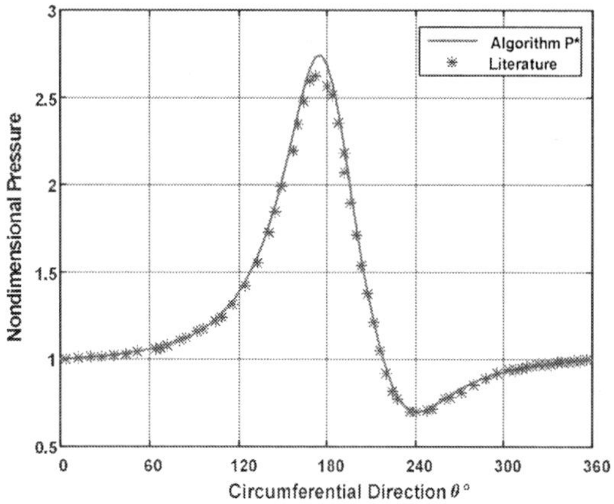

Figure 7. Dimensionless pressure at the axial midsection of rigid bearing compared with reference.

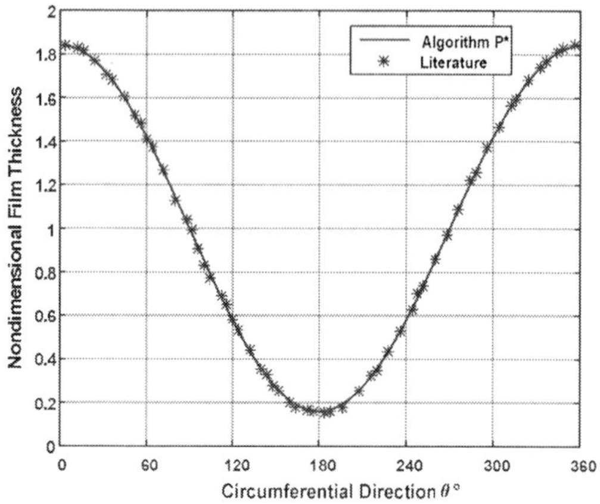

Figure 8. Dimensionless film thickness at the axial midsection of rigid bearing compared with reference.

The parameters in reference [19] are selected for calculation. Shown in figure 9 is the dimensionless pressure for the flexible bearing running at 30 krpm with an assumed minimum film thickness of about 16 μm. The corresponding film thickness for a flexible bearing is also plotted for comparison as figure 10. The results show that the calculated pressure of flexible bearing is in good agreement with the reference pressure [19], and the maximum deviation is about 3%. And the film thickness distribution of flexible bearing with foil surface coincides with the reference data completely.

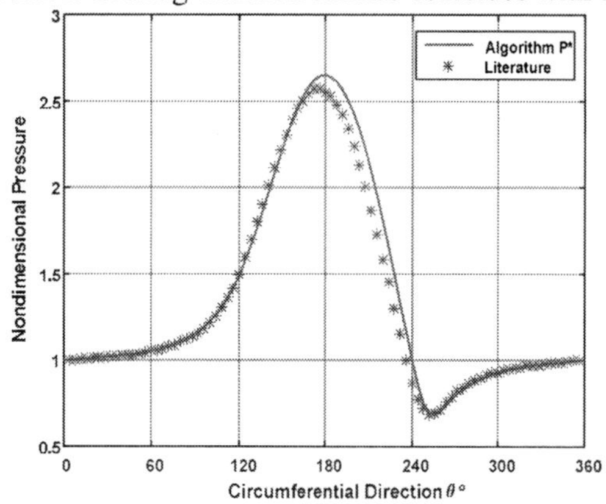

Figure 9. Dimensionless pressure at the axial midsection of flexible bearing compared with reference.

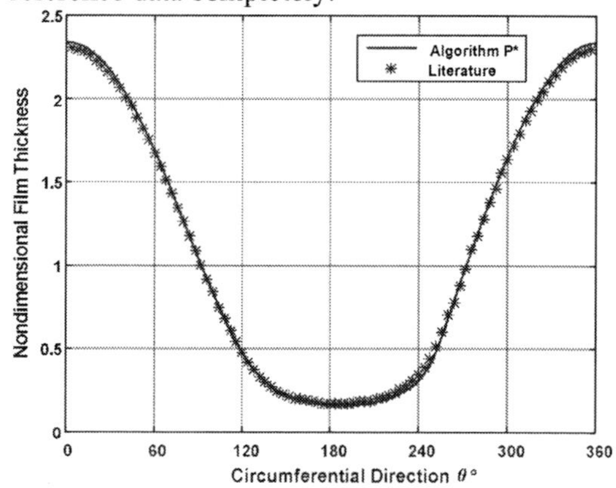

Figure 10. Dimensionless film thickness at the axial midsection of flexible bearing compared with reference.

3.3. Different P algorithms*

In this section, the position of P^* is to be changed to update the algorithm. At the same time, the accuracy of different P* algorithms are compared.

The Reynolds equation of dimensionless steady gas is expressed as follows:

$$\frac{\partial}{\partial\theta}\left(\overline{P}\overline{h}^3\frac{\partial\overline{P}}{\partial\theta}\right)+\left(\frac{2R}{L}\right)^2\frac{\partial}{\partial\xi}\left(\overline{P}\overline{h}^3\frac{\partial\overline{P}}{\partial\xi}\right)=\Lambda_x\frac{\partial\left(\overline{P}\overline{h}\right)}{\partial\theta} \tag{21}$$

Expand the sign of partial derivative. For the convenience of the following description, each item is numbered in the form shown in equation (22).

$$\underbrace{\bar{h}^3\frac{\partial\bar{P}}{\partial\theta}\frac{\partial\bar{P}}{\partial\theta}}_{1}+\underbrace{3\bar{h}^2\bar{P}\frac{\partial\bar{h}}{\partial\theta}\frac{\partial\bar{P}}{\partial\theta}}_{2}+\underbrace{\bar{P}\bar{h}^3\frac{\partial^2\bar{P}}{\partial\theta^2}}_{3}+\underbrace{\bar{h}^3\left(\frac{2R}{L}\right)^2\frac{\partial\bar{P}}{\partial\xi}\frac{\partial\bar{P}}{\partial\xi}}_{4}+\underbrace{3\bar{h}^2\bar{P}\left(\frac{2R}{L}\right)^2\frac{\partial\bar{h}}{\partial\xi}\frac{\partial\bar{P}}{\partial\xi}}_{5}$$
$$+\underbrace{\bar{P}\bar{h}^3\left(\frac{2R}{L}\right)^2\frac{\partial^2\bar{P}}{\partial\xi^2}}_{6}=\underbrace{\Lambda_x\bar{h}\frac{\partial\bar{P}}{\partial\theta}}_{7}+\underbrace{\Lambda_x\bar{P}\frac{\partial\bar{h}}{\partial\theta}}_{8} \quad (22)$$

On the basis of equation (22), algorithm P*₁ changes item 3 into $\bar{P}^*\bar{h}^3\cdot(\partial^2\bar{P}/\partial\theta^2)$, item 2 into $3\bar{h}^2\bar{P}^*\cdot(\partial\bar{h}/\partial\theta)(\partial\bar{P}/\partial\theta)$, and item 5 into $3\bar{h}^2\bar{P}^*(2R/L)^2\cdot(\partial\bar{h}/\partial\xi)(\partial\bar{P}/\partial\xi)$, at the same time changes item 6 into $\bar{P}^*\bar{h}^3(2R/L)^2\cdot(\partial^2\bar{P}/\partial\xi^2)$. In addition to the changes of algorithm P^*_1, P^*_2 also adds another change, which is change item 8 into $\Lambda_x\bar{P}^*\cdot(\partial\bar{h}/\partial\theta)$. P^*_1 is the method introduced in the previous section. Compared with algorithm P^*_1, P^*_2 put P^* on the right end of the equal sign.

On the basis of algorithm P^*_1, algorithm P^*_3 changes item 1 into $\bar{h}^3\cdot\left(\partial\bar{P}^*/\partial\theta\right)\left(\partial\bar{P}/\partial\theta\right)$, item 4 into $\bar{h}^3(2R/L)^2\cdot\left(\partial\bar{P}^*/\partial\xi\right)\left(\partial\bar{P}/\partial\xi\right)$. In addition to the changes of algorithm P^*_3, P^*_4 also change item 8 to $\Lambda_x\bar{P}^*\cdot\left(\partial\bar{h}/\partial\theta\right)$. Compared with the first two algorithms, algorithm P^*_3 and algorithm P^*_4 set P^* before the partial derivative symbol is opened. In this way, when the partial derivative symbol is opened, a partial derivative term for P^* will be generated.

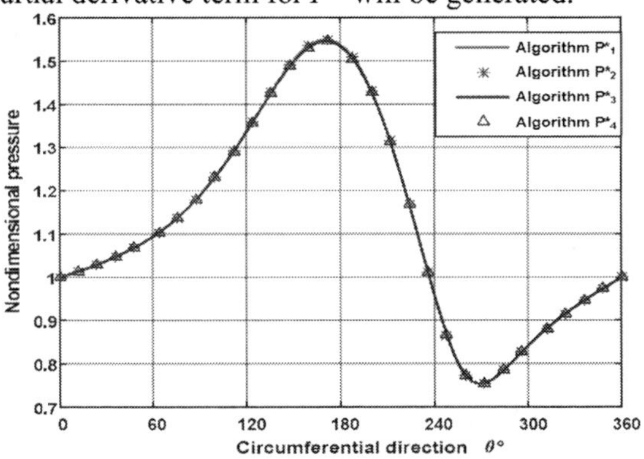

Figure 11. Dimensionless pressure at the axial midsection of different algorithms.

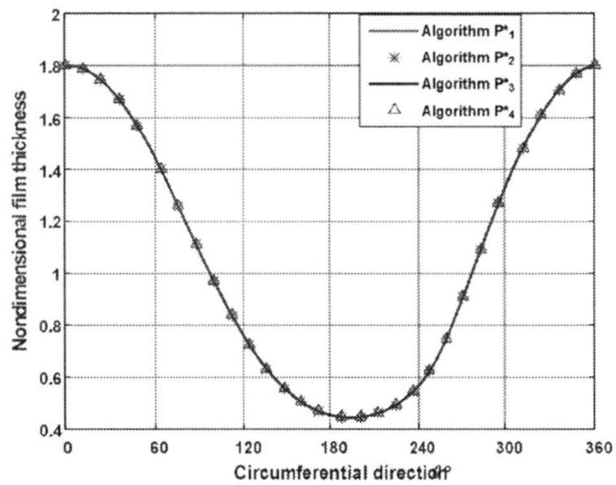

Figure 12. Dimensionless film thickness at the axial midsection of different algorithms.

Figure 11 shows the comparison of dimensionless pressure solution of four algorithms. It can be seen from the figure 11 that the calculation results of algorithms 1 and 2 are exactly the same. The same is true for algorithms 3 and 4, which shows that whether the P^* term is set at the right end of the equal sign has no effect on the calculation results. However, from the comparison of the four algorithms, it can be seen that the results of setting P^* before and after PDE are slightly different. The maximum relative error of the corresponding circumferential position is 0.25%, and the minimum relative error is 0.18%. Figure 12 shows the comparison of the results of dimensionless film thickness calculation of algorithm 1 to algorithm 4, the comparison results are consistent with dimensionless pressure. The maximum relative error of the corresponding circumferential position is 0.39%, and the minimum relative error is 0.17%.

In addition to the four algorithms described above, there are two algorithms. Algorithm P^*_5 change the $\bar{P}_{i,j}$ in the term 3 after the difference into $\overline{Ph}^3_{i,j}\cdot\left(\bar{P}_{i+1,j}-2\bar{P}^*_{i,j}+\bar{P}_{i-1,j}\right)/\Delta\theta^2$,changes item 6 after the central difference into $\left(2R/L\right)^2\overline{Ph}^3_{i,j}\cdot\left(\bar{P}_{i,j+1}-2\bar{P}^*_{i,j}+\bar{P}_{i,j-1}\right)/\Delta\xi^2$. Based on the algorithm P^*_5, algorithm P^*_6 also change item 8 to $\Lambda_x\bar{P}^*\left(\partial\bar{h}/\partial\theta\right)$. Algorithm P^*_5 compared with algorithm P^*_6, the difference is whether P^* is set at the right end of the equal sign. However, some serious things have been found in the numerical calculation by programming. When P is set as P^* in the second-order partial derivative after difference, the calculation cannot achieve convergence. Maybe try to change the iteration method can achieve convergence, but now the two algorithms have not found a solution.

4. Algorithm efficiency comparison

And until now, we can know that the P* algorithm can solve the Reynolds equation accurately. In addition, we also need to pay attention to the efficiency of P* algorithm. Then in the section 4, the efficiency of the P* algorithm and Newton-Raphson algorithm will be compared.

4.1. Compare the Dimension of The Matrix

Figure 13 shows the P* algorithm involved in calculating the matrix dimension, and figure 14 shows the Newton-Raphson algorithm. In the calculation below, set n_θ=151, n_z=26. So, the number of grids participating in the calculation is 150×25. For the P* algorithm, the size of the matrix involved in the calculation is 150×25. As for the Newton-Raphson algorithm, the method of solving the linear equation $Ax=B$ is used. Therefore, the parameter values on each grid node need to be placed at the corresponding position, then A will become a large sparse matrix, and the number of diagonal elements is the number of grid nodes, which should be 150×25. And the matrix size should be (150×25,150×25). Intuitively, the size of the Newton-Raphson algorithm is the square of the P* algorithm.

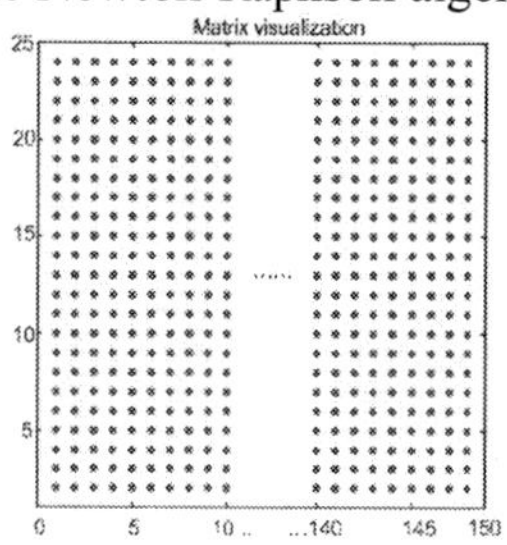

Figure 13. Calculation matrix dimension of P* algorithm.

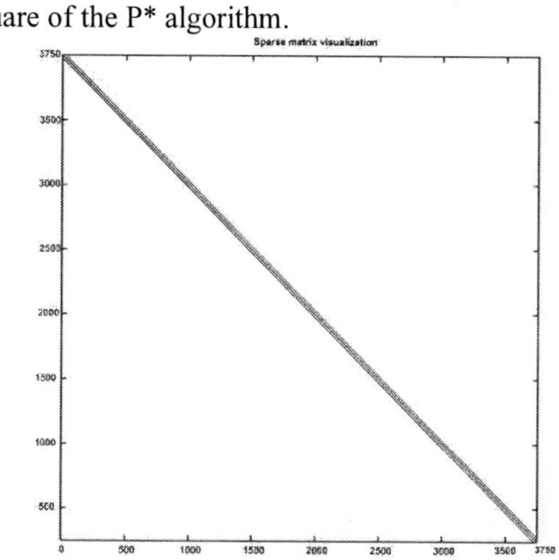

Figure 14. Calculation matrix dimension of Newton-Raphson algorithm.

4.2. Selection of Relaxation Factor

*4.2.1. Algorithm P**. P* algorithm introduces the Successive Over-Relaxation method (SOR)[20-23] at the solution of gas film thickness. $H^{k+1,0}$ is the initial estimate of the k+1 cycle. $H^{k,0}$ is the initial value of the last cycle, and H^k is the calculated result of the last cycle. The relationship between the above values is as follow: $H^{k+1,0} = H^k + SOR \cdot (H^k - H^{k,0})$. It should be noted that the selection of the relaxation factor is very important. If the selection is appropriate, the calculation efficiency can be greatly improved. Otherwise, iterative divergence may occur. In the calculation of flexible gas foil bearing, it is generally considered that the minimum gas film thickness is 4 microns, which is the closest to the engineering practice. In the case of the calculation parameters selected in this paper, if h_{min} of about 4 μm, the eccentricity ratio should reach about 2.5.

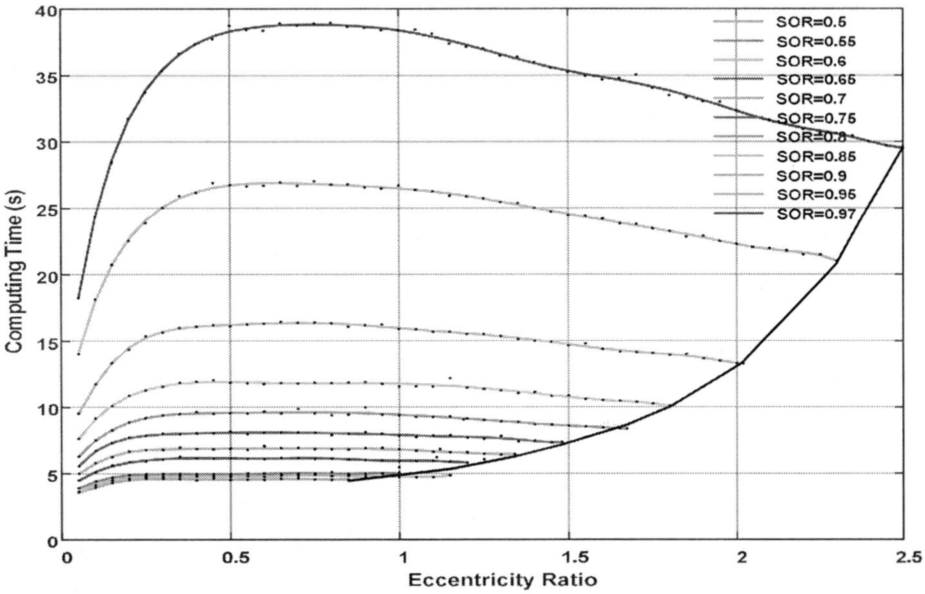

Figure 15. Influence of relaxation factor selection on computational efficiency.

Therefore, the calculation of the relaxation factor is within the range of 0.5 to 1, and the region for normal calculation can be realized. The solution result is shown in figure 15. The x-coordinate is the eccentricity rate, the y-coordinate is the calculation time, and each curve represents the value of a relaxation factor. As can be seen from the figure 15, when the eccentricity ratio is less than 1, the relaxation factor of 0.5 can realize the normal calculation of the whole region, and the calculation speed is the fastest and the efficiency is the highest. However, when the eccentricity ratio is greater than 1, choosing a smaller relaxation factor will lead to iterative divergence. After a lot of calculations, it can be seen that if the relaxation factor is 0.97, the maximum eccentricity can be calculated to 2.5, and the minimum film thickness is 3.81 μm.

4.2.2. Newton-Raphson algorithm. As for the Newton-Raphson algorithm, its general rule is the same as the P* algorithm, there is a calculation field that can do the normal calculation, but the difference is that the boundary line of the Newton-Raphson algorithm moves to the right. That is to say, with the same relaxation factor selected, the maximum eccentricity that can be calculated is larger than the P* algorithm. However, it will bring a long calculation time, and in some cases, it can be found that the convergence process is a run-out process by detecting the residual value.

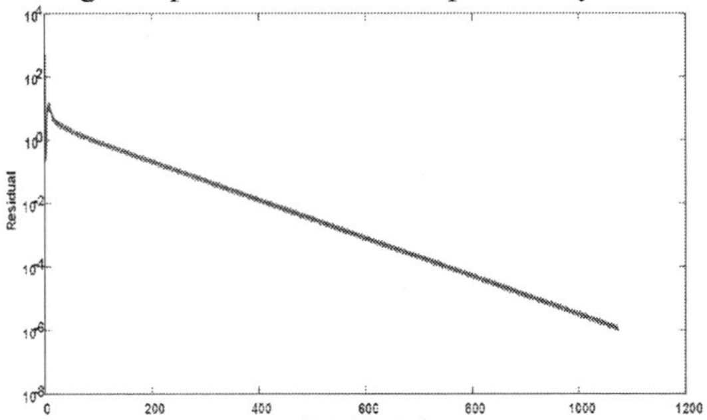

Figure 16. Convergent residual graph with ε=1.3, *SOR*=0.55 of Newton-Raphson algorithm.

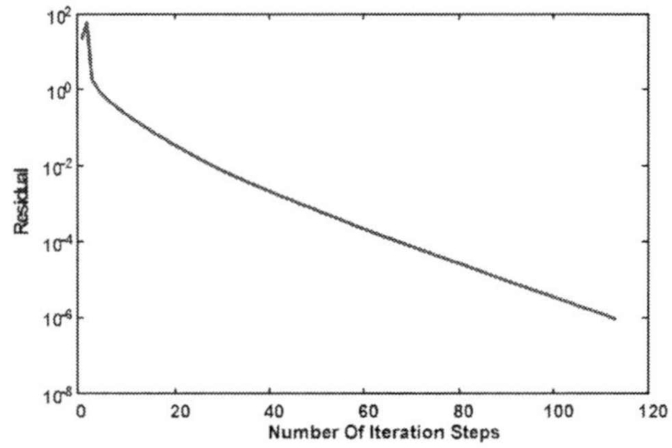

Figure 17. Convergent residual graph with ε=1.12, *SOR*=0.90 of Newton-Raphson algorithm.

For example, select *SOR*= 0.55. The P* algorithm can calculate normally until ε=1.12, and the calculation time is less than 7 seconds. The Newton-Raphson algorithm can calculate the eccentricity to 1.30, the calculation time is 46 seconds, and the iteration steps are 1074 steps. The residual curve starting from the second step is shown in figure 16.

At the same time, if we choose a better segmentation factor and initial value, the convergence effect will be very good. For example, in figure 17, the *SOR*=0.90, and the initial value is 1.1 times the atmospheric pressure. However, it is worth mentioning that in order to adapt to the calculation of all eccentricity ranges, we have made a lot of attempts, and finally choose the initial value of 10 times atmospheric pressure and the *SOR*=0.97.

4.3. Efficiency of different algorithms

In order to compare the calculation time of different algorithms, the same parameters were used for calculation. In this case, the selected parameters are: ε=0.8 and *SOR*=0.98. Firstly, the computation time of different algorithms under different initial values are calculated, and then fit the data points, as shown in figure 18. It is worth mentioning that the selection of the initial value has an important impact on whether the normal iterative calculation can be carried out. Attention should be paid to avoid geometric interference caused by too small initial pressure, that is, to ensure that the gas film thickness value is positive.

It can be seen from the figure 18 that although the coefficient expressions of algorithms 1 and 2 are different, their calculation time is almost the same. Algorithms 3 and 4 are the same too. From the comparison of computing time, it is obvious that the efficiency of the first two algorithms is better than that of the latter two algorithms. And the Newton-Raphson algorithm is the fastest.

From the initial value selection, the latter two P* algorithms are more sensitive to the initial value selection. In contrast, the initial value selection has little effect on the Newton-Raphson algorithm. At the same time, if the four algorithms choose the initial value closer to the calculation results, the speed can be improved. However, when the initial value of iteration pressure is greater than 10 times atmospheric pressure, the effect of initial value increase on calculation efficiency is not obvious. At the same time, if the initial value is too large, it is likely to cause abnormal calculation.

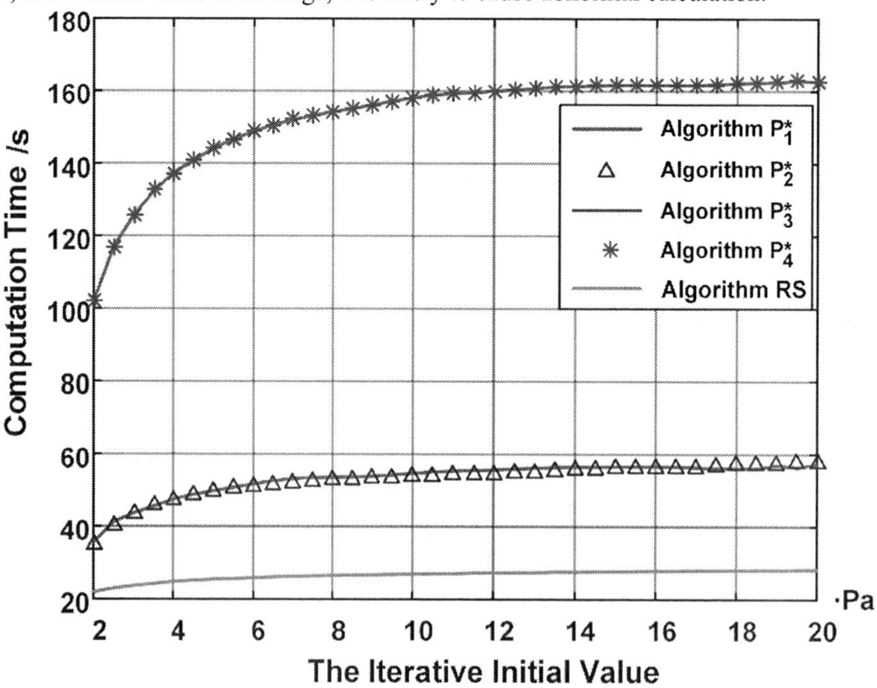

Figure 18. Calculation time of four algorithms under different initial values.

5. Conclusions

In the process of combining P* algorithm with SOR method, the choice of relaxation factor has an important influence on the normal calculation. Under the parameters selected in this paper, when the relaxation factor is 0.5, the calculation speed is fast and the efficiency is high. However, the calculation scope is limited, and the iteration divergence will occur when the eccentricity ratio is greater than 1. When the relaxation factor is 0.97, the calculation speed is slow, but the normal calculation range is wide, and the eccentricity can be calculated to 2.5. At this time, the minimum film thickness is 3.8 μm, which is close to the actual engineering value.

This paper studies different algorithms after changing $P*$ position. From the perspective of calculation accuracy, the maximum relative error of the four algorithms is 0.39%, and the calculation results are consistent. From the perspective of computational efficiency, whether $P*$ is placed on the right side of the equal sign has no influence on the computational speed. The calculation speed of the first two algorithms is better than that of the other two algorithms, and the calculation speed can be increased by about 1.5 times under the same parameter. From the perspective of initial value influence, the latter two P* algorithms are more sensitive to initial value selection. However, when the initial iteration pressure is greater than 10 atmospheres, the sensitivity of the four algorithms to the initial pressure decreases.

Compared with the Newton-Raphson algorithm, the P* algorithm has a distinct mathematical concept and simple programming process. The P* algorithm adopts the form of matrix operation and combines with the SOR method in the calculation process, which can greatly improve the calculation speed. In terms of computing the dimension of the matrix, the matrix dimension of Newton-Simpson algorithm is the square of P* algorithm. Intuitively, the P* algorithm has the advantage of a small matrix dimension. Although the Newton-Raphson algorithm has large dimension, the process of setting initial value for iteration is not involved in the calculation. In terms of computational efficiency, the Newton-Raphson algorithm is superior to the P* algorithm.

Reference

[1] Ju L, Zhang J X, Xiao B and Miao Z 2020 The Analysis and Treatment of HSRL350 Fan Reduction Gearbox[J] *Chinese Journal of Turbomachinery* **62** 59-61

[2] Allaire P Dimond T and Cao J M 2019 Bearing Fix for Large Sub-synchronous Vibration in Steam Turbine-gearbox-generator System[J] *Chinese Journal of Turbomachinery* **61** 4 34-38+4

[3] Jiang L J, Yuan J B, Huo W H and Sun H 2020 Dynamic Characteristics Analysis of Cantilever Rotor Considering Impeller Coupling Effect[J] *Chinese Journal of Turbomachinery* **62** 16-20

[4] Deng Z H 2016 Design and Experimental study of an oil-free high speed Compressor for Fuel Cell[D] *Hunan University*

[5] Shu X J, Xu G, Zheng Y Q, Cui H L and Lan H 2017 Key Technologies for Centrifugal Blowers with Foil Bearings and High Speed Magnet Motor[J] *Chinese Journal of Turbomachinery* **59** 36-42

[6] Zhu B S, Zhang B Z, Rao J Yuan T Z, Niu S T and Lu S C 2020 The Application of Air Bearing in Blower and Turbojet Engine Field[J] *Chinese Journal of Turbomachinery* **62** 25-28

[7] Tian Q Y, Huang W J, Rao J, Zhu X N and Wang X Z 2018 Numerical Calculation and Analysis of Oil Film Characteristics of Hydrodynamic Radial Sliding Bearing[J] *Chinese Journal of Turbomachinery* **60** 44-47

[8] Wang X K, Xu Q, Zhang L X and Yang H 2019 Numerical Analysis of the Static Performance of Aerostatic Journal Bearings[J] *Machinery Design & Manufacture* **339** 14-17

[9] Wu H P, Kong X M and Chen Dong 2000 The Application of GMRES Algorithm in Calculating Reynolds Equation[J] *Lubrication Engineering* **000** 2-4

[10] Zhang D X, Wu G W and Deng H Y 2010 Determination of The Best Relaxation Factor of Symmetric Successive Over Relaxation Iterative Method by Golden Ratio Method[J] *Science and Technology Innovation Herald* **169** 217-219

[11] Xu S L and Fan W X 2013 Multi-grid Method for Solving Reynolds Equation[J] *Mechanical Engineering & Automation* **177** 70-71

[12] Zhang X, Luo H X, Zuo B Q and Li Z Q 2019 Research on Adaptive Step-size Difference Algorithm for Reynolds Equation Based on Hierarchical Model[J] *Lubrication Engineering* **44** 123-130

[13] Wang Q 2020 Influence of Mesh Number on Solving Reynolds Equation by FDM[J] *Lubrication Engineering* **45** 125-129

[14] Heshmat H, Walowit J.A and Pinkus O 1983 Analysis of Gas Lubricated Foil Journal Bearings *ASME Journal of Lubrication Technology* **105** 647– 655

[15] Xu F C and Kim D J 2016 Dynamic performance of foil bearings with a quadratic stiffness model[J] *Neurocomputing* **216** 666-671

[16] Peng Z C and Khonsari M M 2004 On the Limiting Load-Carrying Capacity of Foil Bearings[J] *Journal of Tribology* **126** 817-818

[17] Yan J J 2016 Rotor Dynamic Characteristics of Foil Bearing System with Surface Micro Structure[D] *Harbin Institute of Technology*

[18] Xu F C 2016 Vibration Characteristics of Gas Foil Journal Bearing-Rotor System Based on Effects of Coulomb Friction[D] *Harbin Institute of Technology*

[19] Peng Z C and Khonsari M M 2006 A Thermohydrodynamic Analysis of Foil Journal Bearings[J] *Journal of Tribology 128* 534-541

[20] Liu S M and Liu H M 2016 A Method for Simplifying the Process of Solving Reynolds Equation[J] *Journal of North China University (Natural Science Edition)* **27** 165-167

[21] Lu P 2016 Numerical Study of Static Characteristics of the Foil Hydrodynamic Thrust Gas Bearing with Hemispherical Convex Dots in Turboexpander[D] *Xi'an University of Architecture and Technology*

[22] Wu B Y, Jin Jian and Zhang G X 2002 Solving Reynolds Equation with SOR Method[J] *Journal of Shanghai University(Natural Science)* **8** 130-132,155

[23] Hu F and Yu F X 2006 The Selection Method of the Relaxation Factor ω In Successive Over-Relaxation Iteration Algorithm[J] *Journal of Qinghai Normal University (Natural Science)* **01** 42-45

IOP Conf. Series: Materials Science and Engineering 1081 (2021) 012038 doi:10.1088/1757-899X/1081/1/012038

Numerical analysis of the static characteristic of orifice-type throttled aerostatic journal bearings

X H Zhang [1,2], C L Ke [1], N Peng [1], L Y Xiong [1,2] and L Q Liu [1,2]
[1] Technical Institute of Physics and Chemistry, CAS, Beijing, China
[2] University of Chinese Academy of Sciences, Beijing, China

E-mail: zhangxiaohua15@mails.ucas.edu.cn
pengnan@mail.ipc.ac.cn

Abstract: The static characteristic of aerostatic journal bearing was studied by numerically simulating the state non-linear Reynolds equation. The finite difference method was used to solve Reynolds equation in MATLAB programming. The relation-ship between the static characteristics and hydrodynamic effect and hydrostatic effect were analyzed. The results show that the load capacity of bearing is directly proportional to the eccentricity ratio, the rotating speed, the air supply pressure and the diameter of orifices; increasing eccentricity ratio, the rotating speed is conducive to reduce the rotor attitude angle and improve the stability of the rotor; increasing the length diameter ratio of bearing and selecting the appropriate orifice position are beneficial to reduce the air consumption of bearing.

1. Introduction

Gas lubricated bearing is a kind of sliding bearing with gas as lubricant. It has a series of advantages, such as low friction power consumption, high motion accuracy and high and low temperature resistance. Therefore, it has been widely studied and applied in high-speed turbo expander and precision instruments [1][2].

Han[3] analyzed t influence of bearing air supply pressure on critical speed of aerostatic bearing rotor system. Li[4] carried out a numerical study on the bearing characteristics of aerostatic bearings. Li[5] carried out the simulation and experimental verification of gas bearing characteristics. Yu[6] studies the effects of aspect ratio, width of throttling slot, depth of throttling slot and thickness of gas film on static characteristics of bearing. Wang[7] simulated the flow field of the gas bearing by using the computational fluid dynamics software FLUENT, and solved the pressure distribution of the bearing gas film.

In this paper, the radial aerostatic bearing with small orifice throttling is selected as the research object. We divide it into pure dynamic pressure bearing and pure static pressure bearing. From the angle of hydrodynamic effect and hydrostatic effect, the working principle of orifice throttling hydrostatic gas bearing is clarified. Finally, how the structural parameters of bearing affect the dynamic and static pressure effects to produce static characteristic is studied.

2. Model of Aerostatic Bearing

2.1. Basic Equation of Aerostatic Bearing

For small-hole throttled aerostatic journal bearings, as shown in Figure 1 and Figure 2, the rotor is separated from the bearing under the combined action of dynamic pressure effect and orifice throttling effect, forming a frictionless gas film. Under isothermal conditions, the dimensionless form of Reynolds equation [4] for compressible gas is as follows:

$$\frac{\partial}{\partial \boldsymbol{\Phi}}\left(\boldsymbol{P}\boldsymbol{H}^3\frac{\partial \boldsymbol{P}}{\partial \boldsymbol{\Phi}}\right)+\frac{\partial}{\partial \boldsymbol{\xi}}\left(\boldsymbol{P}\boldsymbol{H}^3\frac{\partial \boldsymbol{P}}{\partial \boldsymbol{\xi}}\right)=2\boldsymbol{\Lambda}\frac{\partial(\boldsymbol{P}\boldsymbol{H})}{\partial \tau}+\boldsymbol{\Lambda}\frac{\partial(\boldsymbol{P}\boldsymbol{H})}{\partial \boldsymbol{\Phi}} \tag{1}$$

In the formula:

$\boldsymbol{P}$ — the dimensionless film pressure

$\boldsymbol{H}$ — the dimensionless film pressure

$\boldsymbol{\Phi}$ — the dimensionless circumferential coordinates

$\boldsymbol{\xi}$ — the dimensionless axial coordinates

$\boldsymbol{\tau}$ — the dimensionless time

$\boldsymbol{\Lambda}$ — the dimensionless velocity

The boundary condition:

(1) The end face of aerostatic bearing is environmental pressure:

$$P(\Phi, \xi=0)=P(\Phi, \xi=\frac{L}{R})=1$$

(2) The aerostatic bearing satisfies periodic boundary conditions:

$$P(\Phi, \xi)=P(\Phi+2\pi, \xi)$$

(3) The outlet pressure of orifice of aerostatic bearing is set as boundary condition:

$$P(\Phi_i, \xi_j)=P_{out}$$

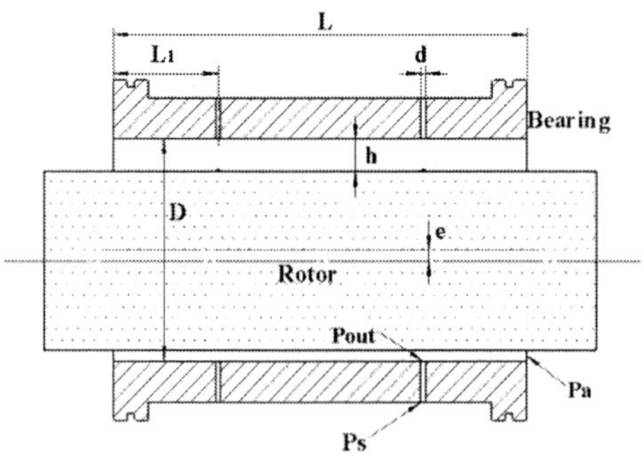

Figure 1. Structural diagram of bearing-rotor system.

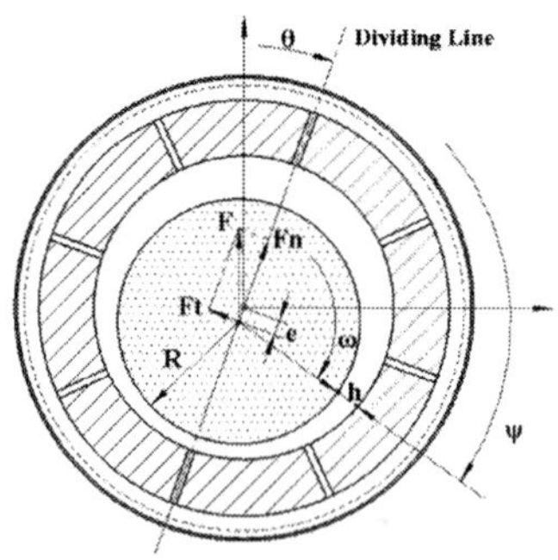

Figure 2. Position of rotor in bearing.

Defining dimensionless parameters:

$$\boldsymbol{P}=\frac{P}{P_a}\ ;\ \boldsymbol{H}=\frac{h}{Cr}=1+\varepsilon\cos\varphi\ ;\ \Phi=\frac{x}{R}$$

$$\xi=\frac{z}{R}\ ;\ \boldsymbol{\tau}=\omega t\ ;\ \Lambda=\frac{6\mu R^2\omega}{P_a Cr^2}\ ;\ \boldsymbol{P_s}=\frac{P_s}{P_a}$$

In the formula:

P — the film pressure
P_s — the air supply pressure
h — the film thickness
C_r — the gas film gap
e — the eccentricity
ε — the eccentricity ratio ($\varepsilon=e/Cr$)
ω — the rotor angular velocity
t — time
μ — the viscosity of lubricant
R — the rotor radius
x — the circumferential coordinates
z — the axial coordinates

2.2. Static Characteristics of Aerostatic Bearing

By solving Reynolds equation (1), the distribution of film pressure in the gas film gap can be obtained. As shown in Figure 2. Through the mechanical equilibrium conditions[8][9], F_n is the bearing loading capacity in the eccentric direction and F_t is the bearing loading capacity in the vertical eccentric direction:

$$\boldsymbol{F_n}=-\int_0^{\frac{L}{R}}\int_0^{2\pi}(\boldsymbol{P}-1)\cos\boldsymbol{\varphi}\,d\boldsymbol{\varphi}d\xi \tag{2}$$

$$\boldsymbol{F_t}=-\int_0^{\frac{L}{R}}\int_0^{2\pi}(\boldsymbol{P}-1)\sin\boldsymbol{\varphi}\,d\boldsymbol{\varphi}d\xi \tag{3}$$

The total bearing capacity of aerostatic bearing:

$$\boldsymbol{F}=(\boldsymbol{F}_n^2+\boldsymbol{F}_t^2)^{1/2} \tag{4}$$

The deflection angle of rotor:

$$\theta=tan\left(\frac{F_t}{F_n}\right) \tag{5}$$

The static stiffness of gas film:

$$k=\frac{dF}{de} \tag{6}$$

3. Solution of Reynolds for Aerostatic Bearing

The outlet pressure $\boldsymbol{P_{i,j}}$ of the orifice is treated as the boundary condition, which is helpful to simplify the Reynolds equation. According to the conservation of mass, the mass flow rate into the orifice is $\boldsymbol{Q_{in}}$ is equal to the mass flow rate $\boldsymbol{Q_{out}}$ out of the orifice. The boundary condition $\boldsymbol{P_{out}}$ can be solved. Finally, the distribution of gas film pressure $\boldsymbol{P_{i,j}}$ is obtained by using the over relaxation iterative method (SOR). The flow chart[10][11] of solving Reynolds equation for gas lubrication is shown in Figure 3.

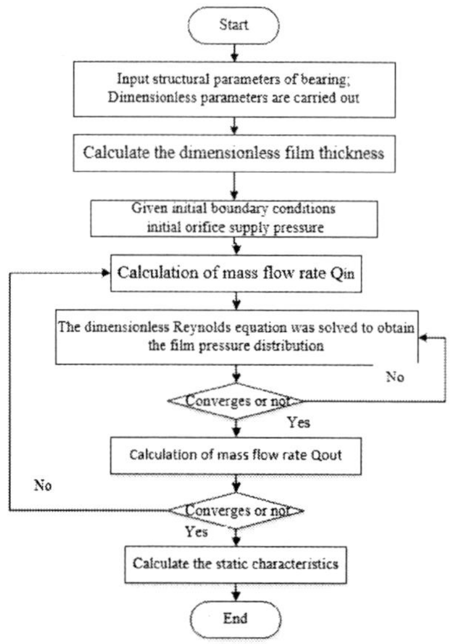

Figure 3. Flow chart of film pressure solution.

Table 1. Basic structural parameters of aerostatic bearing.

Name	Symbol / Unit	Value
Length	L /(mm)	32
Diameter	D /(mm)	16
Gas Film Gap	Cr /(μm)	20
Orifice Position	L_l /(mm)	8
Orifice Diameter	d /(mm)	0.3
Number of Orifices	n	8*2
Air Supply Pressure	P_s /(MPa)	0.5
Environmental Pressure	P_a /(MPa)	0.1
Environmental Temperature	T /(K)	300

4. Operating Principle of Aerostatic Bearing

For pure dynamic pressure gas bearing, under the action of gas viscosity, the high-speed rotating rotor will draw the bearing gas into the wedge-shaped gap, forming a high-pressure area, as shown in Figure 4; the low-pressure area is formed at the position where the bearing gas was continuously transported to the wedge-shaped clearance [12][13], and the bearing capacity is formed under the interaction of high and low film pressure.

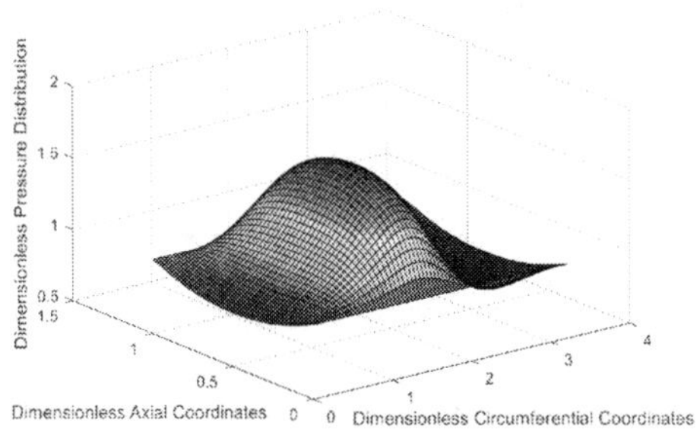

Figure 4. Pressure distribution of pure hydro-dynamic bearing（$\varepsilon=0.4$，$\Lambda=10$）

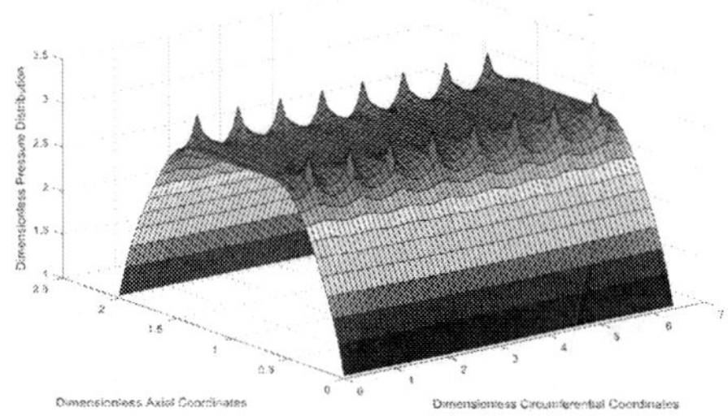

Figure 5. Pressure distribution of aerostatic bearings（$\varepsilon=0$, $\boldsymbol{P_s}=5$）

In the pure hydrostatic gas bearing, the rotor speed is 0, the eccentricity is 0, and the gas film pressure is distributed symmetrically along the radial direction, as shown in Figure 5, the bearing capacity is 0; When the rotor has unbalanced mass, resulting in the rotor eccentricity in the bearing, the gas film pressure has a negative correlation with the gas film thickness, and is axial symmetrically distributed along the dividing line. As shown in Figure 6, the bearing capacity of pure hydrostatic gas bearing is formed under the effect of pressure difference.

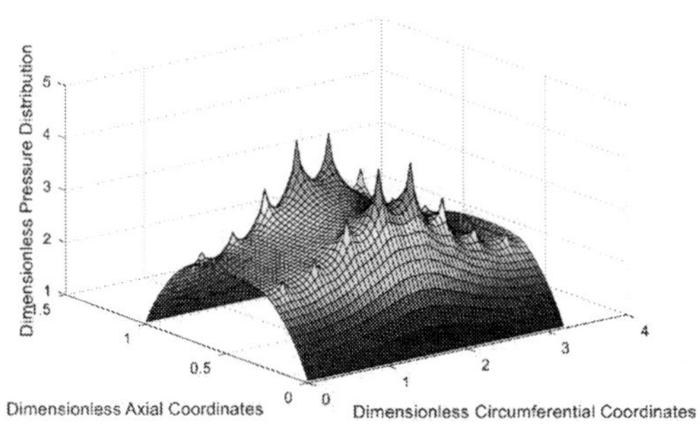

Figure 6. Pressure distribution of aerostatic bearings（$\varepsilon=0.4$，$\boldsymbol{P_s}=5$，$\Lambda=0$）

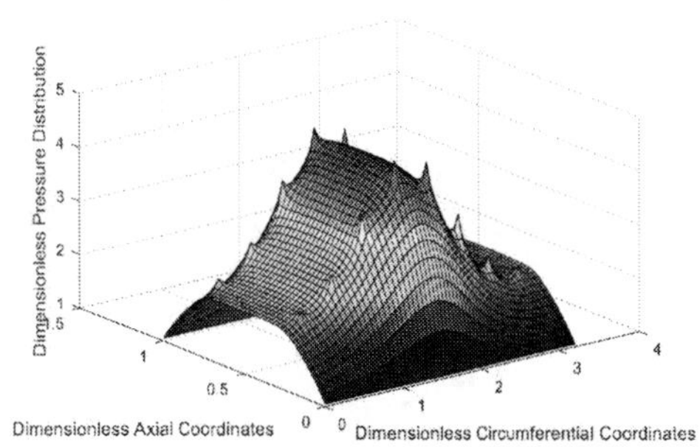

Figure 7. Pressure distribution of aerostatic bearing（$\varepsilon=0.4$，$\Lambda=10$，$\boldsymbol{P_s}=5$）

Combined with Figures 4, 6 and 7, the film pressure produced by the orifice-type radial aerostatic bearing not only has the hydrodynamic pressure effect produced by the hydrodynamic gas bearing, but also has the throttling effect of the hydrostatic gas bearing. As shown in Figure 7, the peak value of film pressure has deviated to the left from the center of symmetry. At this time, the position of the rotor in the bearing can be determined by the eccentricity ε and the attitude angle θ.

The essence of the orifice-type radial aerostatic bearing is the coupling bearing of the pure hydrodynamic pressure bearing and the orifice-type pure hydrostatic gas bearing. Therefore, the structure parameters such as the rotor eccentricity ε, bearing number Λ, air supply pressure $\boldsymbol{P_S}$, orifice diameter d can have an important influence on the static characteristics of the orifice-type aerostatic bearing.

5. Influence of System Parameters on Static Characteristics of Bearing

In order to verify the accuracy of the program results, the calculation results are compared with reference[4]. Under the conditions of rotating speed ω = 0, gas supply pressure of 0.5MPa, and the same bearing structure, the comparison results are shown in Figure 8. The two methods have good trend consistency, and the maximum error is less than 7%, which can inferred the correctness of the calculation results in this paper.

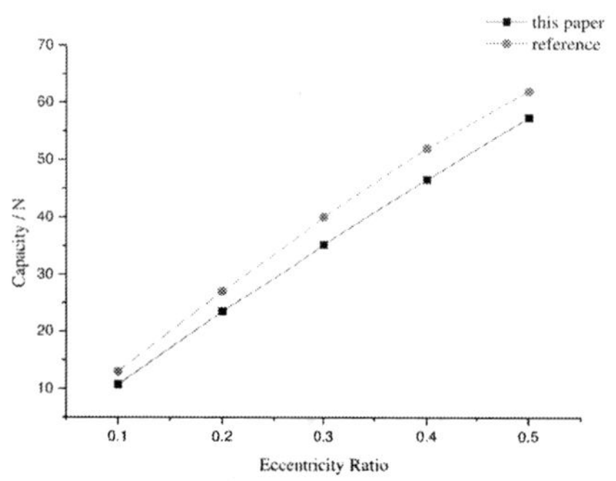

Figure 8. Comparison of calculation results.

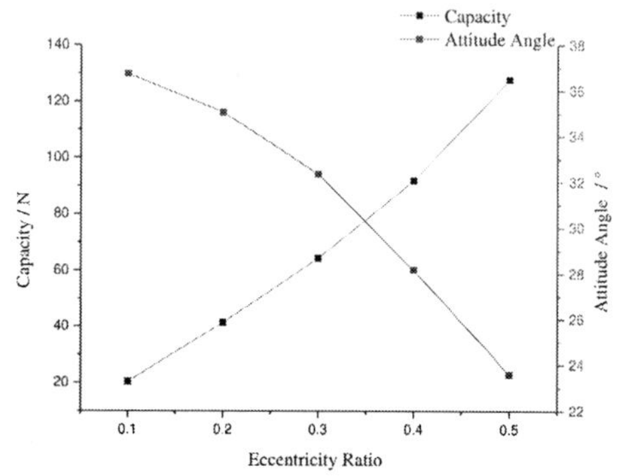

Figure 9. Influence of eccentricity on bearing capacity and attitude angle.

Eccentricity Ratio

The eccentricity ratio represents the ratio of eccentricity to gas film gap. As shown in Figures 9, 10 and 11, the bearing capacity increases with the increase of eccentricity, and the slope $k = dF/de$ of the bearing capacity also increases, which indicates that increasing the eccentricity of the rotor not only increases the bearing capacity, but also improves the static stiffness of the gas film.

As shown in Figure 9, with the increase of eccentricity ratio, the attitude angle of the rotor decreases, the ratio of F_t/F_n decreases. The bearing capacity F_t in the vertical eccentric direction is the main factor causing the rotor whirl[14]. Therefore, increasing the eccentricity ratio increases the bearing loading capacity and improves the stability of the rotor.

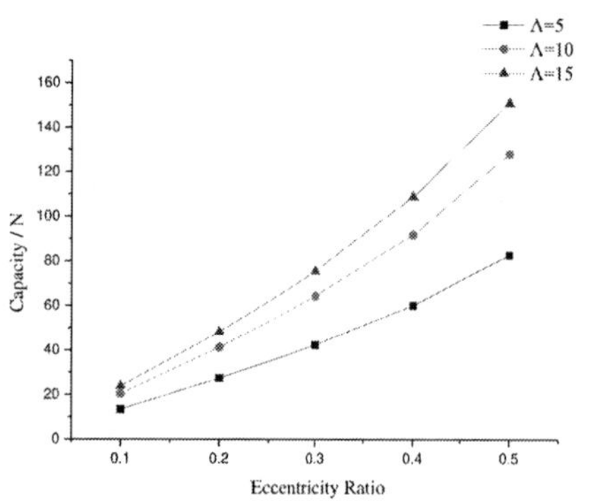

Figure 10. Influence of eccentricity on bearing capacity under different bearing numbers.

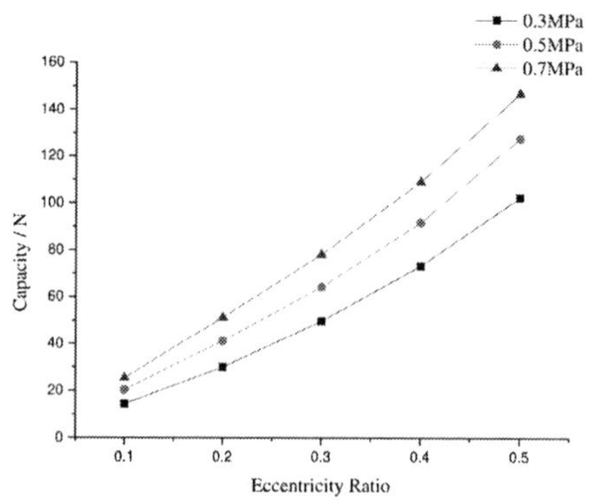

Figure 11. Influence of eccentricity on bearing capacity under different pressures.

Number of bearings

The number of bearings is the dimensionless form of rotor angular velocity. As shown in Figure 12, the bearing capacity increases with the increase of the number of bearings. With the increase of rotor speed, the attitude angle increases first and then decreases, which indicates that the bearing capacity in the vertical eccentric direction increases the instability of the rotor under the condition of small rotating speed. With the increase of rotating speed, the dynamic pressure effect becomes more and more important, and the stability of the rotor is improved continuously.

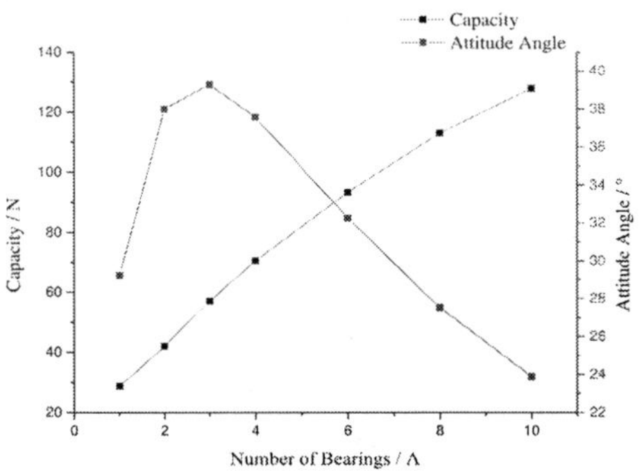

Figure 12. Influence of bearing number on bearing capacity and attitude angle.

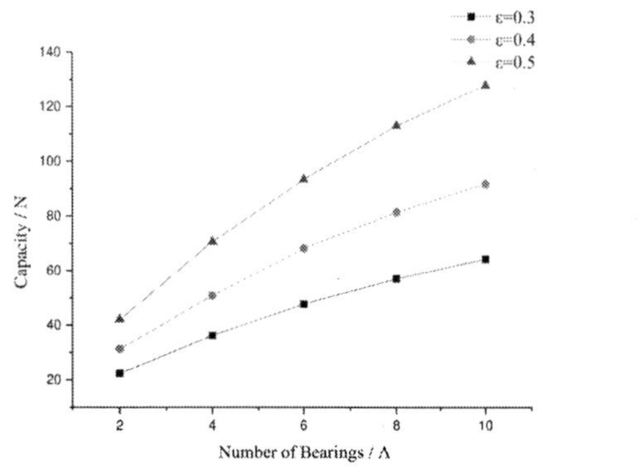

Figure 13. Influence of bearing number on bearing capacity under different eccentricity.

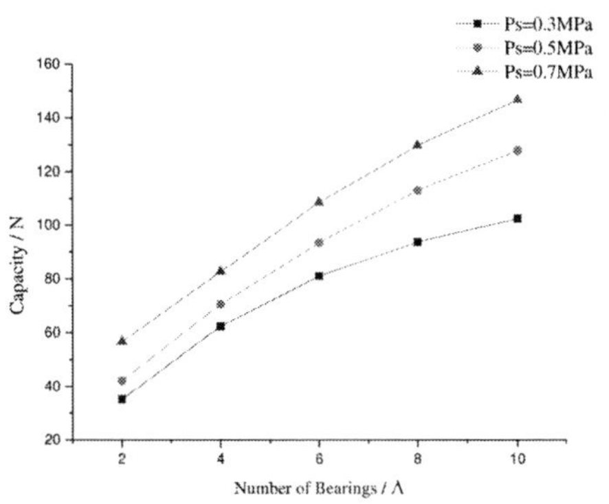

Figure 14. Influence of bearing number on bearing capacity under different air supply pressure.

Figure 13 and Figure 14 show the influence of bearing number on bearing capacity under different eccentricity and gas supply pressure. It can be seen from the figure that the radial bearing capacity of the bearing increases with the number of bearings, but the slope decreases gradually. The results show that to a certain extent, increasing the rotor speed is an effective means to improve the bearing capacity, and the dynamic pressure effect of the bearing is gradually weakened under the condition of too high rotating speed. At this time, the radial bearing capacity of the bearing can be increased by increasing the eccentricity and gas supply pressure.

Air Supply Pressure

Figure 15 shows the effect of orifice supply pressure on bearing loading capacity, rotor attitude and bearing air consumption. With the increase of air supply pressure, the bearing capacity and mass flow rate of the bearing are improved. It shows that increasing the air supply pressure increases the bearing radial bearing capacity and increases the air consumption of the bearing.

It can be seen from the figure that increasing the air supply pressure of the bearing leads to the increase of the rotor attitude angle. When the rotor is eccentric, the static pressure effect makes F_t and F_n increase, and the ratio of F_t/F_n increases, which reduces the stability of the rotor system.

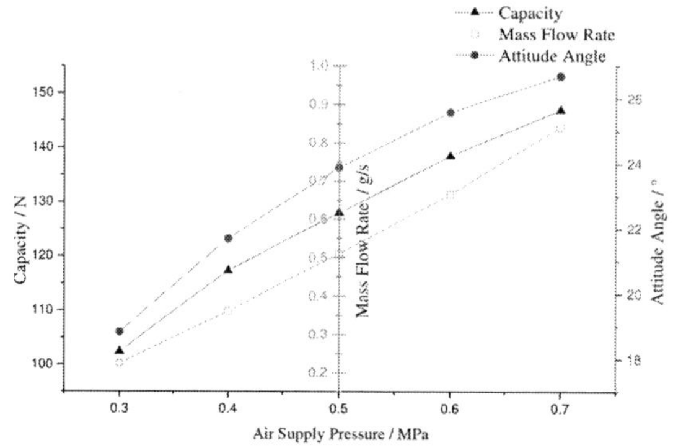

Figure 15. Effect of air supply pressure on bearing capacity, attitude angle and mass flow rate.

Aspect Ratio

It can be seen from Figure 16 that with the increase of length diameter ratio, the bearing capacity increases and the mass flow rate decreases. By increasing the length diameter ratio of the bearing, the high pressure area between the two rows of orifices is increased, and the flow resistance from the orifice to the end face of the bearing is increased. Therefore, the radial bearing capacity of the aerostatic bearing is improved, the air consumption of the bearing is reduced and the cost is saved. Therefore, the bearing with high aspect ratio is more suitable.

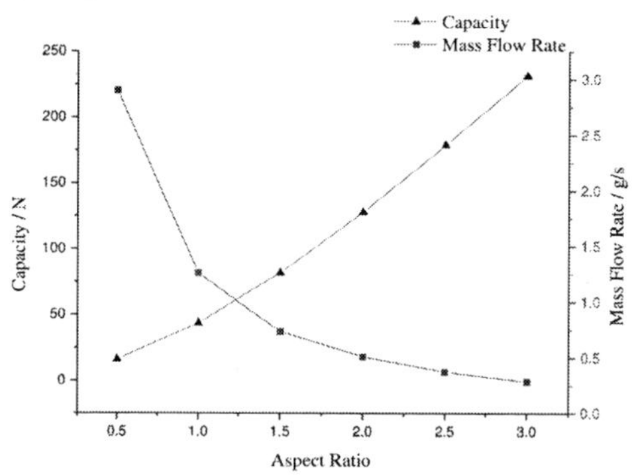

Figure 16. Effect of aspect ratio on bearing capacity, attitude angle and mass flow rate.

Orifice Diameter

Figure 17 shows the effect of orifice diameter on the static characteristics of aerostatic bearings. The bearing capacity is improved by increasing the orifice diameter, but the air consumption and rotor attitude angle are increased at the same time.

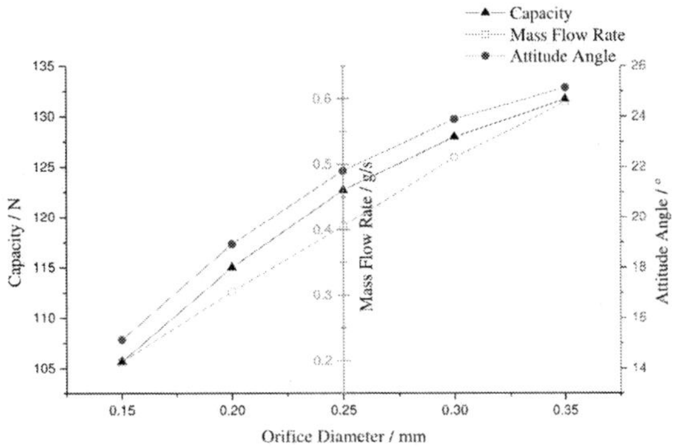

Figure 17. Effect of orifice diameter on bearing capacity, attitude angle and mass flow rate.

Under the condition of keeping the orifice diameter unchanged, the bearing capacity and gas consumption of the bearing can be greatly increased by increasing the air supply pressure of the orifice as shown in Figure 18 and Figure 19. When the orifice diameter is large, the bearing gas consumption is huge, but the bearing capacity increase is small, which is uneconomical. Therefore, the orifice diameter of gas bearing should not be too large.

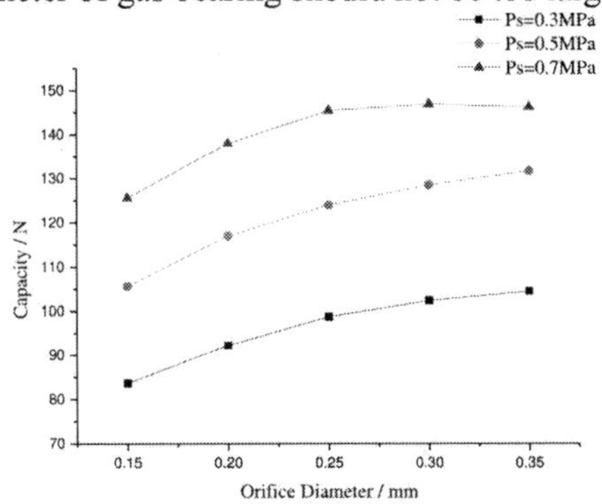

Figure 18. Influence of orifice diameter on bearing capacity under different gas supply pressure.

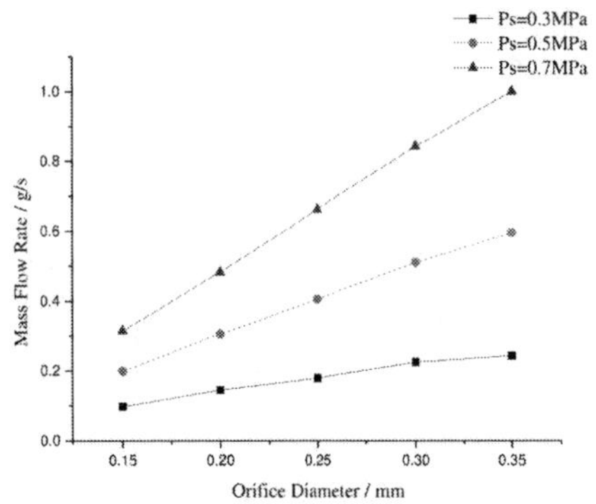

Figure 19. Effect of orifice diameter on mass flow rate under different gas supply pressure.

Orifice Position

The position of the orifice indicates the ratio of the distance L_1 of the orifice to the total length of the bearing L. As shown in Figure 20, the closer the orifice position is to the bearing end face, the more bearing gas leakage, resulting in the smaller bearing capacity; with the increase of the distance from the orifice to the bearing end face, the bearing gas leakage resistance increases, and the bearing gas consumption decreases continuously; as the high pressure area between the exhaust orifices decreases, the bearing capacity begins to decrease at 0.35 of the orifice position.

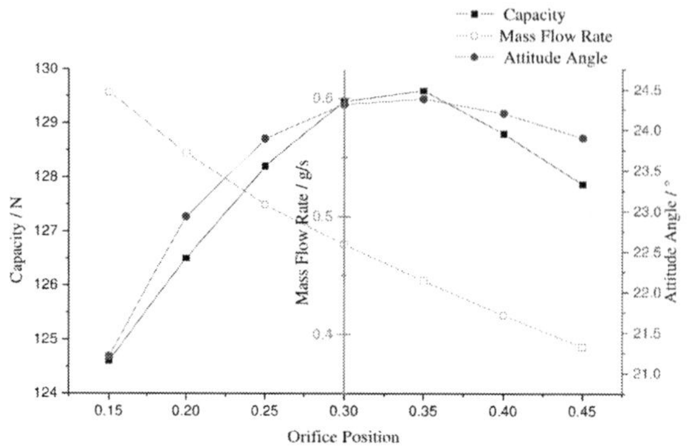

Figure 20. Effect of orifice position on bearing capacity, attitude angle and mass flow rate.

6. Conclusion

1. The orifice-type throttled aerostatic journal bearing is actually a kind of hybrid bearing [15]. The hydrodynamic pressure effect and the hydrostatic throttling effect are coupled together to form the gas film pressure, which makes the rotor rotate at high speed in suspension state. The average pressure produced by hydrostatic throttling effect is greater than that by hydrodynamic pressure effect in the orifice-type aerostatic bearing.

2. The results of numerical analysis show that the bearing capacity of the journal aerostatic bearing with orifice throttling can be improved by increasing the eccentricity of rotor, rotating speed and air supply pressure, adopting large length diameter ratio, large orifice diameter and reasonably arranging orifice position.

3. Bearing capacity in vertical eccentric direction $\boldsymbol{F_t}$ is the main factor causing whirl [14]. The results of numerical analysis show that increasing the eccentricity and rotating speed can reduce the rotor attitude angle. It is not recommended to use too large orifice diameter, and too large orifice position.

4. Gas consumption is one of the important indexes to evaluate the performance of gas bearing. The calculation results show that the air consumption and operation cost will be increased by increasing the air supply pressure and orifice diameter. The leakage of bearing gas can be reduced and the cost can be saved by selecting the proper position of orifice and the ratio of length to diameter of bearing.

Reference

[1] Fan T and Li S S 2004 *J. Forestry Machinery & Woodworking Equipment* **32** 7-8
[2] Wang Z W and Sun A 2006 *J. Journal of Mechanical Engineering* **42** 6-10
[3] Han Z H, Hou D N and Zhao J H 2019 *J.Journal of Vibration and Shock* **38** 246-251
[4] Li J, Li Q and Li X M 2016 *J. Lubrication Engineeing* **41** 1-5
[5] Li Z J, Han D J and Yang J F 2018 *J. Lubrication Engineeing* **43** 52-8
[6] Yu H C, Ma W Q and Wang Z W 2009 *J. Chinese Hydraulics & Pneumatics* **34** 77-81
[7] Wang Y, Wang Q X, Chen A K and Liu N 2012 *J. Machinery* **39** 31-4
[8] Wu L J, Yang C E and Wang W H 2015 *J. Journal of Mechanical & Electrical Engineering* **32** 1201-05
[9] Fan C, Gao S H and Ni Z 2018 *J. International Journal of Fluid Dynamics* **6** 62-8
[10] Zhao S X and Huang H 2008 *J. Journal of Wuhan University of Science and Technology* **33** 69-73
[11] Peng W H, Tao J Z and Xu G 2008 *J. Lubrication Engineeing* **33** 56-9

[12] Wang X M, Du J J and Li S S 2012 *J. Journal of Mechanical Engineering* **48** 121-7
[13] Cheng Z Y, Zhang Y P and Zhang Z F 2017 *J. Lubrication Engineeing* **42** 24-9
[14] Li W, Ma W Q and Qiu X Q 2019 *J. Lubrication Engineeing* **44** 24-8
[15] Li J, Yang S Q and Li X M 2017 *J. Bearing* **000** 42-8

IOP Conf. Series: Materials Science and Engineering 1081 (2021) 012039 doi:10.1088/1757-899X/1081/1/012039

Research the Influence of Hydraulic Model on Vibration of Pump

Feng Zhao[1], He Li[1], Duo Liu[2] and Jun Fu[2]

[1] School of Mechanical Engineering and Automation, Northeastern University, Shenyang 110004, China
[2] Shenyang Blower Works Group CO., LTD. Shenyang 110869, China

Correspondence should be addressed to Feng Zhao; E-mail: 13840424515@163.com

Abstract. In this paper, the cooling water pump of centrifugal equipment is taken as the research object in ship system. The pump set is normally open. And, it is also one of the main vibration sources of ship. On the premise of meeting the requirement of hydraulic performance parameters, several different hydraulic model prototypes of this pump set are manufactured. On the same test bench, the vibration tests of this pump show that the measured values are quite different. Fluid software is used to analyse these hydraulic models under the same boundary conditions. It is found that there are obvious differences in hydraulic fluctuations and fluid vortices among impeller passage, vortex chamber, inlet and outlet passages. In order to verify the inference, vibration test and analysis are carried out on the whole pump set and the foots of installation. The corresponding relationship was found out between the fluid fluctuation of hydraulic components and the vibration characteristics. Based on the corresponding relationship, the current hydraulic model of pump is optimized and improved. The parameters of hydraulic model which can affect the vibration sensitivity of pump are optimized. Finally, the experimental results prove that this method can reduce the vibration effectively.

1. Introduction

Pump set is a widespread general machinery. Where there is liquid flow, there is always a pump at work. With the development of science and technology, it is no longer satisfied with the hydraulic performance of pump. The research on vibration suppression of pump is emphasized gradually. With the increasing requirement of vibration and noise, it is more and more strict in ship equipment system [1-2]. The pump unit researched in this paper belongs to the normally open equipment and is also one of the main vibration sources in the ship system.

At present, the hydraulic components of this pump in China are still using widely which were designed from 1970s to 1990s. In order to adapt to the low vibration performance of ships, the technology of vibration suppression is improved. It is urgent to find out the corresponding rules between different hydraulic models and vibration characteristics of pump set. And the targeted approach is adopted to suppress vibration. In view of such problems, the pulsation and vortex of hydraulic components were analyzed in this paper, such as impeller passage, vortex chamber, inlet and outlet passages. The corresponding relationship with the vibration of this pump set was found out. So, the vibration value of the pump set could be reduced effectively.

2. Hydraulic model calculation

2.1. Theoretical basis of hydraulic modelling

Mathematical models were established that could reflect the nature of engineering and physical problems. Specifically, the differential equation is necessary to be established which can reflect the relationship between variate and corresponding definite solution conditions in the problem. And this is first step for numerical simulation. The basic governing equations of fluid include mass conservation equation, momentum conservation equation, energy conservation equation, and the corresponding definite solution conditions of these equations [3].

According to the hydraulic characteristics of flow field, it can be regarded as an incompressible viscous flow in the pump. The ideal governing equation of conservation type is as follow:

- Mass conservation equation

$$\frac{\partial \rho}{\partial t} + div(\rho \vec{u}) = 0 \tag{1}$$

- Momentum conservation equations which were also known as Naiver-Stokes equations or equations of motion

$$\begin{aligned} &\frac{\partial(\rho u)}{\partial t} + div(\rho u \vec{u}) = div\,(\mu\ grad\ u) - \frac{\partial p}{\partial x} + S_u \\ &\frac{\partial(\rho v)}{\partial t} + div(\rho v \vec{u}) = div\,(\mu\ grad\ v) - \frac{\partial p}{\partial y} + S_v \\ &\frac{\partial(\rho w)}{\partial t} + div(\rho w \vec{u}) = div\,(\mu\ grad\ w) - \frac{\partial p}{\partial z} + S_w \end{aligned} \tag{2}$$

- Energy conservation equation

$$\frac{\partial(\rho T)}{\partial t} + div(\rho \vec{u} T) = div(\frac{k}{C_p}\ grad\ T) + S_T \tag{3}$$

2.2. Design of hydraulic component

The hydraulic performance was calculated and analyzed based on the 3D models of hydraulic components. According to the analysis results, the hydraulic components were optimized. And the parameters which are closely related to the hydraulic performance were adjusted for optimization analysis [4]. The hydraulic components with low vibration were developed and verified by tests. Thus, an excellent hydraulic model would be obtained. And the vibration amplitude could be reduced from the vibration excitation source.

2.2.1. Design of inlet and outlet channel

The axial suction and radial discharge were adopted in the inlet and outlet channel. The suction of straight conical tube was adopted in the inlet channel. And the structure of single vortex chamber was adopted in outlet channel. The diameter of inlet and outlet pipe was adjusted to match the interface of system. The three-dimensional model of the pump set is shown as below.

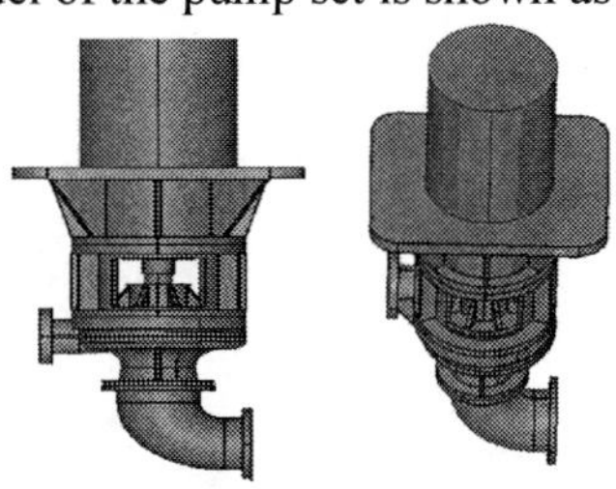

Figure 1. Three-dimensional Model of the Pump Unit

2.2.2. Selection of the impeller

According to the performance of equipment cooling pump: Q=80m^3/h, H=45m, n=2950r/min. It is concluded that ns=82.2. The hydraulic model with similar ns and high efficiency was selected in hydraulic model library. Similar conversion was carried out to ensure that the hydraulic performance of the pump meets the requirements. In order to reduce the vibration caused by hydraulic pulsation, different number of blades (six and seven) were considered.

2.2.3. Design of vortex chamber
From the perspective of vibration, the vortex chamber of different base circles was designed. In order to optimize the appropriate diameter of base circle, the ratio which is the base circle diameter of vortex chamber to the outside diameter of impeller (D3/D2) was changed. And the improvement of the vortex chamber was analysed.

3. Inlet and outlet channel

3.1. Boundary conditions
The first step is to set the boundary conditions. These parameters need to be defined here such as import boundary, export boundary and wall boundary.

3.1.1. Inlet boundary
The inlet boundary is located at the inlet pipe. According to the rated flow of hydraulic parameters, it was set as the velocity inlet boundary. The turbulent kinetic energy and the dissipation rate are calculated according to the actual working parameters [5]. The formula is as follows:

$$k = \frac{3}{2}(v \cdot I)^2 \tag{4}$$

$$\varepsilon = C_u^{3/4} \frac{k^{3/2}}{l} \tag{5}$$

Where $I = 0.16\mathrm{Re}^{(-1/8)}$, and I is turbulence intensity; $\mathrm{Re} = vD\rho / \mu$, Re is Reynolds number, v is fluid velocity, D is hydraulic diameter, ρ is fluid density, μ is fluid viscosity; l is turbulence length scale, $l = 0.07D$; $Cu = 0.0845$.

3.1.2. Outlet boundary
The outlet boundary is located at the end of the outlet pipe. Considering the length of the outlet pipe, the exit condition is set to free outflow.

3.1.3. Wall boundary
The boundary conditions of the wall surface are determined by the properties of the fluid moving in it. In this paper, the non-slip wall condition was selected. All the wall surfaces that move with the impeller were set as moving wall surfaces in the impeller area. And the rotation direction and speed were consistent with the impeller. Such as vortex chamber, inlet and outlet areas, these surfaces were set as static wall which do not change with the rotation of impeller.

3.2. Computational domain setting for hydraulic analysis
Through the analysis of previous data, the unsteady flow structure was the main factor that induces the vibration and noise in the centrifugal pump. The main factors included flow separation on blade surface, vortex, rotating stall, cavitation, dynamic and static interference of impeller-tongue, etc [6]. The fluid analysis on the hydraulic components was carried out. And the hydraulic model of the impeller was established accurately, as shown in figure 2.

In order to determine the influence of inlet and outlet pipelines on the model stress, the pressure chamber was combined with the original impeller, straight cone pipe suction chamber and discharge pipe respectively. In addition, it provided relatively accurate models such as velocity field and pressure field for mechanical research and even vibration analysis. Radial force was analyzed on these four kinds of computational domain grids.

IOP Conf. Series: Materials Science and Engineering 1081 (2021) 012039 doi:10.1088/1757-899X/1081/1/012039

a) Grid Structure 1st b) Grid Structure 2nd c) Grid Structure 3th d) Grid Structure 4th

Figure 2. Models of Four Computational Domains

Through the above four computational domains, the hydraulic components that affect the radial force and the changing factors were obtained. When the impeller was calculated separately in the computational domain 1st, there is basically no radial force. When the inlet pressure changes in computational domain 2nd consisted of impeller and pressure-chamber, the radial force was basically unchanged. When the computational domain connects with the inlet and outlet pipelines, the radial force decreased in last two computational domains. From the calculation results of the third and fourth domains, it could be seen that the position of the elbow at the inlet turning has little effect on the radial force.

3.3. *Vibration test verification*

The vibration acceleration level of the test environment is no more than 80dB, which is lower than 10dB of the device operation. It is considered that the test environment is reasonable. And the vibration test data need not to be modified. This type of vertical water pump unit was tested by acceleration level method.

By adjusting the condition of the turning elbow of the inlet pipe, the vibration test was carried out. It showed that the test results of the last two computational domains were almost the same, which were 121.8db and 122.1db respectively, as shown in figure 3 below. It is proved that the position of the elbow in the inlet has little influence on the radial force and the vibration.

a) Result of Field 3th b) Result of Field 4th

Figure 3. Spectrum Diagram of Different Positions of Inlet Turning Elbow

4. Impeller passage

4.1. *Force analysis of impeller*

There are two kinds of fluid excitation forces on the impeller: One is the steady-state force that causes the eccentricity of the impeller-rotor system. And the other is the transient force that causes the vibration [7-9]. In this paper, the direct integration method was used to calculate the force on the impeller. And the pressure and viscous force on the impeller were integrated respectively by Fluent software. The boundary of the impeller region was selected in software. And the pressure and viscous force were calculated in three directions of X, Y and Z respectively. The resultant force was obtained by adding the two forces together. Then the frequency response function (FRF) of the resultant force on the impeller was obtained by FFT method.

The force of the impeller was analysed. When the transient calculation results converged, the calculation was carried out again. One set of results was recorded for each three degrees rotation of impeller. And the results of impeller rotating were recorded in two entire cycles. The forces on impeller were exported in the three directions. The force condition was shown in figure 4 below.

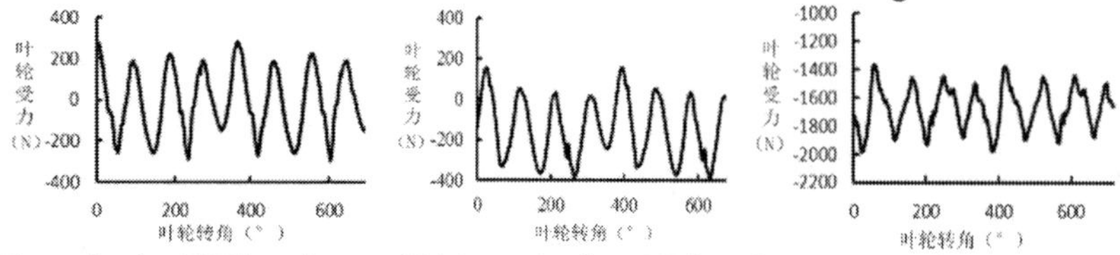

a) Force in the X Direction b) Force in the Y Direction c) Force in the Z Direction

Figure 4. The Forces on Impeller in Three Directions

While the impeller has rotated twice, the change of force in three directions was showed in figure 4. As can be seen from the figure4, the impeller rotated once (i.e. 360°) as a cycle. The force on the impeller was basically a cyclical change after the stable operation of pump. After the curve of force & angle was obtained, it could be expressed in the frequency domain by FFT, as shown in figure 5 below.

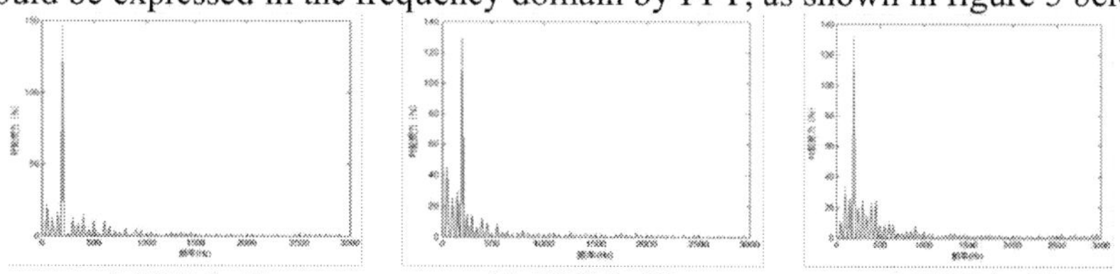

a) FRF in X b) FRF in Y c) FRF in Y

Figure 5. FRF of Impeller Force

According to FRF analysis of pump, the vibration peak existed at the blade frequency. And the generation of blade frequency was related to the hydraulic components. In order to reduce the low frequency vibration from the vibration source, the hydraulic components of the pump set need to be optimized further.

4.2. Impeller fluid analysis

4.2.1. Modelling analysis of hydraulic components

- Steady calculation, full multi-grid accelerated convergence.
- Impeller with full channel technology, rotor parts with the consolidation method.
- Turbulence model: S-A
- Inlet boundary conditions: total temperature, total pressure, speed direction.
- Exit boundary conditions: mass flow, initial pressure.
- The conditions of wall are no slip and adiabatic, impeller given design speed.
- Initial field: uniform field
- The volume loss and mechanical loss were not considered.

According to the design drawing, the grid models of six blades (scheme 1) and seven blades (scheme 2) were generated respectively, as shown in figure 6. The simulation results corresponding to the two hydraulic models of impeller were shown in table 1.

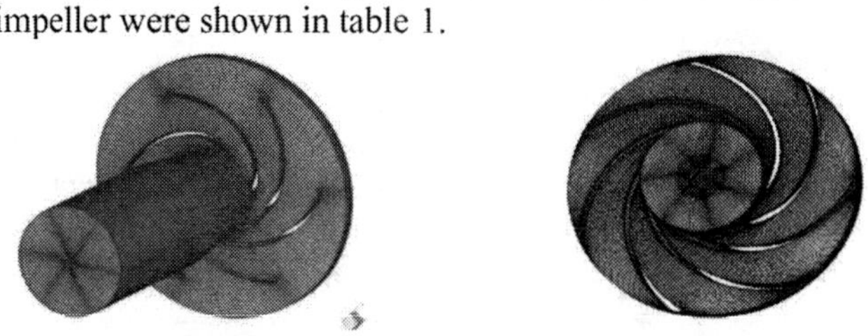

a) Scheme1 with Six Blades b) Scheme2 with Seven Blades

Figure 6. Grid Structure of Impeller

Table 1. Simulation Results of Two Schemes

Scheme	Flow (m^3/h)	Revolving speed (r/min)	Lift (m)	Efficiency (%)
Scheme 1	80	2950	45	80.1
Scheme 2	80	2950	49.6	83.7

4.2.2. Determination of simulation results

The inlet pressure of working point was set as 200000Pa. The calculated results of the two schemes were accurate and reliable through the analysis of y+ value and temperature field. Generally, the range of y+ values of most wall surfaces in the computational domain met the needs of S-A turbulence model as shown in figure 7. It could be found that the temperature changes very small in computational domain from figure 8. And it was consistent with the actual situation. So, the analysis result could be judged to be relatively true and accurate.

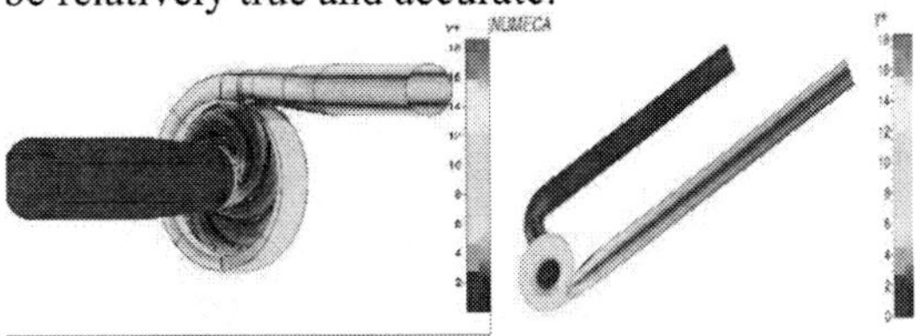

a) Scheme1 b) Scheme2

Figure 7. Distribution of Values y+

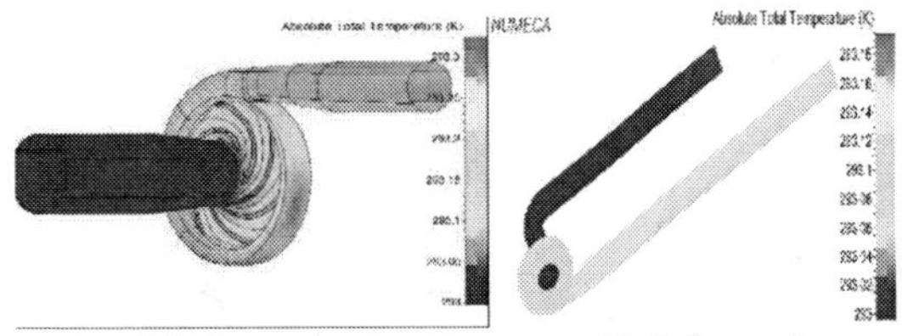

a) Scheme1 b) Scheme2

Figure 8. Distribution of Absolute Total Temperature

4.2.3. Fluid analysis results

The following figure shows that the distribution of flow line inside the impeller is generally uniform. Compared with the two schemes, the impeller with seven blades has reflux zones on the working face.

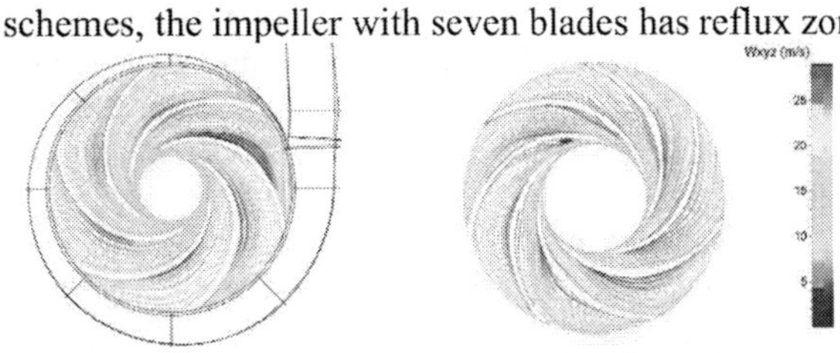

a) Scheme1 b) Scheme2

Figure 9. High Speed Streamline in the Half of Impeller

According to the results of fluid simulation, the hydraulic pulsation of scheme 1 is better than that of scheme 2.

4.2.4. Rotor balance mode for axial force

In order to compare and verify the influence on vibration, two balance methods were adopted. One was to balance the axial force of the rotor by setting front and rear seal rings and balance holes on impeller. The other was that the impeller does not need balance holes. In order to balance axial force, the external balance tube was installed which was connected to the high-pressure chamber behind the impeller and the inlet. The stress state of the rotor was improved. The residual axial force was afforded by the bearing of pump. The guide bearing of water-lubricated was arranged between the impeller and sealing device. As an auxiliary support, it suffered radial force of impeller and centrifugal force of rotor dynamic imbalance. The dynamic stability of rotor system was improved by reducing the cantilever length of

IOP Conf. Series: Materials Science and Engineering 1081 (2021) 012039 doi:10.1088/1757-899X/1081/1/012039

rotor shaft in operation process. And it was conducive to improve the bearing in vibration and noise reduction.

4.3. Vibration measurement

According to the spectrum diagram of FRF obtained from tests, the prominent peak frequency might correspond to the blade frequency of hydraulic pulsation (blade frequency is axial frequency multiplied by blade number).

Compared the influence of scheme 1 (impeller of six blades) and scheme 2 (impeller of seven blades) on vibration of pump, it is found that the vibration value of scheme 1 is 120.7db, which is better than that of scheme 2 (122.8db). After changing the balance mode, the vibration value of scheme 1 is 116dB better than that of scheme 2 (118dB).

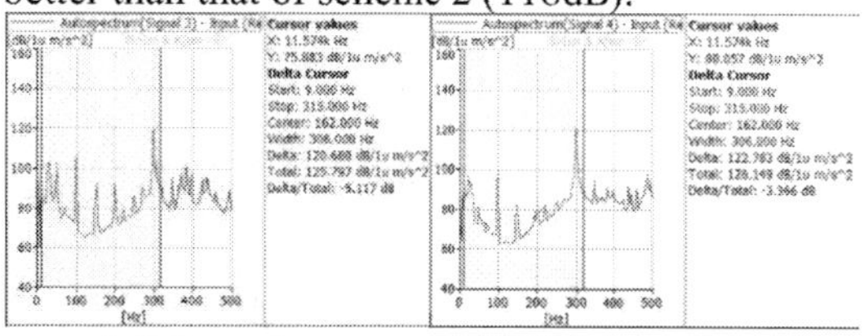

a) Scheme1 b) Scheme2

Figure 10. Results of Tests with Balance Holes

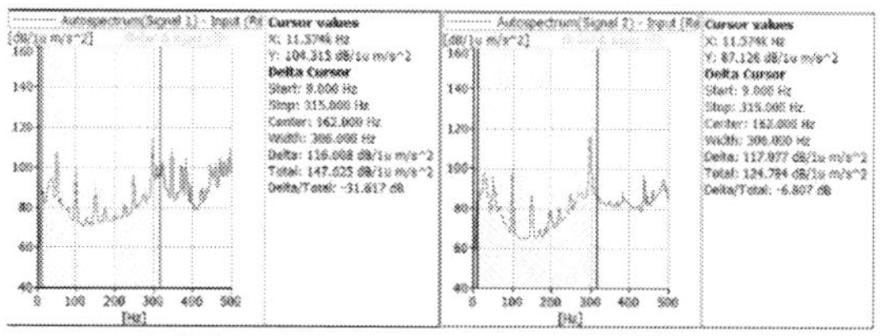

a) Scheme1 b) Scheme2

Figure 11. Results of Tests with Balance Tube

According to the results of vibration test, it could be concluded that the vibration of scheme 1 was better than that of scheme 2 in the same mode of balance holes. When balance holes were closed, the vibration results of both schemes were better than before. After changing balance mode, the vibration of plan 1 was still better than that of plan 2.

4.4. Section of this chapter

According to the results of fluid simulation, it could be obtained that the hydraulic pulsation of six blades was better than that of seven blades. Because the vortices were produced by impeller of seven blades. At the same time, the hydraulic pulsation was affected. And the vibration tests have the corresponding results with simulation. In the frequency range of 10Hz-320Hz, the vibration value of six-blades impeller was better than that of seven-blades impeller.

In addition, the external balance tube was used to connect the high-pressure chamber behind the impeller to the inlet without opening balance holes. From the perspective of theoretical analysis and vibration test verification, it was found that the vibration effect of impellers including six and seven blades was better than that of traditional impeller with balance holes. It could also be concluded that the balance hole produces certain disturbance to hydraulic pulsation. And it could affect hydraulic model on radiating vibration. At the same time, the improvement of balance mode on the impeller was better than the change of the number of blades in vibration reduction.

Therefore, the best choice was six-blades impeller in the comparison of these schemes. And the external balance tube was used to connect the high-pressure chamber behind the impeller with the inlet instead of the previous balance holes.

5. Design of vortex chamber

The vortex chamber with different base circle was designed. And the clearance between impeller and tongue was optimized to reach a better value. The hydraulic performance and hydraulic pulsation were analyzed to make the liquid flow smooth and uniform without vortex. The vibration value was decreased as much as possible from the excitation source caused by hydraulic flow [10-12].

In the past researching on vibration and noise reduction of centrifugal pump, it was proved that the vortex chamber of different base circle had influence on vibration. It means that the different clearance value between impeller and tongue had different vibration value. Therefore, the appropriate base circle of the vortex chamber was researched to match with the impeller of six blades.

5.1. *Calculation of hydraulic performance*

Two kinds of base circle with the limit size of the vortex chamber were provided here. The scheme which the diameter of base circle was 208mm of the minimum limit size, could be defined as scheme 01. And the scheme which the diameter of the base circle was 222mm of the maximum limit size, could be defined as scheme 02. The vortex chambers of two base circles matched the same impeller of six blades were compared and analyzed. At first, the hydraulic performance of the two schemes was calculated.

Table 2. Analysis Results of Variable Working Conditions

Scheme	Flow (m^3/h)	Efficiency (%)	Lift (m)	Scheme	Flow (m^3/h)	Efficiency (%)	Lift (m)
Scheme 01	50	78.5	39.1				
	60	80.0	38.3	Scheme 02	60	74.8	38.3
	70	80.7	37.2		70	80.1	37.0
	80	81.1	45.6		80	80.9	43.9
	90	81.0	43.5		90	76.7	36.6
	100	80.1	40.7				
	110	78.2	37.2				

Through table 2, It could be found that the efficiency of scheme 01 is 81.1% as rated point. The head is 45.6m, and the efficient point is between 80-90m^3/h. The efficiency of scheme 02 is 80.9% as rated points. The head is 43.9m, and the efficient point is around 80m^3/h. There is no significant improvement in the efficiency, and the head is 1.7m lower in Scheme 02.

5.2. *Fluid simulation*

The vortex chamber of two basic circles matched the same impeller, were compared and analyzed [13]. Generally, both two flow diagrams of each scheme showed that the overall flow were well inside the pump, as shown in figure 12. In scheme 01, there exists a backflow area in impeller. In scheme 02, there is no obvious backflow area. And the streamline is basically consistent with the blade profile.

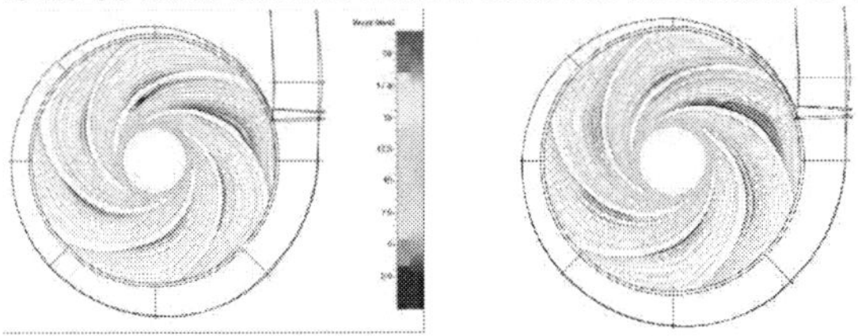

a) Scheme 01 b) Scheme 02

Figure 12. Analysis Diagram of Flow Field

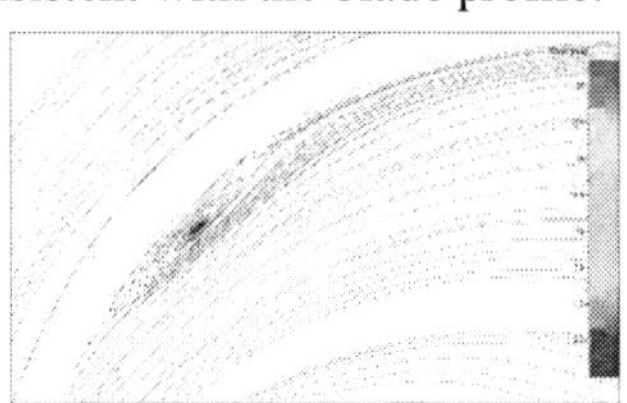

Figure 13. Drawing of Partial Enlargement

The velocity flow diagram of the central surface of the impeller was observed as shown in figure 12(a). The velocity flow from the center to the edge of impeller was relatively uniform. And there was only one section existing abnormal phenomena in the cavity. The detail of vortex was found in this area after magnifying it as shown in figure 13. When the vortex entered the subsequent pressure chamber, it would impact on the fluid and spread. Thus, it caused a large hydraulic pulsation.

The static pressure gradient line of radial distribution was analysed in vortex chamber of scheme 01. Under the influence of the impeller, periodic fluctuations occurred near the inner side. And the gradient line of static pressure formed an arc in the outlet cylinder. Compared with scheme 01, The range of static pressure gradient was small and the hydraulic pulsation is also small in scheme 02.

IOP Conf. Series: Materials Science and Engineering 1081 (2021) 012039 doi:10.1088/1757-899X/1081/1/012039

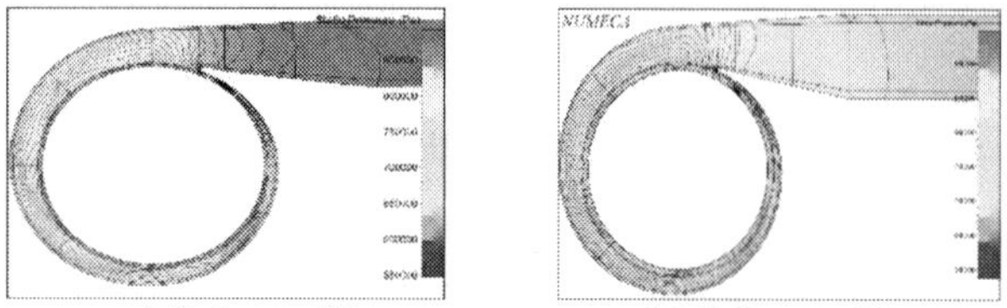

a) Scheme 01 b) Scheme 02

Figure 14. Static Pressure Distribution Diagram of Cross Section in Pump Housing

5.3. Vibration measurement

In the test, the effect of scheme 01 (base circle is 208mm) and scheme 02 (base circle is 222mm) on vibration were compared. It is found that the vibration value of scheme 02 is 110.7db, which is better than that of scheme 01, which is 113.3db.

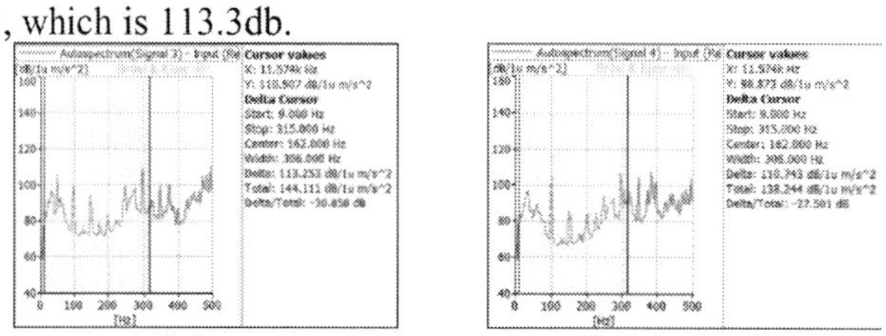

a) Scheme 01 b) Scheme 02

Figure 15. Vibration Test Results of the Pump

5.4. Section of this chapter

By analysing the cloud image of velocity vector and its cross section in the vortex chamber, it could be clearly found that there were obvious velocity gradients in both horizontal and vertical directions near the tongue. The reason for this phenomenon was as follow. Caused by the interference between impeller and baffle tongue, the liquid which was accelerated by centrifugal force of impeller suffered local resistance at the tongue. And the tongue slows down the overall speed. While some of liquid was reaccelerated due to the different pressure between the two sides of baffle tongue. It flowed back to the vortex chamber in the left of baffle tongue and mixed with the high-speed fluid. Because of this part of the fluid backflow on the left side of baffle tongue, the high-speed area was affected. In the common influence area of impeller and baffle tongue, there would be a large alternating gradient of velocity and pressure. And the change of gradient was closely related to the rotation of impeller. This was the reason why the blade frequency of the hydraulic component was obvious in FRF curve [14-16].

From the fluid simulation results, it could be seen that the hydraulic components of this type of pump have such problems as vortex and backflow at the edge of impeller. Therefore, the flow field was not stable. And there also be certain pressure pulsation. This pulsation situation would be more complex after entering the vortex chamber. Based on analysis of pressure pulsation in flow field, it showed that the pressure pulsation was the main contributor to blade frequency peak in FRF.

On the finite element analysis platform, the fluid analysis which include matching different clearance between impeller and tongue was carried out. The structural parameters of hydraulic components with the smallest hydraulic pulsation were found out. So, the liquid flowed into the pump smoothly.

Based on the above, the analysis and calculation results of scheme 01 and scheme 02 were compared. Combined with the test results, scheme 02 (base circle diameter of 222mm) was selected as the base circle size of the new vortex chamber. That is to meet the requirements of hydraulic performance, but also to obtain better hydraulic pulse characteristics.

6. Hydraulic optimization and test

6.1. Calculation basis of hydraulic performance optimization

IOP Conf. Series: Materials Science and Engineering 1081 (2021) 012039 doi:10.1088/1757-899X/1081/1/012039

On the premise of maintaining the performance and efficiency of the pump unit, the hydraulic model was optimized. So, it was necessary to establish a mathematical model for the hydraulic model and performance.

The impeller model with similar specific speed and high efficiency was selected from the data basement of hydraulic model. The design parameters were substituted into formula, and the specific speed is as follows:

$$n_s = \frac{3.65 \times n \times \sqrt{Q}}{H^{3/4}} = 83.4 \tag{6}$$

The similarity conversion is performed by the following formula.

$$\frac{Q_P}{Q_m} = \lambda_Q{}^3 \times \frac{n_P}{n_m} \tag{7}$$

$$\frac{H_P}{H_m} = \lambda_H{}^3 \times \left(\frac{n_P}{n_m}\right)^2 \tag{8}$$

From formula (5) and (6)

$$\lambda_Q = \sqrt[3]{\frac{Q_P}{Q_m} \times \frac{n_m}{n_p}} = 0.85 \tag{9}$$

$$\lambda_H = \sqrt[2]{\frac{H_P}{H_m} \times \frac{n_m}{n_p}} = 0.851 \tag{10}$$

According to λ=0.852, all dimensions were converted. And the performance curve is shown in figure 16.

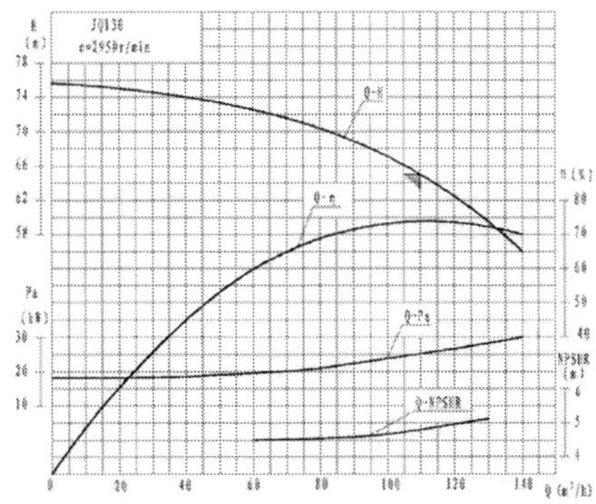

Figure 16. Performance Curves

According to the above analysis of the dynamic characteristics and flow field on this pump unit, reasonable optimization and improvement on the structure of fluid components have carried out. The purpose was to reduce the vibration and improve the efficiency. It included the local optimization and improvement of inlet angle, outlet angle and packing angle in blades [17].

Since optimized and improved impeller without opening balance hole, the vortex chamber of scheme 02 that basic circle is 222mm, was further optimized. So, the clearance between the impeller and the tongue could reached a better value. The liquid flow was stable and uniform without vortex. And the contribution of hydraulic pulsation to the vibration was decreased. So, the source of excitation was reduced greatly.

6.2. *Test and verification*

This new pump set which was optimized and improved was tested. And its vibration level of low-frequency was 108.5db (see figure 17). The influence of hydraulic pulsation on unit vibration was greatly reduced.

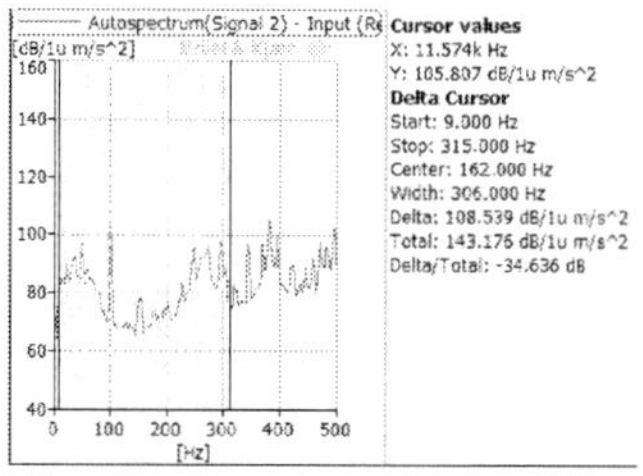

Figure 17. Vibration Energy Spectrum in the 10hz-320hz Frequency Band

7. Conclusion

The range of these parameters was researched in the centrifugal pump, such as inlet form, impeller, tongue and vortex chamber. Under the premise of constant performance, the combination parameters of typical key components were optimized to obtain the minimum pulsating pressure.

The scheme of the new pump set was researched. The hydraulic model of the pump was optimized and improved according to the test and analysis results. This new hydraulic model has broken through the one used since the 1970s. Based on the method of vibration reduction, the design scheme has been successfully completed.

References

[1] Dan Florin NICULESCU, Adrian OLARU, Adrian GHIONEA, Mikulas HAJDUK, Marek VAGAŠ. Assisted Research of the Centrifugal Pump's Vibration Behavior[J]. Mechanics and Materials. ISSN: 1662-7482, Vol. 436, pp 85-91.

[2] A. Albraik, F. Althobiani, F. Gu and A. Ball. Diagnosis of Centrifugal Pump Faults Using Vibration Methods[C]. 25th International Congress on Condition Monitoring and Diagnostic Engineering, Journal of Physics: Conference Series 364 (2012) 012139, 1-12.

[3] W Y Zhao, J G Ge, D Ma, C M Li and S B Bao. Vibration analysis of large centrifugal pump rotors[C]. 6th International Conference on Pumps and Fans with Compressors and Wind Turbines. IOP Conf. Series: Materials Science and Engineering 52 (2013) 022033, 1-8.

[4] Schmitz S. Reducing Pump Noise in Cooling Tower Applications[J]. World pumps. 2004(3): 24-29.

[5] Zhou Yunlong, Zhao Peng.Vibration Fault Diagnosis Method of Centrifugal Pump Based on EMD Complexity Feature and Least Square Support Vector Machine[J]. Energy Procedia, 17 (2012) 939-945.

[6] C.G.Rodriguez, E.Egusguiza, I.F.Santos. Frequencies in the Vibration Induced by the Rotor Stator Interaction in a Centrifugal Pump Turbine[J]. Journal of Fluids Engineering, 2007,129:1428-1435.

[7] Qiang Zhou, Hong-kun Li, Zhen-fang Fan, Hong-jun Niu, Shi-wen Sun. Experimental Investigation on the Unsteady Pressure Pulsation and Structural Vibration of an Engineering Test Reactor Main Pump [J]. Chinese Journal of Turbomachinery, Vol.62, 2020, No.4:75-79.

[8] N.R.Sakthivel, V.Sugumaran, S.Babudevasenapati. Vibration based fault diagnosis of monoblack centrifugal pump using decision tree[J]. Expert Systems with Applications, 37 (2010) 4040 – 4049.

[9] C. G. Rodriguez, E. Egusquiza, I. F. Santos. Frequencies in the Vibration Induced by the Rotor Stator Interaction in a Centrifugal Pump Turbine[J]. Journal of Fluids Engineering, November 2007, Vol. 129 1428- 1435.

[10] William E. Nelson, John W. Dufour. How to Avoid Building Problems into Pumping System[C]. Proceedings of The Eleventh International Pump Users Symposium, 2009, 125-136.

[11] Chun-jun Ji, Xue-wei Zhang, Mao Shi, Chun-yang Li, Jun-yi Fang, Jun-nan Liu. Matching Characteristic Between Impeller and Passage Guide Vane in Multi-Stage Centrifugal Pump [J]. Chinese Journal of Turbomachinery, Vol.61, 2019, No.Z1:7-13.

[12] Senthilkumar M, Yuvaraja M, Metin KOK. Fault Diagnosis of Centrifugal Pump and Vibration Control Using Shape Memory Alloy Based ATDVA[J]. Mechanics and Materials ISSN: 1662-7482, Vol. 787, pp 927-931.
[13] W J Zhou, Y C Yang, G K Xing and L Q Wang. Numerical investigation of nonlinear vibration for rotor-seal system of centrifugal pump[C]. 6th International Conference on Pumps and Fans with Compressors and Wind Turbines, Materials Science and Engineering 52 (2013) 032006, 1-5.
[14] A.Al-Qutub, A.Khalifa, Y.Khulief. Experimental Investigation of the Effect of Radial Gap and Impeller Blade Exit on Flow-Induced Vibration at the Blade-Passing Frequency in Centrifugal Pump[J]. International Journal of Rotating Machinery, Volume 2009, Article ID 704845, 1-9.
[15] Xi He, Sheng-li Xu, Hai-tao Liu, Xiao-fang Wang, Xu-dong Chen. Prediction of the Dynamic Stress of Reactor Coolant Pump Impellers in Water Environments based on a Harmonic Response Analysis [J]. Chinese Journal of Turbomachinery, Vol.60, 2018, No.2:53-59.
[16] Chuan-yue Zhou, Zeng-hong Xue, Xing-fa Liu, Jian-xin Wen, Yuan-zan Yu. The Rotor dynamic Calculation Analysis of Pump Rotor System [J]. Chinese Journal of Turbomachinery, Vol.60, 2018, No.S1:39-48.
[17] Yu Zhang, Sanbao Hu, Yunqing Zhang, Liping Chen. Optimization and Analysis of Centrifugal Pump considering Fluid-Structure Interaction[J]. The Scientific World Journal Volume 2014, Article ID 131802, 1-9.
[18] Fang-ming Zhou 1 Xiao-fang Wang 1 Sheng-li Xun 1 Rong Xie, Hong-chang Yu2 Xing-ping Yong2 Zuo-wen Zhong. The Effects of Staggered Diffuser Vanes on the Hydraulic Characteristics of a Nuclear Reactor Coolant Pump[J]. Chinese Journal of Turbomachinery, Vol.59, 2017, No.1:7-17.

NUMERICAL INVESTIGATION OF THE FLOW FIELD AND AERODYNAMIC LOAD ON IMPELLERS UNDER RADIAL INLET GUIDE VANES WITH TRAILING EDGE BLOWING IN CENTRIFUGAL COMPRESSOR

Jianchi Xin[1], Haitao Liu[2], Zhitao Tian[1], Xiangyang Liu[1], Xiaozhi Kong[1], Xiaofang Wang[2]
(1. Naval Architecture and Ocean Engineering College, Dalian Maritime University;
2. School of energy and power engineering, Dalian University of technology)
E-mail: xinjc@dlmu.edu.cn

Abstract: The radial inlet is a typical upstream component which is widely used in multiple-stage centrifugal compressors or blowers. Adding the guide vane in the radial inlet could improve the flow field along circumference distribution and increase the efficiency of the compressor as well. However, the guide vane in the radial inlet generates the wake which causes the Rotor-Stator-Interaction (RSI) and further raises the issue of impeller fatigue failure. In order to compensate the wake effect caused by the guide vane, the active control of trailing edge blowing (TEB) is adopted. Based on Computational Fluid Dynamics (CFD) simulation, the radial inlets with unevenly distributed guide vanes (UGV) are considered. This study investigates the guide vane wake in a radial inlet with or without TEB, and aims to figure out the flow phenomenon in radial inlet and the aerodynamic load on the downstream impellers. The simulation results show that the trailing edge blowing has effect on the compressor performance and reduces the wake intensity. By conducting unsteady simulations, the aerodynamic load on the impeller's leading edge is obtained for the radial inlets with or without TEB. The dominant frequency and the pulse amplitude are obtained by Fast Fourier Transforms (FFT). Compared with the UGV model, the TEB shows lower excitation at the harmonics of the impeller passing frequency, and has little influence on the amplitude at the impeller rotation frequency. It means the TEB could improve. the safety of the impeller under radial inlet guide vane condition.

Key words: Radial inlet; Trailing edge blowing; aerodynamic load; compressor

1. INTRODUCTION

The radial inlet is a typical upstream component which is widely used in most multiple-stage centrifugal compressors or blowers. It is required to allow for a uniform flow field upstream of the impeller and therefore ensuring an acceptable performance. As the structure of the radial inlet is complex, the flow field in the radial inlet is a complex three-dimensional flow field, the flow can hardly reach a uniform distribution. The distortion flow field in radial inlet has badly influence on performance of impeller [1]. Compared to an axial flow inlet, the radial

[1] Please address all correspondences to dlwxf@dlut.edu.cn.

inlet results in 1 to 4 percent decrease of the efficiency on the performance of the whole centrifugal compressor [2]. Therefore, the study of the radial inlet is meaningful for the compressor.

In order to understand the flow field in a radial inlet, Tan [3] reviewed the radial inlet of the centrifugal compressor, which included the internal flow in radial inlet, the effects of radial inlet on performance, structure parameter design and optimum design method. At presents, most researchers adopted experiments and computation to investigate the flow characteristics. Compared with the radial inlet and axial inlet, Long [4] consider that the uniform at impeller inlet by radial inlet result in 3 percent decrease of the efficiency and 2.4 percent decrease of the total pressure, and impeller rotating has little influence on flow field in radial inlet. Running variable conditions, Dong [5] found that the radial inlet results in the attack angle decreasing close to 1.4°, and the flow field showed stronger uniform distribution along the circumference and radial directions, which is more obviously at off-design conditions. Based on the understanding of the characteristics of inlet volute's structure and interior flow, Wu [6] through analyzed the property of every channel cross- section, provided a parameterized design model, which can simplify radial inlet design process entity's structure and improve design precision and working efficiency. Han [7] measured the detailed flow in radial inlet with five-hole probes. The investigation indicated that there were five main flow separations and vortices, which is the primary cause of the flow loss and the flow distortions. Feng [8] analyzed the flow field of the radial inlet and first stage with CFD, the results indicated that the performance of radial inlet chambers had close relations to geometry shape. In order to improve the bad situation due to the radial inlet, Wang [9] added the splitter vanes in the radial inlet, the results showed that the splitter vanes increased the flow loss, but decreased the distortion of the impeller inlet. By optimizing profile of the vanes, further decreased the separation loss of the flow, which increase the efficiency of the compressor. Using CFD technique, a numerical optimization approach to the radial inlet was developed by Chen [10], the 1.26% improvement of the efficiency was obtained.

As the impeller is a key component of the compressor, the flow field by radial inlet not only affects the performance, but also affects the safety of the impeller. Wang [11] reviewed the centrifugal compressor excited by unsteady pneumatic loads, and pointed out that the flow field by radial inlet should be considered in impeller safety analysis. The research by Han [12] indicated that the impeller has little influence on the radial inlet, but the radial inlet results in aerodynamic load excitation on the impeller. Xin [13] set the guide vane in the radial inlet, the results showed that unevenly distributed guide vanes provided a favorable condition for the impeller, and decreased the aerodynamic load at machine rotating frequency, which improved the safety of the impeller.

All the previous work of radial inlet mainly focused on the fluid analysis and optimization of the structure, which allows for a better flow performance. Adding guide vanes is a good way to improve the flow field in radial inlet. However, the guide vane in the radial inlet generate the wake result in the SRI (Stator rotator interaction) [14], and it is another aerodynamic load excitation on the impeller as well. The trail edge blowing (TEB) is an active control method, which is a way to reduce the wake method. It was proposed by Park [15], and it was widely used in rotating machinery to reduce the SRI. Lewis [16] adopted the TEB method on Francis turbine, the resulted in 43% decrease in global torque variation at the runner passing frequency. Based on the numerical simulations, Xin [17] analyzes the flow field and aerodynamic load under different guide vane angle during the centrifugal compressor operation. It pointed out that the aerodynamic load on impellers has been reduced significantly when the momentumless wake appeared.

In our previous work, although setting guide vane in radial inlet could decrease the aerodynamic load on impeller at machine rotating frequency , it also make the excitation at impeller passing frequency. The aerodynamic load on it is an important issue related to the compressor safety, which ensures a smooth operation of the compressor and therefore a long lifetime. In order to decrease the excitation at impeller passing frequency, based on the best performs UGV model, the TEB method are adopted on the guide vanes. Through the flow field and aerodynamic load analyzed, the effects of TEB method on the performance of centrifugal compressor and aerodynamic load on impeller are studied using the numerical in this paper.

The remainder of this article is organized as follows. Section 2 describes the compressor with its radial inlet model, the computational consideration and the results of the UGV. Section 3 describes the scheme of trailing edge blowing and the momentumless wake criterion. Compared with the UGV model, the flow field and aerodynamic load results in TEB model are shown in Section 4. Finally, section 5 provides some concluding remarks.

2. GEOMETRY DESCRIPTION AND COMPUTATION CONSIDERATIONS

2.1 Radial inlet centrifugal compressor geometry

Fig. 1 gives the model of the centrifugal compressor with radial inlet, which is used to compress the recycle gas in a methanol process. To handle the large volume flow rate at the given working condition, this compressor adopts the double-sided radial inlet with a compact and symmetrical structure. The first stage of the centrifugal compressor is selected in this numerical study is shown in Fig. 1(a). It includes the radial inlet, the first stage impeller, the vaneless diffuser, the curved channel and the return channel. The radial inlet with unevenly distributed guide vanes (UGV) is shown in Fig. 1(b), which includes two splitter vanes in line with the incoming flow, and the number of the guide vans is 18. The stage feature 13 impeller blades, and the return channel has 28 vanes.

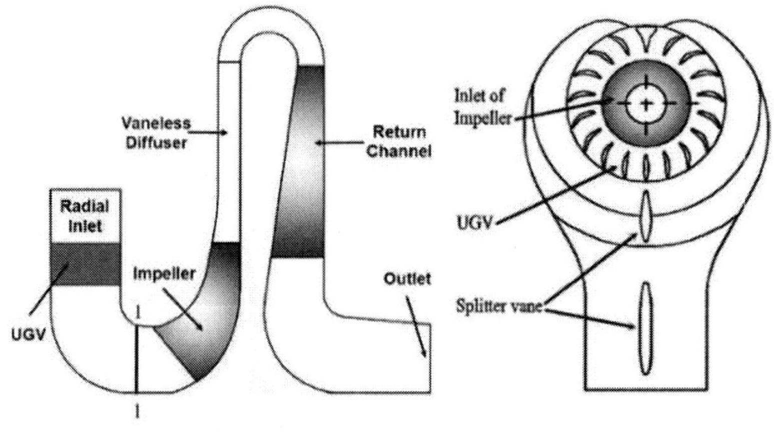

(a) (b)

Figure 1. The first stage with the radial inlet including (a) profile of the compressor meridian plane, and (b) structure of the radial inlet

The compressor performance characteristics at the operation condition are summarized in Table 1. The parameters refer to the compressor first stage without the seal leakage loss.

Table. 1 Compressor performance at operation condition

Parameter	Symbol	Operation Condition
Flow rate (kg/s)	Q_d	60.95
Rotational speed (rpm)	n	7982
Intake total pressure (MPa)	P_{in}	6.53
Intake total temperature (° C)	T_{in}	45
Pressure ratio	ε	1.045
Efficiency (%)	η	89.3

2.2 Computational considerations

In this investigation, all simulations are carried out by ANSYS CFX. In order to improve the CFD computing efficiency, we adopt the radial inlet, the first stage, the vaneless diffuser and the return channel as simulation domain. However, the seal area has been ignored in this simulation. The user-defined recycle gas which adopts the real gas parameters is used as the working medium. Note that the simulation model of the compressor adopts the actual size for processing, ensuring the accuracy of describing the flow region.

In the simulation model, the total energy equation is selected for the heat transfer, the high-resolution advection is used for both steady and unsteady simulations, and the second order backward Euler is used for unsteady simulation. To ensure the convergence, the RMS (Root Mean Square) residual target is set to 1e-5 for pressure and velocity components, and the standard of the convergence should be less than 0.2% for the mass flow relative error between the inlet and outlet. Meanwhile, the efficiency and the pressure ratio have to be kept constant. The steady simulation result is used as the initial condition for the unsteady simulation. During the unsteady simulation, the total computing time is set to 10 circles of the impeller rotation for each radial inlet model in order to get stable and precise aerodynamic data on the impeller blades. The time step size is related to the compressor operation condition, i.e., 180-time steps per impeller rotation cycle are chosen for each case and the data in each time step is saved. The loops of each time step coefficient are set from 3 to 10 to ensure convergence.

As the purpose is to obtain the aerodynamic load on the impeller, the monitoring points are used to save the flow data. Due to the radial inlet in front of the impeller, the monitors are placed at the impeller leading edge from hub to shroud on a single blade equidistantly, in a way that 12 points are on the pressure side of the impeller and the remaining 12 points are on the suction side of the impeller, see Fig. 2. All the monitor points are set on the rotating coordinate system, and the speed of the coordinate system is related to the impeller operation condition. After the unsteady simulation, the aerodynamic load signals of the monitoring points are calculated by means of Fast Fourier Transforms (FFT) to obtain the dominant frequency and pulse amplitude.

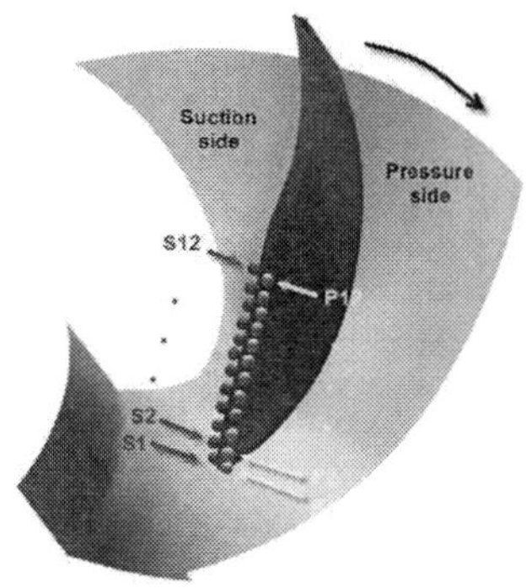

Figure 2. The locations of monitoring points

In this analysis, the $k-\omega$ based SST (Transport of the Turbulent Shear Stress) turbulence model is used. The SST model is based on the equation of turbulence energy and the equation of diffusion rate [18]. This model is designed to give highly accurate predictions of the onset and the amount of flow separation with adverse pressure gradients, and is suitable for the rotating machinery.

In both the steady and unsteady simulations, the boundary conditions are set as follows. The total pressure and total temperature are defined at the inlet, and as the working medium is a compressible fluid, the mass flow rate at the outlet is used. The no-slip condition is set for the blade surfaces and the wall, the rotating and stationary components are connected by the General Grid Interface (GGI). The frozen rotor interface is used for steady simulation because this type of interface does not produce any mixed state in front of or behind the impeller, and the transient rotor-stator interface, which takes into account all the transient flow characteristics and allows for a smooth rotation between components, is used for unsteady.

(a) (b)

Figure 3. 3D meshes of (a) the radial inlet and (b) impeller

In this article, for the blade passage, we use the ANSYS turbo-grid to generate the structured meshes. Then the meshes of a periodic passage are copied to the full wheel model. As it is difficult to generate a structured mesh for the radial inlet, we use the commercial mesh generator ICEM to generate a structured/unstructured mesh in this flow domain, see Fig. 3.

Generally, a mesh independent study should be conducted to investigate the effect of mesh density and to determine which mesh refinement is appropriate for simulation. Fig. 4 shows the simulated results at the operation condition for different mesh numbers.

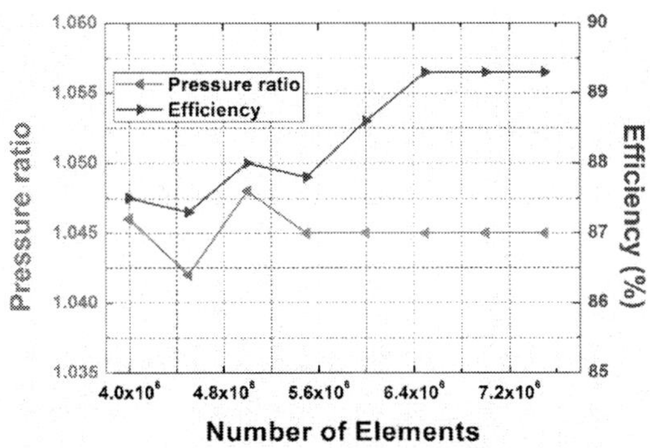

Figure 4. Verification of mesh independence

It is observed that when simulating with more than 6432000 elements, the simulated results are almost the same and in agreement with the data in Table.1. Thus, considering the simulation accuracy and computing time, the 6432000 meshes with a 23.45° minimum orthogonality is used in subsequent calculations. The number of elements for the individual component are as follows. The radial inlet has 3004210, the first impeller has 752895, the vaneless diffuser has 447525, and the return channel has 2227480. Meanwhile, the values of the Y plus on the blade and the wall of the simulation domain are found to satisfy the requirement of the SST turbulence model.

2.3 Results of the UGV model

The steady simulation results offer the overall performance of the compressor and the flow detail in the radial inlet. Fig. 5 gives the contour plots of Mach number, pressure, total pressure and shows the vector plot as well as the streamlines at the outlet of the radial inlet, i.e., the inlet of the first impeller, see section 1-1 in Fig. 1(a).

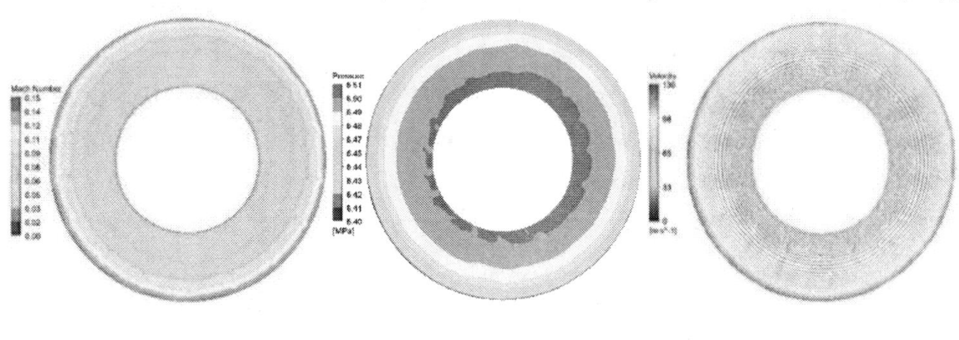

(a) (b) (c)

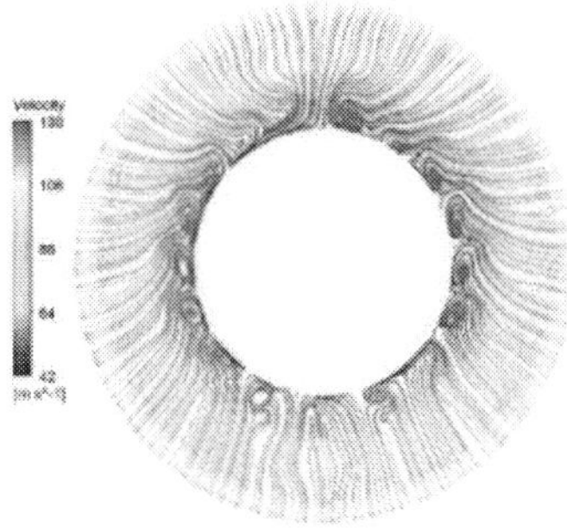

(d)

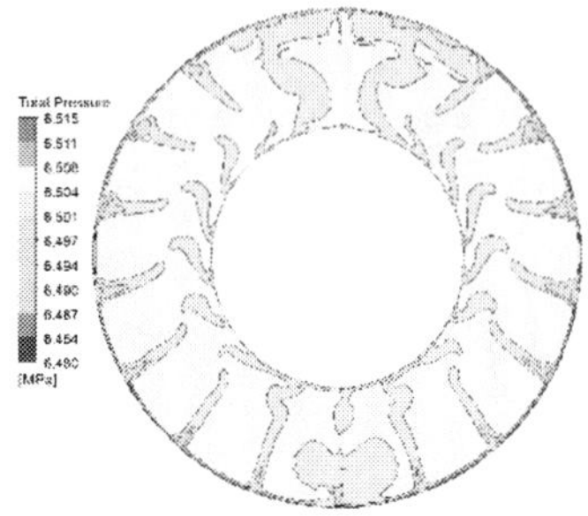

(e)

Figure 5. The distributions of (a) Mach number, (b) pressure, (c) vector, (d) streamline and (e) total pressure at the impeller inlet of the UGV model

After the fluid passes through the splitter vane, the flow direction turns from the radial to the axial. From each contour of the parameters, the flow field of the UGV model shows uniform in the circumference and radial directions. It means that using the guide vane is a good way to reduce the flow losses and improve the flow field distortion in radial inlet. However, the wake of the guide vane leads extra total pressure loss, and makes the impeller generate aerodynamic load at the impeller passing frequency.

Through the unsteady simulation, the frequency domain of the pressure fluctuation of different monitors is calculated by means of the FFT. In this compressor, the main frequencies of the load on impeller leading edge appear at the machine rotating frequency and the impeller passing frequency. For the impeller passing frequency, the 18 guide vanes have to be considered due to the SRI [14]. Table 2 shows the machine rotating frequency and impeller passing frequency of the compressor at operation condition. The parameter f_m represents the machine rotating frequency and the parameter f_i represents the impeller passing frequency, being 18 times higher due to the 18 guide vanes.

Table. 2 Summary of the frequencies of the compressor under the operation condition

Frequency (Hz)	Formula	Value
Machine Rotating Frequency	$f_m = n/60$	133.03
Impeller Passing Frequency	$f_i = f_m \times 18$	2394.6

Fig.6 (a) shows the aerodynamic load at the tip of the impeller leading edge in the UGV model. It is observed that there is an obvious excitation at the 18-frequency harmonic, i.e., the impeller passing frequency. It means that the wake of guide vanes results in the aerodynamic load on the impeller at the impeller passing frequency and further results in the impeller fatigue failure. Fig.6 (b) shows the amplitudes at f_m and f_i distribution at the impeller leading edge. There is no distinct rule in amplitudes distribution at f_m from hub to shroud, and the amplitudes at f_i increase from hub to shroud, which is because that the streamline distance between the guide vane and the impeller is decrease from hub to shroud, and the wake intensity is stronger in shorter distance. Compared with the amplitudes at f_m, the amplitudes at f_i are lower, and the amplitudes at f_m show increasing trend from hub to the shroud, with the highest value at the shroud region. However, the amplitudes at f_m are important aerodynamic load on the impeller, which cannot be ignored by considering the impeller safety.

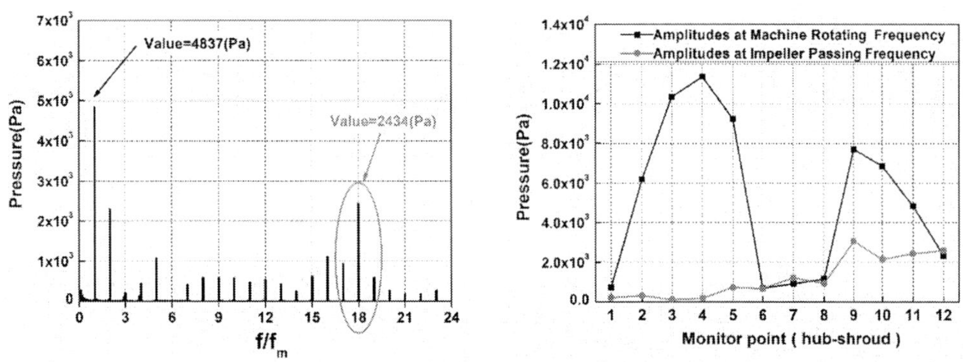

(a) (b)

Figure 6. The aerodynamic load at the impeller leading edge in the UGV model

3. THE SCHEME OF TRAILING EDGE BLOWING AND THE MOMENTUMLESS WAKE CRITERION

As the result shows that the guide vane in the radial inlet will produce wake in front of the impeller, we plan to alleviate it. The active method of trailing edge blowing is a way in theory. Considering the guide vane distribution at radial inlet and the flow field distribution, we set the trailing edge blowing method at the trailing edge of the guide vanes, see Fig. 7. Due to the flow separation near the suction side on left and right guide vanes, the blowing location offsets an area to the suction side.

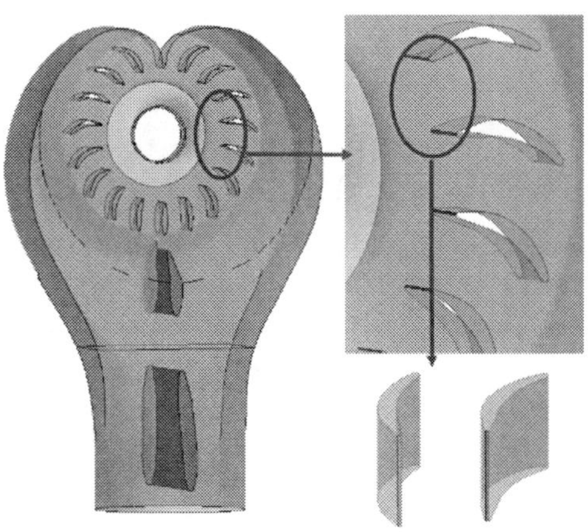

Figure 7. The setting location of the trailing edge blowing at the guide vanes in radial inlet

By using the TEB method, we will compensate the wake deficit, and the goal is to obtain the momentumless wake state. By adjusting the blowing mass flow, we will get four typical wakes. The criterion of momentumless wake is related to the momentum thicknesses θ, which is calculated by the velocity profiles [15], that is

$$\frac{\theta}{d}=\int_{-\infty}^{+\infty}\frac{U}{U_{\infty}}(1-\frac{U}{U_{\infty}})d(\frac{y}{d})d \quad (1)$$

where U_∞ represents the main flow velocity, d represents the guide vane trailing edge width, and y represents the width of the guide vane passage. When $\theta/d > 0$, it represents the pure wake or weak wake, that means there is no blowing or blowing is not enough. When $\theta/d < 0$ it represents for the jet, and it means the blowing flow is overmuch. When $\theta/d \approx 0$, it represents the momentumless wake state, which means that the blowing flow compensates the wake deficit just in time.

4. THE ANALYSIS OF TRAILING EDGE BLOWING

4.1 The compressor performance analysis

During the simulation of the TEB, the main boundary condition and the numerical method are the same as that in the UGV simulation case. Furthermore, the mass flow boundary is set at the guide vane blowing locations, and the extra mass flow has been considered at the outlet. By adjusting the blowing mass flow, the momentumless wakes of the guide vane are selected as the TEB results for the subsequent comparison.

Through the simulations with various running conditions, Table 3 gives the blowing mass flow distribution under various conditions. From 0.8 Q_d to 1.2 Q_d, the needed blowing mass as well as relative flow rate is increased. It is because the high mass flow leads to high main flow velocity, high blowing velocity, high blowing mass flow. Due to the blowing mass flow, the compressor corresponding operation condition has changed, especially in the high flow condition. The TEB method results in 3%—7% mass flow increased at 0.8 Q_d —1.2 Q_d operation condition.

Table. 3 The blowing mass flow distribution under momentumless wake condition

Compressor operation condition	Blowing mass flow (kg/s)	Relative flow rate (%)	Total mass flow corresponding operation condition
0.8 Q_d	2.02	4.13	0.83 Q_d
0.9 Q_d	2.47	4.51	0.94 Q_d
1.0 Q_d	2.87	4.72	1.05 Q_d
1.1 Q_d	3.48	5.18	1.16 Q_d
1.2 Q_d	4.38	6.00	1.27 Q_d

Fig. 8 gives the compressor characteristic curves of the UGV and TEB models. The pressure ratio and efficiency are defined as:

$$\text{Pressure ratio} = P_{out} / P_{in} \tag{2}$$

$$\text{Efficiency} = ((\gamma\text{-}1)\ln(P_{out} / P_{in})) / (\gamma \ln(T_{out} / T_{in})) \tag{3}$$

where γ represents the specific heat ratio, P_{out} and T_{out}, respectively, represent the discharge total pressure, and the discharge total temperature.

From a normalized mass flow of 1.2 down to the 0.8, we find that the guide vane trailing edge blowing in the radial inlet changes the compressor best operation point. When the mass flow is less than that at the design point, the efficiency of the TEB models is higher than that at UGV model. When the operation condition mass flow is equal or higher than the design point, the efficiency of TEB model is lower than that at UGV model. It means that the TEB method makes the efficiency lines left moving.

With the increasing of the mass flow, the pressure ratio shows a decreasing trend. Due to the blowing flow, more mass flow leads to the lower pressure ratio. It is found that the pressure ratio line of the TEB models is lower than that in UGV models from 0.8 Q_d to 1.2 Q_d. It means that the TEB method makes the pressure ratio lines down moving.

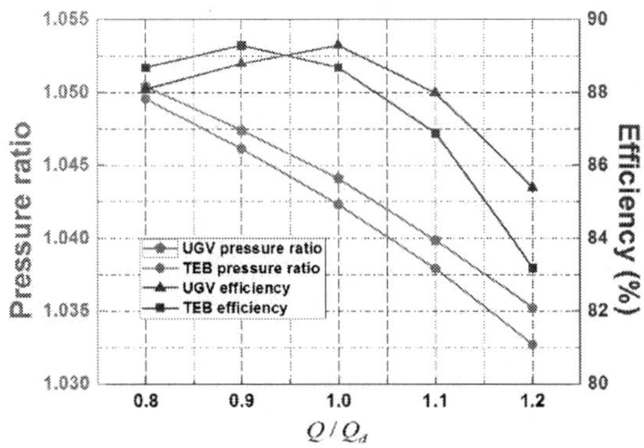

Figure 8. Comparison of the compressor characteristic performance of the UGV and TEB models

4.2 The flow field analysis in radial inlet

All the simulation results of the TEB refer to the 1.0 Q_d operating point. In order to obtain the momentumless wake style, the ring lines are set respectively at 0.1, 0.5 and 0.9 span behind the guide vane, see Fig.9. Calculating the velocity profiles on the ring lines by Eq. (2), we could estimate the wake style and adjust the blowing mass flow distribution.

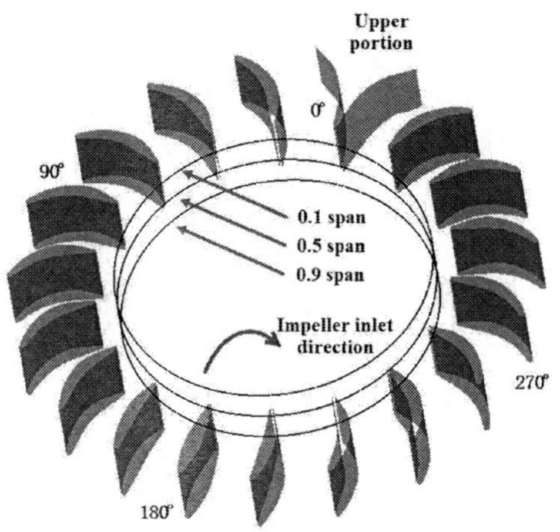

Figure 9. The locations of the ring lines

Fig. 10(a) gives the velocity distribution behind the guide vanes at different spans in the UGV model. According to the Eq. (2), the wake style is significantly influenced by the main flow velocity, and the velocity distributions are different in both circumferential and spanwise directions. The velocity is the highest at 180° because it is the nearest place to the inlet of the compressor. Since the flow channel turns direction from the radial to the axial, the velocity is the highest near the hub. The wake intensity is stronger near 90°and 270° at the 0.5 and 0.9 span, near 180°at the 0.1 span. The wake velocity distribution in radial inlet is significantly different in comparison to the axial inlet.

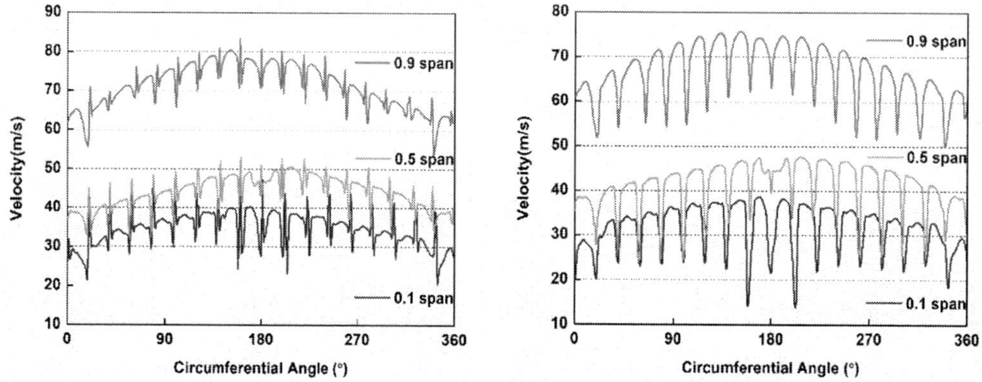

(a) (b)

Figure 10. The velocity distribution behind the guide vane at different spans of (a) the UGV model and (b) the TEB model

According to the velocity profile, if we want to obtain the momentumless wake style, the needed blowing mass flow at each guide vane is different, and the single guide vane that needed the blowing is different from hub to shroud as well. By adjusting the blowing mass flow, Fig.10 (b) gives the velocity distribution behind the guide vane at different spans in TEB model. By using Eq. (2), it is found that the value of the θ/d for each wake approaches zero, which means the momentumless wake of the guide vanes is obtained. Compared to the UGV model, the main flow velocity is higher in the TEB model.

Fig. 11 gives several contour plots at section 1-1 for the TEB model. In comparison to the UGV results in Fig.5, by adding the trailing edge blowing on the guide vanes, the flow field upstream of the impeller has changed a little. Due to the blowing flow, the velocity has increased and the pressure has decreased, the low total pressure region has reduced, especially at the shroud region, which means the wake intensity has decreased.

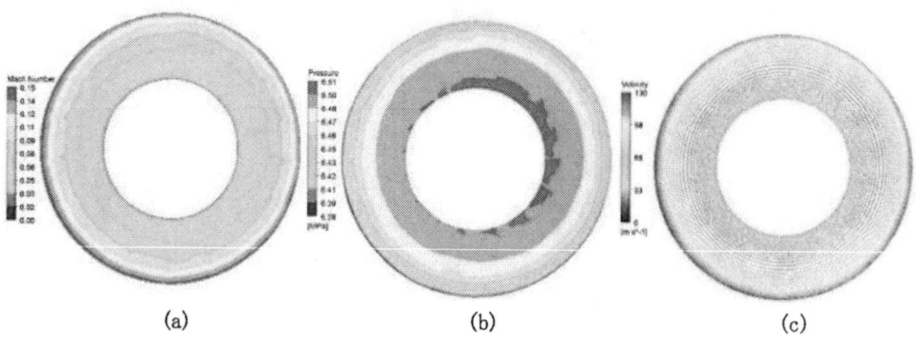

(a) (b) (c)

IOP Conf. Series: Materials Science and Engineering 1081 (2021) 012040 doi:10.1088/1757-899X/1081/1/012040

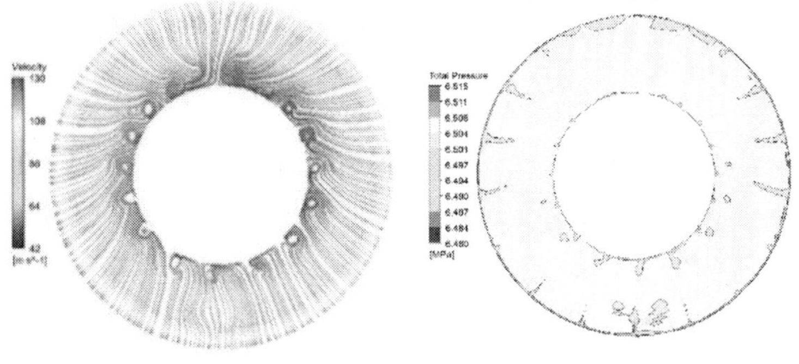

(d) (e)

Figure 11. The distributions of (a) Mach number, (b) pressure, (c) vector, (d) streamline and (e) total pressure at the impeller inlet of the TEB model

Because of the aerodynamic load on the impeller is determined by the total pressure and the flow angle at the outlet of the radial inlet, Fig. 12 gives the flow angle (0°represents the axial direction) and the total pressure ratio (defined as the total pressure divided by the inlet total pressure) distributions of the UGV and the TEB models. The absolute flow angles at the leading edge are close to zero, and total pressure ratios are close to 1, which means the radial inlet performance well. The flow angle and the total pressure ratio show obvious fluctuations at different spans, and the number of fluctuations is equal to the number of guide vanes, which means that the fluctuation is cause by the guide vane wake. There is obviously value change at 0°, which is due to the flow mixing from the left and right sides at top of the radial inlet. Compared with the UGV model, the TEB model shows lower fluctuation intensity due to the trailing edge blowing, especially at the impeller shroud region, which is the region significantly affected by the guide vane wake.

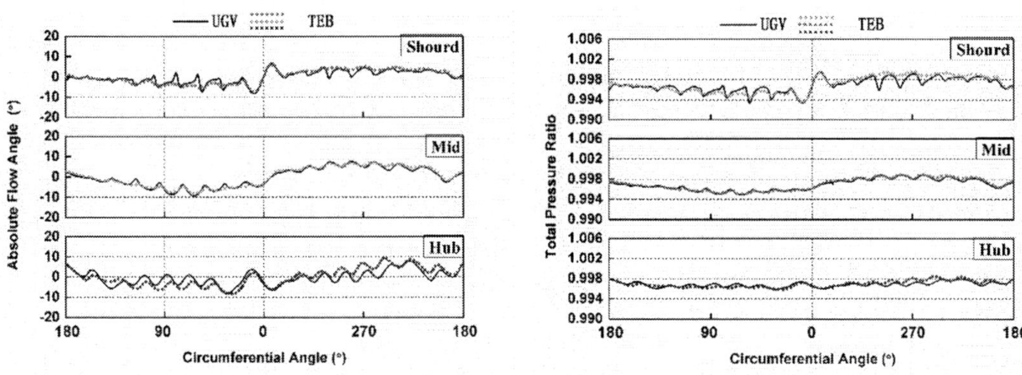

(a) (b)

Figure 12. Comparison of the (a) flow angle and (b) total pressure ratio distribution at the impeller leading edge of the UGV and TEB models

4.3 The aerodynamic load analysis

In order to get accurate and stable monitoring data, all the data are chosen after five revolutions in each unsteady simulation case. Fig. 13 presents the comparison of the time history oscillations (the last two cycles) at the tip monitor 11 on two sides of the impeller of the UGV and TEB models.

IOP Conf. Series: Materials Science and Engineering 1081 (2021) 012040 doi:10.1088/1757-899X/1081/1/012040

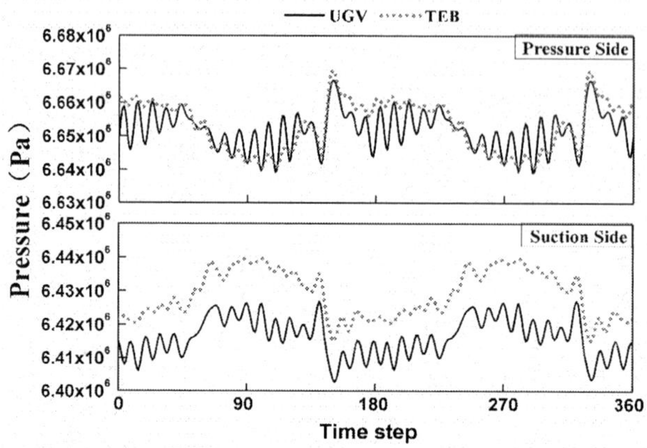

Figure 13. Comparison of the history oscillations at monitor 11 on the two sides of impeller of the UGV and TEB models

The pressure data has an obvious periodicity and stability for two models, and each wavelength corresponds to the time of one impeller revolution. Compared to the suction side, the mean pressure on the pressure side is very high, as this side is the working surface. The trend of the pressure date is almost the same of the two models. Each revolution has 18 fluctuations on both sides on the impeller, which is due to the wake of the guide vanes. The fluctuations in the TEB model are obviously weaker than that in the UGV model, especially on the pressure side. which means that the trailing edge blowing has positive effect on wake.

In order to analyze the aerodynamic data, the frequency domain of the pressure fluctuation of different monitors is calculated by means of the FFT. Fig.14 gives the frequency domain of the aerodynamic load at monitor 11 in two models. Compared to the UGV model, the amplitudes at f_m are a little higher in the TEB model, which is because the extra flow result in the velocity of the main flow has increased, and a little strengthens the non-uniformity along the circumference direction. However, the amplitudes at f_i are significantly lower in the TEB model, and reduction of the value is about 90 percent, which means that the trailing edge blowing has positive effect on the wake, and reduces the aerodynamic load at f_i on the impeller. Meanwhile, the amplitudes at harmonic from 16 to 20 have decreased as well.

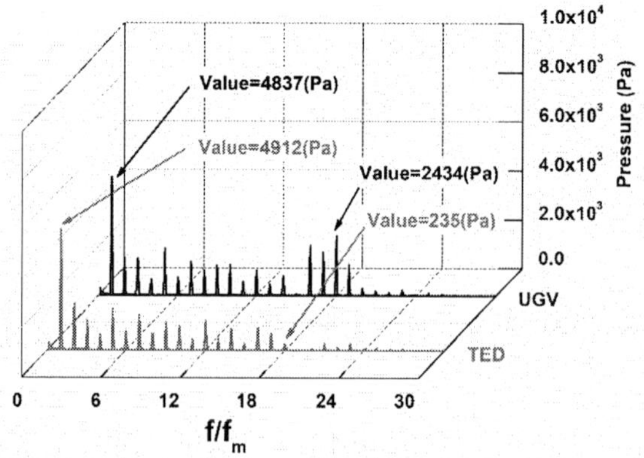

Figure 14. Comparison of the frequency harmonic of pressure fluctuations at monitor 11 of the UGV and TEB models

Fig. 15 gives the aerodynamic load distribution at f_m and f_i from hub to shroud in two models. The amplitude values at f_m have little difference of the two models; however, the amplitude values at f_i in the TEB model are obviously lower than that in the UGV model, especially at the mid and tip region of the impeller. It means that the trailing edge blowing is an effective way to decrease the wake deficit and reduce the aerodynamic load at f_i on the impeller, and further improves the safety of the impeller.

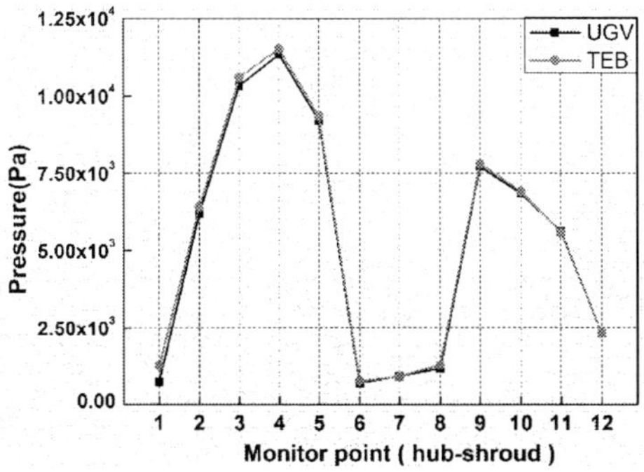

(a)

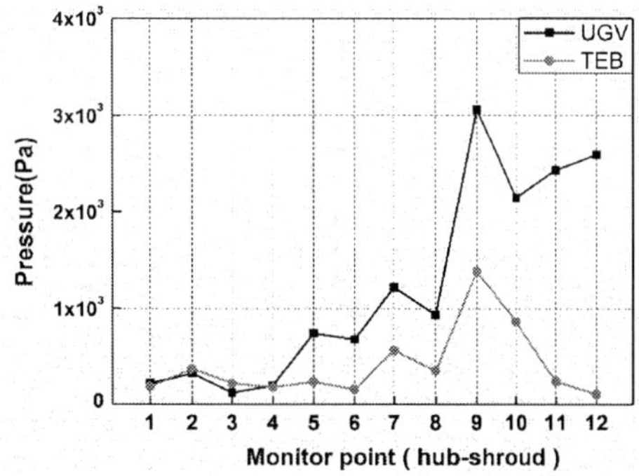

(b)

Figure 15. Comparison of the amplitude values at (a) f_m and (b) f_i at the impeller leading edge from hub to shroud of the UGV and TEB models

5. CONCLUSION

The radial inlet is one of the most crucial parts of a centrifugal compressor. Combined with the trailing edge blowing method, the UGV model of the compressor inlet is analyzed numerically in this paper. Based on the compressor working condition, the characteristic performance and the flow field of two models (UGV and TEB) are analyzed. Unsteady simulations are conducted to obtain the aerodynamic load on the impeller leading edge. The consideration of the aerodynamic load due to the trailing edge blowing method could help the designer to analysis similar questions. According to the simulation results, the conclusions are listed below:

(1) Setting guide vanes in the radial inlet is a good way to improve the flow field, decrease the flow losses and the distortion upstream of the impeller. However, the guide vanes increase the total pressure loss due to the wake, and the wake results in the aerodynamic load excitation at f_i on the impeller, which affects the safety of the impeller.

(2) The trailing edge blowing is an active control method, the wake style is related to the velocity profile behind the guide vane. In the UGV model, the main flow velocity behind the guide vanes is different along the spanwise and circumference directions. Therefore, by adjusting the blowing mass flow in both directions, the momentumless wakes are obtained.

(3) Compared to the UGV model, the characteristic line has changed due to the extra blowing mass flow in the TEB model. It makes the efficiency lines left moving and the pressure ratio lines down moving. The blowing flow leads the main flow increased as well.

(4) The flow field at impeller inlet has improved in the TEB model, the total pressure and the absolute angle fluctuations are lower due to the trailing edge blowing. By the FFT, the TEB model shows a little influence on the aerodynamic load at f_m, but shows obviously lower at f_i, especially at the mid and the tip region of the impeller. It means that the trailing edge blowing is an effective way to improve the safety of the impeller.

6. ACKNOWLEDGMENTS

The authors appreciate the supported by Fundamental Research Funds for the Central Universities (3132020122), National Natural Science Foundation of China (No. 52006021) and Natural Science Foundation of Liaoning Province (No. 2020-BS-069)

REFERENCES

[1] Engeda A, Kim Y, Aungier R. 2003 The Inlet Flow Structure of a Centrifugal Compressor Stage and Its Influence on the Compressor Performance. ASME Journal of Fluids Engineering, **125**: 779- 785.

[2] Tan J, Wang X, Qi D, 2011 et al. The effects of radial inlet with splitters on the performance of variable inlet guide vanes in a centrifugal compressor stage[J]. Proceedings of the Institution of Mechanical Engineers, Part C: Journal of Mechanical Engineering Science, **225**(9):2089-2105

[3] Tan J, Wang X, Liu J, 2014 et al. Review of the Study on the Radial Inlet of Centrifugal Compressor[J]. Compressor, Blower & Fan technology, (**4**):63-70

[4] Long F, Tan J. 2010 The Influence of Radial Inlet Volute on the Stage Performance of Centrifugal Compressor[J]. Compressor, Blower & Fan technology, (**2**):7-11

[5] Dong F, Yu Y, Zhu X. 2012 Influences of the Radial inlet Chamber on the Characteristics of a Centrifugal Compressor [J]. Fluid Machinery, ,**40**(11):15-20

[6] Wu D, Wang L, Chen J, 2010 et al. The Design Method of Radial Inlet Volute in Centrifugal Compressor[J]. Compressor, Blower & Fan technology, (**1**):23-28

[7] Han, F., Qi, D., Tan, J., Wang, L., 2012 et al. Experimental and Numerical Investigation of the Flow Field in the Radial Inlet of a Centrifugal Compressor[C]. Proc. ASME Turbo Expo 2012: Turbine Technical Conference and Exposition, American Society of Mechanical Engineers, 831-845.

[8] Feng Z, Gao Y, Song Q. 2016 Numerical Simulation Research of the Impact of Radial Inlet Chambers on the Performance of Impellers[J]. Compressor, Blower & Fan technology, (**3**):37-41

[9] Wang R, Qi D, Wang X, 2009 et al. Flow loss for the Radial Suction Chamber of Centrifugal Compressor improved Design[J]. Chinses Journal of Applied Mechanics, ,**26**(3):437-443

[10] Chen Z, Gu C, Shu X, **2010** et al. Shape Optimum Design for Centrifugal Compressor Radial Inlet Based on CFD Technique[J]. Journal of Mechanical Engineering, **46**(14): 124-129

[11] Wang Y, Liu Y, Guo T. 2016 Research Progress in Unsteady Flow Characteristics and Dynamic Response of Centrifugal Compressor Impellers[J]. Compressor, Blower & Fan technology, (**5**):81-87

[12] Han F, Mao Y, Tan J. 2017 Effects of Flow Loss and Outlet Distortions Caused by Radial Inlet[J]. Journal of Engineering Thermophysics, (**10**):2125-2129

[13] Xin J, Wang X, Zhou L, 2016 et al. Numerical Investigation of the Flow Field and Aerodynamic Load on Impellers in Centrifugal Compressor with Different Radial Inlets[C] Asme Turbo Expo: Turbine Technical Conference & Exposition. GT**2016**-57180

[14] Dring, R., Joslyn, H., Hardin, L., Wagner, J. 1982 Turbine rotor-stator interaction[J]. Journal of Engineering for Gas Turbines and Power, (**4**): 729-742.

[15] Park W, Cimbala J. 1991 The effect of jet injection geometry on two-dimensional momentumless wakes [J]. Journal of Fluid Mechanics, (**224**):29-47

[16] Lewis B J, Cimbala J M, Wouden A M. 2012: Investigation of distributor vane jets to decrease the unsteady load on hydro turbine runner blades. [C]. Proceedings of the Proc 26th IAHR Symposium on Hydraulic Machinery and Systems, Beijing China. **2006.**

[17] Xin J, Wang X, Chen X, 2019 et al. Numerical investigation of variable inlet guide vane with trailing-edge blowing in centrifugal compressor[J]. Journal of Harbin Unversity,2018,39(6):1006-1011

[18] Ansys, A. F.,, "19.0 Theory Guide," ANSYS inc.

IOP Conf. Series: Materials Science and Engineering 1081 (2021) 012041 doi:10.1088/1757-899X/1081/1/012041

A Study of the Contact Interface for Compressor Blisks with Ring Dampers Using Nonlinear Modal Analysis

Yekai Sun, Jie Yuan, Enora Denimal and Loïc Salles
Department of Mechanical Engineering, Imperial College London
London SW7 2AZ, UK

E-mail: ys5113@imperial.ac.uk

Abstract. The integrally bladed disks, also known as blisks, have been widely used in industrial turbomachinery because of their benefits in aerodynamic performance and mass reduction. Friction damping is considered as the major damping sources in turbomachinery. However, in blisks, the friction damping is negligible due to the lack of the contact interfaces. The friction ring dampers are one of the emerging external damping sources for blisks. In this paper, a full-scale blisk with a friction ring damper is studied, where a 3D contact element is used to compute the contact frictions. The blisk and ring damper is investigated using their damped nonlinear normal modes. The modal damping can be directly calculated and used to quantify the friction damping generated by the ring damper. The contact behaviour within the contact interface is further analysed. The nodes with initial gap show less damping ability. The separations within the contact interface are expected to be avoided to achieve a better damping performance.

1. Introduction

The bladed disk is an important structural component in industrial turbomachinery. The bladed disks are almost designed to their structural limit due to the higher efficiency required. In turbomachinery, the bladed disks are usually working under extreme environment, i.e. high thermal stress, centrifugal stress and vibration loading. The large vibrational stress on the bladed disks leads to the high cycle fatigue which is considered as the major reason for most failures of aircraft engines. In industrial turbomachinery, the material damping and aerodynamic damping are relatively low. Therefore, the external damping source is required to reduce the vibrational amplitude under resonance. The friction damping is a common and effective damping technique used in industrial turbomachinery. The friction dampers have been well-established in literature [5, 12, 18].

However, the use of blisks has become popular in the industry. The traditional bladed disk consists of a single disk and with several individual blades. In blisks, the structure is manufactured by a single-piece component. The blisks usually have a lower mass and better aerodynamic performance compared to the traditional bladed disks. However, due to lack of the contact interfaces, the internal damping of the blisks is extremely low and extra contact surfaces are expected to increase the friction damping. One of the emerging techniques is the use of the friction ring damper. A friction ring is held within the groove of the blisk by the centrifugal

IOP Conf. Series: Materials Science and Engineering 1081 (2021) 012041 doi:10.1088/1757-899X/1081/1/012041

force. The friction ring damper is known as an efficient damping technique in other rotating structures, i.e. labyrinth seals [11] and gears [24, 22].

In industrial turbomachinery, the dynamic analysis of the blisks with friction dampers has been attempted by many researchers. Laxalde et al. proposed a lumped parameter model and the influences of the mass of the ring damper have been studied using forced response analysis [6]. Sun et al. applied the nonlinear modal analysis on the same lumped parameter model [17]. Full-scale structure for the blisk with ring damper has been numerically analysed and studied in literature by many researchers [9, 1, 10, 19]. The experimental works were completed by Laxalde et al. and results are compared between the experiments and numerical simulations [7]. The influence of the geometry on V-shaped ring damper has been studied by Tang and Epureanu [21]. Most researches for the ring damper focused on the damping ability of the ring damper. The contact interfaces between the blisk and ring damper have seldom been studied in literature. In this paper, a full scale blisk with friction ring damper is considered as the academic test case to study the contact behaviours within the contact interfaces between the blisk and ring damper. In addition, based on the results from this work, some guidelines of the design of ring damper are proposed.

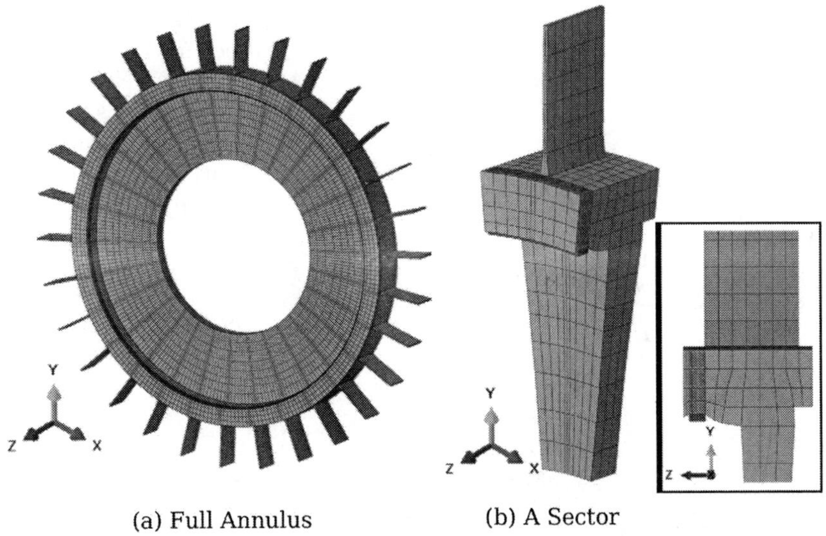

Figure 1. Finite Element Model: (a) Full Annulus of the Blisk and Ring Damper (Red); (b) A Sector of the Blisk and Ring Damper (Red).

The dynamic analysis of the blisk with ring damper is achieved by nonlinear modal analysis. The nonlinear modal analysis has been well established in past several decades. The nonlinear modes were defined by Rosenberg [14] in a conservative nonlinear system. Then, Shaw and Pierre extended the nonlinear modes to non-conservative nonlinear systems using the concept of invariant manifold [16]. The nonlinear modes for a non-conservative nonlinear system were named as damped Nonlinear Normal Mode (dNNM). After that, the extension of concept of complex nonlinear mode was described by Laxalde et al. [8], which can be applied to compute the dNNMs for systems with non-conservative nonlinearities, i.e. system with friction joints. After that, a numerical method, namely Extended Periodic Motion Concept (EPMC), was proposed by Krack [4]. In this method, the dNNMs are numerically calculated based on a periodic concept. In this work, the nonlinear modes of the system are calculated using the EPMC.

IOP Conf. Series: Materials Science and Engineering 1081 (2021) 012041 doi:10.1088/1757-899X/1081/1/012041

The present work aims to study the contact behaviours within the contact interface between the blisk with ring damper using nonlinear modal analysis. The numerical method EPMC is used to calculate the dNNMs, which can be considered as the resonant solutions for forced response [20]. The paper is organised as follows: the Finite Element (FE) model used is firstly given; then, the modelling methodology is described; after that the numerical methods used to compute the dNNMs are briefly described; the contact behaviours within the contact interface are investigated; finally, some guidelines for the design of the ring dampers are proposed.

2. Finite Element Model

A full-scale FE model is used in this work to represent the blisk and ring damper, as shown in Fig.1(a). This FE model is considered as the test case to study the contact behaviours of the ring damper. The model is built in a commercial FE solver ABAQUS. The whole structure is made by a homogenous and isentropic titanium material. The blisk and ring damper are assumed to be perfectly cyclic symmetry and modelled by a sector with cyclic symmetric boundary conditions as shown in Fig.1(b). There are 30 sectors in total. The ring damper is held to the underneath of the blisk.

3. Modelling Methodology

The FE model for the blisk and ring damper has been described in the previous section. The objective of this section is to describe the methodology used in the present work. Firstly, a linear modal analysis of the blisk is completed and a specific mode is selected for the subsequent studies. Then, the rotor dynamic analysis is applied to this selected mode and a resonance case under engine order excitation is obtained. After that, the resonance case is analysed by a nonlinear static analysis. The contact pressure/gap distribution within the contact interface is computed from the nonlinear static analysis and is considered as preloading conditions for the nonlinear modal analysis. The linear modal analysis, rotor dynamic analysis and nonlinear static analysis are completed in commercial FE solver ABAQUS.

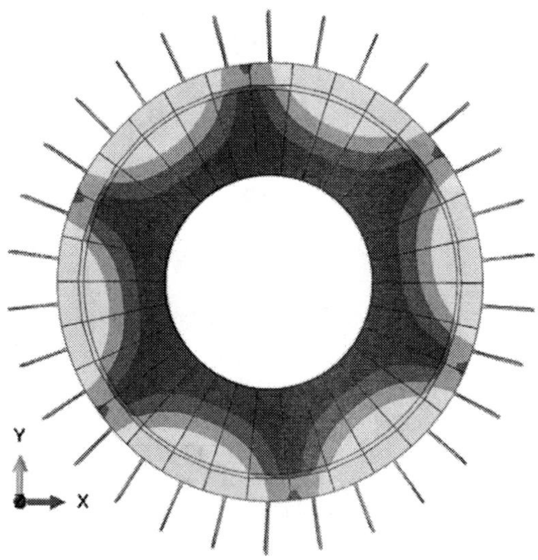

(a) Full Annulus

(b) A Sector

Figure 2. Modeshape of the Mode Studied: (a) Full Annulus; (b) A Sector.

After applying the linear modal analysis, the family of the 1st disk mode with nodal diameter 3 is selected as the test case as shown in Fig.2. The modeshape of the selected mode is highly dominated by the motion of the disk. When the engine rotating speed is 2081 rpm and 27th engine order excitation is applied, then selected mode is excited. This resonance case is considered for the subsequent study. The contact pressure/gap distributions are obtained by the nonlinear static analysis and shown as a map on contact interface in Fig.2. The contact pressure/gap distributions show the initial static contact status for the contact nodes pairs within the contact interface. When the contact pairs show a positive gap, the nodes are separated with the initial gap (i.e. the nodes near the cyclic boundary). The contact pressure is concentrated on the middle of the contact interface. These nodes with positive contact pressure are generally in-contact in a static equilibrium status.

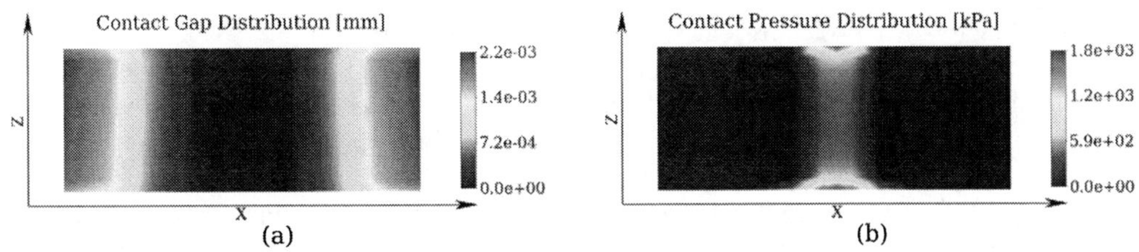

Figure 3. Contact Preloading within the Contact Interface for the Selected Resonance Case: (a) Contact Gap Distribution; (b) Contact Pressure Distribution.

4. Nonlinear Modal Analysis

The description of the FE model and modelling methodology have been given in previous sections. In the present section, the numerical formulations of the nonlinear modal analysis are demonstrated. Firstly, the mass and stiffness matrices of the full-scale FE model is reduced by Craig-Bampton method [3] and cyclic symmetric reduction [13]. The computational expenses can be reduced to an acceptable level. Then, the reduced mass and stiffness matrices are used in nonlinear modal analysis. The contact friction forced is calculated based on a well-established 3D contact element [23], which is a node-to-node contact model. In this contact model, there are normally three contact conditions, which are sticking, separation and siding.

The equation of motion for an autonomous (undamped and unforced) nonlinear system has been given in Eq.(1), where $\underline{\underline{\mathbf{M}}}$ and $\underline{\underline{\mathbf{K}}}$ are the reduced mass and stiffness matrices; $\underline{Q}(t)$ is the displacement of the system; $\underline{F}_{nl}(\underline{\dot{Q}}, \underline{Q})$ is the contact friction calculated from the 3D contact element in [23]. The total energy of the system is dissipated due to the rubbing motions within the contact interface. Therefore, the system is non-conservative and solutions are not periodic. According to the EPMC, an artificial damping term, $\underline{\underline{\bar{\mathbf{C}}}} = -\zeta\omega_0\underline{\underline{\mathbf{M}}}$, is introduced into the system to balance the energy dissipated by the contact frictions, where ω_0 is the resonant frequency and ζ is the modal damping ratio. In addition, based on the concept of nonlinear modes, the dNNMs are energy dependent. In other words, the dNNMs and other modal properties (i.e. resonant frequency and modal damping ratio) vary with the energy level of the system. Therefore, a modal amplitude α is introduced and the solutions of the system can be represented as: $\underline{Q} = \alpha \cdot \underline{Q}_0$. Then, the equation of motion is rewritten and shown in Eq.(2). There are two extra constraints required to solve the equation below which are a mass normalisation and phase normalisation, which can be found in [4].

$$\underline{\underline{\mathbf{M}}} \cdot \underline{\ddot{Q}}(t) + \underline{\underline{\mathbf{K}}} \cdot \underline{Q}(t) + \underline{F}_{nl}(\underline{\dot{Q}}, \underline{Q}) = 0 \tag{1}$$

$$\underline{\underline{\mathbf{M}}} \cdot \alpha \cdot \ddot{\underline{Q}}_0(t) + \underline{\underline{\bar{\mathbf{C}}}} \cdot \alpha \cdot \dot{\underline{Q}}_0(t) + \underline{\underline{\mathbf{K}}} \cdot \alpha \cdot \underline{Q}_0(t) + \underline{F}_{nl}(\alpha, \dot{\underline{Q}}_0, \underline{Q}_0) = 0 \quad (2)$$

The numerical method used to solve the Eq.(2) is the harmonic balance method (HBM) with alternating frequency/time (AFT) method [2]. The HBM with AFT has been widely used to solve nonlinear dynamic problems in frequency domain. In nonlinear modal analysis, the dNNMs are expected to be computed for a range of modal amplitude. Therefore, the well-known continuation method is used to track the evolution of the system against a specific chasing parameter [15].

5. Results

The mode in 1st disk family with nodal diameter 3 is excited by engine order 27 excitation under engine rotational speed 2081 rpm. This resonance case is considered for the nonlinear modal analysis. The dNNMs and other modal properties are computed for a range of modal amplitude. To achieve a physical description of the modal amplitude, the displacement of the tip of the blade is used to represent the vibration amplitude.

The evolution of the resonant frequency ω_0 is demonstrated in Fig.4(a). When the structure is vibrating at low amplitude (the displacement is smaller than 1.5×10^{-2} mm), the ring damper is stuck to the underneath of the blisk. The resonant frequency ω_0 is a constant and equals to the sticking natural frequency of the structure [18]. When the vibration amplitude is greater than 1.5×10^{-2} mm, the contact pairs within the contact interface start to slide leading a softening effect (decrease in resonant frequency ω_0). The evolution of the modal damping ratio ζ is demonstrated in Fig.4(b). At the low vibration amplitude, the contact pairs are either sticking or separating (depending on the initial contact status). Further increasing the vibration amplitude, some contact pairs start to slide and the energy is dissipated by the rubbing motion between the contact nodes in contact pairs. The maximum modal damping ζ is around 0.1%.

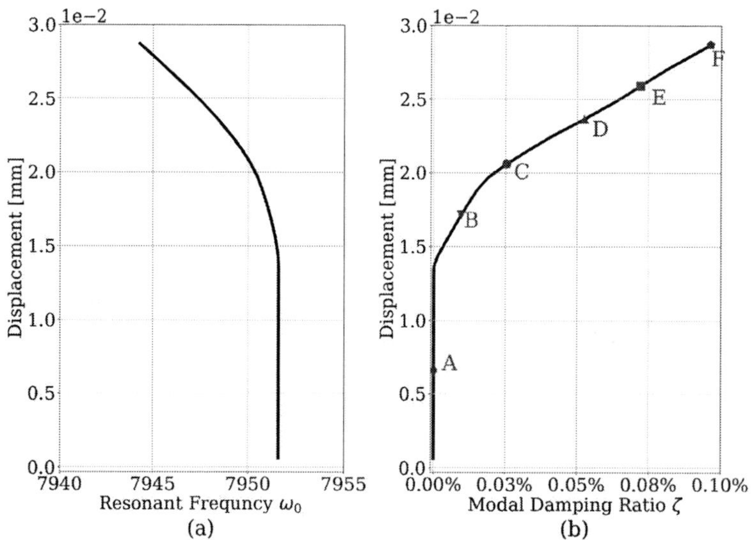

Figure 4. Results from Nonlinear Modal Analysis: (a) Resonant Frequency Plot; (b) Modal Damping Ratio Plot.

To further understand the contact phenomena within the contact interface, six cases with different vibration amplitudes are selected and marked in the Fig.4(b). The distributions of the energy dissipation within the contact interface are computed and shown in Fig.5. For each contact node pair, the energy dissipated by one vibration period is the area enclosed by the hysteresis loop. For Case A, the modal damping ratio is zero and no energy dissipated shown on the map. By considering rest of the cases, the modal damping ratio increases with the vibration displacement (see Fig.4(b)). It is obvious to notice that the nodes in the middle of the contact interface show a higher damping ability. Most of the vibrational energy is dissipated by the nodes in the middle of the contact interface. The nodes near cyclic boundaries have negligible energy dissipated. The nodes in the middle are in-contact at the static equilibrium status whereas the nodes near cyclic boundaries show initial gaps. When the blisk starts to vibrate, these in-contact nodes transit from the sticking condition to the sliding condition. The energy of the system is dissipated due to the rubbing motion of the sliding nodes within the contact interface. However, the contact conditions for these separated nodes are not completely sliding. During one vibration period, the nodes show both separation and sliding. Due to existence of the separation part, the energy dissipated be these nodes are relatively lower than the sliding nodes. The similar results can be found in [18].

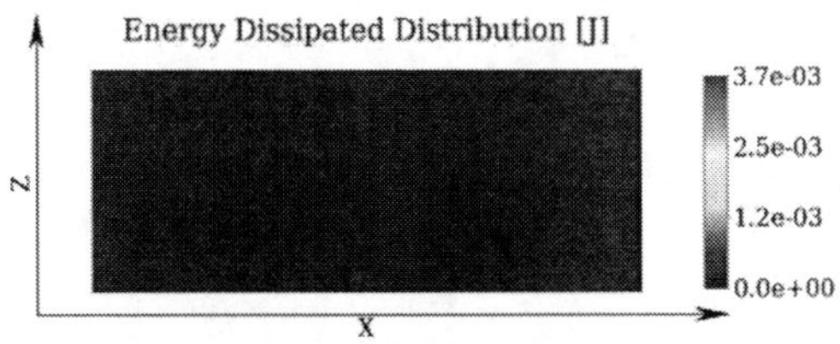

(a) Case A

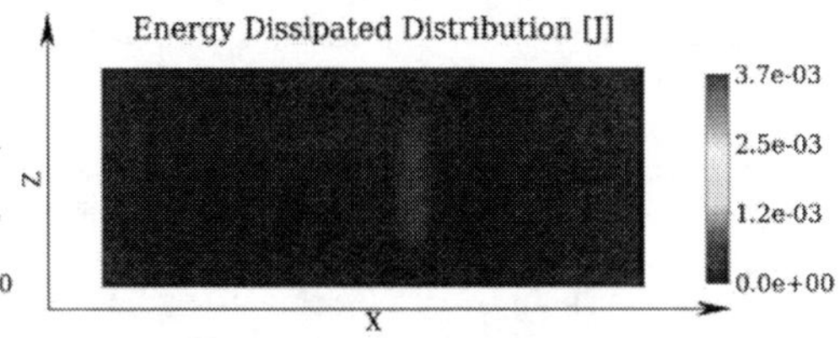

(b) Case B

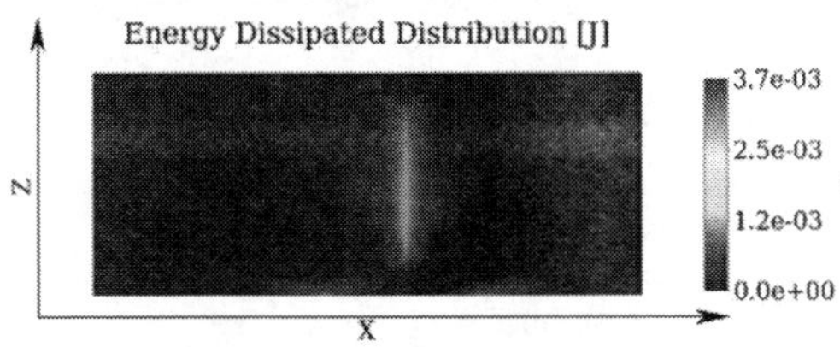

(c) Case C

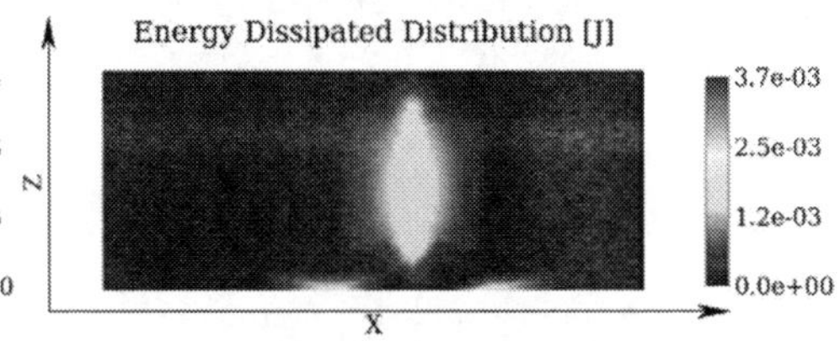

(d) Case D

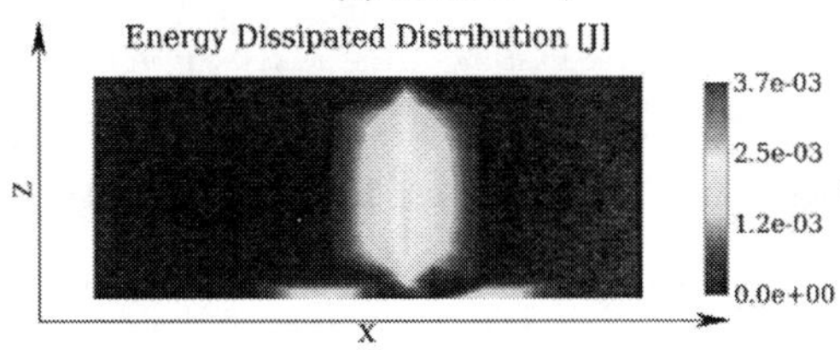

(e) Case E

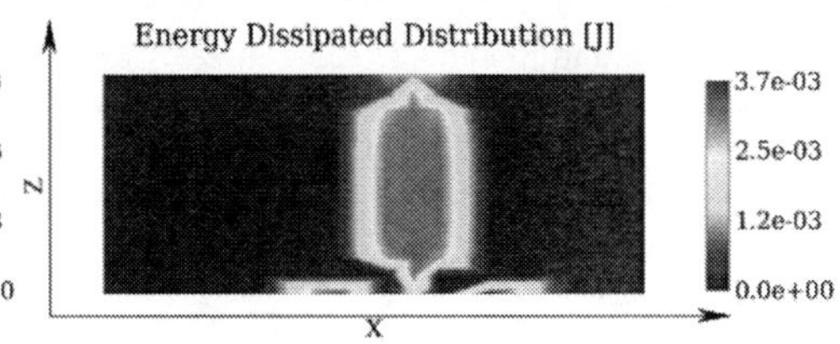

(f) Case F

Figure 5. Energy Dissipation Distributions for Six Cases Shown in Fig.4(b).

6. Conclusions

A full-scale finite element model is used in this paper to represent the blisk and ring damper. This blisk with ring damper is studied using nonlinear modal analysis. In nonlinear modal analysis, the evolutions of the resonant frequency and modal damping ratio are directly computed to understand the dynamic behaviour and damping performance of the ring damper. The contact behaviours and the maps of the energy dissipated within the contact interface are investigated. The results show that the nodes with initial gap has less damping ability. To improve the damping performance of the ring damper, it is important to avoid the separation within the contact interface. The geometric design of the ring damper can be considered as the future work to optimise the damping performance by manipulating the initial contact status.

Acknowledgement

The first author is grateful to China Scholarship Council (File NO. 201708060239) for providing the financial support for this project.

References

[1] Baek S and Epureanu B I, 2017, Reduced-order modeling of bladed disks with friction ring dampers, *Journal of Vibration and Acoustics*, **139**(6), p. 061011.
[2] Cameron T M and Griffin J H, 1989, An alternating frequency/time domain method for calculating the steady-state response of nonlinear dynamic systems, *Journal of Applied Mechanics*, **56**(1), pp. 149–154.
[3] Craig JR R R, and Bampton M C C, 1968, Coupling of substructures for dynamic analyses, *AIAA Journal*, **6**(7), pp. 1313–1319.
[4] Krack M., 2015, Nonlinear modal analysis of nonconservative systems: Extension of the periodic motion concept, *Computers & Structures*, **154**, pp. 59–71.
[5] Krack M, Salles L and Thouverez F, 2016, Vibration Prediction of Bladed Disks Coupled by Friction Joints, *Archives of Computational Methods in Engineering*, **24**, pp. 589–636.
[6] Laxalde D, Sinou J J, Thouverez F, and Lombard J P, 2007, Qualitative analysis of forced response of blisks with friction ring dampers, *European Journal of Mechanics A/Solids*, **26** (4), pp. 676–687.
[7] Laxalde D, Gibert C and Thouverez F, and Lombard J P, 2008, Experimental and numerical investigations of friction rings damping of blisks, *ASME. Paper No.GT2008-50862.*
[8] Laxalde D, Salles L, Blanc L and Thouverez F, and Lombard J P, 2008, Non-Linear Modal Analysis for Bladed Disks With Friction Contact Interfaces, *ASME. Paper No. GT2008-50860.*
[9] Laxalde D, Thouverez F and Lombard J-P, 2010, Forced response analysis of integrally bladed disks with friction ring dampers, *Journal of Vibration and Acoustics*, **132**(1), p.011013.
[10] Lupini A, Mitra M and Epureanu B I, 2019, Application of tuned vibration absorber concept to blisk ring dampers: a nonlinear study, *Journal of Engineering for Gas Turbines and Power*, **141**(10), p.101016.
[11] Niemotka M A and Ziegert J C, 1995, Optimal design of split ring dampers for gas turbine engines, *Journal of Engineering for Gas Turbines and Power*, **117**(3), pp.569–575.
[12] Pesaresi L, Armand J, Schwingshackl C W, Salles L and Wong C, 2018, An advanced underplatform damper modelling approach based on a microslip contact model, *Journal of Sound and Vibration*, **436**, pp.327–340.
[13] Petrov E P, 2004, A method for use of cyclic symmetry properties in analysis of nonlinear multiharmonic vibrations of bladed disks, *Journal of Turbomachinery*, **126**(1), pp.175–183.
[14] Rosenberg R M, 1960, Normal modes of nonlinear dual-mode systems, *Journal of Applied Mechanics*, **27**, pp. 263–268.
[15] Sarrouy E and Sinou J J, 2011, *Advances in Vibration Analysis Research* (Rijeka: nTech), pp. 419–434.
[16] Shaw S W and Pierre C, 1991, Non-linear normal modes and invariant manifolds, *Journal of Sound and Vibration*, **150**(1), pp. 170–173.
[17] Sun Y, Yuan J, Pesaresi L and Salles L, 2018, Nonlinear vibrational analysis for integrally bladed disk using friction ring damper, *Journal of Physics: Conference Series*, **1106**, p.012026.
[18] Sun Y, Yuan J, Pesaresi L, Denimal E and Salles L, 2020, Parametric study and uncertainty quantification of the nonlinear modal properties of friction dampers, *Journal of Vibration and Acoustics*, **142**(5), p.051102.
[19] Sun Y, Yuan J, Denimal E and Salles L, 2020, Nonlinear modal analysis of friction ring damper for compressor blisk, *ASME. Paper No. GT2020-14277.*
[20] Sun Y, Vizzaccaro A, Yuan J and Salles L, 2020, An extended energy balance method for resonance prediction in forced response of systems with non-conservative nonlinearities using damped nonlinear normal mode, *Nonlinear Dynamics.*
[21] Tang W and Epureanu B I, 2019, Geometric optimization of dry friction ring dampers, *International Journal of Non-Linear Mechanics*, **109**, pp.40–49.
[22] Wang Y, Ye H, Jiang X, and Tian A, 2018, A prediction method for the damping effect of ring dampers applied to thin-walled gears based on energy method, *Symmetry*, **10**(12), p.677.
[23] Yang B D and Menq C H, 1998, Characterization of 3d contact kinematics and prediction of resonant response of structures having 3d friction constraint, *Journal of Sound and Vibration*, **217**(5), pp.909–925.
[24] Zucca S, Firrone C M and Facchini M, 2010, A method for the design of ring dampers for gears in aeronautical applications, *ASME. Paper No.IMECE2010-38788.*

Numerical Study on Particle Deposition Characteristics of Impingement Effusion Cooling Structure

Bingran Li[1], Cunliang Liu[1,2], Lin Li[1], Wenbin Chen[3], Yihong He[3]

1. School of Power and Energy, Northwestern Polytechnical University, Xi'an Shaanxi, P. R. China

2. Shaanxi Key Laboratory of Thermal Sciences in Aero-engine System, Xi'an Shaanxi, P. R. China

3. China Aviation Engine Group Co., Ltd. Hunan Power Machinery Research Institute, Zhuzhou, Hunan, P. R. China

Abstract. The paper investigates the effect of impingement velocity and particle diameters on the flow and solid-phase deposition characteristics of the impingement effusion cooling structure by numerical simulation of gas-solid two-phase flow using DPM with UDF. The results show that under the same condition of the particle diameter, both the particle deposition flux and the particle deposition rate become increase with the increasing of the impingement velocity while the dimensionless particle deposition rate increases first and then decreases; Under the same condition of the impingement velocity, with the increase of the particle diameter, the particle deposition flux increases, while the particle deposition rate and the dimensionless deposition rate of the particle both increase first and then decrease. Taken together, Brownian motion of particles, turbulent diffusion, and particle inertia and flow characteristics is the main factors that affect the solid-phase particle deposition behaviour.

Keywords. Impingement effusion structure, Gas-solid two-phase flow, Particle deposition rate, Dimensionless particle deposition rate, Numerical simulation. The first section in your paper.

1. Nomenclature

$D=0.5$	The diameters of impingement hole and film hole [mm]
H	Impingement spacing [mm]
L	Spanwise spacing of impingement chamber [mm]
W	Streamwise spacing of film hole [mm]
P	Spanwise spacing of film hole [mm]
W_i	Spacing between row of impingement holes [mm]
P_i	Spacing between impingement holes in the same row [mm]
St	Particle stokes number
μ	Dynamic viscosity [N·s/m2]
u_∞	The velocity of flow [m/s]

τ_p	Particle relaxation time [s]
τ_f	Gas relaxation time [s]
ρ_d	Density of particle [kg/m3]
D_p	The diameters of particle [μm]
l	Channel characteristic length [mm]

2. Introduction

During the operation of an aviation gas turbine engine, solid particle impurities entrained in the air entering the engine will affect the normal operation of the engine [1]. Especially when working in a dusty geographic environment or encountering a dusty climate, aero engines will inhale a large amount of sand with a particle size ranging from 1 to 1000μm [2-3]. Sand particles intrude into the cooling passages of the turbine blades, which are prone to deposition and blockages in the narrow internal cooling passages, causing the cooling effectiveness of the turbine blades to decrease, thereby increasing the temperature of the turbine blades. These will affect the service life, working performance and safe operation of aero-engines [4].

With the adaptability of aircraft flying in volcanic and desert environments, the deposition of particulate matter on the turbine blade wall has attracted the attention of researchers from various countries, so the more systematic basis and engineering application research have been carried out. Foreign research started early, and the main research methods include experimental research and numerical simulation. Jenson et al. [5] designed a High Temperature Accelerated Deposition Device (TADF) to perform accelerated experiments on particle deposition by increasing the particle concentration in the turbine environment. Ai [6] used accelerated deposition equipment to experimentally study the dust deposition of turbine blades using film cooling, and obtained the dust deposition under different blowing ratios. Shah et al. [7] used large eddy simulation to simulate the impingement of particle diameter on the distribution of particle impingement velocity and angle on the wall surface of ribbed channels and the law of particle energy transmission. Jonathan et al. [8] studied the erosion and deposition characteristics of particles in the channels of a cross-arrangement cylindrical array. The gas-solid two-phase flow model was used to investigate the particle erosion and deposition characteristics in the spoiler channel under different particle Stokes numbers and different softening temperatures. Zhou et al. [9-11] numerically simulated the particle deposition near the turbine cascade and the film hole, studied the particle motion and deposition characteristics, and revealed the flow and heat transfer laws. Yang et al. [12-13] studied the deposition of particulate matter on the surface of turbine blades and cascades and the effect of deposition on film cooling. Pei et al. [14] proposed a new deposition model based on the critical viscosity model and critical velocity model, and studied the effect of mainstream temperature and inlet/outlet pressure ratio on the deposition behavior for the non-film-cooling blade channel model.

From the current point of view, research on the phenomenon of particle deposition is mainly concentrated on the surface of the turbine cascade and the exit of the film hole. In this paper, the particle deposition mechanism of the impingement effusion internal cooling channel structure of the turbine blade is studied, and the deposition laws of different impingement velocities and different particle diameters in the impingement effusion internal cooling structure are analyzed. It has important theoretical guiding significance for avoiding the particle deposition in the cooling channels in turbine blades.

3. Computational physical model and meshing

The computational physical model is based on a simplified impingement effusion cooling structure of a certain type of turbine guide vane structure, as shown in Figure 1. The diameter of the film hole and the diameter of the impingement hole are D, and the impingement spacing is $H / D = 50$. There are two rows of impingement holes, with the spacing between the impingement holes in the same row is $W_i / D = 14$, and the spacing between the impingement holes in each row is $P_i / D = 8$. There are two

IOP Conf. Series: Materials Science and Engineering 1081 (2021) 012042 doi:10.1088/1757-899X/1081/1/012042

rows of film holes. The streamwise spacing of film holes is $W / D = 14$, the spanwise spacing of film holes is $P / D = 8$.

Structural meshing was performed on the computational domain using the ICEM CFD software, as shown in Figures 1. The number of grid cells in the entire computational domain is about 3.6 million. Among them, the grid is encrypted at the wall surface and near the wall of the film hole and the impingement hole. The grid size gradually increases from the wall surface and is enhanced according to the near-wall surface model. The requirement is that the first layer of grids near the wall surface should be below the cohesive sublayer, y+ is about 1.

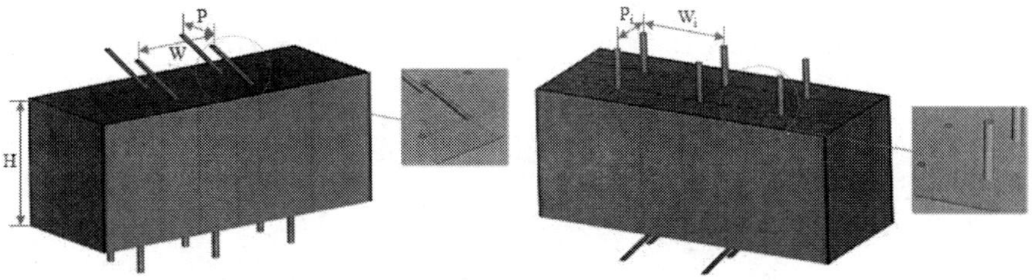

Figure 1. Computational Model and Meshing

4. Numerical simulation method

4.1. EI-Batsh Particle Deposition Impingement Criteria

The impingement model mainly includes two physical processes: one is the interaction between particles and the wall surface, which mainly includes two phenomena of adhesion and rebound; the other is the interaction between the adherent particles and the wall surface and the surrounding flow field to determine whether the particles are finally will deposit.

For the former physical process, the judgment criterion is the critical capture velocity, which is defined as:

$$V_{cr} = \left(\frac{2E}{d_p} \right)^{10/7} \tag{1}$$

In the equation: It is related to the Poisson's ratio of particles and the wall surface of spoiler and Young's modulus.

When the normal velocity of the particles relative to the wall surface is greater than the critical capture velocity, the particles rebound, otherwise they adhere to the wall surface.

Particles adhering to the wall are affected by the flow near the wall, and secondary suspension may occur. Only when the fluid wall friction velocity is less than the critical shear velocity, the particles will eventually deposit on the wall

$$u_{tc}^{\ 2} = \frac{W_A}{\rho d_p} \left[\frac{9\pi^2 W_A \left(k_s + k_p\right)}{8 d_p} \right]^{1/3} \tag{2}$$

4.2. Numerical Simulation Settings and Boundary Conditions

This paper uses the Realizable k-ε model for the numerical turbulence calculation of the two-phase flow in the impingement effusion cooling channel structure. The near-wall area is treated with enhanced wall surface. The particle phase is bidirectionally coupled by DPM and the random walk model is selected. The UDF based on the EI-Batsh impingement criterion is used to perform steady calculations using the SIMPLE algorithm. Momentum, turbulent kinetic energy, turbulent dissipation rate, and energy all uses the second-order QUICK format.

Inlet boundary: inlet gas temperature is 733K; The inlet velocity is matched and calculated by a given engine blade inlet Reynolds number, and the impingement Reynolds number is 300-3000, so we select the inlet velocity(impingement velocities)that meets this Reynolds number range: 5m/s, 20m/s, 32m/s, 50m/s.

Outlet boundary: the outlet is a pressure outlet condition, and the pressure is given by the working atmospheric pressure.

4.3. Mesh independence verification

At the same time, we also conducted a mesh independence verification to eliminate the effect of grid division on the simulation process and results. By adjusting the global mesh parameters and the number of near-wall nodes, we divided three kinds of mesh with different numbers of grids (mesh 1, mesh 2 and mesh 3), the number of mesh are 1609621, 3125567, 7853282 respectively. We selected four different working conditions to achieve four different dimensionless relaxation times and corresponding dimensionless deposition rates as shown in Figure 2. It can be seen that there is no obvious change in the numerical results obtained by the continuous refinement of grid 2, so grid 2 meets the requirements of grid independence.

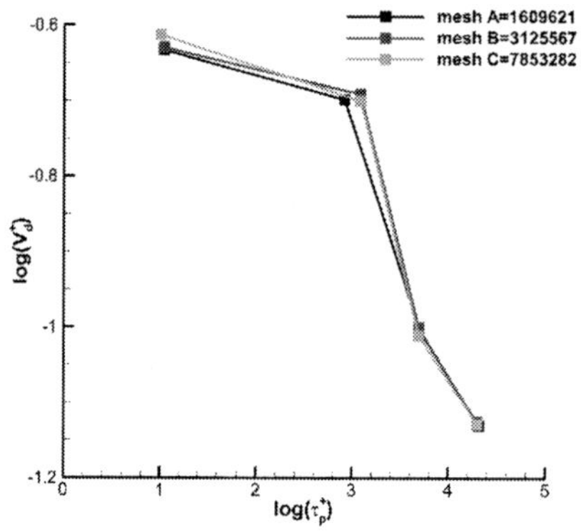

Figure 2. Mesh independence verification

4.4. Verification of Particle Deposition on Two-Dimensional Channel Wall

In order to verify the availability of the numerical simulation method, a two-dimensional channel wall structure is used for modeling, and the above numerical method is used for numerical simulation. The two-dimensional channel model is 800mm long and 12.7mm wide, and still uses structural grid division. Figure 3 shows the channel model and grid division.

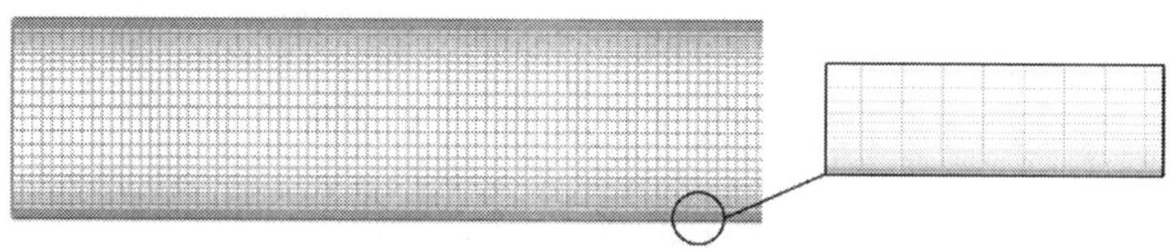

Figure 3. 2D channel meshing

Numerical simulation of particle deposition on the wall with an initial velocity of 5.3 m/s (u_τ= 0.342 m/s) and a diameter in the range of 1-50μm. To verify the rationality of the particle deposition calculation method used in this paper, the numerical simulation results are compared with the experimental results of Liu and Agarwal [15] and Sehmel [16], which have obtained experimental data, are compared. The comparison result is shown in Figure 4. Among them, within the range of $\tau_p^+ < 10$, the dimensionless deposition rate of particles increases with τ_p^+ increasing; while at $\tau_p^+ > 10$, with the τ_p^+ increase, the growth of the dimensionless deposition rate of particles slows down and shows a downward trend, which is the same as the change trend in the literature, indicating that the particle deposition calculation method used in this paper can better reflect the particle deposition trend, thus proving its rationality.

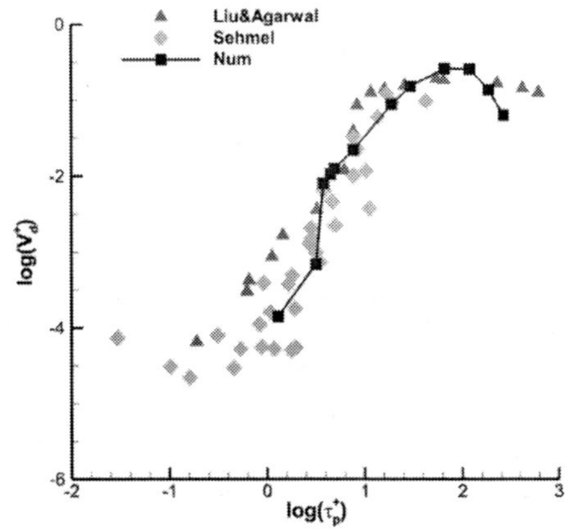

Figure 4. Numerical simulation and verification of 2D channel wall

4.5. Verification of Particle Deposition on Two-Dimensional Channel Wall

Particle deposition flux R_{dep} (kg/s^2) is the total particle deposition per unit area per unit time and is defined as:

$$R_{dep} = \sum_{p=1}^{N} \frac{m_p}{A_{face}} \tag{3}$$

The particle deposition rate V_d (m/s) is a measure of how fast particles are deposited on the wall and is defined as:

$$V_d = \frac{J}{C_0} \tag{4}$$

In the equation 4: J (kg/m^2·s) represents the average deposition flux of particles on the wall surface; C_0 (kg/m^3) represents the average particle concentration.

Considering the shear environment of the near-wall flow field, the ratio of particle deposition rate to wall friction velocity is defined as a dimensionless deposition rate, which is used to measure the deposition capacity of particles in different wall environments. Its expression is:

$$V_d^+ = \frac{V_d}{u_\tau} \tag{5}$$

In the equation 5: u_τ is wall friction velocity.

Considering the interaction of particle deposition and fluid vortex near the wall, the dimensionless relaxation time based on the flow time scale is defined as:

$$\tau_p^+ = \frac{C_c \rho_d d_p^2 u_\tau^2}{18\mu v} \tag{6}$$

In the equation 6: C_c is Cunningham coefficient.

5. Results and analysis.

5.1. Flow Characteristics

In the simplified impingement effusion cooling structure channel, the particle deposition is greatly affected by the channel impingement hole and the film hole structure. During the flow of cooling gas carrying particles in the channel, the impingement on the wall surface will change the fluid streamline in the near-wall area, and vortices will form in the square impingement cavity channel, which will affect the deposition.

Figure 5 shows the streamline distribution and velocity of Y direction distribution of the cooling gas flow field at four different inlet velocities with a particle size of 10μm.

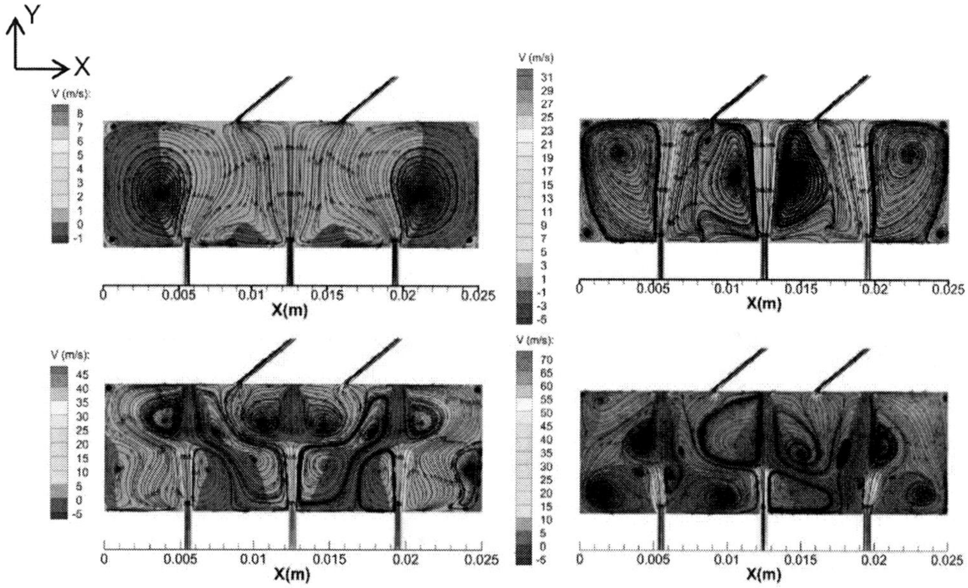

Figure 5. Velocity streamline and velocity of Y direction distribution of gas flow field at different inlet velocities (d_p=15μm)

It can be seen from the velocity streamline distribution that the gas enters from the impingement hole, and the velocity of Y direction distribution pattern of upward impingement is formed after the gas enters the square channel. When the velocity changes from 5m/s to 20m/s, the upward impingement velocity region expands significantly. When the velocity changes to 32m / s, the upper wall recoil velocity region appears, which weakens the impingement velocity region. When the velocity changes to 50m/s, the upper wall recoil velocity region increases, and the impingement velocity region is further weakened. At the same time, it can be seen from the streamline distribution that as the inlet velocity increases, the smaller the size of the eddy current cluster in the square channel, the larger the area of the fluid domain.

Based on the Euler-Lagrange model, and considering both the effect of continuous fluid on particles and the reaction of particles on continuous fluid, the flow direction of gas-solid two-phase fluid in the square channel is opposite to the direction of particle gravity. Therefore, changes in particle gravity caused by changes in particle size will have a certain effect on the fluid velocity flow field.

Figure 6 shows the velocity of Y direction distribution of the cooling gas flow field at four different intake particle diameters and an inlet velocity of 32 m/s.

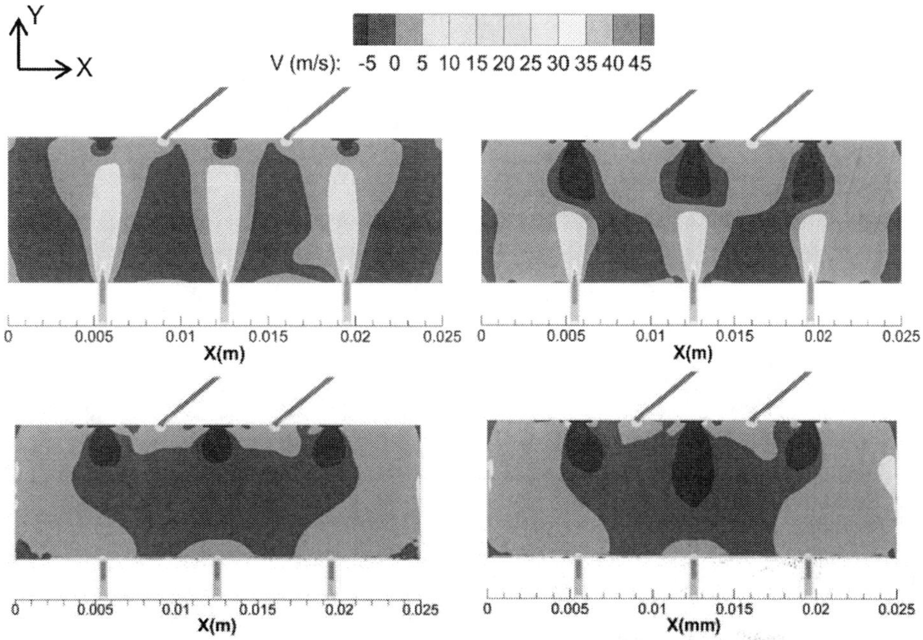

Figure 6. Velocity of Y direction distribution of gas flow field under different particle size conditions (v=32m/s)

It can be obtained from the velocity of Y direction distribution that the particle diameter has a great effect on the gas velocity flow field of the cooling channel. When the particle size is changing, the increase of particle size causes the impingement velocity region continuously decrease, and the impingement of the flow is continuously weakened due to the effect of the particle gravity. When the particle size is greater than 15μm, the gas at the outlet of the impingement hole has almost no ejection, the impingement velocity region almost disappears, and the upper wall recoil velocity region in the negative Y direction increases continuously, the velocity region in the negative Y direction with increase in particle size continuously occupies the middle area of the square channel. And under the condition of larger particle size, the size of the eddy current is increased, and the backflow with large area eddy current suppress the ejection of the impingement hole.

5.2. *Effect of Channel Inlet Velocity of Gas-Solid Two-Phase Flow on Deposition*

The difference in channel inlet velocity of gas-solid two-phase flow greatly affects the fluid flow state in the impingement effusion cooling structure, and then also affects the particle deposition characteristics and deposition morphology in the internal cooling channel structure. At the same time, as the change of the incident velocity of the particles will affect the followability to fluid of the particles and the determination parameters in the El-Batsh deposition model, these will change the probability that the particles are deposited on the wall.

Figure 7 shows the effect of the average gas-solid flow inlet velocity on the particle deposition rate and the dimensionless deposition rate of the cooling channel when the particle diameter is 10μm and the working temperature is 733K. Figure 8 shows the relationship between the frictional velocity of the wall of the channel, the dimensionless relaxation time, and the inlet velocity of the two-phase flow.

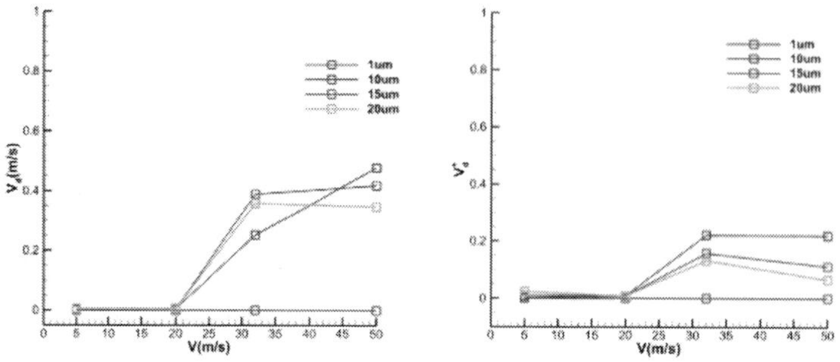

Figure 7. Effect of inlet velocity on deposition rate and dimensionless deposition rate

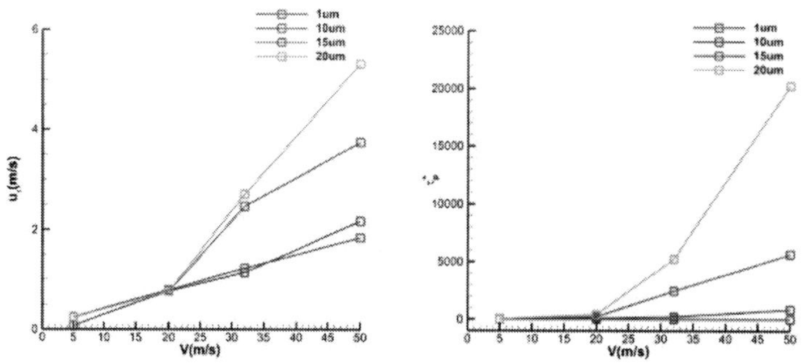

Figure 8. Effect of inlet velocity on wall friction velocity and dimensionless relaxation time

As shown in Figure 7, for the particle deposition rate, as the average velocity of the inlet increases, the deposition rate of particles on the wall as a whole increase, and the dimensionless deposition rate increases first, then flattens and decreases; combined with the change of friction velocity with the average inlet velocity, the increase in the average velocity of gas-solid two-phase flow leads to a constant increase in wall friction velocity, according to the definition of dimensionless deposition rate, the increase in dimensionless deposition rate is continuously slowed until it is flat or even reduced.

For the simplified impingement effusion cooling structure channel, the inlet and outlet are both holes with a relatively small size relative to the rectangular cross section of the channel. As the velocity increases, the dimensionless relaxation time of the particles increases, the Stokes number increases, and the inertial effect of particle continues to increase, which is continuously reduced by the effects of fluid flow lines and eddy currents, and decreases with followability to fluid of the particles, reduced the probability of particles escaping from the channel when the outlet is relatively small compared to the rectangular cross-sectional size of the channel, the probability of interaction between particles and the channel wall surface is increased, and the deposition capacity of the particles on the wall surface is increased, so that the deposition rate increases with the inlet velocity rise. At the same time, due to the particle-wall collision criterion of the EI-Batsh deposition model and the increase in particle velocity, the interaction time of solid particles staying in the channel is reduced, and the deposition capacity of particles on the wall is reduced, making the growth of the dimensionless deposition rate gradually slow down until it starts to decrease.

Figure 9 shows the deposition morphology of the four inlet average velocities for the cooling channel in the case of relatively small particles (d_p=1μm) when the working temperature is 733K.

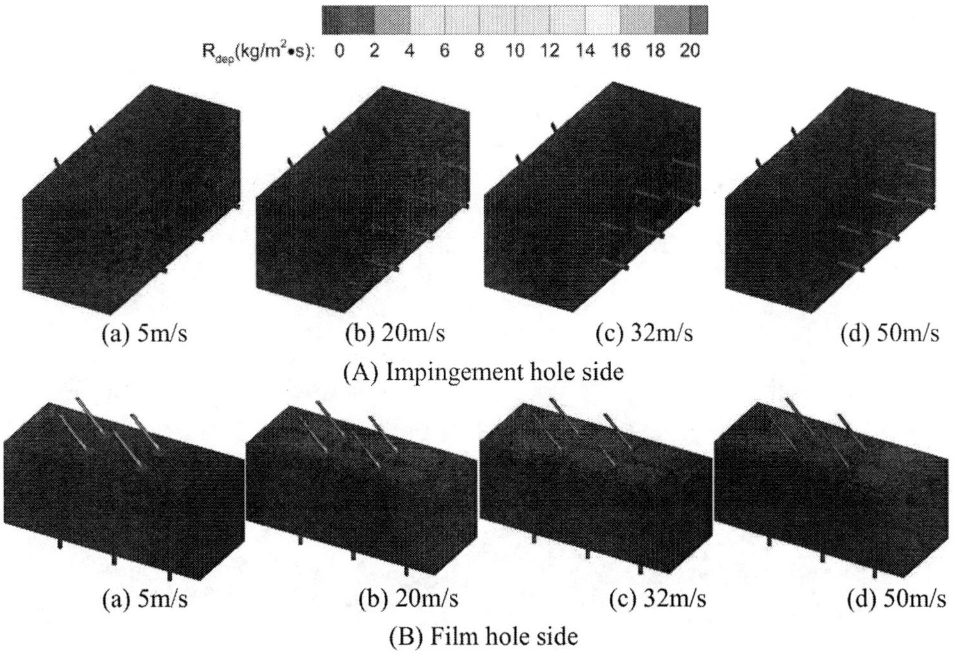

Figure 9. Deposition distribution of cooling channels at different inlet velocities (d_p=1μm)

It can be seen in the results of the deposition morphology of the four different inlet velocities of relatively small particles (d_p=1μm): There is relatively less deposition on the side of the impingement hole and the lower wall of the model; but as the inlet velocity continues to increase, the average deposition flux on the inner wall of the impingement hole is increased, which is due to the relatively high probability of particles contacting the wall surface during the flow of the relatively narrow channel at the inlet impingement hole; according to the channel velocity distribution, the velocity inside the impingement hole is larger, and the increase of the particle velocity increases the particle dimensionless relaxation time, the followability to fluid of particles are weakened, which increases the deposition ability of the inner wall of the impingement hole.

Besides, a certain deposition occurs on the inner wall of the film hole, and as the velocity increases, the inner wall of the film hole decreases in deposition, which is due to an increase in dimensionless relaxation time and an increase in the Stokes number due to an increase in velocity, which reduces the flowability to fluid of the particles and decreases the number of particles transported into the film hole with the gas flow, resulting in a decrease in the average deposition flux on the inner wall surface of the film hole;

As the inlet velocity continues to increase, the average deposition flux on the lower wall surface continues to decrease. According to the definition of the Stokes number, the characteristic velocity of particles has a linear relationship with the Stokes number. An increase in velocity means an increase in the Stokes number of particles, that is, the inertia effect of large particles has continuously played a leading role, causing a large number of particles to be deposited on the upper wall surface during the impingement process, thereby reducing the average deposition flux on the lower wall surface.

The deposition concentration area on the upper wall surface shifts from the entrance of the film hole to the area with the same Y coordinate of the impingement hole as the velocity increases; the same reason is that as the velocity increases, the inertia effect of the particles starts to be greater than the diffusion effect, and particles are less affected by the effects of gas vortices and streamlines; when the velocity is small, the flowability to fluid of the particles is better, and the particles are easy to follow the gas flow and deposited at the inlet of the film hole. When the velocity is large, the particles are easily affected by the impingement velocity and are deposited on the upper wall surface consistent with the Y coordinate of the impingement hole.

Figure 10 shows the deposition morphology of the four inlet velocities for the cooling channel in the case of relatively large particles (d_p=20μm) when the operating temperature is 733K.

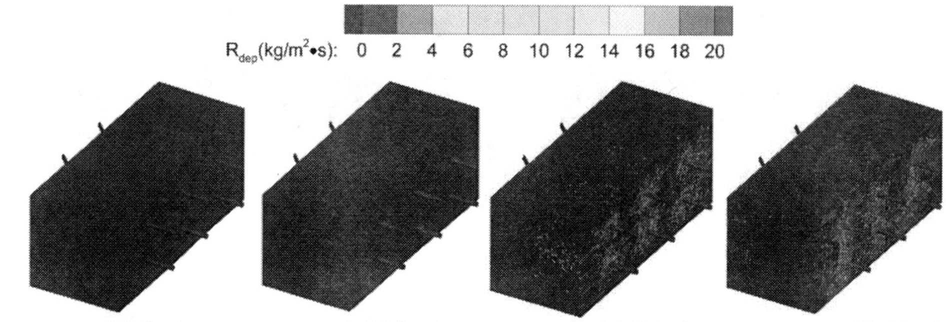

(a) 5m/s (b) 20m/s (c) 32m/s (d) 50m/s

(A) Impingement hole side

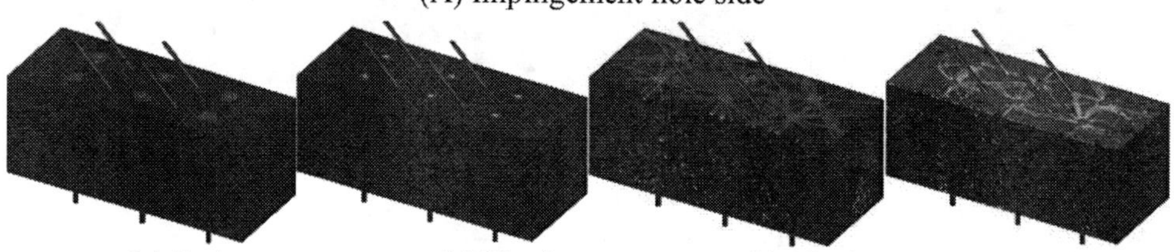

(a) 5m/s (b) 20m/s (c) 32m/s (d) 50m/s

(B) Film hole side

Figure 10. Deposition distribution of cooling channels at different inlet velocities (d_p=20μm)

In the impingement hole side deposition morphology of four different inlet velocities of relatively large particles (d_p=20μm), compare the impingement hole side deposition morphology of small particle diameters and different inlet velocities: under the working conditions of 5m/s, 20m/s, the deposition on side of the impingement hole and the lower wall of the model is relatively small; and under the conditions of 32m/s and 50m/s, the deposition on the lower wall and the surrounding wall increases sharply; on the one hand, in the channel, the gas flow recirculation velocity region along the negative Y direction appears, and the impingement velocity region continues to disappear as the particle size increases. The recoil recirculation velocity region continues to increase, resulting in the case of large particle diameters and high velocity, massive deposition appears on the lower wall surface and surrounding wall surfaces; on the other hand, due to the increase in particle size, the critical capture velocity of the wall surface for particle size decreases, and the critical shear velocity also decreases at the same time, resulting in an increase in the probability of the physical state of particles peeling and rebounding. Part of the deposited particles on the upper wall surface is eventually deposited on the lower wall surface due to vortex peeling or rebound. At the same time, as the velocity increases, the amount of deposition area on the upper wall consistent with the Y-coordinate of the impingement hole increases significantly, which is also the result of the increasing inertia effect of the particles.

5.3. Effect of Particle Size on Deposition

Figure 11 shows the effect of particle diameter on the particle deposition rate and dimensionless deposition rate of cooling channel when the working temperature is 733K and the average inlet velocity of the gas-solid two-phase flow is 32m/s. Figure 12 shows the relationship between dimensionless relaxation time and wall friction velocity with the change of particle diameter.

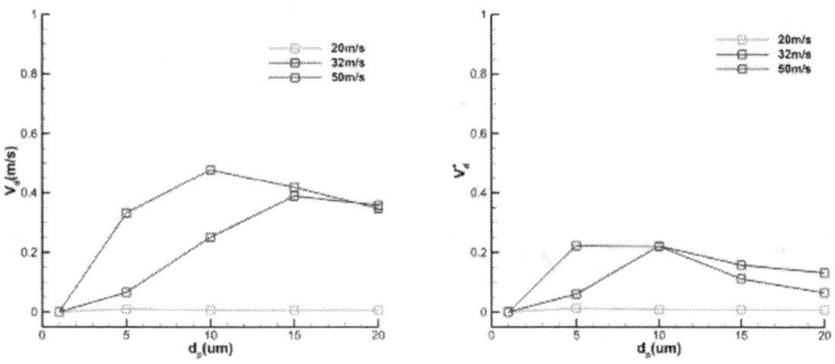

Figure 11. Effect of inlet velocity on deposition rate and dimensionless deposition rate

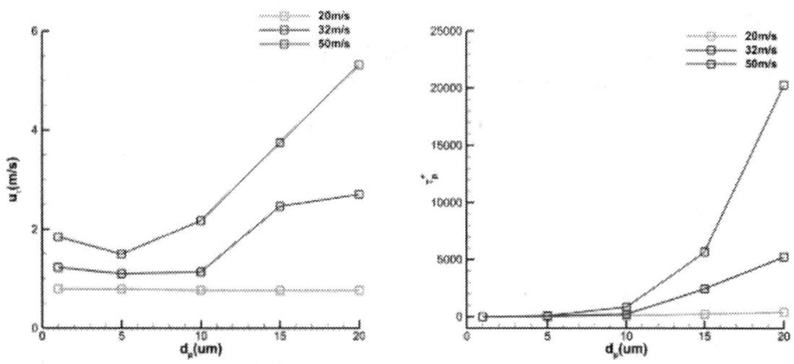

Figure 12. Effect of inlet velocity on wall friction velocity and dimensionless relaxation time

As for the particle deposition rate, as the particle diameter increases, both the particle deposition rate on the wall surface and the dimensionless deposition rate change trend increase first and then decrease, and the peak value of the dimensionless deposition rate takes precedence over the peak value of the deposition rate change. In the case of relatively small particle sizes, the dimensionless relaxation time and particle Stokes number caused by the increase in particle size continue to increase, and the probability of collision between particles in the channel and the wall surface increases due to the increase of inertia of particles, and the decrease in the probability of escaping from the outlet of the film hole leads to an increase in the deposition capacity; at a relatively large particle size according to the particle-wall collision criterion of the EI-Batsh deposition model, the critical capture velocity decreases with increasing particle size, making more particles unable to meet the adhesion conditions and rebounding, resulting in a weakening of the deposition ability. From 12, it can be seen that as the particle size increases, the wall friction velocity increases but the increase is relatively limited, but the critical shear rate gradually decreases with the increase in particle size, making more particles have a wall friction velocity greater than the critical shear velocity makes the deposited particles peel off the wall surface, so that the deposition ability is weakened with the increase of the particle size in the case of a larger particle size.

It is worth noting that in the case of relatively small particle diameters, the wall friction velocity decreases with the increase of the particle diameter, which suppresses the physical peeling of the particles. Therefore, the deposition rate can be maintained and climbed due to the flow shearing environment near the wall, which makes the change peak of dimensionless deposition rate takes precedence over the change peak of deposition rate.

Figure 13 shows the influence of the four particle diameters on the deposition morphology of the impingement hole side and the film hole side when the working temperature is 733K and the inlet velocity is 32m/s.

IOP Conf. Series: Materials Science and Engineering 1081 (2021) 012042 doi:10.1088/1757-899X/1081/1/012042

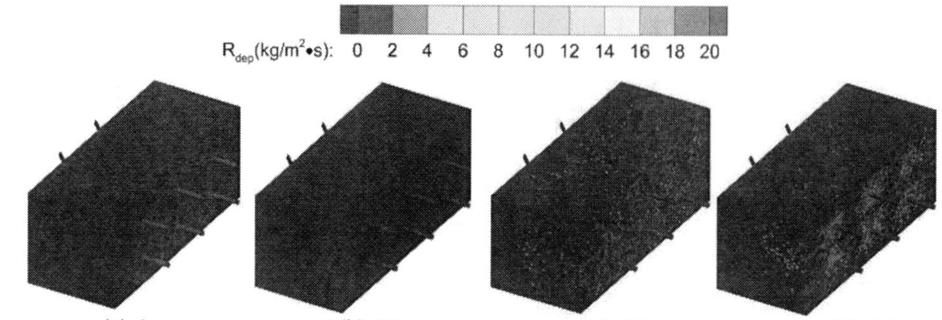

(a) 1μm (b) 10μm (c) 15μm (d) 20μm

(A) Impingement hole side

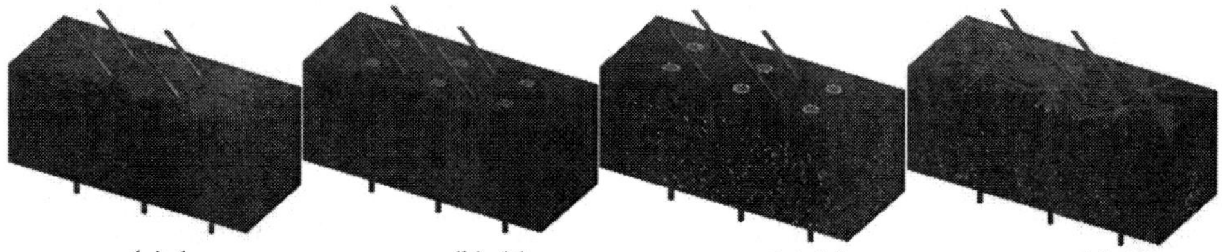

(a) 1μm (b) 10μm (c) 15μm (d) 20μm

(B) Film hole side

Figure 13. Deposition distribution of cooling channels at different particle diameters (d_p=20μm)

In the deposition morphology results of different particle diameters under the condition of 32m/s, it can be seen that the deposition on the inner wall of the impingement hole gradually decreases with increasing particle size; the reason is that the increase in particle diameter reduces the critical adhesion velocity and the critical shear rate makes the physical state of the particles rebound and peel gradually increase, and the deposition ability is limited. At the same time, the inner wall of the film hole is deposited under the condition of the particle size of 1μm, as the particle size increases, the inner wall of the film holes gradually becomes almost non-depositing; the deposition in the film holes is mainly related to the followability to fluid of the particles, which are limited in the case of 32m/s, so the particles appear in the film holes when the particle size is small. With the increase of the particle size, the deposition of the lower wall surface and surrounding wall surface decreases first and then increases, and the deposition morphology changes greatly during the change to 10μm to 15μm. When the particle size is less than 10μm, that is, a relatively small particle size, the impingement of the gas flow is intense, and the reflow effect is weak, the increase in particle size causes more particles to maintain their own movement and deposit on the upper wall surface, resulting in the ability of deposition in lower wall surface gradually decreases. When the particles are larger than 15μm, that is, in the case of a relatively large particle size, according to the definition of the Stokes number, the particle diameter greatly affects the change of the Stokes number, the increase of the particle size means that the Stokes number of the particles increases sharply, that is, the followability to fluid of particles decreases, the larger the particle size, the inertial effect of the particle continues to play a leading role, making a large number of particles unable to escape the channel, and during the effect of eddy current and rebound, the particles constantly weakens its own velocity, so that the particles meet the deposition conditions and finally deposited on the surrounding wall and the lower wall. At the same time, with the increase of the particle size, the deposition of the upper wall area which is consistent with the Y-coordinate of the impingement hole increased significantly and the deposition area became more concentrated, which was also the result of the increasing inertia effect of the particles.

6. Conclusions

In this paper, the Euler-Lagrange method is used to simulate the deposition rule of particles in the impingement effusion cooling structure under different conditions including different inlet velocities and particle diameters of the gas-solid two-phase flow. The rules and mechanisms of deposition in the cooling channel are summarized as follows:

(1) The gravity of particles can affect fluid flow under different working conditions. The impingement of gas flow in the cooling channel is continuously weakened due to the effect of particle gravity.

(2) When changing the inlet velocity and particle diameter of gas-solid two-phase flow, particle deposition still conforms to the deposition mechanism with Brownian motion, turbulent diffusion, and particle inertia as the main reasons for deposition.

(3) The inlet velocity of gas-solid two-phase flow has a great effect on particle deposition. Increasing the inlet velocity on the one hand increases the inertia effect of the particles, thereby reducing the escape of particles from the film holes which are relatively small compared to the rectangular cross section of the channel, allowing more particles to stay inside the channel, and eventually depositing in the channel under the condition of continuous weakening of the velocity of particles, which improves the deposition ability; On the other hand, the wall friction velocity and normal velocity of hitting wall are increased, which weakens the particle deposition ability. In general, for the velocity interval conditions studied in this paper, the increase in inlet velocity enhances the particle deposition ability. At the same time, the velocity affects the flowability to fluid of the particles. The larger the inlet velocity, the worse the flowability to fluid of the particles;

(4) Particle diameter has a great impact on particle deposition. The increase in particle diameter on the one hand increases the particle inertia effect to allow more particles to stay inside the channel, which promotes the particle deposition ability, on the other hand, the critical capture velocity and the critical shear velocity near the wall are reduced, which weakens the particle deposition ability; meanwhile, the particle size affects the flowability to fluid of the particles. The larger the particle size, the worse the flowability to fluid of the particles.

References

[1] Tabakoff W. Review-Turbomachinery Performance Deterioration Exposed to Solid Particulate Environment[J]. ASME Journal of Fluids Engineering,1984,106(7):125-134..
[2] Adel Ghenaiet. Study of sand particle trajectories and erosion into the first fan stage of a turbofan[C]. Proceedings of ASME Turbo Expo 2010, GT2010-22415, Glasgow, UK, 2010.
[3] Adel Ghenaiet. Study of Sand Particle Trajectories and Erosion into the First Compression Stage of a Turbofan[J]. ASME Journal of turbomachinery,2012,134(9):051025.
[4] Zeng Qiang. Numerical Simulation Study of Sand Particle Trajectory in Air Flow Channel [D]. Nanjing: Nanjing University of Aeronautics and Astronautics, 2013.
[5] Jensen J.W., Squire S.W., Bons, J.P., Fletcher T.H. Simulated land-based turbine deposits generated in an accelerated deposition facility[C]. Proceedings of ASME Turbo Expo 2004, GT2004-53324, Vienna, Austria, 2004.
[6] Ai W., Murray N., Fletcher T.H., Harding S., Lewis S., Bons J.P. Deposition Near Film Cooling Holes on a High Pressure Turbine Vane[C]. ASME paper, No.GT2008-50901, 2008.
[7] Shah A, Tafti D. Transport of particulates in an internal cooling ribbed duct[J]. Transactions of the ASME. The Journal of Turbomachinery,2006,129(4):816-825.
[8] Jonathan B C., Danesh K T. Investigation of sand particle deposition and erosion within a short pin fin array[C]. Glasgow, UK: Proceedings of ASME Turbo Expo 2010: Power for Land, Sea and Air, 2010:139-149.
[9] Zhou Junhui, Zhang Jingzhou. Numerical Study on Particle Deposition Characteristics in Turbine Cascades [J]. Journal of Aeronautics, 2013,34(11):2492-2499.
[10] Zhou Junhui, Zhang Jingzhou. Numerical study of particle deposition characteristics near film holes [J]. Journal of Aerospace Dynamics, 2014,29(9):2166-2173.
[11] Zhou Junhui, Zhang Jingzhou. Effect of Local Blockage of Film Holes on the Comprehensive

IOP Conf. Series: Materials Science and Engineering 1081 (2021) 012042 doi:10.1088/1757-899X/1081/1/012042

Cooling Efficiency of Impact-Spoiler-Air Film Structure of Blade Pressure Surface [J] Journal of Aviation, 2016,37(9):2729-2738.
[12] Yang Xiaojun, Zhu Jiaxiong. Numerical Simulation of Particle Deposition Process in Turbine Cascade Channel [J]. Journal of Aeronautics,2017,38(5):31-42.
[13] Yang Xiaojun, Cui Mohan, Liu Zhigang. Experimental Study on Particle Deposition on the Surface of Film Cooling Plate [J / OL]. Propulsion Technology,2018(06):1-8.
[14] Pei Yu. Research on the model of pollutant deposition on the surface of gas turbine blades [D]. China Civil Aviation University, 2015.
[15] Liu B Y H, Agarwal J K. Experimental observation of aerosol deposition in turbulent flow ☆[J]. Journal of Aerosol Science, 1974, 5(2):145, IN1,149-148, IN2,155.
[16] Sehmel G A. Particle sampling bias introduced by anisokinetic sampling and deposition within the sampling line. [J]. American Industrial Hygiene Association Journal, 1970, 31(6):758.

IOP Conf. Series: Materials Science and Engineering 1081 (2021) 012043 doi:10.1088/1757-899X/1081/1/012043

Rapid Prediction Method of Fan Broadband Noise Based on A Three-dimensional Sound Source Model

Yang Ye[1,2], Wang Liangfeng[1,2], Yang Dangguo[1,2*]

[1] State Key Laboratory of aerodynamics, China Aerodynamics Research and Development Center, Mianyang, PR China
[2] High Speed Aerodynamics Research Institute, China Aerodynamics Research and Development Center, Mianyang, PR China

E-mail:yangdg-cardc@163.com

Abstract. The fan/compressor broadband noise is an important noise source of civil large bypass ratio turbofan engine.. The empirical method requires massive amounts of test data for support. Adopting the computational aeroacoustic (CAA) method to predict the fan/compressor broadband noise will cost high amounts of computing resources or time. Thus, this study combines flow field calculation with analytical model to study the prediction method of fan/compressor broadband noise, this method can be used for quick and efficient preliminary evaluation and parameter studies. On the basis of the simulation and modeling of the stator－rotor interference phenomenon of fans, rotor wake flow field and the corresponding parameters are obtained. On this basis, the three-dimensional sound source model is used to calculate the blade surface sound source and its propagation under different working conditions. The distribution of sound energy in the duct is obtained. Then, the broadband noise prediction of fan is further acquired. The broadband noise of a typical single-stage TA36 fan is predicted using this method. Study results suggest that (1) turbulence model and turbulence kinetic energy of inlet boundary condition will have an important effect on the prediction results of broadband noise. (2) When the inlet turbulence kinetic energy is small, different turbulence models slightly influence the prediction results of broadband noise; when the inlet turbulence kinetic energy is large, the sound pressure level of sound field obtained by SST model is approximately 5dB larger than that obtained by model. This method can be used in the early stage of fan/compressor design, and the characteristics of blade broadband noise can be rapidly obtained.

1. Introduction

The noise problem always exists in the development of human society. Acoustic phenomena are also encountered in people's daily lives. Acoustic technology is widely used in the field of natural science and engineering[1-6]. With the increase in people's disposable income, traveling by air has become increasingly popular. This situation brings a strong development vitality to the civil aviation transportation industry. Therefore, the noise problem has gradually attracted people's attention.

Jet noise dominates in early turbojet engines. Jet speed decreases with the increase in bypass ratio. Given that the intensity of jet noise is directly proportional to jet speed to the power of 8, the proportion of jet noise in the total noise level of engine gradually decreases. With the increase in bypass ratio, the proportion of fan/compressor noise becomes increasingly large and becomes an important noise source[7-8]. This situation makes the generation and propagation of fan/compressor noise become one of the main problems in the study of aeroacoustics[9].

The broadband noise of fan is generated from the interaction between nonperiodic air flow and solid surface. According to different generation mechanisms, it is classified as rotor self noise, secondary flow interference noise, and turbulence wake and stator interference noise[10]. Two prediction methods can be used. The first one is the CAA method. According to different calculation modes, it can be divided into hybrid and direct computational methods. In 2005, by using a hybrid computational method, Jacob[11] first obtained the flow field information of cylinder-wing section interference problem through CFD and then acquired its sound field distribution by using the FW-H equation. This hybrid computational method overcomes too much simplification of the semiempirical formula. However, it cannot solve many blade rows. In 2012, Tam & Ju[12] directly numerically simulated wing section turbulence problems by using high-precision CAA method. This direct computational method is not simplified. It is helpful for understanding the mechanism and propagation law of fan noise. However, its computational amount is too much to be used in engineering research at this moment.

The second is to use analytic model. In 1971, Mani[13] and Amiet[14] studied the noise problem of the interference between turbulence and isolated blade and proposed a general idea of establishing a broadband noise sound source model. The response function of the blade was first established to solve the unsteady force on the surface of blade produced by the upwash speed and then to solve the sound field in the duct caused by the unsteady force. On the basis of this idea, Glegg[15] grafted the response function of blade proposed by the former onto the broadband interference noise of fans and achieved a good result in 1998. On this basis, Hanson[16] considered the effects of the nonuniformity of turbulence and swept-curved blade on broadband noise in 2001. In 2002, Evers and Peake[17] established another fan broadband interference noise model by using the response function of blade row developed by Peake and studied the effects of blade shape and nonuniform mainstream. They proved that the blade thickness and small curvature slightly contribute to the broadband noise in high-frequency range. In 2005, Nallasamy[18] improved the fan broadband noise program developed by Ventres and regarded the turbulence flow obtained by the Reynolds average NS equation as the input, including the effects of background turbulence intensity and rotor wake turbulence. In 2011, Grace[19] studied the effects of turbulence model selection, radial wave number, integral scale, and other parameters on fan broadband interference noise. The prediction program of RSI fan broadband noise developed by NASA based on Nallasamy method has achieved a good result, but most of the current prediction models of fan broadband noise still simplify the blade into a two-dimensional rectangular plate and cannot consider the effects of geometric blade parameters. Atassi[20-21] extended the response function of the three-dimensional annular cascade developed by himself to the prediction of broadband noise and proved that the three-dimensional and swirl effects considerably influence the prediction result. However, solving the blade response function at each frequency and wave number by using numerical method will lead to a huge amount of computation.

In summary, the fan/compressor broadband noise is an important component of aeroengine noise, and the proportion is still increasing. Although the prediction method of two-dimensional rectangular plate has achieved some results in the prediction of fan/compressor broadband noise, the three-dimensional geometric effect greatly influences the noise prediction result. Input conditions for the sound propagation design should be provided by noise prediction. The denoise effect will be reduced if the sound propagation design of the sound source is ignored. Zhang Weiguang [22] proposed 3D sound source model, in the algorithm benchmark test, compared with other broadband noise prediction algorithms[21], [23], error is less than 2 dB. Compared with the experimental results, it is also proved that this is a successful broadband noise prediction model. However, the stream input mainly depends on the test data. This increases the overall prediction cycle. CFD computation can rapidly capture the flow field information and can therefore compensate for the shortcomings of the above-mentioned method. By combining the advantages of both methods, the rapid prediction method for the three-dimensional fan/compressor broadband noise established by using CFD rotor wake as the input can further shorten the period and cost of noise prediction. Accordingly, its generation mechanism and denoise method can be studied.

In this work, NUMECA software is used to simulate the flow field of BUAA TA36 single-stage low-speed axial fan. The parameters required by the broadband noise prediction can be obtained by modeling the flow field result. The three-dimensional fan sound source model developed by Zhang Weiguang is used to solve the distribution of the sound pressure level in the duct at different frequencies, and the prediction result of broadband noise is obtained.

2. Broadband noise model

2.1. Nomenclature

ω	frequency
V	blade number of stator
ρ_0	air density
U	front axial mainstream velocity of fan rotor
m	circumferential mode
n	radial mode
k_{mn}	eigenvalues of (m,n) mode
$\psi(k_{mn}r)$	characteristic function of normalized (m,n) order duct
k	wave number
M	Mach number of uniform flow in the main stream inside the duct
$\Im(\vec{K},r',z')$	unsteady load on the blade surface produced by the unit of upwash disturbance $e^{i\vec{K}\cdot\vec{y}}$ with the wave number of $\vec{K}$
$\langle W^*\cdot W\rangle$	upwash velocity spectrum

2.2. Broadband noise model

The general expression of the sound power of fan broadband noise in Zhang Weiguang's three-dimensional sound source model is

$$\wp_\omega^\pm=\frac{\omega V^2}{2\pi\rho_0 U}\sum_{s=-\infty}^{\infty}\sum_{n=1}^{\infty}\left[\frac{m^2}{4\kappa_{n,m}}\left(\frac{\beta^2 M}{\pm M\kappa_{n,m}-k_0}\right)^2\right.$$
$$\iiiint\Im^*\left(\vec{K},r_1',z_1'\right)\Im^*\left(\vec{k},r_2',z_2'\right)$$
$$\frac{\psi\left(k_{mn}r_1'\right)e^{-im\varphi_1'}e^{i\alpha_{1,2}z_1'}}{r_1'}\cdot\frac{\psi\left(k_{mn}r_2'\right)e^{im\varphi_2'}e^{-i\alpha_{1,2}z_2'}}{r_2'}\tag{1}$$
$$\left.\cdot\left\langle W^*\cdot W\right\rangle d\vec{K}d\vec{k}\cdot ds_1ds_2\right]$$

The prediction of broadband noise ensures that the rotor wake turbulence acts on the stator blade to produce sound. Therefore, the upwash velocity spectrum $\langle W^*\cdot W\rangle$ is important. Its expression is

$$
\begin{aligned}
\left\langle W^*\left(\vec{K},\omega,r_1\right)\cdot W\left(\vec{k},\omega',r_2\right)\right\rangle = &\left\langle W^*\left(\vec{K},\omega,r_1\right)\cdot W\left(\vec{k},\omega',r_2\right)\right\rangle_{ww} \\
&+\left\langle W^*\left(\vec{K},\omega,r_1\right)\cdot W\left(\vec{k},\omega',r_2\right)\right\rangle_{bw} \\
&+\left\langle W^*\left(\vec{K},\omega,r_1\right)\cdot W\left(\vec{k},\omega',r_2\right)\right\rangle_{bb}
\end{aligned} \tag{2}
$$

where

$$
\begin{aligned}
\left\langle W^*\left(\vec{K},\omega,r_1\right)\cdot W\left(\vec{k},\omega',r_2\right)\right\rangle_{ww} = &\left(\frac{N_B L_w u'_w}{2\pi\bar{r}}\right)^2\cdot\frac{\pi}{\bar{r}\cdot\left(r_d-r_h\right)}\cdot\sum_m\sum_l\sum_{m_g}\sum_{n_g} \\
&e^{-\frac{1}{4\pi}\left(\frac{mN_BL_w}{\bar{r}}\right)^2}\cdot e^{-\frac{1}{4\pi}\left(\frac{lN_BL_w}{\bar{r}}\right)^2}\cdot e^{i\left(\frac{(m-l)N_B}{\bar{r}\cos x}\right)\vec{D}\cdot\vec{n}} \\
&\cdot\delta\left(\omega-\vec{k}\cdot\vec{U}\right)\cdot\delta\left[\omega'-\omega-\Omega(m-l)N_B\right] \\
&\cdot\Phi\left(\vec{k}-\frac{mN_B}{\bar{r}\cos\chi}n_1,\Delta r\right)\cdot\delta\left[\vec{K}-\vec{k}+\frac{(m-l)N_B}{\bar{r}\cos\chi}\vec{n}\right] \\
&\delta\left(k_2-\frac{\pi n_g}{r_d-r_h}\right)\cdot\delta\left(k_3-\frac{mN_B+m_g}{\bar{r}}\right)
\end{aligned} \tag{3}
$$

$$
\begin{aligned}
\left\langle W^*\left(\vec{K},\omega,r_1\right)\cdot W\left(\vec{k},\omega',r_2\right)\right\rangle_{bw} = &\frac{N_B L_w u'_b u'_w}{2\pi\bar{r}}\cdot\frac{\pi}{\bar{r}\cdot\left(r_d-r_h\right)}\cdot\sum_m\sum_{m_g}\sum_{n_g} \\
&e^{-\frac{1}{4\pi}\left(\frac{mN_BL_w}{\bar{r}}\right)^2}\cdot\delta\left(\omega-\vec{k}\cdot\vec{U}\right)\cdot\delta\left(\omega'-\omega-\Omega mN_B\right) \\
&\cdot\left\{e^{i\frac{mN_B}{\bar{r}\cos\chi}\vec{D}\cdot\vec{n}}\cdot\Phi\left(\vec{k}-\frac{mN_B}{\bar{r}\cos\chi}n_1,\Delta r\right)\right\} \\
&\cdot\delta\left(\vec{K}-\vec{k}+\frac{mN_B}{\bar{r}\cos\chi}\vec{n}\right) \\
&\cdot\delta\left(k_2-\frac{\pi n_g}{r_d-r_h}\right)\cdot\delta\left(k_3-\frac{mN_B+m_g}{\bar{r}}\right) \\
&+e^{-i\frac{mN_B}{\bar{r}\cos\chi}\vec{D}\cdot\vec{n}}\cdot\Phi\left(\vec{k},\Delta r\right)\cdot\left(\vec{K}-\vec{k}-\frac{mN_B}{\bar{r}\cos\chi}\vec{n}\right) \\
&\cdot\delta\left(k_2-\frac{\pi n_g}{r_d-r_h}\right)\cdot\delta\left(k_3-\frac{m_g}{\bar{r}}\right)
\end{aligned} \tag{4}
$$

$$
\begin{aligned}
\left\langle W^*\left(\vec{K},\omega,r_1\right)\cdot W\left(\vec{k},\omega',r_2\right)\right\rangle_{bb} = &u_b'^2\cdot\frac{\pi}{\bar{r}\cdot\left(r_d-r_h\right)} \\
&\cdot\delta\left(\omega-\vec{k}\cdot\vec{U}\right)\cdot\delta\left(\omega'-\omega\right)\cdot\Phi\left(\vec{k},\Delta r\right)\cdot\delta\left(\vec{K}-\vec{k}\right) \\
&\cdot\delta\left(k_2-\frac{\pi n_g}{r_d-r_h}\right)\cdot\delta\left(k_3-\frac{m_g}{\bar{r}}\right)
\end{aligned} \tag{5}
$$

The solution of the sound power in this model is to integrate along the whole stator blade rather than to stack the sound power on each rectangular cascade similar to the two-dimensional strip theory. At the same time, the acoustic radiation problem in the circular duct is solved on the basis of acoustic analogy theory. Strip theory is still adopted to solve unsteady load on the stator blade, but the unsteady load is solved in the radial and chord directions at the same time on the whole stator blade. Therefore, radial influence can be considered. The influence of swept-curved blade design on noise can be also considered.

3. Numerical simulation of rotor wake

3.1. TA36 fan/compressor test bench

In this section, rotor wake will be calculated by means of BUAA TA36 single-stage low-speed axial fan test bench. This test bench is shown in Fig. 1. It mainly consists of three parts, namely, air inlet, a row of rotor blades, and a row of stator blades. The main design parameters of TA36 are shown in Tables 1 and 2.

Figure 1. TA36 compressor test bench.

Table 1. Geometric parameters of TA36 compressor test bench

Geometric parameters	Rotor	Stator
Blade number	20	27
Aspect ratio	1.18	1.40
Tip clearance (mm)	0.6	0.5
Case diameter (mm)	600	600
Hub diameter (mm)	346	401
Installation angle of hub position (°)	45	0

Table 2. Aerodynamic parameters of TA36 compressor test bench

Aerodynamic parameters	
Design rotation rate (rpm)	2930
Design point-mass flow (kg/s)	6.5
Overall pressure ratio	1.026
Efficiency	85%

3.2. Calculation method of three-dimensional fan rotor wake

In this section, a commercial software called NUMECA is used to calculate the flow field of TA36 three-dimensional fan. The result is compared with the experimental result of its efficiency characteristic line, and the reliability of the rotor–stator interference calculation result is proven. Accordingly, the wake modeling can be prepared and the input parameters of the fan noise prediction program can be extracted in the next step.

NUMECA has high-quality automatic grid division, impeller machine design, and other modules. Given its simple operation, good stability, fast calculation, and good convergence, it is suitable for three-dimensional fan blade calculation. The second chapter that we only have to solve the RANS equation based on the statistical mean of turbulence. In this case, the wake produced from the whole row of rotor blades can be regarded as the rotational symmetry of the wake generated by a single blade channel in the circumferential direction. Therefore, only single-channel steady calculation is needed for the TA36 three-dimensional blade, which can reduce huge computational amount.

The turbulent kinetic energy is essential for the prediction of fan broadband noise. Thus, a turbulence model with good stability and convergence, as well the calculation results including the turbulent kinetic energy, should be considered. In this work, two turbulence models are selected, namely, $k-\varepsilon$ model and SST model.

3.2.1 Grid generation

The IGG-Autogrid module in NUMECA is used to generate the grid of the three-dimensional TA36 blade. The final result is shown in Fig. 2. The total number of grids is approximately 2.66 million.

Notably, different from the conventional flow field for solving the mainstream velocity, the turbulent kinetic energy may decay exponentially to a small amount approaching zero as the axial distance increases. The main reason is the large grid spacing. If so, when the turbulence parameters are solved in accordance with the turbulence kinetic energy, the background turbulence intensity in turbulence input parameters is nearly zero, and the three other turbulence parameters seriously deviate. The axial grids should be properly encrypted to avoid such phenomenon, and appropriate numerical settings should be used in FINE solution module. The topology of the rotor grid is shown in Fig. 3. The number of grids is approximately 1.51 million. The topology of the stator grid is shown in Fig. 4. The number of grids is nearly 0.75 million.

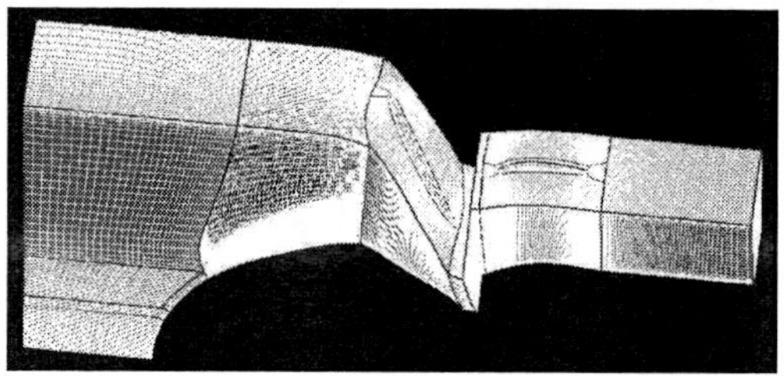

Figure 2. TA36 calculation grid

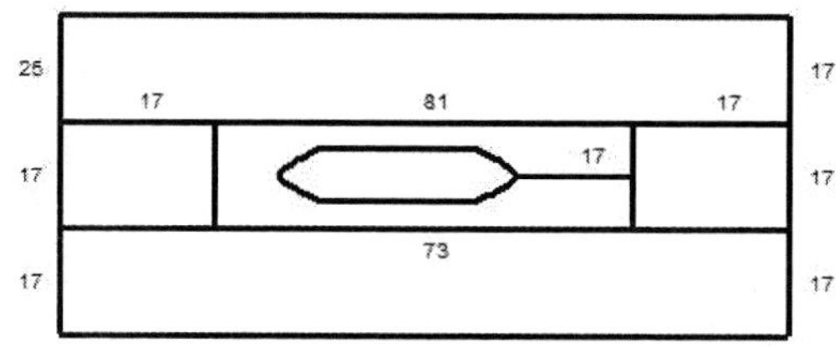

Figure 3. Topology of the rotor grid

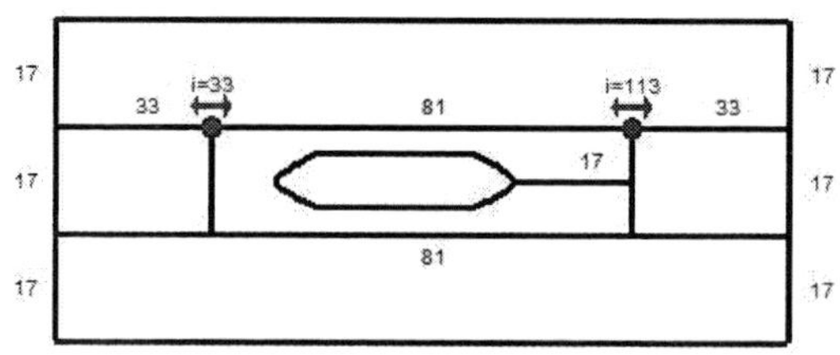

Figure 4. Topology of the stator grid

3.2.2 Boundary condition setting

After grid division is completed, the inlet and outlet parameters are set in accordance with the test conditions. The inlet boundary conditions are set as the total pressure, axial intake, and deflection angle 0; the outlet boundary conditions are set as static parameters; and the static pressure is set at the height of 1/2 blade on the basis of the radial balance equation. To verify the prediction ability of the three-dimensional fan broadband noise prediction program under multiple working conditions, four characteristic lines corresponding to 100%, 90%, 80%, and 70% of rotation rate should be calculated. In accordance with the conditions of characteristic line, the rotation rates are set to 2915, 2624, 2332, and 2041 rpm.

Notably, the turbulence model is $k-\varepsilon$ model or SST model. Thus, the turbulence kinetic energy should be given in the inlet boundary conditions. The intake air in the TA36 single-stage fan test bench is uniform, and the turbulence intensity at the inlet is very small. It is set on the basis of the default value k = 5.

The turbulent kinetic energy determines the turbulent fluctuation velocity but is still smaller than the mainstream velocity. Thus, the setting of the inlet turbulent kinetic energy does not affect the mainstream velocity and compressor efficiency.

3.2.3 Calculation results

The compressor has a corresponding characteristic line at each rotation rate. The abscissa is the flow, and the ordinate is the efficiency. In this example, the change in the blade load is simulated by changing the static pressure at the outlet continuously, to obtain the flow-efficiency characteristic line at each rotation rate.

The characteristic line diagram calculated by NUMECA should be compared with the test results to verify the reliability of the CFD rotor–stator interference calculation. Figs. 5–8 show the flow-efficiency characteristic line diagrams of TA36 compressor at 70%, 80%, 90%, and 100% rotation rates. The red lines represent the results calculated by NUMECA, and the green lines represent the test results. The figures show that the calculated results of the flow field at four rotation rates are same as the test results. This similarity shows that the CFD calculation results are reliable.

However, further improvements are required. At the peak of four rotation rates, the calculated results are larger than the test results. The main reasons for this difference include the following: 1) The measurement point of the outlet static pressure is on the duct wall, and the calculated outlet static pressure is 1/2 of the blade height. Thus, using the static pressure measured on the wall as the outlet boundary condition for calculation will cause differences between the calculated and test boundary conditions. Finally, it leads to the difference in efficiency. 2) TA36 test bench has been used for more than 15 years. The slight difference between the tip clearance in operation and the designed value will lead to the difference in efficiency.

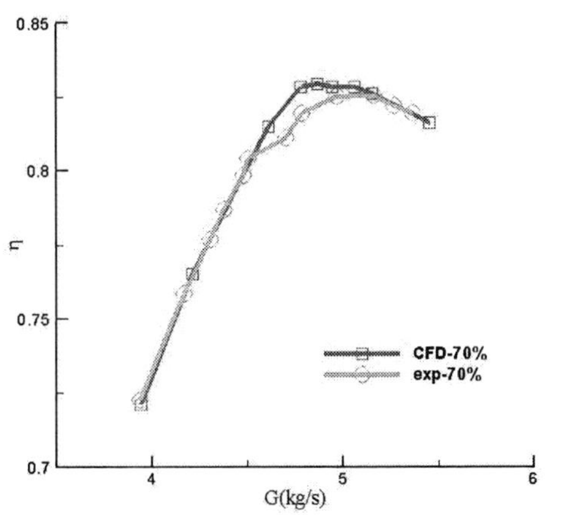

Fig. 5 Comparison result of 70% speed efficiency characteristic lines

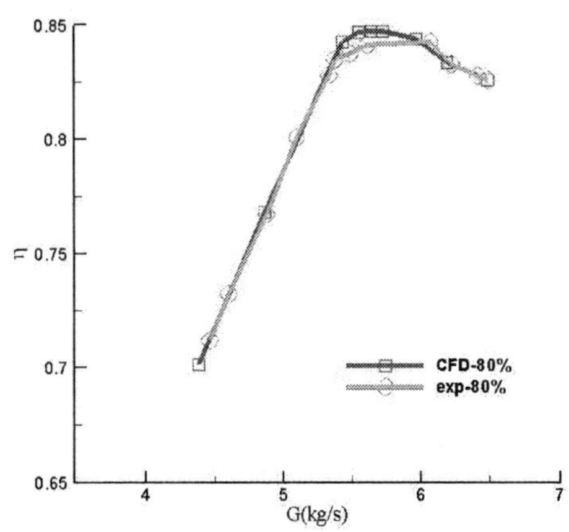

Fig. 6 Comparison result of 80% speed efficiency characteristic lines

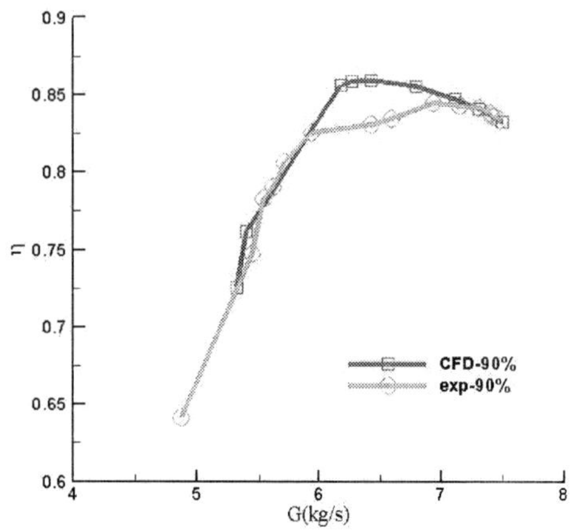

Fig. 7 Comparison result of 90% speed efficiency characteristic lines

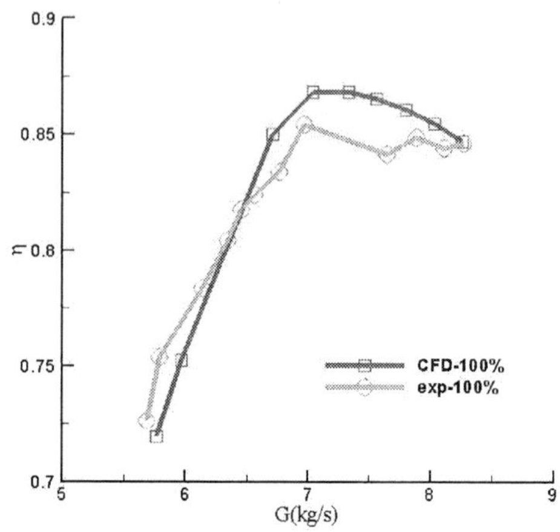

Fig. 8 Comparison result of 100% speed efficiency characteristic lines

4. Rotor wake parameter modeling

4.1. Modeling method

After rotor wake turbulence fluctuation is obtained using CFD, the flow field information cannot be directly used in the broadband noise prediction. Fig. 9 shows the schematic of the change in the two-dimensional plane cascade rotor/stator coordinates. Four turbulence input parameters are required by the fan broadband noise prediction program: turbulence integral scale Λ , background turbulence intensity u_b' , central line turbulence intensity u_w' , and turbulence width L_w . Except for integral scale Λ , the three other variables are determined by turbulence fluctuation speed directly.

The turbulence flow has Gaussian distribution. Background turbulence intensity u_b' is the minimum value of Gaussian distribution; central line turbulence intensity u_w' is the difference between the maximum and minimum values of Gaussian distribution; and turbulence width L_w is the width of Gaussian distribution at semicentral line intensity $u_b' + \frac{u_w'}{2}$.

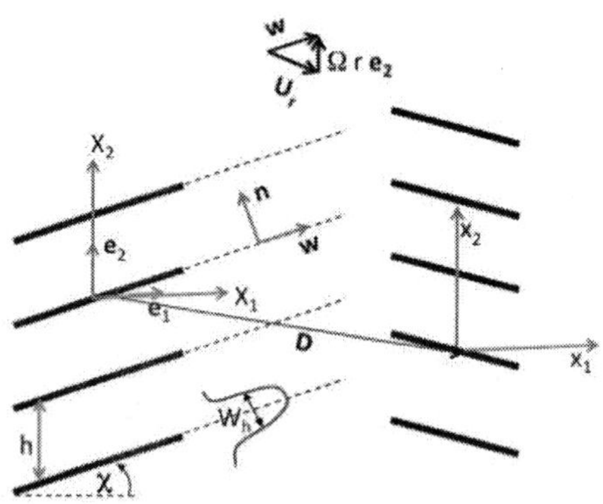

Fig. 9 Rotor/stator coordinate transformation diagram

Based on the isotropic turbulence hypothesis, the relationship between turbulence fluctuation velocity and turbulence kinetic energy is

$$u'_{RANS} = \sqrt{\frac{2}{3}k_{RANS}} \tag{6}$$

where k_{RANS} refers to the turbulence kinetic energy solved using the RANS equation in NUMECA and u'_{RANS} refers to the turbulence fluctuation velocity.

If the turbulence kinetic energy in rotor wake is calculated by means of $k-\varepsilon$ model, then integral scale Λ can be obtained from Eq. (7):

$$\Lambda = C_{\varepsilon}\frac{k_{RANS}^{\frac{3}{2}}}{\varepsilon} \tag{7}$$

where $C_{\varepsilon} = 1$ and ε refers to the dissipation factor. Then, the integral scale Λ from $k-\varepsilon$ is

$$\Lambda = \frac{k_{RANS}^{\frac{3}{2}}}{\varepsilon} \tag{8}$$

If the turbulence kinetic energy in rotor wake is calculated using SST model, which is transformed from $k-\omega$ model, then the following can be obtained:

$$\varepsilon = 0.09k_{RANS}\omega \tag{9}$$

After substituting Formula (9) into Formula (8), the expression of integral scale corresponded by SST model is

$$\Lambda = \frac{k_{RANS}^{\frac{1}{2}}}{0.09\omega} \tag{10}$$

So far, the four turbulence parameters of the two turbulence models have been modeled, and the CFD wake can be transformed into the input parameters of the three-dimensional fan broadband noise prediction program through this transformation relationship.

4.2. *TA36 rotor wake modeling result*

The actual working condition of the fan is complex and diverse. The rotation speed, the inlet-to-outlet pressure ratio, and the inlet distortion will cause different working conditions, which will lead to the changes in the rotor wake flow field. In this work, the fan rotor wake flow field at 70%, 80%, 90%, and 100% rotation velocities is calculated with the working condition for the maximum fan efficiency as the reference.

After acquiring the flow field information of rotor wake by CFD method, turbulence integral scale, background turbulence intensity, and central line turbulence intensity and turbulence width can be obtained on the basis of Eqs. (6), (8), and (10). Geometric and mainstream parameters can be obtained directly from TA36 blade information and flow field information.

On the basis of the selection of the turbulence kinetic energy and turbulence model, the turbulence parameters of the rotor wake are as follows. Tables 3 and 4 show the rotor wake turbulence parameters of SST and $k-\varepsilon$ models at the inlet turbulence kinetic energy $k = 5kg \cdot m/s^2$; Tables 5 and 6 show the rotor wake turbulence parameters of SST and $k-\varepsilon$ models at the inlet turbulence kinetic energy $k = 50kg \cdot m/s^2$.

Table 3 Rotor turbulence parameter of the inlet turbulent energy $k = 5kg \cdot m/s^2$ by SST model

Rpm/ (r/pm)	*Ma*	L_w */m*	u'_w / *(m/s)*	u'_b / *(m/s)*	Λ *(m)*
2915	0.0889	0.02916	5.121	0.2983	0.003554
2624	0.0816	0.02886	4.450	0.2770	0.003400
2332	0.0715	0.02904	3.998	0.2618	0.003512
2041	0.0622	0.02816	3.485	0.2448	0.003630

Table 4 Rotor turbulence parameter of the inlet turbulent energy $k = 5kg \cdot m / s^2$ by $k - \varepsilon$ model

Rpm/ (r/pm)	*Ma*	L_w */m*	u'_w / *(m/s)*	u'_b / *(m/s)*	Λ *(m)*
2915	0.0868	0.02719	5.722	0.002582	0.005747
2624	0.0794	0.02731	5.088	0.002582	0.005471
2332	0.0694	0.02727	4.545	0.002582	0.005657
2041	0.0604	0.02691	3.987	0.002582	0.005769

Table 5 Rotor turbulence parameter of the inlet turbulent energy $k = 50kg \cdot m / s^2$ by SST model

Rpm/ (r/pm)	*Ma*	L_w */m*	u'_w / *(m/s)*	u'_b / *(m/s)*	Λ *(m)*
2915	0.0891	0.02997	5.088	2.281	0.02180
2624	0.0811	0.02869	4.460	2.219	0.02319
2332	0.0712	0.02997	4.030	2.031	0.02399
2041	0.0621	0.02914	3.585	1.882	0.02488

Table 6 Rotor turbulence parameter of the inlet turbulent energy $k = 50kg \cdot m / s^2$ by $k - \varepsilon$ model

Rpm/ (rpm)	*Ma*	L_w */m*	u'_w / *(m/s)*	u'_b / *(m/s)*	Λ *(m)*
2915	0.0834	0.02813	6.041	2.274	0.02680
2624	0.0770	0.02896	5.413	2.143	0.02772
2332	0.0679	0.02848	4.859	1.995	0.02808
2041	0.0578	0.02726	4.243	1.829	0.02806

5. TA36 broadband noise prediction result

The sound power level of the noise in the duct can be obtained by substituting the above-mentioned turbulence, geometric, and mainstream parameters into the three-dimensional fan broadband noise prediction program. The sound wave in the duct is assumed to be plane wave, which can be converted as follows:

$$SPL = SWL - 10\,\lg A \tag{11}$$

The left of Eq. (11) refers to sound pressure level, A refers to the area of duct, and $A = \pi r_d^2, r_d = 0.3m$.

After converting the sound power level to sound pressure level based on the aforementioned relationship, the sound pressure level spectrum of broadband noise, as shown in Figs. 10–13, can be obtained.

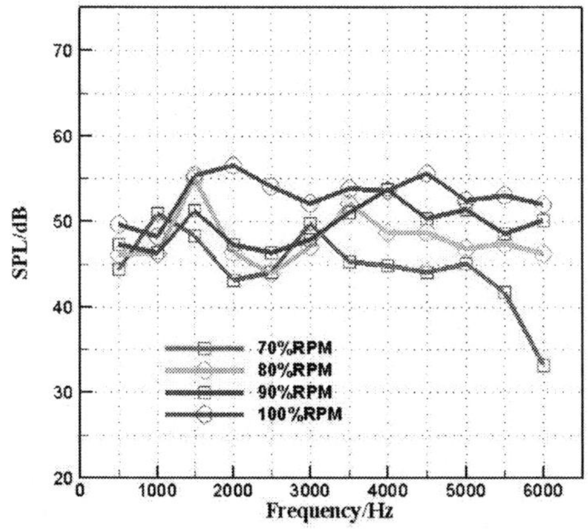

Fig. 10 Broadband noise spectrum of imported turbulent kinetic energyby SST

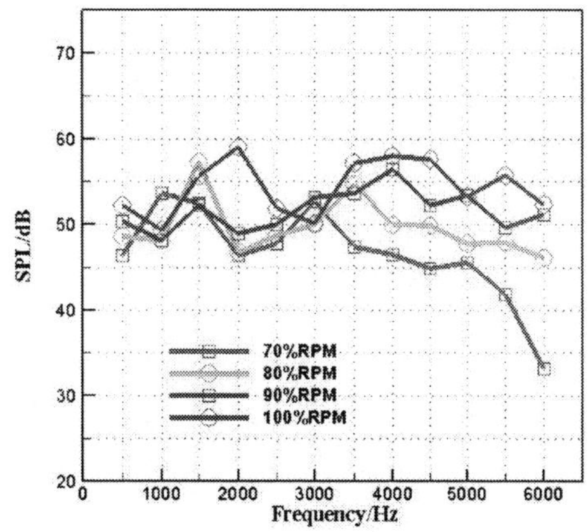

Fig. 11 Broadband noise spectrum of imported turbulent kinetic energy $k = 5kg \cdot m / s^2$ by $k-\varepsilon$ model

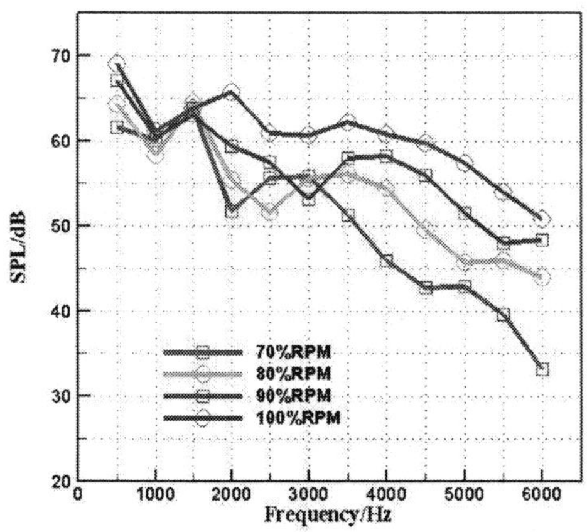

Fig. 12 Broadband noise spectrum of imported turbulent kinetic energy $k = 50kg \cdot m / s^2$ by SST model

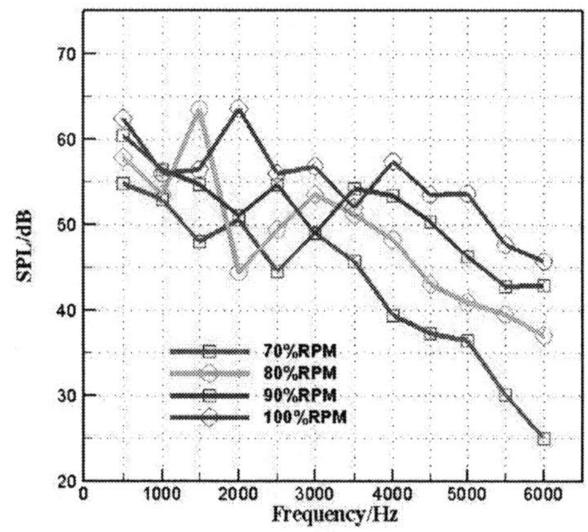

Fig. 13 Broadband noise spectrum of imported turbulent kinetic energy $k = 50kg \cdot m / s^2$ by $k-\varepsilon$ model

Comparing Fig. 10 with Fig. 11 reveals that, except for some differences near 2500 Hz at 100% rotation velocity, the broadband noise spectra obtained from the two turbulence models are nearly the same. This result further verifies the correctness of the previous CFD wake results and indicates the reliability of the three-dimensional fan broadband noise prediction program.

Figs. 12 and 13 show that, when the inlet turbulence kinetic energy is increased, the noise prediction results of different turbulence models are still consistent, but the differences become evident. This observation indicates that the selection of turbulence model and inlet turbulence kinetic energy has very important influence on fan broadband noise prediction.

5.1. Effect of inlet turbulence kinetic energy

The setting of inlet turbulence kinetic energy and the selection of turbulence model will lead to the difference in rotor wake turbulence kinetic energy in the flow field and finally lead to the difference in fan broadband noise prediction.

The reason is that the inlet turbulence kinetic energy and the turbulence model determine the distribution of the turbulent kinetic energy in the wake of the rotor. Eq. (4.1) shows that the turbulence kinetic energy determines the turbulence parameters directly, which is further reflected in the noise prediction.

Comparing Fig. 10 with Fig. 12 shows that the increase in inlet turbulence kinetic energy will lead to the overall increase in the sound pressure level of the sound field by SST model. This result is consistent with the common sense. Comparing Fig. 11 with Fig.13 shows that the increase in turbulence kinetic energy will lead to the increase in low-frequency sound pressure level and the decrease in high-frequency sound pressure level by $k-\varepsilon$ model. Therefore, when the turbulent kinetic energy increases, the noise prediction results obtained by SST model are highly regular.

5.2. Effect of turbulence model

Comparing all parameters in Tables 3 and 4 reveals that the background turbulence intensity by $k-\varepsilon$ model is two orders of magnitude smaller than that by SST model, and the turbulence integral scale is doubled. The results of parameter analysis indicate that the fan broadband noise increases with the increase in background turbulence intensity and decreases with the increase in turbulence integral scale. At $k=5kg\cdot m/s^2$, the sound pressure level of the sound field obtained by $k-\varepsilon$ model is considerably smaller than that by SST model only from the angle of independent change in parameters. However, comparing Fig. 10 with Fig. 11 indicates that, below 3000 Hz, the sound pressure level by $k-\varepsilon$ model is approximately 2dB larger than that by SST model; above 3000 Hz, the sound power levels of both models are nearly the same. This result indicates that $k-\varepsilon$ model will make the turbulence parameters change unevenly, but the total sound pressure level by this model is the same as that by SST model.

Comparing Table 5 with Table 6 shows that the parameters are not very different. From the perspective of the independent change in parameters, the corresponding sound power levels of both models should be similar. However, comparing Fig. 12 with Fig. 13 reveals that, if the inlet turbulence kinetic energy is $k=50kg\cdot m/s^2$, then the sound field by $k-\varepsilon$ model is generally 5 dB smaller than that by SST model.

Observing the background turbulence intensity obtained by $k-\varepsilon$ model in Table 4 indicates that the final prediction result is the same as that by SST model. However, the turbulence fluctuation decays too quickly from the inlet $k=5$ to rotor wake $k=0.002582$. This finding does not conform to the rules from parameter research.

The above-mentioned phenomena show that the turbulence model slightly influences the prediction results of broadband noise when the inlet turbulence kinetic energy is very small. However, when the inlet turbulence kinetic energy is large, the differences in different turbulence models are reflected. SST model conforms to the laws from parametric research for sound field prediction by flow parameters. Therefore, SST model should be preferred.

6. Conclusions

A fast prediction method of broadband noise based on three-dimensional sound source model is proposed, and the application of single-stage TA36 fan is analyzed. The following conclusions are drawn:

(1) This method can be applied to predict fan broadband noise. Noise characteristics can be obtained in the early stage of blade design. Thus, we can effectively save time and cost and shorten the design cycle of blade.

(2) The prediction results of broadband noise are influenced by the turbulence model and inlet turbulence kinetic energy. The result by SST model is more in line with the cognitive law than that by $k-\varepsilon$ model.

Notably, this study only analyzes the influence of the turbulence model preliminarily. In-depth study on the effect of the turbulence model on the turbulence parameters should be conducted to further improve the accuracy of noise prediction.

References

[1] He Lin, Zhu Haichao, Qiu Xiaojun, Du Gonghuan. Acoustic Theory and Engineering Application[M]. Beijing:Science Press. 200605. (in Chinese)

[2] Li Zhonghe; Zhang Ye; Lu Yeming; Liu Haoran; Yu Xi; Wang Xiaofang; Numerical Simulation of the Flow and Acoustic Field of A Cylinder Wake-airfoil Interaction[J]. Chinese Journal of Turbomachinery, 2020,03:52-57

[3] Tian chenye; Wu Liming; Liu Xiaomin; Huang Yutai; Ma lie; Li Jinbo; Study on Aerodynamic Noise Reduction of Multi-Blades Centrifugal Fan by Using Blades with Grooved Inlet End[J]. Chinese Journal of Turbomachinery, 2020,03:58-64

[4] Shen Zheng; He Weilin; Tang Liming; Wang Yihui; Fan Denoise Optimization of Self-ventilated Traction Motor Based on Similar Design Theory[J]. Chinese Journal of Turbomachinery, 2020,03:65-68

[5] Wu Heyuan; Jian Xiaoshu; Chen Qiang; Zhu Zhineng; Noise Analysis and Control of Centrifugal Fan for Space Station[J]. Chinese Journal of Turbomachinery, 2020,02:27-34

[6] Shi Kang; Wu TianChao; Noise Analysis and Control of Centrifugal Fan for Space Station[J]. Chinese Journal of Turbomachinery, 2015,06:32-38

[7] Huff D L. Fan noise prediction: status and needs, NASA TM-97-206533[R]. Cleverland: NASA Lewis Research Center, 1997.

[8] Peake N, Parry A B. Modern challenges facing turbomachinery aeroacoustics[J]. Annu. Rev. Fluid Mech. 2012, 44(1): 227-248.

[9] Xu Yao. The Computation and Optimization of Fan and Bypass Duct Noise of Aircraft Engine[D]. Doctor's degree Dissertation. Shanghai：Shanghai Jiaotong University，2013. (in Chinese)

[10] Ganz U W, Joppa P D, Patten T J, et al. Boeing 18-inch fan rig broadband noise test, NASA-CR-1998-208704[R]. Hampton: NASA Langley Research Center, 1998.

[11] Jacob M, Boudet J, Casalino D. A Rod-Airfoil Experiment as Benchmark for Broadband Noise Modeling[J]. Theoretical and Computational Fluid Dynamics. 2005, 19(3): 171-196.

[12] Tam C K, Ju H. Aerofoil tones at moderate Reynolds number[J]. Journal of Fluid Mechanics. 2012, 690: 536-570.

[13] Mani R. Noise due to interaction of inlet turbulence with isolated stators and rotors[J]. Journal of Sound and Vibration. 1971, 17: 251-560.

[14] Amiet R K. Acoustic radiation from an airfoil in a turbulent stream[J]. Journal of Sound and Vibration. 1975, 41(4): 407-420.

[15] Glegg S A L. Broadband fan noise generated by small scale turbulence, NASA CR-207752[R]. Hampton: NASA Langley Research Center, 1998.

[16] Hanson D B. Theory for broadband noise of rotor and stator cascades with inhomogeneous inflow turbulence including effects of lean and sweep, NASA CR-2001-210762[R]. Cleveland: NASA Glenn Research Center, 2001.

[17] Evers I, Peake N. On sound generation by the interaction between turbulence and a cascade of airfoils with non-uniform mean flow[J]. Journal of Fluid Mechanics. 2002, 463: 25-52.

[18] Nallasamy M, Envia E. Computation of rotor wake turbulence noise[J]. Journal of Sound and Vibration. 2005, 282(3): 649-678.

[19] Grace S M, Maunus J, Sondak D L. Effect of CFD Wake Prediction in a Hybrid Simulation of Fan Broadband Interaction Noise[C]//17th AIAA/CEAS Aeroacoustics Conference, AIAA paper 2011-2875.

[20] Atassi H M, Vinogradov I V. A model for fan broadband interaction noise in nonuniform flow[C]//11th AIAA/CEAS Aeroacoustics Conference, AIAA paper 2005-2880, 2005.
[21] Atassi H M, Vinogradov I V. Modelling broadband fan noise and comparison with experiments[C]//13th AIAA/CEAS Aeroacoustics Conference, AIAA paper 2007-3691, 2007.
[22] Zhang Weiguang, Wang Xiaoyu, Sun Xiaofeng. Prediction of fan broadband noise using lifting surface method[J]. AIAA Journal. 2015, 53(10): 2845-2855.
[23] Posson, H., Moreau, S., and Roger, M., "Fan-OGV Broadband Noise Prediction Using a Cascade Response," 15th AIAA/CEAS Aeroacoustics Conference, AIAA Paper 2009-3150, 2009.

IOP Conf. Series: Materials Science and Engineering 1081 (2021) 012044 doi:10.1088/1757-899X/1081/1/012044

A High-Accuracy and Conservative Unstructured Overset Grid data transfer Method for Numerical Simulation of Turbomachinery

P C Cui[1,2], J Tang[1,2], B Li[1*], H Li[1,2], X J Wu [1] and M S Ma [1]

[1] Computational Aerodynamics Institute, China Aerodynamics Research and Development Center, Mianyang 621000, China
[2] State Key Laboratory of Aerodynamics, China Aerodynamics Research and Development Center, Mianyang 621000, China

[*] pccui@foxmail.com

Abstract. The overset grid method is getting more and more popular in the numerical simulation of turbomachinery, especially in the simulation of wind turbines. Reasonable interpolation between overset grid interfaces is one of the bases to ensure correct calculation of flow fields, non-conservative interpolation is insufficient to guarantee the mass fluxes in the numerical simulation of turbomachinery. This paper develops a conservative interpolation method for hybrid overset mesh based on supermesh technology and centre-based finite volume method. A method to build the local supermesh during overset grid interface by the cell-cut algorithm is presented, and flow field variables could be interpolated conservatively during overset meshes with second-order accuracy via the local supermesh. Numerical results show that the proposed method is strictly conservative for second-order distributed flow variables. Compared with trilinear interpolation method and reverse distance-weighted interpolation method, the method in this paper decreases the numerical errors and the variables are interpolated more accurately. Engineering cases show that the method in this paper is robust and practical for engineering applications.

1. Introduction

For the numerical simulation of turbomachinery, two main approaches exist to transfer the flow through successive bladed rows in relative motion[1]. These are namely the sliding interface method[2-6] and the overset grid method[7-8].The sliding interface method is widely used in the simulation of turbomachinery[9-11] because it's easy to operate than overset grid method. However, the sliding interface method is a simplified computational model, the grid cells move together within an independent domain[12]. The overset grid method separates the complicated flowfield into several simple sections, the grids of each section is generated independently and the flowfield of each section is computed independently, the data of flowfield is transferred between the interfaces of overset grids by interpolation. With respect to the sliding interface method, the overset grid method is known to be more flexible since the size of the overlapping zone can vary and the mesh generation of complex geometries is eased[1], and the overset grid method can achieve more accurate simulation[12].

The overset grid method is getting more and more popular in the simulation of turbomachinery, especially in the simulation of wind turbines and marine current turbines. Li[13] and Zahle[14] simulated the aerodynamic characteristics of NREL Phase VI wind turbine based on overset grid method, the numerical results agree well with experiment results. Chow[15] studied the shape optimization of blades of wind turbine, and researched the aerodynamic characteristics with overset grid method. Tran[16] researched the rotational motion of OC3-Hywind turbine with overset grid method in STAR-CCM+ software.

However, the interpolation between overset mesh interfaces is still a challenge to Computational Fluid Dynamics (CFD). There are usually 2 interpolation methods: conservative interpolation and non-conservative interpolation. The non-conservative interpolation methods are widely used in many non-discontinuity flow field simulations because it is easy to operate. However, it strongly depends on the grids scale in the overset mesh area. When the grid scale matches badly, the non-conservative method is insufficient to guarantee the accuracy, and may cause numerical oscillation in numerical simulation.

In general, conservational interpolation is an important basis to ensure the correct flow field. Farrell[17-18] proposed the concept of "supermesh" and developed an integral conservation interpolation method of two-dimensional mesh, and then he did further research in the conservative interpolation method of three-dimensional hybrid mesh in 2011. Wang and Xu[19-21] proposed the concept of "efficient search" of donor cells in overset grid, which improved the parallel efficiency of interpolation in overset grid, and increased the interpolation accuracy of overlapping grid areas by increasing the number of interpolation cells. Usually in the flow field of strong shock or large separation, the traditional second-order gradient interpolation, distance interpolation and trilinear interpolation could not guarantee the conservation of physical between overset mesh interfaces, which would lead to a wrong shock wave position and a rough flow field[22-25].

Based on supermesh and three dimensional unstructured grids, this paper develops an implicit conservative interpolation method for overset grid, the results show that this method decreases the numerical errors and is strictly conservative for second-order flow variables, and engineering cases show that the method in this paper is robust and practical for engineering applications.

2. Numerical method

The numerical simulations in this paper were conducted by the China Aerodynamics Research & Development Center (CARDC) MFlow-code, which is an unstructured finite volume cell-center CFD solver and has participated in the 5th and 6th CFD drag prediction workshop. Second-order accuracy in space is achieved by linear reconstruction in cells. The vertex-based Gauss method is adopted for gradient computations [26]. Roe scheme is used for inviscid flux computations.

2.1. Governing equations

The unsteady Navier-Stokes equations are discretized by a cell centered finite-volume method. The integral form of unsteady Navier-Stokes equations for a bounded domain V with a boundary $\partial\Omega$ can be expressed as:

$$\frac{\partial}{\partial t}\iiint_V Q dV + \oiint_{\partial\Omega}\left(H(Q)\cdot\hat{n} - Q\vec{v}_g\cdot\hat{n}\right)dS = \oiint_{\partial\Omega} H_v(Q)\cdot\hat{n}dS \tag{1}$$

Where $Q=[\rho\ \rho u\ \rho v\ \rho w\ \rho E]^T$ is conservative vector, $\vec{v}_g$ is wall velocity $H(Q)$ is inviscid flux vector, and $H_v(Q)$ is viscous flux vector. Here, $[\rho\ u\ v\ w\ E]$ denotes the density, velocity of three directions, and specific total energy of the fluid, $\vec{n}$ is outward pointing normal unit vector of boundary. The inviscid flux is replaced by a numerical Riemann flux function Roe schemes, and the viscous flux is discretized with central difference scheme. The Venkatakrishnan limiter is used to restrict oscillation.

2.2. Grid generation

The computational grid is unstructured hybrid grid, including triangular prism, pyramid, and tetrahedral grids. The triangular prisms and tetrahedrons are used to simulate the boundary layer and the isotropic region of the spatial flowfield, respectively, and the pyramid grid is used to transition the triangular prism and the tetrahedron.

3. Conservative interpolation algorithm

3.1. Conservative conditions

Conservative interpolation requires the integration of the physical variables could be kept constant in the overlapping areas of the overlapping grids. Suppose there are two sub-meshes Γ_A, Γ_B, and the overset area is Γ_C. The distribution of physical variable ϕ satisfies the function $\phi^A(x)$ in mesh Γ_A, and the distribution of physical variable ϕ satisfies the function $\phi^B(x)$ in mesh Γ_B. If the physical variable ϕ satisfies the following condition in any area Γ_D in mesh Γ_C:

$$\int_{\Gamma_D} \phi^A(x)dV = \int_{\Gamma_D} \phi^B(x)dV \tag{2}$$

where the physical variable ϕ is strictly conservative in the overset area Γ_C of meshes Γ_A and Γ_B.

3.2. Supermesh

Farrell proposed the concept of "supermesh"[17-18], for any meshes Γ_A and Γ_B, N_A and N_B is the grid nodes, E_A and E_B is the grid edges, supermesh Γ_C is composed with their common nodes and edges, N_C and E_C are the node and edge of grid Γ_C, which satisfy following conditions:

$$N_C \supseteq N_A \cup N_B \quad \Gamma_C \in \{\Gamma_A, \Gamma_B\} \tag{3}$$

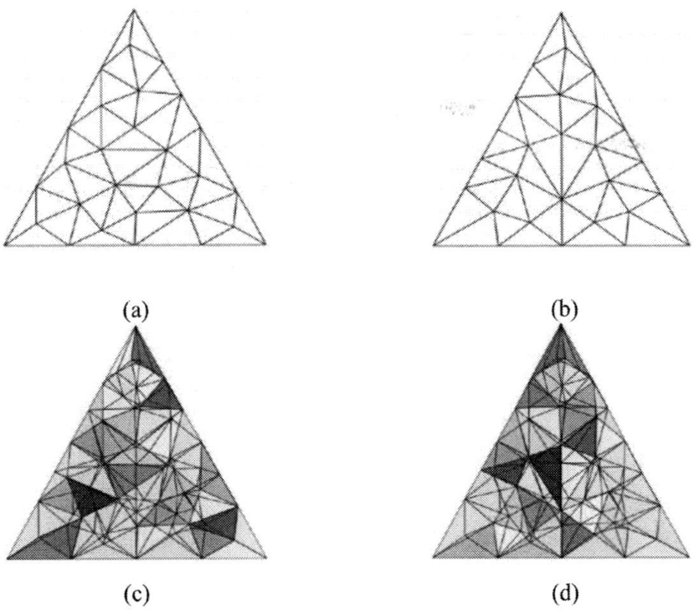

Figure 1. Schematic of supermesh.

In short, all the nodes in the supermesh Γ_C are contained in the parents meshes Γ_A and Γ_B, and all the cells in the supermesh Γ_C are contained in the parents meshes Γ_A and Γ_B. Each grid cell in the supermesh Γ_C is intersected by the grid cells of Γ_A and Γ_B. Cells of supermesh do not intersect with any cells of Γ_A and Γ_B. Obviously, the construction of a supermesh is not unique, but if we treat one of the parents meshes as "target mesh", and treat the other mesh as "donor mesh", let each cell of donor mesh intersect with cells of target mesh, the construction of a supermesh is unique. Fig. 1 shows

IOP Conf. Series: Materials Science and Engineering 1081 (2021) 012044 doi:10.1088/1757-899X/1081/1/012044

the super grid diagram, Fig.1 (a) and (b) show two parents meshes Γ_A and Γ_B, if Γ_A is treated as "target mesh", we can get a supermesh in Fig.1 (c), and the color part of the mesh belong to Γ_A. If Γ_B is treated as "target mesh", we can get a supermesh in Fig.1 (d), and the color part of the grid belong to Γ_B.

For any mesh Γ_B, these exists a supermesh Γ_C, whose grid cell can be found on the specific location and the corresponding physical variables in mesh Γ_A, the variable ϕ has the same distribution in mesh Γ_A and Γ_B, $\phi^A(x)=\phi^B(x)$. The variable of each cell of Γ_B can be summed up by the variable of its supercells, and the variable of each cell of sepermesh can be interpolated by mesh Γ_A. Due to the variable ϕ has the same distribution in mesh Γ_A and Γ_B, so the interpolation function can be write as:

$$\begin{aligned}\int_{\Gamma_C}\phi^A(x)dV&=\sum_{\Gamma_C}\phi^A_i V_i\\&=\sum_{\Gamma_C}\phi^B_i V_i=\int_{\Gamma_C}\phi^B(x)dV\end{aligned} \tag{4}$$

Thus the variable ϕ could be interpolated conservatively from Γ_A to Γ_B through the medium supermesh Γ_C.

Cells of the supermesh Γ_C are constructed by intersection of parents meshes Γ_A and Γ_B. Common three-dimensional polyhedral grid units include tetrahedrons, triangular prisms, pyramids, hexahedrons and so on, of which triangular prisms, pyramids and hexahedrons have quadrilaterals, and in the grid intersection process we can not guarantee the four vertices are in the same plane. Therefore, the polyhedron mesh units are decomposed into multiple tetrahedral units uniformly, and the intersection of polyhedrons is obtained by calculating the intersection of tetrahedral units. The intersection method between tetrahedrons has been mature, and Sutherland-hodgman method is employed in this paper, for example in two dimensions the main idea is to cut the "target polygon" by the edge of "cut polygon" and to get a new polygon. Fig.2 shows a two-dimension case of Sutherland-hodgman method, $A_1A_2A_3$ is "target polygon" and $B_1B_2B_3$ is "cut polygon", $A_1A_2A_3$ is cut by each edge of $B_1B_2B_3$ and we can get a new polygon B_1CDEF.

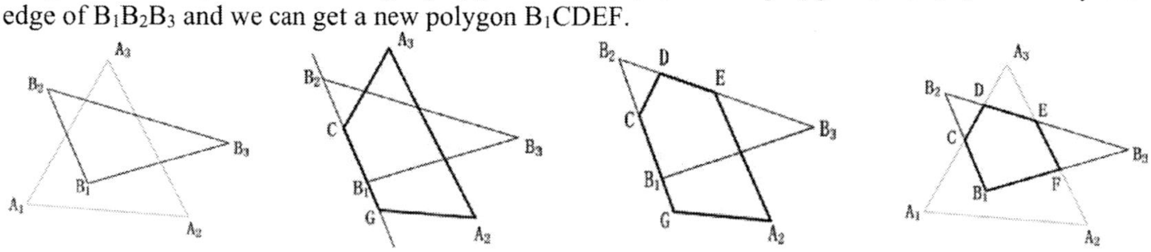

(a) Original polygon $A_1A_2A_3$ and $B_1B_2B_3$ (b) $A_1A_2A_3$ cut by B_1B_2 (c) A_2A_3CG cut by B_2B_3 (d) New polygon B_1CDEF

Figure 2. The intersection of polygons in two dimensions.

Intersection method between three-dimensional tetrahedrons is same as two dimensional polygons, Fig 3 shows the tetrahedral mesh intersection process, in Fig.3 (a) the black tetrahedron is target tetrahedron, the red tetrahedrons are the cutting tetrahedrons, the target tetrahedron is cut by each surface of cutting tetrahedrons to get each unit of the supermesh. Fig.3 (b) shows a supermesh which is constructed by the intersection of target tetrahedron and cutting tetrahedrons.

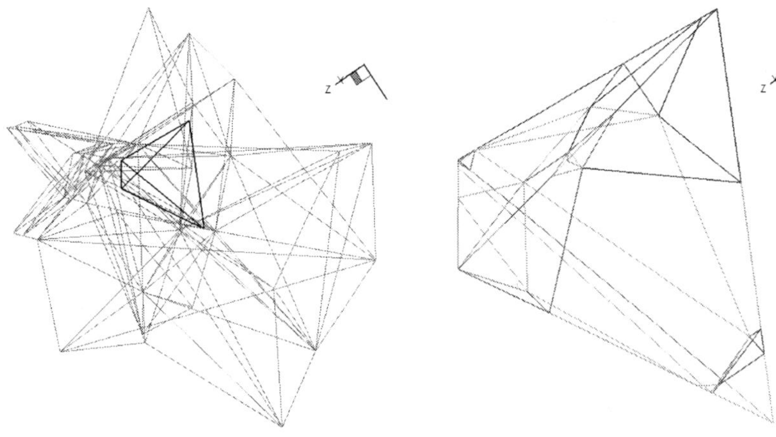

(a) Target tetrahedron and cut tetrahedrons (b) Supermesh in three dimensions

Figure 3. The intersection of tetrahedrons.

An important consideration for the construction of supermesh is run-time efficiency, which depends on the intersections calculations. Menon tested the numerical cost of intersections calculations[27], and it just costs 25.1 seconds to deal with 3.03 million grids using an Intel Core 2 Quad computer running at 2.83GHz with 4Gb of RAM, which is a tiny computational cost compared with normal numerical simulation.

3.3. Interpolation method

Interpolation of variable ϕ from Γ_A to Γ_B is achieved by the medium supermesh Γ_C. K is a grid cell of mesh Γ, the physical variable is firstly interpolated from Γ_A to supermesh Γ_C with a second-order Taylor expansion:

$$\phi(X_{K_C}) = \phi_{K_A} + (\nabla\phi)_{KA} \bullet (X_{K_C} - X_{KA}) \bullet \psi_{KA} \tag{5}$$

where $(\nabla\phi)_{KA}$ is the gradient of the physical variable and ψ_{KA} is the limiter function, and Vencat Limiter is employed in this paper. The physical variables are then interpolated from the supermesh Γ_C to the target mesh Γ_B, In this paper, the second-order Taylor method is adopted in the interpolation method. And the interpolation from source mesh to target mesh is strictly conservative for second-order distributed physical variables.

According to whether the "donate cells" also act as "target cells", the interpolation between overset grids can be divided into implicit interpolation and explicit interpolation. An implicit interpolation method is employed in this paper, the donate cells and target cells are not strictly distinguished, each single overset grid sends data to other overset grids and also receives data from other overset grids during parallel computing. The implicit parallel interpolation is more efficient than explicit interpolation, and can be appropriate for massive grids of complex configurations.

4. Numerical results

4.1. Linear equation

Assume that the physical variables are linearly distributed in the grid area and satisfy the equation:

$$\phi(x, y, z) = 2x + 3y + z \tag{6}$$

Γ_A is a structured mesh and Γ_B is a unstructured tetrahedron mesh, and they share the same grid area. Let the distribution of the physical variables on the mesh Γ_B satisfy equation (6), then interpolate

100 times between Γ_A and Γ_B, and calculate the absolute value of the mean value of the conservation and each cells.

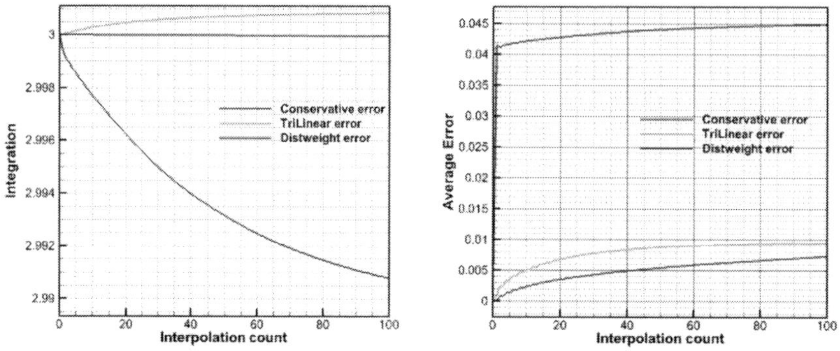

(a) Integration of variable (b) Average error of interpolation

Figure 4. Linear equation interpolation.

Fig.4 shows a comparison of the proposed method in this paper and traditional interpolation methods on accuracy and conservation, Fig.4 (a) and Fig.4 (b) show the integration and average error during the interpolation process. Integration denotes the sum of variable of all cells, and average error denotes the average error of each cell. As is shown in Fig.4 (a), the integral value of the physical variable on the grid shows that the interpolation developed in this paper satisfies the conservation of the integral value of the physical variable in the overlapping area. Fig.4 (b) shows the absolute value of the interpolation error of the physical quantity on the grid. Compared with the distance-weight method and trilinear method, the method in this paper decreases the errors of interpolation.

4.2. Trigonometric function

Assume that the physical variables are trigonometric distributed in the grid area and satisfy the equation:

$$\phi(x,y,z) = 1 + \sin(2px)\,\sin(2py)\,\sin(2pz) \tag{7}$$

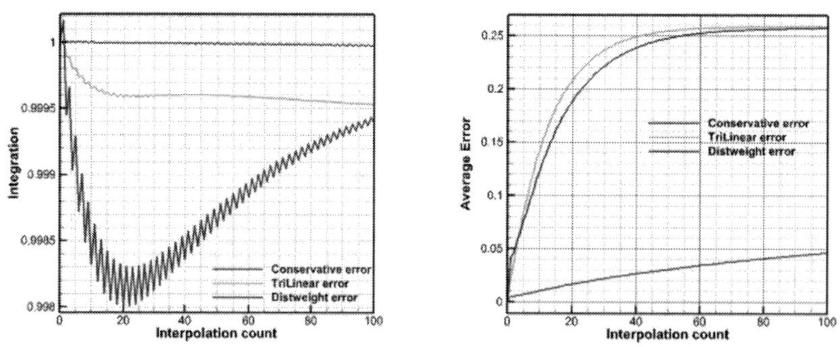

(a) Integration of variable (b) Average error of interpolation

Figure 5. Trigonometric function interpolation.

Fig. 5 also shows a comparison of the proposed method in this paper and traditional unconservative interpolation methods on accuracy and conservation. It can be seen that the average error of proposed interpolation method is much less than other methods, and is almost conservative for the integral of variables.

4.3. Numerical simulation of engineering applications

The overset grid method is getting more and more popular in the simulation of turbomachinery, especially in the simulation of wind turbines[13-16, 28]. An isolated six-bladed propeller at different speeds is numerically simulated to investigate the robustness and engineering practicability of the developed method. Wind tunnel experiments are conducted in 8m×6m Low Speed Wind Tunnel of CARDC. Fig.6 (a) shows the overset grids of propeller, the grid in red color is the sub-grid of blades, and the grid in blue color is the sub-grid of the hub, which is also the background grid and is refined in the zone of slipstream in order to accurately simulate the propeller slipstream. The number of total grids is 18.2 million.

Fig.6 (b) and Fig.6 (c) show the Q-criterion of propeller slipstream at advance ratio 1.58, the vortex is observed clearly because of the mesh refinement behind the propeller. A comparison of numerical simulation and experiment result of torque coefficient is presented in Fig.6 (d), and the abscissa is the propeller advance ratio. The numerical result is well consistent with the experiment result, which verifies that the proposed method is robust and practical for engineering applications.

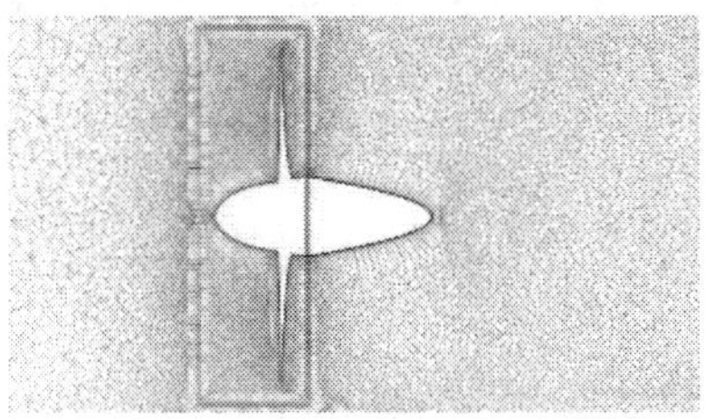

(a) Overset grids

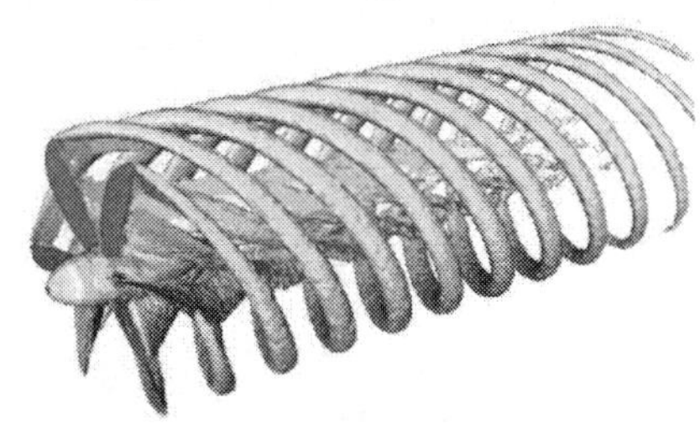

(b) Propeller slipstream

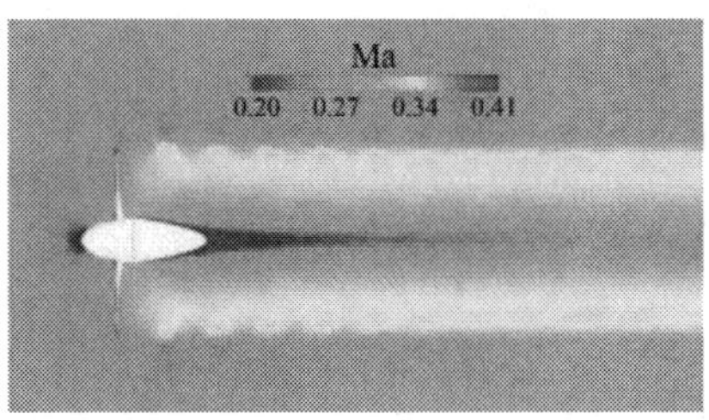

(c)Ma distribution of propeller slipstream

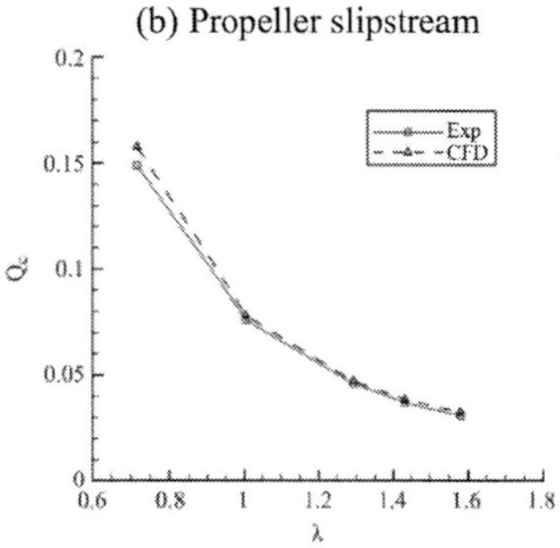

(d) Comparison of the thrust coefficient

Figure 6. Numerical simulation of an isolated propeller.

5. Conclusion

A new conservative data transfer method for overset grid is developed based on unstructured grid and supermesh theory in this paper, which improves the accuracy of data transfer between the overset interfaces, the method is strictly conservative for second-order distributed flow variables, and engineering cases show that the method in this paper is robust and practical for engineering applications.

IOP Conf. Series: Materials Science and Engineering 1081 (2021) 012044 doi:10.1088/1757-899X/1081/1/012044

Acknowledgement
This work is supported by National Numerical Windtunnel (NNW).

References
[1] Jerme D L Duchaine F Gicquel L et al 2018 Journal of Computational Physics **363** 371
[2] Tong Y Yang Z G 2014 Chinese Journal of Turbomachinery **3** 31
[3] Chen J X Lai H X 2016 Chinese Journal of Turbomachinery **1** 24
[4] Sun C D Wang Z Y Yang Z G et al 2016 Chinese Journal of Turbomachinery **1** 8
[5] Liu X M Yang L N 2017 Chinese Journal of Turbomachinery **6** 66
[6] Tian C R Li B K 2018 Chinese Journal of Turbomachinery **1** 33
[7] Brazell M J Sitaraman J Mavriplis D J 2016 Journal of Computational Physics **322** 33
[8] Völkner S Brunswig J Rung T 2017 Computers & fluids **148** 39
[9] Zhao L Chen S Gao T Y et al 2019 Compressor Blower Fan Technology **61**(5) 13
[10] Huang W J Shen Z Z Gu C W et al 2019 Compressor Blower Fan Technology **61**(5) 40
[11] Alexej P Matthias M Wolfgang S 2019 Compressor Blower Fan Technology **61**(4) 1
[12] Liu Y Cai W J Wang Z 2018 Chinese Journal of Ship Research **13**(1) 85
[13] Li Y Paik K Xing T et al 2012 Renewable Energy **37**(1) 285
[14] Zahle F Sørensen N N Johansen J 2009 Wind Energy **12**(6) 594
[15] Chow R Van D C P 2012 Wind Energy **15**(8) 967
[16] Tran T Kim D Song J 2014 Energies **7**(8) 5011
[17] Farrell P E Piggott M D Pain C C et al 2009 Computer Methods in Applied Mechanics and Engineering **198**(33–36) 2632
[18] Farrell P E Maddison J R Computer Methods in Applied Mechanics and Engineering **200**(1–4): 89
[19] Wang W Yan C Yuan W et al 2016 Acta Aeronautica et Astronautica Sinica **37**(10) 2980
[20] Huang Y Yan C Wang W et al 2017 Journal of Beijing University of Areonautics and Astronautics **43**(2) 285
[21] Xu C G Dong H B Liu J 2016 Expolosion and Shock Waves **36**(3) 305
[22] Zhang Y B Zhou N C Cheng J T 2015 Acta Aerodynamica Sinica **33**(3) 279
[23] Journeaux A A Nemitz N Moreau O 2013 Compel International Journal for Computation & Mathematics in Electrical & Electronic Engineering **33**(1/2) 663
[24] Kang Z L Yan C Yu J et al 2013 Acta Mechanica Sinica **29**(2)149
[25] Cui P C Tang J Li B et al 2018 Acta Aeronautica et Astronautica Sinica **39**(3) 121569
[26] Peña E Calvo B Martínez M A et al 2005 Clinical Biomechanics **20**(5) 498
[27] Menon S Schmidt D P 2011 Computer Methods in Applied Mechanics and Engineering, **200**(41) 2797
[28] Gong X Q Ma M S Zhang J et al 2018 Journal of Aerospace Power **33**(2) 345

Volumetric entropy generation rate associated with tip clearance flow in linear cascade

Long Bingxiang[1], Chen Qin[1], Ren Zebin[1], Liao Daxiong[1]

1.Facility Design and Instrumentation Institute of Chinese Aerodynamic Research and Development Centre, No. 6, South Section, Second Ring Road, Mianyang City, Sichuan Province, P.R. China.

bingxiang2011@126.com

Abstract: The tip clearance flow structure and its corresponding entropy generation rate of a linear cascade with 1.85mm radial tip clearance operating under three different incidence angle are investigated. The incidence angle influences the tip leakage vortex position as well as the size of the tip leakage vortex. Larger incidence angle results in larger tip leakage vortex and pushes the tip leakage vortex further away from the suction side. The entropy generation rate related to the tip clearance flow structure is calculated and it is confirmed that tip clearance flow is the main major source of aerodynamic loss in linear cascade. Larger incidence angle leading to higher entropy generation rate. Results also show that it's not the core region of tip leakage vortex but the edge vortex and shear layer of the tip clearance flow which has the highest entropy generation rate. When operating under an incidence angle of 4°, there exist an extra high entropy generation rate region in main flow. It is found that the compression and expansion of fluid due to the existence of tip clearance vortex was the main reason for this extra high entropy generation.

1. Introduction

The radial clearance between stator-vane to end-wall and between rotor-blade to end-wall is indispensable for turbomachine operation and adjusting. The clearance flow, including tip clearance flow and hub clearance flow, due to the existence of radial clearance, however, has been a major source for loss, blockage, instability and noise, and is detrimental to the aerodynamic, aero-acoustic and structural performance of turbomachine. Since clearance flow has significant influences on turbomachinery performance, it attracts numerous researchers' attention.

A number of numerical and experimental investigations have been carried out across the world for the purpose of getting deep insight into the detailed structure of the clearance flow. The investigation of Kang and Hirsch shows that horseshoe vortex and multiple vortex exist in tip region and the existence of tip clearance vortex results in passage vortex close to end wall and suction side, Kang and Hirsch also find that the vorticity of tip clearance vortex increase right after it formed in the leading edge region and then decrease gradually[1][2]. Kang and Hirsch also indentified tip separation vortex and pointed out that leakage flow mixing has a significant influence on the internal loss of compressor cascade[3]. Zhong studied the unsteadiness of tip clearance flow using DDES method and found that smaller tip gap results in weaker anisotropy[4]. The investigation of Kunz shows that Navier-Stokes procedures with k-epsilon turbulence model has satisfactory performance in capturing important physical phenomena of tip clearance flow[5]. Tian and Li found that the intensity and influence area of tip leakage vortex increase with increasing tip leakage clearance[6]. The investigation of Muthanna and Devenport reveals that the dynamics of tip clearance vortex are dominated by streamwise mean-velocity deficit the vortex produces[7]. Alexej studied the tip clearance flow of axial-flow fan using LES method and found that tip clearance flow results in lower turbulence kinetic energy[8]. Some passive methods has been utilized to control the tip clearance flow. Mehdi pointed out that porous treatment can reduce the intensity of tip vortex as well as the shear layer rollup and boundary layer separation[9]. The investigation of Chen demonstrated that circumferential groove can suck low energy tip clearance flow and thus has a significant importance in extending stall margin of centrifugal compressor[10].Wang studied the effects of blade tip profile and found that optimized blade tip profile can reduce tip leakage flow velocity, weakens the strength of mixing, and therefore results in loss reduction[11].

The loss induced by clearance flow is a dominant part of turbomachinery total loss. Clarifying the loss mechanism associated with the complex tip clearance flow structure is beneficial for developing design method to improve the turbomachinery efficiency as high as possible. According to the theory of the second law analysis, the volumetric entropy generation rate is capable to present the local loss intensity and thus providing researchers the possibility of clarifying the loss mechanism[12]. Up to date, three main approaches has been proposed for volumetric entropy generation rate calculation. By adding a turbulent conductivity and turbulent viscosity to the molecular conductivity and viscosity, Moore and Moore proposed the first method to close the volumetric entropy equation[13]. The Moore model has been utilized for studying end-wall sealing flows[14] and tip cavity flows[15] in turbines based on RANS method. Kramer-Bevan[16] proposed a new model for volumetric entropy equation closing and has been used by Kock[17] and Takakura[18] to investigate the entropy generation rate in the tip region of a high-pressure turbine using RANS simulation results. Adeyinka[19] extended past Reynolds averaging techniques for the momentum and scalar transport equations to the second law for turbulent and has been used by Orhan[20] to study loss mechanism of an axial turbine cascade.The applicability of the model proposed by Kramer Bevan in loss mechanism study has been validated and is used in this paper.

Direct and quantitative study of clearance flow in actual turbomachines is challenge due to some safety issues and some technique limitation. Linear cascade with gap can reproduce flow structures quite similar to the tip clearance flow in actual turbomachines and providing researchers an efficient and effective way of studying clearance flow phenomenon. In this paper, a numerical research using RANS is carried out to study the tip clearance flow structure and the loss mechanism associated with tip clearance flow under three different incidence angle (α). The paper is organized as follows:

After the validation of the RANS method, the numerical model is introduced and the numerical results are presented. The tip clearance flow structure and the loss production associated with these flow structures under three different incidence angle is then analyzed in detail to understand the tip clearance flow structure, the corresponding loss-generating mechanism and the influence of incidence angle on clearance flow structure and the corresponding loss mechanism.

2. Model and Numerical Method

2.1. The studied cascade

the linear compressor cascade of LMFA at Ecole Centrale Lyon is selected for the purpose of evaluating the accuracy of numerical methods as there are many detailed and accurate experiments of this cascade were carried out in previous researches. The geometric parameters of the cascade are shown in Table 1. The inlet velocity is 40m/s so that the flow in the cascade can be treated as incompressible. The incident angle is set to 4 degrees, so that the corner separation exist. More detailed information can be found in Reference[21].

In order to study the effects of incidence angle on clearance flow structure and entropy generation rate, a 1.85mm tip clearance gap is added to the LMFA cascade and three different operating condition, which corresponding to an incidence angle of -4°, 0° and 4°, were set in this study.

Table 1 Geometric parameters of the cascade

Parameter	Value
Chord(mm)	150
Camber angle(Degree)	23.22
Stagger Angle(Degree)	42.7
Pitch spacing(mm)	134
Solidity	1.12
Blade span(mm)	370
Aspect ratio	2.47
Design upstream flow angle(Degree)	54.31
Incident angle(Degree)	4
Design downstream flow angle(Degree)	31.09

2.2. Numerical setup

The HOH mesh topology is employed for meshing the studied cascade. To guarantee the accuracy of the numerical simulation, the Y-plus of the first layer near-wall of computational mesh is less than 1. The length of the inlet and outlet region are 1.5 times of the chord length (1.5C) and 2C in order to make these two boundaries less reflective. Half of blade span is used to carry out this research for the purpose of reducing the computational consumption. The mesh dependence was studied and the grid number chosen for this study is about 3.0 million. Mesh information is show in Figure 1.

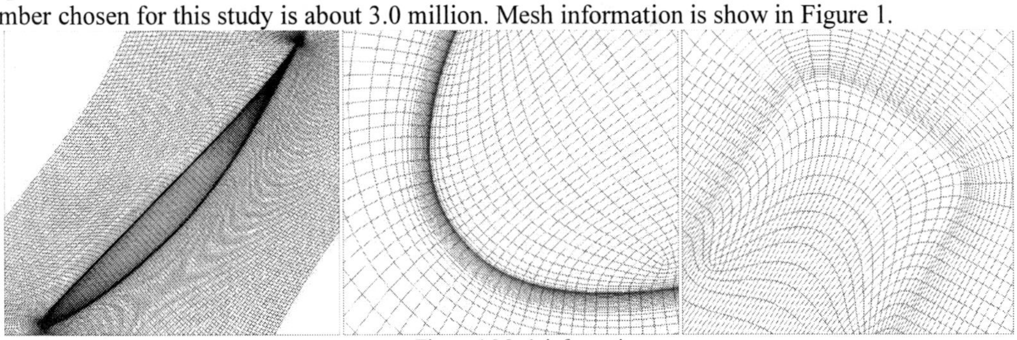

Figure 1 Mesh information

Ansys Fluent 15.0 is used as the computational platform. Realizable k-epsilon turbulent model with enhanced near-wall treatment is employed in this study. Velocity inlet and Outflow boundary condition is set up for inlet and outlet respectively. The inlet velocity distribution is taken from experiments. Other boundary conditions were give in normal ways for internal flow simulation.

2.3. Numerical validation.

Static pressure coefficient is a common parameter to quantify the blade load distribution and Total pressure loss coefficient is a parameter corresponding to loss generation in turbomachine and they are defined as eqn.(1) and eqn.(2) respectively.

$$C_p = \frac{p - p_\infty}{Pt_\infty - p_\infty} \tag{1}$$

$$\omega^* = \frac{\int_0^s C_{pt}(y,z)\rho(y,z)u(y,z)dy}{\int_0^s \rho(y,z)u(y,z)dy} \tag{2}$$

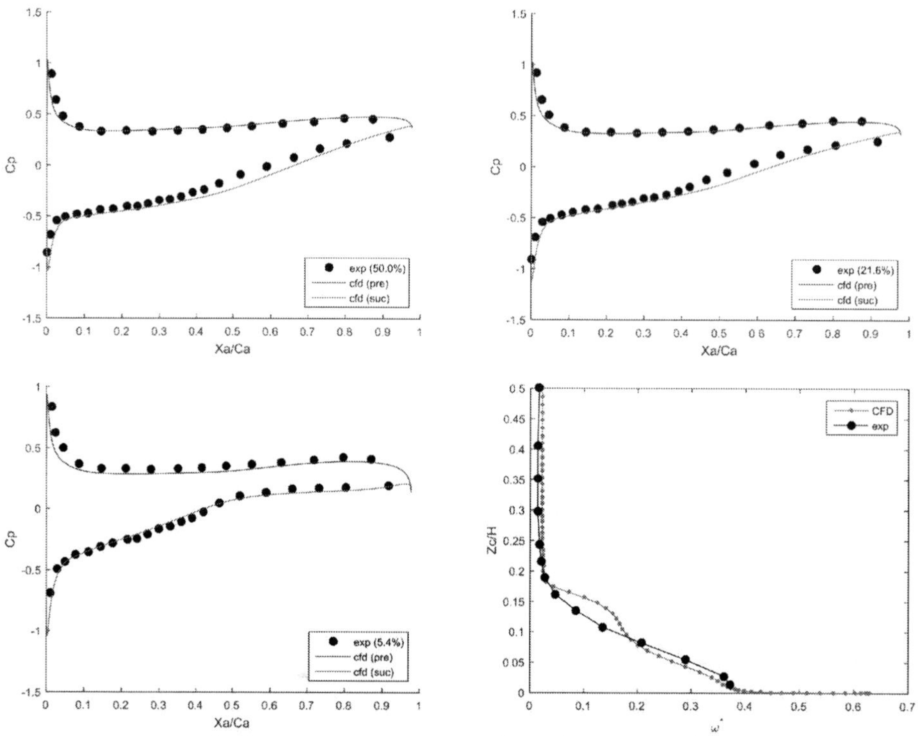

Figure 2 the static pressure coefficient and total pressure loss

Figure 2 depicts the static pressure coefficient on the blade of the experiment and RANS results at the 50%, 21.6% and 5.4% and it also presents the total pressure loss coefficient taken from experiment and CFD results at the plane 0.363 times of the axial chord length away from the trailing edge of the cascade. From Figure 2, it is clear that RANS can predict the separation vortex and the total pressure loss corresponding to the separation vortex with satisfactory accuracy.

3. Results and Analysis

3.1. Clearance flow structure

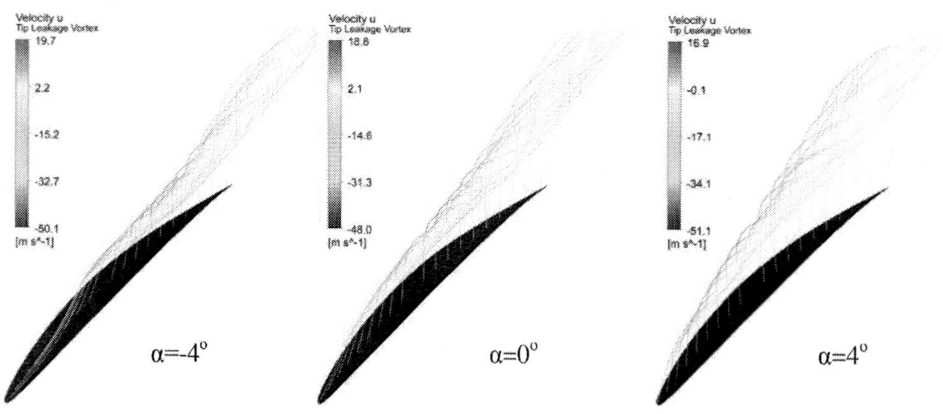

Figure 3 Tip clearance vortex under different incidence angle operating condition

It is clear from Figure 3 that tip leakage vortex originates from the suction side and being pushed away from the suction side to the pressure side of the neighbouring blade as well as lifted away from the casing wall. Figure 3 also presents that the tip leakage vortex size expanded as the tip leakage vortex being convected downstream. The generation and development process of tip leakage vortex revealed is consistent with other previous researches.

Figure 3 also reveals the variation of tip leakage vortex structure with the change of incidence angle. The origin of the tip leakage vortex moves downstream along the blade suction side with decreasing incidence angle. The origin of the tip leakage vortex located around 3%,10% and 30% chord length for incidence angle of 4°, 0° and -4°. Increasing incidence angle pushes the tip leakage vortex further faraway from the suction side. From Figure 3, it is also clear that larger incidence angle results in larger tip leakage vortex.

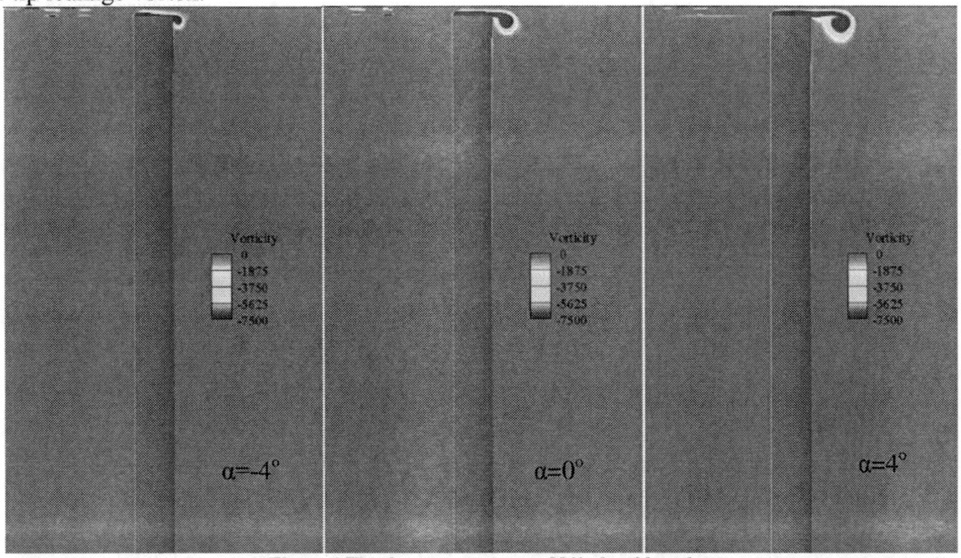

Figure 4 Tip clearance vortex at 50% chord length

Figure 4 demonstrates the strength of tip leakage vortex at 50% chord length. From Figure 4, it is obvious that the strength of the tip leakage vortex increases with the increase of the incidence angel. The streamwise component of tip leakage vortex vorticity is negative, which means tip leakage vortex rotate clockwise. Figure 4 also reveals that tip leakage vortex size and its distance to suction side increase with incidence angle increase. Increasing incidence angle results in large static pressure difference between blade suction side and blade pressure side as well as rapid flow acceleration on the suction side, which is the fundamental reason for tip leakage flow structure and vortex strength variation with incidence angle.

3.2. *Entropy Generation Rate*

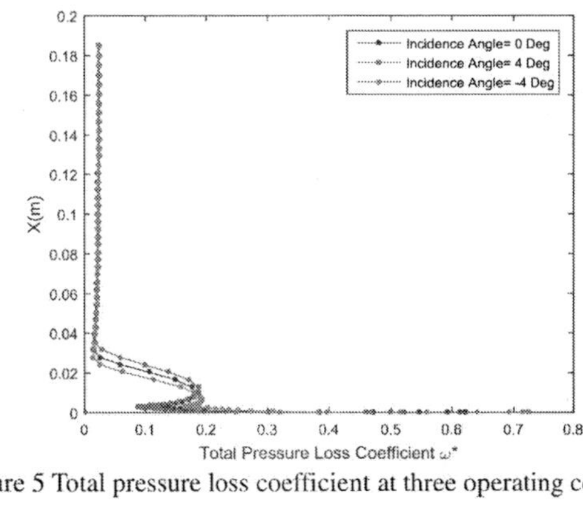

Figure 5 Total pressure loss coefficient at three operating conditions

3.2.1. Total pressure loss coefficient. Figure 5 presents the variation of total pressure loss coefficient with incidence angle. Data is extracted from a plane which is 0.363 times of axial chord length away from the blade trailing edge. Total pressure loss coefficient is calculated according to Eq. (2). From Figure 5, it is clear that the loss magnitude associated with the tip clearance flow increase with incidence angle. It can also be seen from Figure 5 that the region which is influenced by tip clearance flow expanded as incidence angle increases. For operating conditions with an incidence angle of -4°, 0° and 4°, the origin of influence gradually increase from about 0.025m to around 0.035m. In the incidence angle range from -4° to 4°, the influence of incidence angle on the profile loss is relatively small.

3.2.2. Entropy Generation Rate. According to Kramer-Bevan' model, volumetric entropy generation rate (S''') is defined as

$$S''' = \frac{2\mu \bar{s}_{ij}\bar{s}_{ij}}{\bar{T}} + \frac{\varepsilon\bar{\rho}}{\bar{T}} + \left(1+\frac{\alpha_t}{\alpha}\right)\frac{\lambda}{\bar{T}^2}\frac{\partial\bar{T}}{\partial x_i}\frac{\partial\bar{T}}{\partial x_i} \tag{3}$$

Where, $\bar{T}$ is the temperature, ρ is the density, $2\mu\bar{s}_{ij}\bar{s}_{ij}$ is the viscous stress tensor, ε is the turbulence dissipation rate, α_t and α represents the turbulent thermal diffusivity and thermal diffusivity respectively. λ is the heat transfer coefficient. For incompressible flow, $\bar{T}$ is assumed as a constant. In this way, $\left(1+\frac{\alpha_t}{\alpha}\right)\frac{\lambda}{\bar{T}^2}\frac{\partial\bar{T}}{\partial x_i}\frac{\partial\bar{T}}{\partial x_i} \cong 0$. Then, Eq. (3) can be simplified as:

$$S''' = \frac{2\mu \bar{s}_{ij}\bar{s}_{ij}}{\bar{T}} + \frac{\varepsilon\bar{\rho}}{\bar{T}} \tag{4}$$

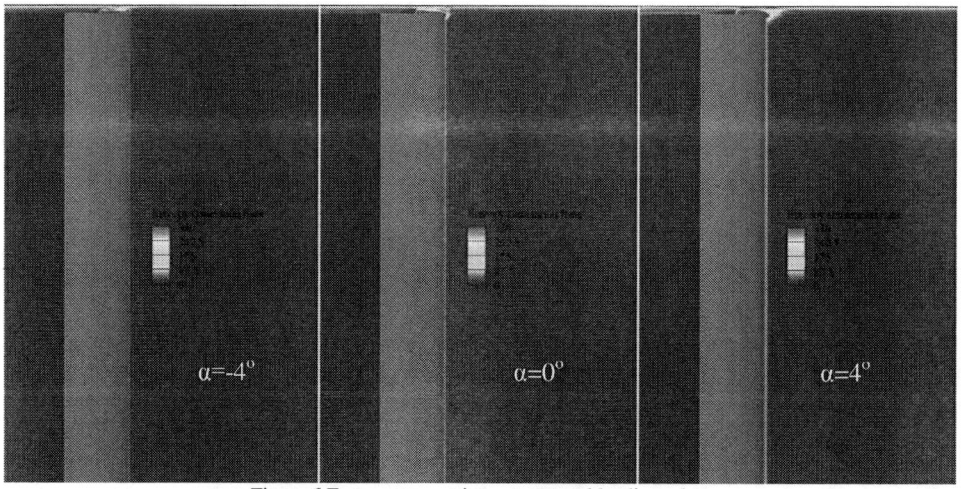

Figure 6 Entropy generation rate around leading edge

Figure 6 and Figure 7 clearly demonstrate the volumetric entropy generation rate associated with tip clearance flow around the leading edge and trailing edge region. Around the leading edge, when incidence angle is -4°, the aerodynamic loss source is confined in the tip clearance region and it is obvious that the highest entropy generation rate occurs around the suction side, which means when cascade operating under incidence angle of -4°, around the leading edge region, part of main flow fluid moves from the suction side to the pressure side, which results in edge vortex exist around the suction side. Increasing incidence angle, entropy generation rate associated with edge vortex located near the pressure side increase and the loss source extend to the main flow close to the suction side. Figure 6 also reveals that, around the leading edge, the magnitude of entropy generation rate of boundary layer is relatively small compared to that of the tip clearance flow.

Figure 7 clearly demonstrates that, around the trailing edge, tip clearance flow is still the main loss source. For tip clearance flow, the highest entropy generation rate associated the edge vortex around the pressure side and shear layer generated by the tip clearance flow and the main flow. The entropy generation rate in the vortex core region, however, is relatively small compared to that of edge vortex and shear layer. From Figure 7, it is obvious that on the pressure side, there exist a region with high entropy generation.

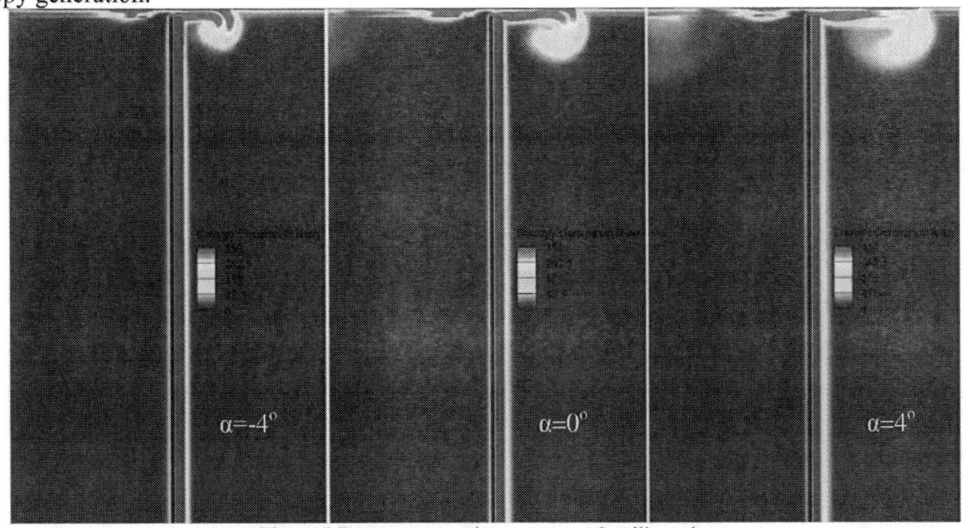

Figure 7 Entropy generation rate around trailing edge

Figure 8 presents the streamlines go through this high entropy generation region. It is clear from Figure 8 that, due to the existence of tip clearance vortex from the neighbouring blade, the streamline undergo quick compression and expansion. The compression and expansion of the fluid is the reason for such an high entropy generation rate region.

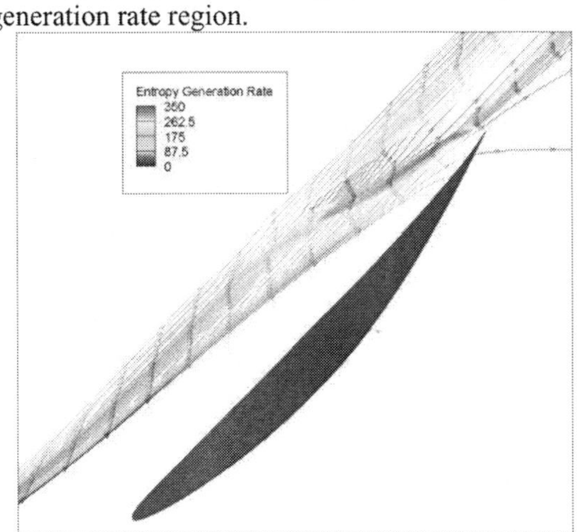

Figure 8 Streamlines pass through high entropy generation rate region

4. Conclusions

Incidence angle plays an important role in deciding the origin, the size and the position of tip leakage vortex. The tip leakage vortex origin moves downstream with the increase of incidence angle, the vortex size increase with the increase of the incidence angle. Increasing incidence angle moves the tip leakage vortex further faraway from the blade suction side.

Tip clearance flow is the main major source of aerodynamic loss in linear cascade. Entropy generation rate associated with tip clearance flow increases with the increase incidence angle.

Edge vortex and shear layer corresponding to the tip clearance flow has the highest entropy generation rate. The entropy generation rate associated with the tip leakage vortex, however, is relatively small when compared with the entropy generation rate of edge vortex and shear layer.

When operating under an incidence angle of 4°, there exist an another high entropy generation rate region in main flow near trailing edge. It is found that the compression and expansion of fluid due to the existence of tip clearance vortex was the main reason for this extra high entropy generation.

References

[1] Kang, S., Hirsch, C., & Fields, Pressure. (1992). Experimental study on the three-dimensional flow within a compressor cascade with tip clearance: part I—velocity and pressure fields. Journal of Turbomachinery,115(3), 1927-1951.
[2] Kang, S., Hirsch. C. (1992). Experimental Study on the Three-Dimensional Flow Within a Compressor Cascade With Tip Clearance: Part II—The Tip Leakage Vortex.
[3] Kang, S., Hirsch, C. (1994). Tip leakage flow in linear compressor cascade.Journal of Turbomachinery,116(4), 657.
[4] Zhong, L. Y. , Liu, Y. W. , & Lu, L. P. . (2019). Unsteady behavior of tip leakage vortex in an axial compressor with different rotor tip-gap sizes using DDES.Chinese Journal of Turbomachinery,061(002), 1-9.
[5] Kunz, R. F. , Lakshminarayana, B. , & Basson, A. H. . (1993). Investigation of tip clearance phenomena in an axial compressor cascade using Euler and Navier-Stokes procedures.Journal of Turbomachinery,115(3), 453.
[6] Tian, C. R. , & Li, B. K. . (2018). Large-eddy simulation study on effect of blade tip clearance on performance of the axial flow fan.Chinese Journal of Turbomachinery.
[7] Muthanna, C. , & Devenport, W. J. . (2004). Wake of a compressor cascade with tip gap, part 1: mean flow and turbulence structure.Aiaa Journal,42(11), p.2320-2331.
[8] Alexej Pogorelov, Matthias Meinke, & Wolfgang Schroder. (2018). Large Eddy Simulation of the Tip Leakage flow in a Axial-fan.Chinese Journal of Turbomachinery,060(001), 1-15.
[9] Mehdi, R., Khorrami, Fei, Li, & Meelan. (2002). Novel approach for reducing rotor tip-clearance-induced noise in turbofan engines.Aiaa Journal.
[10] Chen, X. F, & Qin G. L. (2018). Numerical Simulation of Flow in Centrifugal Compressor with a Single Circumferential Groove. Chinese Journal of Turbomachinery.
[11] Wang, X. , Li, W. , Zhang, X. H. , Zhu, Y. L. , Zuo, Z. T. , & Chen, H. S. . (2019). Tip profile optimization in a low aspect ratio caes radial expander based on orthogonal design.Chinese Journal of Turbomachinery.
[12] C S Tan. (2007). Internal flow - concepts and applications.cambridge university press.
[13] Moore J. & Moore J. G. .(1983). Entropy Generation Rates in Air-Cooled Gas Turbine Nozzles: Part-1-A Turbulent Boundary Layer. ASME paper 83-GT-70.
[14] Zlatinov, M. B. , Tan, C. S. , Montgomery, M. , Islam, T. , & Seco-Soley, M. . (2011). Turbine Hub and Shroud Sealing Flow Loss Mechanisms.Asme Turbo Expo: Turbine Technical Conference & Exposition.
[15] Timothy R. Palmer, Choon S. Tan, Humberto Zuniga, David Little, Matthew Montgomery, & Anthony Malandra. (2016). Quantifying loss mechanisms in turbine tip shroud cavity flows.
[16] Kramer-Bevan J. S. (1992). A Tool for Analysing Fluid Flow Losses. PHD Thesis.
[17] Kock, F. , & Herwig, H. . (2005). Entropy production calculation for turbulent shear flows and their implementation in cfd codes.International Journal of Heat and Fluid Flow,26(4), p.672-680.
[18] Takakura, Tamuto. (2017). Entropy generation in the tip region of a high-pressure turbine. PHD Thesis.
[19] Adeyinka, O. B., & Naterer, G. F. . (2004). Modeling of entropy production in turbulent flows.Journal of Fluids Engineering,126(6), 503-507.
[20] Orhan. O. E. (2014). Investigation of the Effect of Turbulence on Entropy Generation in Turbomachinery. PHD Thesis.
[21] Zambonini, G. , Ottavy, X. , & Kriegseis, J. . (2016). Corner Separation Dynamics in a Linear Compressor Cascade.ASME Turbo Expo 2016: Turbomachinery Technical Conference and Exposition.

IOP Conf. Series: Materials Science and Engineering 1081 (2021) 012046 doi:10.1088/1757-899X/1081/1/012046

Calculation and analysis of interaction flowfield over tiltrotor aircraft in conversion mode

Li Huan, Wu Xiaojun, Jia Hongyin*, Zhou guiyu, Cui Pengcheng, Li Bin

Computational Aerodynamics Institute, China Aerodynamics Research and Development Center, Mianyang,621000, China

email:hongyin@mail.ustc.edu.cn

Abstract. Aiming at solving the complicated aerodynamic interaction problem of conversion mode, the preconditioning method is applied. Combined with the unstructured hybrid technique and the overlapping grid interpolation technique, a numerical simulation method for tiltrotor aircraft in conversion mode is established. On this basis, the validity of the numerical method is verified. Then, with the numerical simulation of isolated rotor, isolated fuselage and tiltrotor aircraft, the laws of the aerodynamic interaction of rotor/fuselage/wing/tail are analyzed with tilting process. The calculation results show that the rule of the influence of the rotor wake on the aerodynamic performance of the wing is different between the inner side and the outer side of the wing. Compared with the isolated rotor, the thrust coefficient of the rotor of the aircraft increases slightly by 0 ~ 5% during the tilting process due to wing and the fuselage.

1. Introduction

Tiltrotor aircraft is a new type of aircraft which combines the functions of vertical take-off and landing and high-speed cruise. It has a wide range of applications in military and civilian applications. At present, the Bell-Boeing V-22 Osprey [1] has been used by the U.S. military, and the AW609 [2] has also been put into full commercial operation. However, it still has serious complex unsteady aerodynamic interaction problems， and the unsteady aerodynamic interaction of its conversion mode is one of the key and difficult points in the research. Conversion mode is the intermediate mode between helicopter mode and fixed wing aircraft mode, which is of great significance for fast, safe and stable takeoff, cruise and landing of tilt rotor aircraft. It can be seen that the prediction of the effect of downwash induced by rotor on the wing and fuselage in the conversion mode is difficult. Therefore, it is of great theoretical and practical significance to study the unsteady aerodynamic interaction of the tiltrotor aircraft.

Due to the remarkable unsteady characteristics of the conversion model, it presents a great challenge to the experiment and numerical simulation. There are few published literatures on this field. Compared with the experimental research [3-5], the numerical simulation research of the tiltrotor aircraft has the advantages with low cost and little shape limitation. Johnson [6-8] in NASA research center calculated the aerodynamic characteristics of tiltrotor with CAMRAD helicopter analysis software based on vortex theory[9], but did not study the aerodynamic interaction of tiltrotor aircraft rotor between wing and fuselage; Yong Su Jung [10] in South Korea carried out the aerodynamic interaction of rotor and tail on conversion mode with CFD method. The result shows that the increase of the sideslip angle causes the amplification of the amplitude of the yaw moment and the roll moment, but the engine nacelle in the conversion mode is only simulated at 80 ° and 90 °. A. Jimenez [11] in the University of Glasgow, UK, used the CFD method of momentum source to simulate the conversion of ERICA, analyzed the aerodynamic interaction of rotor, wing and fuselage, but the engine nacelle only simulated at 30 °, and the accuracy of flow field detail feature capture was not guaranteed. In China, Li Chunhua[12] and others used the free wake method to calculate the conversion mode of the isolated tiltrotor. In reference [13-14], the V-22 conversion mode was simulated by the CFD method based on momentum source[15], but the viscosity is not considered by Euler equation and the details of the flow field are not captured by momentum source method. Li Peng and Zhao Qijun[16] used CFD method to calculate the conversion mode of the isolated tilt rotor, but did not study the aerodynamic interaction of the rotor, wing and fuselage of the tiltrotor aircraft. It can be seen that, there are few published literatures on the study of the conversion mode of the tiltrotor aircraft, and the research methods have gradually shifted from the early experimental research and vortex theory research to CFD numerical simulation research. The research object is mostly a single rotor, and few research of experiment and numerical simulation on the whole tiltrotor aircraft.

In this paper, aiming at solving the tiltrotor aircraft's complicated flow problems, the unsteady preconditioning method and the Spalart Allmaras equation turbulence model [17-18] are used. Combined with the unstructured hybrid grid generation technique and the overlapping grid interpolation technique, which are suitable for the complex configuration and large displacement of tiltrotor aircraft, a numerical simulation method for tiltrotor aircraft is established. With local time step, multigrid technology and large-scale parallel computing technology to accelerate the solution of the subsonic flow field, the interaction flow field between the airframe and the rotor in the conversion mode of the tiltrotor aircraft is analyzed, which has certain guiding significance for the aerodynamic design of the high-performance tiltrotor aircraft in the future.

2. Simulation calculation methodology

2.1. N-S Equation

The flow field solver used in this paper is the self-developed MFlow software [19-20]. The unsteady compressible Navier-Stokes equations for the original variables are as follows:

$$\boldsymbol{M}\frac{\partial}{\partial t}\int_{\Omega}\boldsymbol{Q}d\Omega+\oint_{\partial\Omega}(\boldsymbol{F}_c-\boldsymbol{F}_v)dS=0 \tag{1}$$

Here, Ω and $\partial\Omega$ are the domain and boundary of the control volume, $\boldsymbol{F}_c$ is the inviscid flux, $\boldsymbol{F}_v$ is the viscous flux. The flux defined in reference [21]. $\boldsymbol{Q}$ and $\boldsymbol{M}$ are primitive and conservative variables respectively， $\boldsymbol{M}$ is the transformation matrix for $\boldsymbol{Q}$ and $\boldsymbol{W}$. The forms of $\boldsymbol{Q}$ 、 $\boldsymbol{M}$ and $\boldsymbol{W}$ are follow as:

$$\boldsymbol{M}=\frac{\partial \boldsymbol{W}}{\partial \boldsymbol{Q}}=\begin{bmatrix} \rho_p & 0 & 0 & 0 & \rho_T \\ \rho_p u & \rho & 0 & 0 & \rho_T u \\ \rho_p v & 0 & \rho & 0 & \rho_T v \\ \rho_p w & 0 & 0 & \rho & \rho_T w \\ \rho_p H-1 & \rho u & \rho v & \rho w & \rho_T H+\rho C_p \end{bmatrix} \tag{2}$$

$$\boldsymbol{Q}=[p\ u\ v\ w\ T]^{\mathrm{T}}$$
$$\boldsymbol{W}=[\rho\ \rho u\ \rho v\ \rho w\ \ \rho E]^{\mathrm{T}} \tag{3}$$

Where, u, v, and w represent the components of velocity in the x, y, and z directions. Cp, ρ, T, H and E denote constant pressure specific heat, density, temperature, total energy per unit mass of gas, and total enthalpy, ρ_p and ρ_T are the partial derivative of density versus pressure and temperature.

2.2. Unsteady Preconditioning Equation

Weiss-Smith [22-23] preconditioning matrix replaces the parameters ρ_p of the transformation matrix in (1) with the parameter Θ. Due to the strong unsteady flow field of the tiltrotor aircraft in conversion mode, a two-time step method is introduced to solve the problem. Unsteady preconditioning equations are of the form:

$$\frac{\partial}{\partial t}\int_{\Omega} \boldsymbol{W} d\Omega + \boldsymbol{\Gamma}\frac{\partial}{\partial \tau}\int_{\Omega} \boldsymbol{Q} d\Omega + \oint_{\partial\Omega}(\boldsymbol{F}_c - \boldsymbol{F}_v) dS = 0 \tag{4}$$

Where, t and τ represent the real time and pseudo time respectively. $\boldsymbol{\Gamma}$ is Weiss-Smith preconditioning matrix, the form is as follows:

$$\boldsymbol{\Gamma} = \begin{bmatrix} \Theta & 0 & 0 & 0 & \rho_T \\ \Theta u & \rho & 0 & 0 & \rho_T u \\ \Theta v & 0 & \rho & 0 & \rho_T v \\ \Theta w & 0 & 0 & \rho & \rho_T w \\ \Theta H - 1 & \rho u & \rho v & \rho w & \rho_T H + \rho C_p \end{bmatrix} \tag{5}$$

$$\Theta = \left(\frac{1}{U_p^2} - \frac{\rho_T}{\rho C_p}\right)$$
$$U_p^2 = \min(\max(U^2, k^2 U_r^2), c^2) \tag{6}$$

In equation(6), U_p 、 U_r 、 k and c are the preconditioning speed, rotor tip speed, preconditioning parameter and local sound velocity. The preconditioning parameter is constant. After adding the preconditioning matrix to the time derivative term of the equation, a new convective eigenvalue can be derived. The comparison of eigenvalues is shown in reference [24]. As can be seen from reference [24], when the velocity is low, the eigenvalues of the equation keep at the same order of magnitude, and the numerical rigidity of the system is eliminated. When the velocity exceeds the velocity of sound, the preconditioning equation is transformed into N-S equation.

2.3. *Discretization and Diagonalization*

The right-hand side of the equation is discretized by the Roe scheme, and the dual-time method is used for the time advancement. The LU-SGS method divides the discrete equation into the following three parts:

$$(\boldsymbol{D} + \boldsymbol{L})\boldsymbol{D}^{-1}(\boldsymbol{D} + \boldsymbol{U})\Delta \boldsymbol{W}^n = \boldsymbol{Res} \tag{7}$$

$\boldsymbol{L}$ is the strict lower triangular matrix, $\boldsymbol{U}$ is the strict upper triangular matrix, $\boldsymbol{D}$ is the diagonal matrix. The above equation can be inverted in two steps - the front and back scans. $\boldsymbol{L}$ 、 $\boldsymbol{U}$ and $\boldsymbol{D}$ are in the following form:

$$\boldsymbol{D} = \boldsymbol{S}\boldsymbol{M}^{-1}\left(\frac{Vol}{\Delta\tau} + \frac{1}{2}\sum_i \lambda_i S_i\right) \tag{8}$$

$$\boldsymbol{L} = \frac{1}{2}\sum_{i\subset L} S_i\left(\Delta \boldsymbol{F}_{c,i} - \boldsymbol{S}\boldsymbol{M}^{-1}\lambda_i \Delta \boldsymbol{W}\right) \tag{9}$$

$$\boldsymbol{U} = \frac{1}{2}\sum_{i\subset U} S_i\left(\Delta \boldsymbol{F}_{c,i} - \boldsymbol{S}\boldsymbol{M}^{-1}\lambda_i \Delta \boldsymbol{W}\right) \tag{10}$$

$$\boldsymbol{S} = \frac{(1+\phi)\Delta\tau \boldsymbol{M}}{\Delta t} + \boldsymbol{\Gamma} \tag{11}$$

Where, the concrete forms of $\boldsymbol{Res}$ 、 $\Delta \boldsymbol{F}_{c,i}$ and $\Delta \boldsymbol{W}$ can be seen in reference [24].

2.4. *Overlapping grid interpolation technique and unstructured hybrid grid*

Due to the complex unsteady flow phenomenon of the tiltrotor aircraft, the overlapping grid interpolation technology is used. The overlapping grid allows the sub-grid zones to overlap each

IOP Conf. Series: Materials Science and Engineering 1081 (2021) 012046 doi:10.1088/1757-899X/1081/1/012046

other, and reduces the difficulty of grid generation. Firstly, the wall surface intersection criteria is selected as the determination method of the hole boundary. After determining the hole boundary, alternative digital tree (ADT) method [25] is used to find the corresponding grid cell of the hole boundary in the sub-grid wrapping objects. Finally, the information of overlapping zones is transmitted by hole digging and interpolation. Obviously, the technology which is flexible and suitable for the simulation of tiltrotor aircraft with complex shape and tilting process.

The reasonable design and high quality generation of computational grid are the premise of numerical simulation of tiltrotor aircraft. In this paper, the computational grid is unstructured hybrid grid, including triangular prism, pyramid and tetrahedron grid cells. Triangular prism grid cells are used to simulate the boundary layer, tetrahedron grid cells are used to simulate the isotropic region of space flow field, and pyramid grid cells are used to be transition grid cells of triangular prism grid cells and tetrahedron grid cells. With the unstructured hybrid grid technology, the viscous simulation accuracy in the boundary layer can be guaranteed. Compared with the structured grid, the unstructured hybrid grid is suitable for the simulation of the complex configuration of tiltrotor aircraft with its efficiency and flexibility.

2.5. Grid Generation

The TRAM model [26] is sized to be a 1/4 tiltrotor rotorcraft reduction model. The rotor blades are of positive camber thick airfoils. Because the airfoil data of the model can not be found, some approximations are taken here.

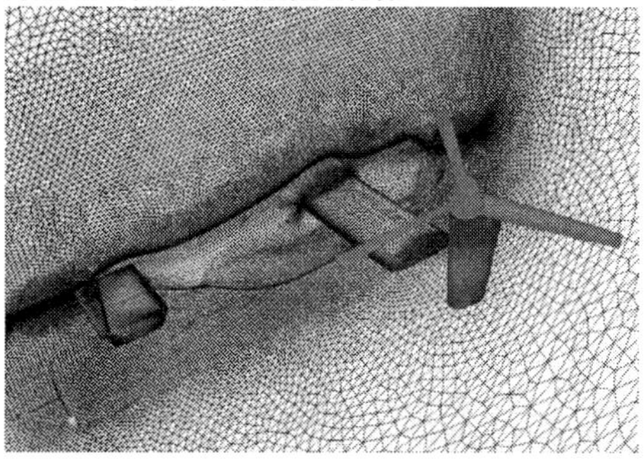

Figure.1 Surface grid of half-span S-TRAM model

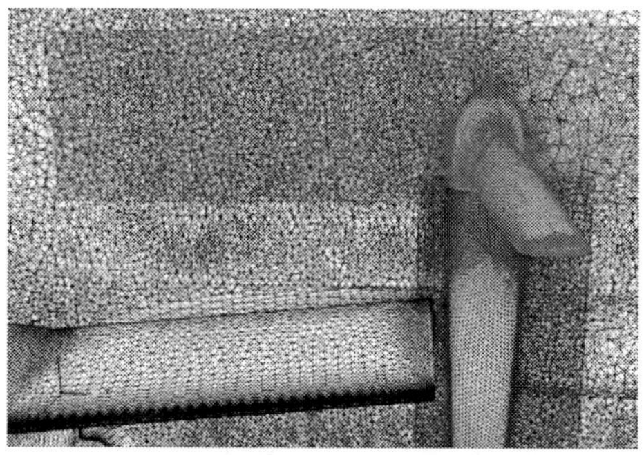

Figure.2 Slice mesh enlargement

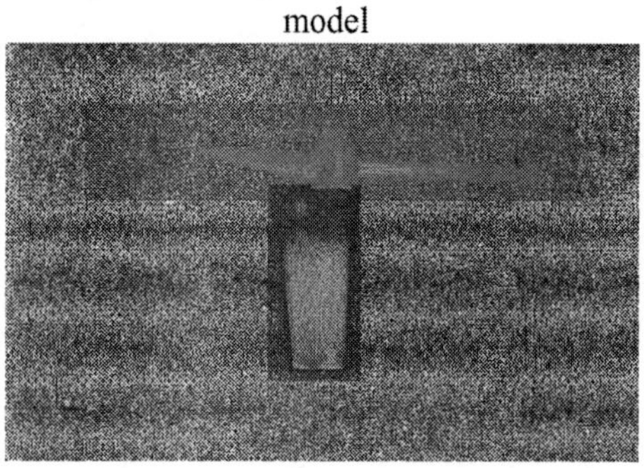

Figure.3 Isolated trotor mesh of I-Tiltrotor model

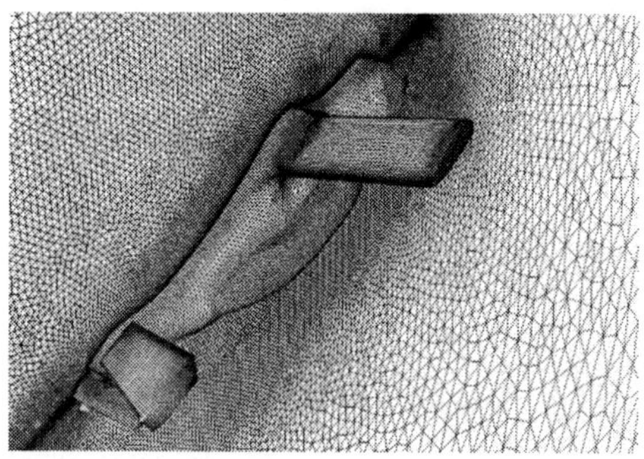

Figure.4 Isolated airframe mesh of S-TRAM model

There generates two models: S-TRAM model and I-Tiltrotor model.The S-TRAM model contains five parts: wings, fuselage, rotors, tails and engine nacelles. The grids of the two models are generated respectively.

Figure 1 shows the surface mesh of the half-span S-TRAM model. Red mesh is of the zone of the blades, blue one represents the zone of the engine nacelle, and black one represents the major-zone.

The $0.20c_{tip}$ is the finest off-body space size and this grid surrounds the blades and extends. The subzone containing rotor blades and that containing the engine nacelle has 4.85 million cells and 0.84 million cells respectively. the other major-zone has 39.37 million cells. Figure 2 is an enlarged view of the overlapping grid sections, showing the mesh size in the interpolated boundary of the three zones is matched. Figure 3 and Figure 4 shows an isolated rotor mesh and an isolated airframe mesh respectively.

3. Test Cases

The numerical simulation ability of the numerical algorithm and grid technology to the forward flight flow field of propeller is examined by this example. The shape parameters of the six blades propeller are defined in reference [27-28], the propeller rotates anticlockwise from the front with the rotating speed n_s = 1020 rpm, and the calculated forward ratios f are 0.716, 1.004, 1.292, 1.432 and 1.580 respectively. The parameters for the case are: calculation height h = 0km, calculation angle of attack α = 0 °, calculation angle of sideslip β = 0 °. The sub-grid containing propeller has 10.18 million cells, including 4.6 million tetrahedral grid cells, 8 million pyramid grid cells and 5.5 million prism grid cells. The major-grid containing shroud has 8 million, among which the numbers of the tetrahedral grid cell, pyramid grid cell and triangular grid cell are 7.03 million, 20 million and 95 million respectively.

Figure.5(a) and (b) show the X-section and Y-section of propeller overlapping grid. Mesh colored red represents the sub-grid of propeller, and mesh colored blue represents the major-grid. The meshes in the interpolated boundary of the two zones and the meshes behind the propeller are refined, and the extrapolated mesh is generated on the interpolated boundary surface.

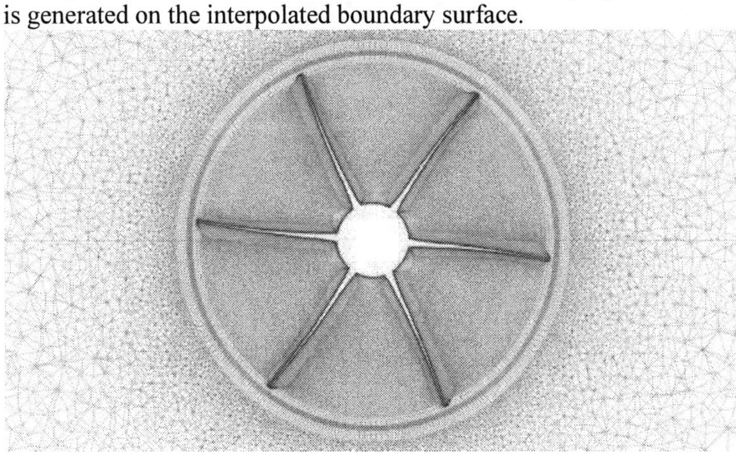

(a) X-section of both zones

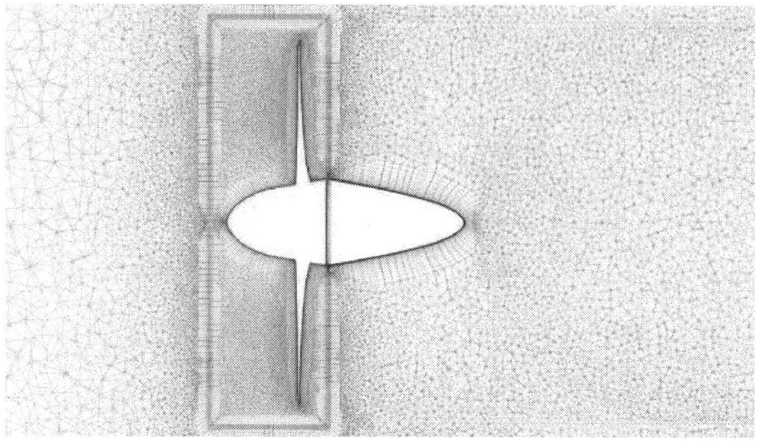

(b) Y-section of both zones

Figure.5 Slice mesh of the both zones

Figure.6 shows the comparison of the experimental and calculated propeller thrust coefficient and torque coefficient with the propeller advance ratio. The numerical results show a good agreement with the experimental data, and the simulation methods for forward flight are verified well.

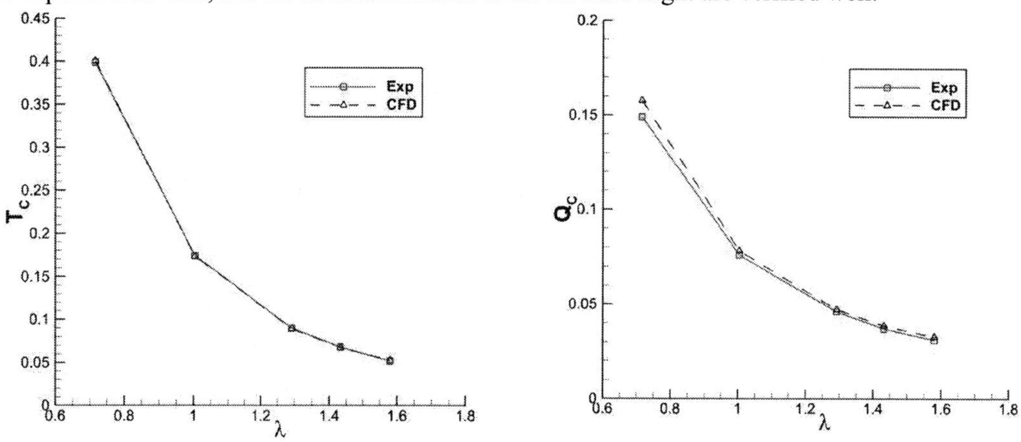

(a) Comparison of the thrust coefficient (b) Comparison of the torque coefficient

Figure.6 Isolated-propeller aerodynamic characteristics

Figure.7 shows the iso-surface of the Q-criterion colored by Mach number, where f=1.292. The spatial vortex caused by rotation of the propeller can be clearly observed because meshes behind the propeller are refined.

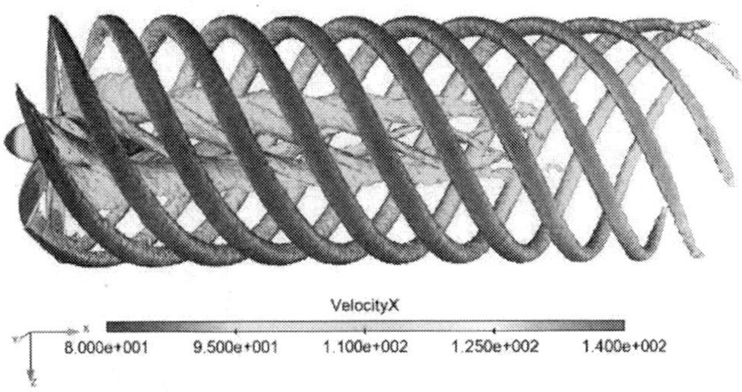

(a) f=1.292

Figure.7 Q-criterion iso-surface of the propeller

4. Tiltrotor Aircraft Conversion Mode

4.1. Flow Field Analysis of Isolated-Rotor

The multigrid method is used to accelerate the computational convergence. The calculation condition for rotor's conversion mode: each time step rotor rotates 2 °, that is, 180 time steps per rotation period. The forward speed is 34m / s and the Mach number of rotor tip is 0.56. The nacelle is fixed at 0 °, 30 °, 60 ° and 90 ° respectively, and the computational grid is the isolated rotor grid of I-Tiltrotor model.

(a) Tilting angle: 0°

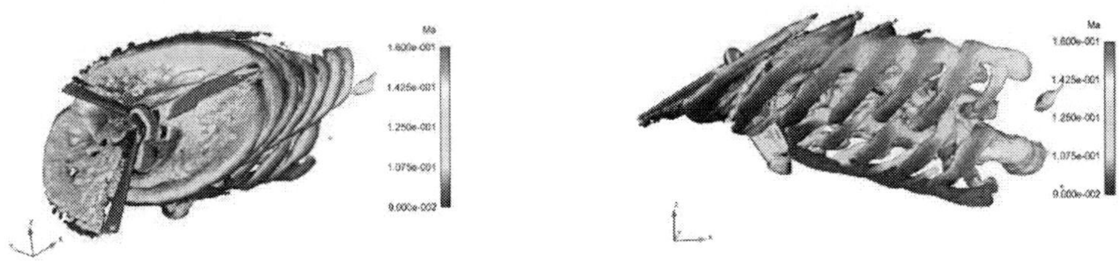

(b) Tilting angle: 30°

IOP Conf. Series: Materials Science and Engineering 1081 (2021) 012046 doi:10.1088/1757-899X/1081/1/012046

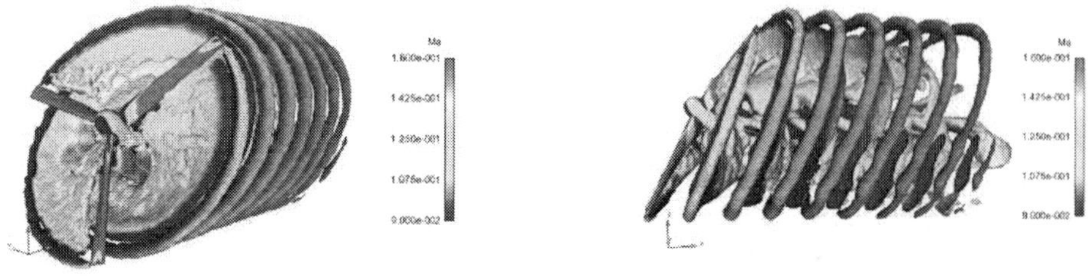

(c) Tilting angle: 60°

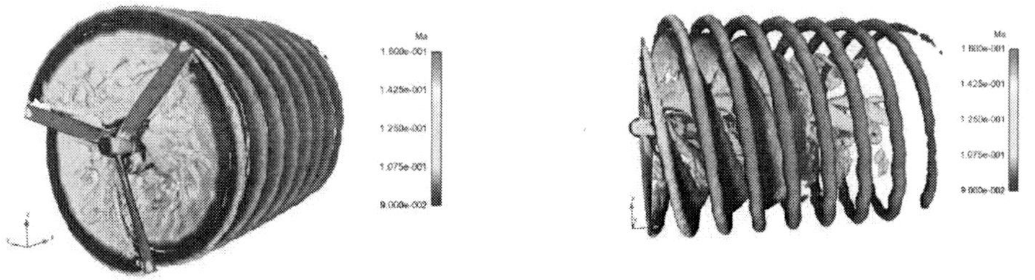

(d) Tilting angle: 90°

Figure.8 Q-criterion iso-surface of the isolated tiltrotor

In Figure.8, the iso-surface of the Q-criterion (colored Mach number) of an isolated rotor at a certain time with different tilting angles is shown. Different from the hovering state of the helicopter, the spiral vortex pulled out by the tip of the tilting rotor is not only transported downward by the induced velocity perpendicular to the rotor disk, but also backward by the incoming flow. When the rotor is not tilted, due to the influence of the incoming flow, the tip vortices gather and move backward in a large amount before developing downward at the front and back sides of the rotor. Therefore, the wake structure is composed of the tip vortex which develops spirally downward and linearly downward. After the rotor is tilted further, the flow induced by the rotor is parallel with the incoming flow gradually, and the tip vortex which develop linearly downward in the wake disappears. The wake structure is mainly composed of the backward spiral tip vortex.

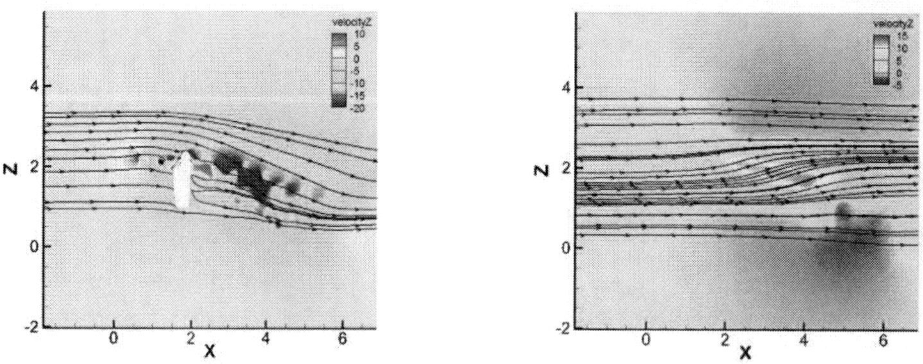

(a) Y=1.83m (b) Y=3.3m

Figure.9 Streamline diagram at 0° tilting angle

IOP Conf. Series: Materials Science and Engineering   1081 (2021) 012046   doi:10.1088/1757-899X/1081/1/012046

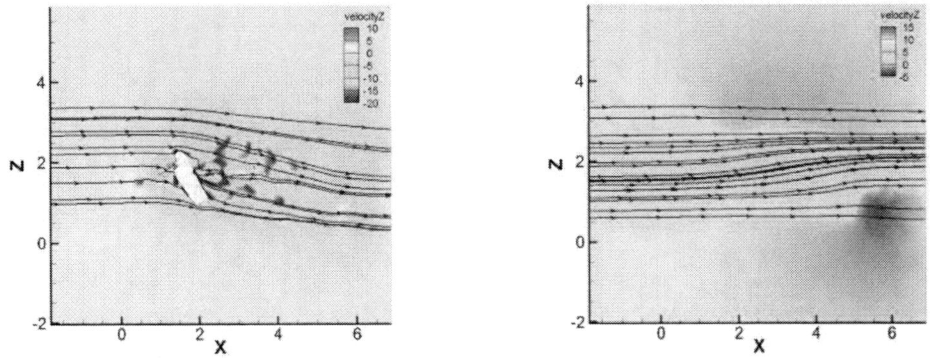

(a) Y=1.83m (b) Y=3.3m

Figure.10 Streamline diagram at 30° tilting angle

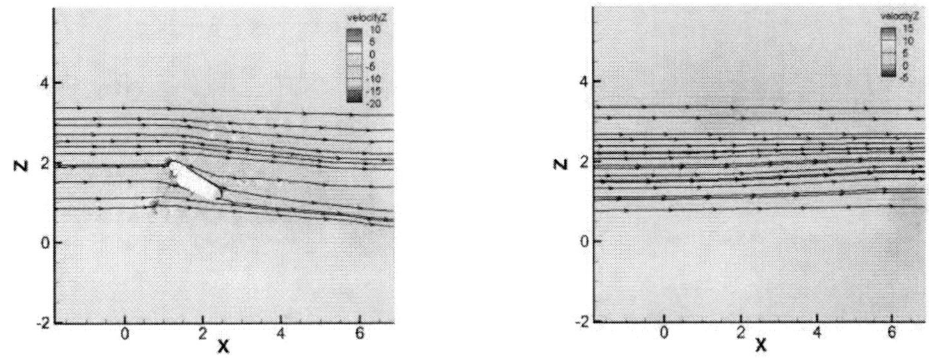

(a) Y=1.83m (b) Y=3.3m

Figure.11 Streamline diagram at 60° tilting angle

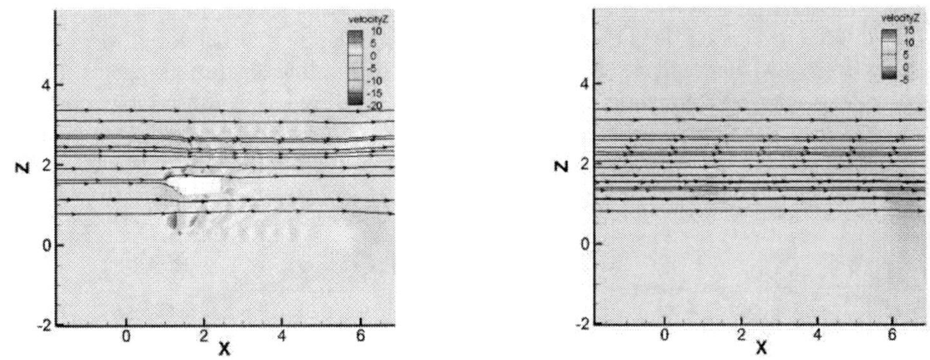

(a) Y=1.83m (b) Y=3.3m

Figure.12 Streamline diagram at 90° tilting angle

Figure.9~ Figure.12 show the streamline diagram of some certain sections of the rotor at each tilting angle. Section y = 1.83m passes through the center of rotor rotation axis, and section y = 3.3m is the outer end of blade front edge blade disk. It can be clearly seen that the air stream through the rotor in the section y = 1.83m flows downward, while the flow in the outer end of the rotor plate flows upward.

IOP Conf. Series: Materials Science and Engineering 1081 (2021) 012046 doi:10.1088/1757-899X/1081/1/012046

With the increase of the tilting angle of the rotor, the deflection effect of the rotor on the flow decreases gradually.

4.2. Flow Field Analysis of Rotor / Airframe

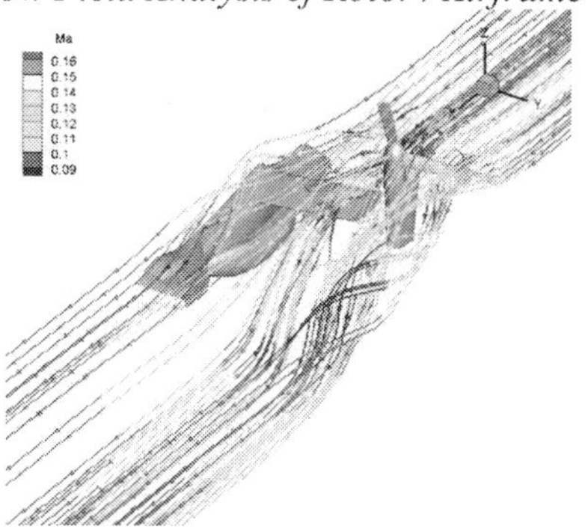
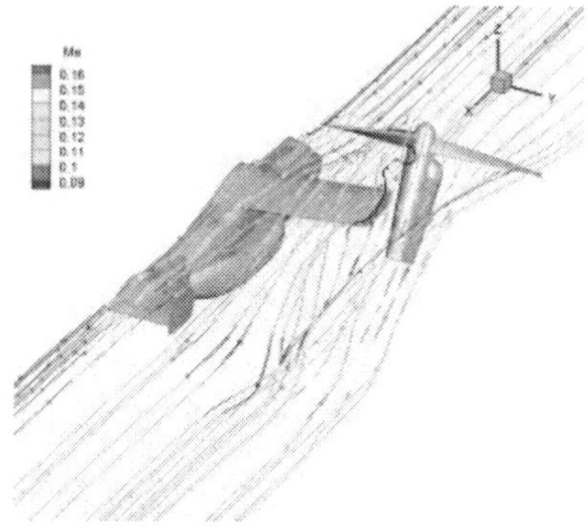

(a) Tilting angle: 0° (b) Tilting angle: 30°

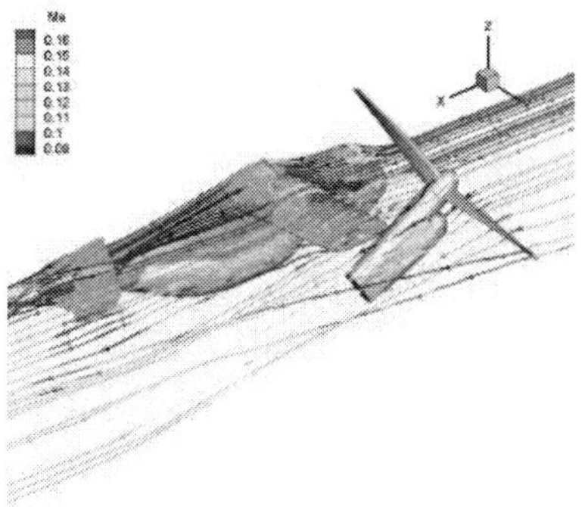
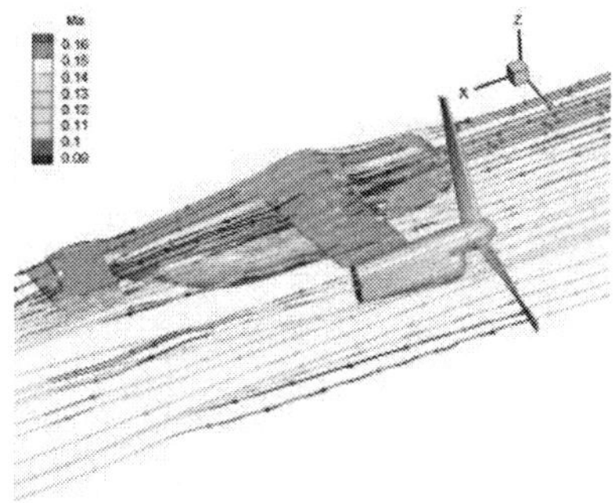

(c) Tilting angle:60° (d) Tilting angle: 90°

Figure.13 Spatial streamline diagram of S-TRAM model

The flow phenomena of the conversion mode of the tiltrotor aircraft is complex. Figure. 13 shows the spatial streamline near the rotor of the S-TRAM model. Figure. 14 shows the spatial Q-criterion iso-surface of the S-TRAM model in the conversion mode (Q-criteria = 500, colored Mach number). It can be seen from the figure that the rotor causes the streamline distortion, which has obvious downward deflection and acceleration effect on the incoming flow. When the rotor is not tilted, the upper surface of the wing is not impacted vertically by the tip vortex wake pulled by the rotor, but the flat tail and vertical tail is impacted directly; When the rotor is tilted further, the aerodynamic interaction of the rotor wake on the wing / horizontal tail / vertical tail is strengthened.

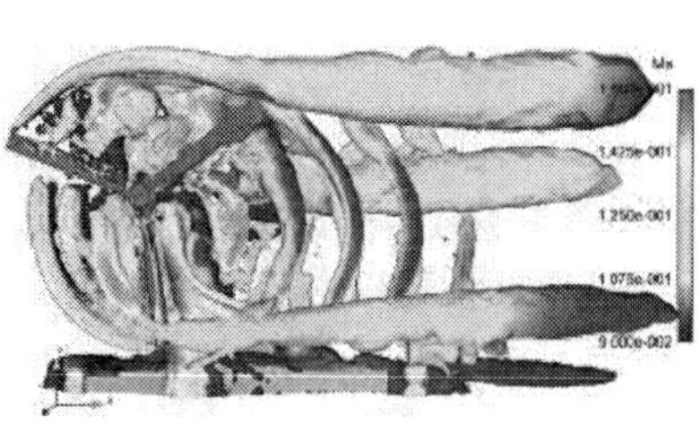
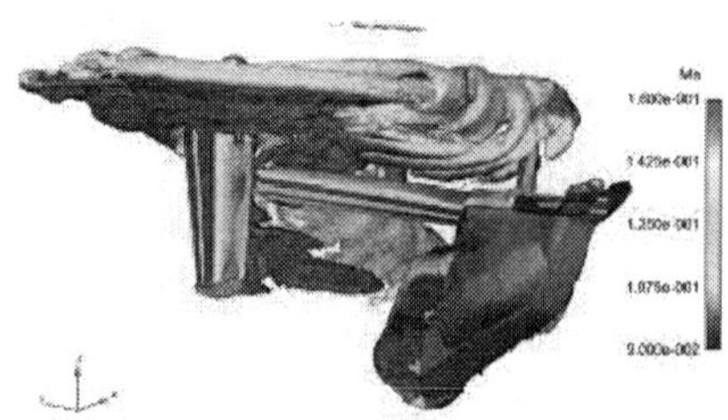

(a) Tilting angle: 0°

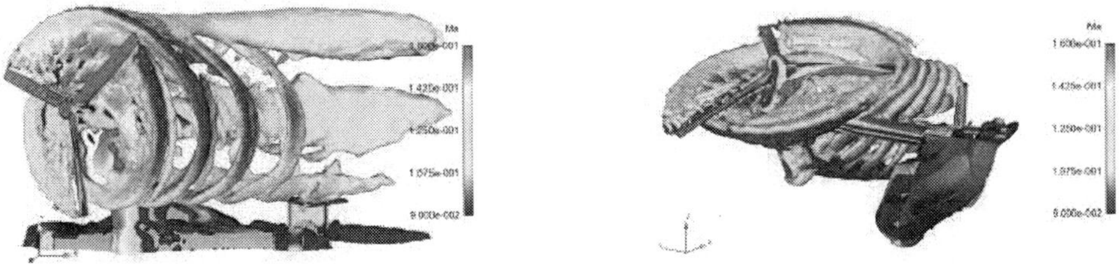

(b) Tilting angle: 30°

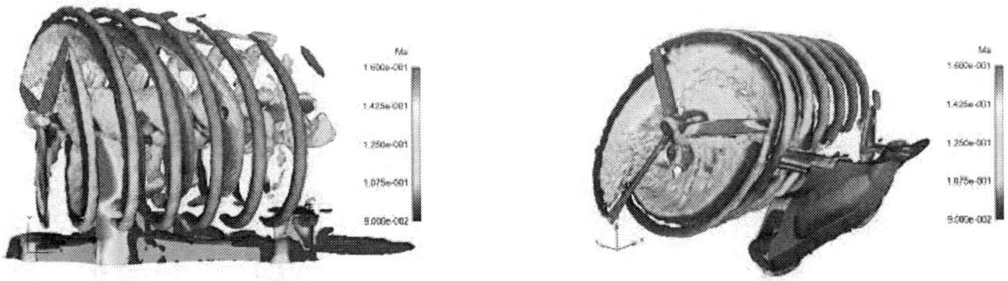

(c) Tilting angle: 60°

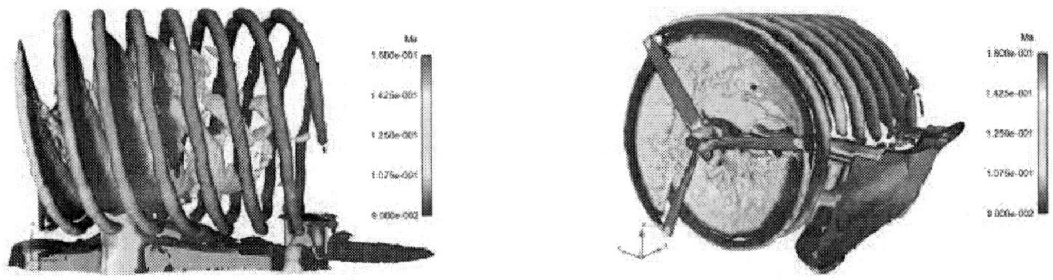

(d) Tilting angle: 90°

Figure.14 Q-criterion iso-surface of S-TRAM model

Figure 15 shows the pressure coefficient distribution of the S-TRAM model in different tilt angles. From the figure, we can see that with the increase of the tilting angle, there is an obvious low-pressure area gradually expanding, outside the upper surface of the wing's leading edge, while the low-pressure area inside the upper surface of the wing's leading edge and the horizontal tail gradually reduces gradually. Obviously, the interaction law of the rotor wake on the wing is completely different in the inner side and the outer side of the wing.

IOP Conf. Series: Materials Science and Engineering 1081 (2021) 012046 doi:10.1088/1757-899X/1081/1/012046

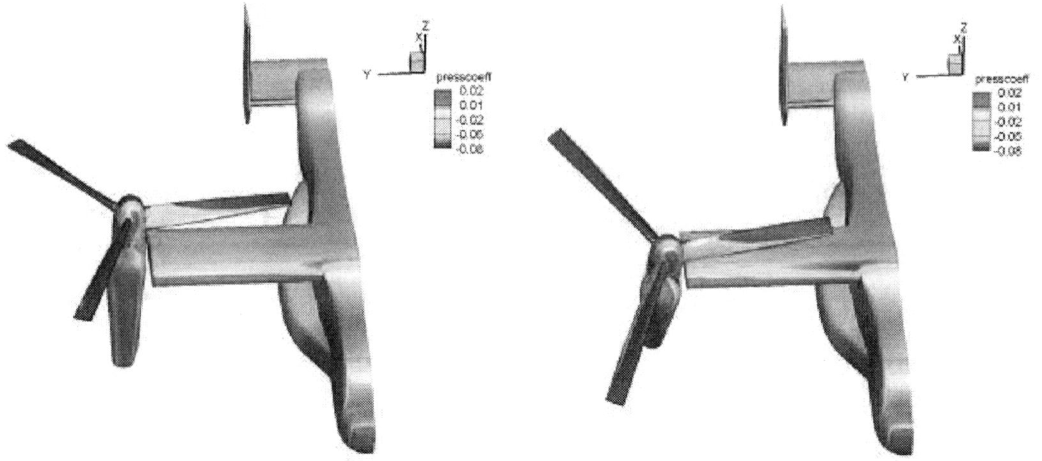

(a) Tilting angle: 0° (b) Tilting angle: 30°

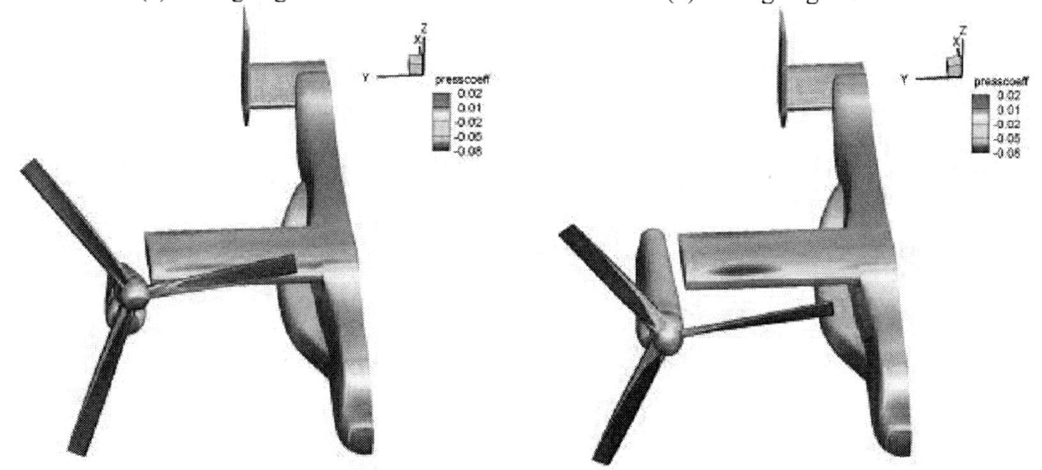

(c) Tilting angle: 60° (d) Tilting angle: 90°

Figure.15 Pressure distribution diagram of S-TRAM model

4.3. Aerodynamic Performance Analysis of Rotor / Airframe

Figure.16 shows the change of the lift coefficient of S-TRAM model and I-Airframe model with time step. Installed_0° represents the S-TRAM model with the rotor tilting angle of 0°. As shown in figure 18, the aerodynamic performance of the S-TRAM model changes periodically with a period of 1/3 rotation, 120 °. When the tilting angle is 90°, the performance coefficient of the airframe changes most dramatically.

Table 1 shows the aerodynamic coefficient of the airframe components of the I-Airframe model and the S-TRAM model at different tilting angles. The results in the table are the periodical average values. From the helicopter mode to the fixed wing mode, with the increase of the tilting angle (0° ~ 90°), the lift coefficients of horizontal tail and fuselage of the S-TRAM model decrease gradually, but are still larger than those of the horizontal tail and fuselage of the I-Airframe model. It shows that the rotor wake has a positive effect on the lift generation of the fuselage and horizontal tail; The lift coefficient of the wing of the S-TRAM model decreases gradually in the tilting process of 30° ~ 90°, and the lift coefficient of the S-TRAM model is the smallest but still greater than that of the I-Airframe model (0°). The result shows that the rotor wake also has a positive effect on the wing lift, and the law dominates the change of lift coefficient of the whole airframe, resulting in the increase of lift coefficient of airframe of the S-TRAM model (0° ~ 30°) and then decrease (30° ~ 90°).

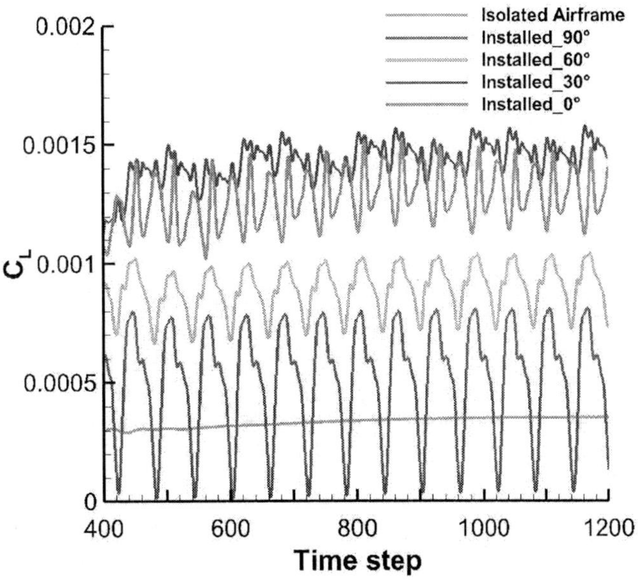

Figure.16 Comparison of the thrust coefficient

Table 1 Aerodynamic characteristics of the components（unit：10^{-5}）

	horizontal tail	fuselage	wing	airframe
Installed_0°	41.16	30.70	51.85	123.71
Installed_30°	25.71	29.85	85.93	141.49
Installed_60°	7.47	7.59	72.57	87.63
Installed_90°	-1.65	-4.58	59.06	49.83
I-Airframe	-3.20	-5.41	43.59	34.98

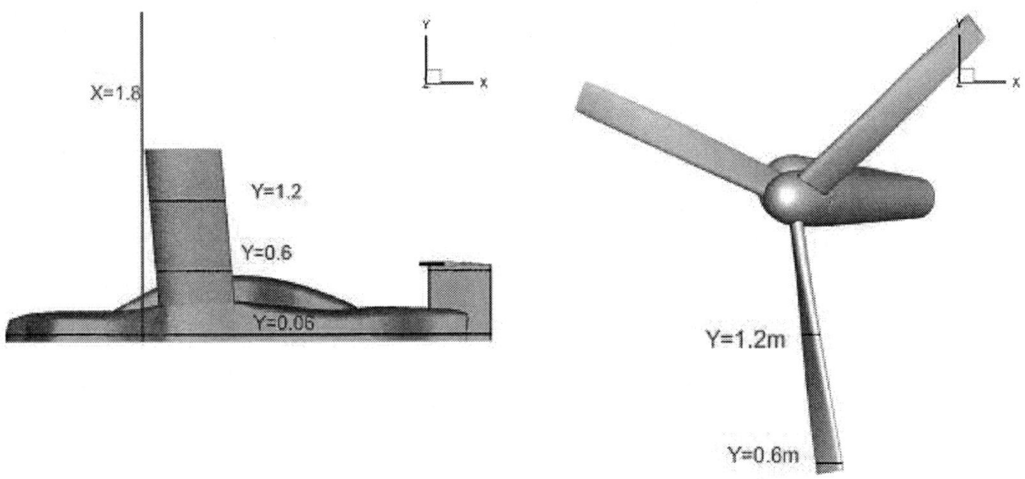

Figure.17 Different sections diagram

Figure 17 shows the location of different sections. Y = 0.06m is the section passing through the fuselage, Y = 0.6m is the section passing through the wing and horizontal tail, Y= 1.2m is the section passing through the outside of the wing. X = 1.8m is a section above the leading edge of the wing.

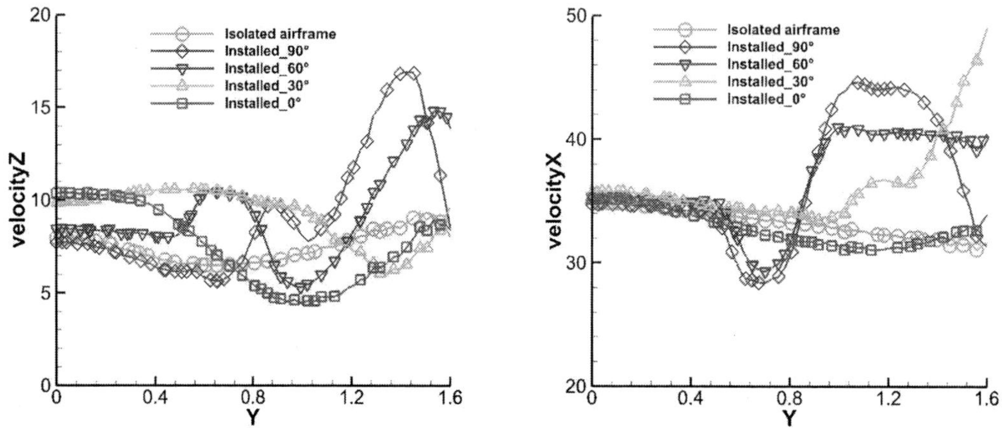

Figure.18 Velocity distribution diagram

Figure 18 shows the velocity in Z direction and X direction distribution in front of the leading edge of the wing (the intersection line of X = 1.8m section and Z = 1.7m section). Y = 0m section passes the inside of the fuselage, Y = 1.6m section passes the wing tip. It can be seen that the velocity X inside the fuselage keeps same basically. the velocity Z near the fuselage and the wing at 0° and 30° tilting angles match well, also at 60° and 90° tilting angles, but the overall trend is that with the increase of the tilting angle, the up wash effect of the rotor wake on the inside of the fuselage decreases. Therefore, the positive effect of rotor wake on the fuselage lift is the improvement of local angle of attack.

From figure 19 showing the streamline and pressure distribution at Y = 0.06m section, the increase of local attack angle of fuselage can be seen more clearly. Figure 22 also shows that the positive effect of rotor wake on the horizontal tail lift is the improvement of local attack angle. With the increase of the tilting angle, the local angle of attack of the airflow at the wing and the horizontal tail decreases gradually. When the tilting angle is 90 °, the local angle of attack of the horizontal tail is negative.

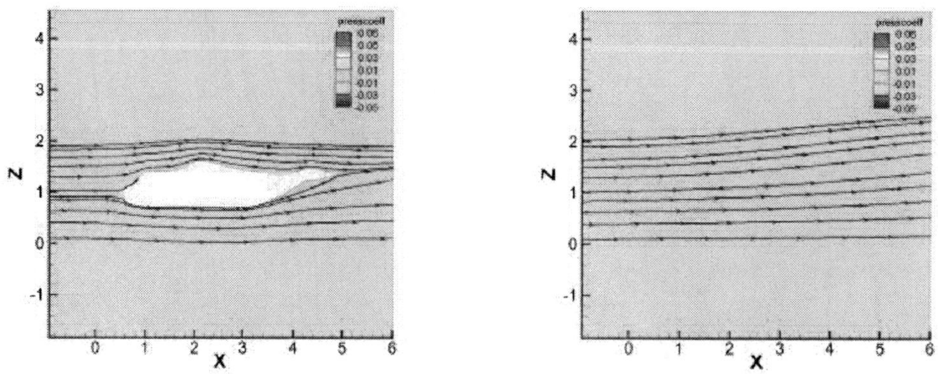

(a) I-Airframe (b) I-Tiltrotor

IOP Conf. Series: Materials Science and Engineering 1081 (2021) 012046 doi:10.1088/1757-899X/1081/1/012046

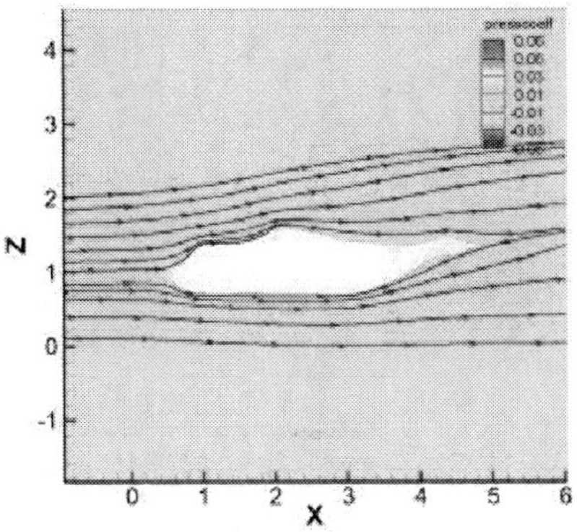

(c) Installed_0°

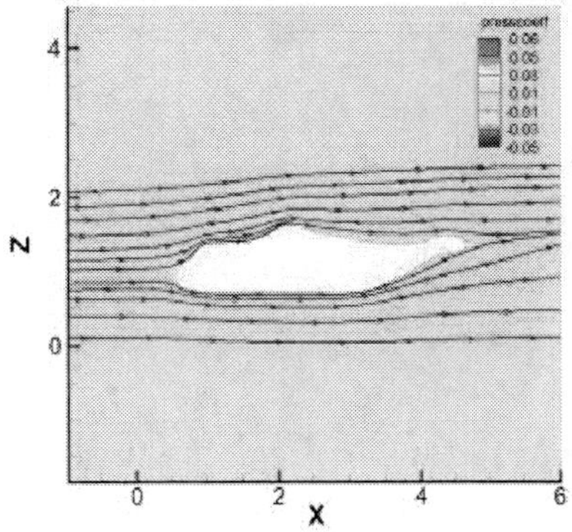

(d) Installed_30°

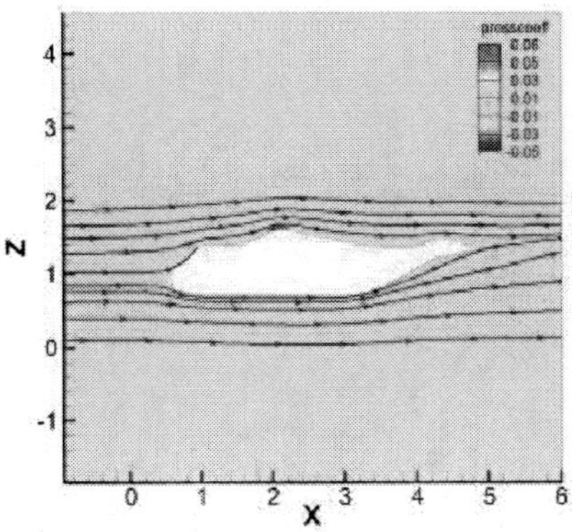

€ Installed_60°

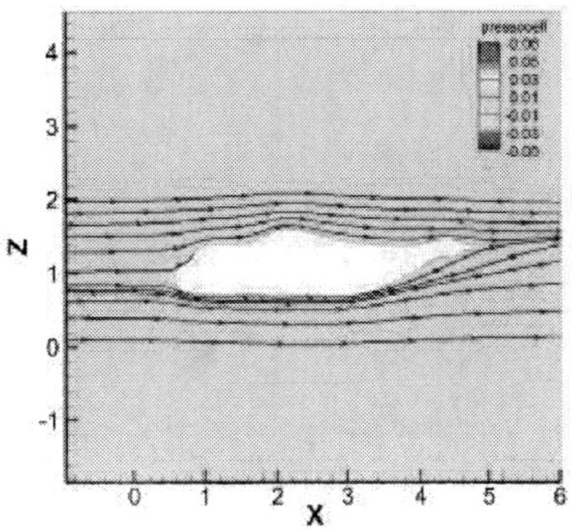

(f) Installed_90°

Figure.19 Streamline diagram at Y=0.06m section

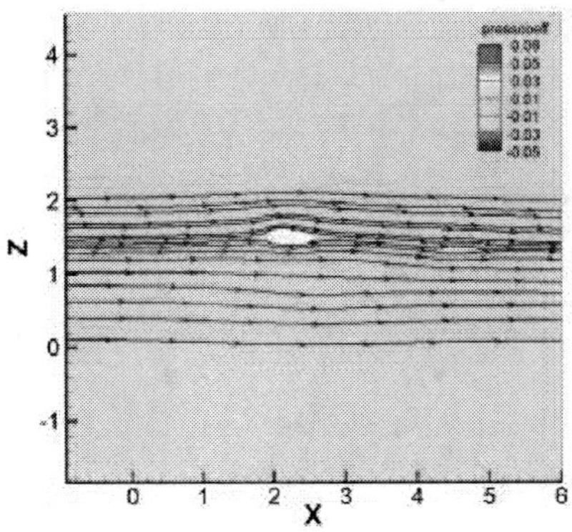

(a) I-Airframe

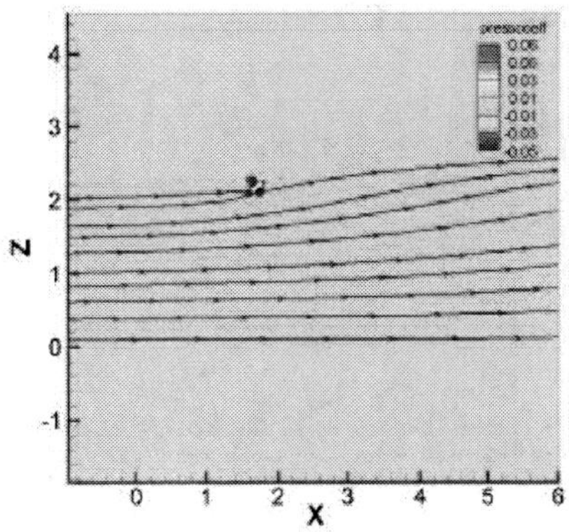

(b) I-Tiltrotor

IOP Conf. Series: Materials Science and Engineering 1081 (2021) 012046 doi:10.1088/1757-899X/1081/1/012046

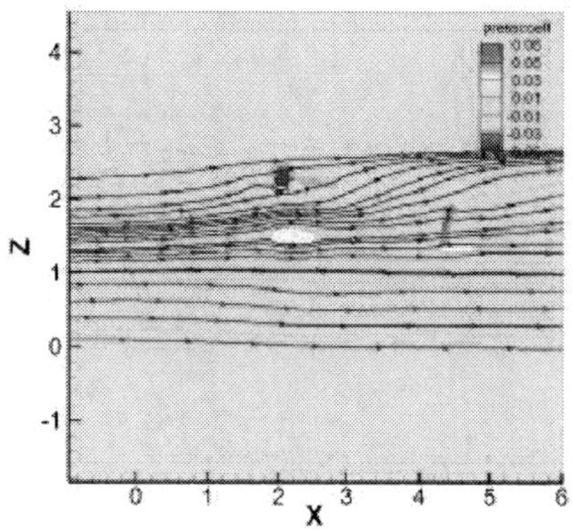

(c) Installed_0°

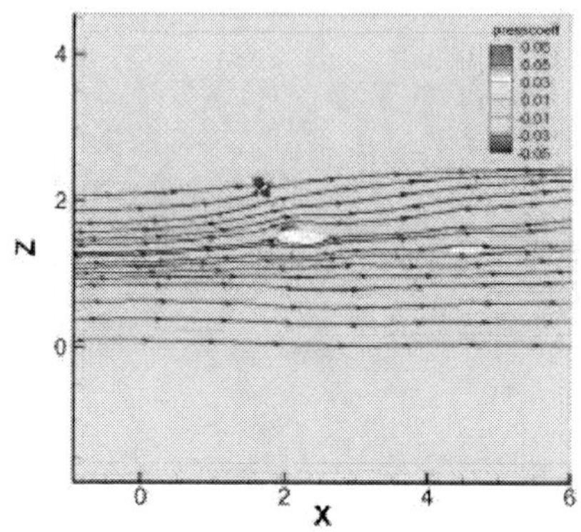

(d) Installed_30°

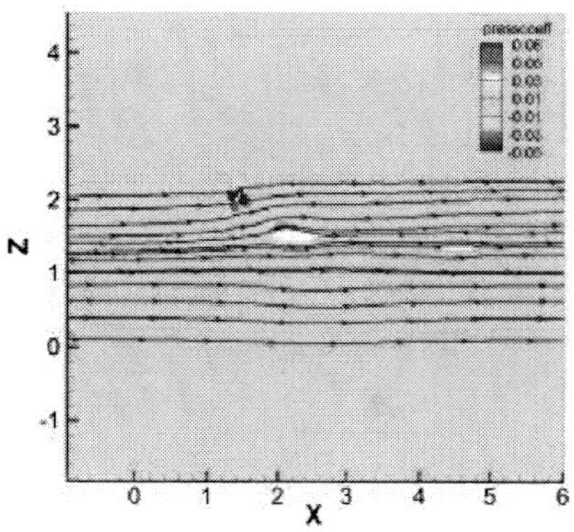

(e) Installed_60°

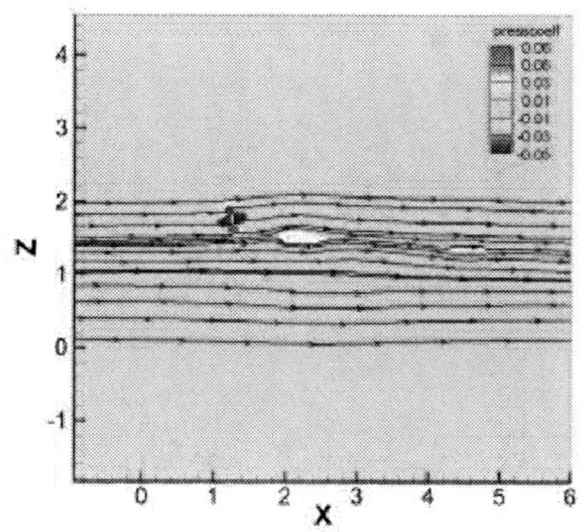

(f) Installed_90°

Figure.20 Streamline diagram at Y=0.6m section

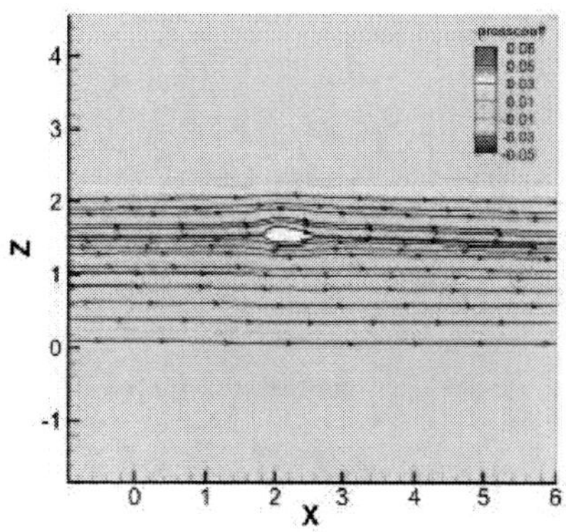

(a) I-Airframe

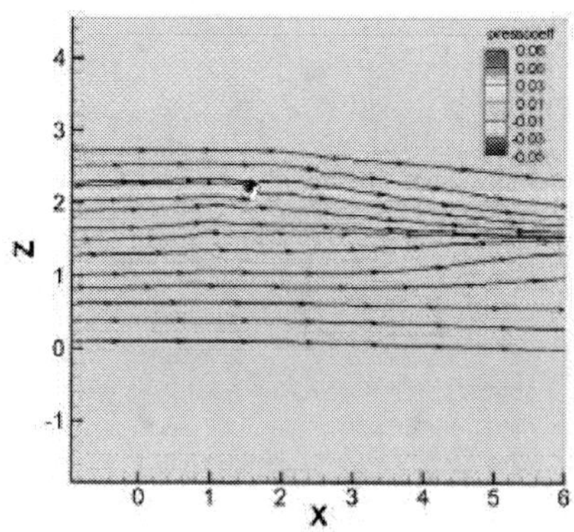

(b) I-Tiltrotor

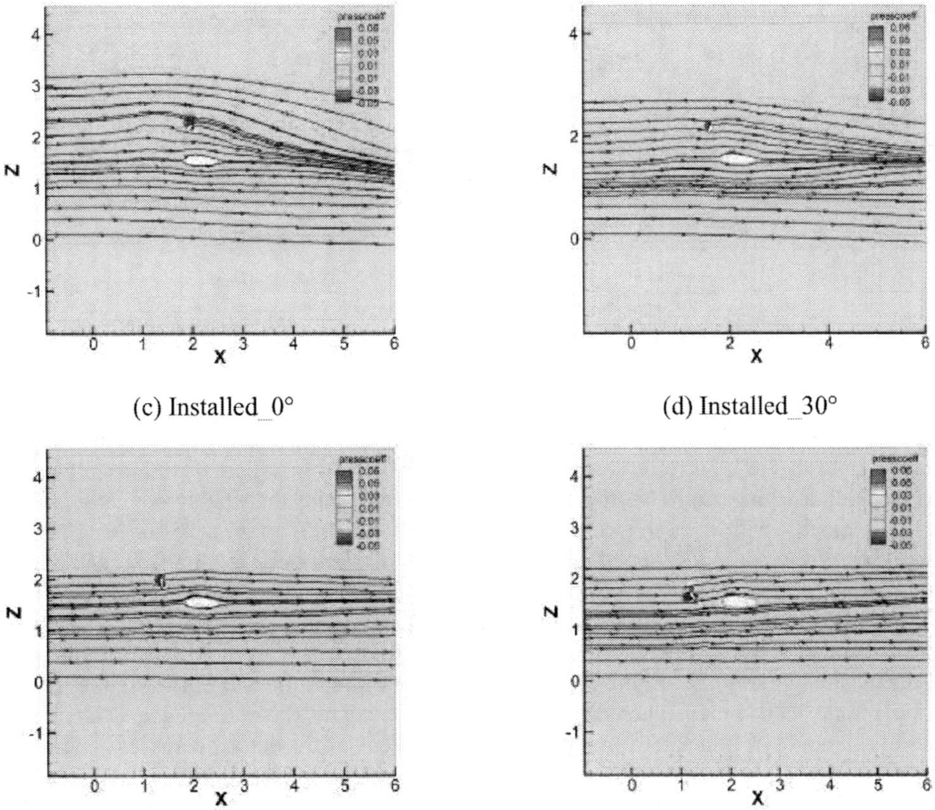

(c) Installed_0° (d) Installed_30°

(e) Installed_60° (f) Installed_90°

Figure.21 Streamline diagram at Y=1.2m section

Figure.21 is the pressure distribution cloud diagram and streamline diagram at section Y = 1.2m. Even when the rotor is not tilted, the downwash flow does not impinge on the upper surface of the wing vertically, which is related to the relative height between the rotor disk and the wing, as well as the downwash flow speed and the incoming flow speed of the rotor. With the increase of the tilting angle, the impact of the downwash flow on the wing is further weakened. Combined with Figure 20, it can be seen that the influence of the rotor on the wing is different at the inner side (Y= 0.6m) and the outer side (Y = 1.2m). At the inner side of the wing, the airstream through the rotor disk flows up and away from the fuselage surface; At the outer side of the wing, the airstream through the rotor disk flows down and near the fuselage surface, but only when the velocity of the incoming flow is small, the downward flow of the airstream can form a certain impact on the upper surface of the fuselage. Therefore, the law of the influence of rotor wake on the wing is not obtained simply.

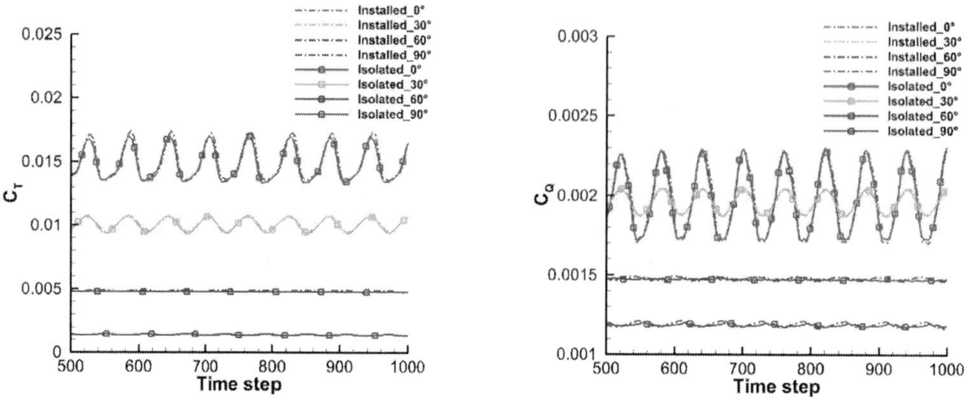

(a) Comparison of the thrust coefficient (b) Comparison of the torque coefficient

Figure.22 History of rotor aerodynamic characteristics

In Figure 22, the unsteady change curve of rotor aerodynamic performance is shown. Isolated_0° refers to the rotor of the I-Tiltrotor model at tilting angle 0°, and Installed_0° refers to the rotor of the S-TRAM model at tilting angle 0°. It can be seen from the figure that the aerodynamic performance changes periodically with a period containing 60 time steps, 1/3 rotation period or 120°. With the increase of the tilting angle, the speed difference of the airstream between the front and back sides of the rotor gradually reduces until the speed is the same, and the amplitude of the aerodynamic cycle change gradually decreases; It can also be seen from the figure that the thrust coefficient and torque coefficient gradually decrease with the increase of the tilting angle. According to the blade element theory, the local angle of attack of the rotor blade decreases due to the rotor's tilt. As the forward flight speed and rotor speed remain unchanged, the lift generated by the rotor blade decreases with the decrease of the angle of attack, so that the thrust coefficient of the rotor decreases. The reduction of the torque coefficient is also related to the reduction of the lift coefficient of the rotor blade. Table 2 shows the comparison of rotor aerodynamic coefficients. In the table, "Isolated rotor" is the isolated rotor, and " Installed rotor" is the rotor of the S-TRAM model. Compared with the isolated rotor, the thrust coefficient and torque coefficient of the rotor of the S-TRAM model increases slightly by 0 ~ 5% during the tilting process.

Table 2 Comparison of rotor aerodynamic characteristics（unit: 10^{-5}）

Tilting angle	CT			CQ		
	Isolated rotor	Installed rotor		Isolated rotor	Installed rotor	
	Calculated value	Calculated value	relative value (%)	Calculated value	Calculated value	relative value (%)
0°	1517.81	1533.49	101.03	203.49	210.98	103.20
30°	953.79	1005.30	105.40	188.42	194.83	103.40
60°	476.72	482.54	101.22	147.19	147.70	100.35
90°	254.67	258.6	101.54	118.43	118.66	100.19

5. Conclusion

In this paper, the numerical simulation of the S-TRAM model of the tiltrotor aircraft is completed. The following conclusions can be obtained:

- The airstream through the rotor in the center section of the isolated rotor flows backward and downward, while the airstream through the rotor at the outer end of the rotor flows backward and upward. With the increase of the tilting angle, the deflection effect of the rotor on the airstream decreases gradually.

- At a fixed tilting angle, the aerodynamic performance of the airframe changes periodically. When the tilt angle is 90°, the lift coefficient of the airframe changes most dramatically, and the rotor wake has a strong unsteady aerodynamic interaction on the airframe.
- The interaction law of rotor wake on wing is totally different in the inner side and the outer side of wing. At the inner side of the wing, the airstream through the rotor flows up and away from the fuselage surface; At the outer side of the wing, the airstream through the rotor flows down and near the fuselage surface.
- The rotor wake increases the local angle of attack of the fuselage and the horizontal tail. With the increase of the tilting angle, the lift coefficients of the horizontal tail and the fuselage of the S-TRAM model decrease gradually, but are still larger than the lift coefficients of the horizontal tail and the fuselage of the I-Airframe model.
- The existence of fuselage / wing improves the thrust coefficient and torque coefficient of rotor. Compared with the isolated rotor, the thrust coefficient and torque coefficient of the rotor of the S-TRAM model increases slightly by 0 ~ 5% during the tilting process. The thrust coefficient and torque coefficient of the rotor decrease with the increase of the tilting angle.

References

[1] Another reference Narducci R, Jiang F, Liu J, Clark R. CFD modeling of tiltrotor shipboard aerodynamics with rotor wake interactions:*AIAA*-2009-3857[R]. Reston: *AIAA*,2009.
[2] Harris J C, Scheidler P F, Hopkins R, Fortenbaugh R L. Initial power-off testing of the BA609 Tiltrotor[C]. *Proceedings of the Sixty-Sixth Annual Forum*, AHS, Phoenix, Arizona, 2000: 1-9.
[3] Chiaramonte J Y, Favier D., Maresca,C. Experimental and numerical study of the proper/fixed wing interaction[J]. *Journal of aircraft* ,1991,28(6):365-373.
[4] Chiaramonte J Y, Favier D., Maresca,C. Unsteady interactional effects between a propeller and a fixed wing:AIAA-1991-3231[R]. Reston: *AIAA*,1991.
[5] Chiaramonte J Y, Favier D. Aerodynamic interaction study of the rotor/wing ensemble in different flow configurations:AIAA-1994-1921[R].Reston: *AIAA*,1994.
[6] Johnson W. Airloads and wake geometry calculations for an isolated tiltrotor model in a wind tunnel[C]. *Proceedings of the 27th European Rotorcraft Forum*, Moscow, 2001: 1-12.
[7] Johnson W. Calculation of tilt rotor aeroacoustic model (TRAM DNW) performance,airloads,and structural loads[C]. *Proceedings of the Aeromechanics Specialists' Meeting*, Atlanta, 2000: 1-65.
[8] Johnson W. Calculation of the aerodynamic behavior of the tiltrotor aeroacoustic model(TRAM) in the DNW[C]. *Proceedings of the 57th Annual Forum*, Washington, D C, 2001: 1-10.
[9] Yincheang Ng, Hai-xin Chen. Rapid Aerodynamic Design of Prop-rotor Blade with Optimization [J]. *Chinese Journal of Turbomachinery*,2019,61(05):19-27.
[10] Jung Y S, You J Y, Kwon O J. Numerical investigation of prop-rotor and tail-wing aerodynamic interference for a tilt-rotor UAV configuration[J]. *Journal of Mechanical Science and Technology*, 2014, 28(7): 2609-2617.
[11] Garcia A J, Barakos G N. CFD simulations on the ERICA tiltrotor using HMB: AIAA-2016-0329[R]. Reston: *AIAA*,2016.
[12] Lichun Hua, Zhang jie, Guo Hua X. Computational analysis on tiltrotor aerodynamic characteristics for transitional flight[J]. *Acta Aerodynamica Sinica*, 2009,27(2):173-179.
[13] Wu dawei, LI Shu. Quasi-steady aerodynamic model of tilt rotor based on optimization algorithm[J]. *Jounal of Aerospace power*, 2017,28(4):0759-0765.
[14] Zhu Qiu Xian, Zhao Qi Jun, et al. Analysis and optimizations on aerodynamic interations of tiltrotor aircraft multi-components on its wing[J]. *Jounal of Aerospace power*, 2017,32(6):1505-1514.
[15] Shuai Yang, Yan Liu, Min Zhang, Jin-guang Yang. Comparison of Aerodynamic Performance Prediction of Winglet Shroud Turbine Cascades Based on Source Terms and Body-Fitted

Grid CFD [J]. *Chinese Journal of Turbomachinery*,2020,62(01):1-10.

[16] Li Peng, Zhaoqijun. CFD analyses aerodynamic char-acterstics of tiltrotor under tipical flight conditions[J]. Journal of Aerospace Power, 2016，31(2): 421-430. Droandi. G, Zanotti. A, Gibertini. G, Grassi D, et al. Experimental investigation of the rotor-wing aerodynam-ic interaction in a tiltwing aircraft in hover[J]. *The Aero-nautical Journal*, 2015, 119(1215): 591-612.

[17] Spalart P R, Allmaras S R. A one-equation turbulence model for aerodynamic flows [R]. *AIAA* 92-0439, 1992.

[18] Yi-kai Zhong, Jin-guang Yang, Min Zhang. Comparison of Turbulence Models of Numerical Simulation in Axial Compressor [J]. *Chinese Journal of Turbomachinery*,2020,62(02):9-18.

[19] Cui Pengcheng, Deng Youqi, Tang Jing, et al. Ad-joint-based grid adaptation and error correction [J]. Acta Aeronoutica et Astronautica Sinica, 2016, 37(10): 2992-3002.

[20] Zhang Y B, Zhou N C, Cheng J T. Numerical investigation of Reynolds number effects on a low-aspect-ratio flying-wing model[J]. *Acta Aerodynamica Sinica*, 2015, 33(3): 279-288

[21] Xiao Tian-hang, Ang Hai-song, Yu Shao-zhi, et al. Us-ing preconditioning method to simulate steady/unsteady flows at all speeds on hybrid grids[J]. *Acta Aerodynami-ca Sinca*, 2007, 25(4): 425-430.

[22] Weiss J M, Smith W A. Preconditioning applied to variable and constant density flows:AIAA-1995-0033[R]. Reston: AIAA, 1995.

[23] Turkel E. Preconditioning methods for solving the incompressible and low speed equations[J]. *Jounal of Compu-tational Physics*, 1987,7(22):277-298.

[24] LI Huan, Chen Jangtao, Ma Mingsheng, et al. Study of the farfield boundary condition for preconditioning sys-tems based on characteristic relations [J]. *Acta Aero-noutica et Astronautica Sinica*, 2017, 37(10): 2992-3002.

[25] BONET J. An alternating digital tree (ADT) algorithm for 3D geometric searching and intersection problems [J]. *International Journal for Numerical Methods in Engineering*, 1991, 31(1):1-17.

[26] Young L A, Lillie D, McCluer M, et al. Insights into airframe aerodynamics and rotor-on-wing interactions from a 0.25-scale tiltrotor wind tunnel model[C]. *Proceedings of the Aerodynamics,Acoustics, and Test and Evaluation Specialists' Conference of the AHS*, San Francisco,2002:1~14.

[27] Nathan Hariharan, Andy Wissink, Michael Steffen, et al. Tip vortex field resolution using an adaptive dual-mesh computational paradigm: AIAA2011-1108[R]. Reston: *AIAA*, 2011.

[28] Gong X Q, Ma M S, Zhang J, et al. Investigation on propeller slipstream by using an unstructured RANS solver based on overlapping grids[J]. *Journal of Mechanics*, 2017:1-13

An Improved Design of a Swirl Distortion Generator

FENG Lu-ning, CHENG Bang-qin, WANG Jia-le, LI Jun, FEI Xiao-wen,

(Aeronautical Engineering College, Air Force Engineering University, Xi'an 710038, China)

E-mail: 1450436728@qq.com

Abstract. In order to study the effect of swirl distortion on the performance and stability of the compressor, this paper optimizes the design of the existing swirl distortion generator, and designs three different structures of pair swirl distortion generator. Using CFD technology, a comparative analysis of the standard pair swirl distortion flow fields and the pair swirl distortion flow fields produced by three newly designed swirl distortion generators and the existing pair swirl distortion generator is performed.The results show that the optimally designed distortion generator reproduces the standard pair swirl distortion flow field to a higher degree than existing distortion generators. On the three measurement rings, the average swirl angle error is reduced by 5.35%, 19.88%, and 8.22%. The total pressure recovery coefficient is increased by 2.2%, which proves that the design method in this paper is more reasonable and achieves the purpose of optimal design.

1. Introduction

The aviation propulsion system must have good reliability indicators and good performance indicators.Besides, due to the interaction between the inlet and the engine, a good aviation propulsion system requires the inlet and the engine to meet certain applicability indicators to ensure that the aviation propulsion system has a good stability margin. With the development of modern warfare, the stealth and mobility of advanced fighters has become the key to improve combat effectiveness. Due to its structural advantages, the S-duct inlet [1] is widely used in the new generation of stealth fighters, and the swirl distortion caused by this inlet [2] [3] has gradually become the focus of attention in recent years.

The F-111A development program was forced to suspend due to stall issues that frequently occurred during flight test. In the late 1970s, the Tornado fighter [4] repeatedly encountered problems such as surge and even engine shutdown during the test flight. This caused a serious flight accident. This series of accidents has caused people to gradually pay attention to the problem of swirl distortion and try to

study the mechanism of swirl distortion on the engine by designing the swirl distortion generator.
There are three main methods for generating swirl distortion: the guide vane method, the delta wing method, and the chamber-type swirl distortion generator.
Govardhan [5] et al deflect the inlet flow through the guide vanes to obtain a specific swirl flow field[6] .This swirl distortion generator is based on the swirl flow field that may occur in actual situations. It can simulate general pair swirl and bulk swirl flow fields. However, it is not accurate enough to make the simulated flow field match the size of the swirl angle of the target flow field at any position.Based on the research of Govardhan, Flitcroft [7] et al used a method of placing a metal net on a load bearing device to generate an bulk swirl. Jiang Jian[8]~[10] of Nanjing University Of Aeronautics And Astronautics and TU Bao-feng[11] et al. also studied the swirl mechanism through a vane-type swirl distortion generator.
Genssler[12] et al used a delta wing method to generate swirl, and adjusted the size of the pair swirl and the position of the swirl core by adjusting the angle of attack and the position of the delta wing. However the delta wing method requires the generator to be placed in front of the engine, which may cause the delta wing to be sucked into the inlet and damage the engine.
Chamber-type swirl distortion generator, by adding an adjustable valve in front of the static pressure chamber, to ensure that only the incoming air from a specific angle can pass through the chamber. This can be adjusted by adjusting the size, position or quantity of the valve to control the direction of airflow to form different swirl pat-terns.
These three types of swirl distortion generators all have a certain degree of limitation, that is, they cannot generate complex pair swirl flow fields, and are not suitable for reproducing the complex pair swirl field patterns generated from real engine / body combinations.
Zhang Lei of Air Force Engineering University has designed a new type of swirl distortion screen, which can design a swirl distortion screen for the target swirl velocity vector field to generate the swirl distortion of the desired structure form. It is possible to reproduce the complex swirl distortion flow field formed by the S-duct inlet and the wing-body fusion aircraft during actual flight. However, the swirl distortion flow field produced by this swirl distortion generator is far from the target flow field. Based on Zhang Lei's research[19], this paper optimizes the pair swirl distortion generator[13] designed by it, and designs three new types of pair swirl distortion generators. Through the simulation calculation method, and then analysis from swirl angle distribution, swirl evaluation index, and total pressure loss coefficient .

2. Design method

In order to simulate the pair swirl distortion flow field, this paper optimizes the existing pair swirl distortion generator and uses a new CDA[14] blade to reduce the flow loss generated when the flow passes through the dis-tortion generator. According to the design method of the swirl distortion screen, three pair swirl distortion gen-erators : constant chord length and forward-swept and backward-swept blades on both sides are designed to compare the recurrence of the three to the standard pair swirl distortion flow field.
The existing pair swirl distortion generators are improved from the following three aspects: (1) The CDA airfoil with better gas flow performance is selected to eliminate or suppress the separation of the surface layer and reduce the pressure loss. (2) Find the best consistency and installation angle, so that

the swirl angle on the aero-dynamic interface (AIP) is closest to the swirl angle of the target flow field.(3) Reduce the interference between the blades by adjusting the blade layout.

2.1 Obtaining the target swirl field

Formula (1)-(3) give the standard pair swirl velocity vector field:·

$$\Omega_0 = \frac{\Gamma_0}{\pi a_0^2} \tag{1}$$

$$u(x,y) = \frac{1}{2}\left(\frac{a_0^2\Omega_0(y-y_1)}{(x-x_1)^2+(y-y_1)^2}\right)\left(1-e^{-((x-x_1)^2+(y-y_1)^2)/a_0^2}\right) - \frac{1}{2}\left(\frac{a_0^2\Omega_0(y-y_2)}{(x-x_2)^2+(y-y_2)^2}\right)\left(1-e^{-((x-x_2)^2+(y-y_2)^2)/a_0^2}\right) \tag{2}$$

$$v(x,y) = -\frac{1}{2}\left(\frac{a_0^2\Omega_0(x-x_1)}{(x-x_1)^2+(y-y_1)^2}\right)\left(1-e^{-((x-x_1)^2+(y-y_1)^2)/a_0^2}\right) - \frac{1}{2}\left(\frac{a_0^2\Omega_0(x-x_2)}{(x-x_2)^2+(y-y_2)^2}\right)\left(1-e^{-((x-x_2)^2+(y-y_2)^2)/a_0^2}\right) \tag{3}$$

Table 1 is the relevant parameter values of the pair swirl in this paper:

Tab.1 Pair swirl parameters

name	value
a_0	0.10
Γ_0	1
x1	-0.15
x2	0.15
y1	0
y2	0
a_0	0.10

According to formula (1)-(3), the circumferential velocity at any point in the plane can be calculated.According to the definition of swirl angle formula (4), the axial velocity can be calculated as 6.31m / s, and the maximum swirl angle is 16 °.

$$\alpha = tan^{-1}\frac{U_\theta}{U_x} \tag{4}$$

When the axial velocity is determined, a VC program can be written according to formula (1)-(4) .The VC program can be used to determine the swirl angle at any position in the plane.

As shown in Figure 1, the initial line (red line) is arranged in the standard pair swirl velocity vector field, so that the direction of the red line is consistent with the swirl side, and then the blade line (blue line) is arranged perpendicular to the red line.In this model, each blade is Select 7 feature points on the line, and obtain the target swirl angle at each position according to the program. The 7 feature points are selected randomly with certain intervals between each point, and the left and right sides are symmetric. Then draw the corresponding blade at each feature point according to the obtained swirl angle.

IOP Conf. Series: Materials Science and Engineering 1081 (2021) 012047 doi:10.1088/1757-899X/1081/1/012047

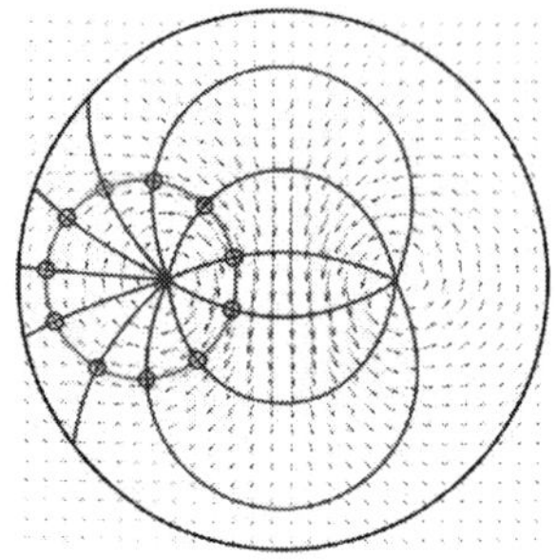

Fig.1 Initial line and blade line

2.2 Blade type selection

After obtaining the layout of the blades, it is necessary to determine the blade parameters at each feature point, mainly including the selection of the blade profile, the appropriate blade consistency, and the adjustment of the installation angle.

2.2.1 Blade type selection

In terms of airfoil, a controllable diffusion airfoil--CDA airfoil[15] with better gas flow performance is used instead of the uniform thickness airfoil which is adopted by existing swirl distortion generator, as shown in Figure 2.

Fig.2 CDA blade type

2.2.2 Consistency selection

In order to select the optimal consistency, an bulk swirl model with a constant chord length of 100 mm and a number of blades of 12 was set. Models were drawn using blades with bend angles of 5 °, 10 °, 15 °, and 20 °, and calculated through simulation. The relationship between the output consistency and the swirl angle is shown in Figure 3.When the consistency is greater than 1.45, the consistency has little effect on the swirl angle, and the swirl angle is similar to the expected target at this time, so 1.45 is selected as the blade consistency.

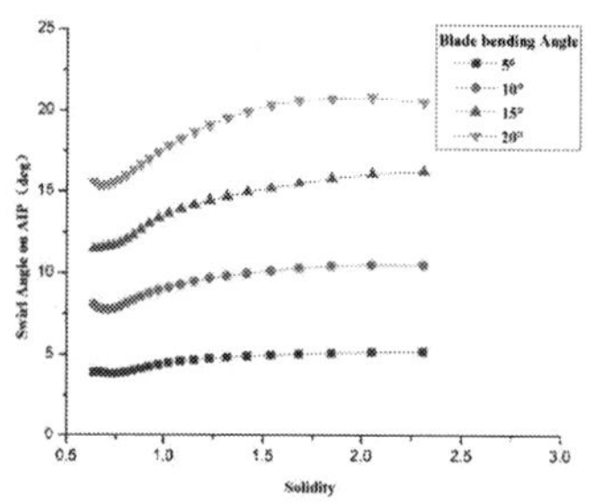

Fig.3 Relationship between consistency and swirl angle

2.2.3 Adjustment of installation angle

Set the bulk swirl model with a blade consistency of 1.45 and a number of 12 blades, and determine the relationship between the swirl angle and the blade bend angle through simulation calculations, as shown in Figure 4:

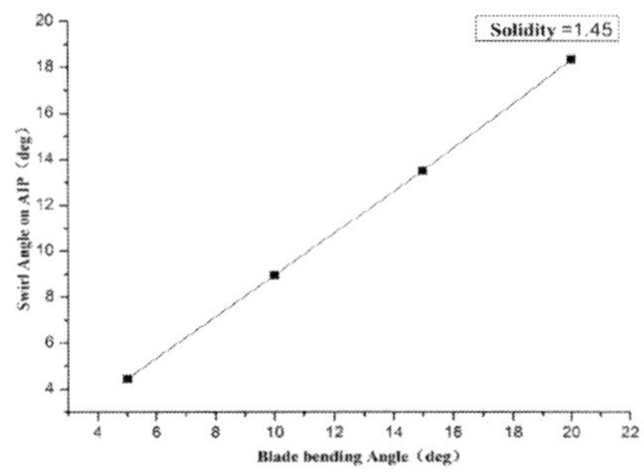

Fig.4 Relationship between blade bend angle and swirl angle

As can be seen from the figure, when the consistency is 1.45, the relationship between the swirl angle α and the blade bend angle β is shown in formula (5). Therefore, add 1.55 ° to the original installation angle of the blade to get the target swirl angle on the AIP surface.

$$\alpha = \beta + 1.55^{\circ} \quad (5)$$

2.3. Drawing of pair swirl distortion generator model

As shown in Figure 5, three new models of pair swirl distortion generators were drawn based on the selected blade shape, consistency, and installation angle. Models with a constant chord length of M1; M2 consistency of 1.45 with blades swept back on both sides; pair swirl distortion generator model with M3 consistency of 1.45 on both sides of the blade swept forward.

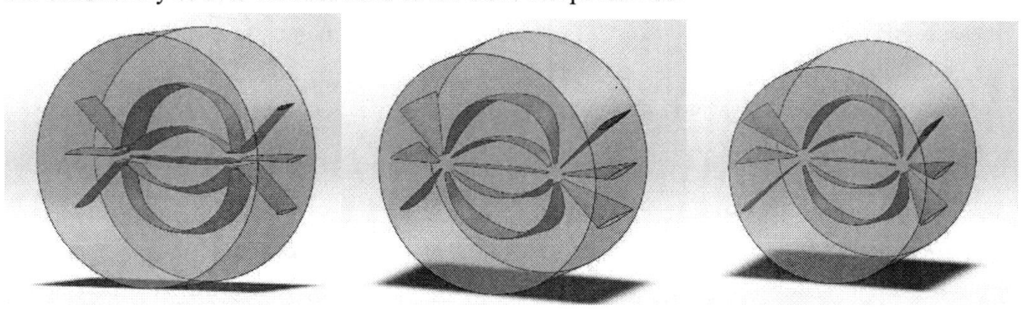

(a) M1 (b) M2 (c) M3

Fig.5 Computational model of pair swirl distortion generator

3. Calculation method

The calculation domain of this paper is set up according to the single-stage low-speed axial-flow compressor test bench of the Plasma Laboratory of Air Force Engineering University. It is divided into three parts: the front pipe, the distortion section and the rear pipe, as shown in Figure 6. It is 1800mm, 300mm, and 800mm, and the diameter of the pipe is 600mm. Considering the actual situation of the test bench, the AIP is set at 400mm at the exit of the distortion section.

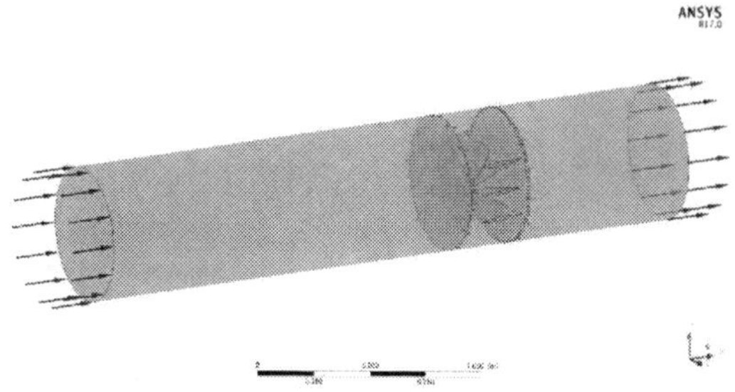

Fig.6 Computational domain

In this paper, the ICEM software is used to divide the three-dimensional computing grid. The front and rear pipes and the distortion section use tetrahedral unstructured grids. After grid independence verification, when the number of grids reaches 7.5 million, the number of grids is calculated. The effect of the results can be ignored, the final distortion segment mesh number is 8.4 million. The numerical simulation calculation uses CFD commercial software ANSYS CFX. The inlet boundary conditions set are total pressure 101325pa, total temperature 288K, outlet boundary condition mass flow 6.5Kg / s, wall surface adopt adiabatic non-slip boundary conditions, the convergence criterion is set to reduce the relative change of each quantity to be solved to 10^{-5}, and monitor the inlet and outlet flow conditions. The total number of grids in the entire calculation domain reaches 11 million. Turbulence model selection k-ω turbulence Model. When studying swirl distortion, the ideal position of AIP is close enough to the cross section of the engine inlet to ensure that the swirl flow field of AIP is the same as the swirl flow field flowing into the engine, and that the gas flowing through AIP can flow into the engine. Therefore, the selection of the AIP surface must meet the requirements of both total pressure distortion and swirl distortion research. Because this article only studies swirl distortion, and the airflow velocity in simulation conditions is very low, the pressure loss is small and the total pressure distortion is negligible, so AIP surface is selected as the compressor inlet. According to the size of the laboratory single-stage low-speed axial flow compressor test bench, AIP is set at the inlet section of the compressor, that is, 400mm away from the distortion section[16][17].

4. Results analysis

In order to compare the reproducibility of the standard pair swirl distortion flow field of the pair swirl distortion generator designed in this paper with existing distortion generator, the swirl angle, the swirl evaluation index, and the total pressure loss are used to analyze the recurrence of constant chord length, consistency 1.45 swept back, consistency 1.45 swept forward pair swirl distortion generator and existing pair swirl distortion generator to the standard pair swirl distortion flow field.

4.1 Swirl angle simulation results

The velocity vector diagram and the swirl angle cloud diagram of the target flow field and the flow field obtained from the pair swirl distortion generator are shown in Figure 7. Among them, the swirl flow field obtained by the existing pair swirl distortion generator is M0, the swirl flow field obtained by constant chord length pair swirl distortion generator is M1, and the swirl flow fields obtained by the pair swirl distortion generator of the blades swept back and swept forward on both sides are recorded

IOP Conf. Series: Materials Science and Engineering 1081 (2021) 012047 doi:10.1088/1757-899X/1081/1/012047

as M2 and M3.

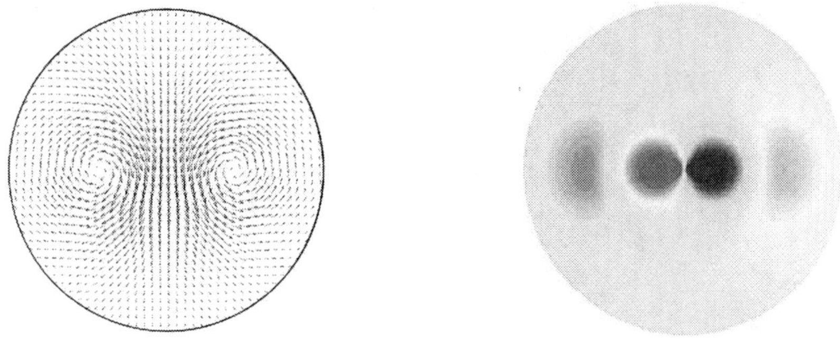

(a) Standard pair swirl flow field

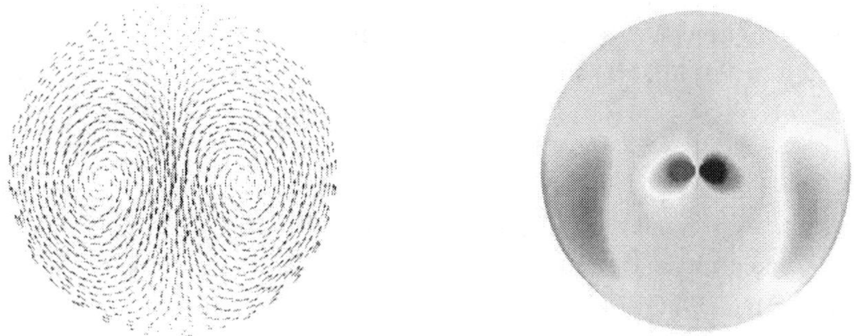

（b）M0

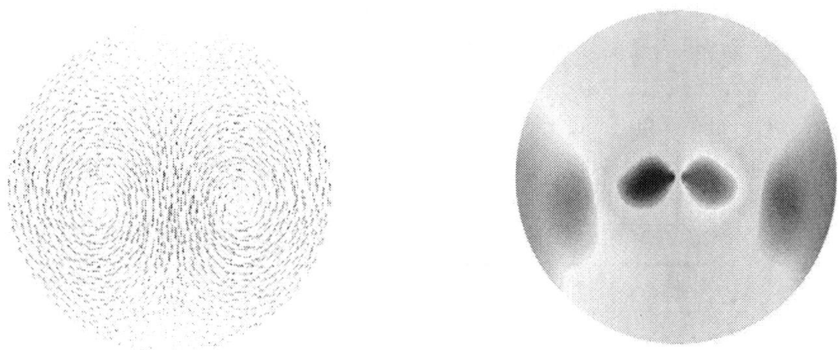

（c）M1

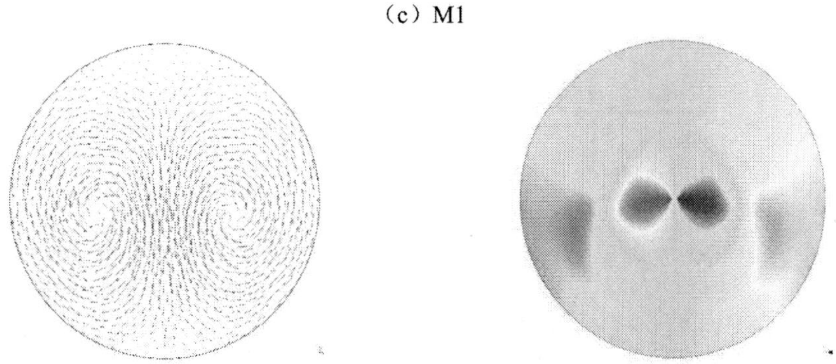

（d）M2

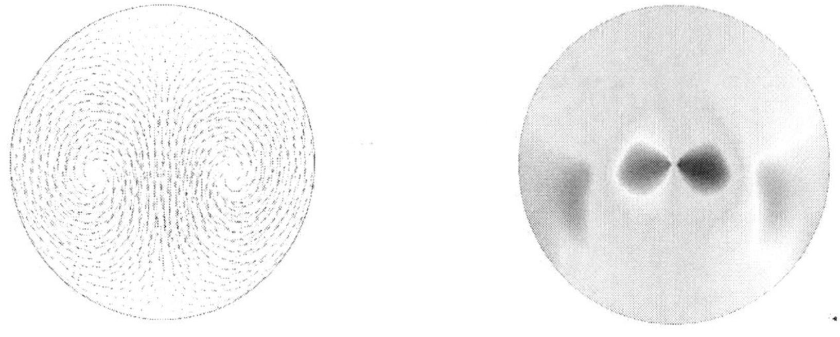

(e) M3

Fig.7 Swirl angle cloud illustration with speed vector

It can be seen from the figure that the swirl flow field M1, M2, M3 obtained after the optimization of the swirl distortion generator is closer to the distribution of the velocity vector or the swirl angle than the swirl flow field M0 obtained from the existing swirl distortion generator . It is particularly obvious at the swirl core.

4.2 Flow field analysis

In order to quantitatively compare the reproducibility of M0, M1, M2, and M3 for the standard pair swirl flow fields, and compare the flow fields M1, M2, and M3 formed by three improved models of swirl distortion generator. In this section the swirl flow field will be analyzed from the swirl angle, the swirl distortion evaluation index and the total pressure loss.

4.2.1 Distribution of swirl angle

In order to specifically compare and analyze the similarity of the distortion flow field, three measurement rings were selected on the AIP. The swirl angle distribution of each model on the measurement ring is shown in Figure 8 (b)-(d). On the three measurement rings R1, R2, and R3, the maximum error and average deviation of M1, M2, and M3 from the target swirl angle as shown in table 2.

Tab.2 Three model error angles (unit: °)

Ring position	M0	M1	M2	M3
R1	1.114	0.910	0.912	0.918
R2	1.647	1.122	0.798	1.003
R3	2.565	3.012	2.558	2.968
average error	1.776	1.681	1.423	1.630

According to the table, it is found that the flow field of M1, M2, and M3 is closer to the target flow field than that of M0 on the R1 ring, and the error of M1, M2 and M3 is similar. On the R2 test ring, the errors are relatively large, but it can be seen that the errors of M2 and M3 are small. On the R3 test ring, the four have a large gap with the target flow field, of which M2 is the best. In summary, we can see that in terms of the swirl angle distribution, the optimal design effect of the distortion generator is obvious, especially on the R1 and R2 rings. Besides, from the average errors we can see the swept forward and swept back models are more accurate than the constant chord length model. The flow field created by the swept forward models is closest to the target flow field.

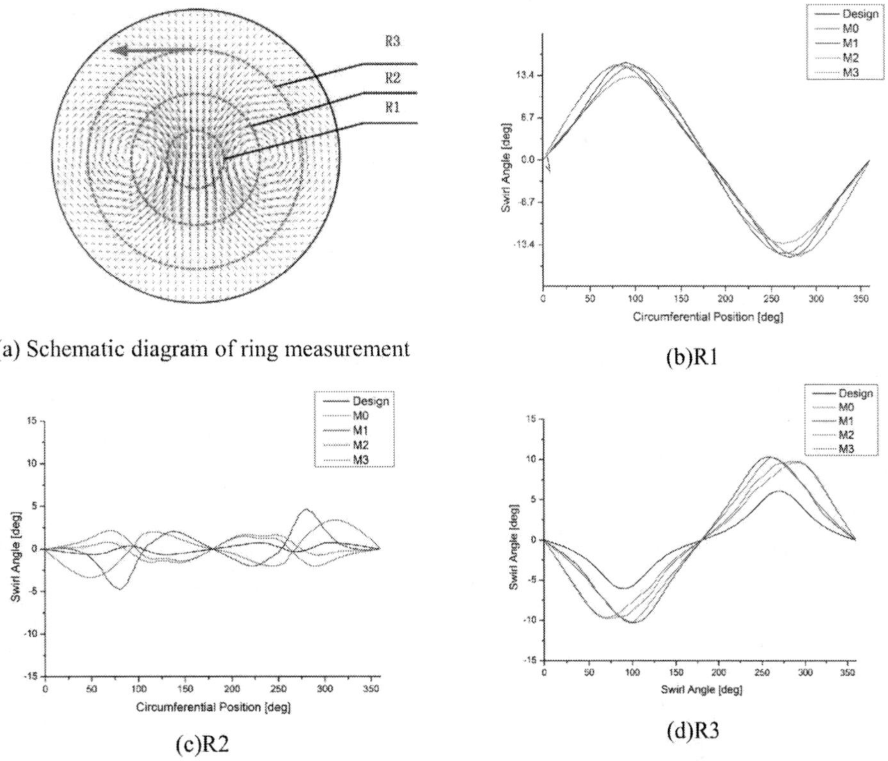

(a) Schematic diagram of ring measurement

(b)R1

(c)R2

(d)R3

Fig.8 Schematic diagram of ring position and swirl angle distribution of different models on the ring

4.2.2 Analysis of simulation results

In order to further compare and analyze the simulation results, based on the relevant standards for the evaluation of swirl distortions air5868 [18] prepared and issued by the S-16 committee, we selected the following four indicators as the evaluation criteria of the swirl distortions we studied this time. They are: Swirl Sector SS, Swirl Intensity SI, Swirl Pair SP, and Swirl Directivity SD.

(1)Sector Swirl SS

The Sector Swirl values on the three measurement rings are shown in Figure 9.

Analyze the Sector Swirl of the three measurement rings: On the R1 measurement ring, M1 is closest to the target flow field, and the error is the smallest, but the error of M1 becomes larger as the distance of the selected measurement ring from the center of the circle increases. However, the overall performance is better than M0. For M2 and M3, the blade chord length at R1 is more likely to cause pair swirl mixing, and the error is slightly larger than M1. On R2 and R3, we can see that M2 is related to the target flow field. The reproduction effect is the best, and the error is the smallest. The degree of reproduction of M3 on the R2 measurement ring is similar to M2, but the error increases rapidly with the increase of the radius.

In general, the three improved models have better reproducibility of the target flow field than the existing model M0, and the combination of the three measurement loops shows that M3 is better than M1 and M2 has the best effect.

IOP Conf. Series: Materials Science and Engineering 1081 (2021) 012047 doi:10.1088/1757-899X/1081/1/012047

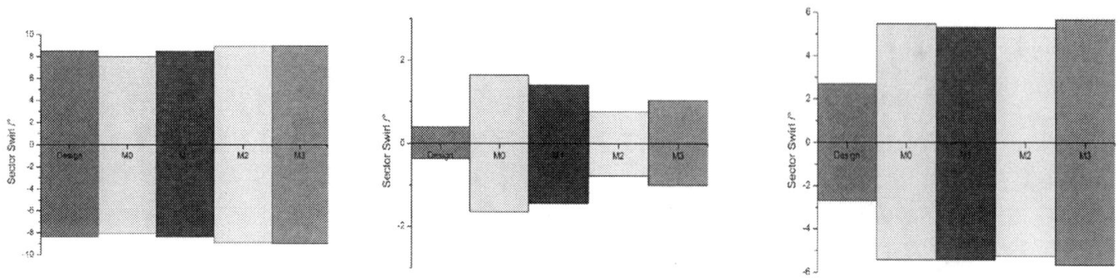

(a) R1 (b) R2 (c) R3

Fig.9 Sector Swirl

(2) Swirl Intensity SI

According to Figure 10, it can be found that, on the three measurement rings, the Swirl Intensity of R1 is the largest, R3 is the second, and R2 is the smallest, and the trend of M0 ~ M3 is consistent with the target flow field. The law of the average error of the Swirl Intensity on each measurement ring. It is similar to the Sector Swirl, so won't go into details.

Combining the evaluation indicators SS and SI, it can be found that M2 is closer to the target flow field in the distribution and intensity of the swirl angle, and the recurrence of M1 and M3 are also better than the flow field M0 generated by the existing distortion generator. On the R1 and R3 rings M1 is similar to M3. On the R2 rings M3 is closer to the target flow field than M1 obviously.

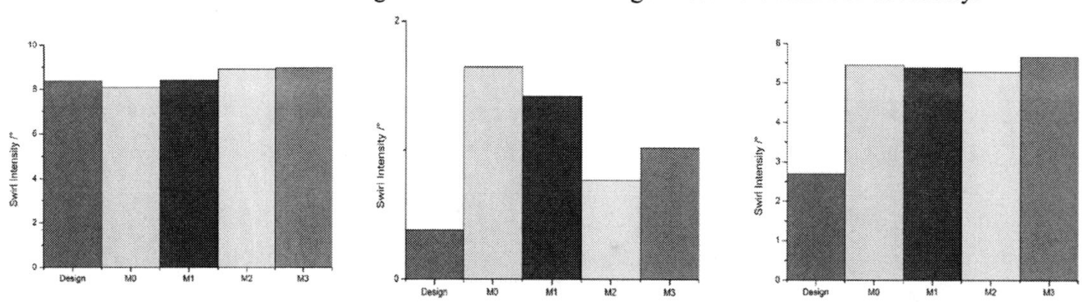

(a) R1 (b) R2 (c) R3

Fig.10 Swirl Intensity

(3) Swirl Direction SD

The standard Swirl Direction of the pair swirl flow field is 0. From the figure 11, we can see that the Swirl Direction values of the M0 ~ M3 flow field on the three measurement rings are approximately 0, and M0 ~ M3 on the three measurement rings. There is a small difference in the value of the swirl direction of the flow field. This indicates that there are equivalent reverse swirl on the three measurement rings, and the four distortion generators all have a good reproduction effect on the Swirl Direction. The difference between the four is little.

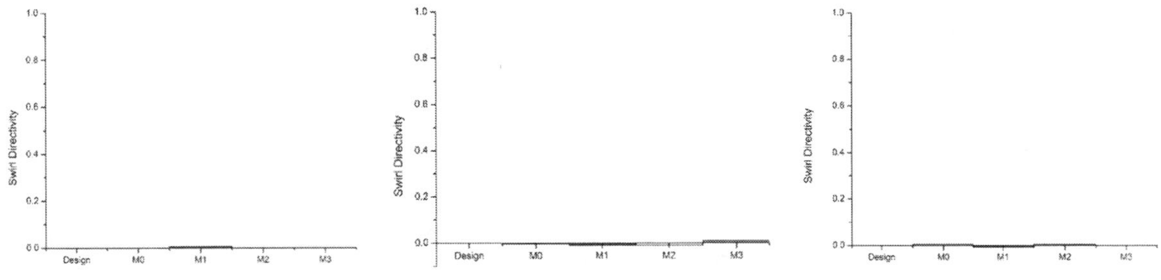

(a) R1 (b) R2 (c) R3

Fig.11 Swirl Directivity

(4) Swirl Pair SP

By analyzing Figure 12: On the R1 and R3 measurement rings, the Swirl Pair in the M0 ~ M3 flow field are close to 1, indicating that there is an equivalent reverse swirl on the measurement ring. On the R2 measurement ring, the target flow field SP is 2.07, because the target flow field has three positive swirl sectors and three negative swirl sectors on the R2 ring, and the intensity of these two sets of positive and negative swirl sectors is larger than the third group, and the intensity of these two groups of positive and negative swirl sectors is similar, so the standard pair swirl flow field at R2 measurement ring has a SP of 2.07. However, M0 ~ M3 only generates two swirl sectors on the R2 measurement ring. The intensity of the two swirl sectors of M3 is similar, with SP of 1.8. M2 is 1.56, M1 is 1.55, M0 is 1.2, and the error of M3 is the smallest, followed by is M2.

It can be seen from the figure that, on the three measurement rings, according to Swirl Pair, M3 is the closest to the target flow field, which proves that the intensity of the swirl sector of M3 is similar on the three measurement rings, and the swirl angle distribution is more uniform; followed by M2, except for R1 measurement ring (the gap between M0 and M2 at R1 measurement ring is small), it is closer to the target flow field in swirl angle distribution, which proves that the swirl angle distribution of M2 flow field is also more uniform; M1 flow under the evaluation index of Swirl Pair, the gap between the target flow field and the target flow field on R1 and R3 is the largest, indicating that the constant-angle distortion generator has a more uneven swirl angle distribution than the constant distortion generator.

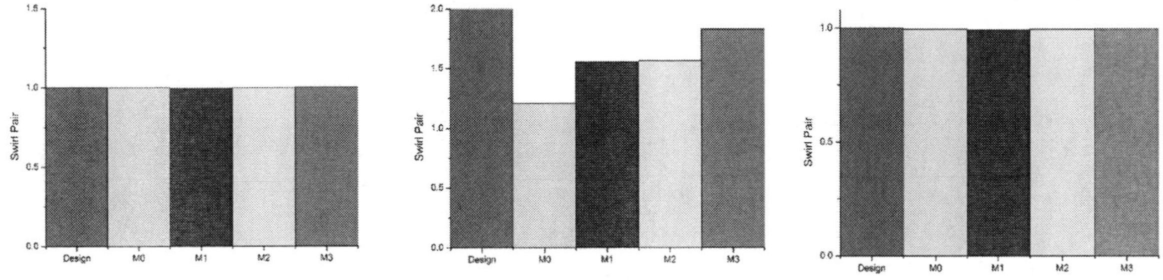

(a) R1 (b) R2 (c) R3

Fig.12 Swirl Pair

4.2.3 Total pressure loss

IOP Conf. Series: Materials Science and Engineering 1081 (2021) 012047 doi:10.1088/1757-899X/1081/1/012047

The M0 model uses a uniform thickness airfoil, which is easy to cause flow separation at the leading and trailing edges of the blade, which will cause a large pressure loss. As can be seen from the figure 13, after using the CDA airfoil (M1 ~ M3 model), the total pressure loss is significantly reduced, of which The total pressure recovery coefficient of the swept back M2 and the swept forward M3 reached 0.999, and the total pressure loss was the smallest. The existing model M0 model had a total pressure recovery coefficient of 0.977, and the total pressure loss was the largest.

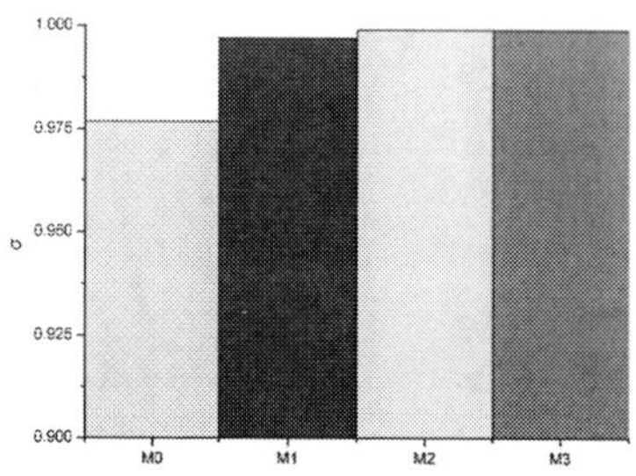

Fig.13 Total pressure recovery factor

Conclusions

In this paper, three optimization models for pair swirl distortion generators are designed--constant chord length pair swirl model, two sides of blades swept back and swept forward pair swirl models. The similarity between the pair swirl flow field obtained by the these models and the standard pair swirl flow field was compared and studied in detail using CFD technology, and compared with the pair swirl flow field obtained by the existing pair swirl model. The conclusions are as follows:

(1) During the design phase of the swirl distortion generator, by studying the bulk swirl model, it was found that when the blade consistency is 1.45 and the installation angle is 1.55 °, the swirl angle on the AIP is closest to the expected.

(2) The CDA airfoil used in this model effectively suppresses air flow separation, improves the accuracy of flow field reproduction, and effectively reduces the total pressure loss, and the total pressure recovery coefficient is increased by 2.2%.

(3) The flow field (M1, M2, and M3) generated by the model designed in this paper are more similar to the standard pair swirl flow field than the flow field (M0) generated by the existing swirl distortion generators. On the three measuring rings, the average swirl angle errors are reduced by 5.35%, 19.88%, and 8.22%. The design method in this paper is more reasonable and effective, and achieves the purpose of optimization.

(4) The swept forward and swept back models are more accurate than the constant chord length model, and the swirl angle distribution position is more accurate. Combined with the swirl evaluation index, the swept back model has the highest reappearance of the target flow field as a whole which proves that the constant consistency swept back layout is the most reasonable.

References

[1] Yu Zheng , Ni Wang. Effects of S-shaped Duct on Fan Blade Vibration [J]. Chinese Journal of Turbomachinery Chinese Journal of Turbomachinery.2019, 61(1):47-49

[2] Walter O B, Anthony F. Progress in Boundary Layer Ingesting Embedded Engine Research[C]. Joint Propulsion Conferences. American Institute of Aeronautics and Astronautics, 2012.

[3] Kawai, R T, Friedman D M, Serrano L. Blended Wing Body (BWB) Boundary Layer Ingestion (BLI) Inlet Configuration and System Studies[R]. NASA Contractor Report, NASA/CR-2006-214534.

[4] M Davis, A Hale , D Beale. An argument for enhancement of the current inlet distortion ground test practice for aircraft gas turbine engines [J].Journal of turbomachinery, 2002,124(2):235–241.

[5] M. Govardhan, K. Viswanath. Investigations on an axial flow fan stage subjected to circumferential inlet flow distortion and swirl [J].Journal of Thermal Science,1997, 6(4):241-250.

[6] Ronald H, Soeder. Effect of Steady-Stat pressure distortion On Flow Characteristic Entering A Turbofan Engine. NASA TM-79134

[7] J E Flitcroft, J Dunham, W A Abbott. Transmission of inlet distortion through a fan. [R],ADP005469 , DTIC Document, 1987.

[8] YE Fei， ZHANG Kun-yuan， JIANG Jian， et al. Experimental Investigation on Simulation and Measurement for Intake Swirl［J］ . Journal of Propulsion Technology, 2009, 30 （3） : 297-281.

[9] JIANG Jian,QU Ji-yun,. Experimental Study on the Interview Carriage of the Inlet Swirl Simulation and Measurement [J]. Gas turbine test and research, 2012, 25(3): 36-39.

[10] JIANG Jian,QU Ji-yun， SHI Jian-bang. Design and Numerical Simulation of Inlet Swirl Generator ［J］ . Gas turbine test and research， 2008， 21 （1） :43-46.

[11] TU Bao-feng, HU Jun, YIN Chao. Numerical Simulation of Vane Swirl Distortion Generator ［J］ . Propulsion technology， 2015， 36 （12） : 1817 -1824.

[12] H P Genssler, W Meyer, L Fottner. Development of intake swirl generators for turbo jet engine test-ing[R].ADP005473, DTIC Document, 1987.

[13] SONG Guo-xing,LI Jun,ZHOU You-tian,NIE Yong-zheng. Experimental and Simulation Study of Inlet Swirl Distortion of Axial Compressor [J]. Chinese Journal of Turbomachinery Chinese Journal of Turbomachin-ery.2017,59(5):2-5

[14] WANG Wen-tao. Numerical Simulation and Experimental Study on Three-Dimensional Corner Separation of Compressor Static Cascade Based on CDA Blade ［D］ . Institute of Engineering Thermophysics, Chinese Academy of Sciences， 2006， 4:15-16.

[15] GAO Li-min,CAI Ming.Wind tunnel test research on compressor airfoil [J]. Chinese Journal of Turbomachinery Chinese Journal of Turbomachinery.2018,60(4):10-15

[16] HUANG Xiu-bao. Shock Tube Simulation and Experiment Comparison of Axial Flow Compressor from Stall to Surge Instantaneous Process [J]. Chinese Journal of Turbomachinery Chinese Journal of Turbomachin-ery.2018,60(3):1-4

[17] QIN Jie,WANG Jian,WANG Yang,LIU Yun,LIU Yan. Effect of Swept Leading Edge Airfoil on Performance of Small Transonic Centrifugal Compressor [J]. Chinese Journal of Turbomachinery Chinese Journal of Tur-bomachinery.2018,60(2):48-50

[18] SAE . A methodology for assessing inlet swirl distortion [R].AIR5686, S16 Turbine Engine Inlet Distortion Committee,2007.

[19] ZHANG Lei, CHENG Bang-qin， WANG Jia-le， QI Yang， LI Jun. Design and Numerical

Simulation for New Type Swirl Distortion Screen [J].JOURNAL OF PROPULSION TECHNOLOGY,2018.39(9):2110-211

IOP Conf. Series: Materials Science and Engineering 1081 (2021) 012048 doi:10.1088/1757-899X/1081/1/012048

Impact of blade leading edge film coolant on the blade tip phantom cooling

F Li, Z Liu* and Z P Feng

School of Energy and Power Engineering, Xi'an Jiaotong University, Xi'an, China
*liuzhao@mail.xjtu.edu.cn

Abstract. Numerical simulations are carried out to predict the tip phantom cooling from blade leading edge (LE) coolant. The impact of spanwise angle of the blade LE film holes on the blade tip phantom cooling are investigated in detail under various the mass flow ratios (*MFR*) of coolant to mainstream from 0.3% to 1.3%. The results demonstrate that the LE film coolant could help cool the tip. These coolants mainly affect the aerothermal performance of the tip forepart and have little impact on that of the tip rear part. Increasing the coolant *MFR* has an obvious improvement on the tip phantom cooling level, the area-averaged cooling effectiveness could increase more than 2 times in comparing the cases with a higher *MFR* and a lower one. The cooling effectiveness on the LE surface first increase then decrease as a result of these coolants lift-off the LE surface with the increasing *MFR*. However, the LE film holes spanwise angle almost has no impact on the tip phantom cooling level at a lower *MFR* and even deteriorate the LE heat transfer procedure, where the cooling effectiveness decreases as the film holes spanwise angle increases.

1. Introduction

The gas turbine operating temperature is extremely high, and exceeds the limited temperature that the metal blade could normally operates. Consequently, many cooling methods has been applied to protect turbine components. However, the blade tip is difficult to cool as a result of some mainstream hot gas would across through the tip gap at a very high speed. The high-speed leakage flow would enhance the heat transfer for that the boundary layer is thin. Furthermore, the flow field structure in the gap is complex and is a three-dimensional problem even for a stationary case. Considering the above factors, the blade tip region is always exposed to a huge thermal load, which may damage the blade material. On the aerodynamic, the leakage flow would form the leakage vortex (LV) at the gap exit, causing the leakage flow loss and reducing the turbine efficiency. Denton [1] revealed that the tip leakage flow could reduce the work done by the mainstream, and could cause the leakage loss.

To better protect the tip, many open literatures have discussed aerothermal characteristics inside the tip gap. Newton and Lock [2], using oil-flow visualization method, found that the leakage flow would form a separation region on the tip near the pressure side (PS), where its heat transfer coefficient is the highest. Xiao and McCarter [3, 4] measured the flow field in the tip gap and in the passage. Their results showed that the highest total pressure loss value in the LV is much higher than that in the passage vortex. O'Dowd et al. [5] experimentally revealed that the tip gap size would greatly affect the size and strength of the tip LV. The blade in real turbine is a rotating component. Thus, it is vital to study the tip aerothermal performance under the rotating condition. Zhou [6], Srinivasan and Goldstein [7] analyzed the impact of relative casing movement on tip aerothermal performance. Zhou and Zhou [8] found that positive in-coming vortex could suppress the LV and reduce the aerodynamic loss. Zhang et al. [9] experimentally found that both the turbulence intensity and the inlet boundary layer thickness don't affect the tip heat transfer. Zhang and He [10], and Jiang et al. [11] demonstrated that the wall thermal boundary condition of the tip has an evident impact on the tip leakage flow structure. Arisi et al. [12] pointed out that higher exit Mach number enhances the tip thermal performance while it does not affect the LV.

What's more, considering the high thermal loads in the tip region, cooling methods for the tip are very important and necessary. Many cooling methods has been applied on the tip to protect this region, such as tip injection, casing injection and PS near-tip injection, or their combined cooling injection. However, the application of cooling method will increase the consumption of coolant and reduce the turbine thermal efficiency. From this point of view, new cooling concept has been introduced and discussed in the 1980s in the turbine industry. Phantom cooling, firstly detailed studied by Roback and Dring [13, 14], is the coolant redistribution on the blade surfaces or the end-wall caused by the upstream coolant injection or the leakage flows. It means that local area surface could be cooled by upstream spent coolant, which is a kind of the second order and unintentional cooling effect. Although the coolant temperature increases a lot after having cooled its target area, it may have the potential to cool other regions. Thus, this kind of unintentional cooling effect could reduce coolant consumption and even there is no need to place film holes on the local surface. For the blade tip, its phantom cooling could be affected by the nozzle shroud cooling discharge, the nozzle surface film cooling, the nozzle trailing edge slot, the leakage flow between tip shoe and nozzle end-wall, and the blade LE and the PS film coolant. Although above factors are not validated by experimental or numerical method in open literatures, they may be considered in industrial design process, especially for the impacts of LE film coolant.

Phantom cooling could save local coolant usage and increase the gas turbines output power and thermal efficiency. Zhang et al. [15], using pressure sensitive paint technology, measured the impact of end-wall hole diameter on the suction side (SS) phantom cooling. Zhang et al. [16] also pointed that decreasing the LE film holes spanwise angle and employing the compound angle would improve the end-wall cooling effectiveness. Liu et al. [17] also carried out similar studies and considered the impact of coolant from the trailing edge slots. Zhang et al. [18], Du et al. [19, 20, 21], Zhang and Yuan [22], and Papa et al. [23] investigated the impact of coolant from leakage slot on phantom cooling at the SS surface near the end-wall. Their studied variables mainly include the swirling jet angle, slot jet angle and blowing ratio. Zhang and Yuan [24], and Li et al. [25] found that both the end-wall and blade SS phantom cooling could be achieved by the coolant from trailing edge slots.

Previous work only focused on the tip aerothermal performance, or the phantom cooling on the blade SS or the end-wall surface. However, no attention was paid to the impacts of coolant from blade LE film hole on the tip phantom cooling effects up till the present moment. In this study, the phantom cooling level on the tip from LE film coolant is studied. The impacts of *MFR* and LE film hole spanwise angle on the tip phantom cooling and aerodynamic performance are detailly analyzed. To investigate these impacts, numerical simulations are performed under five different *MFR* values: 0.3%, 0.5%, 0.7%, 1.0%, and 1.3% at a LE film hole spanwise angle value of 25°; and five different LE film hole spanwise angle values of: 25°, 35°, 45°, 55° and 65° at an *MFR* of 0.5%.

2. Numerical methods

2.1. Geometry configuration

The calculation model derived from the first stage of the GE-E3 engine turbine (Ref. [26]). The blade height is 42.65mm, and the tip gap size is about 1% of the blade height. These dimensions match those in Ref. [26]. To simplify the computational model, the impingement cooling in the LE area is not concluded, and coolant is supplied by a coolant supply plenum (see figure 1(a)). The flat tip is imposed in the present study, while the original one from Ref. [26] is the squealer tip. The three rows LE film hole (Row 1-3) have a spanwise angle ranging from 25° to 65° (see figure 1(b)) and is arranged at a constant pitch with a diameter of 0.36mm. All the holes in Row 1-3 are the cylindrical hole (see figure 1(c)). The location of film hole row is defined according to the percentage of arc length of the airfoil profile. The Row2 is located on the stagnation line of the LE and its percentage value is 0%, while the Row1 and Row3 is located on the SS (the minus value) and the PS (the plus value), respectively. Details about film hole definitions are presented in figure 1, and table 1 shows detailed geometric information.

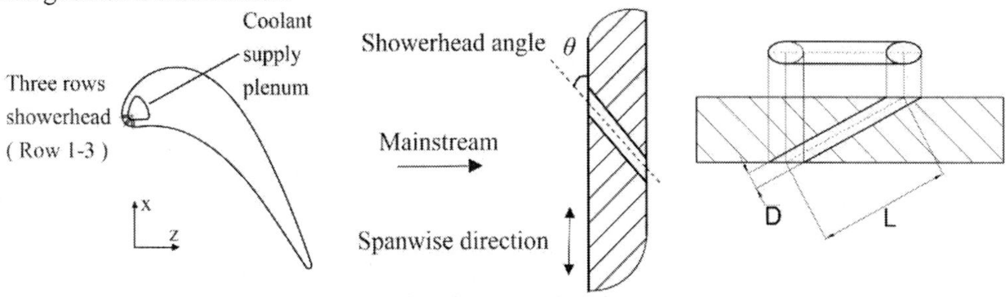

(a) Blade schematic diagram (b) Definition of LE film hole angle (c) Cylindrical hole (CY)

Figure 1. Detailed geometry information and film holes definitions

Table 1. Detailed geometry information of jet configurations

	N	*D*/mm	*P*/*D*	Hole type	*α*/°	Location from LE /%	*θ*/°	*L*/*D*
Row1	11	0.36	10.0	CY	12	-2.2	25,35,45,55,65	3.1~6.6
Row2	10	0.36	10.0	CY	51	0	25,35,45,55,65	3.1~6.7
Row3	11	0.36	10.0	CY	80	3.1	25,35,45,55,65	3.1~6.6

2.2. *Numerical simulations setting*

The boundary conditions come from Ref. [26]. As displayed in figure 2, Total pressure and total temperature are set to the mainstream inlet, and the mainstream outlet is set to the average static pressure. Coolant temperature and its mass flow rate are set to the coolant inlet. The cascade passage either side are set to the rotational periodic boundary condition. and the mixing plane is the stator-rotor interface. The other surfaces are set to the no-slip and adiabatic wall. The blade rotational speed is 8450 rpm. The main fluid and the coolant air are ideal gases. The main fluid's physical properties are functions of temperature. Table 2 presents the boundary conditions in detail.

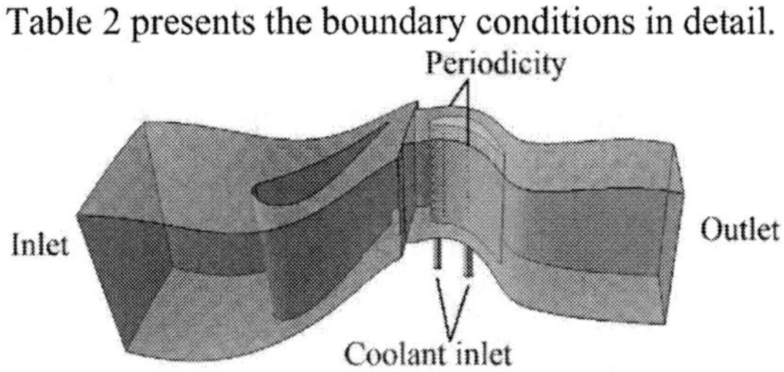

Figure 2. Calculation domain

Table 2. Boundary conditions

	Parameters	Value
Mainstream	T_0/K	709
	P_0/P_2	2.25
	ω/rpm	8450
	Tu_0/%	5
Coolant	T_c/K	344
	MFR/%	0.3, 0.5, 0.7, 1.0, 1.3
	Tu_c/%	1

Present numerical simulations are implied by solving the commercial CFD code ANSYS CFX. This code could solve the steady Reynolds-Averaged Navier-Stokes equations and discretize the conservation equation by applying the finite control volume. The convection term is discretized by employing the high-resolution method with a second order form. The whole calculations are operated under the second order. The convergent target is that the area-averaged η on the tip remains constant with additional iterations.

2.3. Numerical simulations setting

To find out the best turbulence model to carry out this study, a linear cascade experimentally investigated by Kwak [27] is chosen to validate the accuracy of numerical methods with a flat tip case. Figure 3 compared the difference of h distribution between the measurement and predictions. The results reveal that the standard *k-ω* model have similar *h* distribution with experimental data, which achieves the best precise predictions in the separation region near the PS. The SST *k-ω* model also could capture the separation region but underpredicts the high-*h* region. However, the standard *k-ε* model overpredict the high-*h* region and could not catch the high-*h* region near LE. The RNG *k-ε* mis-predicts the separation region and underpredicts the high-h region. It is showed that the standard *k-ω* is the most suitable model and it is employed in the presented simulations.

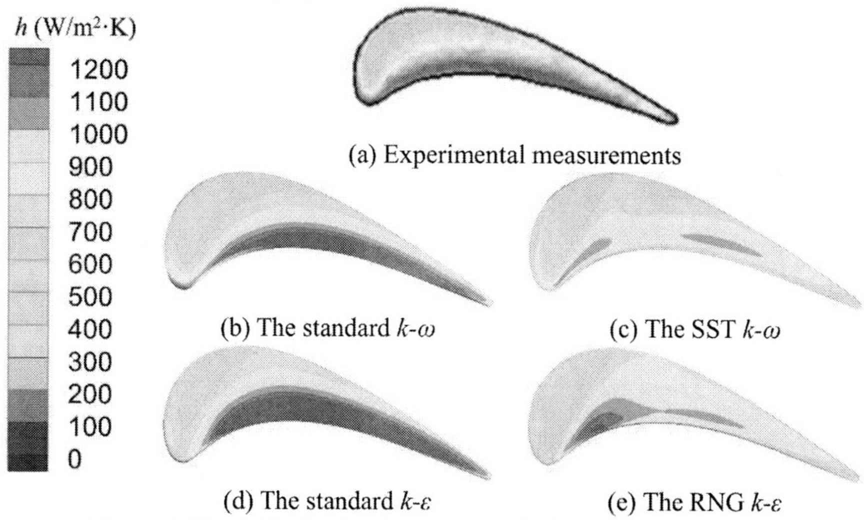

Figure 3. The *h* distribution between predictions and measurement

Three different meshes for the calculation domains with grid numbers of 5.75 million, 8.20 million, and 10.20 million were employed to carry on the grid independence analysis. The unstructured mesh in the calculation domains were generated. The mainstream region consists of tetrahedral mesh and the near-wall regions are full of prism mesh. To ensure that there is no poor-quality mesh, all the mesh elements had been refined. To catch the high velocity gradient near-wall regions, the number of boundary layers is 15, and the first boundary layer height is 0.001mm with a growth factor of 1.1, the wall $y+$ in all cases is about 1.0.

Figure 4 presents the η level on the tip and its pitch-averaged η along the axial-chord direction at three different numbers of nodes for the case with θ=25° and *MFR*=1.0%. The results show that there is almost no difference between the cooling effectiveness with 8.20 million (figure 4(a) (2)) and that with 10.20 million (figure 4(a) (3)). The pitch-averaged η also show the same distributions among the cases with 8.20 million and 10.20 million. It could be concluded that the increasing number of nodes has almost no impact on the predictions. Considering the numerical simulations precision and the calculation cost, the mesh with 8.20 million nodes was employed in present simulations.

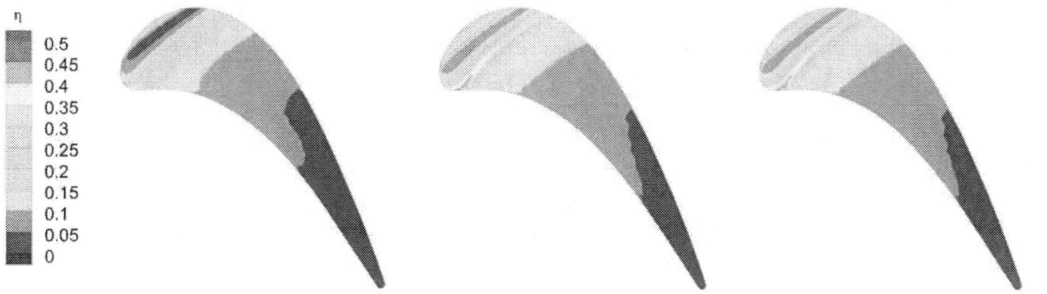

(1)5.75million (2)8.20million (3)10.20million

(a) Contours of η level

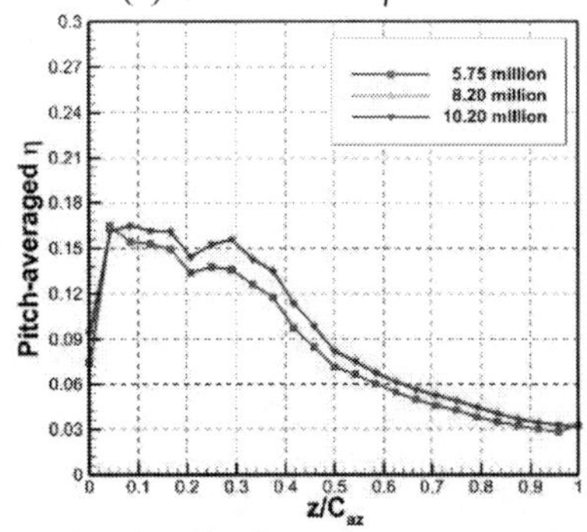

(b) Pitch-averaged η distributions along the axial-chord direction

Figure 4. Gird independence analysis with different nodes for the case with θ=25° and MFR=1.0%

3. Results and discussions

3.1. The tip phantom cooling features

Figure 5 shows the LE film coolant flow path and its phantom cooling effects on the blade tip with a case of θ=25° at an MFR of 0.5%. From the figure 5(a), first, the coolant from the LE film hole cool the LE surface, and form a high-η region on local surface. Then, the coolant, mixing the mainstream, flow into the tip gap. Although the coolant's temperature increases a lot, it still has the potential to cool the tip surface. Thus, it is observed that there are several bar-shaped high-η regions. It is also found that the coolant flowing into tip gap mainly comes from the No. 1, No. 2 and No. 3 hole in Row2 and Row3. These coolant flow into the tip gap for that the pressure difference between the PS and the SS. These coolant from No. 1 in Row1 also form a high-η region on the tip. The coolant is suppressed by the mainstream and become a part of leakage flow. While it is caused by the mainstream incidence angle rather than the pressure difference. Besides, From the figure 5(b), the coolant will form the LV and passage vortex and it will affect the aerodynamic characteristics in the passage.

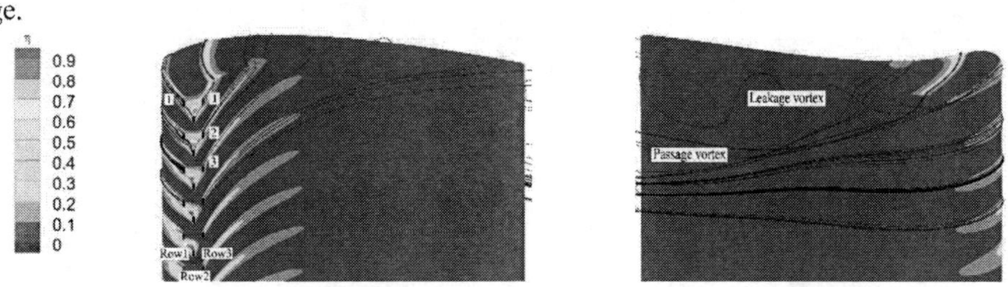

(a) PS view (b) SS view

Figure 5. The LE film injection coolant flow path and its phantom cooling effects on the blade tip

3.2. Impact of MFR

3.2.1. Thermal performance. Figure 6 shows the η level on the tip surface for different *MFR* values at a θ value of 25°. The results show that there are two higher η regions (A and B) at the tip forepart. Region A starts from the z/C_{az}=0 near the SS and ends at the about z/C_{az}=0.25 near the PS. This region is cooled by the coolants from the No.1 hole of Row1 and Row2 of the LE. These coolants are suppressed by the mainstream and directly flow into the tip gap, becoming a part of the leakage flow. Region B starts from the z/C_{az}=0.1~0.2 near the PS and then across the tip gap and ends at the 0.35~0.45 z/C_{az} near the SS. The coolant from the No. 1 hole of Row2 and Row3 cool this area, mixing with the mainstream. It is also observed that two low-η regions (C and D) at the tip central. These regions are cooled by the coolant from the No. 2 and No. 3 hole in Row2 and Row3, respectively. The temperature of coolant rises due to it mixes with mainstream, its second order cooling effect on the tip become weak. Thus, the η in regions C and D is lower than that in region A and B. The η level at other areas on the tip is almost zero except region A, B, C and D. This is because almost no coolant could cover these areas in the studied cases. At a low *MFR* of 0.3%, the area of high-η is small. The highest η level is observed in Region B, and Region D almost cannot be detected. As *MFR* increases to 0.5% and 0.7%, all the high-η regions expand. Regions C and D integrate and the low-η regions shrink. At a high *MFR* of 1.0% and 1.3%, the area that η=0 almost disappear and the results show a high-level phantom cooling performance. At all *MFR* values, the high-η regions are detected in tip forepart (z/C_{az}=0~0.5) and the Region B always shows the highest η value.

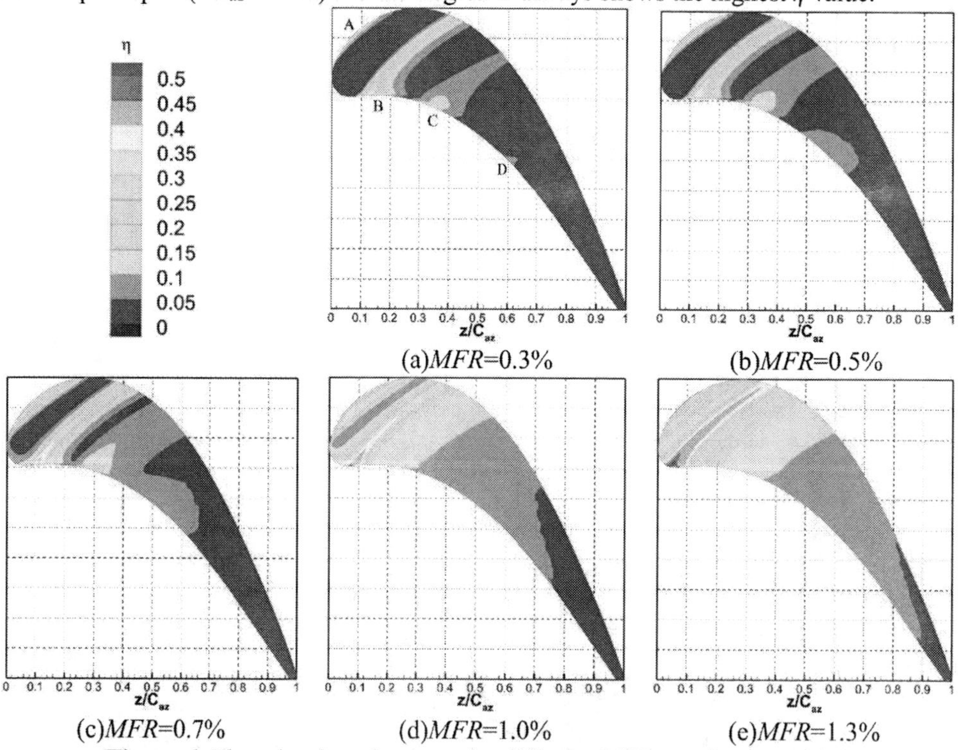

Figure 6. The η level on the tip under different *MFR* at a θ value of 25°

To quantitatively compare the η level for different *MFR* values, figure 7 shows the pitch-averaged η distribution along the axial-chord direction. At lower *MFR*, two peak values of η appear between the z/C_{az}=0.1~0.2 and z/C_{az}=0.3~0.4, which agree with the high η regions in figure 6. While there is no obvious peak value of η at higher *MFR*, and the η level decreases along the axial-chord direction. On the tip rear part(z/C_{az}=0.5~1.0), the η value is much lower than that in the forepart of tip, which means that the LE coolant mainly affects the heat transfer of the tip forepart and have little impact on the tip rear part. In general, the *MFR* has an evident impact on the η level of the tip, particularly in the aera z/C_{az}=0~0.5. The peak value of the pitch-averaged η in the case with the *MFR*=0.3% is 0.1, and this value increases by 190% when *MFR*=1.3%. Combining the figure 6 and figure 7, with an increase of *MFR*, both the higher η regions (A and B) and the lower η regions (C and D) expand (integrate at higher *MFR*).

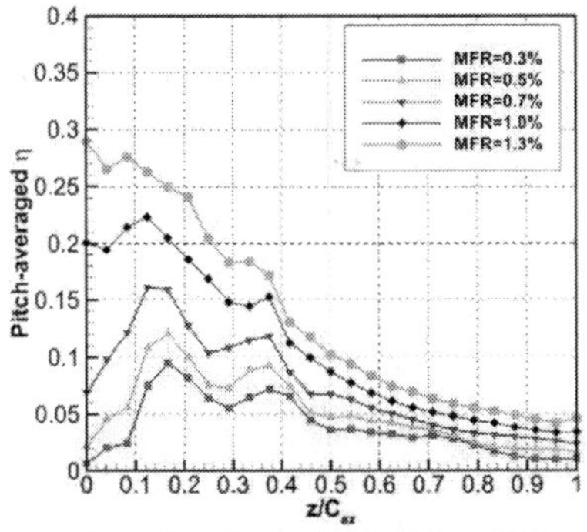

Figure 7. Pitch-averaged η distribution along the axial chord direction on the tip at a θ value of 25°

Figure 8 presents the η level on the LE for different *MFR* values. The left side of each picture is the SS and the right one is the PS. The results show that the high-η regions located on the coolant flow path at any *MFR*, especially on the hole exit the η could reach to 0.9. At a low *MFR* of 0.3%, its η level is relatively high. However, the η value on the hub region of the LE is much lower than other areas and no coolant eject on the stagnation line of the hub region. As increases to 0.5%, the η level both on the PS and the SS enhance, and the hub region also could achieve a better film coverage. However, although the high-η aeras on the tip region expand, its highest η value decrease from about 0.9 to about 0.8. this is owing to that the more coolant discharge in the tip region for that the centrifugal force and the coolant jet off from the LE surface. As the *MFR* continues to increase, the η value in these high-η aeras decreases, especially on the PS and hub region. At a high *MFR* of 1.3%, the high-η aeras on the PS almost disappear, which means that the PS would be difficult to cool under a higher *MFR*. The coolant on the tip region integrate and form a larger high-η aeras, but its level is lower than that in the cases with a low *MFR*. With the increasing *MFR*, the η level on the LE first increases and then decrease, achieving the highest η level with the *MFR* of 0.5%.

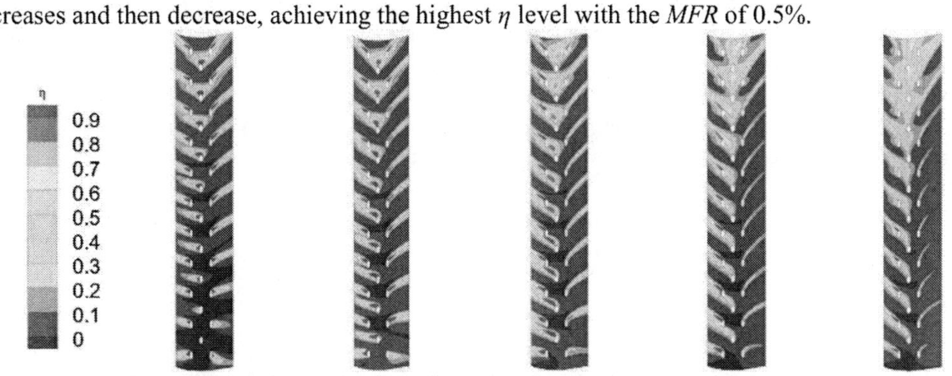

(a)*MFR*=0.3% (b)*MFR*=0.5% (c)*MFR*=0.7% (d)*MFR*=1.0% (e)*MFR*=1.3%

Figure 8. The η level on the LE surface for different *MFR* at a θ value of 25°

To better understand how the *MFR* influence the phantom cooling level on the tip and the film cooling performance on the LE, figure 9 shows the coolant flow structure and the non-dimensional temperature distributions on the cross-section of Row3 near the tip for different *MFR* values. The non-dimensional temperature distributions could reveal the coolant distribution on the LE and in the tip gap. The left side of each picture is the PS, and the right side the SS. The top side is the tip gap. For a low *MFR* of 0.3%, the high-ε region could be detected nearby the hole exit and the ε level in the tip gap is very low. As *MFR* increases to 0.7%, the high ε region on the LE expands and a high-ε region could be found in the tip gap near the SS. As *MFR* continue to increase, some coolant lifts off the LE

IOP Conf. Series: Materials Science and Engineering 1081 (2021) 012048 doi:10.1088/1757-899X/1081/1/012048

surface and become a part of the cross flow. Therefore, the η-level on the LE decrease, which agree with the results in the figure 8. However, more coolant flow into the tip gap and the ε level in the tip gap is relatively higher, it means that increasing the *MFR* would enhance the phantom cooling performance on the tip.

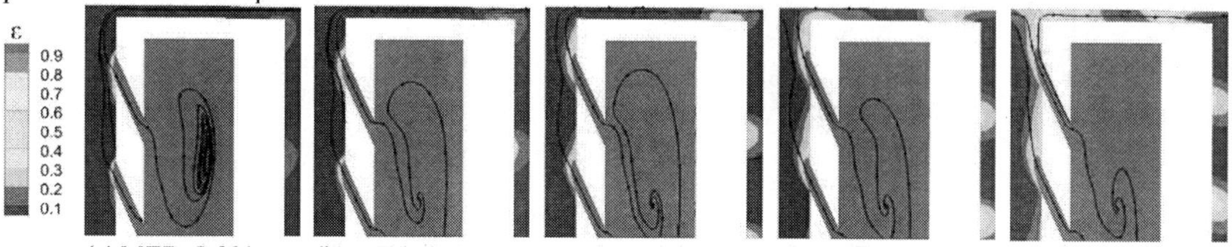

(a)*MFR*=0.3% (b)*MFR*=0.5% (c)*MFR*=0.7% (d)*MFR*=1.0% (e)*MFR*=1.3%

Figure 9. The coolant flow structure and the non-dimensional temperature distributions on the cross-section of row3 near the tip

3.2.2. Aerodynamic performance. Figure 10 displays the normalized tip leakage mass flow rate distribution for different *MFR*. The minus tip leakage mass flow rate values, existing at about z/C_{az}=0~0.13, are caused by the mainstream incidence angle. It means that some mainstream directly flows into the tip gap, mixing with the coolant. With an increase of *MFR*, the tip leakage mass flow rate increase at z/C_{az}=0.25~0.6. While the *MFR* has little impact on it at other parts of tip.

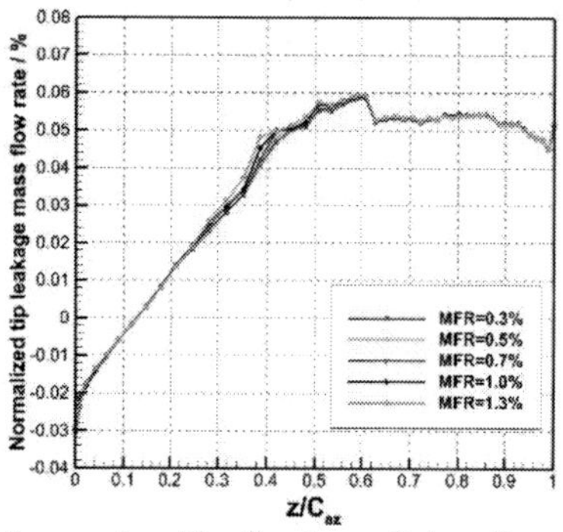

Figure 10. The normalized tip leakage mass flow rate distribution along the chordwise direction at a θ value of 25°

Figure 11 shows the distribution of the dimensionless specific entropy increment at several cut-planes of z/C_{az}=0.2, 0.4, 0.6 and 0.8. The LV begins to appear at the z/C_{az}=0.2 and develops along the streamwise direction. The upper passage vortex also develops along the streamwise and its size is smaller than that the LV. With increasing *MFR*, the size of both the LV and the upper passage vortex almost remains constant, but its values of entropy increment increases, especially at 0.6 and 0.8 z/C_{az} cut-planes.

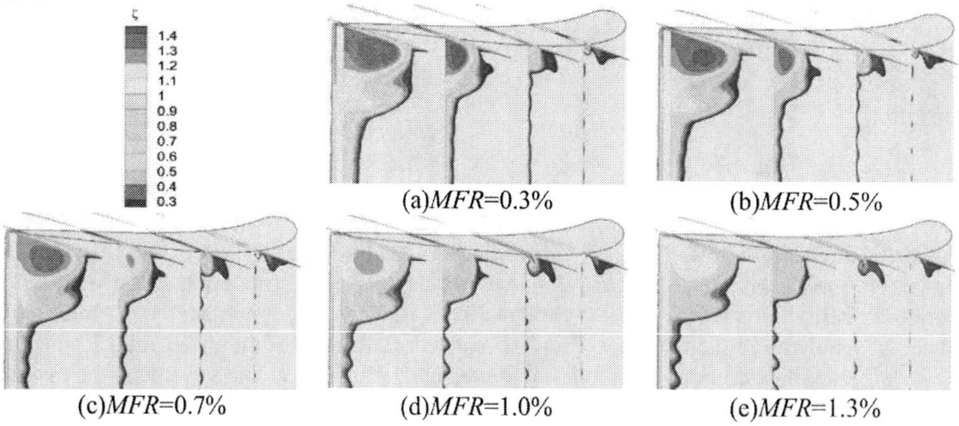

(a)*MFR*=0.3% (b)*MFR*=0.5%

(c)*MFR*=0.7% (d)*MFR*=1.0% (e)*MFR*=1.3%

IOP Conf. Series: Materials Science and Engineering 1081 (2021) 012048 doi:10.1088/1757-899X/1081/1/012048

Figure 11. The distribution of the dimensionless specific entropy increment at different cut-planes at a θ value of 25°

Figure 12 shows the contours of the Cp_s on the z/C_{az}=1.1 cut-plane for different *MFR* at a θ value of 25°. The upper boundary of each figure is casing, and the lower boundary of each figure is hub. There are three obvious higher Cp_s regions corresponding with loss cores, including the LV, the upper passage vortex and the lower passage vortex. Besides, there are boundary layer losses near the hub and shroud regions and profile losses in the middle part of this picture. With the increase of *MFR*, the Cp_s values in the upper vortex decrease, while there is no much difference in other loss cores. To quantitatively compare the Cp_s at different *MFR* values, figure 13 presents the pitch-averaged Cp_s distribution in the spanwise direction. The results show that the *MFR* just have little impact on the Cp_s at 0~0.3 and 0.60~0.96 relative blade height. At the 0.90~0.96 and the 0.60~0.80 relative blade height, corresponding to the LV and upper vortex, the pitch-averaged Cp_s decrease with increasing *MFR*. At the 0~0.3 relative blade height, corresponding to the lower passage vortex, the pitch-averaged Cp_s also decrease with increasing *MFR*.

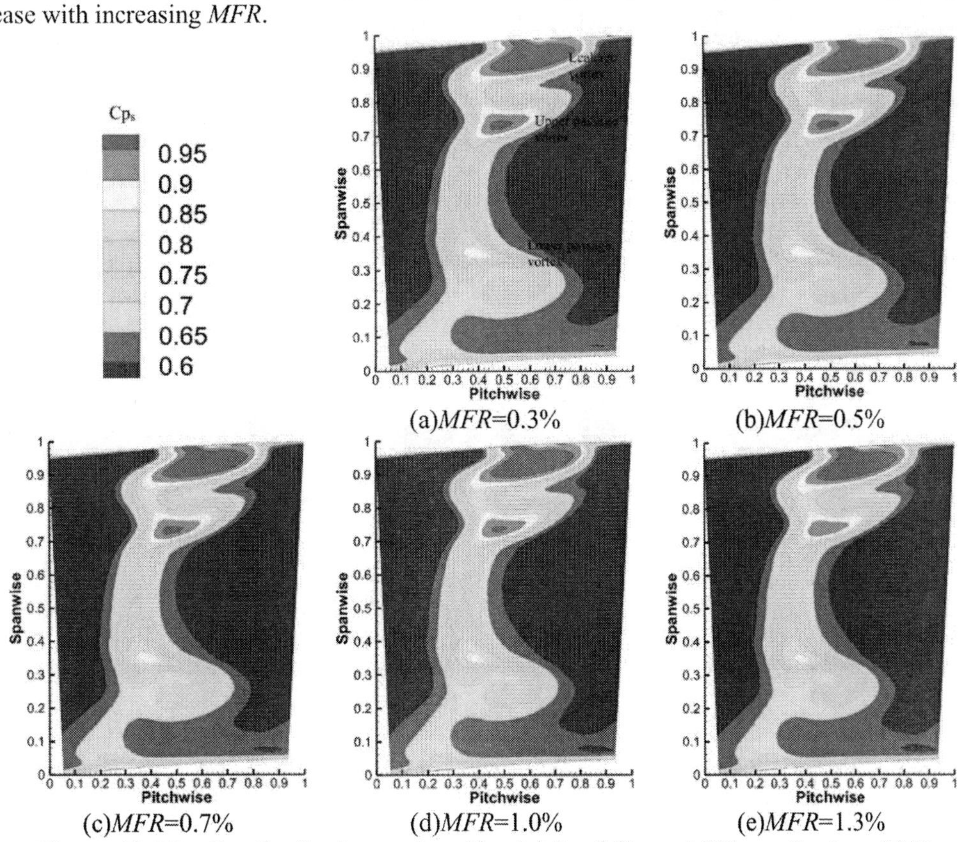

Figure 12. The *Cps* distribution on the z/C_{az}=1.1 for different *MFR* at a θ value of 25°

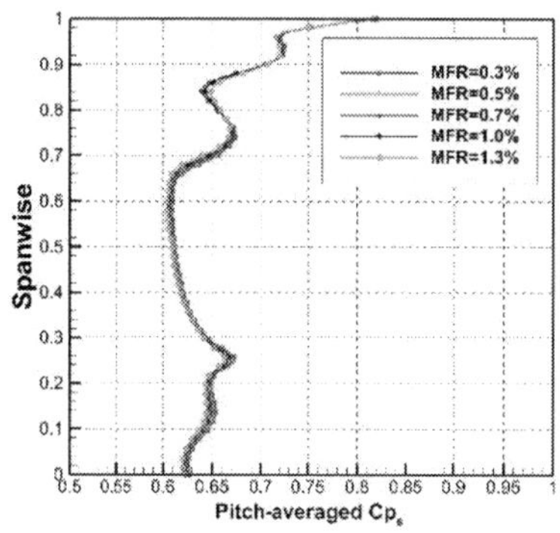

Figure 13. The pitch-averaged Cp_s distribution along the spanwise direction on the z/C_{az}=1.1 cut-plane at a θ value of 25°

3.3. Impact of θ

Thermal performance. Figure 14 shows the η level on the tip for different θ at a constant *MFR* of 0.5%. At a smaller θ value, the radial component of the coolant velocity is relatively large, and under the oppression of the leakage flow, the coolant would cover the tip surface better. With an increase of θ, both the area of all high-η regions and its highest η value decrease. It is obviously found that the Region A nearly vanished in case of θ value of 65° and the Region D disappears in cases of θ values, 55° and 65°. This is because the radial component of coolant velocity decreases, and the axial component increases with an increase θ, then coolant mainly jets into the mainstream hot gas. This will weaken the second order cooling effect on the tip due to mixing with the mainstream hot gas. Figure 15 shows the pitch-averaged η along the axial chord direction for different θ values. There are two peak values of η, located at about z/C_{az}=0.16 and z/C_{az}=0.4. It is observed that the θ has an evident impact in these areas z/C_{az}=0~0.2 and z/C_{az}=0.3~0.4, which are corresponding to the high η values in Regions A, B and C in figure 13. With an increasing θ, the pitch-averaged η decrease in above three regions, while it almost remains constant in other regions.

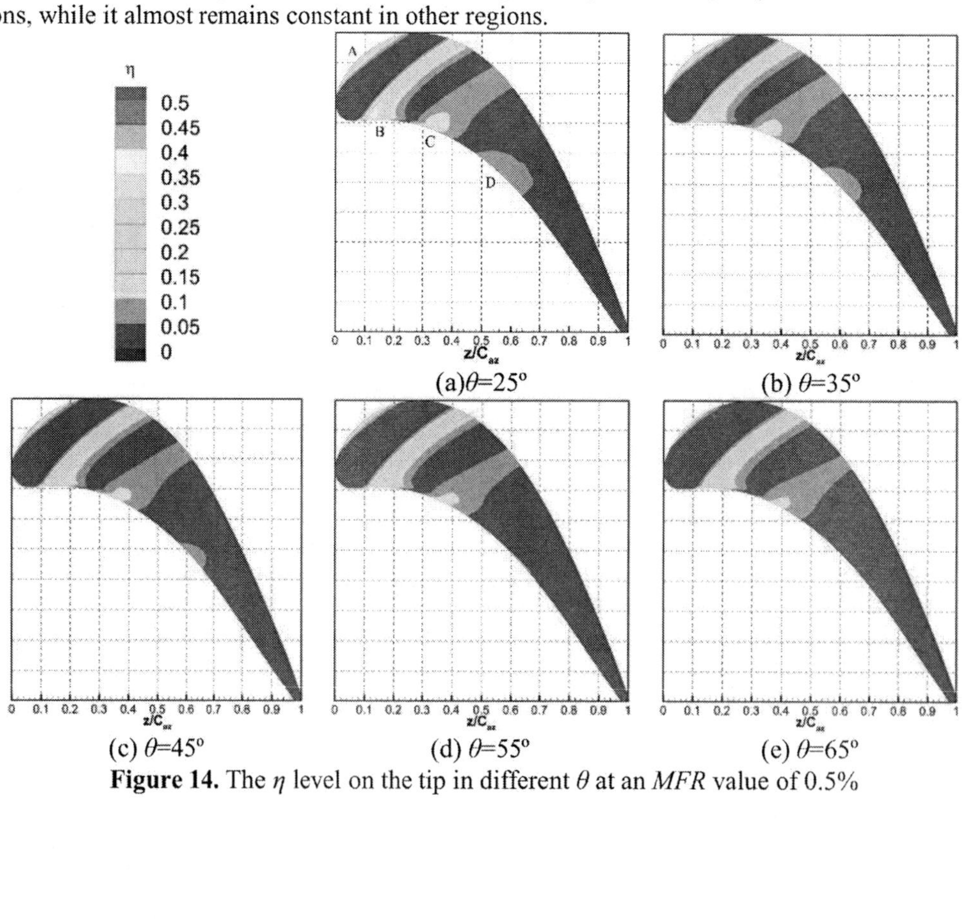

(a)θ=25° (b) θ=35°

(c) θ=45° (d) θ=55° (e) θ=65°

Figure 14. The η level on the tip in different θ at an *MFR* value of 0.5%

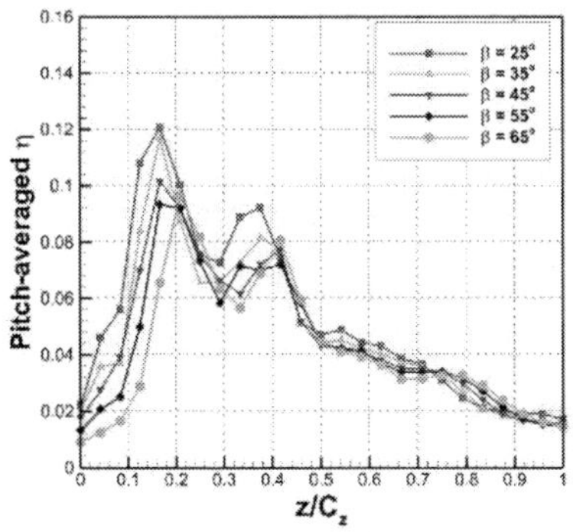

Figure 15. The pitch-averaged η distribution along the axial-chord direction on the tip at an *MFR* value of 0.5%

Figure 16 shows the η level on the LE for different θ values at an *MFR* value of 0.5%. The features of high-η areas in Figure 16(a) are clearly shown and analyzed in Figure 8. The results show that with the increasing θ, both the η value in the high-η regions and the its area decrease. And on the PS, the coolant flow path almost disappears at high θ values. This is because that the radial component of the coolant velocity for the case with a small θ is large, and the coolant would flow along the LE surface under the mainstream suppression. As the θ become larger, the radial component of coolant velocity decreases, and the axial component increases. Thus, the coolant mainly jets into the mainstream hot gas rather than cover the LE surface. Furthermore, there is a low-η region between neighboring film holes in the same row, especially for the Row2 and Row3, for that there is no coolant coverage in these regions.

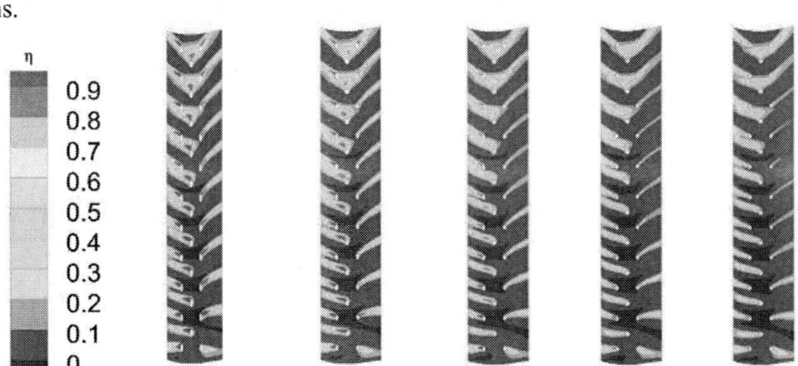

(a)θ=25° (b) θ=35° (c) θ=45° (d) θ=55° (e) θ=65°

Figure 16. The η level on the LE in different θ at an *MFR* value of 0.5%

Figure 17 presents the coolant flow structure and the non-dimensional temperature distributions on the cross-section of Row3 near the tip for different θ values. For a low θ value of 25°, the high-ε region could be detected nearby the hole exit and the ε level in the tip gap is very low. As the θ increase, the ε level near the hole exit still remains high. However, the component of coolant velocity that towards to the tip decreases due to the increasing θ. Therefore, the η-level on the LE decrease, which agree with the results in the figure 16. Meanwhile, the coolant that flows into the tip gap become less, which leads to the phantom cooling performance on the tip become weak. In general, increasing θ would deteriorate the heat transfer process both on the LE and the tip.

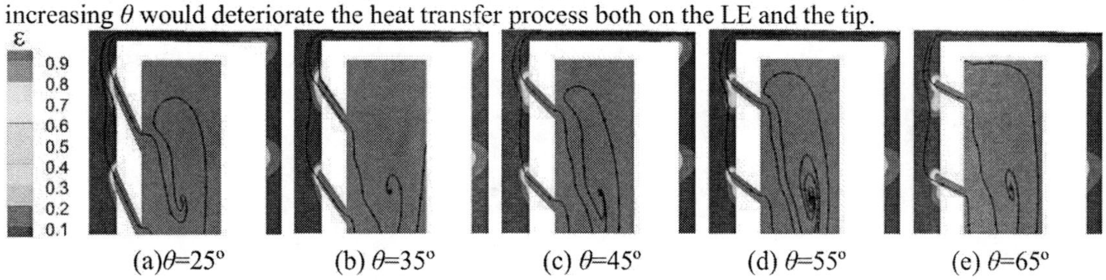

(a)θ=25° (b) θ=35° (c) θ=45° (d) θ=55° (e) θ=65°

Figure 17. The coolant flow structure and the non-dimensional temperature distributions on the cross-section of row3 near the tip

3.3.1. Aerodynamic performance. Figure 18 displays the normalized tip leakage mass flow rate for different θ at an *MFR* value of 0.5%. With the increasing θ, the tip leakage mass flow rate almost remains constant, which means that the θ would not affect the leakage flow in the case with an *MFR* value of 0.5%.

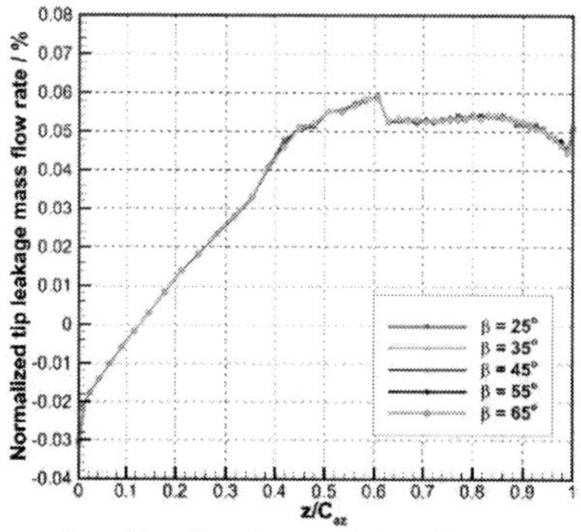

Figure 18. The normalized tip leakage mass flow rate distribution along the chordwise direction at an *MFR* value of 0.5%

Figure 19 shows the distribution of the dimensionless specific entropy increment at several cut-planes of z/C_{az}=0.2, 0.4, 0.6 and 0.8. At any cut-plane, the size of both the LV and the upper passage vortex, and the high values of entropy increment are almost the same with increasing θ. This is because that the entropy increment is mainly affected by the LV, which caused by the leakage flow interacts with the across secondary. However, the θ would not affect the leakage flow as displayed in figure 18. Thus, the dimensionless specific entropy increase distribution at any cut-plane remains the same with increasing θ.

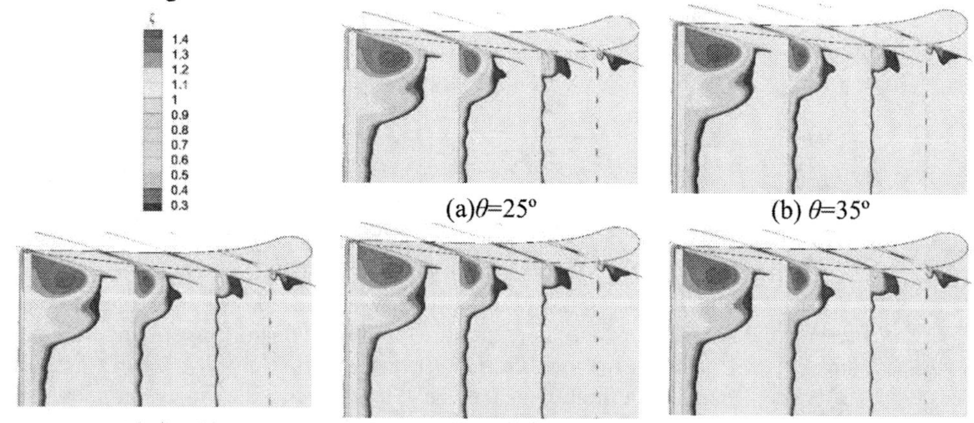

(a)θ=25° (b) θ=35°

(c) θ=45° (d) θ=55° (e) θ=65°

Figure 19. The distribution of the dimensionless specific entropy increment at different cut-planes at an *MFR* value of 0.5%

Figure 20 and Figure 21 show the contours of the Cp_s at the z/C_{az}=1.1 cut-plane and its pitch-averaged values along the spanwise direction for different θ with an *MFR* value of 0.5%, respectively. The loss cores and their corresponding vortex are shown and analysis results are in figure 12. There is no obvious difference in the contours the Cp_s for different θ and the figure 20 reveals that θ has no impact on the Cp_s distributions.

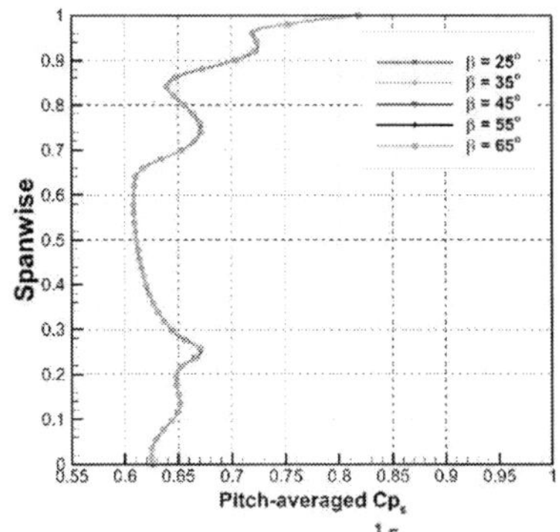

Figure 20. The pitch-averaged cps along the spanwise direction for different θ at an *MFR* value of 0.5%.

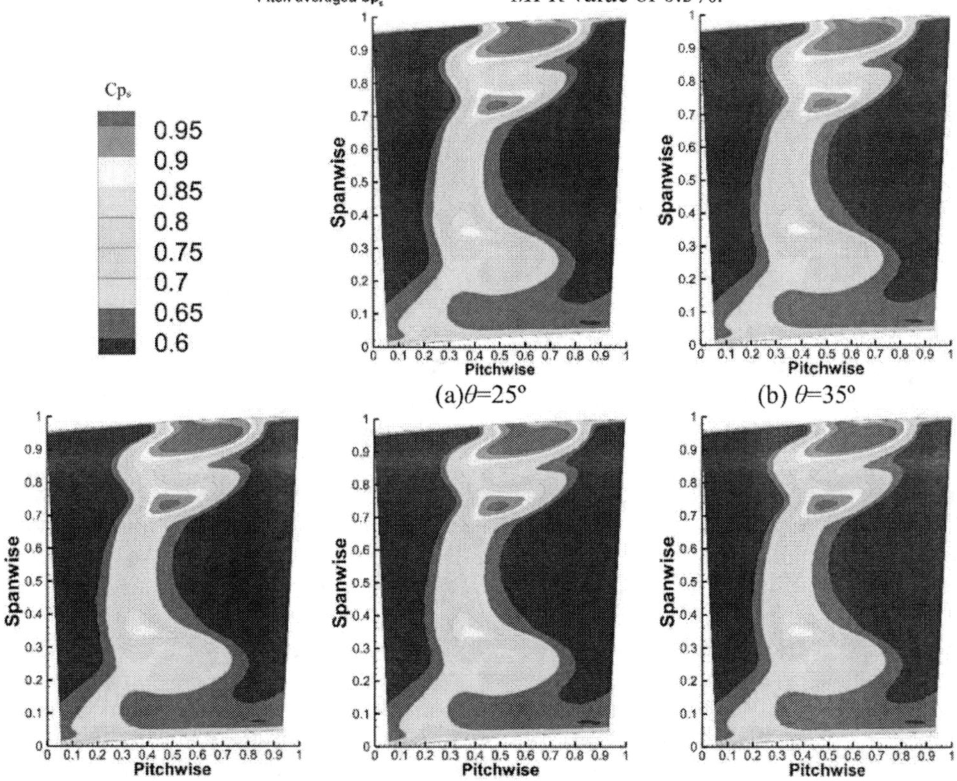

(a)θ=25° (b) θ=35°

(c) θ=45° (d) θ=55° (e) θ=65°

Figure 21. The Cp_s distribution at the z/C_{az}=1.1 cut-plane at an *MFR* value of 0.5%

3.4. Overall performance

Figure 22 shows the area-averaged η on the tip for various θ and *MFR* values. The area-averaged η increases proportionally with the increasing *MFR* under all θ values. For a small θ value of 25°, the area-averaged η in the case of *MFR*=1.3% increase by 220% in comparing with that in the case of *MFR*=0.3%, while for the cases of θ=65°, the rate of increase is 75%. It means that *MFR* could obviously enhance the phantom cooling performance under a smaller θ. Furthermore, the impact of θ on the area-averaged η depends on the value of *MFR*. For a low *MFR* of 0.3%, the θ almost has no impact on the phantom cooling performance. However, as *MFR* increase to 1.3%, the area-averaged η in the case of θ=25° is about 2 times as high as that in the case of θ=65°. It indicates that the impact of θ on the phantom cooling performance of the tip is more obvious at higher *MFR*.

Figure 23 shows the area-averaged η on the LE for various θ and *MFR* values. The results show that the area-averaged η on the LE first increase then decrease with increasing *MFR*. The peak value of the area-averaged η appear at *MFR*=0.5% for all cases, which means that the optimal *MFR* for the LE is 0.5%. When θ increases, the cooling performance on the LE become worse under any *MFR* value. As *MFR* increases from 0.3% to 1.3%, the difference between the area-averaged η of the cases of θ=25° and θ=65° increases from 50% to 172%. It shows that the θ has a more obvious impact on the film cooling performance on the LE at higher *MFR*.

Figure 24 shows the area-averaged stage efficacy on the z/C_{az}=1.1 cut-plane for various θ and *MFR* values. The results indicate that the area-averaged Φ increases proportionally as *MFR* increases at any θ values. As *MFR* increases from lower to a higher one (0.3% to 1.3%), the area-averaged Φ increases by about 2.5%. But, the area-averaged Φ remains constant with the increase of θ, which means that θ has no impact on the stage efficacy.

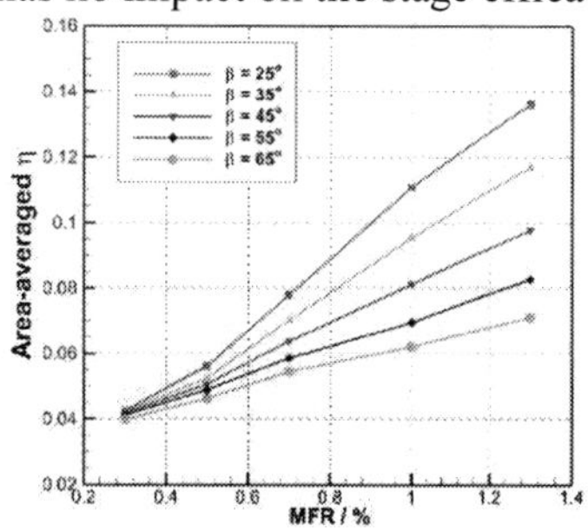

Figure 22. The area-averaged η level on the tip in various θ and *MFR*

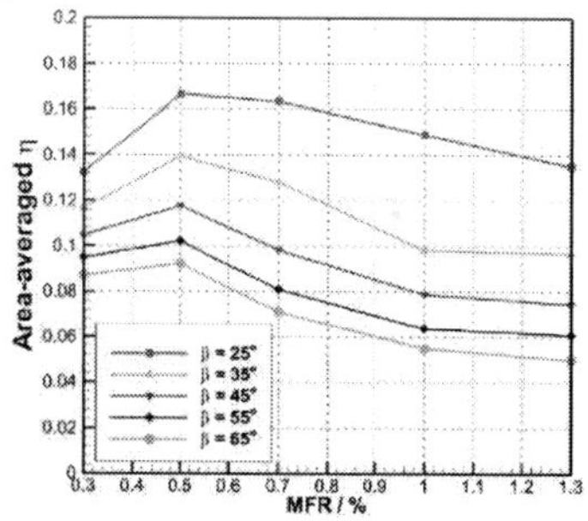

Figure 23. The area-averaged η on the LE for various θ and *MFR*

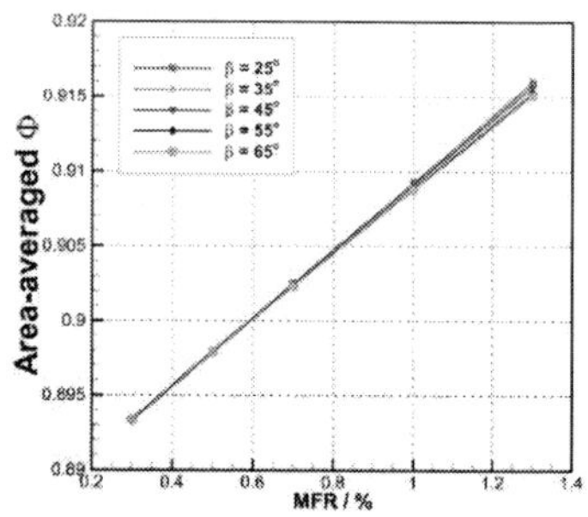

Figure 24. The area-averaged Φ in various θ and *MFR*

4. Conclusions

In present study, numerical simulations are implied to investigate the blade tip phantom cooling performance due to the blade LE film discharge. The features of phantom cooling on the tip are demonstrated. The impacts of coolant *MFR* and LE film hole spanwise angle on the tip phantom cooling are analysed. Turbulence models are validated by contrasting with the experimental measurements in an open literature. Present results reveal that the standard k-ω model is closest to the experimental measurements and it is employed in present studies. A grid independence analysis demonstrates that 8.20 million mesh nodes for the calculation domain could give the reasonable results. The main results can be listed as:

These coolant from blade LE could produce phantom cooling effect on the tip and they mainly affect the cooling performance on the tip forepart and have little impact on the tip rear part. Increasing the *MFR* could not only enhance the tip phantom cooling, but also influence the η distribution on the tip. The peak values of pitch-averaged η mainly exist at the z/C_{az}=0.1~0.2 and z/C_{az}=0.3~0.4 on the tip at low *MFR*. While there is no obvious peak value of pitch-averaged η at high *MFR* cases. The impact of θ on the phantom cooling of the tip is more obvious at higher *MFR*. For a low *MFR* of 0.3%, the θ almost has no impact on the phantom cooling performance. However, as *MFR* increase to 1.3%, the area-averaged η in the case of θ=25° is about 2 times as high as that in the case of θ=65°.

Both the *MFR* and θ have a critical impact on cooling performance of the LE. With the increasing *MFR*, the film coverage on the LE first improve then become worse. The peak value of the area-averaged η on the LE appear at *MFR*=0.5% for all cases. Increasing the θ would worsen the film cooling performance on the LE. This is because that as the θ become larger, the radial component of coolant velocity decreases, and its axial component increases. And the coolant mainly jets into the mainstream hot gas rather than cover the LE surface. At different *MFR* values, the area-averaged η on the LE of the cases of θ=25° and θ=65° decreases from 50% to 172%.

The impact of *MFR* on the aerodynamic performance is greater than that of the LE film hole spanwise angle. The tip leakage flow, corresponding leakage flow loss and the Cp_s increase with increasing *MFR*, while those parameters almost remain constant with increasing θ. As *MFR* increases, the area-averaged Φ increases by about 2.5% at most. the area-averaged Φ also remains constant with the increase of θ, which means that θ has no impact on the stage efficacy.

Acknowledgments
The authors would like to acknowledge the financial support from the National Science and Technology Major Project (2017-III-0009-0035).

Nomenclature

cp	specific heat at constant pressure, [J/(kg·K)]
Cp_s	pressure loss coefficient, $[(p_{s,ref}-p_{s,local})/(p_{s,ref}-p_{local})]$
D	diameter of film hole, [mm]
h	heat transfer coefficient, $[W/(m^2 \cdot K)]$
L	length of film hole, [mm]
LE	leading edge
LV	leakage vortex
m	mainstream, mass flow rate, [kg/s]
MFR	mass flow ratio, $[m_c/m_m]$
N	number of film hole
P	hole pitch in each row, [mm]
PS	pressure side
p	pressure, [kPa]
$p_{s,ref}$	reference total pressure, $[(m_m \cdot P_{s,1}+m_c \cdot P_{s,c})/(m_c+m_m)]$
R	universal gas constant, [J/(mol·K)]
s	entropy, $[\Delta s = cp \ln(T/T_1) - R \ln(P/P_1)]$
SS	suction side
T	temperature, [K]
Tu	turbulence intensity, [%]
V	velocity, [m/s]
$y+$	non-dimensional distance, $[Y\mu_\tau/v]$
Y	transverse coordinates from wall, [mm]
Z	axial chord direction, [mm]
Greek symbols	
α	hole angle in the axial direction, [°]
γ	specific heat ratio
ε	non-dimensional temperature, $[(T_{aw}-T)/(T_{aw}-T_c)]$
ζ	dimensionless specific entropy increase, $[T_2 \cdot \Delta s/0.5 V_2^2]$
η	film cooling effectiveness, $[(T_{aw}-T_w)/(T_{aw}-T_c)]$
θ	LE film hole angle in the spanwise direction, [°]
μ	dynamic viscosity, [Pa·s]
ν	kinematic viscosity, $[m^2/K]$
τ	shear stress
Φ	stage efficiency, $[(T_{s,0}-T_{s,2})/ (T_{s,0} (1-(P_{s,2}/P_{s,0})$^$((\gamma-1)/\gamma)))]$
ω	speed of rotation, [rpm]
Subscripts	
aw	adiabatic wall
az	axial chord direction in Z coordinate, [mm]
c	coolant
is	isentropic
ref	reference
s	total
w	walls

0 stage inlet
1 blade inlet
2 stage outlet

References

[1] Denton J D 1993 Loss Mechanisms in turbomachines *Proceedings of the ASME 1993 International Gas Turbine and Aeroengine Congress and Exposition.* Cincinnati Ohio USA
[2] Newton P J, Lock G D, Krishnababu S K, Hodson H P, Dawes W N, Hannis J and Whitney C 2009 Aerothermal investigations of tip leakage flow in axial flow turbines—part III: tip cooling *J. Turbomach.* **131** 011008
[3] Xiao X, McCarter A A and Lakshminarayana B 2001 Tip clearance effects in a turbine rotor: part I—pressure field and loss *J. Turbomach.* **123** 296-304
[4] McCarter A A, Xiao X and Lakshminarayana B 2001 Tip clearance effects in a turbine rotor: part II—velocity field and flow physics *J. Turbomach.* **123** 305-13
[5] O'Dowd D O, Zhang Q, Usandizaga I, He L and Ligrani P M 2010 Transonic turbine blade tip aero-thermal performance with different tip gaps: part II—tip aerodynamic loss *Proceedings of the ASME Turbo Expo 2010: Power for Land, Sea, and Air* Glasgow UK
[6] Zhou C 2014 Aerothermal performance of different tips in transonic turbine cascade with end-wall motion *J. Propul. Power* **30** 1316-27
[7] Srinivasan V and Goldstein R J 2003 Effect of endwall motion on blade tip heat transfer *J. Turbomach.* **125** 267-73
[8] Zhou K and Zhou C 2018 Aerodynamic interaction between an incoming vortex and tip leakage flow in a turbine cascade *J. Turbomach.* **140** 111004.
[9] Zhang Q, He L and Rawlinson A 2014 Effects of inlet turbulence and end-wall boundary layer on aerothermal performance of a transonic turbine blade tip *J. Eng. Gas Turbines Power* **136** 052603
[10] Zhang Q and He L 2014 Impact of wall temperature on turbine blade tip aerothermal performance, *J. Eng. Gas Turbines Power* **136** 052602
[11] Jiang H, Zhang Q, He L, Lu S, Wang L and Teng J 2018 Experimental evidence of temperature ratio effect on turbine blade tip heat transfer *J. Turbomach.* **140** 121010
[12] Arisi A, Xue S, Ng W F, Moon H Koo and Zhang L 2015 Numerical investigation of aerothermal characteristics of the blade tip and near-tip regions of a transonic turbine blade *J. Turbomach.* **137** 091002
[13] Roback R J and Dring R P 1993 Hot streak and phantom cooling in a turbine rotor passage, part 1–separate effects *J. Turbomach.* **115** 657-66
[14] Roback R J and Dring R P 1993 Hot streak and phantom cooling in a turbine rotor passage, part 2 –combined effects and analytical modeling *J. Turbomach.* **115** 667-74
[15] Zhang L, Yin J, Liu K and Moon H Koo 2015 Effect of hole diameter on nozzle endwall film cooling and associated phantom cooling, *Proceedings of the ASME Turbo Expo 2015: Turbine Technical Conference and Exposition* Montreal Quebec Canada
[16] Zhang L, Yin J and Moon H Koo 2015 The effects of vane showerhead injection angle and film compound angle on nozzle endwall cooling (phantom cooling) *J. Turbomach.* **137** 021003
[17] Liu K, Xu H Z and Fox M 2015 Turbine nozzle endwall phantom cooling with compound angled pressure side injection *Proceedings of the ASME Turbo Expo 2015: Turbine Technical Conference and Exposition* Montreal Quebec Canada
[18] Zhang K, Li J, Li Z and Song L 2019 Effects of simulated swirl purge flow and mid-passage gap leakage on turbine blade platform cooling and suction surface phantom cooling performance *Int. J. Heat and Mass Transfer* **129** 618-34
[19] Du K, Song L, Li J and Sunden B 2017 Effects of the mainstream turbulence intensity and slot injection angle on the endwall cooling and phantom cooling of the vane suction side surface *Int. J. Heat and Mass Transfer* **112** 427-40

[20] Du K, Li Z, Li J and Sunden B 2018 Effects of the swirling coolant jet from the upstream slot on the vane endwall cooling and the vane suction side phantom cooling *Int. J. Heat and Mass Transfer* **121** 952-66
[21] Du K, Li Z, Li J and Sunden B 2017 Influence of the upstream slot geometry on the endwall cooling and phantom cooling of vane suction side surface *Appl. Therm. Eng.* **121** 688-700
[22] Zhang Y and Yuan X 2012 Experimental investigation of turbine phantom cooling on suction side with combustor-turbine leakage gap flow and endwall film cooling *Proceedings of the ASME Turbo Expo 2012: Turbine Technical Conference and Exposition* Copenhagen Denmark
[23] Papa M, Srinivasan V and Goldstein R J 2010 Film cooling effect of rotor-stator purge flow on endwall heat/mass transfer, *Proceedings of the ASME Turbo Expo 2010: Power for Land, Sea, and Air* Glasgow UK
[24] Zhang Y and Yuan X 2014 Experimental investigation of turbine phantom cooling on endwall with trailing edge discharge flow *Proceedings of the ASME Turbo Expo 2014: Turbine Technical Conference and Exposition* Düsseldorf Germany
[25] Li X, Ren J and Jiang H D 2014 Phantom cooling of nozzle trailing edge discharge on blade surface and platform *Proceedings of the ASME Turbo Expo 2014: Turbine Technical Conference and Exposition* Düsseldorf Germany
[26] Timoko LP 1984 *Energy efficient engine high pressure turbine component test performance report*, Cleveland: NASA Lewis Research Center
[27] Kwak J S and Han J C 2003 Heat transfer coefficients of a turbine blade-tip and near-tip regions *J. Thermophys. Heat Transf.* **17** 297–303

IOP Conf. Series: Materials Science and Engineering 1081 (2021) 012049 doi:10.1088/1757-899X/1081/1/012049

Experiment of the Influence of Dynamic Balancing Precision on Rotor Dynamic Behavior of 75KW Oil-free Turbo Blower

Zeda Dong[1] Liukai Hou[2] Bo He[3] Fangcheng Xu[*1]

[1] Dalian University of Technology, School of Control Science and Engineering, Dalian
[2] Beijing Aerospace Propulsion Institute, Beijing
[3] Shijiazhuang Kingston Bearing Technology Co., Ltd, Shijiazhuang

E-mail: fcxu@dlut.edu.cn

Abstract. The principle and structure of the 75kW oil-free turbo blower is briefly introduced and the performance of gas foil bearings (GFBs) is tested. According to the test results, the lift-off speed of the gas foil journal bearings (GFJBs) using in 75kW oil-free turbo blower is about 4000rpm, and the gas foil thrust bearings (GFTBs) can provide at least 140N and 200N load capacity at 7200rpm and 9000rpm, respectively. The rotordynamics behavior of oil-free turbo blower is studied. Two proximity probes and speed sensor are installed to monitor the horizontal vibration displacement, vertical vibration displacement and shaft speed, respectively. And the vibration spectrum and the rotor axis orbit are analyzed. It was found that the maximum vibration amplitude of the rotor with dynamic balancing precision of G-1 was only 3.5μm, and no other low frequency other than low frequency caused by signal interference. The fundamental frequency vibration amplitude of the rotor with dynamic balancing precision of G-2.5 is larger when running stably, but there is no low frequency and high frequency vibration with higher amplitude. Shaft couldn't lift-off with dynamic balancing precision of G-6.3 In general, oil-free turbo blower with two kinds of dynamic balancing precision rotors can operate well.

1. Introduction

A multitude of sewage-treatment plants have been established with the development of modern industrialization. In the process of sewage-treatment, a large part of the power losses is taken up by traditional blower employ oil-lubricated hydrodynamic bearings. Therefore, the oil-free turbo blower emerges as the times required. At present, it mainly includes turbo blower using magnetic bearings and turbo blower using GFBs[1]. Both have many common advantages, such as low friction, no pollution and reliable [2-5]. The GFBs are well received due to the complex control system and high costs of the magnetic bearings.

GFBs have been in use for 50 years, which has been widely applied [6-10]. GFBs are one of the key parts of oil-free turbo blower, mainly including GFJBs and GFTBs [11-14]. Up to now, there is a lot of research on the performance of GFBs. In 1983, Heshmat H. and Walowit J. A. established the foundation model of the bump foil equivalent to spring structure, and gave a calculation model [15, 16]. Later, the model and calculation methods were deeply studied [17-19]. Qi, S. Ho, Y. S. et al. Demonstrated the effects of the load, radial clearance and bearing radius on lift-off speed. The lift-off speed will increase with increasing load and radial clearance and decreasing bearing radius[20]. San

Andrés, L., Chirathadam et al. presented the measurement of the friction torque and rotor lift-off speed. The results proved that the bearing friction torque increases with the static load and rotor speed[21]. Kim, T. H., Park, M. et al. studied the performance in optimizing maximum load capacity of GFTBs. The predictions show that the load capacity of the GFTBs increases with increasing rotor speed and decreasing minimum film thickness. And there is an optimum taper inlet height to maximize load capacity[22]. In 2018, Fu, G., Untaroiu, A. et al. established a 3D CFD model of a six pads GFTB and simulated the static properties. They predicted the relationship between friction torque and applied load and found that the predicted results of 3D CFD model was in good agreement with measurement[23].

Although GFBs has been studied since 1980s, more research on rotordynamics of GFBs has been carried out in recent 20 years. Many research teams began to present their studies on rotordynamic characteristic of oil-free turbomachinery using GFBs. Heshmat, H. , Walton, J. F. et al. developed oil-free turbochargers and small gas turbine engines using GFBs and demonstrated the feasibility of GFBs application in 2000[7]. Walton, J. F., Heshmat, H. et al. presented the design and test program in oil-free turbo blower[24]. Zhou Q., Hou Y. et al. designed a high speed gas turbine test rig and carried out dynamic stability experiments. The relationship between vibration amplitude and rotor speed and rotor axis orbit at 125,413rpm were obtained[25]. Kim, T. H., Lee, Y. B. et al. studied the performance of the 75KW oil-free turbo blower. There were subsynchronous rotor motions at widespread frequencies due to insufficient load capacity of GFTBs. The rotor-GFBs system performance was improved by increasing the out diameter of GFTB[26]. Lee, J., Kim, Y. M. et al. presented rotordynamic performance of three types of journal bearing used in microturbomachinery. The rotordynamic stability characteristics was improved by reduced radial clearance of C-type bearing[27].

The rotor system often exist vibration when the oil-free turbo blower is running. There are many reasons for vibration, such as the imbalance of the rotor, the misalignment, the loosening of foundation fixed structure and so on. More noise, decreasing efficiency, and damaging parts will occur due to excessive vibration. Therefore, it is necessary to study the rotordynamics characteristic of oil-free turbo blower[28-31].

The film between foil and rotor is formed by principle of elastohydrodynamic lubrication. The film pressure isn't enough to counter the gravity of the rotor due to low rotor speed. There will be dry friction between the rotor and bearing foil and the wear is extremely serious if the rotor speed doesn't reach the "lift-off speed". The rotor will float and there is no direct contact between the rotor and GFBs when the rotor speed reaches the "lift-off speed". Therefore, the oil-free turbo blower is very energy-saving and reliable.

2. Description of oil-free turbo blower

The oil-free turbo blower is designed with the integrated design structure of permanent magnet synchronous motor, impeller, GFBs and inverter. As shown in Figure.1, the 75kW oil-free turbo blower is a kind of centrifugal blower in essence, which achieves the purpose of blowing by the continuous rotation of the impeller. The oil-free turbo blower is different from traditional blower using rolling bearing and it has the minimum operation rotor speed.

Figure 1. Physical diagram of oil-free turbo blower.

The film between foil and rotor is formed by principle of elastohydrodynamic lubrication. The film pressure isn't enough to counter the gravity of the rotor due to low rotor speed. There will be dry friction between the rotor and bearing foil and the wear is extremely serious if the rotor speed doesn't reach the "lift-off speed". The rotor will float and there is no direct contact between the rotor and GFBs when the rotor speed reaches the "lift-off speed". Therefore, the oil-free turbo blower is very energy-saving and reliable.

3. Description of test GFBs

The schematic diagram and physical diagram of test rig for GFJBs is shown in Figure.3 and Figure.4, respectively. The motor rotor speed and the friction torque of GFJBs are monitored in real time when testing performance of GFJBs. Simulate the weight of the rotor by loading the bearing with dead weight.

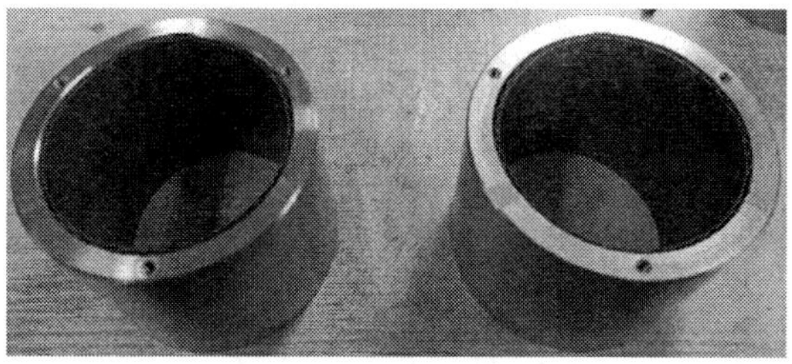

Figure 2. Physical diagram of GFJBs.

The inner diameter of GFJBs used for oil-free turbo blower is 60mm, and axial length is also 60mm, and Figure.2 is the physical diagram of GFJBs. Speed up the motor to 7200rpm, and the test data is shown in the Figure.5. The left Y-axis represents the motor speed and the right Y-axis represents the friction torque. The friction torque increases sharply because the lift-off speed of the GFJBs is not reached when accelerating the motor. At about 4000rpm, the friction torque decreases sharply and tends to be stable. Therefore, it can be judged that around 4000rpm is the lift-off speed of the GFJBs. Similarly, the friction torque will increase sharply when the motor speed reduced.

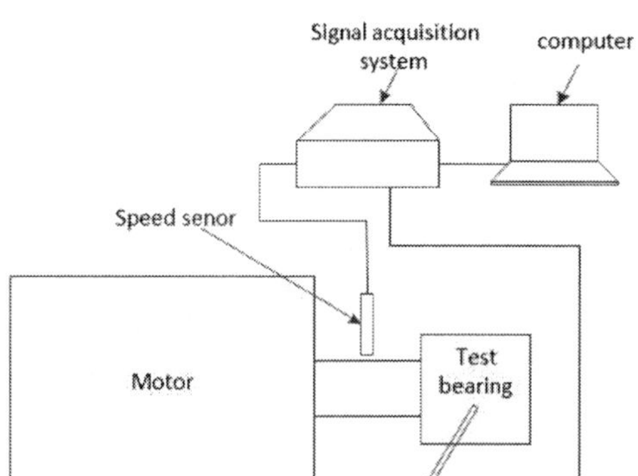

Figure 3. Schematic diagram of test rig for GFJBs. **Figure 4.** Physical diagram of test rig for GFJBs.

The schematic diagram and physical diagram of test rig for GFTBs is shown in Figure.6 and Figure.9, respectively. The motor rotor speed, the load applied to the GFTBs and the friction torque are monitored simultaneously when testing performance of GFTBs. The load applied to GFTBs can be changed by adjusting the opening of valve.

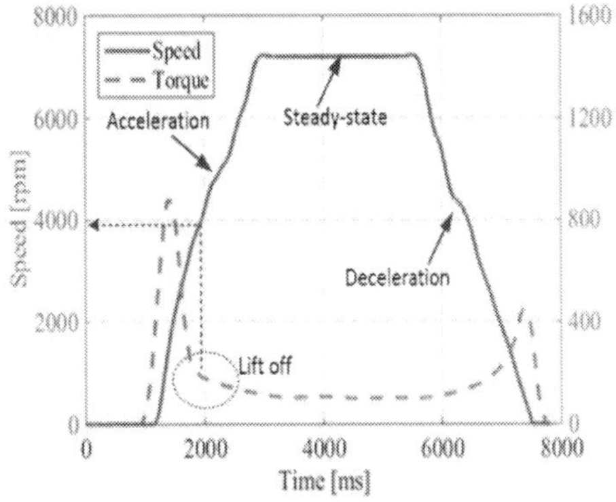

Figure 5. Friction torque versus rotor speed.

Figure 6. Physical diagram of test rig for GFTBs.

The outer diameter and inner diameter of GFTBs used for oil-free turbo blower are 140mm and 69.5mm, respectively. The GFTBs has six pads, the angle of each pad is about 59 degrees and Figure.7 is the physical diagram of GFTBs. The GFTBs were tested at 7200 rpm and 9000 rpm, and the test data is shown in Figure.8. The test results show that the GFTBs can provide at least 140N load capacity at 7200rpm, and can provide at least 200N load capacity at 9000rpm.

Figure 7. Physical diagram of GFTBs.

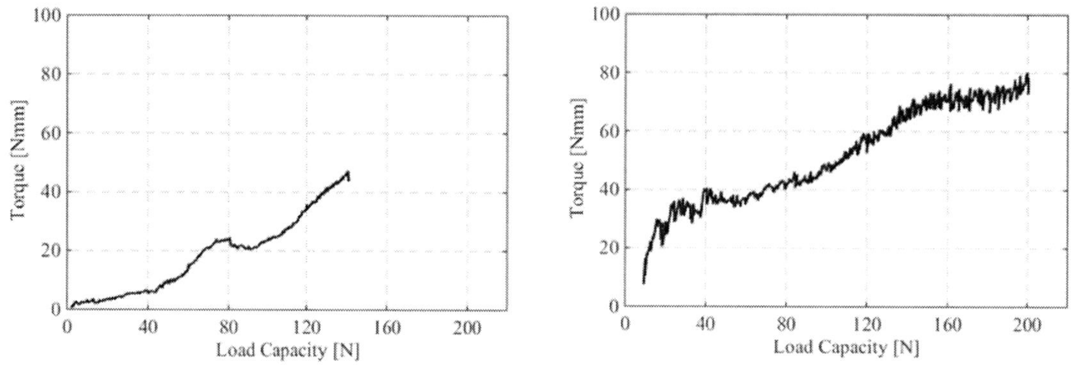

(a) 7200rpm (b) 9000rpm

Figure 8. Friction torque versus load capacity at different speed.

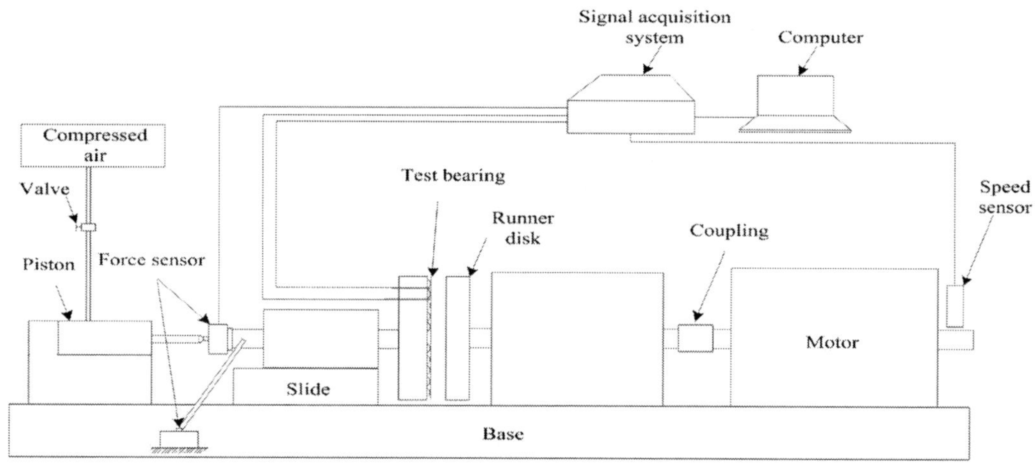

Figure 9. Schematic diagram of text rig for GFTBs.

4. Description of oil-free turbo blower

For 75KW oil-free turbo blower, the rotors with dynamic balancing precision of G-1,G-2.5 and G-6.3 are respectively assembled on the oil-free turbo blower for testing. Two Bently 3300XL proximity probes were installed to measure the horizontal and vertical vibration of rotor. In addition, the speed sensor was used to monitor rotor speed. The bearing parameters used in oil-free turbo blower and the rotor balancing precision are shown in Table 1.

Table 1. Bearing parameters and rotor dynamic balancing precision.

Parameters	value
Inner diameter of GFJBs/mm	60
Axial diameter of GFJBs/mm	60
Inner diameter of GFTBs /mm	70
Outer diameter of GFTBs /mm	140
Rotor dynamic balancing precision	G-1; G-2.5; G-6.3

First, the rotordynamics of oil-free turbo blower with dynamic balancing precision of G-1 is measured. Figure.10 shows the horizontal vibration spectrum, vertical vibration spectrum， and axis orbit of the rotor when the oil-free turbo blower is started and accelerated to 3000rpm. As can be seen from the coordinate values, the rotor axis orbit deviates from centre position. The rotor speed has not reached the "lift-off speed" of GFJBs, and the rotor has not yet floated. The lift-off speed is the indicator to judge whether the rotor and bearing surface enter the state of gas lubrication under specific load. The rotor speed does not reach the take-off speed, and the bearing does not enter the state of gas lubrication, causing the dry friction between the rotor and the bearing foil in direct contact. There are more clutter and higher noise in the spectrum, and the axis orbit deviates from centre position for the same reason. Dry friction exists between rotor and bearing foil. There is more clutter and higher noise in the spectrum.

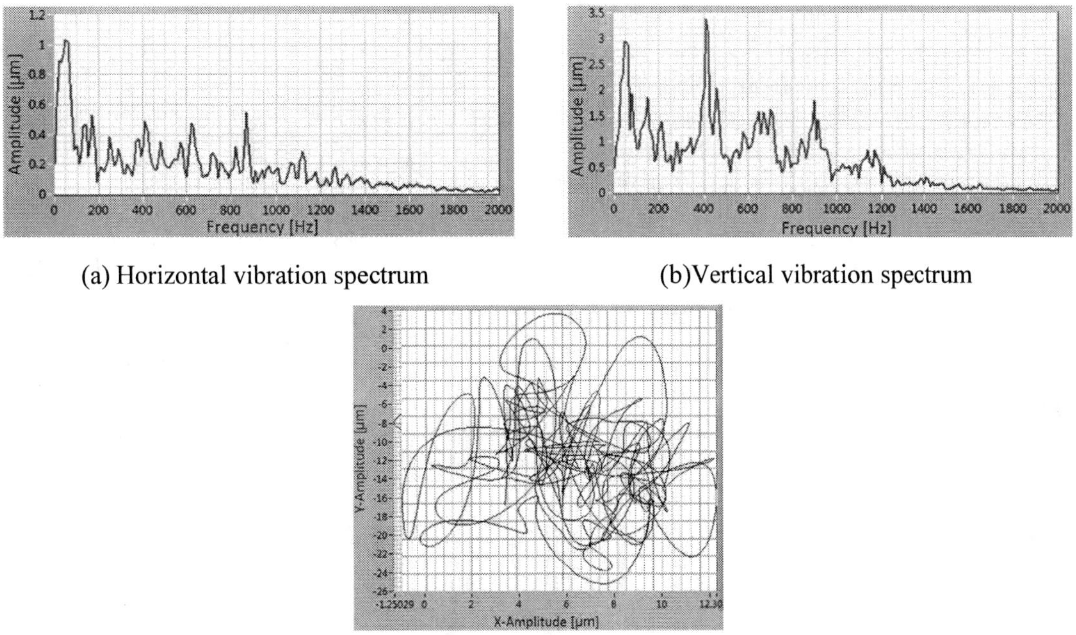

(a) Horizontal vibration spectrum (b)Vertical vibration spectrum

(c) Rotor Axis orbit

Figure 10. Vibration spectrum and rotor axis orbit with dynamic balancing precision of G-1 at 3000rpm.

Then the blower speed is driven to 12000rpm. Figure.11 shows the horizontal vibration spectrum, vertical vibration spectrum， and axis orbit of the rotor at 12000rpm. According to the rotor axis orbit and coordinate values, it can be seen that the axis orbit is basically in the cent centre of the coordinate system. Accordingly, rotor speed has reached the lift-off speed of GFJBs. The rotor is supported by the air film and floats, and hydrodynamic lubrication is formed between the rotor and the bearing foil. In the horizontal and vertical vibration spectrum, there are 2x, 3x and even higher frequency vibration, which are caused by misalignment. The amplitude of 3x frequency in the horizontal direction exceeds the fundamental frequency amplitude. However, it is only 3μm which has little effect on the stable operation of oil-free turbo blower. In addition, the rotor axis orbit is not regular and a low frequency about 50HZ appears on both spectrums.

IOP Conf. Series: Materials Science and Engineering 1081 (2021) 012049 doi:10.1088/1757-899X/1081/1/012049

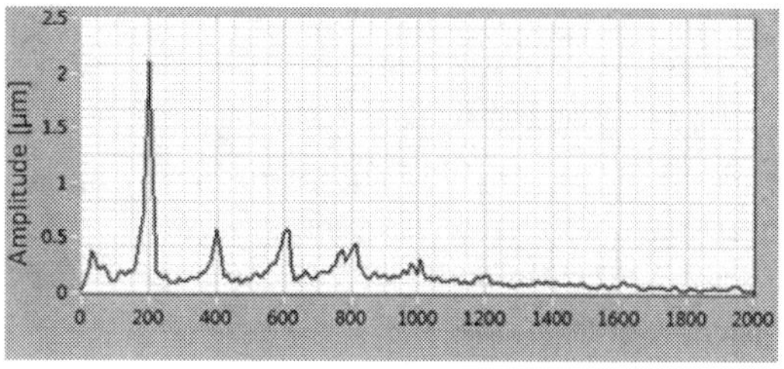

(a) Horizontal vibration spectrum

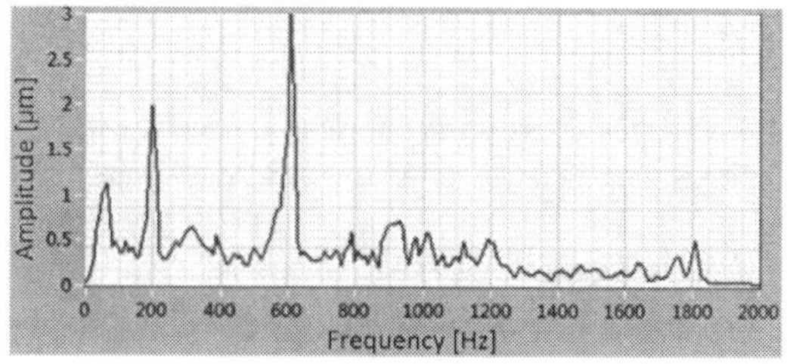

(b)Vertical vibration spectrum

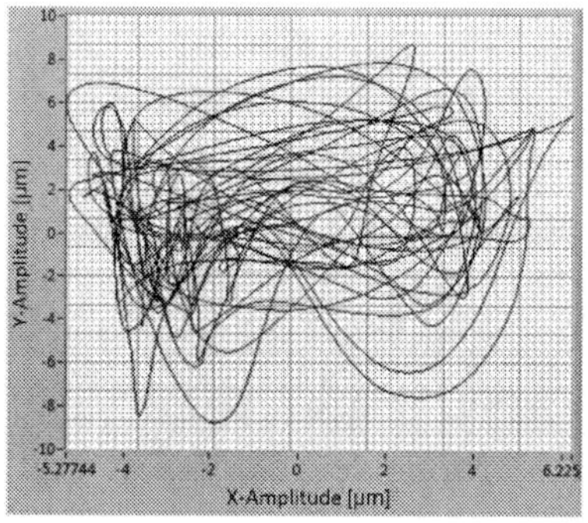

(c) Rotor Axis orbit

Figure 11. Vibration spectrum and rotor axis orbit with dynamic balancing precision of G-1 at 12000rpm.

Figure.12 shows the horizontal vibration spectrum, vertical vibration spectrum and axis orbit of rotor at 19800rpm. Obviously, there are many high frequencies vibration in spectrums, but the 1X vibration amplitude is only 1.6μm. The reason for this phenomenon is that rotor balancing precision is very high, so the high frequency vibration amplitudes are relatively higher caused by misalignment of front and rear journal bearing.

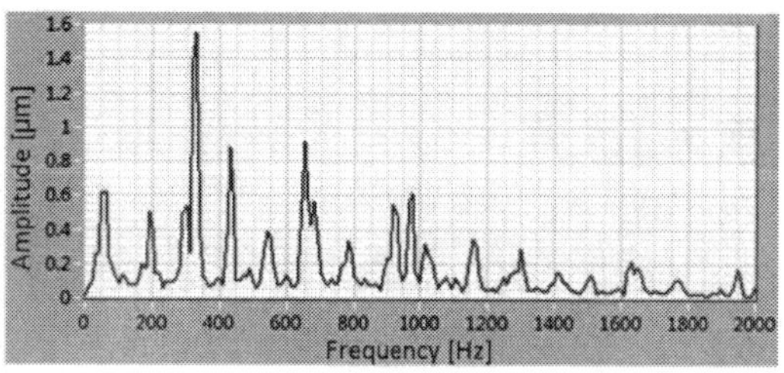

(a) Horizontal vibration spectrum

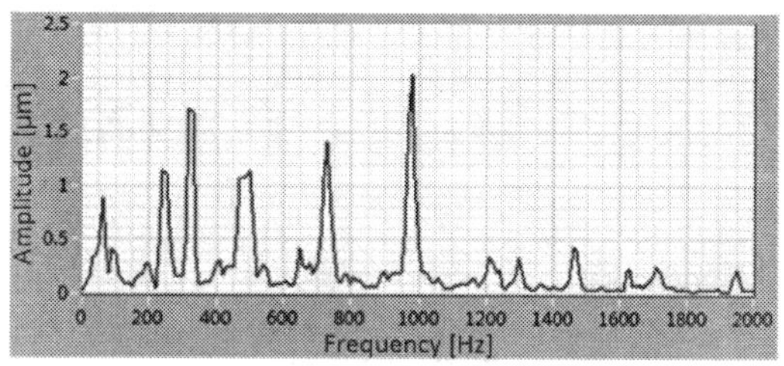

(b)Vertical vibration spectrum

IOP Conf. Series: Materials Science and Engineering 1081 (2021) 012049 doi:10.1088/1757-899X/1081/1/012049

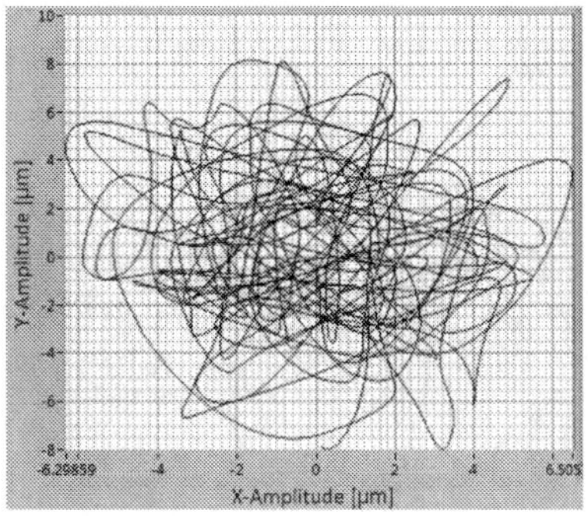

(c) Rotor Axis orbit

Figure 12. Vibration spectrum and rotor axis orbit with dynamic balancing precision of G-1 at 19800rpm.

Figure.13 shows the horizontal vibration spectrum, vertical vibration spectrum, and axis orbit of the rotor when rotor speed is 25200rpm and working stably. According to the rotor Axis orbit and coordinate values, it can be seen that the axis orbit is relatively regular at this time, and the oil-free turbo blower is working well. The 1X vibration amplitude is the largest, about 2.5µm. There are fewer high frequency caused by misalignment and its amplitude is much smaller than 1X vibration amplitude. Additionally, the sub-synchronous, 50HZ and 200HZ is also appeared, which does not change with the increase of rotor speed. Therefore, it can be judged that this sub-synchronous, 50HZ and 200HZ may be caused by electromagnetic interference instead of rotor system problems.

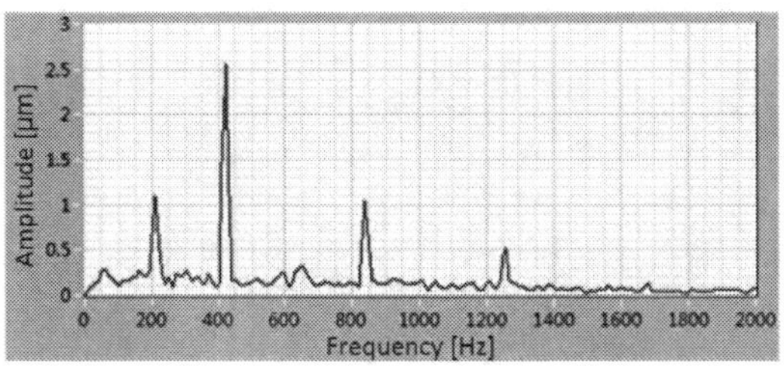

(a) Horizontal vibration spectrum

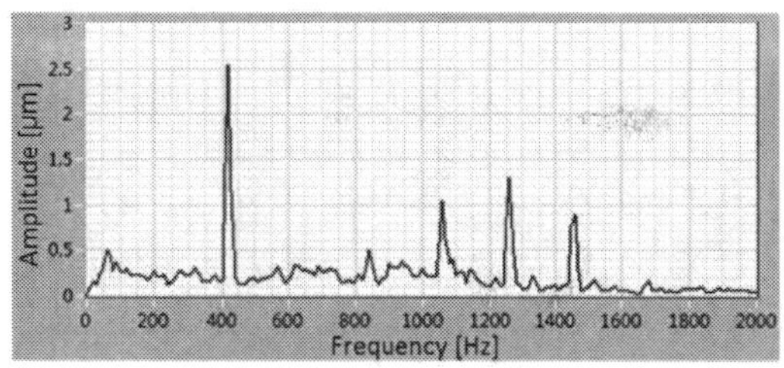

(b)Vertical vibration spectrum

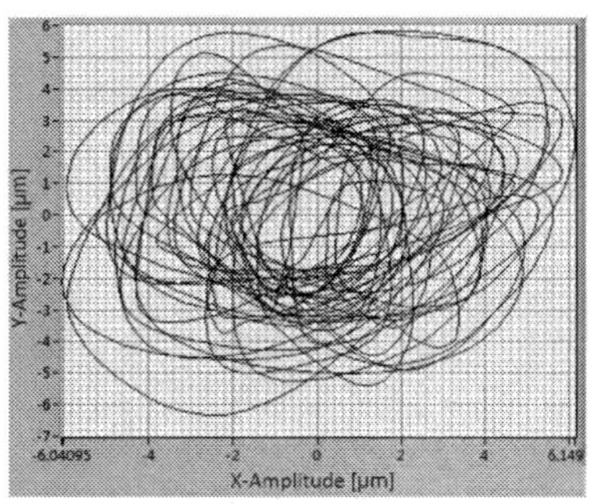

(c) Rotor Axis orbit

Figure 13. Vibration spectrum and rotor axis orbit with dynamic balancing precision of G-1 at 19800rpm.

IOP Conf. Series: Materials Science and Engineering 1081 (2021) 012049 doi:10.1088/1757-899X/1081/1/012049

Figure 14 shows the bearing foil wear after the oil-free turbo blower test. It can be found that the foil coating of GFJBs is slightly worn, and the foil coating of GFTBs is almost not worn. This is because the gravity of rotor is applied to the GFJBs and dry friction is relatively serious during starting and stopping of the oil-free turbo blower.

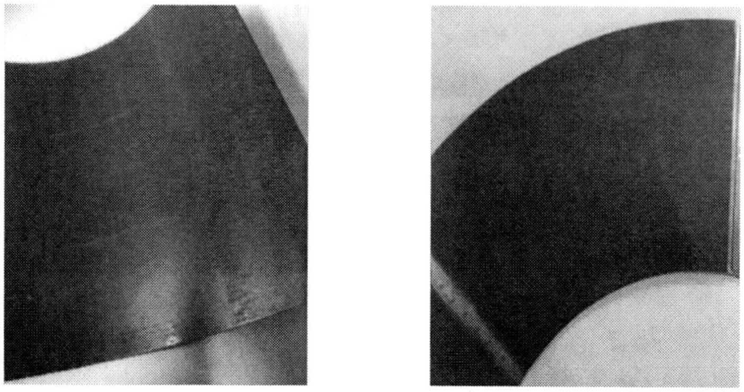

Figure 14. Bearing foils wear after test.

Similarly, the rotordynamics of oil-free turbo blower with dynamic balancing precision of G-2.5 and G-6.3 are measured. Figure.15 shows the horizontal vibration spectrum, vertical vibration spectrum and axis orbit of rotor when rotor speed is 25200rpm and working stably. It can be seen that the fundamental frequency vibration amplitude has reached 12μm, which is much larger than dynamic balancing precision of G-1 because of the mass imbalance of the shaft. The amplitudes of other frequencies in rotor axis orbit and spectrum are not obvious due to the large fundamental frequency amplitude.

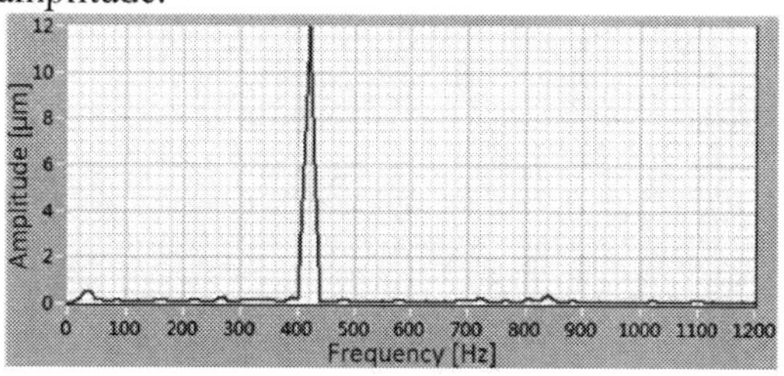

(a) Horizontal vibration spectrum

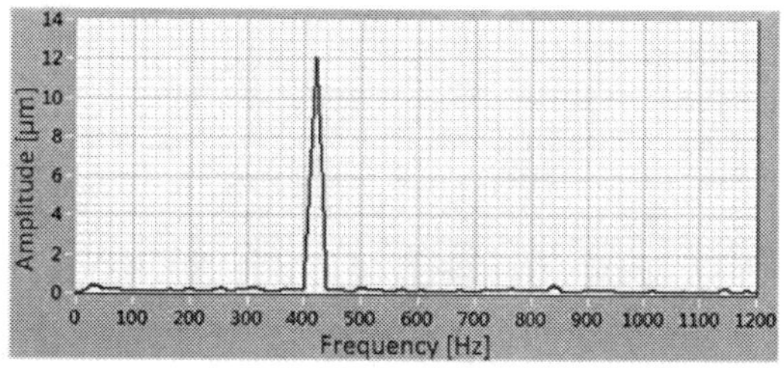

(b)Vertical vibration spectrum

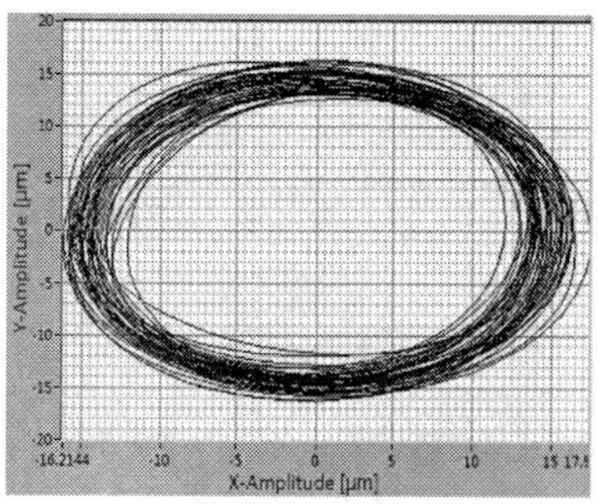

(c) Rotor Axis orbit

Figure 15. Vibration spectrum and rotor axis orbit with dynamic balancing precision of G-2.5 at 25200rpm.

However, when the dynamic balancing precision increases to G-6.3, blower shaft couldn't lift-off because the hydrodynamic film cannot be formed when imbalance force is too large. Top foil coating of both journal bearing and thrust bearing are wear out, as shown in Figure.16.

Figure 16. Journal bearing and thrust bearing wear after test with dynamic balancing precision of G-6.3.

5. Conclusion

The test rig for bearing test was built and bearing performance was tested. The lift-off speed of GFJBs used in the oil-free turbo blower is about 4000rpm. The GFTBs can provide at least 140N and 200N load capacity at 7200rpm and 9000rpm, respectively.

The rotor dynamics of oil-free turbo blower was measured with different dynamic balancing precision. It indicated that the oil-free turbo blower with high dynamic balancing precision can operate very well while shaft cannot lift off with G-6.3 dynamic balancing precision. In general, dynamic balancing precision of G-2.5 can be adopted if cost saving is considered.

Acknowledgement

This research is partially supported by the National Natural Science Foundation of China (NSFC) (Grant No.11702050).

This research is partially supported by the 64th batch of China Post-doctoral Science Foundation (2018M641690).

Reference

[1] Bing-shuo Zhu, Bing-zhe Zhang, Tian-ze Yuan, Shu-tan Niu, Sheng-ce Lu. 2020. " The Application of Air Bearing in Blower and Turbojet Engine Field. " *Chinese Journal of Turbomachinery*, 2020,62(S1):25-28.

[2] Heshmat , H., Chen , H. M., and Walton, J. F. , II 1999. "On the performance of hybrid foil-magnetic bearings." *Journal of Engineering for Gas Turbines and Power*, 122(1), 73-81.

[3] Jeong, S., Jeon, D., and Lee, Y. B. 2016. "Rigid mode vibration control and dynamic behavior of hybrid foil–magnetic bearing turbo blower." *Journal of Engineering for Gas Turbines and Power*, 139(5).

[4] Feng, K., Liu, L. J., Guo, Z., and Zhao, X. 2016. *Parametric study on static and dynamic characteristics of bump-type gas foil thrust bearing for oil-free turbomachinery.*

[5] Park, C. H., Choi, S. K., Hong, D. E., Yoon, T. G., and Lee, S. H. 2013. "Note: Radial-thrust combo metal mesh foil bearing for microturbomachinery." *Rev Sci Instrum*, 84(10), 106102.

[6] Agrawal, G. L. 1997. "Foil air/gas bearing technology — an overview." (78682), V001T004A006.

[7] Heshmat, H., Walton, I. I. J. F., Corte, C. D., and Valco, M. 2000. "Oil-free turbocharger demonstration paves way to gas turbine engine applications." (78545), V001T004A008.

[8] Kim, K.-S., and Lee, I. 2006. "Vibration characteristics of a 75kW turbo machine with air foil bearings." *Journal of Engineering for Gas Turbines and Power*, 129(3), 843-849.

[9] Heshmat, H., Walton, J. F., II, and Nicholson, B. D. "Ultra-high temperature compliant foil bearings: The journey to 870°c and application in gas turbine engines — experiment." *Proc., ASME Turbo Expo 2018: Turbomachinery Technical Conference and Exposition*V07BT34A016.

[10] Park, C. H., Park, J. Y., and Yoon, E. S. "Design and evaluation of hybrid magnetic bearings for turbo compressors." *Proc., ASME Turbo Expo 2018: Turbomachinery Technical Conference and Exposition*V07BT34A008.

[11] Xu, F., Kim, D., and Zamanian Yazdi, B. 2016. "Theoretical study of top foil sagging effect on the performance of air thrust foil bearing." (49842), V07BT31A013.

[12] Ryu, K., and Ashton, Z. 2015. "Bump-type foil bearings and flexure pivot tilting pad bearings for oil-free automotive turbochargers: Highlights in rotordynamic performance." *Journal of Engineering for Gas Turbines and Power*, 138(4).

[13] San Andrés, L., Ryu, K., and Diemer, P. 2014. "Prediction of gas thrust foil bearing performance for oil-free automotive turbochargers." *Journal of Engineering for Gas Turbines and Power*, 137(3), 032502-032502-032510.

[14] Lee, Y.-B., Kim, T. Y., Kim, C. H., and Kim, T. H. 2011. "Thrust bump air foil bearings with variable axial load: Theoretical predictions and experiments." *Tribology Transactions*, 54(6), 902-910.

[15] Heshmat, H., Walowit, J. A., and Pinkus, O. 1983. "Analysis of gas lubricated compliant thrust-bearings." *Journal of Lubrication Technology-Transactions of the Asme*, 105(4), 638-646.

[16] Heshmat, H., Walowit, J. A., and Pinkus, O. 1983. "Analysis of gas-lubricated foil journal bearings." *Journal of Lubrication Technology*, 105(4), 647-655.

[17] Ku, C.-P. R., and Heshmat, H. 1992. "Compliant foil bearing structural stiffness analysis: Part i—theoretical model including strip and variable bump foil geometry." *Journal of Tribology*, 114(2), 394-400.

[18] Iordanoff, I. 1999. "Analysis of an aerodynamic compliant foil thrust bearing: Method for a rapid design." *Journal of Tribology-Transactions of the Asme*, 121(4), 816-822.

[19] Heshmat, C. A., Xu, D. S., and Heshmat, H. 2000. "Analysis of gas lubricated foil thrust bearings using coupled finite element and finite difference methods." *Journal of Tribology-Transactions of the Asme*, 122(1), 199-204.

[20] Qi, S., Ho, Y. S., Geng, H., and Yu, L. "Theoretical prediction of the lift-off speed in aerodynamic compliant foil journal bearings." *Proc., ASME Turbo Expo 2010: Power for Land, Sea, and Air*, 383-388.

[21] San Andrés, L., Chirathadam, T. A., Ryu, K., and Kim, T. H. 2010. "Measurements of drag torque, lift-off journal speed, and temperature in a metal mesh foil bearing." *Journal of Engineering for Gas Turbines and Power*, 132(11).

[22] Kim, T. H., Park, M., and Lee, T. W. 2017. "Design optimization of gas foil thrust bearings for maximum load capacity." *Journal of Tribology-Transactions of the Asme*, 139(3).

[23] Fu, G., Untaroiu, A., and Swanson, E. 2018. "Effect of foil geometry on the static performance of thrust foil bearings." *Journal of Engineering for Gas Turbines and Power-Transactions of the Asme*, 140(8).

[24] Walton, J. F., II, Heshmat, H., and Tomaszewski, M. J. "Design and test program in the development of a 100 hp oil-free, high-speed blower." *Proc., ASME Turbo Expo 2007: Power for Land, Sea, and Air*, 951-957.

[25] Zhou, Q., Hou, Y., and Chen, C. 2009. "Dynamic stability experiments of compliant foil thrust bearing with viscoelastic support." *Tribology International*, 42(5), 662-665.

[26] Kim, T. H., Lee, Y.-B., Kim, T. Y., and Jeong, K. H. 2011. "Rotordynamic performance of an oil-free turbo blower focusing on load capacity of gas foil thrust bearings." *Journal of Engineering for Gas Turbines and Power*, 134(2).

[27] Lee, J., Kim, Y. M., Park, M. S., Kim, T. H., Ha, K. K., Lee, C., and Kim, S. H. 2016. "Rotordynamic performance measurements and predictions of a rotor supported on multilayer gas foil journal bearings for microturbomachinery." (49842), V07BT31A007.
[28] Li-jun Jiang, Jian-bao Yuan, Wen-hao Huo, Hao Sun. 2020. "Dynamic Characteristics Analysis of Cantilever Rotor Considering Impeller Coupling Effect." *Chinese Journal of Turbomachinery* , 62(S1):16-20.
[29] Yi He. 2017. "Discussion on Aerodynamic Design of Magnetic Centrifugal Blower Impeller." *Chinese Journal of Turbomachinery* ,59(S1):13-16+63.
[30] Hong-lei Sha, Tian-ye Yu, Yi He, Zhi-hua Zhang. 2020. "Rotor Dynamics Design and Test of 700kW Magnetic Levitation Turbo Blower." *Chinese Journal of Turbomachinery* , (06),45-47.
[31] Guo-wei Du, You-zhi Wang, Guo-jun Yang, Zhen-gang Shi, Lei Zhao. 2017. "Rotor Dynamics Research of a 165kW Helium Blower with AMBs." *Chinese Journal of Turbomachinery* , 59(04):39-44.

Investigation of noise characteristics of periodic jet impinging on a flat plate

Xin Liu , Jiaxin Hou , Heyuan Huang , Xiaoliang Guo and Junlong Xie*
School of Energy and Power Engineering, Huazhong University of Science & Technology, Wuhan, CHN

***Email:** jlxie@hust.edu.cn

Abstract: Noise characteristics of periodic jet impinging on a flat plate are investigated. The numerical model is established by large eddy simulation and FW-H acoustic analogy method. Compared with experimental data obtained by noise measurement experiment, the effects of acoustic integral surface and different subgrid-scale models on noise simulation results are studied. The accuracy of the optimized numerical model is also validated by verification experiment. Further, the present numerical model is applied to numerical prediction of noise directivity and spectrum distribution with different waveforms and periodic frequency. Compared with the steady impinging jet, noise characteristic of periodic jet impinging on a flat plate are presented and the mechanism of acoustic source is analyzed. The present work provides a reference value for the analysis and control of noise of periodic jet impinging on a flat plate.
Keywords: Periodic impinging jet; Noise characteristic; Waveform; Periodic frequency

1. Introduction

Impinging jet is widely used in many practical problems and industrial fields[1], such as drying of process products, cooling of gas turbine components and electronic equipment, landing of aerospace vehicles[2,3]. The research on impinging jet mostly concentrates on two fields. One is the aerospace field, the other is the heat transfer and cooling field. In the aerospace field, impinging jet noise may lead to fatigue damage of aircraft structure and reduce the safety performance and service life of aircraft. Therefore, noise characteristics and acoustic source mechanism of high-speed impinging jet are the center of various studies. Dhamanekar[4] studied the acoustic and flow variations in jets impinging on inclined plates by experiment. He found that the noise of subsonic impinging jet is equal for all inclinations, however, the noise of supersonic impinging jet increases as the inclination increases because of the presence of impinging tones. Sinibaldi[5] carried out an experiment study on effects of the impinging surface on screech tones of impinging jets. The results validates the coupling between acoustic discrete tones and fluctuating flow velocities. Akamine[6,7] studied the influence of the distance between the flat plate and the jet nozzle on the difference of acoustic field when the supersonic jet impinges on the inclined plate. The experimental results show that the radiation intensity of the sound source of impinging jet mainly depends on the location of the acoustic monitoring point and the pressure ratio at the nozzle outlet. Balakrishnan[8] experimentally investigated the noise spectra and directivity of circular and non-circular impinging jets on flat plate at various nozzle pressure ratios and emission angles. The results reveal that the structure and symmetry of shock cell structures are destroyed for the non-circular jets compared with the circular jet.

In recent decades, the numerical methods has been widely used due to the development of computing technology[9-11]. Romain[12,13] studied the noise characteristics of a supersonic jet impinging on a flat plate using large eddy simulation. The results show that the impinging tones and standing waves are caused by the acoustic feedback loop between the nozzle and the plate. Yao et al.[14] presented the interaction between shock wave and vortex structure when supersonic jet impinges on a flat plate by large eddy simulation. Christoph[15], Lyrintzis[16] and Nonomura[17] numerically investigated the noise characteristics and acoustic source mechanism of supersonic jet impinging on the inclined plate. The effects of inclination on the overall sound pressure level, spectrum and noise directivity are analyzed.

As for the heat transfer and cooling field, numerous researches mainly focus on the heat transfer characteristics of low-speed impinging jet[18-20]. In addition to changing the impinging distance, nozzle parameters and the shape of impinging surface, researchers gradually turn their attention to periodic impinging jets with unsteady velocity to improve heat transfer performance[21,22]. However, there is a paucity of literature on noise generation by periodic impinging jets and most researchers have ignored the accompanying noise problem. In actual industrial applications, noise emission not only affects the work efficiency of the staff, but also hurts the physical and mental health of the staff. Therefore, it is of great significance and innovative value to investigate the noise characteristics of periodic impinging jet.

In this paper, a numerical model for predicting noise generation of impinging jet is established in Section 2. Effects of acoustic integral surface and subgrid-scale models on simulation results are studied and the numerical model is validated by noise measurement experiment in Section 3. The validated numerical model is applied to investigation on noise characteristics and acoustic source mechanism of periodic jet impinging on a flat plate with different waveforms and periodic frequency in Section 4. The present work provides a reference value for the analysis and control of noise of periodic jet impinging on a flat plate.

2. Numerical method

2.1. Numerical model

Fig 1 shows the three-dimensional numerical model of periodic jet impinging on a flat plate. The diameter of the jet nozzle (D) is 10 mm, and the length of the jet nozzle is 25 mm. The average impinging velocity is 30m/s. The size of the flat plate is 80mm×80mm. The distance between nozzle outlet and flat plate (X) is the impinging distance. The computational domain is a hexahedral region with a size of 200mm×200mm×200mm. The size of the acoustic integral surface is defined as L×L×H. Noise monitoring points are arranged on a semicircle with R=1m as shown in Fig 2. Axis is the center line of the nozzle. θ is the angle between axis and the straight line from a monitoring point to the plate center. There are 11 monitoring points in the range of 15° to 165°. 90° direction is perpendicular to the impinging jet.

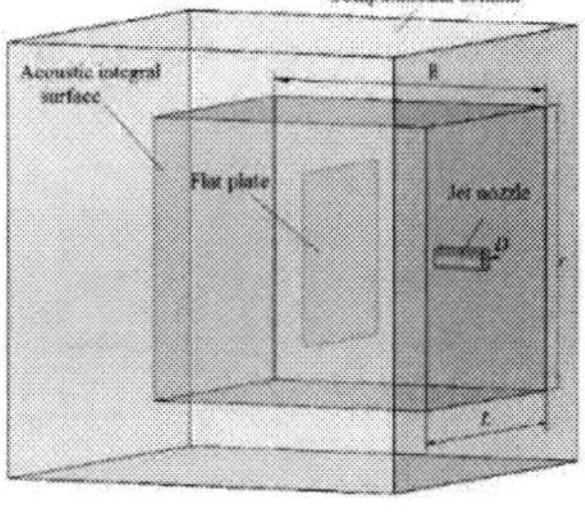

Fig 1. Numerical model of periodic jet impinging on a flat plate

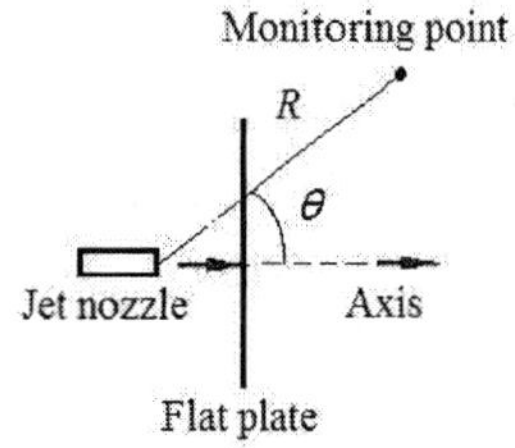

Fig 2. Distribution of Noise Monitoring Points

2.2. Governing equations

The numerical method used in the study of acoustic characteristics is based on large eddy simulation (LES) and Ffowcs Williams-Hawkings (FW-H) acoustic analogy. Firstly, time-accurate solutions of the

IOP Conf. Series: Materials Science and Engineering 1081 (2021) 012050 doi:10.1088/1757-899X/1081/1/012050

flow-field variables can be obtained by LES. For incompressible fluids, governing equations of LES are defined as[23]:

$$\frac{\partial \overline{u}_i}{\partial x_i} = 0 \tag{1}$$

$$\frac{\partial \overline{u}_i}{\partial t} + \frac{\partial}{\partial x_j}(\overline{u}_i \overline{u}_j) = -\frac{1}{\rho}\frac{\partial \overline{p}}{\partial x_i} + \nu \nabla^2 \overline{u}_i - \frac{\partial \tau_{ij}}{\partial x_j} + \frac{F_i}{\rho} \tag{2}$$

$$\overline{\tau_{ij}} = (\overline{u_i u_j} - \overline{u_i u_j}) \tag{3}$$

where τ_{ij} is the subgrid-scale (SGS) stress.

The SGS stress has great influence on LES method and can be solved by SGS models. In order to study the effect of the SGS model on the simulation results, different SGS models are constructed such as Smagorinsky-Lilly (SL) model, Wall-Adapting Local Eddy-Viscosity (WALE) model, Wall-Modeled Large Eddy Simulation (WMLES) model and Kinetic-Energy Transport (KET) model. Secondly, the acoustic field are obtained by solving the FW-H acoustic analogy equation. The method adopts the most general form of Lighthill's acoustic analogy, and is capable of predicting noise generated by equivalent acoustic sources. The FW-H formulation can be written as the following equation[24]:

$$\frac{1}{c^2}\frac{\partial^2 p'}{\partial t^2} - \nabla^2 p' = \frac{\partial}{\partial t}[(\rho_0 U_n)\delta(f)] - \frac{\partial}{\partial x_i}[L_i \delta(f)] + \frac{\partial^2}{\partial x_i \alpha x_j}[T_{ij} H(f)] \tag{4}$$

$$U_i = v_i + \frac{\rho}{\rho_0}(u_i - v_i) \tag{5}$$

$$L_i = P_{ij}\hat{n}_j + \rho u_i (u_n - v_n) \tag{6}$$

$$T_{ij} = \rho u_i u_j + P_{ij} - c^2(\rho - \rho_0)\delta_{ij} \tag{7}$$

p' is the sound pressure at far field, c is the sound velocity, u is the fluid velocity, v is the body surface velocity, n is the normal vector pointing into the fluid, $\delta(f)$ is Heaviside function, $H(f)$ is Dirac delta function, T_{ij} is the Lighthill stress tensor.

The acoustic integral surface has great influence on FW-H acoustic analogy method. In order to study the effect of acoustic integral surface on the simulation results, the size of acoustic integral surface is 100mm≤L≤170m and 95mm≤H≤170mm.

2.3. Mesh strategy

Structured mesh is applied to mesh generation. The computational domain is divided into several parts as shown in Fig 3. Part 1 and Part 2 needs to be encrypted to ensure the simulation accuracy. Then from Part 3 to Part 4, the grid density decreases gradually. The detail view of computational mesh in the cross section of the model can be seen in Fig 4.

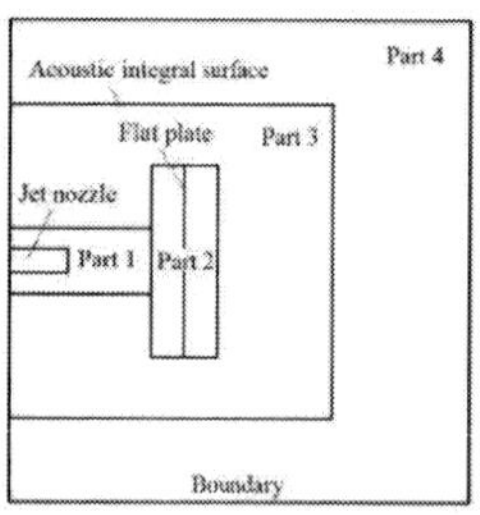

Fig 3. Schematic diagram of model partition

Fig 4. The detail view of computational mesh of the model

Verification of grid independence is carried out. The numerical model is meshed with the mesh number of 2.44×10^6, 3.14×10^6, 3.69×10^6, 4.11×10^6, 4.66×10^6 and 5.26×10^6 respectively when X=5D. As the important parameter of impinging jet, the velocity changes of monitoring points are taken as references to validate the grid independence as shown in Fig 5. Monitoring point A is located on the central axis of the nozzle, and the distance to the center of the nozzle outlet is defined as X1; Monitoring point B is located parallel to the plane, the distance to the plate is 2mm, and the distance to the central axis is defined as X2; Various values are taken for X1 and X2 respectively.The results of independent verification of grids are displayed in Fig 6. One can observe that velocity distribution no longer changes when the mesh number increases to 4.66×10^6. Therefore, 4.66×10^6 grids are employed to discretize the computational domain.

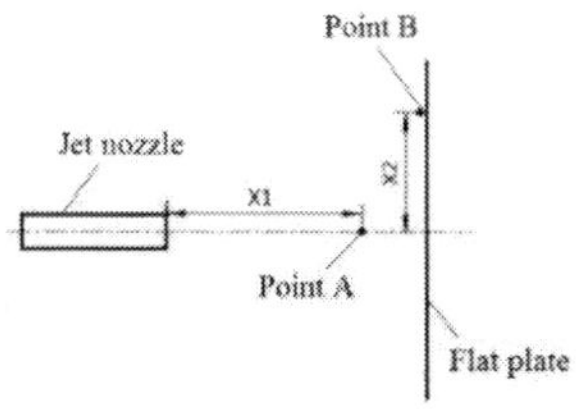

Fig 5. Distribution of velocity monitoring points

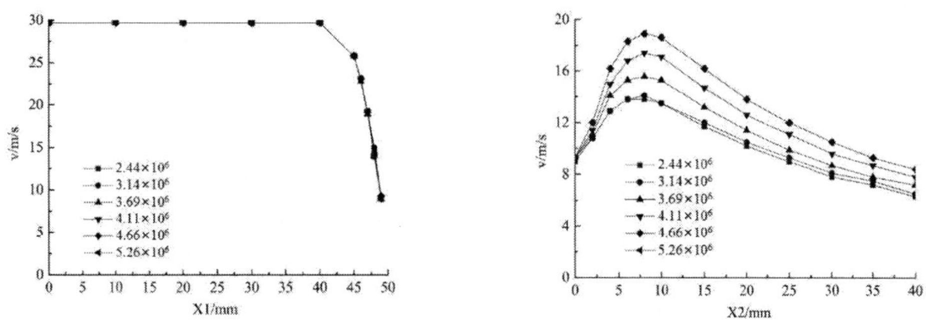

(a) Velocity monitoring point A (b) Velocity monitoring point B

Fig 6. Results of independent verification of grids

2.4. Boundary conditions and solving method

Noise characteristics of periodic impinging jet are compared to that of steady impinging jet with constant velocity. For the steady impinging jet, the boundary condition for the nozzle inlet is velocity-inlet. The left boundary of the computational domain is pressure-inlet. The other boundaries of the computational domain are pressure-outlet. The surface of nozzle and flat plate is provided with non-slip wall boundary conditions. For the periodic impinging jet, most boundary conditions are the same as those of steady case. However, the difference lies in the use of UDF program to achieve the output of periodic velocity signal for the nozzle inlet. Figure 7 shows four typical waveforms of periodic velocity signals used in the paper. The average velocity（$\bar{u}$）is 30m/s. The periodic frequency (1/T) varies from 5Hz to 40Hz. Due to the different instantaneous noise of different periodic impinging jets, equivalent continuous sound pressure level is introduced as an evaluation index of periodic impinging jet noise, as follows.

$$L_{eq} = 10\lg\left(\frac{1}{T}\int_0^T 10^{0.1L_{PA}}\,dt\right) \tag{8}$$

where L_{PA} is the instantaneous sound pressure level at a certain time *t*, and *T* is the specified measurement time. Then the noise source of periodic impinging jet can be equivalent to that of steady impinging jet.

Steady flow field is obtained by standard k-ω model as the initial field of LES. The transient flow field accurately near the noise source is obtained by LES. The PISO scheme is used in pressure-velocity coupling. The PRESTO! is applied to spatial discretization pressure equation and second-order upwind format is applied to momentum equation. Acoustic fields of steady and periodic impinging jets are calculated by FW-H model. Time step size is 2.5×10^{-5}s according to the noise frequency range recognized by the human ear. From the noise source to the receiver, the noise propagation is obtained by solving the acoustic wave equation.

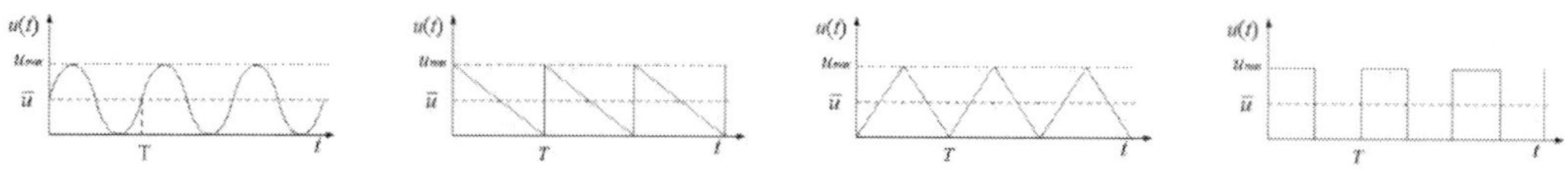

(a) Sinusoidal waveform (b) Triangular waveform (c) Sawtooth waveform (d) Rectangular waveform

Fig 7. Waveform of the periodic impinging jet

3. Experimental verification

Taking the steady impinging jet case as an example, numerical results are compared with experimental data to investigate the effects of acoustic integral surface and SGS model on noise simulation results. The accuracy of numerical model can also be verified based on the comparison results.

3.1. Experimental system

The noise measurement system used for experimental verification is shown as Fig 8. It includes nozzle jet device, control room and anechoic room of size 3.5m × 3.5m × 3m (wedge tip-to tip). In order to reduce the effect of nozzle jet device on experimental results, the gas tank, the control system, gas tank and air compressor of the nozzle jet device are arranged in the control room, while the nozzle is arranged in the anechoic chamber. The impinging velocity is 30m/s. The diameter of the jet nozzle, impinging distance and the size of the flat plate are consistent with the numerical model. The location of the microphone sensor is the same as shown in Fig 8. The microphone sensor is PCB 377B11 with PCB 426E01. Data acquisitions are carried out with NI PXI-1042Q, PXI-4472 and PXI-8106 control system, which has a band width from 20Hz to 20 kHz. The background sound pressure level is 19.9dB. The noise distribution of jet impinging on the flat plate is measured through the above system.

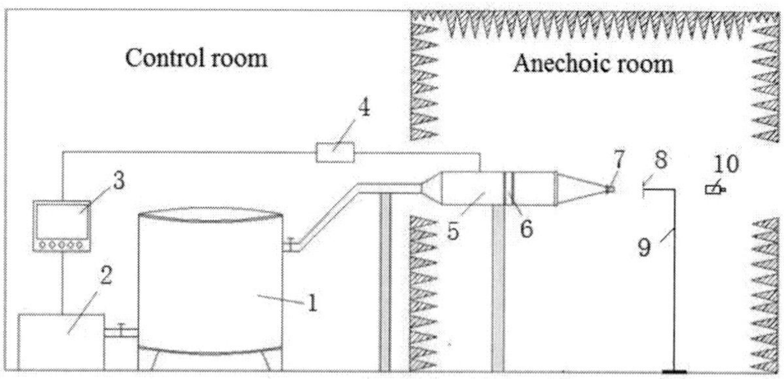

1-Gas tank 2-Compressor 3-PID controller 4-Differential pressure sensor 5-Diffuser section 6-Rectification section 7-Jet nozzle 8-Flat plate 9-Cone support 10-Microphone sensor

(a) Schematic diagram of experimental system

IOP Conf. Series: Materials Science and Engineering 1081 (2021) 012050 doi:10.1088/1757-899X/1081/1/012050

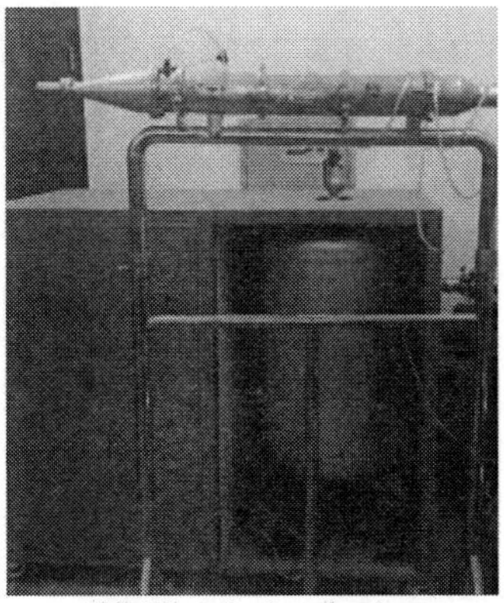

(b) Nozzle jet device

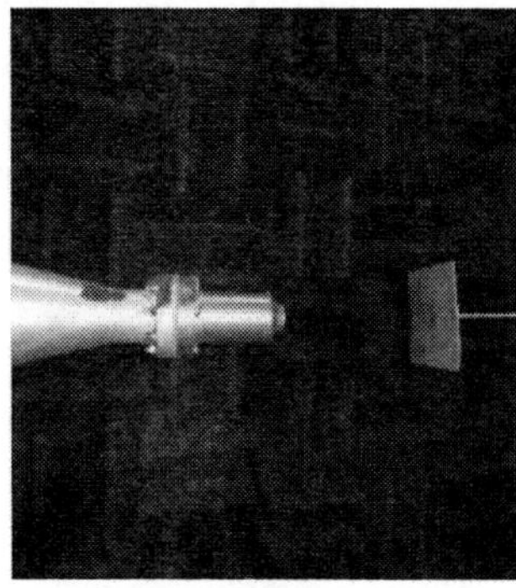

(c) Anechoic room

Fig 8. Device diagram of noise measurement experiment system

3.2. Effects of acoustic integral surface

The influence of size of the acoustic integral surface ($L\times L\times H$) on the calculation results of the acoustic field of impinging jet is studied. Firstly, the KET model is initially selected as the SGS model and the length of acoustic integral surface parallel to the jet direction remains unchanged (H=140mm). The length of acoustic integral surface parallel to jet direction varies from 100mm to 170mm ($10\leqslant L/D\leqslant 17$). Fig 9 shows the variation of overall sound pressure level (OASPL) with L under different noise monitoring points. It can be observed that the impinging jet noise decreases as L increases. When L increases from 100 mm to 130 mm, the maximum difference between the simulation results of impinging jet noise is 2.0dB. When L increases from 130 mm to 170 mm, the maximum difference between the simulation results of impinging jet noise is 0.2dB. Acoustic integral surface has little influence on the simulation results when $L\geqslant$130mm.

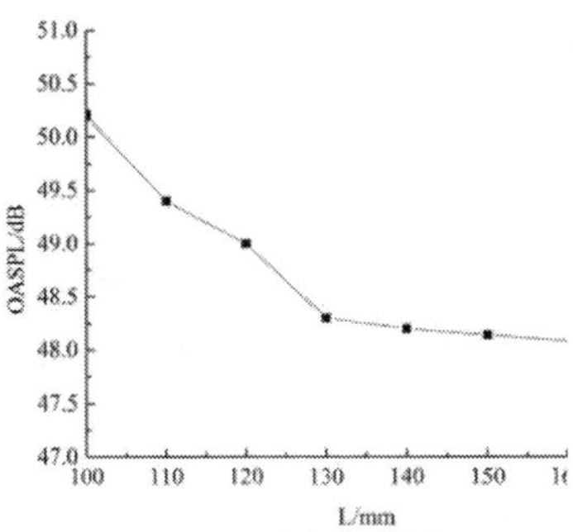

(a) θ=30°

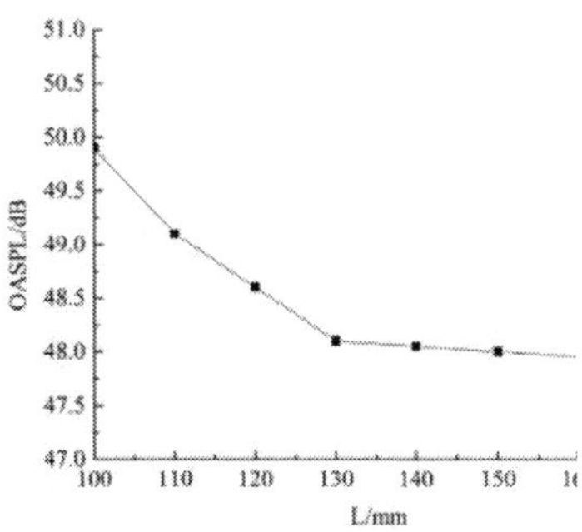

(b) θ=60°

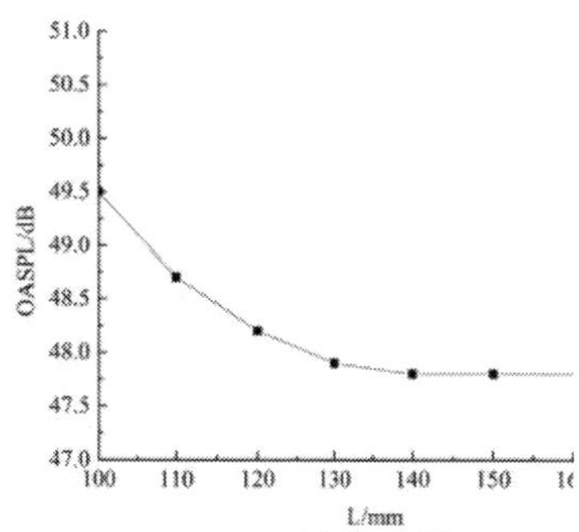

(c) θ=90°

Fig 9. Variation of OASPL with L under different noise monitoring points

In theory, the acoustic integral surface already contains the main acoustic source region. The size of acoustic integration surface should have no effect on the simulation results. However, in the actual simulation progress, the jet impinges on a flat plate and flows through the acoustic integral surface, thus producing a non-physical virtual sound source[25]. The greater the velocity is when flowing through the acoustic integral surface, the greater the influence of the virtual sound source has on the simulation results. Based on this, the variation of velocity through the acoustic integral surface with L is displayed in Fig 10. It can be seen that the attenuation trend of velocity is flat gradually when $L\geqslant$130mm. The velocity decreases to around 3m/s and the simulation results of the impinging jet noise are less affected by the virtual sound sources. The total mesh number will increase as L increases. Considering the

IOP Conf. Series: Materials Science and Engineering 1081 (2021) 012050 doi:10.1088/1757-899X/1081/1/012050

computational efficiency and accuracy of the numerical method, the value of L is chosen to be 130mm (L/D=13).

Secondly, the value of L remain unchanged (L/D=13), the length of acoustic integral surface perpendicular to jet direction varies from 95mm to 170mm (9.5≤H/D≤17). Fig 11 shows the variation of OASPL with H under different noise monitoring points. It can be seen that the impinging jet noise increases as H increases. When H increases from 95 mm to 130 mm, the maximum difference between the simulation results of impinging jet noise is 0.1dB. When H increases from 130 mm to 170 mm, the maximum difference between the simulation results of impinging jet noise is 0.4dB. Acoustic integral surface has little influence on the simulation results when H≥130mm. Since Part 1 and Part 2 should be inside the acoustic integral surface, the value of L is chosen to be 130mm (H/D=13).

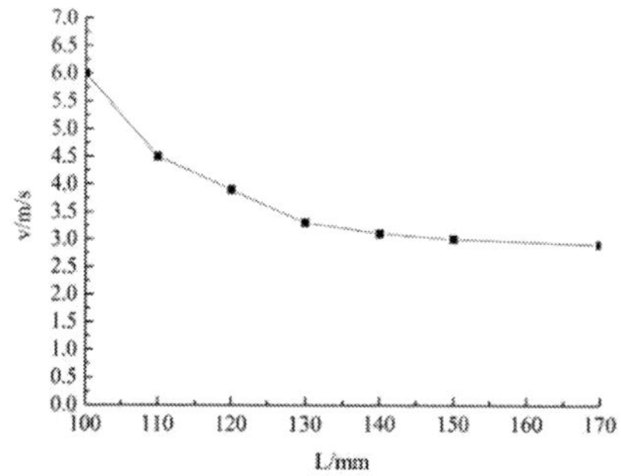

Fig 10. Velocity of jet flowing through different acoustic integral surface

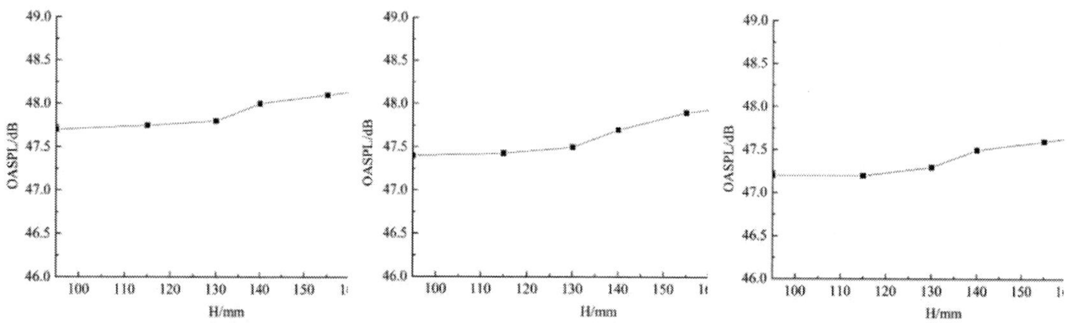

(a) θ=30° (b) θ=60° (c) θ=90°

Fig 11. Variation of OASPL with H under different noise monitoring points

3.3. *Effects of SGS model*

Fig 12 shows the effects of different SGS model on the simulation results of impinging jet noise when the size of the acoustic integral surface is chosen as 130mm×130mm×130mm. Compared with the experimental data, the results of WALE model are smaller and the results of the other three SGS models are larger. The comparison result is quantitatively analyzed. Table.1 shows the noise computation errors for different SGS models. One can observe that the computational errors of SL model and WMLES model are relatively large and the simulation result of KET model is in the best agreement with the experimental data. Therefore, the SGS model of LES is selected as KET model.

3.4. *Validation of numerical model*

The above selected numerical model is applied in the simulation for noise distribution under different impinging distances (3D≤X≤9D) and monitoring positions (15°≤θ≤90°). Compared with the experimental data, the noise calculation errors is shown in Table 2. The maximum calculation error is

1.3dB and the maximum relative error is 2.8%. The numerical results are all in good agreement with the experimental data which validates the accuracy of the numerical model.

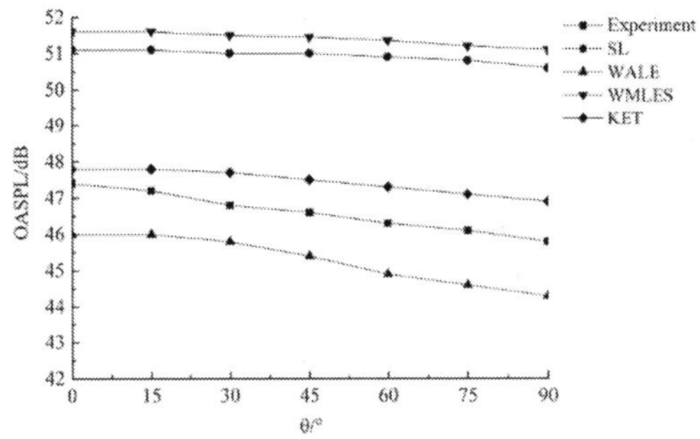

Fig 12. Effects of different SGS models on the simulation results of impinging jet noise

Table 1. Noise computation errors for different SGS models

	Maximum difference (dB)	Average difference (dB)	Average relative error (%)
SL model	4.8	4.3	9.3
WALE model	-1.5	-1.3	2.8
WMLES model	5.3	4.8	10.3
KET model	1.1	0.8	1.8

Table 2. Noise computation errors under various impinging distances and monitoring positions

	Maximum difference (dB)	Maximum relative error (%)
X=3D	1.1	2.4
X= 5D	1.1	2.4
X= 7D	1.2	2.6
X= 9D	1.3	2.8

4. Noise characteristics analysis

The validated numerical model is applied to investigation on noise characteristics of periodic jet impinging on a flat plate with different waveform and periodic frequency.

4.1. Effects of waveform

The noise directivity distribution of periodic impinging jet with different waveform is compared to that of steady impinging jet as shown in Fig 13. Generally speaking, the noise of periodic impinging jet is much larger than that of steady impinging jet and the noise of rectangular case is larger than that of the other cases. The average difference of noise between steady case and rectangular case is 24dB. As for the other three waveforms, the average difference is about 14dB. The noise distribution of rectangular waveform fluctuates greatly in the range of 60° to 120° whereas the noise distribution of the other three waveforms is more uniform.

Specifically, the effect of waveform on noise spectrum is studied as shown in Fig 14. The periodic frequency is 40Hz and the monitoring angle is 90°. For the steady impinging jet, the noise decreases as the frequency increases. As for the periodic impinging jet, the noise decreases first and then increases with the increase of frequency. Compared with the noise spectrum of steady impinging jet, the waveform has a greater influence on high-frequency noise. Especially for the rectangular waveform, the spectrum is of obvious high-frequency characteristic.

IOP Conf. Series: Materials Science and Engineering 1081 (2021) 012050 doi:10.1088/1757-899X/1081/1/012050

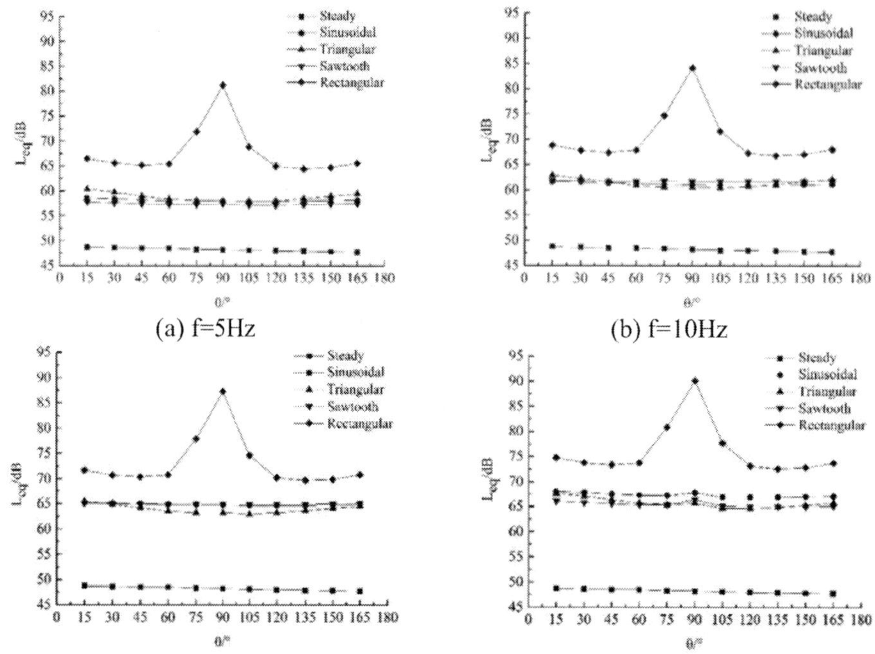

(a) f=5Hz (b) f=10Hz

(c) f=20Hz (d) f=40Hz

Fig 13. Effect of waveform on noise directivity of periodic impinging jet

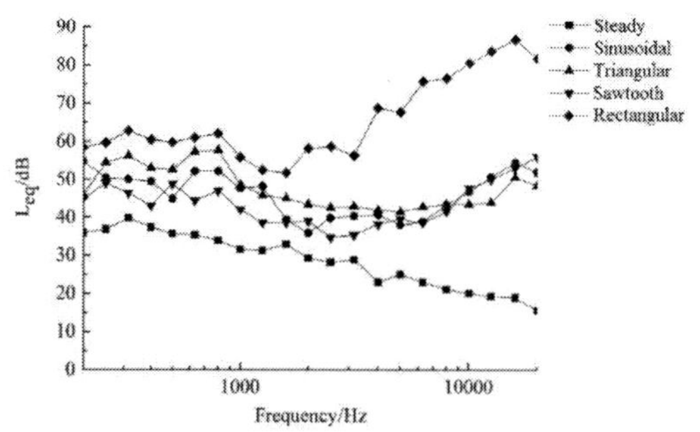

Fig 14. Effect of waveform on noise spectrum of periodic impinging jet (θ=90°)

Acoustic sources of impinging jets mainly contains two factors: one is the quadrupole source caused by the internal stress of fluid in the free jet section, the other is the dipole source caused by the reaction force between fluid and flat plate. Based on our previous conclusions[26], the dipole source leads to the broadband characteristics in medium and high frequency bands and the spectrum of free jet is of low-frequency characteristic. So the quadrupole source plays a more important role on the acoustic source of the steady jet impinging on the flat plate. However, for the periodic impinging jet, another acoustic source is the unstable fluctuation of fluid pressure at nozzle outlet caused by the unsteady velocity. Then the noise of periodic impinging jet is larger than that of steady impinging jet. According to the noise spectrum of the periodic impinging jet, the fluid pressure fluctuation has a greater influence on high-frequency noise. For the rectangular waveform, the fluid pressure fluctuation of impulse type leads to a great increase of noise in high-frequency band. The direction of the acoustic source caused by the fluid pressure fluctuation of impulse type is perpendicular to the jet. So the maximum noise is in the 90° direction. For the other three waveform, the change of velocity is gentle.

IOP Conf. Series: Materials Science and Engineering 1081 (2021) 012050 doi:10.1088/1757-899X/1081/1/012050

The strength of fluid pressure pulsation at nozzle outlet is relatively small. Therefore, the noise distribution is different from that of the rectangular case and the noise is also smaller.

When the periodic variation frequency is 40Hz and the vecolity is 30m/s and the impact distance is 5D, the instantaneous velocity distribution nephogram of the periodic jet flow field of flat plate impact in one cycle (T) is shown in Fig 15 and the end time of one cycle is also the starting time of the next cycle. It also can be seen from the figure that the velocity attenuation rate of rectangular waveform is faster than that of other waveforms, which is also the reason why rectangular waveform has greater impact on high frequency noise.

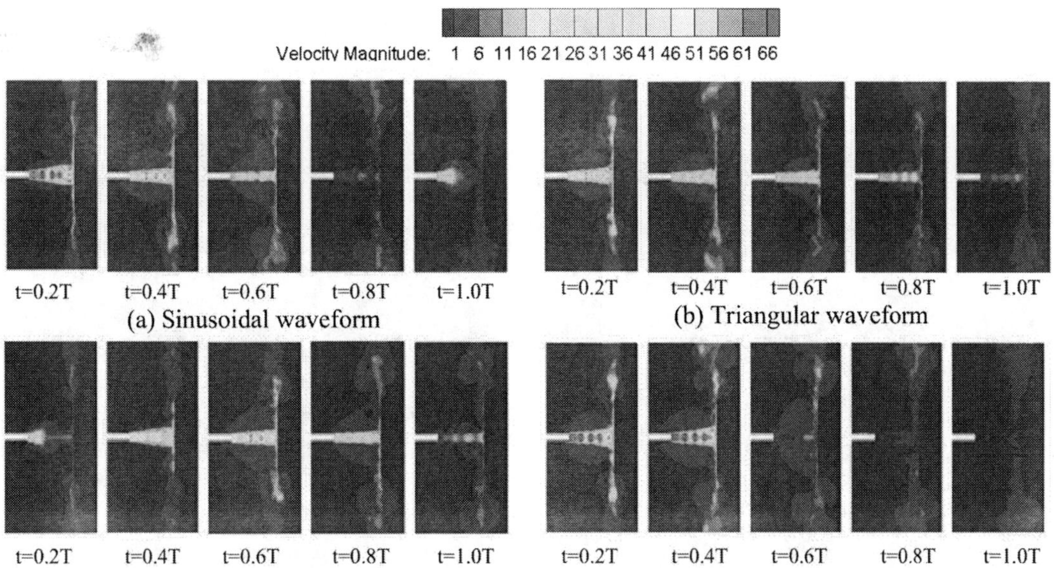

(a) Sinusoidal waveform (b) Triangular waveform
(c) Sawtooth waveform (d) Rectangular waveform

Fig 15. Velocity cloud image of the periodic impinging jet

4.2. Effects of periodic frequency

Fig 16 shows the effects of periodic frequency on noise directivity of periodic impinging jet. The noise of periodic impinging jet increases as the periodic frequency increases. Specifically, when the periodic frequency increases from 5Hz to 40Hz, the average noise increases by 7~9dB.

For the sinusoidal waveform and sawtooth waveform, noise distribution is relatively uniform. But the noise increase of sawtooth waveform is very small when the periodic frequency is greater than 20Hz. For the triangular waveform, noise in the perpendicular direction to the jet is smaller than that in the parallel direction to the jet. As for the rectangular waveform, noise in the perpendicular direction to the jet is larger than that in the parallel direction to the jet.

Fig 17 shows the effects of periodic frequency on noise spectrum of periodic impinging jet (θ=90°). It can be observed that for the rectangular waveform, the rise of periodic frequency only increases the noise value, but has little influence on the spectrum distribution. As for the other three waveforms, the periodic frequency has a relatively small effect on noise in the frequency band from 1000 Hz to 6000 Hz, and the noise in other frequency bands increases with the increase of periodic frequency.

For the rectangular waveform, the main acoustic source is the fluid pressure pulsation at nozzle outlet and the periodic frequency has a greater influence on this factor. As the periodic frequency increases, the strength of the fluid pressure pulsation increases. This is why the periodic frequency only changes the noise value but has relatively small influence on the spectrum characteristics. For the other three waveforms, the main acoustic source is the fluid pressure pulsation and the quadrupole source.

The increase of periodic frequency has a greater influence on the fluid pressure. Therefore, the noise in high frequency band is enhanced more, especially when the periodic frequency increases from 20 Hz to 40 Hz.

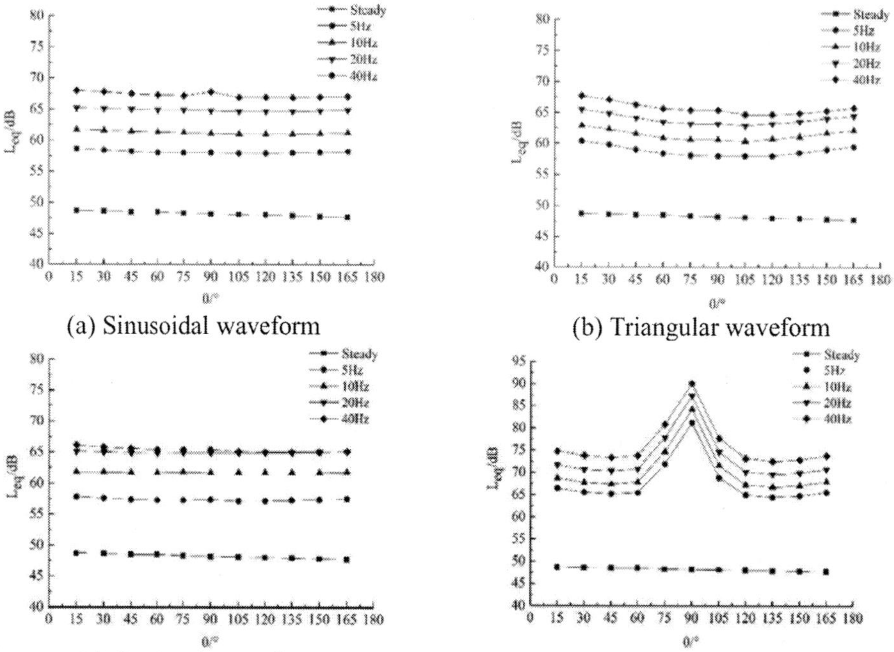

(a) Sinusoidal waveform (b) Triangular waveform

(c) Sawtooth waveform (d) Rectangular waveform

Fig 16. Effect of periodic frequency on noise directivity of periodic impinging jet

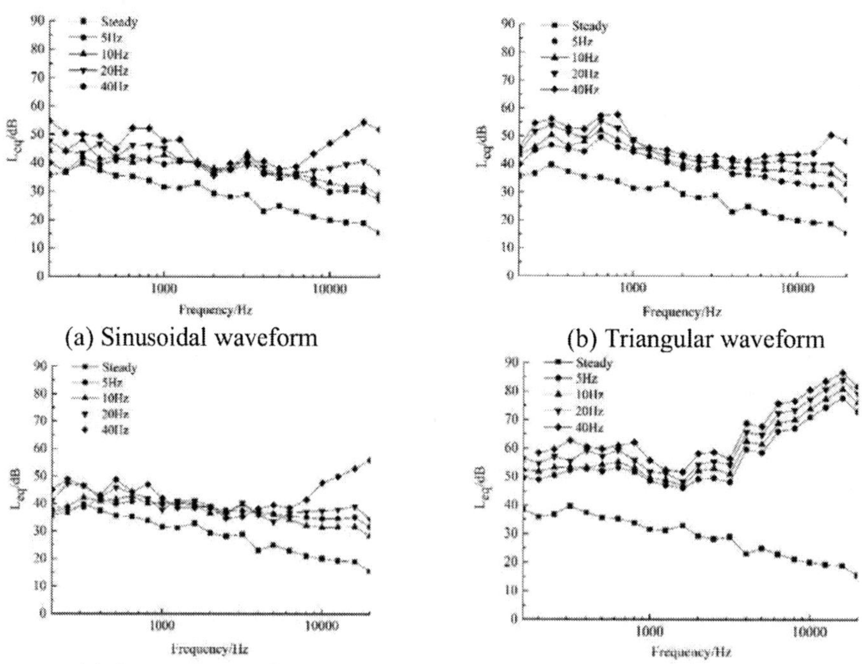

(a) Sinusoidal waveform (b) Triangular waveform

(c) Sawtooth waveform (d) Rectangular waveform

Fig 17. Effect of periodic frequency on noise spectrum of periodic impinging jet

5. Conclusions

In this paper, based on LES and FW-H acoustic analogy, a numerical model for predicting the noise of impinging jet is established and the noise measurement experiment is carried out to validate the numerical model. The comparison between experimental data and numerical results shows that when the size of the acoustic integral surface is chosen as 130mm×130mm×130mm and the SGS model of LES is KET model, the maximum calculation error is 1.3dB and the maximum relative error is 2.8%, which validates the accuracy of the numerical model.

Then the present numerical model is applied to the investigation on noise characteristics of periodic jet impinging on a flat plate. Compared with the steady impinging jet, noise directivity and spectrum distribution of periodic impinging jet are analyzed with different waveforms and periodic frequency. The results show that the average noise of periodic impinging jet is more than 14 dB larger than that of steady impinging jet and the noise of rectangular waveform is the largest. As the periodic frequency increases from 5Hz to 40Hz, the average noise increases by 7~9dB. The differences of noise characteristic between periodic impinging jet and steady impinging jet mainly depend on the unstable fluctuation of fluid pressure at nozzle outlet caused by the unsteady velocity. The drastic change of velocity (such as rectangular waveform) and the increase of periodic frequency lead to the increase of strength of the fluid pressure fluctuation. The fluid pressure fluctuation has a greater influence on high-frequency noise.

References

[1] Xu HH,Zhuang YL and Qiu KM 2016 Research on a thrust test rig for an automatic induction type jet fan *Chinese Journal of Turbomachinery* **58**(5) 54-56.

[2] Dhamanekar A and Srinivasan K 2014 Effect of impingement surface roughness on the noise from impinging jets *Physics of Fluids* **26**(3) 036101.

[3] Abdel-Fattah A 2007 Numerical and experimental study of turbulent impinging twin-jet flow *Experimental Thermal & Fluid Science* **31**(8) 1061-72.

[4] Dhamanekar A and Srinivasan K 2017 Effect of plate inclination on the noise of impinging jets *Applied Acoustics* **127** 354-64.

[5] Sinibaldi G, Lacagnina G, Marino L and Romano G P 2013 Aeroacoustics and aerodynamics of impinging supersonic jets: Analysis of the screech tones *Applied Acoustics* **25**(8) 086104.

[6] Akamine M, Okamoto K, Gee K L,Neilsen TB,Teramoto S,Okunuki T and Tsutsumi S 2018 Effect of nozzle–plate distance on acoustic phenomena from supersonic impinging jet *Aiaa Journal* **56**(5) 1943-52.

[7] Akamine M, Nakanishi Y, Okamoto K,Teramoto S,Okunuki T and Tsutsumi S 2015 Acoustic phenomena from correctly expanded supersonic jet impinging on inclined plate *Aiaa Journal* **53**(7) 2061-67.

[8] Balakrishnan P and Srinivasan K 2019 Impinging jet noise reduction using non-circular jets *Applied Acoustics* **143**(1) 19-30.

[9] Wu HY,Jian XS,Chen Q and Zhu ZN 2020 Noise analysis and control of centrifugal fan for space station *Chinese Journal of Turbomachinery* **62**(2) 27-34.

[10] Zhang XR, Li HY,Zhao S,Tan JF,Xiong JTan JF and Xiong J 2020 Design and analysis of two-stage centrifugal fan *Chinese Journal of Turbomachinery* **62**(3) 22-26.

[11] Wang L,Li JB,Huang YT,Ma L and Liu XM 2020 Study on effects of bionic blade inspired by owl wing on aerodynamic noise of axial fan *Chinese Journal of Turbomachinery* **62**(2) 19-26.

[12] Gojon R, Bogey C and Marsden O 2016 Investigation of tone generation in ideally expanded supersonic planar impinging jets using large-eddy simulation *Journal of Fluid Mechanics* **808** 90-115.

[13] Gojon R and Bogey C 2018 Flow structure oscillations and tone production in underexpanded impinging round jets *Aiaa Journal* **55**(6) 1-14.

[14] Yao Z, Liu H and Zhang X 2012 Interaction of shocks and vortices in a highly underexpanded supersonic impinging jet *Aiaa Journal* **50**(5) 1169-76.

[15] Christoph B, Housman J A and Kiris C C 2016 Noise generation mechanisms for a supersonic jet impinging on an inclined plate *Journal of Fluid Mechanics* **797** 802-50.
[16] Lyrintzis A, Golubev V and Kurbatski K 2014 Detached-Eddy simulations of rocket plume noise at lift-off Journal of the *Acoustical Society of America* **136**(4) 2168.
[17] Nonomura T, Honda H, Nagata Y,Yamamoto M,Morizawa S Obayashi S and Fujii K 2016 Plate-angle effects on acoustic waves from supersonic jets impinging on inclined plates *Aiaa Journal* **54**(3) 816-27.
[18] Alenezi A, Almutairi A, Alhajeri HU,Addali A and Gamil AAA 2018 Flow structure and heat transfer of jet impingement on a rib-roughened flat plate *Energies* **11**(6) 1550.
[19] Vinze R, Chandel S, Limaye M D and Prabhu SV 2015 Influence of jet temperature and nozzle shape on the heat transfer distribution between a smooth plate and impinging air jets International *Journal of Thermal Sciences* **53** 2061-67.
[20] Wang X L, Motala D, Lu T J,Song SJ and KimT 2014 Heat transfer of a circular impinging jet on a circular cylinder in crossflow *International Journal of Thermal Sciences* **78**(3) 1-8.
[21] Hewakandamby B N 2009 A numerical study of heat transfer performance of oscillatory impinging jets International *Journal of Heat & Mass Transfer* **52**(1) 396-406.
[22] Azevedo L F A, Webb B W and Queiroz M 1994 Pulsed air jet impingement heat transfer *Experimental Thermal & Fluid Science* **8**(3) 206-13.
[23] Manchester E L and Xu X Y 2020 The effect of turbulence on transitional flow in theFDA 's benchmark nozzle model using large-eddy simulation International *Journal for Numerical Methods in Biomedical Engineering* **10** e3389.
[24] Liu X, Xie JL, Xie Q and Wang M 2019 Prediction of noise directivity of an impinging jet using les and acoustic analogy method *Chinese Journal of Turbomachinery* **61**(3) 48-52.
[25] Wang M, Lele S K and Moin P 1996 Computation of quadrupole noise using acoustic analogy *Aiaa Journal* **34**(11) 2247-54.
[26] Liu X, Xie JL, Xie Q and Huang HY 2018 Cone angle effects on noise characteristics of jets impinging on conical obstacles *AIP Advances* **8** 115121.

Acknowledgments
This work was financially supported by the National Natural Science Foundation of China (No.51376077).

Research on structure characteristics of the helium-lubricated spiral grooved thrust bearing for helium turbine expander

X H Zhang [1, 2], N Peng [1], B Dong [1], K R Li[1] and L Q Liu [1, 2]
[1]Technical Institute of Physics and Chemistry, CAS, Beijing, China
[2]University of Chinese Academy of Sciences, Beijing, China

E-mail:zhangxiaohua15@mails.ucas.edu.cn
pengnan@mail.ipc.ac.cn

Abstract: The helium-lubricated spiral grooved thrust bearing of helium turbo expander was studied by numerically simulating the steady-state non-linear Reynolds equation, the Reynolds equation of the pressure was solved by finite difference method (FDM), the film pressure distribution is calculated iteratively by MATLAB program, the structural parameters including the spiral angle, the groove length ratio, the groove width ratio influenced on the static characteristics were analyzed. The results show that: The bearing loading capacity of helium lubricated spiral grooved thrust gas bearing is lower than that of air lubricated bearing under the same conditions; The viscosity of the lubricant only affects the gas carrying capacity of the thrust disc when the spiral groove is on the bearing, while the density of the lubricant has a greater influence on the bearing loading capacity; When the spiral groove is designed on the rotor thrust disc, the bearing loading capacity of the spiral grooved thrust bearing increases by 6% on average compared with that of the spiral groove on the bearing.

1. Introduction

Gas lubricated bearing is a kind of sliding bearing with gas as lubricant. It has a series of advantages such as low friction power consumption, high motion accuracy, small vibration, pollution-free, high and low temperature resistance and stable operation in special environment [1]. Therefore, it has been widely applied in high-speed turbine expander, precision instruments and nuclear engineering [2].

The main problem in the development of spiral groove thrust gas bearing is that the working clearance is too small, so it is very difficult to obtain the gas film pressure field by measuring method [2]. Therefore, numerical simulation is a convenient and fast method, which can study the formation mechanism of gas film pressure and help to design excellent structure of spiral grooved thrust gas bearing [3].

IOP Conf. Series: Materials Science and Engineering 1081 (2021) 012051 doi:10.1088/1757-899X/1081/1/012051

In the large-scale cryogenic liquid helium refrigerant system, helium turbo expander is a core part of the system. In order to improve the purity of liquid helium, the bearing gas of helium turbo expander must use helium. Compared with air turbine expander, the physical properties of helium are different from air, for example, the density of helium is less than air. It is a more challenging task to design helium lubricated spiral grooved thrust gas bearing which can produce the same bearing loading capacity as the bearing lubricated by air [4-6].

2.The Model of the Spiral Grooved Thrust Gas Bearing

2.1 Basic equation of spiral grooved thrust gas bearing

The rotor thrust disc will bring the gas into the spiral groove due to the action of viscous lubrication gas, and the viscous gas will be compressed to form the axial bearing loading capacity [7], as shown in Figure 1-2. In polar coordinates, the Reynolds equation of compressible steady-state gas lubricated spiral grooved thrust gas bearings is:

$$\frac{\partial}{\partial r}\left(\frac{rPh^3}{\mu}\frac{\partial P}{\partial r}\right)+\frac{\partial}{r\partial r}\left(\frac{Ph^3}{\mu}\frac{\partial P}{\partial \theta}\right)=6\omega r\frac{\partial Ph}{\partial \theta} \tag{1}$$

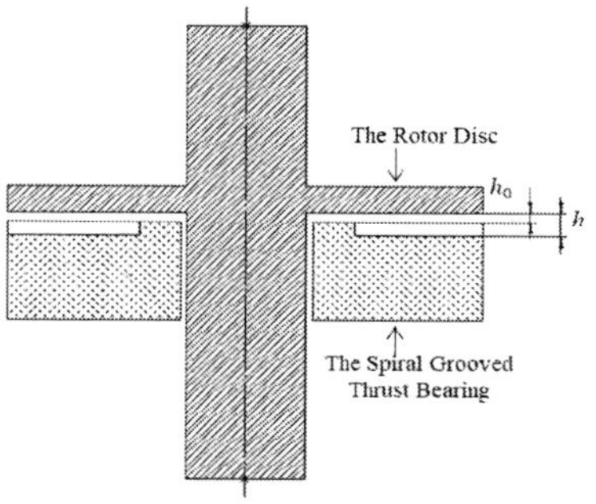

Figure 1. Structural diagram of bearing-rotor.

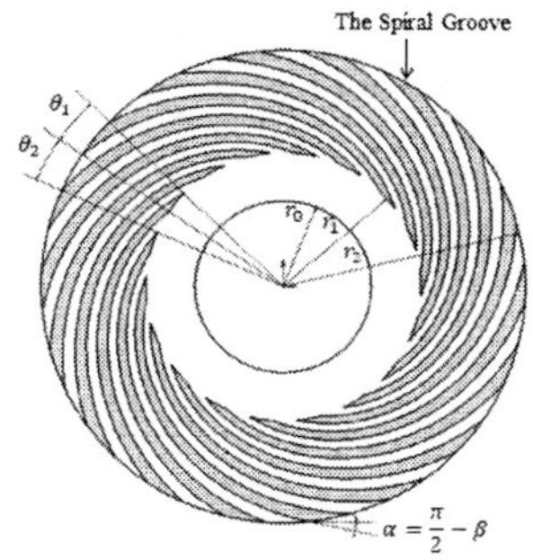

Figure 2. Structural diagram of bearing thrust surface.

Defining dimensionless parameters:

$$\boldsymbol{H}=\frac{h}{h_0},\boldsymbol{R}=\frac{r}{r_0},\boldsymbol{P}=\frac{P}{P_a},\varLambda=\frac{6\mu\omega r^2}{P_a h_0^2}$$

In the formula:

$\boldsymbol{P}$ — the dimensionless film pressure
$\boldsymbol{H}$ — the dimensionless film thickness
ξ — the dimensionless radius coordinates
$\varLambda$ — the number of dimensionless bearings
P — the film pressure
P_a — the environmental pressure
h_0 — the minimum film thickness
ω — the rotor angular velocity
μ — the aerodynamic viscosity
r — the rotor radius
θ — the circumferential coordinates
α/β — the spiral angle
r_0 — the radius of base circle
r_1 — the radius of dam area
r_2 — the outer radius
θ_1 — the width of groove area
θ_1 — the width of platform area

The steady-state nonlinear dimensionless control equation of spiral grooved thrust gas bearing is as follows:

$$\frac{1}{R}\frac{\partial}{\partial R}\left(RH^3\frac{\partial P^2}{\partial R}\right)+\frac{1}{R^2}\frac{\partial}{\partial\theta}\left(H^3\frac{\partial P^2}{\partial\theta}\right)=2\Lambda\frac{\partial(PH)}{\partial\theta} \quad (2)$$

The groove body of the spiral grooved thrust gas bearing is spiral structure, in which the calculation of film pressure is very complicated by finite difference method (FDM). In order to simplify the calculation, conformal transformation is needed for the governing equation and the solution domain. The working surface of the spiral groove thrust gas bearing is expanded into a plane and the conformal transformation method is used to transform the spiral groove into a fan-shaped element [8], as shown in Figure 3.

$$\begin{cases} \zeta=r \\ \eta=\theta-\dfrac{1}{\cot\beta}\ln\left(\dfrac{r}{r_g}\right) \end{cases} \quad (3)$$

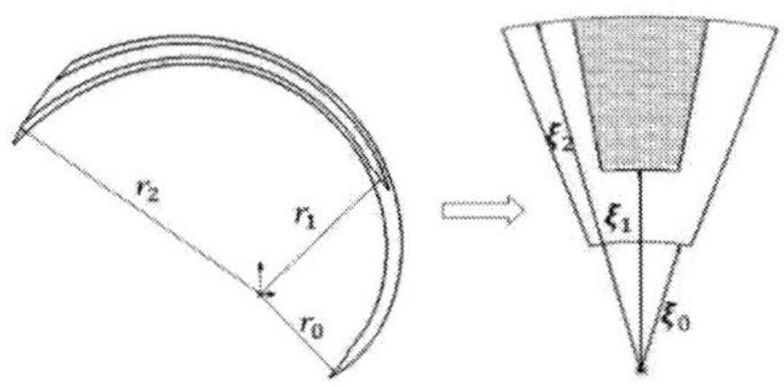

Figure 3. The conformal transformation

From the above two formulas, 'r' is a function of 'ξ' and 'η', and 'Φ' is only a function of 'η'. In the new coordinate system, we derived the new Reynolds equation: the dimensionless Reynolds equation of gas film pressure of spiral grooved thrust gas bearing under steady state condition is:

$$\begin{aligned}&\frac{\partial}{\partial\zeta}\left(\zeta\boldsymbol{H}^3\frac{\partial\boldsymbol{P}^2}{\partial\zeta}\right)+\frac{1}{\zeta}\left(1+\frac{1}{\cot^2\beta}\right)\frac{\partial}{\partial\eta}\left(\boldsymbol{H}^3\frac{\partial\boldsymbol{P}^2}{\partial\eta}\right)-\frac{1}{\cot\beta}\left[\frac{\partial}{\partial\zeta}\left(\boldsymbol{H}^3\frac{\partial\boldsymbol{P}^2}{\partial\eta}\right)+\frac{\partial}{\partial\eta}\left(\boldsymbol{H}^3\frac{\partial\boldsymbol{P}^2}{\partial\zeta}\right)\right]\\&+\boldsymbol{H}^3\frac{\partial\boldsymbol{P}^2}{\partial\zeta}=2\Lambda\zeta\frac{\partial(\boldsymbol{PH})}{\partial\eta}\end{aligned} \quad (4)$$

The Boundary Condition:

(1) The inlet and outlet of spiral grooved thrust gas bearing is environmental pressure:

$$\boldsymbol{P}(\zeta=0,\eta)=\boldsymbol{P}\left(\zeta=\frac{r}{r_0},\eta\right)=1 \quad (5)$$

(2) The spiral grooved thrust gas bearing satisfies periodic boundary conditions:

$$\boldsymbol{P}(\zeta,\eta)=\boldsymbol{P}\left(\zeta,\eta+\frac{2\pi}{N}\right) \quad (6)$$

2.2 Static characteristic of spiral grooved thrust gas bearing

The distribution of gas film pressure of spiral grooved thrust gas bearing can be obtained by solving Reynolds equation (4):

The bearing loading capacity:

IOP Conf. Series: Materials Science and Engineering 1081 (2021) 012051 doi:10.1088/1757-899X/1081/1/012051

$$F = P_a \cdot r_0^2 \cdot \int_{r_0}^{r_2} \int_0^{2\pi} (P-1)\, r dr d\theta \tag{7}$$

The film static stiffness:

$$K = \frac{dF}{dh} \tag{8}$$

3. Solution of Reynolds for spiral grooved thrust gas bearing

The gas film pressure between the rotor disc and the spiral grooved thrust gas bearing is solved by finite difference method (FDM) in this paper. The distribution of the gas film pressure—$P_{(i,j)}$ is obtained by using the over relaxation iterative method (SOR), as shown in Figure 4. The flow chart of solving Reynolds equation of gas lubrication of the spiral grooved thrust gas bearing was shown in Figure 5.

Table. 1 Physical properties of air and helium

Name	**Density**(Kg/m^3)	**Viscosity**($Pa \cdot s$)
Air	1.225	$1.79 * 10^{-5}$
He	0.163	$1.99 * 10^{-5}$

The spiral grooved calculation domain is expanded into a fan-shaped calculation domain, which can speed up the convergence speed and improve the accuracy [9].

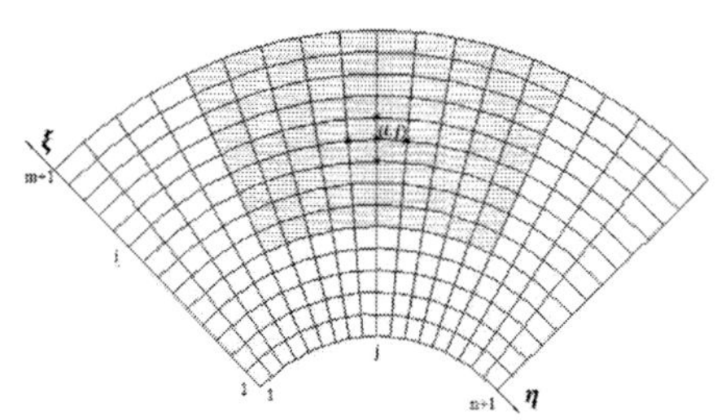

Figure 4. The solution domain of fan-shaped

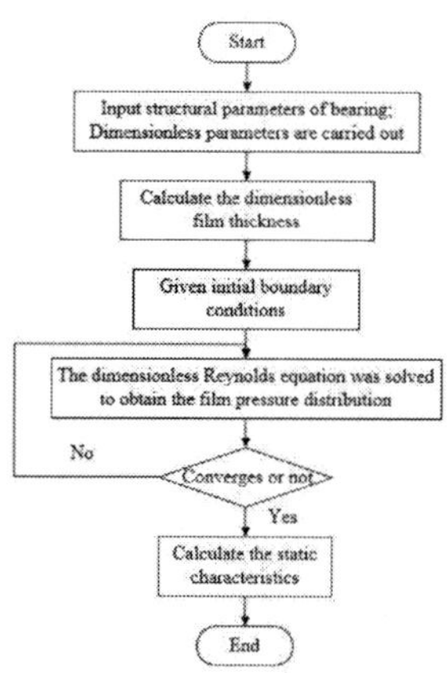

Figure 5. Flow chart of film pressure solution

Table. 2 Basic structural parameters of spiral grooved thrust gas bearing

Structural parameters of bearing	**Symbol / unit**	**Value**
Base circle radius of bearing	r_0/(mm)	8
Radius of bearing dam area	r_1/(mm)	12.5
Outer radius of bearing	r_2/(mm)	22
Spiral angle	α/(°)	14~24
Groove length ratio	$\lambda = (r_2 - r_1)/(r_2 - r_0)$	0.45~0.75
Groove width ratio	$\gamma = \theta_g/(\theta_g + \theta_l)$	0.4~0.8
Minimum film thickness (Dam area)	h_0/(μm)	10~30
Depth of spiral groove	h/(μm)	25~45

IOP Conf. Series: Materials Science and Engineering 1081 (2021) 012051 doi:10.1088/1757-899X/1081/1/012051

Groove depth ratio	$\delta = (h + h_0)/h_0$	3.5~5.5
Number of spiral grooves	N	18
Environmental pressure	$P_a/(\mathrm{MPa})$	0.1

4. The Calculation Results and Theoretical Analysis

4.1 Verification of calculation method

The finite difference method (FDM) is used to calculate the static characteristics of spiral grooved thrust gas bearing. The calculation results are compared with the results of finite element method (FEM) using FLUENT software in this paper.

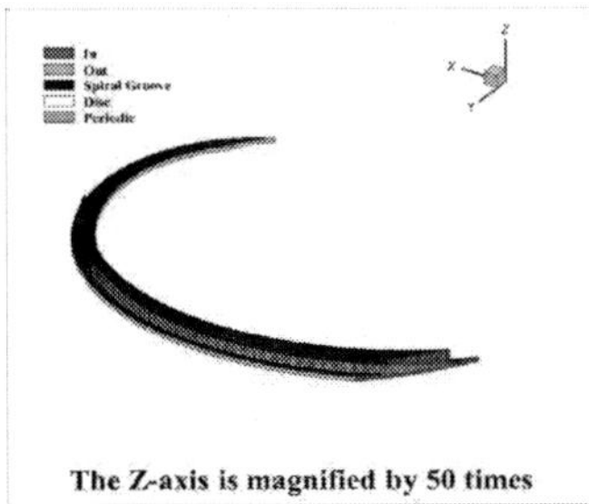

Figure 6. Setting of boundary conditions.

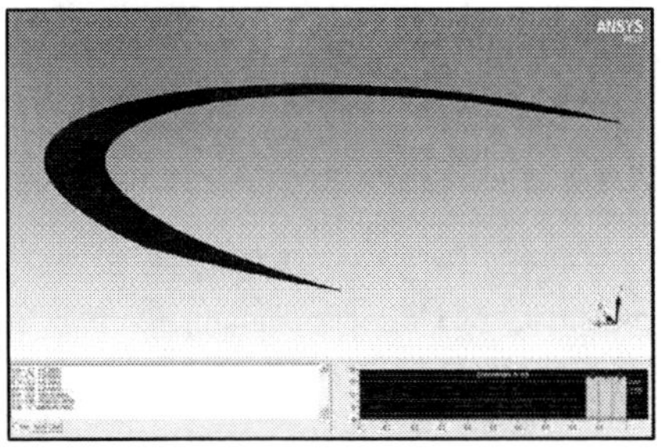

Figure 7. Grid quality.

A period is selected from the fluid domain of spiral groove gas bearing as shown in the Figure 6. According to the actual situation, the corresponding boundary conditions are set. The structured hexahedral mesh is divided by ICEM software, and the grid quality is shown in Figure 7.

The grid is imported into FLUENT fluid calculation software. The inlet and outlet pressures are ambient pressures, and the platform area is set as periodic boundary conditions. When the spiral groove is on the bearing, only the thrust disc rotates at high speed, and the speed of the thrust disc is 60000 rpm.

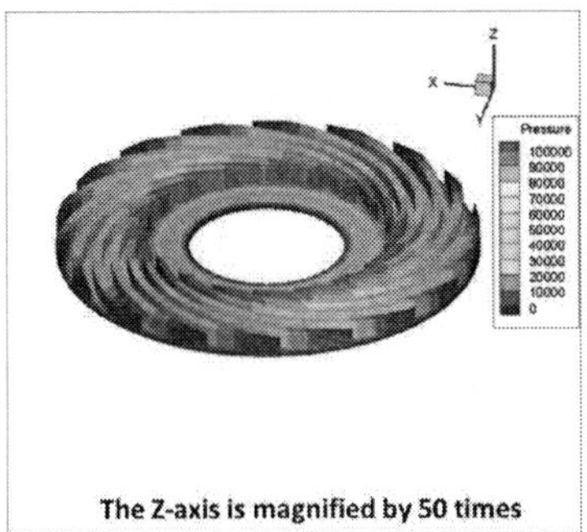

Figure 8. The cloud chart of pressure distribution

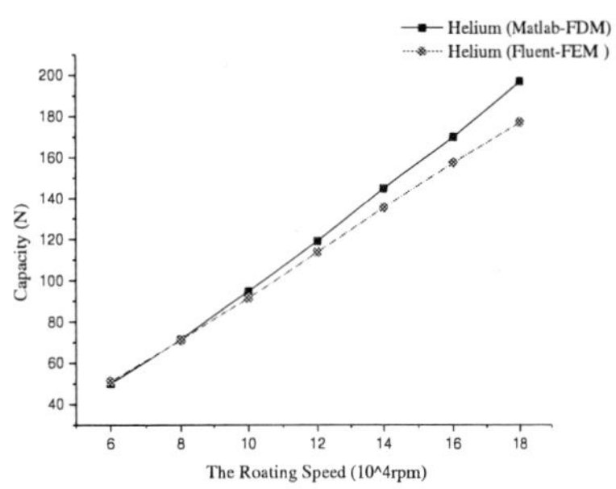

Figure 9. Comparison of calculation results with different methods

After iterative calculation, each parameter reaches the convergence state. The gas film pressure distribution of spiral groove thrust gas bearing is shown in Figure 8. A negative pressure area is formed at the entrance of the spiral groove. Under the action of viscosity, the lubricating gas brought in by the thrust plate flows along the spiral groove, and the lubricating gas is compressed in the dam area to form a high pressure area [10][11].

As shown in Figure 9, the maximum calculation error is less than 10%. The results of the two methods are in good agreement, which verifies the correctness of the MATLAB program. The Figure 10 and Figure 11 show the film pressure distribution of the helium lubricated spiral grooved thrust gas bearing. The Figure 10 is a two-dimensional view from the top, and the Figure 11 is a three-dimensional axonometric drawing.

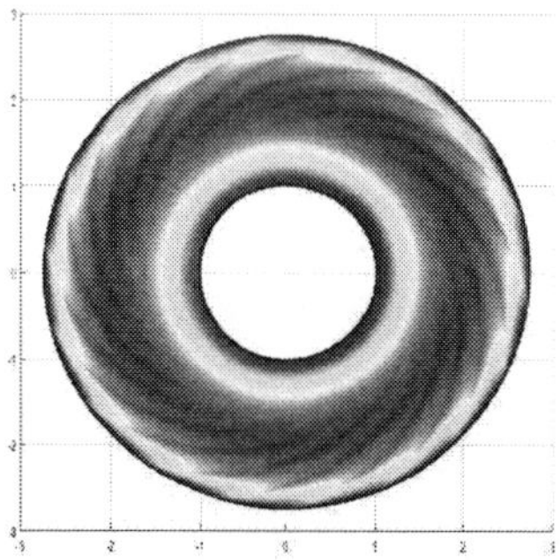

Figure 10. Two dimensional distribution of film pressure

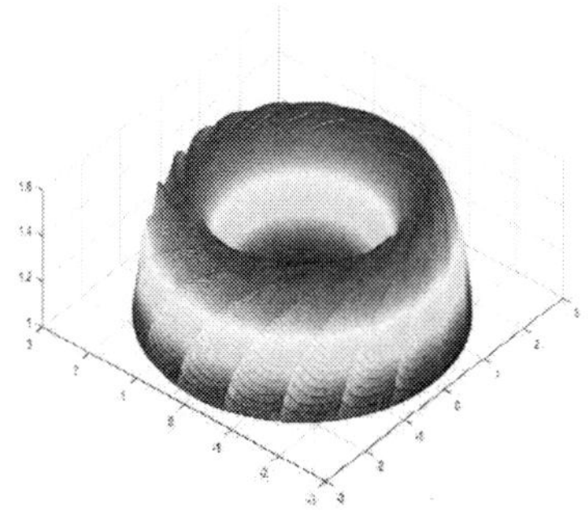

Figure 11. Three dimensional graph of film pressure

The Bernoulli Equation:

$$P_1 + \frac{1}{2}\rho v_1^2 + \rho g h_1 = P_2 + \frac{1}{2}\rho v_2^2 + \rho g h_2 \tag{9}$$

Combined with the physical structure of spiral groove thrust gas bearing, the distribution mechanism of gas film pressure is analyzed: At the entrance of the spiral groove, the pressure can be converted into velocity energy, and the inlet pressure becomes low, which forms a pumping effect on the gas outside the groove from the Bernoulli Equation; most of the viscous gas flows along the spiral groove into the groove, and is compressed in the dam area to form a high-pressure gas film area, producing bearing loading capacity; the bearing gas flows to the environment passing through the dam area finally.

4.2 Influence of different parameters on static characteristics

The spiral groove on the end face of the bearing, it's the thrust disc of the rotor which brings viscous gas into the spiral groove under the action of gas viscosity as shown in Figure 12.

As shown in Figure 13, different spiral angle has different influence on axial loading capacity of the spiral grooved thrust gas bearing. When the spiral grooved thrust bearing is lubricated by air and the optimum spiral angle is 18 degrees, it produces the best bearing loading capacity [14]. However, the optimum spiral angle which can produce maximum loading capacity is 16 degrees under the condition of helium as lubricant.

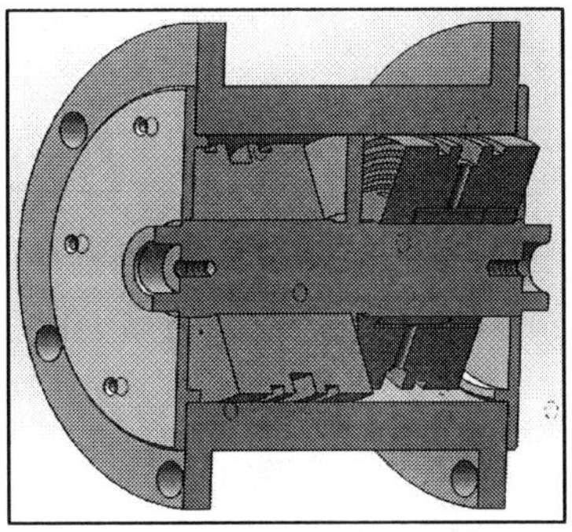

Figure 12. The spiral groove on the bearing

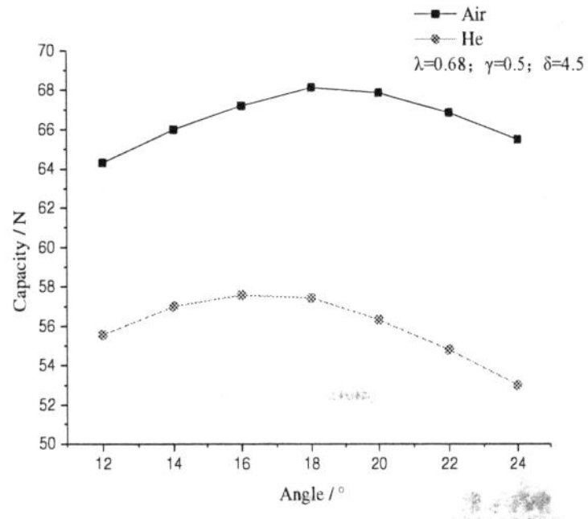

Figure 13. Influence of spiral angle on loading capacity

Due to the spiral angle is small, the spiral groove is close to the annular groove just as the left picture in Figure 14, the circumferential step effect is very weak, and the bearing loading capacity is low [8]; when the spiral angle is larger, the structure of the bearing gradually tends to the radial stepped bearing [9], and the pumping effect of the gas bearing is gradually weakened, which makes the bearing capacity gradually reduce. Therefore, there must be an optimal spiral angle, which makes the spiral grooved thrust bearing have the best pumping effect and circumferential step effect at the same time to produce the best bearing loading capacity.

When the spiral angle is larger, the structure of spiral groove thrust gas bearing is closer to radial stepped bearing as shown in Figure 14. Helium with low density and molecular weight is easier to cross through the platform area and enter the next groove area, resulting in the reduction of the air volume of the bearing that can be compressed. Therefore, compared with the air lubricated spiral groove thrust gas bearing, the spiral angle is need to be designed smaller when helium is used as lubricating gas.

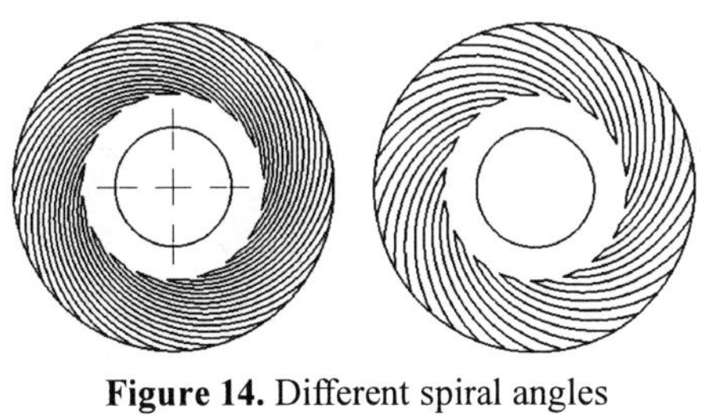

Figure 14. Different spiral angles

In the process of studying the influence of groove length ratio on bearing loading capacity of spiral grooved thrust gas bearing, the groove length ratio has the same influence trend on bearing loading capacity whether the lubricant is air or helium.

The groove length ratio represents the ratio of the length of the spiral groove to the radius of the whole thrust disc, and also reflects the sealing effect of the dam area on bearing gas [15]. When groove length is very small, the combined effect of pumping effect and circumferential step effect is weak, resulting in small loading capacity; when groove length ratio is large, sealing effect of dam area is small with enormous leakage, so the bearing loading capacity is also small. As shown in Figure 15, the optimal groove length ratio is about 0.68-0.7.

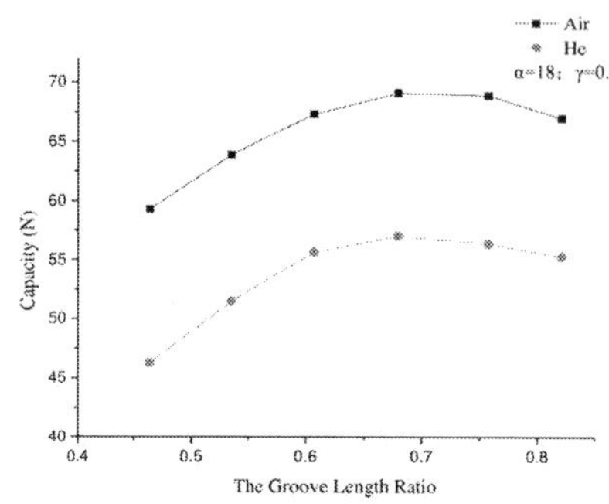

Figure 15. Influence of groove length ratio on loading capacity

The groove width ratio represents the ratio of the width of the spiral groove to the whole thrust disc surface. As shown in the figure 16, the influence of the groove width ratio on the bearing loading capacity showed a trend of increasing and then decreasing. When the groove width ratio is 0.5-0.6, the spiral grooved thrust bearing produce the best bearing loading capacity to the thrust plate.

When the groove width ratio is small, it means that the groove area is relatively narrow, and the gas flow rate pumped into the groove is very small, and there is not enough bearing gas to produce large bearing capacity; when the groove width ratio is large, it means that the platform area is relatively narrow, and the resistance of bearing gas flowing from the spiral groove area to the next spiral groove area becomes smaller, which leads to the large leakage of bearing gas as shown in Figure 17, The results show that the bearing gas which can be compressed in the dam area will be reduced and the bearing loading capacity will be smaller.

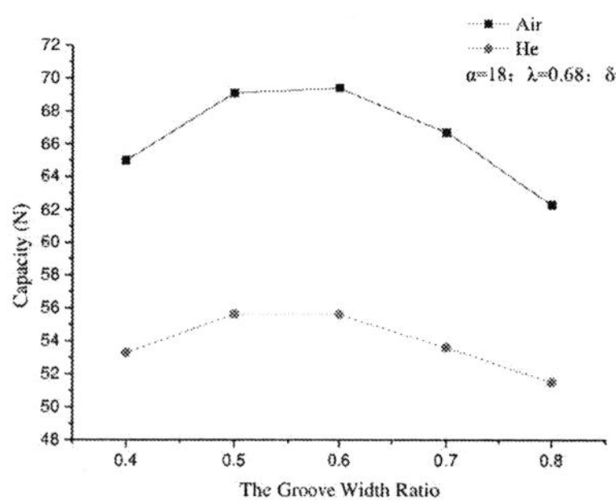

Figure 16. Influence of groove width ratio on loading capacity

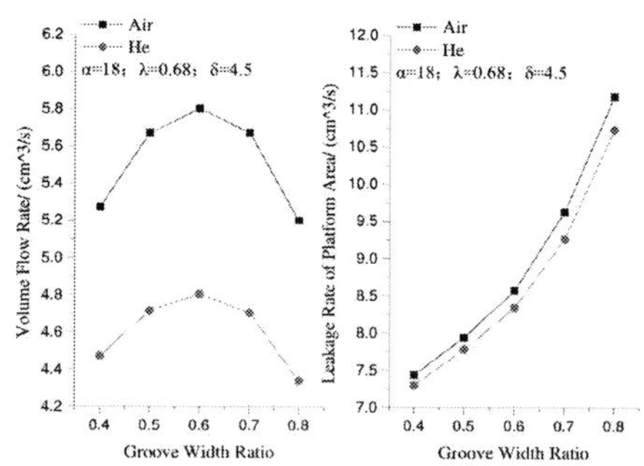

Figure 17. Comparison of volume flow rate and leakage rate

As shown in Figure 18, when air is the lubricant, the optimum groove depth is 35μm, and the gas film thickness ratio is $\delta = 4.5$, it is consistent with the classical 4.0 design value; when helium is used as lubricant, the best film depth ratio and bearing loading capacity are less than those of air.

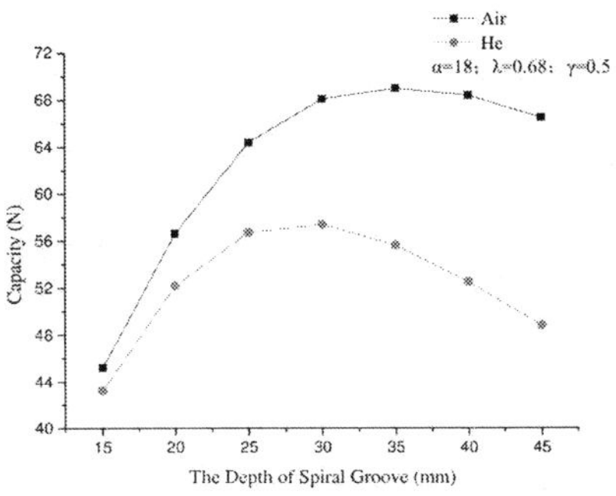

Figure 18. Influence of depth of spiral groove on loading capacity

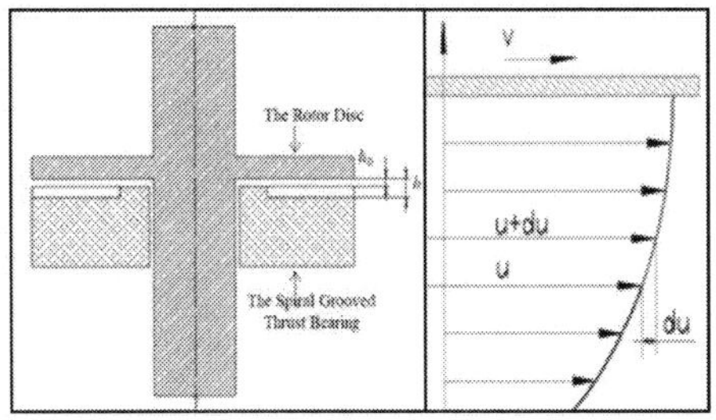

Figure 19. Variation of shear force with depth

For the same lubricant, the thrust disc of the rotor rotates at a high speed, and under the action of viscosity, the gas is brought into the groove and compressed in the dam area; with the increase of groove depth, the shear force gradually decreases, the gas velocity decreases, and the average pressure at the inlet increases. The pressure difference with the atmospheric environment (the pumping pressure) as shown in the figure 20, decreases continuously. Although the inlet area increases, the total volume flow rate increases first and then decreases.

For two different lubricant, air and helium, the viscosity of helium is slightly higher than that of the air, and the carrying capacity of thrust disc to the groove is basically the same. However, the air density is 7.5 times that of helium. According to Bernoulli equation, the greater the gas density is, the smaller the pressure at this point. Therefore, the pumping capacity of air is greater than that of helium. Under the same spiral groove structure, the volume flow rate of air is larger than that of helium, and the bearing loading capacity is also greater than that of helium.

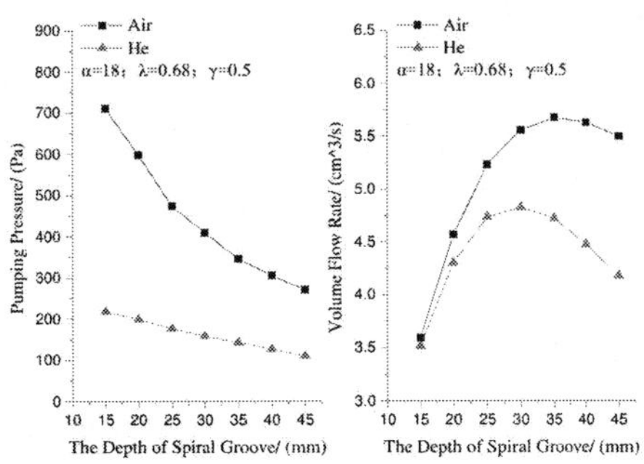

Figure 20. Comparison of pumping effect and mass flow rate

It can be seen from Figure 21 that the gas film thickness in the dam area is a very important factor affecting the bearing loading capacity. With the increase of the gas film thickness in the dam area, the bearing loading capacity characteristics decrease sharply. It cause that the leakage of bearing gas is greatly increased, and the step effect is weakened.

With the decrease of gas film thickness in dam area, the slope of bearing loading capacity curve increases gradually, and the static stiffness of bearing gradually increases as shown in Figure 22.

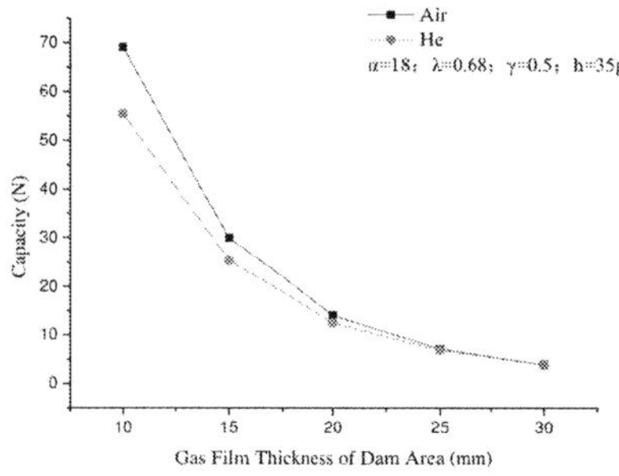

Figure 21. Influence of gas film thickness of dam area on loading capacity

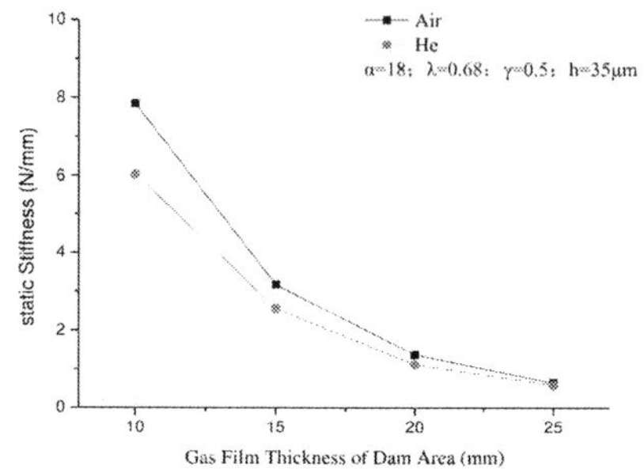

Figure 22. Influence of gas film thickness of dam area on static stiffness

4.3 The static characteristics of the spiral groove on the thrust disc and on the bearing

In this paper, we found that the spiral groove on the end face of the bearing, it's the thrust disc of the rotor which brings viscous gas into the spiral groove, and the lubricating gas is compressed to form a gas film pressure; but there is a big difference when the spiral groove was designed on the thrust disc as shown in figure 23.

The spiral groove is on the thrust disc of the rotor. The spiral groove and the rotor thrust disc have the same speed. The lubricating gas entering the spiral groove is no longer carried by the viscous effect, but has a relative speed with the spiral groove. In this case, the volume flow rate of the lubricating gas is higher than that of the spiral groove on the bearing.

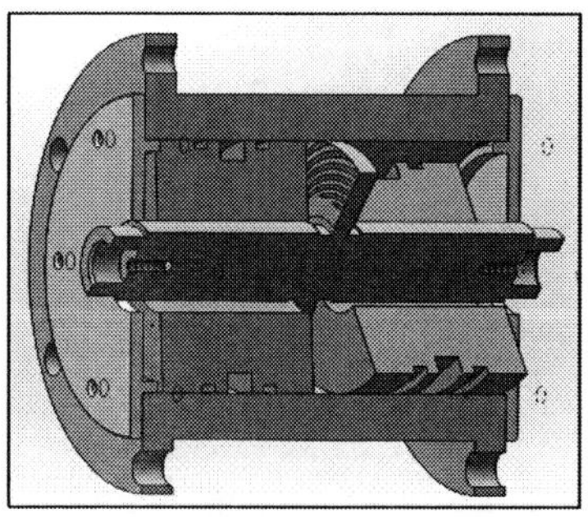

Figure 23. The spiral groove on the disc of rotor.

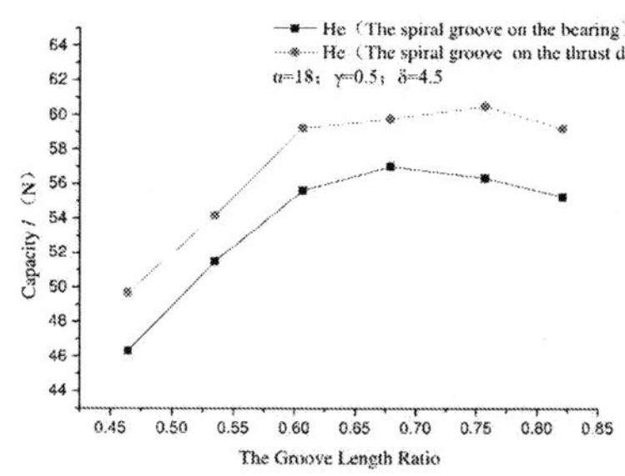

Figure 24. Different groove length ratio.

As shown in figure 24-26, we found that the bearing loading capacity has been improved when the spiral groove was designed on the thrust disc. Under the same structural parameter condition, this kind of design increase the bearing loading capacity by 6 percent on average, when the spiral groove thrust gas bearing lubricated by helium.

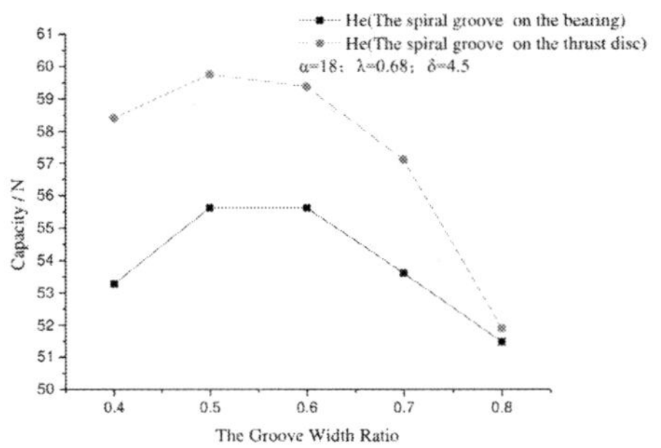

Figure 25. Different groove width ratio.

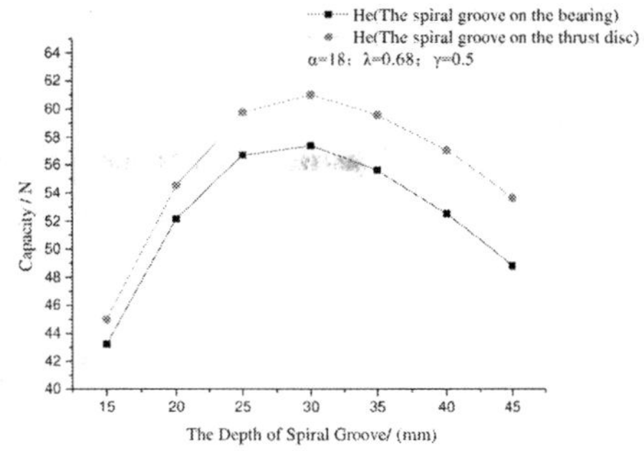

Figure 26. Different groove depth.

To sum up, we calculated the influence of different structural parameters and design styles on the static characteristics of spiral groove thrust gas bearing. According to the design requirements of the bearing loading capacity of helium turbo expander, we can select different and appropriate structural parameters on both sides of the thrust disc of the rotor to meet the design conditions.

5. Conclusion

1. The spiral grooved thrust gas bearing lubricated by helium has the best bearing loading capacity with 16 degree spiral angle. Due to the smaller density and molecular weight of helium, the smaller spiral angle makes it easy for helium to drain into the dam area for compression and to reduce the leakage in the platform area.

2. The optimum ratio of groove width is the result of the joint action of inlet area and leakage of the platform area. The optimal ratio of groove length is only the result of leakage in dam area.

3. For the same kind of lubricating gas, the groove depth increases, the inlet area increases. But the shear force decreases, which makes the average inlet pressure decrease. Therefore, the optimal groove depth ratio is the result of the joint action of the inlet area and the average inlet pressure.

4. The viscosity only affects the gas carrying capacity of the rotor thrust disc, while the density affects the velocity and pressure of the lubricating gas at the inlet, resulting in different pump gas pressure and volume flow rate. Therefore, compared with the viscosity, the density of helium is the more important reason for the smaller bearing loading capacity.

5. Spiral groove on the bearing, under the action of gas viscosity, the thrust disc will bring lubricating gas into the spiral groove, which is a kind of passive compression; when the spiral groove is on the rotor thrust disc, this is a kind of active compression, and the bearing loading capacity increases by 6% on average with the lubricating gas flow rate increases.

Reference

[1] Wang J G, Chen F H, Chen L G and Zhang Y Q 2017 *J. Lubrication and Sealing* **42** 1-7

[2] Zhang C W, Wang Q J and Xie Z J 2018 *J. Tribology* **38** 213-219

[3] Liu R and Wang X L 2010 *J. Transactions of Beijing Institute of Technology* **30** 1400—04

[4] Shan X C, Zhang Q D and Sun Y F *J. Microsystem Technologies* **13** 1501—08

[5] Liu R and Wang X L 2010 *J. Journal of Mechanical Engineering* **46** 113—117

[6] James D D and Potter A F 1967 *J. Journal of Lubrication Technology* **89** 439—443

[7] Hou Y and Chen C Z and Wang B C 2002 *J. Lubrication Engineering* **27** 2—4

[8] Hou Y and Xiong L Y and Wang J 2002 *J. Cryogenic Technology* 6−9
[9] Li J, Yang G W and Yang X W 2020 *J. Lubrication and Sealing* ***45*** 17-24
[10] Jia C H, Fu Y S, Qiu M, Ma W S, Gao J and Zhang H J 2019 *J. Lubrication and Sealing* **44** 56-63
[11] Jia C H, Yang W and Qiu M 2014 *J. Journal of System Simulation* **26** 1763-68
[12] Bai Z, Sun S F, Sun L J, Sun Y, Ren X K and Wang L 2014 *J. Lubrication Engineering* **39** 55−58
[13] Lai T W, You H L, Chen S T, Liu J H and Hou Y 2014 *J. Lubrication Engineering* **39** 1−6
[14] Jia C H, Gao J and Qiu M 2018 *J. Bearing* **000** 22-28
[15] Jia S and Li P K 2019 *J. Journal of Mechanical Science and Technology* **33**

A Study of Stall Condition of a Centrifugal Compressor based on CFD and Experimental Methods

R. Wang[a,1*], H. L. Wang[a,1], S. H. Yang[a,b*], J. J. Tan[a], S. Jiang[a], J.Q. Ren[a]

[a]. Shenyang Blower Works Group Corporation, Shenyang, China
[b]. School of Energy and Power Engineering, Dalian University of Technology, Dalian, China
[1] These authors contributed equally to this work.

[*] wangruiauthor@aliyun.com (R. Wang), yangshuhua@shengu.com.cn (S. H. Yang)

Abstract. The stall point of a centrifugal compressor is obtained by CFD method and FFT method, and an experiment in the stall point is carried out. Firstly, the curves of the compressor efficiency and total pressure ratio with flow rate are obtained by the steady CFD analysis. Secondly, according to the curves, the unsteady CFD analysis and frequency signal analysis are carried out at the operation points where stall may occur. Then the stall point of the compressor is obtained. On the basis of the results of numerical analysis, the model machine is used to carry out the test. By comparing the results of the numerical method and the experimental method, the stall is identified at the similar operation point, which proves the effectiveness of the prediction method in this paper.

1. Introduction

In a compressor, stall can be regarded as the cessation of a continued rise in static pressure recovery in a diffuser or cascade. The blade passages may stall partially or fully. If mass flow continues to reduce, it can turn to the surge and causes audible flow fluctuations[1]. Cheshire[2] found surge in centrifugal compressors. Emmons[3] confirmed that rotating stall was a precursor to surge, and made clear the difference between the two phenomena through experiments. Moore and Greitzer[4-6] proposed a speed disturbance model to predict the rotating stall, and used active control to suppress the disturbance before stall. Fringe[7] found two stalls at different frequencies with experiments, which correspond to impeller stalls and diffuser stalls respectively. Based on the dynamic model proposed in [4-6], Liaw[8] performed the bifurcation analysis and proposed an active control technique in axial flow compression system. Tryfonidis[9] used nine high-speed compressors for research and found that there were traveling waves that developed in the circumferential direction before stall initiation. In literatures [10-11], researchers performed study about the spike type rotating stall and explain the physical mechanism. The rapid development of computer technology has made CFD become one way for studying stall phenomena. Spentzos[12] used 2D and 3D CFD to investigate the dynamic stall of square wings of NACA 0012 and NACA 0015. Sun[13] presented a stall inception model for transonic fan/compressors under some assumptions. By comparing numerical and experimental results, the model can give a reasonable prediction to the stall inception for both subsonic and transonic fan/compressor. In literatures[14-16], CFD and experiments were also applied to study stall at different places of flow channel of centrifugal compressors, respectively. The tests performed by

IOP Conf. Series: Materials Science and Engineering 1081 (2021) 012052 doi:10.1088/1757-899X/1081/1/012052

Day[17] on Cambridge 4-stage axial compressor showed that for the same type of compressor, the rotation speed of the stall cell has a constant proportional relationship with the rotation speed of compressor during stall. Zhao[18] performed the tests for the investigation of blade vibration induced by stall and observed the signal modulation phenomenon. Generally, the eccentricity of the rotor is not taken into consideration in CFD simulation process. Therefore, the pressure pulsation obtained by CFD cannot identify the frequency corresponding to the rotation speed (f_r) through FFT. However, due to the effects of rotation and fluid-structure interaction, the characteristic frequencies corresponding to the stall can still be obtained in the low frequency region and near the blade passing frequency. These frequencies are the frequency of the stall cell (f_s), the rotation frequency of the stall cell (f_{BPF}-f_s or f_{BPF}+f_s) and the blade passing frequency (f_{BPF}). The steady and unsteady CFD simulations of a centrifugal compressor are performed to analysis the operation point where stall may occur in this paper. On the basis of the numerical results, at similar operation point, a series of comparative experiments were carried out using a centrifugal compressor. The results show that both the numerical method and the experimental method obtained the stall phenomena at the similar operation point.

2. CFD Simulation

The research object is a centrifugal compressor with vaned diffuser in this paper. In numerical modeling, the inlet and outlet of the channel were extended appropriately. The calculation domain includes an impeller, a vaned diffuser and a vaned return channel, shown as Fig.2.1.The diameter of impeller is 500mm. The blade numbers of the impeller, diffuser and return channel are 19, 13 and 24, respectively.

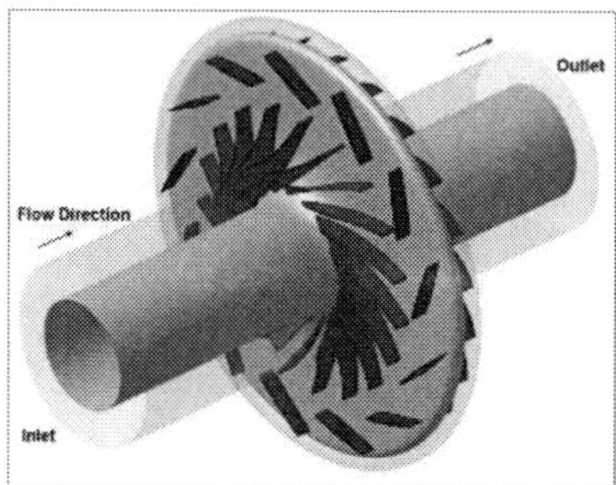

Fig. 2.1. The 3D model of the centrifugal compressor

The IGG module which belongs to the NUMECA software was applied to generate full-channel structured grid. The partial enlarged images of the grids are shown as Fig. 2.2, and the total number of grids is 1.508×10^7. The thickness of the first layer of grid on the wall is 0.01mm to ensure that most of Yplus values are less than 10. The density of the grids on both sides of all moving-static interfaces are properly increased. This can prevent the grid size difference on both sides of the interfaces from becoming too large and causing unnecessary interpolation errors.

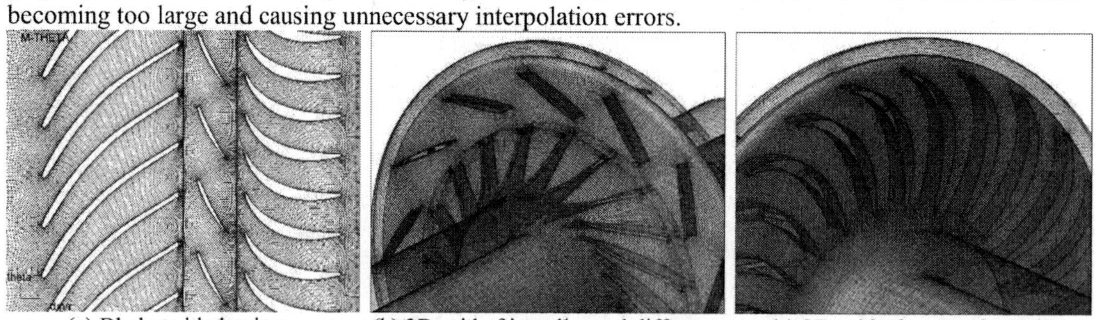

(a) Blade to blade view (b) 3D grid of impeller and diffuser (c) 3D grid of return channel

Fig. 2.2. The mesh of the calculation domain

The FINE/Turbo suite of Numeca software are used to perform the steady and unsteady simulation. The full-channel model are adopted during the CFD simulation. The Spallart-allmaras (SA) turbulence model is employed. The conditions of total pressure and total temperature are set at the inlet, and axial air intake is set to uniform. The condition of mass flow is set at the outlet. The working fluid is ideal air. The boundary conditions at design point are shown as Table 2.1. The total pressure ratio and the polytropic efficiency in this paper are shown as dimensionless values relative to the design points. The area definition method is applied to set the rotating wall in the impeller. The adiabatic and non-slip conditions are applied for the wall. In the steady calculation, the frozen rotor method is selected between the moving and stationary domains for upstream and downstream data transmission. While for the unsteady case, transient rotor-stator interface option is adopted.

Table 2.1. The boundary conditions at design point

Fluid	Total Pressure (Pa)	Total Temperature (K)	Rotation Speed (r/min)	Mass Flow (kg/s)
Ideal Air	101000	293	5911.2	2.55

2.1. Steady Simulation

First, the steady CFD method is used to obtain the centrifugal compressor performance curve, including total pressure ratio and polytropic efficiency versus inlet mass flow. The maximum mass flow is 3.57kg/s. Then reduce the mass flow for calculation, constantly. The minimum mass flow is 1.79kg/s. The dimensionless results of total pressure ratio and polytropic efficiency are shown as Fig. 2.3. The horizontal axis is the mass flow in Fig. 2.3. The two curves in Fig. 2.3 are approximately parabolic with an opening downward. The maximum value of the total pressure ratio appears when the mass flow is 1.98kg/s. While the maximum value of the polytropic efficiency is at the mass flow of 2.3kg/s. As the mass flow decreases, the operating point at which total pressure ratio begins to decrease is selected as the suspected stall point. The mass flow at the suspected stall point is 1.98kg/s. Then the unsteady CFD simulations are performed in the section 2.2 with five mass flow values are 1.79, 1.85, 1.91, 1.98 and 2.04kg/s, respectively.

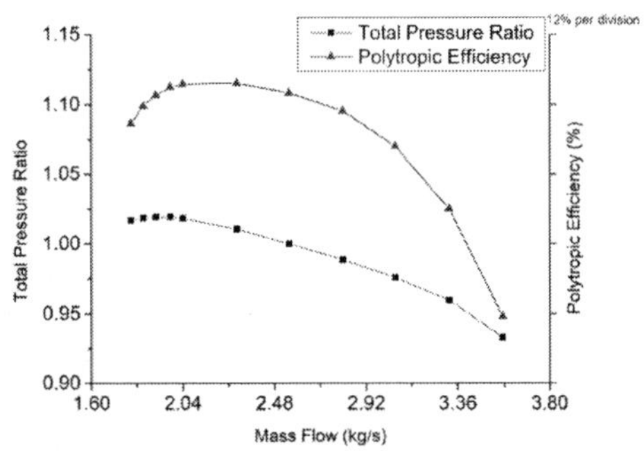

Fig. 2.3. The dimensionless results of total pressure ratio and polytropic efficiency of the steady CFD

2.2. Unsteady Simulation

According to the results of the steady method, five operating points are chosen to implement the unsteady simulations. During the simulation processes, 19 time steps are set for each moving blade channel. Meanwhile, set 50 iterations in each time step for better convergence. The streamline diagrams in the Meridian flow channel with five mass flows are shown in Fig. 2.4. It can be seen that when the mass flow is reduced to 1.98kg/s, two blue separation zones will appear at the impeller inlet and in the diffuser. When the mass flow is further reduced, cells will generate in the above two separation areas, and the volumes of the cells will gradually increase. On the basis of the results of the section 2.3, the rotation stall can be observed firstly when mass flow is 1.91kg/s. The results of the unsteady CFD simulation at mass flow is 1.91kg/s will be taken as an example. The partial enlarged views of the streamline when mass flow is 1.91kg/s are shown as Fig. 2.5.

IOP Conf. Series: Materials Science and Engineering 1081 (2021) 012052 doi:10.1088/1757-899X/1081/1/012052

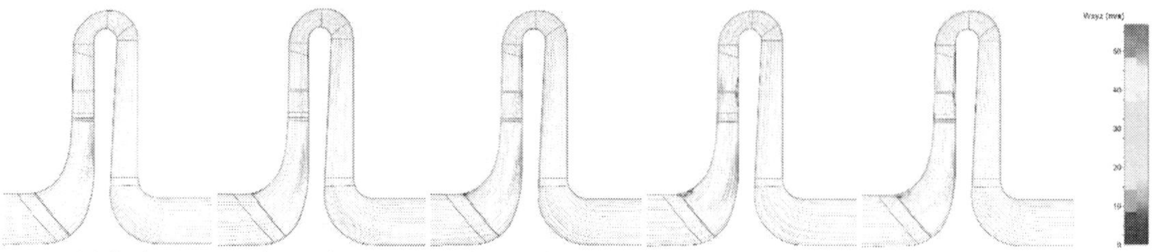

(a) 2.04kg/s (b) 1.98kg/s (c) 1.91kg/s (d) 1.85kg/s (e) 1.79kg/s

Fig. 2.4. The streamline diagrams in the Meridian flow channel

(a) Separation zone at the impeller inlet (a) Separation zone at the diffuser

Fig. 2.5. The partial enlarged views of the streamline when mass flow is 1.91kg/s

According to the positions of the two cells, the velocity and streamline diagrams are taken at the 95% and 5% blade height sections with mass flow is 1.91kg/s, respectively, as show in Fig. 2.6. The meridian velocity and the spatial streamline are shown as Figs 2.6(a, b), and the relative velocity vectors are shown as Figs 2.6(c, d). It can be seen from Figs. 2.6(b) that there are clearly cells near the shroud of the impeller. At this mass flow, the separation of the flow field has occurred, and one location of the stall cells are generated. The location is near the suction surface of each blade channel which shown as Figs. 2.6(a, b). At the location which is at the inlet and the outlet of each diffuser blade which shown as Fig. 2.6(d) in red circle, the attack angle of the flow is not good for working and low speed locations are obviously.

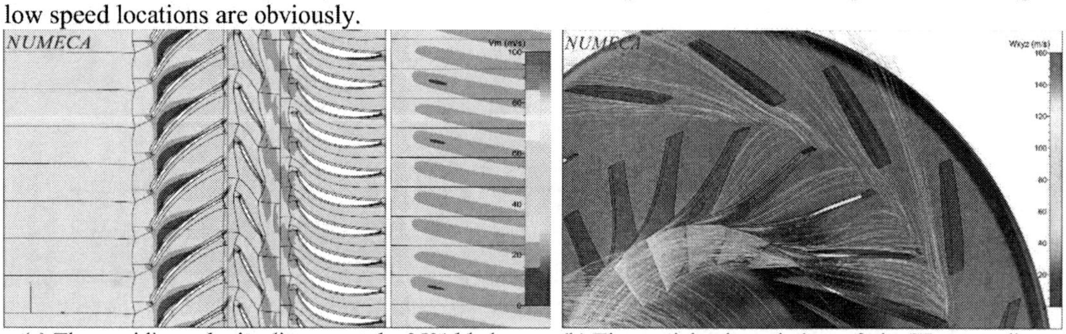

(a) The meridian velocity diagram at the 95% blade height (b) The partial enlarged view of the 3D streamline at the impeller and diffuser

IOP Conf. Series: Materials Science and Engineering 1081 (2021) 012052 doi:10.1088/1757-899X/1081/1/012052

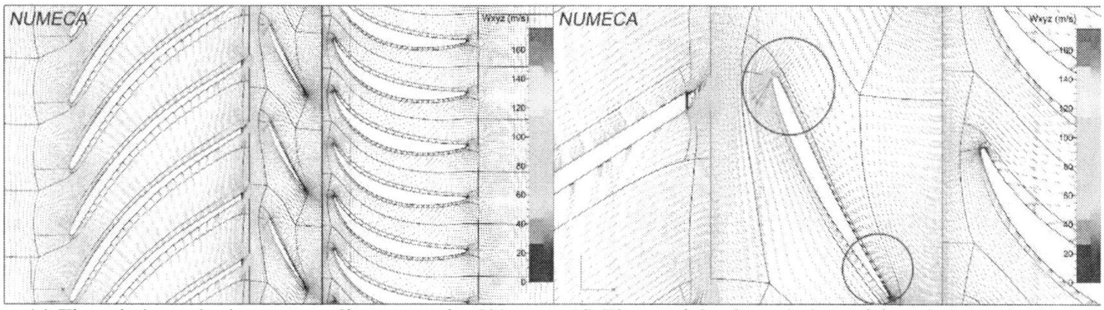

(c) The relative velocity vectors diagram at the 5% blade height (d) The partial enlarged view of the relative velocity vectors at the 5% blade height of the diffuser

Fig. 2.6. The velocity and streamline diagrams when mass flow is 1.91kg/s

In general case, when a rotating stall occurs, the stall cells propagate between different impeller channels. The entropy diagrams at the 95% blade height are shown as Fig. 2.7 when the mass flow is 1.91kg/s. There are four times in the Figs 2.7, respectively. The time range is the time for the impeller to sweep through one blade passage. The four times are evenly distributed. In the section 2.3, the unsteady CFD pressure results at five operation conditions are processed by the FFT method, and the compressor rotating stall phenomenon is studied in the frequency domain.

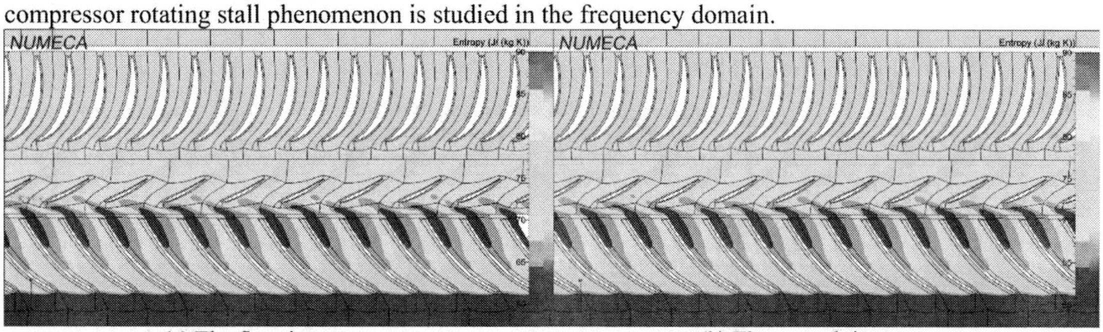

(a) The first time step (b) The second time step

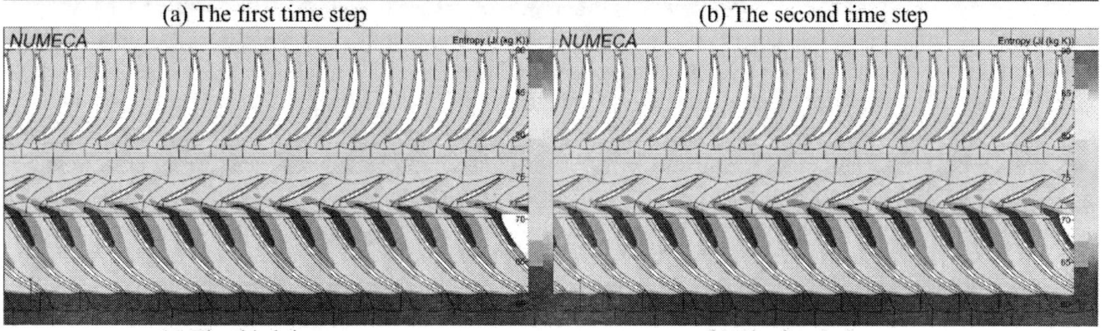

(c) The third time step (b) The fourth time step

Fig. 2.7. The entropy diagrams when mass flow is 1.91kg/s

2.3. The Analysis of the Unsteady CFD Results in Frequent Domain

Seven monitoring sections (S1 to S7) were set along the flow direction in the calculation domain, respectively. The locations of these sections are shown as Table 2.2 and Fig.2.8.

Table 2.2. The locations of monitoring sections

Section	Location
S1	Inlet (in rotating domain)
S2	Diffuser inlet (Impeller outlet)
S3	Diffuser blade

IOP Conf. Series: Materials Science and Engineering 1081 (2021) 012052 doi:10.1088/1757-899X/1081/1/012052

S4	Diffuser outlet
S5	Return channel inlet
S6	Return channel Blade
S7	Return channel outlet

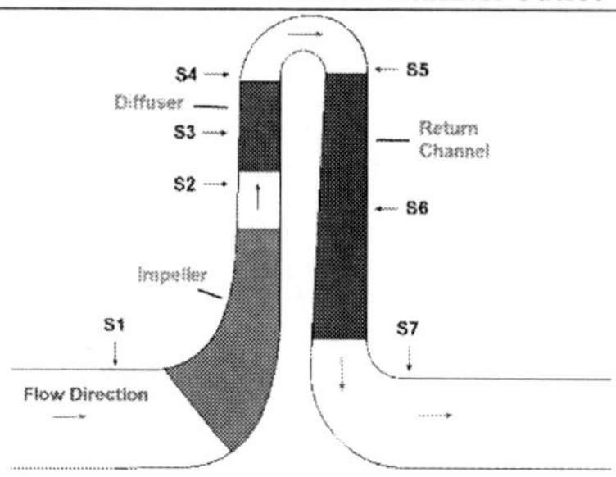

Fig. 2.8. The schematic diagram of monitoring sections

In the calculation process, the convergence or not is determined by monitoring the efficiency, total pressure ratio, inlet and outlet flow, and pressure pulsation of the each monitoring point. The pressure pulsation with mass flow is 1.98kg/s of the S2 section and S7 section in time domain are shown in Fig. 2.9. The pressure pulsation is periodic.

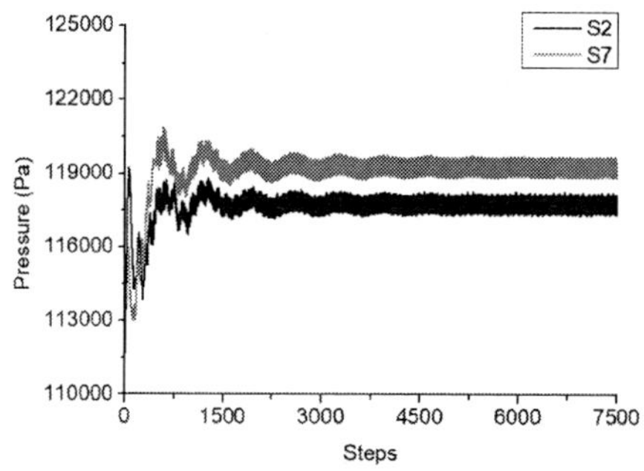

Fig. 2.9. The diagram of pressure pulsation

The FFT method is applied to process the pressure pulsation results of the monitoring points. The results of the last 2048 time steps are chosen as samples for the FFT method in each operation condition. The results of the monitoring points with the five mass flows in frequency domain are shown in Figs. 2.10-2.14, respectively. The abscissa is frequency and the ordinate is pressure in these figures. In general case, the eccentricity of rotor is not taken into account during the CFD simulation. Thus, the frequency of rotation speed and the modulation frequency near it cannot be observed in the results of CFD simulation. The following conclusions can be obtained from Figs. 2.10-2.14:

1) The blade passing frequency (f_{BPF}=1870.3Hz) and its second order frequency (2×f_{BPF}=3470.7Hz) can be observed in all five operation conditions at the Section 2 to Section 7;

2) There are f_{BPF} and modulation frequencies (f_{BPF}-f_s=1818.4Hz or f_{BPF}+f_s=1922.3Hz) in the Figs. 2.12-2.14 when the mass flow is less than or equal to 1.91kg/s;

3) There are same peaks f_s=52Hz.

According to the results of the CFD method and the literatures, the rotational stall can be identified when mass flow is 1.91kg/s.

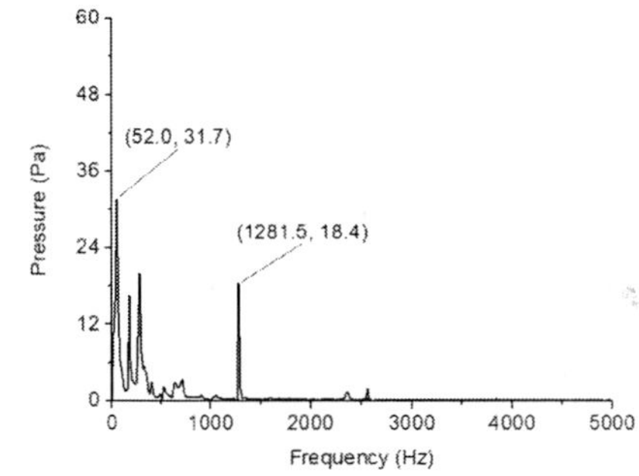

(a) Section 1

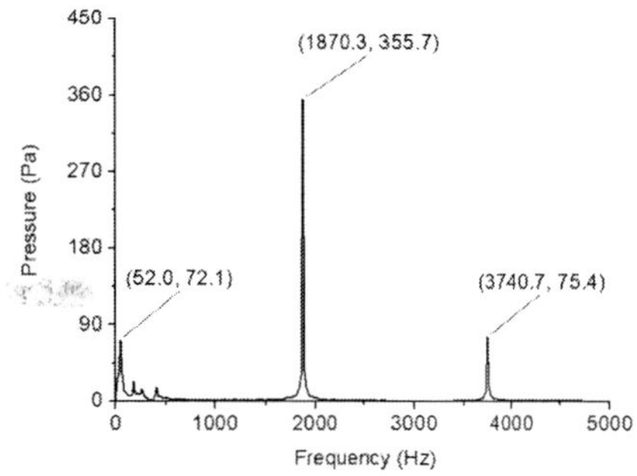

(b) Section 2

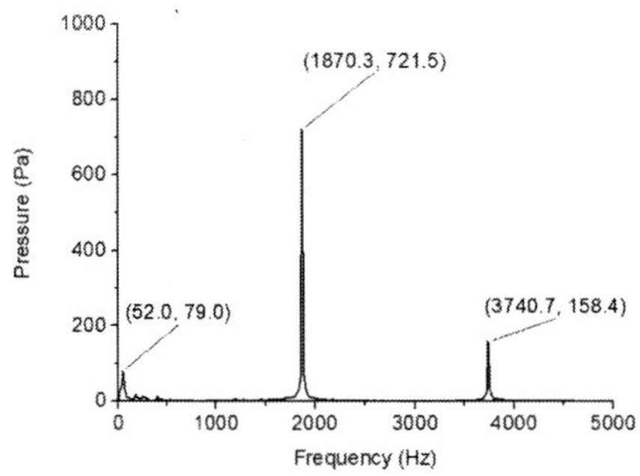

(c) Section 3

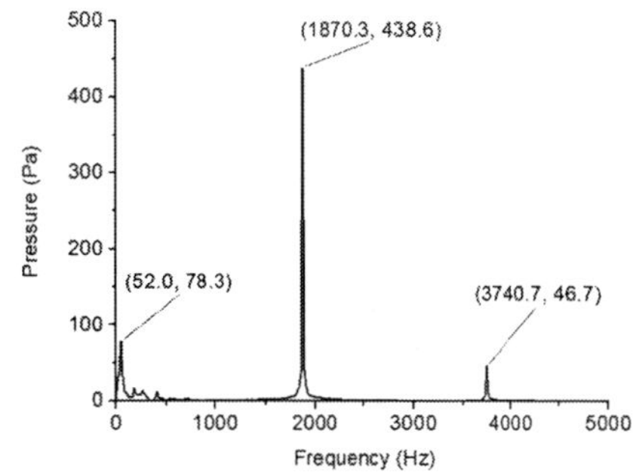

(d) Section 4

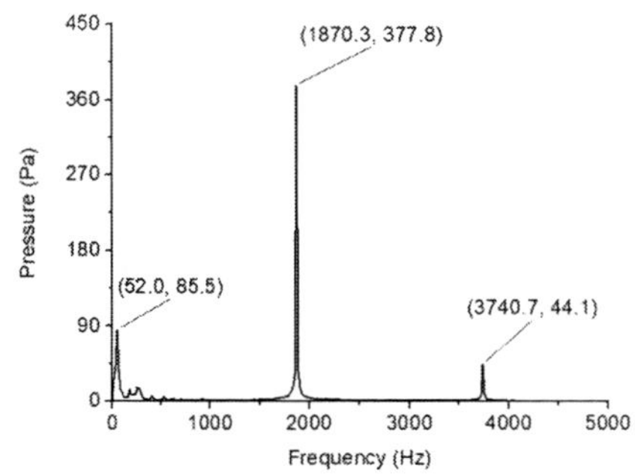

(e) Section 5

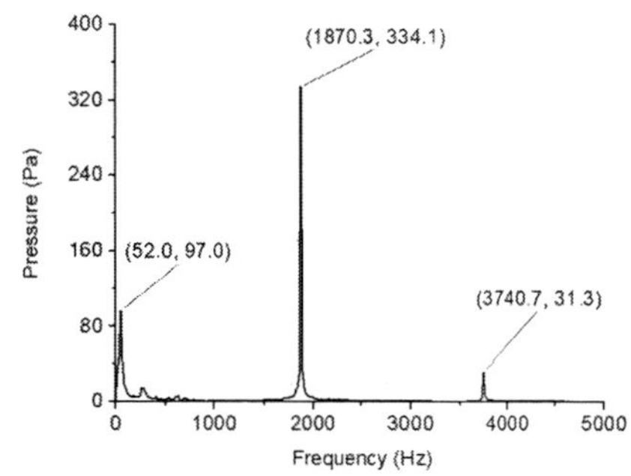

(f) Section 6

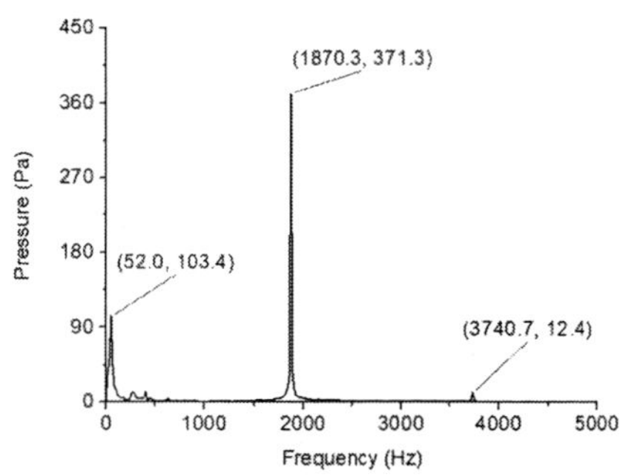

(g) Section 7

Fig. 2.10. The results of FFT at the seven sections with mass flow is 2.04kg/s

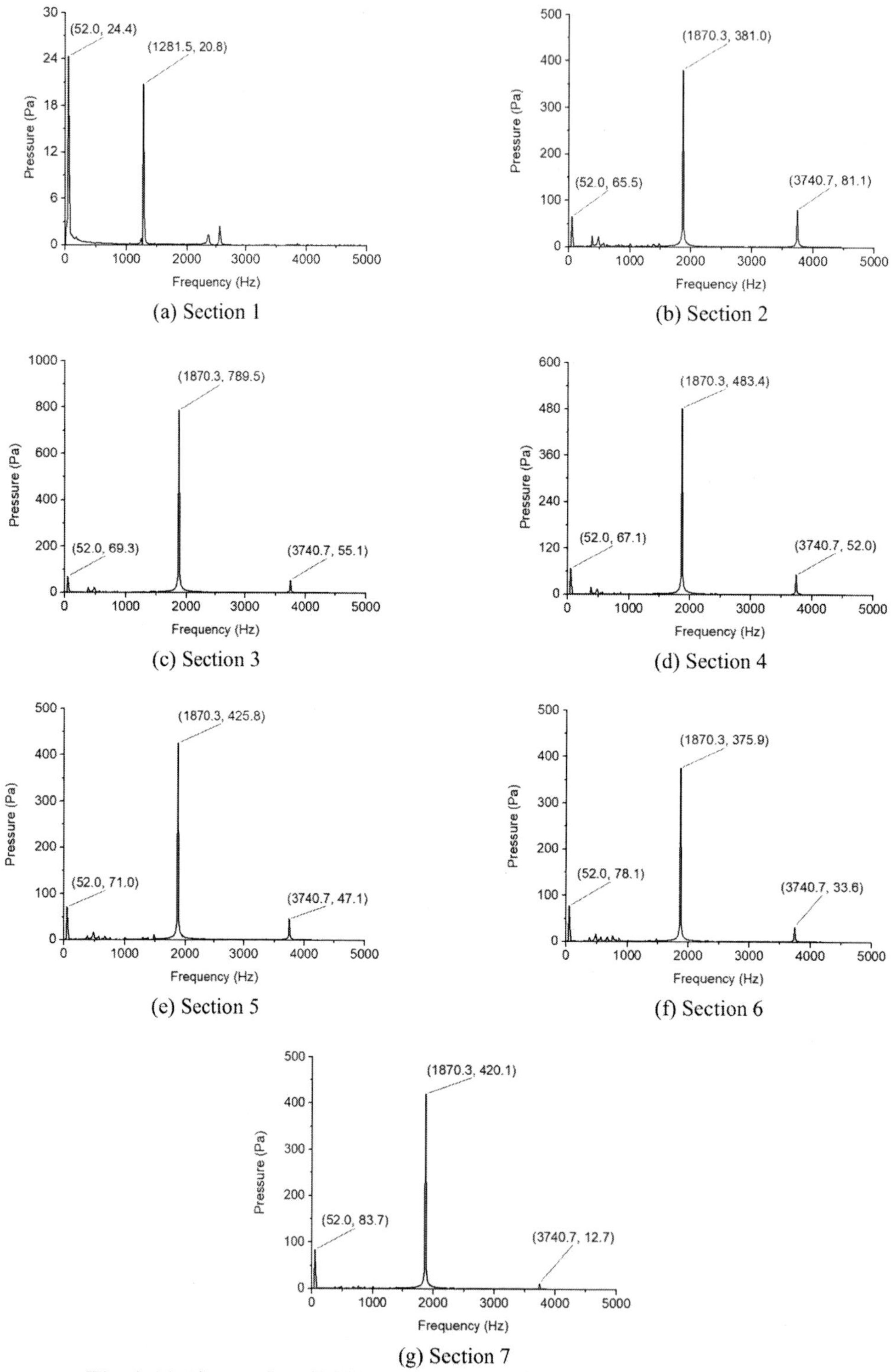

Fig. 2.11. The results of FFT at the seven sections with mass flow is 1.98kg/s

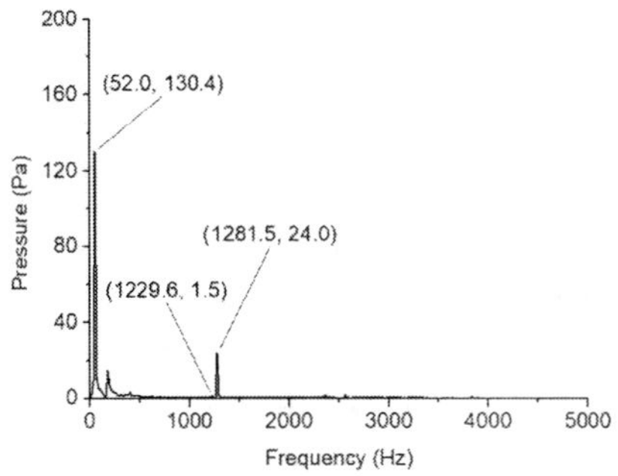

(a) Section 1

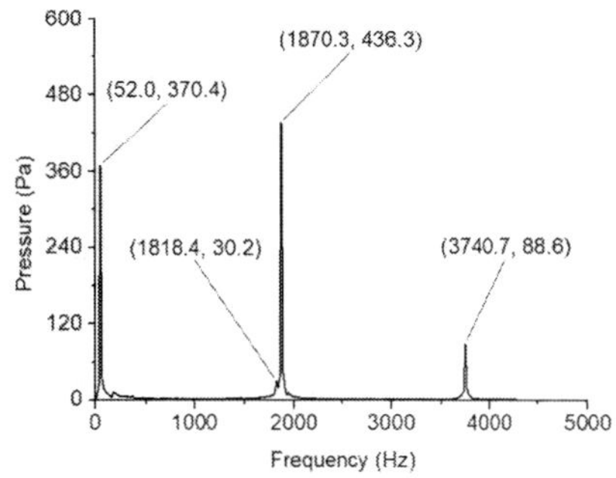

(b) Section 2

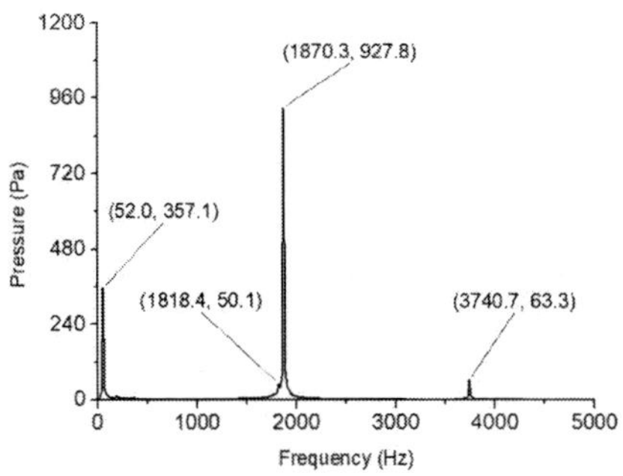

(c) Section 3

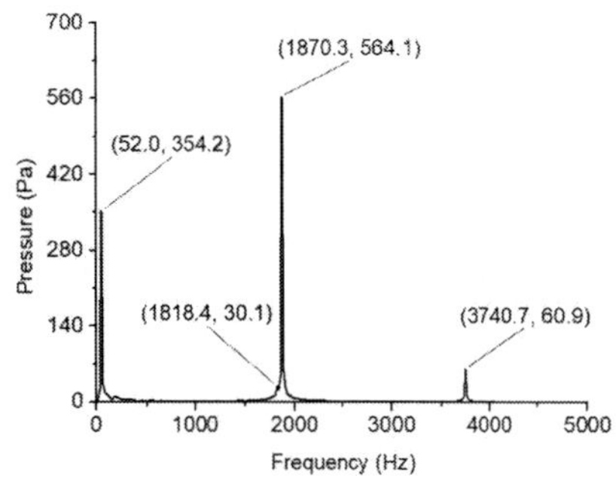

(d) Section 4

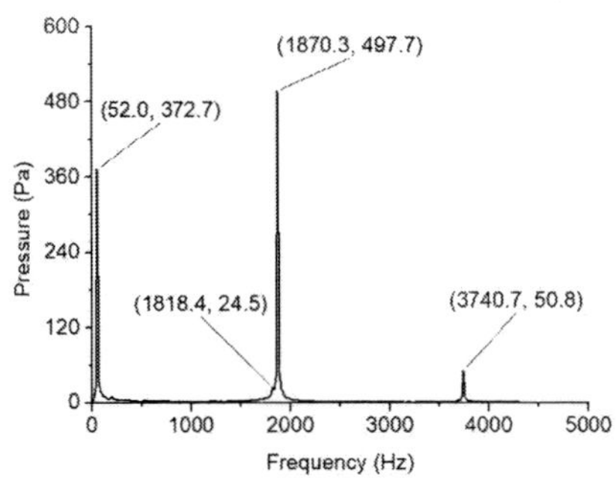

(e) Section 5

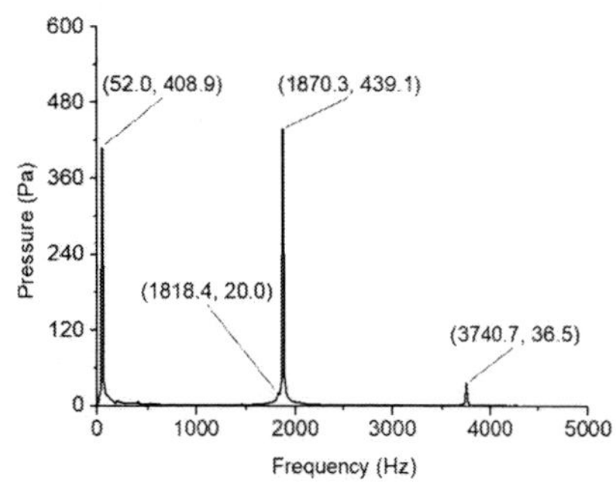

(f) Section 6

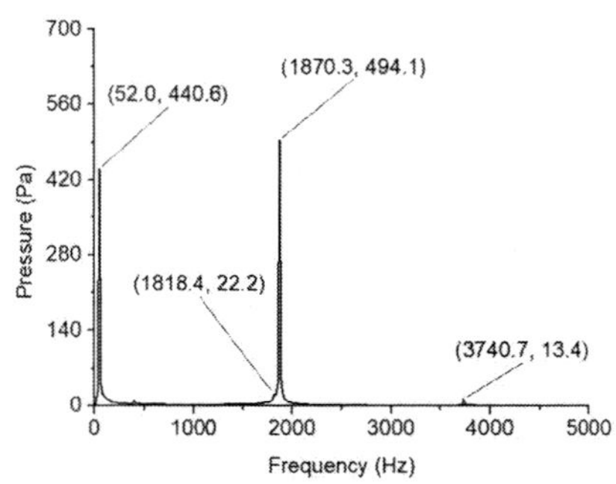

(g) Section 7

Fig. 2.12. The results of FFT at the seven sections with mass flow is 1.91kg/s

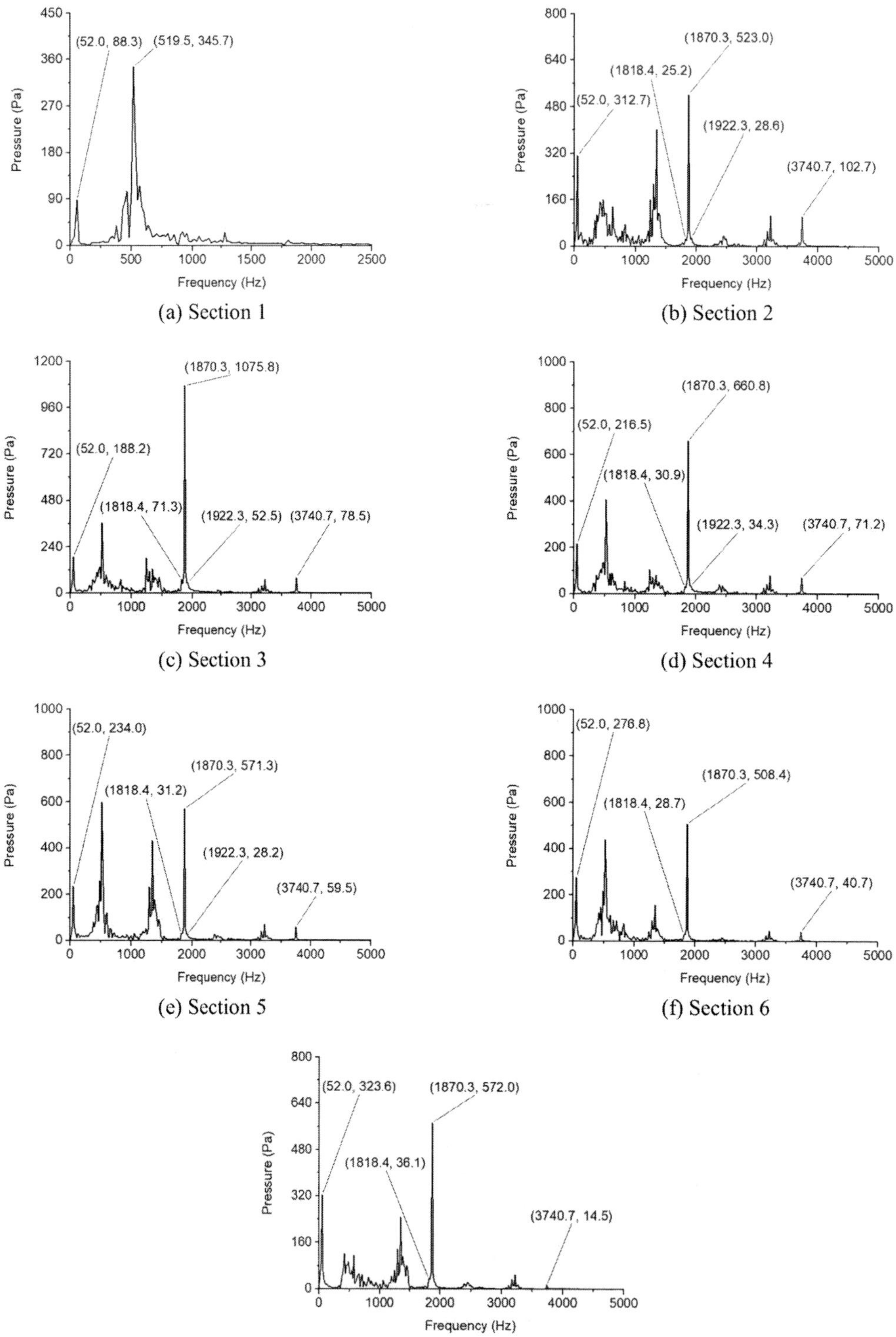

(g) Section 7

Fig. 2.13. The results of FFT at the seven sections with mass flow is 1.85kg/s

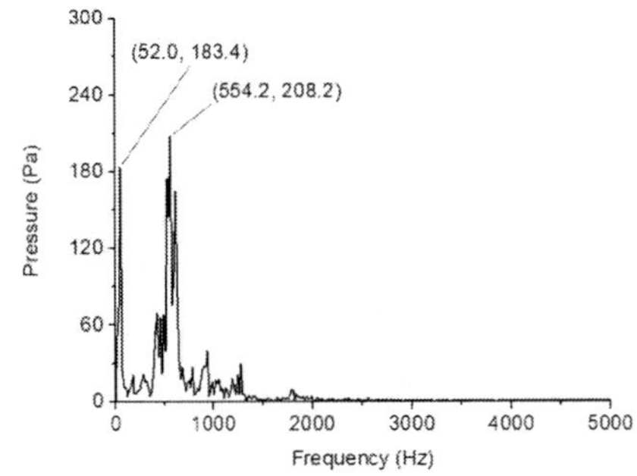

(a) Section 1

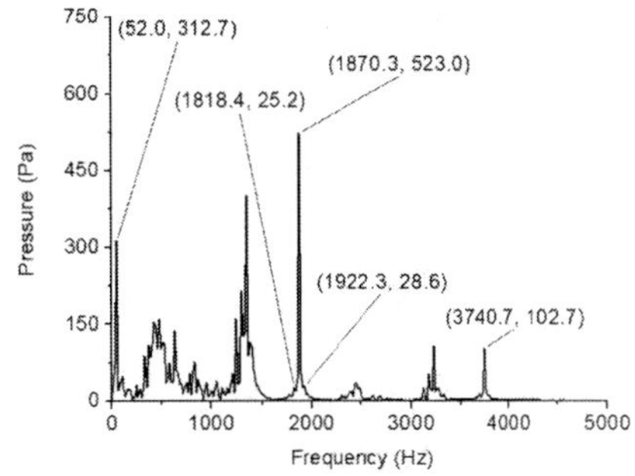

(b) Section 2

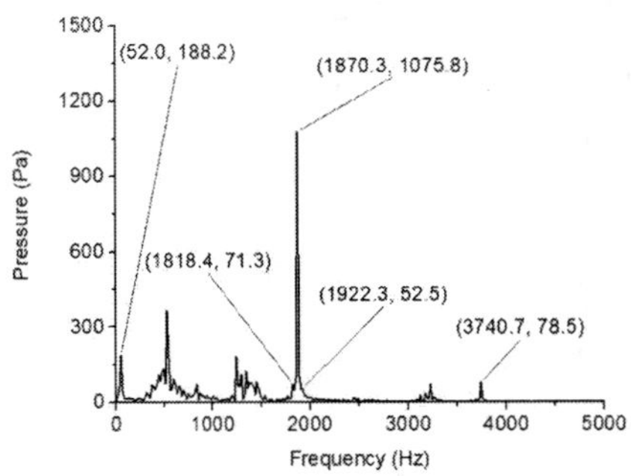

(c) Section 3

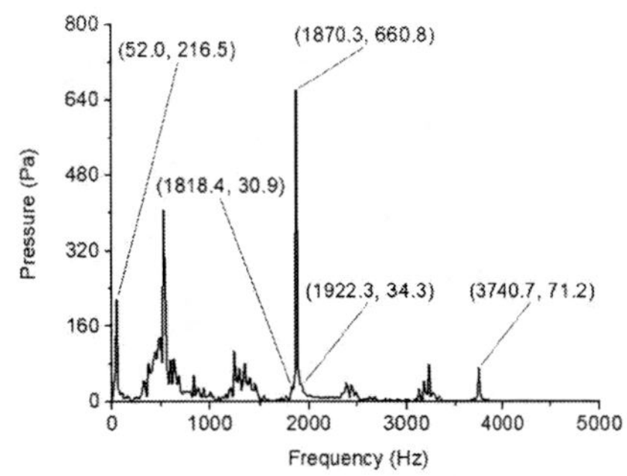

(d) Section 4

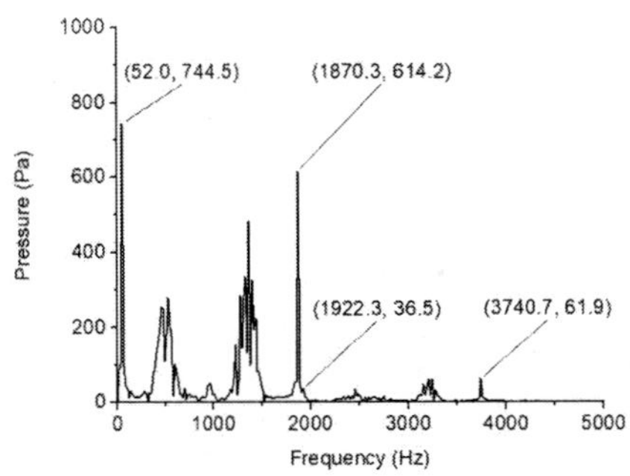

(e) Section 5

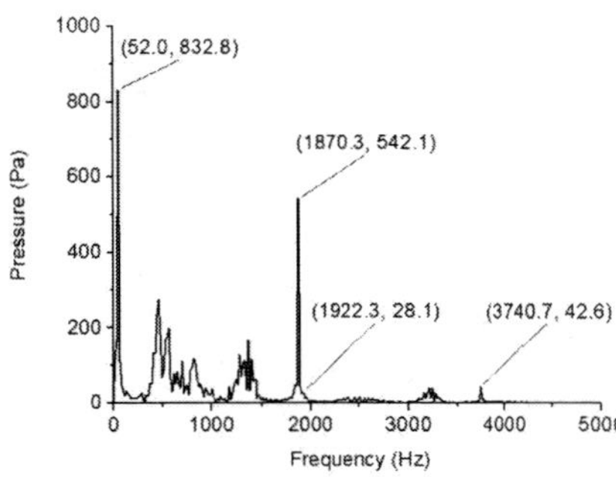

(f) Section 6

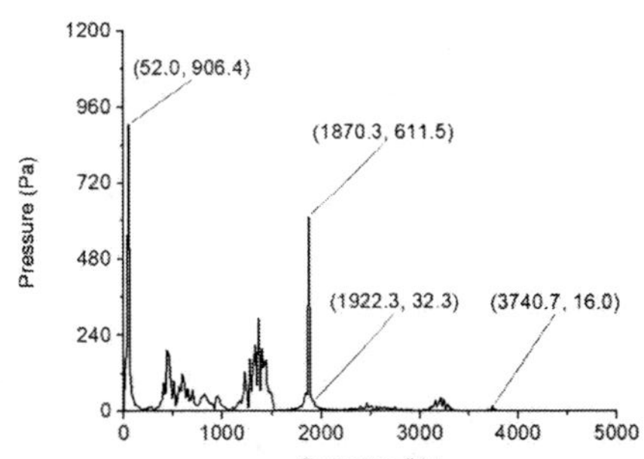

(g) Section 7

Fig. 2.14. The results of FFT at the seven sections with mass flow is 1.79kg/s

3. Experimental Investigation

On the basis of the results of the CFD simulation, a centrifugal compressor was used to carry out the experiments with the range of the mass flow is [1.638, 3.661] kg/s. Experiment details are introduced as the following part.

3.1. Experiment Equipment

The model machine used in the experiments is a one stages centrifugal compressor. The experiment device layout and the model machine are shown as Fig. 3.1, and the arrows in the figures point to the direction of the air flow. At the outlet of the impeller, that is, the inlet of the diffuser, a pressure sensor is installed. The sensor model is 116B pressure sensor of the PCB Piezotronics, Inc. The sampling frequency was set to 51.2kHz. In each operation point, 40s data is collected each time, twice in total. Meanwhile the experiment parameters, including inlet pressure, outlet pressure, inlet temperature, outlet temperature, and mass flow, are monitored. The parameters which have been used in experiments are list in the Table 3.1.

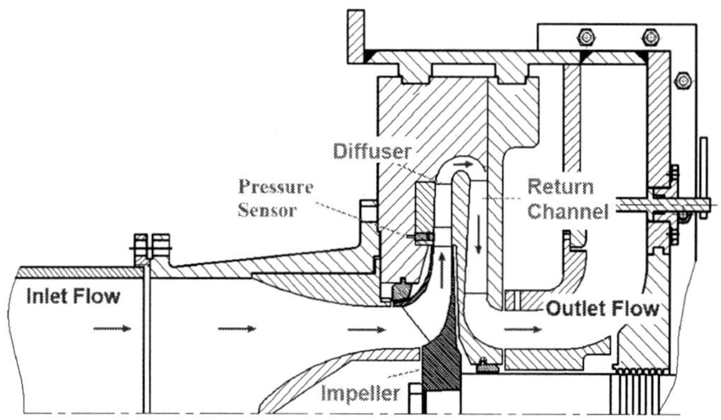

Fig. 3.1. Experiment device

Table 3.1. Experiment Parameters

No.	Inlet Pressure (Pa)	Outlet Pressure (Pa)	Inlet Temp. (K)	Outlet Temp. (K)	Mass Flow (kg/s)
1	98688	105750	295.8	310.3	3.661
2	99316	113904	296.3	312.5	3.001
3	99740	116817	296.5	313.6	2.509
4	99981	118219	296.7	314.7	2.159
5	100048	118573	296.8	315.2	2.040
6	100118	118779	297.0	315.8	1.916
7	100198	118867	297.1	316.4	1.798
8	100298	118643	297.1	317.2	1.638

The dimensionless results of total pressure ratio and polytropic efficiency based on the steady CFD method and experimental method are shown in Fig. 3.2, respectively. It can be seen from Fig. 3.2 that the steady CFD results are in good agreement with the experimental results. It can be proved that the CFD simulation is credible.

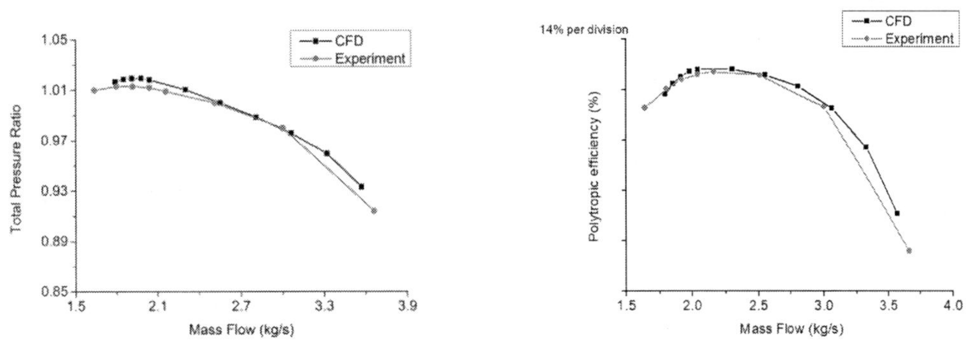

(a) Total pressure ratio (b) Polytropic efficiency

Fig. 3.2. The dimensionless results of the CFD method and the experimental method

3.2. Experiment Results Analysis in Frequent Domain

The results in frequent domain by the FFT method of the eight operation points are shown as Fig. 3.3, respectively.

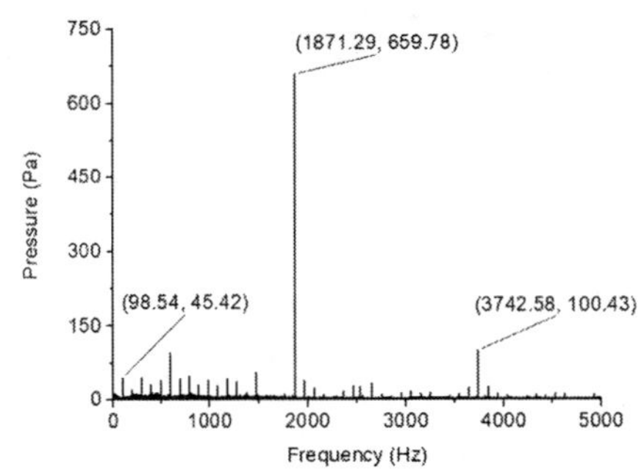

(a) The results of FFT with mass flow is 3.661kg/s

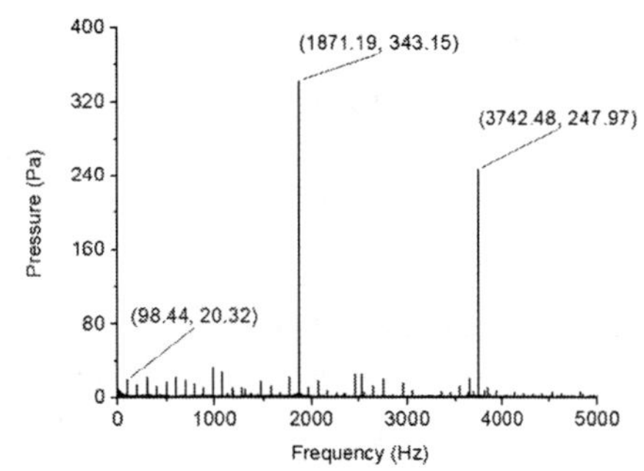

(b) The results of FFT with mass flow is 3.001kg/s

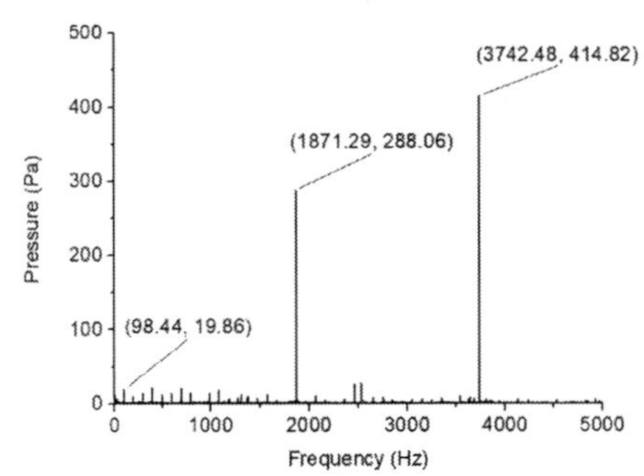

(c) The results of FFT with mass flow is 2.509kg/s

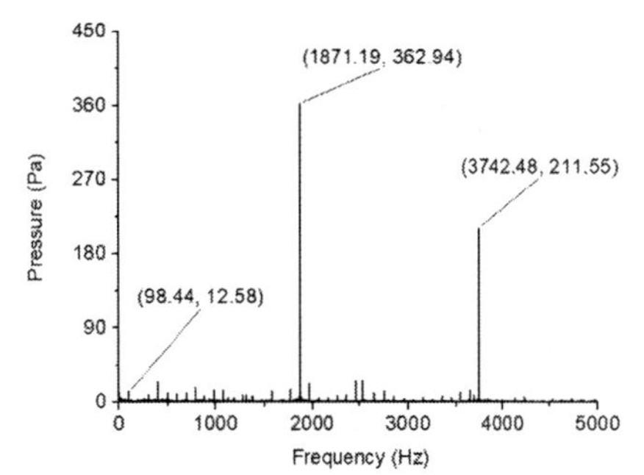

(d) The results of FFT with mass flow is 2.159kg/s

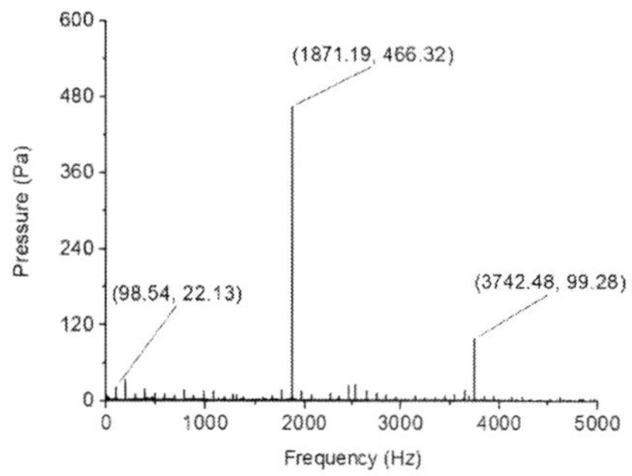

(e) The results of FFT with mass flow is 2.04kg/s

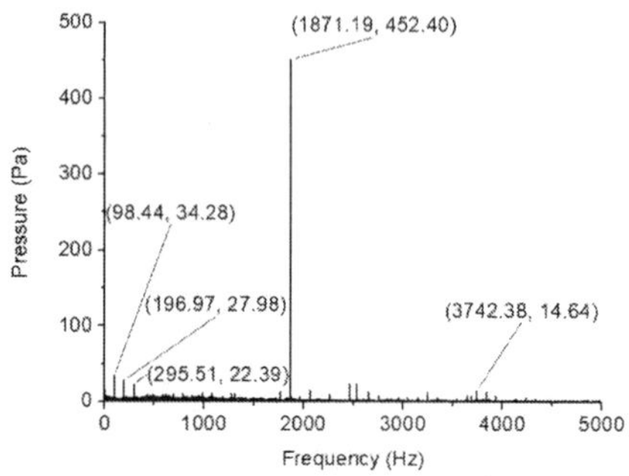

(f) The results of FFT with mass flow is 1.916kg/s

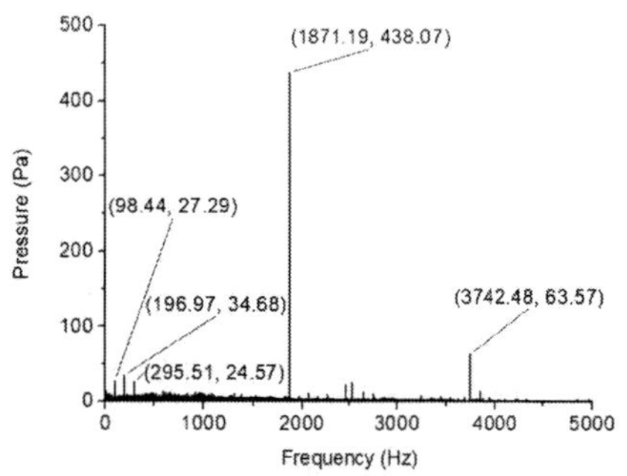

(g) The results of FFT with mass flow is 1.798kg/s

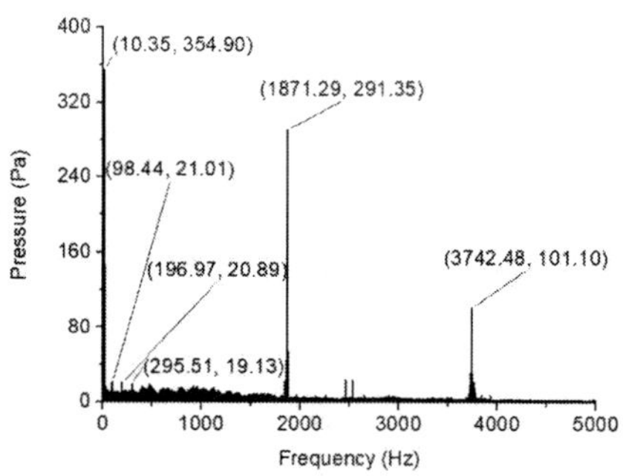

(h) The results of FFT with mass flow is 1.638kg/s

Fig. 3.3. The experimental results in frequent domain

The following conclusions can be obtained from Fig. 3.3:

1) The blade passing frequency (f_{BPF}=1871.19Hz) and its second order frequency (2×f_{BPF}=3742.48Hz) can be observed in all eight operation conditions;

2) The rotational frequency (f_r=98.44) and its higher order frequencies can be observed in each operation condition. Especially when the mass flow is 1.916, 1.798 and 1.638kg/s, the amplitudes of the first three-order rotational frequency are more significant than that of the higher-order frequencies. It can be inferred that the compressor rotor has poor alignment.

In addition to the above conclusions, the part of Fig. 3.3(g) is enlarged, as shown in Fig. 3.4.

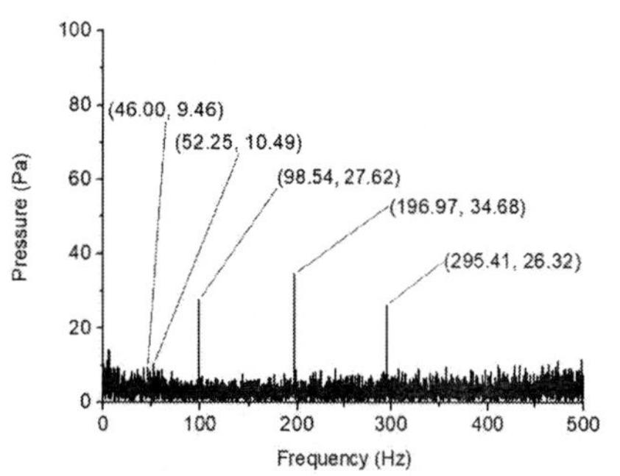

(a) The results in 0-500Hz

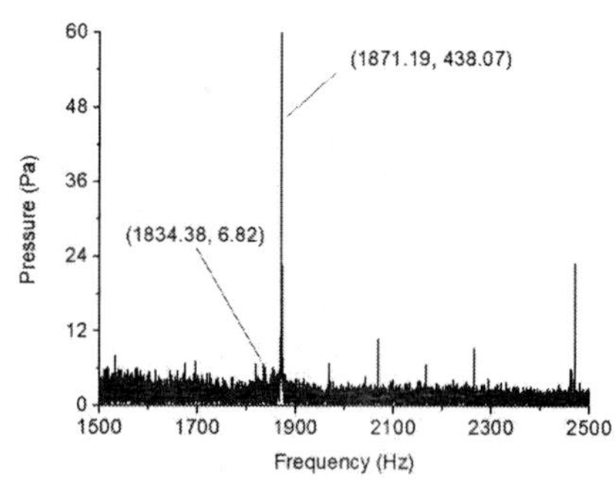

(b) The results in 1500-2500Hz

Fig. 3.4. The results of FFT with mass flow is 1.798kg/s

Although there are many peaks that are difficult to distinguish in the Fig. 3.4, several characteristic peaks can still be identified.

1) There are f_{BPF} and a modulation frequency (f_{BPF}-f_s=1834.38Hz) in the Fig. 3.4(b);

2) There are f_r and a modulation frequency (f_r-f_s=52.25Hz) in the Fig. 3.4(a).

3) There is a peak of f_s=46Hz.

According to the literatures, the rotational stall can be identified when mass flow is 1.798kg/s. Moreover, the part of Fig. 3.3(h) is enlarged, as shown in Fig. 3.5.

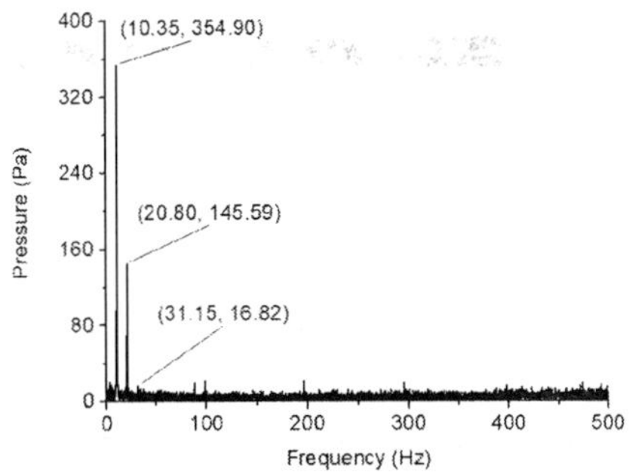

Fig. 3.5. The results of FFT with mass flow is 1.638kg/s

A series of frequencies can be gained. They are 10.35, 20.8 and 31.15Hz, respectively. The 10.35Hz is about 0.1×f_r. The phenomena of the compressor surge has been observed with mass flow is 1.638kg/s. According to the results of the experiments, the compressor surge phenomenon was observed in the frequency domain. The results of the CFD method in this paper cannot observe the surge phenomenon. On the one hand, if continue to reduce the mass flow, which is less than 1.79kg/s, during the unsteady CFD simulation process in this paper, the periodicity of the pressure pulsation cannot be observed. Therefore, it cannot be judged whether it has converged. On the other hand, the compressor surge caused by the flow state of the complete compressor system. The model for the simulation of the CFD method in this paper is focused on the core parts, including impeller, diffuser and return channel. The exhaust volute and downstream piping system have not been taken into account.

4. Conclusions

The stall condition of a centrifugal compressor based on CFD and experimental methods has been investigated. The steady and unsteady CFD simulations has been performed. A series of experiments has been carried out. The following conclusions can be drawn:

1) By comparing the results of the total pressure ratio and polytropic efficiency of the CFD and the experimental methods, it can be proved that the CFD model used in this paper can be performed for the simulation of the model machine.
2) Both the CFD method and the experimental method observe the rotating stall phenomenon in frequency domain. For CFD, the stall appears at flow rate 1.91kg/s, the corresponding f_s is 52Hz. For experiments, the stall appears at flow rate 1.798kg/s, the corresponding f_s is 46Hz. There are some reasons for this deviation. For instance, the exhaust volute, clearances, and surface roughness were not taken into consideration during the process of the numerical simulations.

5. Acknowledgment

The authors are grateful for support from the project of new energy-saving and system intelligent control technology for fluid machinery, the National Key R&D Program of China (Grant No. 2018YFB0606105).

References

[1] R. C. Pampreen (1993). *Compressor Surge and Stall.* Concepts ETI, Inc. Norwich, Vermont, USA.

[2] L. J. Cheshire, M. S. Tech, A. M. I. Mech (1945). *The Design and Development of Centrifugal Compressors for Aircraft Gas Turbines* . Proceedings of the Institution of Mechanical Engineers,

153, pp 426-440.

[3] H. W. Emmons, C. E. Pearson, and H. P. Grant (1955). *Compressor surge and stall propagation* . Transactions of the ASME, 79, pp 455–469.

[4] E. M. Greitzer (1976). *Surge and Rotating Stall in Axial Flow Compressors-Part II: Experimental Results and Comparison with Theory*. Journal of Engineering for Power, 98(2), pp 199-211.

[5] F. K. Moore, E. M. Greitzer (1986). *A Theory of Post-Stall Transients in Axial Compression Systems: Part I-Development of Equations*. Journal of Engineering for Gas Turbines and Power, 108(1), pp 68-76.

[6] F. K. Moore, E. M. Greitzer (1986). *A Theory of Post-Stall Transients in Axial Compression Systems: Part II-Application*. Journal of Engineering for Gas Turbines and Power, 108(2), pp 231-239.

[7] P. Frigne, R. V. D. Braembussche (1984). *Distinction between Different Types of Impeller and Diffuser Rotating Stall in a Centrifugal Compressor with Vaneless Diffuser*. Journal of Engineering for Gas Turbines and Power, 106(2), pp 468-473.

[8] D-C. Liaw, E. H. Abed (1996). *Active control of compressor stall inception: a bifurcation-theoretic approach*. Automatica, 32(1), pp 109-115.

[9] M. Tryfonidis, O. Etchevers, J. D. Paduano, A. H. Epstein, G. J. Hendricks (1995). *Prestall Behavior of Several High-Speed Compressors*. Journal of Turbomachinery, 117(1), pp 62-80.

[10] T. R. Camp, I. J. Day (1998). *A study of spike and modal stall phenomena in a low-speed axial compressor*. Journal of Turbomachinery, 120(3), pp 393-401.

[11] G. Pullan, A. M. Young, I. J. Day, E. M. Greitzer, Z. S. Spakovszky (2015). *Origins and Structure of Spike-Type Rotating Stall*. Journal of Turbomachinery, 137(5), pp 2567-2579.

[12] A. Spentzos, G. Barakos, K. Badcock (2004). *CFD investigation of 2d and 3d dynamic stall*. The AHS 4th Decennial Specialist's Conference on Aeromechanics, San Fransisco, California, January, pp 21-23.

[13] Sun X. F., Sun D. F., Yu. W. W (2011). *A Model to Predict Stall Inception of Transonic Axial Flow Fan/Compressors*. Chinese Journal of Aeronautics, 24(6), pp 687-700.

[14] J.N. Everitt, Z.S. Spakovszky (2013). *An Investigation of Stall Inception in Centrifugal Compressor Vaned Diffuser*. Journal of Turbomachinery, 135(1), 011025.

[15] A. Broatch, J. Galindo, R. Navarro, J. García-Tíscar (2016). *Numerical and experimental analysis of automotive turbocharger compressor aeroacoustics at different operating conditions*. International Journal of Heat and Fluid Flow, 61(B), pp 245-255.

[16] N. Fujisawa, T. Inui, Y. Ohta (2019). *Evolution Process of Diffuser Stall in a Centrifugal Compressor With Vaned Diffuser*. Journal of Turbomachinery, 141(4), 041009.

[17] I. J. Day (1992). *Stall and Surge in Axial Flow Compressor*. Axial Flow Compressor Lecture Series. Von Karman Institute for Fluid Dynamics.

[18] Zhao X. W., Zhou Q., Yang S. H., Li H. K (2109). Rotating Stall Induced Non-Synchronous Blade Vibration Analysis for an Unshrouded Industrial Centrifugal Compressor. Sensor, 19(22), 4995.

Investigation of the shell-side flow in a shell-and-tube exchanger with impingement plate

Yi Wang[1,2], Guoliang Qin[1,*], Yong Zhang[2] Shuhua Yang[2], Yuying Sun[2], Qin Cui[1], Cheng Jia [1]

(1. School of Energy and Power Engineering, Xi'an Jiaotong University, Xi'an 710049, China; 2. Shenyang Blower Works Group Corporation, Shenyang 110869, China)

E-mail: glqin@mail.xjtu.edu.cn

Abstract. For studying the influence of impingement plates on the fluid flow within the shell-side of the shell-and-tube heat exchanger, two-dimensional numerical simulation is carried out using the Fluent software. The different distances between the impingement plate and the tube bundle are taken into consideration. Meanwhile the research investigates the flow field of the non-perforated impingement plate and the perforated impingement plates with various hole diameters. The results show that the fluid behind the non-perforated impingement plate contacts the tube insufficiently. The farther the distance between the non-perforated impingement plate and the tube bundle is, the more tubes contact with the fluid poorly. The flow field behind the perforated impingement plates is more uniform. The protective effect from flow impacting for impingement plates is positive. The perforated impingement plates are recommended in the engineering design of the shell-and-tube heat exchanger.

1. Introduction

Shell-and-tube heat exchangers are widely used in the industrial fields such as the energy and power industry, the chemical industry, the metallurgy, the aerospace and so on. Shell-and-tube heat exchangers have many excellent characteristics, including the simple structure, low cost, easy scale cleaning, resistance quality of high temperature and high pressure. Heat exchanger designers usually put an impingement plate at the entrance of the shell side or the tube side of the heat exchanger according to the provisions of GB-T151-2014 [1]. The impingement plate is a protective auxiliary part. Firstly, it can prevent the tube bundle and the tube sheet of the heat exchanger from directly impacting by the working fluids. Meanwhile, it can prevent the tube bundle vibration caused by the high flow rate working fluid. Finally, it can also prevent the tube bundle and the tube sheet from erosion and corrosion sometimes.

At present, the research on shell-and-tube heat exchangers mainly focuses on heat transfer elements and pressure-bearing elements. There are relatively few studies on impingement plates. Fu Shuangquan [2] developed an Excel program to calculate the using criterion of the impingement plate for heat exchange designers. Liu Xiaoling [3] used empirical formulas to analyze the impingement plate on the shell-side of the heat exchanger from the perspective of the tube bundle vibration. Defan Qing [4] took the experiment on heat transfer and flow resistance of elliptic rib and circular rib. Xiaojun Chen [5] indicated the application of fin tube cooler with bimetallic strip. Bin Xue et al [6] investigated the relationship between resistance and heat transfer of inner finned tube with single or double holes based on measuring and calculating. Feng Su et al [7] used the theory of multi-span beams to study the transverse vibration induced by a tube-side flow. Mingqing Liu [8] summarized the type and character of the heat exchanger used as intercooler of the compressor process.

In recent years, CFD research on impingement plates began to be reported. Wu Yangkuan et al. [9] took the horizontal condenser as the research object, using the Ansys CFX software to study the effect of the slotted impingement plate on the shell side. Liu Xiangbin et al. [10] took the channel of a single-tube-pass heat exchanger as the research object. They simulated the protection effect of impingement plates with the circular flat shape and the angle shape 120° cone by the Ansys CFX software. They found the circular flat impingement plate with holes had the best protection on alleviating the impact of the working fluid within the channel. Salamah S. Al-Anizi et al. [11] developed double perforated impingement plate to stop fouling accumulation on the tubes to prevent under-deposit corrosion.

The flat impingement plate is widely used in shell-and-tube heat exchangers for it is easy to manufacture. The impingement plate at the shell side could be welded on the inner wall of the shell or the spacers of the tube bundle. The installation type is shown in Figure 1. In addition, the flat impingement plate would be made into the non-perforated one or the perforated one, as shown in Figure 2. At present, both the installation type and the perforated structure parameters of the impingement plate depend on the designer's experience. There are no standards or guidelines about the impingement plate.

In a shell-and-tube heat exchanger heat transfers from one fluid in the tube to another fluid in the shell through the tube wall. The effective heat transfer means that the fluid in the shell-side flows fluently through the gaps or the tube bundle. However, few researches have been done about the flow behind the impingement plate in the shell, especially by numerical simulation. Therefore the research is taken into consideration in this paper on how the shell-side flow field of the tube bundle area is effected by the impingement plate.

In this paper, the research object focuses on the fluid flow through the tube bundle area behind the impingement plate at the shell side of the shell-and-tube heat exchanger. Two-dimensional numerical simulations are performed with the Fluent software to investigate the flow field on the different installation type and the various perforated structure of the impingement plate. The conclusions aim to provide references for the designers of shell and tube heat exchangers.

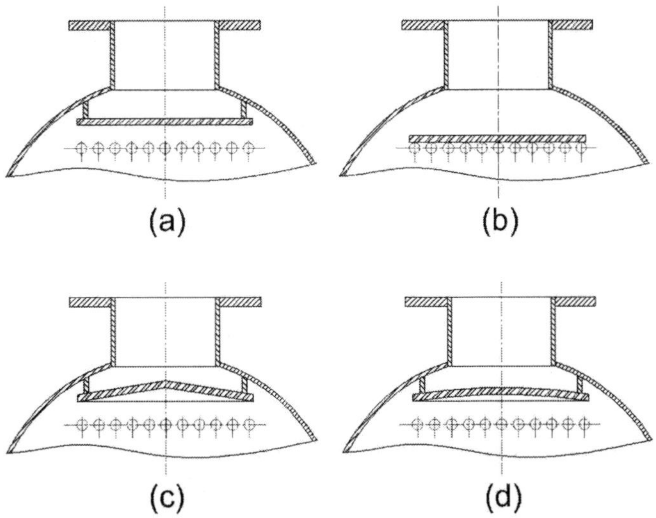

Figure 1. Installation Type of the shell-side impingement plate.

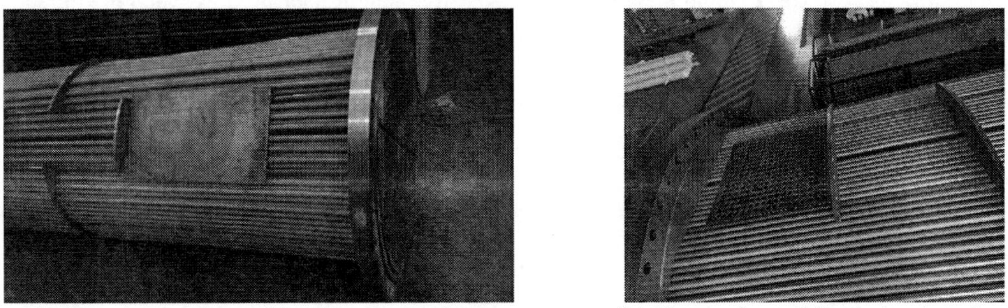

(a) The non-perforated impingement plate. (b) The perforated impingement plate.

Figure 2. The type of the impingement plate.

2. Description of the engineering case

This research is based on a shell-and-tube heat exchanger designed by Shenyang Blower Works Group Corporation. All the simulation data comes from the engineering application. Considering the commercial confidentiality, only basic public data of the shell-and-tube heat exchanger is shown in the paper.

2.1. Calculation of the Reynolds Number

According to the formula in the reference [12], when the tubes layout are designed in the form of equilateral triangle, the equivalent diameter of shell-side flow area is calculated by:

$$d_e = \frac{4\left(\frac{\sqrt{2}}{2}l^2 - \frac{\pi}{4}d_o^2\right)}{\pi d_o} = 17\,(\mathrm{mm}) \tag{1}$$

where the center distance between two adjacent tubes is 25mm, and the tube outer diameter is 19mm.

In the heat exchanger, the baffle spacing is 500mm. The shell diameter is 1600mm. The maximum flow area in the shell-side is calculated by:

$$A_{s,max} = BD_{i,s}\left(1-\frac{d_o}{l}\right) = 0.192\,(m^2) \quad (2)$$

The working fluid in the shell-side is water. The mass flow of water is 243289kg/hr. The water velocity in the shell-side is calculated by:

$$u_s = \frac{Q}{\rho_l A_{s,max}} = 0.35\,(m/s) \quad (3)$$

The Reynolds number of the shell-side is calculated by:

$$Re = \frac{d_e u_s \rho_l}{\mu_l} = 5932 \quad (4)$$

According to the reference [12], the flow pattern at the shell side is determined by the Reynolds number. When the Reynolds number at the shell-side fluid is greater than 100, the flow is turbulent. Therefore, water flows turbulently at the shell-side in the sample.

2.2. *Criterion of the Impingement Plate*

According to GB/T151-2014[1], the minimum flow area of the shell-side inlet nozzle is calculate by:

$$A_{s,min} = \pi D_{i,n} H_{tl} + F_1\left(\frac{\pi D_{i,n}^2}{4}\right)\frac{(l-d_o)}{F_2 l} = 42461.95\,(mm^2) \quad (5)$$

where $D_{i,n}$ is the inner diameter of the shell-side inlet nozzle. In the case, the inner diameter of the shell-side inlet nozzle is 255mm. H_{tl} is the averaged free height measured between the tube layout and the inlet nozzle, H_{tl} =50mm. The coefficients F_1 and F_2 are dimensionless coefficients, which are taken as recommended in GB/T151-2014[1], F_1=1 and F_2=1.

It has been stated in GB/T151-2014[1] that an impingement plate is needed when the value of $\rho_l v^2$ is greater than 2230kg/(m·s²). In the sampled heat exchanger, the value of $\rho_l v^2$ is calculated by:

$$\rho_l v^2 = \rho_l\left(\frac{Q/3600\rho_l}{A_{s,min}}\right) = 2523.55\,(kg/(m\cdot s^2)) \quad (6)$$

Then, an impingement plate is needed at the shell-side of the heat exchanger.

2.3. *Averaged Velocity at the Shell-side Inlet Nozzle*

According to the fluid flow data in the shell-side, the average velocity at the shell-side inlet nozzle is calculated by:

$$u_{in} = \frac{Q}{3600\rho_l\left(\frac{\pi D_{i,s}^2}{4}\right)} = 1.33\,(m/s) \quad (7)$$

Various velocities are studied in the simulation for investigating its influence on the flow field. All values of the velocity in the study is based on the values calculated above.

3. Numerical simulation

3.1. *Geometric Model*

Flow change is neglected along the longitudinal direction of the shell in the research. The tube layout is simplified for simulation. Figure 3 shows the geometric model of the simplified numerical calculation area. The tube bundle is quite far from the inlet and the outlet to reduce the influence of the inlet and outlet boundary on the flow field. The tubes are arranged in the form of a regular triangle as stated in GB/T151-2014[1]. The outer diameter of the tube is 19mm. The spacing of the tube is 25mm.

In order to investigate the effect of the impingement plate on the flow field in the tube bundle area, there are two kinds of geometric models- with or without the impingement plate. Many variables are considered in the model with the impingement plate. The different distances between the impingement plate and the tube bundle are investigated first. Then, the research displays the flow field behind the impingement plates perforated or not. In addition, flow field behind the perforated impingement plates with various diameters is studied.

Assuming that the holes on the perforated impingement plate are round, the distance between the two holes center is 2 times the hole diameter. The cross-section is selected as the calculation area where the center of the hole is located. The variables for simulation are shown in Table 1.

It is recommended that the tubes should be as many as possible in the heat exchanger when designing. Therefore, the vertical distance is basically not more than 120mm between the center line of the tubes in the first row and the intersecting line of the shell and the inlet nozzle. Meanwhile, the distance between the impingement plate welded on the shell and the intersecting line of the tubes in the first row is about 60~70mm based on the design experience. Based on the above statement, the maximum distance between the first row of the tubes and the impingement plate is 51mm in the research, which can cover the actual distances of most engineering cases.

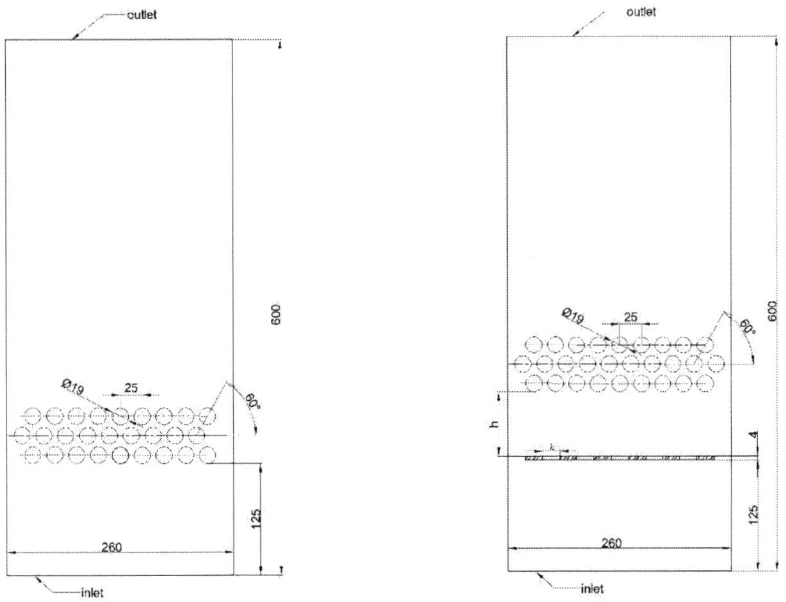

(a) the case without the impingement plate. (b) the case with the impingement plate.
Figure 3. Geometric Model (mm).

Table 1 The variables for simulation(mm).

name	value	denotation
	1	The impingement plate is welded on the spacers
h	21	The impingement plate is welded on the shell. The distance between the impingement plate and the tube bundle is 21mm.
	51	The impingement plate is welded on the shell. The distance between the impingement plate and the tube bundle is 51mm.
	0	The non-perforated impingement plate
λ	10	The perforated impingement plate with the hole diameter 10mm
	20	The perforated impingement plate with the hole diameter 20mm

3.2. *Mathematical model*

In this paper, the numerical simulations of the above geometric models are carried out with the Fluent software. The working fluid is assumed as water. The flow is assumed to be a two-dimensional steady-state incompressible flow. The mathematical models used in the research are the continuity equation, the momentum conservation equation, and the standard k-ε turbulence model. The second-order upwind difference format is used to discretize the above-mentioned equations. The COUPLED algorithm is used to deal with the coupling of pressure and velocity. The convergence condition is the equation residual less than 10^{-5}.

The boundary conditions are imposed as follows. The inlet boundary is the velocity inlet. The outlet boundary is the free flow. The fixed and non-slip wall surface is used on the surface of the tubes and the impingement plate, as well as the two side walls.

3.3. *Meshing*

Unstructured grids are used in the entire calculation area. Mesh encryption is carried out in the tube bundle and the impingement plate area for the calculation accuracy. The sparse mesh is used in the area of inlet and outlet where little influence of these areas is imposed on the result.

In this study, the grid independence is verified by different geometric models. Taking the geometric model in Figure 3(a) as an example, the relationship is studied between the nodes and the averaged outlet velocity on the y-direction, as shown in Figure 4. In order to save the calculation cost, this case adopts a meshing scheme with 53674 nodes for calculation.

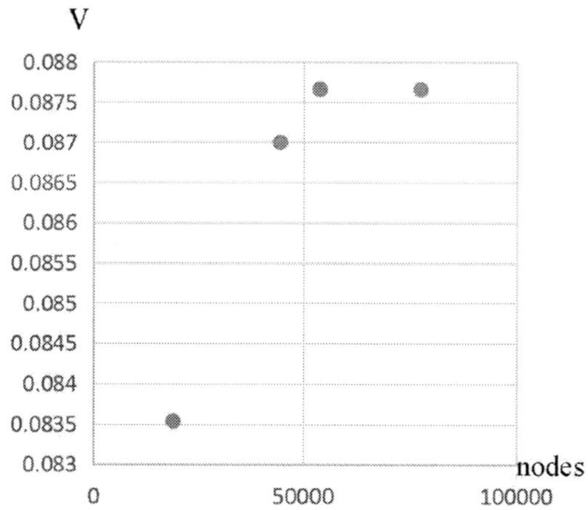

Figure 4. Grid independence verification.

4. Results and discussions

4.1. *The Flow With or Without the Impingement Plate*

Two cases are modeled in order to investigate the influence of the impingement on the flow field of the tube bundle area. Case 1 is without the impingement plate. Case 2 is with the non-perforated impingement plate welded on the spacers. The water inlet velocity is 0.09m/s. The flow velocity contour is shown in Figure 5.

When the water flows through the tube bundle, the flow field shows a symmetrical distribution along the vertical centerline of the model's area in both cases. Vortexes appear between the tubes in the last row and the outlet. Actually, the gap between the tube bundle and the shell wall is very small in the design of the heat exchanger, so that there is no space for large vortex generation.

The main purpose of this research is to study the fluid flow in the tube bundle area, not too much blank area. Therefore, in the subsequent discussions, the appropriate space around the tube bundle is

IOP Conf. Series: Materials Science and Engineering 1081 (2021) 012053 doi:10.1088/1757-899X/1081/1/012053

intercepted for flow field analysis. Meanwhile, half of the area along the vertical centerline is cut for the symmetry. Then the streamline and velocity contour of these two cases are shown in Figure 6.

The fluid can smoothly pass through the gap between the tubes in case 1as shown in Figure6 (a). In case 2 as shown in Figure6 (b), an obvious vortex is generated around the right tube of the first row after the impingement plate. Affected by the vortex, the fluid bypassing the impingement plate cannot smoothly flow into the windward side of the tubes in the first row. The fluid flows from the leeward side of the tubes in the first row into the gaps of the tubes in the second row and the third row. The non-perforated impingement plate welded on the spacer will prevent the first row of tubes from convection with the fluid.

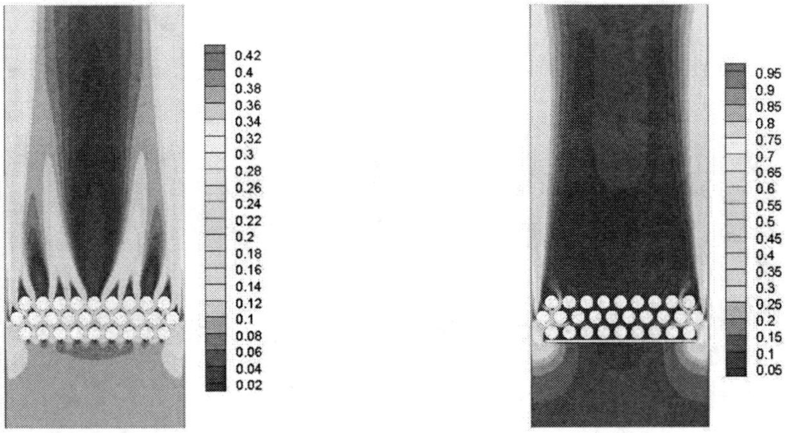

(a) Case 1 without the impingement plate. (b) Case 2 with the non-perforated impingement plate.
Figure 5. Velocity contour at inlet velocity 0.09m/s.

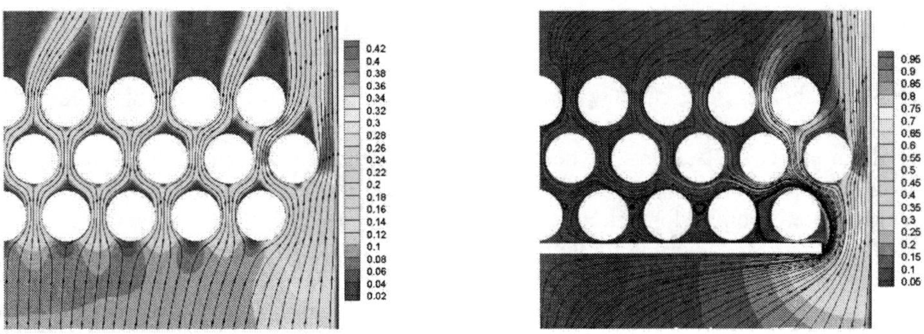

(a) Case 1 without the impingement plate. (b) Case 2 with the non-perforated impingement plate.
Figure 6 Streamline and velocity contour at inlet velocity 0.09m/s.

4.2. The Distance between the Impingement Plate and the Tube Bundle

Three cases are studied with various distances between the impingement plate and the tube bundle in this section. The distance between the impingement plate and the tube bundle is 1mm in case 2. The distance between the impingement plate and the tube bundle is 21mm in case 3. The distance between the impingement plate and the tube bundle is 51mm in case 4. The impingement plate is non-perforated. The water inlet velocity is 0.09m/s for three cases. The streamline and velocity contours of these three cases are shown in Figure 7.

The flow field of case 2 shown in Figure7 (a) is discussed in the section 3.1 when analyzing the flow behind the non-perforated impingement plate welded on the spacer shown in Figure6 (b).

When the distance between the impingement plate and the tube bundle is 21mm as shown in Figure7 (b), a large vortex is formed between the impingement plate and the first tube on the right side of the first row. A number of small vortices are formed between the impingement plate and the other tubes of the first row. These vortices make most of the streamlines pass through the leeward side of the tubes in the first row, flowing into the area of the second row and the third row. Similar to the flow of case 2, the fluid does not fully contact with the first row bypassing the impingement plate.

When the distance between the impingement plate and the tube bundle is 51mm as shown in Figure7 (c), two large vortices are formed between the first row of tubes and the impingement plate. These two vortices cause many small vortices in the gaps between the tubes in the first row. Affected by these vortices, the fluid basically does not contact the tubes in the first row. Meanwhile, the fluid does not contact some tubes in the second row near the vertical centerline for these vortices.

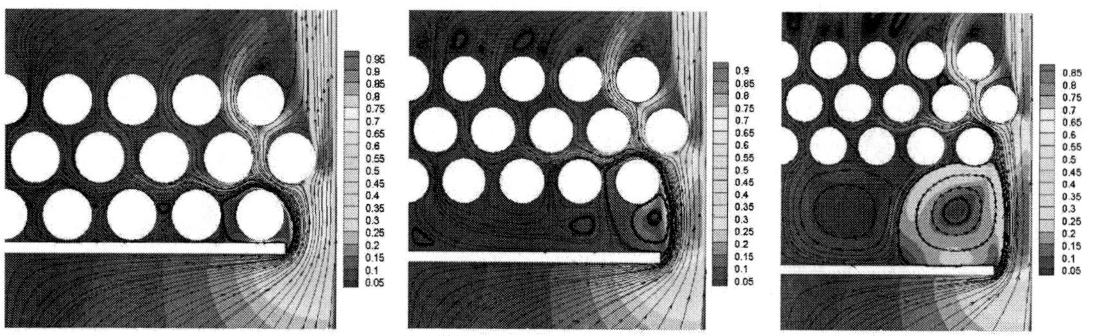

(a) case 2 (h=1mm). (b) case 3 (h=21mm). (c) case4 (h=51mm).

Figure 7 Streamline and velocity contour at inlet velocity 0.09m/s.

4.3. *The Inlet Velocity of the Shell-side Fluid*

When the inlet velocity of the water is 0.35m/s, the streamline and velocity contours behind the non-perforated are shown in Figure8. Three cases are studied with various distance between the impingement plate and the tube bundle. The distance between the impingement plate and the tube bundle is 1mm in case 5. The distance between the impingement plate and the tube bundle is 21mm in case 6. The distance between the impingement plate and the tube bundle is 51mm in case 7. The impingement plate is non-perforated.

The flow field after the non-perforated impingement plate welded on the spacer still shows that the first row tubes are in insufficient contact with the fluid, as shown in Figure8 (a). For case 6 and case 7, the vortices become larger than that in case 5, as shown in the Figure8 (b) and (c). Not only the tubes in the first row, but also some tubes in the second and third row are in the vortex zone. That means more tubes cannot effectively contact the fluid for the cases that the non-perforated plate is welded on the shell when the inlet velocity of the water is 0.35m/s.

When the inlet velocity of the water is 1.33m/s, the vortices fill the flow area regardless of the distance between the impingement plate and the tube bundle, as shown in Figure9. All tubes in Figure9 contact the fluid insufficiently.

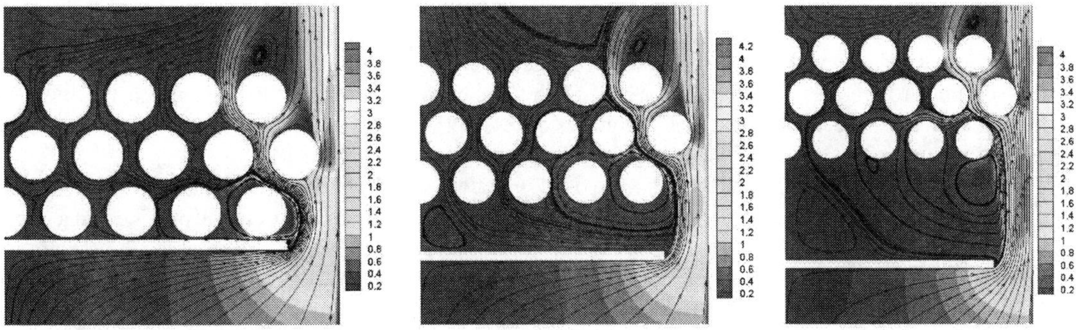

(a) case 5 (h=1mm). (b) case 6 (h=21mm). (c) case 7 (h=51mm).

Figure 8. Streamline and velocity contour at inlet velocity 0.35m/s.

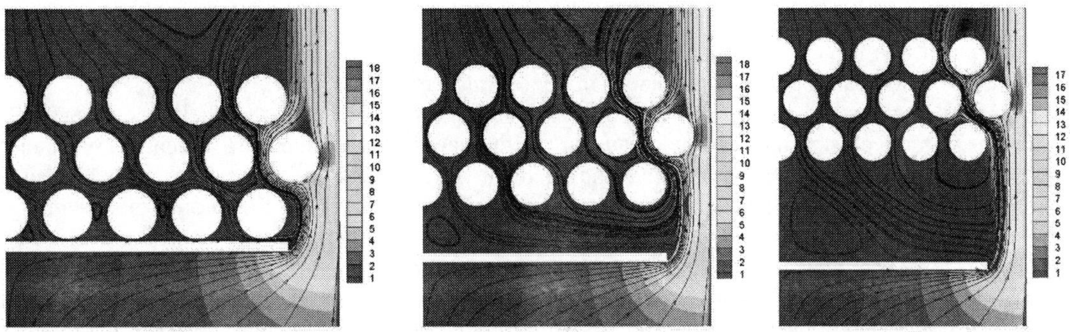

(a) case 8 (h=1mm). (b) case 9 (h=21mm). (c) case 10(h=51mm).

Figure 9. Streamline and velocity contour at inlet velocity 1.33m/s.

Although the non-perforated impingement plate can prevent the tube bundle from fluid impacting, it will also somewhat affect the flow behind it causing the fluid cannot fully contact the tubes. By comparing the flow field of the various inlet velocities in Figure 7, Figure 8 and Figure 9, it can be seen that the flow field behind the non-perforated impingement plate welded on the spacer is relatively fluent in a wide inlet velocity range. The fluid bypassing the non-perforated impingement plate welded on the spacer could contact more tubes.

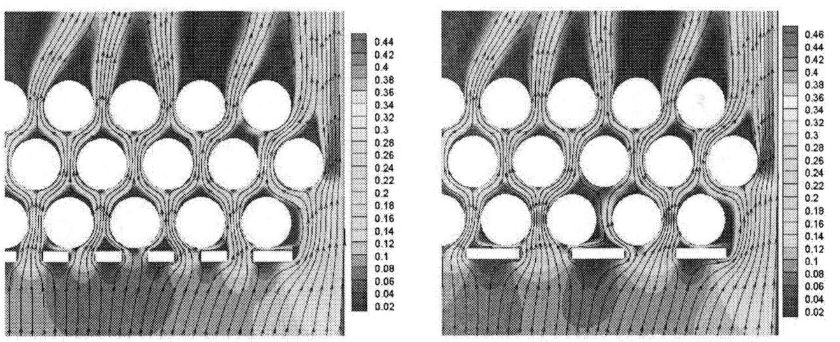

(a) case 11 (λ=10mm). (b) case 12 (λ=20mm).

IOP Conf. Series: Materials Science and Engineering 1081 (2021) 012053 doi:10.1088/1757-899X/1081/1/012053

Figure 10. Streamline and velocity contour at inlet velocity 0.09m/s.

4.4. The Hole Diameter of the Impingement Plate

A common situation in engineering design of the impingement is to punch holes when the impingement plate is welded on the spacers of the tube bundle. Based on the engineering foundation, this section investigates the flow field behind the perforated impingement plate when h=1mm. The diameter of the hole is specified as 10mm in case 11 and 20 mm in case 12. The streamline and velocity contours when the inlet velocity is 0.09m/s are shown in Figure 10.

It can be seen in Figure 10 that the fluid velocity on the windward side of the perforated impingement plate is reduced. It indicates that even if the impingement plate is punched, it can still play a role in protecting the tube bundle from the fluid impacting. It can be seen that the streamlines are uniform and the low-velocity area at the tube leeward is very small in Figure 10 (a). However, the streamlines are in the uneven distribution somewhere between the gaps of the tubes in case 12. The low-velocity area in the tube bundle extends from the leeward of the tubes to the gaps of the tubes in Figure10 (b). Based on the discussion between these two cases, the hole diameter is recommended to be 10mm when the perforated impingement plate is designed.

5. Conclusion

In this paper, the fluid flow behind the impingement plate in the shell-side of the shell-and-tube heat exchanger is studied. The main conclusions obtained are as follows:

(1) The flow field behind the non-perforated impingement plate on the shell-side is inevitably uneven. Vortices exist in the flow field causing the fluid contacts the tubes insufficiently. By changing the distance between the non-perforated impingement plate and the tube bundle, it is found that the smaller the distance is, the smaller the vortices are and the more tubes the fluid contacts. According to the above analysis, for an impingement plate is necessary for a heat exchanger according to GB/T151-2014[1], it is recommended to weld on the tube bundle spacers.

(2) Not only can the perforated impingement plate welded on the tube bundle spacers prevent the tube bundle from the fluid impacting, but also it improves the effect of contact between the tubes and the fluid. Meanwhile, the flow field of the perforated impingement plate is more uniform than that behind the non-perforated impingement plate.

(3) Heat exchangers with perforated impingement plates are not suitable for the engineering limitations sometimes. The designer should take into consideration the case of the fluid contact the tubes behind the non-perforated impingement plate insufficiently and the possibility of the insufficient heat transfer capacity. Increasing the heat transfer margin of the heat exchanger is a recommended way to avoid the failure of the heat exchanger caused by insufficient heat transfer capacity.

References

[1] China Standardization Committee on Boilers and Pressure Vessels. Heat exchanger: GB/T151-2014.Beijing: China Standard Press, 2015.

[2] Fu Shuangquan. A Program of EXCEL For Impingement Plate on Tube-shell Heat Exchanger, 2013, 30(4):35-37.

[3] LIU Xiao-ling. Flow-induced vibration analysis on the shell side entrance impingement plate setting of shell and tube heat exchanger. Polyester Industry, 2015, 28(3):1-4.

[4] Defan Qing. The experimental investigation on heat transfer and flow resistance of elliptic rib tube for compressor intercooler. Chinese Journal of Turbomachinary, 2002(5):18-20.

[5] Xiaojun Chen. Application of fin tube cooler with bimetallic strip. Chinese Journal of Turbomachinary, 2003(1):24.

[6] Bin Xue, Wang Shangfeng, Yang Ming. Influence of inner finned tube with single or double holes in assembled compressor on resistance and heat transfer. Chinese Journal of Turbomachinary, 2011(5):8-10.

[7] Feng Su, Yue-fang Wang. Evaluation of transverse vibration characteristics for a tube induced by tube-side flow in a heat exchanger. Chinese Journal of Turbomachinary, 2018,60(6):70-74.
[8] Mingqing Liu. Summarization of intercooler with centrifugal air compressor. Chinese Journal of Turbomachinary, 2009 (3):33-34.
[9] WU Yangkuan, LIU Bohan, HAN Dong, PU Wenhao, DAI Tadong, MA Aiyun. Numerical simulation and experimental study on heat transfer enhancement of horizontal condenser with perforated impingement plate. Energy Chemical Industry., 2017, 38(6):42-47.
[10] LIU Xiangbin, Wang Yu, CHANG Yang, HUANG Weichang, DAI Yang, XU Dalu, REN Wenyuan, SUN Zhitao, QIAN Jiang. Simulation and Analysis of Impingement Plate in Tube Channel in One Tube Side Heat Exchanger. Process Equipment and Piping, 2016, 53(4):41-43.
[11] Salamah S. Al-Anizi, Abdullah M. AI-Otaubi. Double Perforated Impingement Plate in Shell-and-Tube Heat Exchanger. Heat Transfer Engineering, 2009, 30(10-11):885-894.
[12] Chen minheng, Cong Dezi, Fang Tunan, Qi Mingzhai, Pan Helin. Units Operation. Beijing, Chemical Industry Press, 2015.

Acknowledgments
This work is financially supported by the National Key Research and Development Program No. 2018YFB0606104, which is highly acknowledged by the authors.

The influence of wall roughness on centrifugal pump performance

Von-Hao Wang, Kun-Hang Li, Jing-Yin Li,

(School of Energy and Power Engineering, Xi'an Jiaotong University, Xi'an, Shaanxi, China, 710049)

E-mail: jyli@mail.xjtu.edu.cn

Abstract. Wall roughness is an important factor in the pump head and efficiency. Influence of wall roughness on the performance of a centrifugal pump is conducted with numerical approach. In the present study, a critical value of wall roughness is found. The performance of the pump decreases significantly while the wall roughness is less than critical value and decreases slightly while exceeding it. Influence of wall roughness of each domain is compared; the result revealed that the influence of wall roughness was decreased for volute, impeller, inlet and outlet respectively. Blade wall roughness of impeller has the greatest effect on pump performance, and pressure surface roughness has a greater influence than that for suction surface. This study could provide a reference for the manufacture and design of this type of centrifugal pump.

1. Introduction

Pumps have great applications in industry and agricultural production [1-4]. The washing tower circulating pump, as a centrifugal pump, would transport water mixed with corrosive media during the work process, and corrosion is an important reason for increasing the roughness of the wall surface. The wall roughness has an important influence on the pump efficiency. Therefore, surface treatments of the pump are particularly important.

In the past years, many scholars have studied the effect of wall roughness on flow. The research methods of flow field wall surface roughness are mainly based on experiments and models. Syverud et al. [5] and Kind et al. [6] proposed two calculation models to fit the wall roughness, but their application is still inadequate. Boyle [7] observed the surface microstructure and determined the detailed value of roughness with the instrument test, and then estimated the relationship between *Ra* and *hs*. Stripf et al. [8] proposed an external heat transfer measurement method for high-load turbine blade surface roughness changes. The results showed that the roughness has a great influence on the transition and the heat transfer coefficient of the turbulent boundary layer. In the area affected by the secondary flow, the roughness has the most significant effect on the heat transfer, reaching twice the two-dimensional turbulent boundary layer. Montomoli et al. [9] studied the formula for calculating the inner boundary layer thickness theoretically and the results were verified by experiments. Walker et al. [10] studied the influence of surface roughness on the boundary layer on the suction surface of turbine blades. It was concluded that reducing roughness can delay separation of the boundary layer and reduce separation loss while increasing roughness can induce the transition of boundary layer in advance. Yamazaki et al. [11] took the effect of roughness into account when the pump is studied. The results showed that the wall roughness near the throat inlet had the greatest influence on the efficiency of the jet pump. Aldas et al. [12] used numerical calculation to determine the effect of pump absolute roughness and relative roughness on pump efficiency. It was found that when the wall roughness was considered, the efficiency could be significantly improved to a certain degree. Yun and Kim [13] used numerical calculation to determine the effect of pump absolute roughness and relative roughness on pump efficiency. It was found that when the wall roughness was considered, the efficiency could be significantly improved to a certain degree. Zhu et al. [14] studied the influence of wall roughness on flow field characteristics of axial flow pump and the performance difference of axial flow pump under the influence of different wall roughness was obtained through simulation calculation. Lim and Sohn [15] studied the effect of impeller surface roughness on centrifugal pump efficiency by experiment. It was found that the sensitivity of impeller blades to roughness was relatively high. Gu et al. [16] applied an experimental method to investigate the effect of wall roughness on the pump head efficiency and shaft power. The results showed that head and efficiency decreased but shaft power is opposite as roughness increased. Deshmukh and Samad [17] analyzed the influence of turbulent flow energy and eddy viscosity characteristics of electric submersible pumps with different roughness and showed the effect on the pressure distribution and velocity field. It was found that the effect of roughness on performance was different from different medium and Reynolds numbers.

To sum up, wall roughness has a great influence on pump performance. However, there are few studies comparing the effects of the wall roughness of each part of the pump. Due to the limitation of manufacturing method and production cost, high wall roughness of the whole system could not be considered in the actual manufacturing. Therefore, it has great significance to study the effect of wall roughness of each part on the performance of the whole pump.

The paper is organized as follows. Section II provides numerical calculations, including Physics model, Grid validation and Turbulence model. Roughness model is explained in Sec. III. The results of the influence of wall roughness are presented and discussed in detail in Sec. IV. The paper ends with concluding remarks given in Sec. V. The purpose of the article is to provide support for the production and maintenance of this type of pump.

2. Numeral calculations

2.1. Physics model

The presented model is a special centrifugal pump designed for washing tower. The pump is simplified into four domains: inlet, impeller, volute, and outlet. The parameters and schematic diagram of the model are shown in Table 1 and Figure 1.

The design flow rate Q is 18m³/h, and the head H at the design flow rate is 10 m. The impeller inlet diameter D_1 is 54mm and outlet diameter D_2 is 102mm. The average outlet blade angle β_{2A} is 40.5°. The impeller outlet width b is 10mm. The nominal rotation speed is 2900 r/min. The blade number is 6 and the volute diameter D is 110mm.

Table 1. Pump design parameters.

Parameters	Value
Q (m³/h)	18
H (m)	10
Rotating speed (r/min)	2900
D_1 (mm)	54
D_2 (mm)	102
B (mm)	10
Blade number	6
β_{2A}	40.5°
D (mm)	110

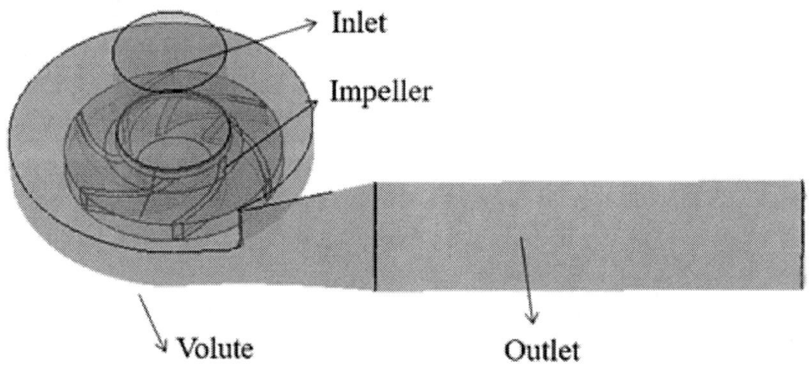

Figure 1. General view of the pump model.

2.2. Grid validation

The hexahedral structured grid was generated based on the topology of the pump. To determine the appropriate number of grids, under the same settings, selected 5 sets of different numbers of grids for grid-independent analysis, as shown in Figure 2. It could be seen that when the number of grids reached about 3.5 million, the head was stable, and the calculation requirements have been met. Considering the coordination of calculation accuracy and time, 3.5 million grids were selected as the calculation standard.

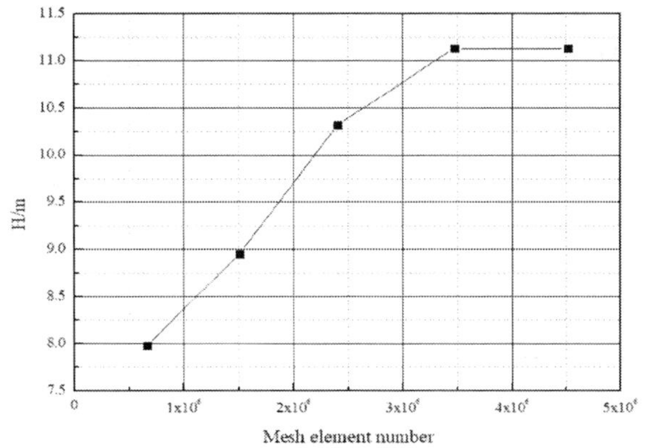

Figure 2. Grid-independence validation for the pump model.

2.3. Turbulence model

The selection of the turbulence model is very important in numerical simulation. [18] Generally, different models would have different results, which need to be selected according to the actual situation. In this paper, the calculation results of these models were compared with the experimental values, as shown in Table 2. The results showed that the SST model was in good agreement with the experimental results, so the subsequent numerical calculations in this paper used the SST model.

Table 2. Comparison of different turbulence models.

Turbulent Models	η	H (m)
k-ε	0.734	10.66
RNG k-ε	0.723	10.45
SST	0.712	10.39
Experimental value	0.698	10.34

3. Roughness model

3.1. Equivalent sand grain roughness model

In actual production, the roughness has peaks and valleys, and the shapes and sizes are different. The technical roughness of peaks and valleys with different shapes and sizes could be described by the equivalent sand roughness (*hs*). As shown in Figure 3, a rough surface could be equivalent to a sphere with an average height *hs* densely spread on a smooth surface.

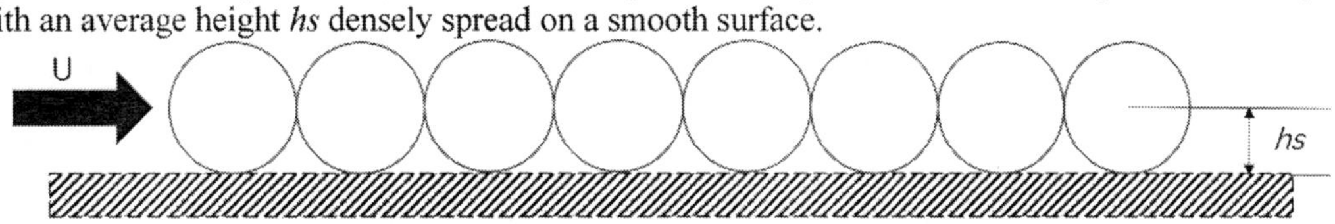

Figure 3. Schematic diagram of equivalent sand model

3.2. Wall roughness parameter setting

In previous study, *hs* represent the average height of the uniformly laid sand grains in the equivalent sand grain model and this was different from the actual wall roughness (Ra). Moreover, *hs* would have a different definite relationship with Ra according to the processing technology and processing method. Adams et al. [19]

proposed an algorithm to convert the measured wall roughness parameters to equivalent sand grain roughness, and used this algorithm to convert the wall roughness to equivalent sand grain roughness and the model was verified by experiments. According to his research, this article determined the relationship as:

$$hs = 5.863Ra$$

For convenience of explanation, the actual wall roughness Ra would be replaced by the equivalent sand grain roughness *hs*. Table 3 is the specific conversion between *Ra* and *hs*.

Table 3. Conversion between *Ra* and *hs*

hs (μm)	*Ra* (μm)
0	0
100	17.06
200	34.11
300	51.17
400	68.22
500	85.28
600	102.34
700	119.39
800	136.45

4. Results and discussion

4.1. Influence of wall roughness on pump performance

To study the influence of wall roughness on the performance of the pump, a total of 9 roughness of 0μm, 100μm, 200μm, 300μm, 400μm, 500μm, 600μm, 700μm and 800μm were selected for numerical calculation. The results are shown in Figure 3 and Table. 4. It could be seen that there was a critical wall roughness (hs = 600μm) for the influence of roughness on the performance of the pump. When $0<hs<600$, as *hs* increased, the head and efficiency gradually decreased; when $600<hs<800$, as *hs* increases, the head and efficiency are basically unchanged. When *hs* increases from 0 to 600μm, the head decreased by 11.9% and the efficiency decreased by 13.1%. This showed that when the wall roughness was less than the critical wall roughness, the wall roughness had a greater impact on the performance of the pump. While the wall roughness was greater than the critical wall roughness, the wall roughness has little impact on the performance of the pump. It could be concluded that in actual industrial production, the reduction of wall roughness through the transformation of processing methods or the improvement of process level is of great significance to the performance improvement of this type of pump.

Table 4. Pump performance with different wall roughness under the rated flow condition

hs (μm)	0	100	200	300	400	500	600	700	800
η	0.782	0.722	0.713	0.703	0.696	0.689	0.651	0.647	0.646
H (m)	11.39	10.61	10.47	10.37	10.32	10.24	10.03	9.99	9.97

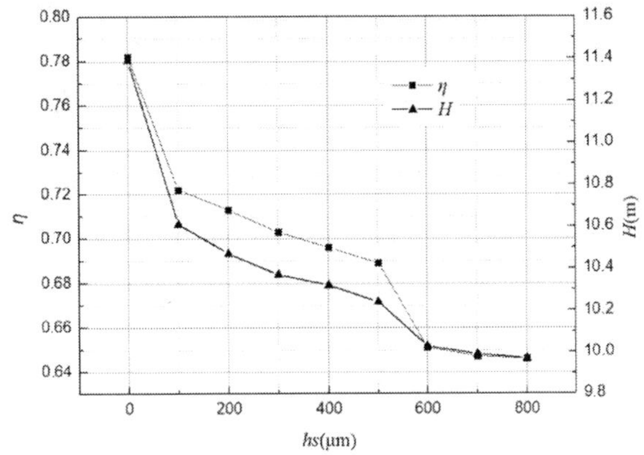

Figure 4. Pump performance with different wall roughness under the rated flow condition

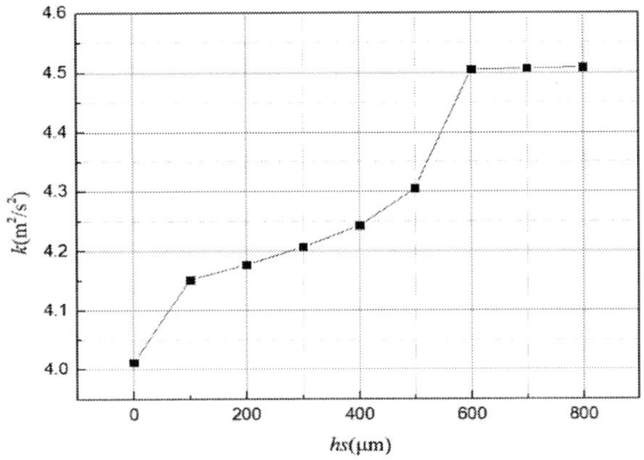

Figure 5. Turbulent kinetic energy with different wall roughness under the rated flow condition

It was important to study the reasons of the pump performance variation with different wall roughness and the significance of the critical value. Figure 5 shows turbulent kinetic energy with different wall roughness under the rated flow condition. It could be seen that the turbulent kinetic energy increased gradually with the increase of roughness. The increase of turbulent kinetic energy means that the flow state was more violent and the flow dissipation was larger. Therefore, turbulent kinetic energy increased when the wall roughness increased. When the critical value is reached, the turbulent kinetic energy basically does not change. This indicated that the fluid state was basically unchanged and flow losses increased little when the roughness was more than the critical value.

4.2. Influence of roughness of different parts on pump performance

The decrease in wall roughness means a lower price. In order to decrease price in actual production, only a certain part or some parts are usually smoothed. Therefore, it was necessary to study the influence of wall roughness of different parts on pump performance. In the process of exploring the influence of wall roughness of different parts on the pump performance, this paper controlled *hs* of other walls to 600μm and the wall surfaces of inlet, impeller, volute and outlet are smoothed (hs = 0) respectively. The results are shown in Table 5. It could be seen that the smooth treatment of each part of the wall surface promoted the efficiency of the whole pump, but the roughness of the wall surface of different parts improved the performance of the centrifugal pump differently. The influence of wall roughness was decreased for volute, impeller, inlet and outlet respectively. Among them, the efficiency and the head increased by 11.1% and 10.2% while the volute wall was smoothed and the efficiency and the head increased by 6.1% and 1.8% when the impeller wall was smoothed. The efficiency and the head increased by 4.4% and 0.59% while the inlet wall was smoothed and the efficiency and the head increased by 3.3% and 0.40% while the outlet wall was smoothed.

Table 5. Pump performance with wall roughness of different part

Part	Inlet smooth (hs=0)	Impeller smooth (hs=0)	Volute smooth (hs=0)	Outlet smooth (hs=0)	All rough (hs=600μm)
η	0.695	0.712	0.766	0.684	0.651
H (m)	10.09	10.21	11.05	10.07	10.03

Table 6 shows the pressure difference of different parts. Figure 6 illustrates the variation of pressure difference of different parts. It could be seen that the pressure difference between the inlet and outlet of the volute was relatively small, but the increased amplitude was the largest under the influence of wall roughness. Therefore, the wall roughness of the volute had the greatest influence on the pump performance. The impeller has a characteristic of high-speed rotation and the main energy is converted. So the pressure difference was relatively large. However, the variation of pressure difference is relatively small under the influence of wall roughness. Therefore, the effect of the wall roughness of impeller on the pump performance was not as good as that of volute. This indicated that the effect of roughness on pump performance was mainly related to the flow area.

Table 6. Pressure difference of different part

	Inlet	Impeller	Volute	Outlet
rough(hs=600μm)	-2.562e+03	1.336e+05	-3.396e+04	-1.125e+04
smooth(hs=0)	-1.125e+03	1.362e+05	-2.082e+04	-9.902e+03
variation	1437	2600	13140	1348

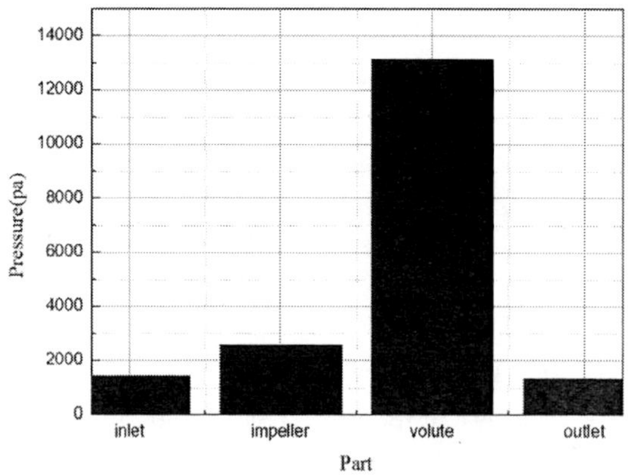

Figure 6. The variation of pressure difference of different part

4.3. Influence of volute wall roughness on pump performance

The surface roughness of the volute had the greatest influence on the performance of the pump. To study the influence mechanism of the surface roughness of the volute on the performance of the centrifugal pump, the presented study compared and analysed the internal flow of volute between the rough wall and the smooth wall. The central section of the volute was selected as the research object. Figure 7 is the velocity contours of the central section of the volute between the rough wall and the smooth wall. Figure 8 is the total pressure contours of the central section of the volute between the rough wall and the smooth wall.

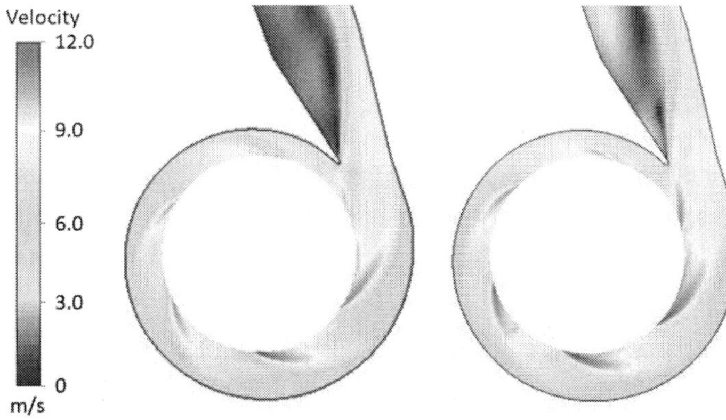

Figure 7. Velocity contours of the central section of the volute between the rough wall and the smooth wall

Figure 8. Total pressure contours of the central section of the volute between the rough wall and the smooth wall

Figure 7 and Figure 8 showed that the total pressure at the outlet of the volute after smooth treatment was higher than that before, which indicated that the existence of volute wall roughness increased the hydraulic loss of the volute. But the velocity field and total pressure field at the outlet of the impeller were almost unchanged, which indicated that the roughness of the volute wall has almost no influence on the flow field of the impeller. Therefore, the roughness of the volute wall increased the frictional resistance loss caused by the fluid flow in the volute and it could be inferred that part to part would not affect each other by the wall roughness. If the volute wall surface was smoothed and the roughness was reduced, the performance of the pump could be greatly improved without worrying about the flow situation in the impeller.

4.4. Influence of impeller wall roughness on pump performance

The impeller of the article was composed of hub, shroud and blade, so it was of practical significance to study the effect of wall roughness of each part on pump performance. Table 7 shows pump performance with wall roughness of different parts of the impeller. It could be seen that the blade wall roughness had the greatest effect on the performance of the centrifugal pump.

Table 7. Pump performance with wall roughness of different part of impeller

Part	Hub smooth	Shroud smooth	Blade smooth	All rough
η	0.682	0.680	0.699	0.651
H (m)	10.08	10.07	10.18	10.03

For blades, different parts had different effects on flows and the performance of the pump on different working conditions would be different. To further study the influence of pressure surface and suction surface roughness on pump performance, the pump head and efficiency at low flow rate(0.8Q), rated flow(Q)and large flow rate(1.2Q) were calculated respectively under the design scheme in Table 8.

Table 8. Blade surface roughness design scheme

Scheme	Pressure Surface hs (μm)	Suction Surface hs (μm)
1	600	0
2	0	600
3	600	600
4	0	0

As shown in Figure 9 and Figure 10, both the smooth treatment of the suction surface and the pressure surface could improve the characteristics of the pump, but the smooth treatment of the pressure surface was obviously better. At the rated flow, the head and efficiency increased by about 1.2% and 3.4% when the pressure surface was smoothed while the head and efficiency increased by 0.6% and 1.4% with the smooth suction surface. The former was almost twice as effective as the latter.

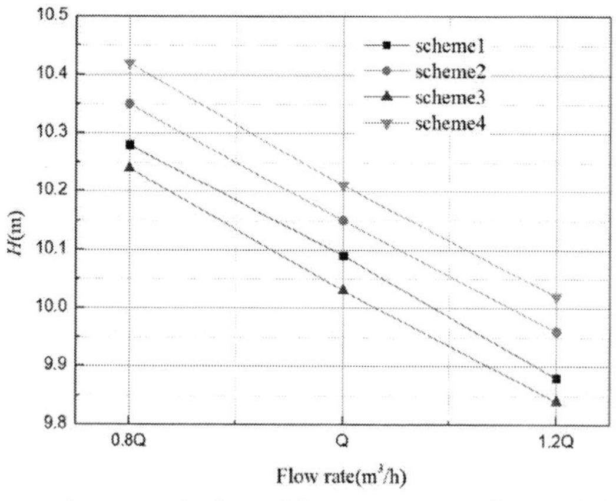

Figure 9. Influence of the roughness variation of the pressure surface and the suction pressure surface on the head under different working conditions

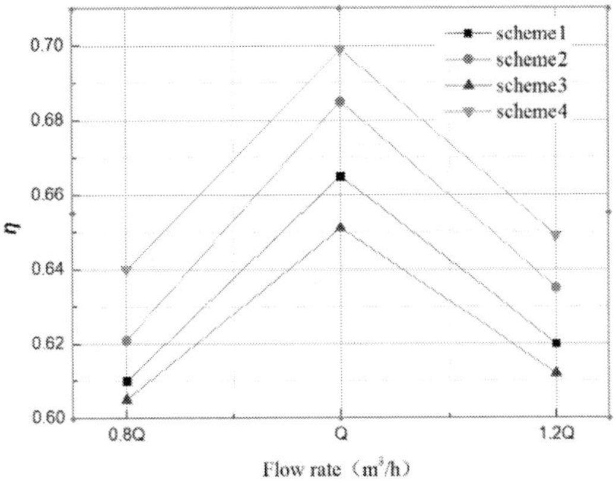

Figure 10. Influence of the roughness variation of the pressure surface and the suction pressure surface on the efficiency under different working conditions

Figure 11 is the velocity and total pressure contours of the central section of the impeller of the blade with wall roughness. From Figure 11, it could be seen that the fluid velocity on the suction surface was not high and the fluid pressure was small. When the roughness increased, the shear stress on the surface wall of the blade increased and the velocity decreased, the kinetic energy loss increased, the pressure decreased, and the efficiency decreased. However, due to more lager velocity of pressure surface, so the efficiency dropped more. And the suction surface did little work on the fluid, which could not directly affect the fluid velocity loss. It indicated that the fluid velocity could affect the influence of wall roughness on the pump performance.

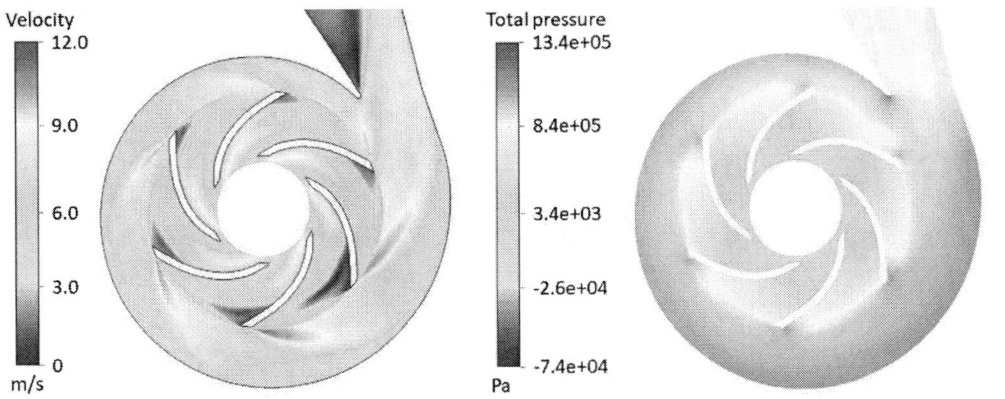

Figure 11. Velocity and Total pressure contours of the central section of the impeller of the blade with wall roughness

Conclusions

A numerical study is carried out to investigate the influence of wall roughness on the centrifugal pump performance. The characteristics of pump performance with different wall roughness were studied. The wall roughness influence for each domain is compared separately and the corresponding influence on the pump performance was analysed. The conclusions can be drawn as follows:

(1) Pump head was decreased with the wall roughness increasing, and a critical wall roughness was found. When the wall roughness is less than the critical value, the head and efficiency are greatly decreased. While the roughness exceeds the critical value, the wall roughness process slightly influences on the pump performance.

(2) Different domain with the same critical wall roughness shows a different pressure difference. The influence of each part of the pump is decreased from volute, impeller, and inlet to outlet. The efficiency and the head increase by 11.1% and 10.2% when the volute wall is smoothed. The efficiency and the head increase by 6.1% and 1.8% with the impeller wall smoothed.

(3) Compare with the shroud and hub wall roughness, the blade wall roughness has the greatest impact on the pump head. More specifically, the pressure surface roughness has a greater impact on pump performance than the suction surface roughness and the former is almost twice as effective as the latter.

(4) This study could provide a reference for the manufacture and design of this type of centrifugal pump. In actual processing and production, the performance and corrosion resistance of the pump can be improved by smoothing the wall surface of the volute and the pressure surface of the blade, which can effectively save costs.

References

[1] He, X. , Xu, S. L. , Liu, H. T. , Wang, X. F. , & Chen, X. D. . (2018). *Prediction of the dynamic stress of reactor coolant pump impellers in water environments based on a harmonic response analysis*. Chinese Journal of Turbomachinery, 60(2):53-59..

[2] Guan, X. , Wang, Y. F. , Wang, R. , & Guo, T. . (2017). *Design optimization of a thick cylindrical pump for minimum structural weight*. Chinese Journal of Turbomachinery, 59(5):36-39.

[3] Patil, A. , Delgado, A. , Chen, Y. , Steck, D. , & Morrison, G. L. . (2018). *Mechanical reliability of electrical submersible pumps*. Chinese Journal of Turbomachinery, 60(5):69-77.

[4] Zhou, C. Y. , & Schmied, J. . (2017). *Design of magnetic bearing controller using madyn 2000*. Chinese Journal of Turbomachinery, 59(4):51-78.

[5] Syverud, E. , Brekke, O. , & Bakken, L. E. . (2007). *Axial compressor deterioration caused by saltwater ingestion*. Journal of Turbomachinery, 129(1), 119-126.

[6] Kind, R. J. , Serjak, P. J. , & Abbott, M. W. P. . (1998). *Measurements and prediction of the effects of surface roughness on profile losses and deviation in a turbine cascade*. Journal of Turbomachinery, 120(1):20-27.

[7] Bons, & J. (2005). *A critical assessment of reynolds analogy for turbine flows*. Journal of Heat Transfer, 127(5):472-485.

[8] M. Stripf, S. Wittig, & A. Schulz. (2005). *Surface roughness effects on external heat transfer of a hp turbine vane*. Journal of Turbomachinery, 127(1): 200-208.

[9] Montomoli, F., Hodson, H., & Haselbach, F. (2010). *Effect of roughness and unsteadiness on the performance of a new low pressure turbine blade at low Reynolds numbers*. Journal of turbomachinery, 132(3): 0310181-0310189.

[10] Walker, J. M., Flack, K. A., Lust, E. E., Schultz, M. P., & Luznik, L. (2014). *Experimental and numerical studies of blade roughness and fouling on marine current turbine performance*. Renewable Energy, 66: 257-267.

[11] Yamazaki, Y., Nakayama, T., Narabayashi, T., Kobayashi, H., & Shakouchi, T. (2006). *Effect of surface roughness on jet pump performance*. JSME International Journal Series B Fluids and Thermal Engineering, 49(4): 928-932.

[12] Aldaş, K., & Yapıcı, R. (2014). *Investigation of effects of scale and surface roughness on efficiency of water jet pumps using CFD*. Engineering Applications of Computational Fluid Mechanics, 8(1): 14-25.

[13] Yun, J. E., & Kim, J. H. (2014). *Effect of surface roughness on performance analysis of centrifugal pump for wastewater transport*. Transactions of the Korean Society of Mechanical Engineers B, 38(2): 147-153.

[14] ZHU, H. G., YAN, B. P., & ZHOU, J. R. (2006). *Study on the Influence of Wall Roughness on the Hydraulic Performance of Axial-flow Pumps* [J]. Journal of Irrigation and Drainage, 153:1455-1464.

[15] Lim, S. E., & Sohn, C. H. (2018). *CFD analysis of performance change in accordance with inner surface roughness of a double-entry centrifugal pump*. Journal of Mechanical Science and Technology, 32(2): 697-702.

[16] Gu¨ lich, J. F. (2003). *Effect of Reynolds number and surface roughness on the efficiency of centrifugal pumps*. J. Fluids Eng, 125(4): 670-679.

[17] Deshmukh, D., & Samad, A. (2019). *CFD-based analysis for finding critical wall roughness on centrifugal pump at design and off-design conditions*. Journal of the Brazilian Society of Mechanical Sciences and Engineering, 41(1): 58.

[18] Li, H. , Wang, J. , Zhou, S. Q. , Hu, X. B. , Xiong, G. Z. , & Xu, T. C. (2017). *The effect of the number of blades on the performance of a multi-blade centrifugal fan*. Chinese Journal of Turbomachinery, 59 (02): 19-22.

[19] Adams, T., Grant, C., & Watson, H. (2012). *A simple algorithm to relate measured surface roughness to equivalent sand-grain roughness*. International Journal of Mechanical Engineering and Mechatronics, 1(2), 66-71.

Acknowledgments

This work was financially supported by the National Science and Technology Major Project（2017-II-0007-0021）.

IOP Conf. Series: Materials Science and Engineering 1081 (2021) 012055 doi:10.1088/1757-899X/1081/1/012055

Effect of taper parameters on mechanical seal performance of taper/pores composite mechanical seal

X P Cheng, Q Liu, Y L Zhang, J Wei
Institute of Applied Physics, Jiangxi Academy of Sciences, 330029, Nanchang, China

E-mail:654268480@qq.com

Abstract. a new combination fluid seal with the cone/pores is given, and the 3D fluid/solid coupling theory model is constructed, considering the interaction between the liquid pressure field and the stress deformation of the seal rings, led to seal clearance unceasingly change. Through the corresponding numerical calculation, film pressure distribution and end face deformation situation are obtained between the seal end faces, the influence rule of taper on sealing performance under the high and low pressure conditions is analyzed. The results show that the dynamic pressure effect caused by rhomboid pores can produce circumferential and radial wavy deformation of the seal end faces, while the static pressure effect also changes with the taper change of the end faces. For the low pressure and medium speed equipment, the conical end face which the taper is 0-2.4 should be chosen priority. For higher pressure and rotational speed device, the convergence end face seal which the taper is 0-2.4 should first be chosen.

1. Introduction

In 1980 Etsion studied the performance parameters of taper end face seal in detail and first presented the feasibility of taper end face seal theoretically[1,2]. Subsequently, scholars from various countries have carried out researches on the relatively axial, radial, angular syndication and end faces deformation of mechanical seals during operating[3,4]. By controlling the end faces clearance, the design of the seal structure and working conditions, the relative deflection of the end faces is reduced and the end faces deformation is neutralized to prevent the contact between the two end faces[5,6].Metcalfe[7] adopts the influence coefficient method and considers the constraints and forces between the components of the sealing system to attribute the seal end faces deformation to the comprehensive influence of various parameters, and calculates the deformation value of each point of the sealing rings. Mayer[8] believed that force and temperature were the factors affecting the deformation and calculated the deformation of the mechanical seal end faces by using the ring theory. Salant[9] analyzed the seal end faces deformation of different structures and obtained experimental verification. And it is found that the experimental values are in good agreement with the calculated ones. Metcalfe[10] classified the deformation of the seal end faces into the comprehensive influence of various parameters by considering the constraints between components and the stress of the seal structure, and calculated the deformation of end faces. Blasisk[11] used displacement sensor and piezoelectric brake device to measure the end faces deformation of non-contact fluid mechanical seal, and learned that angular velocity and ring material have similar effects on the end faces temperature rise. Peng Xudong[12] systematically explored the influence of seal rings deformation on performance under various structures and constraints. Meng Xiangkai[13] calculated the end faces deformation

IOP Conf. Series: Materials Science and Engineering 1081 (2021) 012055 doi:10.1088/1757-899X/1081/1/012055

behavior of mechanical seals under different pressures by adopting the integral contact finite element method and considering the influence of auxiliary sealing rings.

2. Model

2.1. Geometric model

Fig.1 shows the schematic model of the local end faces geometric structure of cone-diamond pores composite mechanical seal. There is a certain taper on the static torus, among them: φ-cone angle; r_m-starting radius of the cone; r_i、r_o- inner and outer diameter of end face Seal ring; h_0- basis of film thickness; the taper is defined as ϕ=100×(r_o-r_m)×tanφ/h_0[14] .The rhombic pores in the rotary ring are distributed symmetrically in the center along the circumferential direction, then the rhombic pores are equally spaced in the radial direction, And the diamond pores is equal depth, their depth is h_1. Diamond cut faces are directional pores, two structure parameters are used to indicate its geometric characteristics, and is defined the direction factor ξ for:

$$\xi = b/a \qquad (1)$$

Among them: a-semi-major axis of diamond pores symmetry axis, b-semi-minor axis.

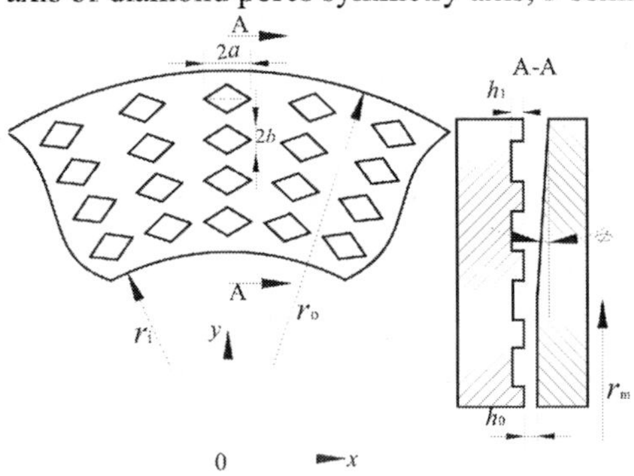

Fig.1 Schematic diagram of diamond pores distribution and conical surface structure

2.2. Mathematical model

The fluid pressure between the end faces is constant along the direction of film thickness by assuming. The sealing fluid is Newtonian fluid whose viscosity remains constant, then the governing equation used to describe the liquid film pressure on the end faces can be expressed by two-dimensional Reynolds Equation:

$$\frac{\partial}{\partial x}\left(\frac{h^3}{6\mu}\frac{\partial p}{\partial x}\right)+\frac{\partial}{\partial y}\left(\frac{h^3}{6\mu}\frac{\partial p}{\partial y}\right)=U\frac{\partial h}{\partial x}+V\frac{\partial h}{\partial y} \qquad (2)$$

Among them: U, V-average linear velocity of end face x and y, h-the thickness of liquid film, p-liquid film pressure, μ-viscosity coefficient of sealing medium. The end surface liquid film thickness equation for:

$$h=\begin{cases} h_0-h_i+h_j & \textit{No cone area outside the pores} \\ h_0-h_i+h_j+h_k & \textit{Cone area outside the pores} \\ h_0+h_1-h_i+h_j & \textit{No cone area inside the pores} \\ h_0+h_1-h_i+h_j+h_k & \textit{Cone area intside the pores} \end{cases} \qquad (3)$$

Among them: h_i-the amount of deformation at a point on the rotary ring; h_j-the amount of deformation at a point on the stationary ring; h_k-the cone height at a point on the cone; When the mechanical seal operates stably, the axial force of the rotary ring assembly is balanced, then:

Closing force F_c= opening force F_o (4)

Among them:

$$F_c=\pi(r_o^2-r_i^2)[p_{sp}+B(p_o-p_i)] \qquad (5)$$

$$F_o = p_a A_a N \iint p dx dy \quad (6)$$

In the formula: p_{sp}-spring pressure, B-seal end face balance ratio, A_a-the cross-sectional area of a single diamond pore, p_i、p_o-inner and outer diameter pressure of sealing ring, p_a-environmental pressure or cavitation pressure, N-period number of rhomboid pores.
Formula (2) is discretized by the finite difference method, then combined with formula (3)-(6) is solved to establish a three-dimensional mathematical analysis model of fluid-solid coupling for hydrodynamic mechanical seals. The deformation of the seal rings is calculated by commercial finite element software programming, and the p distribution is solved again with film thickness deformed, and the iterative solution is carried out repeatedly until the convergence criterion meets the program setting, then the iteration stops. At fast, seal performance parameters such as liquid film stiffness, opening force and leakage rate are calculated from the final iterated pressure. See the references for detailed algorithms.[15]

3. Calculation results and discussion

The influence of the taper of the seal end face on seal performance of the cone-diamond pores combination seal under the condition of high and low pressure was studied. Structural parameters: r_i=127.25mm，r_o=141.75mm，a=2mm，h_1=1.25μm，Number of diamond pores in single row n_s=4，ξ=0.67. Elastic modulus of rotary and static rings: E_1=617GPa、E_2=23GPa，Poisson's ratio of rotary and static rings: v_1=0.28、v_2=0.25，the thickness of rotary and static sealing rings: h_{b1}=10mm、h_{b2}=15mm，B=0.85，p_{sp}=0.1，N=150. Operating parameters: p_a=p_i=0.1MPa，low pressure p_o=0.3 MPa, high pressure p_o=6.6MPa，μ=0.001Pa.s.
When calculating and studying the influence of certain parameters on sealing performance parameters, the other parameters remain unchanged except for explanation. On this basis, the change rule of sealing performance under high and low pressure conditions and different taper conditions would be analyzed.

3.1. Influence of taper on sealing performance at low pressure (=0.3MPa)

Under the working condition of low pressure, ω=800r/min, and different tapers, figure 2 shows that changes of film thickness caused by deformation of end faces during stable operation. The results show that the liquid flow in the pores area increases due to the presence of rhomboid pores on the end face, and the strong dynamic pressure effect is produced due to the block and extrusion of the pores boundary, so the pressure field of the liquid film between the end faces has regular and alternating trough and peak area. In the circumferential direction, the high pressure region appears on the side blocked by the boundary of the rhomboid pores; on the contrary, the trough appears in the divergent region of the liquid film in the rhomboid pores. In the radial direction, the highest pressure peak occurs near the outside diameter when the ϕ is 0, then the pressure peak decreases in turn to the inside diameter, while The highest pressure peak occurs near the inside diameter side when the ϕ is not zero, the pressure peak decreases in turn to the outer diameter side. At the same time, the peak pressure varies with the tapers different, and the highest pressure peak has the following rule: $p_{h(\phi=0)} > p_{h(\phi=2.4)} > p_{h(\phi=4)}$，so the change of the maximum pressure difference($\Delta p_{(\phi=0)}$) between the end faces is the largest when ϕ is zero, while $\Delta p_{(\phi=2.4)}$ and $\Delta p_{(\phi=4)}$ is progressively smaller. The range of high and low pressure areas is also different when the ϕ is different, the low pressure area of the liquid film between the end faces is very large, while the high pressure area is very small when ϕ is 4. Therefore, the flow pressure field can easily cause the unbalanced operation of the seal end faces. By contrast, the range of high and low pressure of the liquid film pressure field is relatively balanced when ϕ is zero, so the two sealing end faces in this state will be very stable in the process of operation, not easy to contact, it is difficult to wear, And mechanical seal fluid film stability in the three when ϕ is 2.4.
It can also be seen that the hydrostatic pressure action gradually enhanced with the change of taper along the radial direction from the outside diameter to the inside diameter, the dynamic pressure effect is corresponding change, resulting in the sealing ring surface compression deformation to form uneven surface, the film thickness changes accordingly, no longer is the same thickness, with a certain

waveform and taper. With the increase of taper, the film thickness on the inside diameter side gradually decreases, while that on the outside diameter side gradually increases. In the radial direction, the film thickness increases from inside diameter side to outside diameter side, and the order of maximum and small film thickness is as follows: $h_{max(\phi=4)}>h_{max(\phi=2.4)}>h_{max(\phi=0)}$, $h_{min(\phi=0)}>h_{min(\phi=2.4)}>h_{min(\phi=4)}$. The reasons are as follows: with the increase of the surface taper, the scope which is influenced by the static effect of liquid film between end faces began to grow and increase, and the range which is influenced by the dynamic pressure effect gradually become smaller and less. To maintain the balance of axial force, mechanical seal will automatically adjust the overall film thickness between end faces to control the membrane pressure in the operation of the sealing device, so if the film thickness increases, membrane pressure will decrease accordingly.

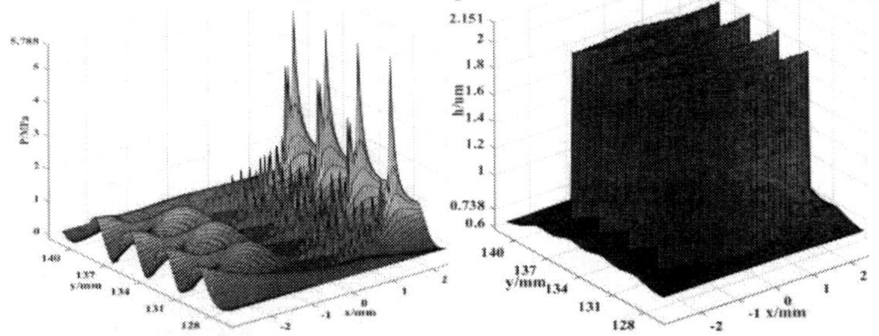

(a) ϕ=0

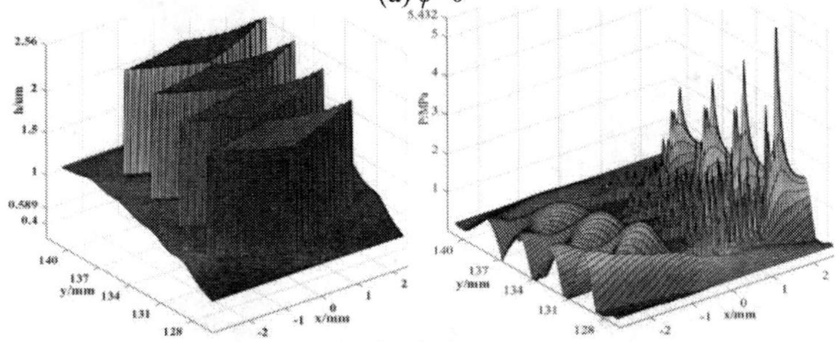

(b) ϕ=2.4

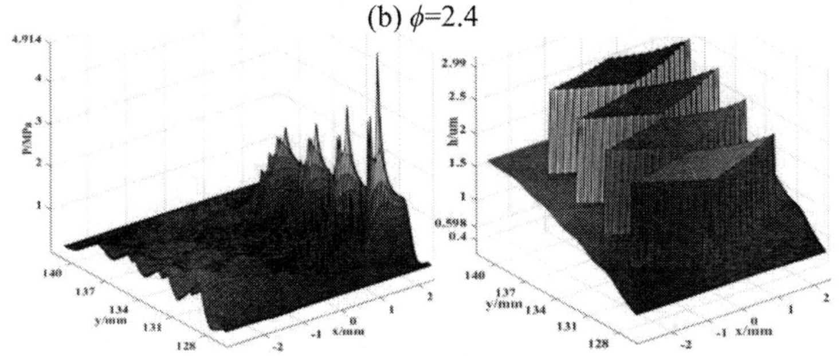

(c) ϕ=4

Fig.2 Film pressure and film thickness under different taper conditions

Fig.3 shows the influence rule of taper on sealing performance when p_o is 0.3MPa, and ω is 800r/min. The left vertical axis corresponds to the gray, black and yellow bar chart, and the right vertical axis corresponds to the colored bar chart. The results show that: the opening forces are not very different when ϕ is 0, 2.4, 4 and 6.4, while the liquid film stiffness varies greatly, sort for: $K_{z(\phi=0)}>K_{z(\phi=2.4)}>K_{z(\phi=4)}>K_{z(\phi=6.4)}$, this indicates that the liquid film stability between the sealing end faces is greatly affected by the uniformity and strength of the liquid film pressure fluctuation. The more balanced the liquid film pressure is, the greater the strength is, the better the corresponding liquid film stability is, and the greater the liquid film stiffness is. The more stable the sealing surface is in the process of operation, the more difficult it is to contact, and the longer the life cycle is. It can also be seen that the law of minimum film thickness and leakage rate in the color histogram on the right: $h_{min(\phi=0)}>h_{min(\phi=6.4)}>h_{min(\phi=4)}>h_{min(\phi=2.4)}$ 和 $Q_{(\phi=6.4)}>Q_{(\phi=0)}>Q_{(\phi=4)}>Q_{(\phi=2.4)}$， that is, the minimum film thickness of the seal face basically determines the change of its leakage rate, the smaller the minimum film thickness is, the better the sealing is, but the smaller the minimum film thickness is, the sealing face is more easy to contact wear, the life will be shortened.

To sum up, considering the stability of liquid film and sealing reliability, the low pressure and medium speed equipment, convergence taper face which the taper is 2.4 should be chosen, the leakage rate is lesser, the opening force and liquid film stiffness are better.

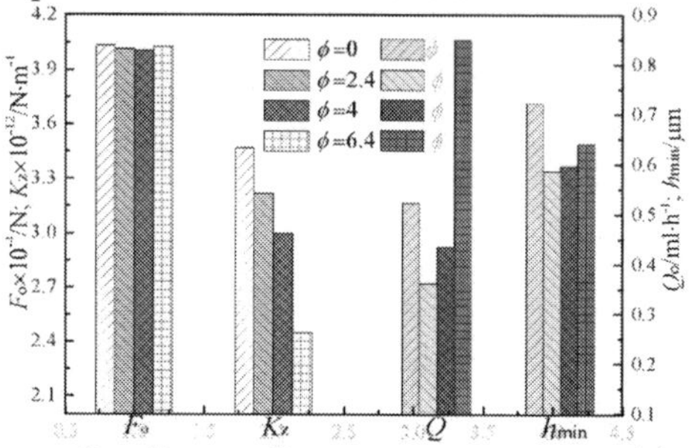

Fig.3 Comparison of sealing performance parameters with different tapers

3.2. *Influence of taper on sealing performance at high pressure (=6.6* MPa*)*

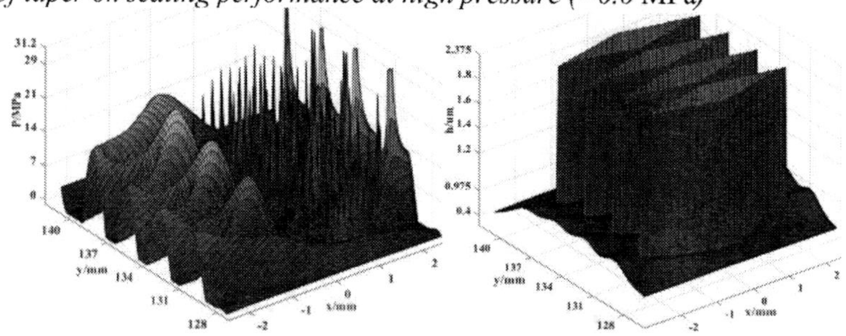

(a) ϕ=0

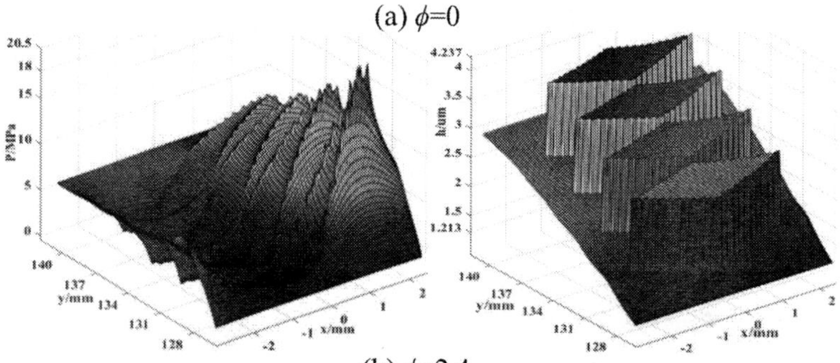

(b) ϕ=2.4

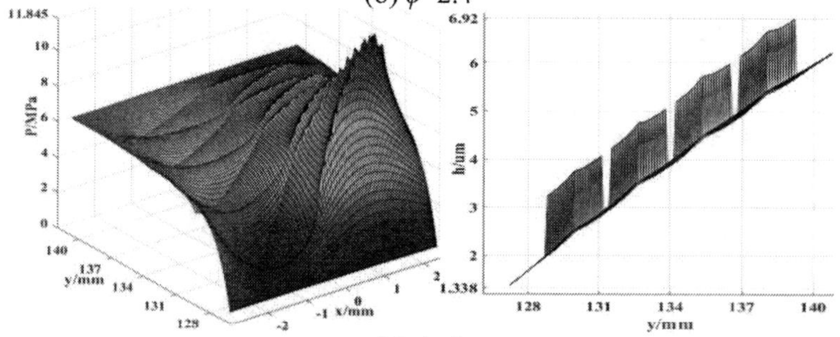

(c) ϕ=4

Fig. 4 The film thickness and film pressure diagram under the different tapers

It can be seen that the changing rule of the pressure field and the liquid film thickness between end faces under the working condition of high pressure, high speed (ω=4000r/min) and different tapers in

Fig.4. It can be seen from the figure that the static pressure effects of the liquid film on the end faces play an extremely important role under the conditions of high speed and high pressure, and the peak pressure varies greatly with the difference of taper. And the sorting size of the highest pressure peak is $p_{high\,(\phi=0)}>p_{high\,(\phi=2.4)}>p_{high\,(\phi=4)}$, the variation range of pressure difference ($\Delta p_{(\phi=0)}$) between the end faces is the largest, then $\Delta p_{(\phi=2.4)}$ and $\Delta p_{(\phi=4)}$ go down. The local scope of ultrahigh pressure liquid membrane area is far less than the range of low pressure area when the taper is zero, and the pressure difference is very big, the difference between high and low pressure area is smaller when the taper is 2.4 and 4, so the stability of liquid film is better. In addition, with the increase of taper, the influence range of internal and external static pressure difference on the whole seal end face is larger, the action range of static pressure effect is much larger than that of dynamic pressure effect.

It can also be seen that the film thickness gradually decreases from the outside diameter to the inside diameter, forming a linear radial convergence cone, and the size of the taper increases with the increase of the taper. From the perspective of the overall film thickness, the larger the taper is, the larger the film thickness is, while the maximum and minimum film thickness also increase with the increase of the taper. The reason is as follows: with the increase of the taper, the influence range of the static pressure effect on the liquid film between the seal end faces starts to increase and strengthen, while the influence range of the dynamic pressure effect gradually decreases and weakens. In order to maintain the balance of axial forces under the action of fixed closing force, the sealing device will automatically adjust the overall film thickness between end faces to control the film pressure of end faces during operation.

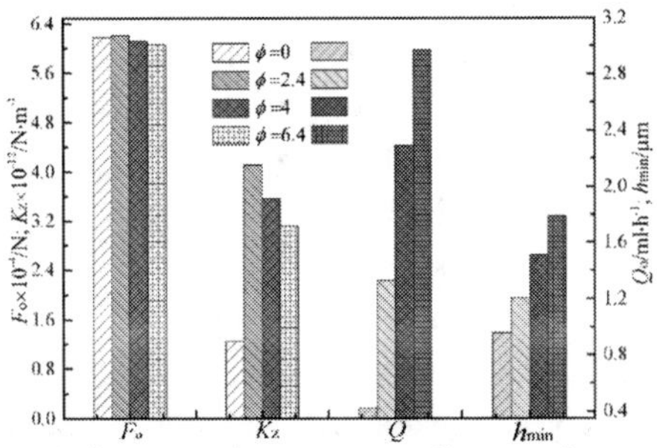

Fig.5 Influence of taper on sealing performance

The influence law of taper on sealing performance parameters under high pressure and high speed (ω=4000 r/min) conditions in Fig.5. When the taper is 0, 2.4, 4 and 6.4, the opening forces in the gray and yellow bar on the left are not very different by the figure, but the liquid film stiffness sequence is $K_{z(\phi=2.4)}>K_{z(\phi=4)}>K_{z(\phi=6.4)}>K_{z(\phi=0)}$. The order of the leakage rate and minimum film thickness in the color histogram on the right are $h_{min\,(\phi=6.4)}>h_{min\,(\phi=4)}>h_{min\,(\phi=2.4)}>h_{min\,(\phi=0)}$ 和 $Q_{(\phi=6.4)}>Q_{(\phi=4)}>Q_{(\phi=2.4)}>Q_{(\phi=0)}$. Thus, mechanical seal which the taper is 2.4 has the best stability and good sealing, Mechanical seal of which the tapers are 0 and 2.4. The liquid film stability and sealing are common, the effect of ϕ=6.4 is the worst, so it can be selected that the appropriate end face seal according to the different requirements of the equipment.

4. Conclusion

1) Dynamic pressure effect caused by the diamond pores can make faces circumferential and radial wave distortion under the conditions given, and with the change of taper, the static pressure effect also changes accordingly in the area of end faces. The membrane stability is greatly affected by the uniformity and strength of the membrane pressure fluctuation. The more balanced the pressure changes, the greater the strength is, the better the stability of the corresponding liquid film is, and the greater the liquid film stiffness is. The operation is the smoother, the harder the contact is and the longer the life cycle is.

2) For low pressure and medium speed equipment, in order to improve the stability of liquid film and seal reliability, the convergence taper (ϕ=0-2.4) seal face should be chosen. For devices with higher pressure and speed, the convergence taper seal face should be chosen as 2.4-4.

References

[1] Etsion I. Performance of end-face seals with diametral tilt and coning-hydrostatic effects [J]. ASLE Trans., 1980, 23 (3): 279-288.

[2] Etsion I. Dynamic analysis of non-contacting face seals [J]. ASME J. of Lubr. Tech., 1982, 44(4):460-468.

[3] Griskin E N. Study of face seals with eccentric rotation [M]. Soviet Journal of Friction and Wear, Allerton Press, New York, 1987.

[4] Wileman J. Dynamic response of eccentric face seals to synchronous shaft whirl [J]. ASME J. Tribol., 2004, 126(4): 301-309.

[5] Zou M, Dayan J, Green I. Parametric analysis for contact control of a non-contacting mechanical face seal[C]. In Proceedings of Vibration, Noise and Structural Dynamics 99, Venice, Italy, 1999: 493-499.

[6] Zou M, Dayan J, Green I. Sensitity analysis for the design and operation of a non-contacting mechanical face seal[C]. Proc Instn Mech Engrs, 2000, 214, Part C.

[7] Metcalfe R. End-face seal deflection effect–the problem of two-component stationary or rotating assemblies [J].
ASLE Transactions, 1980, 23(4): 393-399.

[8] Mayer E. Mechanical seal [M]. Beijing: Chemical Industry Press, 1981.

[9] Salant R, Key W E. Development of an analytical model for use in mechanical seal design [C]. Proceedings of 11th International Conference on Fluid Sealing, Insbruck, 1984, 325-343.

[10] Metcalfe R. End-face seal deflection effect–the problem of two-component stationary or rotating assemblies [J]. ASLE Transactions, 1980, 23(4): 393-399.

[11] Blasisk S, Laski P A, Takosoglu J E. Parametric analysis of heat transfer in non-contacting face seals[J]. International Journal of Heat and Mass Transfer, 2013, 57(1): 22-31.

[12] Peng X D, Feng X Z, etc.. Numerical analysis of deformation of a non-contacting gas lubricated Seal[J]. Tribology, 2004, 24(6):536-540.

[13] Meng X K, Wu D Z, Wang L Q. Study on deformation behavior of mechanical seal faces influenced by secondary seal [J]. Lubrication engineering, 2008, 33(1): 101-103.

[14] Peng X D，Liu X，Meng X K, etc.. Performance of mechanical face seals with both surface micro-pores and convergent coning[J].CIESC JOURNA, 2011, 62(12): 3463-3470.

[15] Peng X D, Du D B, Sheng S E, etc.. Effect of face asperity geometry on performance of a liquid lubricated face seal [J]. Tribology, 2007, 27(4): 352-356.

Acknowledgments

The National Regional Science foundation of China (Grant No. 51865019); 2020 Major R & D Project of Jiangxi Academy of Sciences (Grant No. 20201BBE51015).

AUTHOR INDEX